MFPT Conference 2014

Technology Solutions for Affordable Sustainment

Virginia Beach, Virginia, USA
20-22 May 2014

Volume 1 of 2

ISBN: 978-1-5108-8842-5

Printed from e-media with permission by:

Curran Associates, Inc.
57 Morehouse Lane
Red Hook, NY 12571

Some format issues inherent in the e-media version may also appear in this print version.

Printed by Curran Associates, Inc. (2019)

For permission requests, please contact Society for Machinery Failure Prevention Technology at the address below.

MFPT Society
715 Twining Road
Suite 135
Dresher PA 19025

Phone: (215) 346-2870
Fax: (215) 352-6902

https://mfpt.org

Additional copies of this publication are available from:

Curran Associates, Inc.
57 Morehouse Lane
Red Hook, NY 12571 USA
Phone: 845-758-0400
Fax: 845-758-2633
Email: curran@proceedings.com
Web: www.proceedings.com

TABLE OF CONTENTS

VOLUME 1

VOLUME 2

Technical Program for

MFPT 2014 CONFERENCE

Technology Solutions for Affordable Sustainment

Sheraton Virginia Beach Oceanfront Hotel

May 20–22, 2014

MFPT

The Society for Machinery Failure Prevention Technology

Pursuing prognostics and health management capabilities for military and commercial systems

MFPT

MFPT 2014 CONFERENCE

Technology Solutions for Affordable Sustainment

Sheraton Virginia Beach Oceanfront Hotel
May 20–22, 2014

MFPT
The Society for Machinery Failure Prevention Technology

Welcome ...

Dear Distinguished MFPT Conference Attendees,

I welcome you to Virginia Beach, Virginia on behalf of the Society for Machinery Failure Prevention Technology for the Society's 68th conference, Technology Solutions for Affordable Sustainment.

As last year, with Government travel restrictions in effect, the program committee members and others have worked hard to provide you with yet another outstanding program for the MFPT Conference. This event provides an excellent resource for our technical community as we explore the new possibilities available to us with the increased use of complex systems. In this economic climate we must keep in mind that with new technology we must consider affordable sustainment.

This year we are pleased to offer nearly 60 presentations and 4 tutorials. This is a great opportunity for you to interact with experts in Order Tracking, Model Based Maintenance Decision Making, Human Performance Monitoring, and Strain Gage Technology.

Also, with the increased demands on a shrinking workforce and turnover from experienced to young engineers, we must develop ways to sustain the corporate knowledge as well. I encourage you to embrace the younger community and foster an environment for future engineers to excel in. As we did last year, we are including a Student Paper/Presentation Prize to honor their great work. Last year's top papers were separated by a tenth of a point each!

I sincerely thank all of you for attending the MFPT 2014 Conference and I hope you find this an engaging and valuable event. Please tell your colleagues if you do!

John M Lucero
Chairman, MFPT Board of Directors

Note:

Antitrust and Competition Law: MFPT operates and conducts business in a manner that complies with the spirit and letter of federal, state and international antitrust and competition laws.

MFPT

KEYNOTE SPEAKERS

John S. Thackrah is the Executive Director of the U.S. Navy's Military Sealift Command - the leading provider of ocean transportation for the Navy and the rest of the Department of Defense - operating approximately 110 ships daily around the globe needed to sustain the USA's warfighting forces and deliver specialized maritime services in support of national security objectives in peace and war. In his current role, he is responsible for providing legal and business advice on the full spectrum of issues at Military Sealift Command. Mr. Thackrah is a graduate of the University of Delaware where he earned his Bachelor of Mechanical Aerospace Engineering degree and a graduate of Rensselaer Polytechnic Institute where he earned a Master of Business Administration. He began his career in 1979 with Pratt & Whitney (PW) in project and test engineering, followed by program management and model management. After working in commercial flight test activities and as general manager of a joint venture start-up for PW, he held positions of increasing responsibility until being promoted to Vice President (VP) of all commercial products. After a period as VP of Engineering for the Otis Elevator division of UTC, he returned to PW as VP of Aftermarket Services, and then VP/general manager for one of Pratt's strategic business units. In 2005, Mr. Thackrah was appointed to the Senior Executive Service and became Deputy Assistant Secretary for Management and Budget, responsible to the Assistant Secretary of the Navy (Research, Development and Acquisition). From November 2007 to July 2008, he served as Acting Assistant Secretary of the Navy (Research, Development and Acquisition). In December 2008, he became principal deputy, Strategic Systems Programs before assuming his current position in February 2011.

Jamie Smith is the director of product marketing for embedded systems at National Instruments. He leads the product marketing for the company's industrial and embedded products, including NI CompactRIO, Vision, Motion, and Wireless. Since joining NI in 1996 as an application engineer, Mr Smith has held key leadership positions in sales, engineering, product strategy, corporate development, and marketing. He has helped drive numerous strategic product and operational innovations during his time at National Instruments. In 2012, he was recognized as Top Embedded Innovator by Embedded Computing Design and received an R&D 100 Award. Jamie has served on the Board of Directors of the ISPO Alliance and as a member of the Embedded Vision Alliance. He holds a bachelor's of science degree in physics from UC Santa Barbara and a masters in applied physics from Stanford University

Professor Mike Lipsett is a Professor in the Department of Mechanical Engineering and Ernest & Gertrude Poole Chair in Management for Engineers at the University of Alberta. He holds a PhD in robotics from Queen's University, and studied applied systems science at Kyoto University in Japan. Prior to 2006, he held technical and leadership roles with Syncrude Canada Ltd, the world's largest crude oil producer from oil sands. He was also a senior research engineer at Atomic Energy of Canada Ltd. His research interests are the reliability and sustainability of integrated oil sands operations, rail systems, and other complex systems. This area includes modeling maintenance processes to understand the risk and value of process changes, developing advanced inspection and diagnostic methods for oil sands process systems & distributed fleet equipment, and developing new technologies for bitumen production processes and tailings (such as robotics for industrial environmental monitoring). Mike teaches engineering finance, senior mechanical engineering design, and modeling and simulation of engineering systems. Mike has been a Vice-Chair of the International Society for Condition Monitoring; and he is an Associate Editor of the International Journal of Condition Monitoring and Diagnostic Engineering Management. Mike remains closely linked to industrial issues through his consulting activities; and he has served on the boards of two not-for-profit innovation organizations. Mike Lipsett is a registered professional engineer in Alberta, Canada.

MFPT

CONFERENCE THEME

The theme of MFPT 2014 is:

"Technology Solutions for Affordable Sustainment"

to highlight how failure prevention, condition monitoring and PHM technologies can make a significant impact in reducing through-life costs and greatly enhance the viability of an enterprise.

CONFERENCE ATTIRE

Business casual, or business, is acceptable attire for the Conference.

TECHNICAL PROGRAM

Pages 7 to 10 show the three parallel tracks for the Conference/Symposium after the opening plenary session (shown on Page 6). Please see the diagram below for the rooms assigned to each track:

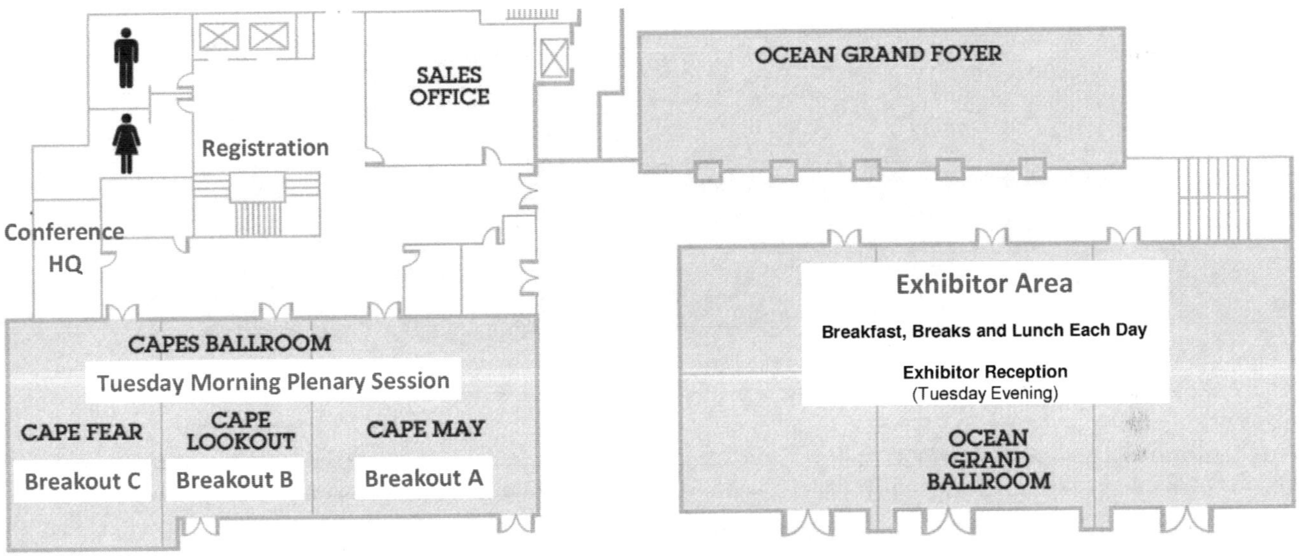

REGISTRATION AND INFORMATION DESK

The information and message center will be located in the registration area where telephone messages left via Hotel Reception on +1 (757) 425-9000 and special notices will be posted. Callers leaving messages should ask for the MFPT 2014 Registration Desk.

Dates and Times Registration Desk is Open

Monday	19 May	6:00 pm – 9:00 pm
Tuesday	20 May	7:30 am – 5:00 pm
Wednesday	21 May	7:30 am – 5:30 pm
Thursday	22 May	7:30 am – 2:00 pm

MFPT

REFRESHMENTS, SOCIAL AND NETWORKING ACTIVITIES

Hot breakfast (available from 7:00 am), lunch, and mid-session breaks will be available on Tuesday, Wednesday and Thursday. Refreshments will be situated in the Exhibitor Area (in the Ocean Grand Ballroom). And don't forget the *Exhibitor Reception on Tuesday evening*, and the *Wednesday evening reception sponsored by The GBS Group*, where there will be a wide range of hors d'oeuvres and a cash bar – see below – both excellent opportunities for networking with your failure prevention colleagues!

Exhibitors' Reception

5:00-8:00 pm
Tuesday, 20 May 2014

Hors d'oeuvres and a Cash Bar

In the *Exhibitor Hall*
(Ocean Grand Ballroom)
The Sheraton Oceanfront Hotel

Reception Sponsored by The GBS Group

5:30-7:00 pm
Wednesday, 21 May 2014

Hors d'oeuvres and a Cash Bar

In the *Cabanas Seaside Bar and Grille*
The Sheraton Oceanfront Hotel

THE GBS GROUP

EXHIBITOR AREA

The Exhibitor Area will be open as follows:

Exhibitor Setup:	Tuesday, 20 May 2014	3:00-5:00 pm
Principal Exhibit Hours:	Tuesday, 20 May 2014	5:00-8:00 pm[1] Exhibitors' Reception
Principal Exhibit Hours:	Wednesday, 21 May 2014	7:30-8:30 am[1] Breakfast 10:00-10:30 am[1] Morning Break Noon-2:00 pm[1] Lunch 3:30-4:00 pm[1] Afternoon Break
Principal Exhibit Hours:	Thursday, 22 May 2014	7:30-8:30 am[1] Breakfast 10:00-10:30 am[1] Morning Break Noon-1:30 pm[1] Lunch
Exhibitor Tear Down:	Thursday, 22 May 2014	From 1:30 pm

[1] These are the principal exhibitor traffic times; however, the Exhibitor Area will be open throughout the day should exhibitors wish to man their booths or meet with attendees at other times.

MFPT

Tuesday May 20 - Morning

From 7 AM — **Breakfast** (Ocean Grand Ballroom) - sponsored by PCB PIEZOTRONICS, A PCB GROUP COMPANY

Opening Plenary Session - Capes Ballroom

8:30 AM

Welcome	**John Lucero** - Chairman, MFPT Society Board of Directors
Welcome	**Dave Corelli** - President, The Vibration Institute
Introduction to MFPT	**Chris Pomfret** - Executive Director, MFPT Society
Introduction to BINDT	**Cameron Sinclair** - CEO, British Institute of Non-Destructive Testing
Introduction of MFPT Board of Directors and Focus Group Chairs	**Chris Pomfret** and **BoD Members**

9:15 AM

First Keynote — **John S. Thackrah, Executive Director, US Navy Military Sealift Command**, ***"MSC - Sustaining a Global Fleet"***

10:00 AM — **Break** (Ocean Grand Ballroom)

10:30 AM

Second Keynote — **Jamie Smith**, Director, Embedded Systems Product Management and Marketing at National Instruments: ***"Big Analog Data Solutions are Keeping Machines Running"***

11:15 AM

Third Keynote — **Professor Mike Lipsett**, University of Alberta: ***"Extending Machinery Failure Prevention Technologies to Enable Improved Environmental Performance"***

12:00 PM — **Presentation of Jack Frarey Award** and **MFPT Fellowship**

Introduced by Awards Chair - **Mike Roemer, PhD**
Presented by MFPT Chairman - **John Lucero**

12:15 PM — **Lunch** (Ocean Grand Ballroom)

Special Presentation by **Brian Weiss, PhD**, NIST
"Prognostics and Health Management for Smart Manufacturing Systems – PHM4SMS"

Tuesday May 20 - Afternoon

Time	Session A1 - Cape May	Time	Session B1 - Cape Lookout	Time	Session C1 - Cape Fear
	Paper Session **Signal Analysis 1**		Paper Session **Diagnostics 1**		Paper Session **Sensors**
	Chair - **Suri Ganeriwala**, SpectraQuest Inc		Chair - **Simon Jessop**, Impact Technologies, A Sikorsky Innovations Company		CoChairs - **Harvey Niska**, Honeywell Aerospace Co and **Chris Nemarich**, US Navy Military Sealift Command HQ
1:30 PM	*Data Reduction Approach – Acoustic Emission (AE) Based Defect Analysis: Development of Burst Detector Algorithms,* **Domenic Boos**, RWTH Aachen	1:30 PM	*Implementing an Efficient SHM System into a Steel Model Bridge,* **Mohammad Azarbayejani**, University of Texas-Pan American, Civil Engineering Department	1:30 PM	*Mechanical Waveguide Vibration and Temperature Sensors for Monitoring Gas Turbine Engines,* **Chris Larsen**, Etegent Technologies; presented by **Richard Roth**
2:00 PM	*A Computational and Accuracy Based Evaluation of Automated Algorithms for Mechanical System Health Assessment and Fault Classification,* **Behrad Bagheri**, Center for Intelligent Maintenance Systems (IMS), University of Cincinnati	2:00 PM	*A New Acoustic Emission Sensor Based Bearing Fault Diagnostic Technique,* **Brandon Van Hecke**, The University of Illinois at Chicago	2:00 PM	*Smart Sensors for Technical Health Assessment of Subsea Oil and Gas Equipment,* **Pratichi Vaidya**, NTNU Norway
2:30 PM	*Order Tracking under Run-up and Run-down Conditions,* **Bob Randall**, University of New South Wales	2:30 PM	*Calibration of Industrial Vibration Protection Systems,* **Mike Scott**, The Modal Shop	2:30 PM	*Measurement Systems and Sensors for Gas Turbine Engine Development,* **Harvey Niska**, Honeywell Aerospace Co
3:00 PM	**Break** (Ocean Grand Foyer) - sponsored by SAE International				
	Paper Session **Strategic Perspectives and Business Case Analysis**		Paper Session **Diagnostics 1 (Continued)**		Tutorial
	Chair - **Tom Mooney**, SOAR Engineering LLC				**Professor Bob Randall**, University of New South Wales
3:30 PM	*Identifying the True Cost of No Fault Found (NFF) within the United Kingdom Ministry of Defence (MOD),* **Chris Hockley**, Cranfield University, EPSRC Centre for Through-Life Engineering Services	3:30 PM	*Application of Information Theory's Channel Capacity as an Industry Machine Health and Diagnostic Metric,* **Ted Costuros**, Applied Materials, Inc.; presented by **Mike Bryant**	3:30 PM	**Phase Demodulation Based Computed Order Tracking** *- for a full description see Page 11*
4:00 PM	*Integrated Health and Contingency Management for Flight Critical Systems,* **Mike Roemer**, Impact Technologies, A Sikorsky Innovations Company	4:00 PM	*Machinery Fault Diagnosis Using Polar Wavelet Analysis,* **Lim Meng Hee**, Universiti Teknologi Malaysia		
4:30 PM	*Some Reflections on Machinery Vibrations during Procurement and Acceptance,* **Ron Eshleman**, Vibration Institute	4:30 PM	*Statistical Analysis of the Vibration of Mistuned Centrifugal Compressor Impellers with Crack Damage,* **Shuai Wang**, State Key Laboratory for Manufacturing System Engineering		
5:00 PM - 8:00 PM	**Exhibitors' Reception** (Ocean Grand Ballroom)				

MFPT

Wednesday May 21 - Morning

From 7 AM — **Breakfast** (Ocean Grand Ballroom) - sponsored by Impact Technologies, A Sikorsky Innovations Company

	Session A2 - Cape May		Session B2 - Cape Lookout		Session C2 - Cape Fear
	Paper Session **Prognostics 1**		Paper Session **Data Management and Analysis**		Tutorial
	Chair - **Brian Weiss**, National Institute of Standards and Technology		Chair - **Preston Johnson**, National Instruments		**Professor Michael Lipsett**, University of Alberta
8:30 AM	*Intelligent Maintenance Systems: Vehicle Health and Usage Tracking,* **John Lacontora**, Drexel University	8:30 AM	*Big Analog Data™ Lessons Learned in Fleet Wide Asset Monitoring of Gas Turbines and Supporting Equipment in Power Generation Applications,* **Preston Johnson**, National Instruments	8:30 AM	**Using Reliability Analysis and Maintenance Process Simulation for Improved Operational Decision Making** *- for a full description see Page 12*
9:00 AM	*Effective Accuracy Control for CBM Applications,* **Thomas Lagö**, QirraSound Technologies Europe AB	9:00 AM	*The Merger of Field Sensory Data (Big Analog Data™) with Control and Semi-Structured and Un-Structured Data Breeds New Opportunities for Data Management including Hadoop™ and Application Specific File Formats,* **Tik Hei Sze**, Dell		
9:30 AM	*An Efficient Prognostic Estimator,* **Jae Yoon**, University of Illinois-Chicago	9:30 AM	Panel Discussion: *The Six "V's" of Big Data Volume, Velocity, Variety, Value, Veracity, Visibility*		

10:00 AM — **Break** (Ocean Grand Ballroom)

	Paper Session **Prognostics 1 (Continued)**		Paper Session **Systems Engineering**		Paper Session **Failure Prevention for Materials and Structures**
			CoChairs - **John Lucero**, NASA John H Glenn Research Center and **Michael Lipsett**, University of Alberta		Chair - **Tom Considine**, US Army Research Laboratory
10:30 AM	*New E-learning Systems for CBM Applications,* **Thomas Lagö**, QirraSound Technologies Europe AB	10:30 AM	*Design of a Modular, Cost-Effective Embedded Wireless Condition Monitoring System for Asset Management of Distributed Fleets of Light Commercial Vehicles,* **Michael Lipsett**, University of Alberta	10:30 AM	*Mitigating Technical Risk when Applying Novel Materials & Processes,* **Bob Ware**, US Air Force Research Laboratory
11:00 AM	*Significance of Uncertainty Quantification in Prognostics and Health Monitoring,* **Shankar Sankararaman**, SGT Inc., NASA Ames Research Center	11:00 AM	*Practical Heat Pumps in Cold Climates – An Enabling Technology,* **Tom Walter**, Mechanical Solutions; presented by **Juan Gamarra**	11:00 AM	*Bending Fatigue Life Comparison between DMLS and Cold-Rolled Nickel Alloy 718,* **Onome Scott-Emuakpor**, US Air Force Research Laboratory
11:30 AM	*PHM Development and Implementation Challenges for the Off-Road Engine Market,* **Mark Friedrich**, John Deere	11:30 AM	*Formal Validation and Verification of Flight Critical System Software using Modeling and Simulation,* **Mike Roemer**, Impact Technologies, A Sikorsky Innovations Company	11:30 AM	*Cursory Development of an NDI Method for Internal Coatings on Army Gun Barrels,* **Reinhold Ludwig**, Worcester Polytechnic Institute; presented by **Dan Kaplowitz**, US Army Research Laboratory

12:00 PM — **Lunch** (Ocean Grand Ballroom)
Lunchtime discussion - **12:30-1:45 pm**
"End Users' Perspectives"

Wednesday May 21 - Afternoon

	Session A3 - Cape May		**Session B3** - Cape Lookout		**Session C3** - Cape Fear
	Paper Session **Prognostics 2**		Paper Session **Health Management Tools and Capabilities 1**		Paper Session **Failure Analysis**
	Chair - **Mike Roemer**, Impact Technologies, A Sikorsky Innovations Company		Chair - **Paula Dempsey**, NASA Glenn Research Center		CoChairs - **Bob Ware**, US Air Force Research Laboratory and **Bruce Wallin**, Hamilton Sundstrand
2:00 PM	*Integrated Prognostics and Control Provides Enhanced Asset Utilization,* **Fred Discenzo**, Rockwell Automation, Advanced Technology	2:00 PM	*Data Fusion Tool for Spiral Bevel Gear Condition Indicator Data,* **Paula Dempsey**, NASA Glenn Research Center	2:00 PM	*Characterization of Tape Adhesion to Chemical Agent Resistant Coatings,* **Tom Considine**, US Army Research Laboratory
2:30 PM	*Health Monitoring and Prognostic Assessment in a Fleet Context,* **Alexandre Voisin**, Predict; presented by **Jean-Baptiste Léger**	2:30 PM	*Non-intrusive Load Monitoring of Electric Motor Driven Machine Equipment,* **Jake Siegel**, Luna Inc.	2:30 PM	*A Measurement Validation Algorithm Used in Test Cell,* **Aurélie Gouby**, Snecma
3:00 PM	*Investigations on Spindle Bearings Health Prognostics Using a Data Mining Approach,* **Divya Sardana**, University of Cincinnati; presented by **Raj Bhatnagar**	3:00 PM	*Fluid-Based Portable Machinery Assessment Tools,* **Pat Henning**, Spectro Inc, QNA	3:00 PM	*Failure Analysis of Fan Mill in Thermal Power Plant,* **Chao Liu**, Department of Thermal Engineering, Tsinghua University

3:30 PM — **Break** (Ocean Grand Ballroom) - sponsored by GASTOPS

	Paper Session **Prognostics 2 (Continued)**		Paper Session **Health Management Tools and Capabilities 1 (Continued)**		Tutorial
					Dragan Djurdjanovic PhD, University of Texas
4:00 PM	*A Generic Bayesian Approach Using Laplace Approximation for Model-Based Failure Prognosis,* **Liang Xu**, SpectraQuest, Inc.	4:00 PM	*Dynamic Condition Based Feature Extraction Strategy for Machine Health Monitoring Applications,* **Ann Kao**, University of Cincinnati	4:00 PM	**Human Performance Monitoring:** **A System-Based Approach to Monitoring the Neuromusculoskeletal System** *- for a full description see Page 13*
4:30 PM	*Wireless HART: An Enabling Technology for Essential Asset Monitoring,* **Steven Moore**, Emerson Process Management	4:30 PM	*Investigation and Evaluation of Condition Indicators, Variable Selection, and Health Indication Methods and Algorithms for Rotorcraft Gear Components,* **David Siegel**, University of Cincinnati - Center for Intelligent Maintenance Systems		
		5:00 PM	*Rotordynamic Model Development of a Spiral Bevel Gear Test Rig Drive Train,* **Mike Hurrell**, Sierra Lobo, Inc.		

5:30 - 7:00 PM — **Reception** (Cabanas Seaside Bar and Grille) - sponsored by THE GBS GROUP

MFPT

Thursday May 22 - Morning

From 7 AM — **Breakfast** (Ocean Grand Ballroom)

	Session A4 - Cape May		Session B4 - Cape Lookout		Session C4 - Cape Fear
	Paper Session **Condition Based Maintenance**		Paper Session **Diagnostics 2**		Paper Session **Signal Analysis 2**
	Chair - **Chris Sautter**, University of Alabama in Huntsville		Chair - **Bill Marscher**, Mechanical Solutions Inc		Chair - **Chris Nemarich**, US Navy Military Sealift Command HQ
8:30 AM	*Actionable Maintenance Information from HUMS,* **Rodrigo Teixeira**, University of Alabama in Huntsville; presented by **Chris Sautter**	8:30 AM	*A Study on Comparing Acoustic Emission and Vibration Sensors for Gearbox Fault Diagnostics,* **Yongzhi Qu**, University of Illinois at Chicago	8:30 AM	*A Comparison of Different Techniques for Rolling Element Bearing Fault Diagnosis,* **Suri Ganeriwala**, SpectraQuest Inc
9:00 AM	*Comparison of state of the art SCADA and High Sampled Data Condition Monitoring for Wind Turbines,* **Marc Hilbert**, RWTH Aachen	9:00 AM	*Fault Detection of Rotating Machinery Using a Hybrid Hidden Markov Model,* **Ayan Sadhu**, University of Waterloo	9:00 AM	*A Comprehensive Study of Shaft/Coupling Misalignment Signature using Vibration Analysis,* **Mate Toth**, SpectraQuest, Inc.
9:30 AM	*Wind Turbine Lubrication Oil Condition Monitoring, the Impact of Water Contamination and Oxidation,* **Junda Zhu**, Renewable NRG Systems	9:30 AM	*Performing Diagnostics on the Spallation Neutron Source Vision Beam Line to Eliminate High Vibration Levels and Provide Sustainable Operation,* **Blake Van Hoy**, Oak Ridge National Laboratory	9:30 AM	*Differential gearbox diagnostics using Adaptive Wavelet Decomposition,* **Budhaditya Hazra**, SpectraQuest, Inc

10:00 AM — **Break** (Ocean Grand Ballroom)

	Paper Session **Health Management Tools and Capabilities 2**		Paper Session **Diagnostics 2 (Continued)**		Tutorial
	Chair - **Andy Hess**, The Hess PHM Group, Inc				**Don Thompson**, The GBS Group
10:30 AM	*Failure Cause Determination for a Field-Deployed Solenoid Valve,* **Jordan Jameson**, CALCE University of Maryland	10:30 AM	*Intrinsic Method for Monitoring Industrial Applications for Virtual Diagnostics,* **Markus Schütz**, RWTH Aachen; presented by **Marc Hilbert**	10:30 AM	**Use of Strain Gauge Technology for Main Propulsion Shafting Alignment** - *for a full description see Page 13*
11:00 AM	*Using a Systems Engineering Method to Provide a Particle Imaging Velocimetry Capability for a Supersonic Wind Tunnel,* **John Lucero**, NASA John H Glenn Research Center	11:00 AM	*Experimental Investigation and Analysis of Some Common Faults for Wind Turbine,* **Wenguang Yang**, Tsinghua University		
		11:30 AM	*Adaptive Tunable Q-factor Wavelet Transform for Fault Feature Extraction of Gearbox Based on Vibration Signals,* **Wangpeng He**, State Key Laboratory for Manufacturing Systems		

12:00 PM — **Lunch** (Ocean Grand Ballroom)

Presentation of **Best Paper Awards**

Introduced by Awards Chair - **Mike Roemer, PhD,** Presented by MFPT Chairman - **John Lucero**

followed by Exhibitor Prize Draw - Must be Present to Win!

From 1:30 pm — **Conference Close**

PAPER AND TUTORIAL ABSTRACTS

Full abstracts for the MFPT Conference papers (in session order) can be found on the MFPT website at www.mfpt.org. The Tutorial abstracts are also on the website but are reproduced below.

TUTORIALS

Session C1 (Tuesday, 3:30 PM) – Cape Lookout

Tutorial
Phase Demodulation Based Computed Order Tracking

Professor Bob Randall,
University of New South Wales

Order tracking has made possible the analysis of vibration signals from variable speed machines as though they were exactly constant speed, for example establishing that certain frequency components are exact harmonics of a particular shaft speed. The amount of speed variation may be quite small, for example that given by an induction motor or generator over its normal load range, but still having to be removed before synchronous averaging of for example gear signals can be achieved. Alternatively, many machines operate with larger speed variations, for example wind turbines for which it can be as much as ±30%, and yet another application is to the analysis of complete run-ups and/or run-downs between zero and full speed. A number of methods have been proposed, including direct sampling at a rate proportional to the machine speed, but this means that all information about time is lost. If a speed reference signal is recorded at the same time as the vibration signals, eg, a once-per-rev tacho or a shaft encoder signal, then this can be used to resample the signals to give a constant number of samples per rotation, meaning that the resulting spectra are then on a harmonic order basis, and are known as order spectra. In some situations the speed reference signal can be extracted from the response signal itself. The process of resampling in the angle domain is known as computed order tracking. A very powerful method of order tracking is to perform phase demodulation of the reference signal to obtain a map of phase vs time, which can then be used to resample from equal intervals of time to equal intervals of phase. The advantage of the phase demodulation method is that, provided the spectrum of the particular reference component to be demodulated does not overlap with any others in the frequency domain, all information is available about the continuous function of phase vs time and can be obtained by suitable interpolation. This tutorial first describes the basics of amplitude and phase demodulation using Hilbert transform methods, then how to extract the phase-time map from the reference signal. A number of restrictions are made clear, including the limitations given by the range and rate of speed variation, and the extra problems in using the response signal as a reference. These include the response time of the system itself when the speed change is too rapid, and the problems with amplitude modulation as the dominant harmonics pass through resonances. A large number of different examples are used to illustrate the different application areas and the corresponding choices of analysis parameters.

MFPT

Session C2 (Wednesday, 8:30 AM) – Cape Lookout

Tutorial

Using Reliability Analysis and Maintenance Process Simulation for Improved Operational Decision Making

Professor Michael Lipsett,
University of Alberta

Many condition monitoring systems and predictive maintenance programs fail to deliver their full value, partly because it is so difficult to assess what their value is. There are several reasons for this. New technologies and programs are added to existing technologies and organizational systems. Incremental value is thus hard to calculate, let alone measure.

This tutorial introduces simple methods for structured analysis of work flow, based primarily on concepts used in lean manufacturing, for understanding the value of failure prevention technologies in investment analysis and operational decision-making. An overview of reliability analysis concepts sets the stage for calculating the probable costs and benefits of implementing a new reliability program or changing maintenance methods. Techniques for identifying important events include failure modes and effects analysis and HAZOP.

Facilitated risk assessments allow qualitative comparisons of how risk is mitigated when a change is made; but this approach relies on expert opinion and group dynamics that can sway the assessment. Instead, a bottom-up analysis of actions that occur when a failure event occurs, coupled with value-stream mapping, can yield a model of how an organization is affected by a failure. Techniques will be described for identifying and connecting how events take place when failure occurs, with worked examples. General methods are then developed for producing discrete-event simulations from a model. The costs and benefits are often not well estimated, and so sensitivity analysis can aid the analyst in discovering what parts of the organization are most cost sensitive, and thus which costs need to be known most clearly.

Ideally, model verification and validation are done, so that the model is a good predictor of the value of a technology or process. Then, examples are developed to show how a model can be modified to compare the value of one solution over time with a different solution. The importance of appropriate financial assumptions is also discussed, as well as how organizational dynamics can be as important as technical aspects. Finally, the model is only as good as the means by how much a decision-maker wants to use it. Some simple human factors examples are presented to show good and poor ways to visualize how a system behaves.

Session C3 (Wednesday, 4:00 PM) – Cape Lookout

Tutorial
Human Performance Monitoring: A System-Based Approach to Monitoring the Neuromusculoskeletal System

Dragan Djurdjanovic PhD,
University of Texas

A system-based diagnostic methodology for monitoring the neuromusculoskeletal system is presented in this paper. The algorithm is based on autoregressive models with exogenous inputs, which link surface electromyographic signals and joint kinematic variables in order to detect changes in system dynamics as well as to assess joint level and muscle level contributions to those changes. Instantaneous energy and mean frequency of electromyographic signals were extracted from their reduced interference time frequency distributions. These features were used as inputs into the model, while angular velocities of the monitored joints formed the vector of outputs. Anomalous behavior of the whole monitored system or individual joint models was detected by analyzing changes in one-step ahead prediction errors of the corresponding models over time. Finally, analysis of the recursively updated models tracking the behavior of the neuromusculoskeletal system over time were used to characterize changes at the joint and muscular levels. The methodology is demonstrated on data recorded from 12 human subjects completing a repetitive sawing motion until voluntary exhaustion. Statistically significant decreasing trends in EMG frequencies, as well as in the similarities of the NMS models to those observed in the rested state could be observed in all subjects as each subject fatigued. In addition, decreased joint response to muscle output, as well as changes in the coordination and motion planning have been detected with all subjects considered in this study and can all be seen as indicative of fatigue.

Session C4 (Thursday, 10:30 AM) – Cape Lookout

Tutorial
Use of Strain Gauge Technology for Main Propulsion Shafting Alignment

Don Thompson,
The GBS Group

The training will cover an overview of main propulsion shafting components. We will then review shaft alignment criteria and when alignments should be performed. We will cover the typical alignment methods. These include Strain Gage Method, Jack/load Cell Method, Optical Method and other methods. We will concentrate on the strain gauge method and the benefits for complete system alignment. We will cover shafting analysis, where the following factors are addressed: Thermal Growth, Bearing wear, Vertical/Horizontal Alignment. We will provide examples and case studies of actual alignment tasks and lessons learned from these.

MFPT

EXHIBITOR PRIZE DRAW

Be sure to enter the Exhibitor Prize Draw to have a chance to win a Pebble Smart Watch, a Google Chromecast or an MFPT Polo Shirt

How to participate:

1. Visit all of the exhibitors and have them mark their company slot on your Exhibitor Prize Draw Entry Form

2. Bring the completed Entry Form to the Registration Desk before the end of the morning break on Thursday

3. Attend the draw in the Exhibitor Area on Thursday lunchtime

You must be present at the draw to win!

MFPT

AN INTRODUCTION TO MFPT

MFPT is a non-profit professional society with a 47-year legacy of promoting failure prevention technology. An interdisciplinary technical organization, MFPT is strongly oriented towards the practical application of health management across every engineering sphere. The MFPT community includes professional scientists, engineers, failure analysts, maintenance specialists and others who represent a wide variety of disciplines from government agencies, universities, research institutes and industry. For a comprehensive history and a library of excellent information go to www.mfpt.org.

MFPT THANKS ITS CORPORATE MEMBERS

Silver

SAE INTERNATIONAL

PCB PIEZOTRONICS INC. A PCB GROUP COMPANY

Impact Technologies A Sikorsky Innovations Company

GASTOPS

Bronze

MEGGITT smart engineering for extreme environments

NATIONAL INSTRUMENTS

Mechanical Solutions, Inc. Engineering Analysis, Test & Technology

DYTRAN INSTRUMENTS, INC.

SQi SpectraQuest, Inc.

MFPT THANKS ITS CONFERENCE TECHNICAL CO-SPONSORS

BINDT THE BRITISH INSTITUTE OF NON-DESTRUCTIVE TESTING

SAE INTERNATIONAL

iMS CENTER FOR INTELLIGENT MAINTENANCE SYSTEMS

phmsociety

COMADEM

MFPT

MFPT HONORS AND AWARDS

Fellows of the MFPT Society

Elected 2013 - Bill Marscher PE

Elected 2012 - Michael J Roemer PhD

Elected 2011 - Marc Pepi

Elected 2010 - The late Professor C. James Li

Elected 2010 - Ronald L. Eshleman PhD

Elected 2009 - Henry C Pusey

Elected 2009 - Christopher P Nemarich

Elected 2001 - Paul L Howard

Elected 1998 - Howard A Gaberson PhD

Elected 1997 - Henry R (Hank) Hegner

Elected 1996 - The late John L (Jack) Frarey

Jack Frarey Award for Excellence in the Field of Machinery Failure Prevention Technology

May 2013 - Ron Eshleman PhD PE

April 2012 - Vic Champagne

May 2011 - Bill Nickerson

April 2010 - John E Judd

April 2009 - Michael J Roemer PhD

May 2008 - Howard A Gaberson PhD

Henry and Sallie Pusey "Best Paper" Award - the winner of the 2014 Henry and Sallie Pusey Award will be announced at Thursday lunchtime.

The Saab Defense and Security USA Award - the winner of the 2014 Saab Defense and Security USA Best Student Paper Award will be announced at the Thursday lunchtime.

IN MEMORIAM

MFPT recognizes the following member of our community who sadly passed away in the last 12 months.

Dr Howard A 'Howie' Gaberson (9 June 2013)

Member of the MFPT Board of Directors from 'before records began', including Diagnostics & Signal Analysis Chair for well over 10 years, and inaugural winner of the Jack Frarey Memorial Award for Excellence

MFPT

MFPT BOARD OF DIRECTORS

MFPT

NOTES

MFPT 2014 CONFERENCE

Technology Solutions for Affordable Sustainment

Sheraton Virginia Beach Oceanfront Hotel
May 20–22, 2014

MFPT
The Society for Machinery Failure Prevention Technology
Pursuing prognostics and health management capabilities for military and commercial systems

MFPT 2014

20-22 May 2014

Virginia Beach, VA

"Technology Solutions for Affordable Sustainment"

MFPT 2014 CONFERENCE

Technology Solutions for Affordable Sustainment

Sheraton Virginia Beach Oceanfront Hotel
May 20–22, 2014

MFPT
The Society for Machinery Failure Prevention Technology
Pursuing prognostics and health management capabilities for military and commercial systems

Welcome from

John Lucero

Chairman

MFPT Society Board of Directors

MFPT 2014 CONFERENCE

Technology Solutions for Affordable Sustainment

Sheraton Virginia Beach Oceanfront Hotel
May 20–22, 2014

MFPT
The Society for Machinery Failure Prevention Technology

Welcome from

Dave Corelli

President

The Vibration Institute

MFPT 2014 CONFERENCE

Technology Solutions for Affordable Sustainment

Sheraton Virginia Beach Oceanfront Hotel
May 20–22, 2014

MFPT
The Society for Machinery Failure Prevention Technology

An Introduction to MFPT

Chris Pomfret

Executive Director

MFPT

MFPT

MFPT: the Beginning

What started as **Mechanical Failure Prevention Group (MFPG)**

became **Machinery Failure Prevention Technology (MFPT)**

In April 1967, 42 engineers and scientists concerned with the troublesome and costly problems of mechanical failures met at the Office of Naval Research to form the Mechanical Failures Prevention Group (MFPG). The participants, who represented several government agencies and industries, met to develop a forum to exchange information and innovative ideas on methods to avoid or predict mechanical failures in a wide variety of vehicles, equipment and structures. This meeting was the genesis of a broad interdisciplinary society that has become and serves today as a powerful resource in the field of machinery failure prevention, analysis, diagnostics and prognostics.

MFPT 2014 CONFERENCE

Technology Solutions for Affordable Sustainment

Sheraton Virginia Beach Oceanfront Hotel
May 20–22, 2014

MFPT
The Society for Machinery Failure Prevention Technology
Pursuing prognostics and health management capabilities for military and commercial systems

Conference Theme

"Technology Solutions for Affordable Sustainment"

MFPT 2014 CONFERENCE

Technology Solutions for Affordable Sustainment

Sheraton Virginia Beach Oceanfront Hotel
May 20–22, 2014

MFPT
The Society for Machinery Failure Prevention Technology
Pursuing prognostics and health management capabilities for military and commercial systems

MFPT 2014 CONFERENCE

Technology Solutions for Affordable Sustainment

Sheraton Virginia Beach Oceanfront Hotel
May 20–22, 2014

MFPT
The Society for Machinery Failure Prevention Technology
Pursuing prognostics and health management capabilities for military and commercial systems

MFPT 2014 CONFERENCE

Technology Solutions for Affordable Sustainment

Sheraton Virginia Beach Oceanfront Hotel
May 20–22, 2014

MFPT
The Society for Machinery Failure Prevention Technology
Pursuing prognostics and health management capabilities for military and commercial systems

Conference Theme

"Technology Solutions for Affordable Sustainment"

MFPT 2014 CONFERENCE

Technology Solutions for Affordable Sustainment

Sheraton Virginia Beach Oceanfront Hotel
May 20–22, 2014

MFPT
The Society for Machinery Failure Prevention Technology
Pursuing prognostics and health management capabilities for military and commercial systems

MFPT 2014 CONFERENCE

Technology Solutions for Affordable Sustainment

Sheraton Virginia Beach Oceanfront Hotel
May 20–22, 2014

MFPT
The Society for Machinery Failure Prevention Technology
Pursuing prognostics and health management capabilities for military and commercial systems

MFPT 2014 CONFERENCE

Technology Solutions for Affordable Sustainment

Sheraton Virginia Beach Oceanfront Hotel
May 20–22, 2014

MFPT
The Society for Machinery Failure Prevention Technology
Pursuing prognostics and health management capabilities for military and commercial systems

Dr Brian Weiss

National Institute of Standards and Technology (NIST)

MFPT 2014 CONFERENCE

Technology Solutions for Affordable Sustainment

Sheraton Virginia Beach Oceanfront Hotel
May 20–22, 2014

MFPT
The Society for Machinery Failure Prevention Technology

MFPT 2014 CONFERENCE

Technology Solutions for Affordable Sustainment

Sheraton Virginia Beach Oceanfront Hotel
May 20–22, 2014

MFPT
The Society for Machinery Failure Prevention Technology

2014 Best Student Paper Award

Sponsored by Saab Defense and Security USA

SAAB TECHNOLOGIES

SAAB
Defense and Security

MFPT 2014 CONFERENCE

Technology Solutions for Affordable Sustainment

Sheraton Virginia Beach Oceanfront Hotel
May 20–22, 2014

MFPT
The Society for Machinery Failure Prevention Technology

MFPT 2014 CONFERENCE

Technology Solutions for Affordable Sustainment

Sheraton Virginia Beach Oceanfront Hotel
May 20–22, 2014

MFPT
The Society for Machinery Failure Prevention Technology

MFPT 2014 CONFERENCE

Technology Solutions for Affordable Sustainment

Sheraton Virginia Beach Oceanfront Hotel
May 20–22, 2014

MFPT
The Society for Machinery Failure Prevention Technology

DILBERT

MFPT Essential Needs

- Keep sessions running on time – *and please **<u>do not</u>** switch or change the schedule for any reason!*
- Some Program changes unavoidable – see list of changes at Registration
- Session chairs and presenters please meet at breakfast on the morning of your session (special tables will be marked) to:
  - We can check presentations are loaded on the computer and hand bio sheets to the Session CoChairs
- Please fill in your questionnaires – we *really* want to know what you think…!

MFPT

Exhibitors

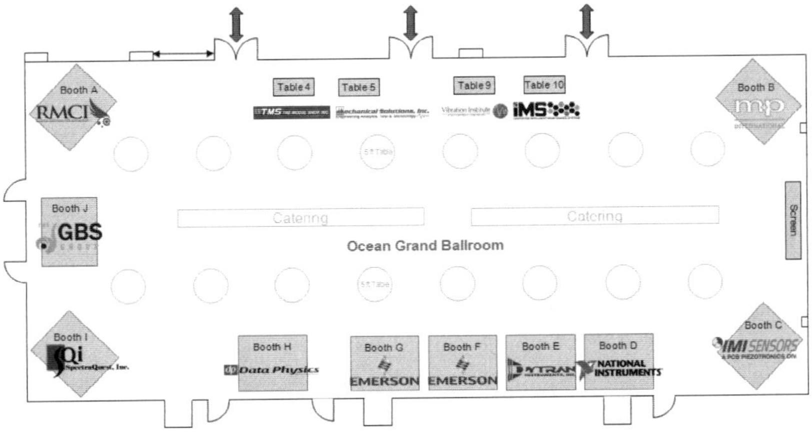

MFPT 2014 CONFERENCE

Technology Solutions for Affordable Sustainment

Sheraton Virginia Beach Oceanfront Hotel
May 20–22, 2014

MFPT
The Society for Machinery Failure Prevention Technology

Wednesday Evening Reception Sponsored by

THE GBS GROUP

MFPT 2014 CONFERENCE

Technology Solutions for Affordable Sustainment

Sheraton Virginia Beach Oceanfront Hotel
May 20–22, 2014

MFPT
The Society for Machinery Failure Prevention Technology
Pursuing prognostics and health management capabilities for military and commercial systems

Australia
Canada
China
France
Germany
Malaysia
Norway
China
Sweden
UK
USA

MFPT 2014 CONFERENCE

Technology Solutions for Affordable Sustainment

Sheraton Virginia Beach Oceanfront Hotel
May 20–22, 2014

MFPT
The Society for Machinery Failure Prevention Technology
Pursuing prognostics and health management capabilities for military and commercial systems

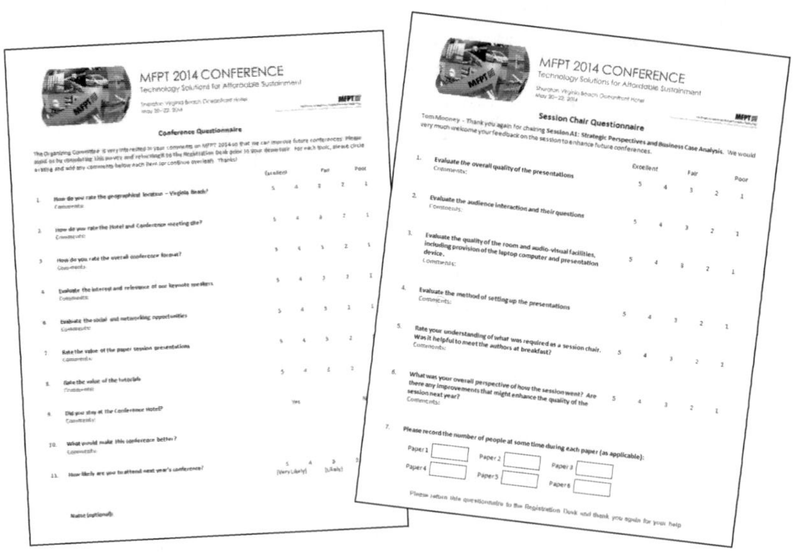

MFPT

STOP PRESS!

This information and updates posted at Registration Desk

Tuesday 20 May - Afternoon		
Session C1 - **Sensors**	**2:00 PM**	*Smart Sensors for Technical Health Assessment of Subsea Oil and Gas Equipment,* **Pratichi Vaidya**, NTNU Norway – **presentation cancelled (paper still in Proceedings)**

MFPT 2014 CONFERENCE
Technology Solutions for Affordable Sustainment

Sheraton Virginia Beach Oceanfront Hotel
May 20–22, 2014

MFPT
The Society for Machinery Failure Prevention Technology

Cameron Sinclair

CEO
British Institute of Non-Destructive Testing

MFPT 2014 CONFERENCE
Technology Solutions for Affordable Sustainment

Sheraton Virginia Beach Oceanfront Hotel
May 20–22, 2014

MFPT
The Society for Machinery Failure Prevention Technology

An Introduction to BINDT

Cameron Sinclair

CEO

British Institute of Non-Destructive Testing

MFPT 2014 CONFERENCE

Technology Solutions for Affordable Sustainment

Sheraton Virginia Beach Oceanfront Hotel
May 20–22, 2014

MFPT
The Society for Machinery Failure Prevention Technology
Pursuing prognostics and health management capabilities for military and commercial systems

MFPT 2014 CONFERENCE

Technology Solutions for Affordable Sustainment

Sheraton Virginia Beach Oceanfront Hotel
May 20–22, 2014

MFPT
The Society for Machinery Failure Prevention Technology
Pursuing prognostics and health management capabilities for military and commercial systems

MFPT 2014

20-22 May 2014

Virginia Beach, VA

"Technology Solutions for Affordable Sustainment"

MFPT 2014 CONFERENCE

Technology Solutions for Affordable Sustainment

Sheraton Virginia Beach Oceanfront Hotel
May 20–22, 2014

MFPT
The Society for Machinery Failure Prevention Technology

An Introduction to MFPT Board of Directors and Focus Group Chairs

Chris Pomfret and BoD Members

MFPT MFPT Board of Directors

- John Lucero **Board Chairman, Systems Engineering Focus Group**, NASA/John H. Glenn Research Ctr
- Chris Nemarich **Vice Chairman, Sensors Technology Focus Group**, US Navy Military Sealift Command
- Fred Discenzo **Immediate Past Chairman**, Rockwell Automation
- Dave Corelli **President, Vibration Institute** IMI Sensors
- Mike Roemer **Awards Chairman**, Impact Technologies, A Sikorsky Innovations Company
- Preston Johnson **Data Management Focus Group**, National Instruments
- Suri Ganeriwala **Signal Analysis Focus Group**, SpectraQuest, Inc
- Simon Jessop **Diagnostics & Prognostics Focus Group**, Impact Technologies, A Sikorsky Innovations Company
- Rachel Moss **Member-at-Large** Saab Defense and Security USA
- Bill Marscher Mechanical Solutions, Inc
- Ron Eshleman **Technical Director, Vibration Institute**
- Hank Hegner
- Rick Wade **Administrative Officer, MFPT**

MFPT 2014 CONFERENCE

Technology Solutions for Affordable Sustainment

Sheraton Virginia Beach Oceanfront Hotel
May 20–22, 2014

MFPT
The Society for Machinery Failure Prevention Technology
Pursuing prognostics and health management capabilities for military and commercial systems

MFPT 2014

20-22 May 2014

Virginia Beach, VA

"Technology Solutions for Affordable Sustainment"

MFPT 2014 CONFERENCE

Technology Solutions for Affordable Sustainment

Sheraton Virginia Beach Oceanfront Hotel
May 20–22, 2014

MFPT
The Society for Machinery Failure Prevention Technology
Pursuing prognostics and health management capabilities for military and commercial systems

John S. Thackrah

Executive Director, US Navy Military Sealift Command

MFPT 2014 CONFERENCE

Technology Solutions for Affordable Sustainment

Sheraton Virginia Beach Oceanfront Hotel
May 20–22, 2014

MFPT
The Society for Machinery Failure Prevention Technology

Keynote

John S. Thackrah

Executive Director, US Navy Military Sealift Command

"MSC - Sustaining a Global Fleet"

MFPT 2014 CONFERENCE

Technology Solutions for Affordable Sustainment

Sheraton Virginia Beach Oceanfront Hotel
May 20–22, 2014

MFPT
The Society for Machinery Failure Prevention Technology

MFPT 2014

20-22 May 2014
Virginia Beach, VA

"Technology Solutions for Affordable Sustainment"

MFPT 2014 CONFERENCE

Technology Solutions for Affordable Sustainment

Sheraton Virginia Beach Oceanfront Hotel
May 20–22, 2014

MFPT
The Society for Machinery Failure Prevention Technology

Jamie Smith

Director, Embedded Systems Product Management and Marketing at National Instruments

MFPT 2014 CONFERENCE

Technology Solutions for Affordable Sustainment

Sheraton Virginia Beach Oceanfront Hotel
May 20–22, 2014

MFPT
The Society for Machinery Failure Prevention Technology

Keynote

Jamie Smith

Director, Embedded Systems Product Management and Marketing at National Instruments

"Big Analog Data Solutions are Keeping Machines Running"

MFPT 2014 CONFERENCE
Technology Solutions for Affordable Sustainment

Sheraton Virginia Beach Oceanfront Hotel
May 20–22, 2014

MFPT
The Society for Machinery Failure Prevention Technology

MFPT 2014

20-22 May 2014

Virginia Beach, VA

"Technology Solutions for Affordable Sustainment"

MFPT 2014 CONFERENCE
Technology Solutions for Affordable Sustainment

Sheraton Virginia Beach Oceanfront Hotel
May 20–22, 2014

MFPT
The Society for Machinery Failure Prevention Technology

**Professor
Mike Lipsett**

University of Alberta

MFPT 2014 CONFERENCE

Technology Solutions for Affordable Sustainment

Sheraton Virginia Beach Oceanfront Hotel
May 20–22, 2014

MFPT
The Society for Machinery Failure Prevention Technology

Keynote

Professor Mike Lipsett

University of Alberta

"Extending Machinery Failure Prevention Technologies to Enable Improved Environmental Performance"

MFPT 2014 CONFERENCE

Technology Solutions for Affordable Sustainment

Sheraton Virginia Beach Oceanfront Hotel
May 20–22, 2014

MFPT
The Society for Machinery Failure Prevention Technology

MFPT 2014

20-22 May 2014

Virginia Beach, VA

"Technology Solutions for Affordable Sustainment"

MFPT 2014 CONFERENCE

Technology Solutions for Affordable Sustainment

Sheraton Virginia Beach Oceanfront Hotel
May 20–22, 2014

MFPT
The Society for Machinery Failure Prevention Technology

Jack Frarey Memorial Award for Excellence

Professor Jay Lee

in recognition of outstanding contributions to the field of health management for manufacturing

MFPT 2014 CONFERENCE

Technology Solutions for Affordable Sustainment

Sheraton Virginia Beach Oceanfront Hotel
May 20–22, 2014

MFPT
The Society for Machinery Failure Prevention Technology

MFPT Fellowship 2014

Fred Discenzo, PhD

Rockwell Automation

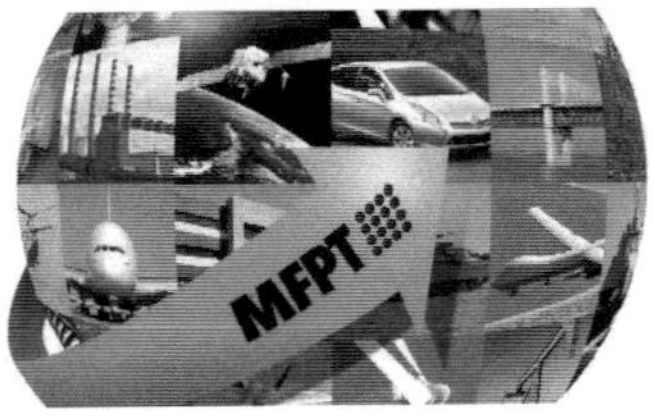

MFPT 2014 CONFERENCE

Technology Solutions for Affordable Sustainment

Sheraton Virginia Beach Oceanfront Hotel
May 20–22, 2014

MFPT

The Society for Machinery Failure Prevention Technology

MFPT 2014

20-22 May 2014

Virginia Beach, VA

"Technology Solutions for Affordable Sustainment"

MFPT 2014 CONFERENCE

Technology Solutions for Affordable Sustainment

Sheraton Virginia Beach Oceanfront Hotel
May 20–22, 2014

MFPT

The Society for Machinery Failure Prevention Technology

MFPT 2014

20-22 May 2014

Virginia Beach, VA

"Technology Solutions for Affordable Sustainment"

MFPT 2014; LUNCHTIME DISCUSSION

"What's really important to end-users of Condition Monitoring"

Several attendees from the end-user world started by providing their perspectives; this generated many follow-on comments from other conference attendees. A summary of the remarks (all non-attributable) follows:

- Create tools that are simple to use; production people love end-to-end simplicity and user friendliness
    - End-users need to be able to look at a screen and figure out within 8 seconds what's going on
        - And in no more than 3 clicks of the mouse
    - Always remember the difference in skill sets and interests between the person that created the tool and the person that uses it in an operational world
- Need to address/pursue Workforce Optimization aspects
    - i.e. tools and capabilities that are less human-dependent
    - Automating Diagnostics and Prognostics
    - Take advantage of the communication revolution
- Embrace the convergence of technologies to create flexible systems and solutions that learn, adapt to meet changing usage and that can communicate
    - But, consider/identify where the weakest link is
        - The airline industry loses $8B/year in delays that affect 50,000 flights
            - And put customer inconvenience on top of that
- Some end users are still in their infancy with embracing failure prevention capabilities, be it what they use to manufacture a product and/or what they deploy on the end product that customers then operate/use
    - Customers change their needs/wants and so have to be adaptable
- The traditional insurance aspects need to be more closely aligned with the technologies as CM or failure prevention tools are a type of insurance policy. Insurance could actually be a barrier to implementing CM technologies
    - Maybe there should be more attractive credits from traditional insurers if an asset is fitted with failure prevention capabilities
        - It's a cradle-to-grave/life-cycle perspective
- Need to get a greater realization that maintenance is part of the cost of doing business, not a cost center that needs to be minimized without due consideration of the effects
    - Try to dispel the mindset of accountants that maintenance is a cost center
- Sharing of data is a thorny issue that leads to "lose-lose" instead of "win-win" between OEM and user

- Sharing of *information* should be the line to take
    - End users can pass valuable operating data to OEMs
    - OEMs can turn data into information to the betterment of all end-users, without any of the original data being attributable to specific users
- 90% of industrial practices for collecting data are archaic; need to get the workforce to transfer to newer designed equipment
    - In oil refineries, people are walking 22 miles/month (210 hrs a week) collecting data on foot. Data then passed to an SME. But who is the SME? Someone with an ear for a problem (the past) who does the diagnosis on the hoof or a machine/electronic system (the current)
    - If a machine is to be the new SME, still need a human who knows how to interpret what the machine tells them and doesn't necessarily believe everything it says
    - But newer data collection methods need to be backward compatible

- Decision makers live and die by metrics; be honest and put the data into numbers that the decision makers can relate to
    - E.g. consider a quad redundant system on an airplane with 1 in 10^6 chance of failure but finding occurrences of 3 "failures" (i.e. 1 in 10^{18}) were supposedly occurring!
- End users will help themselves by giving feedback on how to do diagnostics
    - Sensing is generally a wired system
    - Diagnostics is wireless
- Watch the dangers from the perception that smart systems mean that lower skilled people are needed for ops and maintenance
    - A balance is needed
- Also watch the dangers of overselling; it happened in the NDE community and put the capabilities back 10 years. But, conversely, we can't wait until everything is, for sure, perfect otherwise we'll never introduce anything

BOTTOM LINE(S):

- Need to communicate the value of CM/HM to stakeholders, especially skeptics
    - Great technologists are often poor sellers/advocates for their creations
    - Convey enthusiasm and conviction to a wider group of stakeholders
    - Keep the explanations simple; high technical content isn't always wanted!
- We need to show that the state of the art HM tools are a significant improvement on the current (more human dependent) capabilities, not that they are forever perfect
- **Make the systems end-user friendly; intuitive, quick, synergistic with the operator**

MFPT

STOP PRESS!

This information and updates posted at Registration Desk

Wednesday 21 May - Afternoon		
Session A3 - **Prognostics 2**	**2:30 PM**	*Health Monitoring and Prognostic Assessment in a Fleet Context*, **Alexandre Voisin**, Predict; presented by **Jean-Baptiste Léger** – **presentation cancelled (paper still in Proceedings)** **Replaced by** *Intrinsic Method for Monitoring Industrial Applications for Virtual Diagnostics*, **Markus Schütz**, RWTH Aachen; presented by **Marc Hilbert** (from Session B4, Thursday 10:30 AM)

MFPT 2014 CONFERENCE

Technology Solutions for Affordable Sustainment

Sheraton Virginia Beach Oceanfront Hotel
May 20–22, 2014

MFPT
The Society for Machinery Failure Prevention Technology

Wednesday Evening 5:30-7:00 pm Reception Sponsored by

THE GBS GROUP

Take elevator down **two** floors to the **Basement** (**B**) and follow the signs

MFPT 2014 CONFERENCE

Technology Solutions for Affordable Sustainment

Sheraton Virginia Beach Oceanfront Hotel
May 20–22, 2014

MFPT
The Society for Machinery Failure Prevention Technology

MFPT 2014

20-22 May 2014

Virginia Beach, VA

"Technology Solutions for Affordable Sustainment"

MFPT 2014 CONFERENCE

Technology Solutions for Affordable Sustainment

Sheraton Virginia Beach Oceanfront Hotel
May 20–22, 2014

MFPT
The Society for Machinery Failure Prevention Technology

Lunchtime Discussion

"End Users' Perspectives"

led by **Chris Pomfret**, Executive Director, MFPT

MFPT

"End Users' Perspectives"

Led by **Chris Pomfret**, Executive Director, MFPT

Introductory Speakers

Tom Mooney – SOAR Engineering LLC

Tony Barlow – Chevron Products Company

Mark Friedrich – John Deere

Steven Tibbetts – The GBS Group

Preston Johnson – National Instruments

MFPT 2014 CONFERENCE

Technology Solutions for Affordable Sustainment

Sheraton Virginia Beach Oceanfront Hotel

May 20–22, 2014

MFPT

The Society for Machinery Failure Prevention Technology

Pursuing prognostics and health management capabilities for military and commercial systems

MFPT 2014

20-22 May 2014

Virginia Beach, VA

"Technology Solutions for Affordable Sustainment"

MFPT 2014 CONFERENCE

Technology Solutions for Affordable Sustainment

Sheraton Virginia Beach Oceanfront Hotel
May 20–22, 2014

MFPT
The Society for Machinery Failure Prevention Technology
Pursuing prognostics and health management capabilities for military and commercial systems

MFPT 2014

20-22 May 2014

Virginia Beach, VA

"Technology Solutions for Affordable Sustainment"

MFPT 2014 CONFERENCE

Technology Solutions for Affordable Sustainment

Sheraton Virginia Beach Oceanfront Hotel
May 20–22, 2014

MFPT
The Society for Machinery Failure Prevention Technology
Pursuing prognostics and health management capabilities for military and commercial systems

Please fill out and hand in all your Questionnaires – Conference, Session, Tutor, Exhibitor: We *really* want to know!

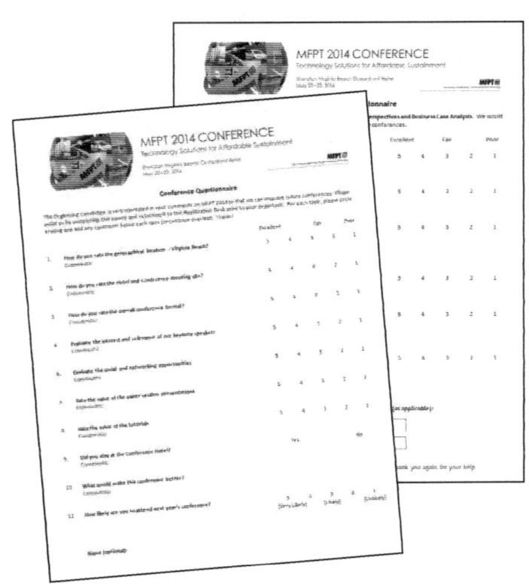

MFPT 2014 CONFERENCE

Technology Solutions for Affordable Sustainment

Sheraton Virginia Beach Oceanfront Hotel
May 20–22, 2014

MFPT
The Society for Machinery Failure Prevention Technology
Pursuing prognostics and health management capabilities for military and commercial systems

In the interests of economy and recycling...

Please drop off your name badge before you leave!

MFPT 2014 CONFERENCE

Technology Solutions for Affordable Sustainment

Sheraton Virginia Beach Oceanfront Hotel
May 20–22, 2014

MFPT
The Society for Machinery Failure Prevention Technology
Pursuing prognostics and health management capabilities for military and commercial systems

- I liked MFPT 2014 *because*…
- I would come to MFPT again *because*…

MFPT 2014 CONFERENCE

Technology Solutions for Affordable Sustainment

Sheraton Virginia Beach Oceanfront Hotel
May 20–22, 2014

MFPT
The Society for Machinery Failure Prevention Technology
Pursuing prognostics and health management capabilities for military and commercial systems

MFPT Honors and Awards

2014 Saab Defense and Security USA, Best Student Paper Award

Henry and Sallie Pusey, Best Paper Award

MFPT 2014 CONFERENCE

Technology Solutions for Affordable Sustainment

Sheraton Virginia Beach Oceanfront Hotel
May 20–22, 2014

MFPT
The Society for Machinery Failure Prevention Technology
Pursuing prognostics and health management capabilities for military and commercial systems

MFPT Honors and Awards

2014 Saab Defense and Security USA, Best Student Paper Award

Henry and Sallie Pusey, Best Paper Award

MFPT 2014 CONFERENCE

Technology Solutions for Affordable Sustainment

Sheraton Virginia Beach Oceanfront Hotel
May 20–22, 2014

MFPT
The Society for Machinery Failure Prevention Technology

2014 Saab Defense and Security USA, Best Student Paper Award

Yongzhi Qu

"A Study on Comparing Acoustic Emission and Vibration Sensors for Gearbox Fault Diagnostics"

MFPT 2014 CONFERENCE

Technology Solutions for Affordable Sustainment

Sheraton Virginia Beach Oceanfront Hotel
May 20–22, 2014

MFPT
The Society for Machinery Failure Prevention Technology

Henry and Sallie Pusey, Best Paper Award

David Siegel and **Jay Lee,** Center for Intelligent Maintenance Systems, University of Cincinnati
and
Paula Dempsey, NASA Glenn Research Center

"Investigation and Evaluation of Condition Indicators, Variable Selection, and Health Indication Methods and Algorithms for Rotorcraft Gear Components"

MFPT 2014 CONFERENCE

Technology Solutions for Affordable Sustainment

Sheraton Virginia Beach Oceanfront Hotel
May 20–22, 2014

MFPT
The Society for Machinery Failure Prevention Technology

MFPT 2014

20-22 May 2014

Virginia Beach, VA

"Technology Solutions for Affordable Sustainment"

MFPT 2014 CONFERENCE

Technology Solutions for Affordable Sustainment

Sheraton Virginia Beach Oceanfront Hotel
May 20–22, 2014

MFPT
The Society for Machinery Failure Prevention Technology

EXHIBITOR PRIZE DRAW

Known As (First Name)	Last Name	Organization	Position	Address Line 1	Address Line 2	City	US State/CA Province	Zip/Postal Code	Other State/ Province	Country	Work Phone	E-mail
Ken	**Anderson**	Universal Synaptics Corp	Director of Sales & Marketing	4066 S. 1900 W. Suite B		Roy	UT	84067			(801) 731 8508	Ken.Anderson@usynaptics.com
Mohammad	**Azarbayejani**	University of Texas-Pan American, Civil Engineering Department		1201 W. University Dr., ASB 1.203		Edinburg	TX	78539			(956)-665-3103	azarbayejanim@utpa.edu
Behrad	**Bagheri**	Center for Intelligent Maintenance Systems (IMS), University of		3242 Whitfield Ave. #E3		Cincinnati	OH	45220			513 407 0696	bagherbd@mail.uc.edu
Tony	**Barlow**	Chevron Products Company	RER Trainer / Condition Monitoring Analyst	Pascagoula Refinery	250 Industrial Road, RM 1021	Pascagoula	MS	39581			228-938-4306	WBDN@chevron.com
Raj	**Bhatnagar**	University of Cincinnati		ML0030 University of Cincinnati		Cincinnati	OH	45221			(513)556-4932	Raj.Bhatnagar@uc.edu
Rob	**Bloomquist**	Data Physics										
Domenic	**Boos**	RWTH Aachen		Wuellnerstr. 12		Aachen		52064	NRW	Germany	+492418090748	dboos@imr.rwth-aachen.de
Eric	**Breen**	The GBS Group										
Brian	**Bricker**	The Timken Company		1737 Hiram St		Lousiville	OH	44641			330-471-2000	brian.bricker@timken.com
Michael	**Bryant**	University of Texas at Austin		Mechanical Engineering	1 University Station C2200	Austin	TX	78712-0292			512 288-5191	mbryant@mail.utexas.edu
Corey	**Bufi**	General Electric Power & Water		1 River Rd		Schenectady	NY	12345			518 385 8367	bufi@ge.com
Bryan	**Butch**	The Modal Shop		3149 E Kemper Rd		Cincinnati	OH	45241			513-351-9919	bbutch@modalshop.com
Randy	**Byrum**	RE Mason Company	Reliability and Wireless Specialist	201 Wylderose Drive		Midlothian	VA	23113			804-858-5807	Randy.byrum@remasonco.com
Mark	**Chaffee**	Renewable NRG Systems		110 Riggs Road		Hinesburg	VT	05461			18024822255 x509	mrc@renewablenrgsystems.com
Mukesh Kumar	**Chippa**	University of Akron		195 Wheeler street	Apt 204	Akron	OH	44304			3307805085	mc80@zips.uakron.edu
Tom	**Considine**	US Army Research Laboratory		4600 Deer Creek Loop		Aberdeen Proving Ground	MD	21005			410-306-2564	thomas.a.considine.ctr@mail.mil
David	**Corelli**	IMI Sensors	Director of Application Engineering	3425 Walden Avenue		Depew	NY	14043-0987			(716) 684-0001, also 800.959.4464	dcorelli@pcb.com
Paula	**Dempsey**	NASA Glenn Research Center		21000 Brookpark Road, MS 23-3		Cleveland	OH	44135			216 433 3398	paula.j.dempsey@nasa.gov
Fred	**Discenzo**	Rockwell Automation, Advanced Technology	Manager, R&D, Rockwell Fellow	1 Allen-Bradley Drive		Mayfield Heights	OH	44124			(440) 646-7325	fmdiscenzo@ra.rockwell.com
Dragan	**Djurdjanovic**	University of Texas		204 E. Dean Keeton Street		Austin	TX	78712			734-260-2525	dragand@me.utexas.edu
Doug	**Dubowski**	GasTOPS Ltd.		1011 Polytek Street		Ottawa		K1J9J3		Canada	613 744 3530	ddubowski@gastops.com
Hiroaki	**Endo**	Test Devices Inc	Chief Technology Officer	571 Main Street		Hudson	MA	01749-3035			978 562 4946	hendo@testdevices.com
Ron	**Eshleman**	Vibration Institute	Technical Director	6262 S Kingery Hwy	Suite 212	Willowbrook	IL	60527			(630) 654-2254, -2053	reshleman@vi-institute.org
Mark	**Friedrich**	John Deere		John Deere Engine Works	3801 W. Ridgeway Ave.	Waterloo	IA	50704			319 292 4723	friedrichmarka@johndeere.com
Juan	**Gamarra**	Mechanical Solutions Inc		11 Apollo Drive		Whippany	NJ	07981			(973) 326-9920	jdg@mechsol.com
Suri	**Ganeriwala**	SpectraQuest Inc		8227 Hermitage Road		Richmond	VA	23228-3031			(804) 261-3300	suri@spectraquest.com
Bob	**Golden**	The GBS Group										bobgolden@thegbsgroup.us
Sarah	**Golden**	The GBS Group										
Aurélie	**Gouby**	Snecma		Rond-point René Ravaud - Réau		Moissy-Cramayel		77550	Seine et Marne	France	+33 1 60 59 42 53	aurelie.gouby@snecma.fr
Christer	**Hallberg**	Saab AB				Linkoping		SE-58188		Sweden	+46734181014	christer.hallberg@saabgroup.com
Nathan	**Haywood**	The GBS Group		397 Little Neck Road	Building 3300 South, Suite 204	Virginia Beach	VA	23452			757 897 2740	nathanhaywood@thegbsgroup.us
Wangpeng	**He**	State Key Laboratory for Manufacturing Systems Engineering		Xi'an Jiaotong University	No.28, Xianning West Road	Xi'an		710049	Shaanxi	P R China	15829301428	hewangpeng2007@stu.xjtu.edu.cn
Hank	**Hegner**	Engineering Consultant		812 Lakemount Drive		Moneta	VA	24121-5776			(540) 721-2114	hrhegner@embarqmail.com
Pat	**Henning**	Spectro Scientific, Inc		One Executive Drive	Suite 101	Chelmsford	MA	01824			978 431 1118	phenning@spectrosci.com

Andy	**Hess**	The Hess PHM Group, Inc	President and Guru	2010 Dundee Road		Rockville	MD	20850			301 605 7625	andrew_hess@comcast.net
Marc	**Hilbert**	RWTH Aachen		Wuellnerstrasse 2		Aachen		52056	NRW	Germany	49 2418096878	mhilbert@imr.rwth-aachen.de
Chris	**Hockley**	Cranfield University, EPSRC Centre for Through-Life Engineering Services		Centre for Defence Acquisition, The Vincent Centre	The Defence Academy	Shrivenham		SN6 8LA	Wiltshire	UK	01793 783284	c.hockley@cranfield.ac.uk
Casey	**Holycross**	US Air Force Research Laboratory		AFRL/RQTS, 1950 Fifth St	Building 18D, Rm D122	WPAFB	OH	45433			937-656-5530	casey.holycross@us.af.mil
Ed	**Hulick**	Military Sealift Command		471 East C Street		Norfolk	VA	23511			7574430992	william.hulick@navy.mil
Mike	**Hurrell**	Sierra Lobo, Inc.		102 Pinnacle Drive		Fremont	OH	43420			216 433 9335	michael.j.hurrell@nasa.gov
Jordan	**Jameson**	Center for Advanced Life Cycle Engineering	Student	Department of Mechanical Engineering	University of Maryland	College Park	MD	20742				njordan.jameson@gmail.com
Simon	**Jessop**	Impact Technologies, A Sikorsky Innovations Company		300 Canal View Boulevard		Rochester	NY	14623			585-424-1990 ext 172	simon.jessop@impact-tek.com
Steven	**Johnson**	Data Physics		2187 Oak Grove Dr		Temperance	MI	48182			408 216 8415	Steven.johnson@dataphysics.com
Preston	**Johnson**	National Instruments		11500 North Mopac Blvd		Austin	TX	78759-3504			(512) 683-5444	preston.johnson@ni.com
Ann	**Kao**	Center for Intelligent Maintenance Systems, University of Cincinnati		51 Goodman Dr		Cincinnati	OH	45219			513 556 3412	kaohn@mail.uc.edu
Dan	**Kaplowitz**	US Army Research Laboratory		Aberdeen Proving Ground		Aberdeen	MD	21005			410 306 0751	daniel.a.kaplowitz.ctr@mail.mil
Magnus	**Kempe**	Saab AB				Linköping		SE-58188		Sweden	+46734186176	magnus.kempe@saabgroup.com
John	**Lacontora**	Drexel University	Associate Research Professor, Dept of Mech Eng & Mechanics	3141 Chestnut Street	MEM - Randell 115	Philadelphia	PA	19104-2884			(215) 895-1752	john.m.lacontora@drexel.edu
Thomas	**Lagö**	QirraSound Technologies Europe AB		Slagetorp 2		Sävsjö		57692	Småland	Sweden	(479) 571 0035	thomas.lago@qirrasound.com
Jay	**Lee**	IMS Center, University of Cincinnati	Ohio Eminent Scholar & L.W. Scott Alter Chair Professor in Advanced	PO Box 210072, 2600 Clifton Ave		Cincinnati	OH	45221-0072			(513) 556-2493	jay.lee@uc.edu
Jean-Baptiste	**Léger**	PREDICT & DIAG 21	President	19 Avenue de la Forêt de Haye		Vandœuvre Lès Nancy		54500	Meurthe et Moselle	France	00-33-677328244	jean-baptiste.leger@predict.fr
Ruoyu	**Li**	SKF USA RDC		3443 N. Sam Houston Pkwy		Houston	TX	77047			312-608-3828	ruoyu.li@skf.com
Michael	**Lipsett**	University of Alberta	Professor	5-1J Mechanical Engineering		Edmonton	Alberta	T6G 2G8		Canada	1 780 492 9494	mlipsett@ualberta.ca
Chao	**Liu**	Department of Thermal Engineering, Tsinghua University				Beijing		100084		P R China	0086 10 62794046	chaoliu13@tsinghua.edu.cn
John	**Lucero**	NASA John H Glenn Research Center		MS6-2	21000 Brookpark Road	Brook Park	OH	44135			(216) 433-2684	John.M.Lucero@nasa.gov
Adbelrhman	**Mahamadi**											
Bill	**Marscher**	Mechanical Solutions Inc		11 Apollo Drive		Whippany	NJ	07981			(973) 326-9920 Ext 111	wdm@mechsol.com
Pete	**Martinez**	The GBS Group		397 Little Neck Road	Building 3300 South, Suite 204	Virginia Beach	VA	23452			757.965.4274	petemartinez@thegbsgroup.us
Melvin	**Mathew**	Drexel University		3141 Chestnut St		Philadelphia	PA	19104			210 350 2098	mdm359@drexel.edu
Kenneth	**Maynard**	Penn State Univ/ARL		301B Applied Sciences Building		State College	PA	16804-0030			(814) 863-4077	kpm128@psu.edu
Skip	**McCusker**	Morgan Borszcz Consulting		2601 Commons Blvd	Suite 130	Beavercreek	OH	45431				skip.mccusker@mbc360.com
Lim	**Meng Hee**	Universiti Teknologi Malaysia		Institute of Noise & Vibration, Universiti Teknologi Malaysia		Kuala Lumpur		54100		Malaysia	60326154368	mhlim@ic.utm.my
Tom	**Mooney**	SOAR Engineering LLC	President & CEO	10, Schooner Ridge		Gloucester	MA	01930			(781) 258-1453	tmooney@soarengineeringllc.com
Steven	**Moore**	Emerson Process Management		835 Innovation Drive		Knoxville	TN	37932			8652530490	steven.moore@emerson.com
Rachel	**Moss**	Saab Defense and Security USA LLC	Vice President, Business Development, Electronic Warfare	2101 L Street, NW, Suite 325		Washington	DC	20037			+1 (703) 406-7238	rachel.moss@saabusa.com
Steve	**Nelson**	Emerson Process Management		2560 Cedar Ridge Lane		Charlottesville	VA	22901			202-262-0572	stephen.nelson@emerson.com
Chris	**Nemarich**	US Navy Military Sealift Command HQ	Senior Electrical Engineer	Engineering, N723	914 Charles Morris Ct SE	Washington Navy Yard	DC	20398-5540			(202) 685-5782	christopher.nemarich@navy.mil
Lang	**Nguyen**	The GBS Group	VP Business Development & Recruitment	397 Little Neck Road	Building 3300 South, Suite 204	Virginia Beach	VA	23452			757-965-4274	LangNguyen@TheGBSGroup.us
Thong	**Nguyen**	Military Sealift Command		471 East C St., bldg Sp-30		Norfolk	VA	23511			7574433949	thong.n.nguyen@navy.mil
Harvey	**Niska**	Honeywell Aerospace Co	Senior Technology Manager	111 S 34th Street MS 302-2		Phoenix	AZ	85034			602-231-7584	h.niska@honeywell.com

Thomas	**Nostrand**	Renewable NRG Systems		110 Riggs Road		Hinesburg	VT	05461			(802) 482-2255	tjn@renewablenrgsystems.com
Chris	**Pomfret**	MFPT Society	Executive Director	5100 Springfield Street	Suite 420	Dayton	OH	45431-1264			(937) 256-2285	chris@mfpt.org
Henry	**Pusey**	SAVIAC	Manager of Technical Services	1877 Rosser Lane		Winchester	VA	22601			(540) 678-8678	hcp1@comcast.net
Yongzhi	**Qu**	University of Illinois at Chicago		842 W Taylor Street	Rm. 2039 ERF, MC 251	Chicago	IL	60607			312-996-3410	yqu5@uic.edu
Bob	**Randall**	University of New South Wales	School of Mechanical and Manufacturing Engineering	School of Mechanical and Manufacturing Engineering		Sydney		2052	NSW	Australia	+61 4 2222 8214	b.randall@unsw.edu.au
Mike	**Roemer**	Impact Technologies, A Sikorsky Innovations Company	Director of Engineering	300 Canal View Blvd	3rd floor	Rochester	NY	14623			(585) 424-1990	mike.roemer@impact-tek.com
Richard	**Roth**	Etegent Technologies		5050 Section Ave, Suite 110		Cincnnati	OH	45212			513-631-0579	richard.roth@etegent.com
Jeff	**Rybak**	PCB Southeast Office		3425 Walden Avenue		Depew	NY	14043-2495			1800 381 2721	jrybak@pcb.com
Ayan	**Sadhu**	University of Waterloo		200 University Avenue West		Waterloo	ON	N2L 3X4		Canada	5198884567	ayansadhu.civil@gmail.com
Shankar	**Sankararaman**	SGT Inc., NASA Ames Research Center		NASA Ames Research Center		Moffett Field	CA	94035				shankar.sankararaman@nasa.gov
Chris	**Sautter**	University of Alabama in Huntsville	Interim Director, CMR Reliability and Failure Analysis Lab,	301 Sparkman Drive	VBRH D-27B	Huntsville	AL	35899			(256) 824 6147 and (256) 684 2032	chris.sautter@uah.edu
JB	**Schroeder**	MFPT Society		5100 Springfield St, Ste 420		Dayton	OH	45431				
Martha	**Schroeder**	MFPT Society		5100 Springfield St, Ste 420		Dayton	OH	45431				
Mike	**Scott**	The Modal Shop		3149 E Kemper Rd		Cincinnati	OH	45241			513-351-9919	mscott@modalshop.com
Onome	**Scott-Emuakpor**	Air Force Research Laboratory		AFRL/RQT, 1950 Fifth St	Building 18D	WPAFB	OH	45433			937-255-6810	onome.scott-emuakpor.1@us.af.mil
Chris	**Sensor**	m+p international inc.		271 Grove Avenue, Building G		Verona	NJ	07044		USA	973 239 3005	chris.sensor@mpina.com
Jake	**Siegel**	Luna Inc.		706 Forest St.	Suite A	Charlottesville	VA	22903			(434) 220-7682	siegelj@lunainc.com
David	**Siegel**	University of Cincinnati - Center for Intelligent Maintenance Systems		PO Box 210072		Cincinnati	OH	45221			(513) 390-8163	siegeldn@mail.uc.edu
Cameron	**Sinclair**	BINDT	CEO	Newton Building	St George's Avenue	Northampton		NN2 6JB		UK	01144-1604-893817	cameron.sinclair@bindt.org
Jamie	**Smith**	National Instruments	Director, Embedded Systems Product Management and			Austin	TX					
Kenneth	**Speaks**	RMCI, Inc.	Program Manager	110 Wandering Ln.		Harvest	AL	35749			256-313-9040	kspeaks@rmci-inc.com
Tik	**Sze**	Dell		2300 W Plano Parkway		Plano	TX	75075			214 415 6126	Tik_Sze@Dell.com

John	**Thackrah**	US Navy Military Sealift Command	Executive Director									202 685 5011	john.thackrah@navy.mil
Josiah	**Thomas**	RMCI, Inc		1525 Perimeter Pkwy		Huntsville	AL	35806				256-975-3532	josiah.thomas@rmci-inc.com
Don	**Thompson**	The GBS Group		397 Little Neck Road	Building 3300 South, Suite 204	Virginia Beach	VA	23452				757-965-4274	donthompson@thegbsgroup.us
Steven	**Tibbetts**	The GBS Group	Vice President, Chief Operating Officer	397 Little Neck Road	Building 3300 South, Suite 204	Virginia Beach	VA	23452				757 965 4274	steventibbetts@thegbsgroup.us
Randy	**Torfin**	US Navy Military Sealift Command HQ	Head, Engineering Maintenance Management Branch N711	914 Charles Morris Ct SE		Washington Navy Yard	DC	20398-5440				(202) 685-5744	randel.torfin@navy.mil
Mate	**Toth**	SpectraQuest, Inc.		8227 Hermitage Road		Richmond		23228				804-261-3300	mate.toth@spectraquest.com
Bernard	**Truesdale**	IMI Sensor, a division of PCB Piezotronics		3425 Walden Avenue		Depew	NY	14043				716-684-0002	btruesdale@pcb.com
Pratichi	**Vaidya**	NTNU Norway		S.P Andersens vei 5	Valgrinda	Trondheim		7491	Sor Trondhelog	Norway		+47 45490287	pratichivaidya@yahoo.com
Brandon	**Van Hecke**	The University of Illinois at Chicago		Dept. of Mechanical & Industrial Engineering	842 West Taylor Street	Chicago	IL	60607				312-996-3410	bvanhe2@uic.edu
Blake	**Van Hoy**	Oak Ridge National Laboratory	Senior R&D staff	Box 2008	MS-6174	Oak Ridge	TN	37831-6174				865-250-0495	vanhoybw@ornl.gov
Håvard	**Vold**	ATA Engineering, Inc.		688 Deepwood Drive		Charleston	SC	29412				(703) 581-5657	hvold@ata-e.com
Rick	**Wade**	MFPT Society	Administrative Officer	5100 Springfield Street	Suite 420	Dayton	OH	45431-1264				(937) 256-2285 x2	rick@mfpt.org
Bruce	**Wallin**	UTC Aerospace Systems		11618 Tyson Drive		Rockford	IL	61114				(815) 966-8193	bruce.wallin@utas.utc.com
Shuai	**Wang**	State Key Laboratory for Manufacturing System Engineering		No.28, Xianning West Road		Xi'an		710049	Shaanxi	P R China		+86 82667963	wangsh_106551@stu.xjtu.edu.cn
Bob	**Ware**	US Air Force Research Laboratory		2977 Hobson Way		Wright-Patterson AFB	OH	45433 -7733				937 656 9146	robert.ware.4@us.af.mil
David	**Weeda**	Chase Supply, Inc		307 E Street		Hampton	VA	23661				757 325 6340	David.Weeda@chasesupplyinc.com
Brian	**Weiss**	National Institute of Standards and Technology	Associate Program Manager, Standards for Smart Manufacturing	Intelligent Systems Division	100 Bureau Drive, Stop 8230	Gaithersburg	MD	20899				301 975 4373	brian.weiss@nist.gov
Frank	**Wells**	Military Sealift Command / USNS Grasp		9290 Kings Landing Ln		Smithfield	VA	23430				410 963 5306	frank.t.wells.civ@msc.navy.mil
Kevin	**Westhora**	Dytran Instruments Inc	Director of Sales	21592 Marilla St		Chatsworth	CA	91311-4137				614 417 1542, (818) 700 7818	kwesthora@dytran.com
Chris	**Wilcox**	m+p international inc.		271 Grove Avenue, Building G		Verona	NJ	07044		USA		973 239 3005	chris.wilcox@mpina.com
Liang	**Xu**	SpectraQuest, Inc.		8227 Hermitage Road		Richmond		23228				804-261-3300	liang.xu@spectraquest.com
Wenguang	**Yang**	Tsinghua University		Dept. of Thermal Engineering		Beijing		100084		China		861062794046	ywg10@mails.tsinghua.edu.cn
Jae Myung	**Yoon**	University of Illinois at Chicago		842 W Taylor Street	Rm. 2039 ERF, MC 251	Chicago	IL	60607				312-996-2387	jyoon52@uic.edu
Junda	**Zhu**	Renewable NRG Systems		110 Riggs Road		Hinesburg	VT	05461				1 802 482 2255 x502	jz@renewablenrgsystems.com

Broadening Relevance:

Extending Machinery Failure Prevention Technologies to Enable Improved Environmental Performance

Mike Lipsett
University of Alberta
May 20, 2014
MFPT 2014, Virginia Beach

1

Faculty of ENGINEERING
University of Alberta

May 21, 2014

Knowing About Failure

- Defining Failure
- Knowing what its implications are
- Knowing what to do about it

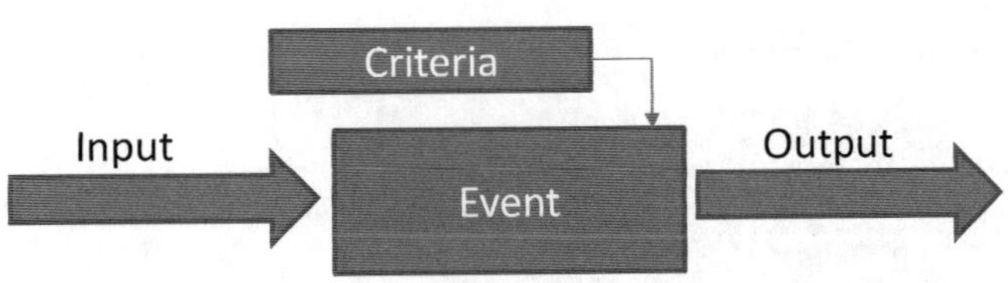

- Implications can be broad
- Our input/output observations may not be enough

2

Faculty of ENGINEERING
University of Alberta

May 21, 2014

Knowing About Failure Locally

- Failure in technical terms
- Implications are pertinent to component life & function
- Maintenance and possibly change in production

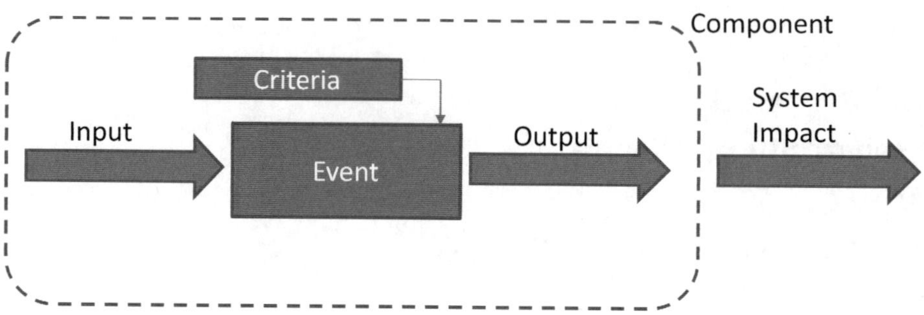

3

May 21, 2014

Faculty of ENGINEERING
University of Alberta

Knowing About Failure Locally

- Failure in technical terms
- Implications are pertinent to component life & function
- Maintenance and possibly change in production

Pump impeller damage

4

May 21, 2014

Faculty of ENGINEERING
University of Alberta

Cascading Failure

- We can see the progression of technological failure
- From local effects to company-wide impacts
- Small failures can have big consequences

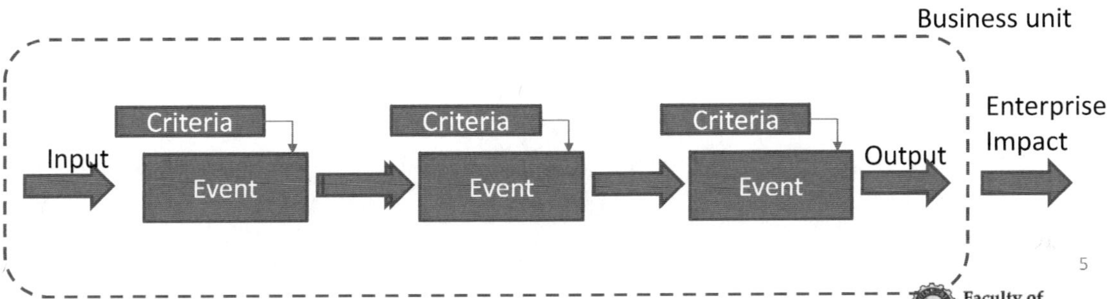

Cascading Failure

For want of a nail
the shoe was lost.
For want of a shoe
the horse was lost.
For want of a horse
the rider was lost.
For want of a rider
the battle was lost.
For want of a battle
the kingdom was lost.
And all for the want
of a horseshoe nail.

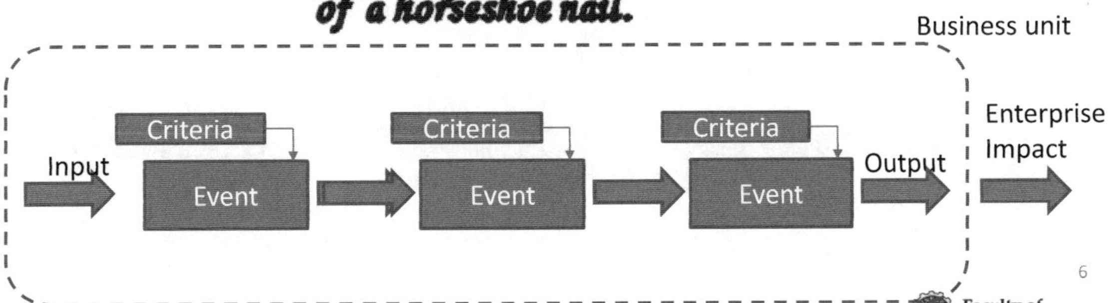

Complicated Failure

- Technologies
- Human error
- Inappropriate procedures

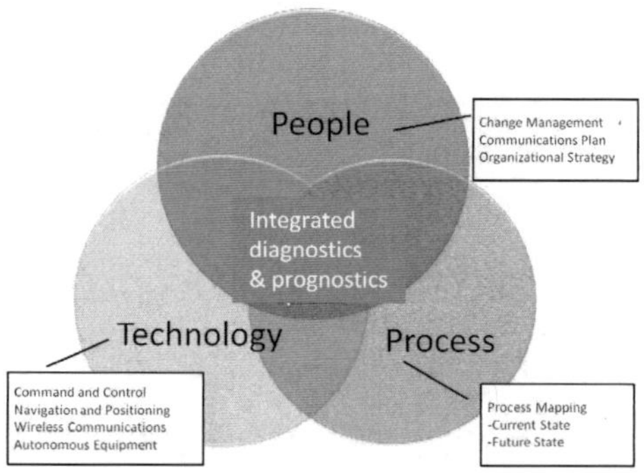

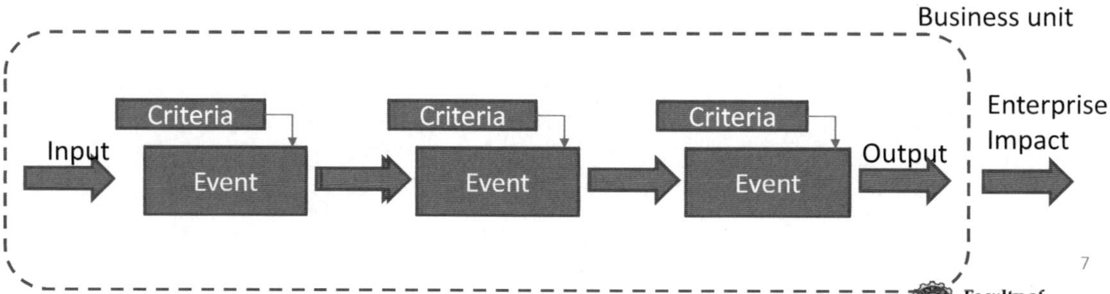

Faculty of ENGINEERING
University of Alberta

Other Impacts of Failure

- Safety may be compromised
- The reality is that safety is good business

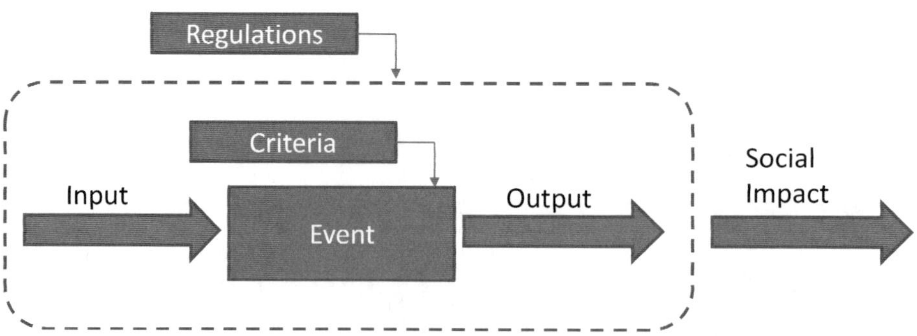

Faculty of ENGINEERING
University of Alberta

Other Impacts of Failure

- Environmental damage may occur
- The social licence to operate is becoming very sensitive to environmental performance

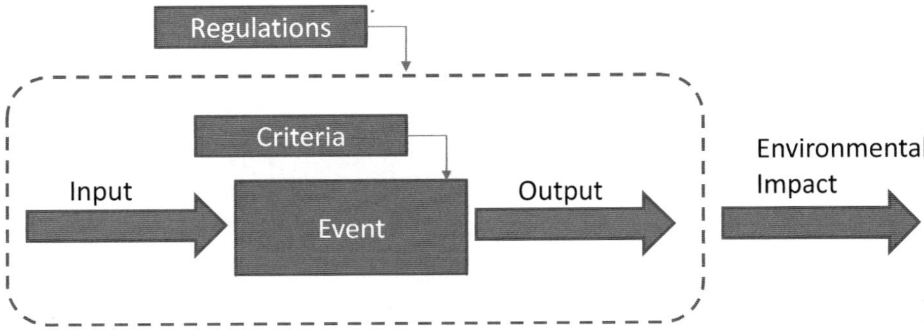

Faculty of ENGINEERING
University of Alberta

Knowing About Failure at the Right Level

- Defining failure at the enterprise level
  - Economic
  - Social
  - Environmental
- Knowing what to measure (inside and outside the box)

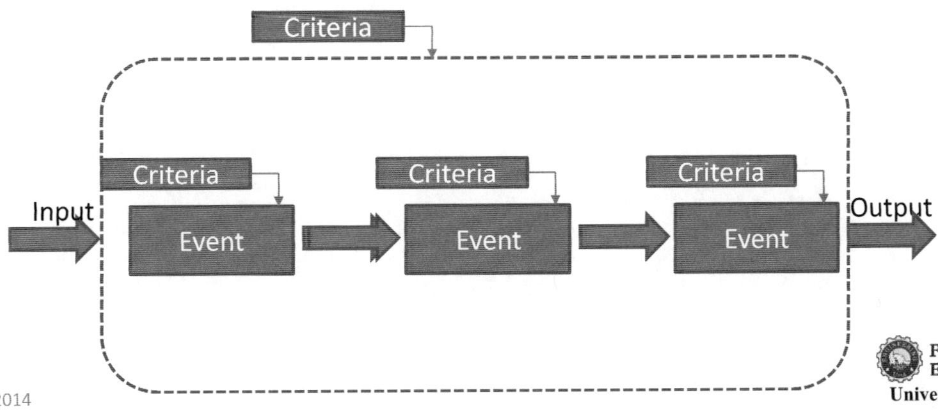

Faculty of ENGINEERING
University of Alberta

May 21, 2014

Going Beyond Machinery Failure Prevention to Enterprise Failure Prevention

- Component level failure prevention is no longer good enough to ensure that an organization delivers its triple bottom line

Going Beyond Machinery Failure Prevention to Enterprise Failure Prevention

- Component level failure prevention is no longer good enough to ensure that an organization delivers its triple bottom line
- Our systems are ever more complicated
  - Bigger, more components, more interactions, more control systems

Going Beyond Machinery Failure Prevention to Enterprise Failure Prevention

- Component level failure prevention is no longer good enough to ensure that an organization delivers its triple bottom line
- Our systems are ever more complicated
  - Bigger, more components, more interactions, more control systems
- Our systems are more complex
  - Not just technology, but processes, and people too

13

Faculty of ENGINEERING
University of Alberta

May 21, 2014

Going Beyond Machinery Failure Prevention to Enterprise Failure Prevention

- Component level failure prevention is no longer good enough to ensure that an organization delivers its triple bottom line
- Our systems are ever more complicated
  - Bigger, more components, more interactions, more control systems
- Our systems are more complex
  - Not just technology, but processes, and people too
- Systems are more connected
  - including global supply chains and labour markets

14

Faculty of ENGINEERING
University of Alberta

May 21, 2014

Going Beyond Machinery Failure Prevention to Enterprise Failure Prevention

- Component level failure prevention is no longer good enough to ensure that an organization delivers its triple bottom line
- Our systems are ever more complicated
  - Bigger, more components, more interactions, more control systems
- Our systems are more complex
  - Not just technology, but processes, and people too
- Systems are more connected
  - including global supply chains and labour markets
- Cascading failures have big nonlinear consequences

15

Faculty of ENGINEERING
University of Alberta

May 21, 2014

Going Beyond Machinery Failure Prevention to Enterprise Failure Prevention

- Component level failure prevention is no longer good enough to ensure that an organization delivers its triple bottom line
- Our systems are ever more complicated
  - Bigger, more components, more interactions, more control systems
- Our systems are more complex
  - Not just technology, but processes, and people too
- Systems are more connected
  - including global supply chains and labour markets
- Cascading failures have big nonlinear consequences
- Social licence to operate is a moving target

16

Faculty of ENGINEERING
University of Alberta

May 21, 2014

Limited Observability

- Sometimes we are not measuring the right things

Or looking in the right place

Expanding the Definition of Reliable Systems: Outputs

- Conventional definition based on production
- More stakeholders increases expectations
- Economic
- Health and safety
- Environmental (direct and indirect)
- Social Licence
- Sustainability

Expanding the Definition of Reliable Systems: Inputs

- Typical inputs for machinery failure diagnostics and prognostics are based on anomalies with respect to the physics of the operating machine
- More sophisticated prognostics are based on physics of failure (but how many failure modes really need to be considered?)
- Damage accumulation models are of little use for prognostics unless the initiation time is known
- Diagnostics and prognostics are prone to errors due to component variability, disturbances to the system, and noise in measurement

19

Faculty of ENGINEERING
University of Alberta

May 21, 2014

Disturbance

- Signal that is "outside the box"
- May affect outputs, but can be differentiated from change in the system
- Leading indicators are more useful for operational control than outputs

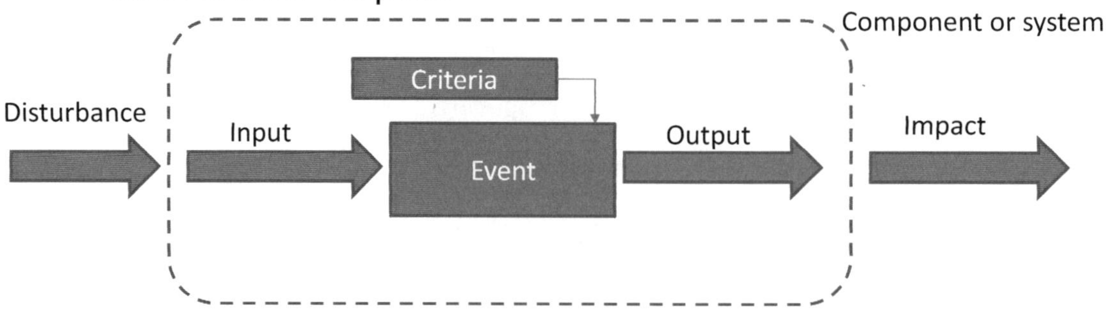

- Some things outside the system are easier to measure

20

Faculty of ENGINEERING
University of Alberta

May 21, 2014

Component Thinking

- Machine monitoring: component-level state variables
- Damage accumulation: correlation to observable variables
- Data-driven or Physics-of-failure prognostics ignoring external factors

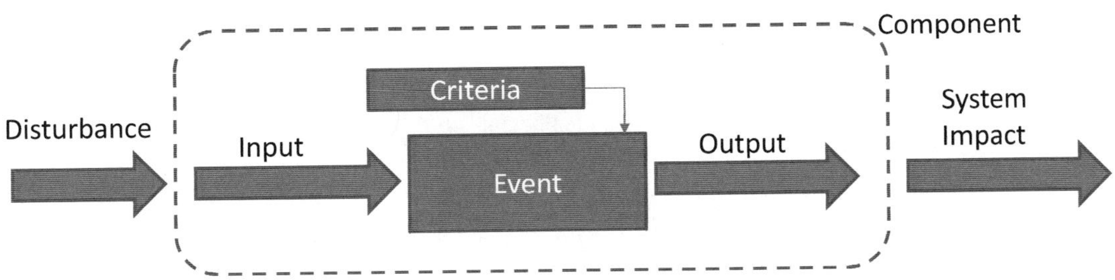

21

Faculty of ENGINEERING
University of Alberta

May 21, 2014

Systems Thinking

- Critical thinking about the context of machine operating conditions, damage accumulation & consequences of failure
- Measuring inputs allows us to track leading indicators
- Measuring outputs not only directly relevant to component
- Enterprise-wide condition indicators

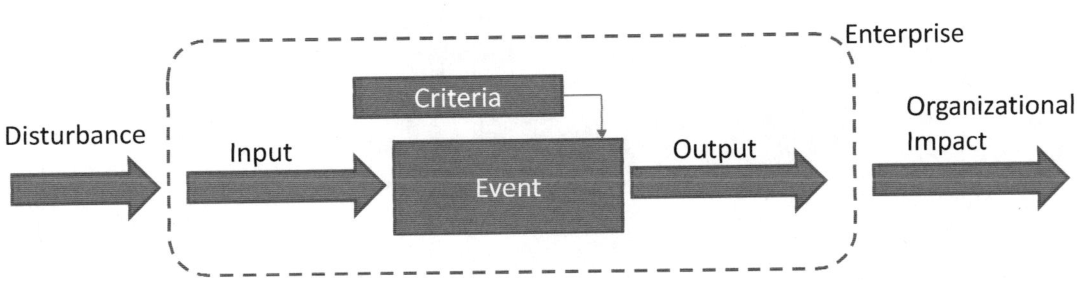

22

Faculty of ENGINEERING
University of Alberta

May 21, 2014

Appropriate Levels of Diagnostics

- Component level failure understanding
- Conditional probabilities of failure
- Enterprise-wide understanding of consequences

Stuck Relief valve

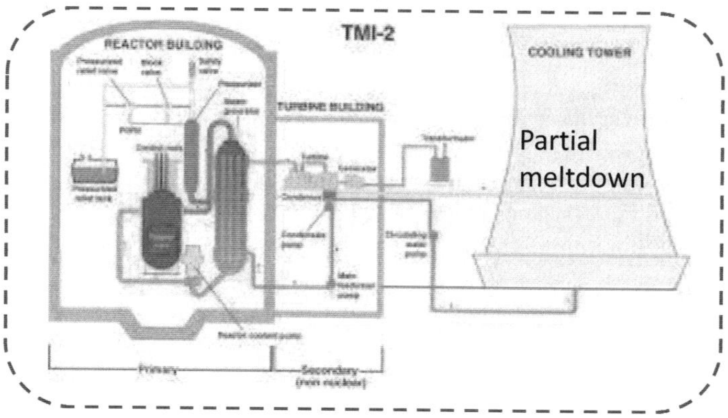

Thirty-year setback in US nuclear power

Faculty of ENGINEERING
University of Alberta

I Know What You're Thinking

- "I am still trying to figure out how to detect a particular failure mode – how am I supposed to be a holistic thinker?"

Faculty of ENGINEERING
University of Alberta

Well, You're Right

- But…
- One of our challenges is demonstrating our relevance
- Machinery failure prevention becomes an integral part of an enterprise when there are clear links to how reliability
  - improves the bottom line
  - helps companies to keep the public trust
- We have to broaden our observations and understanding to explain the value to the enterprise

25

Faculty of ENGINEERING
University of Alberta

May 21, 2014

Examples:

- Enbridge leak at Kalamazoo
- Recently purchased pipeline
- Integrity not fully understood
- Estimated costs > $100M (not including reputation)

Imprecise Leak detection

Marshall Pipeline leak

Hundreds of intervenors protesting new pipelines

26

Faculty of ENGINEERING
University of Alberta

May 21, 2014

Examples:

- BP Deepwater Horizon (Macondo)
- Completion & gas control
- Regulations on blowout preventors
- Contractor oversight
- 11 dead; estimated costs > $30B (not including reputation)

Technology
Process
People

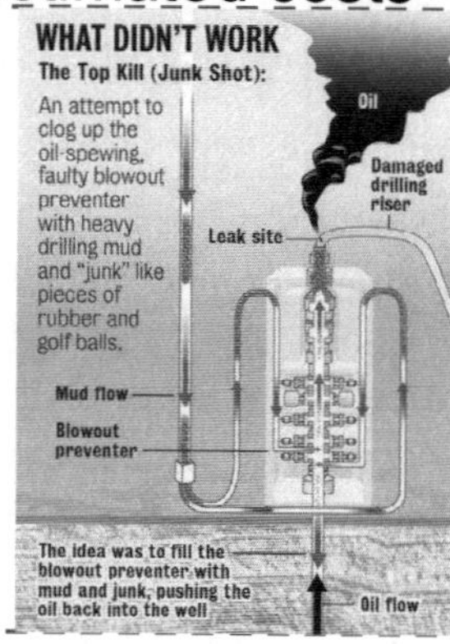

Widespread
environmental
& economic impact

27

Faculty of ENGINEERING
University of Alberta

May 21, 2014

Examples

- But these alarming examples are only half the story
- Commercial aircraft are extremely reliable
  - complex and well managed suite of technologies
  - effective business and technical processes
  - well trained and motivated people

28

Faculty of ENGINEERING
University of Alberta

May 21, 2014

FMECA Is Critical to Risk Management

- Focus on what failure modes have consequences that are key to the organization

- Any contributing technology
- Any process (production process, supply chain, maintenance)
- Measure whatever may affect the state of the machine
- And outputs affected by the machine
- Part of the value proposition

Managing Distributed Assets with Embedded Systems and Mobile Sensors

- Inspection of industrial plants and transportation systems
  - improve system reliability
  - reduce monitoring costs
  - Timely assessment of asset condition & environmental impacts
- Minimize risk through low cost remote sensing platforms
  - Remove workers from hazardous environments
- Monitor periodic changes
- Quick response for situational awareness

General System Description

- Inspection using airborne robots for preliminary visual surveys & applicable remote sensing
- Ground-based sensors: detailed measurements & calibration
- Mobile robots or fixed sensors *as appropriate*
  - Aerial mapping & direct measurements on ground

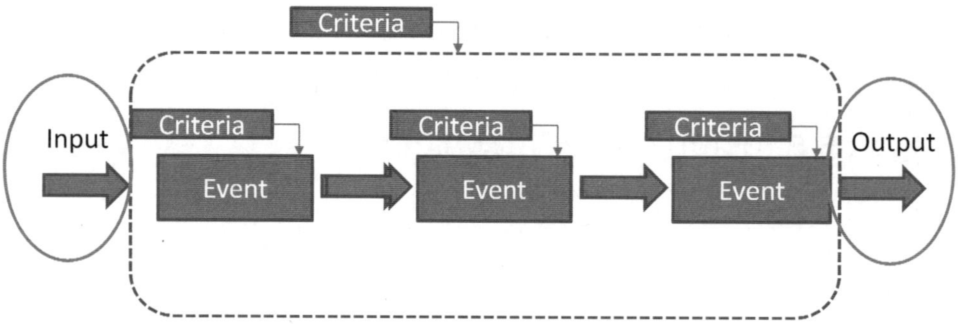

- Multipurpose environmental monitoring, capturing what is just outside the box of the enterprise

31

Faculty of ENGINEERING
University of Alberta

May 21, 2014

System Architecture

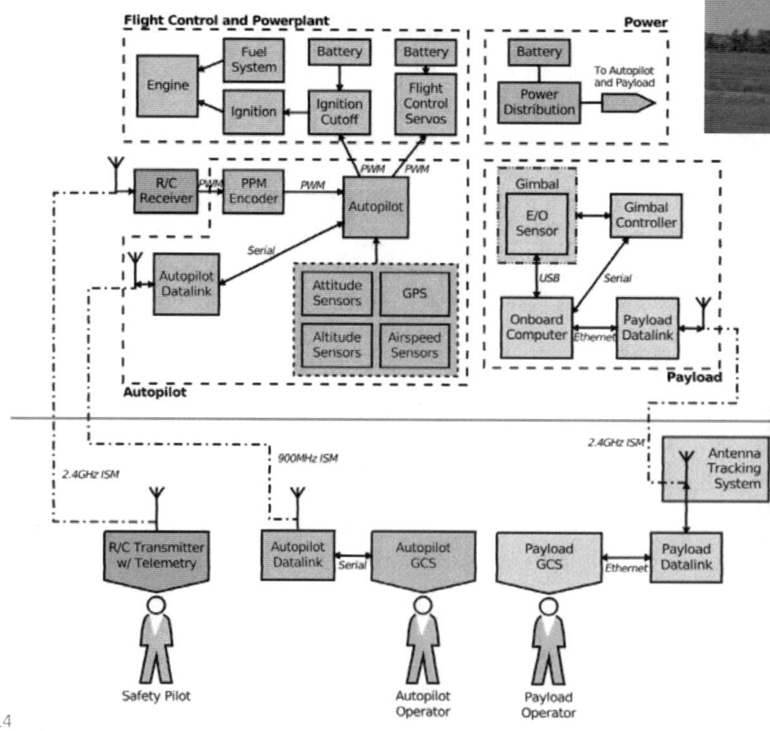

Airborne

Ground-vehicle

2

Faculty of ENGINEERING
University of Alberta

May 21, 2014

Retrofitting to Plant Equipment

Haul truck struts: detect strut fault or road conditions

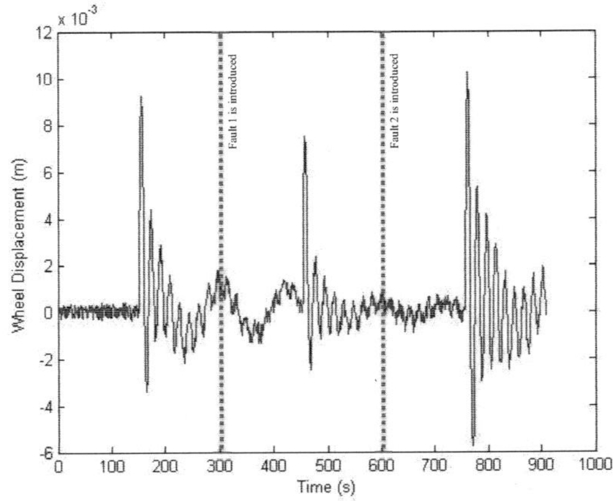

Front wheel displacement

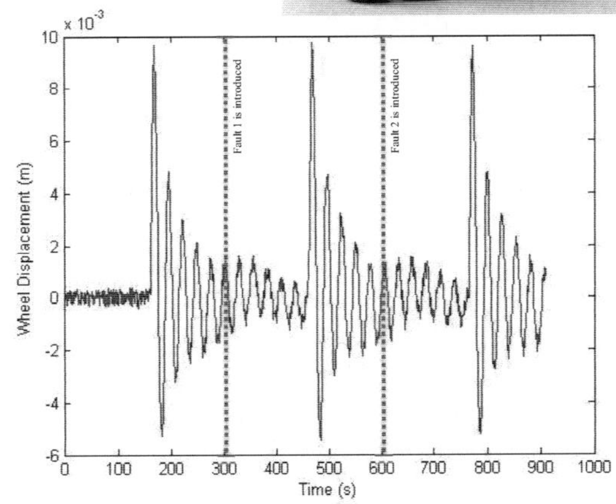

Rear wheel displacement

Car moves over three bumps in different operating condition

Two different faults are considered →

- Internal leakage at t=303 s
- Broken valve t=606 s

Expanding Our Observations and Our Understanding of Fault Effects

- Understand the value proposition
- Present a credible business case

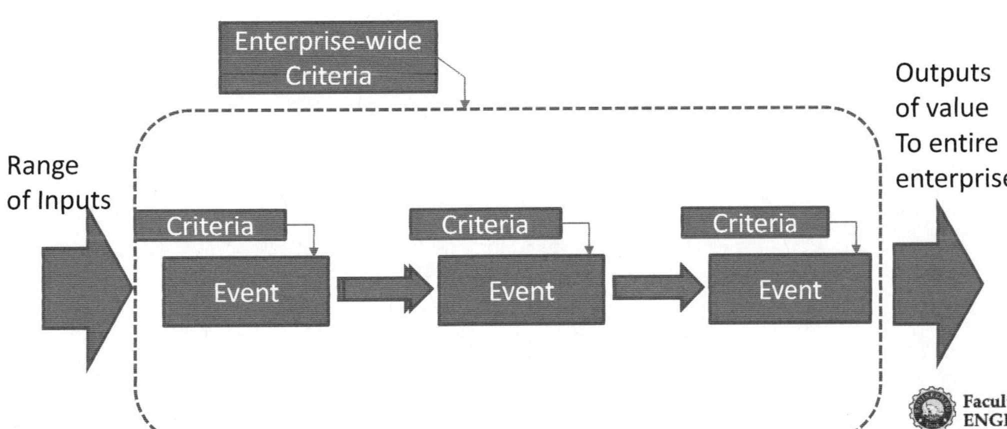

Broadening Relevance:

Machinery Failure Prevention Technologies to Enable Improved Environmental Performance

Use FMECA to select likely important fault modes

Observe inputs and outputs related to corporate objectives

Relate faults to key performance indicators

Improve performance – and show your relevance

35

Faculty of ENGINEERING
University of Alberta

May 21, 2014

Big Analog Data™ Keeping Machines Operational

Jamie Smith, Director of Embedded Systems
National Instruments

ni.com

NATIONAL INSTRUMENTS

Science, Discovery, and Innovation ...

Experimental Science | Theoretical Science | Computational Science | Data-Intensive Science

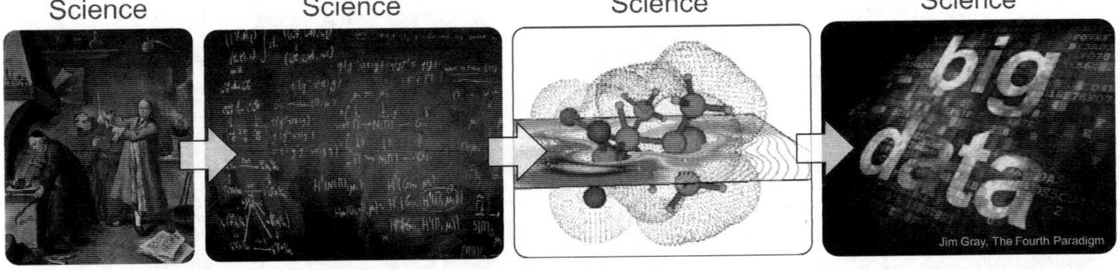

- Internet of Things
- Intelligent Systems
- Big Analog Data™ Solutions
- Programmable World
- M2M
- Industrial Internet
- Smarter Planet
- Cyber-physical Systems
- Industrie 4.0

ni.com 2

NATIONAL INSTRUMENTS

Big Data Characterized

Variety

Mix of structure and format

Volume

Large amounts

Velocity

High speed, high sample rates

Value

Importance of analysis, which was previously limited by technology

The fifth "V" . . .

Visibility

Access from disparate geographic locations

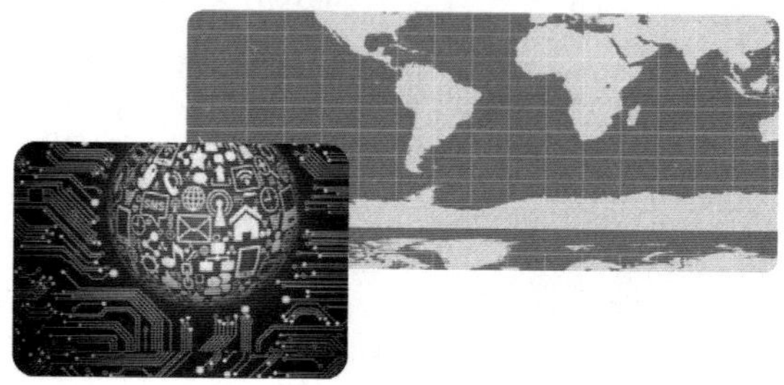

Big Data Characterized

Sources of Big Data:

Industry/ IT sources

- Enterprise apps: ERM, CRM, HR
- IT data: events, logs, inventories
- Process and control

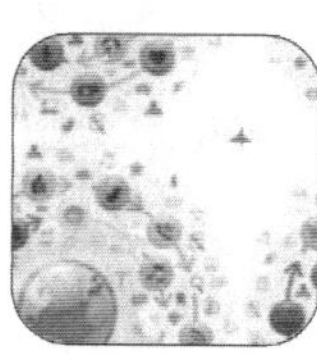

New/ emerging data sources

- Social data, behaviors, sentiments
- Tweets, posts, comments

Big Analog Data™ Sources

Engineering/scientific data sources

- Physical phenomenon
- Life science/ genomics
- High energy physics
- Transportation and energy

Big Data Characterized

Big Analog Data™

is Big Data derived from the physical analog world

is sourced from nature, people, and electrical and mechanical devices and machines

is digitized via A/D conversion

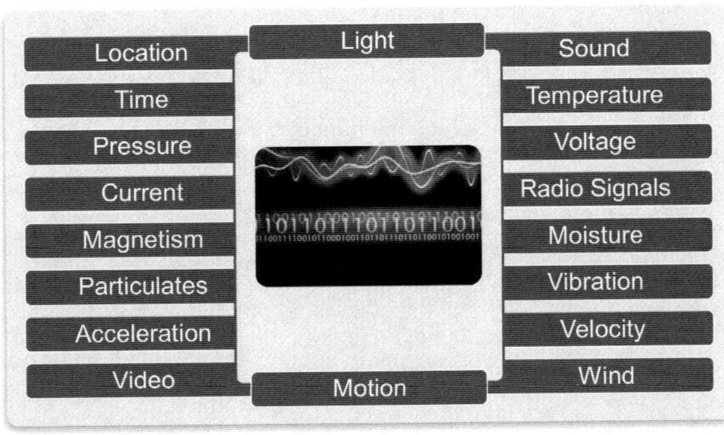

Big Analog Data™ Applications Characterized

Sunday	Monday	Tuesday	Wednesday	Thursday	Friday	Saturday
		1	2	3	4	5
6	7	8	9	10	11	12
13	14	15	16	17	18	19
20	21	22	23	24	25	26
27	28	29	30			

Smart Power Grid Measurements:
5TB / Month

Turbine Test Measurements:
10TB / Day

Jet Engine Measurements:
20TB / Hour

Physics Experiment Measurements:
40TB / Second

Big Analog Data™ Sources

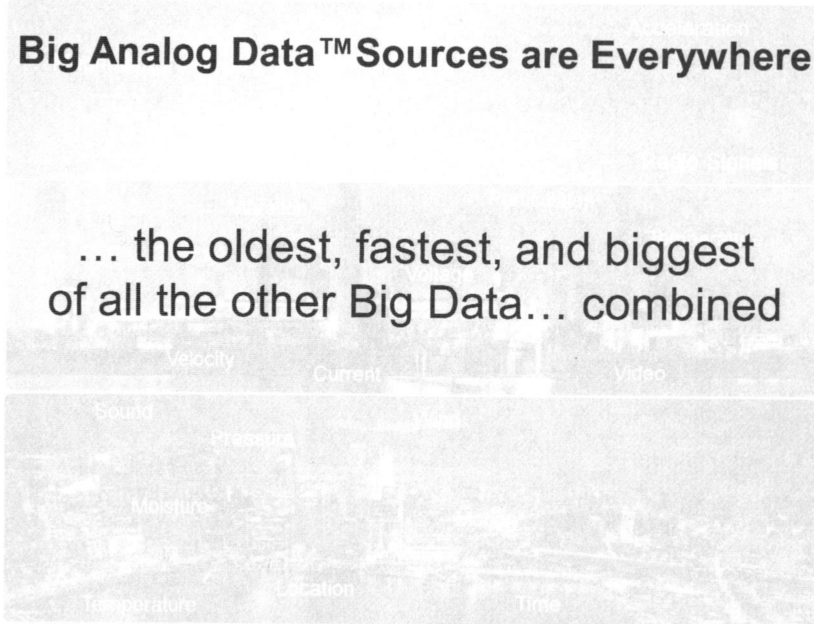

ni.com

NATIONAL INSTRUMENTS

Wind Farm in Ardrossan, Scotland

How The Problem is Primarily Being Solved

Current

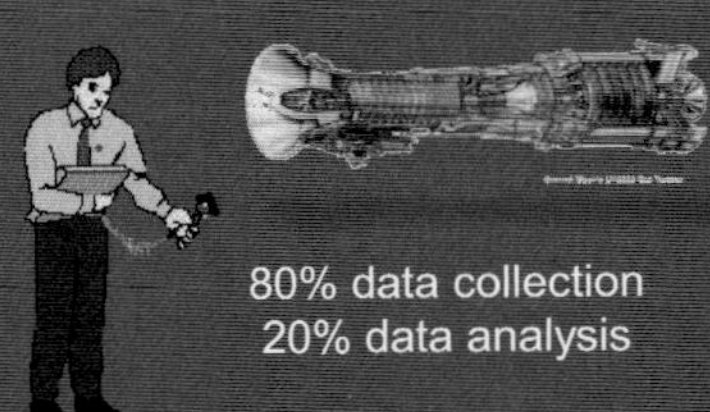

80% data collection
20% data analysis

- Smaller asset coverage
- Periodic manual collection
- 20% analytics
- Limited instrumentation
- Inconsistent diagnostics
- Limited risk capabilities
- Poor integration & visuals
- M&D center – process data
- Workforce – low value tasks

How The Solution is Shifting

Current

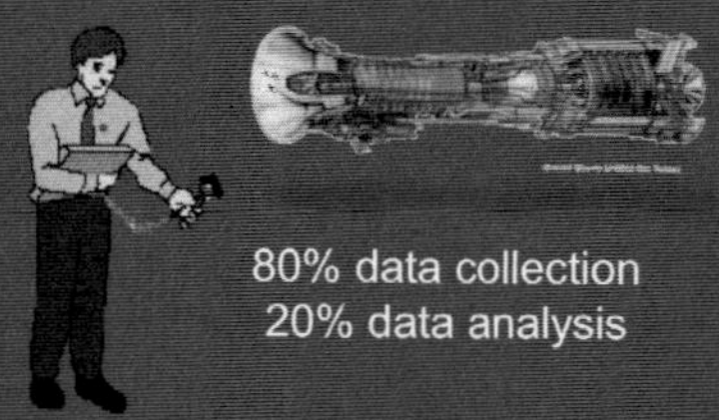

80% data collection
20% data analysis

- Smaller asset coverage
- Periodic manual collection
- 20% analytics
- Limited instrumentation
- Inconsistent diagnostics
- Limited risk capabilities
- Poor integration & visuals
- M&D center – process data
- Workforce – low value tasks

Future

Automated data
collection & analysis

- Greater asset coverage
- Continuous monitoring
- 80% analytics
- Fleet-wide system & sensors
- Diagnostic Advisors
- Risk Advisor- failure predictions
- Data fusion & federated visualization
- M&D center – integrating all info
- Workforce – high value analytics

Big Analog Data™ End-To-End Solution Architecture

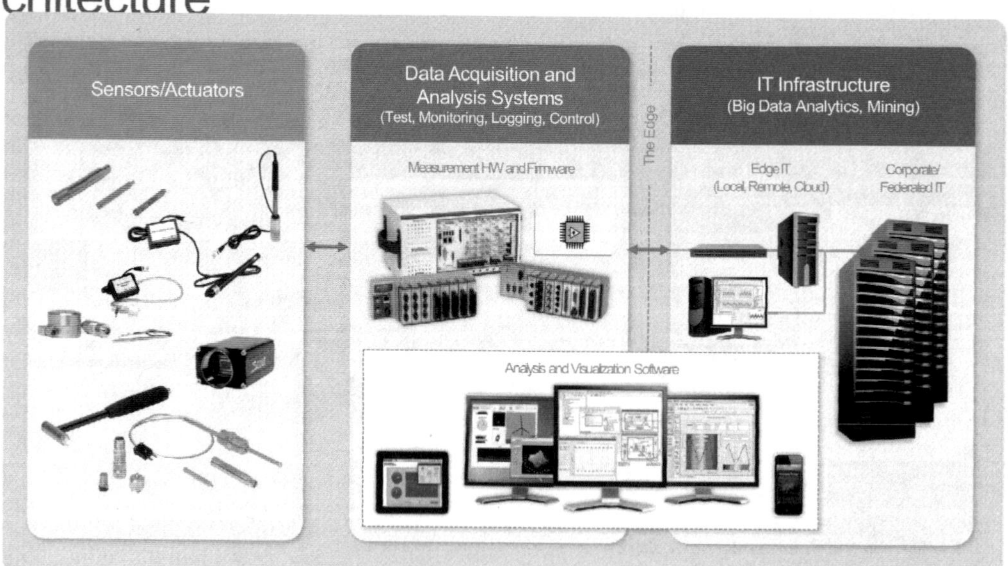

Big Analog Data™ End-To-End Solution Architecture

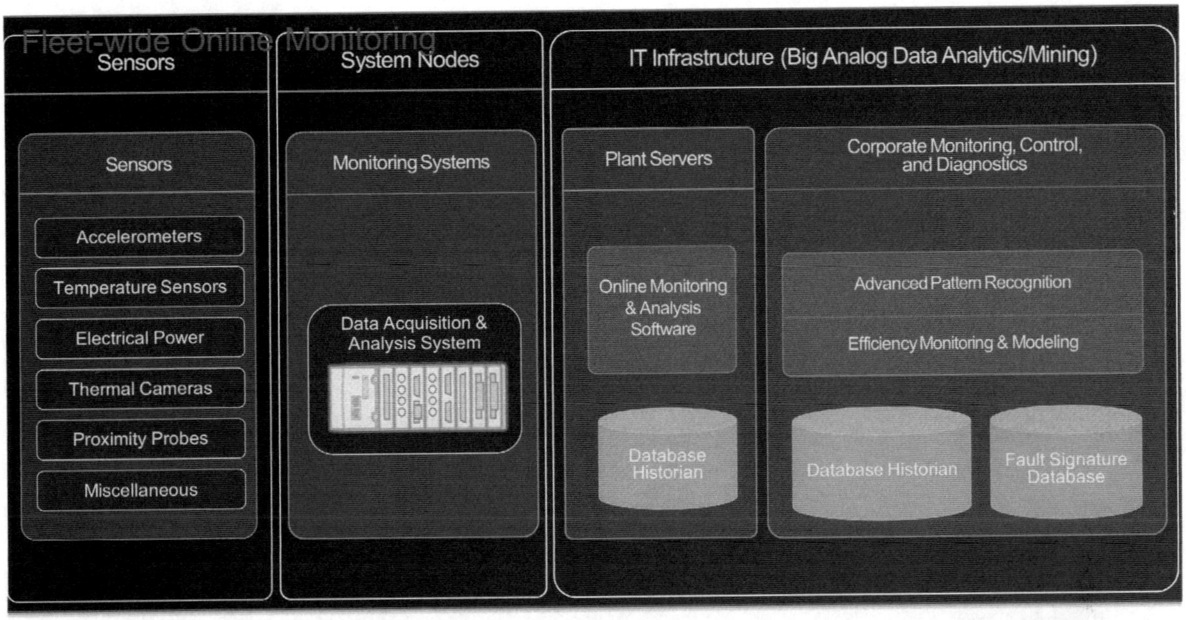

Example Big Analog Data™ End-to-End Solution

Fleet-wide Online Asset Condition Monitoring

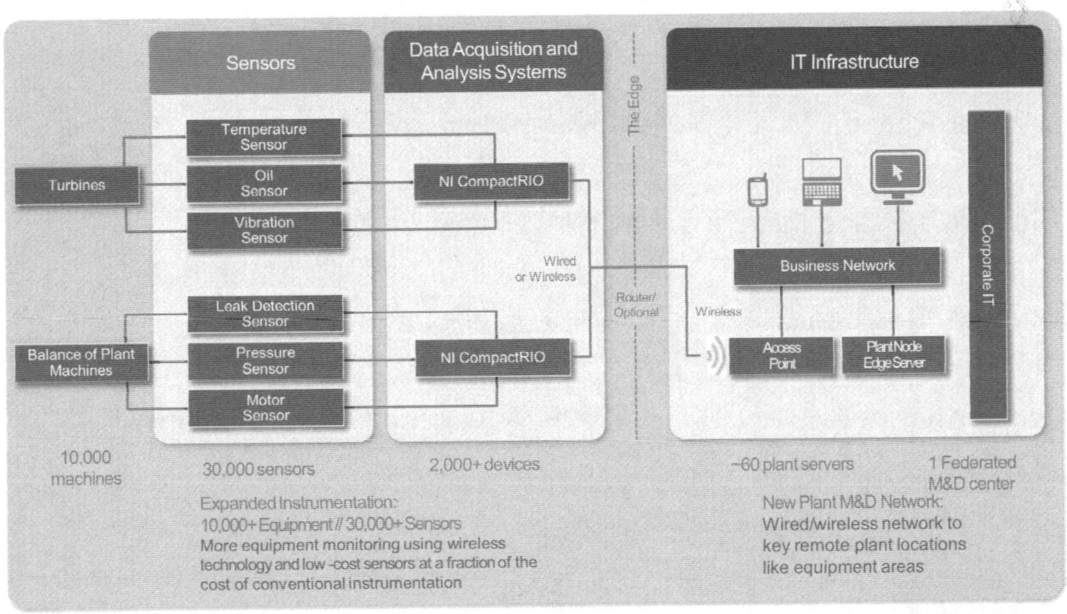

Example Big Analog Data™ Solution Architecture

Smart Grid - Synchrophasor/Phasor Measurement Units (PMUs)

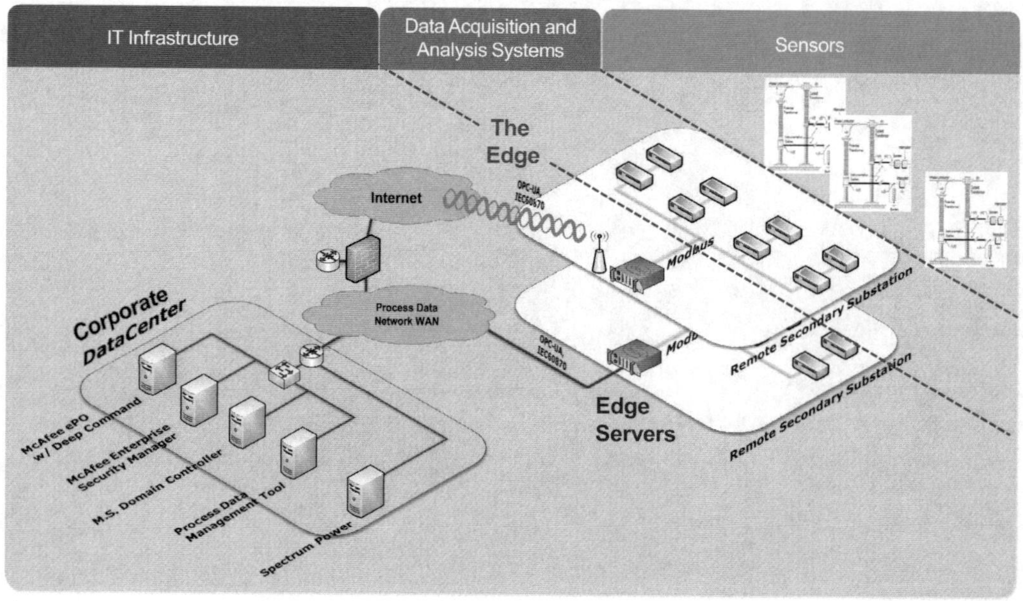

ni.com 16 NATIONAL INSTRUMENTS

Industry Wants, Needs and Wishes Today

Proprietary	→ Open Architecture
Manual Rounds	→ Continuous Monitoring
Schedule Based Maintenance	→ Predictive Maintenance
Human "Databases"	→ Intelligent Advisers
Limited Visibility	→ Advanced Sensor Fusion

ni.com 19 NATIONAL INSTRUMENTS

Enabling Technology Vectors to Consider in Building Big Analog Data™ Solutions

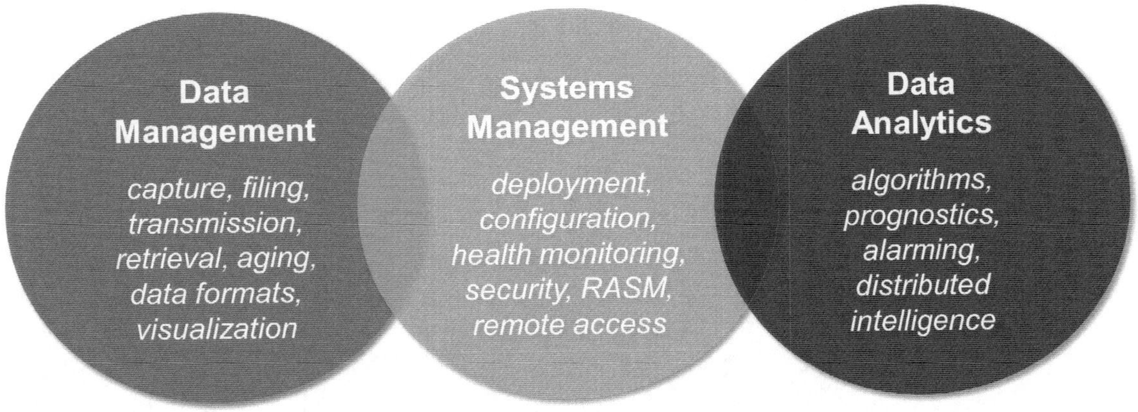

Technology Disruptors Tomorrow

ni.com 22 NATIONAL INSTRUMENTS

CYBER-PHYSICAL SYSTEMS
BIG ANALOG DATA
FUTURE OF STEM
RF
MOBILE
MODELS OF COMPUTATION
NI Trend Watch 2014
Technology Trends That Accelerate Your Productivity
ni.com/trendwatch
ni.com
23
NATIONAL INSTRUMENTS

Sustaining a Global Fleet

Our U.S. Navy's

Military Sealift Command

John Thackrah

Executive Director

May 20, 2014

MSC Global Laydown

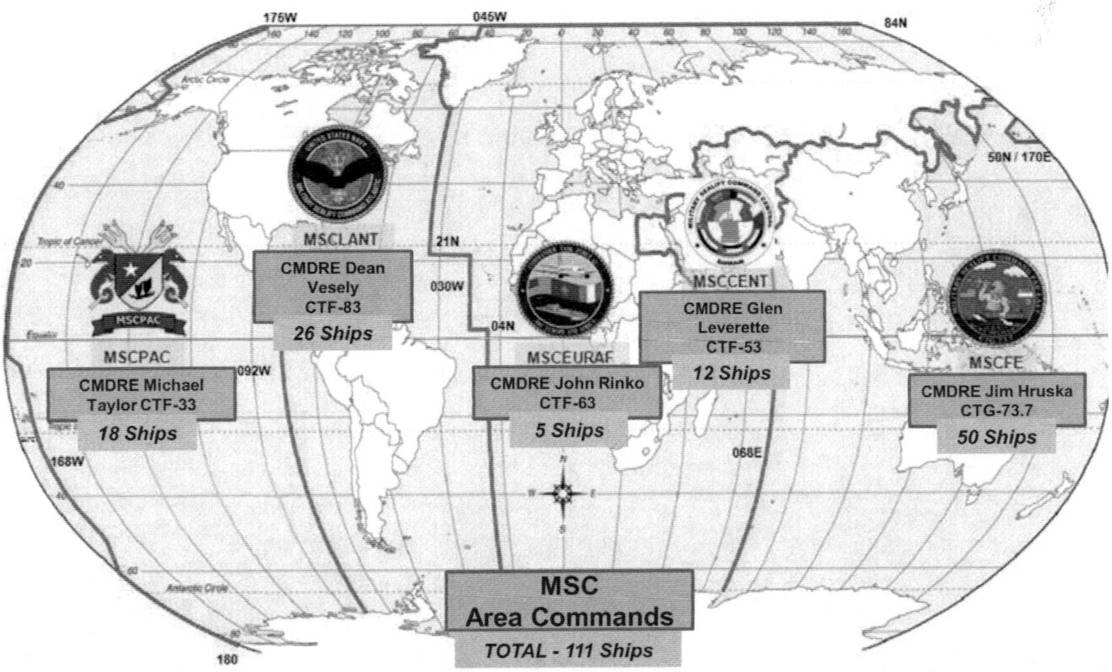

As of 14 Jan 2014

Building Trust

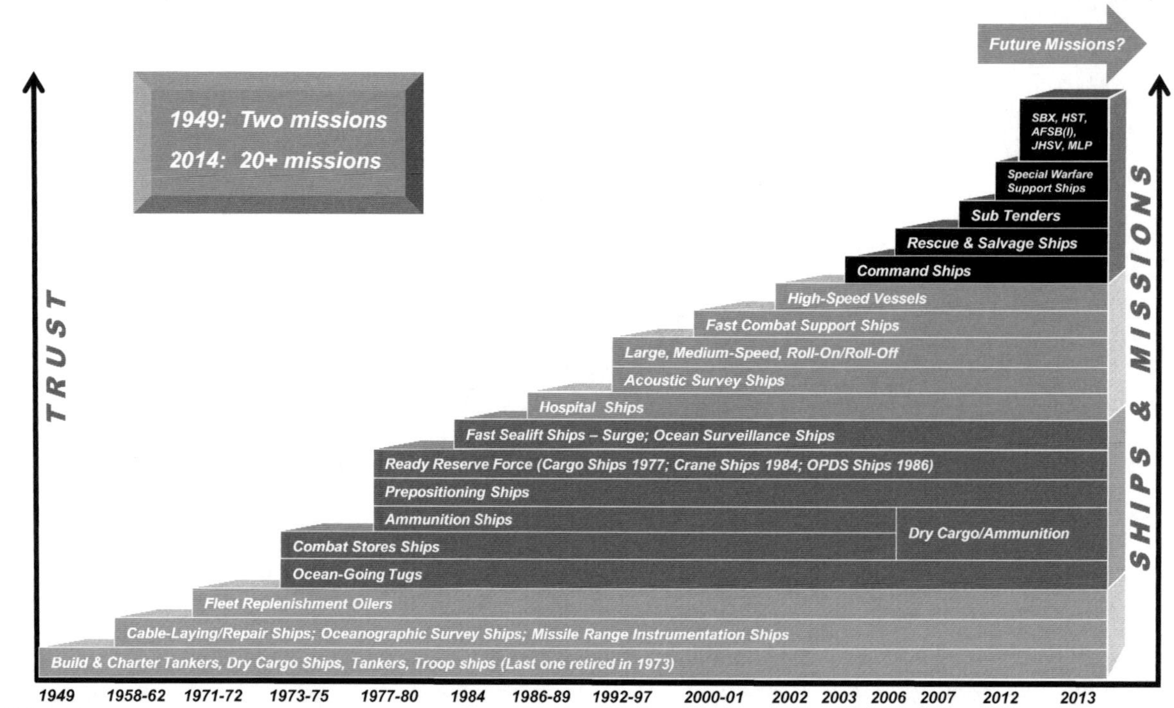

UNCLAS

Military Sealift Command — We Deliver!

Combat Logistics Force

Military Sealift Command — We Deliver!

Special Mission

UNCLAS 5

Military Sealift Command We Deliver!

Prepositioning

UNCLAS

Military Sealift Command We Deliver!

Service Support

UNCLAS

Military Sealift Command

We Deliver!

Strategic Sealift

Military Sealift Command

We Deliver!

MV Cape Ray

Military Sealift Command We Deliver!

Maintenance Technology Adoption

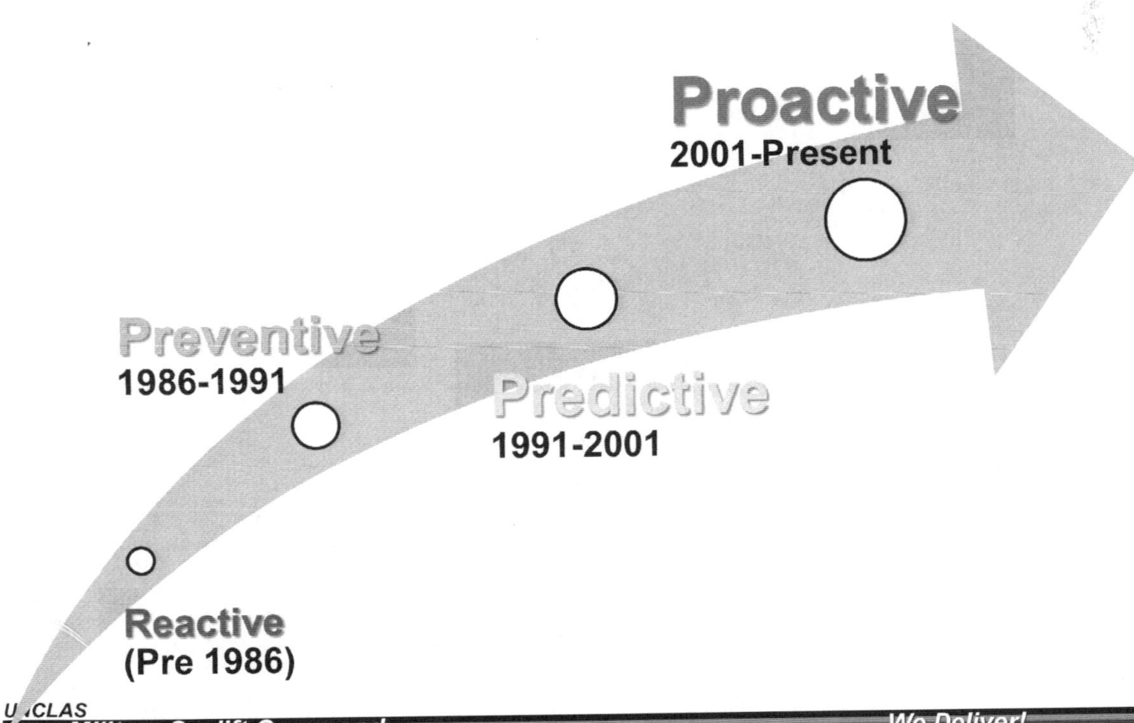

UNCLAS

10

Military Sealift Command We Deliver!

Maintenance Philosophy

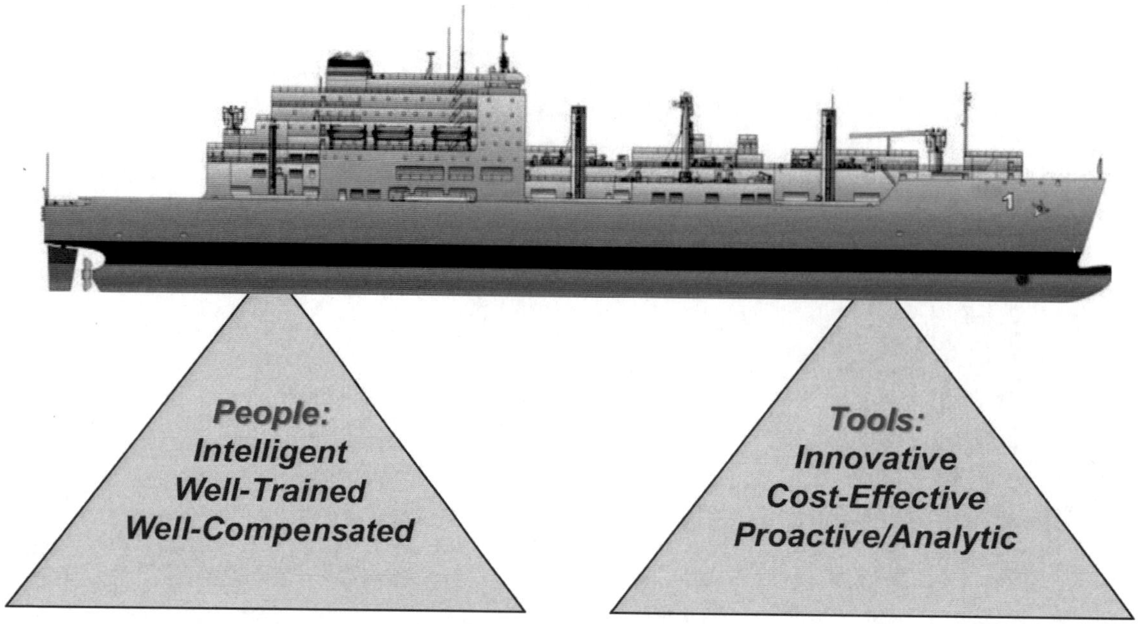

UNCLAS 11

Military Sealift Command We Deliver!

Shipboard Automated Maintenance Management

Vibration, Fuel, Oil, Chemical & Diesel Engine Combustion Analysis

UNCLAS 12

Military Sealift Command We Deliver!

Hand-held Tools We Use At MSC

ViPRe - vibration spectra on rotating machinery

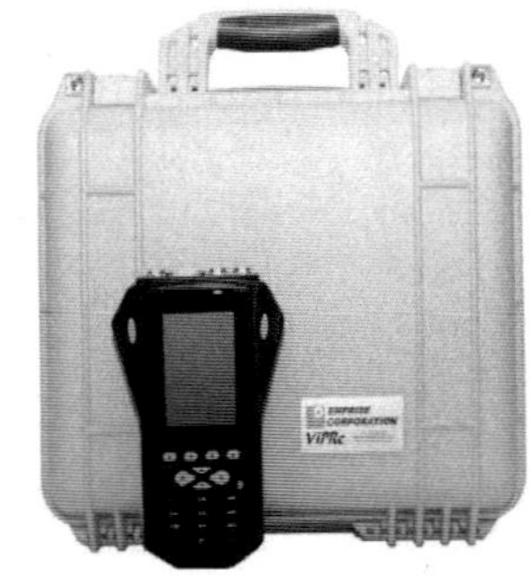

DOCTOR - combustion analysis on diesel engines

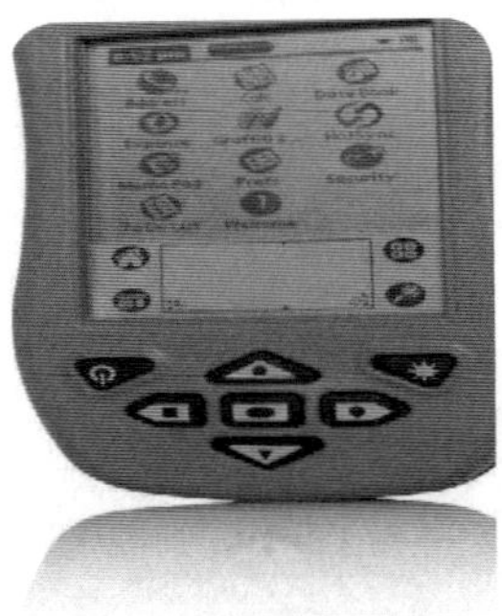

Logbook RDA - engineering plant parametric data

Hand-held Tools We Use At MSC

Fluid Scan - lubrication and hydraulic oil

Thermal Imager - electrical switch gear and mechanical connections

MSC Key Energy Initiatives

UNCLAS 15

Military Sealift Command We Deliver!

Artificial Intelligence – Future Tool?

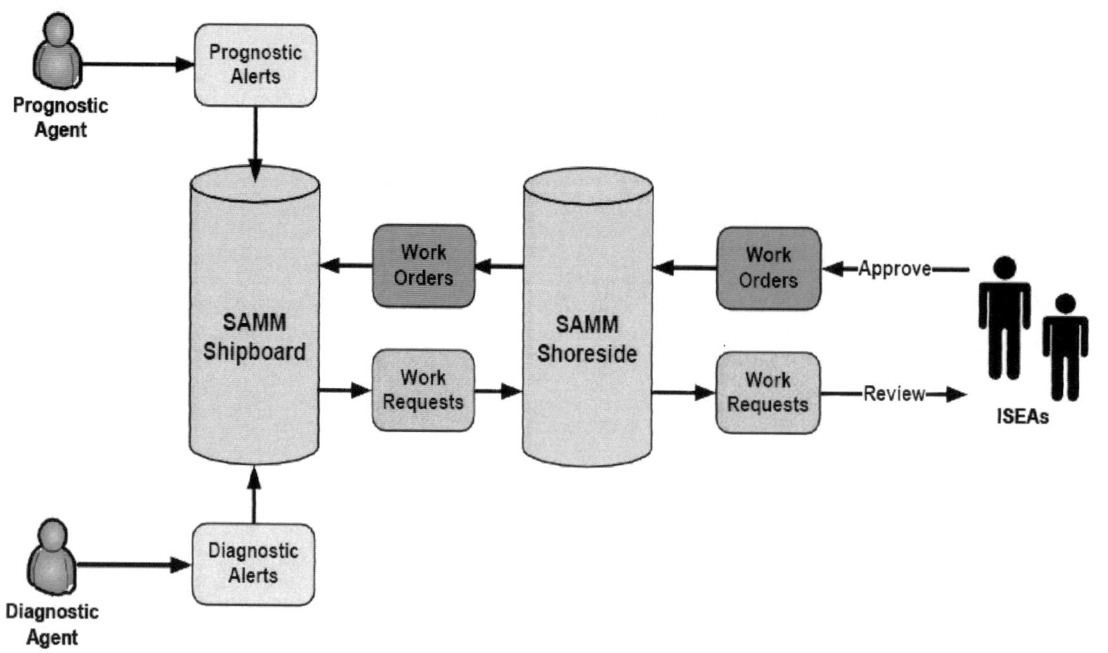

UNCLAS 16

Military Sealift Command We Deliver!

The Challenges We All Face

BIG DATA

Higher energy cost$

Smaller budgets

17

UNCLAS

Military Sealift Command *We Deliver!*

MSC Delivers!

Join the MSC conversation:

www.facebook.com/MSCdelivers

www.twitter.com/MSCsealift

MSCSealift.dodlive.mil

www.msc.navy.mil

18

UNCLAS

Military Sealift Command *We Deliver!*

A COMPUTATIONAL AND ACCURACY BASED EVALUATION OF AUTOMATED ALGORITHMS FOR MECHANICAL SYSTEM HEALTH ASSESSMENT AND FAULT CLASSIFICATION

Behrad Bagheri, David Siegel, Jay Lee
Center for Intelligent Maintenance Systems
2600 Clifton Avenue
Cincinnati, OH 45221-0072
Telephone: (513) 556-3412
bagherbd@mail.uc.edu

Abstract: The increasing demand of using automated algorithms for fault diagnosis and health assessment in current industry is undeniable. Consequently, implementing emerging technologies such as cyber physical systems, cloud computing in prognostics and health management algorithms is inevitable. Due to these improvements and the emerging Big Data environment, optimizing algorithms for consuming less process and network bandwidth but generating high accuracy is important. Considering these factors, in this research, we propose a generic fault diagnosis framework for rotating machineries with variant methods and parameters for different steps of the algorithm. Methods, such as Fast Fourier Transform and Wavelet Transform for signal processing and Artificial Neural Networks and K-Nearest Neighbor for fault classification are considered in this study. This framework is applied to a gearbox and bearing system in order to evaluate the effect of simplicity in the overall performance of prognostic and health management algorithms. Finally, results of this paper shows that precise selection of methods and parameters will eliminate the requirement of using more complex and computationally expensive methods.

Key words: Diagnostics; Rotating Machinery; Signal Processing; Wavelet; Nearest Neighbor; Artificial Neural Network;

Introduction: While the use of sophisticated systems and machinery is increasing at a sharp rate in industry, a new methodology called Prognostic Health Management (PHM) has been developed and growing since the last decade [1]. In addition to complex systems such as aircrafts, military vehicles, and automotive vehicles, PHM has a broad set of applications in various industrial sectors. Manufacturing systems are another industrial application where applying prognostic methods can be used to improve the production and prevent unexpected downtime [2]. In today's industrial environment, manufacturers are trying to gain their highest production rate with very good product quality, and this goal is difficult to achieve without considering an automated intelligent system. PHM algorithms can be used to automatically recognize and monitor minor defects in various industrial equipment and machines.

Bearings and Gearboxes are two key components in most of current industrial machineries [3], [4]. In recognition to this broad usage, lots of researches have been focused on monitoring and diagnosing the operation condition and faults of these two components by implementing different algorithms and most of the results were focused on the accuracy of fault detection [4]-[9]. In addition to having accurate results, the speed of calculation plays a significant role in the possibility of embedding the developed algorithm in real-time and online tasks. Based on this concept, the major focus of current study is on evaluating a balanced automated algorithm in order to have the best accuracy with the shortest calculation duration.

In most cases, implementing more sophisticated algorithms increases the accuracy of the final outcome but complex methods are more process consuming which results in a longer calculation duration. Despite recent advances in computer and networking industry with emerging technologies such as cloud computing [10], process and data size efficiency are still two critical factors for efficient integration of new paradigms such as cyber physical systems [11] in PHM methodologies. Data size for data types with low sampling frequency might not be an important issue but continuously transferring gigabytes of high sampling rate vibration data for real-time online monitoring of assets is not an efficient way of designing the PHM algorithm. The best solution to tackle this issue is to locally extract enough and relevant features out of the available data. Since only extracted features will be passed to the computational core of the algorithm, extracting and selecting appropriate features to investigate all the important correlations and signatures in the data is important to prevent accuracy loss.

In this current paper, the same approach has been applied to two sets of vibration data, one from a gearbox and the other one from a bearing test-rig, in order to study the effect of good and efficient feature extraction and feature selection on the final results of a PHM algorithm. Signals are processed using Fast Fourier Transform (FFT) and wavelets along with an Improved Distance Evaluation (IDE) method to select appropriate features that are extracted by various statistical parameters. The processed data has been fed into K-Nearest Neighbor (KNN) and Artificial Neural Networks (ANN) classifiers in order to classify the faults of gearbox and bearing case studies. Results have revealed that by extracting good and significant features, less complex classifiers and signal processing methods could provide the highest possible accuracy with considerably less processing duration.

Methodology: In this current paper, in order to make a comparison possible, different algorithms and/or variant parameters are used in each step of data analysis. The initial assumption is to have high sampling frequency data, such as vibration or acoustic signals, available from the target machinery. Although in both case studies we have used vibration signals, but this methodology is extendable into different signal types e.g. acoustic emission as it is shown in [9].

Figure 1 shows the schematic overview of the data analysis process in this research. As one can observe, in the signal processing and fault classification stages, two algorithms are applied on the same data set, where one of the algorithms is more computationally expensive than the other one. Additionally, the parameters in the wavelet transform and feature selection method are varied in order to distinguish the effect of changing algorithm parameters on the final accuracy of the fault diagnosis method.

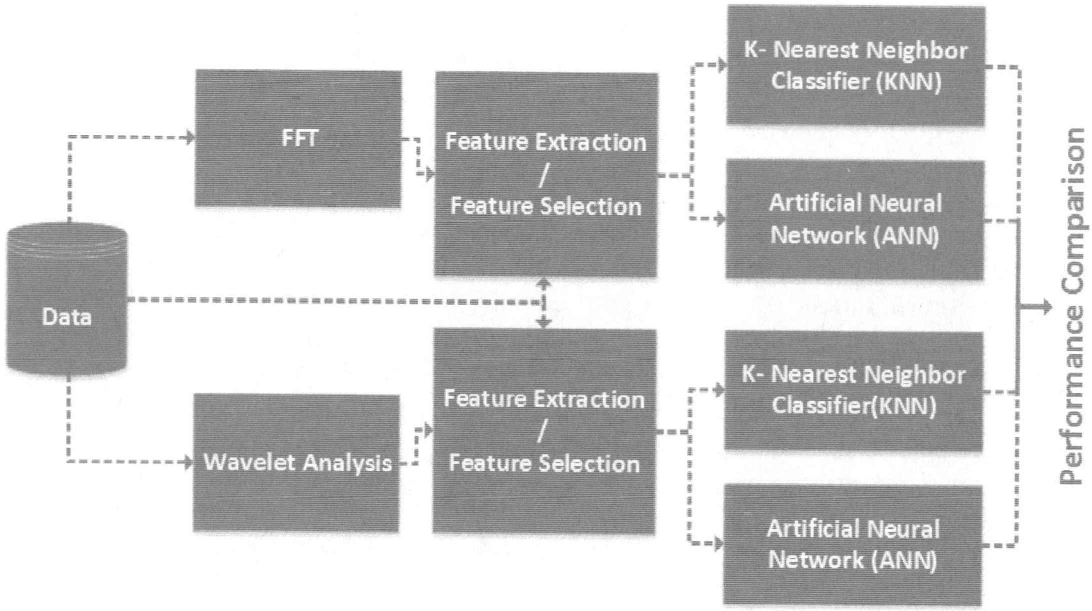

Figure 1: The Schematic View of PHM Framework

The beginning step in processing signals is to divide them into signal blocks where each of which will be processed separately. The division process requires a block size variable where this variable can be selected arbitrarily or based on tachometer signals to include integer multipliers of a machine revolution in each block. In case of the unavailability of a tachometer or any angular reference signal, the block size has to be selected carefully and overlap and averaging is recommended for decreasing leakage. In this research we arbitrarily chose block size and averaged four blocks with 50% overlap to segment the raw data. Lastly, a Hanning window [12] is applied on blocks to reduce the leakage effect.

Fast Fourier Transform (FFT) is a conventional frequency domain algorithm for analyzing time domain signals. Several researchers used the FFT as their main signal-processing algorithm to reveal useful information in the frequency spectrum [7], [13], [14]. On the other hand, the more complex time-frequency signal processing algorithms, such as Wavelet Transform, are also useful in fault diagnosis of rotary machineries, especially in presence of transient signals [9], [15], [16]. As aforementioned, to make a complete comparison, we varied some parameters in each algorithm, such as the decomposition level and mother wavelet in the Discreet Wavelet Transform (DWT), which is used in this research study. Using 8 mother wavelets and 6 decomposition levels, 48 combinations of wavelet parameters are applied to the data. Although Daubechies wavelet family are well-known in fault recognition of rotating machineries some recently published work by Bagheri [17] revealed that in some cases Coiflet wavelet family is capable of performing better than others. The complete list of used wavelets are as follow: "Db1", "Db2", "Db3", "Db4", "Coif1", "Coif2", "Coif3", "Coif4".

Feature Extraction: Processed signals are not directly useful in classifiers and it is required to investigate useful information by extracting relevant features. To do so, 29 functions, which are mostly statistical parameters, have been applied to processed signals. Features have been extracted from either time domain blocks, frequency spectrum and wavelet approximation and details. In total, 58 features have been extracted from data that has been analyzed with FFT (29 from Frequency spectrum and 29 from time domain) or similarly, a data block that is processed with two levels of decomposition with discreet wavelet transform will result in 116 features (29 Time Domain, 29 Approximations, 2x29 details). Table 1 shows these 29 features named from T1 to T29. These features will be used for classifying the faults in the rotating machinery, and after performing this feature calculation, the raw data is not further processed.

Table 1: Feature Parameters That Are Used For Feature Extraction

$$T_1 = \frac{\sum_{n=1}^{N} x(n)}{N} \qquad T_2 = \sqrt{\frac{\sum_{n=1}^{N}(x(n) - T_1)^2}{N}}$$

$$T_3 = \left(\frac{\sum_{n=1}^{N} \sqrt{|x(n)|}}{N}\right)^2 \qquad T_4 = \sqrt{\frac{\sum_{n=1}^{N}(x(n))^2}{N}}$$

$$T_5 = \max(|x(n)|) \qquad T_6 = \frac{\sum_{n=1}^{N}(x(n) - T_1)^3}{N.{T_2}^3}$$

$$T_7 = \frac{\sum_{n=1}^{N}(x(n) - T_1)^4}{N.{T_2}^4} \qquad T_8 = \frac{T_5}{T_4}$$

$$T_9 = \frac{T_5}{T_3} \qquad T_{10} = \frac{N.T_4}{\sum_{n=1}^{N}|x(n)|}$$

$$T_{11} = \frac{N.T_5}{\sum_{n=1}^{N}|x(n)|} \qquad T_{12} = \frac{\sum_{n=1}^{N}(x(n) - T_1)^2}{N}$$

$$T_{13} = \frac{N}{\sum_{n=1}^{N}\frac{1}{x(n)}} \qquad T_{14} = \sqrt[N]{\prod_{n=1}^{N} x(n)}$$

$$T_{15} = \frac{T_2}{T_1} \times 100 \qquad T_{16} = \frac{\sum_{n=1}^{N}|x(n) - T_1|}{N}$$

$$T_{17} = \frac{\frac{1}{N}\sum_{n=1}^{N}(x(n) - T_1)^3}{(\sqrt{\frac{1}{N}\sum_{n=1}^{N}(x(n) - T_1)^2})^3}$$

$$T_{18} = \sqrt{\frac{N-1}{N}} \times \frac{1}{(N-2).{T_{12}}^{1.5}} \times \sum_{n=1}^{N}(x(n) - T_1)^3$$

$$T_{19} = \frac{(N-1)\times(N+1)}{(N-3)\times(N-2)\times N\times {T_{12}}^2} \times \sum_{n=1}^{N}(x(n) - T_1)^4 - \frac{3\times(N-1)^2}{(N-2)(N-3)} + 3$$

$$T_{20} = \frac{\sum_{n=1}^{N}(x(n) - T_1)^3}{N}$$

$$T_{21} = \frac{\sum_{n=1}^{N}(x(n) - T_1)^4}{N}$$

$$T_{22} = \frac{\sum_{n=1}^{N}(x(n) - T_1)^5}{N}$$

$$T_{23} = \frac{\sum_{n=1}^{N}(x(n) - T_1)^6}{N}$$

$$T_{24} = \frac{3\times(T_1 - \hat{X})}{T_2}$$

$$T_{25} = \frac{T_{21}}{(T_{12})^2}$$

$$T_{26} = \min(x(n))$$

$$T_{27} = \sum_{n=1}^{N} x(n)^2$$

$$T_{28} = \sum_{n=1}^{N} log(x(n)^2)$$

$$T_{29} = \sum_{n=1}^{N} x(n)^2 \times log(x(n)^2)$$

Feature Selection The entire set of extracted features might not be related to the status of the system. To reduce the computational load and also improve the accuracy of final results, it is required to use the most dominant features that convey related information to target task of the study that is being considered. Therefore, feature selection algorithms have to be used to accomplish this task. Feature selection, on the one hand, reduces the computational load on classifier and helps in accuracy improvement. On the other hand, once the most dominant features are selected, for final implementation of the PHM algorithm, only those features will be required to be extracted and in this manner, less process will be required for feature extraction stage and consequently, the data size and network usage will be reduced.

In the current study we have used Improved Distance Evaluation (IDE), which grades features based on their clustering/classification ability between zero and one. As it is described in the article by Lei and colleagues[18], the IDE algorithm ranks features based on separation caused by the feature between data clusters of different classes and the less variance between data points in same class. In other words, a feature that creates higher distance between conditions and at the same time has less variance inside each condition is rated higher than others. Once features are ranked, the most dominant ones can be selected by defining a score threshold. In the current work, we selected different combination of features by varying the threshold between 0 and 1 with steps of 0.1.

Fault Classifier: In the classification step, there are various options for a PHM solution designer to choose from. Depending on the type of application, some algorithms have advantages and disadvantages. In this stage, we have selected the powerful Artificial Neural Network as one of the classifiers. ANN has been successfully used in several PHM applications and in particular for fault diagnosis of gearbox and bearing [2], [17], [19].

In contrast with ANN, which is an eager learning method, we chose K-Nearest Neighbor (KNN) classifier as a lazy learning method to compare its classification rate and computational load. The reason for this comparison is to reveal the processing difference between a simple and complex classifier along with their classification performance, and highlight the possibility of using simpler methods with the support of better feature extraction.

Evaluation: Using different methods and parameters will result in various combinations where their classification rate and process duration are important factors for selecting the most optimized combination for each case study. We will compare the effect of signal processing and classification method on process duration and finally among all combinations, the most accurate yet computationally optimized combination will be selected.

Case Studies: In this paper, we applied the same fault diagnosis method to two case studies. The first case study is the gearbox test bed data that was used in The 2009 Prognostics and Health Management Data Challenge (PHM 2009 Data Challenge). The gearbox contains several mechanical elements such as 3 shafts, 4 gears and 6 bearings. In the original dataset the gearbox had been run under 5 different speeds, 2 different loads and 2 different gear types [20]. In this research, we chose helical gears with revolution speed of 50Hz (3000 rpm) and high load (the amount of torque load is not specified).

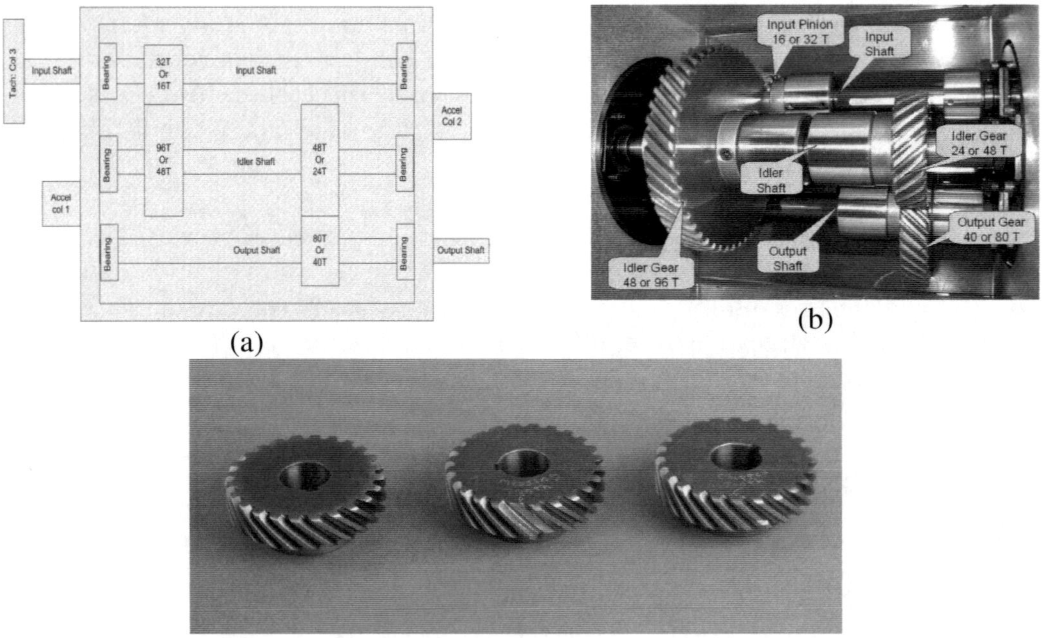

(a) (b)

(c)

Figure 2: Schematic (a) and Actual Inside View (b) of Gearbox Used in PHM 2009 Challenge Data (PHM 2009 Data Challenge)
(c) Example of Gear Faults. Left to Right: Normal, Broken Tooth, Chipped Tooth

Table 2: Gearbox Settings For Six Health Classes

Case	Gears				Bearings						Shafts	
	16T	48T	24T	40T	IS:IS	ID:IS	OS:IS	IS:OS	ID:OS	OS:OS	Input	Output
1	N	N	N	N	N	N	N	N	N	N	N	N
2	N	N	Chp	N	N	N	N	N	N	N	N	N
3	N	N	Brk	N	N	N	N	Cmb	Inner	N	Bent	N
4	N	N	N	N	N	N	N	Cmb	ball	N	Imbal	N
5	N	N	Brk	N	N	N	N	N	Inner	N	N	N
6	N	N	N	N	N	N	N	N	N	N	Bent	N

Chp: Chipped Tooth   Brk: Broken Tooth   Inner: Inner Race Fault   Ball: Ball Fault
Imbal: Imbalance   Cmb: Combined   ID: Idler Shaft   IS: Input Shaft
:IS : Inner Side   :OS : Outer Side

The second case study consists of data acquired from a rolling element bearing test bed. One channel of vibration data has been acquired with a sampling frequency of 50 kHz, while the shaft was rotating at a constant speed of 800 rpm. Conventional bearing faults (damage on the inner race, outer race, on the rolling element) have been seeded into the test bearing and data has been collected from a normal bearing, bearing with only one fault and bearings with a combination of faults. The dataset can be divided into 8 health conditions where class name represents the location and type of fault(s) in bearing: "Normal", "Inner Race", "Outer race", "Roller", "Outer race and Inner race", "Outer race and Roller", "Inner race and Roller", "Outer race, Inner race and Roller".

Results and Discussion: Variations in signal processing method, mother wavelet, decomposition level, feature selection threshold and classifier resulted in 1078 combination of parameters for each case study. Precious information lies inside these results that can be represented in various ways. First and most importantly, the result for a classification problem is the accuracy of classification. Although, the maximum possible classification rate, 100%, has been achieved by several combinations; our focus is to investigate the effective factors in obtaining such good results as well as finding the most accurate and efficient combination of methods and settings for similar cases. The availability of two case studies from different assets helps to prove and evaluate the capabilities of the chosen combinations for accurate and process efficient fault diagnosis of similar rotating machines. Figure 3 shows the average accuracy achieved by different signal processing settings. Due to presence of all sets of selected features by different feature selection thresholds, the average accuracy is lower that the best set of features but it helps to emphasize the relevancy of classification performance with respect to the signal processing methods used. In both case studies, ANN has better overall performance comparing to KNN, which is in agreement with the prior expectations when comparing these two algorithms. Although this accuracy difference represents the higher vulnerability of KKN to input data there are still more factors to consider when determining which algorithm has the best overall performance.

From the signal processing point of view, despite the capability of wavelet analysis in capturing transient behavior of signals, there is not a significant difference between the FFT performance and various wavelet settings. This result shows the unnecessity of using more complex signal processing methods when the FFT can provide reasonable results.

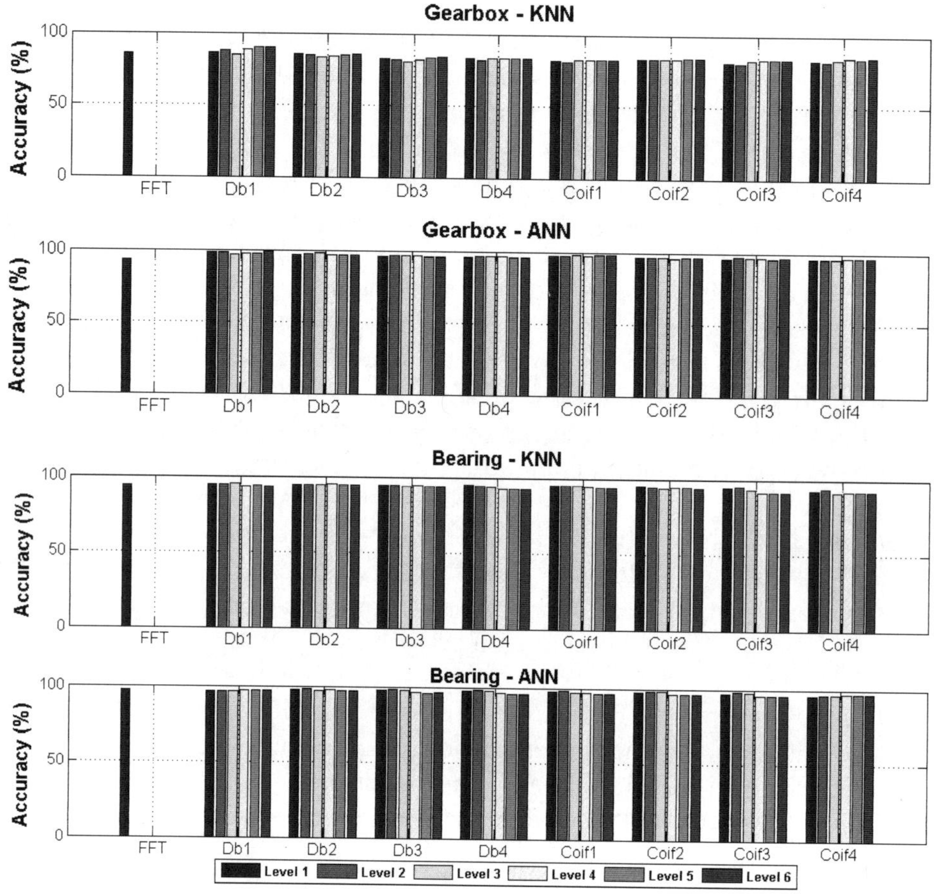

Figure 3: Average Performance Of ANN and KNN With Various Signal Processing Settings

The second considerable metric when we care about process load is the running duration. This metric calculates the amount of time required for performing a certain task on a particular computer hardware. To guarantee the comparability of results, all calculations have been done on same computer running Mac OS X with 2.53GHz core2due processer, 6MB cache and 4GB DDR3 memory. Since minor interruptions cause significant changes in process duration, values have been calculated using average of several runs to eliminate error. For our comparison purpose, we calculated process duration for signal processing (FFT, Wavelet) and classification steps (ANN, KNN). It is worth being mentioned that for the ANN, the duration of simulating the network has been acquired because the training stage is a one-time job that is not required to be done simultaneously. Figure 4 shows the process durations of signal processing methods that have been studied in this research work. Based on the results, the simplest and fastest wavelet setting which is Db1 (Haar) mother wavelet with 1 level of decomposition is more than 5 times slower than FFT where this significant difference turns FFT into the best candidate for the final diagnostic method. On the other hand, the effect of decomposition level on process duration of wavelets matches the expectations where higher decomposition levels have more complexity and require more processing time.

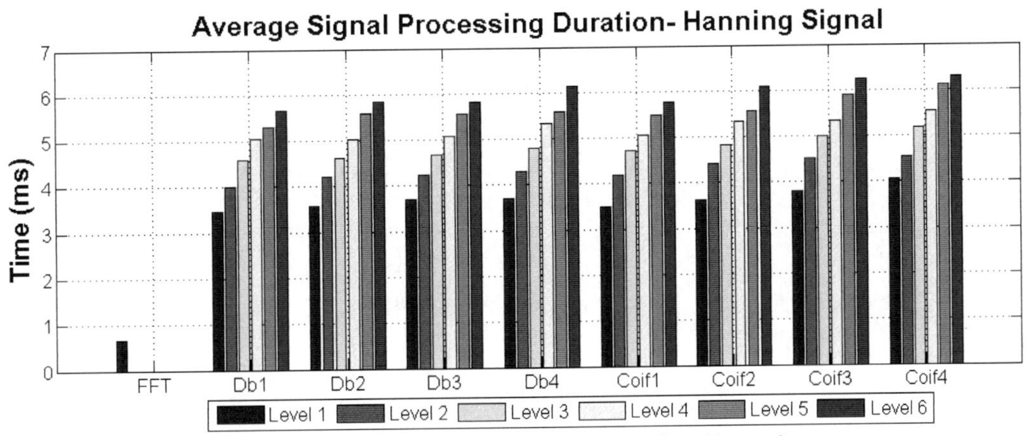

Figure 4: Average Signal Processing Duration

The other most time consuming part of the PHM algorithm is the classification step. ANN is required to simulate the previously generated neural network with the test data and for KNN, it needs to only compare the incoming data with the training set. Although simulating pre-produced neural networks might appear to be faster, the mathematical complexity of neural networks increases the processing time. With an average of 11.84 milliseconds for KNN and 73.70 milliseconds for ANN, the classification of faults with ANN is more than 6 times longer than KNN.

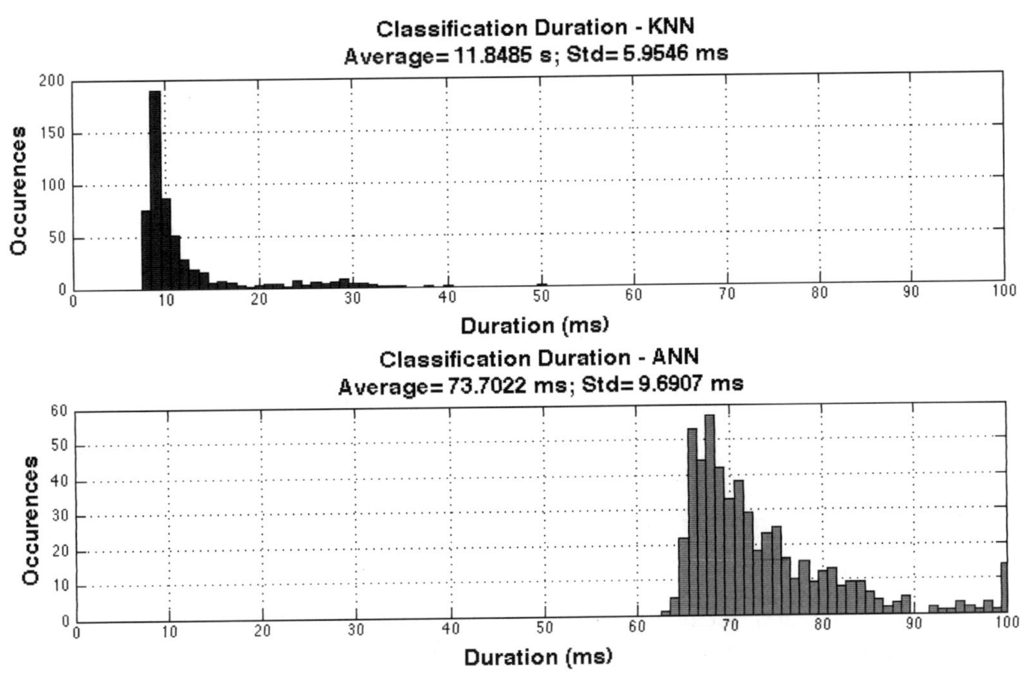

Figure 5: Classification Duration For KNN (Top) And ANN (Bottom)

The process duration comparisons revealed the major differences between our complex and simple methods for signal processing and classification. Therefore to generate a precise yet computationally optimized PHM method, both accuracy and process duration should be considered. As aforementioned, using various settings and methods resulted in 1078 different combinations for each case study where among those we can select the most accurate and process efficient set. Table 3 lists the fastest and slowest combinations with similar output for both case studies where DSP stands for Digital Signal Processing method, FST stands for Feature selection Threshold, #F stands for number of selected features, T_{SDP} presents process duration for signal processing and T_C represents classification duration.

Table 3: Fastest and Slowest Combinations With 100% Classification Rate

Case	Classifier	DSP		FST	#F	$T_{DSP\,(ms)}$	$T_{C\,(ms)}$	Selected
Gearbox	KNN	FFT	-	0.5	29	0.675	9.04	X
	ANN	Wavelet	Db1 Level 5	0.1	75	5.31	28.89	
Bearing	KNN	FFT	-	0.1	6	0.55	5.53	X
	ANN	Wavelet	Coif3-Level 6	0	239	6.93	20.476	

As one can observe, for both case studies there are computationally efficient combinations with exactly the same classification rate as much more complex combinations. e.g. for bearing case study, without considering the number of selected features, the represented ANN-Wavelet combination is more than 4 times slower than the simpler combination. In addition, the number of selected features significantly reduce the data size. In the same example, the set of selected features is 40 times smaller for the efficient combination.

Conclusion: In this article we used a generic and powerful framework along with two tough fault classification case studies to evaluate the importance of the correct selection of methods in each step of fault diagnosis framework. Results of this study revealed the importance of avoiding complexity in solving problems. In particular, it showed the fact that more complex methods will not necessarily improve the final results. Based on the outcome of current research, in designing PHM solutions, particularly those with cloud technology and cyber physical implementations, prioritizing and considering more abstract methods is necessary to make the whole algorithm more efficient. Additionally, excellent classification rates prove the power of the proposed feature extraction parameters despite the independency of these parameters from each case study and application domain.

Ultimately, although the risk in using simple methods is undeniable, this conservative point of view should not prevent PHM algorithms from being computationally optimized. Results of this current effort showed that this framework for algorithm design can result in a much more optimized PHM framework that considers both accuracy and computational requirements. More focus on this subject is required for proposing design metrics to help researchers and engineers select the most efficient PHM combination for their application.

Acknowledgments: The work is supported by the NSF Industry/University Cooperative Research Center on Intelligent Maintenance Systems (IMS) at the University of Cincinnati as well as its company members.

References:

[1] Y. Chen, F. Zhu, and J. Lee, "Data quality evaluation and improvement for prognostic modeling using visual assessment based data partitioning method," *Computers in Industry*, vol. 64, no. 3, pp. 214–225, Apr. 2013.

[2] J. Lee, F. Wu, W. Zhao, M. Ghaffari, and L. Liao, "Prognostics and health management design for rotary machinery systems—Reviews, methodology and applications," *Mechanical Systems and*, vol. 42, no. 1, pp. 314–334, Jan. 2013.

[3] B. Bagheri, "Application of data mining and feature extraction on intelligent fault diagnosis by Artificial Neural Network and k-nearest neighbor," presented at the The XIX International Conference on Electrical Machines - ICEM 2010, 2010, pp. 1–7.

[4] O. D. Snell, O. D. Snell, I. R. C. M. 2. 4. I. I. C. O. Nairne, and I. R. C. M. 2. 4. I. I. C. O. Nairne, "Acoustic bearing monitoring — The future RCM 2008," presented at the Railway Condition Monitoring, 2008 4th IET International Conference on, 2008, pp. 1–5.

[5] Z. Y. Han, Y. L. Liu, H. Y. Jin, and H. Y. Fu, "A Review of Methodologies Used for Fault Diagnosis of Gearbox," *Applied Mechanics and Materials*, vol. 415, pp. 510–514, Dec. 2013.

[6] W. Bartelmus and R. Zimroz, "A new feature for monitoring the condition of gearboxes in non-stationary operating conditions," *Mechanical Systems and Signal Processing*, vol. 23, no. 5, pp. 1528–1534, 2009.

[7] Y. Zhan, V. Makis, and A. Jardine, "Adaptive state detection of gearboxes under varying load conditions based on parametric modelling," 2006.

[8] D. He and E. Bechhoefer, "Development and Validation of Bearing Diagnostic and Prognostic Tools using HUMS Condition Indicators," *He, D*, pp. 1–8, 2008.

[9] B. Bagheri, H. Ahmadi, and R. Labbafi, "Implementing discrete wavelet transform and artificial neural networks for acoustic condition monitoring of gearbox ," 2011.

[10] *Cloud Computing*.

[11] L. Sha, S. Gopalakrishnan, X. Liu, and Q. Wang, "Cyber-physical systems: A new frontier," *Machine Learning in Cyber*, pp. 1–11, Jan. 2009.

[12] R. W. Potter, "Compilation of Time Windoes and Line Shapes for Fourier Analysis," Hewlett Packard, 1985.

[13] B. Lu, Y. Li, X. Wu, and Z. Yang, "2009 IEEE Power Electronics and Machines in Wind Applications," presented at the 2009 IEEE Power Electronics and Machines in Wind Applications (PENWA), 2009, pp. 1–7.

[14] H. Al-Atat, D. Siegel, and J. Lee, "A Systematic Methodology for Gearbox Health Assessment and Fault Classification," International Journal of Prognostics and Health Management, 2011.

[15] K. Zhu, Y. S. Wong, and G. S. Hong, "Wavelet analysis of sensor signals for tool condition monitoring: a review and some new results," *International Journal of Machine Tools and …*, 2009.

[16] J. D. Wu and C. C. Hsu, "Fault gear identification using vibration signal with discrete wavelet transform technique and fuzzy–logic inference," *Expert Systems with Applications*, 2009.
[17] B. Bagheri, "Acoustic Condition Monitoring of MF285 Gearbox Using Wavelet Transform and Artificial Neural Networks ," University of Tehran, Tehran, 2011.
[18] Y. Lei, Z. He, and Y. Zi, "A new approach to intelligent fault diagnosis of rotating machinery," 2008.
[19] R. Isermann, "Model-based fault-detection and diagnosis – status and applications," *Annual Reviews in Control*, vol. 29, no. 1, pp. 71–85, Jan. 2005.
[20] "PHM 2009 Data Challenge Competition," 2009. [Online]. Available: http://www.phmsociety.org/references/datasets.

DATA REDUCTION APPROACH – ACOUSTIC EMISSION (AE) BASED DEFECT ANALYSIS: DEVELOPMENT OF BURST DETECTOR ALGORITHMS

F. D. Boos, R. Baltes, F. Toutounchi and K. Nienhaus
RWTH Aachen
Wuellnerstr. 2
Aachen, Germany 52062
Telephone: (+49) 241-8094522
dboos@imr.rwth-aachen.de

S. Denault
MIT
3 Ames St W215
Cambridge, MA 02142

Abstract: Acoustic emission (AE) techniques offer the potential for new techniques of condition monitoring of bearings, especially for very large roller bearings with slow rates of rotation. Acoustic emission data processing must be carried out by a computer, necessitating algorithms to process recorded signals. One such algorithm is a burst detector, used to identify emission events. This paper deals with the design of a burst detector with the goal of improving upon existing algorithms. The algorithm has the benefit of detecting small amplitude and overlapping bursts, as well as supplying additional information about the decay portion of the burst.

Key words: Acoustic Emission; Burst Detection; Diagnostics; Signal Processing;

Introduction: Acoustic Emission, once developed as non-destructive testing method for steel design, has become commonly known in machine health, bearing diagnostics and diagnostics for mechanical structures. Especially for slow rotating or oscillating bearings such as in wind turbines or grinding mills. Since the early 2000 years publication activity regarding Acoustic Emission analysis has risen to a significant number in these topics. International publications show how to describe a failure of a system such as bearing or other mechanical components by analyzing AE data in time- and frequency-domain. Often, publications show a development of features in time-domain based on the detection of an Acoustic Emission burst. Normally this is done by using a static threshold for the detection of an AE burst. This is associated with problems. For example:

- Because of the static threshold changes in noise can't be monitored → This leads to a problem regarding the detection of bursts
- Burst which run into each other can't be detected
- If the static threshold is too high no bursts are detected.

In order to solve these problems new methods for the detection of bursts are necessary. Two new methods are presented in this paper. Based on these methods, the same AE features as described in [1] or [2] can be used. The following detection methods will be presented.

- Static Threshold
- Edge Detection
- Shifting Windows

Burst detection using a static threshold: One of the main challenges in the field of acoustic emissions is handling and analyzing the data which is collected, as many applications involve a plurality of sensors in different positions with a very high sampling rate (1 MHz or greater). Thus, all analysis of the data must be automated, leading to the need to develop algorithms which can recognize Acoustic Emission events present in data and extract important parameters from these events.

One established burst detection method, with roots in the analog electronic techniques used during early development, makes use of a static threshold which can be compared with the signal level. In Figure 1 the red line indicates a static threshold V_{th} at 0.5V. If the signal rises above this threshold, a burst is counted. In this case – the small-amplitude burst D is not counted.

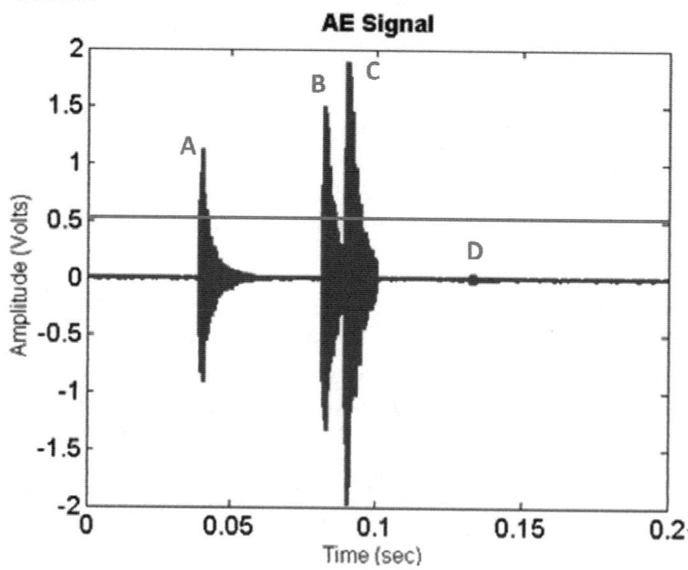

Figure 1: Sample AE signal containing bursts and static threshold level V_{th} of 0.5 V , note single burst (A), overlapping bursts (B and C) and small-amplitude burst (D).

This method has the advantage of being very simple with regard for implementation; however, issues occur if bursts are of a small amplitude (Figure 1 (D)) relative to underlying noise and if bursts occur in quick succession. This can lead to an overlapping (Figure 1 (B and C)) and can result in the fact that only one burst is counted if the amplitude does not fall back below the threshold value used to determine the burst event.

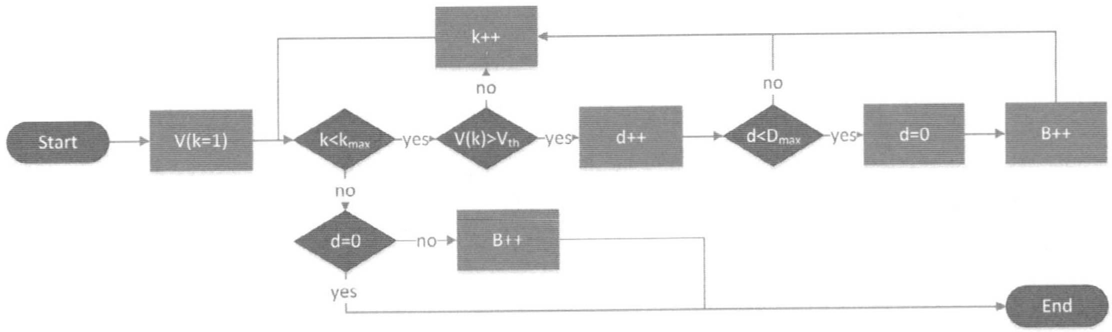

Figure 2: Possible implementation of a burst detector algorithm with static threshold.

A possible implementation of a burst detecting algorithm with static threshold is shown in Figure 2, where k is the index of every sampled AE signal value V(k). V_{th} represents the threshold value, d is a counting index for the number of sampled values V(k) greater than the threshold V_{th}. D_{max} represents the defined maximum value of burst duration and B represents the count of detected AE bursts. This described algorithm can detect the burst A from Figure 1. But the correct detection of burst B and C as two separate bursts depends on the right choice of the threshold value and maximum duration count D_{max}. Burst D can't be detected since the maximum amplitude is far below the threshold value.

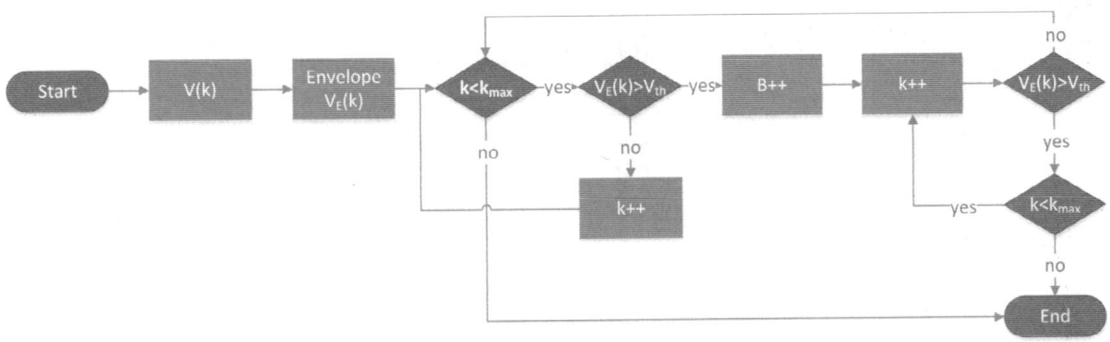

Figure 3: Possible implementation of a burst detector algorithm with static threshold.

Figure 3 shows another possible algorithm for static threshold burst detection. Based on the AE signal V an envelope V_E of the burst containing AE signal is calculated. As long as the value $V_E(k)$ of the enveloped signal is greater than the threshold value, only one burst is counted. If the level of the envelope signal drops below the threshold value, the next burst can be detected. This implementation needs less parameters than the algorithm from Figure 2. There is no need for using the maximum duration time. The detection of burst A from Fig 1 is possible. Burst B and C can be separated if the threshold value is chosen right, but burst D can't be detected since the maximum peak value is still far below the threshold value. An Implementation of this method can be found in [3] and [4].

Burst Detection using the concept of edge detection: The following algorithm searches for burst events based on the archetypal characteristics of acoustic emission bursts from common sources as concrete and steel; these signals have a sudden rise in signal intensity at the onset of the acoustic emission event (see Figure. 4). Regions with this characteristic are recognized as the onset region of the hit. Thus, as mentioned above, two shortcomings of the static threshold method are overcome:

1) small amplitude bursts, which might otherwise not exceed a threshold value, can be recognized solely based on their onset.
2) consecutive bursts which overlap each other in time can be recognized, even when the signal does not fall below a threshold value in the period between the onset of the two bursts.

This method can be thought of as analogous to the edge detection method used in digital image processing, albeit far simpler, since a digital image has an additional dimension compared to an acoustic emission signal. The algorithm essentially searches for regions of the signal where the signal envelope climbs steeply "rising edges" of the envelope.

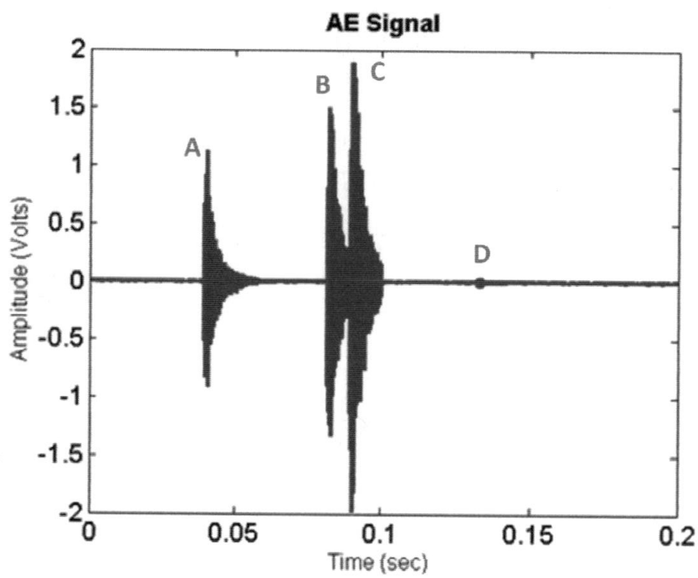

Figure 4: Sample AE signal containing bursts, note single burst (A), overlapping bursts (B and C) and small-amplitude burst (D).

The algorithm consists of the following steps: the envelope of the data from the acoustic emission sensor is first calculated by rectifying the signal and applying a low-pass filter (Figure 4 shows a typical acoustic emissions signal with bursts, the envelope is displayed in Figure 5).

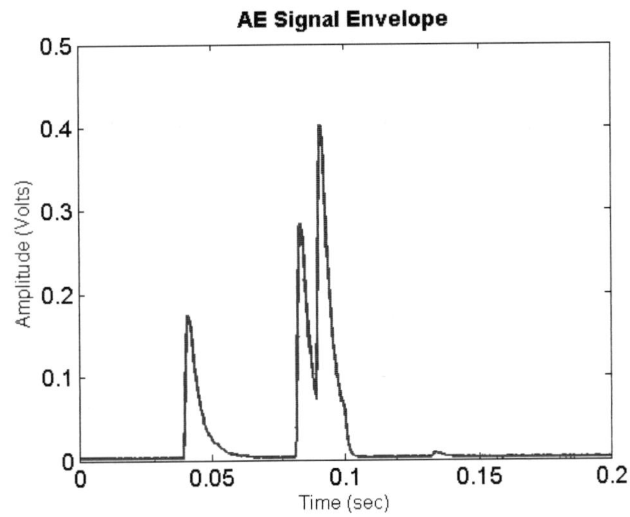

Figure 5: Envelope of sample signal. Characteristic shape of burst envelope is demonstrated.

The envelope is divided into small zones, and the average derivative of the envelope for each zone is computed (as shown in Figure 6).

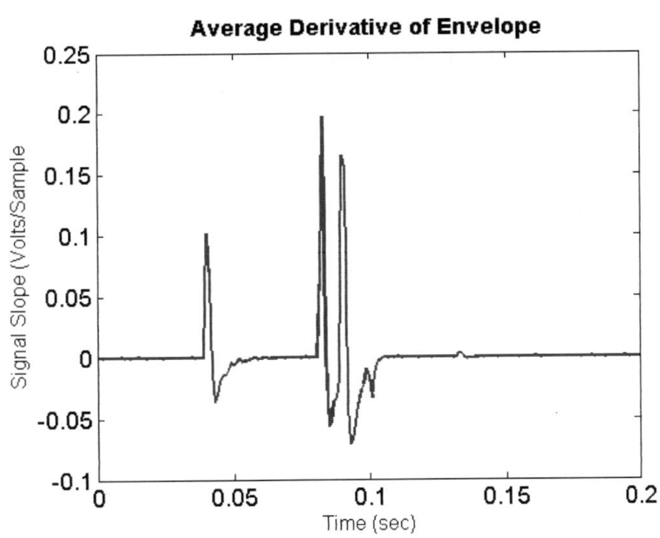

Figure 6: Average derivative of envelope over small regions

Note that it is unnecessary to compute the continuous derivative of the signal; the average derivative is given by the following equation:

$$\frac{\Delta V_e}{\Delta t} = \frac{(V_{e2} - V_{e1})}{(t_2 - t_1)} \tag{1}$$

where Ve_1 and t_1 are the envelope value and time at the beginning of the region, and Ve_2 and t_2 are the envelope value and time at the end of the region. Thus, only two data points are needed per region. If this derivative exceeds a certain threshold then the region is recognized as being a part of a steeply rising portion of the signal. Consecutive regions recognized as such are taken together to represent the onset of a burst (see Figure 7).

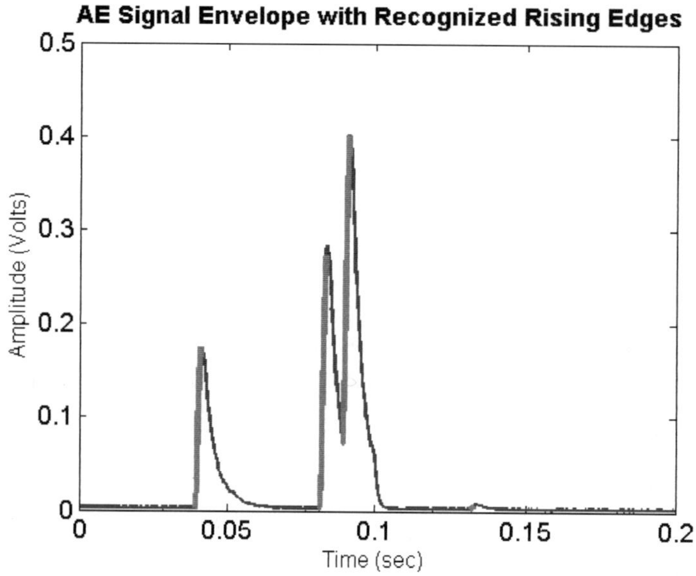

Figure 7: Signal envelope with highlighted regions (in red) that were detected by the algorithm as having a high average derivative. Note that the overlapping and small-amplitude bursts were also recognized.

From here, the algorithm must determine where the detected burst ends. This can be problematic, especially for bursts with very low amplitude; the current method uses a fixed threshold to determine where the signal falls below a level close to the underlying signal noise. In the case of a small-amplitude burst, there is a possibility that the signal was not above the threshold to begin with. In this case, it may be practical to assume a decay time equal to the onset time, or use another minimal decay time depending on the application.

Burst Detection by using sliding windows: The following method describes a third principle how to detect an Acoustic Emission burst. This algorithm uses sliding windows and detects local maxima in order to describe bursts. For detecting bursts in a discrete data set, it is important to find the local maxima of the data set. A burst happens whenever an extreme change of amplitude occurs in the data set. Therefore this change of amplitude can be represented as a local extremum or a local maximum.

By windowing we can limit the amount of data being examined. The maximum value and its position can be found within this data set and marked as a burst candidate. The window starts examining the data set from the beginning and then moves along the time axis with a specific step size.

Window definition and length considerations:
A window is referred to a block set of data samples to be analyzed together at a time. When this block moves along the time axis the so-called sliding window is constructed. While moving the window, an overlap between the ith data set and $(i+1)^{th}$ occurs. When a greater maximum in the overlapping part of the following window is detected the maximum index will be replaced by this greater maximum.

In order to avoid detecting regular maxima, and labeling them as bursts, the window length should be sufficiently large. On the other hand, in order to avoid missing the important bursts, the step size for moving the window should be sufficiently small. After testing the algorithm for different types of data, a window length of 2 sec, which is equivalent to 2 M data points, looks sufficient for a sampling rate of 1 Mhz. Furthermore the step size of 0.2 sec, equivalent to 200 K symbols, looks also reasonable. The choice of the window length and the step size is a very critical part of the algorithm, which may differ for different sets of data. The overall length of the data set is also important, when choosing the window length and the step size.

Burst detection algorithm step by step:
In the following figures the execution of the algorithm can be seen step by step for a sample data set. As shown in Figure 8, the data set length is 3 seconds and the window length is 2 seconds long and the data set has been extracted with a sampling rate of 1 MHz. The window begins by examining the sample values of the data set starting from left and tries to find the local maximum. The marked sample (colored red) is a candidate burst within the dataset (Figure 8 - 1 – 2). The window starts sliding with a step sizes of 0.2 sec. The procedure continues until (Figure 8 - 1 – 6) the first candidate happens to be out of the window. Then another maximum is being chosen as the candidate burst and the window keeps sliding until it reaches the end of the dataset.

To get a more accurate measure at the beginning and ending part of the data set, it is possible to extend the dataset at the start and end to have a bigger range in the windowing and get more precise burst detection.

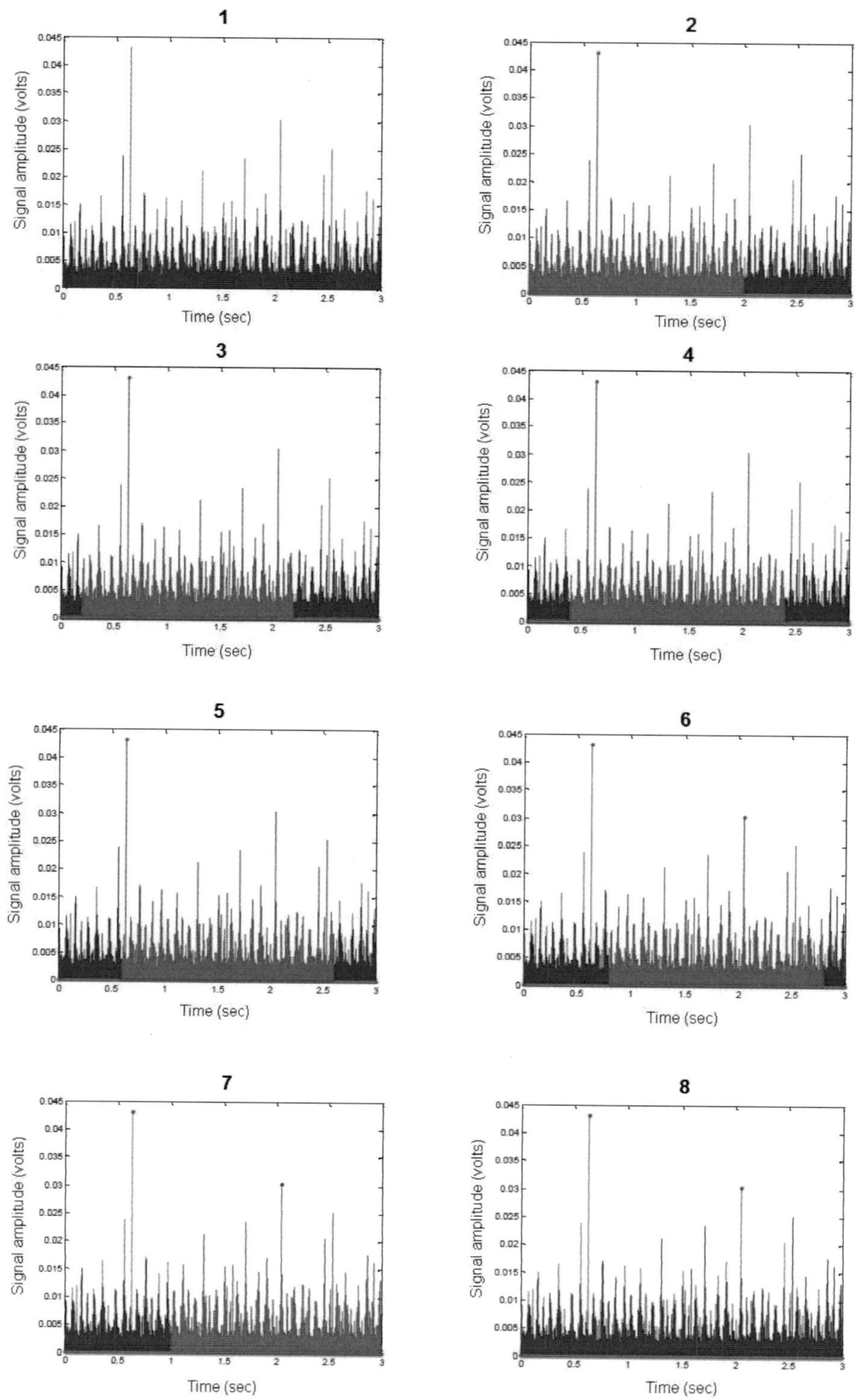

Figure 8: Step by step algorithm from 1 upper-left to 8 lower-right)

The advantages of the method:
There is no need to define a threshold like in the conventional methods and the algorithm doesn't need a parameter as an input. Although the window size and the step size should be chosen, they are independent of the signal level. For these parameter only the length of the data and the type of the data, i.e. how many bursts and how much noise is expected, needs to be considered. Since there is no continuous approach in this method and the maximum values are used for finding the final bursts, the positions of bursts are completely precise as there is no approximation used in finding them. In noisy data sets, wherein it is hard to distinguish a real burst from noise, the algorithm works quite robust and shows bursts which can be achieved by selecting long window lengths (Figure 9).

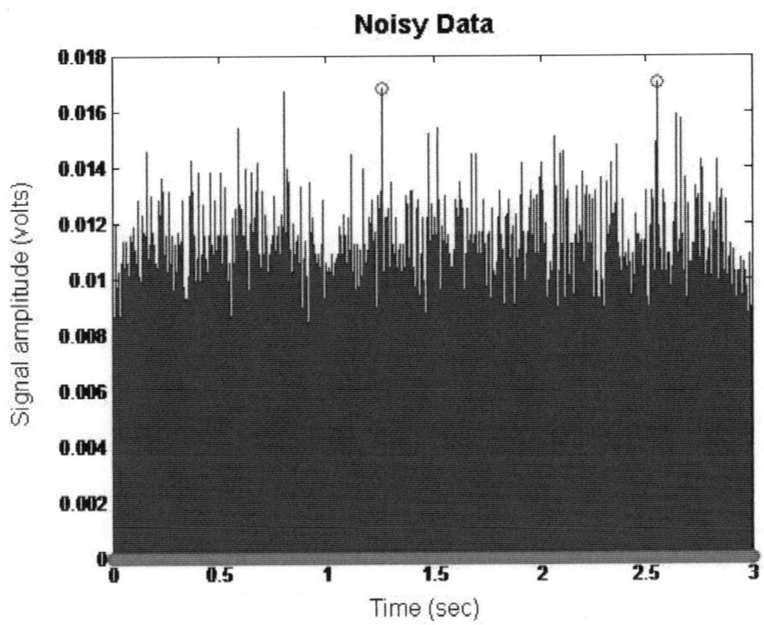

Figure 9: Bursts detected by the algorithm from a noisy set of data

Another advantage of the method is when multiple bursts are present and it is necessary to identify the dominant burst. By using this method, the window does not let the smaller burst appear in the final results and only the dominant burst appear to be real burst and an output of the algorithm(Fig 10).

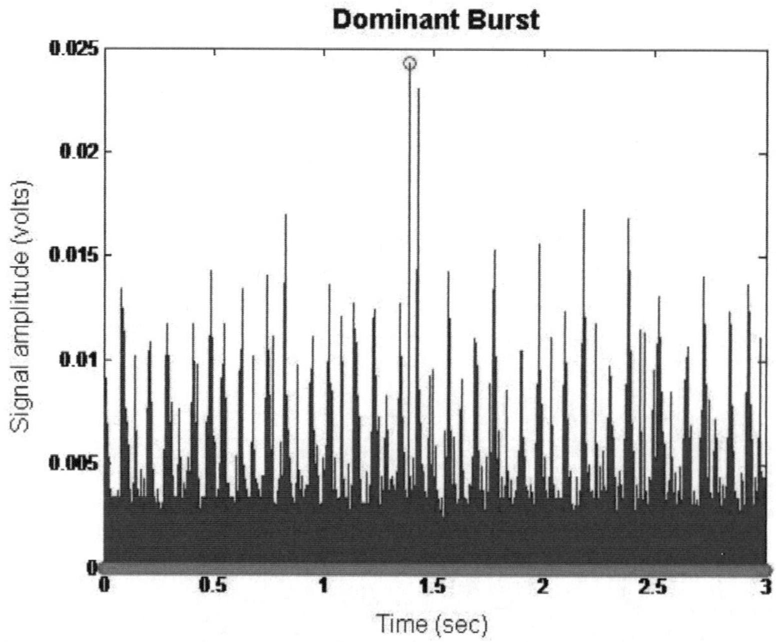

Figure 10: Keeping the dominant burst and rejecting the nearby ripples

Results: Both presented algorithms proved satisfactory to detect bursts in a test signal containing acoustic emission events. Results were especially gratifying for the case of overlapping bursts, where the issues surrounding the decay threshold were not present. Small-amplitude events could also be identified. The results appear to correspond quite closely to the results of manual burst detection, however a statistical test to determine efficacy is suggested to obtain a quantitative measure of the algorithm's performance compared to other methods.

If the algorithm is applied to data in which the bursts do not have a well-defined envelope form with a distinct rising region, the algorithm does not give adequate results. This might be the case for the acoustic emissions of a friction test or emissions from a material with highly nonlinear wave propagation, such as wood. The static threshold based burst detector would likely function better for this case.

The edge detection algorithm operates with very little computational overhead, as it only performs as very simple operation on a small amount of the envelope data. This method could be implemented with embedded systems and run in real time.

Conclusion: The algorithms laid out here are readily applied to analysis of acoustic emission data and should prove useful in the field of condition monitoring. The method of detecting steep increases in signal intensity improves on the static threshold for certain cases, without demanding an increase in complexity. The method of shifting windows can be implemented in modern AE equipment as well. The presented methods can be implemented in many different AE diagnostic fields such as [5] and [6].

Further work is needed to implement the algorithms in completed form, either as real-time software or post-processing. Tests on established Acoustic Emission sources, such as tensile tests and Hsu - Nielsen tests, may help to optimize the algorithm and make its strengths and weaknesses clearer.

References:

[1] D. Mba, MFPT Acoustic Emissions and monitoring bearing health, Tribology Transactions, 46 (3) 2003, Pages 447-451.

[2] J. Miettinen, P. Pataniitty, Acoustic Emission in Monitoring Extremely Slowly Rotating Rolling Bearing 1999, Pages 289-297.

[3] K. Nienhaus, F.D. Boos, C. Kuech, R. Baltes, Data Reduction Approch - A Study for Simplifying Acoustic Emission (AE) Data, Progress in Acoustic Emission XVI Proceedings of the 21st International Acoustic Emission Symposium, Japan 2012, Pages 31-36.

[4] K. Nienhaus, F. D. Boos, K. Garate, R. Baltes, Development of Acoustic Emission (AE) based defect parameters for slow rotating roller bearings, 25th International Congress on Condition Monitoring and Diagnostic Engineering 2012, Huddersfield, UK - Journal of physics / Conference series, 364, Pages 1 - 10.

[5] K. Nienhaus, C. Klein, F. D. Boos, R. Baltes Acoustic Emission zur Materialerkennung im Schneidprozess. AKIDA / Aachener Kolloquium für Instandhaltung, Diagnose und Anlagenüberwachung 2012, Pages 45 - 58.

[6] A. Keshtgar, M. Modarres, Detecting Crack Initiation Based on Acoustic Emission, Chemical Engineering Transactions Vol. 33 2013 Pages 547-551.

ORDER TRACKING UNDER RUN-UP AND RUN-DOWN CONDITIONS

Michael D. Coats and Robert B. Randall
School of Mechanical and Manufacturing Engineering
University of New South Wales
Sydney, NSW2052, Australia
b.randall@unsw.edu.au

Abstract: In recent years there has been considerable interest in the analysis of machine vibration signals in the presence of speed variations, sometimes over wide ranges. A typical application is to wind turbines, where the most efficient systems allow the speed to vary by ±30%, which can just be accommodated by phase demodulation based order tracking techniques, limited to a maximum speed range of 2:1. This paper shows how analysis over a much greater speed range, for example a run-up or run-down, can be achieved by dividing the signal into overlapping segments, in each of which the speed range is less than 2:1. In the order domain, the highest frequency to be retained is proportional to the speed, and so at lower speeds the required sampling frequency can be reduced accordingly, as long as higher frequency components are first removed to avoid aliasing. The paper explains how this can be done in the most efficient way. The speed reference signal can be a once-per-rev tacho, a multi-pulse-per-rev encoder signal, or can be extracted from the signal itself under certain conditions which are explained in the paper. The paper discusses the pros and cons of the different approaches, with respect to the possible applications of the results.

Key words: Computed order tracking; large speed range; phase demodulation; run-downs; run-ups.

Introduction: Order tracking has made possible the analysis of vibration signals from variable speed machines as though they were exactly constant speed, for example establishing that certain frequency components are exact harmonics of a particular shaft speed [1]. If a signal is sampled with a fixed number of samples per revolution of a shaft, events that are phase-locked to the rotation of that shaft, such as unbalance or gearmeshing, will be approximately periodic in the rotation angle domain, even if the speed varies. Any deviation from periodicity will come from amplitude variations; for example a faulty gear tooth will always mesh at the same point in each rotation, but the response amplitude may vary as resonance frequencies are traversed, in particular over a wide speed range.

Order tracking has long been a requirement as a precursor to TSA (time synchronous averaging). However, this usually required that the speed did not vary greatly along the total record, for example the case of gearboxes driven by an induction motor. In recent years, there has been an increasing interest in the analysis of machines with more widely varying speed, for example pitch controlled wind turbines with doubly fed generators, which can vary in speed by as much as ±30%.

Order tracking is sometimes of use in situations where speed varies over an even wider range, for example a complete run-up or run-down between near zero and full speed. Such order tracked signals would in general not be suitable for TSA, but there can be interest in seeing how the various harmonics change with speed, for example as they traverse fixed resonance frequencies. This can be studied to some extent by using a waterfall array (spectrogram) of a series of normal spectra at different speeds, but the width (and thus amplitude) of the spectral components is sensitive to how much the speed changes along each segmental record. With order tracked records, there can be some broadening of spectral peaks from amplitude variation (modulation), but this is an intrinsic property of the signal itself, and not of the analysis method.

There are a number of ways of achieving order tracking, of which the earliest was to simply sample the signal at a rate proportional to the speed (using a sampling clock controlled by a phase-locked-loop), but this gave problems in ensuring that the signal was lowpass filtered at a variable cut-off frequency proportional to speed, and all information about the timing of events was lost. An alternative is given by computed order tracking, where a reference signal, typically a once-per-rev tachometer signal from the reference shaft, is recorded at the same time as the actual signals to be analysed, and then used to divide each tacho period into a fixed number of resample periods. Interpolation using various orders of polynomials was initially used to determine both the time sampling points within each tacho period, and the values of the resampled signals at those times. One of the earliest methods was by Fyfe and Munck [2], whose method implied constant angular acceleration (steps) between tacho pulses, and thus quadratic variation in angular displacement. McFadden [3] showed that cubic spline interpolation gave much lower errors than zero, first or second order polynomials.

If the spectrum of a phase modulated signal is bandlimited, then it can be windowed by a rectangular bandpass filter, and this corresponds to sinc(x) interpolation in the time domain, in principle giving the true phase/time curve to any degree of resolution. However, the sinc(x) function is infinitely long in the time domain and would have to be truncated if applied there, whereas inverse transformation from the frequency domain gives the true interpolated curve. In [4] the authors have shown that for practical purposes, "bandlimited" can be taken as separated down to -40 dB (with respect to the largest component in the band) from adjacent unrelated frequency components. In [4] it is also shown that this 40 dB separation can just be achieved between the first and second harmonics of a pulse tacho signal, for a maximum speed variation of ±33% of the fundamental component, in other words from 2/3 to 4/3 of it, or a maximum possible speed range of 2:1. This applies only for very low modulation frequencies and the range is even lower if the rate of frequency change is higher [4]. The amount of maximum frequency deviation against modulation frequency, both expressed as a percentage of the carrier frequency, is given in Figure 1 for three different criteria for what constitutes significant sidebands. The most restrictive is the 40 dB dynamic range (DR) criterion mentioned above, but curves for 20 dB DR and Carson's rule are also given [4]. When the modulation signal is not sinusoidal, equivalent parameters can be used as described below. Figure 1 is for the first harmonic, but an equivalent diagram can be made for higher harmonics [4].

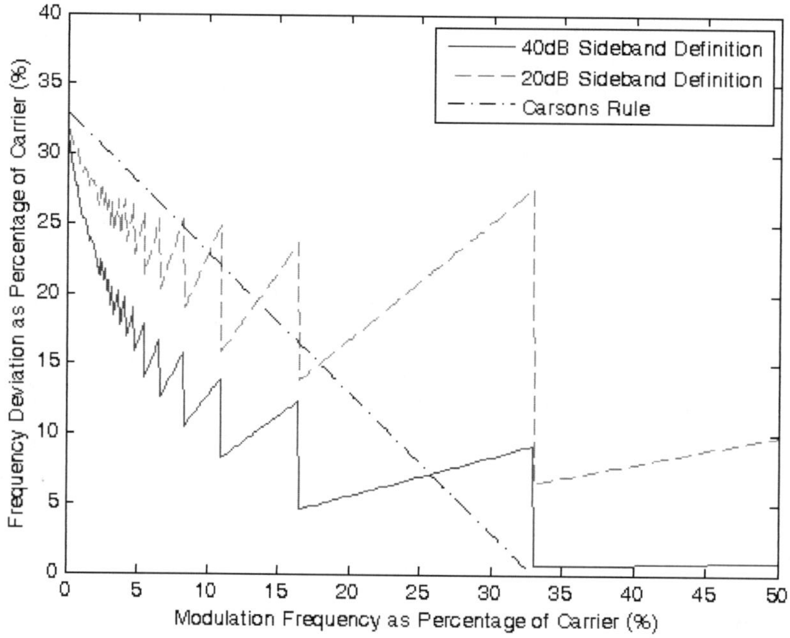

Figure 1: Envelope of permissible combinations of frequency deviation and modulation frequency

In a run-up or run-down situation, the overall speed range is generally greater than 2:1, but this paper proposes a method whereby the overall signal can be broken up into segments in each of which the speed range is < 2:1. If these segments are overlapped, by use of appropriate windowing, a smooth transition can be made between the segments to obtain the order-tracked version of the whole signal. Since the segments are defined by a maximum speed ratio, it is not possible to extend down to zero frequency, but four or five segments will normally extend down to a speed below which there is nothing of interest in the signal.

Segmentation of the Signal: This section describes how the signal can be segmented, including choice of a suitable window in the overlap areas, such that the individually processed segments can simply be added together after processing to give a continuous signal with no apparent disruption in the overlap areas.

Four segments of 2:1 (not considering the overlap) will mean that the centre of the lowest frequency band will be a factor of 8 below that of the highest frequency band (75% of the maximum frequency), and the lowest frequency will then be 1/16 (6.25%) of the highest frequency. Five segments will normally cover the same frequency range in more restrictive situations, including allowance for overlap. For a constant rate of frequency sweep, the length of each segment will be proportional to the centre frequency of the band, and thus proportionally shorter at lower frequencies.

There are two reasons why the ends of each segment should be weighted by an appropriate window before processing, apart from simply ensuring that the total recombined signal has a smooth transition with a weighting of 1 in the overlap areas:

1. Since the processing is done using the FFT, there are always wraparound errors near the ends of any processed signal after transformation to the frequency domain, editing in the frequency domain and transformation back to the time domain, and the window minimizes this effect.
2. The final interpolation of the signals at the resample positions is done by cubic spline interpolation, with errors near the ends because slope and curvature are not specified there (internally they are implicit from the values of the two samples on each side).

To maintain a uniform weighting of the recombined signal after treatment, the weighting of each signal must decrease to zero towards each end of each segment (where the wraparound errors are greatest) at the same time as the weighting on the same part of the signal in the adjacent segment increases correspondingly. This is illustrated in Figure 2.

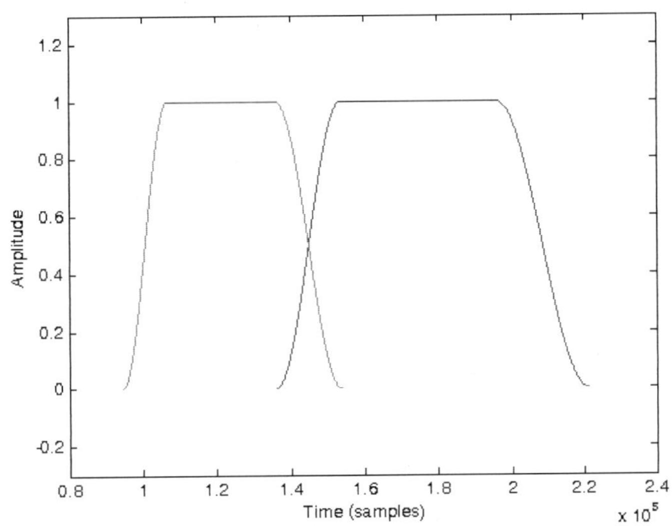

Figure 2: Windowing Functions for Two Adjacent Segments

The windowing functions illustrated there are half Hanning functions where a full Hanning window is variously described as a "raised cosine" or "cosine squared" when centred on zero time. A half Hanning weighting is used to taper each end as shown. They overlap at the point where each equals 0.5, and add to 1 everywhere in the overlap zone. Even though the transformation from equal time to equal phase intervals distorts the x-axis of the windows, this distortion is identical for both segments so that they still add to 1 everywhere, and each equals 0.5 at the intersection point originally in the centre of the overlap region in the time domain. It is suggested that these (break) points should be made the reference points for lining up the phase axis for the whole record after resampling.

It is desirable to keep the overlapping of segments to the minimum needed to suppress errors due to the wraparound errors from FFT processing, and cubic spline end effects, since this will minimise the total number of segments required to adequately span the original signal. This will not necessarily be the case if the range of each segment is taken as the range between break points rather than the total range including overlaps, and it can be argued that this is a reasonable compromise since the overlap region should be dominated by the part with weighting > 0.5 which is within the breakpoints, in particular if the overlap region is kept small such as the 5% used in the example given here.

In some cases it has been found critical that the reference signal used for order-tracking is not windowed. Windowing should have no effect on the phase of a reference signal, which is its only property of interest in the order-tracking process, and the phase can become indeterminate when the amplitude gets too small. If it is desired to have an order tracked version of the reference signal, it is recommended that a copy should be made and treated in the same way as other measured signals. It is not necessary to window the extreme ends of the highest and lowest speed segments, as the ends should be discarded in any case. For this reason, a section of the signal at constant speed should be included at the upper end so that a portion can be discarded finally without loss of information, and the same for a portion at the lowest speed end where the vibrations are not of much interest. The whole order-tracked signal after re-assembly can be weighted with a new window function for further processing, such as order spectrum analysis.

It is suggested that the overlap should be expressed as a percentage of the length of the lower speed segment, for example the one on the left in Fig. 1. This means that the overlap will in general be different at the two ends of each segment, and be smaller than the nominal percentage at the upper end of each segment. The nominal overlap used for the results given here was ±5%.

Prevention of aliasing: In general it will be desired to retain a given number of orders in the order tracked signal, typically the highest valid order in the spectrum at the highest speed. This will normally be determined by the lowpass antialiasing filters used for the original data acquisition, typically at about 40% of the sampling frequency. The frequency of this highest order goes down proportionally to the speed, and since the resampling needs a minimum of two samples per period of this highest order, the required sampling frequency should reduce proportionally. However, to avoid aliasing, the sampling frequency must be more than twice the highest retained frequency even at the lowest speed in each segment. Since the maximum speed ratio for any segment is 2:1, it is suggested that the sampling frequency should initially be doubled, before any further processing, so that the sampling theorem will still be satisfied at the lowest speed. For lower speed segments, the frequency content should be reduced correspondingly by lowpass filtering the signal at a cutoff frequency corresponding to the highest order at the highest speed in that segment.

This lowpass filtering is done most efficiently in the frequency domain, by simply setting spectrum components above this frequency to zero before transforming back to the time domain. Since this effectively gives an ideal filter, the final resampling frequency only has to be just > twice the cutoff frequency at all speeds in the segment to avoid aliasing. It is important to distinguish between sampling in the time domain and in the angle domain. For optimum efficiency it would be possible to reduce the sampling frequency in the time domain as long as it is still at least twice the lowpass filter frequency everywhere in each segment. It is easiest to do this reduction in factors of two (by dropping every second sample), so it would not be possible for every successive segment unless the speed range were always 2:1. If the time sampling rate is different for adjacent segments, it should be kept in mind that the length of the half Hanning windows in time must be the same in the overlap section, so the length in samples must have a ratio of 2:1 and thus be an even number for the higher sampling frequency. The resampling frequency in samples per period must be the same for the whole signal, and is thus the same for adjacent segments. Since the ratio of sampling frequency to highest valid order in the original digitized signal is typically 2.5 (though increasing to 5 when doubled), the number of samples per period of the fundamental should typically be a minimum of 2.5 times the highest order, but it may be of advantage to increase it to give a better representation of waveforms.

Experimental data: The general procedure will be demonstrated using a signal generated to resemble the response on the casing of a machine such as a gearbox during a run-up from low speed at a linearly increasing rate. It is in fact a gearbox casing (without internals) but excited by a force signal, applied through a shaker, with ten harmonics of equal amplitude (at least in the control signal sent to the shaker amplifier) with fundamental frequency increasing from zero to 200 Hz over 30s. (sweep rate 6.67 Hz/s). Thus, the highest excitation was 2000 Hz, but because of distortion, response harmonics are present over the full valid frequency range of 6,400 Hz (order 32 at the highest speed). The sampling frequency was 2.56 × the highest valid frequency or 16,384 Hz. Figure 3 shows spectrograms of the force and response acceleration signals.

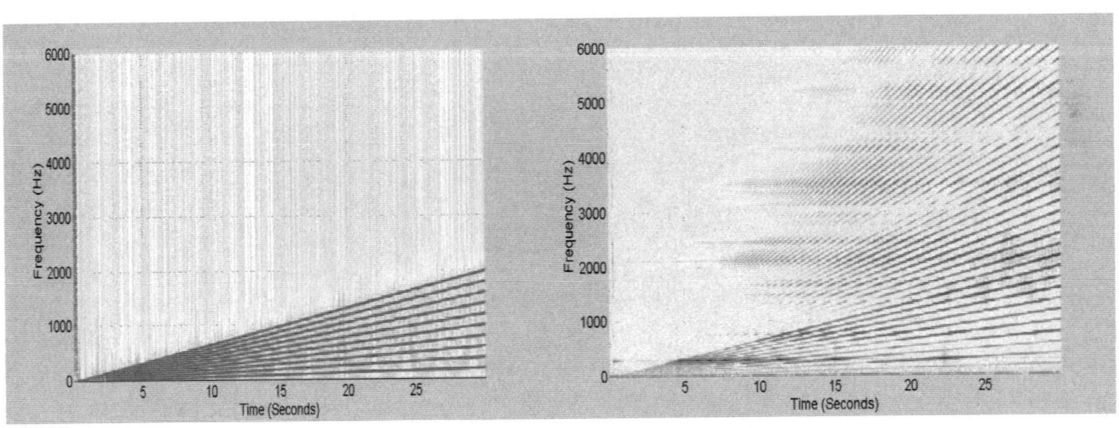

Figure 3: Spectrograms (a) Force signal (b) Response acceleration

The radial lines emanating from the bottom left origin represent the various harmonic orders, and horizontal bands in (b) represent constant resonance frequencies.

Typical procedure: The first step is to decide the break points of the segments, based on a speed vs time diagram. This can be extracted from the signal itself, or from a separate speed sensor on the machine, which cannot be used directly for order tracking in itself.

One way of extracting the speed information is using a spectrogram such as Fig. 3(a) or (b), and tracking the peak value of the most prominent harmonic, as described in [5]. Another is to identify the zero crossings of such a harmonic, when filtered out from the rest of the signal as in [6], since the instantaneous angular speed can be calculated from the reciprocal of the time intervals between zero crossings.

In this case, the speed vs time diagram was taken from the signal sent to the shaker controller. The speed vs time diagram should be resampled to have the same number of samples as the signals (including tacho) to be processed. The speed vs time diagram does not have to be completely accurate, but any uncertainty should be taken into account when choosing break points.

The diagram of Fig. 1 can then be used to choose the maximum allowable frequency deviation f_d for each segment, starting from the highest speed. This may require a little trial and error, since the carrier frequency f_c (used to normalize the deviation and modulation frequencies) will be the mean of the upper and lower frequencies of the segment, and the deviation itself half the difference between these frequencies. The equivalent modulation frequency f_m should be determined from the rate of frequency sweep.

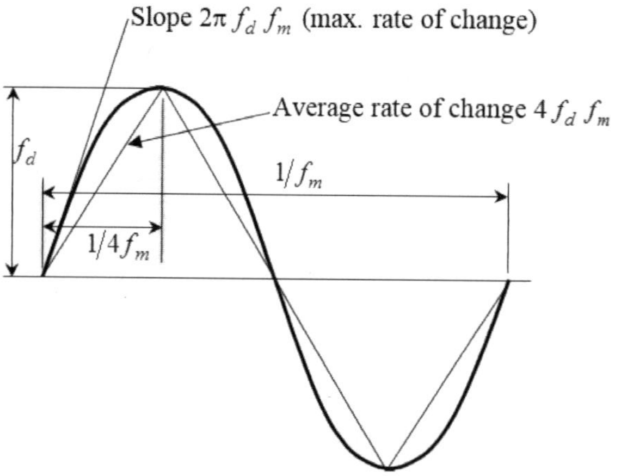

Figure 4: Equivalent modulation frequency for sweep rate W

Figure 4 shows that the maximum sweep rate for modulation frequency f_m is $2\pi f_d f_m$, but an average sweep rate can be found from the line joining a zero crossing to the next peak value, and is given by $4 f_d f_m$. That has been used here and can be used to find the equivalent modulation frequency f_{meq} as follows:

$$W = f_d \times 4 \times f_{meq} \quad \text{from which} \quad f_{meq} = \frac{W}{f_d \times 4} \quad \text{and} \quad \frac{f_{meq}}{f_c} = \frac{W}{f_c^2 \times \left(\frac{f_d}{f_c}\right) \times 4} \tag{1}$$

In this case, assuming $f_d/f_c = 0.3$, then

$$f_c = \frac{f_{\max}}{1.3} = \frac{200}{1.3} = 153.8 \text{ Hz and } \frac{f_{meq}}{f_c} = \frac{6.67}{153.8^2 \times 0.3 \times 4} = 0.023\%$$

so from Fig. 1, $f_d/f_c = 0.3$ is conservative. For constant sweep rate *W*, as here, the ratio f_{meq}/f_c will increase inversely with the square of the speed, so with f_c reducing by 8:1 the value of f_{meq}/f_c would be 1.5%, and f_d/f_c should be 0.25 for the 40 dB criterion. The value of f_d/f_c actually used here for demonstration is 0.18 or 18%.

Figure 5 shows the resulting break points and overlap zones used here, with overlap ±5%.

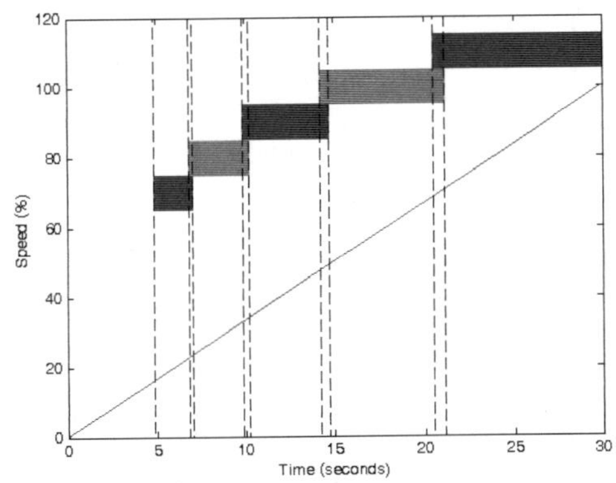

Figure 5: Segment locations with overlap

Figure 6 shows the windowed time records for segment four (from the left). Note that the one for the tacho signal was used for reconstructing the order tracked tacho signal, but not for generating the phase/time map (non-windowed version used). The passage of the response signal through several resonances is seen clearly.

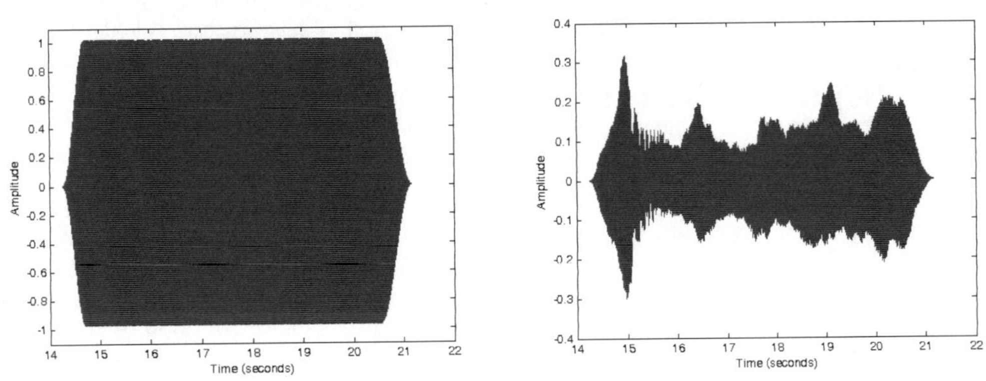

Figure 6: Windowed segment number four (a) Tacho (b) Acceleration

The phase vs time curve is next determined for each segment by phase demodulation of the first order of the non-windowed tacho signal, as described in [4]. Figure 7 compares the spectra of these segments for the lowest (segment 1) and highest (segment 5) speeds. Note that even though they are numbered from lowest to highest, the order of processing

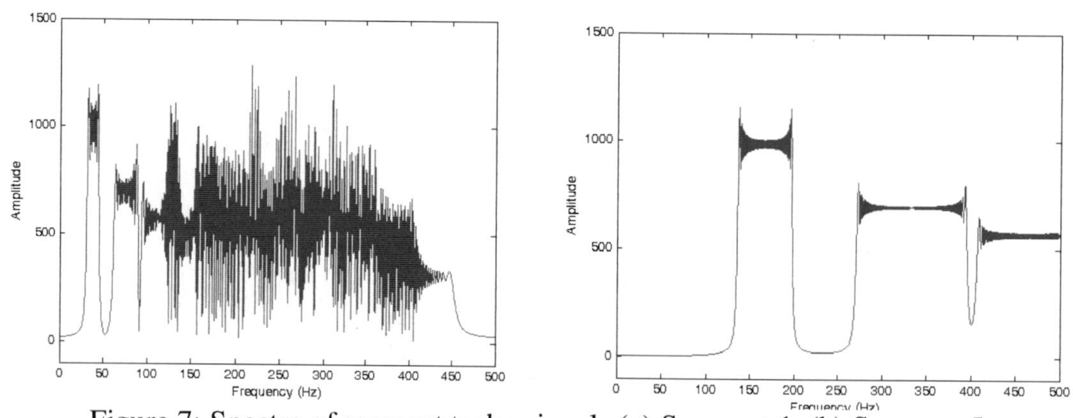

Figure 7: Spectra of segment tacho signals (a) Segment 1 (b) Segment 5

is in the inverse order. It can be seen that because of the small deviation of 18%, the separation of the lowest order from those higher is very distinct, and in fact the second order is almost separated.

As explained in [4] the first order is windowed by a rectangular window and phase demodulated. To aid in the demodulation and unwrapping of the resulting phase curve, the carrier frequency f_c is removed (shifted to zero) as part of the demodulation process, but it is known precisely, and can be added back in to obtain the total phase $2\pi f_c t + \phi(t)$, where $\phi(t)$ is the variation in phase (around the carrier) found by the demodulation process. For each segment, the initial phase at the beginning of the record is typically near zero, but must be raised to match the phase of the adjacent segment at the common breakpoint. It is suggested that the matching should be done at these breakpoints, where there is equal accuracy in both segments. Thus, even though the demodulation is done from right to left, the phase at the lowest frequency of the lowest segment is set to zero, and corrections added to each higher segment so as to attain a match at each breakpoint. Figure 8 shows the phase vs time map obtained by this process for the data in this example.

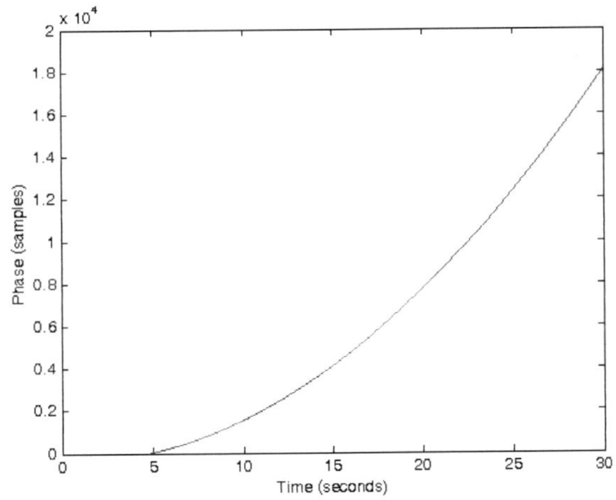

Figure 8: Assembled phase vs time curve from the five segments

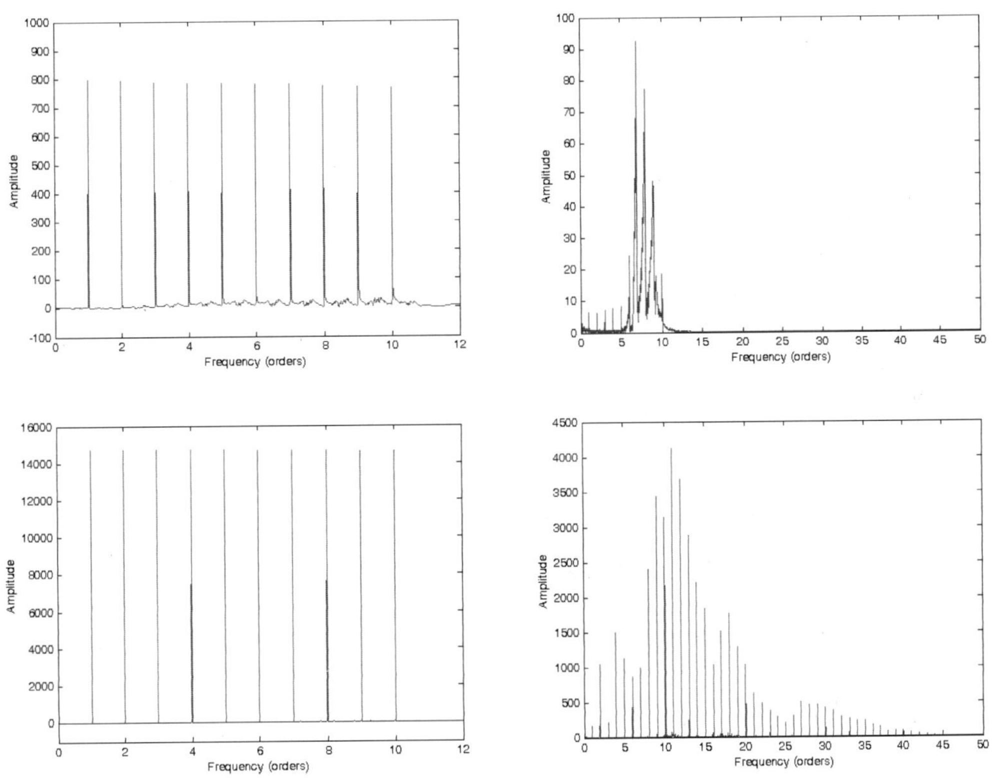

Figure 9: Typical order spectra after order tracking; (a, b) 1st segment; (c, d) 5th segment (a, c) Tacho signal (b, d) response signal

Figure 9 shows order spectra for the lowest and highest speed segments. Note that the smearing at the base of the response spectra (primarily at the low speed) must come from amplitude modulation, since it is not present in the tacho signal spectra.

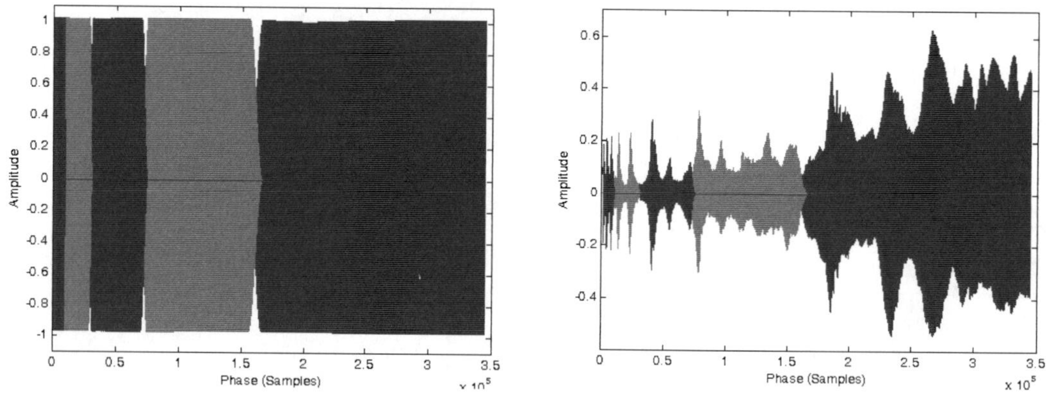

Figure 10: Overlaid order tracked signals for the five segments

Figure 10 shows the overlaid order tracked signals for the five segments after processing. When added together, there were no discontinuities in the records.

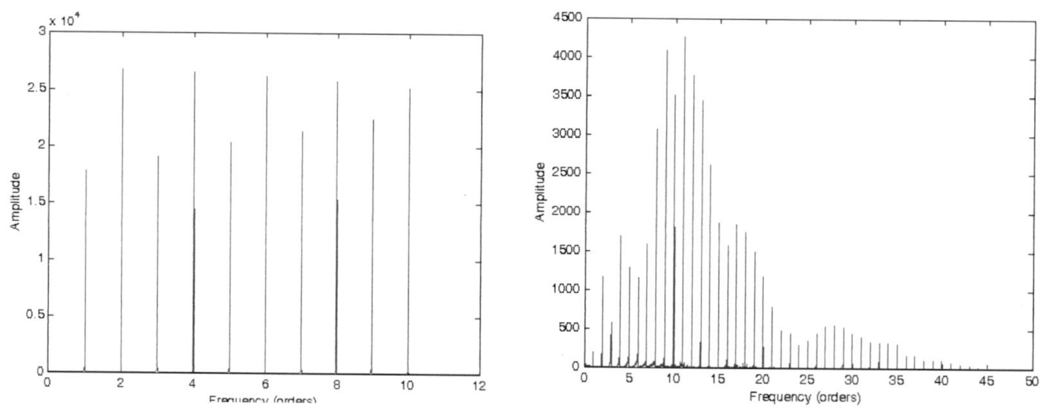

Figure 11: Order spectra of the re-combined signals (a) Tacho (b) Response

Finally, Figure 11 shows the order spectra for the total combined signal. There is still a small amplitude modulation effect at the base of the low orders in the response spectrum, but much less than in Fig. 9(b) when taken over the whole signal with uniform scaling.

Figure 12 shows spectrograms for the order tracked tacho and response signals, for comparison with Fig. 3. The number of samples per period of the fundamental harmonic was chosen to be 120, so spectra are shown up to order 60, although as can be seen, the whole diagram is only valid to the highest valid order at full speed, viz 32. Orders up to 35 can be seen there, but are affected by the lowpass antialiasing filter. Harmonic orders are now horizontal lines, while constant resonance bands follow a 1/speed trajectory.

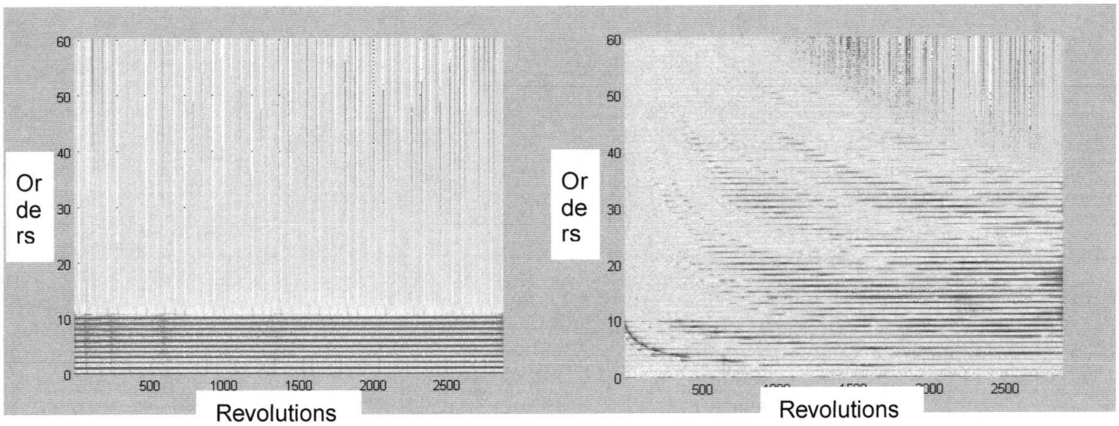

Figure 12: Spectrograms for the order tracked signals (a) Tacho (b) Response signal.

Reference signal: The reference signal used here for demodulation purposes was a tacho signal with 10 harmonics, of which the first was used for generating the phase/time map. However, the first harmonic of the response signal itself can often be used if it is strong enough (with respect to adjacent noise) throughout the whole record, and if the sweep rate is very slow with respect to the response time of the system [4]. If the latter is not true, the response at a given time (and corresponding speed) has been excited a short time previously when the speed was different. This is discussed by Borghesani [7]. The other extra limitation given by using the response signal is that it is often amplitude modulated as well as frequency modulated, and thus the total bandwidth of each order is larger, further restricting the possible range of deviation which could be achieved in each segment [4].

Figure 13 (from [4]) shows a case where the third harmonic of a response signal could be used for order tracking where the machine speed varied by ±10%. In this case the first harmonic was not strong enough, so this response signal would not have been very suitable for a ramp analysis (it would have meant many more segments, each with a speed range of 1.2:1). Inspection of the spectra in (b) and (d) shows that the amplitude modulation of the response signal has not significantly increased the bandwidth of the third harmonic, even though it has made it non-symmetrical. Ref. [4] discusses how an estimate of this effect can be made before using the diagram of Fig. 1.

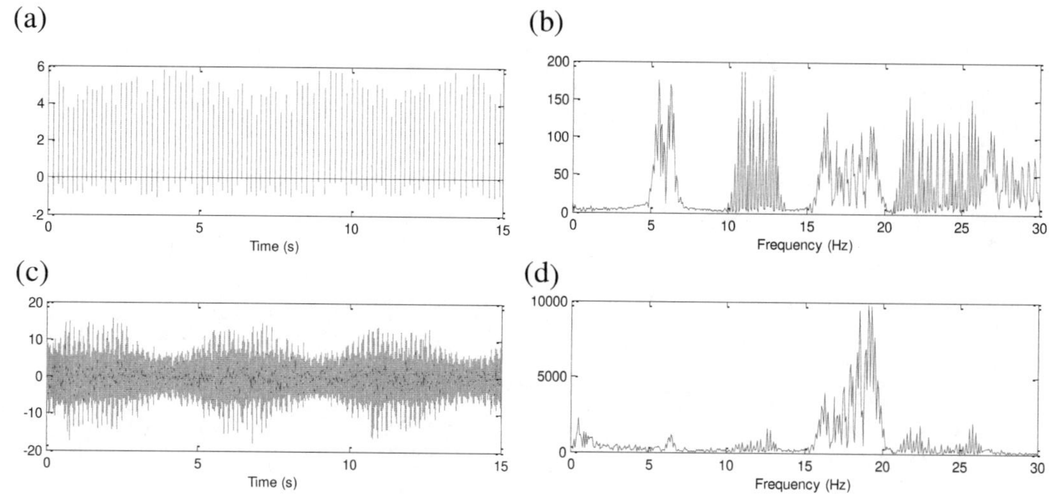

Figure 13: Signals with ±10% speed variation (a, b) tacho (c, d) acceleration
(a, c) time signals (b, d) spectra

Conclusion: This paper shows how the phase demodulation based method of order tracking, normally restricted to a maximum speed range of 2:1, can be extended to analyze run-up or run-down signals over a much wider frequency range. In general, such an order tracked signal could not be used for some purposes, such as time synchronous averaging of gear signals, but it can be useful to track the amplitude of each order as a function of speed. This can be displayed as a spectrogram, as in Figure 12(b), but alternatively it would be simple to filter out each harmonic in the order domain, and inverse transform it back to a time (actually revolution-based) signal (to relate it to speed and time via the speed vs time curve, since rotation angle is the integral of speed).

Order analysis over a full speed sweep might also be useful to determine at which speed certain phenomena show up best (for example a gear or bearing fault) after which a more detailed analysis could be carried out in a more restricted frequency range. Thus, the criteria for the wide speed range analysis, such as the choice of dynamic range criterion in Fig. 1, would most likely be less critical than for a subsequent more detailed analysis.

Acknowledgements: This research has been supported by the Australian Research Council under Discovery Project DP0878266, and by Australia's Defence Science and Technology Organisation.

References:

[1] N Sawalhi, RB Randall, Gear parameter identification in a wind turbine gearbox using vibration signals, 2014, Mechanical Systems and Signal Processing, 42(1-2) pp. 368-376.

[2] KR Fyfe, EDS Munck, Analysis of computed order tracking, 1997, Mechanical Systems and Signal Processing 11(2) pp. 187-205.

[3] PD McFadden, Interpolation Techniques For Time Domain Averaging Of Gear Vibration, 1989, Mechanical Systems and Signal Processing 3(1), pp. 87-97.

[4] MD Coats, RB Randall, Single and multi-stage phase demodulation based order-tracking, 2014, Mechanical Systems and Signal Processing, 44(1-2), pp.86-117.
[5] J Urbanek, T Barszcz, J Antoni, A two-step procedure for estimation of instantaneous rotational speed with large fluctuations, 2013, Mechanical Systems and Signal Processing, 38(1) pp. 96-102.
[6] N Sawalhi, RB Randall, D Forrester, Separation and enhancement of gear and bearing signals for the diagnosis of wind turbine transmission systems, 2014, Wind Energy 17(5) pp.729-743.
[7] P Borghesani, P Pennacchi, RB Randall, R Ricci, Order tracking for discrete-random separation in variable speed conditions, 2012, Mechanical Systems and Signal Processing 30, pp. 1-22.

SOME REFLECTIONS ON MACHINERY VIBRATIONS DURING PROCUREMENT AND ACCEPTANCE

Ronald L. Eshleman, Ph.D., P.E.
Technical Director
Vibration Institute
6262 South Kingery Hwy.
Willowbrook, IL 60527
Telephone: (630) 654-2053
reshleman@vi-institute.org

Abstract: The procurement process including selection, specification, and testing for acceptance for rotating machinery is described with emphasis on machine vibration. Several situations where things often go wrong are described.

Key Words: Design, functionality, installation, reliability, testing, specifications, standards.

Introduction: The procurement of costly equipment for manufacturing, process, transportation, or other functional purposes is a serious endeavor and should be viewed as an activity that requires professional attention. When this process goes wrong many hours and dollars can be expended on frustrating resolution. Issues such as cost versus quality must be thoroughly addressed. This means realistic expectations on function and reliability must be observed. The process of procurement involves the functionality and performance of the machine system including its foundation and environment. The design, specification, and testing of the system will be addressed.

Design: Basically, this early step in the process involves the selection of machine components that will perform the task or process in a specified manner. The design needs to consider the operational sequence of the system as well as its environment. This will not only affect the machine but its foundation, piping, and other peripheral influencing structures and process interactions. For example, if one wants to operate a pump at shut-off, a recirculation line is needed. If a wide range of operational speed is required using a variable frequency drive the threat of resonance is going to have to be seriously considered. If a change in speed is required to effect efficiency, gearbox design must be considered. At this stage the whole issue of cost versus quality and performance needs to be addressed. Planning for future expansion may be wise but it may backfire by procurement of a machine that does not operate at an efficient state.

Specification: The design is complete and now it is time for procurement. There is always the concern about finances; however, the owner needs to face up to performance and reliability. Owners do not want to spend money frivolously however this is not the time to go cheap. Beware of over specifying it may cost you dearly in the end when the vendor cannot make the unrealistic specification. Therefore this is a very serious step and should be handled as such. This is a situation where you may want to consult a person with background and experience. There are a number of basic factors to be considered and specified.

1. **Functionality**
2. **Reliability**
3. **The environment**
4. **Compatibility**

This paper is concerned mainly about reliability. Usually the functionality portion of the procurement is straight forward; however, there may be alternative methods of ending at the same goal. If that is the case all factors may be considered in alternate procurement paths. Since the author has witnessed a number of procurements that went wrong because of inadequate or misunderstood specifications this paper is basically concerned with the reliability path although in some applications such as printing and medical equipment vibration can be a very real performance factor.

Therefore vibration needs to be addressed in terms of functionality and reliability. There are a number of ways to address the vibration specification –past experience, vibration standards, and procurement specifications like API—all credible. The writer needs to be realistic. A standard written excessively tight may be the ticket to a lot of trouble without a good ending. The specification writer needs to know what is impossible or excessively costly to build as well as what is too loose. The writer needs to know the threats- short term failure due to manufacturing, installation, or resonance; long term fatigue; inadequate performance.

A good place to start is to review ISO machine standards [1] and [2] and/or machine specific association standards such as AGMA, NEMA, and HI. Compare the levels provided in these documents. However, sometimes it is impossible to compare documents because of differing processing. The writer will at least get a feel for what will work in a cost-effective way.

Also, the use of previous experience is a good way to define acceptable limits. The writer must be careful that limits are based on similar situations—process and environment. For very specialized procurements users like API do have purchase specifications where decisions are minimized.

Whatever levels are selected for acceptance, acquisition and processing of the data must be specified as well as the operational conditions during test. It is best to stay away from unrealistic operating conditions e.g. operating pumps far from the BEP. Freedom from bearing defects, location of natural frequencies, balancing limits, and alignment limits are sometimes specified. These factors may be difficult to effectively specify so in general it is recommended to refrain from locking down manufacturing and installation limits. If there is a problem the vibration symptoms will point to the offending parameter.

It is recommended that the specification provide some *guidance* on foundation design, piping restraint, and balancing and alignment limits. In complicated systems with wide frequency ranges provided by variable frequency drives resonance is always a threat. The specification writer needs to have viable options. There are situations where the vendor cannot provide a resonance free system- particularly with vertical pumps.

Acceptance Testing: Acceptance testing needs to be planned based on the specification. In other words a test plan is a requirement [3]. A good field test plan will involve types of data, machine operating speeds, measurement parameters, measurement locations, and process conditions. The types of data will depend on the design of the machine –normally bearing type. Relative displacement is normally used for large machines (greater than 500 HP) with sleeve bearings and integrated accelerometers for lower speed (less than 1,800 RPM) rolling element bearing machines. However, gearboxes may require acceleration and relative displacement.

Normally the system will be operated at a number of conditions pertinent to the life-time operation of the machine. It will be expected that the specified parameters will be within the limits selected at all test conditions. It is important to set up instruments so that relevant data are collected. There is a wide difference in the values of vibration measures depending on how the data are processed and there is generally no way to convert from one measure to another. For example RMS values will be lower than peak values -- commonly by a factor of 2.5 depending on what is causing the vibration [4]. Even RMS values calculated in the spectrum can be vastly different than those measured from the waveform. In most cases only overall levels are specified but the owner may have special needs that require spectrum and time waveform analysis. In some cases critical speeds and natural frequencies may be measured with or without specification. The moral of the story is to understand what is specified and set up the instrument to process that measure properly.

Case Histories: A few case histories that reflect some of the things that can go wrong follow.

No 1: Inadequate measurement A large gearbox was properly specified and built for a motor driven extruder. After installation the customer was not happy with the measured vibration velocity levels (over 2. IPS- peak) obtained from the monitoring system. Upon review of the acquisition system and the data it was determined that the data were not processed properly. A large unfiltered low frequency ski-slope was vastly distorting the vibration levels. Detailed vibration measurements revealed that the vibration levels were not as high as reported but still did not meet the specification. In the absence of viable fixes the owner was given an extended warranty.

Often I see data acquired that are not relevant to the specified specification. For example, Hydraulic Institute requires unfiltered RMS processed vibration velocity. Peak measurements will not work unless one wants to estimate the RMS energy from the spectrum.

No 2: Unrealistic Specifications: A specification for a group of piston pumps was set at 0.025 IPSrms. The owner complained that the vendor could not make the specification. I inquired about how the acceptance level was developed. The owner said they measured existing piston pumps. This is a good idea but obviously something was wrong because this is not a credible number.

I have seen specifications that rule out critical speeds from vertical pumps that have a wide operating speed range. This becomes a huge problem leading to many unsuccessful "fixes". Resonance is probably the worst enemy in getting machines to make vibration specifications.

There are times when the manufactures specifications do not match the customer's expectations. For example the ISO balancing specification ISO 1940 [5] may not provide the owners expectations. However sometimes the customer expects too much like specifying 2W/N (W=rotor weight and N=RPM). More is not always better.

No 3: Careless Application: A group of Diesel driven firewater pumps were installed to protect a manufacturing plant in case of power outage during a fire. Unfortunately plant personnel wanted to wash down the facility with these pumps which meant operating the pumps near shut-off (way off the curve). This caused catastrophic vibration that could have been avoided with a re-circulation line.

No 4: Inadequate Installation: Inadequate foundations are often encountered- from bad grouting to unsupported piping. In one case a stack of shims caused the mounted natural frequency to be lowered into the operating speed range. Happily this was one of the easier problems to fix. Needless to say these situations, without correction, will lead to non-acceptable machines.

Conclusion: In many cases procurement is not handled seriously, clouded by past bad experience, or done by non-professionals that do not have a clue about the technology. We constantly hear about the bad and the ugly but not much about the good. I must give credit to my diseased friend Charlie Jackson who took this job seriously and always had positive results in field tests and moreover had machinery that ran at 99% reliability.

References:

[1] ISO 10816, "Mechanical Vibration—Evaluation of machine vibration by measurements on non-rotating parts- Parts1-8, International Standards Organization, Geneva, Switzerland.
[2] ISO 7919, Mechanical vibration – Evaluation of machine vibration by measurements on rotating shafts- Parts 1-5, International Standards Organization, Geneva, Switzerland.
[3] Eshleman, R.L. and Nagle-Eshleman, J. *Basic Machinery Vibrations,* VI Press, Inc., Clarendon Hills, IL 1999.
[4] Eshleman, R. L. Machine Condition Evaluation: Using Seismic Measurements, Proceedings of the 37th Meeting of the Vibration Institute, Jacksonville, FL, 2013.

[5] ISO 1940-1, Mechanical vibration—Balance quality requirements of rigid rotors—Part 1: Specification and verification of balance tolerances, International Standards Organization, Geneva, Switzerland.

IDENTIFYING THE TRUE COST OF NO FAULT FOUND (NFF) WITHIN THE UNITED KINGDOM MINISTRY OF DEFENCE (MOD)

Wg Cdr CJ Hockley OBE, CEng, MRAeS, RAF(Rtd)
EPSRC Centre for Innovative Manufacturing in Through-life Engineering Services, United Kingdom, Cranfield University, Bedford, MK43 0AL
Telephone: -01234 750111 ext2281
E-mail: c.hockley@cranfield.ac.uk

Abstract: Availability is a key requirement for the Armed Forces yet the occurrence of faults where no attributable reason can be established still threatens to reduce the required availability and still causes nugatory maintenance effort. Such faults with no apparent cause are usually described in the UK as No Fault Found (NFF) and in USA as Retest OK (RTOK) or Can Not Duplicate (CND); their occurrence remains a huge and disruptive effect on the successful delivery of through-life customer support. Whilst commercially it causes huge loss of revenue, in the military environment, costs are often less obvious and hidden, yet may be much higher. This lack of visibility of the true cost is a worrying trend that must be tackled as the UK Armed Forces reduce in size as it will have a more significant operational and financial impact. A study of MOD's NFF occurrences has been conducted by the EPSRC Centre for Through-Life Engineering Services in order to establish more accurate costs and to establish the key reasons and causes. Whilst the study has so far concentrated on the air environment where information is most readily available, the research and lessons learned are applicable to other industries.

Key words: No fault found; NFF; Maintenance; Availability;

Introduction: The Royal Air Force has undergone considerable change over the last 15 years with a large proportion of its maintenance and support being contracted out to civilian repair organisations. The size of the RAF has reduced from 55,800 in 1998 to 37,030 in April 2013 and is planned to reduce further to around 33500 by 2015.[1] Maintenance at the aircraft squadron level is still done by RAF personnel at what is called the "Forward" support level, where daily flying operations are carried out or wherever the squadron is deployed. All remaining maintenance, at what used to be known as second and third line, is now carried out mainly by contractors, at what is known as "Depth"; the Depth level combines what was second and third line servicing which were mostly manned by service personnel. Such change and reductions in uniformed personnel have necessarily caused a great deal of change in the maintenance support organisation, but there is now greater visibility of any loss of availability and of the downtime due to faults and maintenance. The support organization is now therefore much more aware of the actual cost of maintenance downtime and is thus incentivised to reduce the cost of any nugatory maintenance. Faults that cannot be replicated or diagnosed successfully are a waste of time and resources and may even be penalised in some contracts where the contract is one for defined levels of support or availability, which may then be reduced as a result.

As part of one of the major projects undertaken by the EPSRC Centre for Innovative Manufacturing in Engineering Through-Life Services, visits to 5 RAF aircraft fleets were made to research the NFF rates that each fleet was experiencing. The fleets included fast jet, helicopters and larger aircraft. Previous research by an RAF colleague for the Harrier aircraft fleet had identified that the information available on the Logistic Information System (LITS) was not always able to produce a comprehensive picture of the true NFF rate. When detailed research is made of the LITS data, the true rates of NFF are far greater and it is evident that the true cost of NFF is therefore not fully appreciated by management and the MOD. There are a number of reasons for this ranging from cultural and human factor issues, to operational imperatives. Indeed the research of the 5 MOD fleets showed that the reasons were not dissimilar to those found in the commercial world. The broad and general reasons identified from interviews and observations can be summarised as follows:

- Testing and rectification procedures are either overly complicated or overly simplified, or did not even exist for certain faults.
- Reliance on Built-in-Test and test equipments has produced a culture of reliance on their results rather than a questioning attitude and an admission that a fault exists but has not been found.
- Testing in most cases could not replicate the operational environment.
- There is under-utilisation of the available data particularly the feedback of fault and repair data from workshops and suppliers.
- Information on fault history of aircraft systems or LRUs was not easily available to the technicians who had the diagnostic task.
- Whilst significant amounts of data exist within LITS, there is no consensus or formal feedback of how this data could be used more effectively to reduce NFF. Changes to LITS are expensive to achieve so are rarely funded.
- A lack of commonality in policy, understanding and acceptance that there is a significant cost involved from high levels of NFF.
- Where a NFF register existed to record NFF history, there was no clear point of contact for maintenance of the register or responsibility identified for monitoring it and analysing the records.
- Awareness of NFF is low and its implications are not covered in training programmes.

Because the NFF rates are much higher than would appear from a routine search for NFF in the LITS data, solutions and mitigation have hitherto perhaps not been given sufficient priority. However, support contracts and contracts for availability are proving a fertile opportunity now to tackle the issue as there is a visible commercial incentive to reduce inappropriate maintenance and improve fleet availability. Somewhat worryingly there is also no recognition that NFF can easily be a flight safety issue, yet there are examples such as an AAIB investigated incident in 2010 where a NFF issue nearly caused a crash of a Dash-8 Q400 aircraft.[2] Operational risks must occur if reported faults cannot be diagnosed or replicated and the research discovered much anecdotal evidence that flight safety is indeed often at risk.

The Mod Maintenance Environment: It is important when considering fault resolution to understand the maintenance environment where it has to be resolved. Maintenance itself is well understood in all environments as being concerned with inspection and servicing equipment so as to make sure it can be put back into operation in a fit state to get to the next planned servicing intervention. It is also generally accepted that maintenance is about finding failures and fixing them. Definitions in the aircraft environment also make the necessary and vital link with flight safety and the need for maintenance to be carried out to ensure, or restore, "aircraft integrity". Jack Hessburg though, who had complete control over the design of the Boeing 777 in order to influence the design for ease of maintenance, says maintenance is therefore "nothing more than the management of failures".[3] Such management though must consider making the most effective use of personnel and resources in order to understand where effort must be applied to improve availability.

However, in the aircraft environment, more so than in some others, the management of failures is determined by the consequences of failures. These are:

- The safety impact
- The operational availability impact

In the military environment both are equally important. Whilst safety is of course paramount, in a wartime situation availability is equally vital and may at times supersede safety in the interests of winning the battle. In the commercial air environment, clearly safety is paramount, but availability is usually subservient, particularly if safety is affected. However a commercial operator loses revenue and reputation if they accept delays and cancellations; this can encourage a culture and process of changing anything and everything in order to solve a fault that might cause a delay or cancellation. Safety is preserved but such activity will almost certainly generate NFFs in the components removed at the next level of repair as some of them will have been removed speculatively, in a 'blanket' act of maintenance to ensure the fault is removed. Maintenance in the MOD environment in peacetime though should have more time to enable thorough diagnosis to be performed; the fact is however, that a similar attitude of operational pressure is often prevalent and possibly understandable as it replicates the operational attitudes that the military must train for in peace. There is also a natural facet of human nature for those who command the resources which mean that they want to produce the most number of serviceable assets in the shortest possible time in order to maintain the highest availability. However, the inability to replicate faults on the ground, incorrect procedures and lack of tracking fault and NFF history, have resulted in NFF rates that should be worthy of effort and which on deeper investigation, are shown to be much higher.

In the MOD environment maintenance and fault rectification is now done on a more centralised basis for each particular fleet. At the Squadron level now called "Forward", the rectification of faults is still carried out exclusively by military personnel. As soon as a deeper investigation is required, the aircraft is transferred to "Depth," which is mainly staffed by contractors who then provide the deeper rectification activity and also the majority of scheduled servicing. Depth is usually a contractor led support organisation that will serve several squadrons, perhaps from different airfields on a centralised regional basis. This increased level of contractorisation where costs are more in evidence, suggests an environment that should be more responsive to reducing NFF costs whatever level they are.

Availability & NFF Cost: When trying to understand the cost impact of NFF on availability we must consider what the constituent parts of availability are. As we are considering the impact and cost on the operator or user we must define operational availability. It is generally accepted that operational availability consists of a ratio of uptime and standby time over the total time:

$$A = \frac{Uptime}{\text{Total Time}} \tag{1}$$

The total time of course is made up of the uptime and the downtime and then the uptime is also made up of the operational usage time and the standby time; similarly the downtime is also made up from the time taken for preventative maintenance, corrective maintenance and the admin and logistic delay time. So for Operational Availability we need to define Equation (1) a lot more precisely as in Equation (3):

$$\mathrm{A_o} = \frac{Uptime}{Uptime + Downtime} \tag{2}$$

$$\mathrm{A_o} = \frac{OT + ST}{OT + ST + TPM + TCM + ALDT} \tag{3}$$

Where OT = operational time
ST = standby time
TPM = Total Preventative Maintenance Time
TCM = Total Corrective Maintenance Time
ALDT = Administrative and Logistic Delay Time

We can now consider where NFF might have a contribution and it is in fact to any one of the contributors to the downtime, depending on where and when the fault occurs; in other words it may occur as part of some preventative maintenance or of corrective maintenance but wherever it occurs, it will certainly have an effect on ALDT. It could be argued that as part of preventative maintenance, a fault may occur perhaps as part of the inspection or the functional tests and it is now a fault requiring just corrective maintenance; this would then mean any NFF hours should be attributed to the corrective maintenance downtime. Another view, however would be from the point of view of the need to attribute the manhours to the particular maintenance level, Forward or Depth. If our interest is to identify where, or at what level, the costs are occurring - as it needs to be for the MOD - then it is useful to accept that some NFF manhour costs will occur as a result or as a consequence of scheduled servicing and will need to be shown as a cost to that level of maintenance.

So in order to identify costs and impact we need to further expand Equation (3) to highlight the specific time spent on NFF in each of the areas of downtime.

$$A_o = \frac{OT + ST}{OT + ST + TPM + TCM + ALDT + NFFpm + NFFcm + NFFaldt} \qquad (4)$$

Where, ***NFF****pm* = Total time spent on NFF within (or associated with) preventive maintenance

NFF*cm* = Total time spent on NFF within corrective maintenance

NFF*aldt* = Total time of admin and logistic delay attributable to NFF

Having established a simple analysis of how the cost of NFF might be highlighted more easily, the problem however, is to obtain the data in sufficient detail in order to isolate the individual impacts and costs in preventative maintenance, corrective maintenance and administrative and logistic areas. The LITS data can provide the analysis for the ***NFFpm*** and ***NFF****cm* but not the ***NFF****aldt*. For this, significant analysis is needed item by item, LRU by LRU and repair contract by repair contract. It is perhaps here that costs are significant for items repaired under contract yet which had no fault in the first place. As the LITS data has been made available by the MOD for this study, only the rates and levels of NFFcm and NFF pm have been analysed so far. For ***NFF****aldt* further research with individual repair contracts is needed.

RAF Fleet Analysis: Whilst five fleets have been analysed for NFF rates, the data in this paper is only from the Harrier fleet which is now out of service and is therefore unclassified. Nevertheless, the initial results from research on the other fleets show very similar rates and levels of NFF once the LITS data is analysed in the same way.

Harrier Fleet Analysis: In looking at the Harrier data an earlier study by an RAF engineer, Stewart Roke, has been used as it had made a very detailed analysis of the data from 2005 to 2008 for an MSc thesis at Coventry University.[4] A similar process of analysis was used therefore for this research and the lessons learnt during that earlier thesis analysis were re-evaluated and confirmed. Harrier data from 2008 until the end of 2010 when the aircraft was retired from service was used to update the previous results and formed the basis for analysis of the other fleets.

LITS collects data from both Forward and Depth via Maintenance Work Orders (MWOs). If any work is required to be done on the aircraft a MWO is raised and completed by the maintenance personnel and entered into the LITS system. LITS Data is then able to be viewed by both Forward and Depth organisations and by the Harrier Project Team as well as industry partners. The MWOs have 12 data fields which include information such as aircraft tail number, dates, description of work required, work carried out and manhours. In the work carried out field there are standard default phrases which LITS uses and the LITS default phrase "01 – No fault found after check/test" is the one that covers NFF. Consequently when analysis of LITS data is used in statistical returns, NFF manhour rates will be determined from the numbers against this code. Roke found from analysis of the LITS data that the initial manhour rates for NFF from the 01 code showed a reducing level from 3.8% to 2% over the period, caused by a period of improved input coding.[5] However, human nature is such that a technician will very often put a much more varied response in the 'work carried out' field as a natural reaction that suggests the work was carried out satisfactorily, for example, fault cleared itself during investigation. This will then not generally be coded into the 01 code field. A list of similar workaround phrases are shown in table 1.

Table 1 – No Fault Found Variations in Reporting

Ser	Description	Acronym
1	No Fault Found	NFF or nff
2	Fault cleared itself during investigation	
3	Fault Not Found	FNF
4	No fault apparent	
5	Tested satisfactorily	
6	No fault indications found	
7	Cannot reproduce fault	
8	No fault codes, or No failure codes	
9	Cannot duplicate	CND
10	No fault symptoms apparent	
11	No damage found	
12	Unable to replicate fault or symptoms	
13	Unable to reproduce fault	

The problem of taxonomy for NFF has been confusing and hiding the rates and levels of NFF across many industries for a very long time and is certainly not particular to the RAF. It is one of the aims of the EPSRC Centre to rationalise the taxonomy as part of its research programme and an initial paper has already been completed by Khan et al.[6]

For the Harrier LITS data therefore, a more thorough analysis was required in order to discover the true levels of NFF manhour expenditure. By analysing the 'work carried out' field in more depth and recognising the myriad of different descriptions such as those shown in table 1, other phrases particular to Harrier engineering practices were recognised and needed to be included in the count of NFF. Some of the Harrier specific ones are shown in table 2.

Table 2. Harrier specific engineering NFF descriptions

	Work Carried Out
1	49 – Fault cleared during investigation
2	OFP re-load carried out
3	PFM re-load carried out
4	Inertia switch reset
5	Functionals carried out, unable to reproduce fault
6	Unable to reproduce fault on ground
7	INAS align carried out satis, no fault codes recorded
8	49 – rcu re-seated satis
9	49 - fault cleared itself on BIT

Once all the possible NFF work had been assessed, the manhours were shown by Roke to have increased substantially to nearly 12% and reducing to 10% at the end of the period with an average of 10.75%.[7] The true NFF rate was therefore some 3.7 times higher than the usual LITS data analysis showed. The true NFF rates actually produced a total of 35,950 manhours for the 4 years 2005-08. At the 2008 capitation rates for a corporal technician which equated to £55 per hour [8], this gives a total of £494,313 per year or nearly £2M over the 4 years. Clearly if the usual LITS data capture is only showing a little more than a quarter of this figure and a level of 3% of all corrective maintenance being NFF, it is unsurprising that it had not received greater attention. However, a cost that is nearly four times greater and a level that is over 10% of all corrective maintenance at the Forward area, should have been a matter of grave concern. In our Equation (4) this gives us the cost of $\boldsymbol{NFF}cm$.

However, this result was only for the Forward domain where operational pressure has most influence to keep aircraft serviceable and available. Whilst this may cause further NFF occurrences at the subsequent levels of repair for Line Replaceable Units (LRUs) and components, the NFF expenditure in the Depth domain might be expected to be similar or less than at the Forward domain. In Depth where most of the maintenance work is preventative servicing, there may be less pressure and more time for diagnosis, so one might expect a lower manhour expenditure on NFF. However, when the LITS data was analysed in the normal way, an average 12.5% of all corrective maintenance manhours was found to be NFF and when the deeper analysis of all 'work carried out' fields was made, there was 28.4% spent on NFF. Actual manhours in the four year period were 19,670 manhours. This equates to £270,462 per year. There is of course a logical explanation for the higher percentage in that most of the work at Depth is scheduled maintenance; consequently there is less corrective maintenance work to do and NFF will be a much bigger proportion of the smaller amount of corrective maintenance. No doubt as well, some of the problems that are difficult to diagnose will have been handed over to

Depth and these are the ones most likely to generate NFF. Hence the ***NFFpm*** cost can be identified. By analysing the subsequent data from 2008 until the end of the Harrier service in 2010, the previous results from the Roke study were confirmed.

One problem with the LITS analysis is that the subsequent cost of NFF with components and LRUs being sent for repair at the next deeper level is hidden. The information is only partially available within specific repair contracts and has to be individually collected for each one in order to assess the overall cost of ***NFFaldt***. Nevertheless, using this approach allows management to identify more easily some of the true costs of NFF.

Other RAF Fleets: The EPSRC Centre has been given excellent access to other fleets of RAF aircraft and their LITS data in order to research and establish whether there are similar findings for the manhour cost of NFF. Two helicopter fleets, one fast jet fleet and one large multi-engine aircraft fleet are being investigated. The Harrier fleet has now finished its service with the RAF and thus the detail reported above has been possible without compromising security. A similar depth of investigation is still underway for the current in-service fleets and it will be for MOD to decide what detail can be publicly released in due course once the analysis is complete and MOD has considered and accepted the final report that they commissioned. However, initial indications are, that whilst some fleets have improved their reporting processes, the indication is that the results will be broadly similar.

Lessons Learned: From the research done with maintenance personnel with each of these RAF fleets, it is clear that there are some lessons to be learned and some best practice to share. Any fault starts with a fault symptom or reported fault state and a large majority of these will be described by the user; the user in the RAF is generally the aircrew. If the fault is not carefully described and in such a way that the correct diagnosis can be set in motion, then the chance of a NFF result is greatly increased. One fast jet fleet has changed the aircrew de-brief reporting such that the pilot is only able to classify a fault into a number of defined categories. This has resulted in the need to change the LITS software – at some cost – but with a reported reduction in NFF occurrences of about 25%. Another fleet had created a NFF cell that monitored faults on each squadron on a daily basis and advised on fault and LRU history and helped to isolate rogue units and to make decisions on the most cost-effective place for LRU test and repair. The experience gained with this dedicated team has shown that the most cost-effective place to solve the level of NFF occurrences is at the aircraft rather than in the Depth domain or at a contractor's repair facility.

Conclusion:

It is clear from the research that the MOD is under-estimating the levels of NFF in all their fleets of aircraft. How this can be best resolved is difficult as a major contribution would be to make changes to the LITS system which would almost certainly be very expensive. The major benefit though of this research has been to prove what the true levels of NFF really are and to make both MOD and support providers aware that there is a serious cost in both availability and maintenance effort that needs to be reduced. The research conducted by the EPSRC Centre has also established a number of recommendations for the MOD to procedures and processes which would contribute. It is also clear that many of the findings that were discovered in this research of MOD aircraft fleets are similar to those found in other industries which have been discovered by the research carried out by the EPSRC Centre for Innovative Manufacturing in Engineering Through-Life Services.

References:

[1] MOD, 1998. Defence Statistics available at: http://www.dasa.mod.uk/publications/UK-defence-statistics-compendium/1998/1998.pdf. P33

[2] AAIB Bulletin: 6/2012 G-JECF EW/C2010/09/04 available at http://www.aaib.gov.uk/publications/bulletins/june_2012/dhc_8_402_dash_8__g_jecf.cfm

[3] Hessburg J, 2013, Functionability Management – A tribute to Jack Hessburg at the 23rd MIRCE international Symposium, 3-5 Dec 2013.

[4] Roke S, 2009. Harrier No Fault Found Reduction, MSc Dissertation in Engineering Business Management, Coventry University, September.

[5] Roke S, 2009. Harrier No Fault Found Reduction, MSc Dissertation in Engineering Business Management, Coventry University, September. P79

[6] Khan, S., Phillips, P., Hockley, C., Jennions, I. (2012), "No Fault Found, Retest OK, Cannot Duplicate or Fault Not Found? –Towards a Standardised Taxonomy" *The 1st International Conference for Through-life Engineering Services,* Shrivenham, (Nov 2012)

[7] Roke S, 2009. Harrier No Fault Found Reduction, MSc Dissertation in Engineering Business Management, Coventry University, September. P82

[8] MOD 2013, Defence Statistics available at: http://www.dasa.mod.uk/publications/personnel/military/annual-personnel-report/2013/1-april-2013-revised.xls, P8.

Impact Technologies

A Sikorsky Innovations Company

Integrated Health and Contingency Management for Flight Critical Systems

Michael Roemer, Jianhua Ge, and Sudarshan Bharadwaj

May 22, 2014

OVERVIEW

- Integrated PHM and fault adaptive control
- Real-time system identification
- Technical approaches for automated contingency management
- Modeling and simulations of the hex-rotor
- Fault adaptive control with motor fault
- Conclusions

PHM AND ACM INTEGRATION

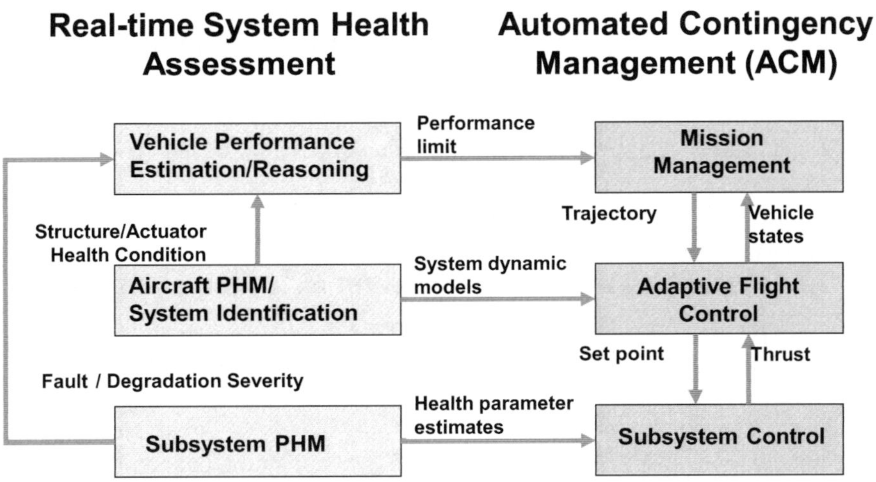

OPTIMAL PERFORMANCE OBJECTIVE

Optimize the utility and performance of the air vehicle with impaired capabilities to accomplish assigned mission:

Mission adaptation (M) allows the control architecture to pursue relaxed mission objectives *M (instead of Mcom)* in order to achieve greater vehicle usefulness U.

Mission Level: $J(M) = \max_{M} U(P_e, M, M_{com})$

Control Level: $J(R) = \max_{R} P_e(F_m, R, M)$

Optimize vehicle performance *Pe* with consideration of mission requirements, through restructuring and reconfiguration, R.

Fm (Failure Mode) is a vector of indicator variables that characterizes the fault modes detected on the aircraft; R is a vector of indicator variables that characterizes all restructuring applicable to the system.

SYSTEM ID AND CONTROL

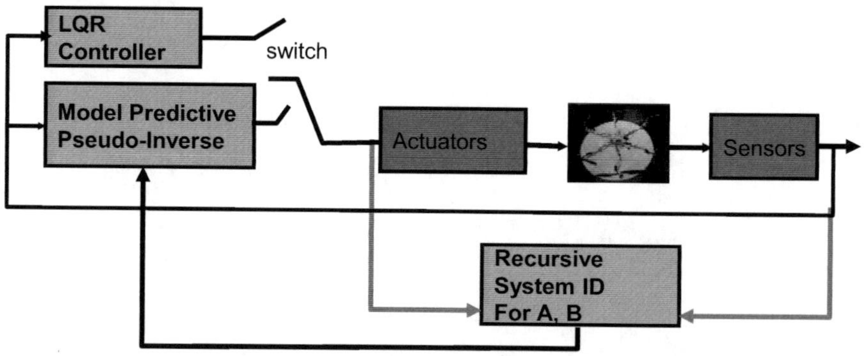

- The capability of detecting in real-time the dynamics of the system is a significant advantage to advanced FCM functionality.
  - Severe system faults can be detected by analyzing the A & B matrices;
  - System controllability can be analyzed;
  - New trim conditions can be derived;
  - System's performance limits can be evaluated;
  - Use of advanced control algorithms such as Model Predictive Control;

SYSTEM IDENTIFICATION

Recursive Least Square (RLS) Algorithm

Cost Function: $J(\theta(n)) = \frac{1}{2} \sum_{k=n-N+1}^{n} \lambda^{n-k} \,||y(k) - \theta(n)^T \phi(k)||^2 + \alpha \,||\theta(n) - \theta(n-1)||^2$

Recursive solution: $\theta(n) = \theta(n-1) + P(n)\phi(n)[y(n) - \phi(n)^T \theta(n-1)] + \alpha\lambda P(n)[\theta(n-1) - \theta(n-2)]$

Covariance matrix: $P(n) = (1/\lambda)P(n-1) - (1/\lambda)P(n-1)C(n)[\lambda I + C(n)^T P(n-1)C(n)]^{-1} C(n)^T P(n-1)$

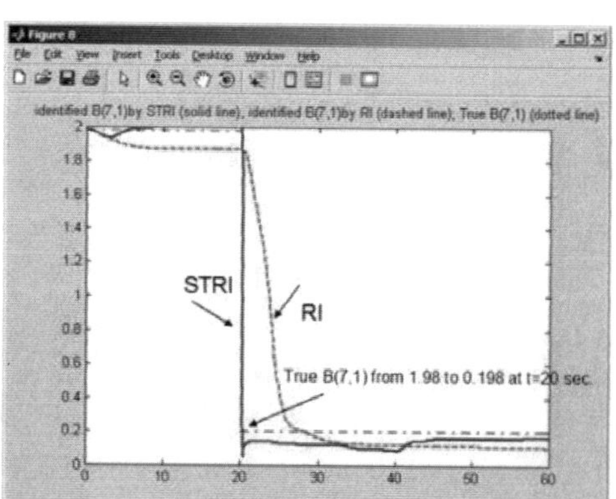

2 x 2 matrix inverse, good for real-time execution!

$$C(n) = [\phi(n), \sqrt{n_p \alpha (1-\lambda)} \cdot e(n)]$$

$$e(1) = \begin{bmatrix} 1 \\ 0 \\ \vdots \\ 0 \end{bmatrix}, \quad e(2) = \begin{bmatrix} 0 \\ 1 \\ \vdots \\ 0 \end{bmatrix}, \quad e(n_p) = \begin{bmatrix} 0 \\ 0 \\ \vdots \\ 1 \end{bmatrix}, \quad e(n_p+1) = \begin{bmatrix} 1 \\ 0 \\ \vdots \\ 0 \end{bmatrix},$$

PSEUDO-INVERSE CONTROL

A linear system example:

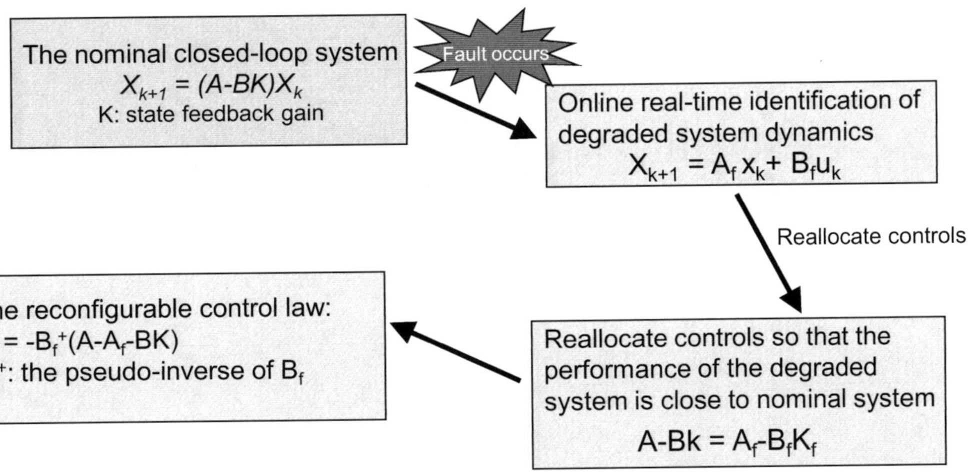

The same concept can be applied to non-linear system.

Impact Technologies
A Sikorsky Innovations Company

This page contains no technical data controlled by the ITAR or EAR

MODEL PREDICTIVE CONTROL

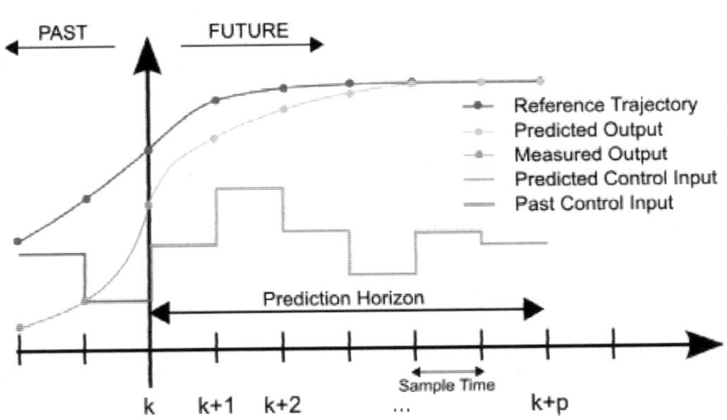

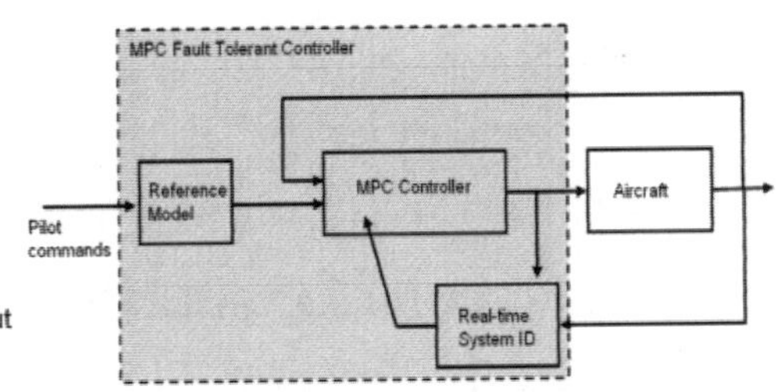

$$J = \sum_{i=1}^{N} w_{x_i}(r_i - x_i)^2 + \sum_{i=1}^{N} w_{u_i}\Delta u_i^{\,2}$$

Referenced **Controlled** **Input**

- Based on iterative, finite horizon optimization of plant model
- Rely on dynamic models of the plant that can be obtained by system identification methods
- Uses plant measurements, current state of plant, MPC models and reference targets to calculate future changes in independent variables.

Impact Technologies
A Sikorsky Innovations Company

This page contains no technical data controlled by the ITAR or EAR

Application Applied to Hex-Rotor UAV

Impact Technologies
A Sikorsky Innovations Company

This page contains no technical data controlled by the ITAR or EAR

NASA CHALLENGE PROBLEM

- NASA motivated by concern over asymmetric thrust condition being masked by automatic control in multi-engine fixed wing aircraft
- Selection of test platforms with closed loop control systems and redundant control effectors:

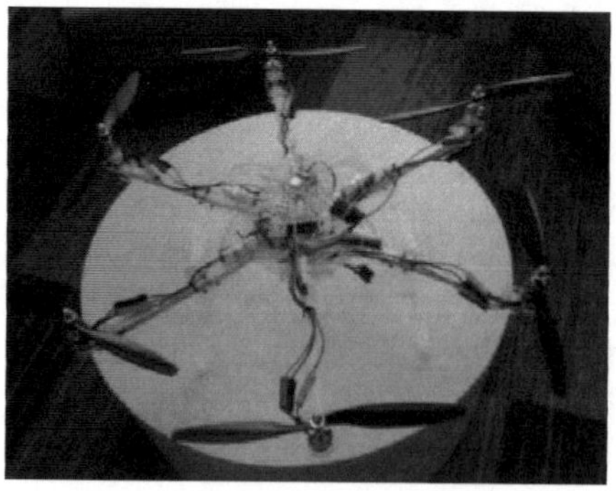

Impact Technologies
A Sikorsky Innovations Company

This page contains no technical data controlled by the ITAR or EAR

10

UNMANNED HEX-ROTOR HELICOPTER

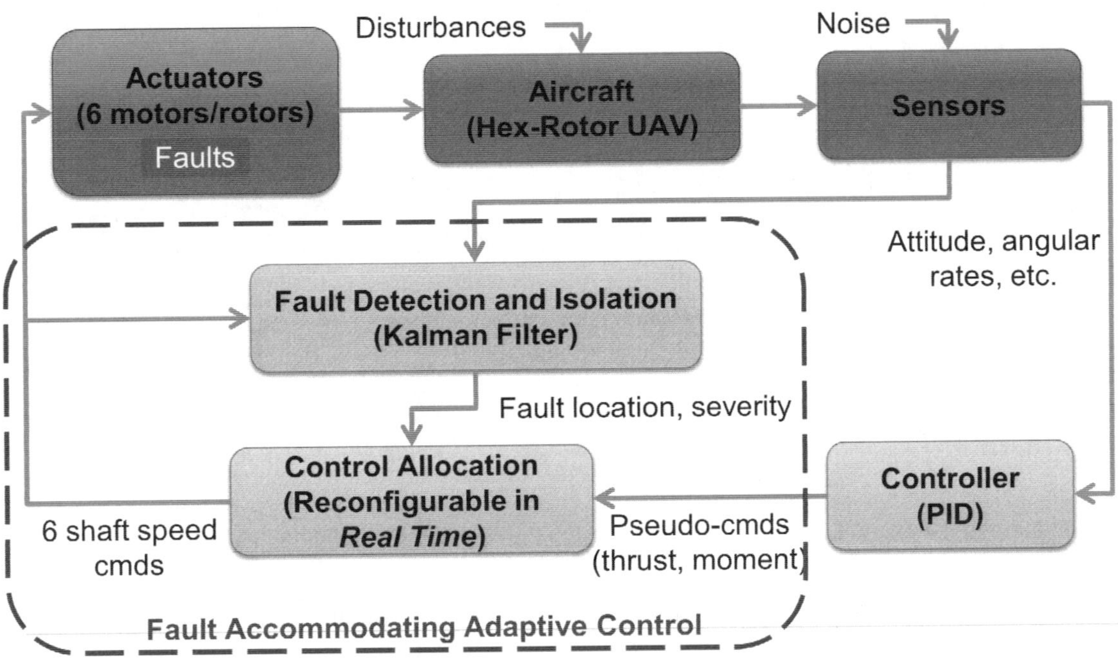

HEX-ROTOR RIGID BODY MODEL

Geometry & Coordinate System Definition

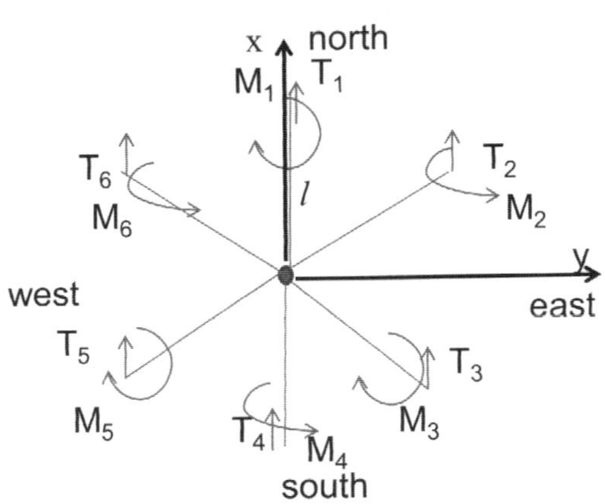

Rotor Geometry

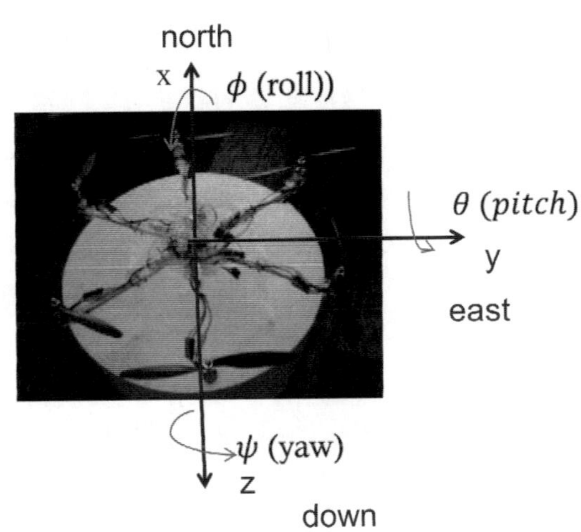

Motors 1, 3, 5 rotate clockwise, and motors 2, 4, 6 rotate counter-clockwise.

HEX-ROTOR DYNAMIC STATES

- Translational velocity (body frame): $v^b = \begin{bmatrix} u \\ v \\ w \end{bmatrix}$
- Rotational velocity (body frame): $\omega^b = \begin{bmatrix} p \\ q \\ r \end{bmatrix}$
- Attitude (inertial frame): $\Phi^i = \begin{bmatrix} \phi \\ \theta \\ \psi \end{bmatrix}$
- Position (inertial frame): $P^i = \begin{bmatrix} x \\ y \\ z \end{bmatrix}$
- State vector: $X = [v^b \quad \omega^b \quad \Phi^i \quad p^i]^T$
- PID controllers (qty. 4) regulate roll angle, pitch angle, yaw angle, altitude

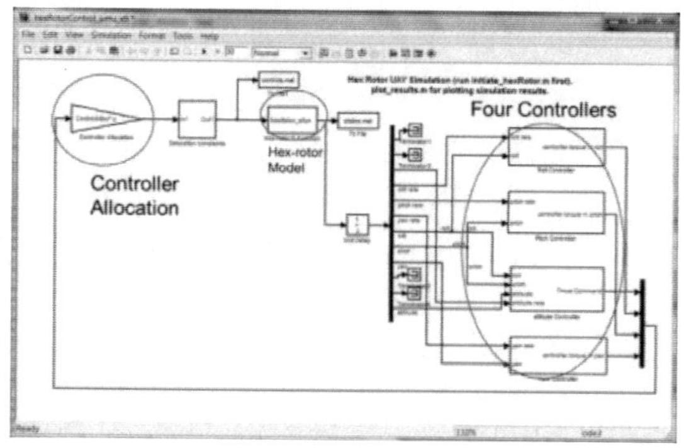

Impact Technologies
A Sikorsky Innovations Company

This page contains no technical data controlled by the ITAR or EAR

FDI VIA KALMAN FILTERING

Modeling Loss of Effectiveness (LoE)...

We model the vehicle dynamics in linear discrete-time state variable form as given by:

$$x_{k+1} = A_d x_k + B_d u_k + w_k^X \tag{19}$$

$$y_k = C x_k + v_k \tag{20}$$

w_k^X and v_k are zero mean uncorrelated random sequences with covariance matrices Q^x and R, respectively. We assume that the initial conditions x_0 is a random Gaussian vector with $E\{x_0\} = \bar{x}_0$ and $E\{(x_0 - \bar{x}_0)(x_0 - \bar{x}_0)^T\} = P_0^x > 0$.

The loss of control effectiveness can be modeled as follows:

$$\gamma_k = \begin{bmatrix} \gamma_{1_k} \\ \gamma_{2_k} \\ \vdots \\ \gamma_{6_k} \end{bmatrix}, 0 \le \gamma_{i_k} \le 1; \text{ i=1,...,6}$$

= 1 – (input/output$_{actual}$)/ (input/output$_{healthy}$)

Where $\gamma_{i_k} = 0$ and $\gamma_{i_k} = 1$ means that the *i*th actuator is completely healthy or fully damaged, respectively.

Impact Technologies
A Sikorsky Innovations Company

This page contains no technical data controlled by the ITAR or EAR

FDI VIA KALMAN FILTERING

Modeling Loss of Effectiveness…

To demonstrate actuator faults explicitly for the convenience of fault parameters estimation, the state equation in Eq. (19) is modified as

$$x_{k+1} = A_d x_k + B_d u_k - B_d U_k \gamma_k + w_k^x \tag{22}$$

where $U_k = diag\begin{bmatrix} u_k^1 & u_k^2 & \dots & u_k^6 \end{bmatrix}$. We want to determine the amount of loss in the effectiveness, γ_k, in a recursive form, so that the control allocation routine can be reconfigured. The effectiveness factors are modeled as the following form:

$$\gamma_{k+1} = \gamma_k + w_k^\gamma \tag{23}$$

where w_k^γ is zero-mean white Gaussian noise vectors with covariance Q^γ. The γ_k term can also be thought of as a health parameter related to the health condition of the corresponding actuators.

FAULT-TOLERANT CONTROL

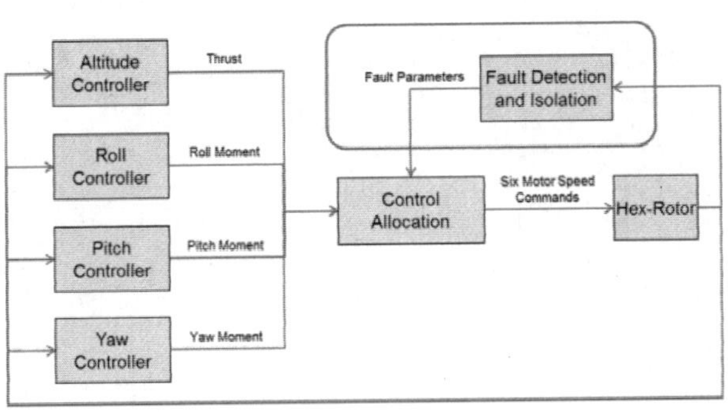

The PD controllers for the hex-rotor are designed below:

$$T = \left(g + K_{z,D}(\dot{z}_d - \dot{z}) + K_{z,P}(z_d - z)\right)\frac{m}{cos\phi cos\theta}$$

$$L = \left(K_{\phi,D}(\dot{\phi}_d - \dot{\phi}) + K_{\phi,P}(\phi_d - \phi)\right)I_{xx}$$

$$M = \left(K_{\theta,D}(\dot{\theta}_d - \dot{\theta}) + K_{\theta,P}(\theta_d - \theta)\right)I_{yy}$$

$$N = \left(K_{\psi,D}(\dot{\psi}_d - \dot{\psi}) + K_{\psi,P}(\psi_d - \psi)\right)I_{zz}$$

$$\begin{bmatrix} T_1 \\ T_2 \\ T_3 \\ T_4 \\ T_5 \\ T_6 \end{bmatrix} = pinv\left(\begin{bmatrix} 1 & 1 & 1 & 1 & 1 & 1 \\ 0 & -lsin\left(\frac{\pi}{3}\right) & -lsin\left(\frac{\pi}{3}\right) & 0 & lsin\left(\frac{\pi}{3}\right) & lsin\left(\frac{\pi}{3}\right) \\ l & lcos\left(\frac{\pi}{3}\right) & -lcos\left(\frac{\pi}{3}\right) & -l & -lcos\left(\frac{\pi}{3}\right) & lcos\left(\frac{\pi}{3}\right) \\ -K_{TM} & K_{TM} & -K_{TM} & K_{TM} & -K_{TM} & K_{TM} \end{bmatrix}\right)\begin{bmatrix} T \\ L \\ M \\ N \end{bmatrix}$$

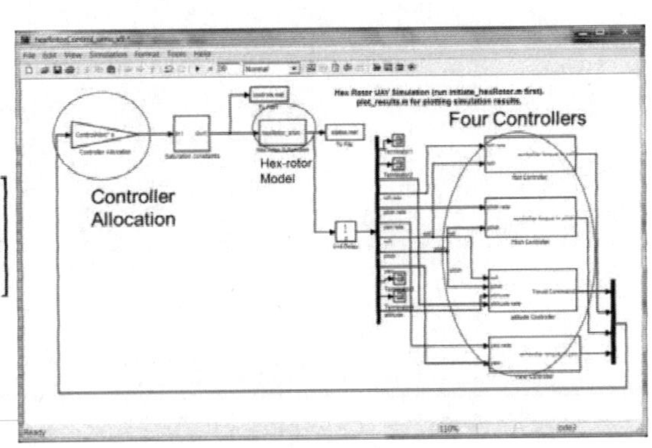

CONTROL ALLOCATION

- 1st principles physics-based approach
- Reconfigurable, based on LoE estimates from Kalman filter:

LoE factors (1 - γ_i) scale corresponding column

PID Output → **Motor Thrust Cmds**

$$\begin{bmatrix} T \\ L \\ M \\ N \end{bmatrix} = \begin{bmatrix} 1 & 1 & 1 & 1 & 1 & 1 \\ 0 & -l\sin\left(\frac{\pi}{3}\right) & -l\sin\left(\frac{\pi}{3}\right) & 0 & l\sin\left(\frac{\pi}{3}\right) & l\sin\left(\frac{\pi}{3}\right) \\ l & l\cos\left(\frac{\pi}{3}\right) & -l\cos\left(\frac{\pi}{3}\right) & -l & -l\cos\left(\frac{\pi}{3}\right) & l\cos\left(\frac{\pi}{3}\right) \\ -K_{TM} & K_{TM} & -K_{TM} & K_{TM} & -K_{TM} & K_{TM} \end{bmatrix} \begin{bmatrix} T_1 \\ T_2 \\ T_3 \\ T_4 \\ T_5 \\ T_6 \end{bmatrix}$$

And angular shaft speed is $\omega_i = \sqrt{\frac{T_i}{K_T}}$

- Solving control allocation problem is 4x6 matrix inversion
- Process can be performed offline, and "gain scheduled" if necessary

Impact Technologies
A Sikorsky Innovations Company

This page contains no technical data controlled by the ITAR or EAR

17

CLOSED-LOOP SIMULATIONS

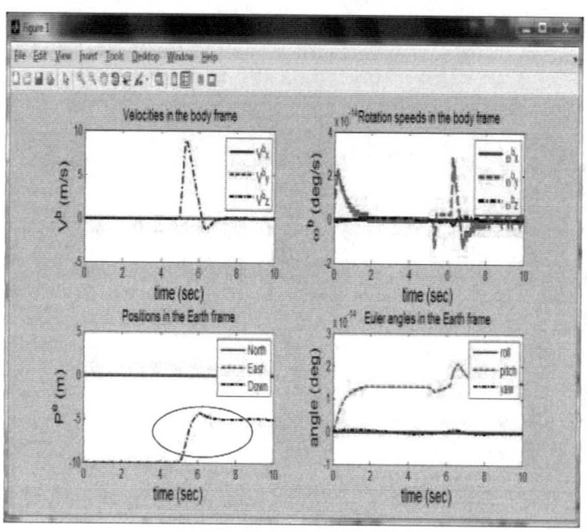

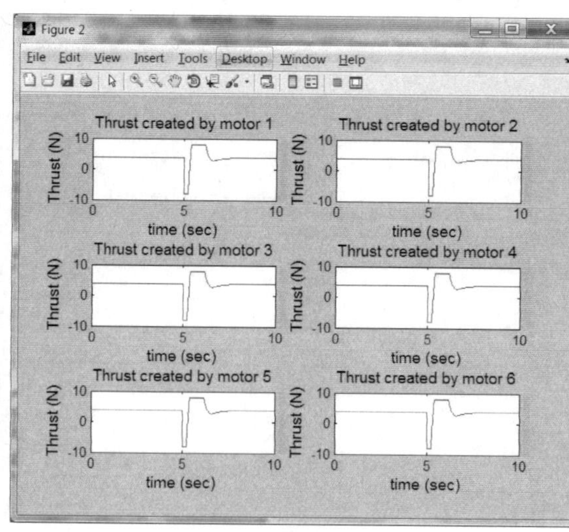

states controls

Hex-rotor reduces altitude from 10 meters to 5 meters at t=5 sec, and hovers at 5 meters.

Impact Technologies
A Sikorsky Innovations Company

This page contains no technical data controlled by the ITAR or EAR

RESPONSE: MOTOR 1 FAILURE

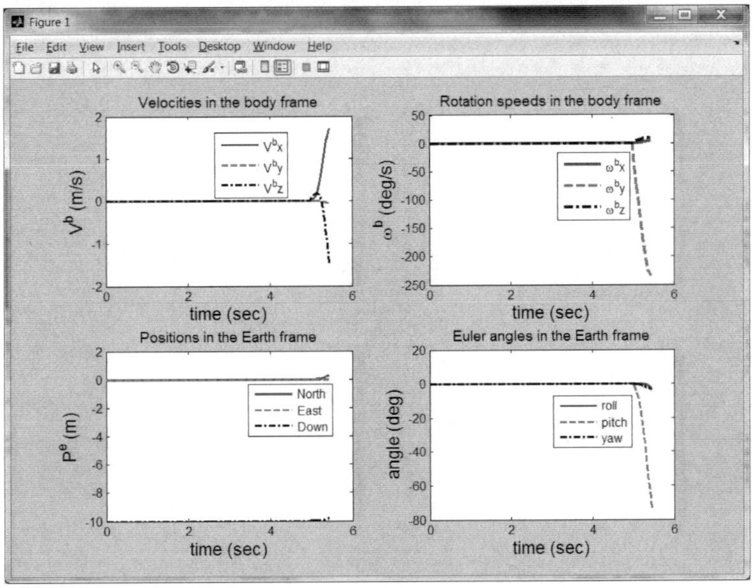

Hex-rotor control system becomes unstable because of motor 1 failure at t=5 s.

CONTROL FOR MOTOR 1 FAILURE

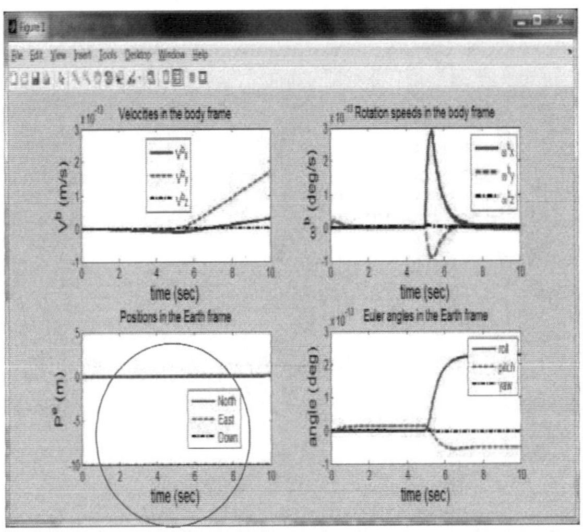

states

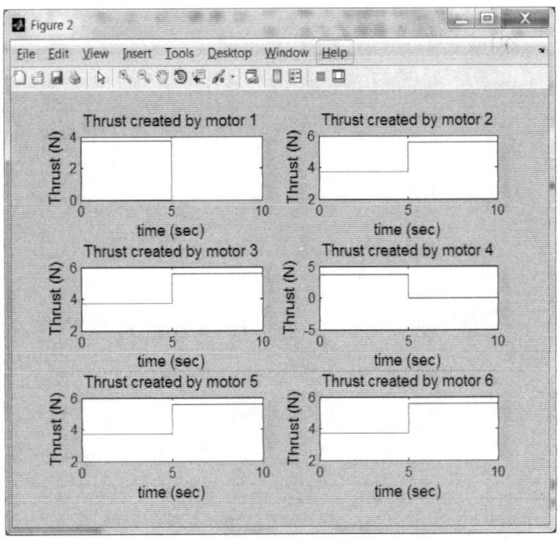

controls

The hexrotor with the remaining five motors is stable after the controller reallocation.

Simulation Scenario: hexrotor hovers at10 meters. Turning off motor 1 at t=5 s, and new controller allocation is applied at t=5 s. No time delay between fault and reconfigurable control

SAFE LANDING MOTOR 1 FAILURE

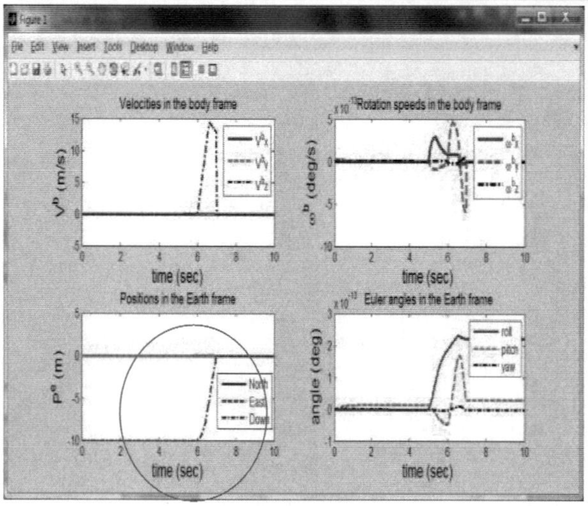

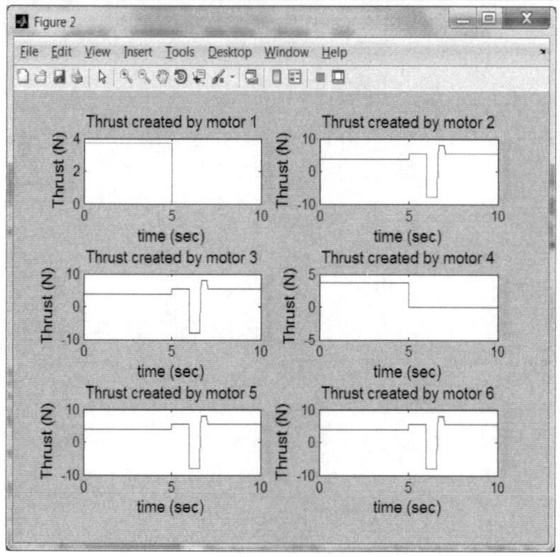

states

controls

The hexrotor with the remaining five motors can follow the altitude command.

Simulation Scenario: hex-rotor hovers at altitude=10 meters from t=0 s to 6s, then lands at t=6 s. Turning off motor 1 at t=5 seconds, and new controller allocation is applied at t=5 s.

CONCLUSIONS

- Developed fault adaptive control method based on continuous health parameters monitoring and LOE estimation.
- Developed real-time system identification for use in model-based control reconfiguration
- Optimized passive fault tolerance based on controller parameters tuned with Particle Swarm Optimization
- Applied to small scale hex-rotor UAV
- Simulations show the proposed controller reallocation against motor failure can effectively keep hex-rotor stable, and has acceptable performance with delay time less than 0.2 seconds.

INTELLIGENT MAINTENANCE SYSTEMS: VEHICLE HEALTH AND USAGE TRACKING

John M. Lacontora and Melvin D. Mathew
Drexel University
3141 Chestnut Street
Philadelphia, PA 19104-2884
john.m.lacontora@drexel.edu
mdm359@drexel.edu
215-895-1752

Abstract: Over the last two decades, Intelligent Maintenance System (IMS) has emerged as an enabler for manufacturers and owners of complex systems to significantly increase their systems reliability and lower the overall cost of ownership. IMS health monitoring and analysis tools have the potential to determine the remaining useful life of components and subsystems resulting in a considerable reduction of operation and maintenance costs.

Key words: Health and usage monitoring system (HUMS); Intelligent Maintenance System (IMS); Performance support

Introduction:

Intelligent Maintenance Systems (IMS) are gaining in popularity based on its ability to decrease downtime and cost of maintenance of engineered systems. On modern rotorcraft, Health and Usage Monitoring Systems (HUMS) are being used to monitor critical components and track aircraft usage. These systems serve to capture, store and analyze data of specific aircrafts and can also be used to analyze a fleet of aircraft in order to monitor aircraft status, decrease downtime, and modify maintenance processes and schedules.

The rotorcraft industry is very interested in developing a validated process that would extend the life of life-limited components through credit granting. This process would need to satisfy the United States Federal Aviation Administration's (FAA) regulatory requirements, maintain safe operations of the aircraft, and provide a means to extend the life of life-limited rotorcraft components. For this, usage and condition monitoring are both viewed as an interrelated opportunity to determine the current state of a component, predict its remaining life and enable a support activity to maintain an aircraft based on that state/condition of the component.

Recent studies and publications show the value of using HUMS and IMS tools to determine the current state of components, enable semi-automated and automated diagnostics, and provide prognostics and interactive technical manuals for performance support.

This paper represents a work in progress related to IMS and Condition-Based Maintenance (CBM) for vehicle maintenance and repair. It addresses emerging tools and techniques related to CBM for granting credits to extend the useful life of life-limited components. This paper also presents a framework for a systems approach to vehicle maintenance.

Intelligent Maintenance System:

IMS strive to provide a comprehensive set of tools that provide degradation assessment and failure prediction. In looking at a systems approach to vehicle maintenance, IMS tools can serve to enhance on-board HUMS or within an off-board data analysis capability.

The Watchdog Agent™ toolkit developed by the IMS Center, headed by the University of Cincinnati is capable of providing degradation assessments based on readings from multiple sensors measuring critical values for the system under consideration [10]. The toolkit's software prognostic capability lies in analyzing data trends and subsequent statistical modeling allowing it to predict future behavior and forecast health of the system. In addition, the Watchdog Agent possesses diagnostic capabilities from its ability to catalog significant signature patterns in order to recognize a situation that have been observed in the past or to aware the users of a situation that was never observed before. The Watchdog Agent has elements of intelligent behavior that enable it to answer the following questions:

- When the observed process, or equipment is going to fail, or degrade to the point when its performance becomes unacceptable
- Why the performance of the observed process, or equipment is degrading, i.e.
- What is the cause of the observed process or machinery degradation

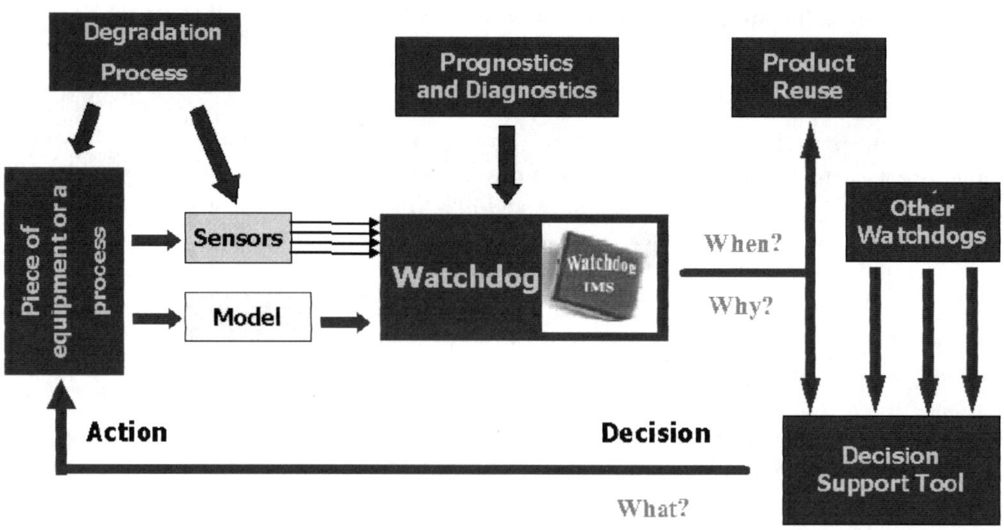

Figure 1. Schematic Representation of an Intelligent Maintenance System

Health and Usage Monitoring System:

Rotorcraft HUMS typically performs vibration monitoring, exceedance monitoring, and rotor track and balance to detect mechanical failures [13]. The next phase in HUMS usage is to become a key enabler for condition-based maintenance programs in extending the life of life-limited components.

Sikorsky recently conducted a research study for the Federal Aviation Administration (FAA), in support of the U.S. Department of Transportation (DOT), that addresses usage monitoring as a means to grant credits to life-limited rotorcraft components. This work mainly focused on usage monitoring and Usage-Based Maintenance (UBM) [1]. The Boeing Company recently conducted a study in support of the US Army CBM program that combined usage monitoring and the condition of components to grant credits to life-limited rotorcraft components [12].

HUMS for rotorcraft derived out of a need to transition rotorcraft to a CBM system within the US Army. It has since become apparent in the US Army CBM program that usage monitoring is one of the key tenants of the program and a subset of wider reaching System Health Management (SHM) program [12]. Aircraft SHM programs enable (a) efficient fault detection, isolation, and recovery, (b) prediction of impending failures or functional degradation, (c) increased reliability and availability, (d) enhanced situational awareness for crew and support personnel, (e) condition-based and just-in-time maintenance practices and (f) efficient ground processing and increased asset visibility and availability. The US Army's investment in HUMS showed a 12-22% decrease in parts cost per flight hour for HUMS-equipped helicopters between 2007 and 2009 [13].

Rotorcraft HUMS capture component degradation, as well as aircraft flight profiles or regimes. This combination serves to provide both component condition analyses alongside flight regimes load data thereby allowing us to predict current and future component state based on component health and flight capability limits.

Condition-Based Maintenance:

CBM is based on the theory that accurate wear and failure of a component is predicted through analysis [4]. The development of CBM systems is directed towards the ability to diagnose any abnormalities and to calculate the Remaining Useful Life (RUL) using Artificial Intelligence (AI) techniques [8]. For rotorcrafts, CBM programs call for the monitoring of health and usage information of individual components as well as the ability to keep track of a vehicle's overall condition. The development of CBM was created due to a desire to measure each component individually for wear and tear, predict when failure is approaching, and subsequently, the replacement of only the individual component before it becomes fully worn out and causes damage to adjacent systems. This idea of maintenance is expected to save costs by extending the operational usage of an aircraft and decrease maintenance burden, both in an economical and human way, while still retaining or enhancing the safety. CBM has a significant advantage compared to traditional maintenance methods such as phase-based maintenance and periodic inspections, thanks to the vibration analysis [9]. CBM development aims for the ability to diagnose deviations and calculate RUL and to enhance the early identification of degraded or damaged components [3,6].

The US Army ADS-79C-HDBK [4] has set up four goals regarding CBM implementation: reduce current maintenance burden required for assurance of airworthiness, increase availability of the aircraft, improve flight safety, and reduce overall maintenance costs. In addition, a set of guidance-lines for a functional CBM system is that it shall include monitoring of: engine, dynamic system components, structure (framework), and usage. Along with these goals, it is also intended to include exceedance recording electronics and an electronic logbook interface. These goals would apply to all Army aircraft systems. It also encourages incorporation of CBM as a maintenance approach, when developing a new aircraft or an Unmanned Aerial System (UAS).

ADS-79C [4] describes elements that enable the possibility to issue CBM credits by measuring condition and actual usage of a component, in order to change inspection intervals and removal criteria. CBM depends on data collection from the various sensors that is processed, analyzed and correlated to set material conditions that requires maintenance. These maintenance actions are needed to ensure continued airworthiness. Since CBM is dependent on data, the main goal of the requirements and standards are focused onto data collection, transmission, storage, and processing. CBM is now widely used in many applications globally and an international standard has been developed; ISO 13374:2003. Palmer, et al [11] states that to maintain a robust health management system, it is critical to ensure the integrity of the data before using it for diagnostic and prognostic purposes. The practice of CBM is enabled through three basic methodologies (1) embedded diagnostics/prognostics for components, (2) usage monitoring and (3) fatigue life management.

Usage-Based Maintenance:

A comprehensive knowledge of actual aircraft usage helps to assure safe operational procedures and usage profiles and could potentially reduce operating costs via UBM credits. Ultimately, the goal is to transition from a time-based method to CBM. Usage monitoring via regime recognition is a means to accomplish this [1].

UBM credits are used to change the maintenance interval of a specific end item or component. Algorithms that utilize Regime Recognition (RR) based maintenance data are being used to conservatively estimate the current state of an item based on actual usage. Changes in inspection timelines and maintenance schedules should keep in mind safety and reliability factors [7]. Clustering algorithms can be employed to analyze the output of onboard RR Algorithms (RRA). The RRA's output would be a listing of regimes flown along with the time spent in those regimes. All event based maneuver occurrences for the flight are also identified [12]. UBM processes are being perceived to be an acceptable means for assessing the remaining useful life of a component. It takes into consideration base reliability and the HUMS RR data.

The clustering algorithm concept illustrated in Figure 2 maps the RR output directly to Composite Worst Case (CWC) regime events [7]. The algorithm is driven by a semi-automated process to ensure engineering experts are part of the process and keeps in mind the FAA regulatory requirements.

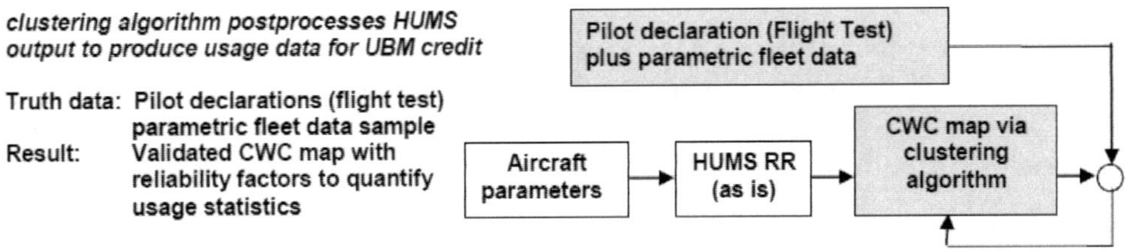

Figure 2: Clustering Algorithm Concept

The concept maintains the premise the a pilot or human input is an important attribute and must remain in the loop to substantiate the accuracy of the data and to ensure that the clustering algorithm parameters are compatible with known physics of flight and aircraft performance characteristics [2].

CBM IMS Notional Framework:

From past experience and literature, a notional CBM IMS framework emerged. A representation of this framework is presented in Figure 3. This framework is based on a systems level approach to maintenance operations. The effectiveness of the framework relies on the initial and later versions of the Failure Mode Effects and Criticality Analyses of the system and system components.

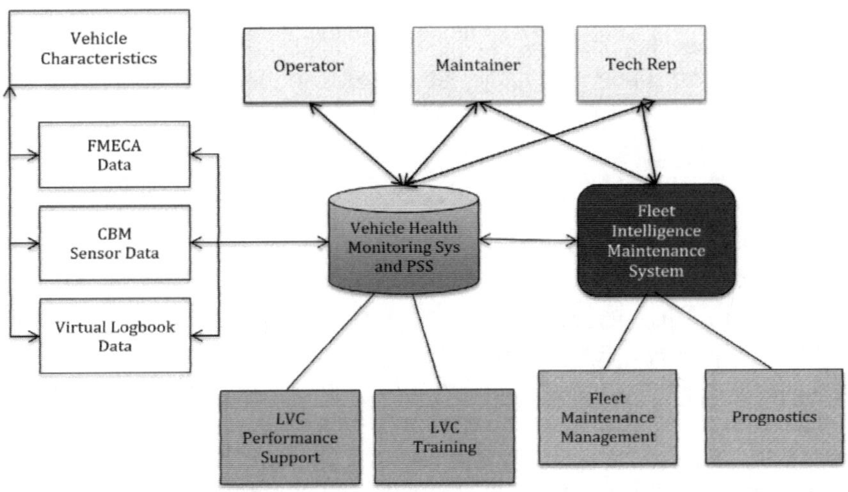

Figure 3: CBM IMS Notional Framework

The framework is perceived to have tenets that are useful for both manned and unmanned vehicles and systems. With the coupling of robust onboard HUMS and an IMS connection to a useful PSS, a near zero unscheduled maintenance program maybe achievable.

Conclusion

Many of the system attributes associated with an intelligent maintenance system make it a good candidate for inclusion in CBM programs. One of the main concerns of HUMS, CBM, and UBM developers is data validity. For these extensive validity criteria, regulatory approval and back up by system redundancy must be made available. One of the most notable facts that emerged during a US Army review is the need for UBM to be a subset of CBM.

The following questions are a sample of possible research foci: Can Health and Usage Monitoring Systems be coupled to provide a comprehensive condition-based maintenance program? Are regime based clustering algorithms appropriate for predicting remaining useful life of components? Can intelligent maintenance systems also serve as a performance support system for logistics support personnel and maintenance technicians? Can a condition-based maintenance intelligent maintenance system serve to enable autonomous systems safe operations?

REFERENCES

1. P. Bates, M. Davis and P. Sadegh, "Structural Usage Monitoring and Flight Regime Recognition", SAC, September 2012.
2. P. R. Bates, M. Davis and P. Sadegh, "Rotorcraft Dynamic Component Usage Based Maintenance Process", presented at the American Helicopter Society 66th Annual Forum, May 2010.
3. T. D. Batzel, D. C. Swanson, "Prognostic Health Management of Aircraft Power Generators", February 2008.
4. D. B. Cripps, "Aeronautical Design Standard Handbook for Condition Based Maintenance Systems for US Army Aircraft Systems" ADS-79C-HDBK, January 2012.
5. K. Daly, "Extending HUMS Reach," Flight-International, 175(5176), pg 29, February 17-23, 2009.
6. P. J. Dempsey, "AH64D Nose Gearbox Spiral Bevel Gear Maintenance Credit Evaluation," NASA Glenn Research Center.
7. Federal Aviation Administration Advisory Circular 29-2C MG, "Structural Usage Monitoring and Flight Regime Recognition- Compliance Validation and Demonstration Using the Health Monitoring and Usage Monitoring System, " U.S. Department of Transportation, September 2012.
8. P. Funk and M. Jackson, "Experience Based Diagnostics and Condition Based Maintenance within Production systems", August 2005.
9. M. P. Gaguzis, "Effectiveness of Condition Based Maintenance in Army Aviation", June 2009.
10. J. Ni, J. Lee and D. Djurdjanovic, "Watchdog – Information Technology for Proactive Product Maintenance and Its Implications to Ecological Product Re-Use".

11. C. A. Palmer, N. A. Mackos and M. J. Roemer, "Approach to Monitor and Assess the Quality of Sensor Data in Support of Calibration and Condition Based Maintenance for Turbine Powered Navy Vessels", May 2007.
12. P. Shanthakumaran, T. Larchuk, R. Christ, D. Mittleider and E. Hitchcock, "Usage Based Fatigue Damage Calculation for AH-64 Apache Dynamic Components", May 2010.
13. US Joint Helicopter Safety Implementation Team, HFDM Working Group, "Health and Usage Monitoring Systems Toolkit," International Helicopter Safety Team, February 2013.

EFFECTIVE ACCURACY CONTROL FOR CBM APPLICATIONS.

Thomas L. Lagö and Alan Boyer
Qirra Sound Technologies Europe AB
Stora Torget
Jonas Bronck Center
Sävsjö, SE 57692 Sweden
Telephone: +46 708 357337
thomas.lago@qirrasound.com

Valdi Ivancic
Medison University
Stora Torget
Jonas Bronck Center
Sävsjö, SE 57692 Sweden

Abstract: Controlling amplitude accuracy is an old concept. It is common to calibrate sensors to 2% and specify temperature drift and many other parameters for sound and vibration measurements. However, the sensor is typically not the main "error contributor." Instead, measurement practices, algorithm errors and faulty data collection principles are the main error sources. One unfortunate thing with these errors is that they are deterministic and hence averaging does not make them go away or decrease. We strive for success concepts in CBM and we strive for a better understanding and clever strategies. However, the basic assumptions for the measurements and analysis principles used is sometimes forgotten or "ignored." This can then lead to large amplitude errors and hence mislead the following analysis, classification and prognosis. This paper describes some of these errors and how incorrect conclusions and results may occur. By understanding these errors and why they exist, it is possible to mitigate and/or minimize their impact on the end results. Common challenges could be improper selection of anti-aliasing filters and/or sampling rate versus the filter bandwidth and roll-off rate. Another challenge is related to the fact that analyzed signals not obeying the required properties, may result in large amplitude errors. The background to the challenges is discussed, and typical solutions or strategies to avoid such errors are presented.

Key words: Diagnostics; prognostics; health management; e-learning; virtual labs; education; signal processing; aliasing; FFT; calibration; amplitude errors

Introduction: From vast empiric research, it has been found that common errors in sound and vibration measurements are made by: 1. Human "error" based on measurement practices. 2. Measurement systems that do not comply to known signal processing theory (e.g. sampling and reconstruction). 3. Faulty algorithm implementations. 4. Incorrect use of the signal class (random, stationary, tonal etc.)

Category "1" is, to a large extent, an educational topic. Often, standards state how to make the measurements but these recommendations are not always followed since it is not believed that it will make such a big difference in the end result. However, if the end result should be in the range of the 2% error that we often set for the sensors, all human errors must be considered in full. The user often repeats these "mistakes" and hence, there is a bias in the measurement that is user induced.

Category "2 and 4" will be covered in the next few sections. This is a common area for errors and many misconceptions exist. It could lead to very large errors! The two categories will be covered in more detail and examples presented.

Category "3" is a challenge. For SLM (Sound Level Meters), multiple standardization organizations (like ANSI and ISO) specify what should be done and why (including the human operation and interaction). This does not mean that all of the instruments are "good." Recently, a noise level situation in a nightclub in a large US city stimulated an OSHA investigation. The results showed that the levels were above the allowed legal limit. The owner asked another team to make a measurement with certified instrumentation and expert engineers, hoping that this would show lower values. The expert SLM measurement gave 4.5 dB higher levels than what OSHA had found with their SLM. After a full investigation of all possible error sources, it was concluded that the SLM (Type-2 instrument) that OSHA used had a faulty averaging method and that caused the large error. The SLM that the expert team used gave the correct answer with a Type-1 accuracy. This is a good example in regards to the faulty and non-tested algorithms. In our vast empiric research, we have found this type of error many times.

Sampling: Today, a digital representation of the world around us becomes almost mandatory. This makes the conversion of an analog signal into a digital representation frequent. When performing this conversion, it is of interest to make sure that it is possible to correctly reproduce the analog signal from the record of collected digital data samples. The sampling procedure has been covered in many textbooks, especially in electrical engineering and for analog time data. However, the sampling process is general and valid for any sampling i.e. spatial sampling, data points from a medical study etc. Thus, it is important to remember that the strict rules described in early papers by i.e. Nyquist and Shannon, [8][9], must be followed, and if they are violated, it is not possible to find the real world signal in the sampled data set. The sampling rules will be covered in this article, and the result may surprise some groups, especially non-electrical engineers. In figure 1 below, a typical ADC system is depicted, and suggested sample points are indicated along the analog time signal axis.

Figure 1: Example of an ADC and the samples taken at equidistant points along the time axis, creating a series of digital "data points" that should represent the analog signal.

It is obvious that if one takes an infinite number of samples, this will be easy, but if too few are collected, it is not possible to reconstruct the analog signal from the sample points. In figure 2, an illustration is made showing that it is possible to draw an infinite number of other signals through the collected data points, and thus it is not possible to be certain that a reconstruction of our original signal is valid, unless some limitations to the analog signal and the corresponding sampling rate is defined. This sampling technique or requirement is often referred to as "The Sampling Criterion," "The Sampling Theorem", "Nyquist Sampling Criteria", or something similar. There are many names and descriptions for this process and this article will discuss the validity of these criterions and their assumptions.

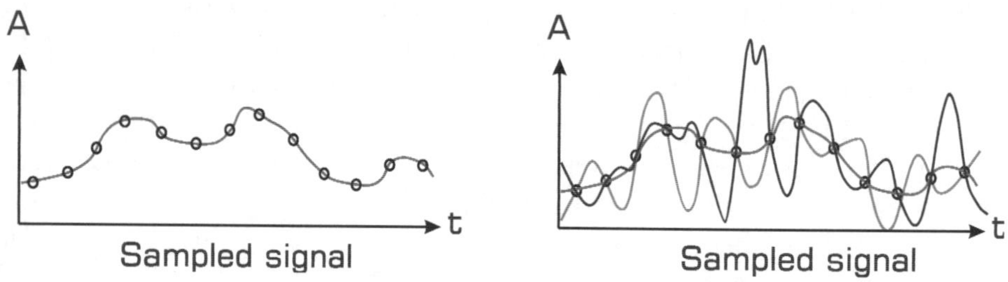

Figure 2: Illustration of a sampled signal and how it is possible to superimpose many different analog signals that will go through the exact same sample points. This fact makes the sampling process ubiquitous and that is not good. By adding some restrictions to the analog signal and making sure the sampling rate is in harmony with these restrictions, we avoid this unfortunate situation.

By studying figure 2, it is intuitive to differentiate the three signals by frequency content. The one that "moves the most" between the data points must have a larger bandwidth. The one that "moves the least," must have the lowest bandwidth or frequency content. This leads us to believe that the sampling rate should be coupled to the frequency bandwidth or frequency content of the signal.

In many textbooks and articles, especially books in Mechanical Engineering, it is common that the sampling theorem is stated as: the sampling frequency must be larger than twice the maximum frequency, [1][2][3][4]. There are many more books that could have been included in this list, but these books were selected as examples only, and there is no intent to discredit their otherwise excellent content, but it is important to highlight the scope of this misconception. It is also common to refer to Shannon when making this sampling selection statement, thus making it sound even more correct. The sampling formulation above is unfortunate and could easily lead to an incorrect understanding or interpretation of how a correct sampling rate should be selected. The proposed approach does not lead to a fully correct sampling rate selection and this common misconception will be discussed in more detail in this article.

Both Nyquist and Shannon understood how to select the sampling rate and demonstrate the reasons why they were appropriate. However, they talked about different types of signal, general versus so-called baseband, and Shannon explained this difference clearly before stating his well known Theorem 1, which will be presented later. This means that Shannon assumed that the signals started at zero Hertz and contained signal information content up to f_{max} Hz, while Nyquist did not. Nyquist talked about the general definition using the bandwidth W and never discussed the restriction using baseband signals. Figure 3 illustrates the difference in spectral content for the two cases, in which the sampling rate is the same, but the maximum frequency is different.

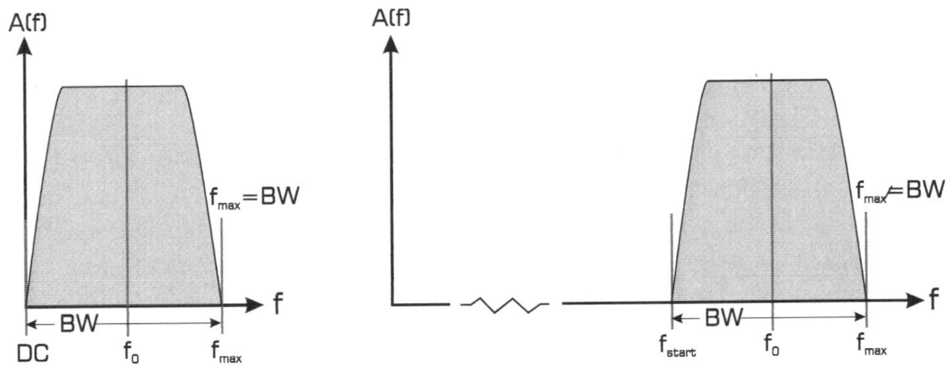

Figure 3: Illustration of the difference between a baseband signal and a general signal with frequency content in a band, and the starting frequency separated from DC. The bandwidth BW, is the same in these two cases, but the maximum frequency can be totally different. The bandwidth could be 1 kHz, but the maximum frequency, given that we use an ultrasonic transducer, could be MHz. Please note that the maximum frequency notation used is not equal to the one commonly used for filter applications.

Shannon and Nyquist: In many textbooks as well as in articles, it is common that the sampling theorem is stated as: the sampling frequency must be larger than twice the maximum frequency, [1][2][3][4]. This is not a fully correct sampling rate selection and this misconception will be discussed in this article. It is also common to refer to Shannon when making this statement, thus making it sound even more correct. In an article in Sensors, July 1999, [5], this issue was highlighted and some misconceptions were discussed in order to clarify the matter. Unfortunately, some explanations did not fully reveal the process behind a correct selection of the sampling rate. In the text, a discussion why Shannon used the terminology "maximum frequency" could easily lead to wrong conclusions. The article states: "If you need to measure a 1 kHz component, Shannon's theorem would demand that you sample at >2 x the component (see the sidebar)." The sidebar then states, and refers to Shannon: "if a function $f(t)$ contains no frequencies higher than W cps (Hz), it is completely determined by giving its ordinates at a series of points spaced ½W seconds apart. This has come to be known as Shannon's sampling theorem, or simply the sampling theorem, often misnamed the Nyquist theorem."

The formulation in this article is unfortunate and could easily lead to a wrong understanding or interpretation of how a correct sampling rate should be selected. Shannon certainly understood which sampling rates should be used and why, but he did express it as the quote above states, [9]. However, he assumes in his article, probably for ease of understanding, that the signals are of a so-called baseband type, and he explains this clearly before stating Theorem 1. This means that the signals start at zero Hertz and contain a signal information content up to f_{max} Hz. Shannon mentions Nyquist in "A Mathematical Theory of Communication" from 1948 [9], but is referencing article [8] for a more thorough discussion. Article [9] discusses bandwidth, but sometimes hides this by using a wording like "components up to." This is unfortunate, since given that the signal contains frequency components within the dynamic range of interest, the sampling rate as discussed earlier could be correct, given that we assume that "up to" implies that we start at zero Hz. The problem with most of those statements is that the "zero to f_{max}" discussion as a definition of the bandwidth is very often hidden by improper wording. Both Nyquist and Shannon stated the sampling theorem in a similar manner but used different wording,, so there are no conflicts and/or differences between them. This will be discussed in more detail later in this article. Several books are explicit and thus correct, most of them being more mathematically based signal processing books. It is rare to find a more practical book to be correct, but Hewlett-Packard's [6] is a good example. However, many application notes uses a less correct sampling rate statement, unless we assume baseband signals. It is interesting to note that the products the application notes are written for do not use the maximum frequency for sampling determination, but the bandwidth. Thus, there seems to be a major confusion in textbooks and articles if maximum frequency or bandwidth should be used when determining the sampling frequency. Shannon said the following [9]:

II. The Sampling Theorem
Let us suppose that the channel has a certain bandwidth W in cps starting at zero frequency, and that we are allowed to use this channel for a certain period of time T. Without any further restrictions this would mean that we can use as signal functions and functions of time whose spectra lie within the band W and whose time functions lie within the interval T. Although it is not possible to fulfill both of these conditions exactly, it is possible to keep the spectrum within the band W, and to have the time function very small outside the interval T. Can we describe in a more useful way the functions which satisfies these conditions? One answer is the following:

Theorem 1: If a function *f(t)* contains no frequencies higher than W cps, it is completely determined by giving its ordinates at a series of points spaced ½ W seconds apart.

It is possible to interpret The Sampling Theorem and Theorem 1 incorrectly. Shannon talks about the "band W". He assumes, as stated in the beginning, that the starting frequency is zero Hz. Thus, the bandwidth W is equal to the maximum frequency denoted f_{max} and given by W. This is the case for baseband signals only, and not a more general statement where the band W must be used. He explains this very clearly, except in the theorem, and thus many readers have had difficulty in understanding the difference between f_{max} and W. Maximum frequency f_{max} and the bandwidth W may imply the same sampling frequency f_s for baseband signals, but only for baseband signals. Thus, there is no difference between Nyquist's explanation and Shannon's. They are both based on the bandwidth W, which is the correct way of doing it.

The use of maximum frequency does not apply for a bandlimited signal around some carrier frequency fc. The continuous time signal $x_r(t)$ representing e.g. an ultrasonic signal, is typically bandlimited. However, there is often a center frequency e. g. 1 MHz. It is only the signal information centered at the carrier frequency that is of interest. This provides a very good opportunity to "under sample" relative to the carrier frequency, but not relative to the sampling theorem. Therefore, it is possible to fulfill the Nyquist sampling theorem for the narrowband signal that represents the ultrasonic signal, yet sample with only $2*W$, where W is the bandwidth of the signal around the center frequency f_c. It is not a contradiction, due to the limited bandwidth. Also, a correct sampling based on the bandwidth W is sparse versus the frequency content, which is important when performing FFT analysis. If this is not the case, very large block lengths must be used for the FFT analysis and/or a resampling of the data must be performed before the FFT calculation can take place. Both methods require computer power and complicate the analysis. A common selection of sampling rate in FFT analyzers is 2.56 times the bandwidth (not frequency if they can do real zoom analysis), [11]. In this case, only 20% of the frequency lines are wasted whereas in an analyzer with 10:1 in sampling rate, 80% of the lines are wasted.

Let's assume that we have an ultrasonic transducer with a 1 MHz center frequency and a bandwidth of 1 kHz. Using f_{max} as the rule for selecting the sampling rate will lead to $f_s >$ 2 MHz, but using the bandwidth W, f_s is only > 2 kHz. There is a 1000:1 decrease in sampling rate without loss of information. If we are now interested in the frequency information within the 1 kHz band, an FFT size of 2.097.152 (1024x2048) samples must be used for $f_s = 2 * f_{max}$, but $f_s = 2*W$ only leads to 2048 samples, a great saving of storage and complexity. The computation time for the 2 M sample FFT is enormous in comparison to the 2048 FFT. Most of the frequency lines when using f_{max} sampling must be disregarded for a situation like this. This is why most FFT analyzers sample against the bandwidth and not f_{max}, despite what their application notes declare.

There is one disadvantage with selecting the sampling-rate versus the bandwidth W for the example above. The time signal does not look "right" since the "carrier frequency" is gone due to the bandwidth sampling. The time signal can be reconstructed and thus having the original signal reconstructed, but this implies interpolation and filtering. For example, with oscilloscopes, it is common to display the time samples only, without interpolation, even for bandlimited signals, and in this case one must sample versus f_{max} and not the bandwidth. This has nothing to do with Nyquist and/or Shannon. It is simply one method to avoid a more complex handling of the data, and if you know that the signal will be "looked at" in time domain, this could be a good strategy.

If the signals have been sampled correctly and follow the above rules, they can be completely reconstructed, despite the sparse sampling method. All samples, between the 2:1 (two samples per period for the maximum frequency (baseband signal or otherwise bandwidth)), can be completely found. The data series or sequence contains all the information in regards to which data points should be inserted in between the actual sample points. There is a common misconception that this is not possible, but it is! [17].

Frequency Domain Analysis: When performing FFT analysis, it is not always fully understood that the assumption for this analysis is based on deterministic signals as described by Figure 7 below (green boxes). If signals belong to different class than the deterministic one, large amplitude scaling errors are likely to happen. This can be avoided given that we know that we have e.g. ergodic data or transients. This is rarely the case. In CBM applications, a combination of signals is the most common case and all levels (different peaks in the data) cannot be properly scaled unless they belong to the same category. This is typically not considered since all data (all frequency values) are scaled using the same principle.

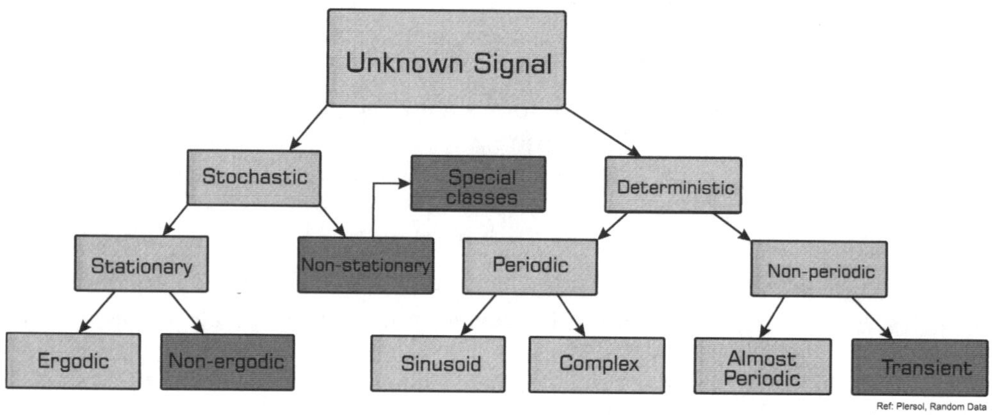

Figure 4: An example of different signal classes. The "Green" category is what most textbooks base their theory on. The "pink" is a complication but "possible" and the red requires more "thinking and handling." All are possible given that we understand the properties of the classes.

In Figure 5 below, two tones are visible but only one exists in reality. The "tone" to the right is about -25 dB according to the display. However, the one to the right is really not a tone. If we try to zoom in on the signal, it will "disappear" despite being clearly visible in this figure. Since we often base diagnosis on signal and/or component amplitudes, it is essential to scale them properly.

In Figure 6, a zoom has been performed, covering the "peak" that was visible before. The peak is "gone" when performing a zoom analysis and levels have decreased to almost -51 dB. This compares to a large amplitude error and in this case it is around 100,000 percent. If we make 100,000 percent amplitude errors, this is bad, especially if we use this as the foundation for prognosis.

This 100,000% error is substantial and it could be argued that this is not the "every time situation." From empiric research, it has been shown that this situation happens multiple times, but the more common situation is only a few 100% amplitude error. This is still substantial when compared to the 2% for sensor calibration.

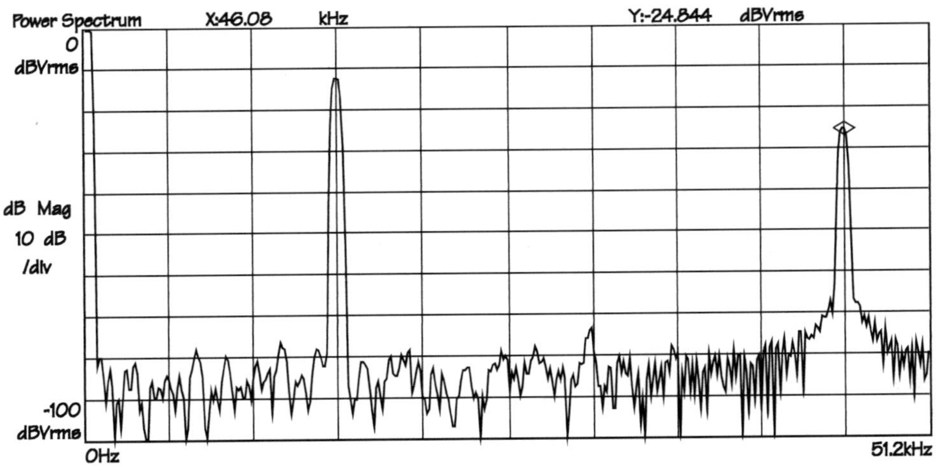

Figure 5: An example of what appears to be two frequency components but only one actually exists. There is only one tonal frequency component in the data.

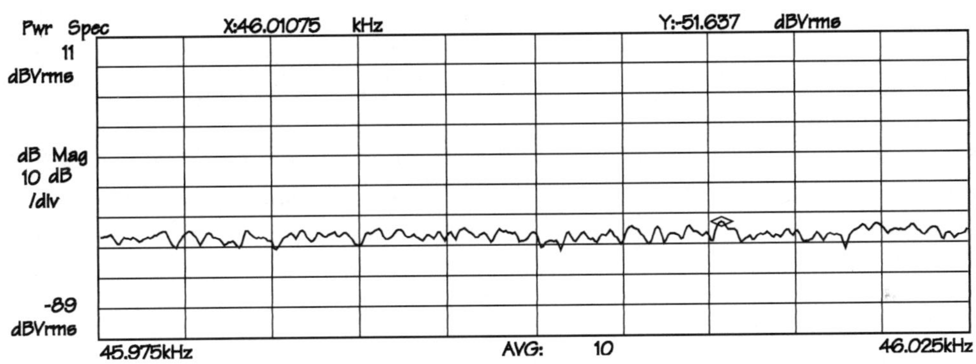

Figure 6: An example of what appears to be two frequency components but only one actually exists. There is only one tonal frequency component in the data.

There are methods, where it is possible to quickly determine if signals belong to the proper category. This is a typical example:

- Calculate three FFTs with resolutions: 400, 2400, 9600 lines.
- Overlay the data in the plot using different colors for the three plots.
- Assume "tonal" or deterministic. If all three overlay properly on all components of interest, the assumption was correct.
- If instead, the signal is "random" or ergodic, it will drop in amplitude every time the resolution is increased.

All of this can be automated in a MATLAB® routine that allows any signal to be automatically checked and have the proper scaling method applied. There are more topics and areas that can be "checked" for amplitude accuracy by using a proper understanding of what foundation the signal processing is based upon and a software routine to help,

[18][19][20][21][22]. Unfortunately, this has rarely been implemented in analyzers and hence, the user must use his own judgment and knowledge base to determine the correct answer. Hence, these challenges often become a "human error" that causes errors in the amplitude estimates. Large amplitude errors are often the end result, much larger than the 2% that has been the target for the sensors. Making sure all errors are in the same "range" makes a difference and should be the goal. Having one error at 2% and another one at 120% does not make sense. ALL errors must be controlled and we believe this is a systematic approach including all parts of the measurement chain plus proper training of the user, [23]. Everything that can be utilized by a computer or DSP (Digital Signal Processor) should be implemented in order to avoid user errors.

Conclusion: This paper has discussed some common errors that will have a strong impact on the amplitude accuracy accomplished. Since amplitude is often one of the more important parameters, it is essential to find strategies and approaches that lead to a good amplitude accuracy. Two common challenges have been discussed: 1. Sampling together with anti-aliasing filters and 2. Scaling of amplitude data from the FFT. For baseband signals, it is correct to interpret the sampling theorem as $2*f_{max}$ since f_{max} and the bandwidth W, will be the same as the maximum frequency, but for this case only. Both Nyquist and Shannon understood this and both used the bandwidth as the key parameter in their mathematical analysis of sampling rate selection. However, Shannon talked about a bandwidth *W* starting at zero Hertz, and that could be interpreted as the maximum frequency. Without the restriction, starting at zero Hertz, it would be a different statement. For FFT analysis, it is essential to have a good understanding of the signal class. If not, large amplitude errors are prone to happen. Some examples of errors have been presented where 100,000 percent scaling error is the end result. Strategies and rules of thumb to deal with these challenges have been presented and a more holistic approach to accuracy verification and control is suggested. Balancing errors to about the same level is essential and good measurement practice. Using a computer or DSP to help control the errors is possible and should be used.

References:

[1] Broesch, James D., Digital Signal Processing Demystified, HighText Publications, 1997.

[2] Zölzer, Udo, Digital Audio Signal Processing, John Wiley & Sons, 1995.

[3] M. P. Norton, Fundamentals of Noise and Vibration Analysis for Engineers, Cambridge University, 1990.

[4] Daniel J. Inman, Engineering Vibrations, John Wiley & Sons, 2007.

[5] National Instruments, "Data Acquisition," Sensors Magazine, Page 45, July 1999.

[6] Application note AN-243-1, Effective Machinery Measurements Using Dynamic Signal Analyzers," Hewlett-Packard Company, 1991.

[7] Dorf, Richard C., Electrical Engineering Handbook, CRC Press, 1993.

[8] Nyquist, H. "Certain Topics in Telegraph Transmission Theory." AIEE Trans., vol. 47, pp. 617-644, 1928.

[9] Shannon, C.E. "Communication in the Presence of Noise." Proc. IRE, vol 37, pp. 10-21, 1949.

[10] White, Robert A., Electronic Test Instruments, Theory and Applications, Hewlett-Packard Professional Books, Prentice Hall, 1993.

[11] Application note AN-243, "Fundamentals of Signal Processing," Hewlett-Packard Company, 1981.

[12] Proakis, John G. and Manolakis, Dimitris G., Introduction to Digital Signal Processing, Macmillan, New York, 1998.

[13] Lagö Thomas L., "A Novel Signal Model and Estimation Method for Acoustic Doppler Current Meters," Doctoral thesis, Lund University, LTH, Civil Engineering, department of Engineering Acoustics, LUTVDG/TVBA–99/1007–SE, April 1999.

[14] Thomas Lagö and Ingvar Claesson, "Spectral Estimation Errors When Using FFT Analyzers," International Journal of Acoustics and Vibration, Proceedings of the Fifth International Congress on Sound and Vibration, ICSV5, Adelaide, December 1997.

[15] Thomas L. Lagö, "Time and Frequency Consideration using Heisenberg's Uncertainty Principle Applied in Mechanical and Acoustic Applications," International Institute of Acoustics and Vibration, Proceedings of the Ninth International Congress on Sound and Vibration, ICSV9, Orlando, USA, Invited Paper, July 2002.

[16] Bendat and Piersol, Random Data, Wiley-Interscience 2nd Edition, 1986.

[17] Anders Brandt, Thomas L. Lagö and Kjell Ahlin, "Interpolation and Resampling Methods in Time and Frequency Domain and Their Applications," Society for Experimental Mechanics, Proceedings for the 22nd International Modal Analysis Conference, IMAC XXII, Dearborn, USA, January 2004.

[18] A.H. Brandt, T. L. Lagö, K. Ahlin, "Summary of Spectrum Estimators for Vibration Signals and Their Errors," Society for Experimental Mechanics, Proceedings for the 22nd International Modal Analysis Conference, IMAC XXII, Dearborn, USA, January 2004.Thomas L. Lagö and Anders Brandt, "Typical Accuracy in Data Collection and Analysis," Society for Experimental Mechanics, Proceedings of the 23rd International Modal Analysis Conference, 2006, St. Louis, Missouri USA.

[19] Thomas L. Lagö, "Condition Monitoring Solutions with Prognosis Capabilities," CM 2011 and MFPT 2011, The Eighth International Conference on Condition Monitoring and Machinery Failure Prevention Technologies, 20-22 June 2011, St David's Hotel, Cardiff, Wales, UK.

[20] Thomas L. Lagö, "Proper Amplitude Estimation Strategies for CBM Applications," MFPT, The Prognostics and Health Management Solutions Conference. 24-26 April 2012 – Crowne Plaza Hotel, Dayton, OH, USA.

[21] Thomas L. Lagö, "Condition Monitoring Requirements with a Focus on Prognosis Success," Proceedings of the Nordic Maintenance World Congress, 14-15 March 2012, Gothenburg, Sweden, Invited Keynote Address.

[22] Thomas L. Lagö, "Automatic Amplitude Estimation Strategies for CBM Applications," 18th World Conference on Nondestructive Testing, 16-20 April 2012, Durban, South Africa. Voted "Best Session Paper."

Bibliography:

[1] **Thomas L. Lagö**, Ph.D, Chief Scientist. Dr. Thomas Lagö is globally accomplished in the upper echelons of the sound and vibration world. Dr. Lagö has "4" Post Doctoral degrees in conflicting scientific disciplines and consults with organizations such as NASA, Rolls Royce, ABB, Volvo, SAAB, Scania, Boeing, Atlas Copco, TATA, Indian Navy, and many other world leading organizations. Dr. Lagö has implemented paradigm shifting technologies around the world and presents advanced classes to these companies to assist them in solving problems in advanced acoustic and vibration applications. He has more than 40 years of industrial experience and has worked for companies including Hewlett-Packard (today Agilent Technologies). Dr Lagö is a creative, enthusiastic, high-achieving executive with outstanding performance based on a solid education. Formerly, President and Chief Scientist of a $9 million leading USA sales company in the Sound & Vibration business. Prior responsibility reflects outstanding career growth and progress through various high-tech businesses, including European Product Line Manager for Hewlett Packard's Dynamic Signal Analyzer Products, as well as multi functional responsibilities including sales and marketing, strategic planning, engineering, but also finance and human resource leadership. Dr. Lagö has a vast practical experience. Lagö teaches sound and vibration courses and applicable mitigation methods on an international basis and has given courses for and many international universities and expert groups. His practical experience is vast and he is very practically oriented despite his thorough theoretical education and experience. His solutions and courses have helped many teams better their environment and solve challenging problems without spending huge money. Dr Lagö has become a BINDT Fellow, (British Institute for Non-Destructive Testing) and he is the Industrial Chair for ISCM - The International Society for Condition Monitoring. Lagö is also involved in a number of other scientific organizations like IIAV, International Institute for Acoustics and Vibration and is also serving as a Board of Directors member for (current and past): SEM-IMAC (Society for Experimental Acoustics), SAE (Society for Automotive Engineering, NVC), ICSV (International Institute for Sound and Vibration), SVIB (Scandinavian Vibration Society). He is also an AES member and participates in multiple standardization groups for sound in concert halls, movie theatres and speaker systems. Lagö is author of more than 290 scientific papers, written multiple chapters in international textbooks and holds more than 25 patent/ patent applications and has received multiple awards. He has 1900 international contacts on Linkedin and 10,000+ personal contacts in his CRM database.

[2] **Alan Boyer**, Ph.D, MBA, LTCL, ALCM, President and CEO of Qirra Sound Technologies LLC,. Dr. Alan Boyer's career in sound and vibration spans almost thirty years. He has been Executive VP of Sales and Marketing, and COO for a leading design and manufacturing company of sound and vibration instrumentation. He has also owned his own international company and during his career, has been awarded many high level contracts with organizations including; London's Heathrow, Gatwick, and Stansted airports, NASA, Boeing, Land Rover, Jaguar, US Airforce, and others. Dr. Boyer is also an accomplished musician and performed as a teenage virtuoso trumpeter and professional conductor. The combination of many years as a musician,

acoustics/vibration executive and entrepreneur give Qirra Sound Technologies and outstanding chance for success in the three target market categories of Industrial, Pro-Audio, and Audiophile. Dr. Boyer was born in England and graduated from Sheffield University, England, studying education, psychology and music. In 1998, after heading the development of the Larson Davis, Inc. European Center, Dr. Boyer moved to the USA to become manager of the corporate marketing department and was subsequently appointed to the position of Senior Vice-President and COO. One of his most important projects was the formation of a strategic alliance between leading international companies in the area of sound and vibration. This successful alliance drew great international attention and began to change the way that collaborative relationships were formed and nurtured in the industry. Other projects successfully led by Dr. Boyer include sound detection systems for the airports of London Heathrow, Gatwick and Stansted. He has also delivered test and measurement systems to Land Rover, Honda, Toyota, Rolls Royce, Jaguar Cars, Lotus Cars, Ford Motor Company, Vauxhall (GM), and others. Boyer has also carried out projects regarding ISO 9002 (for the European office) and ISO 9001 certification for the corporate office in the USA. Dr.Boyer implemented and oversaw projects in relation to product catalog, data sheet, instruction manual design drawing on his background in education, instructional psychology and sales/marketing.
Dr. Boyer is much sought after as a consultant on all areas of business and enjoys working with start-up companies and home based business owners. His teaching, presenting, seminar, and public speaking abilities are extremely effective and his combination of presenting complex materials combined with his English sense of humor, enable him to teach in a very effective manner. Boyer's consulting activities have been to various USA and European based businesses (established and start-up) in diverse industries such as Neutraceutical, Nano Science, Cosmetics, Acoustics/Vibration, and others. Boyer is an accomplished musician and conducts medium to large music ensembles. He is a Licentiate of Trinity College of Music and an Associate of the London College of Music. As a teenager, he was an accomplished trumpet virtuoso. He began to conduct at the age of fifteen and has directed many ensembles in Europe and the USA. In 2006, Dr. Boyer was presented with the National Arts and Humanities Award by Provo Arts Council for services to the arts in Utah and throughout the USA. Dr. Boyer sits on a number of for profit and not for profit boards in the areas of business and music/arts. As CEO of Qirra Sound Technologies LLC, Tech Fuzion, and Embla Global, LLC, Dr. Boyer brings important leadership skills and high levels of success and expertise to the organization both in the USA and European teams.

[3] **Valdi Ivancic**, founder of Medison University, a global and international university. Medison University belongs to the Medison company group with offices in London and California. Valdi is a visionary, serial entrepreneur, keynote speaker, innovator and PM Candidate. A person with ability to inspire other people, working for a better world and improve people's lives. He is the founder, co-founder, president and speaker in several start-ups, educational organizations, research groups, university courses and other non-profit organizations. He is a natural leader who makes co-workers grow in responsibility and as individuals. He creates and finds innovative solutions with the ability of connecting ideas and solutions from different areas into a bigger working picture. He has been traveling and living on several continents and speaks 7+ languages.

He is multi-cultural and has a vast understanding of different business cultures and political views. He has been attending 50+ courses at different universities worldwide (MOOCs) and been a key Lecturer/Speaker at several Swedish universities. He has raised over $25M to projects, enterprises, cities and universities. He is also the head of a venture partner company and a crowd funding organizations, helping SMEs getting funded. Ivancic has recently started SMENET™, powered by EquityNet™ from the USA. SMENET™ is a system implementation for Europe and Latin America of the successful EquityNet™ system for North America. Ivancic is globally well connected in several areas such as manufacturing, assembling, research, education, medicine, Internet companies, media and venture capital among others.

NEW E-LEARNING SYSTEMS FOR CBM APPLICATIONS

Thomas L. Lagö and Valdi Ivancic
Medison University
Stora Torget
Jonas Bronck Center
Sävsjö, SE 57692 Sweden
Telephone: +46 708 357337
thomas.lago@medisonuniversity.org

Alan Boyer
Qirra Sound Technologies LLC
9550 South Eastern Avenue
Las Vegas, NV 89123

Abstract: Today, there are multiple educational platforms for CBM applications like the simulators from e.g. SpectraQuest that can be used by individuals. For many students, it would be an advantage to include both sound and vibration like "touch and feel" in these simulator platforms. The Swedish company SenseGraphics [1] offer an open software platform that enables "touch and feel." This is often referred to as "haptics" originated from the Greek words *haptikos* and *hapthesthai*, which means to grasp or to touch. They have already, together with partners, implemented simulators for medical, dental and multiple other applications. Qirra Sound has developed a platform for reinforced sound that is able to make sound so realistic that it does not feel like there are any loudspeakers. This technology would be very useful in educational simulators where even the sound and vibration aspect would be included. In Sweden, another project known as VISIR™ (Virtual Instrument Systems in Reality) was initiated together with National Instruments in the USA, to disseminate an online laboratory concept created using open source technologies in collaboration with other universities and organizations. The concept is about adding a remote operation option to traditional instructional laboratories to make them more accessible, irrespective of whether the students are on campus or mainly off campus. This paper will discuss an educational concept combining these technologies for CBM applications and how that would benefit students worldwide.

Key words: Diagnostics; prognostics; health management; e-Learning; virtual labs; education; haptics; VISIR, MOOC

Introduction: Open educational platforms are needed to help student understand the theoretical material better. Today, there are educational platforms for CBM applications. Some of them can be bought and others have been designed specifically for a certain university or company education. The simulators from e.g. SpectraQuest are one example of a more generic and openly available platform, [2]. The Machinery Fault Simulator and the Wind Turbine Simulator are two such examples for training and diagnosis in Machine

Vibration Analysis, Rotor Balancing and Shaft/Coupling Alignment. One missing element is "touch and feel" where students can hear and feel the vibration signature, not only measure it via an instrument.

For many students, it would be an advantage to have the ability for "touch and feel" included in such educational platforms. The Swedish company SenseGraphics [1], offer an open software platform that enables "touch and feel." The brainchild of SenseGraphics, H3D API is fast becoming the standard for high performance haptics application development. This open source, hardware independent platform is designed for anyone interested in building haptics applications from scratch. The H3D API uses the open standards X3D, OpenGL and SenseGraphics haptics. It offers a unified scenegraph which simplifies haptic and graphic rendering. They have already, together with partners, implemented such simulators for a vast range of applications including medical and dental.

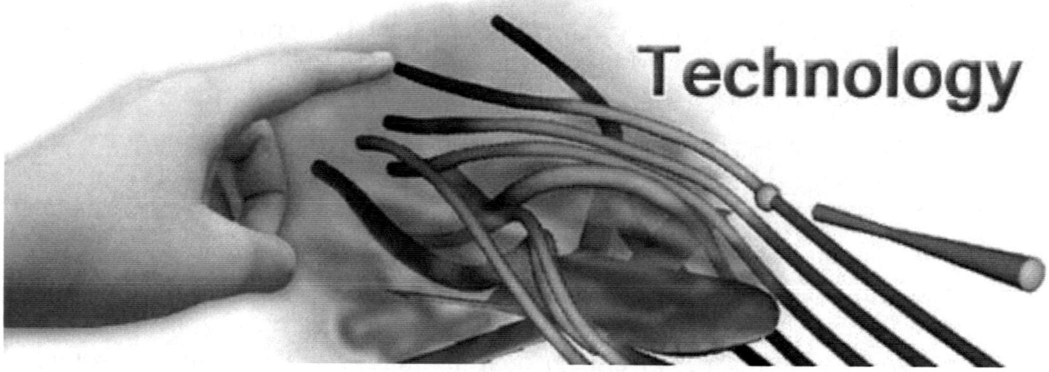

Figure 1: The haptics platform from SenseGraphics in Sweden is base on open source software. Many applications exist today where this software has been embedded as the haptics engine.

Qirra Sound Technologies: Qirra Sound™ is a paradigm-shifting technology in reproduced and reinforced sound for placement in the areas of Pro-Sound, Home Theater, and Industrial market segments. The sound quality, clarity and dynamic range is unique and these loudspeaker platforms would enable a simulator to reproduce even the tiniest sounds. It is a must to experience Qirra Sound™ to fully understand how it performs and how its sounds penetrate your soul. This groundbreaking technology has already received tremendous accolades from performers, audiences, sound engineers, and globally recognized scientists in the field of acoustics and vibration. The frequency range of this speaker technology is capable of handling sounds from 1 Hz up to 50 kHz. This sound technology would help such educational platforms to produce sounds like if they were real. Even low frequency Wind Turbine sounds could be simulated in a very realistic manner including the high frequency components simultaneously!

Collaboration: One intent with this paper is to attract interested partners to collaborate in bringing different expertise and knowledge to the table and create a unique and powerful educational platform for CBM applications. Collaboration is key and the author was part of the creation of the book on the importance of collaboration: COLLABORATE, The Art of WE, [4]. This is also the goal for Medison University, a global and international university. The VISIR consortium has already come a long way but haptics is missing. Several of the existing educational platforms for CBM are good but lack haptics and e-Learning capabilities.

In the year from the fall of 2012 to the summer of 2013, the first 17 HarvardX and MITx courses launched on the edX platform were vast. In that year, 43,196 registrants earned certificates of completion. Another 35,937 registrants explored half or more of course content without certification. An additional 469,702 registrants viewed less than half of the content. And 292,852 registrants never engaged with the online content. In total, there were 841,687 registrations from 597,692 unique users across the first year of HarvardX and MITx courses. ß

A 2013 study by Pennsylvania State University examined the behavior of one million enrollees (Coursera has got over 5 million enrollees by October 2013) in 16 Coursera offerings, and found that completion rates averaged just 4%. Rates were higher for courses with fewer assignments, but even then only reached about 6%. Up to now, however, the success of MOOCs in fulfilling the hopes — and earning back the money — invested in them has been decidedly mixed. Other questions include the viability of pay models. The scale and open nature of MOOCs definitely create challenges — the courses examined in the Penn State study had enrolled anywhere from 13,000 to 110,000 students. Across all courses, about half of those who registered viewed at least one lecture within their selected course. The share of registrants viewing at least one lecture ranged from [27% to 68%].

By October 2013, Coursera enrollment continued to surge, surpassing 5 million, while edX had independently reached 1.3 million. (The main organizations). As of 2013, the range of students registered appears to be broad, diverse and non-traditional, but concentrated among English-speakers in rich countries.

Coursera enrollees

Country	Percentage
United States	27.7%
India	8.8%
Brazil	5.1%
United Kingdom	4.4%
Spain	4.0%
Canada	3.6%
Australia	2.3%
Russia	2.2%
Rest of world	41.9%

We believe that with a strategic collaboration in e.g. an SBIR project or similar would be of interest and we the team would make a major history mark. The author could be the PI, coordinating efforts from all the different groups since work has been performed with all of them in the past.

VISIR: Cooperation between universities is likely to favor their competitiveness and improve their ability to handle the large issues of the society of today. Mankind must live up to the principles of nature and utilize our environment with a focus on sustainability. The engineers who design the new technologies required in the modern society must during their education study how to "communicate with nature" to learn its principles. Engineers' method of communicate with nature is since centuries named physical experiments. However, during recent decades the amount of laboratory work in engineering education has been reduced. The prime cause is clearly the task of coping with greatly increased student numbers, while staff and funding resources have scarcely changed. Contributory factors include the seductive appeal of simulating experiments on computers, where there are no unexpected or unpleasant clashes between theory and simulation [5].

Using Internet for long distance learning, e-Learning, is a great complement and helps many students extend their knowledge base and degrees. This need is expanding worldwide. The mission from Medison University is to do exactly that. Key partners in this quest are sought, and several have already signed up. By combining existing platforms with haptics and also enable remote and real virtual labs, it is believed that the educational expansion can be greater.

Reducing the number of lab sessions is easy because the laboratory work is seldom evaluated and the cost reduction obtained is often considerable. However, for example, ABET (Accreditation Board for Engineering and Technology) in USA has pointed out that learning objectives for laboratory work must exist and be evaluated [6]. Thus, the amount of laboratory work in a course dealing with physical phenomena must be correlated to the learning objectives of the course. Unfortunately, few well-informed persons believe in a substantial rise of course funding resources despite the fact that mankind today need to penetrate deeper and deeper into nature in order to progress. Universities must produce new great experimenters such as Michael Faraday did in the nineteenth century.

Flexible education means providing students with extended accessibility to learning resources and increased freedom to organize their learning activities. From a technological perspective, flexible education corresponds to an adequate exploitation of information and communication devices and infrastructures, especially the Internet. Today, many academic institutions offer a variety of web-based experimentation environments so called remote laboratories that support remotely operated physical experiments [7]-[17]. Such remote experiments entail remote operation of "distant" physical equipment offering students more time for laboratory work. This is one way to compensate for the reduction of lab sessions with face-to-face supervision without significant increase of cost per student. Remote experiments are adapted to the flexible environment of the students of today and permit low cost methods for lab work evaluation.

Web-based experimentation is an excellent supplement to traditional lab sessions. The students can access lab stations outside the laboratory and perform experiments around the clock. It is possible to design virtual instructors in software, which will protect the equipment from careless use; also theft of equipment will not be a problem. Interfaces enabling students to recognize on their own computer screen the instruments and other equipment in the local laboratory may easily be created. Apart from the fact that each student or team of students works remotely in a virtual environment with no face-to-face contact with an instructor or other students in the laboratory, the main difference between a lab session in the remote laboratory described here and a session in a local laboratory is that it is not possible for students to manipulate physical equipment e.g. wires and electronic components with their fingers in a remote laboratory. However, students can, for example, sit at home in peace and quiet and learn how to use a function generator and an oscilloscope in the same way as in a noisy local laboratory where time is limited.

The VISIR™ initiative emanates from a remote laboratory project started at BTH in 1999 to ascertain that it is feasible to design a remote electronics laboratory comprising standard equipment to supplement local instructional laboratories and provide free access to the experimental equipment to students enrolled in circuit analysis and electronics courses. Today BTH has two online laboratories one electronics lab and one signal processing lab for mechanical vibration experiments based on the BTH Open Laboratory concept. The concept is about providing new possibilities for students to do laboratory work and become experimenters by adding a remote operation option to traditional instructional laboratories to make them more accessible for students, irrespective of whether they are on campus or mainly off campus. This option is equipped with a unique interface enabling students to recognize on their own computer screen the instruments and other equipment most of them have previously used in the local laboratory. The software produced is released as open source.

The VISIR™ project started in late 2006 at the Department of Signal Processing, BTH, in cooperation with National Instruments with financial support from VINNOVA (Swedish Governmental Agency for Innovation Systems) to disseminate the online laboratory concept using open source technologies. Other universities and organizations have been invited to participate and several team worldwide have joined.

The first remote control option implemented was for an electronics laboratory and the second one was for a signal processing laboratory with emphasis on mechanical vibration experiments. The electronics' lab option features remote circuit wiring using a virtual breadboard and a relay switching matrix combination. The goal is an international standard, enabling teams worldwide to expand and develop jointly this powerful approach into distributed online laboratories by using standardized software such as IVI (Interchangeable Virtual Instruments) and equipment platforms such as PXI (PCI eXtensions for Instrumentation) and LXI (LAN eXtensions for Instrumentation).

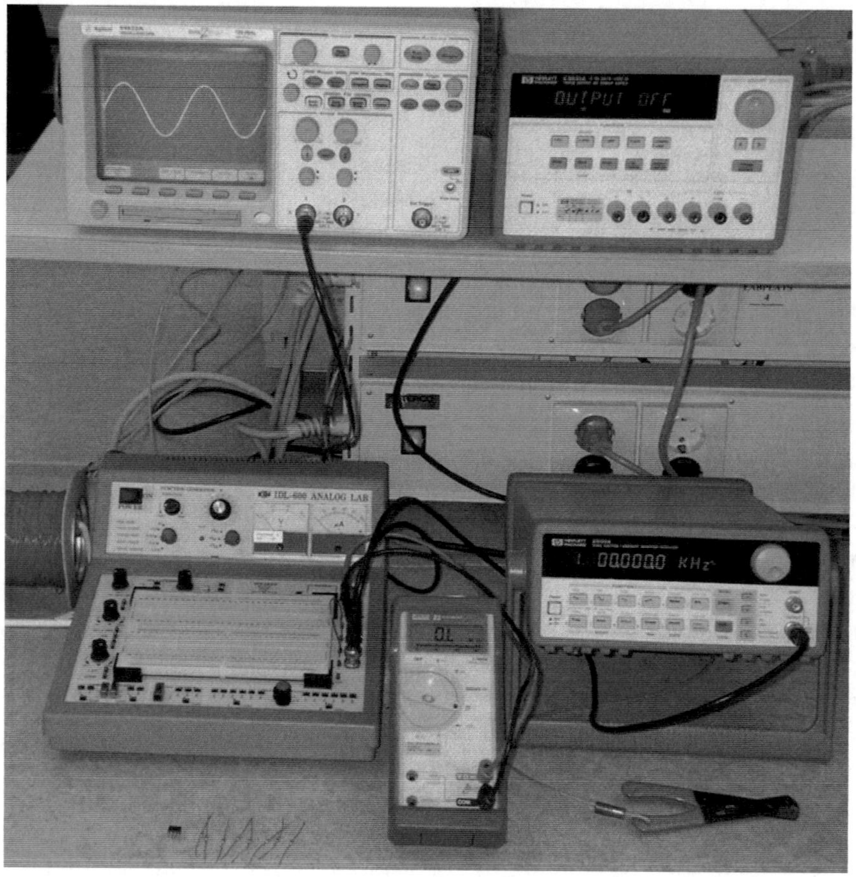

Figure 1: "Virtual lab station" in a local electronics laboratory at BTH in Sweden.

Conclusion: Efficient educational platforms for CBM applications are needed due to increased interest and need in predicting failures and handling maintenance. Today's educational platforms, looking in a vast range of segments outside of CBM has come further. Medical and dental have implemented haptics via SenseGraphics and a fully blown "virtual world" can be used for education or training. Qirra Sound would bring a sound experience that could cover even Wind Turbine sounds. The VISIR consortium has successfully demonstrated that it is possible to create a virtual instrument system application with real life data – **in reality**! The authors propose a strategic collaboration in between core teams in the creation of an educational platform for CBM that will include e-learning, virtual reality, haptics and more. Medison University is poised and ready to become a PI for e.g. a SBIR initiative. It is believed that this is much needed and it would greatly help this industry. Learning is key, and excellent e-Learning systems are internationally sought after. A combined educational platform including all these elements would open for a new educational strategy that we believe would be very good.

References:

[1] http://www.sensegraphics.com
[2] http://www.spectraquest.com

[3] http://www.qirrasound.com
[4] Dan Sanker and Jossey-Bass. *"COLLABORATE, The Art of WE, combining capabilities to create new opportunities for success,"* An Imprint of John Wiley & Sons, ISBN 978-1-118- 11472-8, 2012.
[5] D. Magin and S. Kanapathipillai, 'Engineering Students' Understanding of the Role of Experimentation', European Journal of Engineering Education, 2000, Vol. 25, no. 4, pp. 351-358.
[6] L. D. Feisel and A. J. Rosa, 'The Role of the Laboratory in Undergraduate Engineering Education', Journal of Engineering Education, January 2005, pp 121-130.
[7] Gillet, A. V. N. Ngoc, Y. Rekik, 'Collaborative Web-Based Experimentation in Flexible Engineering Education', IEEE Transactions on Education, Vol. 48, No. 4, November 2005.
[8] Z. Nedic, J. Machotka, and A. Nafalski, 'Remote Laboratories Versus Virtual and Real Laboratories', Proceedings of the 33rd ASEE/IEEE Frontiers in Education Conference, Bolder, USA, November 5 – 8, 2003.
[9] M. Auer, A. Pester, D. Ursutiu, and C. Samoila, 'Distributed Virtual and Remote Labs in Engineering', Proceedings of the ICIT'03 International Conference on Industrial Technology, Maribor, Slovenia, December 10 - 12, 2003.
[10] A. A. Asumadu et al.,'A Web-Based Electrical and Electronics Remote Wiring and Measurement Laboratory (RwmLAB) Instrument', IEEE Transactions on Instrumentation and Measurement, Vol. 54, No. 1 February 2005.
[11] A. M. Scapolla, A. Bagnasco, D. Ponta, and G. Parodi, 'A Modular and Extensible Remote Electronic Laboratory', International Journal of Online Engineering, Vol. 1 No. 1 (2005).
[12] http://www.pxisa.org/, 2010.
[13] http://www.lxistandard.org/home, 2010.
[14] http://www.ivifoundation.org/, 2010.
[15] I. Gustavsson et al., 'An Instructional Electronics Laboratory Opened for Remote Operation and Control', Proceedings of the ICEE 2006 Conference, San Juan, Puerto Rico, July 23 - 28, 2006.
[16] Åkesson, H. et al., 'Vibration Analysis of Mechanical Structures over the Internet Integrated into Engineering Education', Proceedings of the 2006 ASEE Annual Conference, Chicago, USA, June 18 - 21, 2006.
[17] M. E. Auer, V. J. Harward and A. K. M. Azad (ed.). *The VISIR Open Lab Platform accepted for publication in Internet Accessible Remote Laboratories*: Scalable E-learning Tools for Engineering and Science Disciplines, IGI Global, 2011.

Bibliography:

[1] **Thomas L. Lagö**, Ph.D, Chief Scientist. Dr. Thomas Lagö is globally accomplished in the upper echelons of the sound and vibration area. Dr. Lagö has "4" Post Doctoral degrees in conflicting scientific disciplines and consults with organizations such as NASA, Rolls Royce, ABB, Volvo, SAAB, Scania, Boeing, Atlas Copco, TATA, Indian Navy, and many other world leading organizations. Dr. Lagö has implemented

paradigm shifting technologies around the world and presents advanced classes to these companies to assist them in solving problems in advanced acoustic and vibration applications. He has more than 40 years of industrial experience and has worked for companies including Hewlett-Packard (today Agilent Technologies). Dr Lagö is a creative, enthusiastic, high-achieving executive with outstanding performance based on a solid education. Formerly, President and Chief Scientist of a $9 million leading USA sales company in the Sound & Vibration business. Prior responsibility reflects outstanding career growth and progress through various high-tech businesses, including European Product Line Manager for Hewlett Packard's Dynamic Signal Analyzer Products, as well as multi functional responsibilities including sales and marketing, strategic planning, engineering, but also finance and human resource leadership. Dr. Lagö has a vast practical experience. Lagö teaches sound and vibration courses and applicable mitigation methods on an international basis and has given courses for and many international universities and expert groups. His practical experience is vast and he is very practically oriented despite his thorough theoretical education and experience. His solutions and courses have helped many teams better their environment and solve challenging problems without spending huge money. Dr Lagö has become a BINDT Fellow, (British Institute for Non-Destructive Testing) and he is the Industrial Chair for ISCM - The International Society for Condition Monitoring. Lagö is also involved in a number of other scientific organizations like IIAV, International Institute for Acoustics and Vibration and is also serving as a Board of Directors member for (current and past): SEM-IMAC (Society for Experimental Acoustics), SAE (Society for Automotive Engineering, NVC), ICSV (International Institute for Sound and Vibration), SVIB (Scandinavian Vibration Society). He is also an AES member and participates in multiple standardization groups for sound in concert halls, movie theatres and speaker systems. Lagö is author of more than 290 scientific papers, written multiple chapters in international textbooks and holds more than 25 patent/ patent applications and has received multiple awards. He has 1900 international contacts on Linkedin and 10,000+ personal contacts in his CRM database. Dr. Lagö was the Chairman of the VISIR initiative and founded the name.

[2] **Valdi Ivancic**, founder of Medison University, a global and international university. Medison University belongs to the Medison company group with offices in London and California. Valdi is a visionary, serial entrepreneur, keynote speaker, innovator and PM Candidate. A person with ability to inspire other people, working for a better world and improve people's lives. He is the founder, co-founder, president and speaker in several start-ups, educational organizations, research groups, university courses and other non-profit organizations. He is a natural leader who makes co-workers grow in responsibility and as individuals. He creates and finds innovative solutions with the ability of connecting ideas and solutions from different areas into a bigger working picture. He has been traveling and living on several continents and speaks 7+ languages. He is multi-cultural and has a vast understanding of different business cultures and political views. He has been attending 50+ courses at different universities worldwide (MOOCs) and been a key Lecturer/Speaker at several Swedish universities. He has raised over $25M to projects, enterprises, cities and universities. He is also the head of a venture partner company and a crowd funding organizations, helping SMEs getting funded.

Ivancic has recently started SMENET™, powered by EquityNet™ from the USA. SMENET™ is a system implementation for Europe and Latin America of the successful EquityNet™ system for North America. Ivancic is globally well connected in several areas such as manufacturing, assembling, research, education, medicine, Internet companies, media and venture capital among others.

[3] **Alan Boyer**, Ph.D, MBA, LTCL, ALCM, President and CEO. Dr. Alan Boyer's career is sound and vibration spans almost thirty years. He has been Executive VP of Sales and Marketing, and COO for a leading design and manufacturing company of sound and vibration instrumentation. He has also owned his own international company and during his career, has been awarded many high level contracts with organizations including; London's Heathrow, Gatwick, and Stansted airports, NASA, Boeing, Land Rover, Jaguar, US Airforce, and others. Dr. Boyer is also an accomplished musician and performed as a teenage virtuoso trumpeter and professional conductor. The combination of many years as a musician, acoustics/vibration executive and entrepreneur give Qirra Sound Technologies and outstanding chance for success in the three target market categories of Industrial, Pro-Audio, and Audiophile. Dr. Alan Boyer was born in England and graduated from Sheffield University, England, studying education, psychology and music. Boyer has more than 25 years of experience in the sound and vibration industry and has worked at all levels of the organization. In 1998, after heading the development of the Larson Davis, Inc. European Office, Boyer moved to the USA to become manager of the corporate marketing department and was subsequently appointed to the position of Senior Vice-President and COO. One of his most important projects was the formation of a strategic alliance between leading international companies in the area of sound and vibration. This successful alliance drew great international attention and began to change the way that collaborative relationships were formed and nurtured in the industry. Other projects successfully led by Dr. Boyer include sound detection systems for the airports of London Heathrow, Gatwick and Stansted. He has also delivered test and measurement systems to Land Rover, Honda, Toyota, Rolls Royce, Jaguar Cars, Lotus Cars, Ford Motor Company, Vauxhall (GM), and others. Boyer has also carried out projects regarding ISO 9002 (for the European office) and ISO 9001 certification for the corporate office in the USA. Dr. Boyer implemented and oversaw projects in relation to product catalog, data sheet, instruction manual design drawing on his background in education, instructional psychology and sales/marketing. Dr. Boyer is much sought after as a consultant on all areas of business and enjoys working with start-up companies and home based business owners. His teaching, presenting, seminar, and public speaking abilities are extremely effective and his combination of presenting thorough material combined with his English sense of humor, enables him to teach in a very effective manner. Boyer's consulting activities have been to various USA and European based businesses (established and start-up) in diverse industries such as Neutraceutical, Nano Science, Cosmetics, Acoustics/Vibration, and others. Boyer is an accomplished musician and conducts medium to large music ensembles. He is a Licentiate of Trinity College of Music and an Associate of the London College of Music. As a teenager, he was an accomplished virtuoso trumpeter. He began to conduct at the age of fifteen and has directed many ensembles in Europe and the USA. In 2006, Dr. Boyer was presented with the National Arts and Humanities Award by Provo Arts Council for services to the arts in

Utah and throughout the USA. Dr. Boyer sits on a number of for profit and not for profit boards in the areas of business and music/arts. As CEO of Qirra Sound Technologies LLC, Tech Fuzion, and Embla Global, LLC, Dr. Boyer brings important leadership skills and high levels of success and expertise to the organization both in the USA and European teams.

SIGNIFICANCE OF UNCERTAINTY QUANTIFICATION IN PROGNOSTICS AND HEALTH MONITORING

Shankar Sankararaman
SGT Inc., NASA Ames Research Center
Moffett Field, CA 94035
Telephone: (650) 604-0552
shankar.sankararaman@nasa.gov

Kai Goebel
NASA Ames Research Center
Moffett Field, CA 94035
Telephone: (650) 604-4204
kai.goebel@nasa.gov

Abstract: This paper discusses the significance of uncertainty quantification in prognostics and health monitoring, which are important for failure prevention and risk-informed decision-making with respect to the operation of engineering systems. Since there are several sources of uncertainty that affect the future loading and operating conditions of these engineering systems, it is important characterize and quantify these sources of uncertainty, and quantify their combined effect on future prediction. Failure threshold levels are defined and used to quantify the remaining useful life, which is also uncertain due to the aforementioned sources of uncertainty. It is important to quantify the overall uncertainty in the remaining useful life using principles of probability, since this information is directly useful for risk-informed decision-making. Methods for such uncertainty quantification are discussed and illustrated using a numerical example.

Key words: Prognostics; health management; uncertainty quantification; remaining useful life; probability distribution.

1. Introduction: Prognostics and health management (PHM) deals with continuous monitoring of the performance of an engineering system, checking for the presence of faults, predicting future failures, estimating the remaining useful life, and aiding decision-making activities. An accurate PHM system is essential for failure prevention and fault mitigation in different types of structural, mechanical, and aerospace engineering applications.

In order to meaningfully predict the failure of engineering systems, it is necessary to adjudge the future loading, operating, and environmental conditions of the system. In practical applications, it is almost impossible to precisely estimate these quantities due to limited knowledge regarding the future. However, it may be practically possible to understand the variations in these quantities and characterize the uncertainty in them. Additionally, there are also several other sources of uncertainty, that affect prognostics and failure prediction. For example, measurement data available through sensors, that are a part of the PHM system, may be uncertain due to the presence of noise, bias, and gain. When such measurements are used for state estimation, and in turn for quantifying the extent of damage in the system, the resultant estimates are also uncertain. Finally, it is necessary to predict how the aforementioned damage will degrade in the future, based on the future loading, operating, and environmental conditions. Such forecasting is usually performed using a data-driven or physics-based model, which may also be uncertain. When the amount of predicted damage increases beyond a specified threshold level, failure is said to occur and this indicates the end-of-life (EOL). Due to the aforementioned sources of uncertainty, the prediction of EOL is rendered uncertain. The remaining useful life (RUL) is defined as the amount time of left until EOL is attained, and the computation of RUL is an important goal in PHM [1].

While the topic of prognostics and health management has received considerable attention in the literature during the past ten years, only recent research efforts have been focusing on the effects of uncertainty in prognostics. It is important to fully understand the significance and impact of uncertainty, and develop methods for rigorous uncertainty quantification and management before computational methods for PHM can be applied to practical problems.

This paper focuses on explaining the significance of uncertainty in prognostics and health monitoring. While uncertainties have been discussed from representation, quantification, and management points of view [2 - 6], these three are different processes, and are often confused with each other and interchangeably used. In this paper, the different tasks related to uncertainty quantification and management are classified into four, as explained below. These four tasks need to be performed in order to accurately estimate the uncertainty in the RUL prediction and inform the decision-maker regarding such uncertainty.

a. Uncertainty Representation and Interpretation
b. Uncertainty Quantification
c. Uncertainty Propagation
d. Uncertainty Management

All of these activities are explained in detail, with an emphasis on using probabilistic methods for the treatment of uncertainty. The interpretation of uncertainty is explained from a statistical point of view, and then, methods for uncertainty quantification are explained in detail. Finally, the arguments presented in the paper are illustrated using numerical examples.

2. Uncertainty Representation and Interpretation: The first step is uncertainty representation and interpretation, which in many practical applications, is guided by the choice of modeling and simulation frameworks. There are several methods for uncertainty representation that vary in the level of granularity and detail. Some common theories include classical set theory, probability theory, fuzzy set theory, fuzzy measure (plausibility and belief) theory, rough set (upper and lower approximations) theory, etc.

Amongst these theories, probability theory has been widely used in the PHM domain [7]. Even within the context of probabilistic methods, uncertainty can be interpreted and perceived in two different ways: physical probabilities versus subjective probabilities. It is essential to understand the difference between these two interpretations before attempting to interpret the uncertainty in prognostics and RUL prediction.

Frequentist Interpretation: Physical probabilities [8] also referred to objective or frequentist probabilities, are related to random physical experiments such as rolling dice, tossing coins, roulette wheels, etc. Relative frequencies of occurrence are equal to physical probabilities, which are defined only in the context of random experiments. The theory of classical statistics is based on physical probabilities. In the context of physical probabilities, randomness arises only due to the presence of physical probabilities. If the true value of any particular quantity is deterministic, then physical probabilities cannot be used to represent this type of uncertainty. Further, the uncertainty in the parameter estimate cannot be used for further uncertainty quantification. For example, if the model parameters of a battery model are estimated under a particular loading condition, then this uncertainty cannot be used for quantifying the battery-response for a similar loading condition. This is a serious limitation, since it is not possible to propagate uncertainty after parameter estimation, which is often necessary in system-level uncertainty quantification [9].

Subjective Interpretation: Subjective probabilities [10] can be assigned to any "statement". It is not necessary that the concerned statement is in regard to an event which is a possible outcome of a random experiment. In fact, subjective probabilities can be assigned even in the absence of random experiments. The Bayesian methodology is based on subjective probabilities, which are simply considered to be degrees of belief and quantify the extent to which the "statement" is supported by existing knowledge and available evidence. In this approach, even deterministic quantities can be represented using probability distributions. As a result, probability distributions can be assigned to parameters that need to be estimated, and therefore, this interpretation facilitates uncertainty propagation after parameter estimation. The concept of likelihood and its use in Bayes' theorem rely on the theory of subjective probability. Today, Bayesian methods are used to solve a variety of problems in engineering. Filtering techniques such as particle filtering, Kalman filtering, etc. are also primarily based on the use of Bayes theorem, and sequential sampling.

Summary: It is important to understand the implications of the above interpretations of probability on prognostics and health monitoring. When multiple nominally identical components are tested for failure, in the context of testing-based prognostics, there is true variability from one specimen to another, and hence, frequentist interpretation and the theory of classical probability are applicable. However, in the context of condition-based monitoring, the focus is on the operation of one particular engineering component/system, there is no frequentist-related variability in this context, and the Bayesian interpretation of uncertainty is suitable in this case. For example, when a filtering approach calculates the uncertainty in a state estimate, the "true" state is actually deterministic, and the estimated uncertainty reflects the analyst's belief or knowledge regarding the state uncertainty. Therefore, filtering approaches such as particle filtering, Kalman filtering, etc. are popularly called as Bayesian tracking approaches, not only because they use Bayes' theorem recursively, but also interpret the resultant uncertainty from a Bayesian perspective.

4. Uncertainty Quantification: Sources of Uncertainty: It has been conventional to classify the different sources of uncertainty into aleatory (physical variability) and epistemic (lack of knowledge), where epistemic uncertainty consists of data uncertainty and model uncertainty. However, in the case of condition-based monitoring, there is only one particular system being monitored and not multiple realizations of a population, and therefore, it is not meaningful to discuss variability. For example, the system is at a particular state at any time instant and there is nothing variable about it. As mentioned earlier, variability would need to be accounted for, only in the case of reliability-testing methods where multiple components/systems are tested. Therefore, the sources of uncertainty in prognostics are classified into:

Modeling uncertainty: Predicting the future is the most important aspect of prognostics, and typically, a physics-based model or a data-driven model is used for predicting future behavior. This model is usually represented using state-space equations. Modeling uncertainty represents the difference between the predicted response and the true response (which can neither be known nor measured accurately), and comprises of several parts such as model parameters, model form, process noise, etc.

Present uncertainty: The first step of prognostics is to estimate the condition/state of the component/system at any time instant. Output data (usually collected through sensors) is used to estimate the state and many filtering approaches are able to estimate the state, and calculate the uncertainty associated with the state estimate.

Future uncertainty: The most important source of uncertainty in the context of prognostics is due to the fact that the future is unknown, i.e. both the loading and operating conditions are not known precisely. The future behavior (i.e., the response of the system to the loading and operating conditions) needs to be estimated using a model; the usage of a model imparts additional uncertainty as explained earlier.

5. Uncertainty Propagation: The third step is uncertainty propagation and is most relevant to prognostics, since it accounts for all the previously quantified uncertainties and uses this information to predict (1) future states and the associated uncertainty; and (2) remaining useful life and the associated uncertainty. The former is computed by propagating the various sources of uncertainty (discussed earlier in Section 4) through the prediction model. The latter is computed using the estimated uncertainty in the future states along with a Boolean threshold function which is used to indicate end-of-life. In this step, it is important to understand that the future states and remaining useful life predictions are simply dependent upon the various uncertainties characterized in the previous step, and therefore, the distribution type and distribution parameters of future states and remaining useful life should not be arbitrarily chosen. Sometimes, a normal (Gaussian) distribution has been assigned to the remaining useful life prediction; such an assignment is erroneous. Even if the prediction model and the threshold function are both linear, their combination, which is used to predict the remaining useful life, becomes non-linear. The probability distribution of RUL needs to be estimated though rigorous uncertainty propagation of the various sources of uncertainty through the state space model and the EOL threshold function, both of which may be non-linear in practice.

6. Uncertainty Management: The fourth and final step is uncertainty management, and it is unfortunate that, in several articles, the term "Uncertainty Management" has been used instead of uncertainty quantification and/or propagation. Uncertainty management is a general term used to refer to different activities which aid in managing uncertainty in condition-based maintenance during real-time operation. There are several aspects of uncertainty management. One aspect of uncertainty management attempts to answer the question: "Is it possible to improve the uncertainty estimates?" The answer to this question lies in identifying which sources of uncertainty are significant contributors to the uncertainty in the RUL prediction. For example, if the quality of the sensors can be improved, then it may be possible to obtain a better state estimate (with lesser uncertainty) during Kalman filtering, which may in turn lead to a less uncertain RUL prediction. Another aspect of uncertainty management deals with how uncertainty-related information can be used in the decision-making process.

6. Computational Framework for Uncertainty Quantification in Prognostics: While the previous sections discussed the different aspects of uncertainty representation, interpretation, quantification, propagation, and management in prognostics, this section presents a comprehensive computational framework for the treatment of uncertainty in prognostics, particularly in the context of condition-based prognostics and health management. A model-based architecture for prognostics is discussed for this purpose, and then computational methods for quantifying prediction uncertainty are investigated.

Architecture for Prognostics: The goal in prognostics is to predict the future behavior of a component/system at any generic time instant, denoted by t_P. This is accomplished by estimating the states of the system at all time instants $t > t_P$. The inputs ($\boldsymbol{u}(t)$) and outputs ($\boldsymbol{y}(t)$) to the system are known until the prediction time $t = t_P$, and in order to perform prediction the future inputs, i.e., $\boldsymbol{u}(t)\ \forall\ t > t_P$ also need to be available. A generalized architecture for prognostics is shown in Fig. 1.

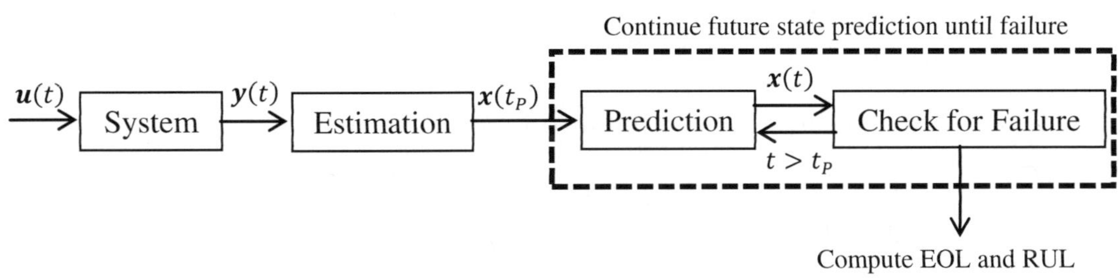

Figure 1. Model-Based Prognostics Architecture

The first step in prognostics is to estimate the state at time t_P, which is indicative of the extent of damage in the system at t_P. Using the estimated state, the second step is to predict future states until failure, and thereby, compute the end-of-life (EOL) and the remaining useful life (RUL) of the system.

Modeling: State-space models are developed needed for both estimation and prediction. Consider a generic state space model which is used to continuously predict the state of the system, as:

$$\dot{\boldsymbol{x}}(t) = f\big(t, \boldsymbol{x}(t), \boldsymbol{\theta}(t), \boldsymbol{u}(t), \boldsymbol{v}(t)\big) \tag{1}$$

where $\boldsymbol{x}(t) \in \mathbb{R}^{n_x}$ is the state vector, $\boldsymbol{\theta}(t) \in \mathbb{R}^{n_\theta}$ is the parameter vector, $\boldsymbol{u}(t) \in \mathbb{R}^{n_u}$ is the input vector, $\boldsymbol{v}(t) \in \mathbb{R}^{n_v}$ is the process noise vector, and f is the state-space equation, that represents the evolution of the states of the engineering system. Typically, one state, or a collection states is representative of the damage in the system. This equation can be constructed using physics-based principles or using data-driven techniques.

While Eq. 1 is used for state prediction, actual sensor measurements (which are available until time $t = t_P$) are used for state estimation. The sensor measurements are modeled using a generic output equation, such as:

$$y(t) = h\big(t, \boldsymbol{x}(t), \boldsymbol{\theta}(t), \boldsymbol{u}(t), \boldsymbol{n}(t)\big) \tag{1}$$

where $\boldsymbol{y}(t) \in \mathbb{R}^{n_y}$ is the output vector, $\boldsymbol{n}(t) \in \mathbb{R}^{n_n}$ is the measurement noise vector, and h is the output equation. Since the output measurements are available only until time $t = t_P$, the output equation is used only in the estimation stage and not for prediction purposes.

State Estimation: Bayesian tracking approaches as Kalman filtering, particle filtering, etc. can be used for estimation purpose. These methods use Bayes theorem to update the uncertainty in the states continuously as a function of time, as and when new measurements are available. While particle filtering is the most general method that can account for different types of probability distributions and varying degrees of non-linearity, Kalman filtering can be used only when Eq. 1 is linear and all the uncertain quantities are Gaussian. When the uncertain quantities are Gaussian, the extended Kalman filter can be applied by linearizing Eq. 1.

Daigle et al. [11] used unscented Kalman filters to approximate the mean and variance of the system states, and this approach is applicable even for non-linear models. This method is based on generating weighted samples and continuously updating the probability distributions of the state variables using Bayesian filtering.

Prediction: Having estimated the uncertainty in the states at time t_P, this uncertainty, along with the other uncertainties associated with the model and the future prediction, need to be propagated through Eq. 1 until failure occurs. This dependence of RUL on the different sources of uncertainty can be graphically represented as shown in Fig. 2.

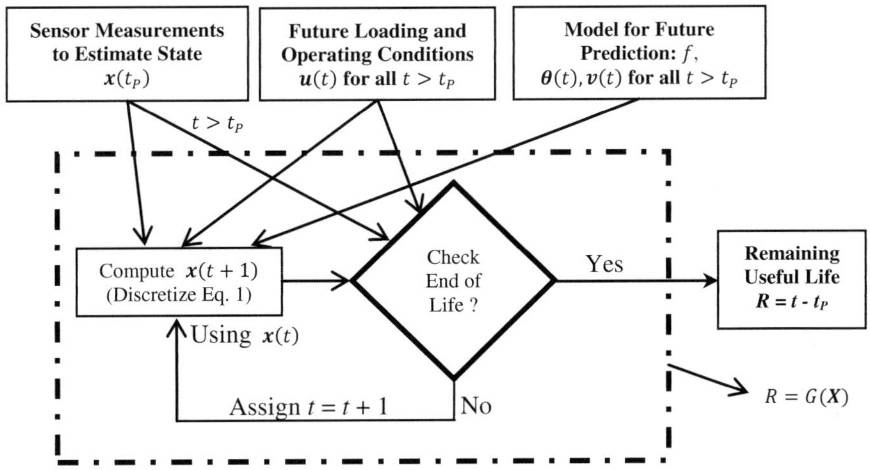

Figure 2. Computing RUL through Uncertainty Propagation

In Fig. 2, "*G*" denotes the function through the different sources of uncertainty (denoted by the vector ***X***) need to be propagated; note that this function is a combination of the state space model in Eq. 1 and the failure threshold criterion. The threshold function is used to check for failure at every time-instant. If the threshold indicates failure, then RUL is compute; if not, future state prediction is continued until failure. For a given set of realizations of ***X***, there exists one value of RUL. Since ***X*** is uncertain, RUL is also uncertain.

The problem of computing the uncertainty in the RUL can be viewed as an uncertainty propagation problem, i.e. propagate the uncertainty in ***X*** through "*G*" and compute the probability distribution of RUL. The result can be expressed, either in terms of the probability density function "$f_R(r)$" or the cumulative distribution function "$F_R(r)$". Such an uncertainty propagation problem can be solved using a variety of statistical methods. Such methods can be classified into two categories: sampling-based methods and analytical methods. Some of the commonly used sampling-based methods are:

Importance Sampling: This algorithm does not generate random realizations of ***X*** from the original distribution. Instead, random realizations are generated from a proposal density function, the statistics of *R* are estimated and then corrected based on the original density values and proposal density values.

Stratified Sampling: In this sampling approach, the overall domain of ***X*** is divided into multiple sub-domains and samples are drawn from each sub-domain independently. The process of dividing the overall domain into multiple sub-domains is referred to as stratification. This method is applicable when subpopulations within the overall population are significantly different.

Latin Hypercube Sampling: This is a sampling method commonly used in design of computer experiments. When sampling a function of N variables, the range of each variable is divided into M equally probable intervals, thereby forming a rectangular grid. Then, sample positions are chosen such that there is exactly one sample in each row and exactly one sample in each column of this grid. Each resultant sample is then used to compute a corresponding realization of R, and thereby the PDF $f_R(r)$ can be calculated.

Unscented Transform Sampling: Unscented transform sampling is a sampling approach which focuses on estimating the mean and variance of R accurately, instead of the entire probability distribution of R. Certain pre-determined sigma points are selected in the $\boldsymbol{X}$ — space and these sigma points are used to generate corresponding realizations of R. Using weighted averaging principles, the mean and variance of R are calculated. While this approach has been predominantly used in the context of state estimation, it can also be used for prediction [11].

Some of the commonly used analytical methods are:

First-Order Second Moment Method: This method uses only the mean and variance of all the uncertain quantities and the first-order Taylor's series expansion of "G", and estimates the mean and variance of the response quantity "R".

First-Order Reliability Method: This method calculates the CDF function "$F_R(r)$" by linearizing "G" around the so-called most probable point [12]. While this approach is an approximation, it can estimate the CDF with reasonable accuracy in many practical applications.

Inverse First-Order Reliability Method: This method is the inverse of the first-order reliability method, i.e., it calculates the value of "r", such that such that $F_R(r)$ is equal to ψ, for a given value of ψ (which lies between 0 and 1). By repeating this approach for several values of ψ, the entire cumulative distribution function can be easily calculated, thereby estimating the uncertainty in R.

In addition to the aforementioned methods, there are several surrogate modeling approaches that have been commonly used by researchers for the purpose of uncertainty quantification. Such surrogate modeling approaches include regression techniques [12], polynomial chaos expansion [13], kriging [14], etc. Each of these methods use different types of basis functions and one may approximate "G" better than the other; therefore, it may be useful to study such surrogate modeling approaches for uncertainty quantification in prognostics. Even though there are several such uncertainty propagation methods, it is important to investigate their applicability to prognostics. While some methods like Monte Carlo sampling can be very accurate, they may not be readily suitable for online health monitoring due to the computational cost involved. On the hand, analytical methods, though based on approximations, can obtain quick results that can be processed during the operation of the engineering system. Even after selecting a suitable approach for uncertainty propagation, there are several practical challenges involved during implementation. For example, it is challenging to characterize the different sources of uncertainty that affect the RUL prediction. Another challenge is related to dimensionality; if a fine level of discretization is chosen to evolve the state-space models, then the number of elements in $\boldsymbol{X}$ may be very large, thereby rendering uncertainty propagation difficult. Some of these challenges are still being addressed by researchers and may be considered in future work.

7. Numerical Illustration: This section discusses the application of the proposed computational framework for condition-based monitoring to the power system of a planetary rover. The power system consists of several lithium batteries, and is based on the empirical model used by Daigle et al. [11].

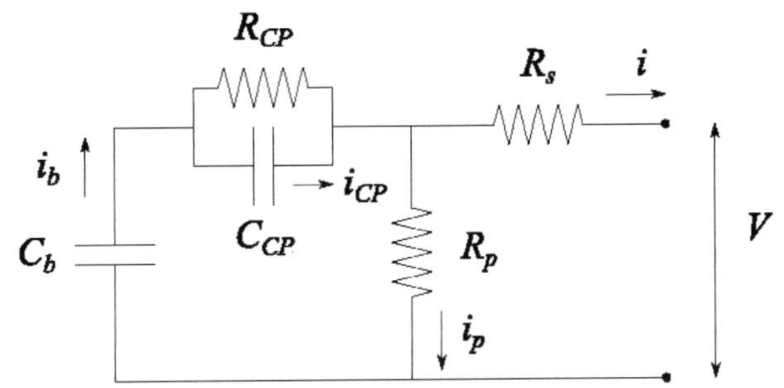

Figure 3. Equivalent Circuit Model of the Lithium-Ion Battery

The battery model is based on an electrical circuit equivalent as shown in Fig. 3. The large capacitance C_b holds the charge q_b of the battery. The nonlinear C_b captures the open-circuit potential and concentration over-potential. The R_{CP}-C_{CP} pair captures the major nonlinear voltage drop due to surface over-potential, R_s captures the so-called Ohmic drop, and R_p models the parasitic resistance that accounts for self-discharge. This empirical battery model is sufficient to capture the major dynamics of the battery while ignoring temperature effects and additional minor processes. The governing equations for the battery model are presented in continuous time below. The implementation of the proposed methodology considers a discrete-time version with a discrete time-step of 1 second. The governing equations and the numerical values of this lithium ion battery model can be found in the authors' earlier publications [15]. This section discusses certain key results, in the context of uncertainty quantification, by considering two types of loading conditions and presenting the results of RUL estimation.

The current in the battery is chosen to be the input and two cases of input loading are considered: constant amplitude loading and variable amplitude loading. In constant amplitude loading, the future current demand is assumed to be constant, but the constant value is a random variable that follows normal distribution (N(1.375, 1/6)), but is truncated at a specified lower bound (0.75) and an upper bound (2.00). In the variable amplitude loading scenario, constant amplitude loading is assumed for a particular time segment; the length of the time segment and the constant amplitude for that time segment are both chosen at random. This numerical example considers consecutive time segments; the time-length of each segment is assumed to be uniformly distributed (with lower bound T_L = 500 seconds and standard deviation T_U = 1000 seconds) and the constant amplitude for that time segment is assumed to be normally distributed (with mean I_μ = 1.375 amps and standard deviation I_σ = 1/6 amps). Multiple realizations of the loading trajectory (current versus time) are indicated in Fig. 4.

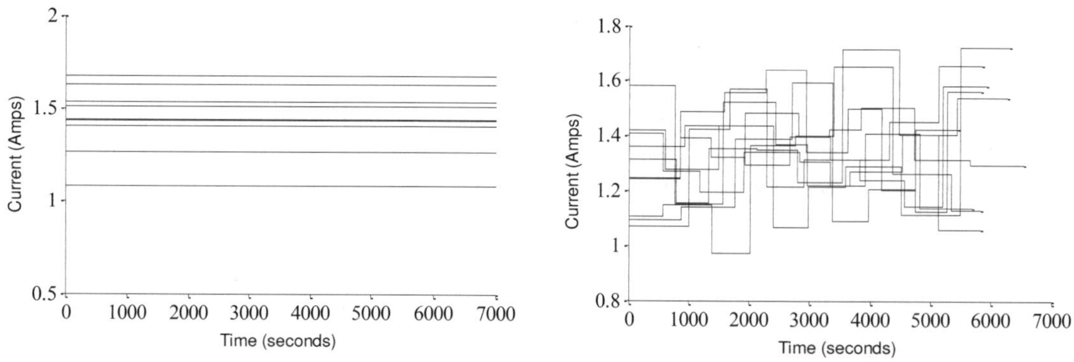

(a) Constant Amplitude Loading (b) Variable Amplitude Loading
Figure 4. Uncertainty in Loading Conditions

The results of quantifying the uncertainty in the end-of-life prediction (which corresponds to the end-of-discharge of the battery), continuously as a function of time, are shown in Fig. 5. It can be seen that the uncertainty in the end-of-discharge prediction decreases with time. Initially, it is necessary to assume loading conditions for a longer period of time and therefore, there is a large amount of uncertainty. In later stages of monitoring, loading conditions need to be assumed for a shorter period of time, and therefore, results in decreased uncertainty in the end-of-discharge prediction.

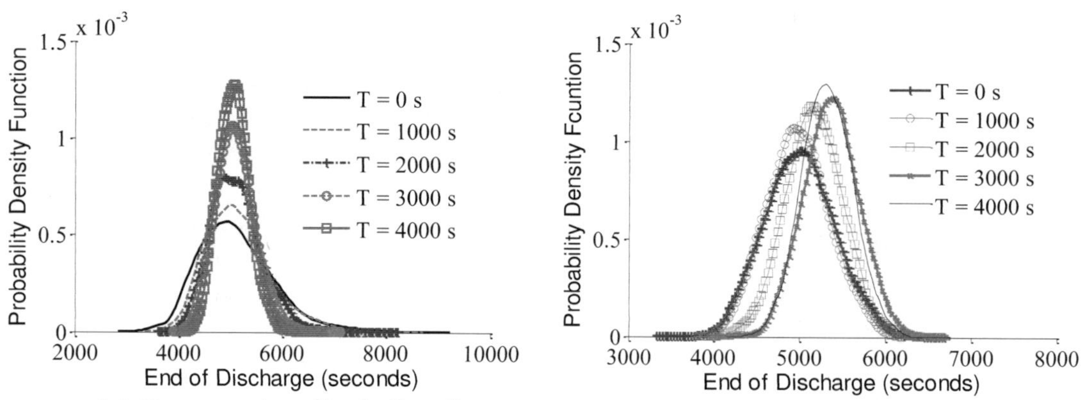

(a) Constant Amplitude Loading (b) Variable Amplitude Loading
Figure 5. Uncertainty in End of Discharge Prediction

The most important goal during uncertainty quantification is to be able to accurately capture the entire probability distribution. Sometimes, the probability distribution of end-of-life may be multi-modal in nature; sometimes, the shape of the probability distribution may significantly change shape over the course of time [16]. Therefore, computational methods need to be able to address such issues. For this purpose, the methods discussed in Section 6 need to be studied in detail and their applicability to prognostics and condition-based monitoring needs to be investigated.

8. Conclusion: This paper discussed the significance, interpretation, and quantification of uncertainty the context of prognostics and health management. The prediction of remaining useful life in engineering systems is affected by several sources of uncertainty, and it is important to correctly interpret this uncertainty in order to facilitate meaningful decision-making. Uncertainty can be interpreted in two ways, either in terms of physical probabilities from a frequentist point of view or in terms of subjective probabilities from a Bayesian point of view. While a frequentist interpretation may be suitable for testing-based prognostics, there are no physical probabilities in the context of condition-based prognostics. Therefore, uncertainty in the context of condition-based monitoring needs to be interpreted subjectively, and hence, a Bayesian approach is more suitable for this purpose. The various sources of uncertainty were discussed in detail, and it was explained that it is necessary to compute the overall uncertainty in the remaining useful life prediction by rigorously accounting for these sources of uncertainty. To this end, this paper presented a computational framework for uncertainty quantification in prognostics, in the context of condition-based monitoring. This framework consists of two steps; the first step involves estimating the state of the system using Bayesian tracking methods and the second step involves predicting the future behavior by propagating the uncertainty in the state estimates, model uncertainty, and uncertainty in the future loading (and usage) conditions through the system model until the failure threshold is reached. This indicates the end of life (EOL), and the uncertainty in EOL and RUL can be calculated using different types of uncertainty propagation techniques.

It is important to accurately estimate the entire probability distribution of EOL and RUL while being able to perform such computation in real time so that the method can be suitable for online health monitoring. While some methods (like Monte Carlo sampling) are accurate, they may be computationally intensive and therefore, not suitable for online health monitoring. Some other analytical methods may be approximate but suitable for online health monitoring. Future work will investigate the applicability of different types of uncertainty quantification methods to prognostics and identify those methods which can be implemented in online health monitoring and condition-based maintenance.

9. Acknowledgment: The work reported herein was in part funded by the NASA System-wide Safety Assurance Technologies (SSAT) project under the Aviation Safety (AvSafe) Program of the Aeronautics Research Mission Directorate (ARMD), and by the NASA Automated Cryogenic Loading Operations (ACLO) project under the Office of the Chief Technologist (OCT) of Advanced Exploration Systems (AES).

10. References

[1] B. Sun, S. Zeng, R. Kang, M. G. Pecht, Benefits and Challenges of System Prognostics, Reliability, IEEE Transactions on 61 (2) (2012) 323–335.

[2] R. deNeufville, Uncertainty Management for Engineering Systems Planning and Design, in Engineering Systems Symposium, MIT, Cambridge, MA, 2004.

[3] D. Hastings, and H. McManus, A Framework for Understanding Uncertainty and its Mitigation and Exploitation in Complex Systems, in Engineering Systems Symposium, MIT Cambridge MA, 2004.

[4] K.C. Ng, B. Abramson, Uncertainty Management in Expert Systems, IEEE Expert Systems, 1990.

[5] M. Orchard, G. Kacprzynski, K. Goebel, B. Saha, G. Vachtsevanos, Advances in Uncertainty Representation and Management for Particle Filtering Applied to Prognostics, in: Prognostics and Health Management, 2008. PHM 2008. International Conference on, 2008, pp. 1 –6.

[6] L. Tang, G. Kacprzynski, K. Goebel, G. Vachtsevanos, Methodologies for Uncertainty Management in Prognostics, in: Aerospace conference, 2009 IEEE, 2009, pp. 1 –12.

[7] J. Celaya, A. Saxena, K. Goebel, Uncertainty Representation and Interpretation in Model-based Prognostics Algorithms based on Kalman Filter Estimation, in: Proceedings of the Annual Conference of the PHM Society, 2012.

[8] L. Szab´o, Objective Probability-like Things With and Without Objective Indeterminism, Studies In History and Philosophy of Science Part B: Studies In History and Philosophy of Modern Physics, Vol. 38, No. 3, pp. 626–634, 2007.

[9] S. Sankararaman, Uncertainty Quantification and Integration in Engineering Systems, Ph.D. Dissertation, Vanderbilt University, 2012.

[10] B. de Finetti, Theory of Probability, Volumes I and II, Bull. Amer. Math. Soc. 83, 94-97, 1977, 0002–9904.

[11] M. Daigle, A. Saxena, and K. Goebel, An Efficient Deterministic Approach to Model-based Prediction Uncertainty Estimation, Annual Conference of the Prognostics and Health Management Society, 2012, pp. 326–335.

[12] A. Haldar, and S. Mahadevan, Probability, Statistics and Reliability in Engineering Design, John Wiley & Sons, New York, 2000.

[13] H.N. Najm, Uncertainty Quantification and Polynomial Chaos Techniques in Computational Fluid Dynamics, Annual Review of Fluid Mechanics, Vol. 41, pp. 35–52, 2009.

[14] R.P. Dwight, and Z.H. Han, Efficient Uncertainty Quantification using Gradient-Enhanced Kriging, 11th AIAA Non-Deterministic Approaches Conference, Palm Springs, CA, Apr 2009.

[15] S. Sankararaman, K. Goebel. An Uncertainty Quantification Framework for Prognostics and Condition-Based Monitoring, In the 16th AIAA Non-Deterministic Approaches Conference, National Harbor, MD, Jan 2014.

[16] S. Sankararaman, K. Goebel, Why is the Remaining Useful Life Prediction Uncertain?, in the Proceedings of the Annual Conference of the Prognostics and Health Management Society, New Orleans, LA, Oct 2013.

AN EFFICIENT PROGNOSTIC ESTIMATOR

Jae Yoon and David He
Department of Mechanical and Industrial Engineering
University of Illinois at Chicago
Chicago, IL, 60607, USA
davidhe@uic.edu

Paula Dempsey
NASA Glenn Research Center
Cleveland, OH, 44135, USA

Abstract: In this paper, a new prognostic estimation technique for online gear health management system is proposed and demonstrated with real spiral bevel gear run-to-failure test data. Unlike conventional particle filter based prognostic estimation methods, the prognostic technique presented in this paper is a hybrid of the unscented Kalman filter and particle filter. It is designed to improve the processing efficiency whilst the state estimation accuracy is maintained. The unscented Kalman filter is utilized to obtain the "best estimate" of the states of a degrading nonlinear component and the particle filter *l*-step ahead prediction technique is employed to obtain the remaining useful life of the component. In addition, data mining techniques are applied to efficiently define the system dynamics model, observation model, and predicted measurement information for the prognostic estimator. At last, the feasibility of the presented prognostic estimator is demonstrated with satisfactory results using the actual oil debris mass and health index data obtained from a spiral bevel gear test rig at the National Aeronautics and Space Administration (NASA) Glenn spiral bevel gear test facility.

Key words: bevel gear; CBM; Condition based management; health management; particle filter; prognostic estimator; prognostics; unscented Kalman filter; data mining.

Introduction: In any industry, cost reduction by increasing the availability of equipment (and parts) while reducing the breakdown rate is an important point of management. Unlike the breakdown maintenance (the earliest type of maintenance) and planned maintenance (a later type of maintenance), condition based maintenance (CBM) can significantly reduce maintenance costs by reducing the number of unnecessary maintenance actions if properly established and effectively implemented [1]. In the related field of machine prognostics and health monitoring, prognosticating remaining useful life (RUL) is a critical technique required to carry out CBM.

State estimation (aka. filtering) techniques are one of applicable strategies in the RUL prediction tasks. Typically, RUL is computed by propagating the current machine (or component) state up to its pre-specified future breakage state. Even though there exists a wide variety of state estimation algorithms that can be utilized for industry applications based on the desired accuracy and processing time, the Kalman filter and particle filter

(PF) variants are amongst the most utilized strategies. Kalman and particle filters are tools used to map the state of a dynamic system. To use them requires a model of how the system changes with time and a model of observations in specific system states. Predict/update time increments based on probabilities of observations of a state at a specific time are used to estimate current and future system states. The Kalman filter generates an optimal estimate of the system state, given a set of measurements and a single Gaussian probability density function (pdf). Particle filters can represent any pdf as a set of samples or particles.

Since both the Kalman filter and particle filter have pros and cons for tracking the state of dynamic systems, their utilization has always been a matter of choice between processing efficiency and accuracy. The Kalman filter requires less computational power and provides the optimal solution for linear systems [2], but its utilization in nonlinear systems is limited and dependent on the complexity of the models. This is because the Kalman filter performs poorly in the following situations [3]: 1) if the system dynamics or observation models have high nonlinearity; 2) if the Gaussian or normality assumption doesn't hold. On the other hand, the particle filter is known as the most accurate state estimation technique which captures almost every statistical feature of the time evolving system of interest. From this perspective, the particle filtering based *l*-step ahead prediction technique has attracted the researchers' attention.

Many different methods of implementing particle filters were found in the literature. An on-line PF-based framework for fault diagnosis and failure prognosis was proposed by Orchard and Vachtsevanos [4] in order to estimate axial crack growth of a UH-60 planetary gear plate. Two PF based RUL estimating modules were attested: 1) sequential importance sampling (SIS) filter based RUL estimator, 2) regularized PF based RUL estimator. In the framework, if any anomalous condition is detected, then prognostic routines are activated to predict the evolution of the fault indicator and RUL of the faulty subsystem. Butler *et al.* [5] addressed the potential ability of the PF based RUL estimator for use in a CBM strategy. In this study, a standard PF based RUL estimator and Gaussian mixture models combination is applied to a degradation model of a thermal processing unit in semiconductor manufacturing. Saha *et al.* [6] proposed the relevance vector machines (RVMs) that combined the standard PF RUL estimator and the Rao-Blackwellized PF (RBPF) RUL estimator with an application of lithium-ion battery capacity depletion. In the study, the RBPF has shown the efficiency of reducing uncertainty in a RUL estimator. In [7], a dynamic fatigue crack evolution of a mechanical component was studied. In [8], a deterministic but efficient unscented transform (UT)) based prognostic prediction framework is proposed. In long-term uncertainty propagation perspective, UT based sampling has a limitation in capturing higher order statistics (*e.g.* skewness, kurtosis, etc.). Thus, in this study, a standard PF based *l*-step prediction technique is maintained to estimate RUL of a crack growing mechanical component model. The Paris law, which is the most popular crack growth model describing the relation of the stress intensity to crack growth under a fatigue stress regime, was utilized as a state transitional function. In [9] an integrated approach for gear health prognostics using the standard PF based RUL estimator was developed. In this study, data mining based techniques are utilized to define the gear degradation model and health index (*i.e.* Cholesky decomposition based whitening transform) utilized to obtain the measurement model. In [10] [11] [12], physics based water and particle contaminated lubrication oil models were developed to quantify the relationship between water and particle contamination level and the readily available commercial dielectric constant and viscosity sensor outputs. Likewise, a standard PF based RUL estimation technique is exploited to estimate the RUL of this particle contaminated lubrication oil. The aforementioned application results have shown that particle filters represent a potentially powerful prognostics tool due to its capability in handling non-linear dynamic systems and non-Gaussian noise.

However, to maintain the higher state estimation accuracy, larger samples or particle populations should be maintained than those presented in the literature. Also, as the state dimension of the system increases, an exponentially increased particle population is needed. To overcome this problem, this paper presents a new hybrid prognostic approach. Two nonlinear state estimation techniques, *i.e.*, unscented Kalman filter (UKF) and particle filter, are combined to strengthen their advantages and reduce their weaknesses. The unscented Kalman filter defines the general state estimation and the particle filter defines the l-step ahead prediction using the resulting statistics of the unscented Kalman filter. The efficiency of the proposed prognostic estimator is demonstrated using the actual oil debris mass (ODM) generated and a health index (HI) calculated from vibration data during damage progression tests on spiral bevel gear sets (pinion/gear). These tests were performed in the Spiral Bevel Gear Fatigue Rig at the National Aeronautics and Space Administration (NASA) Glenn Research Center (GRC).

The remainder of this paper is organized as follows. In the first section, Experiment, the test facility and experimental setup for the data used to validate the proposed prognostic framework is explained. In the next section, Theoretical Methods, a brief theoretical background of UKF, a standard PF, and the RUL prediction with an l-step prediction method are provided. It is then followed by the presentation of the on-line hybrid UKF and PF framework for fault diagnosis and failure prognosis. The validation results are provided in the Model Development section. Finally, the Conclusions section concludes the paper and gives a direction for future work.

Experiment:

(1) Test Facility Description:

Tests were performed in the Spiral Bevel Gear Fatigue Test Rig at NASA GRC. A detailed description of this test facility, as illustrated in Figure 1(a), can be found in References [13] and [14]. In addition to developing gear health monitoring tools, the test facility has been used to study the effects of gear material, gear tooth design, and lubrication on the fatigue strength of gears. Two sets of spiral bevel gears are installed in the test rig and tested simultaneously. Facing the gearboxes per
Figure 1(a), the left gear set (pinion/gear) is referenced as 'left' and the right gear set (pinion/gear) is referenced as 'right' within this paper. A photograph of the spiral bevel gear set is shown in
Figure 1(b).

Both gear sets are lubricated with oil jets pumped from an oil reservoir. The lubrication from the gearbox then exits the gearbox and flows through an inductance type in-line oil debris sensor. The inductance type oil debris sensor was used to measure the ferrous debris generated during fatigue damage to the gear teeth. The sensor measures the number of metal particles and their approximate size based on user defined particle size ranges or bins. Based on the bin configuration, the average metal particle size for each bin is used to calculate the cumulative mass for the experiment by assuming the average metal particle size as a diameter spherical in shape and multiplying it by the density of steel. Health indexes were calculated from vibration data collected using accelerometers mounted on the left and right housing of the test rig. Speed was measured with an optical tachometers mounted on the left gear shaft.

(2) Experimental setup:

The failure observed on this gear set at test completion was damage on one tooth of the left side pinion as shown in
Figure 1(c). This failure mode is defined by class (contact fatigue), general mode (macropitting) and degree (progressive) per American gear manufacturers association (AGMA) standards for gear wear terminology outlined in reference [15]. Pitting was observed on greater than 1mm of the tooth contact surface. In this experiment, two different information sources provide the degrading state and observation data: oil debris mass (ODM) and health index (HI). The ODM and corresponding vibration sensor data were collected when the gear was in the healthy state and progressed to pitting damage on the left pinion tooth. The HI was developed by Bechhoefer *et al.* in [16] fusing several gear vibration condition indicators. For a detailed description of the test procedure, see [16] and [17] . The ODM and HI data are utilized to validate the feasibility of the presented prognostic estimator. The general framework of the validation procedure, identifying the inputs and outputs of the state estimator, is provided in
Figure **2**. The next section will discuss the methods used within this framework.

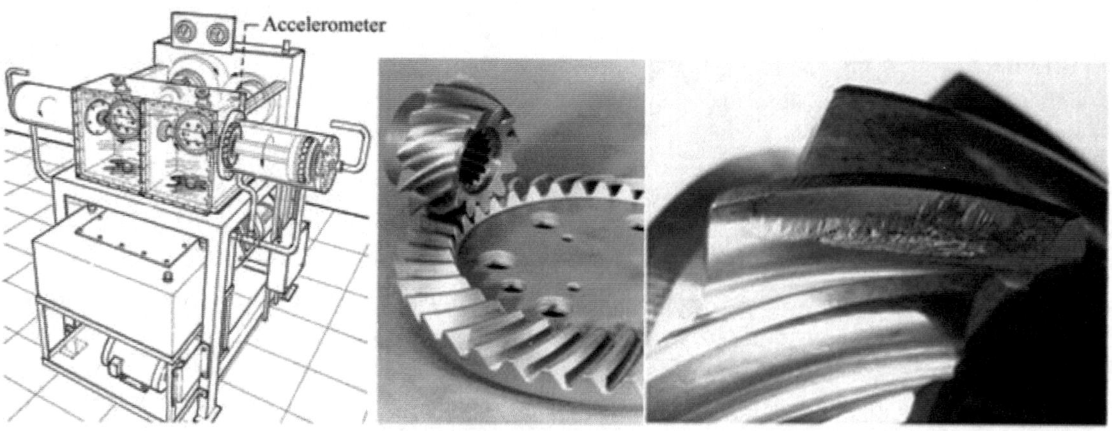

(a) The bevel gear test rig (b) spiral bevel gear (c) defective gear component

Figure 1: The spiral bevel gear test rig at the NASA Glenn test facility [17].

Theoretical Methods: To apply any state estimation method, the underlying problem should be formulated first as stated in [3], and [18]. The problem of state estimation (aka. filtering) is to find the best estimate of the dynamic system in terms of the posterior probability density function (pdf) per Bayesian statistics, based on all available information. This includes the physical laws believed to govern dynamical behavior of the system and sequence of measurements up to the current time step *k*. A complete model of the geared system requires an in depth understanding of the dynamic characteristics unique to this system, under varying operational conditions and failure modes. For this analysis it was simplified to a causal model, the rate of debris generated as gear tooth damaged progressed and gear health degraded. The oil debris mass generated during test run-time was the basis of the system state model. Physical inputs to the system such as oil temperatures, applied torque and operating speed were held constant. End of gear useful life was defined as macropitting on one or more gear or pinion teeth.

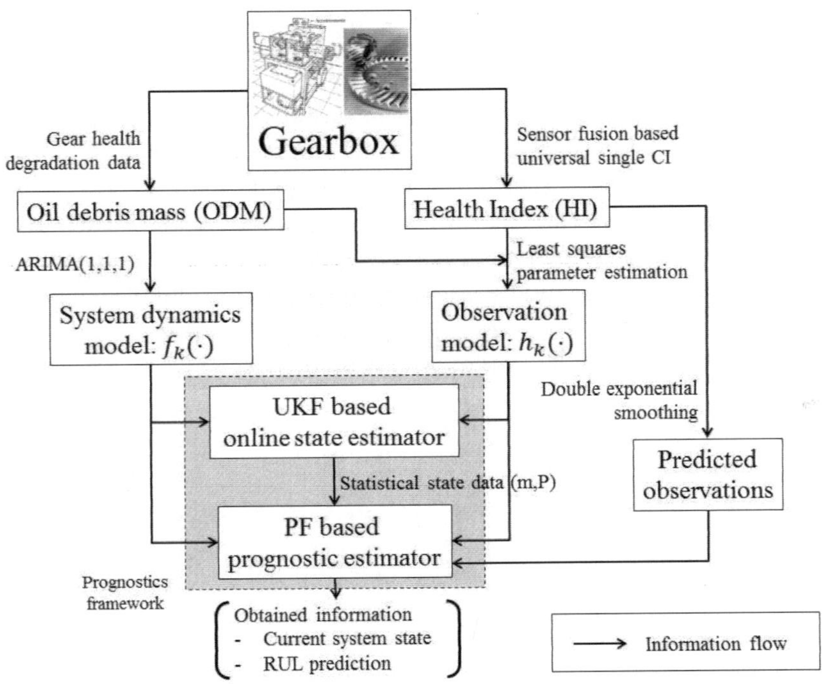

Figure 2: The integrated gear prognostics framework.

(1) State estimation problem formulation:

The first step in the analysis is to outline the methods used within the integrated prognostics framework outlined in Figure 2. The state transition function (f_k), referred to as the system dynamics model in the framework, represents the degradation evolution in time of the component. The measurement function (h_k), referred to as the observation model in the framework, represents the relationship between the degradation state of the component and health indicator. The UKF and PF based state estimators are used to define an effective time step ahead remaining useful life (RUL) estimator. The system dynamics model and observation model are inputs into the UKF and PF state based estimators per Figure 2.

Consider the following discrete-time probabilistic model (aka. System state transition (*f*) and observation (*h*) model) where probability of x_k given x_{k-1} is known:

$$f(x_k|x_{k-1}) \triangleq p(x_k|x_{k-1}) \tag{1}$$

$$h(z_k|x_k) \triangleq p(z_k|x_k) \tag{2}$$

The state transition and the state observation models can be rewritten in functional forms as follows:

$$x_k = f_{k-1}(x_{k-1}, q_{k-1}) \tag{3}$$

$$z_k = h_k(x_k, r_k) \tag{4}$$

Note that x_k is the state at time k, z_k is the observation at time k, q is the process noise and r is the observation noise.

The pdf of the system state of interest can be recursively obtained in two steps: prediction and update. In the prediction step, suppose the state estimate at the time $k-1$ ($i.e. p(x_{k-1}|z_{0:k-1})$) is known. Note $z_{0:k} = \{z_0, z_1, \dots, z_k\}$ is the sequence of observation that is available up to the kth time step. Then the one-step prediction (or prior) of the time evolving system model is obtained via the *Chapman-Kolmogorov* equation as:

$$p(x_k|z_{0:k-1}) = \int p(x_k|x_{k-1})p(x_{k-1}|z_{0:k-1})dx_{k-1} \tag{5}$$

In the update step, the new measurement z_k becomes available and the posterior can be obtained via the *Bayes* rule as follows per the probability of the current state given the current observation:

$$p(x_k|z_{0:k}) = \frac{p(z_k|x_k)p(x_k|z_{0:k-1})}{p(z_k|z_{0:k-1})} \tag{6}$$

where the normalizing constant is:

$$\mathrm{p}(\mathrm{z_k}|z_{0:k-1}) = \int p(x_k|z_{0:k-1})\, p(z_k|x_k)dx_k \tag{7}$$

However, the above recursive propagation of the posterior pdf is a conceptual solution; it cannot be analytically determined [2]. In this paper, the system dynamic function (*i.e.*, Equation (1)) describes a mathematical representation of the data driven bevel gear state degradation in time. Also, the observation model (*i.e.*, Equation (2)) represents the data driven health monitoring sensor outputs indicating degradation of the bevel gear. For this reason, methods to estimate the future states are required. UKF and PF based estimators were selected for further analysis.

(2) Unscented transform (UT) and unscented Kalman filter (UKF):

Extended Kalman Filter (EKF), a non-linear version of the KF has been one of the most efficient state estimation techniques for many decades. However, it assumes that nonlinear functions are represented with local linearization through Taylor series expansion and the higher order terms become negligible. This assumption breaks down when the models are highly nonlinear or non-Gaussian. As a result, UKF is introduced as a solution of nonlinear problem by Julier and Uhlmann, [19]. UKF is normally used when the models are highly nonlinear or the importance of the higher order terms of Taylor series expansion is big. The first advantage of UKF is that it uses actual nonlinear models rather than approximated dynamic model and observation model via *Jacobian*. The next is that UKF accurately captures system nonlinearity up to second order (*i.e.*, mean and covariance) with deterministic but smartly chosen sample points so called sigma points.

(2.1) Unscented transform:

The unscented transform (UT) is a method to capture the statistics of a nonlinear transform of a random variable (*e.g.* $\mathbf{y} = \mathrm{f_N}(\mathbf{x})$). The first two moments, the true mean and covariance, of the nonlinear transform are computed using sigma points, a minimal number of deterministically chosen weighted sample points $(\mathbf{x}^{(i)}, \mathrm{w}^{(i)})$. The number of sigma points (*i.e.* $2n_x + 1$) is dependent upon the dimension of the state vector. Unscented transform is performed in the following steps outlined in [20] [21] [22]. The matrix form of the UT transform in [22] was fed into the unscented Kalman filter (UKF)

(2.2) Unscented Kalman Filter (UKF):

Using the matrix form of the UT transform, the prediction and update stages of the UKF were outlined in several steps presented in [22]. From these equations, predicted state estimation variables mean $\mathbf{m}_k$ and covariance $\mathbf{P}_k$ were calculated.

(3) RUL Prediction with Particle filtering based l-step prediction:

Particle filter (PF, aka. sequential *Monte-Carlo* method) were highlighted as an effective on-line state estimation tool in [23], [24], and [3]. In order to estimate the remaining useful life (RUL) using PF, an *l*-step ahead estimator is required. An *l*-step ahead estimator is a method to update and predict a future system state at next time step given current system state. An *l*-step ahead estimator will provide a long term prediction of the state pdf $p(x_{k+l}|z_{0:k})$, for $l = 1, \dots, T_\lambda$, where T_λ is the breakage or damage instance of mechanical components; λ is a prespecified failure state at time T_λ. In making an *l*-step ahead prediction, it should be remembered that no measurement information is available for updating the state lying along the future time span (*i.e.* $x_{k+1:k+l}$). In other words, the future measurements $z_{k+1:k+l}$ cannot be utilized for updating the prediction in general circumstances. Thus, the desired state pdf before its failure, $p(x_{k+l}|z_{0:k})$, should be factorized with the current posterior pdf $p(x_k|z_{0:k})$ and the system dynamic model, $p(x_k|x_{k-1})$.

PF based approximation procedure of the l-step-ahead estimator is introduced in [7] and [9] as:

$$p(x_{k+l}|z_{0:k}) \approx \sum_{i=1}^{N} w_{k+l-1}^{i}\ p\left(x_{k+l}|x_{k+l-1}^{i}\right) \tag{8}$$

where any weight update in RUL estimation stage would be available with assistance of data mining technique known as double exponential smoothing (DES). Assume that the state x_k represents the spiral bevel gear degradation level at the current time k; the gear degradation level increases and, conversely, RUL decreases over time. If an l-step ahead state (*i.e.* x_{k+l}) goes across λ (*i.e.* $x_{k+l} \geq \lambda$), then the object's RUL can be computed as $\mathrm{RUL}_k = (k+l) - k = l$. At each time step before its failure the state x_{k+l} would be projected up to T_λ. In this manner, estimating $\mathrm{RUL} \leq l$ is equivalent to estimating $x_{k+l} \geq \lambda$, rewriting as:

$$\begin{aligned} \hat{p}(RUL \leq l|z_{0:k}) &= \hat{p}(x_{k+l} \geq \lambda|z_{0:k}) \\ &\approx \hat{p}(\{(x_{k+l+\tau 1}^{1}, \dots, x_{k+l+\tau i}^{i}, \dots, x_{k+l+\tau N}^{N})\} \geq \lambda|z_{0:k}) \end{aligned} \tag{9}$$

where $k+l+\tau_1,\ldots,k+l+\tau_N$ are the time indices of the particles crossed the critical state λ; τ_i is the time index difference of each particle from $k+l$, for $\tau_i \in \mathbb{Z}$; i is a particle index, e.g. $i = 1, \dots, \mathrm{N}$; Then RUL can be estimated as:

$$\widehat{RUL_k} \approx \frac{1}{N}\sum_{i=1}^{N} k + l + \tau_i \tag{10}$$

In the current PF based prognostics framework, PF state estimation and PF based RUL estimator is performed in a same frame whilst the particle population is maintained. For example, if one thousand particles are used in the PF state estimation, the same amounts of particles would be used in the RUL estimation. A graphical representation of the current PF based RUL is shown in Figure 3(a). Generally speaking, in less complex models, achieving high levels of state estimation accuracy does not require the computationally resource intensive technique PF. Because estimating RUL requires more computational resources than estimating the current state, investing the same amount of computational power for both is a waste of resources.

In the presented hybrid prognostic estimator, two estimation techniques are utilized for different purposes: State estimation and RUL estimation are performed via UKF and PF, respectively. A graphical representation of the proposed RUL estimator is shown in 3(b). At each time evolution, the best state estimation, statistic information (*i.e.* $(\mathbf{m}_k, \mathbf{P}_k)$) of the posterior information, becomes available via UKF. Then, particle distribution for l-step prediction is elastically constructed based on $(\mathbf{m}_k, \mathbf{P}_k)$. Those constructed particles are propagated up to end of life (EOL) in RUL estimation stage. Less sample points are required to predict an accurate distribution of the system state per this approach.

Since the processing time of state estimation stage could potentially be optimized by using UKF, investing redundant computational resources to the RUL estimation stage becomes available. The hybrid RUL estimation method could be strategically used in two ways: 1) Eliminating redundant processing time while the same level of RUL accuracy is maintained, 2) Increasing RUL accuracy level while the same processing time is maintained.

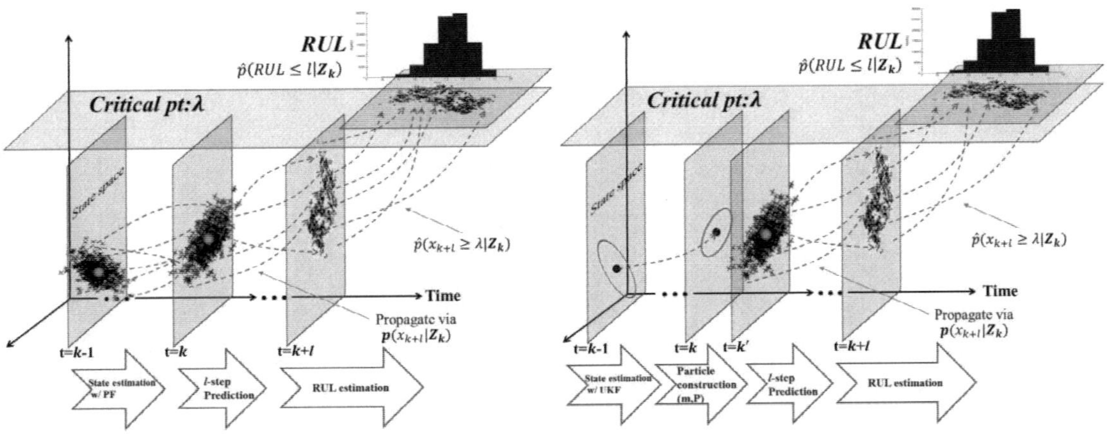

(a) PF based RUL estimation framework (b) hybrid RUL estimation framework

Figure 3: Graphical representation of (a) the current PF based RUL estimation framework (b) the hybrid RUL estimation framework.

Model Development: In this study, the gear degradation model was derived using data mining approach. Autoregressive integrated moving average (ARIMA) was utilized to obtain a predictive formula in the form of state transition function by identifying the patterns in the ODM data. ARIMA(p,d,q) applies statistical regression analysis to predict future trends. The three parameters refer to the degree or order of the autoregressive (p), integration (d) and moving average (q) terms of the model. After testing varied ARIMA models over the ODM data, ARIMA(1,1,1) model was selected from the perspective of the minimum root mean squared errors. The ODM and HI data plots are presented in

Figure 4. In the figure, the solid line and dashed line represent the true ODM data and the HI data, respectively.

The selected system dynamics model using ARIMA(1,1,1) has a form of the following **[25]**:

$$\hat{x}_{k+1} = 0.0165 + 1.1415x_k - 0.1415x_{k-1} - 0.1032(x_k - \hat{x}_k) + q_k \tag{11}$$

Where x is the state of the system, process noise q_k is zero-mean white Gaussian noise. Unlike the typical Markovian state transition model, where future state depends on present state only, ARIMA(1,1,1) model contains previous state (*i.e.* x_{k-1}) and error correction term (*i.e.* $x_k - \hat{x}_k$). Figure 5 shows the verification of ARIMA(1,1,1) model under noisy conditions. In the figure, the solid line represents verification of ARIMA model with process noise and dashed line represents true the ODM data.

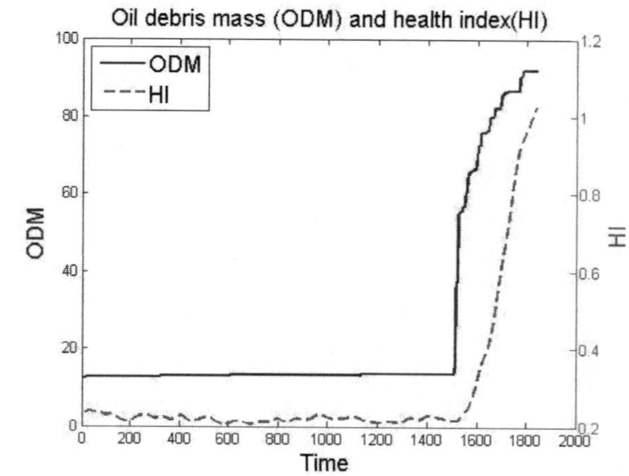

Figure 4: Raw ODM and HI data.

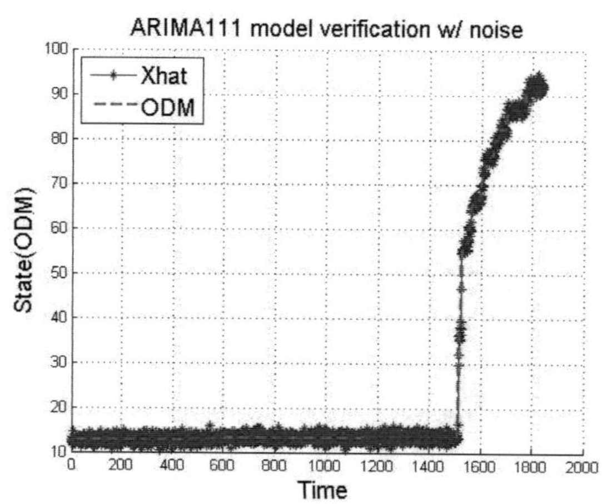

Figure 5: ARIMA(1,1,1) model verification with process noise.

The observation model was derived from the functional relationship between ODM and HI using a least squares parameter estimation method. After testing degrees of x from 1 through 10, a degree of 4 was selected. From the actual data set, the following observation model was obtained:

$$z_k = h(x_k, r_k) = \boldsymbol{x}_k^T \cdot \boldsymbol{p} + r_k \quad (12)$$

where,

$$\boldsymbol{x}_k^T = [x^4\ x^3\ x^2\ x\ 1];$$

$$\boldsymbol{p} = [9.7201\mathrm{e}-08 \quad -1.4502\mathrm{e}-05 \quad 7.7864\mathrm{e}-04 \quad -0.0174 \quad 0.3439]^T.$$

In Equation (12), superscript T represents vector transpose; the observation noise r_k is zero-mean white Gaussian noise. The relationship between oil debris mass (ODM) and health index (HI) is shown in Figure 6. In the figure, the solid line and dashed line represent the true relationship and functional relationship, respectively. The verification of the obtained observation model under noisy circumstance is shown in Figure 7. In the figure, the solid line and dashed line represent the actual HI and the observation model output, respectively.

The initial conditions assumed by the filter were: $\mathrm{x}_0 = \mathrm{x}_{-1} = 12.457$, and P=0.7. In the PF based RUL prediction stage, the error correction term in the ARIMA(1,1,1) becomes eliminated as shown in Equations.(13) - (15). Since the current ODM is always equal or greater than the past ODM ($e.g.\ \hat{x}_k \geq \hat{x}_{k-1}$) at any time, the predicted l-step ahead state only increases exponentially.

$$\hat{x}_{k+1} \approx 0.0165 + 1.1415\hat{x}_k - 0.1415\hat{x}_{k-1} - 0.1032(\hat{x}_k - \hat{x}_k) + q_k \tag{13}$$

$$= 0.0165 + 1.1415\hat{x}_k - 0.1415\hat{x}_{k-1} \quad + q_k \tag{14}$$

$$\geq 0.0165 + \hat{x}_k + q_k \tag{15}$$

Therefore, another data mining technique, double exponential smoothing (DES), was applied to obtain predicted observation ($\tilde{z}_{k+l}$). Double exponential smoothing weighs current observations higher than past observations and takes into consideration trends in the data set. Based on all of the available observations up to the current time, DES captured the trend and level from the noisy information. Using the level and trend, forecasting observation became available. With the predicted observations, the error correction terms in the RUL prediction stage were computed using the following inverse function of the observation model:

$$\tilde{x}_{k+l} = h^{-1}(\tilde{z}_{k+l}) \tag{16}$$

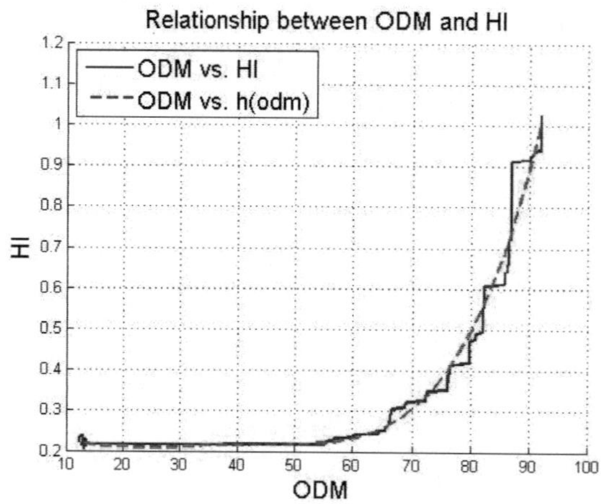

Figure 6: The relationship between ODM and HI.

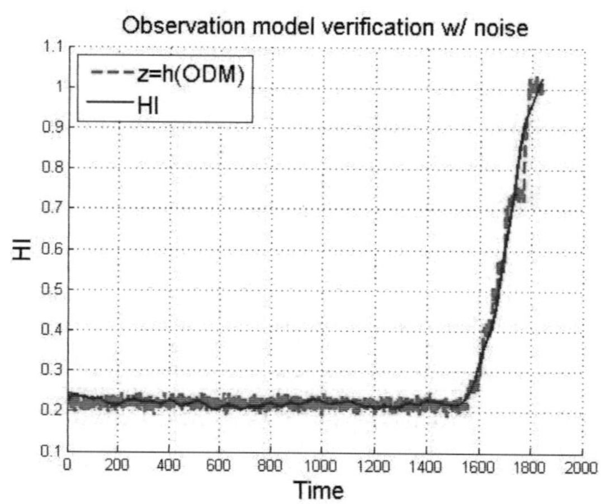

Figure 7: Least squared observation model verification w/ observation noise.

An example of the DES prediction is shown in Figure 8. In this example, the DES was performed at the 1673th time step. By using the available information up to the current time (*i.e.* $z_{0:1673}$), HI prediction up to the failure time was obtained. In the figure, solid line and dashed line represent the predicted HI and actual HI, respectively.

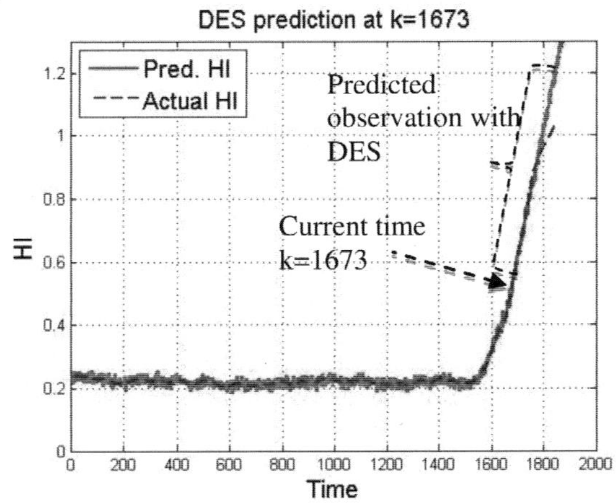

Figure 8: HI prediction at k=1673 with double exponential smoothing (DES).

Then the error correlation term in the l-step ahead prediction becomes the following:

$$\hat{x}_{k+l+1} \approx 0.0165 + 1.1415\hat{x}_{k+l} - 0.1415\hat{x}_{k+l-1} - 0.1032(\tilde{x}_k - \hat{x}_k) + q_k \quad (17)$$

$$= 0.0165 + 1.1415\hat{x}_{k+l} - 0.1415\hat{x}_{k+l-1} - 0.1032(h^{-1}(\tilde{z}_{k+l}) - \hat{x}_{k+l}) + q_{k+l} \quad (18)$$

In the experimental setup, HI value of 0.9 and its corresponding ODM state value of 90 were set as the failure point of the spiral bevel gear component ($\lambda_{HI} = 0.9$, $\lambda_{ODM} = h^{-1}(0.9) = 90$)). That is, $\hat{p}(x_{k+l} \geq (\lambda_{ODM} = 90)|z_{0:k})$ was considered as the remaining useful life of the component while RUL estimator was performed.

Analysis results: The performance of the presented method is provided in Figure 9, Figure 10, Table 1, and Table 2. First, the state estimation accuracy of each state estimation method is investigated by measuring their mean square error (MSE). In the Figure 9, the MSE results are presented.

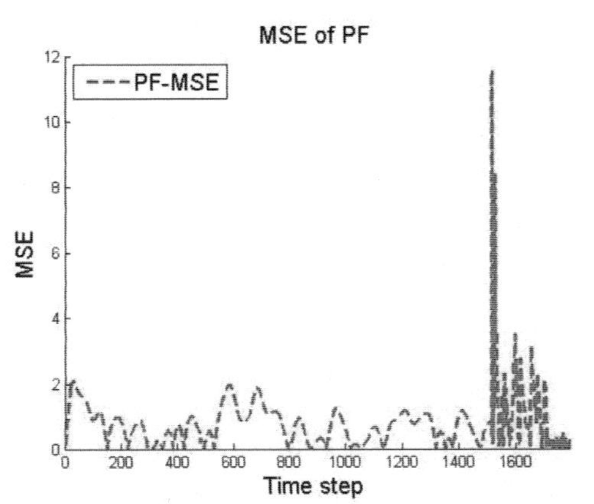

a) Mean square error (MSE) of PF (N=200)

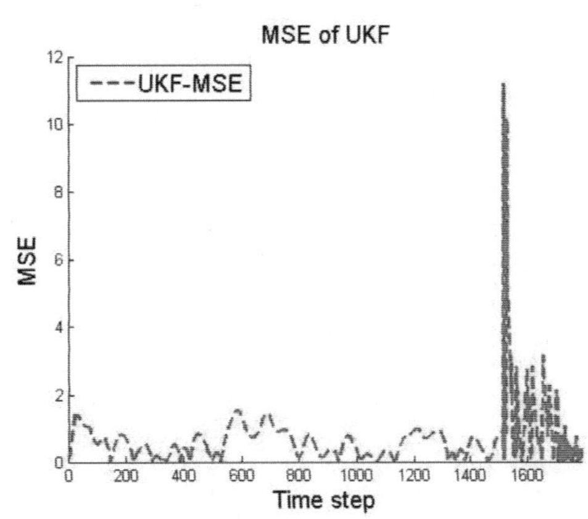

b) Mean square error (MSE) of UKF

Figure 9: Mean square error (MSE) of state estimation methods

The left and right figures represent MSE plots of the PF method and the UKF method in the state estimation stage, respectively. Even though it is well known that the performance of PF is largely dependent on the particle population, one could confirm that the state estimation accuracy between UKF and PF is not significantly different under comparable conditions.

However, in the processing time perspective, the difference between the two state estimation methods became significant. The resulting statistics of the processing time at the state estimation stage are provided in Table 1. UKF was about 8 times faster than PF with N=200. Note that these statistic results were obtained based on a set of 20 runs or iterations.

Table 1: Processing time summary in state estimation stage.

State estimation methods		Average (μ) (sec.)	Std. Dev. (σ) (sec.)
PF with (N=)	50	1.6003	0.0739
	100	3.2422	0.0183
	200	6.7676	0.0243
UKF estimation		0.8364	0.0202

In addition, the time efficiency of the presented method was validated by measuring the processing time of the entire RUL process (*i.e.* the state estimation and RUL prediction stages combined.) The resulting statistics of the experiment is provided in Table 2. In comparison with the PF based method, the presented method is about 20% more time efficient. Also, when N=200 particles were used, the standard PF based RUL estimator could not complete its RUL computation (each time step is 1 minute). Even though the presented method achieved about 20% time efficiency increment with the spiral bevel gear data, one could expect even higher time efficiency as the system model becomes more complicated.

Table 2: Processing time summary in the entire RUL estimation process.

State estimation methods		Average (μ) (sec.)	Std. Dev. (σ) (sec.)
PF with (N=)	50	20.0452	1.1864
	100	36.1533	1.4163
	200	(62.2412)	4.2356
Presented method with (N=)	50	18.2015	1.1599
	100	31.7836	1.4236
	200	52.4478	3.2372

Finally, in Figure 10, the actual RUL from the test data is compared to the estimated remaining useful life (RUL) and corresponding 90% confidence interval for the hybrid RUL method. The vertical lines along the actual RUL plot indicate the width of the distribution at that time step.

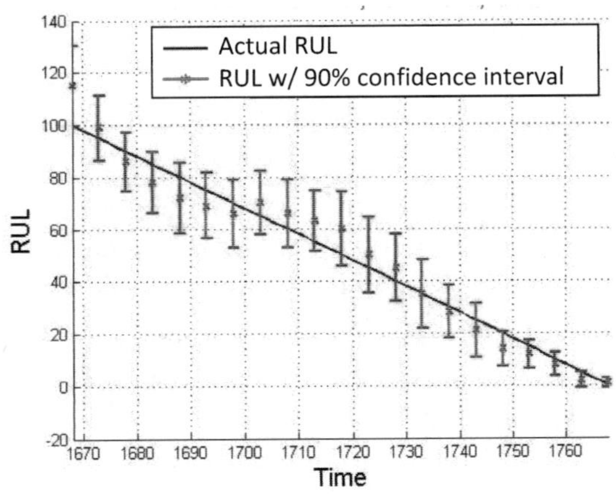

Figure 10: RUL estimation result for spiral bevel gear application.

Conclusions: In the machine prognostics and health monitoring (PHM) field, prognosticating exact remaining useful life (RUL) is important for condition based maintenance (CBM). In this paper, a hybrid prognostic estimation method was proposed and demonstrated with spiral bevel gear run-to-failure test data. In the proposed method, the particle filter (PF) and unscented Kalman filter (UKF) was combined to improve the processing efficiency whilst the state estimation accuracy was maintained. The UKF was mainly used to estimate the system state. The PF based RUL estimator was committed to predict the RUL of a degrading component at a future time. The resulting computational efficiency of the proposed method was compared with that of the traditional PF based RUL estimator. The comparison results have shown that a 20% efficiency improvement could be achieved using the proposed method on the bevel gear degradation data.

The method presented in this paper can be exploited tactically in the following two ways: 1) eliminating redundant processing time while the same level of RUL accuracy is maintained, 2) Increasing RUL accuracy level while the same processing time is maintained. In addition, there exist avenues for future research investigations. One area of investigation is to substitute the UKF state estimation stage with another state estimator to determine if the overall estimation performance could be improved in both time efficiency and accuracy. A second area of investigation is to substitute double exponential smoothing (DES) with other forecasting methods, such as parameter estimation, time series, and other data mining techniques, to also determine if overall performance could be improved.

References:

[1] A.K.A Jardine, D. Lin, and D. Banjevic, "A review on machinery diagnostics and prognostics implementing condition-based maintenance," *Mechanical Systems and Signal Processing,* no. 20, pp. 1483-1510, 2006.

[2] B. Ristic, S. Arulampalam and N.l Gordon, Beyond the Kalman Filter, Artech House Publishers, 2004.

[3] A. Doucet and A. M. Johansen, "A Tutorial on Particle Filtering and Smoothing: Fifteen years later," 2008.

[4] M. E. Orchard and G. J. Vachtsevanos, "A particle-filtering approach for on-line fault diagnosis and failure prognosis," *Transaction of the Institute of Measurement and Control,* vol. 31, no. 3/4, pp. 221-246, 2009.

[5] S. Butler and J. Ringwood, "Particle filters for remaining useful life estimation of abatement equipment used in semiconductor manufacturing," in *Proceedings of the First Conference on Control and Fault-Tolerant Systems*, Nice, France, October 6-8, 2010.

[6] B. Saha, K. Goebel, S. Poll, and J. Christophersen, "Prognostics Methods for Battery Health Monitoring Using a Bayesian Framework," *IEEE Transactions on Instrumentation and Measurement,* vol. 58, no. 2, pp. 291-296, 2009.

[7] E. Zio and G. Peloni, "Particle filtering prognostic estimation of the remaining useful life of nonlinear components," *Reliability Engineering and System Safety,* no. 96, pp. 403-409, 2011.

[8] M. Daigle, A. Saxena, and K. Goebel, "An Efficient Deterministic Approach to Model-based Prediction," in *Annual Conference of the Prognostics and Health Management Society*, Minneapolis, Minnesota, USA, 2012.

[9] D. He, E. Bechhoefer, J. Ma and J. Zhu, "A particle filtering based approach for gear prognostics," in *Diagnostics and Prognostics of Engineering Systems: Methods and Techniques, Chap 13*, 2012.

[10] J. Zhu, J. Yoon, D. He, B. Qiu, and E. Bechhoefer, "Online Condition Monitoring and Remaining Useful Life Prediction of Particle Contaminated Lubrication Oil," in *IEEE PHM conference*, Washington DC, 2013.

[11] J. Zhu, J. Yoon, B. Qiu and D. He, "Lubrication Oil Condition Monitoring and Remaining Useful Life Prediction with Particle Filtering," *International Journal of PHM,* vol. 4, no. 020, pp. 1-15, 2013.

[12] J. Zhu, D. He, J. Yoon, and E. Bechhoefer, "Online Particle Contaminated Lubrication Oil Condition Monitoring and Remaining Useful Life Prediction for Wind Turbines," *Wind Energy,* 2014.

[13] R. F. Handschuh, "Thermal Behavior of Spiral Bevel Gears," NASA TM-106518, 1995..

[14] R. F. Handschuh, "Testing of Face-Milled Spiral Bevel Gears at High-Speed and Load," NASA/TM—2001-210743, 2001.

[15] ANSI/AGMA, *Appearance of Gear Teeth – Terminology of Wear and Failure.,* ANSI/AGMA 1010-E95.

[16] E. Bechhoefer, D. He, and P. Dempsey, "Gear Health Threshold Setting Based On a Probability of False Alarm," in *Annual Conference of the Prognostics and Health Management Society*, Quebec, Canada, 2011.

[17] P. J. Dempsey, R. F. Handschuh, and A. A. Afjeh, "Spiral bevel gear damage detection using decision fusion analysis," in *Proceedings of the Fifth International Conference on Information Fusion*, Annapolis, MD, USA, 2002.

[18] R. V. D. Merwe, A. Doucet, N. D. Freitas, and E. Wan , "The unscented particle filter," *Proceedings of Neural and Information Processing Systems,* 2000.

[19] S.J. Julier and J.K. Uhlmann, "A new extension of the Kalman filter to nonlinear systems," *Proceedings of Aerospace,* Vols. Multi Sensor Fusion, Tracking and Resource Management II, 1997.

[20] S. J. Julier, J. K. Uhlmann, and H. F. Durrant-Whyte, "A new approach for filtering nonlinear systems," in *In Proceedings of the 1995 American Control*, Seattle, Washington, 1995.

[21] S.J. Julier and J.K. Uhlmann, "Unscented Filtering and Nonlinear Estimation," *Proceedings of the IEEE,* vol. 92, no. 3, pp. 401-422, 2004.

[22] S. Sarkka, Recursive Bayesian Inference on Stochastic Differential Equations, Helsinki University of Technology, 2006.

[23] N.J. Gordon, D. J. Salmond, and A. F. M. Smith, "Novel approach to nonlinear/non-Gaussian Bayesian state estimation," *IEEE Proceedings-F,* vol. 140, no. 2, pp. 107-113, 1993.

[24] M. Sanjeev Arulampalam and Simon Maskell and Neil Gordon, "A tutorial on particle filters for online nonlinear/non-Gaussian Bayesian tracking," *IEEE TRANSACTIONS ON SIGNAL PROCESSING,* vol. 50, pp. 174-188, 2002.

[25] D. He, E. Bechhoefer, P. Dempsey, and J. Ma, "An Integrated Approach for Gear Health Prognostics," in *The American Helicopter Society 68th Forum, May 1-3*, Fort Worth, Texas, 2012.

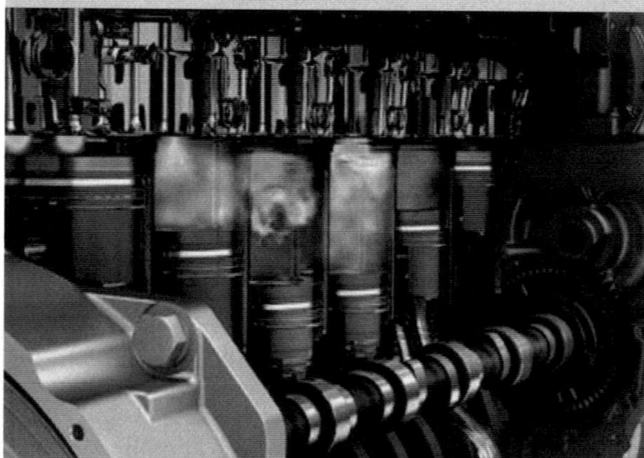

PHM Development & Implementation Challenges for the Off-Road Engine Market

21 May 2014

Mark Friedrich – John Deere Power Systems

MFPT 2014 CONFERENCE
Technology Solutions for Affordable Sustainment
Sheraton Virginia Beach Oceanfront Hotel • May 20–22, 2014

JOHN DEERE

Agenda

- Introduction
- Review of Off-Road Engine Market
- Off-Road Market Culture & Focus
- Challenges to PHM
- Current Remote Capabilities
- Implementation
  - Applying technologies

The Past

JOHN DEERE

The Present

JOHN DEERE

The Future

JOHN DEERE

What is the Off-Road Engine Market?

Agricultural Equipment

7 | MFPT 2014 Conference - Virginia Beach, VA | 21 May 2014

JOHN DEERE

Construction Equipment

8 | MFPT 2014 Conference - Virginia Beach, VA | 21 May 2014

JOHN DEERE

Forestry Equipment

9 | MFPT 2014 Conference - Virginia Beach, VA | 21 May 2014

JOHN DEERE

JDPS External Markets

Thousands of Applications

10 | MFPT 2014 Conference - Virginia Beach, VA | 21 May 2014

JOHN DEERE

Ag Specialty & Irrigation Applications

11 | MFPT 2014 Conference - Virginia Beach, VA | 21 May 2014

JOHN DEERE

Industrial Applications

12 | MFPT 2014 Conference - Virginia Beach, VA | 21 May 2014

JOHN DEERE

Generator Set Applications

Marine Applications

Off-Road Market Culture & Focus

Off-Road Market Focus & Culture

- Focus on fleet management & machine uptime
  - Change to proactive instead of reactive
  - Planned machine repairs during normal downtime
- Culture varies depending on application
  - Construction Market
    - Mostly continuous operation
  - Agriculture Market
    - Intense periodic use
  - Customer Base
    - Business vs. Individual Customer

JOHN DEERE

Challenges to PHM

- Huge amount of data – how do we use it effectively?

JOHN DEERE

Challenges to PHM

JOHN DEERE

Challenges to PHM

- Huge amount of data – how do we use it effectively?
- Data channels & frequency vs. cost to get the data
- Data quality
- Data privacy laws
  - Business vs. Individual
  - Owner vs. Operator
- Mobile applications
  - Repairs in the field
  - On-board & Off-Board Analysis
- Machine Alerts
  - Go to customer or dealer or both?
  - How to get from alert to action

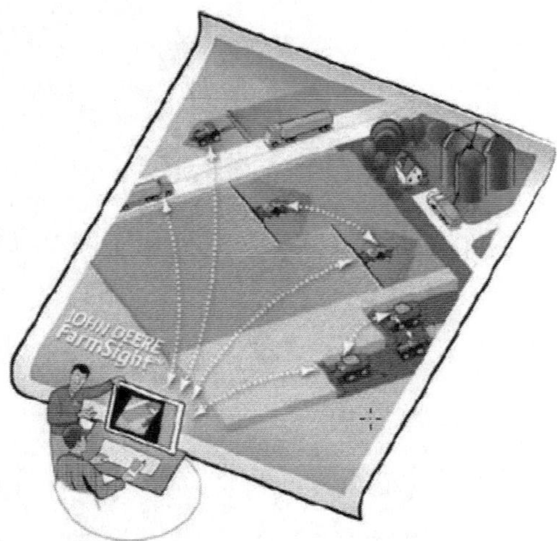

JOHN DEERE

Challenges to PHM

- Building infrastructure to support *automated* prognostics
  - No human in the loop
- Cellular coverage limited in some areas / cost of satellite $$$
- Product & customer variations
- Competing systems and resources

Available data

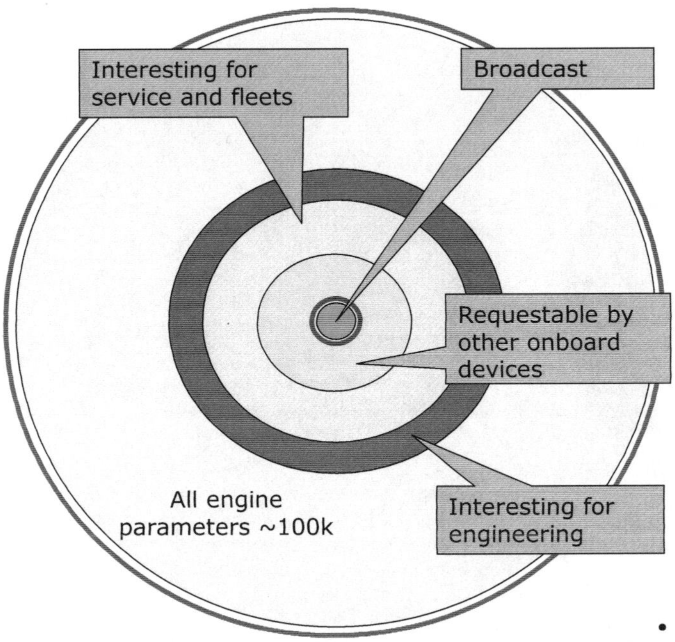

Acquiring Data Remotely

- Service Advisor Remote™
  - Request ***any*** data from ***one*** machine
  - Request ***any*** data from ***many*** machines
- Service Information Monitor
  - SIM
  - Machine(s) send data to Deere

- **Remote Programming**

JOHN DEERE

Implementation - How we are using it

- Remote diagnostics confirmed EGR sensor was de-rate root cause
  - **Correct part sent to field first trip** and half the normal downtime
- Overheating hydraulics
  - Diagnosed and **resolved remotely** with software
  - Trip saved & uptime for customer
- Engine derate and *many* DTC's
  - Diagnosed remotely as plugged fuel filters
  - **Customer changed filters and resolved issue**
- Load profiles retrieved to **validate** FT4 soot accumulation models
- Transient data gathered for EGR sensor
  - Approx. 8000 tractors over 3 months
  - Will be used to **build** a model

25 | MFPT 2014 Conference - Virginia Beach, VA | 21 May 2014

JOHN DEERE

Bottom Line for our Customers

26 | MFPT 2014 Conference - Virginia Beach, VA | 21 May 2014

JOHN DEERE

Telematics enabled machines

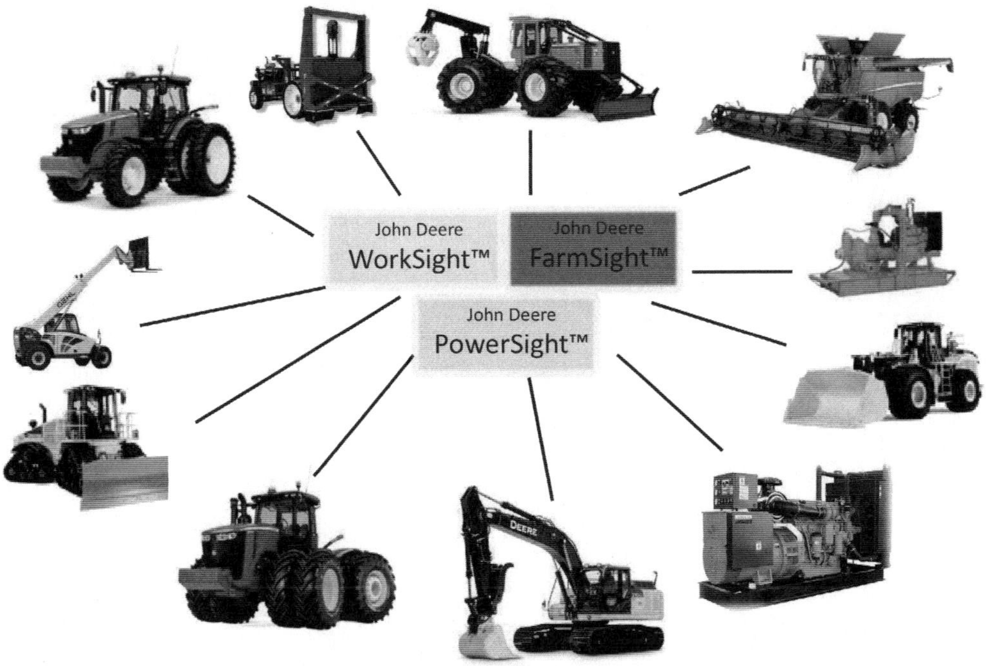

John Deere productivity tools working together

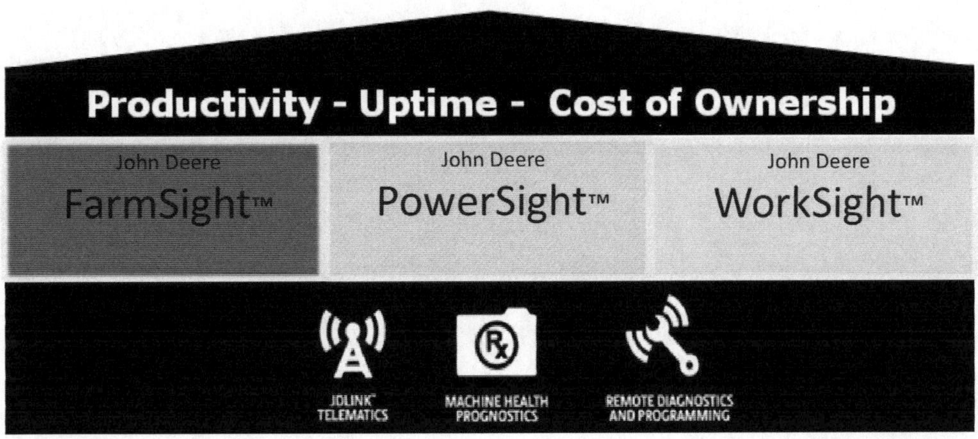

Resources

More information available on the web …

- John Deere WorkSight™
  - www.deere.com/worksight
- John Deere FarmSight™
  - www.deere.com/farmsight
- John Deere PowerSight™
  - www.deere.com/powersight

29 | MFPT 2014 Conference - Virginia Beach, VA | 21 May 2014 JOHN DEERE

JOHN DEERE

INVESTIGATIONS ON SPINDLE BEARINGS HEALTH PROGNOSTICS USING A DATA MINING APPROACH

Divya Sardana and Raj Bhatnagar
School of Electronic and Computing Systems
University of Cincinnati
2600 Clifton Avenue
Cincinnati, OH 45220
sardanda@mail.uc.edu

Radu Pavel and Jonathan Iverson
TechSolve, Inc.
6705 Steger Drive
Cincinnati, OH 45237
pavel@techsolve.org

Abstract: In the metalworking industry, the machine tool is a key component of the manufacturing chain. Unmanaged downtime is directly related to the performance characteristics of the machine tool and may result not only in high repair costs, but also in production delays, customer dissatisfaction, and lower potential sales. Therefore, maintaining the machines in good working condition and avoiding unexpected break-downs is of utmost importance. One of the machine tool systems that fail most often is the spindle. Spindle breakdowns are commonly caused by bearing failures that could be generated by a number of causes such as corrosion, wear, and improper lubrication. Although many attempts to predict time to failure of spindle bearings have been reported in the literature, there are still challenges related to data availability and methodology. In addition, there is significant variation from case to case due to the complexity of system usage and failure modes. This paper presents the results of an investigation focused on developing a new data-driven methodology for predicting the time to failure of the bearings of a spindle. Data collected from over five run-to-failure tests of a spindle test-bed has been reviewed and processed using a number of pattern discovery tools. A Fourier Transform-based analysis showed that episodes of bursts of energy were observed with monotonically increasing frequency and intensity as the spindle bearings' health deteriorated. The predictions made based on the observed pattern of energy bursts have provided encouraging results. The methodology developed and the findings to date are presented in the paper along with the experimental setup and the degradation evolution recorded for each case.

Key words: Data mining; prognostics; spindle bearings health characterization

Background and Motivation: In manufacturing, maintenance methodologies have evolved in response to the necessity to improve machine performance and decrease downtime. Early detection of emerging faults and degradation trends can prevent downtime, target maintenance efforts, increase productivity, and save costs. Condition-based maintenance (CBM) systems in manufacturing plants continuously deliver data related to the machine's status and performance; however, the challenge for field engineers and management staff is making effective use of the huge amount of data to accurately detect equipment degradation and further predict the remaining useful life.

CBM can reduce the cost of total maintenance while keeping the reliability high. This is achieved by monitoring critical parts inside the machine and performing the maintenance only when necessary. Individual machine systems can be analyzed on a case-by-case basis. Prognostics-based maintenance (PHM) links studies of failure mechanisms to system lifecycle management. The goal is to predict the future condition of a component and/or system of components with respect to remaining useful life or time to failure. Using prognostic technologies, smart products and systems with embedded and remote intelligence can be enabled to predict and forecast their performance proactively, as well as to synchronize their service needs with business support systems [1].

This paper presents the results of an investigation focused on developing a new data-driven methodology for predicting the time to failure of the bearings of a spindle. The spindle is a key component of machine tools. Its failure renders the machine unusable until a replacement is installed. Comparing to the other subsystems of a machine such as feed axes, fluid tanks, and tool magazines, the spindle has one of the highest rates of failure and the ability to predict its end of life within a few weeks before the event is highly desirable.

Many studies have been conducted with respect to characterization of the condition of bearings and ability to predict their end of life. A selection of such studies is briefly summarized below. Various algorithms have been developed in an effort to identify robust methods for predicting the end of life of bearings, in particular, and other machinery components, in general. However, the area of prognostics is still considered to be in its early development stage as the robustness needed for an industrial implementation has yet to be achieved.

A comprehensive survey of the anomaly detection methodologies was compiled by Chandola et al [2]. The authors reviewed different categories of anomaly detection technologies and their assumptions as well as the computational complexity. Other surveys are focused on specific methods or approaches such as statistical [3] and neural network methods [4].

Many anomaly detection, diagnostics, and prognostics models can be seen in various application areas. Medjaher et al [5] apply a mixture of Gaussian hidden Markov models in a dynamic Bayesian network for bearing wear prognostics and remaining useful life estimation. Mohanty et al. [6] propose a multivariate Gaussian process approach using online training for fatigue crack length estimation in aluminum specimens. Liu et al [7] propose an advanced autoregressive time series model with a nonlinear accelerated degradation factor for remaining useful life estimation of lithium-ion batteries. Sikorska J.Z. et al [8] described remaining useful life prediction models, including knowledge-based, life expectancy, artificial neural networks and physical models, and presented a classification table to assist model selection for various industries.

These and other publications can give a good perspective of the interest allocated to the health characterization and prognostics for safety and maintenance purposes, and the amount of research already conducted in this field. However, despite the progress to date, there are still many practical issues that have been insufficiently addressed. For example, a combination of signal features may appear to capture well the trend of the fault progression through a system's lifetime. However, there are cases when features extracted by well-known signal processing techniques may not be able to correlate well with the fault progression due to an atypical failure mode for example [9].

The research described in this paper approached different bearing failure cases from a data mining perspective. The varying conditions leading to the spindle/bearing failure created different responses of the degrading system which challenged the ability to determine a generally applicable predictive solution. This research is still in progress; however, several notable findings that were considered of potential interest for the Machinery Failure Prevention Technology community were selected for presentation and discussion.

Methods and Technical Solution:

The prognostic tool being developed can be essentially divided into two units as described below. Also, Figure 1 presents a block diagram of the prognostic tool.

A. Data Monitoring Unit

This unit involves monitoring the health of the spindle using different sensors to obtain a record of spindle health over its entire lifetime. This unit was set up at TechSolve to collect a total of five spindle health datasets. A description of the features of this unit, as well as the collected data, is described in the next section.

B. Prognostic Unit

This unit is based upon patterns discovered in the frequency domain representation of the signatures of a spindle's operation. This analysis helps identify those intervals of operation during which the energy inherent in some frequency ranges of the signatures significantly exceeds the levels of average operation. Episodes of disturbed and calm operation of the spindle are defined in terms of the energy characteristics of the signatures. A scoring scheme is developed based upon the features of these calm and disturbed periods of operation. These scores are then used to develop prognostic measures to issue multiple warnings to predict potential breakdown of the bearings.

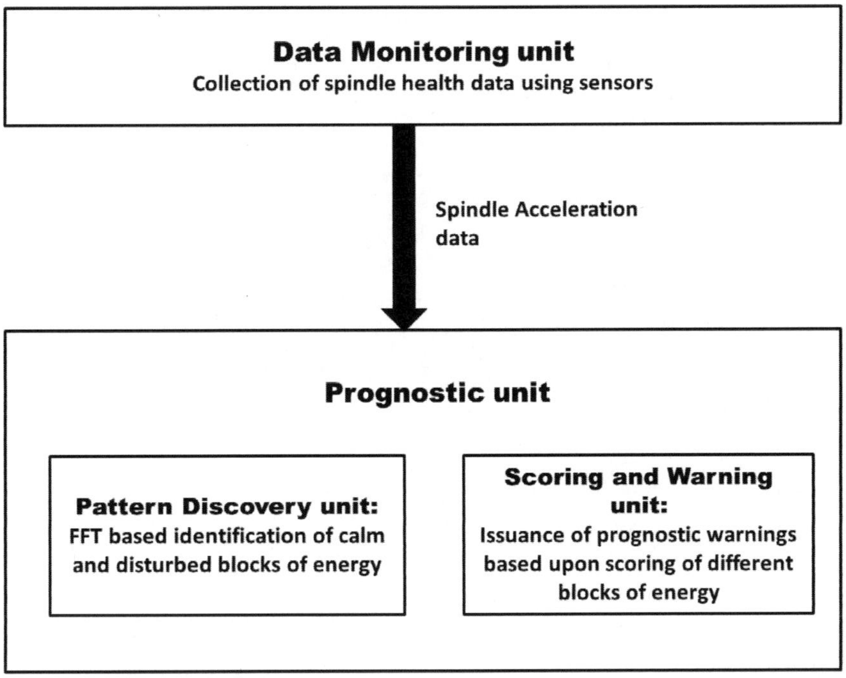

Figure 1. Block Diagram describing the prognostic tool

Data Monitoring Unit

The essential methodology used in the data monitoring unit is based upon monitoring the health of the spindle over its entire life with sensors to capture different features of the spindle operation. The essential properties of this unit are described below.

A. Spindle Test Bed

A spindle test-bed was built by TechSolve using a frequency drive, a motor, a poly-V belt transmission, and a simplified spindle using two bearings identical to the ones used in the horizontal machining center. Figure 2 presents a picture of the spindle test-bed showing the motor, the belt transmission, and the actual spindle. A loading mechanism pulling on the nose of the spindle was added to accelerate the degradation. The force pushing down on the spindle nose was kept approximately constant at 35 lbs throughout all tests. Figure 3 presents a view of the section of the loading mechanism located under the supporting stand. With the exception of one case (Case 4 below), the bearings were manufactured by NSK and can be identified by their code: 7014CSN24TRDULP4Y.

Figure 2: Spindle test-bed at TechSolve

Figure 3. Bottom view of the loading mechanism

B. Data Acquisition Strategy

The spindle test-bed was instrumented with a number of sensors. A uniaxial accelerometer type 607A11 (manufactured by PCB Piezotronics) was placed on the spindle housing, on top of the back bearing. A thermocouple was inserted in a hole drilled into the spindle housing, close to the back bearing's outer surface. The amperage drawn at the motor was monitored using a current sensor type SCT 050CX5 (manufactured by Ohio Semitronics, Inc.), and the temperature of the motor was captured using a thermocouple. The ambient temperature was also acquired using a thermocouple placed in the vicinity of the spindle test-bed. All thermocouples were J-type, Teflon covered for insulation (manufactured by Watlow). The data channels are schematically represented in Figure 4.

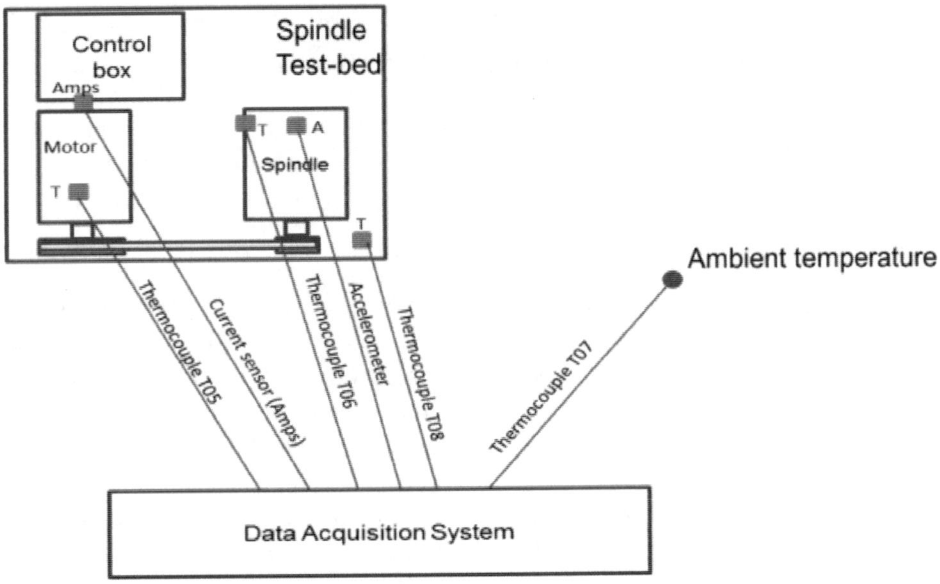

Figure 4. Data channels diagram for the spindle test-bed

C. Experimental Design

For all tests, the spindle was run at the same speed (approx. 9100 rpm), while a load of 35 lbs. was applied perpendicular on the nose of the spindle. For this paper, the following cases have been considered:

Case 1: Lightly greased bearing – the back bearing was greased with only 0.50 ml grease (versus the normal 5 ml); the test was accelerated after 20 days by wiping the races of the back bearing with a cotton swab dipped in isopropyl alcohol. The spindle was run for approximately eight hours every workday, for 25 days, with accelerated degradation induced for the last 5 days.

Case 2: Lightly rusted / greased bearing – the back bearing was first cleaned of its packaging oil; then the races were wiped with a cotton swab dipped in salted water. The bearing was left overnight to develop a very fine layer of rust on its races. Next day, the back bearing was greased with only 0.50 ml grease. The spindle was run for approximately eight hours every workday, for 25 days, with accelerated degradation induced after 21 days. The test was accelerated by wiping the races with cotton swabs dipped in isopropyl alcohol and then with another swab moist with salted water.

Case 3: Lightly greased bearing – only 0.25 ml of grease was added to the back bearing in this case and the spindle test-bed was run non-stop until failure. Similar to the previous cases, the front bearing had what was considered the normal amount of grease (5 ml). The spindle was run non-stop for approximately 5 months.

Case 4: This failure test was run with bearings Barden C114HCUL extracted from the original spindle of a machine tool. The bearings were re-greased with approx. 0.25 ml grease each and were re-installed in the spindle test-bed. The spindle was run non-stop until close to end of life (126 days). Because of the fire hazard concerns, for the last 3-4 days of its life, the spindle was stopped at the end of each workday.

Case 5: This case was a replica of Case 3. Same type of bearings and same quantities of grease were used. Aside from very brief interruptions due to power failures, the spindle was run non-stop to the end of life (spindle locked due to back bearing failure). The duration of the test was approximately 93 days.

For all cases, the back bearing had an induced anomaly (e.g., less lubrication, cleaning with cotton swabs) to create the conditions for accelerated wear. In all cases, it appeared that the cage was the element that had the most advanced wear -- the gradual degradation of the cage was considered to be the main cause for increased vibration and friction.

Prognostic Unit

The data collected using the Data Monitoring unit needs to be analyzed to make predictions about any changes in the health of the spindle. The Prognostic unit performs this task of data analysis in order to identify patterns and make predictions based upon these patterns. This unit is essentially composed of two subunits, A. Pattern Discovery Unit, and B. Scoring and Warning Unit. In the first pattern discovery subunit, a Fourier transform based analysis is performed on the spindle acceleration data in order to separate episodes of bursts of energy from the relatively calm periods of energy. The second scoring and warning subunit involves building a scoring scheme to rate or score different bursts of energy based upon their energy levels and hence their impact on the spindle health. Finally, these scores are used to develop formal prognostic measures to predict spindle failure time. Next, the two subunits are described in detail.

A. Pattern Discovery Subunit

The data collected consists of readings of spindle vibration acceleration, electrical current, and temperature of the spindle during its long-term operation. The spindle vibration acceleration measurements are used as input to the Pattern Discovery unit. A Fourier Transform-based analysis revealed that these sensor readings correlated the most with the health and the extent of deterioration of the bearings. An example is provided for spindle dataset 3 (Case 3) using FFT plots of spindle vibration acceleration. A contiguous period of about 0.4 seconds containing about 10,000 discrete readings of the acceleration was randomly selected. This signature is analyzed using an FFT function to determine the frequency components of the signature. Figure 5 shows the Discrete Fourier Transform (DFT) of an acceleration signature from the spindle during its beginning life (9/7/2011). Figure 6 shows the frequency distributions just before its failure (3/3/2012). The x-axis in these plots is the normalized frequency ranging from –pi/2 to +pi/2. The absolute frequency can be obtained by multiplying the normalized frequency by the sampling rate (25,000 samples / second). The y-axis represents the amplitude of each frequency component as it exists in the underlying signatures. An investigation of a large number of signatures was performed to determine distinct regions within which significant changes in amplitude values are observed as the disturbances take place in a spindle's operation. The three distinct regions in the frequency range that appear to have interesting features are shown in Figure 5 below. These three frequency regions were decided upon as follows. The frequency range was divided into two, three, four, five, and six equally sized intervals and then the average entropy of the energy distributions (squares of amplitude values) across the multiple regions was measured. The entropy value was the smallest for three partitions, indicating that the maximum information about energy distribution can be obtained by dividing the range into three equal parts. Having divided the DFT plot into three zones, we can see that near the end of life of the spindle in Case 3, the energy concentrated in the frequency regions corresponding to zone-1 increases considerably comparing to its early life.

In the tests #3, #4, and #5 a signature was selected at regular interval of every few minutes for analysis and measured the energy in each of the three energy zones and also the total energy in all the three zones. An energy-burst event was defined to be formed by a contiguous set of signatures when: (i) any five contiguous signatures within the burst region have an average total energy higher than a set threshold, and (ii) the total width of the burst regions is more one hour of contiguous signatures. The analysis methodology processes signatures and looks for bursts of energy having structures as described above. Intuitively, a burst of energy corresponds to some disturbance episodes that are experienced by the spindle. Within each burst the total energy is measured based upon the maximum energy level of the signatures contained in it as well as the burst duration. The burst energy calculation is described in detail in the next section.

In general, the analysis conducted for all the test cases shows that throughout the life of the spindle, episodes of bursts of energy are observed with monotonically increasing intensity (amount of energy in the episode) and also the number of burst episodes increases as the time elapses. Figure 7 shows the observed bursts of energy along time by enclosing them within red rectangles. In the beginning, when the spindle is healthy, there are fewer burst episodes and, towards the end, when the health of the spindle has deteriorated, there are more frequent and more intense episodes of energy bursts.

These bursts of energy, in all the tests, are especially pronounced as the health of a spindle deteriorates and it nears its end of life. Also, these bursts of energy were observed in all the three ranges of frequencies independently in the spindle acceleration signatures. A historical record of the observed bursts of energy and the energies and duration in each frequency range for each burst are recorded. These energy levels can help in setting the thresholds to determine when the calamitous bursts might occur that result in a stalled or broken spindle. Figure 7 shows the observed historical record of bursts of energy for dataset of Case 3. A similar pattern and behavior of burst episodes was observed consistently in each of the six datasets examined by us. The predictions about the imminence of spindle failure made based on the observed patterns of energy bursts have turned out to be correct in all the datasets.

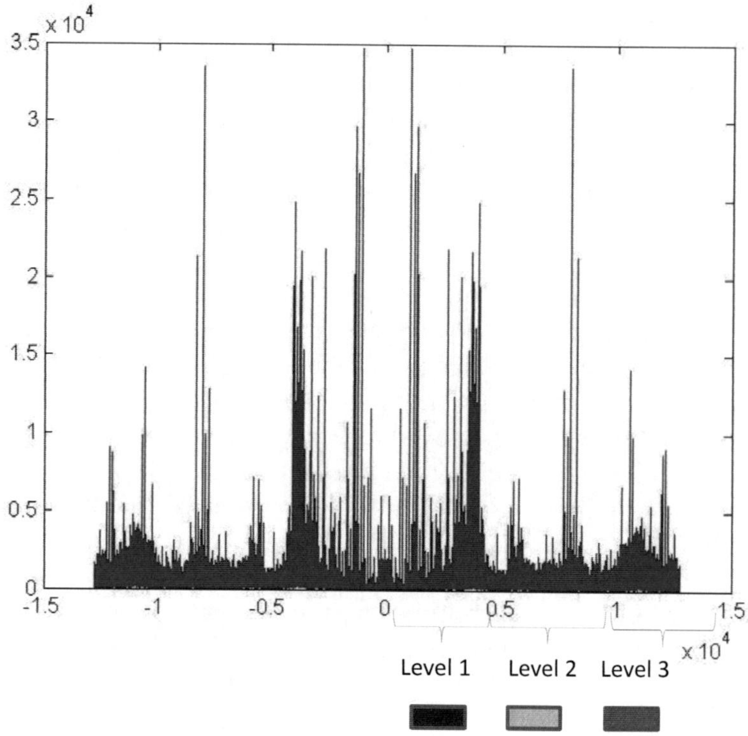

Figure 5. FFT for Spindle Vibration Acceleration values for spindle dataset 3 during its beginning life

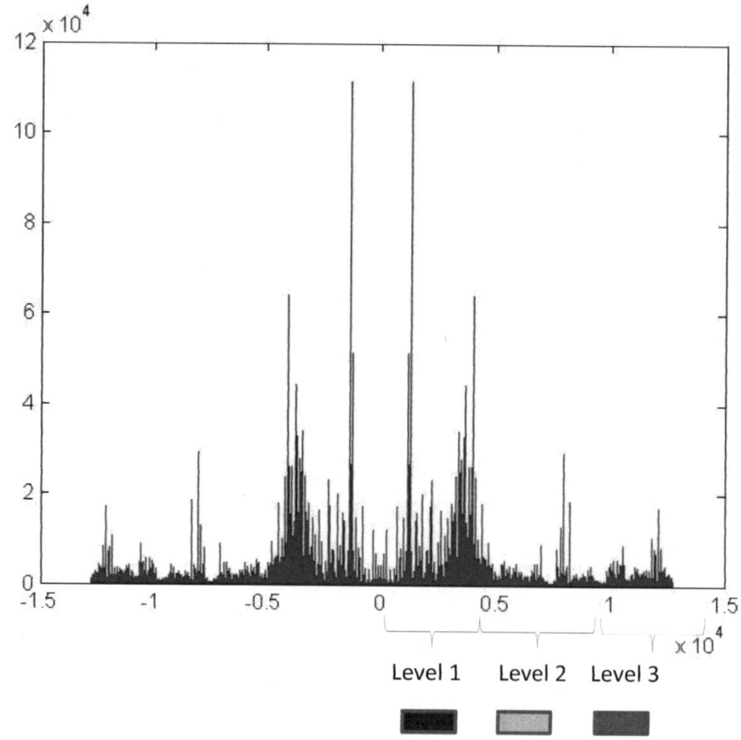

Figure 6. FFT for Spindle Vibration Acceleration values for spindle dataset 3 during its life just before failure

B. Scoring and Warning Subunit

The frequency analysis based assessment of spindles over their life helped to identify properties of energy-burst episodes that could be used to predict the remaining life of the spindles. A prognostic tool was built to study these different bursts of energy in order to make predictions about the impending failure of the spindle. More specifically, a scoring scheme was built to rank each burst of energy based upon its level of energy concentration and duration. Further, a warning issuance scheme was developed on top of this scoring scheme in order to deliver short-term and long-term warnings about the impending failure of the spindle. Next, a more detailed discussion of both the scoring, as well as the warning scheme, is provided.

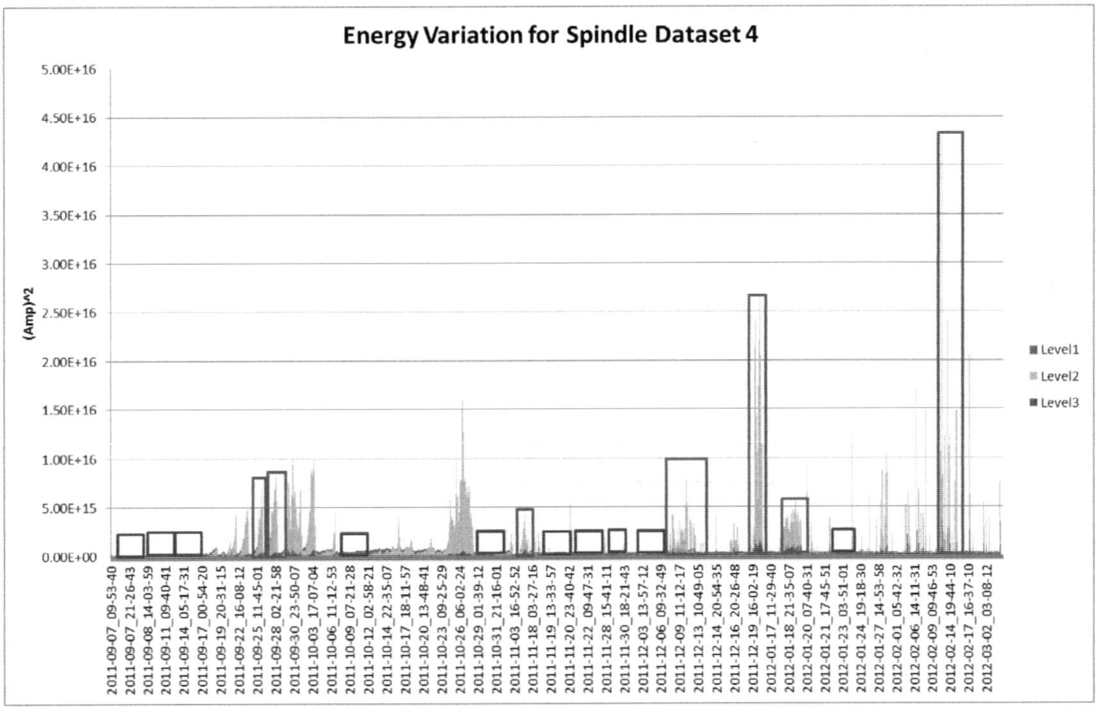

Figure 7. A date wise historical record of energy variation obtained for spindle dataset 3 with some bursts and calm regions highlighted by red and blue rectangles respectively

<u>Scoring Scheme</u>

The pattern discovery subunit involves plotting a historical record of energy variation in a spindle's functioning to identify bursts and calm periods of energy. The scoring scheme tries to analyze this energy variation in detail by studying contiguous signatures in order to separate the burst regions of energy from the calm periods. It assigns scores to different bursts of energy based upon their maximum energy level as well as their duration. The definition of a burst of energy is provided next.

A burst of energy is composed of contiguous FFT signatures such that:

- Number of contiguous signatures with a width of more than one hour and includes all such contiguous signatures

- Total energy of at least five signatures in the burst region is > θ1 and <= θ2. θ1 and θ2 are energy thresholds defined based upon the observed variation of energy levels in the spindle datasets. Further, based upon the study of the six spindle datasets, a minimum bound of 10^15 on the value of θ1 was set.

Different bursts of energy can vary in their intensity based upon their duration, as well as the energy of signatures contained in them. In order to quantify this intensity of each burst, the scoring scheme involves assigning a BurstScore to each individual burst of energy based upon the maximum energy signature contained in it and the time duration of the block. Further, separate BurstScore values are calculated for each level of energy. A Richter Scale-based scoring strategy is used to score all the burst regions. The intuition behind choosing such a scoring strategy for burst periods is that as the duration and intensity of the burst period increase, the more worsening effect it has on the health of the spindle. This in turn has a direct effect on the impending lifetime of the spindle. As a default case, all calm periods are assigned a score of 0, 0.5 or 1 depending upon their duration. This scoring criterion has been described in detail in Table 1.

Table 1. Description of scoring scheme for energy bursts observed in the spindle energy variation plots

Energy	Duration	Burst Score
<10^14	Any	0
>=10^14 & <10^15	<= 1 hour	0.5
	> 1 hour	1
>=10^15 & <10^16	<= 1 hour	1
	> 1 hour	2
>=10^16 & <10^17	<= 1 hour	2
	> 1 hour	4
>=10^17	<= 1 hour	4
	> 1 hour	8

The burst scores are further summed up to provide an idea of the history of spindle health. Two types of history scores are defined below.

Short-term history score (STH score) =

$$STH\ score = Burst\ Score\ of\ the\ current\ burst\ of\ energy$$

Long-term history score (LTH score) =

$$LTH\ Score = \sum Burst\ score\ for\ each\ burst\ before\ the\ current\ signature + Burst\ score\ of\ the\ current\ burst$$

The STH score is used to provide an assessment of the current status of the spindle, without any consideration for previous Burstscore values. On the other hand, the LTH score uses all the Burstscore values assigned to bursts prior to the current burst to provide an assessment of the current health of the spindle. The intuition behind using prior Burstscore values in calculating LTH score is that any wear and tear in the spindle bearings over a period of time leads to the ultimate failure of the spindle in the end. Therefore, taking into account the effect of prior Burstscore values becomes very important for making any prognostics about spindle health. Further, it is to be noted that separate history scores are calculated for each level of energy.

Warning Scheme

The short and long-term history scores calculated by the scoring scheme are used to give warnings about the impending health failure of the spindle. Two types of warnings are issued, the first ones are based upon the STH score, and the second ones are based upon the LTH scores. A detailed description of these two types of warnings is provided below.

Warnings based upon STH score

These types of warnings are issued when the short-term history score for any energy level becomes >= 4. These types of warning are helpful in the cases where the history data for the functioning of the spindle may be unavailable.

Warnings based upon LTH score

Three warnings are based upon the LTH scores as below:

- Warning 1: Issue Warning 1 when:
  LTH Score > 50 for any one level of energy
- Warning 2: Issue Warning 2 when:
  LTH Score > 50 for any two levels of energy
- Warning 3: Issue Warning 3 when:
  LTH Score >50 for all three levels of energy

The above three warnings provide an assessment of the gradual decline in the spindle's health. Figures 8 and 9 show the trend of how long-term and short-term history scores vary over time for spindle dataset 4. The three long- term and short-term warnings are marked on the graphs.

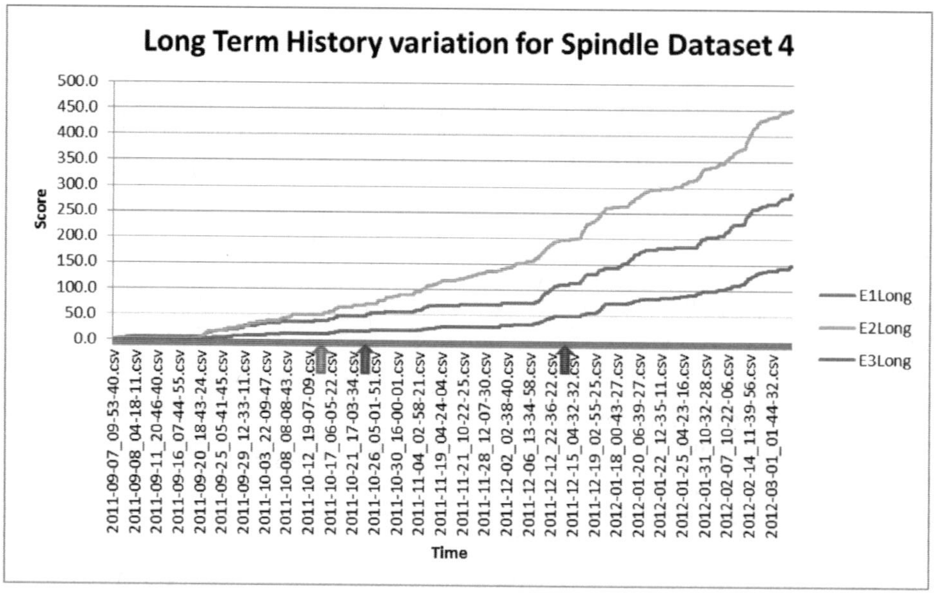

Figure 8. A date wise variation of long-term history scores for spindle dataset 3. The green, blue, and red arrows show the 1st, 2nd and 3rd warnings based upon the long-term history scores.

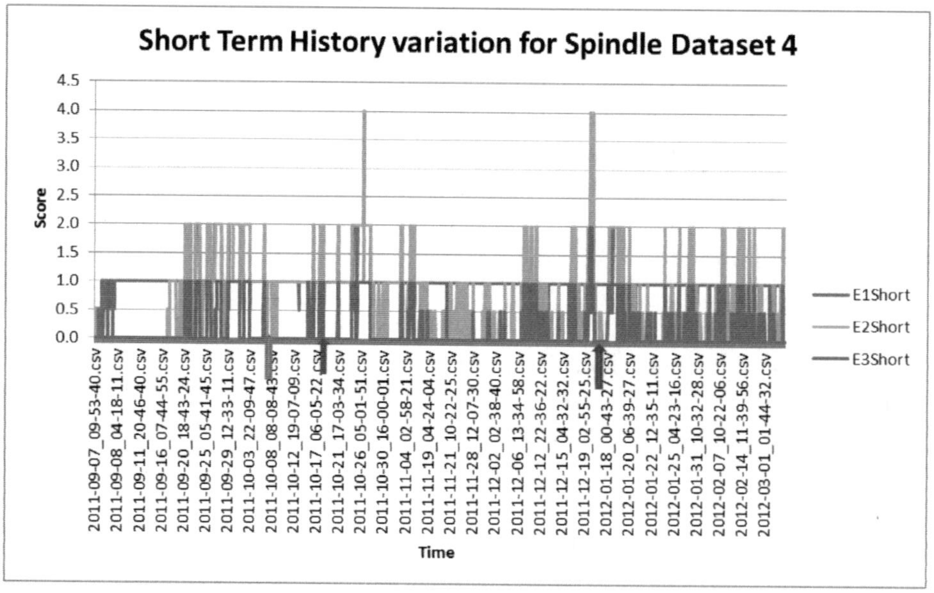

Figure 9. A date wise variation of short term history scores for spindle dataset 3. The green, blue and red arrows show the 1st, 2nd and 3rd warnings based upon the long-term history scores superimposed on this graph.

Experimental Evaluation:

In this section, the graphs obtained for Spindle energy variation, as well as the long-term history variation for spindle datasets 1 to 5, are presented. For spindle datasets 1 and 2, the system generates no warnings as both these experiments involved artificially inducing failure of the spindle using alcohol. Spindle datasets 3 to 5 were generated by letting the spindle run unto failure. The warnings issued for these datasets are marked on the graphs.

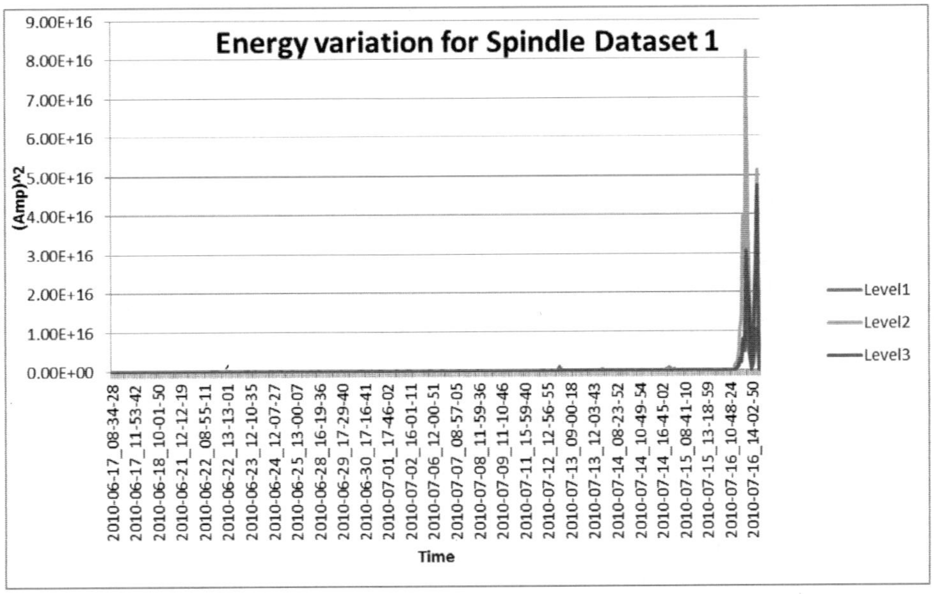

Figure 10. A date wise historical record of energy variation obtained for spindle dataset 1. No warnings were issued for this dataset

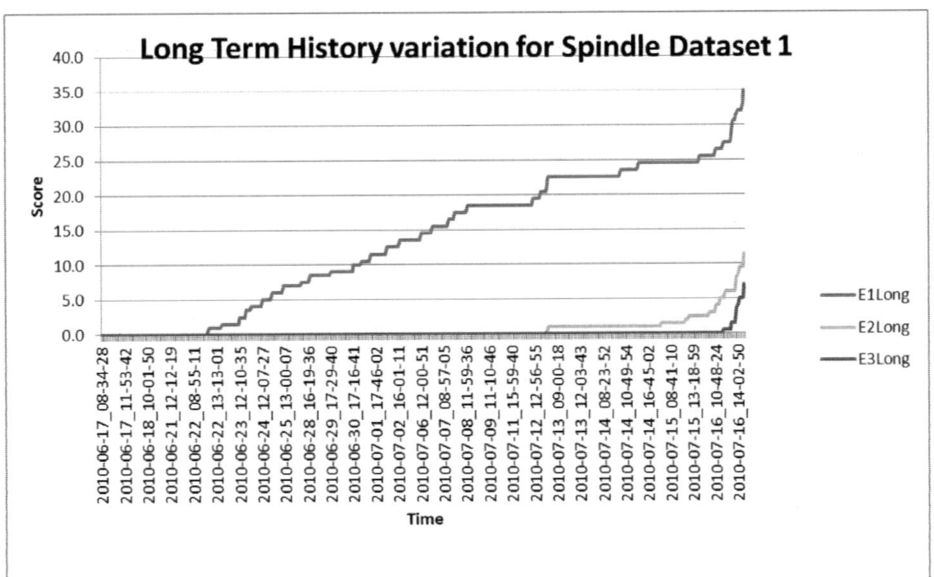

Figure 11. A date wise variation of long-term history scores for spindle dataset 1. No warnings were issued for this dataset as the long-term history scores remain below 50.

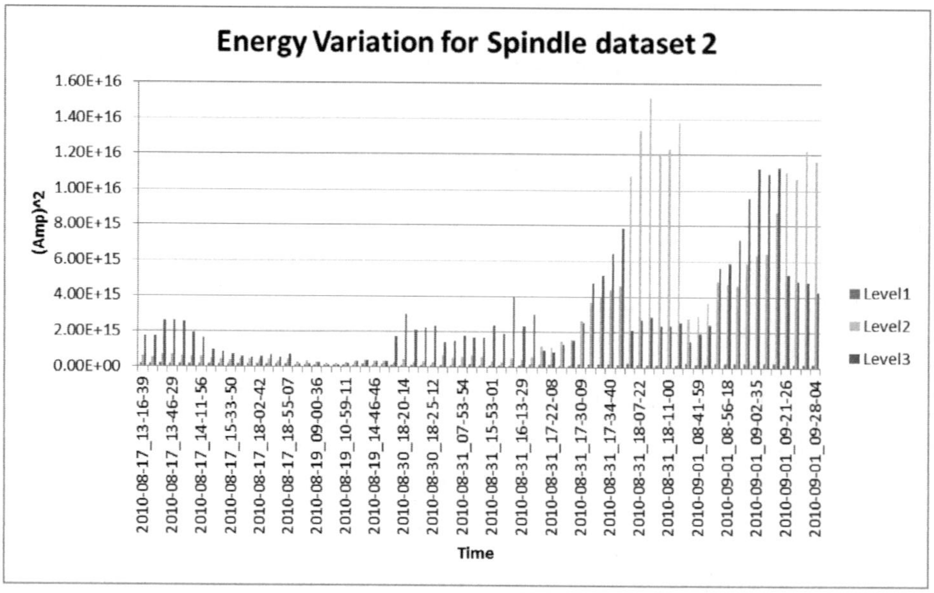

Figure 12. A date wise historical record of energy variation obtained for spindle dataset 2. No warnings were issued for this dataset

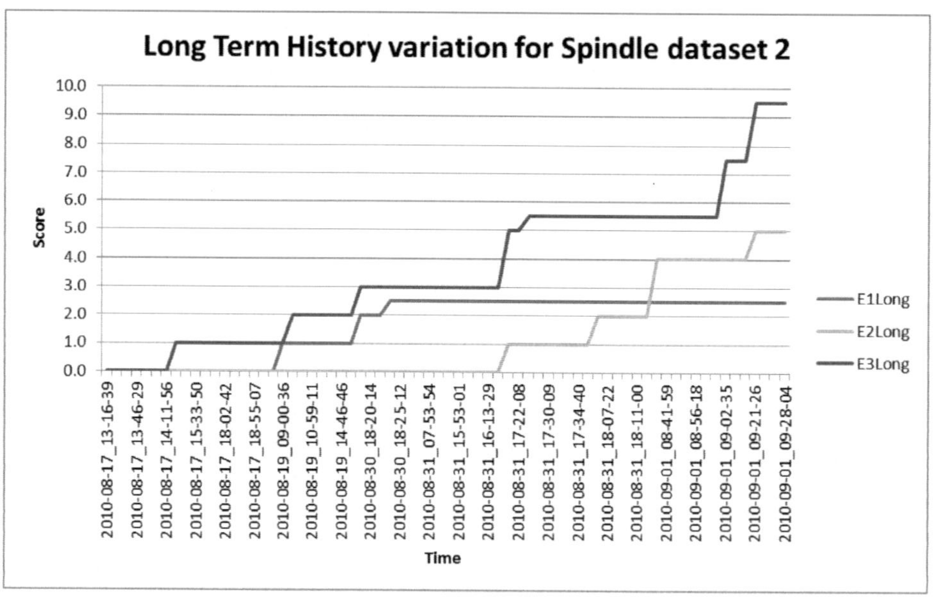

Figure 13. A date wise variation of long-term history scores for spindle dataset 2. No warnings were issued for this dataset as the long-term history scores remain below 50.

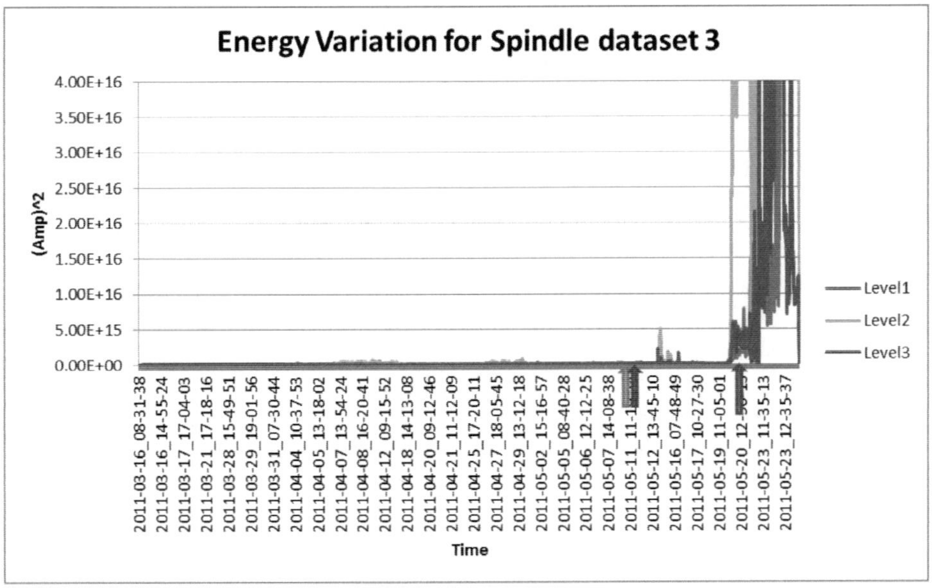

Figure 14. A date wise variation of long-term history scores for spindle dataset 3. The green, blue, and red arrows show the 1st, 2nd and 3rd warnings respectively issued by the tool based upon the long-term history scores.

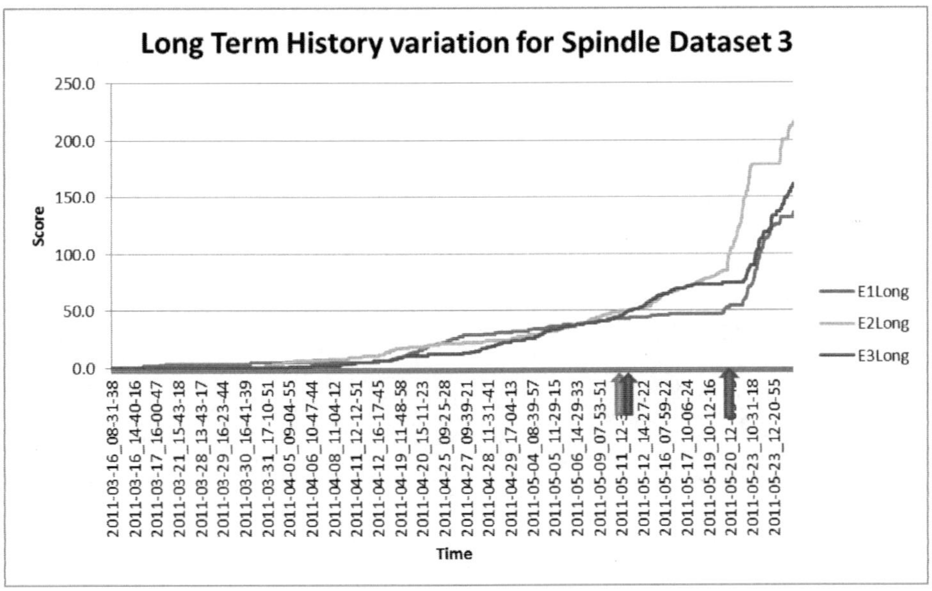

Figure 15. A date wise variation of long-term history scores for spindle dataset 3. The green, red, and blue arrows show the 1st, 2nd and 3rd warnings respectively issued by the tool based upon the long-term history scores.

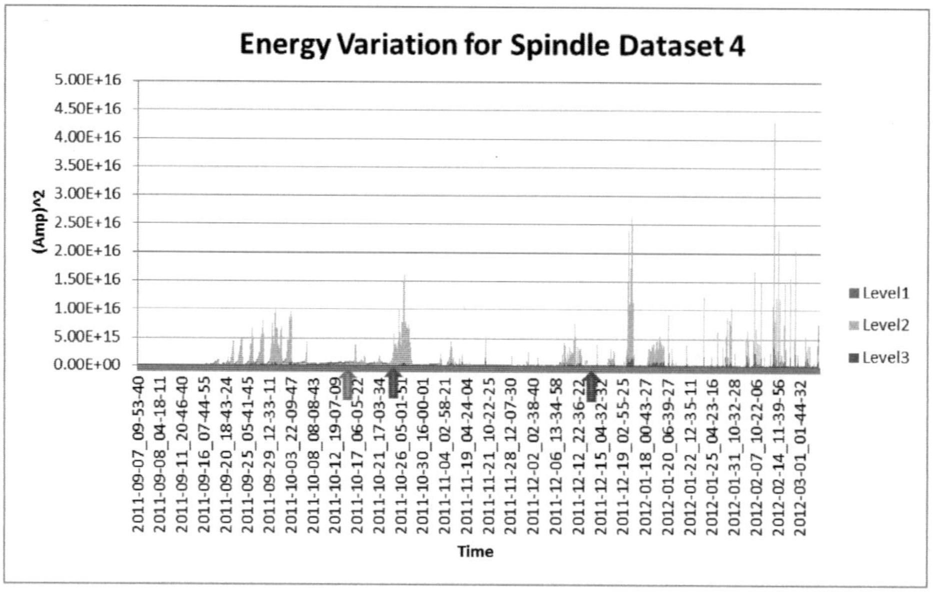

Figure 16. A date wise variation of long-term history scores for spindle dataset 4. The green, blue, and red arrows show the 1st, 2nd and 3rd warnings respectively issued by the tool based upon the long-term history scores.

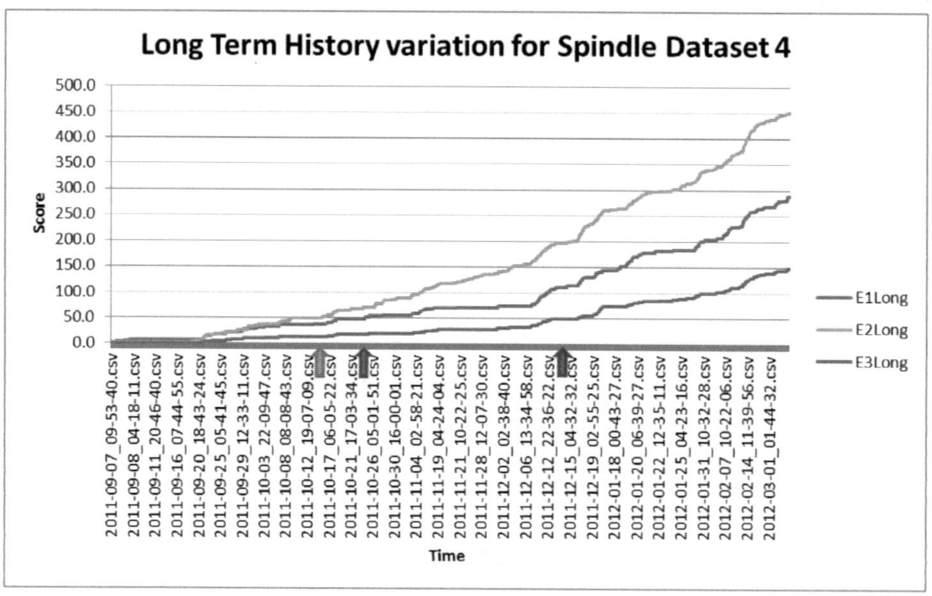

Figure 17. A date wise variation of long-term history scores for spindle dataset 4. The green, blue, and red arrows show the 1st, 2nd and 3rd warnings respectively issued by the tool based upon the long-term history scores.

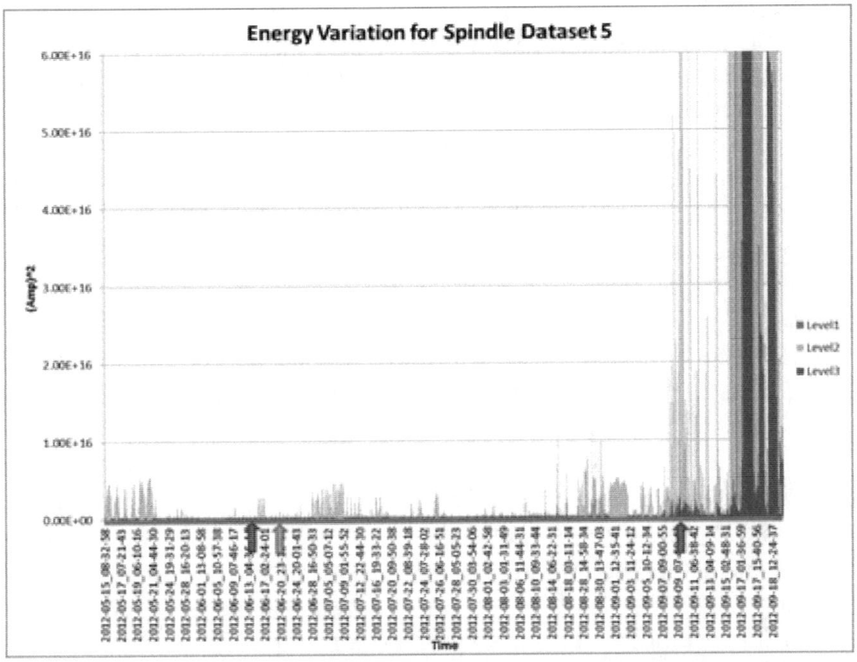

Figure 18. A date wise variation of long-term history scores for spindle dataset 5. The red, green, and blue arrows show the 1st, 2nd and 3rd warnings respectively issued by the tool based upon the long-term history scores.

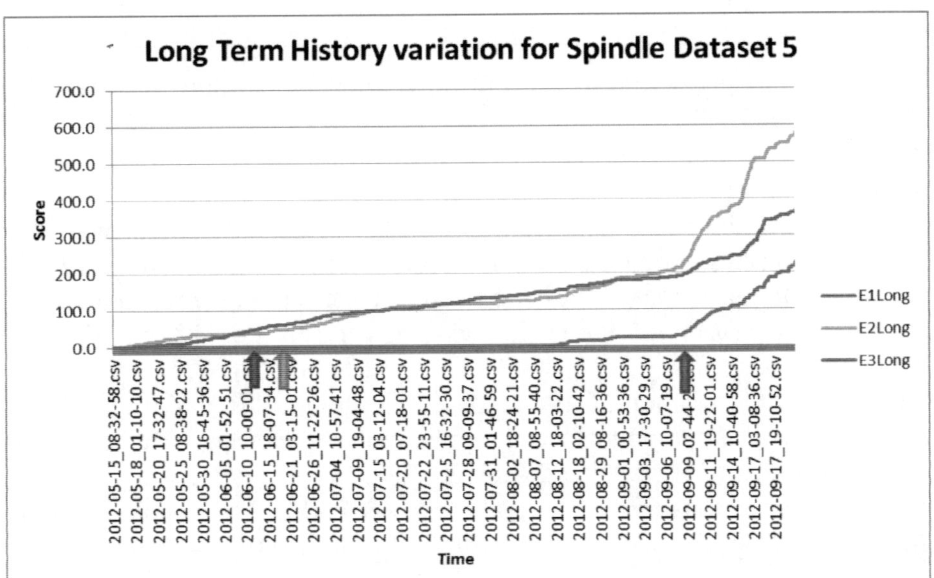

Figure 19. A date wise variation of long-term history scores for spindle dataset 5. The red, green, and blue arrows show the 1st, 2nd and 3rd warnings respectively issued by the tool based upon the long-term history scores.

Significance and Impact:

A purely data-driven approach for understanding the characteristics of a spindle as it goes through its slow aging process and deterioration in its operational signatures was presented. A detailed frequency domain analysis of the monitored data has revealed that numerous bursts of energy occur in the acceleration of spindle and the intensity of these bursts is a very strong indicator of the health of the spindle. Predictions about the fatal crash of the bearings can be made based on the intensity of the energy bursts. This has been demonstrated by the test cases presented in the preceding discussion.

Acknowledgment

The authors would like to thank SETCO, from Cincinnati, Ohio, the spindle manufacturing company, for support with spindle test-bed maintenance and bearing life tests related discussions.

Part of the run-to-failure tests conducted at TechSolve and by TechSolve representatives was sponsored by the U.S. Army Benet Laboratories and was accomplished under Cooperative Agreement Number W15QKN-06-2-0100. The views and conclusions contained in this document are those of the authors and should not be interpreted as representing the official policies, either expressed or implied, of U.S. Army Benet Laboratories or the U.S. Government. The U.S. Government is authorized to reproduce and distribute reprints for Government purposes notwithstanding any copyright notation heron.

References:

[1] J Lee, J Ni, Smart prediction to prevent downtime, Innovation Magazine Vol. 4 No. 3 (2004), pp. 41-42.

[2] V. Chandola, A. Banerjee, and V. Kumar, "Anomaly Detection: A Survey," ACM Computing Surveys, vol. 41, no. 3, July 2009.

[3] Markos Markou and Sameer Singh, "Novelty detection: a review - part 1: Statistical approaches," Signal Processing, vol. 83, no. 12, pp. 2481-2497, 2003.

[4] Markos Markou and Sameer Singh, "Novelty detection: a review - part 2: neural network based approaches," Signal Processing, vol. 83, no. 12, pp. 2499-2521, 2003.

[5] K. Medjaher, D.A. Tobon-Mejia, and N. Zerhouni. "Remaining useful life estimation of critical components with application to bearings," Reliability, IEEE Transactions on, vol. 61(2), pp. 292-302, 2012.

[6] S. Mohanty, A. Chattopadhyay, P. Peralta, S. Das, and C. Willhauck, "Fatigue life prediction using ultivariate Gaussian process," Technical report, DTIC Document, 2008.

[7] D. Liu, Y. Luo, Y. Peng, X. Peng, and M. Pecht, "Lithium-ion battery remaining useful life estimation based on nonlinear ar model combined with degradation feature," in Annual Conference of the PHM Society 2012, vol. 3, 2012.

[8] J.Z. Sikorska, M. Hodkiewicz, and L. Ma, "Prognostic modeling options for remaining useful life estimation by industry," Mechanical Systems and Signal Processing, vol. 25, pp. 1803-1836, 2011.

[9] L. Liao and R. Pavel, "Machinery Time to Failure Prediction – Case Study and Lesson Learned for a Spindle Bearing Application", Prognostics and Health Management (PHM), 2013 IEEE Conference, 24-27 June 2013, Gaithersburg, MD (print ISBN 978-1-4673-5722-7).

Bibliography:

[1] Goh, K.M., Tjahjono, B., Baines, T. & Subramaniam, S., "A review of research in manufacturing prognostics", IEEE International Conference on Industrial Informatics, 2006, pages 417-422.

[2] A Heng, S Zhang, A C C Tan, J Mathew, "Rotating Machinery prognostics: State of the art, challenges and opportunities", Mechanical Systems and Signal Processing, 23, pp. 724-739, 2009.

[3] Jardine, A.K.S., Lin, D. & Banjevic, D., "A review on machinery diagnostics and prognostics implementing condition-based maintenance", Mechanical Systems and Signal Processing 20, 2006, pages 1483-1510.

[4] Kothamasu, R., Huang, S. & VerDuin, W., "System health monitoring and prognostics – a review of current paradigms and practices", International Journal of Advanced Manufacturing Technology 28, 2006, pages 1012-1024.

[5] Lee, J., Ni, J., Djurdjanovic, D., Qui, H. & Liao, H., "Intelligent prognostic tools and e-maintenance", Computers in Industry 57, 2006, pages 476-489.

[6] Liao, L. and Pavel, R., "Machine Anomaly Detection and Diagnosis Incorporating Operational Data Incorporating Operational Data Applied to Feed Axis Health Monitoring", Proceedings of ASME 2011 International Manufacturing Science and Engineering Conference, Corvallis, OR, USA, 2011.

[7] K P Logan, K.P., "Prognostic Software Agents for Machinery Health Monitoring", IEEE 2003, IEEEAC paper # 1311, page 1 to 13.

[9] Luo, J., Namburu, M., Pattipati, K., Qiao, L., Kawamoto, M. & Chiqusa, S., Model-based prognostic techniques, IEEE Systems Readiness Technology Conference Proceedings, 2003, pages 330-340.

[10] Pavel, R., Snyder, J., Frankle, N., Key, G., Miller, L., "Machine Tool Health Monitoring using Prognostic Health Monitoring Software", MFPT Conference, Huntsville, Alabama, April 13-15, 2010.

[11] Schwabacher, M., "A survey of data-driven prognostics, American Institute of Aeronautics and Astronautics", Infotech@Aerospace, 26 - 29 September 2005, Arlington, Virginia, AIAA 2005-7002.

[12] S Uckun, K Goebel, P J F Lucas, "Standardizing Research Methods for Prognostics", 2008 International Conference on Prognostics and Health Management, 10 pages.

[13] Umeda, Y., Tomiyama, T. & Yoshikawa, H., "A design methodology for a self-maintenance machine", IEEE First International Conference on Intelligent Systems Engineering 360, 1992, pages 179-184.

[14] Y. Li, T. R. Kurfess, S. Y. Liang, "Stochastic prognostics for rolling element bearings," Mechanical Systems and Signal Processing, vol. 14, no. 5, pp. 747-762, 2000.

HEALTH MONITORING AND PROGNOSTIC ASSESSMENT IN A FLEET CONTEXT

Alexandre Voisin[1], Gabriela Medina-Oliva[2],
Maxime Monnin[2], Jean-Baptiste Leger[2], Benoit Iung[1]
[1] Centre de Recherche en Automatique de Nancy (CRAN), Université de Lorraine, UMR 7039 CNRS-UHP-INPL, Faculté des Sciences-1er Cycle - BP239, 54506 Vandœuvre-Lès-Nancy Cedex – France
alexandre.voisin@univ-lorraine.fr
benoit.iung@univ-lorraine.fr

[2] PREDICT 19, Avenue de la Forêt de Haye, CS 10508, 54519 Vandœuvre-Lès-Nancy, FRANCE
gabriela.medina-oliva@predict.fr
maxime.monnin@predict.fr
jean-baptiste.leger@predict.fr

Abstract: Prognostics aims at estimating the remaining useful life in order to plan a maintenance action before unit performances are affected. However, such goal is hard to reach since many parameters affect system's behaviour. Many approaches have been proposed to performed prognostics each of them with strength and weakness. In the present paper we propose to follow historical based prognostics together with the notion of fleet. The originality of the paper lies in the use of "similar" system historical instead of identical system historical as usually done in such approaches. In the present paper a fleet is composed of heterogeneous units (mainly components but could be systems or sub-systems) that are grouped together considering some similarities. Hence, the fleet can provide capitalized data and information coming from other members of the fleet for the prognostics. In order to achieve such a goal within a fleet-wide dimension, it is thus necessary to manage relevant knowledge arising from the fleet taking into account heterogeneities and similarities amongst components, operational context, behaviours, etc. This paper will focus mainly in the formalization of a data-driven prognostic model considering a fleet-wide approach. The model is based on a prognostic approach of the system health using Relevant Vector Machine. The proposed model is based on historical data coming from similar units of a fleet. The heterogeneity of the monitored data is treated by assessing a global health index of the units. The proposed approach is shown on a case study in the marine domain.

Key words: health assessment, fleet-wide, ontology, prognostics.

1. INTRODUCTION

PHM involves the following processes: monitoring the process variables of a current situation, assessing the health of the system, prognosticating the Remaining Useful Life

(RUL) of the system and making decision for maintenance action. In that sense, the data coming from the different variables of the process is also used for evaluating and monitoring a global indicator representative of the health state of a unit. The health state allows to supervise easily the degradation behaviour and to detect early enough drifts in operations (Rizzolo et al., 2011). If the health state is not satisfying, then predictive diagnostic could be performed. This process allows to identify the causes of a degradation before a failure occurs. Based on the potential degradation modes producing a drift in operation, the representative variables of this degraded component are used to predict the degradation trajectory and to assess the remaining time to reach a threshold, for instance a specified performance loss level. This time is called Remaining Useful Life. Finally the results are used for decision-making in order to select the maintenance actions to be performed in order to solve the drift.

Finally, prognostic requires some specific model for each degradation, each operational condition, and each material part. Such number of dimension implies that the efforts (according to the type of model, number of data, laboratory tests...) needed for the definition of the model have to be important. Moreover, prognostic deals with the estimation of the future and thus uncertainty appears. However uncertainty could be reduced when more efforts are made (Pecht, 2010).

However to improve PHM processes for large and complex systems such as power plants, ships and aircrafts, one possible approach is to take advantage of the fleet dimension. This dimension can provide knowledge and data to improve prognostic by sharing relevant data/information arising from several system (Medina-Oliva et al., 2013).

A fleet shall be viewed as a set of systems, sub-systems and equipment. In this paper, the naval domain is addressed but the proposed approach can be broadened to other domains. Hence, in the following a unit of a fleet will be considered as a system (e.g. ship), a sub-system (e.g. propulsion or electric power generation) or equipment (e.g. diesel engine, shaft...) depending on the nature of the study. To be in accordance with the need of improving prognostics at the fleet level, an original methodology is proposed in this paper wherein individual knowledge (of each unit) is capitalized for reuse purpose in order to improve prognostics. To take advantage of the individual knowledge at the fleet level, a semantic model is used for the PHM activities in the naval domain. Such a semantic model enables to reuse particular data, such as maintenance history, reliability analysis, failure analysis, data analysis at a fleet level in order to elicit knowledge from each individual database. As data become available, prognostic process could benefit from more contextual information.

Authors have already proposed a predictive diagnostic and prognostics fleet wide approach. The present paper focuses more specifically on prognostics, on the software/interface developed for the application and on the potential use of the prognostics results.

The paper is organised in 6 sections. It starts with the problem statement in section 2. Then in section 3, a bibliographical overview of the PHM with a fleet approach is presented. Next in section 4, our proposition of fleet-wide prognostic is presented. This proposition is applied to a case study in section 5. And finally the paper finish with conclusions and perspectives.

2. PROBLEM STATEMENT

Prognostic process aims at determining the Remaining Useful Life (RUL) of a unit on which a degradation is running. Some literature review, such as (Byington et al., 2002; Jardine et al., 2006; Heng et al., 2009), propose an overview of this domain and consider classification among the prognostic models. (Byington et al., 2002) propose the first classification and classify the prognostic model into three categories:

- Model based approach issued from physical laws of the degradation,
- Data based approach issued from data or indicator monitored on the system,
- Experience based approach mainly issued from reliability model.

In this paper the aim is to benefit from the knowledge acquired during the operation of every unit of the fleet, i.e. events that occurred and have been solved, in order to solve the event occurring on the present unit (Medina-Oliva et al., 2012b). The objective to benefit from the stored knowledge is subject to these constraints:

- Units are heterogeneous (e.g. technically, structure, mission, environment…) since in the naval domain every ship is highly customized. Moreover, even if units are of the same kind (same technical features), the mission they have to fulfil as well as the environment in which they are evolving have a significant impact on the degradation behaviour.

- Signals are heterogeneous. Indeed, signals are heterogeneous in two ways. The first one is for the same kind of unit, since they are evolving in different environment, with different mission… monitored signals show some significant variations. The second deals with the technical differences among units. In that sense, units could have different number of sensors since they are not technically identical. For instance, if engines have different number of cylinders hence, the monitoring of cylinder temperature means that the number of signals is also different.

Moreover, in order to benefit from the latest information, since units are on-line monitored the proposed approach aims at integrating all the available information as soon as it is available through its integration at the fleet level. Such integration can be performed almost "on-line" through communication channel such as satellite or with some delay through USB hard disk for instance.

The proposed approach is dedicated not to work as a single tool but together with some experts of the corresponding field. For instance, in (Medina-Oliva et al., 2013), the

authors show how experts can perform predictive diagnosis using the fleet knowledge. For this goal, the experts are using an iterative process in order to select a target sub-fleet that contains the proper information to solve the case under study (Medina-Oliva et al., 2012b).
Furthermore, one has to consider some constraint arising from the industrial context. Those constraints will help in the choice of well-fitted tools to support the fleet wide approach:

- The nature of the monitoring systems embedded in the ships. As for industrial systems, there exist several systems, which do not share a common conceptual data model (Umiliacchi et al., 2011).

- Monitored data is real data. The signal embeds a part of randomness due to, for instance, measurement noise, singular behaviour…

- Due to unique service life of every unit, there exist some heterogeneity between the measured signals. Hence, the sensor signals of a degradation processes can be captured by the probabilistic nature of the prognostic tool.

3. PHM AT THE FLEET LEVEL

For PHM activities, one of the industrial realities is the lack of capitalization of knowledge and model reuse which represents high costs and efforts for the enterprises (Weber et al., 2012) (Medina-Oliva et al., 2012a). In some fields such as the naval one, units are very customized leading to heterogeneous units. This fact limits mainly data and knowledge capitalization and exploitation.

To tackle this issue the fleet dimension can provide enough information and data to improve/perform PHM activities. In that sense, when searching non-identical but similar units a higher volume of data becomes available to reduce. However, most of the existing fleet-wide approaches treat identical units either for the definition of thresholds based on the data of the fleet (Patrick et al., 2010), technical solution capitalization (Reymonet et al., 2009) or RUL estimation based on a similarity-based approach (Wang et al., 2008). The fact of comparing similar units has rarely been addressed as a whole in the literature. To deal with this issue, this paper is based on a methodology that leads to search non-identical but similar units. To do it, knowledge about different and general characteristics of units was formalized within an ontology (Medina-Oliva et al., 2012b), (Monnin et al., 2011a). This knowledge allows to group heterogeneous units based on shared common characteristics that are relevant for a given situation. Indeed, an expert determines the criteria (i.e. characteristics) to be matched in an iterative process. These criteria depend on the partial knowledge of the current situation, the unit under study, the goal of the expertise (here, predictive diagnosis and prognostic), and the expert itself.

Regarding the prognostic techniques to be used, it has to be defined according the constraints previously defined. First, as the units are on-line monitored, time series data of either sensors or indicators are available for processing. Hence, data-based techniques are well fitted. A review of these techniques has been proposed by (Jardine et al., 2006). Among them, we chose techniques that process the past degradation time series. This choice has been guided since the definition of the unit population used to solve the current case is iterative, i.e. some units are dynamically removed or added. Hence the chosen prognostic model has to be able to integrate quickly new information, in our case time series, to perform its computation. Two example of such techniques are available in (Liu et al., 2007; Wang et al., 2008). (Lui et al., 2007) propose to compute match matrixes that are images of the fitting between the multidimensional time series, the current and the past one, for every past time series. Then, in every image, the best similarity indexes are selected and an Auto Regressive Moving Average model (ARMA) (Box & Jenkins, 1976) predicts the time to failure. Finally, the global RUL is computed by combining the ARMA models according to their degree of similarity. (Wang at al., 2008) propose to compute, first, a health index from the multi dimensional current time series in order to get a mono-dimensional time series. Then, they use Relevance Vector Machine (RVM) and Sparse Bayes Learning (SBL) techniques (Tipping, 2001) in order to synthetize the mono-dimensional signal in a few numbers of kernels. The online prediction process employs the background health information for the health prognostic using the Similarity Based Interpolation (SBI) technique. Moreover, (Wang et al., 2008) mention: "This framework also enables the continuous update of the background health information through offline Sparse Bayes Learning and continuous update of the prognostic results in real-time with new sensory signals through SBI… The SBL process can be carried out individually for different training unit which enables the background health knowledge to be built sequentially without complicated retraining process and updated as more offline training units are gradually available."

Secondly, the online update of the available knowledge can be satisfied. For that purpose, the operator computing the health index has to be determined for every unit in the fleet. Then, once a new event makes new knowledge available, the RVM learning through SBL make knowledge available for prognostic. The resulting information is very sparse and does not require too much space to be stored. Moreover, its use in the on-line phase does not require a huge amount of computing facilities.

Thirdly, since the considered units are heterogeneous, for instance they do not have the same number of cylinder, the monitored signals cannot be handle in the multi-dimensional signal space. Hence, the match matrix cannot handle this aspect. On the contrary, as (Wang et al., 2008) compute a mono-dimensional health index from the multidimensional time series, it is possible to compare the evolution of 2 health indexes even if the underlying units have different number of signals/indicators.

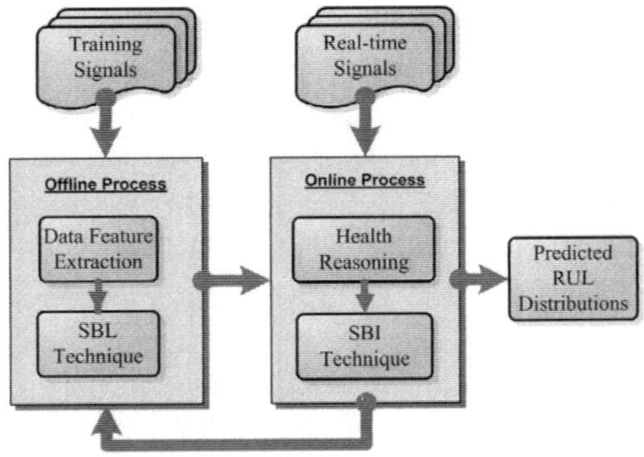

Figure 1. A generic framework for structural health prognostic (Wang et al., 2008)

4. FLEET-WIDE PROGNOSTICS APPROACH

The prognostic algorithm conforms to the one proposed by (Wang et al., 2008) using RVM prognostics on historical data. The fleet-wide dimension is provided through the selection of historical data within the fleet (Figure 2). Such elicitation process could be based for instance on experts knowledge or in relation with predictive diagnostic. In the present paper, an expert in interaction with the ontology performs the elicitation. In the following we detail the selection and the prognostics process.

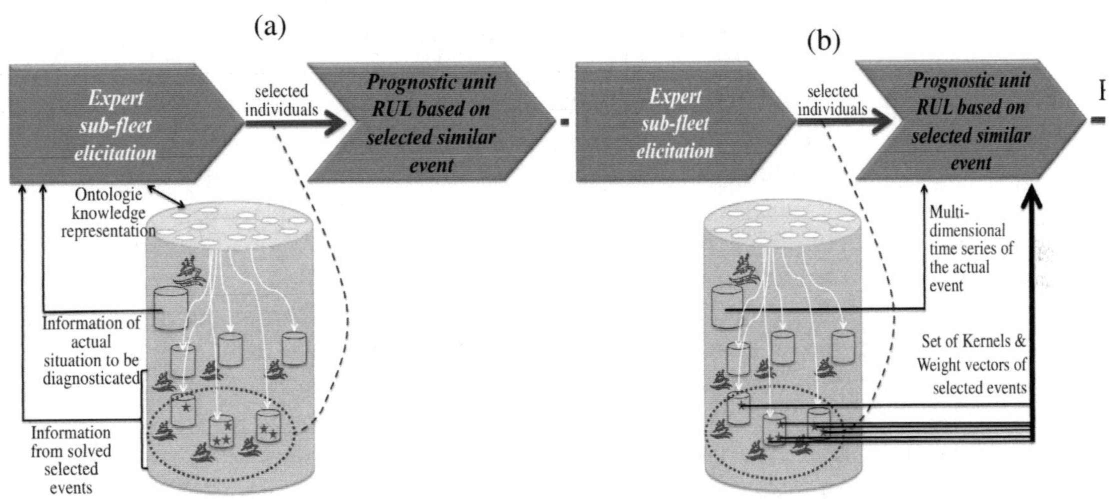

Figure 2. PHM fleet wide diagnostic (a) and prognostic (b) proposed approach process

4.1. Selection process

To facilitate the selection of the relevant information among the fleet, the objective of our proposition is to create an iterative investigation process that will allow the definition of a sub-fleet for prognostics purpose. The sub-fleet is defined by grouping a set of units (i.e. systems, sub-systems or equipment) based on "similar characteristics". Figure 3 presents the main steps of this process. Situation characterization consists in the description of the unit subject for which the prognostics has to be performed (e.g. type, age, usage, operating environment) as well as data on which analysis will be carried out (Figure 3-A). To guide this process, the fleet-wide application proposes different criteria based on the technical features of units as well as on the mission and on the environment description. According to this, a targeted population is defined within the whole fleet (Figure 3-B). The potential historical data from the selected units are investigated in order to decide to use them in the prognostic (Figure 3-C). In an iterative process of steps B and C. Targeted fleet can be refined (closed loop by restricting criteria) if results (i.e. potential similar events) are too far from the current situation. Conversely, targeted fleet can be extended (closed loop by loosening criteria) if it points out too few similar situations.

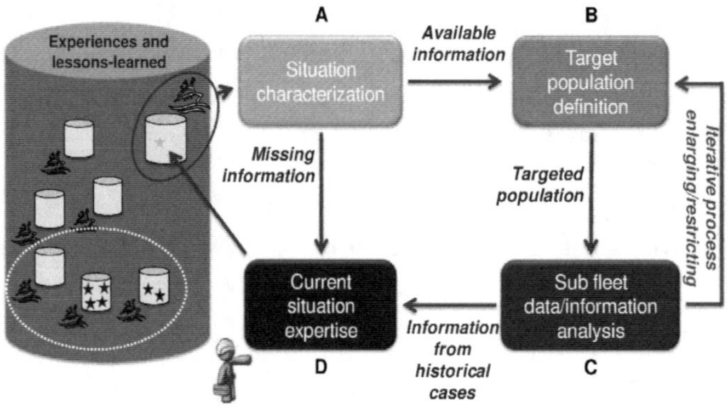

Figure 3. Main steps for fleet case re-use

Toward this goal, the knowledge corresponding to the fleet domain must be well formalized and structured in order to facilitate the manipulation of the multidimensional aspects of the fleet and the heterogeneous data among all fleet units' databases.

Nowadays, fleet managers/engineers query separately ship databases for identical units to obtain data/information. However, units (e.g. diesel engine) in the naval domain are very specific and customized. This fact leads to dispose of few identical units. For this reason the proposed methodology leads to search non-identical units but similar ones. Furthermore, another issue emerges. At a fleet level, engineers must treat different databases of different units. These databases might be heterogeneous, in the sense that the database structure might be different; they might have different names for the tables, the primary keys and foreign keys, etc. To tackle these issues, common semantic is used (Figure 4) (Medina-Oliva et al., 2012), (Monnin et al., 2011a). The semantic model provides a high level definition of terms that is common to all databases allowing to query them. It allows the definition of characteristics of similarities among units and contexts, for instance, by defining common characteristics in the technical, operational and contextual domains. The semantic model is specified by means of ontologies.

Key elements to perform diagnostic/prognostic such as technical characteristics of the system/sub-system/equipment, degradation modes, degradation indicators, the mission and the environment were integrated in the semantic model (i.e. ontology). In that sense the semantic model formalize these pieces of knowledge in order to define "similarities" among units.

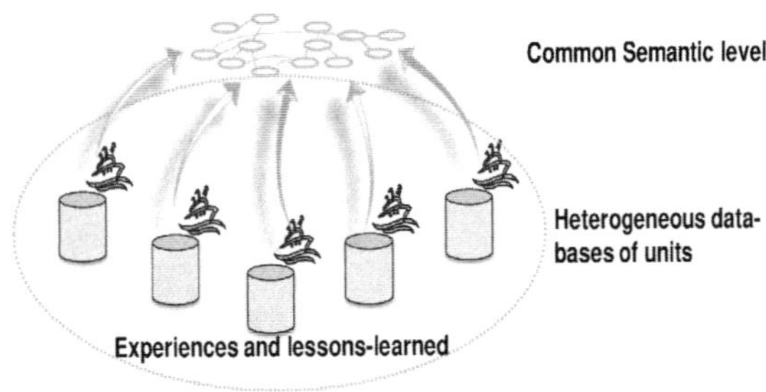

Figure 4. Semantic level to query heterogeneous databases

Based on (Medina-Oliva et al., 2012), (Monnin et al., 2011a) several contexts were defined: a technical context (e.g. characteristics of the system/sub-system/equipment), a dysfunctional context (e.g. degradation modes), an operational context (e.g. mission and environment), a service context (e.g. usage of units) and an application context (e.g. degradation indicators). Once the ontology model was formalized comparisons of heterogeneous units shall be performed on the basis of context similarity.

4.1.1. Ontology for providing semantic

An ontology determines formal specifications of knowledge in a domain by defining the terms (vocabulary) and relations among them (Gruber, 2009). Ontologies are composed of classes, properties of the classes and instances. These elements are explained as follow: "Classes" to describe concepts in the domain, "Properties" which contain in a class definition and describe relationships among the classes, and "Instances" which are the set of specific individuals of classes.

Ontologies seem to be a suitable modeling method to provide common semantic and to query heterogeneous databases. Some of the capabilities provided consist in (Noy and McGuinness, 2001): sharing common understanding of the structure of information among people or software agents, making domain assumptions explicit, defining concepts and knowledge and making domain inferences to obtain non-explicit knowledge.

To represent knowledge and to explicit semantic (vocabulary and relations), the ontology is coded in Web Ontology Language (OWL) supported by Protégé[1] ontology editor. In that sense, OWL embeds all the features previously mentioned. For instance, in our application, a taxonomy is used to represent systems, subsystems and equipments. In addition, the characteristics of languages such as OWL facilitate the definition of generic conceptualizations that can be used in multiple domains, facilitating the creation of Web-based applications, such as a module of an e-maintenance platform in the PHM domain (Garcia-Crespo et al., 2011).

Moreover, OWL provides inference capabilities with plugged reasoners. OWL reasoners shall performed consistency checking. Hence, one shall guarantee that the ontology has been built correctly in the sense that no syntactic error and inconsistency remain in the

1 http://protege.stanford.edu/

ontology. For example, if a fuel-engine is associated to an electrical degradation, an inconsistency will be point out by the reasoner. Moreover, explicit and manually constructed classes that belong to taxonomy constitute an asserted hierarchy. But thanks to OWL reasoners, an inferred hierarchy is automatically computed allowing to emerge new knowledge. For example if an engine has an internal electrical degradation, the ontology can induce that it is an electrical engine.

The proposed approach based on PHM ontology is not a goal itself. This ontology is a support for PHM software applications through the KASEM® software platform[2]. For the purpose of the underlying software application, the ontology model is integrated by means of an SQL-backed storage. The java framework JENA allows the manipulation of the ontology stored knowledge and then it uses this knowledge to search individuals (that correspond to the described criteria in the ontology) in the different and distributed databases. The JENA inference engine allows semantic queries and inference rules to be solved within the platform. The information issue from the ontology as well as the occurrences (e.g. individuals) found in the different databases are shown through the KASEM® platform via JENA (Leger, 2004), (Monnin, 2011c). KASEM® platform provides the user a web portal that allows benefiting of the fleet-wide knowledge capitalization. Relevant contextual information can be retrieved and gathered for the purpose of, for instance, failure anticipation, investigation or expertise sharing.

4.2. Fleet-wide ontology application

Based on the ontologies and these web-semantic applications, an ontology based application is presented. This application shows how the fleet-wide approach is useful for experts during the decision making process for PHM domain. It allows to query different and even heterogeneous databases in order to capitalize knowledge arising from the fleet. The application is composed of four main sections (Figure 5): taxonomy structured knowledge, linked knowledge, semantic query definition and the inferred statistics about the obtained population.

The taxonomy structured knowledge section contains an arrangement of the technical components (e.g. diesel engine, ship, propulsion installation) organized in the form of a hierarchical classification or taxonomy. General concepts are represented as classes and they are linked by a subsumption relation, i.e. "is a" relation, allowing a taxonomy to be defined. Hence, a hierarchy of classes is established, from general classes to specific ones (i.e. subclasses).

[2] http://www.predict.fr

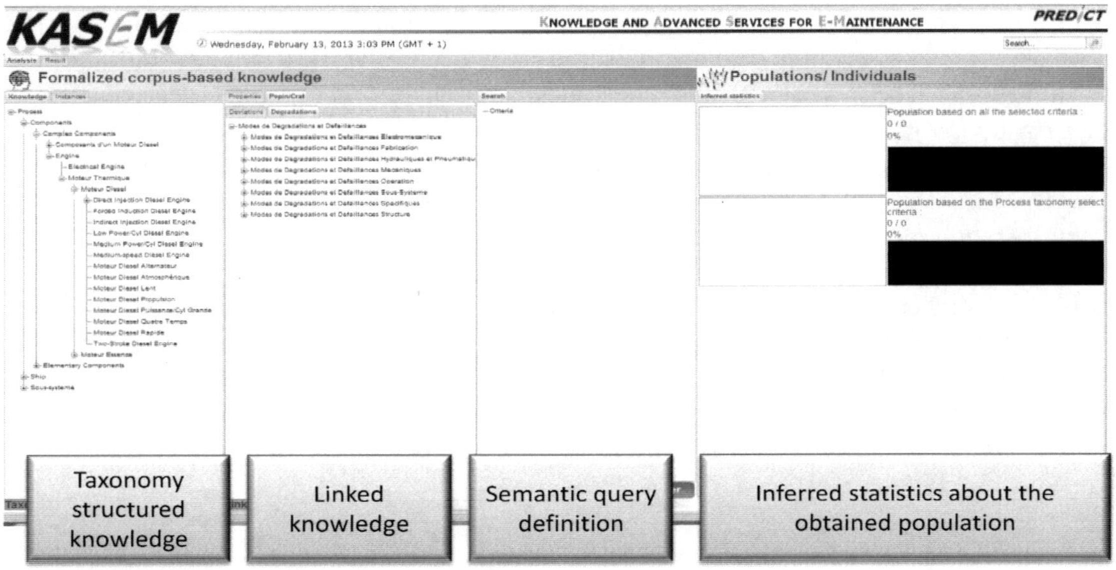

Figure 5. Fleet-wide predictive diagnostic software application

For instance, in this section, one can find the taxonomy of "Equipment" as illustrated in Figure 6. "Equipment" is a general concept or general class. It is then possible to find more specific concepts of equipments such as "Engine", "Pump", "Propeller", and so on. "Engine" has two subclasses, "Electrical Engine" and "Heat Engine". Both of these concepts are "Engines". Then "Heat Engine" has as well two subclasses "Diesel Engine" and "Gas Engine". This way knowledge is structured in order to obtain a classification of the different kind of equipments which allows the user to perform comparisons of heterogeneous units for instance when seeking common and relevant characteristics such as "4-strokes engines" or "propulsion engines" among a heterogeneous fleet.

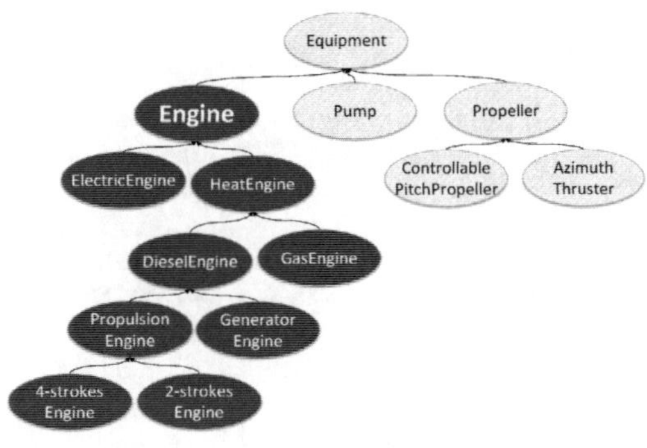

Figure 6. Part of the equipment taxonomy

The linked knowledge section encodes knowledge about the relations among the different concepts. For instance, an equipment has a relation or is linked to a concept degradation mode/deviation. These relations allow to guide the query of the user to search for example the causes or the associated data of the abnormal behavior. As a matter of fact, the user has the possibility to browse the associated knowledge with the equipment in order to define the objective of the query and to contextualize of the situation. For example, the user (e.g. maintenance engineer) might want to investigate the evolution of some indicators. Then, the ontology encodes knowledge about the degradation modes/deviations that are monitored with a given indicator (i.e. objective of the query). Indicators are also linked to units (units is a term used to denote equipments/installations and ships). Moreover, units are exposed to an environment and they are also exposed to different operation modes (i.e. contextualization of the situation).

The software application embeds this formalized knowledge, thus it could guide the expert to enable the identification of a degradation/deviation mode linked to an indicator and linked to one unit (e.g. component) with a given environment for example. This way, knowledge is encoded in the ontology and the expert can use it to see which criteria are more appropriated to choose in order to retrieve similar cases and data to perform fleet-wide prognostic (Figure 3-A (situation characterization)).

The semantic query section contains the set of criteria chosen by the user in order to search the individuals of the population (i.e. fleet) that correspond to the given criteria. It is called semantic query since it has the capabilities of querying within a semantic model (i.e. ontology). This kind of queries benefits of a reasoner engine. This means that even if an individual has not been directly specified, the semantic relations and the taxonomy model allow to find individuals even if they have not been labeled in the searched class. For instance, if the user selects the criteria "Diesel Engines" the query will find the individuals labeled as "Diesel Engines" as well as the individuals not labeled but present in their subclasses, "Propulsion Engine", "Generator Engine", and so on.

The last section, the inferred statistics about the obtained population, contains statistics about the "use" or "exploitation" of the past experiences found in the fleet. This information is useful for the iterative process allowing to refine/extend the population definition (Figure 3-between the steps B and C).

4.3. Prognostics process

4.3.1. Off line stage

The off line stage is composed (Figure 7) of (a) the determination of the aggregation function for health index and of (b) Sparse Bayes Learning (SBL) for health time series.

(a) Health Index computation

(Wang et al. 2008) use a linear data transformation matrix T such that:

$$T = (Q^T Q)^{-1} Q S_H \qquad (1)$$

Where, Q is composed of both faulty (degraded) and nominal multi-dimensional signals and SH is a {0,1} matrix corresponding to every element of Q according to its state, i.e. 0 for degraded and 1 for healthy. T is able to transform any set of multi-dimensional signal into a mono dimensional signal of health index.

For the purpose of our approach, T has to be determined for every unit, event type and operating condition. On one hand, the computation of the matrix is not time consuming neither required complex data selection (2 sets of data: normal and degraded). On the other hand, this job has to be performed for every unit and every event type since degradation signals changes according to these 2 features. Both of them are easily identifiable. Moreover, operational conditions mode are influencing degradation signals as well, but are more hardly identifiable. Hence, some work is required for such a purpose. Then, normal and degraded signal are extracted as well as operational mode identification, and T matrixes are computed.

(b) off training scheme with SBL

For the sake of conciseness, we do not present the SBL. For more details, one can refer to (Wang et al. 2008) or to the original paper of Tipping (2001). The SBL is a generalized linear model in a Bayesian form and it shares the same functional form of the Support Vector Machine (SVM). Tipping has formulated this generalized linear model in a Bayesian form, named the Relevant Vector Machine (RVM). It achieves comparable machine learning accuracy to the SVM but provides a full predictive distribution with substantially fewer kernel functions.

The RVM is a special case of a sparse linear model:

$$h(t) = \sum_{i=1}^{N} \omega_i \phi(t, t_i) + \varepsilon(t) \qquad (2)$$

where $\varepsilon(t)$ is the measurement noise, $\omega = [\omega_1, \omega_2, \cdots, \omega_N]$ a weight vector and basis functions are formed of kernel functions $\phi(t, t_i)$ centered at the training point t_i. The sparseness property enables the automatic selection of a proper kernel at each location by pruning all irrelevant kernels. A sparse weight prior distribution can be assigned, in such a way that a different variance parameter is assigned to each weight. Moreover, SBL allows to integrate the uncertainty contained in the health index time series by using the statistics of the coefficients ω of the RVs.

In the fleet repository, the degradation time series, associated to a solved event, are summarized by the Gaussian kernels and the weight vectors (mean and covariance matrix). It represents available knowledge for prognostic purpose. SBL is performed every time new event has been solved and a degradation time series has to be integrated.

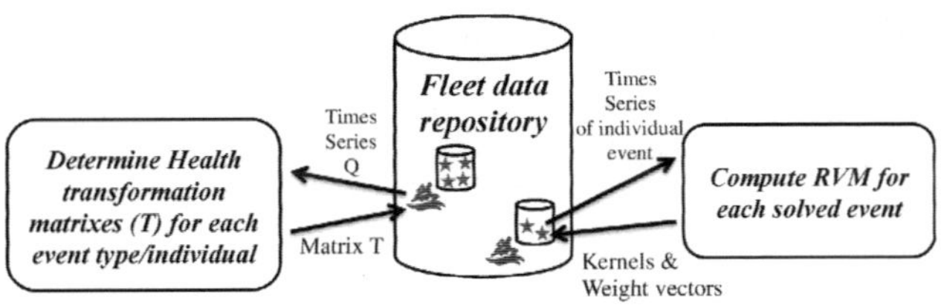

Figure 7. Off line stages of prognostics process

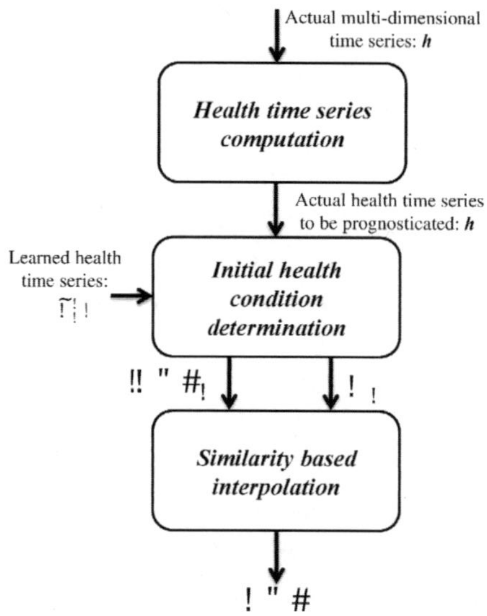

Figure 8. On line stages of prognostics process for one learned time series

4.3.2. On-line stage

The on line stage (Figure 8) is split into 3 steps: (a) Actual health time series computation, (b) initial health condition and (c) Similarity Based Interpolation

(a) Actual health time series computation

Based on the transformation matrix T (see eq. 1), the actual multi-dimensional time series is transformed into mono-dimension health time series h.

(b) Initial health condition determination

Lets consider $\tilde{h}$ a time series of health degradation from learned units with a length of $\tilde{l}$. $\tilde{h}$ and h the current one transformed using T with a length of l. Then, if both series represents the same degradation, first $l < \tilde{l}$ and second we supposed that h can be found in $\tilde{h}$. Indeed, as (Wang et al. 2008) explained h and $\tilde{h}$ may have different initial health index at the beginning of the time series due for instance to manufacturing variability or different service life. Hence the RUL estimation (see Figure 9), according to that single $\tilde{h}$, is:

$$RUL = \tilde{l} - l - T_0 \qquad (3)$$

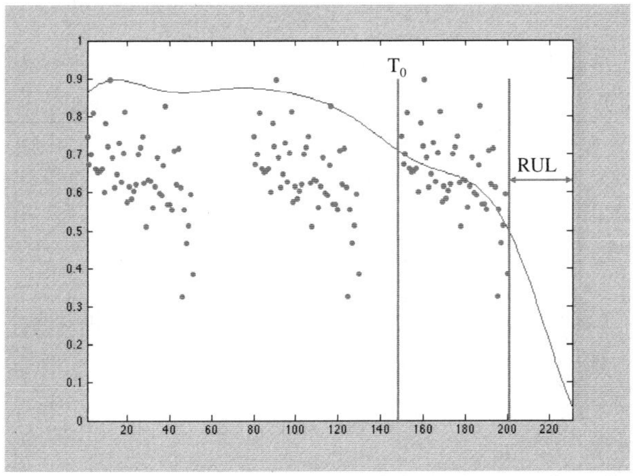

Figure 9. Initial health condition determination

with T_0 the initial time of matching between h and $\tilde{h}$ determined as:

$$T_0 = min_{T_0 \in [0, \tilde{l}-l]} \sum_{i=1}^{l} \left(h(t_i) - \tilde{h}(T_0 + t_i) \right)^2 \quad (4)$$

(c) Similarity based interpolation

Indeed, the computation of the RUL using step (a) is performed for every degradation time series selected using the fleet wide capability, i.e. $\tilde{h}^u$. Similarity based interpolation aims at combining these several RULs. Obviously, h does not match every $\tilde{h}^u$ with the same level of similarity. Hence, the combination of the RUL depends on the degree of matching. For a single unit u, its weight is issued from the matching step:

$$W_u = \left[\sum_{i=1}^{l} \left(h(t_i) - \tilde{h}(T_0^u + t_i) \right)^2 \right]^{-1} \quad (5)$$

The final RUL is computed as:

$$RUL = \frac{1}{W} \sum_u (W_u RUL_u) \quad where\ W = \sum_u W_u \quad (6)$$

4.3.3. Uncertainty management

As explained earlier, the RVM approach allows to capture uncertainty contained in the data by means of the vector ω and the associated covariance matrix. Hence, for every unit u, instead of having only one $\tilde{h}^u$, for instance the mean curve, one has several $\tilde{h}_k^u$ corresponding to random realization of the weight ω^u. Hence eqs. (3) and (4) have to be computed for all the random realizations of the weight ω^u.

Finally, eq. (6) is re-written as:

$$RUL = \frac{1}{W} \sum_u \left(\frac{1}{W_u} \sum_k (W_k^u RUL_k^u) \right)$$
$$where\ W = \sum_u W_u, and\ W_u = \sum_k W_k^u \quad (7)$$

5. CASE-STUDY

To illustrate the feasibility of the proposed approach as well as the added-value, a scenario was developed. This scenario shows how the fleet-wide approach is useful for prognostics purposes. The scenario is developed using an ontology-based fleet-wide software application (Medina-Oliva et al., 2012 & 2013) and KASEM® (Knowledge and Advanced Service for E-Monitoring) e-maintenance software platform.

We consider Volvo Penta D16 MG diesel engine suffering degradation on the poppet valve of the outtake gas to be prognosticate.

5.1. Events fleet-wide retrieval

The air-intake system of this machine is a turbocharged system. A turbo-compressor consists of a turbine and a compressor connected by a shaft. The compressor draws in ambient air and compresses it. The compressor is connected to the turbine by a shaft and its outlet is routed to the engine cylinder air intake. Exhaust gas from the engine cylinders enters the turbine and expands, performing work on the turbine. The turbine spins the shaft connected to the compressor (Figure 10).

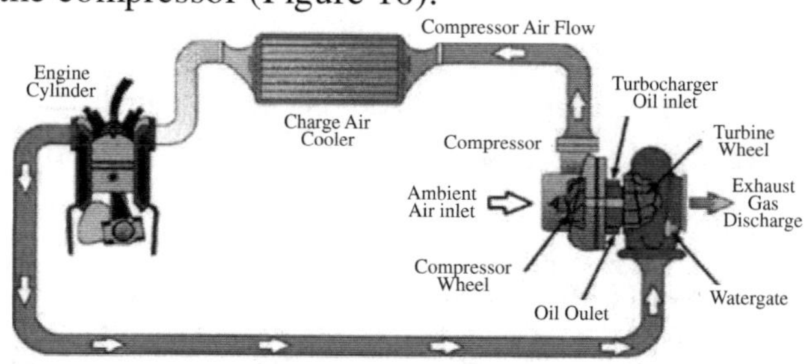

Figure 10. Turbo-charge system

The objective is first to help an expert to extract/retrieve data coming from the fleet in order to solve the diagnostic of this situation. In that sense, the expert should identify which are the most probable root-causes of degradation in the air-intake system either internal or external causes to the turbo-compressor (Muller et al., 2008).

For the purpose of this example, the fleet is limited to diesel engines. Two hundred eighteen (218) events that occurred on diesel engines are considered.

Table 1 presents an extract of the engine units of fleets and their technical features. In the following we present 3 scenario an expert could follow for elicitating a set of indivuals from the fleet.

Table 1. Extract of engine fleet technical features stored in the data bases

Engine Ref	Output power (kW)	Nb. of Cylinder	Configuratio n	Engine Speed (rpm)	Tag related to the ontology	Engine cycle	Air admission	Total	Installation
Wärtsilä 12V38	8700	12	V	600	Fuel engine	4	Turbocharged	2	Propulsion engine
Wärtsilä RT-flex50	13960	8	L	124	Fuel oil	2	Turbocharged	2	Propulsion engine
Wärtsilä RT-FLEX82T	40680	9	L	80	Fuel oil	2	Turbocharged	1	Propulsion engine
Baudouin 12M26P1FR	357,94	12	V	1800	Fuel engine	4	Naturally-Aspirated	5	Generator engine
Wärtsilä 16V38	11600	16	V	600	Fuel engine	4	Turbocharged	3	Propulsion engine
Wärtsilä 9L38	6525	9	L	600	Fuel engine	4	Turbocharged	1	Propulsion engine
Wärtsilä 8L38	5800	8	L	600	Fuel engine	4	Turbocharged	1	Propulsion engine
Volvo Penta D16C – AMG	500	6	Ligne	1800	Fuel engine	4	Turbocharged	2	Generator engine
ABC 12VDZC	2652	12	V	1000	Fuel engine	4	Turbocharged	2	Propulsion engine
Baudouin 6 M26 SR P1	331	6	Ligne	1800	Fuel engine	4	Turbocharged	3	Generator engine
Baudouin 12 M26 SR	662	12	V	1800		4	Turbocharged	2	Propulsion engine

In the first intuitive scenario, the expert considers identical units to Volvo Penta D16 MG diesel engine. In that sense, the software application gives the possibility to start the query directly with the same features than the given individual. If the user clicks on the individual, then the application is able to provide all the information that it is known about the instance such as the classification of the individual in the different classes of the taxonomy. Moreover, thanks to ontology reasoner engine, an inferred hierarchy (i.e. not directly specified in the ontology model) is automatically computed allowing the emersion of new knowledge (

Figure **11**). For example if an engine has an internal electrical degradation, the ontology can induce that it is an electrical engine.

However, in this scenario, only three events are retrieved if one considers the same kind of units with the same symptom and working in a power-generation installation as well. This is not enough historical data for performing similarity-based prognostic. For this reason, other approaches are investigated.

Figure 11. Fleet-wide predictive diagnostic software application: selection of identical units

In the second scenario, to take advantage of the fleet dimension, the expert considers all the fuel engines in the fleet. In that sense, the expert has the possibility to make a query in the proposed application about the past events that produced degradations on the air-intake system (Figure 12). In this case 218 events are retrieved.

This scenario allows the capitalization (re-use) of data. However, the fleet is composed of a wide variety of different engines. Having a look to Table 1, one notices the fleet is composed of big size, high power, two-stroke engines such as the Wärtsilä RT-FLEX82T engine and of other engines with lower output power and four-stroke cycles such as the Volvo Penta D16 MG engine. These units are very different (from an engine expert point of view) and thus their degradation behavior is very different as well. For this reason considering all the fuel engines is not suited to study the degradation behavior of the Volvo Penta D16 MG engine.

Finally, in the third scenario, since units are heterogeneous, they have different technical features, the expert shall benefit of the fleet-wide application. This application aims at helping the expert in the search of similar cases among a heterogeneous fleet of engines that allows the identification of the root-causes. This way, to search the causes of the degradation the application guides the user and proposes different criteria such as the properties or technical features of units. For instance, since there is a degradation in the air-intake system for a turbo-charged engine, the embedded knowledge in the ontology (i.e. classification of engines) allows to select only turbo-charged engines. This criterion is essential to analyze the same type of degradation, for this reason it is necessary to integrate this criterion in the query. This kind of cluster could be relevant for the user since this criterion allows the definition of common and similar characteristics of engines behavior even though they are not identical (Figure 13).

Once an ontology-based query is performed among the fleet, the user might be able to investigate the past events that have occurred in the fleet in order to reuse this information for example for predictive diagnostic purposes (Medina-Oliva et al., 2013).

This way, information/data of past events is capitalized. The application allows to reuse all the historic data about the real causes producing the abnormal behavior found among the selected population. In this scenario, the cause producing the symptom is a degradation on the poppet valve of the outtake gas, which is delayed to open.

Then, based on the retrieved events, a prognostic will be performed using the health state trajectory of the resulting 74 events that are presenting a problem with the poppet valve. This way it will be possible to estimate the remaining useful life of the Volvo Penta D16 MG diesel engine in order to plan a maintenance action before the engine performances are affected.

The on-line prognostic process in performed on the Volvo Penta D16 MG diesel engine. The first step is to compute the health time series of the engine.

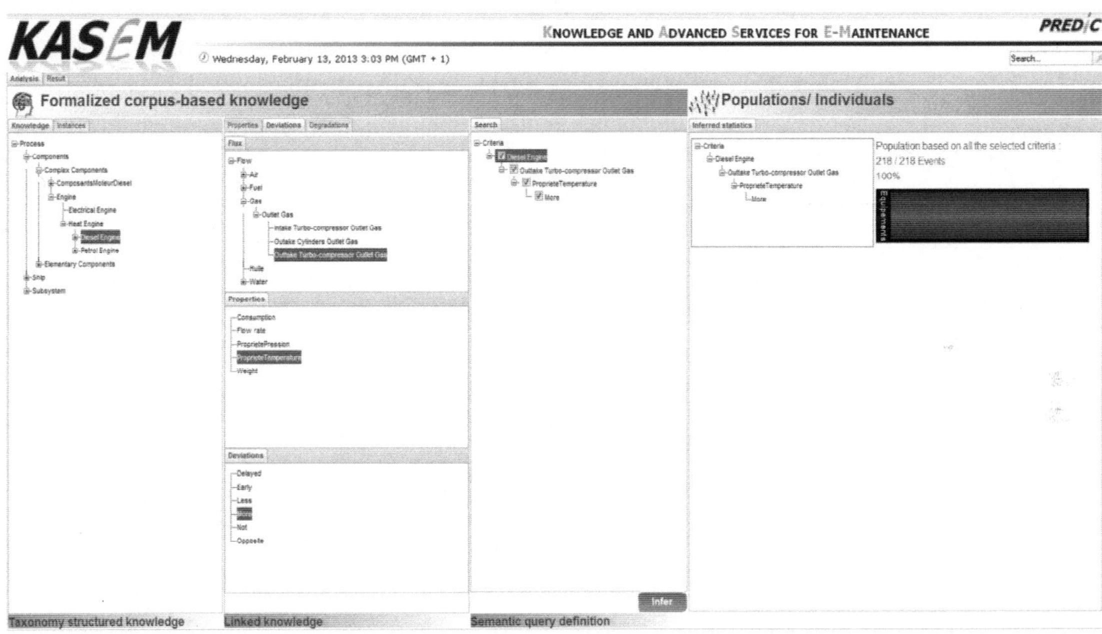

Figure 12. Fleet-wide predictive diagnostic software application: selection of all the units of the fleet

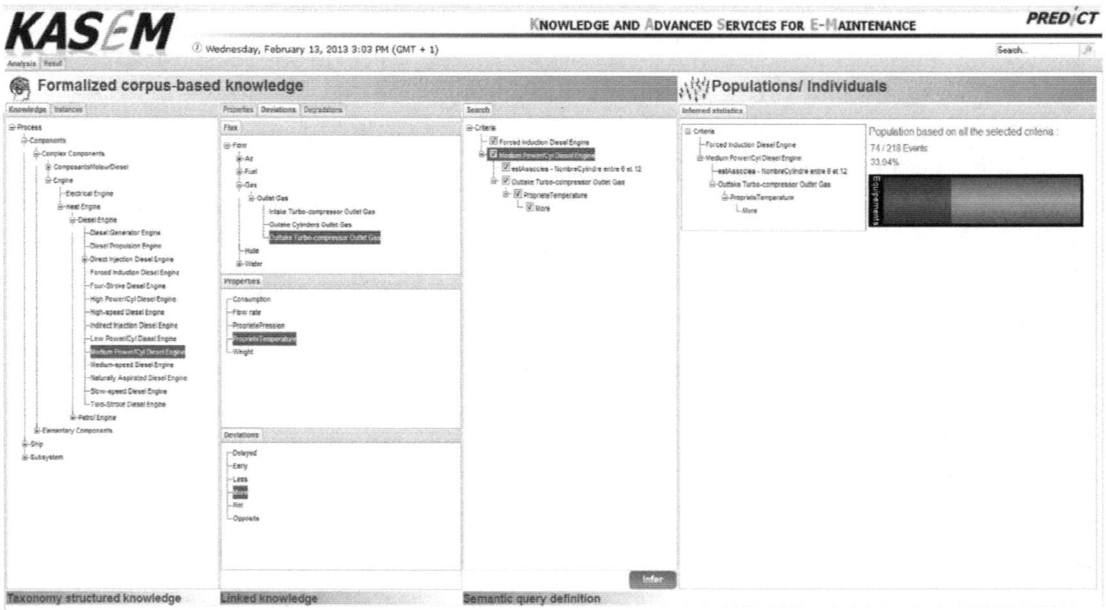

Figure 13. Fleet-wide predictive diagnostic software application: semantic-based selection according to the engine classification

5.2. Prognostic based on the obtained fleet-wide population

The second step is the computation of the RULs for every event time series. This part requires to get their degradation background knowledge, i.e. kernel vector, weights and covariance matrix. Then, several $\tilde{h}_k^u$ curves are generated. We show some curves in order to show different level of uncertainties capture by the RVs for k in {1…100} (

Figure 15).

For every event u and $\tilde{h}_k^u$ curves, a RUL_k^u is computed.

Figure 16 shows the histograms RULs for k in {1,…,10000}. One can notice that the relative dispersion of the histograms do not always correspond to a larger uncertainty in $\tilde{h}_k^u$. For instance, for event 41, $\tilde{h}_k^{41}$'s show some uncertainty (

Figure 15c) while the RUL_k^{41}'s do not since a single value has been found. In the same way, RUL_k^{26} and RUL_k^{33} (

Figure 16b and d) exhibit the same dispersion while $\tilde{h}_k^{26}$ and $\tilde{h}_k^{33}$ (

Figure 15 b and d) do not. Moreover, the contribution of every RUL_k^u in the final RUL through the weight W_k^u, eq. (7), allows to draw histograms for every event as well (

Figure 17).

Over all the events, only 68 gave a proper result to be used in the computation of the overall RUL. In such application, the computation of a single RUL does not seem of great relevance. Instead, the analysis of the histograms of the RUL^u would give better information. Such analysis could be performed with different number of histogram class (

Figure 18). We take 4 numbers of class, N, between 6 and 13. One can notice between only one mode (

Figure 18 b) for N=6, three modes ((

Figure 18 a and c) for N=8 and 11 or even four modes ((

Figure 18d) for N=13. Such differences could be further investigated by going down the cause-tree, investigating the population homogeneity according their service life, mission… and with the help of engine experts.

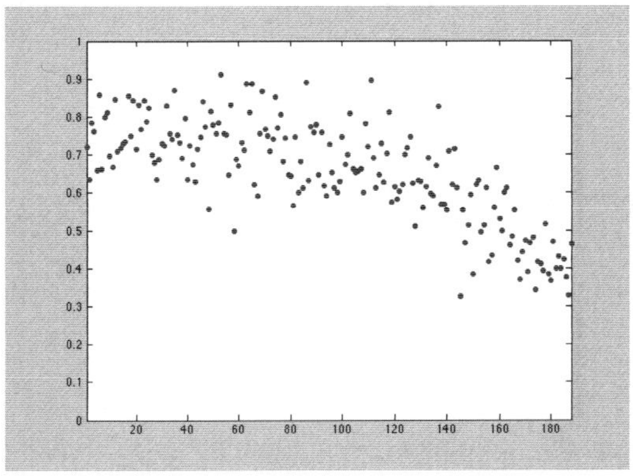

Figure 14: Health time series of the Volvo Penta D16 MG diesel engine with ill-defined running degradation

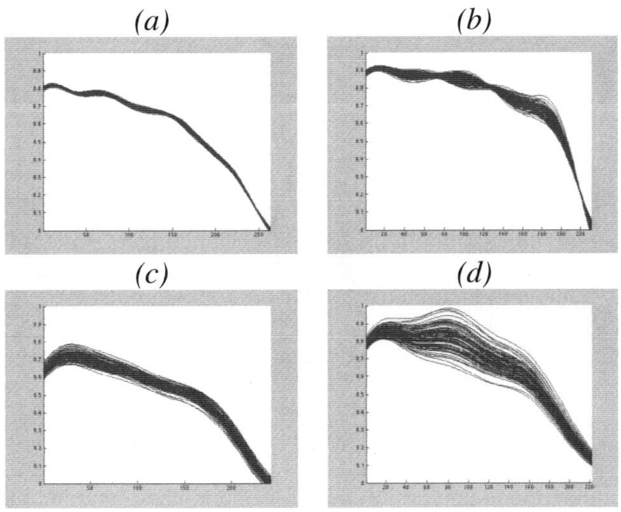

Figure 15: Several health curves for event 74 (a), 26 (b), 41 (c), 33 (d).

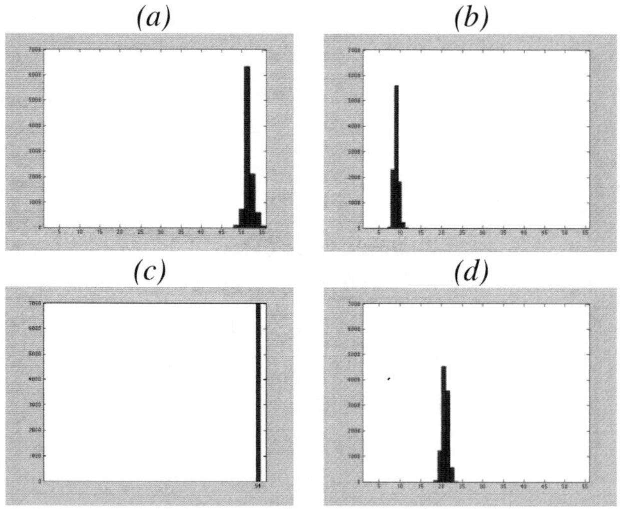

Figure 16: Histograms of RUL_k^u for the event 74 (a), 26 (b), 41 (c), 33 (d).

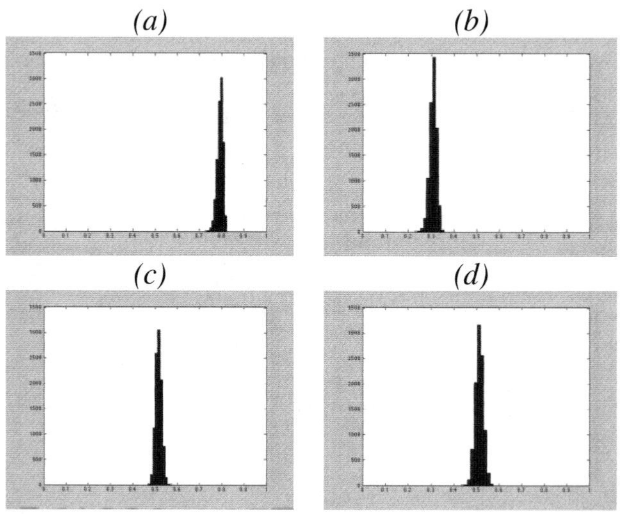

Figure 17: Histograms of W_k^u for the event 74 (a), 26 (b), 41 (c), 33 (d).

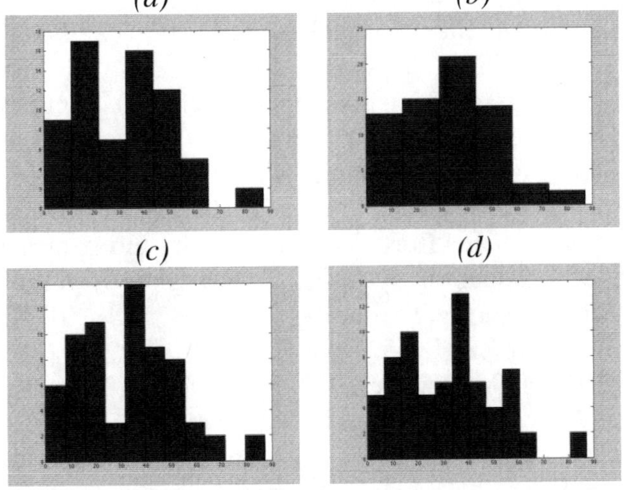

Figure 18: Histogram of the RUL^u of the Volvo Penta D16 MG diesel engine with ill-defined running degradation with different numbers of class: 8 (a), 6 (b), 11 (c) and 13 (d).

6. Conclusions

In this paper we proposed an approach taking advantage of all available knowledge at a fleet level for prognosis based on historical data series. The originality of this work lies in the ability to make dynamically chose the historical series contributing to the prognosis.

Despite the presented case study belongs to the naval domain, the proposed approach is general and can be applied to fields such as wind turbine farms, vehicle fleets…

The presented work is a first step in that direction. Further steps can improve the usefulness of the proposed approach in several directions such as the computation of the units health which has to be homogenous over the entire fleet. An investigation on the operators to be used could be performed. The computation of the histograms of the RUL^u could be performed at every stage of the cause-tree.

References

[1]Bonissone, P.P., Varma, A. (2005). Predicting the Best Unit within a Fleet: Prognostic Capabilities Enabled by Peer Learning, Fuzzy Similarity, and Evolutionary Design Process. In Proceedings of the 14th IEEE International Conference on Fuzzy Systems, IEEE, pp. 312-318.

[2]Byington C.S., M.J. Roemer, G.J. Kacprzynski et T. Galie (2002). Prognostic Enhancements to Diagnostic Systems for Improved Condition-Based Maintenance. 2002 IEEE Aerospace Conference, Big Sky, USA.

[3]G.E.P. Box, G.M. Jenkins (1976), and Time Series Analysis: Forecasting and Control, Holden-Day, San Francisco.

[4]Heng A., S. Zhang, A.C.C. Tan et J. Mathew. (2009). Rotating machinery prognostic: State of the art, challenges and opportunities, Mechanical Systems and Signal Processing, vol. 23 (3), pp. 724-739.

[5]Jardine A.K.S., D. Lin et D. Banjevic (2006). A review on machinery diagnostics and prognostic implementing condition-based maintenance. Mechanical Systems and Signal Processing, vol. 20, pp.1483-1510.

[6]Liu J., Djurdjanovic D., Ni J., Casoetto N., Lee J. (2007) Similarity based method for manufacturing process performance prediction and diagnosis, Computers in Industry, 58, 558–566

[7]Medina-Oliva G., Léger J-B, Voisin A., Monnin M., (2012) Predictive Diagnostic based on a Fleet-wide Ontology Approach, MFPT 2013, 13-17 May 2013 - Cleveland, Ohio, USA.

[8]Medina-Oliva G., Weber P., Levrat E., Iung B. (2012a) Using probabilistic relational models for knowledge representation of production systems: A new approach to assessing maintenance strategies. CIRP Annals - Manufacturing Technology. in press. DOI: 10.1016/j.cirp.2012.03.059

[9]Medina-Oliva G., Voisin A., Monnin M., Peysson F., Leger JB. (2012b). Prognostic Assessment Using Fleet-wide Ontology. PHM Conference 2012, Minneapolis, Minnesota, USA.

[10]Medina-Oliva G., Peysson F., Voisin A., Monnin M., Léger J-B. (2013). Ships and marine diesel engines fleet-wide predictive diagnostic based on ontology, improvement feedback loop and continuous analytics. Proceedings of 25th International Congress on Condition Monitoring and Diagnostics Engineering Management, 11–13 June, 2013, Helsinki, Finland.

[11]Medina-Oliva G., Voisin A., Monnin M., Léger J-B. (2013). Predictive diagnosis based on a fleet-wide ontology approach. Knowledge Based Systems. In Press, Corrected Proof. http://dx.doi.org/10.1016/j.knosys.2013.12.020

[12]Monnin M., Voisin A., Leger JB., Iung B. (2011a). Fleet-wide health management architecture. Annual Conference of the Prognostic and Health Management Society. Montreal, Quebec, Canada.

[13]Monnin M., Abichou B., Voisin A., Mozzati C. (2011b). Fleet historical cases for predictive maintenance. The International Conference Surveillance 6. October 25-26. Compiègne, France.

[14]Muller A., Suhner M-C., Iung B. (2008). Formalisation of a new prognosis model for supporting proactive maintenance implementation on industrial system. Reliability Engineering and System Safety. 93(2) 234-253.

[15]Patrick, R., Smith, M J., Byington, C S., Vachtsevanos, G J., Tom, K., Ly, C. (2010). Integrated Software Platform for Fleet Data Analysis, Enhanced Diagnostics, and Safe Transition to Prognostic for Helicopter Component CBM, in Proceedings of Annual Conference of the Prognostic and Health Management Society, October 10-16, Portland, Oregon.

[16]Reymonet, A., Thomas, J., Aussenac-Gilles, N. (2009). Ontology Based Information Retrieval: an application to automotive diagnosis, in Proceedings of International Workshop on Principles of Diagnosis, June 14-17, Stockholm, Sweden, pp. 9-14.

[17]Rizzolo L., Abichou B., Voisin A., Kosayyer N. (2011), Aggregation of Health Assessment Indicators of Industrial Systems. In Proceedings of the 7th conference of the European Society for Fuzzy Logic and Technology, EUSFLAT-2011, Aix-Les-Bains, France.

[18]Tipping M.E. (2001) Sparse Bayesian Learning and the Relevance Vector Machine, Journal of Machine Learning Research 1, 211 244

[19]Verma, A. K. and Srividya, A. and Ramesh, P. (2010). A systemic approach to integrated E-maintenance of large engineering plants, International Journal of Automation and Computing, vol. 7, pp. 173-179.

[20]Wang P., Youn B., Hu C. (2012) A generic probabilistic framework for structural health prognostic and uncertainty management. Mechanical Systems and Signal Processing. 28, Pages 622–637

[21] Wang T., Yu J., Siegel D., Lee J. (2008). A similarity-based prognostic approach for Remaining Useful Life estimation of engineered systems. International Conference on Prognostic and Health Management. Denver, USA.

[22] Weber P., Medina-Oliva G., Simon C., Iung B. (2012). Overview on Bayesian networks Applications for Dependability, Risk Analysis and Maintenance areas. Engineering Applications of Artificial Intelligence, vol. 25 (4), (671-682).

Integrated Prognostics and Control Provides Enhanced Asset Utilization

Fred M. Discenzo, Ph.D.
Mgr. R&D, Rockwell Fellow
Rockwell Automation
May 21, 2014

MFPT

Rockwell Automation

Discussion Topics

1. Background & Motivation
2. Framework
3. Examples (Pumps, Motor Windings)
5. Gaps & Opportunities
6. Conclusion

Background & Motivation

In some cases machinery failure can be an inconvenience ...

In many cases machinery failure can be catastrophic!

Trend Toward More Complex Coupled Systems

Unexpected Failures Often Have Significant Consequences

Failures Typically Are Preceded by Cues

- With early warning of impending fault can avoid failure, delay the failure or mitigate damage and loss

Prognostics & Control: Key Assumptions

- **Avoiding machinery failure is critical for many applications**
  reduced scrap, less damage, reduced downtime, safer operation…
- **Machines rarely fail suddenly with no prior indication**
  Temperature rise, noisy operation, increased vibration, greater current consumption, slow response to control action …
- **The way a machine is controlled can affect the time to failure**
  Control can effect electrical and mechanical losses - heat generation, bearing load, stimulate resonant frequencies …
- **Altering the control appropriately can reduce component stress, wear, or fatigue & serve to manage the remaining useful life of machines - avoid or defer failure - maximize asset value**
  Reduce controller gains, limit motor torque, reduce pump speed, …

Active Control of Machinery Lifetime

- **Prognostics**
  - Detect incipient fault signatures
  - Estimate remaining lifetime (usually short-term life prediction)
    - "Something is starting to break; how much longer will it last?"
- **Lifetime Control**
  - Correlate operating condition with life expectancy (long-term life prediction)
  - Control operating condition to extend lifetime
    - "How do I make it last 20 years, or 2 hours, or through 100 missions?"
  - Trade-off between machine performance & life expectancy
  - Establish suite of possible future operating conditions – estimate life expectancy
- **Benefits of lifetime control**
  - Avoid untimely failures & reduce maintenance cost
  - Improve mission readiness
  - Support logistic & procurement planning
  - Optimize asset utilization

Integrated Prognostics & Control Framework

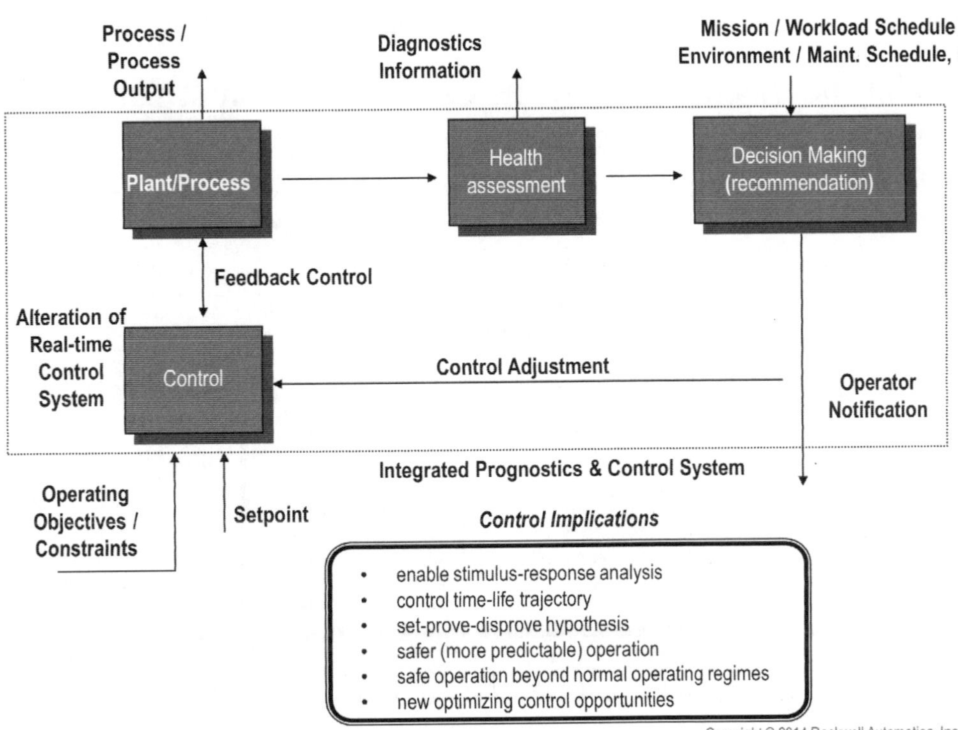

Traditional Automation & PHM System

Control and Diagnostic Typically Separate Systems
No Information Exchange Between the Systems
(reflects organizational boundaries)

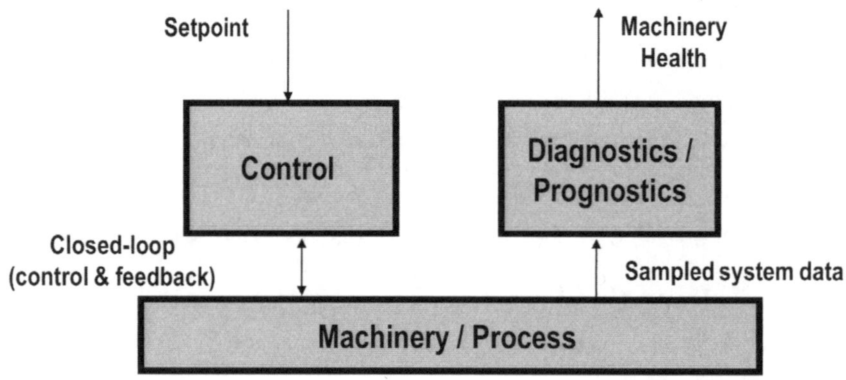

Integrated Automation & PHM System

Weak Coupling – Information Exchange

Automation and Diagnostic Systems Can Exchange information to

a) Alter the control to reduce machinery stress, and
b) Analyze machinery operation in response to control actions (stimulus-response analysis) – enhanced diagnostics & prognostics

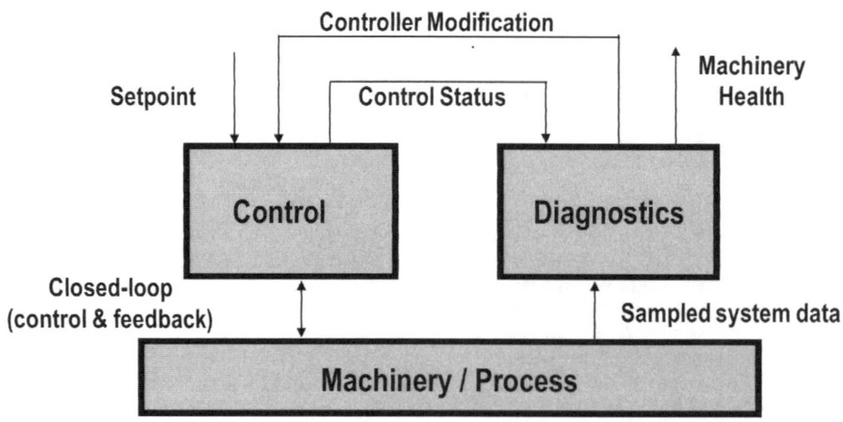

Evolution of Integrated Automation & PHM System

Tight Coupling – Functional Integration

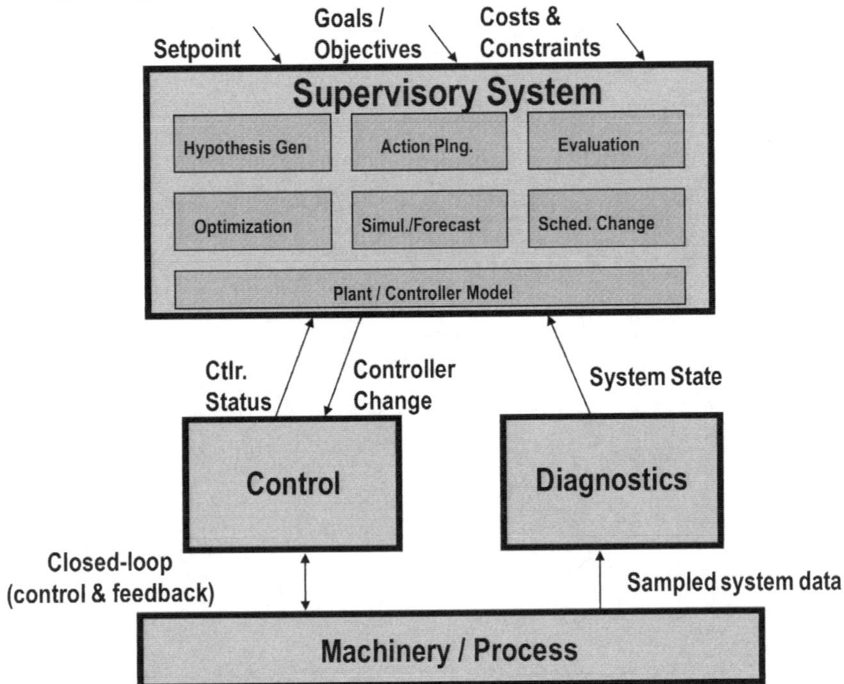

Realization of Integrated Prognostics & Control

General Linear Control State Space Representation

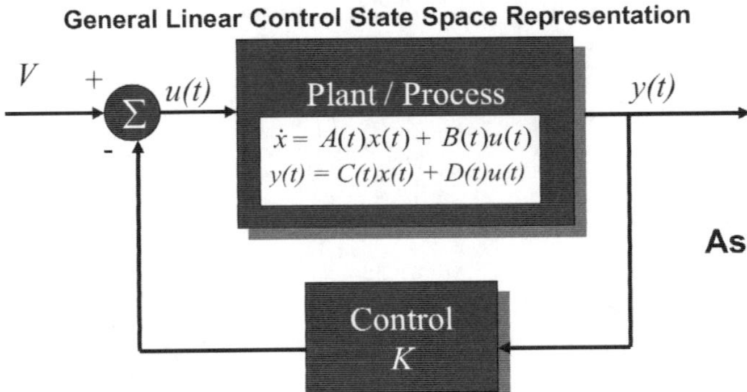

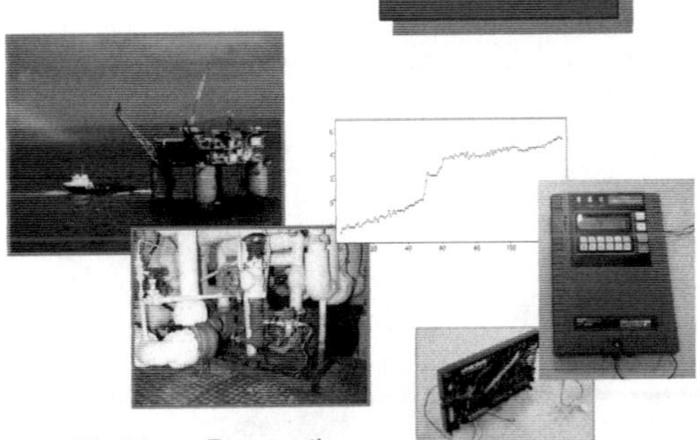

Machinery Prognostics (ARMA(10,10) Model of vibration trajectory)

Objective:

Alter the Control or Configuration to drive specific state variable(s) on a more desirable trajectory subject to input constraints and process requirements

Asset Optimization:

Unique opportunity to optimize key performance metric(s) (e.g. energy, lifetime, throughput, LCC ...)

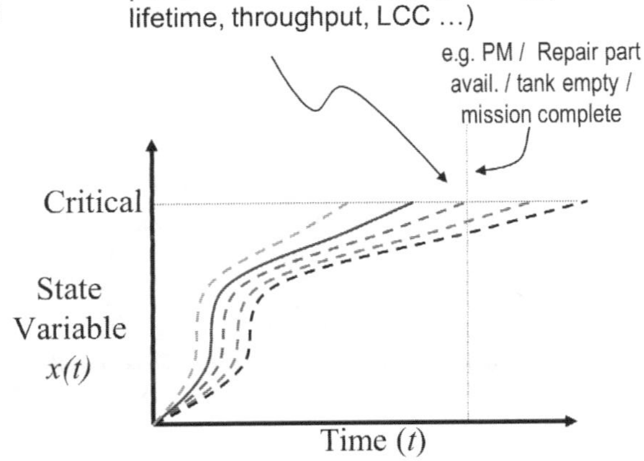

Example: Pump Prognostics & Control

- Cavitation is the generation and subsequent collapse of vapor bubbles – harmful condition – erodes impellers, induces vibration, accelerates failure
- Can detect the degree of cavitation & dynamically alter cavitation state
- Commercial drive (VFD) modified with:
  1. Diagnostic board with pump cavitation algorithms
  2. Board has a link to regulator board in the VFD to control drive parameters (e.g. speed)

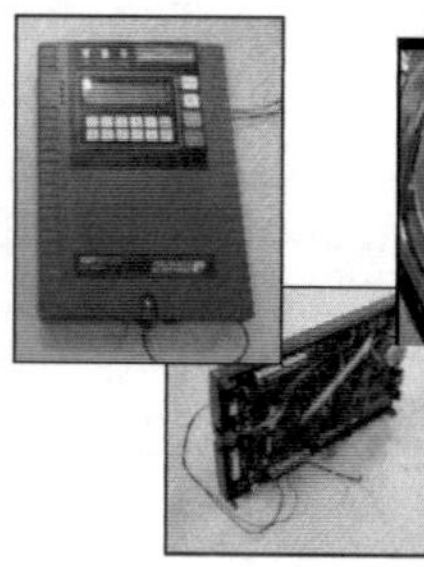

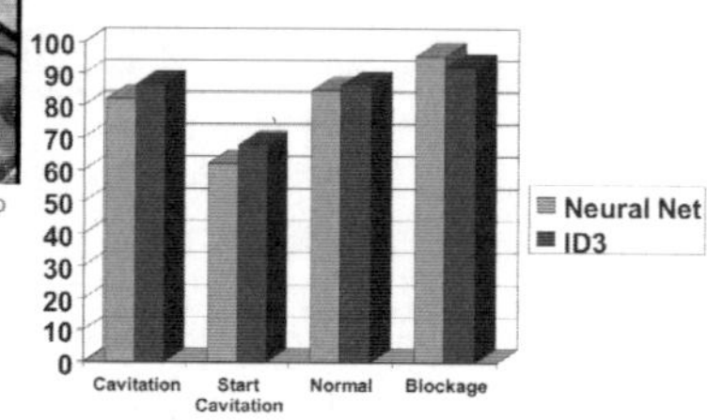

Laboratory Pump Loop and Prototype VFD with Integrated Diagnostics & Control

From Next Generation Pump Systems Enable New Opportunities for Asset Management and Economic Optimization, F. Discenzo, et.al., Rockwell Automation, Pump Symposium, 2002

Example: Pump Prognostics & Control

- Models exist to predict remaining impeller life based on cavitation bubble length (estimated from cavitation severity, impeller geometry, ...)
- A representative model of impeller damage is given by Cooper[a]:

$$\text{Erosion rate (mm/h)} = f\,[(\text{bubble length})^n*(\text{NPSHa flow factor})^3*(\text{flow factor})^6*\text{rho}^3*(I_{tensile})^{-1}]$$

- The severity of cavitation affects impeller damage rate in a non-linear / non-monotonic manner: i.e. increased cavitation severity does not always correspond to increased impeller erosion rate!
- Diagnostics board dynamically controls cavitation via VFD input

[a]Pump Cavitation – Various NPSHR Criteria, NPSHA, Margins, and Impeller Life Expectancy, B.Shiavello, F.Visser, Flowserve Pump Division, Proceedings of the Twenty-fifth International Pump Users Symposium, 2009

Example: Control of Motor Winding Lifetime

- Motors break down primarily due to winding insulation failure & bearing failure
- Life expectancy of winding insulation is typically a function of winding temperature
  - Winding lifetime reduces roughly ½ for every 10°C rise in temperature
  - Can control winding lifetime by controlling winding temperature
  - Can control winding temperature by controlling motor current (or carrier freq. for inverter-driven motors)
- Can track level of motor current over time to continually estimate remaining winding life
- Dynamically alter control to manage residual winding life

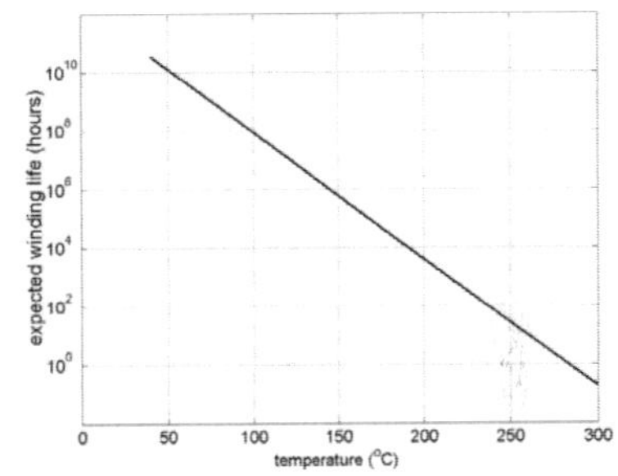

Control of Motor Winding Lifetime: Approach

- Target application is a brushless DC motor-driven actuator
- Controller estimates motor remaining lifetime based on temperature history
  - Uses heuristic method (assumes damage is linearly additive – Miner's Rule)
- Controller reduces motor current (which reduces winding temperature) when estimated remaining lifetime is less than the desired lifetime

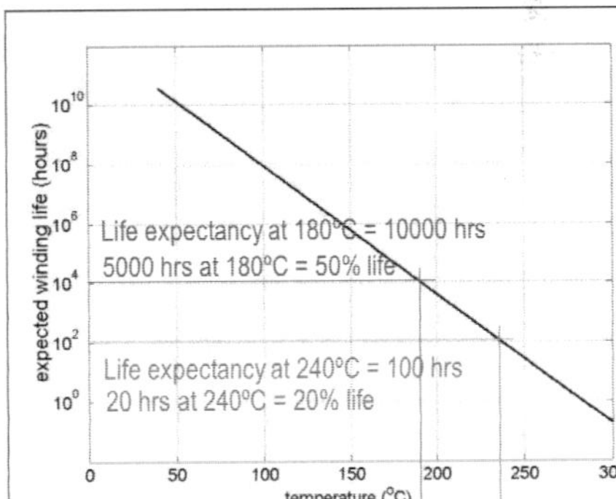

Calculation of Damage Effects:

Motor winding spends 5,000 hours at 190°C, where lifetime of new winding is 10,000 hours

Damage effect = consumed 50% lifetime of new winding

Assume % life consumed at each temperature level is additive

Keep track of sum of % life consumed – no need to store historical data

BLDC Actuator

Motor Winding Lifetime Control: Methodology

- Standard linear control law

$$u = -K[X(t) - X_{cmd}(t)] \qquad X = \begin{bmatrix} \theta \\ \omega \end{bmatrix}$$

Control effort (motor current) → u; Feedback gains → K; Actual motor position & speed → $X(t)$; Commanded motor position & speed → $X_{cmd}(t)$

- Standard linear optimal control computation

Compute K to minimize $\mathbf{J}(t_0) = \int_{t_0}^{\infty} \left[[X(t) - X_{cmd}(t)]^T Q [X(t) - X_{cmd}(t)] + Ru^2 \right] dt$

Weighting factors to minimize position & velocity error → Q; Weighting factor to minimize control effort → R

- Choose large Q to reduce tracking error at the expense of higher motor current
- Choose large R to reduces motor current at the expense of tracking error

From Adaptive Control of Actuator Lifetime, L.Gokdere, et.al., Rockwell Scientific, Boeing Company, IEEEAC paper #1337, 2006

Motor Winding Lifetime Control: Methodology

- Modify optimal control performance measure by including a scaling factor ρ

$$\mathbf{J}(t_0) = \int_{t_0}^{\infty} \left[[X(t) - X_{cmd}(t)]^T Q [X(t) - X_{cmd}(t)] + Ru^2 \right] dt$$

⇓

$$\mathbf{J}(t_0) = \int_{t_0}^{\infty} \left[[X(t) - X_{cmd}(t)]^T \frac{1}{\rho} Q [X(t) - X_{cmd}(t)] + \rho R u^2 \right] dt$$

- Increasing ρ
  - Increases importance of minimizing motor current (increases winding lifetime)
  - Reduces importance of minimizing tracking error (reduces performance)
- Adjust ρ using the linear control law

$$\dot{\rho} = -K_{Tp}\left(Life - Life^*\right) - K_{Td}\frac{d}{dt}\left(Life - Life^*\right)$$

Residual lifetime (or winding temp) → $Life$; Desired lifetime (or equivalent desired winding temperature) → $Life^*$

From Adaptive Control of Actuator Lifetime, L.Gokdere, et.al., Rockwell Scientific, Boeing Company, IEEEAC paper #1337, 2006

Adaptive Linear Optimal Control Simulations

- Motor under constant load
- Note that low desired lifetime cannot always be realized due to upper bandwidth limit (performance limit) of the digital control

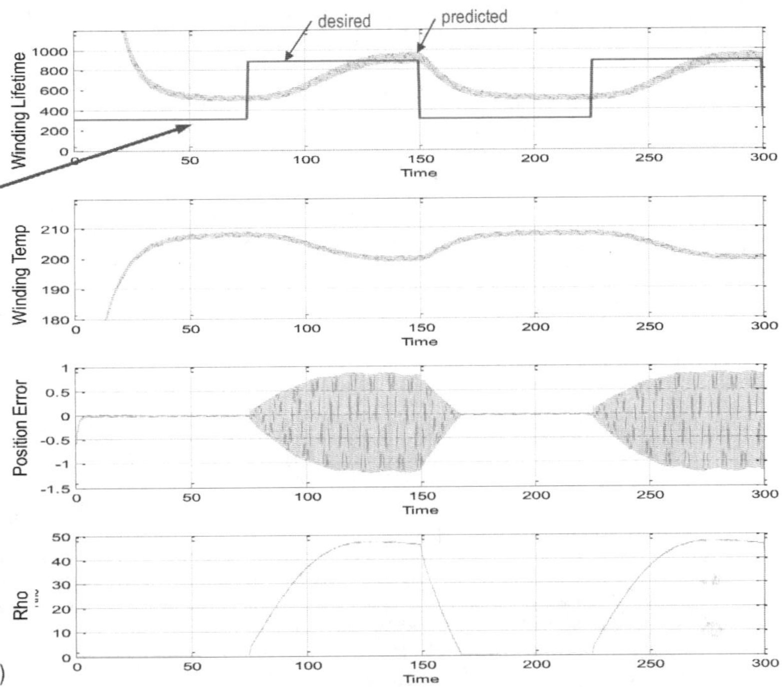

Tabletop Actuator Test Bed
(servomotor with embedded temp sensors)

From Adaptive Control of Actuator Lifetime, L.Gokdere, et.al., Rockwell Scientific, Boeing Company, IEEEAC paper #1337, 2006

Performance Comparison - Expanded View

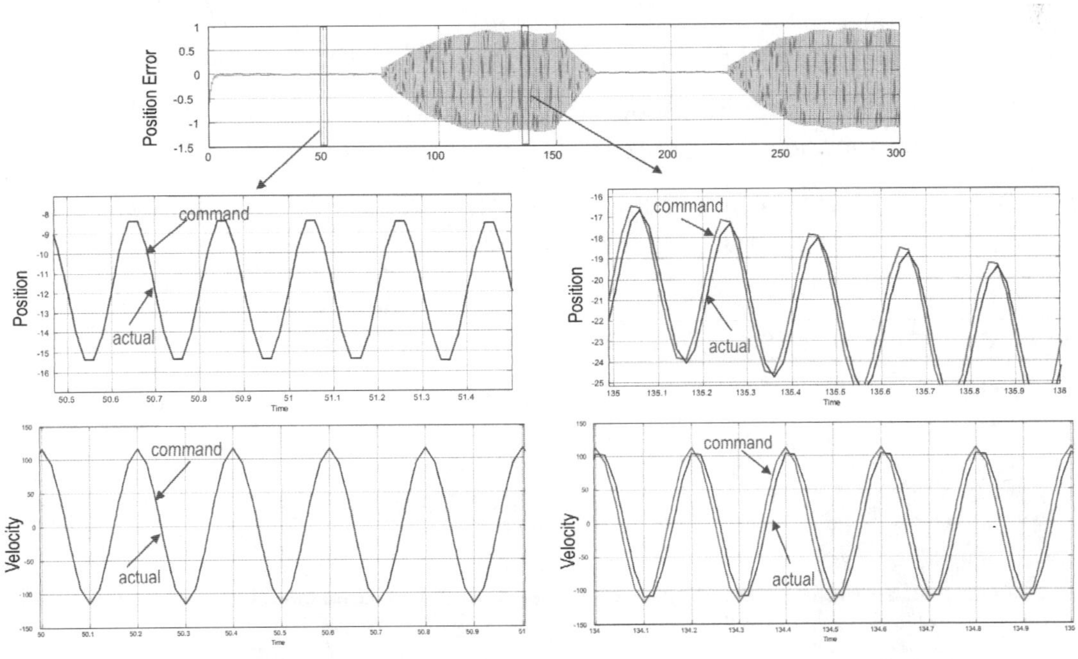

From Adaptive Control of Actuator Lifetime, L.Gokdere, et.al., Rockwell Scientific, Boeing Company, IEEEAC paper #1337, 2006

Gaps & Opportunities

- *Ad hoc* prognostics & control solutions exist – ***component focus***
- Broadscale integrated prognostics & control requires:
  1. ***Reliable prognostics*** (context specific) – state prediction
  2. ***Robust control*** (compensating) and/or reconfiguration
  3. **Systems Approach – not just component** (system-of-systems)
  4. ***Integrated decision support system*** – mission focus
- Open collaboration among the prognostics and control community needed to insure success

Operational Benefits

1. Opportunity to automatically take action rather than just issuing an alert
2. Supports lights-out unattended operation - extended operation of USV
3. System response to control action provides a unique mechanism for enhanced diagnostics (i.e. stimulus-action response analysis)
4. May avoid a shutdown by permitting continued reliable operation at reduced levels, less efficient, noisier, or slower response
5. Permits operation in unusual, unexpected, or extreme conditions (e.g. elevated temperatures, high torque levels, exceed ratings!)
6. Convergence of PHM & control provides a powerful tool for protecting critical assets and sustaining operation while avoiding failure

Concluding Remarks

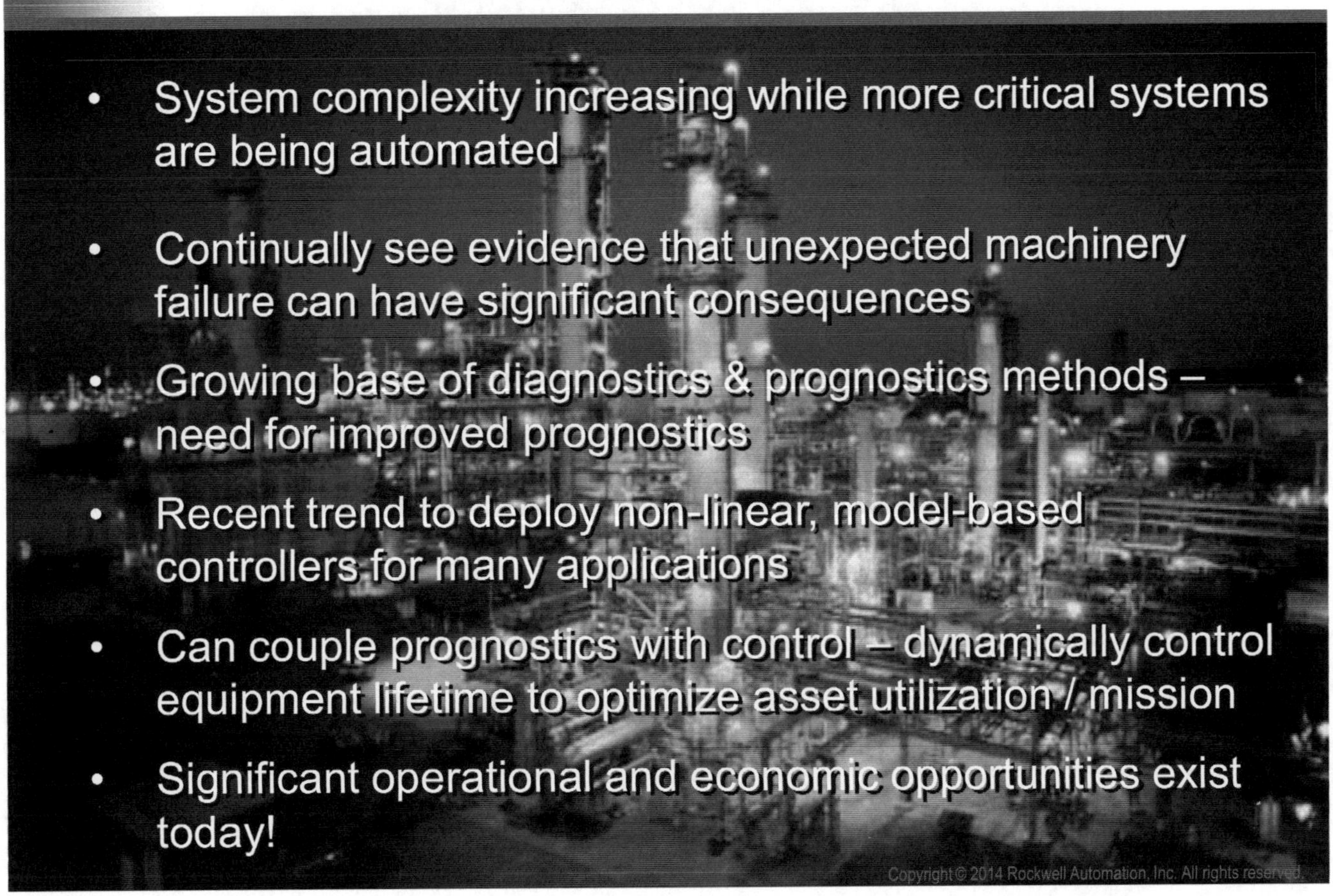

- System complexity increasing while more critical systems are being automated
- Continually see evidence that unexpected machinery failure can have significant consequences
- Growing base of diagnostics & prognostics methods – need for improved prognostics
- Recent trend to deploy non-linear, model-based controllers for many applications
- Can couple prognostics with control – dynamically control equipment lifetime to optimize asset utilization / mission
- Significant operational and economic opportunities exist today!

IMR Mining and Metallurgy Machinery

INTRINSIC METHOD FOR MONITORING INDUSTRIAL APPLICATION FOR VIRTUAL DIAGNOSTICS

M. Schütz, M. Hilbert, F. D. Boos, C. Bernet, R. Baltes, K. Nienhaus, C. Sequeira
Presenter: **M. Hilbert**

MFPT, Mai 2014

RWTH AACHEN UNIVERSITY

RAG

IMR Mining and Metallurgy Machinery
RWTH AACHEN UNIVERSITY

Nordex SE

Content

1. Concept of Virtual Counterparts
2. i-MaSS Measurement System
3. Application Chain Conveyors
4. Application Wind Turbine Gearbox

2

Concept of Virtual Counterparts

IMR Mining and Metallurgy Machinery
RWTH AACHEN UNIVERSITY

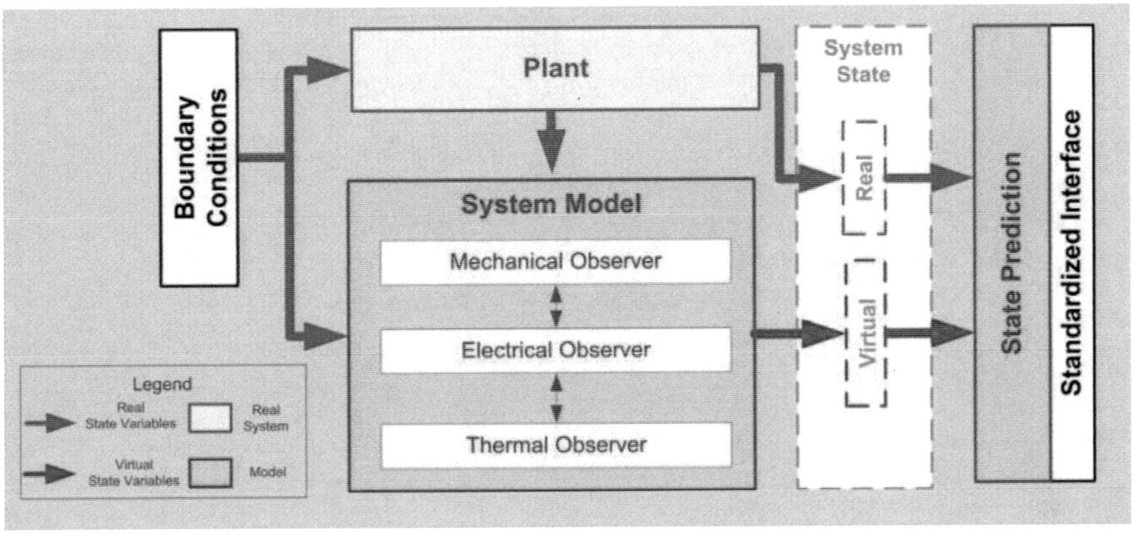

3

i-MaSS Measurement System

IMR Mining and Metallurgy Machinery
RWTH AACHEN UNIVERSITY

- Easy application on **industrial environment**
- **Multi sensor system:** monitoring of vibrations, temperature and force
- Use of **standardised sensing elements**
- **Wireless** data transmission
- **Period of application** of several months without maintenance
- Distinction from telemetry systems by data pre-processing

4

Application Chain Conveyors

5

Application Chain Conveyors

- Applied on **Underground Longwall Mining Operations**
- Transports the coal from the coal face to the main gate - **not redundant**
- Up to 470m long, up to 5000t coal and rock per hour
- Typical malfunction: chain break, jumped out scrapper, **wear and fatigue due to wrong adjusted preload force**

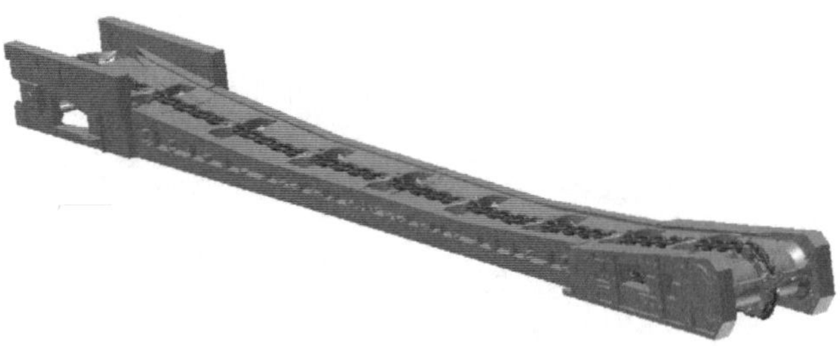

6

Application Chain Conveyors

- Round chain links, wire diameter of 30mm
- Special chain link including a **strain gauges**, power supply, transmitter and the sensor

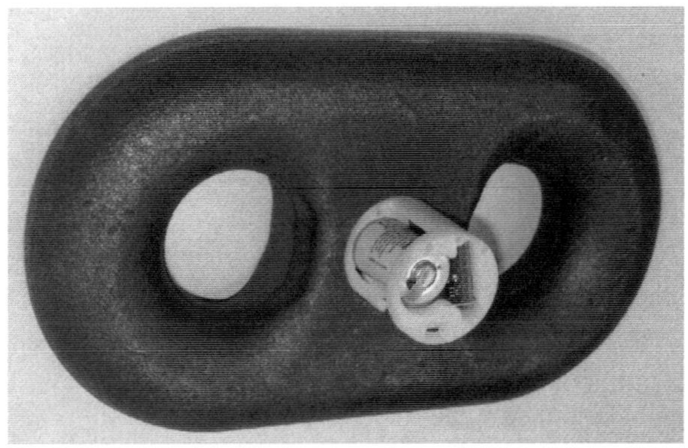

7

Application Chain Conveyors

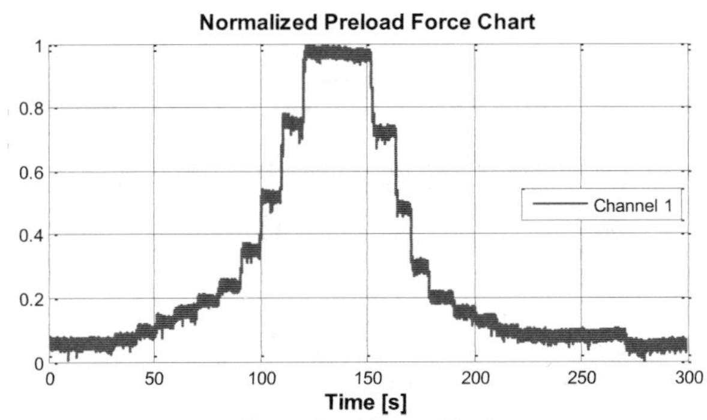

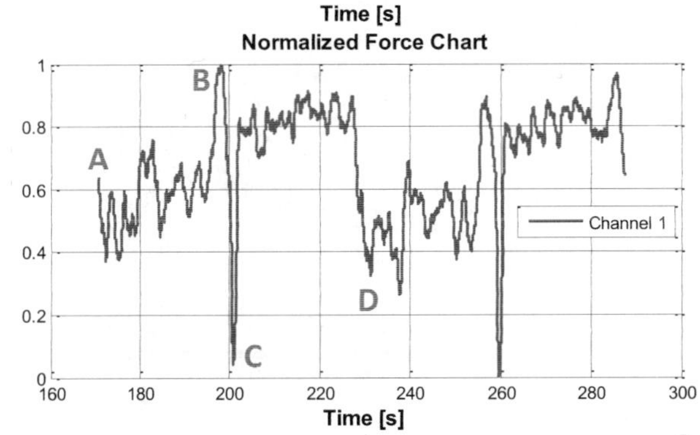

- Test in a **experimental chain conveyor** without coal

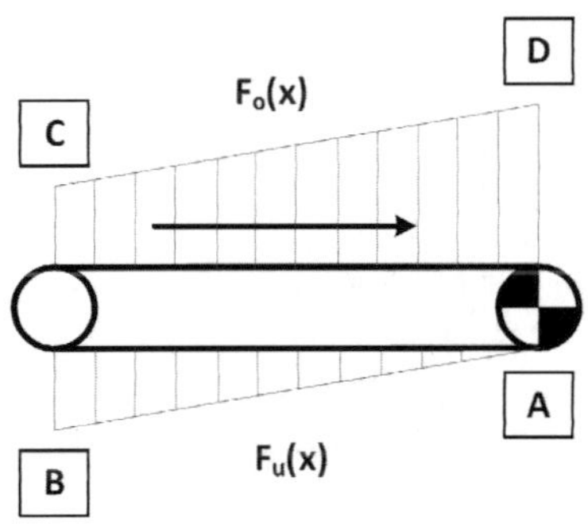

8

Application Wind Turbine Gearbox

9

Wind Turbine Gearbox

Classic Drive Train

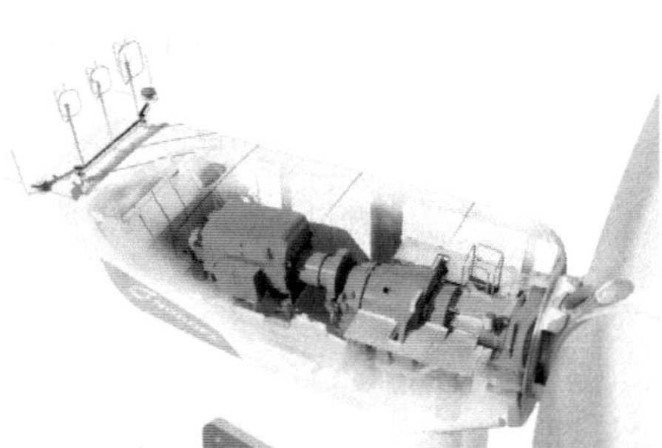

Nordex SE

Eickhoff

Integrated Drive Train

AREVA

10

Wind Turbine Gearbox

IMR Mining and Metallurgy Machinery
RWTH AACHEN UNIVERSITY

State of the Art

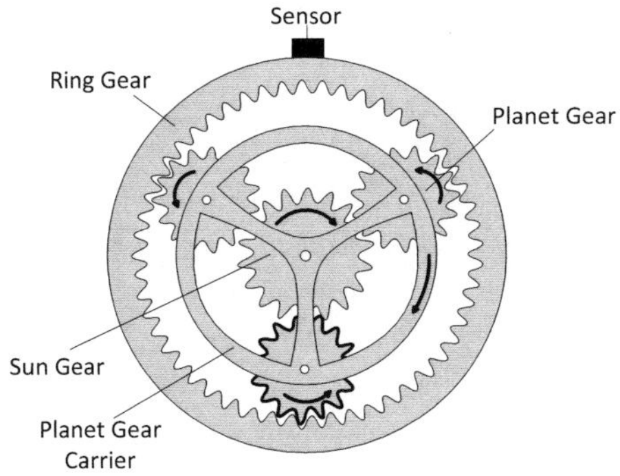

- Installation of the acceleration sensor on the **gearbox housing**

i-MaSS Approach

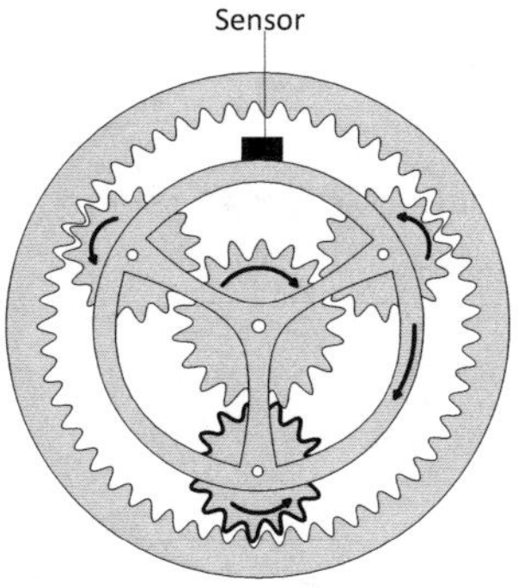

- Installation of the acceleration sensor on the **carrier plate** of the planetary stage

11

Wind Turbine Gearbox

IMR Mining and Metallurgy Machinery
RWTH AACHEN UNIVERSITY

State of the Art

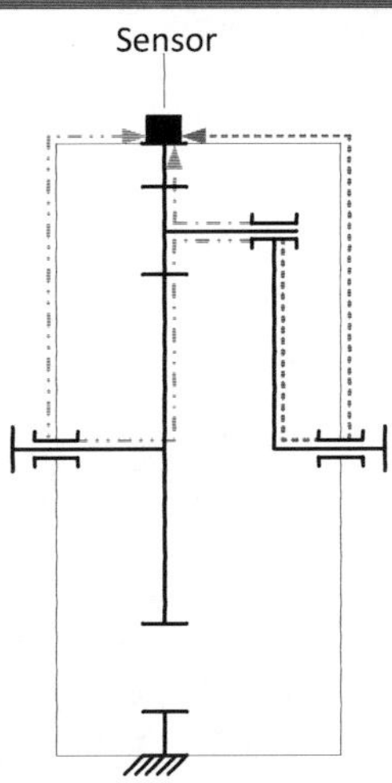

$G_1(t)$
$G_2(t)$
$G_3(t)$

i-MaSS Approach

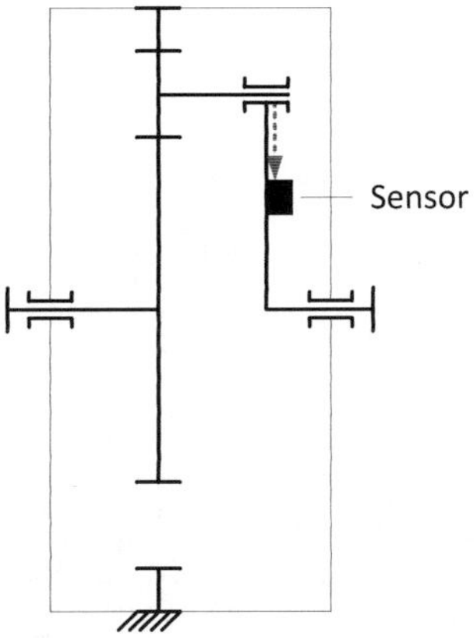

$G_4(t)$

12

Wind Turbine Gearbox

State of the Art

- **Lubrication** of the planet bearing and of the gearing
- **Clearance** of the planet bearing and the gearing
- **Stiffness** of the gearing and the transmission structure
- **Position** of the planetary wheel in relation to the sensor
- Application of particular analysis methods

Time variant damping of the vibration signal

i-MaSS Approach

- **Defined** transfer function
- Application of conventional analysis methods

Time invariant damping of the vibration signal

13

Wind Turbine Gearbox

- No direct visual contact between transmitter and receiver
- Attenuation by lubricant which may contain ferromagnetic particles

14

Wind Turbine Gearbox

IMR Mining and Metallurgy Machinery
RWTH AACHEN UNIVERSITY

State of the Art

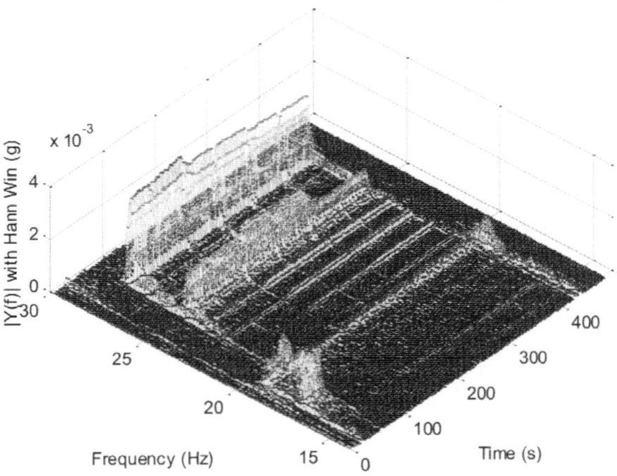

i-MaSS Approach

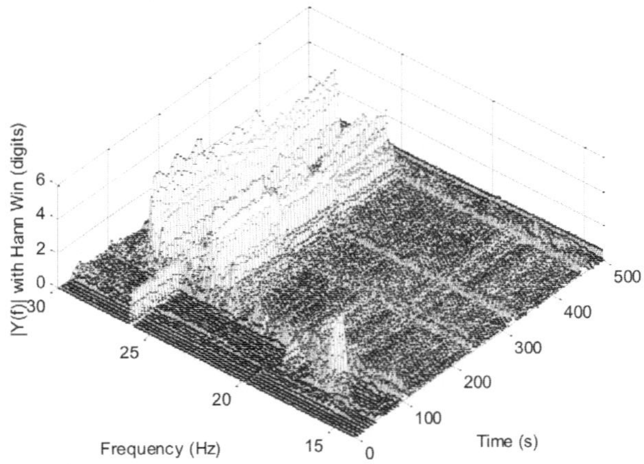

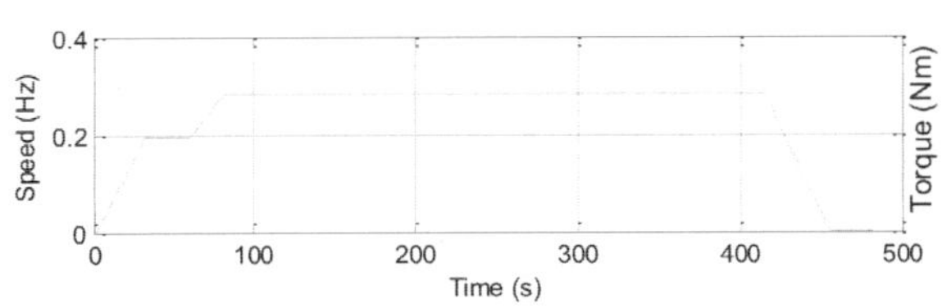

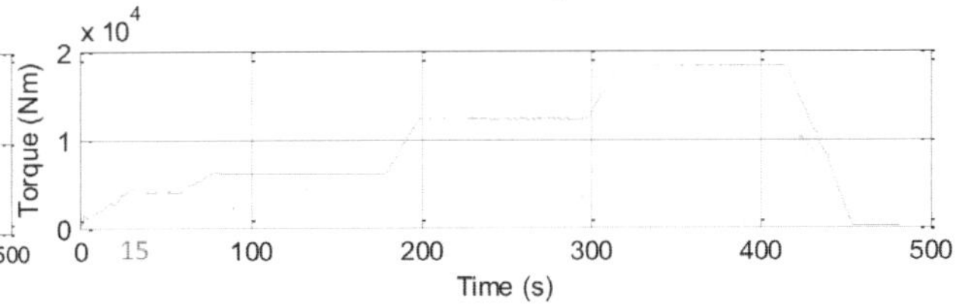

Wind Turbine Gearbox

IMR Mining and Metallurgy Machinery
RWTH AACHEN UNIVERSITY

State of the Art

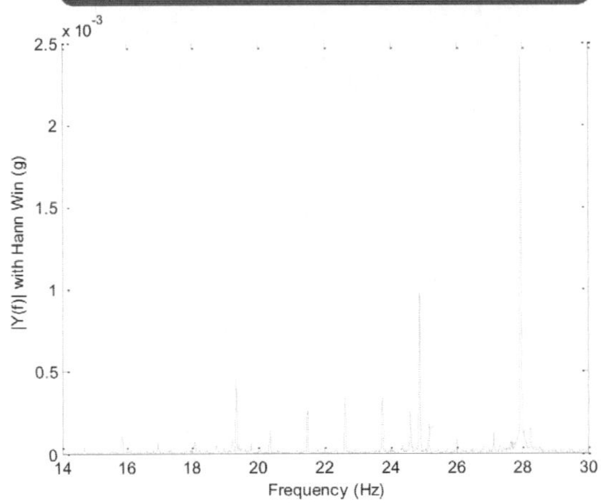

i-MaSS Approach

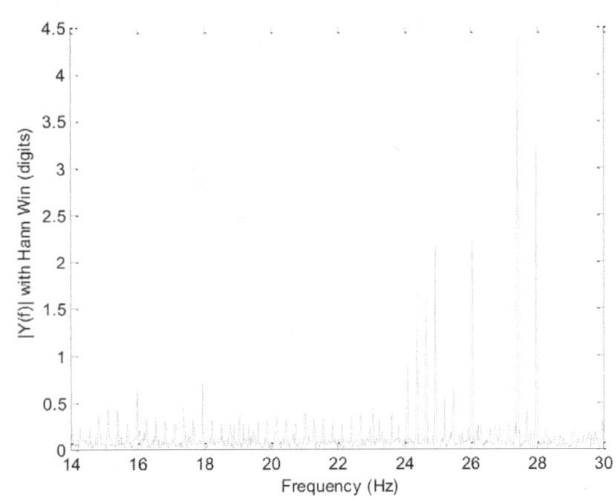

- Gear mesh frequency – 34.9 Hz
- Carrier rotational frequency – 0.28 Hz
- Number of planets - 4

16

Summary

IMR Institut für Maschinentechnik der Rohstoffindustrie
RWTH AACHEN UNIVERSITY

- **Acquiring information in industrial applications** to:
  - **Improve knowledge** of the system
  - Support the **virtual counterpart** with operational data of machine elements
- **Increase machine reliability** by condition prognostics based on data of critical machine elements
- **Improve the design** of machine elements

17

IMR Mining and Metallurgy Machinery
RWTH AACHEN UNIVERSITY

Thank you for your attention

Marc Hilbert
mhilbert@imr.rwth-aachen.de

Institute for Mining and Metallurgy Machinery (IMR)
RWTH Aachen University

18

WIRELESS HART: AN ENABLING TECHNOLOGY

STEVEN A. MOORE

EMERSON PROCESS MANAGEMENT

EMERSON.
Process Management

WIRELESS HART: WHAT IS IT?

- A WORLDWIDE STANDARD: IEC62591
- TOTALLY DIGITAL COMMUNICATION
- AN OPEN STANDARD: MORE THAN A DOZEN MANUFACTURERS
- SAME AS HART WITHOUT THE WIRES

EMERSON.
Process Management

SMART WIRELESS VISION

- SEAMLESSLY INTEGRATE BOTH WIRED AND WIRELESS PLANT AND
  FIELD NETWORKS TO ADDRESS PROCESS INDUSTRY NEEDS
- BASED ON OPEN STANDARDS
  - 802.11
  - 802.15.4
  - WIRELESSHART
  - IEC 62591
- NETWORK CHARACTERISTICS
  - SCALABLE
  - SECURE
  - RELIABLE
  - EASY

*Wireless*HART

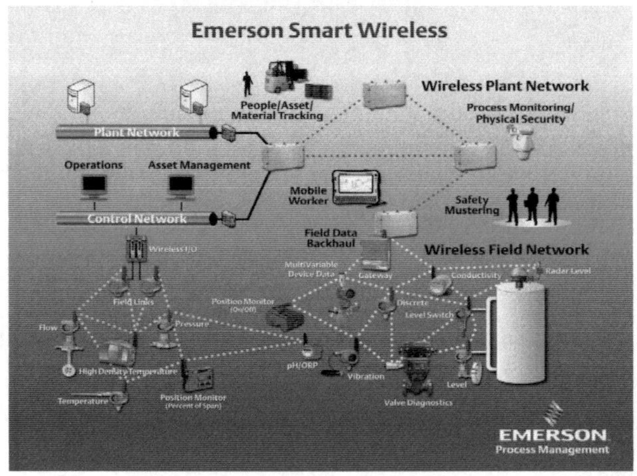

EMERSON
Process Management

IEC 62591 (WIRELESSHART) IS HART WITHOUT THE WIRES

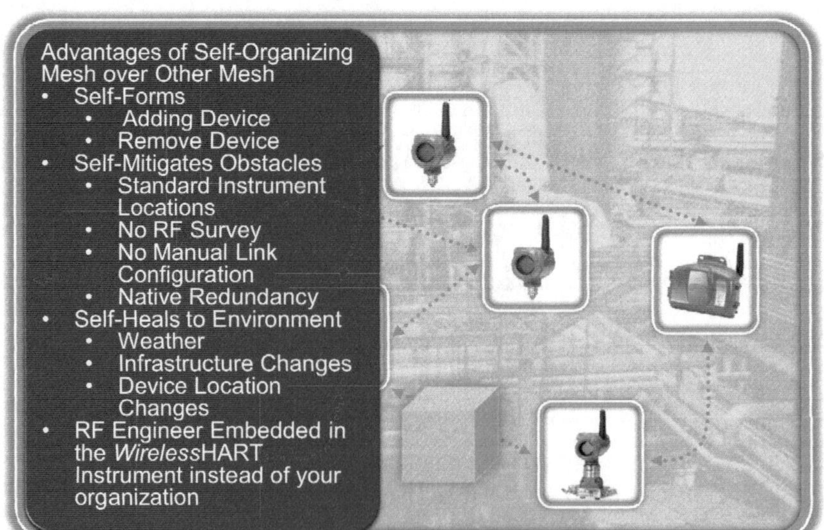

EMERSON
Process Management

ABB – E+H – Emerson – MacTek – Pepperl+Fuchs – Rohrback Coscaso – Siemens – Softing – Tyco – 200+ Vendors

HART Communication Foundation 7.1
IEC 62951

End-to-End Protection
- Encryption
- Authentication
- Verification
- Anti-Jamming
- Key Management

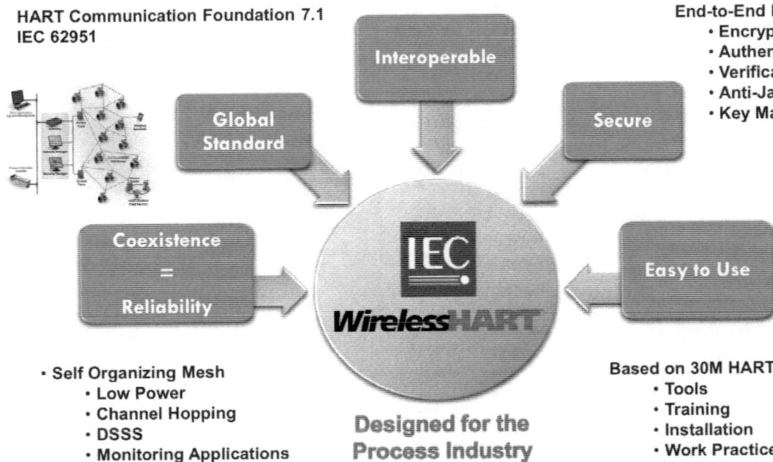

- Self Organizing Mesh
  - Low Power
  - Channel Hopping
  - DSSS
  - Monitoring Applications
  - Control Applications
- RF Engineer Embedded in Technology to ensure reliability

Designed for the Process Industry

Based on 30M HART Devices
- Tools
- Training
- Installation
- Work Practices
- Asset Management
- System Integration

EMERSON
Process Management

THINKING THROUGH WIRELESS: SECURITY

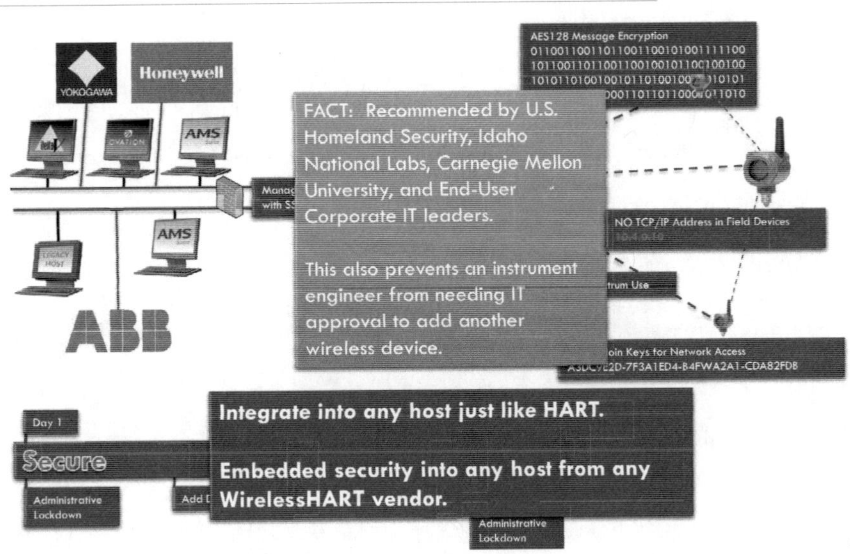

EMERSON
Process Management

THINKING THROUGH WIRELESS: INTEGRATION

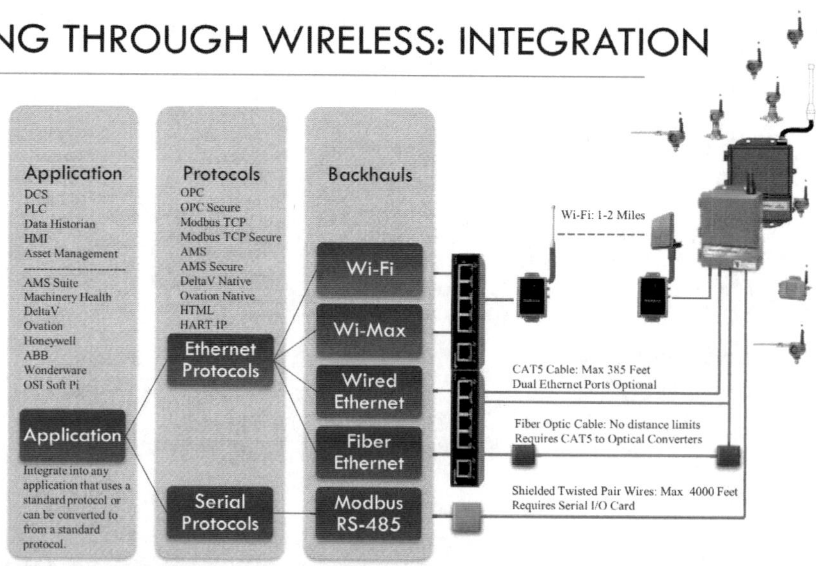

Goal: Connect applications with data coming to and from the WirelessHART Network. Do what is easiest.

EMERSON. Process Management

EVER EXPANDING WIRELESS PORTFOLIO

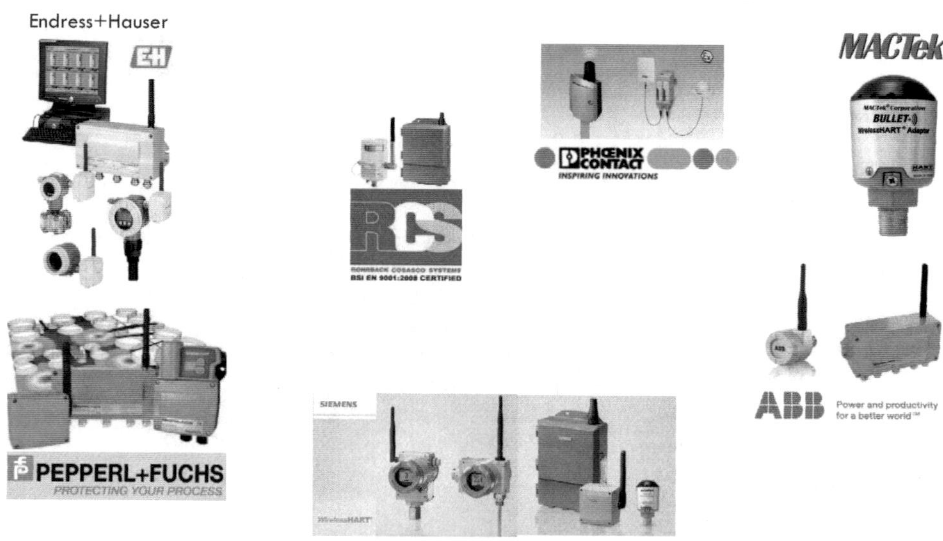

EMERSON. Process Management

EMERSON'S EXPANDING WIRELESS PORTFOLIO

EMERSON. Process Management

INTEROPERABILITY

The HART Communications Foundation has in excess of 240 Members

EMERSON. Process Management

WIRELESS APPLICATIONS ARE EVERYWHERE

Process / Asset Reliability

- Motor, pump, blower monitoring
- Valve monitoring
- Filter monitoring

Personnel Productivity

- Manual gauge replacement
- Reduced operator rounds
- Mobile workers

Efficiency / Throughput

- Tank level monitoring
- Heat exchanger monitoring
- Rotating kiln / moving equipment monitoring
- Steam trap monitoring
- Oil & Gas / Steam injection

Health, Safety, Environment

- Emissions / discharge monitoring
- Eye wash & safety shower monitoring
- Pressure relief valve monitoring
- Secondary level measurements
- Video

EMERSON
Process Management

WATER TREATMENT / EFFLUENT APPLICATIONS

Inlet Flow Rate for Regulatory Requirements: Rosemount 8700 Mag Flow with THUM adapter

Inlet Temperature for Regulatory Requirements: Rosemount 648 Wireless Temperature

pH/ ORP Monitoring for Water Treatment: Rosemount 6081 – Wireless PH/ORP

Vibration on Pumps: CSI 9420 Wireless Vibration

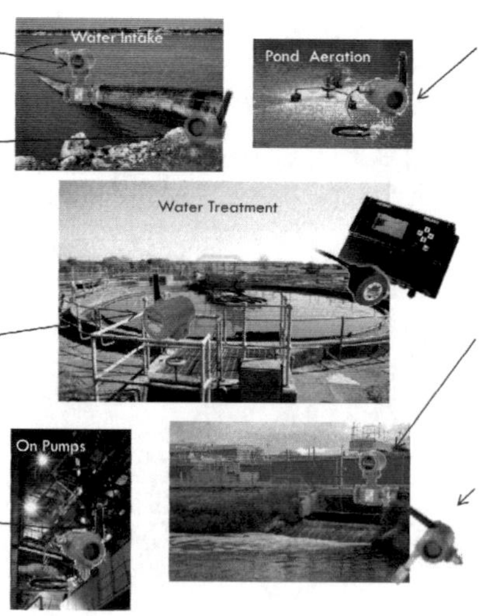

Aerator Vibration Monitoring: CSI 9420 Wireless Vibration

Sludge Blanket Level Monitoring For Sediments: Mobrey MSL600 with 4-20 output to a 648 Wireless

Outlet Flow Rate for Regulatory Requirements: Rosemount 8700 Mag Flow with THUM adapter

Outlet Temperature for Regulatory Requirements: Rosemount 648 Wireless Temperature

EMERSON
Process Management

Emerson Smart Wireless Solutions Help Power Plants Move to the Next Level

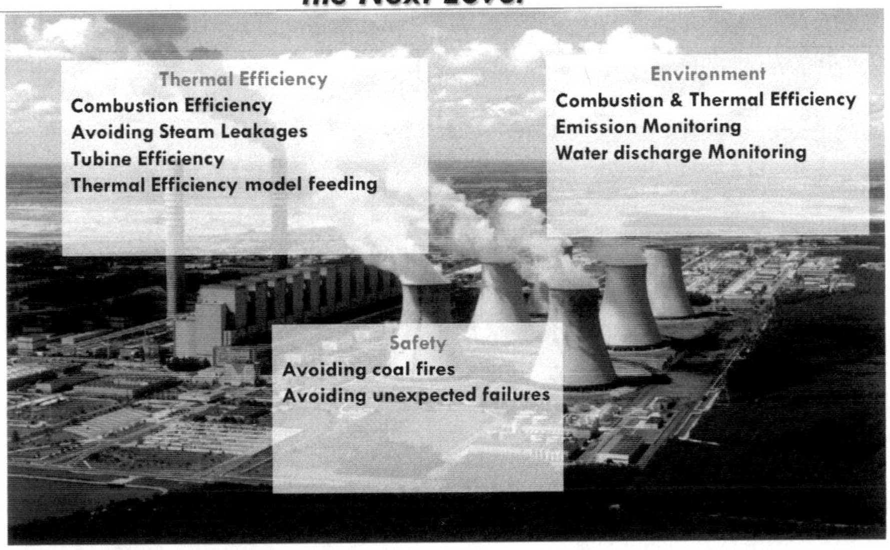

EMERSON. Process Management

WIRELESS MONITORING OF EYEWASH AND SAFETY SHOWER

BENEFITS

- IMPROVE SAFETY TEAM RESPONSE TO A PERSON IN TROUBLE
- PROVIDE TIMESTAMP DATA ON SHOWER USE FOR INCIDENT REPORTING AND SAFETY COMPLIANCE AUDITS

EMERSON. Process Management

WIRELESS VALVE POSITION MONITOR

Features:
- Linkage-less sensor
- Rotary or linear movement
- Two power module options
- Easy Installation - attach & calibrate
- Suitable for use in any Zone (intrinsic safe)

Applications:
- Incremental monitoring
- Automated valves – throttling and on/off
- Manual valves

Emerson Advantage
- Valve health issues uncovered wirelessly before they affect the process
- Reduce operator rounds in hazardous areas
- Increase monitoring options in hard to reach locations

EMERSON.
Process Management

4300 PNEUMATIC OUTPUT OPTION

- SAME BRANDING– TOPWORX 4310 AND FISHER 4320
  - SAME RADIO AND ANTENNA
  - SAME POSITION SENSING TECHNOLOGY
  - SAME MOUNTINGS
  - SAME PRODUCT SUPPORT STRUCTURE
- ADDED PNEUMATIC OUTPUT OPTION

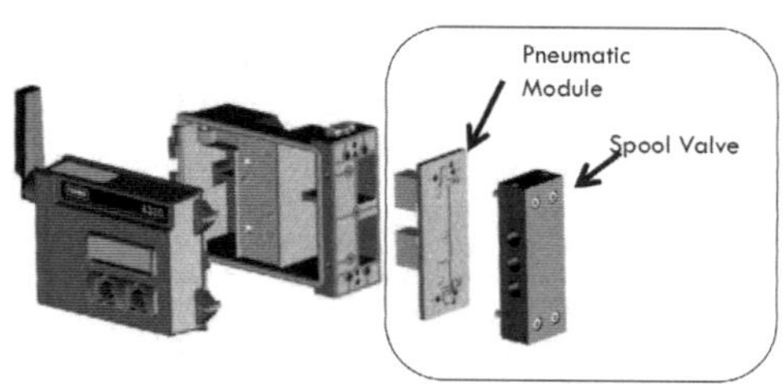

Pneumatic
Output
Option

EMERSON.
Process Management

CSI 9420 WIRELESS VIBRATION TRANSMITTER BASE FUNCTIONALITY

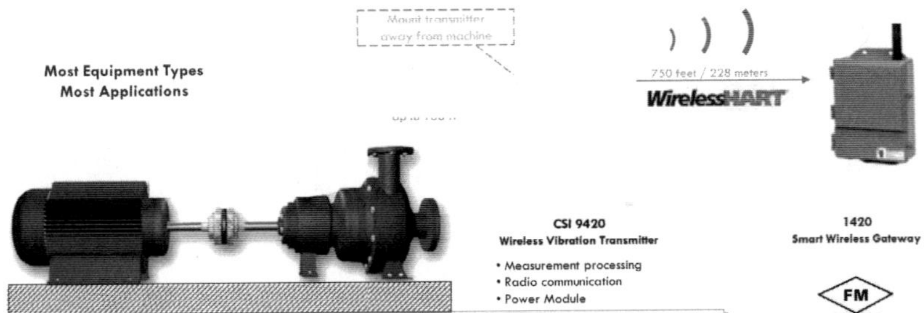

Outputs (scalar values)

1. **Overall Vibration** Units: Velocity (in/sec or mm/sec)
   Range: 2 Hz – 1000 Hz (per ISO 10816 standard).
   Comment: Identifies problems like imbalance and looseness

2. **PeakVue® Vibration** Units: G's Peak
   Range: 1,000 Hz – 20,000 Hz
   Comment: Quantify early bearing & gearbox defects

3. **Temperature** Range: -30 to +250 F / -34 to +120 C
   Comment: When sensor with embedded temperature used

EMERSON. Process Management

Early Detection

PEAKVUE™ FOR EARLIEST BEARING FAILURE AND PUMP CAVITATION WARNING

- GAME CHANGER IN THE CONTROL ROOM
- MEASURE OF THE PEAK VALUE OF IMPACTS WHEN METAL IMPACTS METAL INSIDE THE MACHINE
- DETECTS KEY MECHANICAL FAULTS
  - CAVITATION (PUMPS)
  - LUBRICATION
  - BEARING DEFECT
- PROVIDES EARLY AND ACCURATE WARNING
- PROVIDES SEVERITY ASSESSMENT

EMERSON. Process Management

EARLIER IS BETTER FOR CONDITION MONITORING

PEAKVUE DETECTS PROBLEMS

Stage	Bearing Life	Vibration (in/sec)	PeakVue (g's)
0	20-100%	0.15	0
1	<20%	0.15	4
2	<10%	0.15	8
3	<5%	0.16	12
4	<1%	0.18	25
Failure	0%	>0.30	>40

Traditional Vibration Measurement

Minimal change in classic vibration measurement until failure is imminent or has occurred (traditional analysis method)

PeakVue Vibration Measurement

Significant change in patented PeakVue value as we near failure

EMERSON. Process Management

TRACKING BEARING FAILURE WITH OVERALL VIBRATION VS. PEAKVUE EARLY WARNING

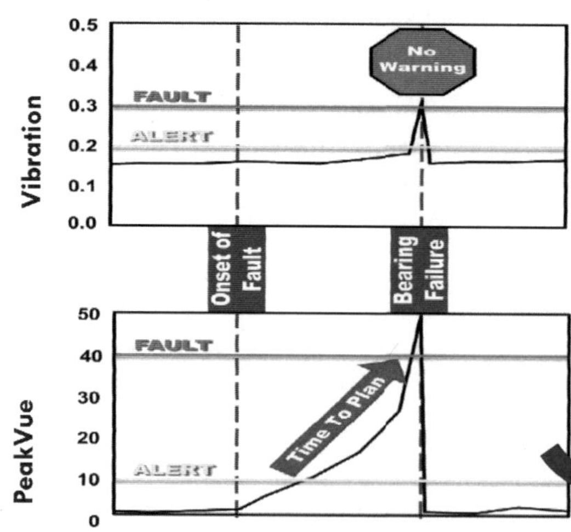

Earlier indication of impending defects:
- More time to plan corrective maintenance
- Avoid lost production from downtime
- Reduce overall maintenance costs
- Achieve greater OEE

EMERSON. Process Management

Essential Asset Monitoring Suite

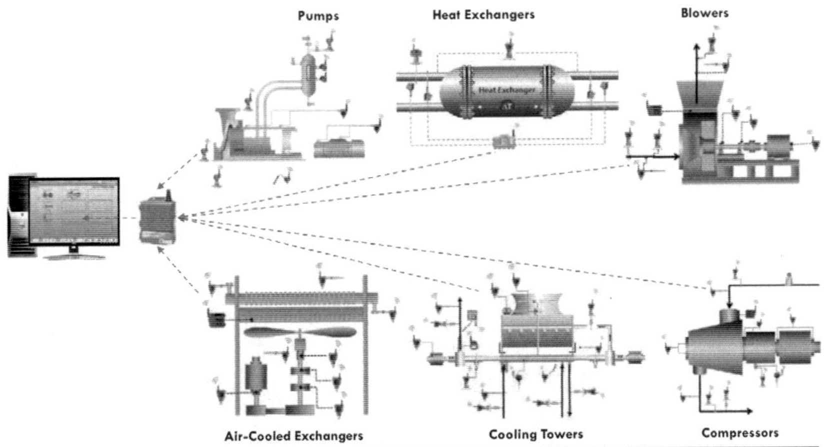

Preconfigured applications built on AMS Asset Graphics software which combines process and asset data obtained from the field to report Overall Asset Health

EMERSON.
Process Management

WIRELESS INSTRUMENTATION ALLOWS YOU TO CAPTURE MISSING DATA COST EFFECTIVELY – COOLING TOWERS

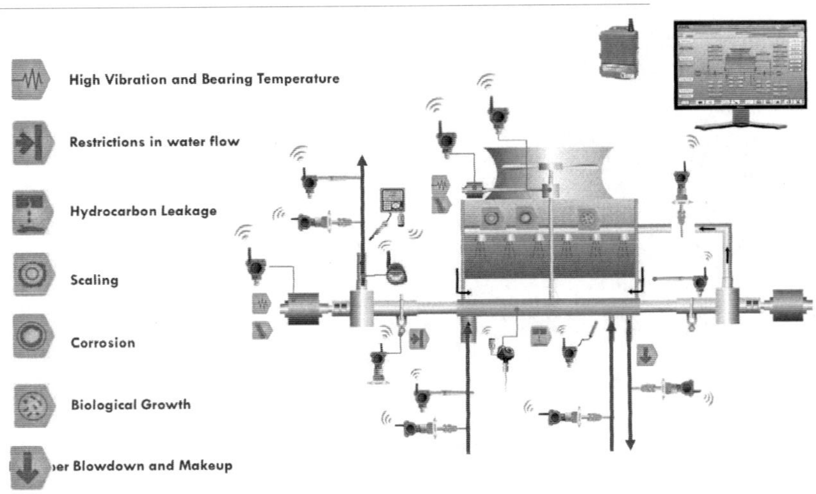

EMERSON.
Process Management

Cooling Tower Monitoring Algorithms

The following KPI variables will be calculated if sufficient inputs are provided:

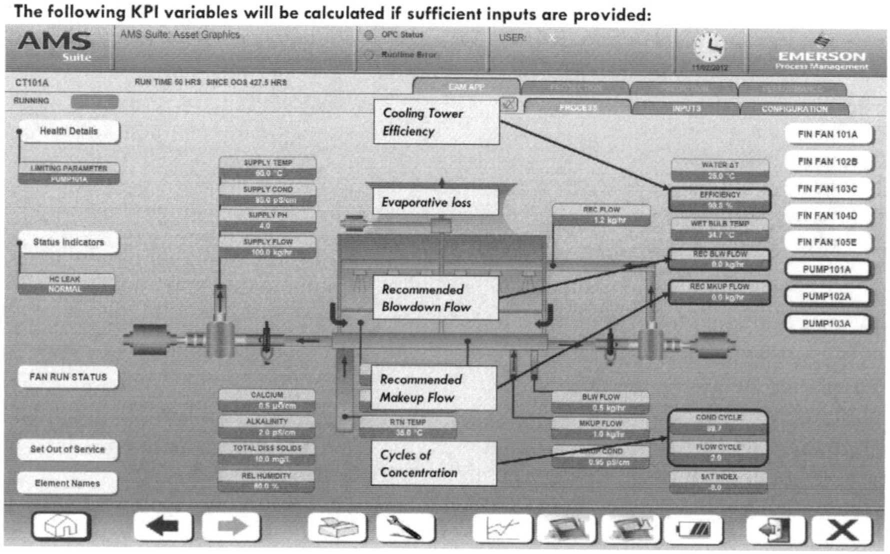

EMERSON
Process Management

PUMP APPLICATIONS

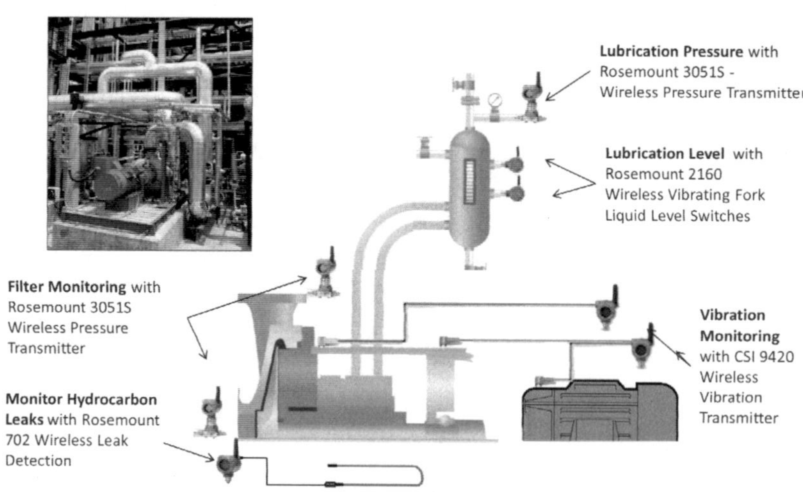

EMERSON
Process Management

MULTIPLE SOLUTIONS WITHIN ONE NETWORK

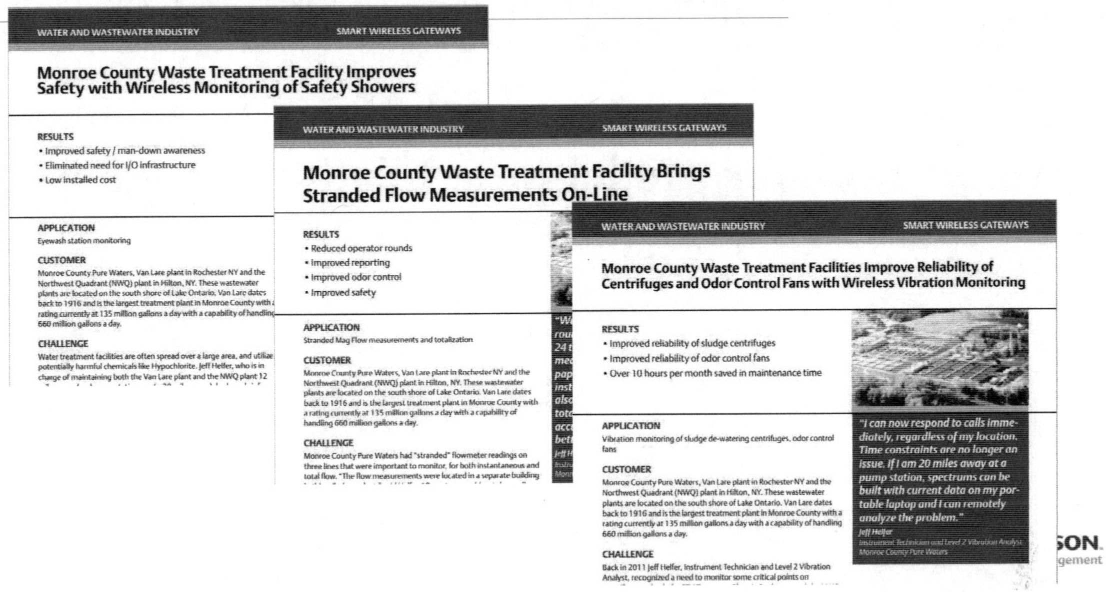

WHY USE WIRELESS HART: SUMMARY

- SMART WIRELESS REDUCES COSTS AND COMPLEXITY OVER WIRED
- WIHART IS SIMPLE, SCALABLE, RELIABLE, SECURE, AND STANDARDS-BASED
- PROVEN IN THOUSANDS OF CHALLENGING APPLICATIONS AROUND THE WORLD
- BROADEST RANGE OF WIRELESS PRODUCTS AVAILABLE
- YOU CAN START ANYWHERE
- EMERSON HAS THE EXPERIENCE AND EXPERTISE YOU CAN RELY ON

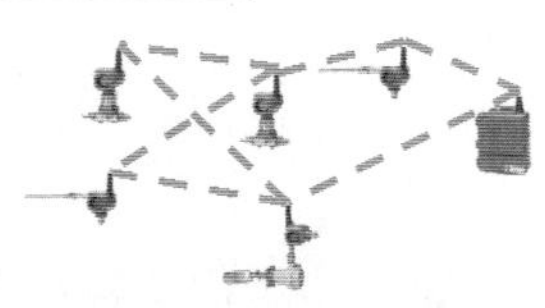

www.EmersonSmartWireless.com

EMERSON. Process Management

05-23-2014

A Generic Bayesian Approach Using Laplace Approximation for Model-based Failure Prognosis

Liang Xu

SpectraQuerst. Inc

Content

- Introduction
- Model-Based Failure Prognosis
- Laplace Approximation
- Prognosis Application and Case Study
- Closing Remarks

Introduction: Motivation

- Safe and reliable operation of complex engineered systems requires health monitoring and prognosis using real time sensory signals.
- Effective failure prognosis provides tremendous benefits such as improved safety, reliability, and reduced costs.
- Challenges: accurately assess the current health state of system degradation and precisely evaluate the remaining useful life.

SpectraQuest. Inc

Failure Prognosis

Essential Tasks

- **Health sensing function**
  To ensure high damage detectability by designing an optimal sensor network (Λ)
- **Health reasoning function**
  To extract system health relevant information in real-time and to diagnose system health condition (κ)
- **Health prognosis function**
  To predict remaining useful lives (RULs) of engineered systems in real-time (κ)
- **Health management function**
  To enable optimal decision making on maintenance of engineered systems (κ)

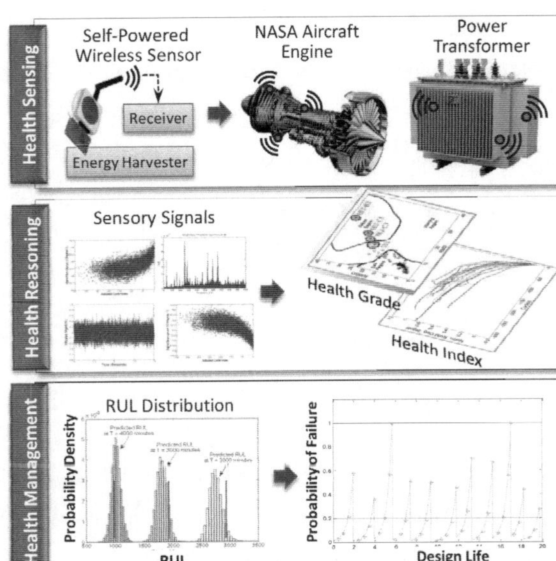

SpectraQuest. Inc

Model-Based Failure Prognosis

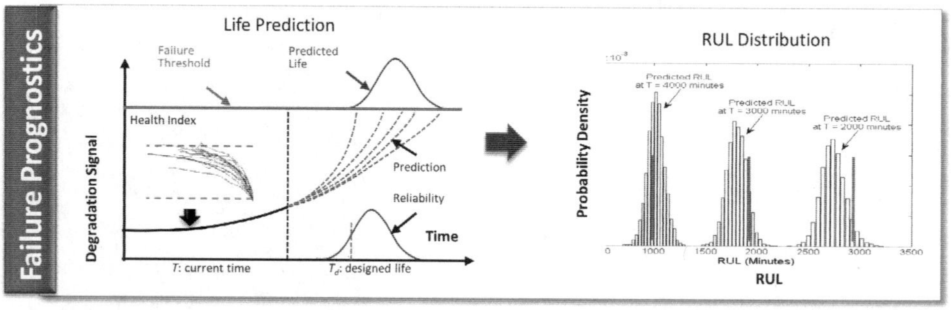

- Failure Prognosis: Model-based Approach [1-4];
  Data-Driven Approach [5] .
- A valid system performance degradation model;
- An effective model updating approach to update the model parameters based on real time sensory signals.
- Bayesian Updating [6-8].
- Objective: a general Bayesian inference mechanism and employs an efficient Bayesian updating approach.

SpectraQuest. Inc

Bayesian Inference Model

Bayesian theorem

Based on the Bayes' theorem, the posterior distribution of□ given new observation x can be expressed as:

Posterior

Likelihood

Prior

$$f_{\square|X}(\square|x)=\frac{f_{X,\square}(x,\square)}{f_X(x)}=\frac{f_{X|\square}(x|\square)\times f_{\square}(\square)}{f_X(x)}$$

The Bayesian approach is used for updating information about the degradation model parameters: □ , given the sensory measurements, x

SpectraQuest. Inc

A Generic Bayesian Prognosis Platform

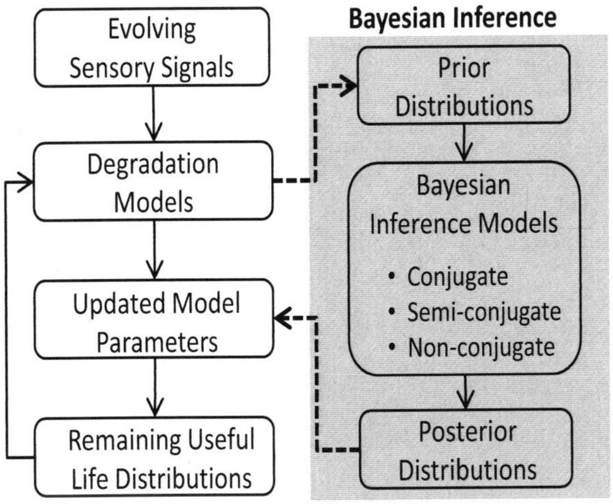

Step 1 : selecting an appropriate degradation model, and prior.

Step 2 : building a likelihood function, with the new data.

Step 3 : updating the joint probability distributions of the model parameters.

Step 4 : updating the lifetime distribution based on the updated degradation model parameters.

Laplace Approximation

Laplace Approximation approximates an unknown distribution with a Gaussian or multivariate Gaussian distribution.

According to natural logarithm to the second order of Taylor series expansion at a point x^*

$$\ln p(\mathrm{x} \mid L) = \ln p(\mathrm{x}^* \mid L) + (\mathrm{x} - \mathrm{x}^*)^T \cdot J(\mathrm{x}^* \mid L)$$
$$+ \frac{1}{2}(\mathrm{x} - \mathrm{x}^*)^T \cdot H(\mathrm{x}^* \mid L) \cdot (\mathrm{x} - \mathrm{x}^*) + O(\mathrm{x}^* \mid L)$$
$$where\ J(\mathrm{x}^* \mid L) = \nabla \ln p(\mathrm{x}^* \mid L)\ \ and\ \ H(\mathrm{x}^* \mid L) = \nabla^2 \ln p(\mathrm{x}^* \mid L)$$

where J(.) and H (.) are the Jacobian and Hessian matrixes.

Laplace Approximation

Let the expansion point $\mathbf{x}^*$ to be a local maximum, the Jacobian matrix J(.) will be vanished:

$$\ln p(\mathrm{x} \mid L) = \ln p(\mathrm{x}^* \mid L) + (\mathrm{x}$$ [redacted] $$\mid L) + \frac{1}{2!}(\mathrm{x}-\mathrm{x}^*)^T \cdot H(\mathrm{x}^* \mid L)\cdot(\mathrm{x}-\mathrm{x}^*) + O($$ [redacted] $$)$$

$$\ln p(\mathrm{x} \mid L) \approx \ln p(\mathrm{x}^* \mid L) + \frac{1}{2}(\mathrm{x}-\mathrm{x}^*)^T \cdot H(\mathrm{x}^* \mid L)\cdot(\mathrm{x}-\mathrm{x}^*)$$

So the posterior distribution $p\,(x|L)$ can be transformed to:

$$p(\mathrm{x} \mid L) = \exp(\ln p(\mathrm{x} \mid L)) \approx p(\mathrm{x}^* \mid L)\cdot \exp\left\{\frac{1}{2}(\mathrm{x}-\mathrm{x}^*)^T \cdot H(\mathrm{x}^* \mid L)\cdot(\mathrm{x}-\mathrm{x}^*)\right\}$$

$$\propto \exp\left\{-\frac{1}{2}(\mathrm{x}-\mathrm{x}^*)^T \cdot \Sigma^{-1} \cdot (\mathrm{x}-\mathrm{x}^*)\right\} \qquad \Sigma = [-H(\mathrm{x}^* \mid L)]^{-1}$$

the poster distribution can be approximated with a multivariate normal distribution with a mean value vector x^* and a covariance matrix of $[-H\,(x^*|L)]^{-1}$

SpectraQuest. Inc

Laplace Approximation

- Step 1: Transform the posterior distribution to a natural logarithm
- Step 2: Find the local maxima point $\mathbf{x}^*$
- Step 3: Evaluate the Hessian matrix H(.) at the local maxima point $\mathbf{x}^*$
- Step 4: Use the point $\mathbf{x}^*$ and Hessian matrix to get the reasonable approximation

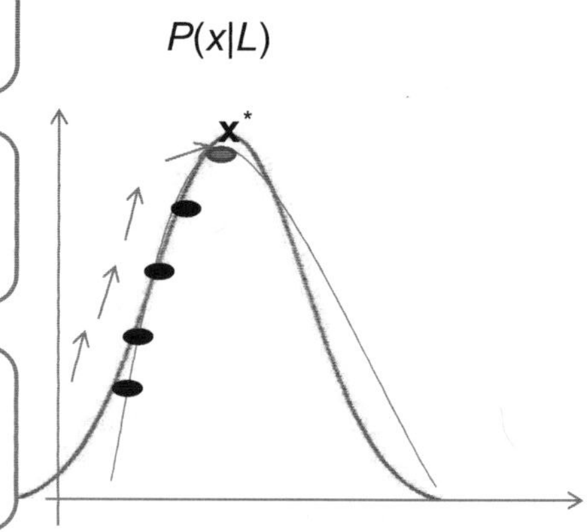

SpectraQuest. Inc

Prognosis Application

Degradation Model

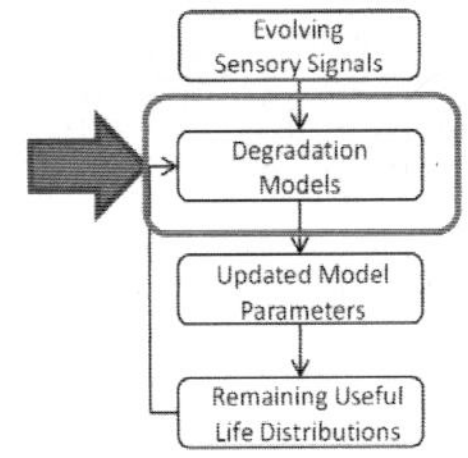

A quadratic exponential degradation model is used in this study [7,8], as :

$$S(t_i) = S_0 + \square \times \exp(\square t_i^2 + \square t_i + \square(t_i) \square \frac{\square^2}{2})$$

$S(t_i)$ represents the degradation signal at time t_i;

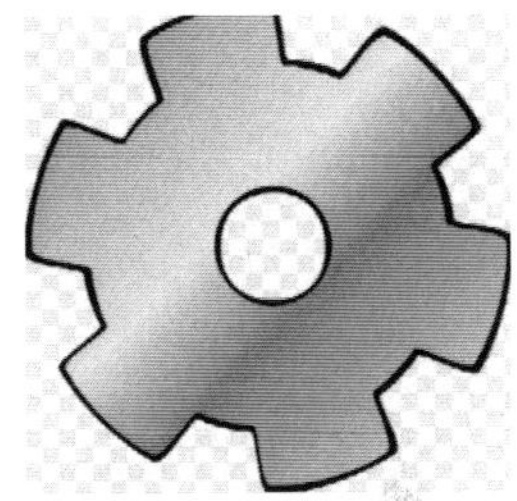

S_0: is a constant;

$\square$, $\square$ and $\square$ are stochastic model parameters;

$\square$ represents the random error term which follows normal distribution with zero mean and $\square^2$ deviation.

SpectraQuest. Inc

Prognosis Application

Likelihood Function

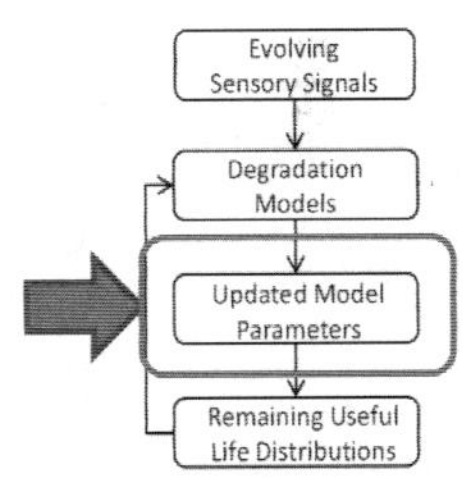

It is more convenient to work with the logged signal at time t_i which will denoted as:

$$L(t_i) = \ln(S(t_i) \square S_0) = \ln \square \square \frac{\square^2}{2} + \square t_i^2 + \square t_i + \square(t_i)$$

$$= \square' + \square t_i^2 + \square t_i + \square(t_i)$$

Where $\square' = \ln \square - \square^2/2$.

Since the error terms are iid normal random variables the following likelihood function can be obtained :

$$p\left(L_1, L_2, \cdots, L_k \mid \delta', \alpha, \beta\right) \sim \left(\frac{1}{\sqrt{2\pi\sigma}}\right)^k \cdot \exp\left\{-\sum_{i=1}^{k}\left(\frac{\left(L_i - \delta' - \beta t_i - \alpha t_i^2\right)^2}{2\sigma^2}\right)\right\}$$

SpectraQuest. Inc

Prognosis Application

Bayesian Updating

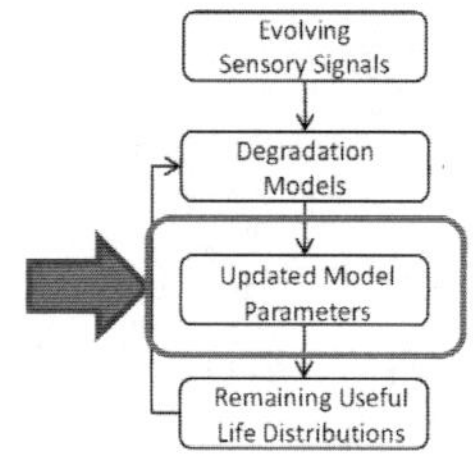

If the prior distributions for □′, □ and □ are provided, for example, □$_0$(□ ′, □ , □), the joint posterior distribution for these parameters can be expressed as:

$$p\left(\delta',\alpha,\beta \mid L_1,L_2,\cdots,L_k\right)\sim$$

$$\left(\frac{1}{\sqrt{2\pi\sigma}}\right)^k\cdot\exp\left\{-\sum_{i=1}^{k}\left(\frac{\left(L_i-\delta'-\beta t_i-\alpha t_i^2\right)^2}{2\sigma^2}\right)\right\}\times\pi_0\left(\delta',\alpha,\beta\right)$$

The research challenge is how to update the posterior efficiently, without using the simulation methods. In this research the Laplace Approximation has been used.

Prognosis Application

Remaining Useful Life

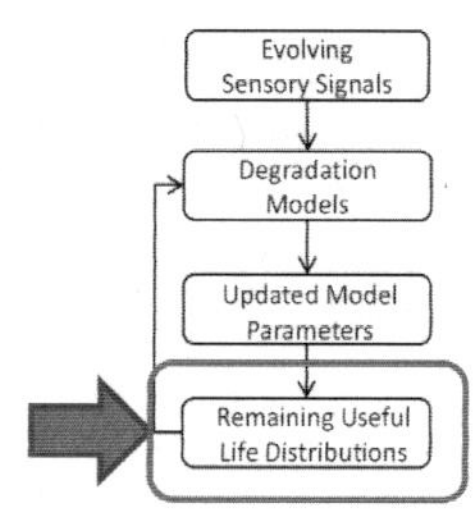

The life distribution information can be obtained when a failure threshold D is specified. The life data can be determined by solving the N sets of quadratic equations as:

$$T_i=\text{roots}\left(\alpha T^2+\beta_i T+\delta'_i+\varepsilon_i-D=0\right),\quad where\ i=1,2,\ldots,N.$$

the life T_i is the positive real root of the quadratic equation. The sample size, N, can be increased because solving the quadratic equation N times is trivial.

Prognosis Case Study

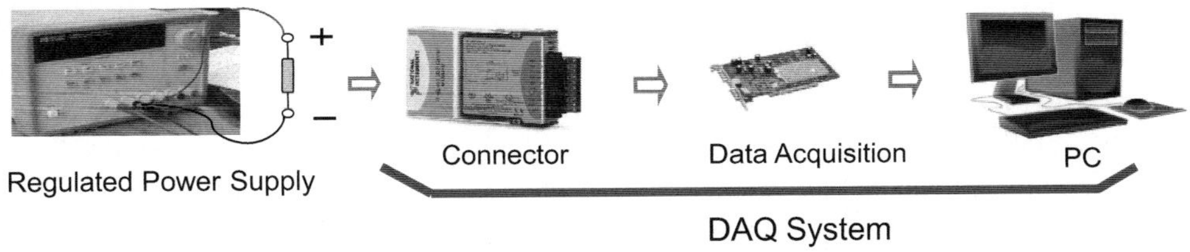

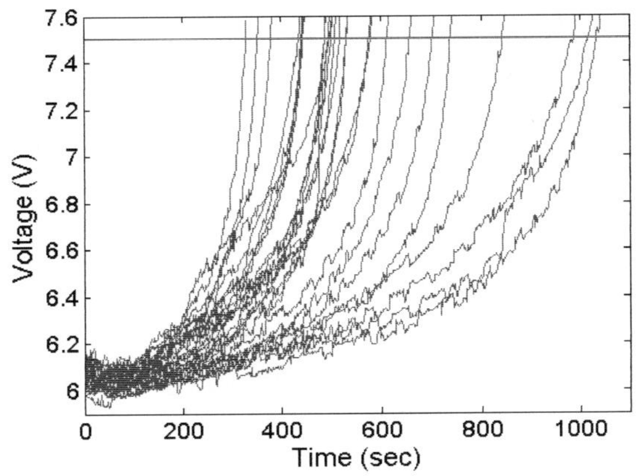

Prognosis Case Study

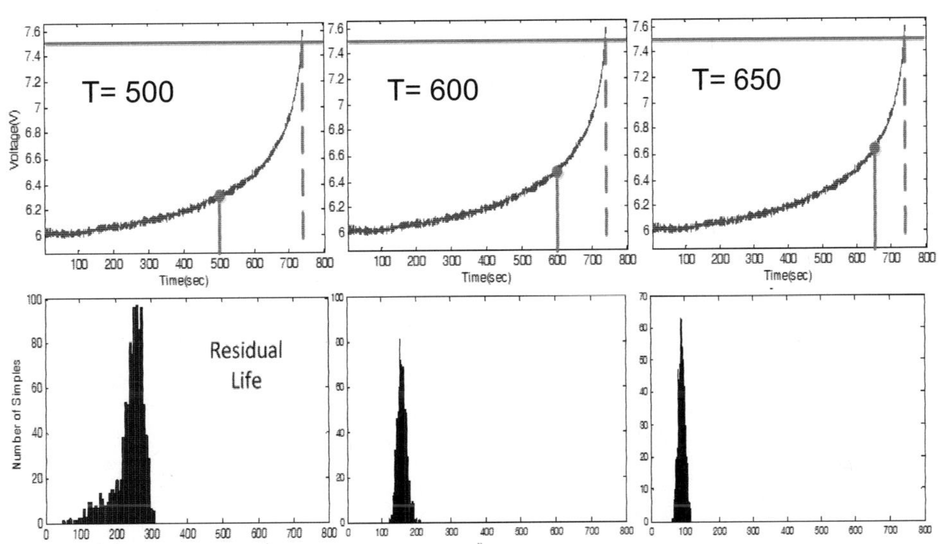

Prognosis Case Study

Resistor	Prediction Time	Actual Life	Predicted Life		Resistor	Prediction Time	Actual Life	Predicted Life	
			Mean	STD				Mean	STD
1	400	506.74	470.89	11.38	14	400	485.24	471.42	6.51
2	400	432.76	435.36	3.71	15	400	494.35	490.07	4.66
3	800	1016.22	1082.74	19.01	16	400	485.38	483.79	10.98
4	250	351.53	322.64	4.48	17	400	441.53	437.16	10.28
5	400	494.29	479.75	10.37	18	400	440.44	457.23	10.44
6	500	738.53	690.37	11.17	19	250	327.23	325.15	1.42
7	400	529.95	498.65	5.76	20	700	837.90	8.55	5.98
8	200	256.36	246.08	3.65	21	400	443.14	4.48	11.54
9	400	574.35	569.19	20.09	22	400	499.55	517.63	13.72
10	400	611.74	609.11	11.53	23	800	1032.94	1041.37	8.73
11	200	243.66	240.45	1.02	24	400	531.35	484.99	3.35
12	800	983.07	933.06	19.11	25	400	576.22	535.13	6.29
13	250	378.06	353.65	7.52	26	400	655.03	643.27	4.34

Prognosis Case Study

Comparison with MCMC

Time	Method	NoFE[a]	Time[b]
400	Laplace Approximation	76	0.116s
	MCMC	50000	33.64s
600	Laplace Approximation	95	0.138s
	MCMC	50000	34.70s

[a] Number of Function Evaluation [b] Based on a 2.40 GHz CPU

Time	Laplace Approximation		MCMC		True RUL
	Mean	STD	Mean	STD	
400	258.16	14.74	275.28	13.19	383.53
600	115.03	10.97	130.32	10.65	183.53

Closing Remarks

- A generic Bayesian framework using Laplace approximation is presented for model-based remaining useful life prognosis.
- The developed Bayesian prognosis approach can eliminate the dependency of evolutionary sampling process or a selection of distribution types for the degradation model parameters.
- The Bayesian updating process can be carried out efficiently with the Laplace approximation method, which enables the proposed approach to be applied for real-time prognosis applications.

Future Work

- Future research can be conducted to address the challenges of potentially multi-mode posterior distribution types.

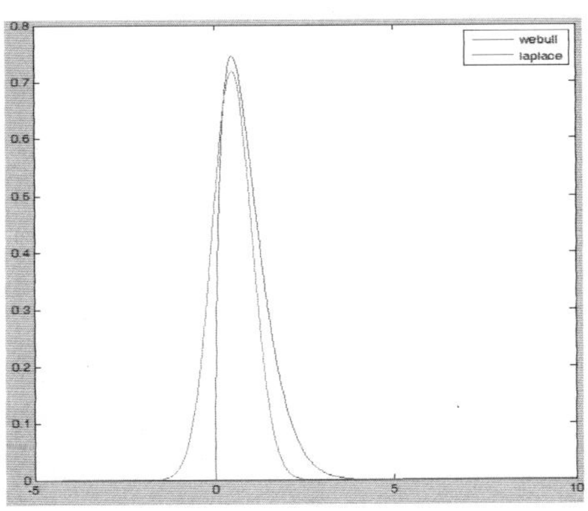

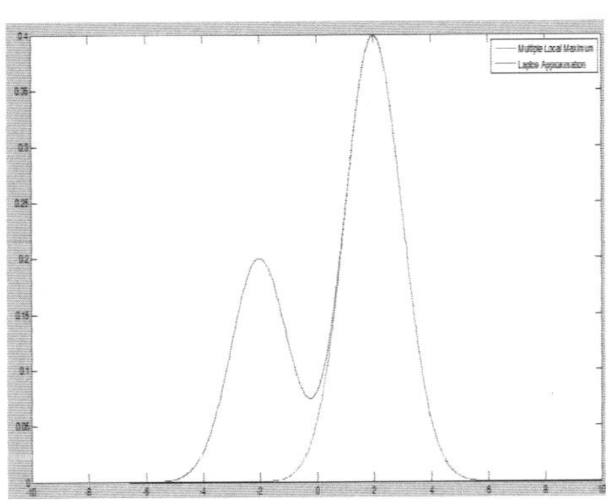

References

1. Doksum, Kjell A., and Arnljot Hbyland., 1992."Models for Variable-Stress Accelerated Life Testing Experiments Based on Wiener Processes and the Inverse Gaussian Distribution." *Technometrics* 34, no. 1: 74-82.
2. C. J. Lu., 1995,"Degradation processes and related reliability models," Ph.D. thesis, McGill University, Montreal, Canada.
3. Whitmore, G. A., M. J. Crowder, and J. F. Lawless., 1998,"Failure inference from a marker process based on a bivariate Wiener model." *Lifetime Data Analysis* 4, no. 3: 229-251..
4. Bagdonavicius, Vilijandas, and Mikhail S. Nikulin., 2001,"Estimation in degradation models with explanatory variables." *Lifetime Data Analysis* 7, no. 1: 85-103..
5. Alguindigue, I. E., Anna Loskiewicz-Buczak, and Robert E. Uhrig., 1993, "Monitoring and diagnosis of rolling element bearings using artificial neural networks." *Industrial Electronics, IEEE Trans.* 40, no. 2: 209-217.
6. Youn, Byeng D., and Pingfeng Wang., 2008,"Bayesian reliability-based design optimization using eigenvector dimension reduction (EDR) method." *Structural and Multidisciplinary Optimization* 36, no. 2: 107-123.
7. Gebraeel, Nagi Z., Mark A. Lawley, L. I. Rong, and K. JENNIFER., 2005, "Residual-life distributions from component degradation signals: A Bayesian approach." *IIE Transactions* 37, no. 6: 543-557.
8. Gebraeel, N., M. Lawley, and R. Liu., 2002,"Vibration-based condition monitoring of thrust bearings for maintenance management." *Proc. ANNIE 2002 Smart Engineering System Design: Neural Networks, Fuzzy Logic, Evolutionary Programming, Artificial Life and Data Mining*: 543-551.
9. Li, Y., S. Billington, C. Zhang, T. Kurfess, S. Danyluk, and S. Liang., 1999,"Dynamic prognostic prediction of defect propagation on rolling element bearings."*Tribology transactions* 42, no. 2: 385-392.
10. Martin, K. F., 1994, "A review by discussion of condition monitoring and fault diagnosis in machine tools." *International Journal of Machine Tools and Manufacture* 34, no. 4: 527-551.
11. Booth, C., and J. R. McDonald., 1998,"The use of artificial neural networks for condition monitoring of electrical power transformers." *Neurocomputing* 23, no. 1: 97-109.
12. Macian, V., B. Tormos, P. Olmeda, and L. Montoro., 2003, "Analytical approach to wear rate determination for internal combustion engine condition monitoring based on oil analysis." *Tribology International* 36, no. 10 : 771-776.
13. Chen, Wei, B. CYRUS, and Farrokh Mistree., 1994, "COMPROMISE: an effective approach for condition-based maintenance management of gas turbines."*Engineering Optimization* 22, no. 3: 185-201.
14. Lee, Jay., 1996,"Measurement of machine performance degradation using a neural network model." *Computers in Industry* 30, no. 3: 193-209.
15. Gebraeel, Nagi Z., Mark A. Lawley, L. I. Rong, and K. JENNIFER., 2005, "Residual-life distributions from component degradation signals: A Bayesian approach." *IIE Transactions* 37, no. 6: 543-557.

SpectraQuest. Inc

Questions?

Thank You!

SpectraQuest. Inc

ACTIONABLE MAINTENANCE INFORMATION FROM HUMS

Rodrigo E. Teixeira, Kari E. Morris and F. Christian Sautter
UAH Research Institute
Reliability and Failure Analysis Laboratory
University of Alabama in Huntsville
Huntsville, AL 35899
Tel: 256.824.6147
Email: chris.sautter@uah.edu

Abstract: Condition Based Maintenance (CBM) of military helicopters is tracked by Condition Indicators (CI) calculated from Health Usage and Monitoring Systems (HUMS) vibration sensors. Even though many CIs have been proposed and implemented, they remain highly variable and difficult to interpret, leading maintainers to become desensitized. Here we show that a sequential Monte Carlo algorithm operating a stochastic non-linear model that includes a description of fault evolution can circumvent the fundamental shortcoming of the CI approach. We estimate fault probabilities from vibration spectra time-histories, showing excellent artifact rejection and accurate fault detections several months prior to all other existing warnings. We expect this approach may eventually allow scheduled maintenance to substitute unscheduled downtime, reducing maintenance cost.
Key words: Vibration; Stochastic modeling; Machine learning; Fault probability; Fault detection; Diagnostics; Condition Based Maintenance (CBM); Health Usage and Monitoring Systems (HUMS).

Introduction: From the outset, the Health Usage and Monitoring Systems (HUMS) program was predicated on reducing maintenance burdens by detecting faults (cracks) before failures, and substituting unscheduled with scheduled maintenance. By substituting reactive with proactive actions, aircraft downtime would be minimized, resulting in increased availability at reduced cost.

HUMS hardware includes sensor banks that are mechanically coupled to various aircraft components. For instance, the Modernized Signal Processing Unit (MSPU) includes sensing for engines and transmissions, fans, thrust bearings, and hanger bearings. Some components, such as the combining transmission, carry more than one sensor. Sensor suites monitor vibration (acceleration vector), but can also include flight parameters such as temperature, pressure, humidity and altitude.

During post flight operations, HUMS data is transferred and analyzed by an algorithm that calculates the health state of the monitored mechanical component. Following Condition Based Maintenance (CBM) practice, this health measure is termed a *Condition Indicator*, or CI. The underlying premise is that a fault alters the dynamic response of a component, and this alteration can be detected in the vibration signature. Others have shown, for instance, that the presence of any fault alters the response of an intact steel beam from harmonic to non-linear, creating vibrational amplitudes at sub- and super-harmonic frequencies [1]. As those vibrations propagate through a component, they become manifest through attached sensors. As such, we surmise that most faults are detectable from their vibrational signatures alone.

Even though faults should be detectable by HUMS, there is no consensus on which CI to use. By and large, CIs in current practice have not achieved their intended goals of pinpointing hidden conditions and reducing unnecessary inspections. For instance, it is not uncommon for a CI to exceed permissible thresholds multiple times, confounding the maintainer (Figure 1). As a result, maintainers have become desensitized, and CIs have been relegated to Aviation and Missile Command (AMCOM) engineers for analysis. Knowledge gained from comparing CI output to inspections has been used to adjust current CIs.

CIs are derived from features in the *current* vibration Power Spectral Density (PSD). In a typical flight regime, vibration time-trajectories are recorded for a time frame of a few seconds to minutes. This data is downloaded and the PSD is estimated using Fourier transform methods. Then, a CI is calculated from this power spectrum. For instance, a CI can reflect the total power contained within a (fixed or variable) frequency range of interest. Other CIs use the skewness, kurtosis, or some other feature of the PSD. Subsequently, another vibration measurement is recorded and a new PSD and CI are calculated. The maintainer utilizes the software in the ground station to plot CIs along a time axis and makes decisions based in part on observed trends and exceedances (values above a predetermined threshold.)

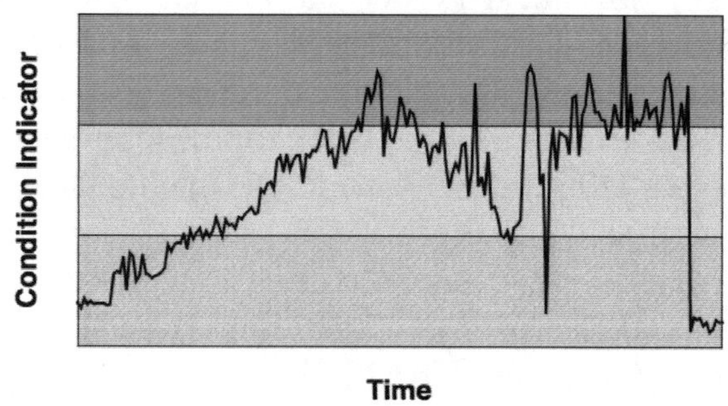

Figure 1. Condition indicator exceedances. *The high frequency bearing energy CI for an engine transmission is plotted in chronological sequence. The multiple exceedances into the red (warning) zone makes it difficult for the maintainer to determine the right course of action.*

Importantly, the PSD estimation step assumes that all vibrations picked up by the sensor are *stationary* stochastic processes. That is, all observed dynamics can be described by probability distribution functions (PDF) that do not change with time (*i.e.*, noises). On the other hand, *non-stationary* stochastic processes are artifacts, and their PDFs change with time. An example of an artifact is a load such as a heavy cargo or a tight maneuver: they are transient and may be gone (or intensified) during the next recording. Therefore, the presence of artifacts, whether they are known and quantifiable or not, distort the PSD and the CI. In theory, artifacts could be minimized by recording under controlled conditions (*e.g.*, grounded, unladen, recent track and balance, calm weather, etc.). In practice, these conditions cannot always be controlled for a given aircraft, and even less so for the entire fleet. Consequently, even if a well-designed CI extracts a fault-sensitive spectral feature, the resulting CI time-history would become obscured.

To remedy this situation one must find a strategy to reject the artifacts that inevitably distort the PSD. However, artifacts are generally unknown externalities that are difficult or impossible to model. Therefore, we focus instead on developing sufficiently accurate stochastic models for the known processes, such that any signature that does not conform to known physics will be, in essence, disregarded as an artifact. Here, an adequate description of fault evolution: its appearance and progression as a result of external stresses should be of paramount importance. Because the CI approach does not include a description of fault evolution, or even take into account past information in some way (*i.e.*, has no "memory"), it cannot do an adequate job of rejecting artifacts. As a result, its successes are limited to a few specific cases.

Numerical Method: As seen previously, fault evolution is a non-linear stochastic process. Hence, a new numerical method is required that estimates health states from vibration in the context of a non-linear stochastic model of fault. More specifically, given only noise and artifact-corrupted vibration (or its PSD), we seek good estimates of some internal state variable - a *fault magnitude*, for instance - which we can model with reasonable accuracy. Fortunately, such a numerical method already exists.

The well-known *Kalman* filter was perhaps the first Probabilistic Inference numerical method. Developed in the late 1950s, the Kalman filter was used in the Apollo program as part of the spacecraft's navigation computer. It estimated, recursively and in real time, course corrections from sensor data that was corrupted with noise and artifact. This initial application, which was limited to linear models, has since been expanded to accommodate non-linearities. This research track saw the creation of the *extended Kalman filter* and later the *unscented Kalman filter*. These two families of algorithms used different strategies to improve accuracy in non-linear processes, with the unscented Kalman being generally more accurate and also easier to implement. Yet, some assumptions and approximations still persisted. It was not until recently, in the 1990s, that *Sequential Monte Carlo* (SMC) methods were developed. SMC are much more powerful and general than preceding algorithms because they do not make any assumptions about the stochastic model of the system: nonlinearities and noises can take any form. SMC methods have resulted in several engineering breakthroughs during the last decade. Examples include tracking and navigation algorithms for autonomous robots, voice recognition, and "quant" financial algorithms for stock trading optimization. More broadly, since internal variables are recursively updated given information about the external world, SMC methods represent a form of machine learning.

Briefly, machine learning algorithms based on SMC are the only known numerical methods that can estimate the states of non-linear systems, as is the case with any fault process, with high accuracy. SMC is based on the representation of PDFs of any form as a collection of independent and identically distributed (i.i.d.) probability particles. As each particle is individually propagated through the model (*a.k.a.* dynamic state-space model, or DSSM), a time-updated *prior* PDF is estimated. The prior is convolved with new noisy samples (measurements) by use of the Bayesian update rule to yield the *posterior* PDF. As the number of particles grows large, estimates converge to true distributions and the algorithm reaches optimality; that is, it makes the best possible use of all available information. In the practical implementation of SMC to real world problems, the only remaining possible source of inaccuracy is the model itself. The rewriting of deterministic physical models to stochastic forms is not straightforward, and a rather heavy burden is placed on the modeler.

In practical terms, we seek to develop a DSSM containing, at the very least, an accurate non-linear model of fault evolution from an intact material. Also, due to the mathematical structure of the method, the DSSM should be a Markov process of 1st order. The DSSM takes on the general form:

$$\mathbf{x}_k = \mathbf{f}(\mathbf{x}_{k-1}, \mathbf{v}_{k-1}) \text{ and } \mathbf{y}_k = \mathbf{h}(\mathbf{x}_k, \mathbf{n}_k) \quad (1)$$

That is, the hidden system state, $\mathbf{x}_k$ (*e.g.*, fault magnitude), propagates over time index k, according to the system model **f** (component), and system noise $\mathbf{v}_{k-1}$ (external stresses). In the second equation, noisy observations, $\mathbf{y}_k$ (vibration) about the hidden states are given by the observation model **h** (sensor), and observation noise $\mathbf{n}_k$ (measurement noises). Both **f** and **h** can be any nonlinear map. Also, $\mathbf{v}_k$ and $\mathbf{n}_k$ are not necessarily additive or Gaussian. The problem is to successively estimate the hidden states given only noisy samples. This task can be performed by various proprietary and open-source SMC software [2].

Results: During the first year of this project our team sought to demonstrate the feasibility of the approach just described. To this end, we focused on the development of a DSSM that imparts the algorithm with an accurate mathematical context for learning and interpreting the vibration signal. The model can be conceptually divided into three parts. The first is a description of fault creation and progression as a result of external stresses. Here, we use pseudo-random stressors applied to a part and stochastically create a microscopic fault that subsequently grows with accumulated strain. This is coupled to a vibration model, which describes how the vibration signature is altered due to the properties of the fault. Lastly, we have a description of the transduction process, or measurement of vibration. Here, we include all known noises associated with the non-idealities of the sensor, its digital readout, and electronic interferences from the surrounding environment. This is illustrated conceptually in Figure 2.

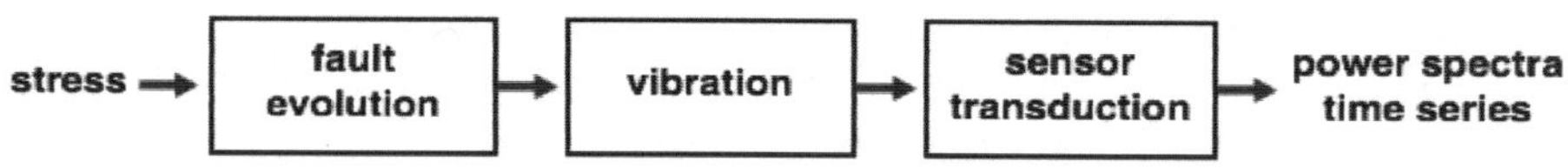

Figure 2. Block diagram representation of the DSSM. *The dynamic state-space model functional structure is represented schematically, starting from pseudo-random external stresses and ending with power spectral density time-series.*

The results are reported not as the fault magnitude, but as the *fault probability*. In the context of the present DSSM, the fault probability is simply the probability that a fault is present. Since microscopic faults may not be of concern to maintainers, we introduce an arbitrary minimum magnitude, or threshold, above which a fault is said to be present. Because the SMC algorithm is probabilistic, the fault magnitude is available not as a scalar, but as a vector representing the PDF. Therefore, the fault probability is estimated as the area under the fault magnitude PDF and to the right of the threshold (Figure 3). Accordingly, a fault probability of zero indicates a normal component, while a probability of 0.95 indicates a 95% chance a significant fault is present, a high level of certainty. Our research indicates that this is the most direct and useful representation for the health state of a component.

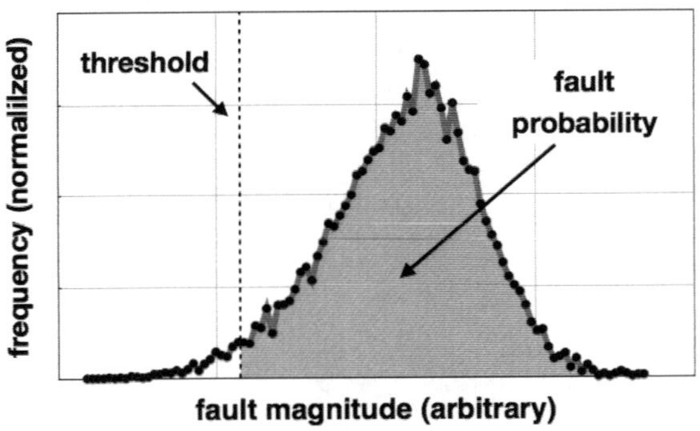

Figure 3. Fault probability calculation. *A fault magnitude probability distribution function is plotted. Also, an arbitrary minimum fault magnitude threshold is depicted. The fault probability is the area under the curve above the threshold (**blue area**). This distribution was estimated with a sequential Monte Carlo algorithm operating 10^4 probability particles.*

A Helicopter Example: A US Army Aviation Fleet Management Office (FMO) provided our team with data from their HUMS that could be utilized in this new approach to provide actionable maintenance information. The FMO was primarily concerned with two issues: *i*) they wanted a more definitive alert for the need take maintenance actions (*i.e.*, true positives), *ii*) while also eliminating No Evidence of Failure (NEOF) cases (*i.e.*, false positives). Based on the data provided by the FMO, the algorithm proved successful in meeting both goals.

The components chosen for this initial study were the number 1 and 2 engine transmissions since they have the largest set of teardown validations (ground truth data). In the results below, probabilities appear as black squares, and currently employed condition indicators as light and dark gray circles. The algorithm was identical for all number 1 and 2 engine transmissions. That is, there were no changes or adjustments between runs.

The first plot (Figure 4) represents data from a number 1 engine transmission. Existing CIs increased in magnitude, but in this case did not cause a removal. Instead, the component was removed due to a cockpit chip light. Upon inspection, the transmission revealed a bearing inner race that broke in multiple fragments. Those fragments caused secondary damage within the housing and the illumination of a chip warning light. The fault probability increased from zero to one about 705 flight hours prior to the chip light and the subsequent removal. Providing an early alert of this type would have allowed the maintenance crew to replace the transmission during a scheduled maintenance activity, reducing downtime and costs, as well as preventing the secondary damage to the component.

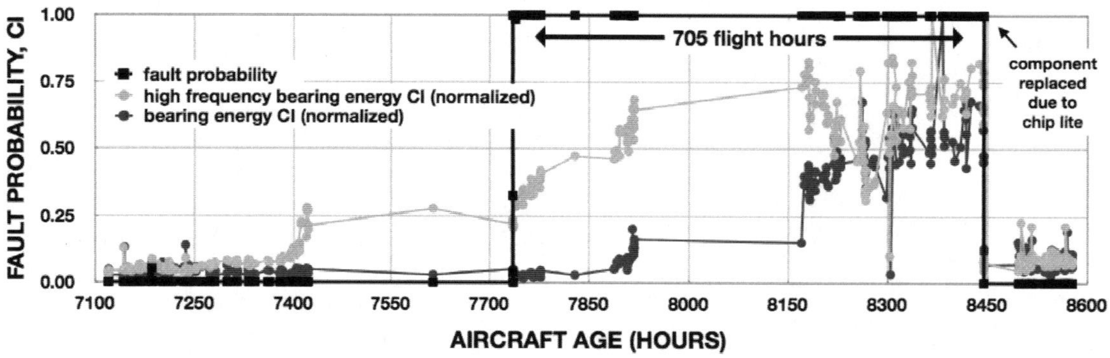

***Figure 4. Results for a #1 engine transmission.** The algorithm shows a high fault probability 705 flight hours before a cockpit chip light turned on, causing the grounding of the aircraft and the replacement of the transmission. Inspection revealed a bearing inner race failure and secondary damage.*

In the next case, condition indicators for another number 1 engine transmission caused a removal of that component (Figure 5). Upon a teardown inspection, the component was declared a NEOF (false positive). The algorithm estimated a fault probability of zero during the time in question, indicating correctly that the component was normal. Provided to the maintenance crew, this information would have eliminated significant maintenance actions, resulting in down time and the cost of a replacement component.

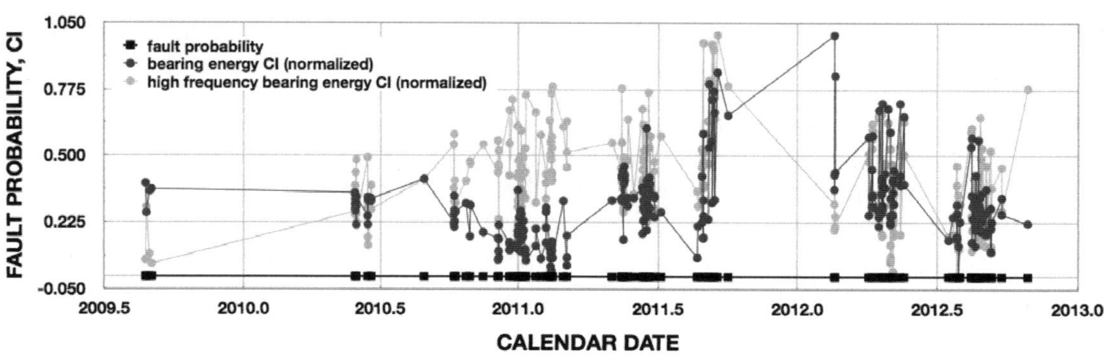

***Figure 5. NEOF in a #1 engine transmission.** Currently used CIs caused a removal, which was declared a NEOF. The algorithm estimated fault probability of zero throughout.*

In another aircraft, a number 1 engine transmission was removed due to the bearing energy CI (Figure 6). Inspection revealed abnormal wear patterns on both the input pinion and output bevel gear teeth with abnormal staining on the inner bearing race. Also, another bearing exhibited false brinelling on both inner and outer bearing races. These damages were determined by the airworthiness authority to be not significant. As a result, the component condition was set to green and the removal was considered a NEOF. The fault probability for this component is seen to rise gradually from zero by mid-year 2012 to ~0.23 at the time of replacement about 5 months later. The low fault probability would not indicate a removal is necessary thus supporting the operational status and eliminating the NEOF.

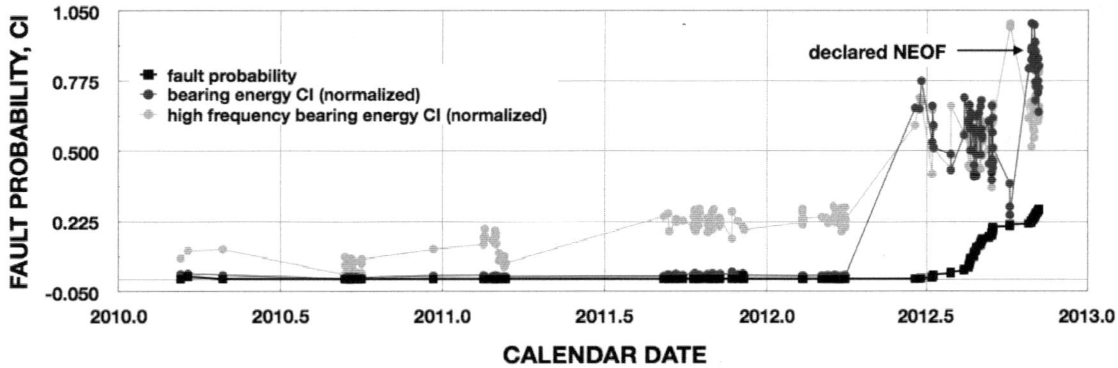

Figure 6. Low damage severity in a #1 engine transmission. *The bearing energy CI caused a removal by the end of the plotted time history. Inspection found some, but insignificant damage and the component condition was set to green. The fault probability at the time of inspection was ~0.23, which is consistent with the level of damage found.*

In this next case, a number 2 engine transmission was replaced. Even though the HUMS data was not complete, the algorithm indicates a fault probability of ~1 about 19 months prior (Figure 7).

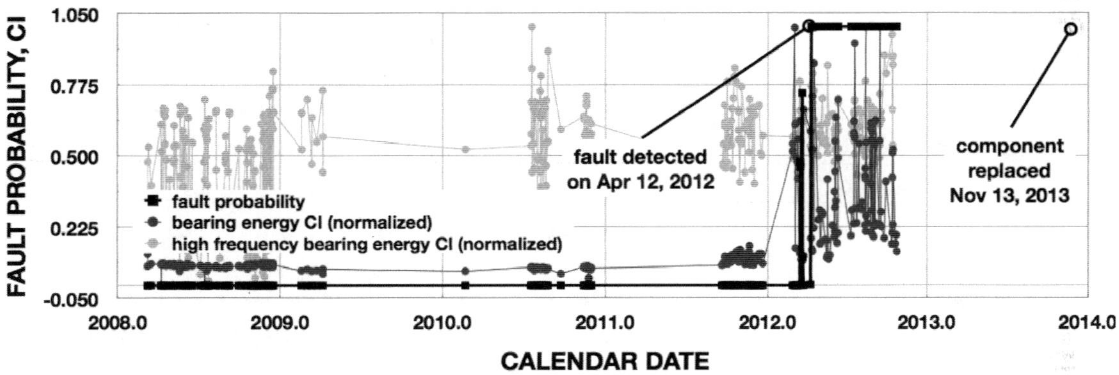

Figure 7. Results for a #2 engine transmission. *Here we detect a fault 19 months prior to its replacement. Shortly before that, the fault probability rose to about 0.75, but only momentarily.*

In the next case, another number 2 engine transmission was removed, this time due to noisy chattering. The component was inspected and declared a NEOF. Even though this time history is rather short with only 25 time samples, the algorithm was able to correctly estimate a fault probability of zero throughout (Figure 8).

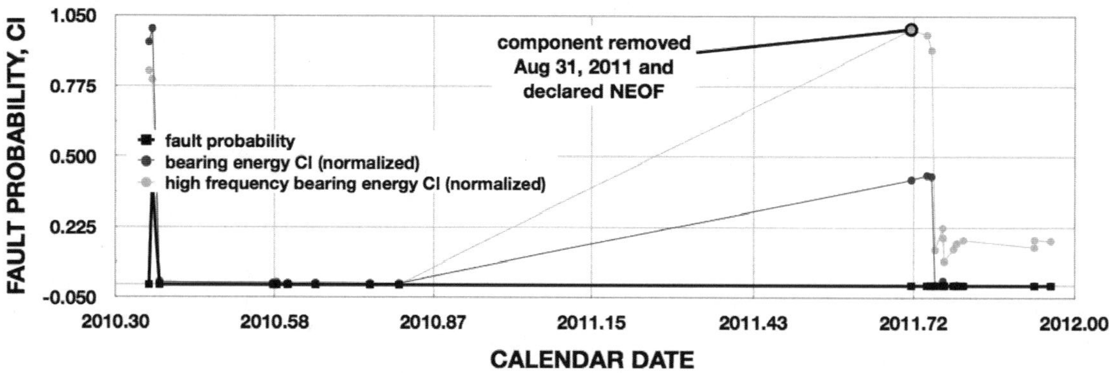

Figure 8. NEOF case for the #2 engine transmission. *This short time history depicts a removal which resulted in a NEOF. The fault probability remains at zero.*

In this next number 2 engine transmission (Figure 9), the component was replaced a few days prior to the available data. Both CIs and the fault probability show a "green" component after the replacement, except in the first 2 data points. In this case, both CIs and fault probability were consistent with the known health state, assuming the component remained normal throughout the duration.

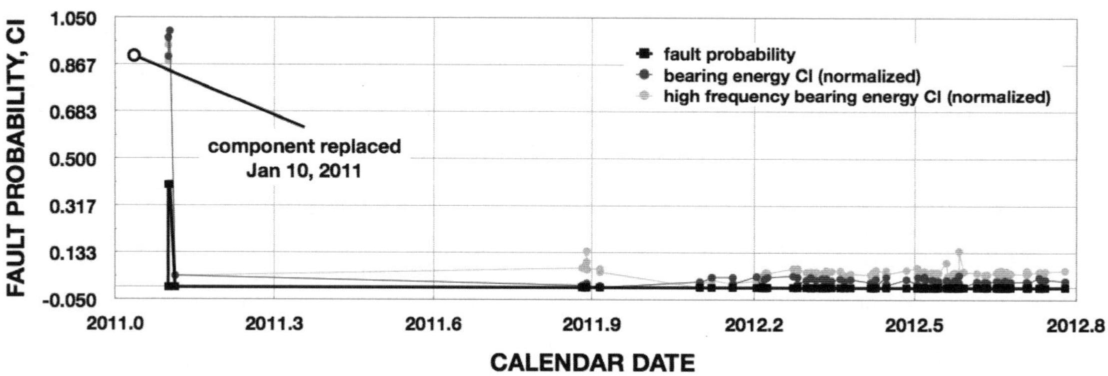

Figure 9. Known "green" #2 engine transmission. *A replacement right before this time-history established the condition as "green". Both CIs and the fault probability were consistent with the known health state.*

A Second Airframe Type: Following initial results with the first airframe, the team was challenged to test the algorithm on two sets of vibration data for a gearbox on a second aircraft type. This was a "blind" study: we did not know the origins of the data. Our only information was that one of the two datasets was normal and the other was faulty.

The first task was to recalibrate the algorithm for the baseline and white noise levels of the gearbox sensor. We also adjusted the covariance for one of the random variables in the process model in order to stabilize fault magnitude estimates. Then we ran the algorithm on both cases. Figure 10 shows the faulty case. The CIs for this component did not indicate a problem. As noted, the algorithm detected a fault was present 230 flight hours - 6 calendar months - prior to the incident.

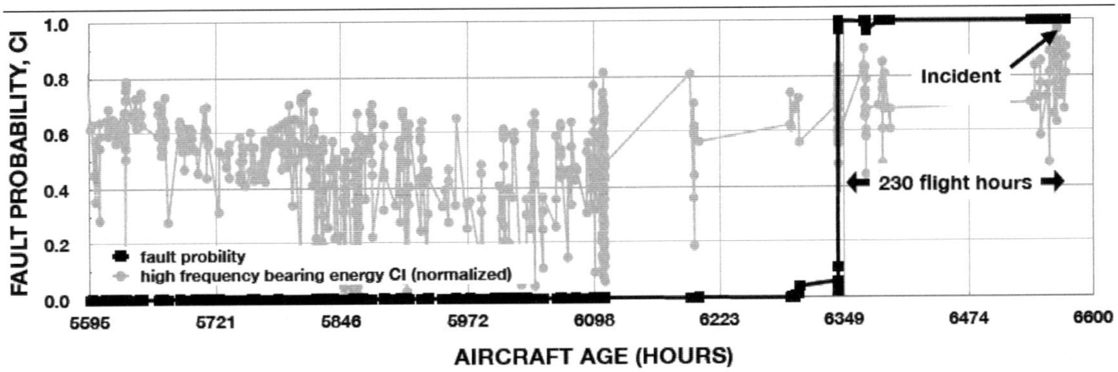

Figure 10. Gearbox. *This dataset ends in an incident, which was caused by a failure adjacent to a gearbox. The algorithm detected a fault at a gearbox 230 flight hours prior to the incident.*

Aligning with the previous work, the results are only considered valid if the same algorithm estimates a fault probability of zero for the known operational (normal) gearbox. As is shown in Figure 11, the algorithm estimates correctly a zero fault probability.

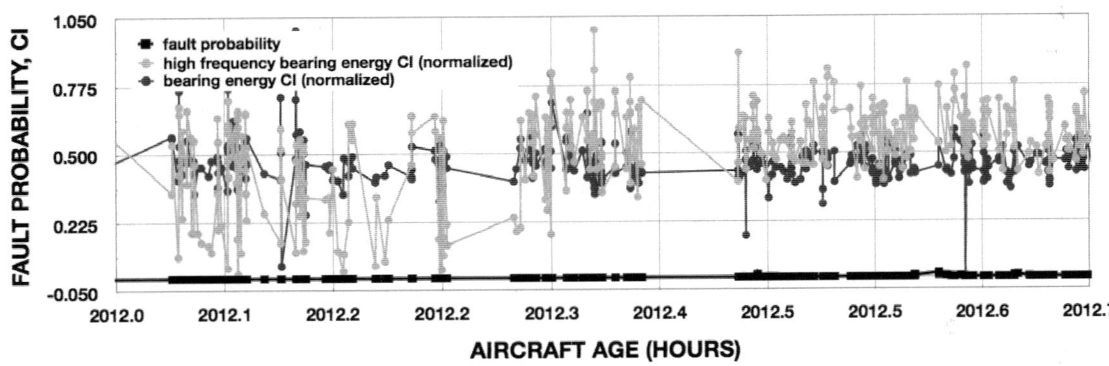

Figure 11. Normal gearbox. *For the operational gearbox, the same algorithm indicates a fault probability of zero.*

Conclusions: Our work has demonstrated that a sequential Monte Carlo algorithm operating a dynamic state-space model of stochastic non-linear fault evolution, mechanical vibration and vibration transduction, can circumvent the fundamental shortcoming of the CI approach. In essence, this is a physics-based machine learning algorithm that interprets noise- and artifact-corrupted vibration spectra time histories in the context of possible underlying fault evolutions. Since the algorithm learns and adapts to the dynamics of individual components, it can reject the artifacts which inevitably corrupt vibration measurements. This approach also allows us to estimate actual fault probabilities, a more direct and useful metric for the maintainer.

We ran the algorithm on several datasets encompassing three mechanical components and two airframes. Fault probability estimates fluctuate very little, demonstrating robust artifact rejection. For all cases thus far, which included both true positives (significant faults or failures) and false positives (NEOFs), our results were consistent with direct visual inspections during teardown. Moreover, our fault detections reached near 100%

certainty several months prior to any other warnings, demonstrating the value of the algorithm for substituting unscheduled with scheduled maintenance, reducing downtime and cost.

The initial successes of this new approach shows that actionable maintenance information can be obtained from HUMS data, offering an opportunity that has not been demonstrated to date. Thanks to our collaboration with experts at AMRDEC and the FMO, we have gained access to large databases, and have analyzed mechanical components on ~240 aircraft. However, the number of datasets with both removals and teardown analysis is limited. We continue to work with all stakeholders to maximize the number of validated cases, as well as expand our approach to more components and airframes.

Bibliography

1. Andreaus, U. and P. Bragatti, Experimental damage detection of cracked beams by using nonlinear characteristics of forced response. Mechanical Systems and Signal Processing, 2012. 31.
2. de Freitas, N. Software for Research Purposes. Available from: http://www.cs.ubc.ca/~nando/software.html.

WIND TURBINE LUBRICATION OIL CONDITION MONITORING, THE IMPACT OF WATER CONTAMINATION AND OXIDATION

Junda Zhu, Tom Nostrand, and Cody Spiegel
Renewable NRG Systems
Hinesburg, VT 05461, USA
Phone: (802) 482-2255
jz@renewablenrgsystems.com

Abstract: The safety and reliability of a functioning wind turbine depend largely on the protective properties of the lubrication oil for its drive train subassemblies such as the gearbox and thus highlights the important role of lubrication oil condition monitoring and degradation detection on a wind turbine. The purpose of lubrication oil condition monitoring and degradation detection is to determine whether the oil has deteriorated to such a degree that it no longer fulfills its function. Lubrication oil degradation is classified into three categories: particle contamination, water contamination and oxidation which are defined as three basic degradation features. In this paper, the impact of water contamination and oxidation on the health condition of turbine gearbox oil are investigated. Water contamination can cause oxidation to increase tenfold, which leads to premature aging of the lubricant. Therefore, these two basic degradation features play a significant role in turbine failure prevention. In this research, gear oil oxidation and water contamination were simulated in a controlled laboratory environment. Lubrication oil performance parameters including dynamic/kinematic viscosity, density, dielectric constant and water content were closely monitored and trended against increasing water contamination and oxidation level. The effectiveness and sensitivity of each performance parameter against water contamination and oxidation level will be discussed.

Key words: Dielectric constant; oxidation; lubrication oil; water contamination; viscosity;

Introduction: According to the United States Department of Energy (DOE), wind energy is the most rapidly advancing source of energy worldwide. United States wind power produces as much clean electricity as nearly 10 nuclear power plants without environmental pollution and consumes virtually no water. However, to increase the wind energy production rate, there is a strong need to improve the wind turbine availability and reduce the operational and maintenance costs. The reliability and availability of a functioning wind turbine depends largely on the protective properties of the lubrication oil for its drive train subassemblies such as the gearbox and means for lubrication oil condition monitoring and degradation detection. It was estimated that one standard oil change of a mega watt class turbine can consume up to 100 gallon of lubricant, 3 technicians, and 8 hours turbine down time. And the total cost can reach up to 150,000 dollars (Lubrizol Corporation, 2013). The reason for the high volume consumption of gearbox oil is that traditional draining method in the wind turbine gearbox will result in a large portion (over 20%) of the used oil remaining in the gearbox (Windpower Development & Engineering, 2009). To overcome this barrier, the gearboxes need to be flushed by a large amount of the new oil. The flushing process consumes a great amount of lubricant, which in turn needs to be recycled properly to avoid considerable environmental impact. This recycling will cost extra labor and equipment. Moreover, a mega watt class turbine generates approximately 1200 dollars worth of electricity per hour of operation. Production interruption can be very expensive and even more costly when it is unscheduled.

The wind industry currently uses offline lubrication oil sample analysis or particle counters to detect gearbox and bearing wear but these methods cannot detect the functional failures of the lubrication oils. The main purpose of lubrication oil condition monitoring and degradation detection is to determine whether the oils have deteriorated to such a degree that they no longer fulfill their functions. By means of lubrication oil condition monitoring in an online fashion, wind turbine fleet maintenance cost can be reduced as well as the environmental impact caused by excessive waste of high viscosity lubricant.

Lubrication oil is an important information source for early machine failure detection just like the role of the human blood sample testing in order to perform disease detection. In modern industries, lubrication oil plays a critical part in condition based maintenance of complicated machineries such as wind turbines. The high speed side of the gearbox is vulnerable to faults by nature because of the universal lubrication within the gearbox (320 Gear lube is not optimum for the high speed section of the gearbox). A great portion of wind turbine gearbox failures are caused by bearing failures on the high speed shaft and intermediate speed shaft. Bearings failure rates are strongly connected to the lubrication oil water contamination level. It was reported by Timken that water content in the lubricant leads to bearing premature failure. 1000 ppm of water reduces the bearing life by 70%. At the same time, based on the study done by Cantley (1977) which is also mentioned in the AWEA recommended practice 2013, a bearing with 25 ppm of dissolved water has a fatigue life 5 times that of a bearing with 400 ppm of dissolved water as shown in Figure 1. As one can observe in Figure 1, 100 ppm of water results in a standard 100% bearing rated life, increasing water contamination level will decrease the bearing life exponentially. For the wind energy industry, water content should be kept under at least 300 ppm. 300 to 600 ppm is the watch region and the oil should be changed or reconditioned when the water contamination level is over 600 ppm. Water contamination can also cause the oxidation rate to increase tenfold, which leads to premature aging of the lubricant. Therefore, water contamination level is considered one of the most important degradation features of wind turbine lubricant.

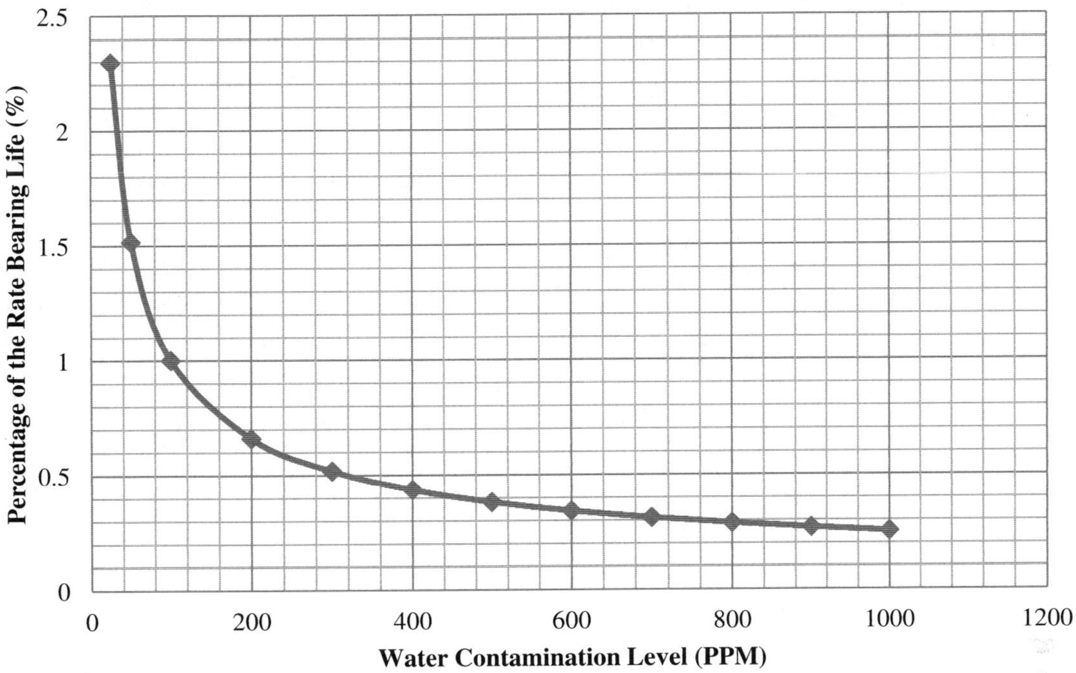

Figure 1. The relationship between percentages of rated bearing life and water contamination level

Oxidation is another basic degradation feature of turbine gearbox lubrication oil as stated by Zhu et al. (2013). Lubrication oil basic degradation features include water contamination, oxidation and particle contamination. Among which, water contamination and oxidation are considered oil condition monitoring while particle contamination is recognized as means of machine health monitoring. Lubricant oxidation can build up acid products that thicken the oil and may lead to a gummy substance that will coat and insulate heat exchanging surfaces. Both particle contamination (mainly metallic) and water contamination can induce a faster rate of oxidation. It was reported that water can accelerate oxidation as rapid as copper. Moreover, based on a previous study, when water is present alongside copper or iron, the oxidation rate can be increased up to 120 times. Hence, oxidation is another aspect of oil condition monitoring that is crucial to operations and maintenance personnel. The unit to evaluate the level of oxidation is mg/KOH which is a performance parameter called Total Acid Number (TAN). Depending on different suppliers of the wind turbine gearbox lubricant, the base line for TAN can vary. Normally maintenance personnel consider 1.5 over the baseline as the watch region while 2.0 over the new oil will trigger the oil changing process assuming the other performance parameters agree with the TAN reading.

The lubrication oil basic degradation features can be described by a selection of parameters. These parameters that are used to describe the basic degradation features are called performance parameters. For water contamination, these parameters include viscosity, humidity, dielectric constant and so on. Over the years, scientists and experts have developed considerable sensors and systems to monitor one or more of the performance parameters in order to monitor the oil health condition effectively. The relationship between the basic degradation feature, performance parameters and available oil condition sensors is shown in Figure 2.

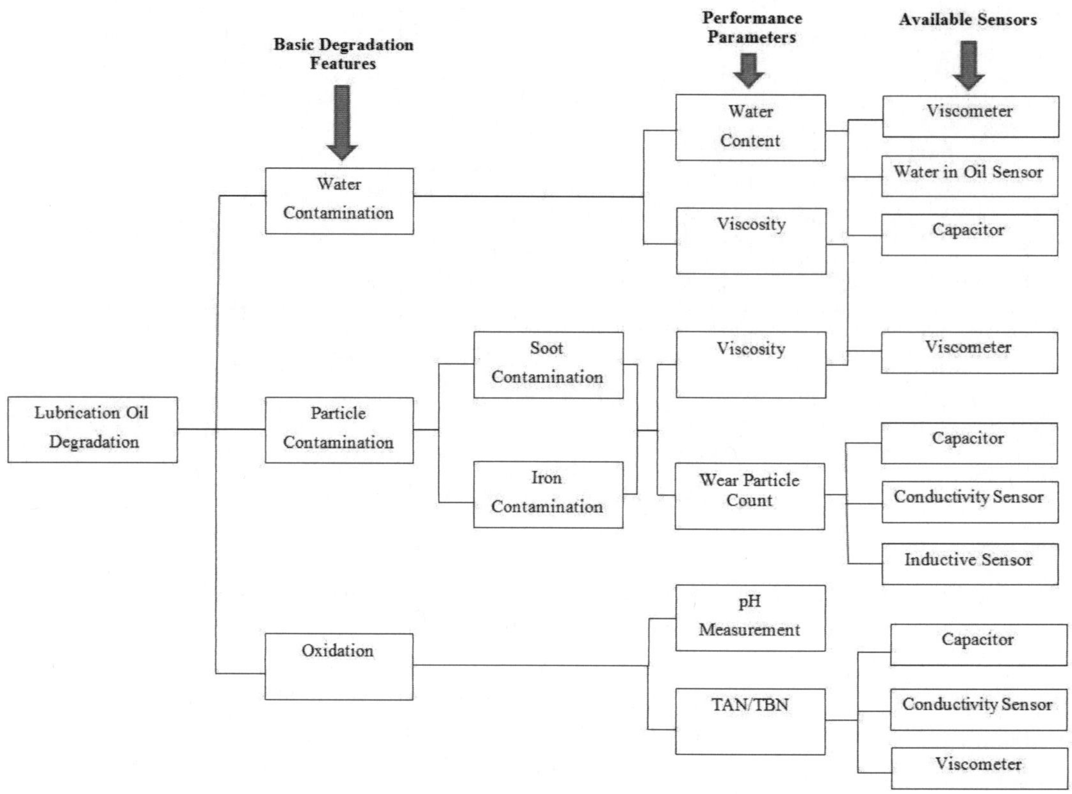

Figure 2. The relationship among the basic degradation features, performance parameters, and available oil condition sensors

Since water contamination and oxidation play a significant role in turbine failure prevention, this paper will focus on these two basic degradation features along with their impact on a selection of performance parameters. Limited publications have reported the ability to accurately simulate the oxidation process of high viscosity turbine gearbox oil. In this research, oxidation and water contamination of lubrication oil were both simulated in a controlled laboratory environment. Lubrication oil performance parameters including dynamic/kinematic viscosity, density, dielectric constant and water content were closely monitored and trended against increasing water contamination and oxidation level. The effectiveness and sensitivity of each performance parameter against water contamination and oxidation level will be discussed.

Experimental Setup: The system used to monitor the performance parameters during the gearbox lubrication oil water and oxidation test was developed by Renewable NRG Systems. The developed system is capable of measuring water content, dielectric constant, viscosity, density and temperature of the test subject which in this case is Castrol A320. Castrol A320 is a common lubricant that is currently deployed in a great portion of turbines in the wind energy industry. 4000mL of Castrol A320 was placed in a cylindrical temperature controlled chamber with an adjustable height mixing blade in the center. The heating surface was installed in the bottom of the chamber. The testing temperature was dynamical adjusted at the temperature of interest. The experimental setup is shown in Figure 3. The sensing system was designed to interface with the TurbinePHD condition monitoring system from Renewable NRG Systems.

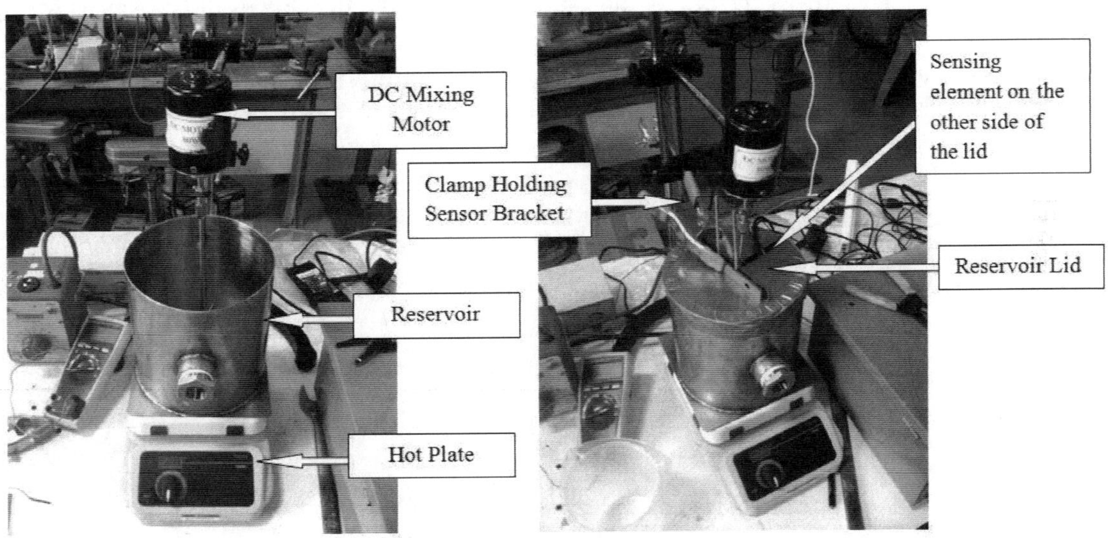

Figure 3. Experiment setup.

The developed sensing system was capable of logging test data into text form for instant post processing. The output unit for viscosity is absolute viscosity, which is Cp. The output unit for Density is gm/cc. Therefore the dynamic viscosity can also be calculated from absolute viscosity and density as shown in Equation (1). The output unit for water and temperature are Volts and Degrees Celsius respectively.

$$Kinematic\ Viscosity = \frac{Absolute\ Viscosity}{Density} \tag{1}$$

For the water contamination test, water was injected into the lubricant with a syringe during the high speed mixing process. After a pre defined mixing time interval, data was logged. Distilled water was used for the water contamination test. From past experiences, it is clear that tap water contains too many particles that potentially have a negative effect on the sensor characteristic test results.

Likewise, for the oxidation test, a special fabricate high concentration acid was added into the lubricant with a non acid reactive syringe. Again, data was logged after a certain time interval. The tests were repeated for multiple cycles. After each cycle of the water contamination test, the sensing system was thoroughly cleaned with a standard cleaning procedure to eliminate oil acid or water residuals on the sensors.

Water Contamination Test: There were two purposes to the water contamination test. The first purpose was to verify if the developed system can measure water contamination variation under 1500 ppm with a resolution of 250 ppm. The second purpose was to see how all the measured performance parameters react against an increasing water contamination level.

Three cycles of the water contamination test were conducted starting from 250 ppm to 1500 ppm with 250 ppm as the step increase. Three cycles of the healthy oil benchmark test were also tested. However, the benchmark test can be integrated into the water contamination test starting with healthy oil on each cycle and increment 250 ppm of water as a step increase. Each cycle of the water contamination test required 4000mL of lubrication oil. Therefore, the total Castrol A320 oil needed in terms of volume was 4.5 Gallon including the loss associated with the transfer of oil between containers. After each cycle of the water contamination test, the test subjects were sampled and shipped to an offsite oil analysis lab to compare the sensor readings with the lab report. The readings for the three cycles of the water contamination test were averaged and are shown in the figures below. The x axis is the level of water contamination.

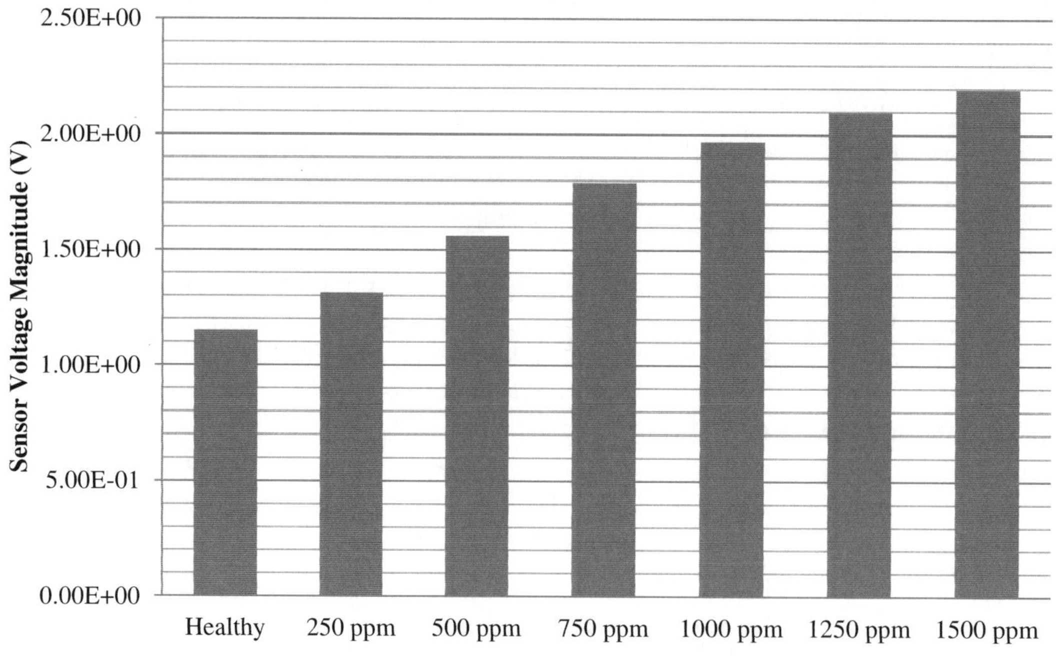

Figure 4. Water content sensing unit reading (water contamination test)

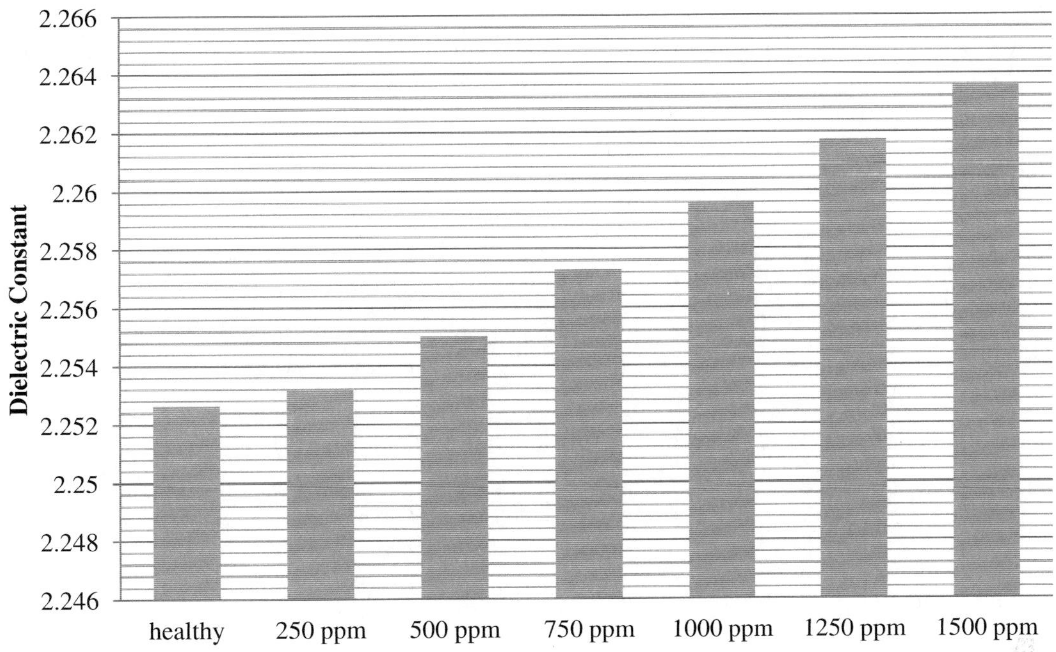

Figure 5. Dielectric sensing unit reading (water contamination test)

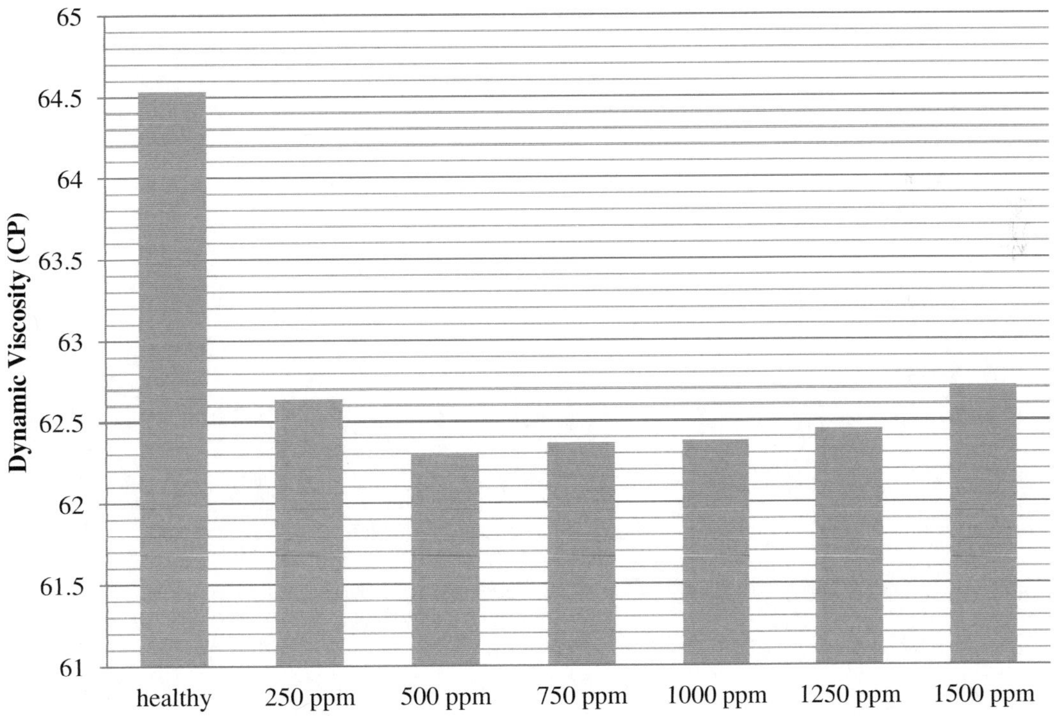

Figure 6. Dynamic viscosity sensing unit reading (water contamination test)

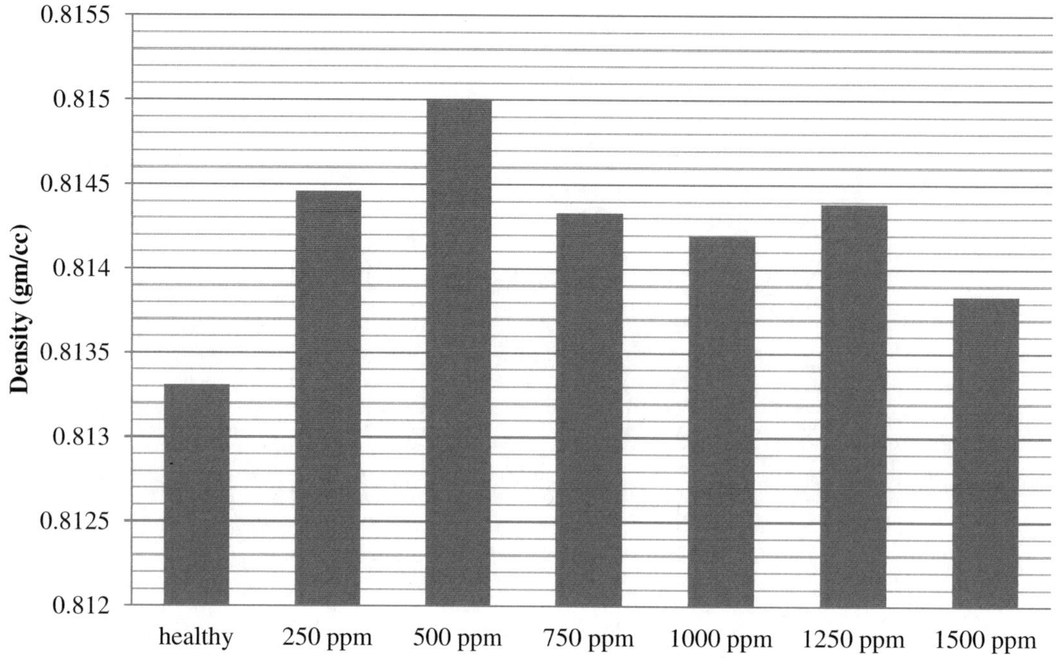

Figure 7. Density sensing unit reading (water contamination test)

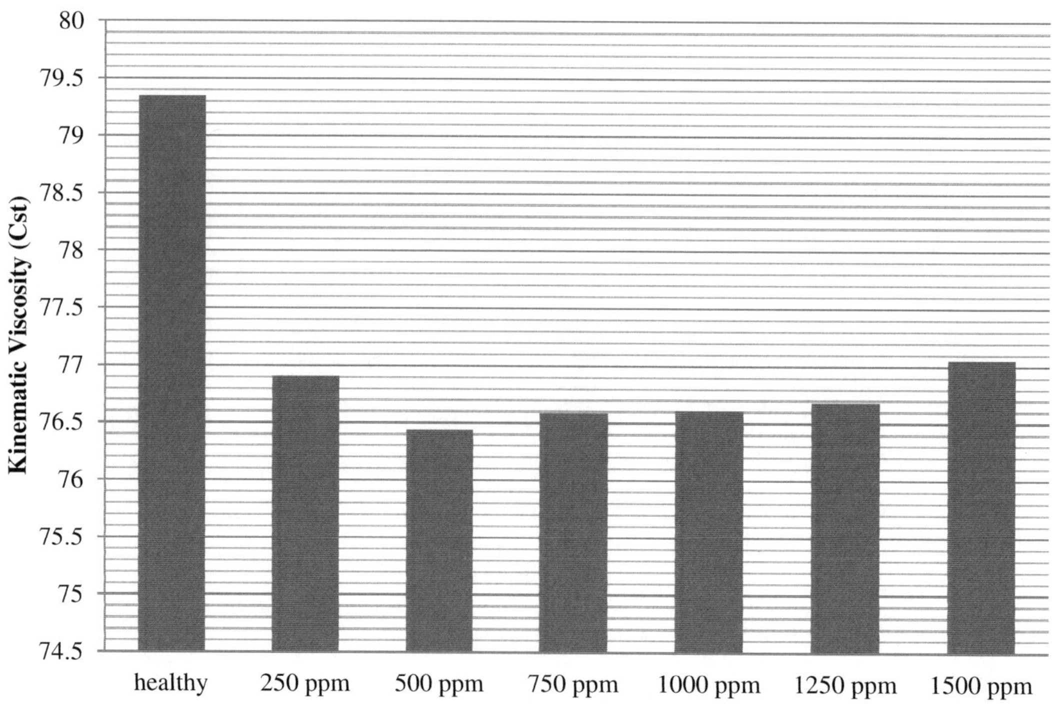

Figure 8. Kinematic Viscosity calculated from the density and dynamic viscosity reading (water contamination test)

From the test results, it is evident that the oil condition monitoring unit from Renewable NRG Systems is capable of water contamination detection under 1500 ppm with a resolution of 250 ppm. The readings of the dielectric sensing unit agreed with the water content sensing unit. However, both dynamic viscosity and kinematic viscosity are unable to detect increasing water contamination level under 1500 ppm. Moreover, based on the previous research on dielectric constant variation based on water contamination level published by Zhu *et al* (2013), the dielectric constant measured by the developed system matched the theoretical dielectric constant progression pattern as shown in Figure 9.

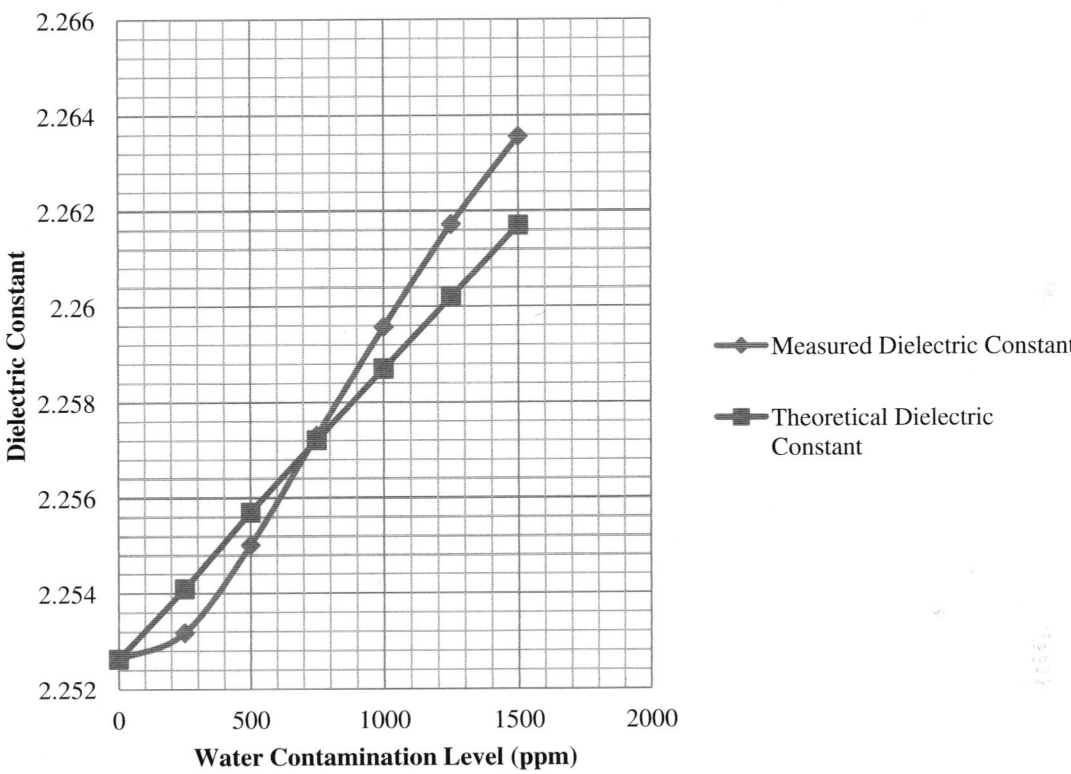

Figure 9. Water contamination progression pattern, measured dielectric constant vs. mathematically modeled theoretical dielectric constant.

The equation used for the dielectric constant model for water contamination was the Maxwell Garnett equation as mentioned by Zhu *et al* (2013). The Maxwell Garnett dielectric formula is the most widely used mixing rule to calculate the average dielectric constant of a mixture consisting of multiple components. The dielectric constant of a water and oil mixture at a certain temperature $\varepsilon_{M,T}$ can be computed as:

$$\varepsilon_{M,T} = \varepsilon_{oil,T} \times (1 + 3 \times P \times \frac{\varepsilon_{water,T} - \varepsilon_{oil,T}}{\varepsilon_{water,T} + 2 \times \varepsilon_{oil,T} - P \times (\varepsilon_{water,T} - \varepsilon_{oil,T})}) \quad (2)$$

where:

$$\varepsilon_{water,T} = 80 - 0.4 \times ((T + 273) - 293) \tag{3}$$

Note that in Equation (2), $\varepsilon_{oil,T}$ is defined as the healthy lubrication oil information and is extracted from our initial test while $\varepsilon_{water,T}$ is defined as the water physical attribute which can be considered known factors as stated in Equation (3). Considering, Equation (2) is developed and validated under an ideal or highly controlled laboratory environment, the minor difference in Figure 8 should be considered acceptable.

To verify the accuracy of the water content sensor, three unknown samples were tested with the Renewable NRG developed oil condition monitoring system. After that, the data was logged and the samples were sent to an offsite oil analysis lab for result comparison.

Table 1. Test result comparison between the developed system and the offsite test report

	Sample 1	Sample 2	Sample 3
Water Content (Measured with the developed system)	284 ppm	232 ppm	194 ppm
Water Content (Measured from the offsite lab)	203 ppm	228 ppm	191 ppm

The conclusion for the water contamination test section is that for water contamination under 1500 ppm, the performance parameters that can keep track of the water progression are water content and dielectric constant. These two measurements can achieve a resolution of at least 250 ppm. The dielectric constant data correlates with that of the water constant very well. Moreover, the dielectric constant reading agrees with that of the theoretical value calculated by the oil water mixture mathematical model.

Oil Oxidation Test: The purpose of the oxidation testing is twofold, one to verify if the developed system can measure oxidation variation and two to see how all the different measured performance parameters react against an increasing oxidation level.

A specially fabricated high concentration acid was added. The correct amount of acid was carefully calculated before it was added to the lubricant to achieve a specific oil and acid mixture. Three cycles of the oxidation tests were conducted with the TAN starting from 0.1 mg KOH/g to 0.6 mg KOH/g with 0.1 mg KOH/g as the step increase. Three cycles of the healthy oil benchmark test were also performed. The benchmark tests can be integrated into the oxidation tests by starting with healthy oil on each cycle and adding 0.1 mg KOH/g as a step increase. Each cycle of the oxidation test consumed 4000 mL of lubrication oil. Therefore, the total Castrol A320 oil needed in terms of volume was 4.5 gallons including the losses associated with the transfer of oil between containers. The readings for the three cycles of the oxidation test were averaged and are shown in the figures below.

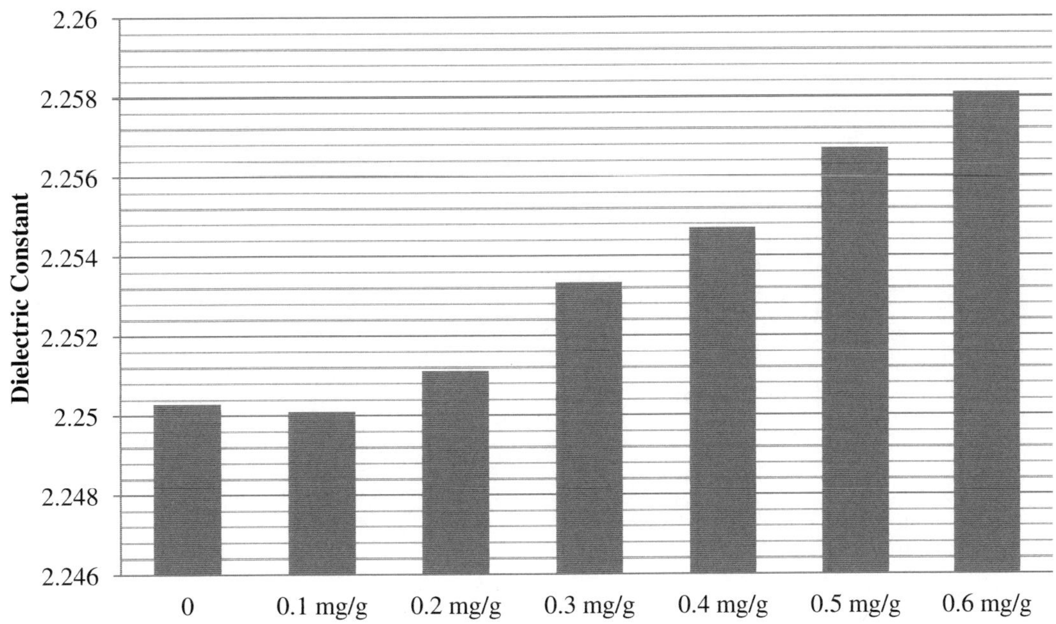

Figure 10. Dielectric sensing unit reading (oxidation test)

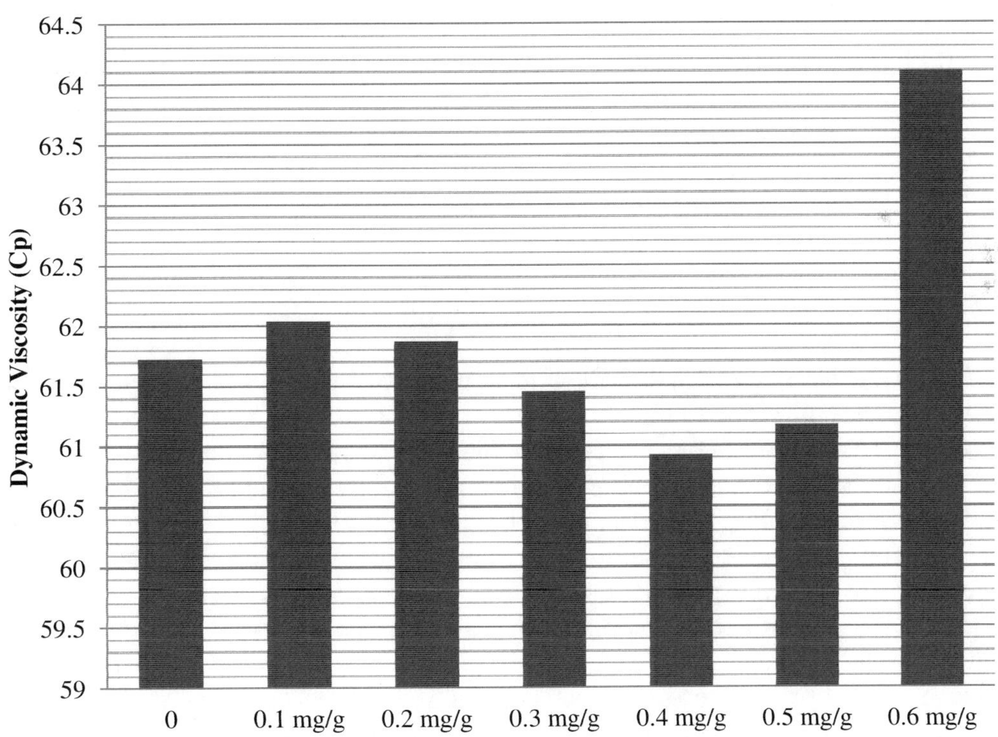

Figure 11. Dynamic viscosity sensing unit reading (oxidation test)

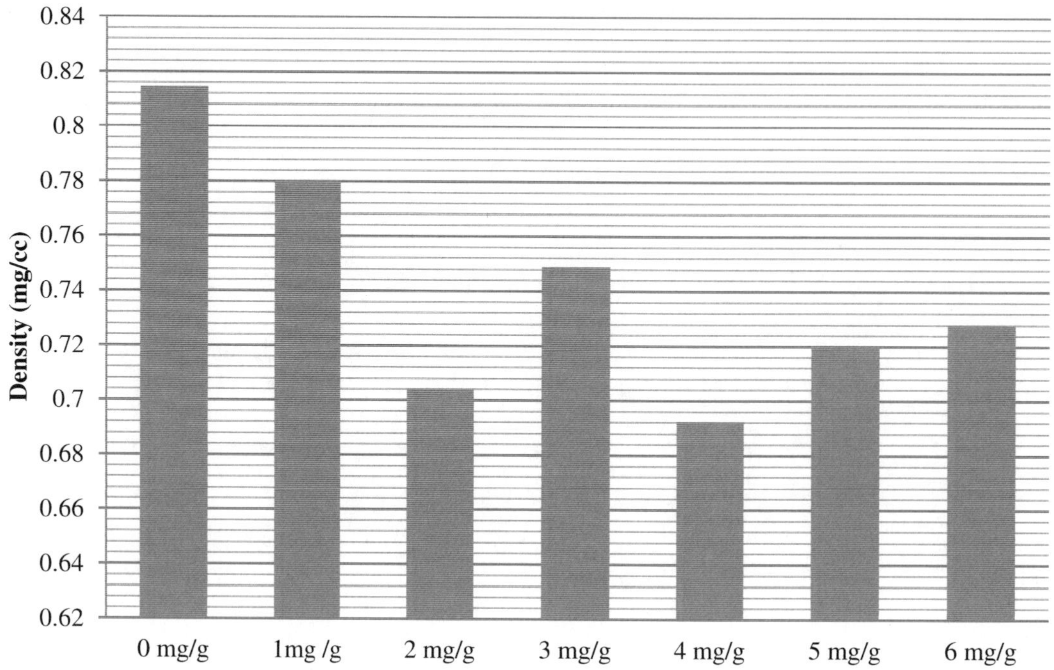

Figure 12. Density sensing unit reading (oxidation test)

From the test results it is evident that the oil condition monitoring unit from Renewable NRG Systems is capable of oxidation detection with a resolution of 0.1 mg KOH/g. However, dynamic viscosity cannot detect an increasing oxidation level. At the start of the acid progression process, the dielectric constant stays relatively stable. This may be caused by the antioxidant starting to act against the incoming acid. After the antioxidant wears out, the oxidation rate started to trend as more acid was added. Therefore, one can conclude that dielectric constant is capable of tracking lubrication oxidation progression with a resolution of 0.1mg KOH/g/.

Conclusion: Over the years, many online wind turbine gearbox oil condition monitoring techniques have been developed and implemented. These techniques focus on different aspects of the gearbox oil deterioration process. The oil degradation can be summarized into 3 categories. They are water contamination, particle contamination and oxidation respectively which are defined as lubrication oil basic degradation features by Zhu *et al* (2013). A selection of parameters was used to describe the basic degradation features. They are called performance parameters. Simulation tests were conducted to test the effectiveness of some common performance parameters for the wind turbine industry which included water content, dielectric constant, viscosity temperature and density. Two major basic degradations were investigated which were water contamination and oxidation. Based on the research results, water content and dielectric constant were capable of water contamination detection under 1500 ppm with a resolution of 250 ppm. For the oxidation, the dielectric constant was able to catch the TAN increase with a resolution of 0.1 mg KOH/g. For water contamination and oxidation simulation tests, dynamic viscosity, density and kinematic viscosity were not capable of detection of the increasing water or oxidation level.

References:

[1] American Wind Energy Association Operation and Maintenance Working Group, AWEA Operation and Maintenance Recommended Practices (Version 1.0), May 6, 2013, Washington DC.

[2] R E Cantley, The Effect of Water in Lubricating Oil on Bearing Fatigue Life, ASLE Transactions, American Society of Lubrication Engineers, 1977, Volume 20, No. 3, pp. 244 – 248.

[3] Lubrizol Corporation, Gear Oils and Oil Sensing Technology, 2013, http://www.lubrizol.com/OurCompany/WindEnergy/GearAdditive.html

[4] Windpower Engineering & Development, How to Keep Them Working 20 Years and Longer, November 25, 2009, Windpower Engineering, http://www.windpowerengineering.com/maintenance/how-to-keep-them-working-20-years-and-longer/

[5] J Zhu, J M Yoon, D He, Y Qu and E Bechhoefer, Lubrication oil condition monitoring and remaining useful life prediction with particle filtering, 2013, International Journal of Prognostics and Health Management , Vol. 4, No.20, pp. 27 - 35.

Bibliography:

[1] Junda Zhu received his B.S. degree in Mechanical Engineering from Northeastern University, Shenyang, China, and M.S. degree in Mechanical Engineering from The University of Illinois at Chicago in 2009, and Ph.D. degree in Industrial Engineering and Operational Research from The University of Illinois at Chicago in 2013. Dr. Junda Zhu is a Systems Engineer in the Turbine Health Monitoring Group in Renewable NRG Systems. His current research interests include rotational machinery health monitoring, diagnosis and prognosis with vibration or acoustic emission based signal processing techniques, lubrication oil condition monitoring and degradation simulation and analysis, physics/data driven based machine failure modeling.

[2] Tom Nostrand received his BS in Engineering from the University of New Hampshire. He is currently an Engineering Manager at Renewable NRG Systems, Turbine Products Group. He has been working in the wind power industry for 6 years. Prior to this he spent 20 years in the aerospace electronics field working on many different commercial aircraft platforms and systems.

[3] Cody Spiegel received his B.S. in Mechanical Engineering from The University of Vermont in 2013. He began a mechanical engineering internship at Renewable NRG Systems in 2011 and is now an Associate Mechanical Engineer working in the Turbine Health Monitoring Group. His current work includes condition monitoring diagnostics and support as well as R&D of new condition monitoring systems.

IMR Mining and Metallurgy Machinery

COMPARISON OF STATE OF THE ART SCADA AND HIGH SAMPLED DATA CONDITION MONITORING TECHNECHIQUES FOR WIND TURBINES

M. Hilbert, K. Nienhaus

MFPT, Mai 2014

RWTH AACHEN UNIVERSITY

The Institute

IMR Mining and Metallurgy Machinery
RWTH AACHEN UNIVERSITY

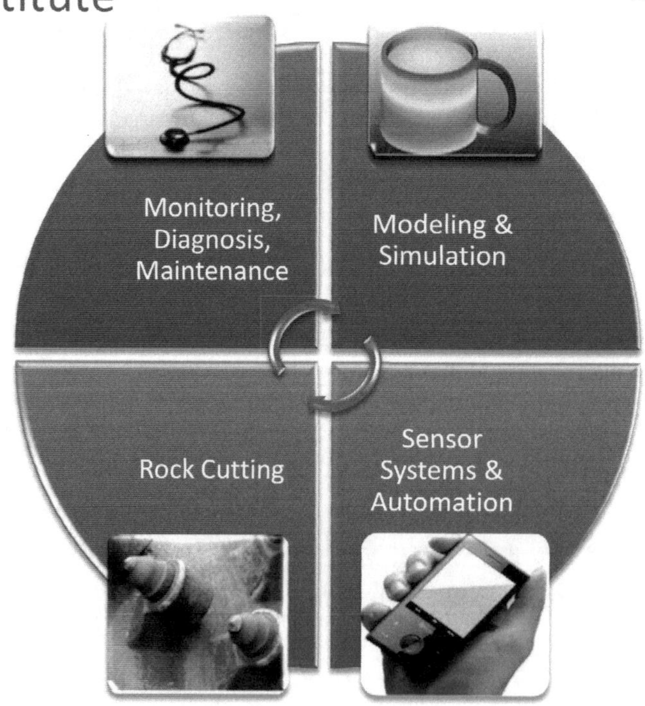

2

The Institute

85 Employees

- State Budget: 7 Scientists, 9 Staff, 8 Apprentices
- Third Party R&D: 20 R&D-Engineers, 2 Staff + 5 R&D Engineers "Spin-off"
- 30 part time Student Research Assistants

Teaching

- 26 Courses (12 Summer, 14 Winter)
- 50 Hours per Semester-Week

Finances

- 9% Budget State Financing: approx. 130.000 €
- 91% Third Party R&D: approx. 1.300.000€

Nordex SE

Questions

1. What are the **differences** between Condition Monitoring using CMS and SCADA?
2. What are the **advantages** and **disadvantages** of each system?
3. What are the **future developments**?

Nordex SE

Terminology

IMR Mining and Metallurgy Machinery
RWTHAACHEN UNIVERSITY

"Condition monitoring is the process of monitoring a parameter of condition in machinery"

- Condition Monitoring
- Fault Diagnosis and Identification
- Condition Prognostics

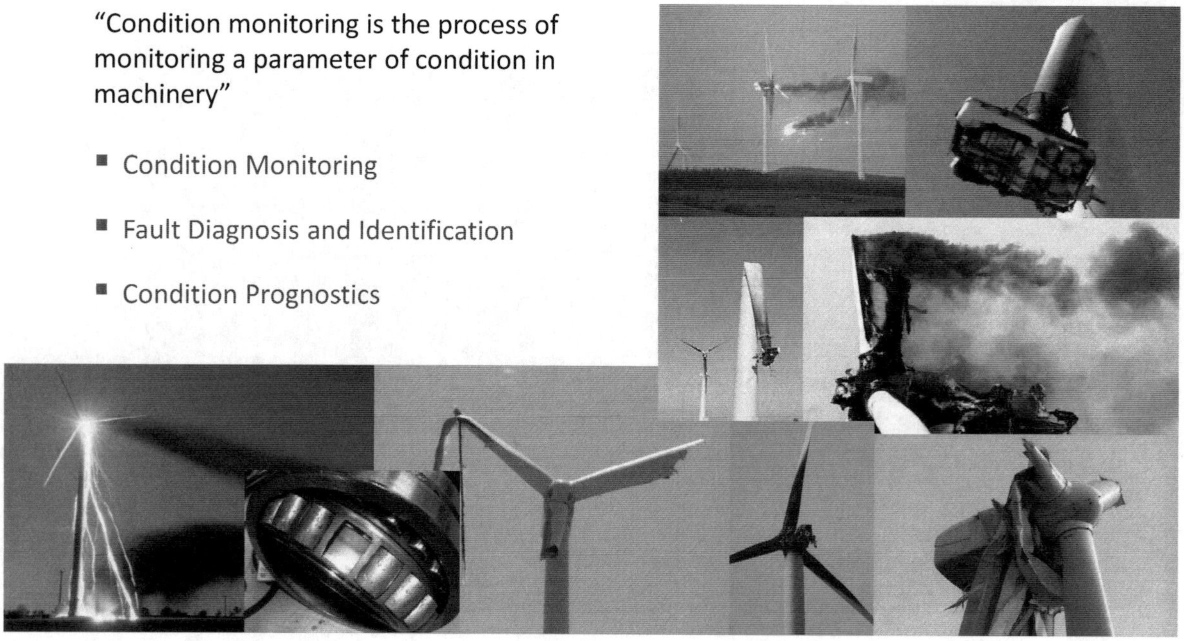

5

Challenges

IMR Mining and Metallurgy Machinery
RWTHAACHEN UNIVERSITY

- **Sites**
  - More remote onshore and offshore sites for new turbines
- **Diversity**
  - Type of machine like Hybrid Drive
  - Age and diversity of wind turbine fleet increases
- **Size**
  - Experience of fault behavior for increasing size
- **Complexity**
  - New technologies and materials are used
- **Cost**
  - Pressure for competitive CoE

6

Methods

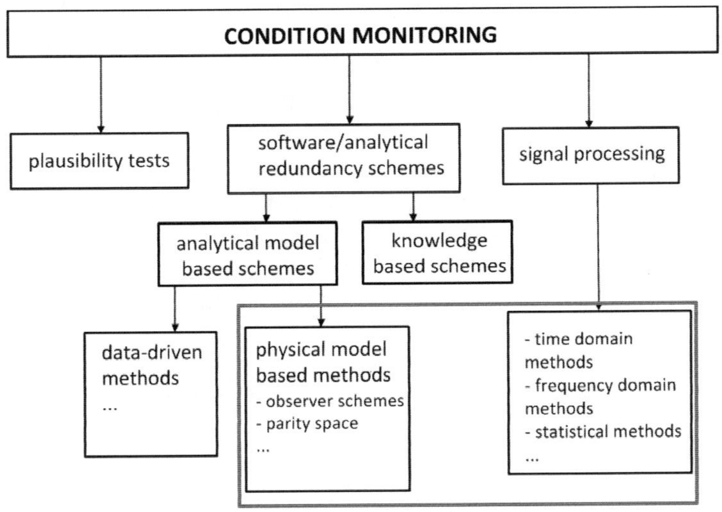

Ding S. X. (2008). Model-based Fault Diagnosis Techniques, Springer-Verlag, Berlin Heidelberg,

7

Methods

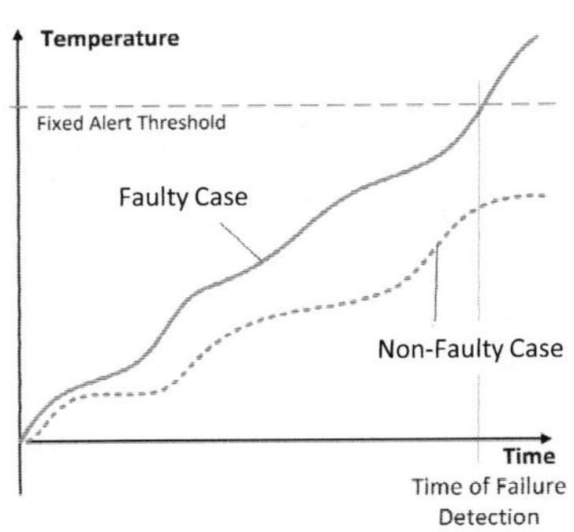

Signal-based:

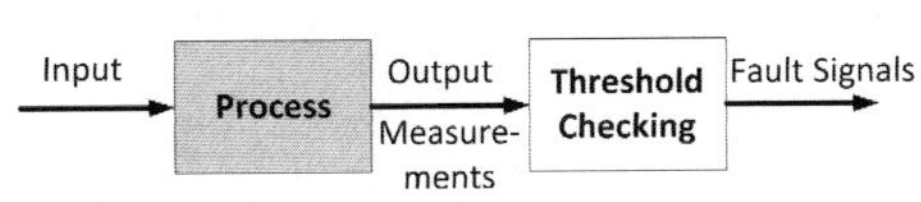

8

Methods

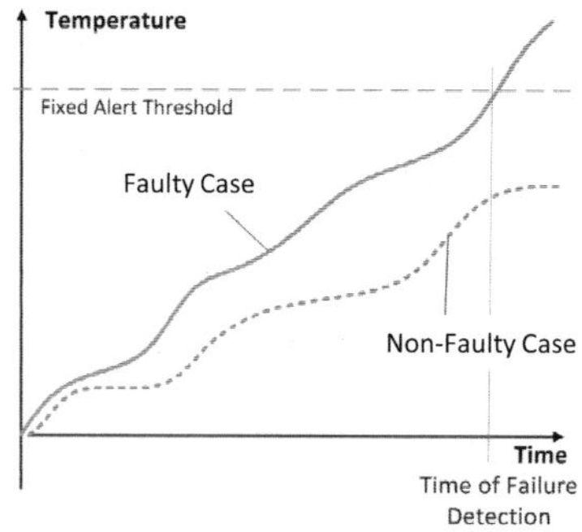

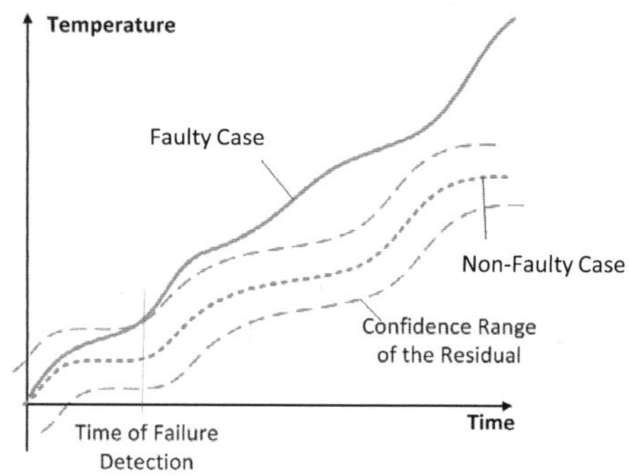

Signal-based:

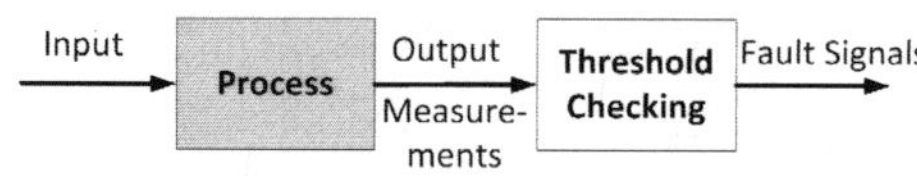

Physical model-based:

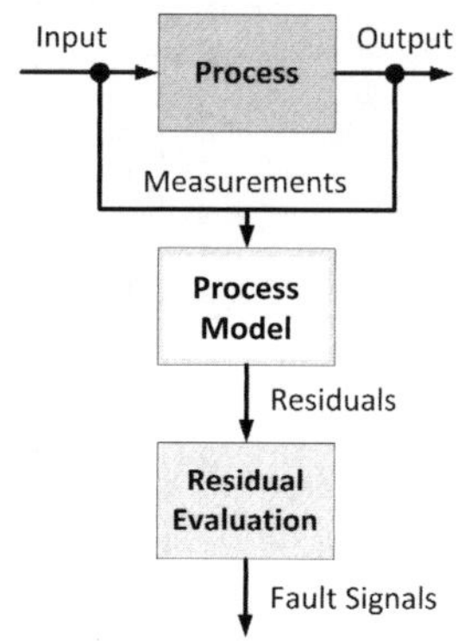

9

Methods

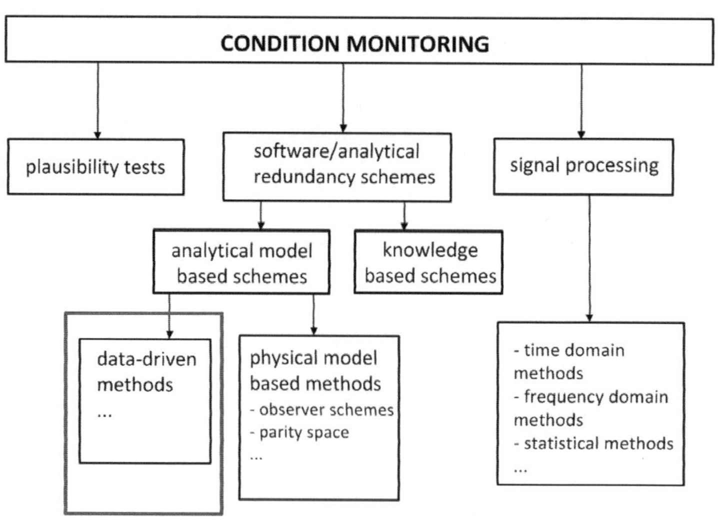

Ding S. X. (2008). Model-based Fault Diagnosis Techniques, Springer-Verlag, Berlin Heidelberg,

10

Methods

- Discovering patterns in data
- Minimum use of machine design knowledge

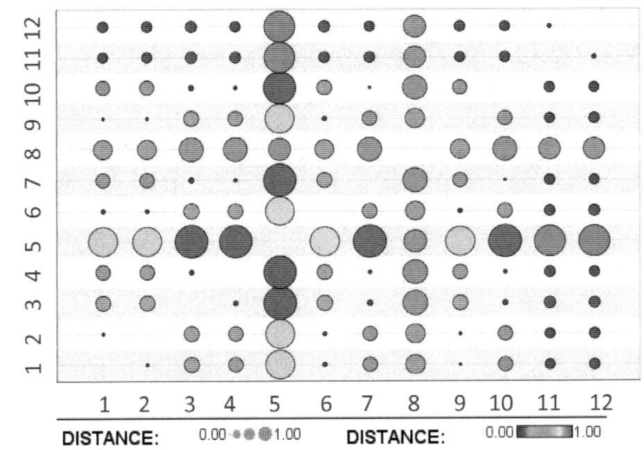

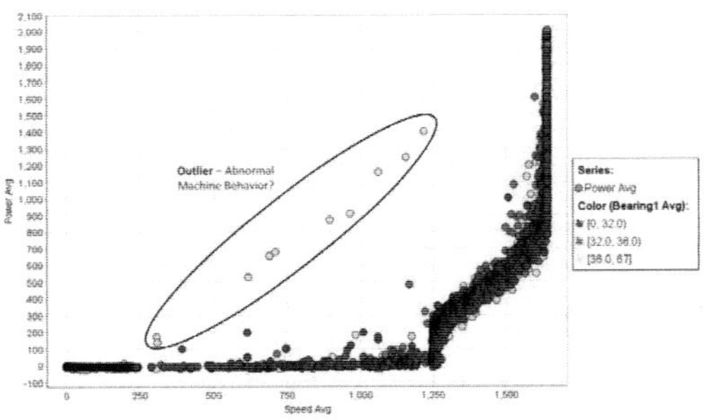

11

SCADA

- **S**upervisory **C**ontrol **A**nd **D**ata **A**cquisition (SCADA)
- Data is collected mostly trough the **wind turbine controller** and transferred to the back office database
- Processing of the data is done in the back office and offline
- Parameters in **10min intervals**:

Active Power Output

Generator Currents and Voltages

Nacelle Temperature

Generator Bearing Temperatures

Power Factor

Gearbox Lubrication Oil Temperature

Turbine and Generator Shaft Speed

Alarm Status

Reactive Power

Anemometer Measured Wind Speed

Generator Winding Temperatures

Gearbox Bearing Temperatures

12

SCADA

ADVANTAGE	DISADVANTAGE
▪ Data is being **collected** and stored ▪ Highly **automated** data analysis is possible	▪ Only few essential **CM signals** are available through the SCADA System ▪ Very low **sampling rate** compared to the changing conditions ▪ Missing suitable software **solutions** for fault diagnosis ▪ Missing/wrong **data**

13

CMS

- **C**ondition **M**onitoring **S**ystems (CMS)
- Data **is collected and pre-processed** at the turbine
- Most commen measurements **are structure born sound** and **oil analysis**
- Limited **interface** to the wind turbine controller
- Data is sent, analyzed and stored at a 3rd CMS provider back office
- **Reports** are generated offline

14

CMS

ADVANTAGE

- **High probability** of fault detection

DISADVANTAGE

- Data is stored and analyzed in a **complex** way
- Analysis cannot be fully **automated**
- **3rd party** hardware & software needs to be installed
- Detailed **machine information** is needed
- **Costs**

15

Conclusion

	ADVANTAGE	DISADVANTAGE
SCADA	**Highly automated** data analysis is possible	Very low **sampling rate** compared to the changing conditions
CMS	**High probability** of fault detection	Detailed **machine information** is needed

16

Future Development s

- CM for techniques for **sub-assemblies**, specifically electrical components
- **Automation and Integration** of CMS and SCADA
- **Fleet Monitoring** for a large amount of wind turbines
- **Combination** with other systems like SAP and energy demand
- Reduced **evaluation complexity**

Condition **B**ased **M**aintenance

17

Thank you for your attention

Marc Hilbert
mhilbert@imr.rwth-aachen.de

Institute for Mining and Metallurgy Machinery (IMR)
RWTH Aachen University

18

Field-Deployed Solenoid Valve Failure Determination

May 22, 2014

CALCE Team: N. Jordan Jameson,
Dr. Michael H. Azarian, Prof. Michael Pecht

Introduction to Solenoid Valve

- A solenoid-operated valve (SOV) is an electromechanical device used to control the flow of gas or liquid by passing an electric current through a coiled wire, thereby altering the valve position and opening or closing the valve orifice.
  - Can be utilized to shut off, discharge, dose, allocate, or combine fluids.

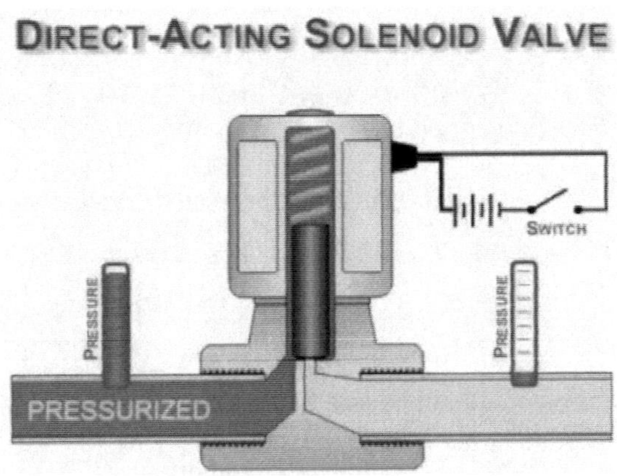

•Two-way normally closed solenoid valve (Ref. http://www.solenoid-valve-info.com/)

Solenoid Valve Nomenclature

- Solenoid operated valve design can be distinguished by:
  - Two-way, three-way, or four-way
  - Direct-acting, pilot operated
  - Normally open, normally closed

Direct-Acting vs. Pilot-Controlled

- A direct-acting valve is one where all flow passes through an orifice that is opened directly by an electromagnet and plunger.
- Pilot-controlled: solenoid valve that operates by means of a minimum and maximum pressure differential and uses an electromagnet and plunger to open or close a small orifice to control the opening and closing of a piston or diaphragm.

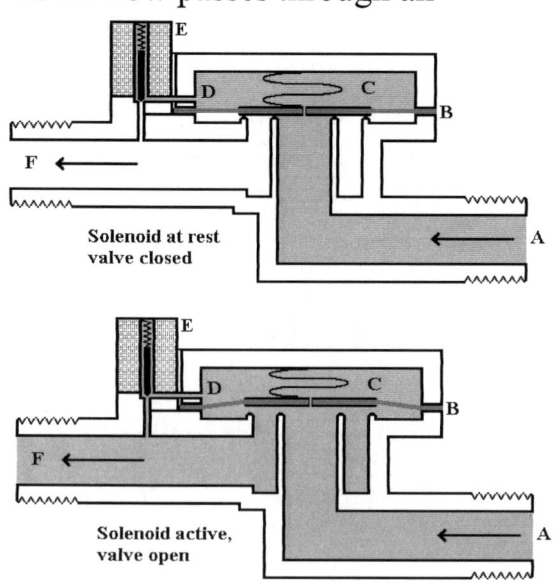

Pilot-controlled two-way SOV

Three-Way Valve

- Built with three ports and two orifices
- Can operate with: an inlet port, an outlet port, and an exhaust port for operating a single acting cylinder; one inlet port and two outlet ports for selecting or diverting flow; or two inlet ports and one outlet port for mixing fluids.

Three-way solenoid valves (Ref. Norgren Fluid Controls)

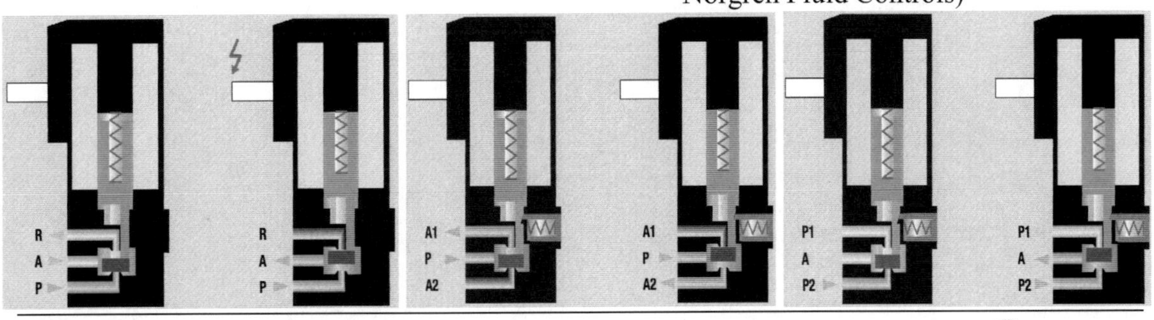

Common SOV Failure Modes

- Decrease in electrical efficiency
  - Electrical coil open
  - Electrical coil short
- Improper media flow
  - Seal leakage
  - Body leakage
  - Incorrect metering

Common SOV Failure Mechanisms

- Corrosion
  - Affecting: plunger, seal, plunger return spring, electrical coil
- Fatigue
  - Affecting: plunger spring, seal
- Debris buildup
  - Affecting: plunger, plunger tube, seal, orifice
- Residual magnetism
  - Affecting: plunger
- Wear
  - Affecting: plunger (frictional), plunger tube (frictional)
- Dielectric breakdown
  - Affecting: electrical coil

Case Study

- CALCE received 3 pilot-operated, 3-way solenoid valves from a customer
  - One valve was new
  - Two valves were failed while deployed
  - Reported operating environment: 0-60°C, 60%-90%RH
- Asked to perform failure analysis and determine the cause of valve failure

Initial Examination

- Performed visual inspection of the valves
  - Noticed significant discoloration on the outside of one failed valve, but not on the other
- X-ray examination was conducted
  - Electrical terminals in one of the failed valves appeared to be disconnected from electrical input
  - In other failed valve, electrical terminals appeared to be rotated from expected position
  - Plunger appeared to be undamaged

New and Failed Valve (1)

Top view of electrical coil

New valve

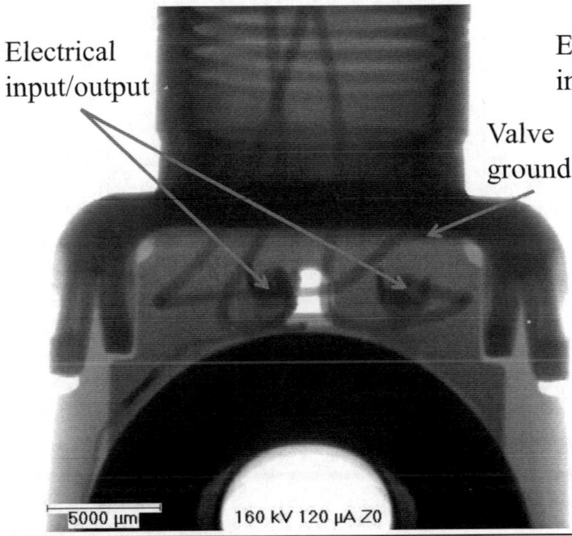

Failed valve (1)

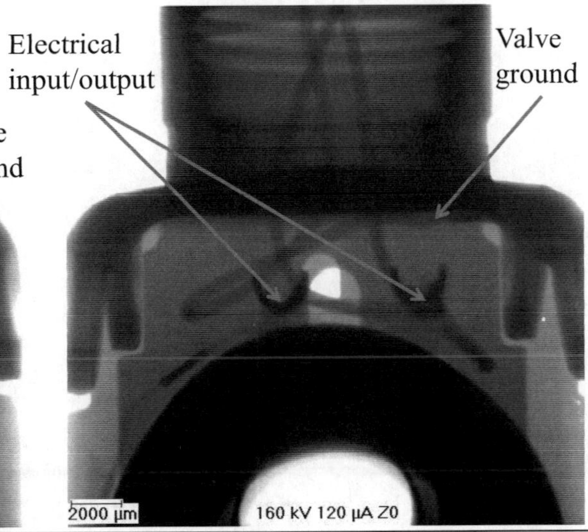

New and Failed Valve (2)

Top view of electrical coil

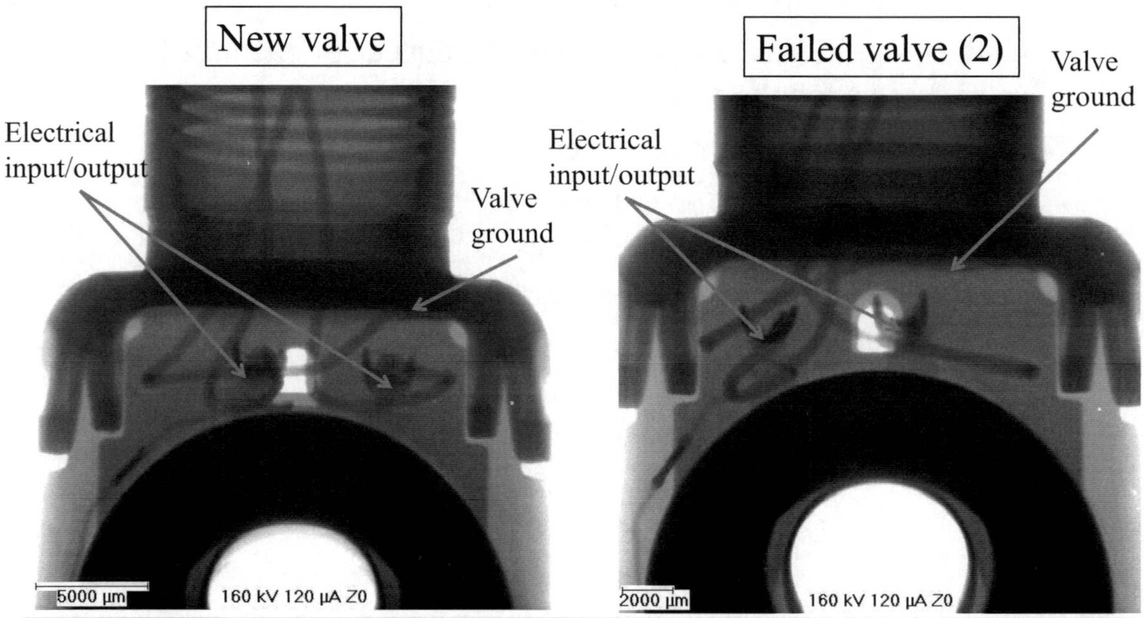

Voltage – Current Testing

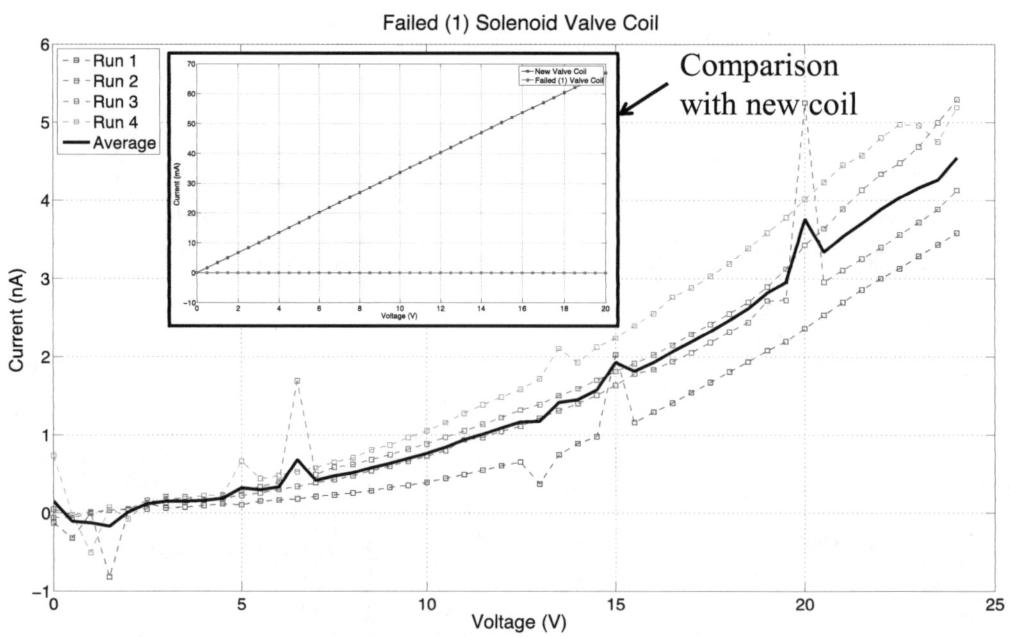

Failed SOV (1)

Voltage – Current Testing

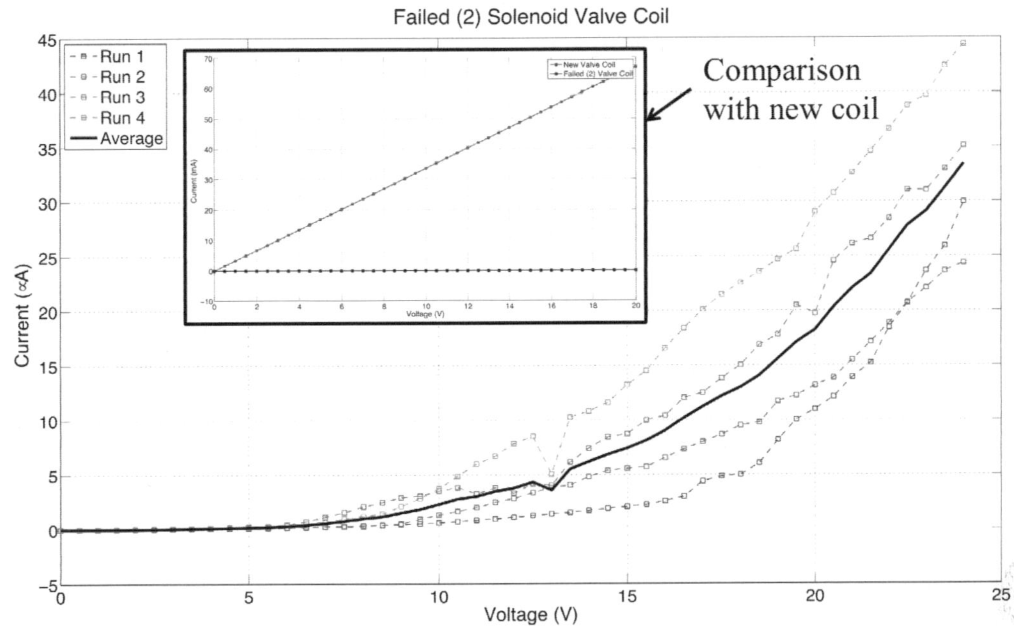

Failed SOV (2)

Impedance Analysis

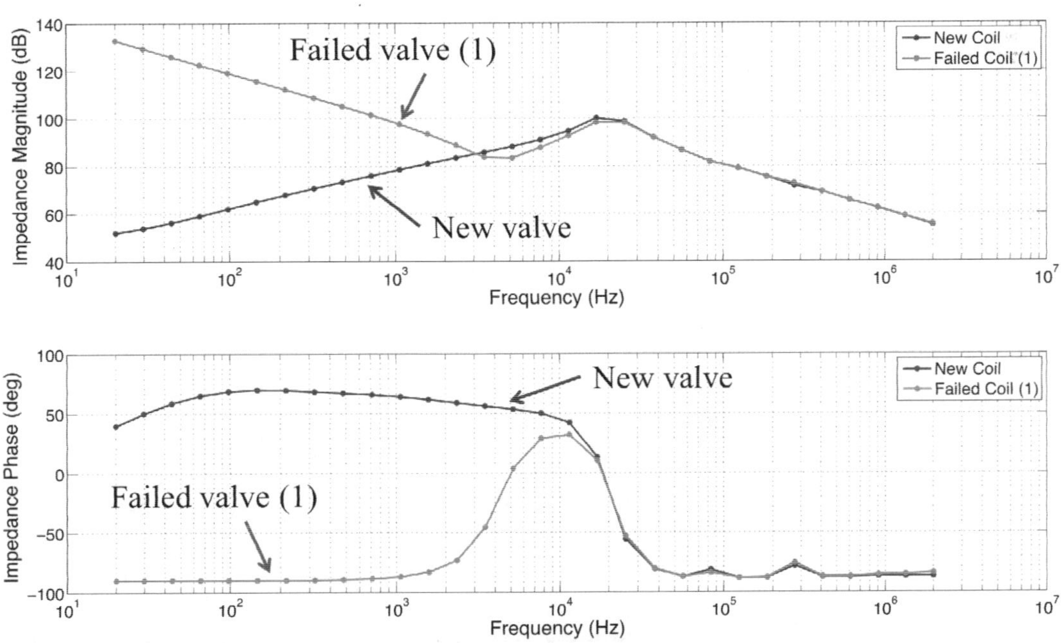

Impedance Comparison for Valve (1)

Impedance Analysis

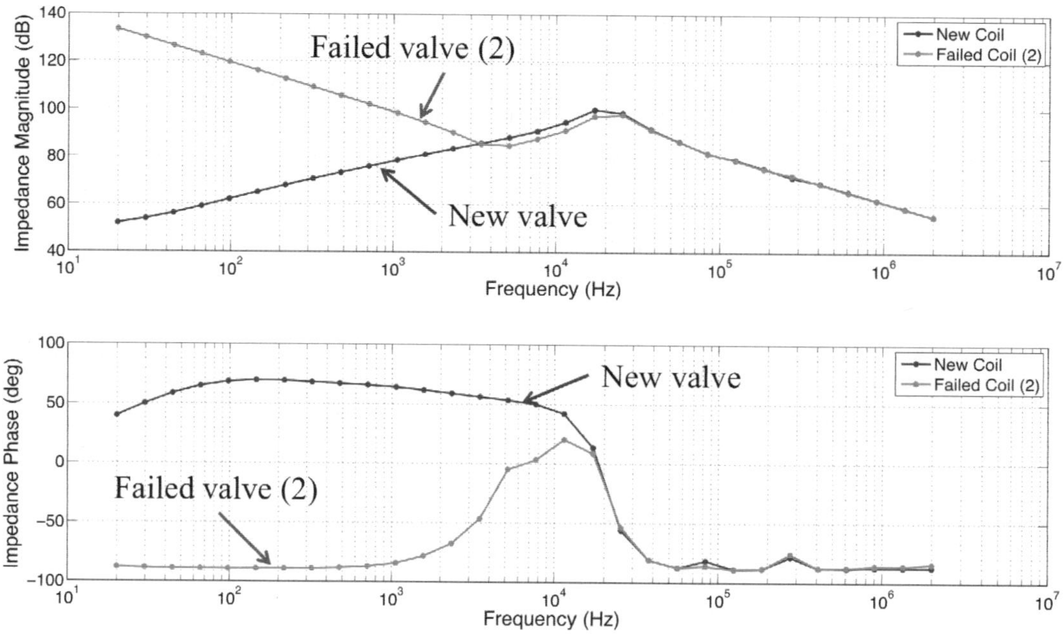

Impedance Comparison for Valve (2)

calce Center for Advanced Life Cycle Engineering 15 A. JAMES CLARK SCHOOL OF ENGINEERING

Impedance Nyquist Plot

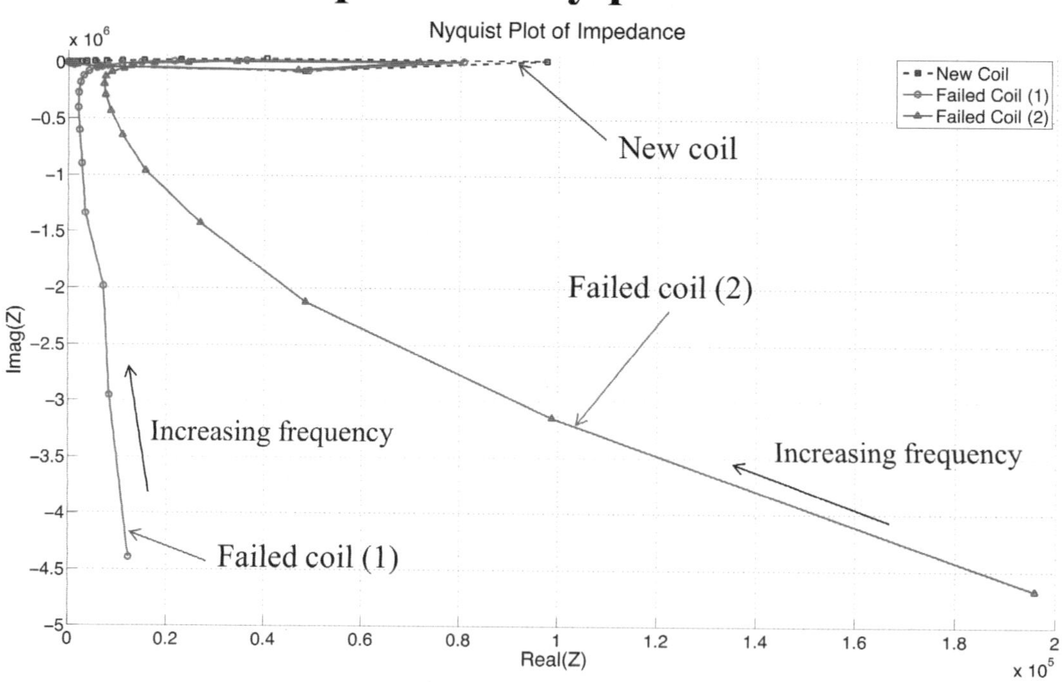

calce Center for Advanced Life Cycle Engineering 16 A. JAMES CLARK SCHOOL OF ENGINEERING

Impedance Nyquist Plot (Zoomed)

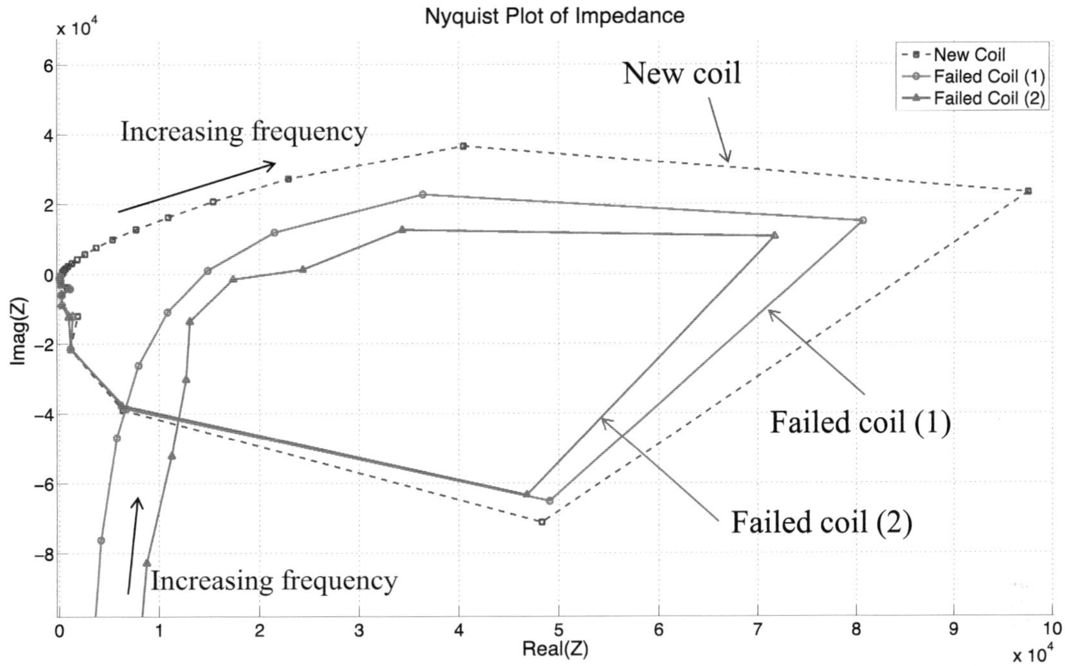

Electrical Characterization Observations

- Electrical performance in the failed valves is significantly different from that of the healthy valve.
  - Much higher resistance
    - However, the failed valves show relative orders of magnitude difference in their resistances
  - Impedance magnitude response reveals an additional point of inflection in the failed valves
  - Impedance phase response shows an approximate –180° phase shift in the healthy valve
    - Failed valves experience an approximate +90° phase shift at resonance (~10^4 Hz) but then return to initial frequency phase response
  - In Nyquist plot, the failed valves have a relatively large imaginary impedance (i.e. electrical reactance)

Electrical Terminal and Coil Inspection

- Terminals in both valves were tested separated from the coil
  - Terminals were separated from coil, shorted, and then the resistance was measured across the input wires
  - Resistance measured approximately 0.15 Ohms
  - Failure from damaged coil terminals was eliminated as the root cause
- Coil resistance was measured when the coil was separated from the terminals
  - Measured high resistance
  - **Failure site lies inside the coil**

Electrical Coil Examination

- Electrical coils were opened and the magnet wire unwound
- Visual inspection of the coil in the valves revealed a break in the magnet wire
  - Burns and discoloration are evidence of a thermal event leading to wire breakage and an open circuit
  - Failure site was located deep in the coil

SEM Inspection
Failed Valve (1)

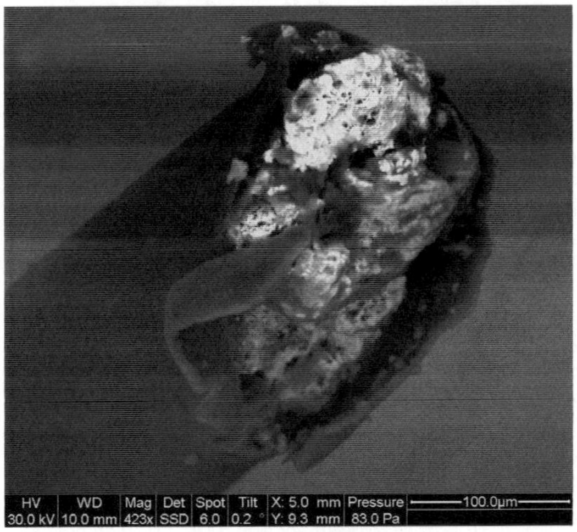

Side of break with greater damage
(Failure Site 1)

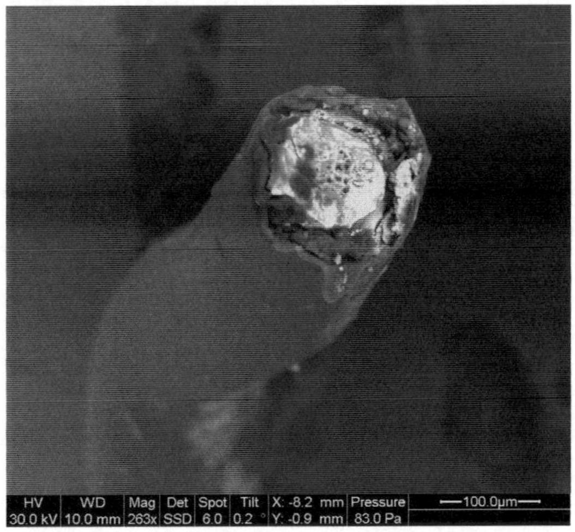

Side of break with lesser damage
(Failure Site 2)

SEM Inspection
Failed Valve (2)

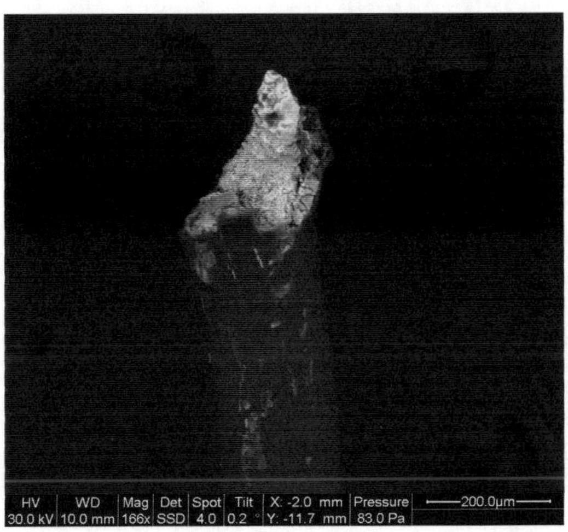

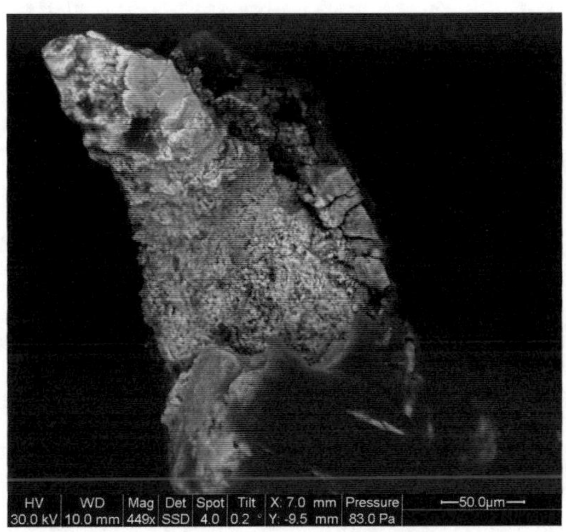

Failure Site 3

Investigations of Chlorine Contamination

- Chlorine was found to be present at the breakage points of the magnet wires in failed valves. It was reasoned that the chlorine may have infiltrated the valve in one of four possible ways:
  1. Ingress from outside environment
  2. Migration from coil epoxy (which contains chlorine) to wire
  3. Contamination due to disassembly procedure
  4. Manufacturing defect
- In the course of the investigation, the first three avenues of ingress were mostly eliminated due to the lack of chlorine present on the magnet wire in locations other than the breakage points.
  – The evidence for this is presented in the tables that follow.

EDS Analysis

	Percent Weight/Sample				
Element	Failure Site 1 Valve (1)	Failure Site 2 Valve (1)	Failure Site 3 Valve (2)	Cleaned Wire Tip	Un-cleaned Wire Tip
Carbon	--	--	--	53.58	70.15
Oxygen	35.79	45.54	30.95	12.44	18.84
Copper	53.59	50.72	62.94	33.49	10.82
Chlorine	5.13	1.72	4.42	--	--
Sulfur	2.75	0.33	--	--	--
Other	2.74	1.69	1.69	0.49	0.19

EDS Analysis

Element	Percent Weight/Sample				
	Enamel Near Failure Site 1	Enamel Near Failure Site 2	Enamel Near Failure Site 3	Cleaned Enamel	Un-cleaned Enamel
Carbon	73.57	73.37	72.61	67.05	69.83
Oxygen	24.85	25.13	24.95	24.09	29.78
Copper	1.36	1.29	2.35	8.76	0.39
Chlorine	--	--	--	--	--
Sulfur	--	--	--	--	--
Other	0.22		0.09	0.49	0.19

Conclusions from Failure Analysis

- Both valves examined by CALCE failed due to magnet wire breakage in the electrical coil
  - Failure mode reported in other literature [1-3]
- Chlorine was present at all break sites
  - Sulfur was found at the break site of one valve electrical coil
  - Not common in solenoid valve failure literature
- Root cause suspected as manufacturing defect whereby chlorine was deposited on the magnet wire enamel causing corrosion and conductor diameter reduction

[1] V. P. Bacanskas, G. C. Roberts, and G. J. Toman, *Aging and Service Wear of Solenoid-operated Valves Used in Safety Systems of Nuclear Power Plants. Volume 1: Operating Experience and Failure Identification*. Oak Ridge National Laboratory, 1987.

[2] Angadi, S.V., R.L. Jackson, Song-yul Choe, G.T. Flowers, J.C. Suhling, Young-Kwon Chang, Jung-Keol Ham, and Jae-il Bae, "Reliability and Life Study of Hydraulic Solenoid Valve. Part 1: A Multi-Physics Finite Element Model," *Engineering Failure Analysis* 16 (2009) 874–887.

[3] Angadi, S.V., R.L. Jackson, Song-yul Choe, G.T. Flowers, J.C. Suhling, Young-Kwon Chang, Jung-Keol Ham, and Jae-il Bae, "Reliability and Life Study of Hydraulic Solenoid Valve. Part 2: Experimental Study," *Engineering Failure Analysis* 16 (2009) 944–963.

Recommendations

- Further investigation into possible sources of contamination is necessary.
  - Possible points of contamination in the manufacturing process:
    - Magnet wire manufacturer
      - Most likely enamel composition: polyurethane (proprietary information)
    - Electrical coil manufacturer
      - Epoxy molding contains chlorine
- Implications of chlorine contamination originating in the manufacturing process:
  - Process control must be strengthened
  - Qualification testing for end users must be developed
  - In-situ health monitoring techniques for the electrical coil must be developed and deployed

Further Magnet Wire Testing at CALCE

- Magnet wire is being tested at CALCE
  - Healthy wire at nominal current
  - Wire with chlorine contamination at nominal current
  - Wire with enamel damage at nominal current
  - Wire with enamel damage and chlorine contamination at nominal current
- Want to determine the effect of chlorine and enamel damage on the performance of magnet wire

Possible Health Monitoring Technique

- Observations from the Bode plot of a failed valve

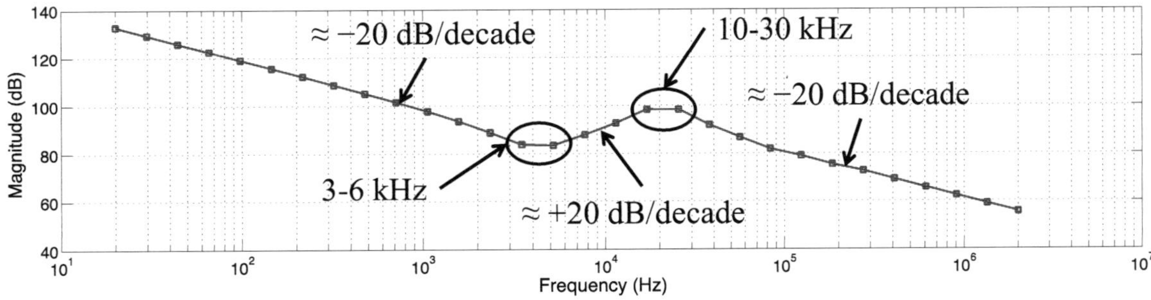

- There is an initial slope of −20 dB/decade and four subsequent changes in slope with increments of ±20 dB/decade.
  - This suggests a model with 2 zeros and 3 poles

$$Z(s) = \frac{b_2 s^2 + b_1 s + b_0}{s^3 + a_2 s^2 + a_1 s + a_0}$$

Parameter Estimation

- The impedance parameters can be fit using non-linear least squares optimization
- Model parameters can be used as features for health monitoring
- When fit to an equivalent circuit model, the parameters could provide physical insight into how different electrical coil components are degrading over time

Third Order Model Results – Failed Coil

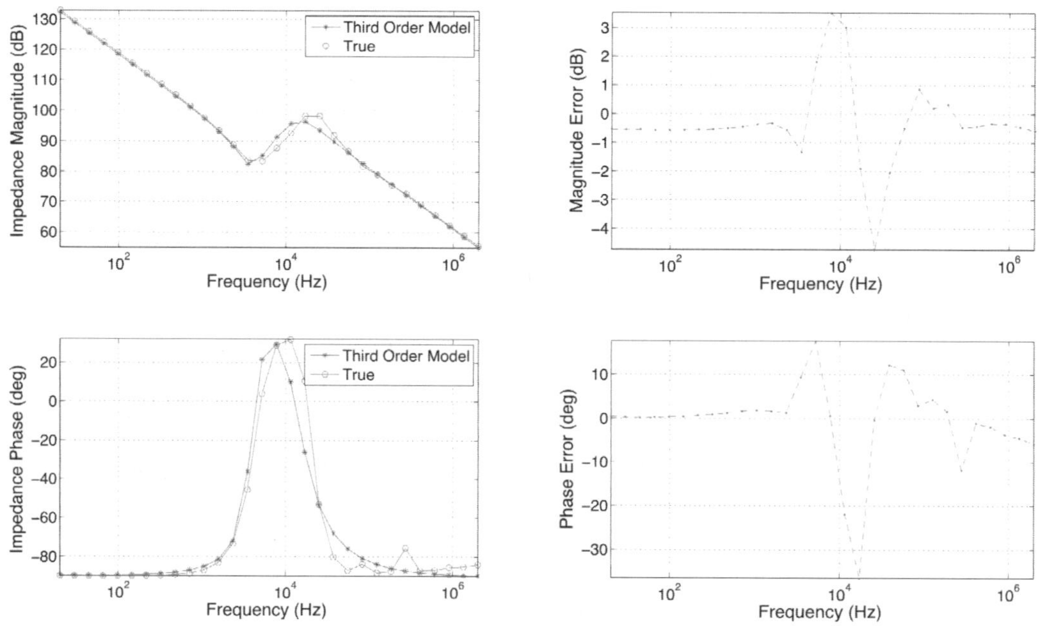

Third order model with non-linear least-squares optimization

Magnitude $||e||_2 = 7.9$, Phase $||e||_2 = 52.3$

Third Order Model Results – New Coil

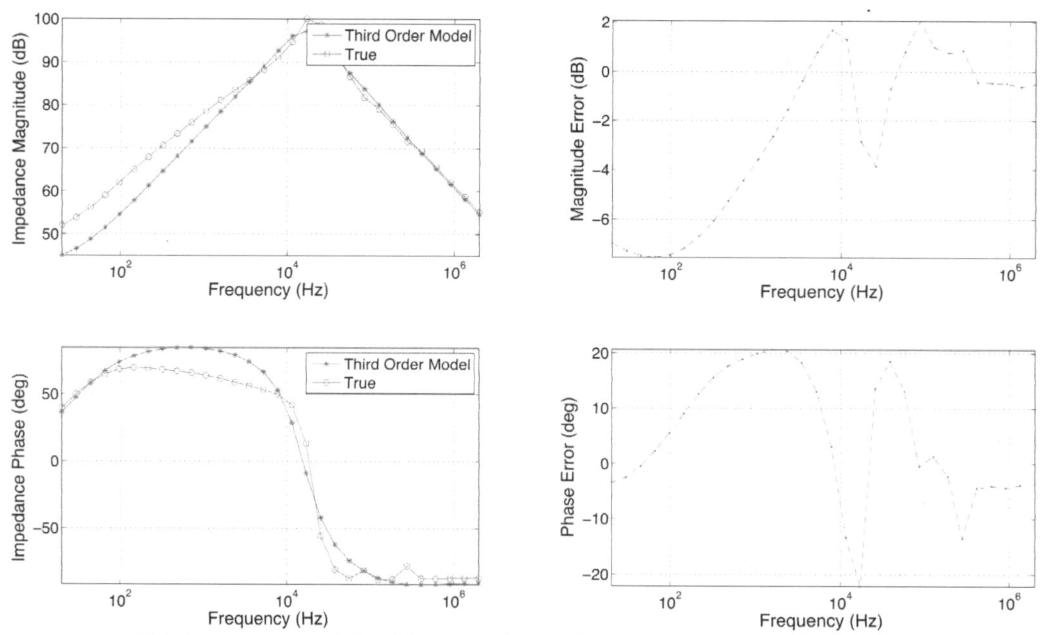

Third order model with non-linear least-squares optimization

Magnitude $||e||_2 = 22.6$, Phase $||e||_2 = 67.9$

Model Parameter Migration

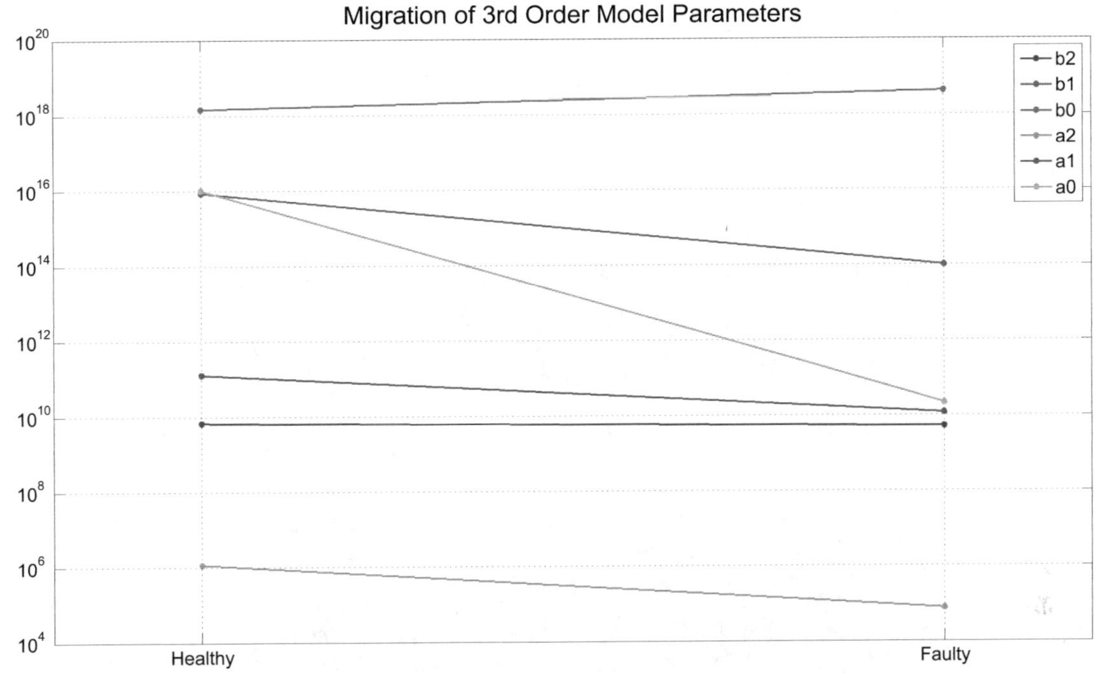

Future Work in SOV Electrical Coil Health Monitoring

- Determine equivalent circuit model structure from the third order system model
  - Evaluate circuit parameters based upon optimized impedance model
- Determine how equivalent circuit model parameters respond to coil degradation and can be updated using obtained measurements
  - Can parameter drift be used as an indication of coil degradation?
  - Is the drift of different parameters indicative of different degradation mechanisms?

National Aeronautics and Space Administration

Developing a Particle Imaging Velocimetry (PIV) Capability for a Supersonic Wind Tunnel

John Lucero
NASA Glenn Research Center
5/21/2014

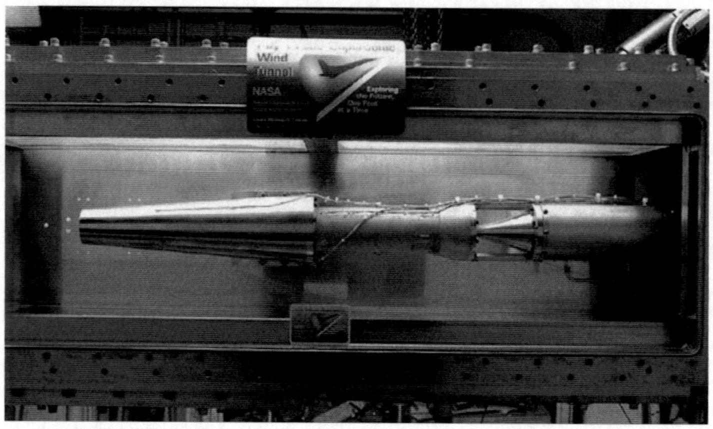

www.nasa.gov 1

National Aeronautics and Space Administration

Objective: Develop a PIV capability for flow visualization in the 1x1 Foot Supersonic Wind Tunnel.

- (I) **Background**
  - **(Current facility capabilities/ Rationale for PIV)**
- (I) **PIV System Design**
  - **(Model Seeding system/Tunnel Seeding System)**
- (II) **Conclusions**

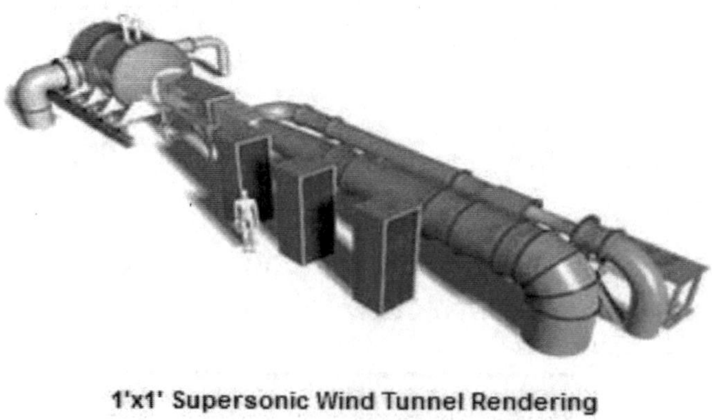

1'x1' Supersonic Wind Tunnel Rendering

www.nasa.gov 2

National Aeronautics and Space Administration

Background

(rationale / current facilities / current fans)

www.nasa.gov 3

National Aeronautics and Space Administration

BACKGROUND:

- Originally built around 1949, used to conduct fundamental vehicle focused research in supersonic fluid mechanics. Mach = 1.3 – 4.0
- In 1989, an additional piping circuit with an electric heater was installed allowing the tunnel's speed range to be extended to Mach 6.
- Specialized support systems such as auxiliary bleed, model hydraulics, and probe actuation systems.
- Flow visualized with Schlieren, or blue Zyglo and black lights.

www.nasa.gov

National Aeronautics and Space Administration

CONCERN:

- High cost of running larger SWT with PIV capability limits the parametric investigations required for detailed understanding of the physics necessary for successful technology development.
- Lack of quantitative flow visualization capability at the 1x1 SWT for proof of concept tests.
- Quantitative optical diagnostic technique, LDV, was attempted in 1991, yielding mixed results due to low level of flow seeding achieved.

www.nasa.gov

National Aeronautics and Space Administration

Rationale

NEED:

A new capability to quantitatively measure the velocity profile around a test article in supersonic flow.

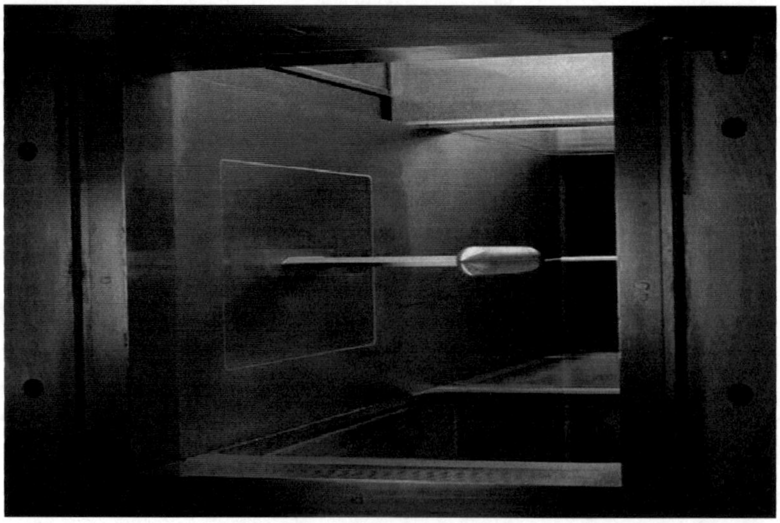

www.nasa.gov

Systems Engineering Approach

Traditional Systems Engineering involves:

1. Understanding how to develop and engineer a system
   - Concepts, terminology, techniques
2. Understanding the importance of:
   - Teams, communication, requirements definition, interface control, focus on deliverables
3. Useful tools and techniques
   - ConOps, Functional Analysis, WBS, Reviews, V&V, CM, Risk Management

SE Dilemma

MUST BALANCE

Risk (R)! Performance (P)! Cost (C)!

C*R = P

1. Reduce C at constant R = P drops
2. Reduce R with constant C = P drops
3. Reduce P at constant C = Risk drops
4. Reduce P at constant R = Cost drops
5. Reduce C at constant P = Risk increase
6. Reducc R at constant P = Cost increase
7. Etc.

National Aeronautics and Space Administration

Design, Develop, Verify and Validate

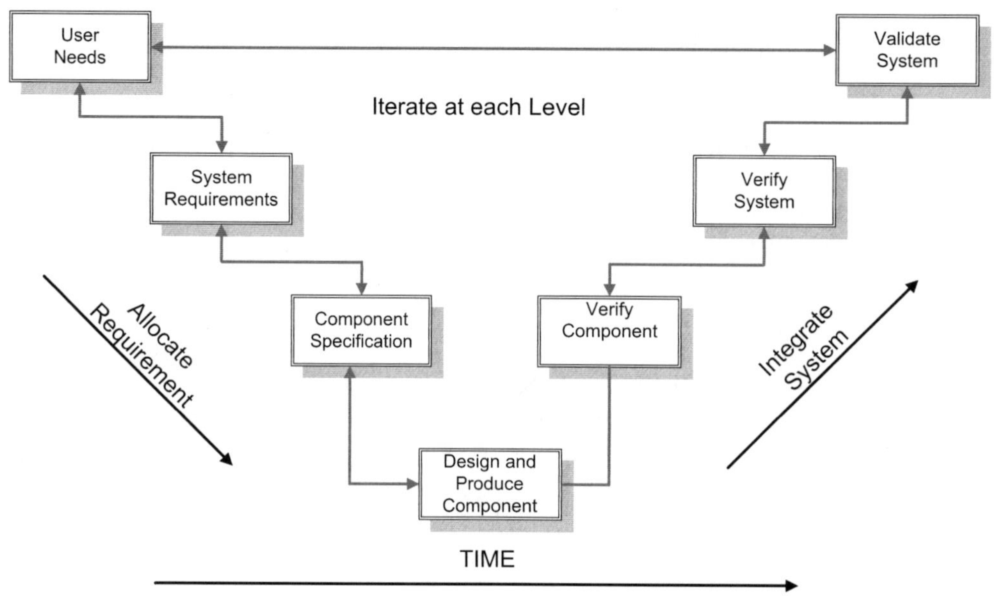

National Aeronautics and Space Administration

Systems Engineering Approach

Requirements Definition:

Stakeholder Expectation

- Understand User Needs
- User Vision to develop ConOps
- Cost/Risk
- Educate Diplomatically

Requirements Development (High Level)

- System Requirements
- Technical Requirements
  Key Performance Parameters (KPPs)

National Aeronautics and Space Administration

Requirements Lessons Learned

- Achievable
- Verifiable (VCRM)
- Crystal clear
- Completeness
- Appropriate level
- Reflects need not solution
  (no design solution at high level)

www.nasa.gov 11

National Aeronautics and Space Administration

Concept of Operations

What would it look like?

(High level design requirements)

- Existing optical access ports at test section will provide adequate access for PIV and/or stereo PIV.

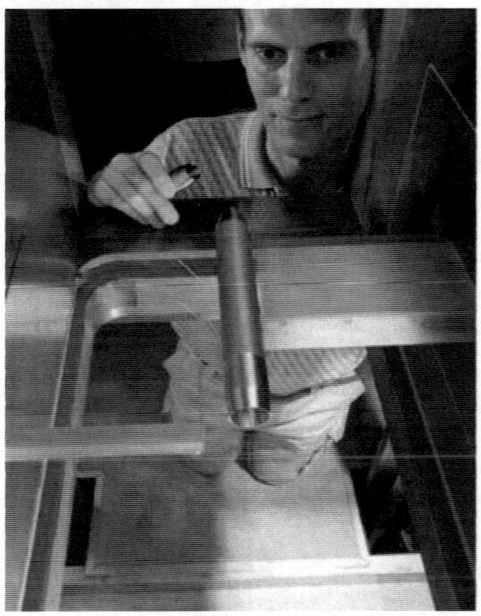

www.nasa.gov

National Aeronautics and Space Administration

Concept of Operations

What would it look like?

(High level design requirements)

- Seed material injected through the nozzle mounting port.

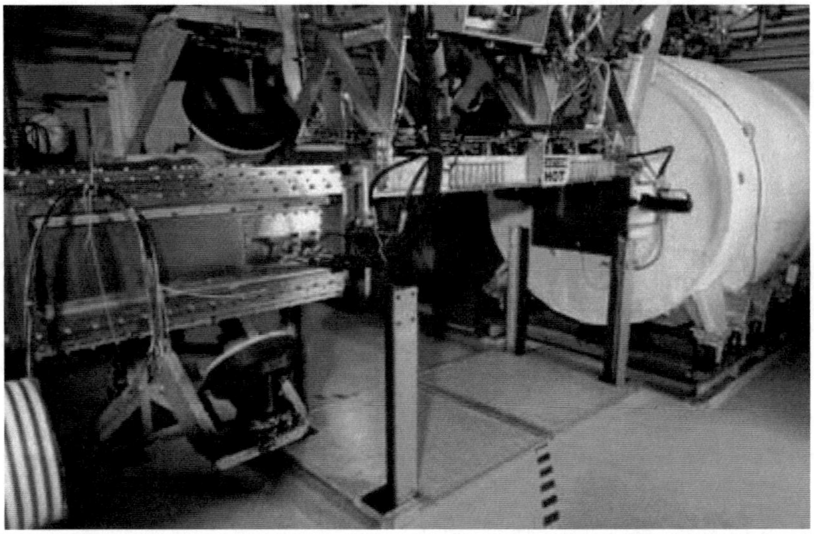

www.nasa.gov

National Aeronautics and Space Administration

- Seed material injected into 40 psi Combustion Air supply.

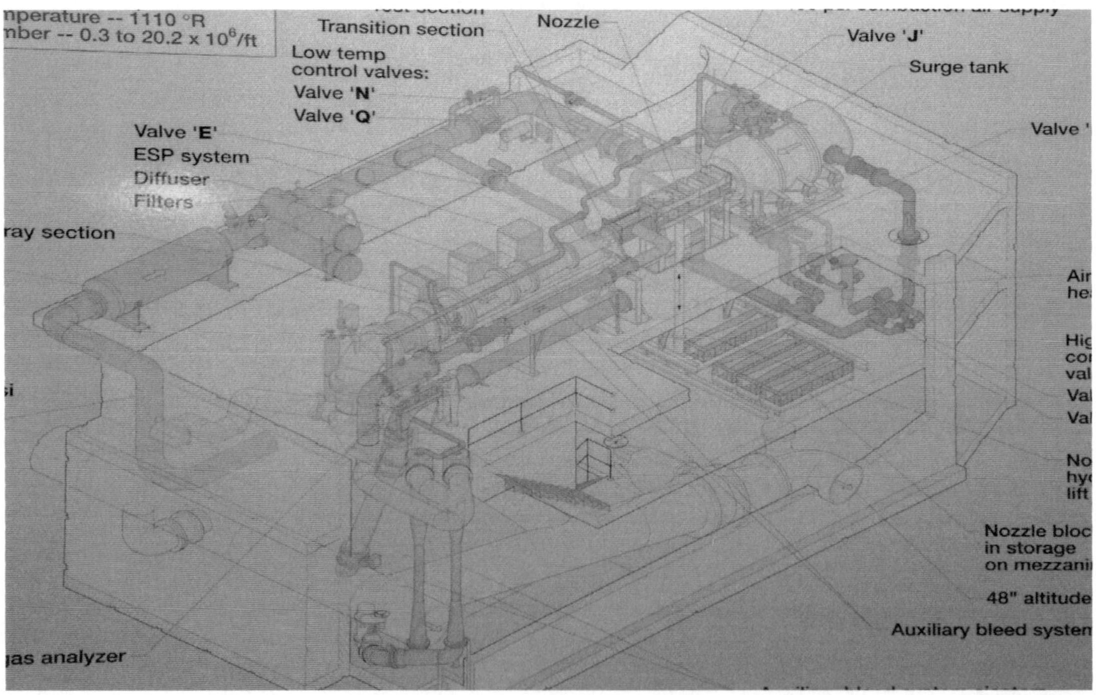

www.nasa.gov

National Aeronautics and Space Administration

Concept of Operations

Proposed PIV configurations:

1) 2-D Streamwise PIV

2) 3-D Streamwise PIV

3) 3-D Cross-Stream PIV

Note: Based on the limitation of 3 optical access ports

www.nasa.gov

National Aeronautics and Space Administration

Concept of Operations

2-D PIV, Streamwise Plane Above Model

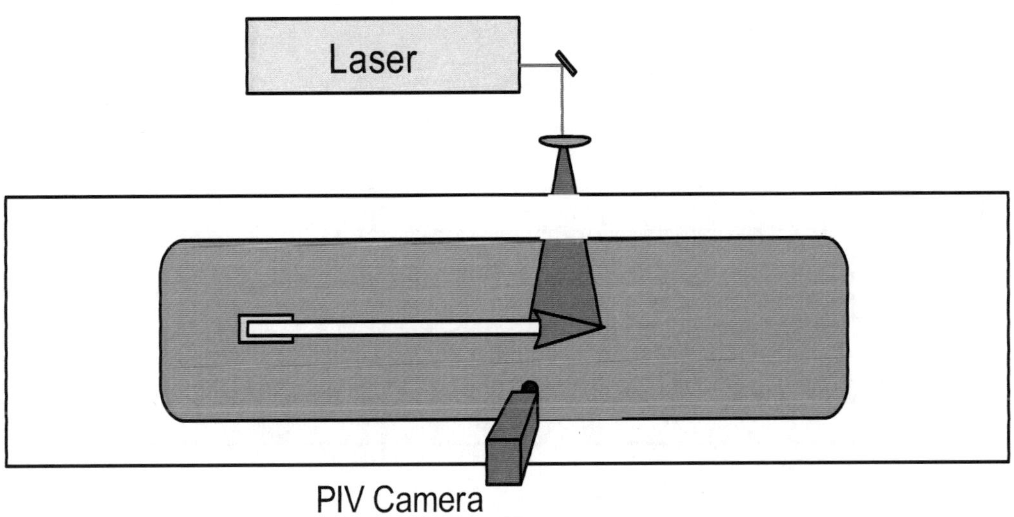

www.nasa.gov

National Aeronautics and Space Administration

Concept of Operations

Stereo PIV, Streamwise Plane Above Model

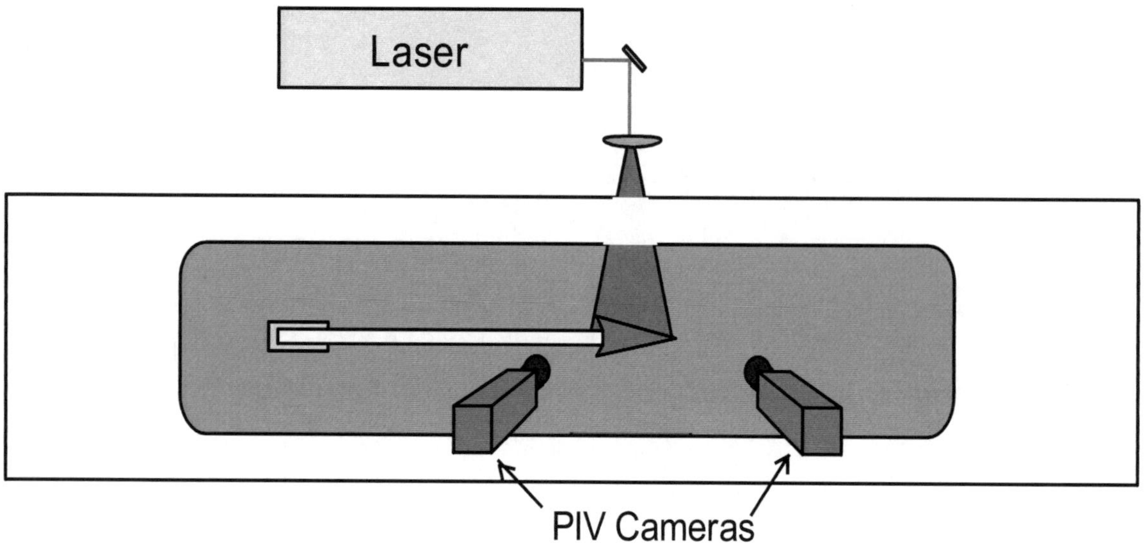

www.nasa.gov

National Aeronautics and Space Administration

Concept of Operations

Stereo PIV, Cross-Stream Plane

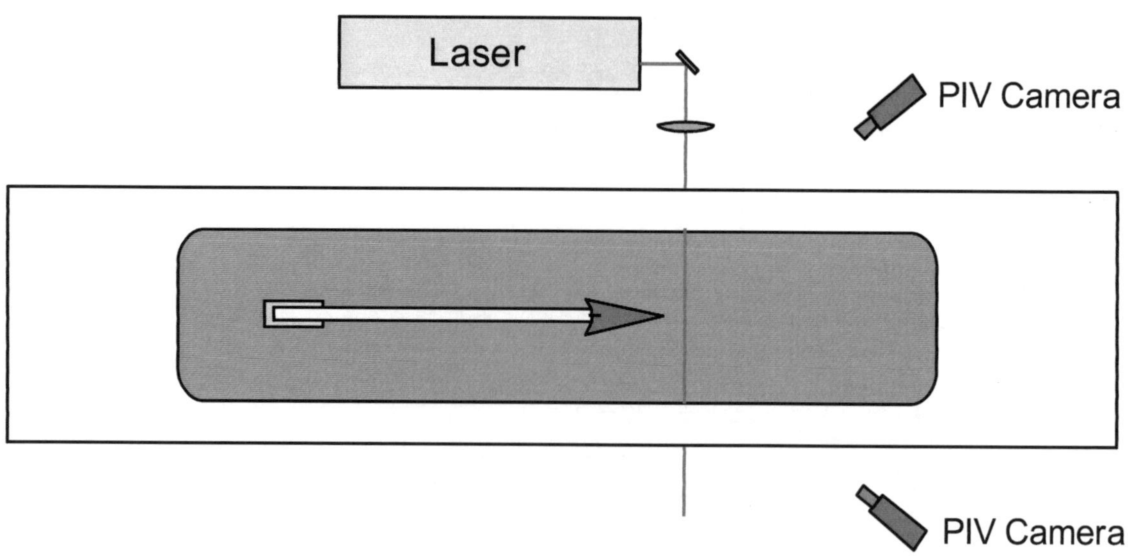

www.nasa.gov

National Aeronautics and Space Administration

Functional Analysis Tools

- Functional Architecture
- System Block Diagram
- Functional Interface Analysis (N squared diagram)
- Timeline analysis (sequential/simultaneous operations)

www.nasa.gov

National Aeronautics and Space Administration

System Block Diagram

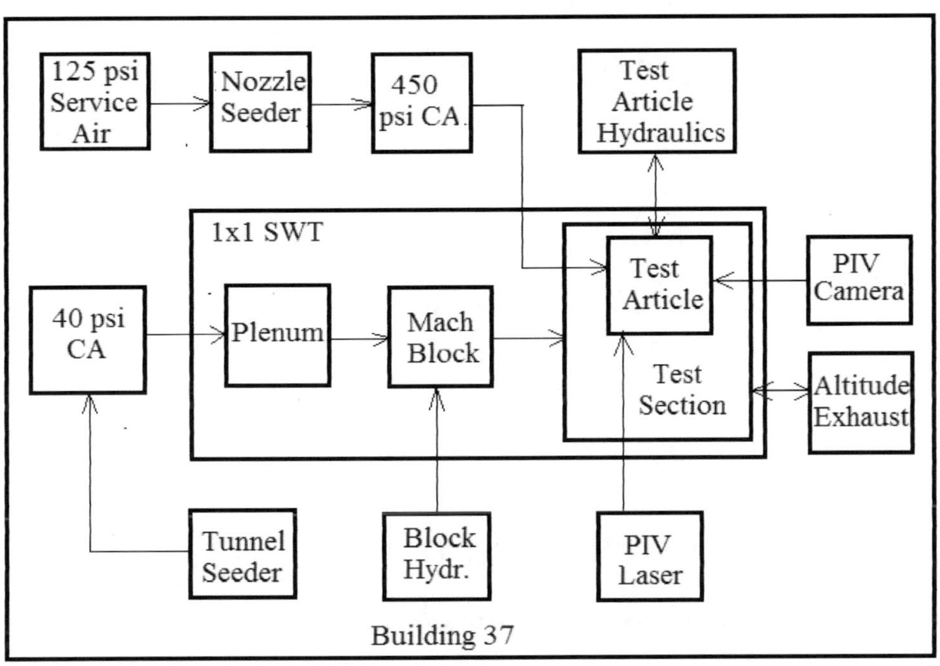

www.nasa.gov

National Aeronautics and Space Administration

Functional Interface Analysis N Squared Diagram

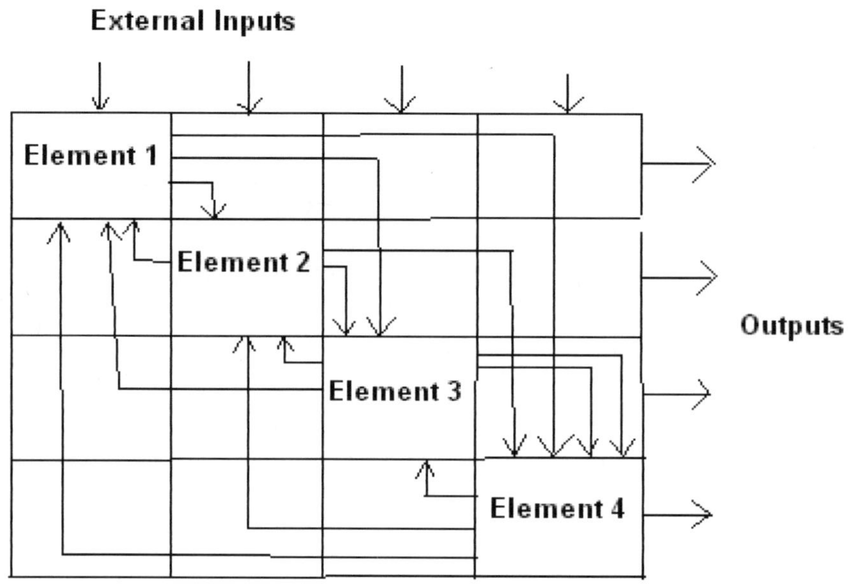

www.nasa.gov

National Aeronautics and Space Administration

Technical Solution Definition

- Logical Decomposition
  - Breakdown from the Functional Analysis
- Design Solution
  - Solution to the problem (actual numbers and hardware)

www.nasa.gov

Requirements Definition

1:17:52 PMRevision: Draft	Document No: PIV Pressure Vessel
Release Date:	Page 2 of 34
Title: System Requirements for the PIV System Pressure Vessel Project	

National Aeronautics and Space Administration

PIV System Pressure Vessel

(Second DRAFT)

RELEASE DATE: TBD

FTB0 Aviations Environment Test Engineering Branch

Draft Maturity 95%
June 3, 2010

Seeder Pressure Vessel SOW

SOW

Pressure Vessel for 1NW Seed System

Purpose

The purpose of this document is to provide a statement of work to fabricate a U stamped pressure vessel (PV) to be used in the test cell 1NW in Building 37. The scope of this work is to fabricate and certify the PV per the Pressure System Office (PSO) and Safety Office (SO) at Glenn Research Center. A picture of the desired PV is shown in Figure 1a. The MAWP is required to be 220 psig. A mobile cradle is needed for the PV to sit on. A representative cradle is shown in Figure 1b.

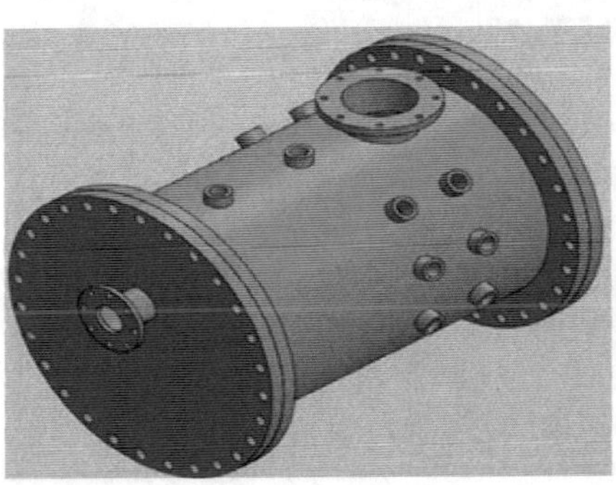

Figure 1a: PV for 1x1 Seed System

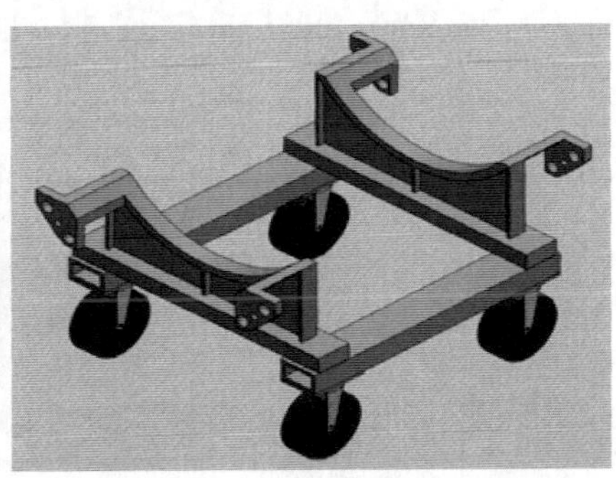

Figure 1b: Mobile Support Cradle for PV

Seed Feasibility Analysis

	# of Screens	Screen Mesh	Seed Material
NATR	3	16	Vicount/Al_2O_3
SHAR	0	-	Vicount/Al_2O_3
10x10	3	12x12	Vicount
9x15	4	10x11.5	Vicount
1x1	6	30	-
15x15cm	2	25	-

National Aeronautics and Space Administration

Seed Material Feasibility Analysis

Compare the behavior of three different seed materials:

1. Di-ethyl-hexyl-sebacat (DEHS),
2. CFC-180 Smoke Fluid, common name white mineral oil
3. Olive oil under identical conditions.

www.nasa.gov

National Aeronautics and Space Administration

Seed Feasibility Analysis

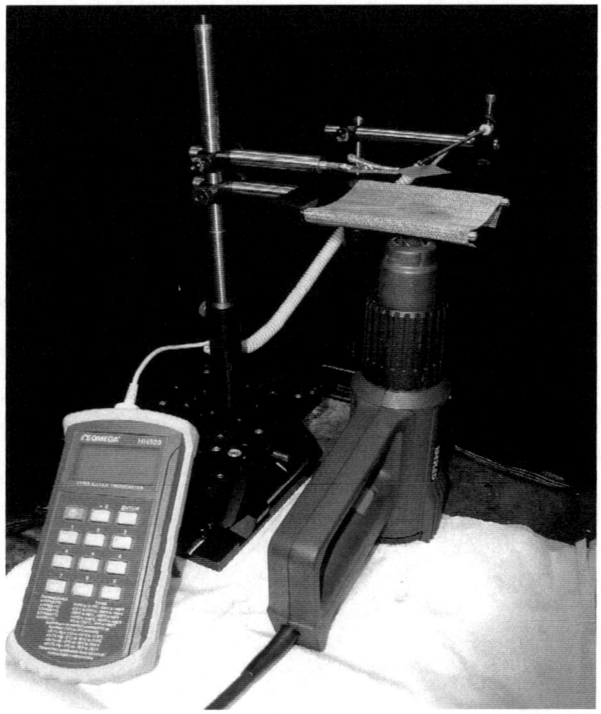

Experimental set-up for performing DEHS/CFC comparison

www.nasa.gov

National Aeronautics and Space Administration

PIV Laser Window

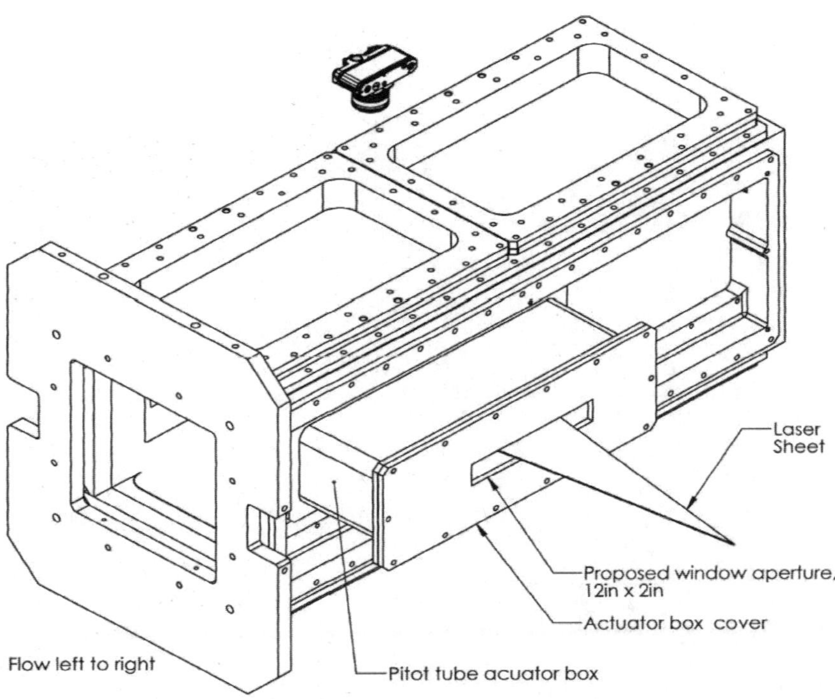

www.nasa.gov

National Aeronautics and Space Administration

Analysis - Nominal Wind Tunnel Operation Results

<u>Glass Window</u>
Max Stress = 159 psi
Allowable = 7,600 psi
FS = 47.8

<u>Requirement</u>
FS = 10 (Pass)

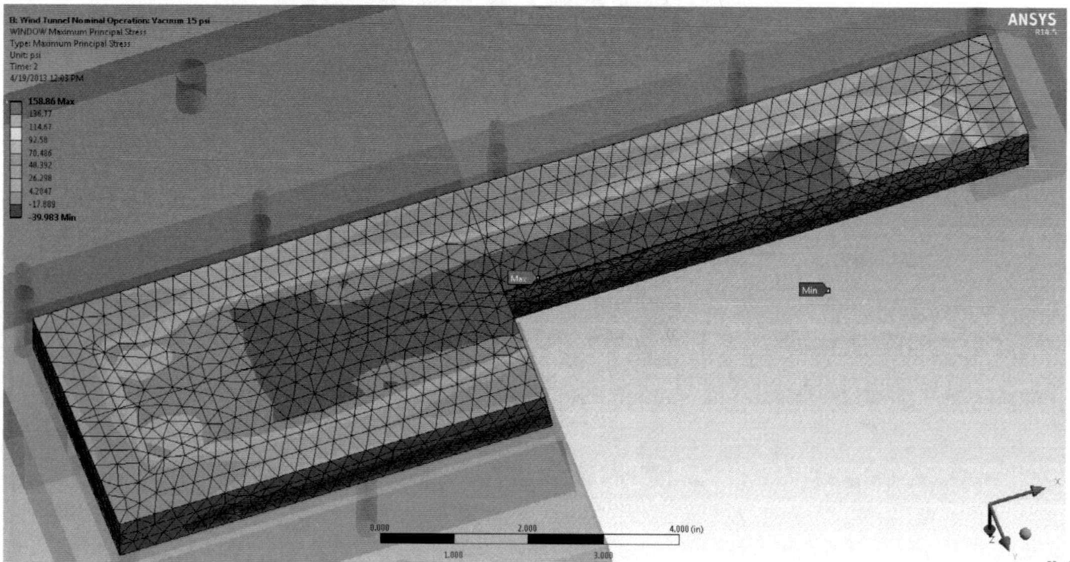

1x1 PIV Window

National Aeronautics and Space Administration

Analysis - Nominal Wind Tunnel Operation Results

<u>Frame</u>
Max Stress = 5,107 psi
Allowable = 55,000 psi
FS = 10.8

<u>Requirement</u>
FS = 5 (Pass)

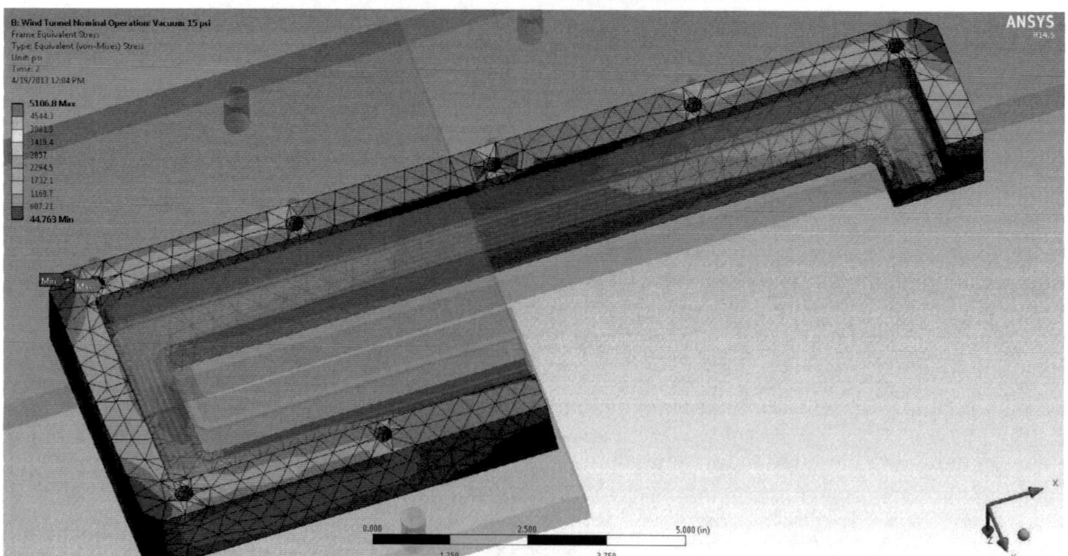

1x1 PIV Window

National Aeronautics and Space Administration

Analysis - Nominal Wind Tunnel Operation Results

<u>Cover</u>
Max Stress = 4,714 psi
Allowable = 55,000 psi
FS = 11.7

<u>Requirement</u>
FS = 5 (Pass)

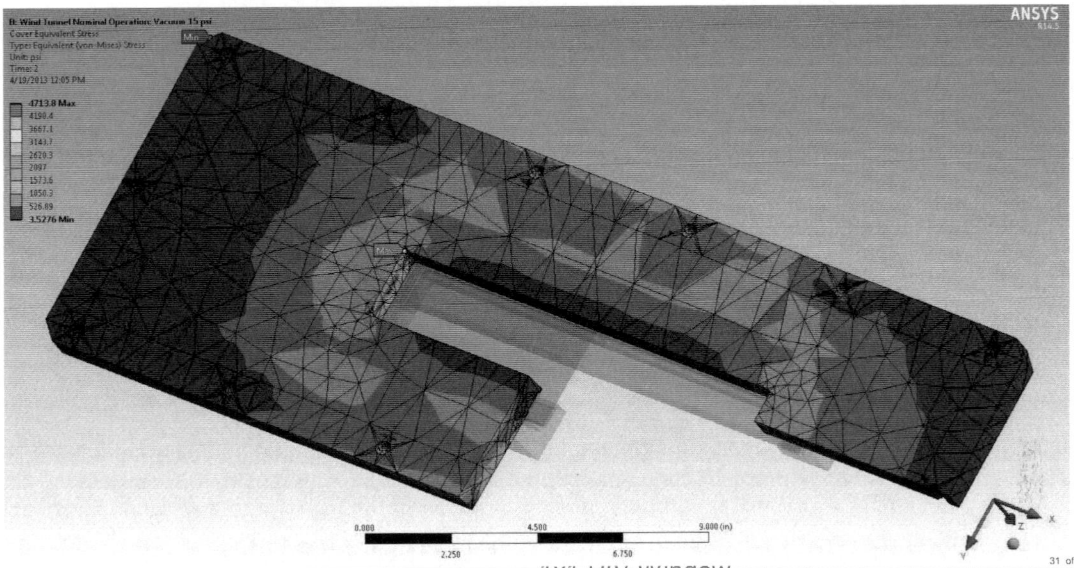

National Aeronautics and Space Administration

Laser Window (Quartz)

National Aeronautics and Space Administration

Pressure Certification

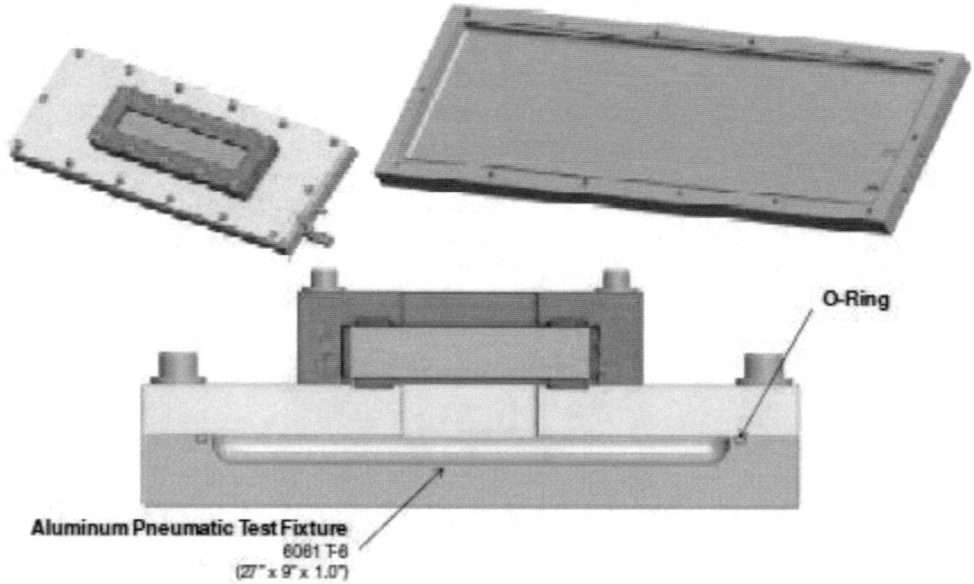

Pneumatic Test Fixture Setup

www.nasa.gov

National Aeronautics and Space Administration

Laser Window Pneumatic Test

www.nasa.gov

National Aeronautics and Space Administration

Nozzle Seeder Schematic

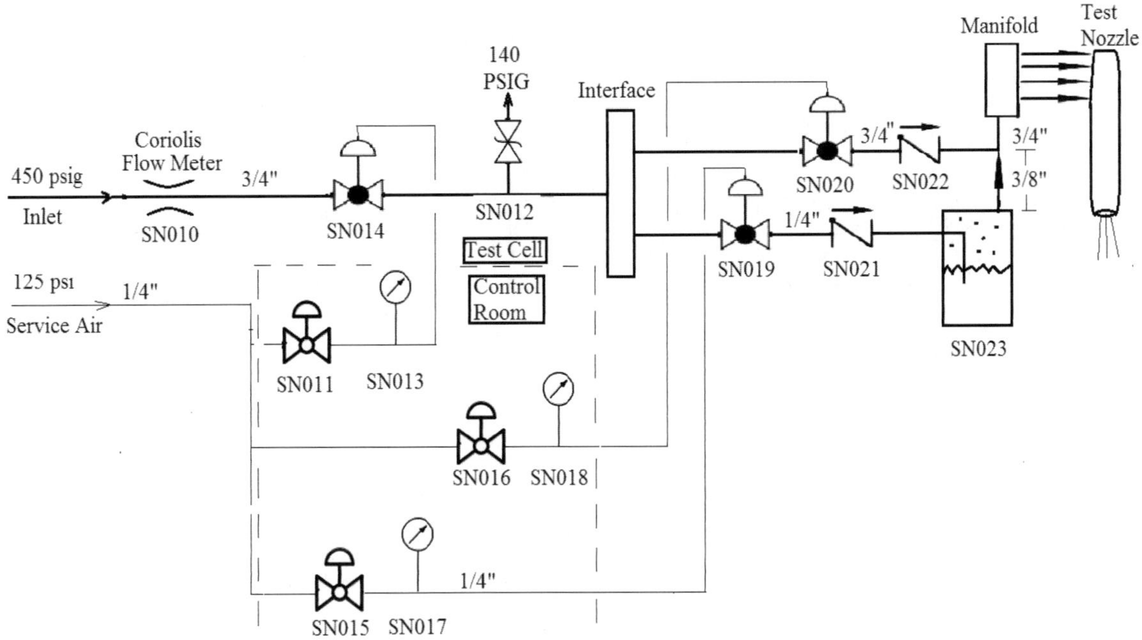

www.nasa.gov

National Aeronautics and Space Administration

Tunnel Seeder Schematic

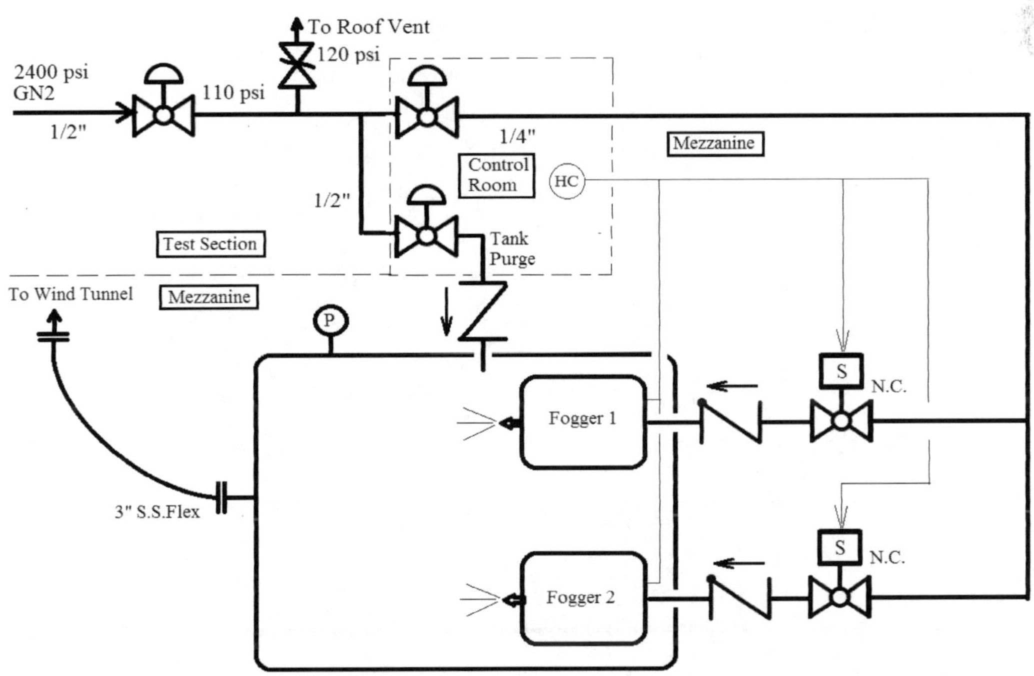

www.nasa.gov

National Aeronautics and Space Administration

Tunnel Seed Tank

www.nasa.gov

National Aeronautics and Space Administration

Fog Generators in Seed Tank

www.nasa.gov

National Aeronautics and Space Administration

Fog Generator

www.nasa.gov

National Aeronautics and Space Administration

Tunnel Seed Line

www.nasa.gov

Summary

National Aeronautics and Space Administration

Summary

1) Testing was delayed due to model test hardware problems

2) Schlieren Configurations completed

3) PIV Testing to be started June 2014

National Aeronautics and Space Administration

Acknowledgements
1x1 SWT Project Team

Research:

Raymond Castner	Randy Locke
Mark Wernet	Adam Wroblewski

Facilities Testing:

Robert Clark	John Lucero
Brent Seifert	Rick Senyitko
Erhart Hartman	Robert McCluskey

Design Engineering:

Allen Guzik

www.nasa.gov 43

IMPLEMENTING AN EFFICIENT SHM SYSTEM INTO A STEEL MODEL BRIDGE

Mohammad Azarbayejani, Sebastian Salinas, David Renteria and Francisco Cantu
Civil Engineering Program, University of Texas-Pan American
1201 W. University Dr.
ASB. 1.203
Edinburg, TX 78539
(956) 665-3103
azarbayejanim@utpa.edu

Abstract: Structure Heal Monitoring (SHM) systems have been implemented over the past decades to detect early damage caused by external forces and environmental effects acting on large-scale structures. An effective decision can be taken based on the evaluation of the damage in order to ensure a good performance of the current structure. In this paper, a SHM system is implemented on a model steel truss bridge; piezoelectric accelerometers are sending signal vibrations due to the traffic on the bridge to module NI 9234 of a NI Compact Rio - 9022 (cRIO). Field-programmable gate array (FPGA) housed in the cRIO process all the data information. Optical Sensing Interrogator sm125 system is connected to the cRIO to obtain strain measurement, Fiber optic Bragg Grating (FBG) which is connected to it facilitates this process. Three different scenarios are introduced on this investigation to present damage situations similar in a real bridge. Based on the data obtained at different states of the bridge and using the finite element model of the brisge, Monte Carlo simulation is implemented as the decision-maker with a range of possible outcomes and the probabilities they will occur for any choice of action. The method used on the model bridge can be implemented on real structures to enhance a damage detection process.

Key words: Structural Health Monitoring (SHM); Monte Carlo Simulation; Field Programmable Gate Array (FPGA); Fiber Optic Bragg Grating (FBG); Finite element model

Introduction and Background: Structural Health Monitoring (SHM) technologies have had a significant role in different areas such as aviation and civil engineering infrastructure for the past two decades. The development of this global system technology that increases the reliability and safety of a structure has become a topic of serious consideration. Being able to sense damage with SHM on a structure, especially bridges with a lot of life's aboard, and then processing that response into a warning for maintenance, is why this topic continues to be examined. A warning that not only alert's of the danger, but that locates the danger and its severity. Bridges around the world, including other structures, continue to fail before their service life. Many bridges are reaching their age limits, and yet most continue to be time-based inspected. More attention to SHM occurs, when not only there is a need of monitoring aged bridges, but new ones as well. Figure 1 shows the Yangmingtan Bridge in Harman, opened in 2012. Less than a year after being opened, a long section of the bridge failed, resulting in some fatalities. It is believed that either the material, construction, or overload problems caused this accident. Regardless of what caused it, it is clear that there was no warning of this failure. Any damage detection that could have been obtained, could have at least serve as a safety measure of closing down the section. This example is only one of the many possibly preventable failures that have occurred throughout the bridge failure history. These exemplars show the importance of SHM systems and how they can be utilized to recognize the current state of any infrastructure around the world.

Figure 1: Failure of Yangmitan Bridge section showing trucks flipped over.

A new vision of structural health monitoring is has the ultimate goal that SHM systems are not only limited to damage identification, but also to describe the structure by a probabilistic model [1]. Implementing Bayesian method will perform all four levels of structural damage detection and assessment: damage indication, its location and severity, and its impact on the structural reliability [2]. This particular method is well known to be a successful tool for implementing it in truss structures. A new perspective for addressing uncertainties in truss bridges associated with truss joint configurations in damage identification based on Bayesian probabilistic model updating and model class selection. [3]. A typical SHM system consists of a data acquisition device, data processing algorithms and different type of sensors including: thermocouples, piezoelectric accelerometers, fiber optic strain gauges etc. Each sensor has its own data acquisition platform and is used for different purposes for example: cracks, strain, change in temperature and excessive vibrations make by the traffic on the bridge.

In this research the principal objective is to implement an effective SHM system on a model steel truss bridge to monitor the health of the bridge on-need and on-demand. Two piezoelectric accelerometers are placed strategically in the bridge to gather signals from the traffic loading on the bridge, placement strategies for traditional sensor systems for SHM has been a well-studied subject [4]. These sensors are connected to the data acquisition device NI Compact Rio - 9022. Another system is the Optical Sensing Interrogator sm125, which uses a Fiber optic Bragg Grating (FBG) to measure strain. The system is designed to work to detect different states of the steel model truss bridge. Three different scenarios are introduced to see the difference to each other.

Experimental procedure: In this paper an SHM system is used to provide real time data to detect damage in the structure [5]. This SHM system is implemented on a steel model truss bridge located in the Civil Engineering building at UTPA. Trying to make this experiment closer to reality, this model steel truss is very similar to a Howe truss used in real bridges. The SHM system includes two accelerometers sensors and a fiber optic strain sensor. These sensors are used to gather data from the different tests made on the bridge. The size of the truss bridge is 12 feet long, 3 feet wide and 3 feet tall. The deck of this truss bridge was made of half of an inch Plexiglas and it was connected from the bottom with quarter inch bolts. Over the Plexiglas deck, a rough carpet was placed to represent the asphalt in a real bridge. Cars and toy trucks move over the carpet from end to end in order to simulate random traffic loading on the bridge. It is important to say that light and heavy traffic are implemented to make this experiment closer to a real situation. The truss bridge was placed on top of two tables that allow easy access for symbolic changes in bolts and members and provide a simply support boundary conditions to the structure. Figure 2 shows the model steel truss bridge implemented in the research laboratory of The University of Texas Pan-American.

Figure 2: Model steel truss bridge located at UTPA

Two piezoelectric accelerometers with the sensitivity of 1000 mv/g are installed at the top center of each member of the truss. When light or heavy traffic is applied to the bridge, the structure experiences some vibrations that need to be measured. These vibrations are measured with the piezoelectric accelerometers. Figure 3a shows a piezoelectric accelerometer implemented in the bridge. The accelerometers are connected to a NI data acquisition device, which gathers every signal made from traffic. A fiber optic strain gauge is connected under the bridge. This strain gauge sensor determines the displacement made on the bridge when the cars are pulled through the bridge from end to end. Figure 3b shows the fiber optic strain gauge connected under the bridge.

Figure 3: a) piezoelectric accelerometer, b) Fiber optic strain gauge Implemented in the bridge.

More over thermocouples sensors are installed in the bridge. Since the bridge is located inside a laboratory with constant temperature, it is impossible to see a difference in temperature. However the program is made with the purpose of implementing this SHM system in a real bridge, in which thermocouples sensors are needed to measure the change in temperature. The system works as a parallel system. Two devices are needed to interpret the data acquired by the sensors. The fiber optic strain sensor is connected to a Micron Optics sm125, which is connected by an Ethernet cable to the NI data acquisition device and those are connected to a portable computer. Figure 4 shows the parallel system connected to a portable computer.

Figure 4: Structural Health Monitoring system connected to a Portable computer.

The data collected from the accelerometers and the strain gauge is sent to a web domain in real time. The user has the ability to put security access codes to the web domain. The program is constantly sending information of the activity made in the bridge to the web domain. This method allows the user to read the information at any hour any time. This SHM system would detect any possible damage in the bridge.

Implementation of the SHM system in the bridge: In order to visualize the information transmitted by the accelerometers and the strain gauges installed in the bridge, a system-designed platform to develop the environment for a visual programming language was needed. A LabVIEW (Laboratory Virtual Instrument Engineering Workbench) interface was chosen as the designer program in this experiment. The principal feature used by this program in this SHM system is data acquisition. The programmer connects different function-nodes by drawing wires which certain variables, which can be executed by any node as soon as the input data become available. Implementing a start and stop buttons, the user is allowed to get information at a desired amount of time. In this study, the program is started before the cars are pulled through the bridge, and it ends after this action is completed. The cars are pulled with an average velocity of 3ft/sec, taking 4 seconds to finish the desired trajectory. Figure 5 shows the control panel of the LabVIEW interface used in this SHM system. Two charts are visualized in the control panel interface, one for the accelerometers, and one for the strain gauges. A box chart containing information from the thermocouples is located in the interface.

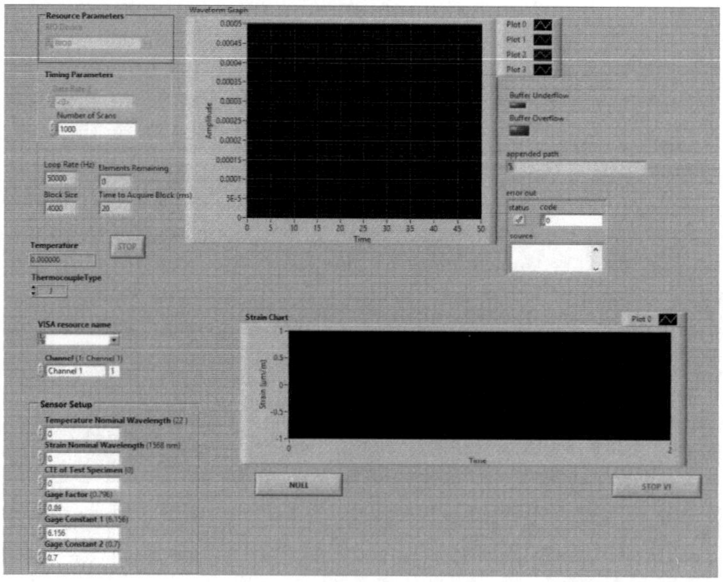

Figure 5: LabVIEW interface control panel showing the accelerometers and strain gauges charts

Constant signals emitted by the sensors are sent to the program. The program writes the information into different TDMS files (Test Data Exchange Stream), this type of files allows the user to open the information in different locations trough a web domain. The web domain is protected with a password and identified with an IP address. Because of the high sensibility of the accelerometers, the TDMS files contains an immeasurable amount of data points, making impossible the analysis of them, for this reason a

MATLAB code is introduced to the SHM system. The MATLAB code analyzes the information, getting rid of the noise signals emitted by the sensors. Thermocouples were implemented in this SHM system, since the tests were conducted in a closed laboratory the bridge is maintained at a constant temperature, for this reason the temperature does not affect in the strain readings. Is important to make a statement that in a real bridge the temperature will affect the properties of the materials, this is why thermocouples sensors were applied in this SHM system.

Different damage scenarios were implemented in the bridge. The Healthy damage simulated a healthy bridge, in which all the bolts were tight and all the members were connected properly. After the Healthy damage, the Medium damage was introduced. Medium damage simulated a bridge which is still serviceable, but not in the best conditions. To represent this damage, bolts were loosed making the beams untighten so the bridge made more signals of vibrations when the cars were pulled. Figure 6 introduces the Medium damage of the bridge. The last stage was the heavy damage. In the heavy damage beams were disconnected, simulating a bridge with ugly conditions, in which a collapse will happen at anytime, and need to be repaired. For each damage state, 25 tests were generated with random traffic and weight, simulating the scenarios of a real bridge. Figure 7 represents the Heavy damage inducted in the steel model truss bridge.

Figure 6: Medium damage state represented in the bridge

A two dimensional finite element (FE) model was presented in this experiment. This 2D model allows the user to damage and represent different distortions into the bridge. In order to generate the model in a FE image, properties of the bridge were deducted. The A36 standard steel was composed of different types of cross-sections. A 2X1 inches tube cross-section with a thickness of 1/16 inches was selected in most members of the bridge. The diagonal members were steel plates with a thickness of 1/8 inch. The model was implemented in SAP2000; this program allows the user to implement featured types of damage into the desired model. In this model, unit loads were displaced, measuring the deflections of each member. Figure 8 shows the FE model realized in SAP2000 simulating the steel model truss bridge located at the University of Texas Pan-American.

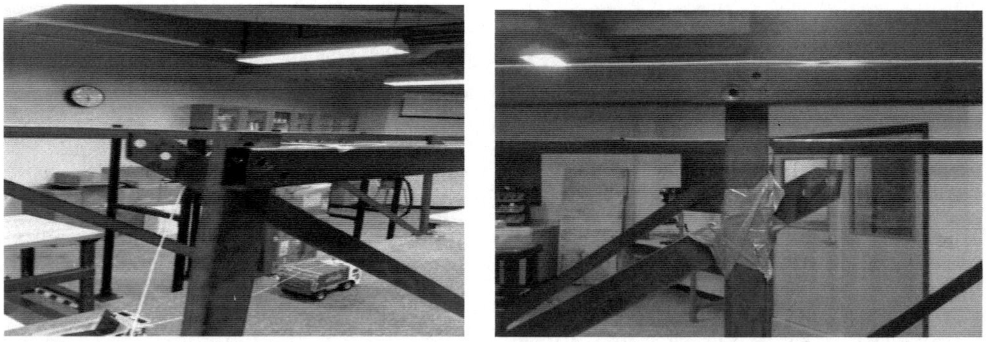

Figure 7: Heavy damage introduced into the bridge by loosing the braces

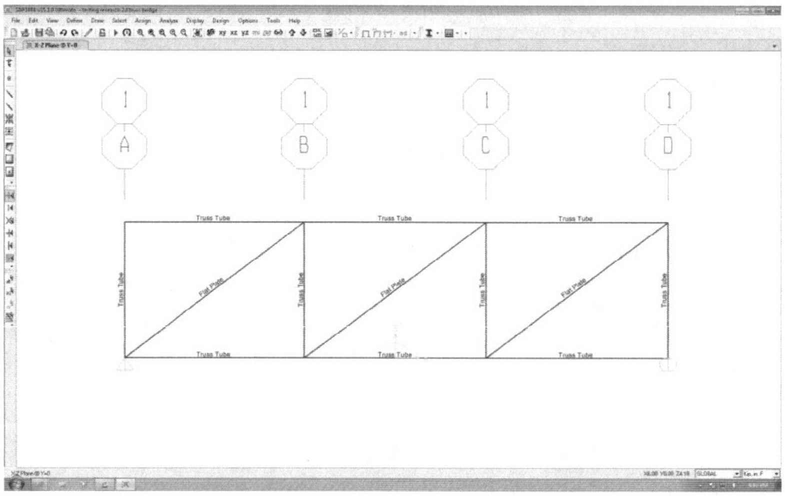

Figure 8: FE two-dimensional model implemented.

Results and Discussions: A MATLAB code was necessary in this experiment to interpret all the damage signals emitted by the sensors. This code generates a vector for each accelerometer, in which the data is analyzed using a Wavelet function. Wavelet [5] is a wave oscillation that begins with zero, increases, and then goes back to zero. As a mathematical tool, the wavelets are used to extract information from many different data, in this case the noise. Using the Wavelet function, the noise emitted by the accelerometers can be eliminated from the data points acquired by the SHM system. The elimination of the noise is a very important step in this SHM system, because it allows the user to obtain certainty in the data acquired and reduce the hassle of large data analysis. Wavelet has the ability to provide accurate information in frequency and time domains. A very useful tool to decompose the original signal to its original components is by using Wavelet Multi Resolution Analysis (WMRA). The wavelet decomposition allows breaking down the signals into groups with different frequency level as shown in Figure 9, approximations is used as a tool to reduce the noise in the SHM system.

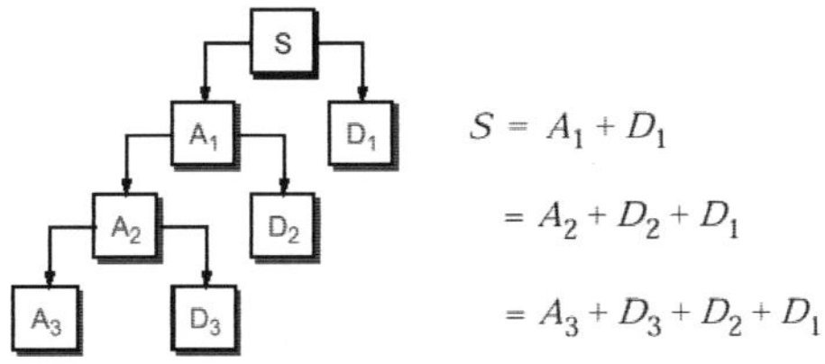

Figure 9: Decomposition of the signals transmitted by the accelerometers using three approximations and details.

In turn to compute the decomposition signals of $x(n)$ at level k the given equation (1) is used:

$$x'(n) = \sum_{w=-\infty}^{\infty} e_{h,w} .\phi_{h,w}(n) + \sum_{g=1}^{h} \sum_{w=-\infty}^{\infty} u_{h,w} .\varphi_{h,w}(n) \qquad (1)$$

Where $\phi(n)$, $\varphi(n)$ are the scaling and wavelet basic functions, respectively and $x'(n)$ is the decomposed signal. In this experiment, the scaling and wavelet functions are selected for the Daubechies db4 mother wavelet. The main purpose of it selection was to obtain the desired decomposition. The approximation and detail coefficients $e_{h,w}$ and $u_{h;w}$ are calculated in equation (2) and (3):

$$e_{h,w} = 2^{(-h/2)} \sum_{n} x(n)\phi(2^{-h} n - w) \qquad (2)$$

$$u_{h,w} = \sum_{n} x(n)\varphi_{h,w}(n) \qquad (3)$$

Using equation (4), the energy of the signals was calculated in order to extract damage features from the third approximation of signals.

$$\lambda = \sum_{f} (e_{2,f})^2 \qquad (4)$$

The approximation signal that was selected and used for this truss bridge was approximation two in order to extract damage features. It was necessary to compute the second approximation signal for accelerometer A and B to observe the difference between the raw signals. Wavelet Multi Resolution Analysis [6] was crucial for this project because it is capable of removing the noise from the signals and providing accurate information from the original signal. As shown in Figure 10, the WMRA removed the noise and the difference between the two signals is now visible. Signals with noise and without noise provide a better interpretation of the data.

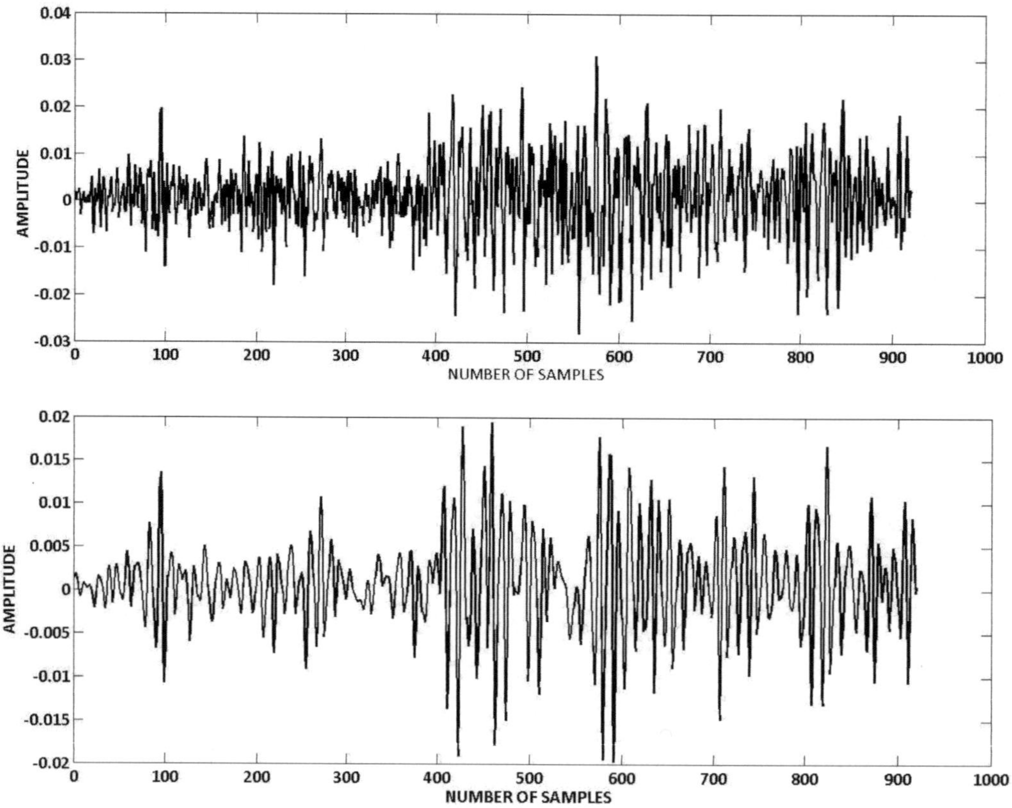

Figure 10: Noise removed using WMRA in the signals of the accelerometers

A total of 75 tests were performed in this experiment (25 tests per Damage scenario). The tests were realized with random traffic and specified amount of weight. Each car was weighted before the test was started and a chronometer took the precise time of each realization. Fast Fourier Transformation function (FFT) was introduced to the experiment. FFT function along with the FE model implemented, allowed the SHM system to determine the exact location of the damage in the bridge. Figure 11 represents the analysis of the data for a heavy damage test. After the tests were performed, the different damage scenarios were plotted in a line graph for both accelerometers, in this case A and B. The data registered in the different damage determined the performance of the bridge. The difference between the different states can be visualized with the plots. In order to get a better understanding of the realizations, standard deviation and mean of all the tests for the different damage scenarios were calculated. A difference of at least 2 figures was conducted for each damage state. The information obtained let us to determine that the SHM system implemented in the bridge is truly efficient, and the prediction of the next damage scenario is needed in order to prevent future catastrophes, for this reason Monte Carlo simulation was introduced to this SHM system. Monte Carlo simulation is a probabilistic method that allows determining through computational algorithms that rely on repeated random samplings to obtain desired numerical results. For this experiment, Monte Carlo simulation was implemented after the Medium damage state was introduced. Normalized random points with mean of 0 and standard deviation of 1 were calculated in order to determine the number of cycles needed to translate from one state of damage into another. Figure 12 represents the Monte Carlo simulation introduced in the bridge.

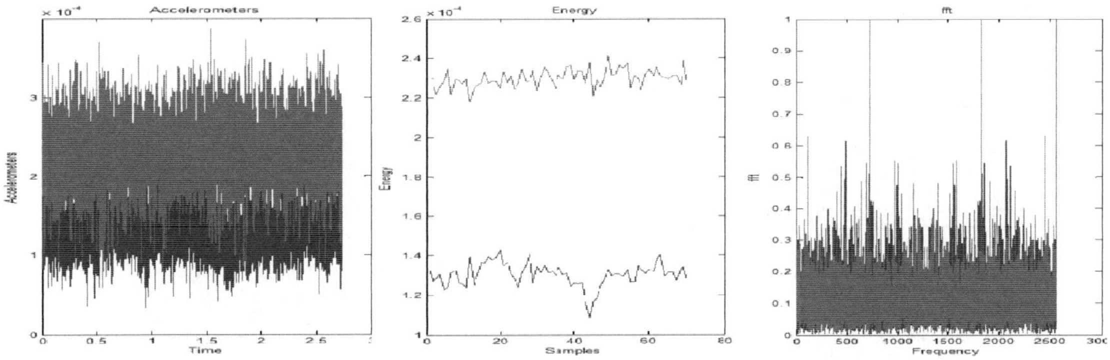

Figure 10: a) Represents the signals from the accelerometers; b) Represents the energy, which was calculated with WMRA, c) Represents the Frequency, acquired with a FFT function.

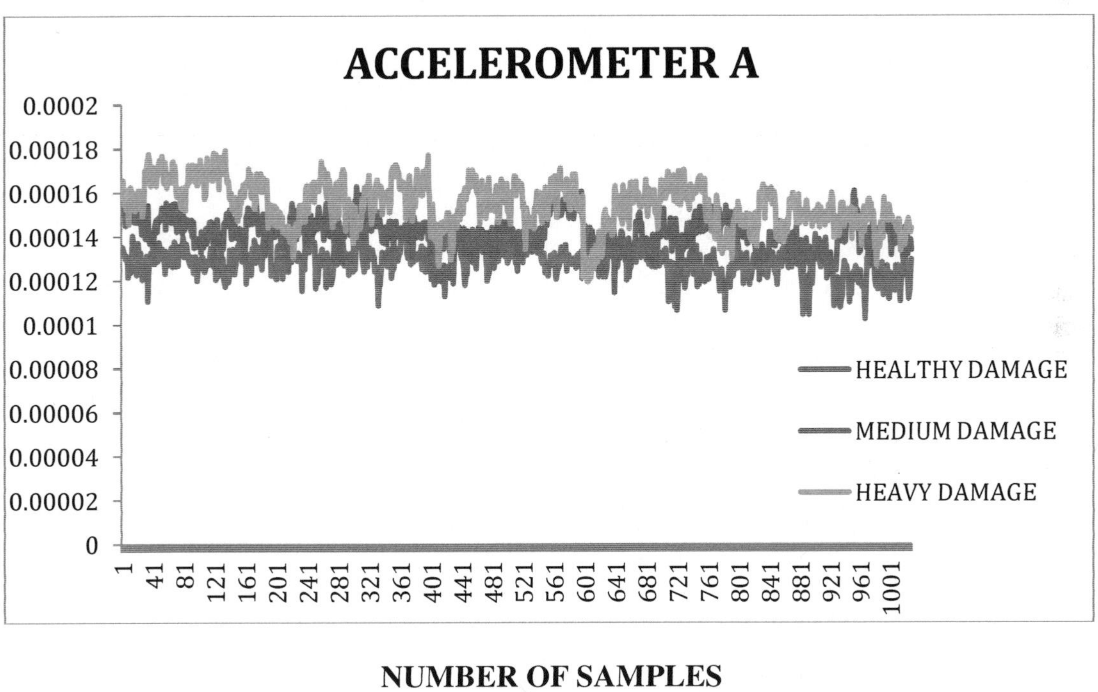

Figure 11: Represents the difference in accelerometer A between the damage scenarios.

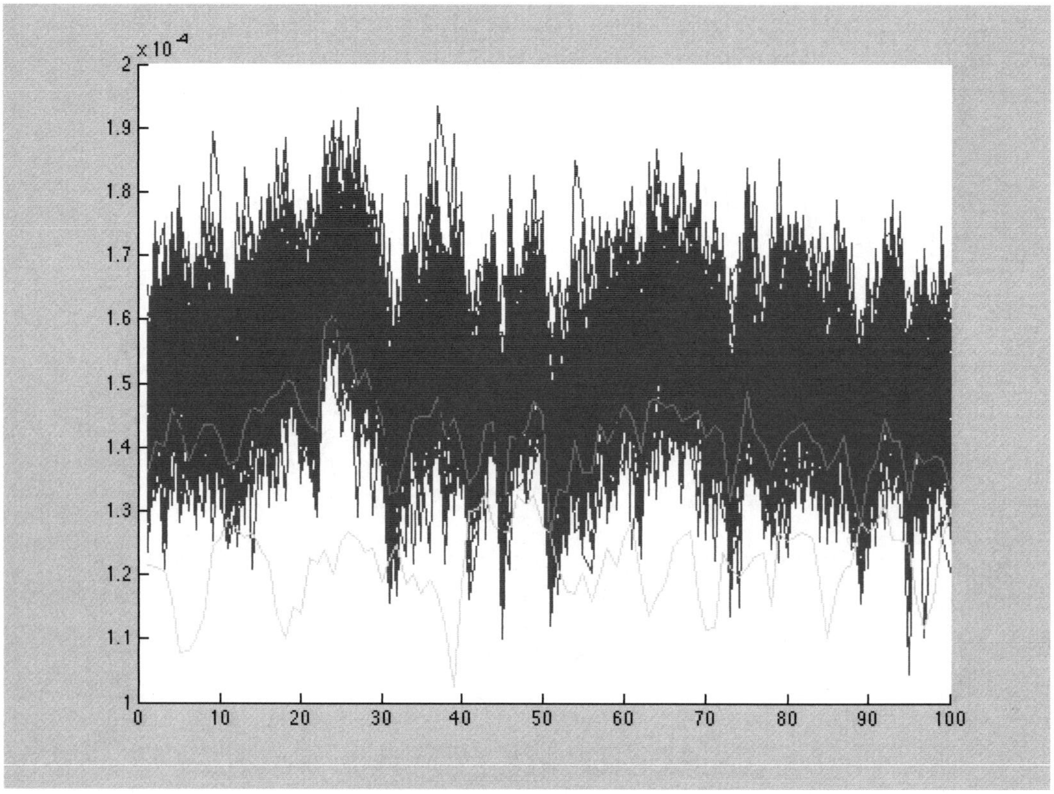

Figure 12: Monte Carlo simulation implemented after the Medium damage was introduced into the bridge.

The strain is a very important component of this SHM efficient system; Table 1 presents the data acquired by the strain gauges signals. Since the strain gauges are from a Fiber Optic interrogator, the noise is reduced in comparison with the electro-mechanical strain gauges. In this case a MATLAB code will not be needed to interpret the information.

Table 1: Strain data obtained from the Fiber Optic Interrogator.

HEALTHY (nε)	MEDIUM (nε)	HEAVY (nε)
30412	61498	74895
28867	49166	73125
27603	58598	71229
30202	66708	75398
30871	52967	74735

Conclusion: An effective SHM system is installed in a steel model truss bridge. Two piezoelectric accelerometers and a strain gauge are installed in the bridge to measure vibrations and displacement respectively. To detect the difference between the different damage case scenarios bolts and trusses were loosed. A wavelet function was used to obtain the energy of the accelerometers. This energy is needed in order to determine the damage of the different states of the bridge. The location of the damage area is determining by using a frequency method. The difference between the healthy, medium and heavy damage can be recognized by plotting all damage scenarios of the bridge. This

SHM system can be used to monitoring the behavior of a certain structure; they include bridges, tower, tunnels dams etc. These types of systems are needed in order to prevent catastrophes and most important save human lives.

References

[1] Yuen, Ka-Veng, James L. Beck, and Siu Kui Au, Structural damage detection and assessment by adaptive Markov chain Monte Carlo simulation, Structural Control and Health Monitoring 11, Pages 327-47.

[2] Chen, Yangbo, and Maria Q. Feng, Structural Health Monitoring by Recursive Bayesian Filtering, Journal of Engineering Mechanics 135 (2009), Page 231.

[3] Zheng, Wei, and Yi Yu, Bayesian Probabilistic Framework for Damage Identification of Steel Truss Bridges under Joint Uncertainties, Advances in Civil Engineering 2013, Pages 1-13.

[4] Flynn, Eric B., and Michael D. Todd, A Bayesian approach to optimal sensor placement for structural health monitoring with application to active sensing, Mechanical Systems and Signal Processing 24 (2010), Pages 891-903.

[5] Karbhari, Vistasp M., and Farhad Ansari, Structural health monitoring of civil infrastructure systems, UKWoodhead, 2009,Cambridge.

[6] Weeks, Michael, Digital signal processing using MATLAB and wavelets, Infinity Science P, 2007 Hingham, MA.

APPLICATION OF INFORMATION THEORY'S CHANNEL CAPACITY AS AN INDUSTRY MACHINE HEALTH AND DIAGNOSTIC METRIC

Ted Costuros
Applied Material Inc.
9700 US HWY 290 East
Austin, TX 78724-1199
Telephone: (512) 272-2264
ted_v_costuros@amat.com

Michael D. Bryant
The University of Texas at Austin
1 University Station C2200
Austin, TX 78712-0292

Abstract:
A metric for degradation quantification, and assessment of machine health was constructed from an analogy between a machine and a communications channel. The metric is based on Shannon's channel capacity concept applied to the difference between measured signals (induced by faults) and baseline signals of a machine system response. Machine faults that hinder operation also alter signal flow through a machine system; the effects appear as a residual signal riding on a baseline signal. Similar to noise hampering data transmission, this residual signal can be viewed as noise within the context of information theory, permitting application of information theory to machines, resulting in a health measure for assessment of machine health.

The technique successfully predicted failures in a closed loop chain robot. The technique is favorable for industry, because it applies to measurement data, and calculations are done in the time domain without need of expensive frequency transformations. The technique can be used in the semi-conducting industry as a tool to monitor system performance during production, and to lower fab operating cost by extending component use and scheduling maintenance as needed.

Key words: Diagnostics; Prognostics; Health management; Information theory; Shannon's channel capacity; Machine health.

1.0 Introduction: Industry uses robots to perform prescribed tasks at given frequencies with minimum variation. Technology barriers are pushed further with more precise measurements and better assembly capabilities. Human labor in industries such as nanotechnology cannot perform tasks with precise repeatability in high volume production environments, without the assistance of robots. In a fab facility, microchips are made on wafers, where electronic circuits are developed in a nano scale, handled by robots. A wafer is exposed to different processes where each step involves precise wafer handling which can only be accomplished with robots.

A single robot failure causes the entire production line to fail. A line failure costs millions of dollars in lost revenue. Since robots are vital to the production line, management has implemented preventive maintenance schedules. In the semiconductor world, depending on the robot type and usage, preventive maintenance can occur anywhere from a few weeks up to a few months. The robot is replaced during preventive maintenance, and because no robot failure has been reported, preventive maintenance replaces a good robot. Each robot averages $250K. For a fab operating 300+ robots, a method to assess robot health can be profitable.

A critical problem in a semi-conduction fab is particle monitoring. Particles in a system will ruin entire wafer boats, resulting in tremendous losses. This necessitates wafer production facilities to monitor particles in a system. Robots with moving parts require lubrication. Few lubrication agents are cleanroom approved. Even fewer prevent out-gassing in vacuum. These lubricants tend to breakdown faster, causing the robot to run dry and generate wear particles. However, particle issues are not all robot dependent and there can be other sources of contaminants in the system such as negative pressure spots drawing ambient air with high particle contamination inside the system. Most chip manufacturing facilities replace a robot when it registers high particle count and wipe the inside of the module to remove particles. Nonetheless, this does not address particle generation nor provide evidence that the particles are generated from a source other than the robot. A method to identify robot degradation will help identify the root cause of particle generation, and lower fab operating cost by extending the robot use, and scheduling maintenance on an as needed basis.

1.1 General Methods of System Diagnostics: System diagnosis develops and implements rules which, based on measurements, state the operating condition of a system. Literature on fault diagnosis ranging from methods requiring extensive modeling to methods based on experience and fault knowledge [1] [2] can be classified into four categories: i) Model and measurement unavailable, ii) Model unavailable, measurement available, iii) Model available, measurement unavailable, and iv) Model and measurement available.

Without model or measurements, a diagnosis method must rely on fault experience. Such "knowledge-based" or "expert systems", popular in industry for forty years, involves a very long flow chart with many decision blocks. The flow chart ties a knowledge database correlating corrective actions to root causes on known problems. The database is constantly updated as new problems get reported and logged. Expert system diagnostic methods take much time, require database maintenance, can only diagnose specific problems, and is unsuitable for analysis of dynamic phenomena. Scheduled preventive maintenance is the only way to avoid unexpected faults that cause production downtime.

When measurement is only available, deviations can be tracked and trended from a measured nominal value(s). Characteristic applications of such method are Statistical Process Control (SPC), and time frequency or wavelet analysis.

When a mathematical model is only available, faults can only be simulated to elucidate the physics behind a fault. Design of Experiments (DOE) with the popular Monte Carlo method can computationally determine expected system output based on parameter variation.

When both mathematical model and measurement are available system health assessment and fault diagnosis are improved by noting the difference between simulated and measured outputs. Appropriate statistical bounds of the model output serve as a real-time system health check. Venkatasubramanian et al [3] identified three classes of failure: 1) Gross parameter changes, 2) Structural changes, and 3) Malfunctioning sensors and actuators.

In the first failure class, independent variables cause a disturbance which affects system performance, for example, a change in the heat transfer coefficient due to fouling in a heat exchanger. The second failure class is due to hard malfunctions in equipment. Problems with sensors and actuators attributed to instrument or actual sensor failure comprise the last class. A failed instrument with an erroneous signal to the sensor results in a system fault. Electronic Control System diagnosis, which evaluates residual information between model and measurement, forms a logical array to isolate faults. Observer Based fault detection compares actual measurement data to model predictions, to identify changes in system behavior.

1.2 Desirable Characteristics of Diagnostic System: Desirable characteristics of a diagnostic system include: 1) Ability to Capture General Faults, 2) Ability to Measure and Quantify System Degradation, 3) Ability to Assess Health, 4) Adaptability, 5) Capture Intermittent Changes in System Performance, 6) Fault Isolation, 7) Low Computational Demand, 8) Low Maintenance, 9) Low Storage Demand, 10) Quick Detection and Diagnosis, and 11) Robustness.

Table 1 compares different diagnostic techniques against the recommended desirable characteristics [4]. Based on personal experience and reference consideration, techniques were scored at each characteristic: 1 - Low, 2- Medium, and 3 - High. The total score is the product of all numbers along the technique column. A higher score indicates a better technique. From the table, evaluation of residual information works best, followed by SPC, and time/frequency analysis.

Despite strengths and weaknesses, the techniques described are labor intensive, considering measured or simulated system response. No single technique has all the desired characteristics. Multiple techniques must complement each other when applied in a diagnostic system. Important response parameters like shape, load, and sampling rate are not considered in diagnosis.

Table 1: Comparison of diagnostic techniques against diagnostic characteristics.

	Knowledge Base	SPC	Wavelet Transformations	Monte Carlo	Residual	Observer
Ability to Capture General Faults	1	3	3	2	3	3
Ability to Measure System Degradation	1	2	2	1	1	1
Ability to Perform Health Assessment	1	2	2	1	1	1
Adaptability	1	3	2	1	2	2
Capture Intermittent Changes	1	1	2	1	1	1
Fault Isolation	3	2	2	1	3	2
Low Computational Demand	3	3	1	2	3	2
Low Maintenance	1	2	2	3	2	3
Low Storage Demand	1	2	3	3	3	2
Quick detection & diagnosis	3	1	1	2	3	3
Robustness	3	2	2	1	3	2
total score:	81	1728	1152	72	2916	864

Bryant et al [5], proposed an analogy between machine and communication channel to estimate fault severity. This work systematically studied how variations in mathematical model parameters affect system performance. By tuning the model parameters to match actual measurements faulty components were identified. Based on actual sensor measurements, the mathematical model can predict future system behavior. Choi's [6] work combined communication theory and mathematical models to detect, identify, and predict faults in motor and pumps. Bryant's work blends communication theory fundamentals with machine response and noise levels, accounting for important parameters. Viewing a machine as a communications channel enables application of Shannon's theorems in assessing performance. Power flowing through a machine changes due to faults. The suggested method considers the difference between the measured power of an output signal from its nominal performance, over a production cycle, to determine a channel capacity value used to quantify degradation and detect faulty components. Models that capture the physics behind system performance usually require extensive time to develop, debug and validate. To industry, this can delay product introduction to markets, and is not cost effective. Because of this, industry usually avoids building models, but will take measurements. A simple diagnostic method must be developed, capable of assessing system degradation directly from measurements. Costuros [7] developed criteria so communication theory can be directly applied to measurements to detect, identify, and predict faults. The method calculates the channel capacity of a mechanical system derived from measurements, simulations, or combinations. The channel capacity technique is best between Statistical Process Control (SPC) and residual evaluation, with high ratings to better measure system degradation and capture intermittent changes. In Table 1, channel capacity scored a high 3 on all characteristics except in the ability to perform health assessment and low maintenance where it maintains a medium 2. For the attributes listed in Table 1, channel capacity scored about 40x better than either SPC or residual comparison.

2.0 Introduction to Communication Theory: In 1948 Shannon`s "Communication in the Presence of Noise" [8] established information theory. Communication conveys information between a sender and receiver over a channel contaminated with noise, see Figure 1. A sender can send many possible messages. Shannon quantified the amount of information communicated about an event A as a function of the probability of occurrence P_A of the event, using the negative logarithm as a measure [9]:

$$I = -\log_2(P_A) = \log_2\left(\frac{1}{P_A}\right). \tag{1}$$

From Equation (1), communication about an event of high probability has less information than an event of low probability. A message about an event certain to happen ($P_A \longrightarrow 1$) contains little information ($I \longrightarrow 0$); in contrast, a message about an event that rarely occurs ($P_A \longrightarrow 0$) contains much information ($I \longrightarrow \infty$).

A binary signal that assumes values of '1' or '0' (called a bit) can communicate knowledge of two equi-probable events such as heads or tails of a coin flip. Four equi-probable events such as flipping two coins require two bits. Three bits convey information on $2^3 = 8$ equi-probable events. Based on this analogy, a message about q equi-probable events, contains $\log_2(q)$ bits of information.

$$2^m = q \xrightarrow{yields} m = \log_2(q). \tag{2}$$

If messages containing m-bits sent serially through a channel are generated at rate r (messages/second), then the rate of information, R, is mr bits/sec (bps). Figure 1 shows a typical communications channel, wherein a transmitter encodes and sends original message with information x_1 as signal $x(t)$ over a channel media corrupted by noise, $n(t)$. For additive channels, the receiver receives signal

$$y(t) = x(t) + n(t), \tag{3}$$

consisting of original signal $x(t)$ plus noise $n(t)$. The goal is to have the information in the received message $\overline{x}_1$ identical to that in the original message x_1, despite the uncertainty imposed by the presence of noise.

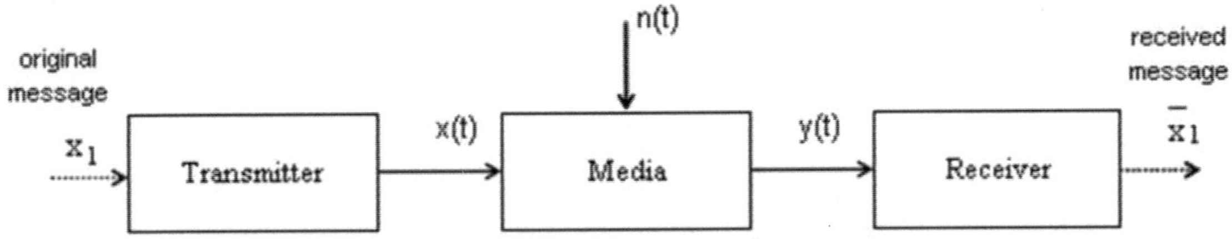

Figure 1: A communications channel.

The field of diagnostics defines a residual signal as the difference between an actual signal and an expected baseline signal [10], [11]. Shannon's information theory is based on noise, generally defined as the difference between an actual (received) signal $y(t)$ and a desired (transmitted) signal $x(t)$. Electric communications systems mostly involve intrinsic noise generated by natural phenomena. Besides intrinsic noise embedded in the measurement apparatus, here noise signal $n(t)$ can include the residual signal due to the fault, deemed "fault noise". Shannon`s channel capacity used as a health metric wiil be applied to a measured signal containing both intrinsic noise and signals generated by faults.

Shannon applied the concept of information entropy, a measure of randomness and uncertainty in a signal, to formulate the information capacity C of a channel of bandwidth W, perturbed by white Gaussian noise of average power N, and received signal, S_y, having a white noise Gaussian density:

$$C = W \log_2 \left(\frac{S_y}{N}\right) \qquad (4)$$

The log in Equation (4) ultimately arises from Equation (1). The capacity C of a noiseless channel $N = 0$ is infinite, since without noise there is no uncertainty in the received signal $y(t)$. Costuros [7] interpreted channel capacity C as a degradation measure, assessed machine channel health via Equation (4), and redefined noise

$$n(t) = y(t) - y_o(t), \qquad (5)$$

as the difference between an actual measured signal $y(t)$ and a desired exemplar baseline signal $y_o(t)$ that replaced $x(t)$ in Equation (3). For example, a robotic system can perform poorly because bearing failure increases system friction, which increases the motor torque needed to drive the robot. Here measured motor torque signal $y(t)$ of the degraded system exceeds the baseline motor torque profile $y_o(t)$ characteristic of the healthy system because of an additional "fault noise" signal component in $n(t)$ induced by the bearing fault. A more intense fault manifests as a larger fault noise, which by Equation (4) results in a lower C value. System degradation can be determined by monitoring the change of C through time.

3.0 Machine Health Assessment and Degradation Measurement of Robot Assembly Motors. Channel capacity health metric concept successfully evaluated performance of a wafer handling robot system via degradation measurements through time. Values of C were determined via Equation (4), using measured signals $y(t)$, baseline signals $y_o(t)$ exemplary of good health, and noise signals $n(t)$ via Equation (5). The change in C was monitored to assess system degradation.

3.1 Procedure for Calculating Channel Capacity. For the channel capacity technique to quantify system degradation directly from measurement data under industry conditions, a signal baseline $y_o(t)$ must be defined, via the following procedure:

1. Operate system at nominal conditions and measure an ensemble of the baseline signals $\{ y_o(t)\}$, where brackets denote ensemble. Statistically the more signals in the ensemble, the better the baseline accuracy. A minimum 5 signals in the ensemble is suggested.
2. Sample all ensemble signals at the same times t_k , require all ensemble signals to have the same number of sample points, and align all ensemble signals in time.
3. For each time sample point t_k , average all ensemble signals at that point to obtain an average system response. This generates the baseline signal y_o.

After the system baseline $y_o(t)$ has been defined, evaluate system degradation via the following procedure:

1. Use Equation (5) to extract noise $n(t)$ from each signal $y(t)$, by subtracting the baseline signal $y_o(t)$.
2. Calculate the ratio of power S_y of signal $y(t)$ to power N of noise $n(t)$ by dividing the sum of the square values of the signal $\sum_k[y(t_k)]^2$ by the corresponding sum of the square values of the noise response $\sum_k[n(t_k)]^2$.
3. Calculate the channel capacity of the corresponding signal by taking the base 2 logarithm of the signal to noise power, and then multiply that result by the signal bandwidth, which here was equated to the sampling frequency.
4. When there are multiple collections of signal $y(t)$, average the channel capacity values from each of the samples to obtain the channel capacity for diagnostics.

3.2 The Robot Design. To demonstrate application of the proposed health metric concept, channel capacity C was calculated for eight servos in 4 identical robots subjected to reliability tests in an industry environment. Each robot had two motors, designated A and B, which drove a robot mechanism. A motor system schematic is shown in Figure 2. A handy terminal FHT11 through an RS-232 interface sent user commands to the ELA driver. The maximum specified motor torque of 10 N-m corresponded to a maximum current of 3A per phase.

The robot, which handles wafers IN and OUT of process chambers, has dual transparent blades, arranged 180° apart to minimize process cycle time. The blades carry wafers. Each blade attaches to two bar linkages which comprise left and right wing assemblies. Coordinated rotation of the hubs extend the wing assemblies, which translate the blades through space. Figure 3 depicts the robot assembly with blades, wing assemblies, hub assemblies, a base ring for mounting, and an enclosure over the motor hub assembly to maintain vacuum. Inside the enclosure is a motor hub assembly containing motors A and B of Figure 2 that drive the hubs. The robot installs on the base ring where an o-ring seals vacuum and atmospheric environments.

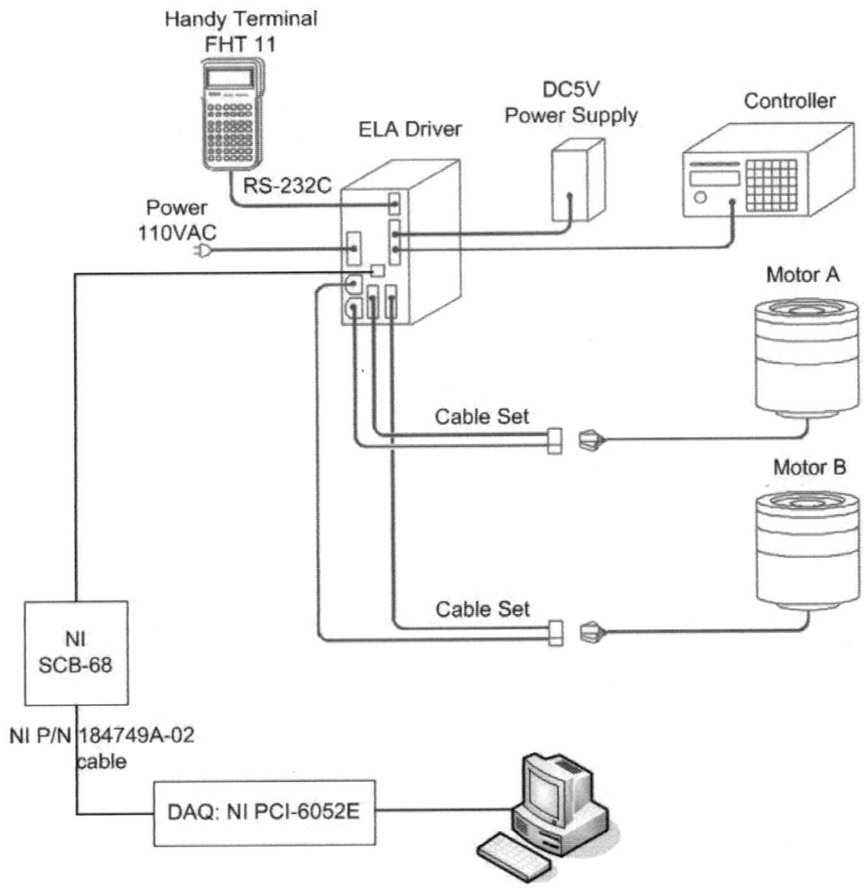

Figure 2: Motor system outline [12].

In the motor hub assembly in Figure 4 are identical 3 phase variable reluctance stepper motors. Motors A and B attach to and move the upper and lower atmospheric hub assemblies. Motor A drives the upper atmospheric hub, while motor B drives the lower atmospheric hub. Each motor phase consists of 6 stator coils daisy-chained on a switch. Rotor speed and position are controlled by three switches which control the three phases. Because this assembly is used in vacuum, a stainless steel barrier bowl separates the atmosphere from the vacuum environment. Each hub assembly has a magnetic coupling which transmits power across the barrier bowl to the vacuum side. The assembly has 5 bearings: one on each motor, one bearing between the upper and the lower atmospheric hubs to allow rotation between hubs, one bearing on the vacuum side to allow the lower vacuum hub to rotate relative to the steel vacuum barrier, and one bearing to allow rotation between the upper and lower vacuum hubs.

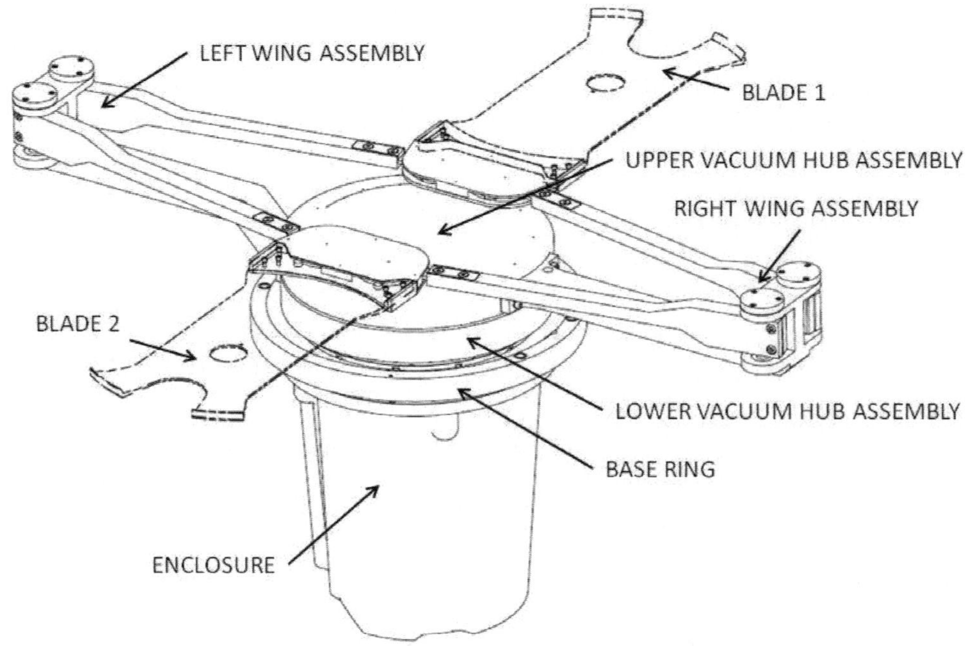

Figure 3: Robot assembly.

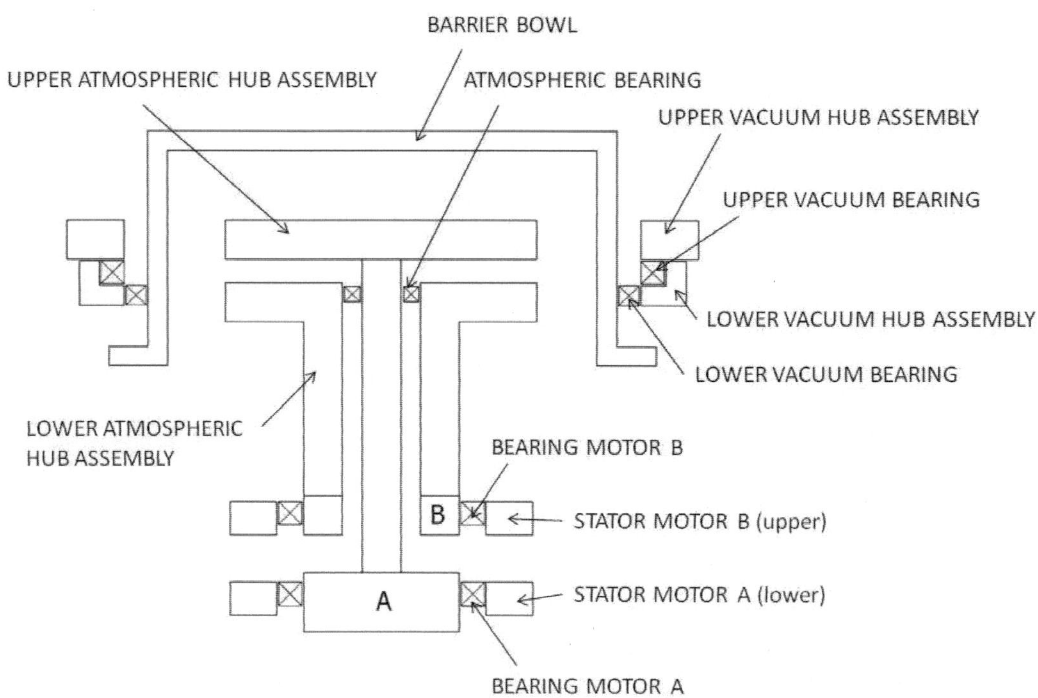

Figure 4: Schematic of robot hub assembly.

3.3 Test Setup and Apparatus. Figure 2, depicts the robotic system motors and data acquisition. Two motors connect to the ELA driver. Each megatorque stepper motor has a rotor mass moment of inertia of 0.013 kg-m3, and 491,520 steps comprise one revolution. A NSK Handy Terminal via the ELA driver directs motor torque. Motor torque signals in the ELA driver directed by the National Instruments SCB-68 I/O connector block transmit into the National Instruments PCI-6052E Data Acquisition card via National Instruments cable 184749A-02. For data collection of motor torque signals, the sampling rate was 1,000 Hz. Signals were recorded via National Instruments Labview 8.5 and analyzed via Mathworks Matlab R2009.

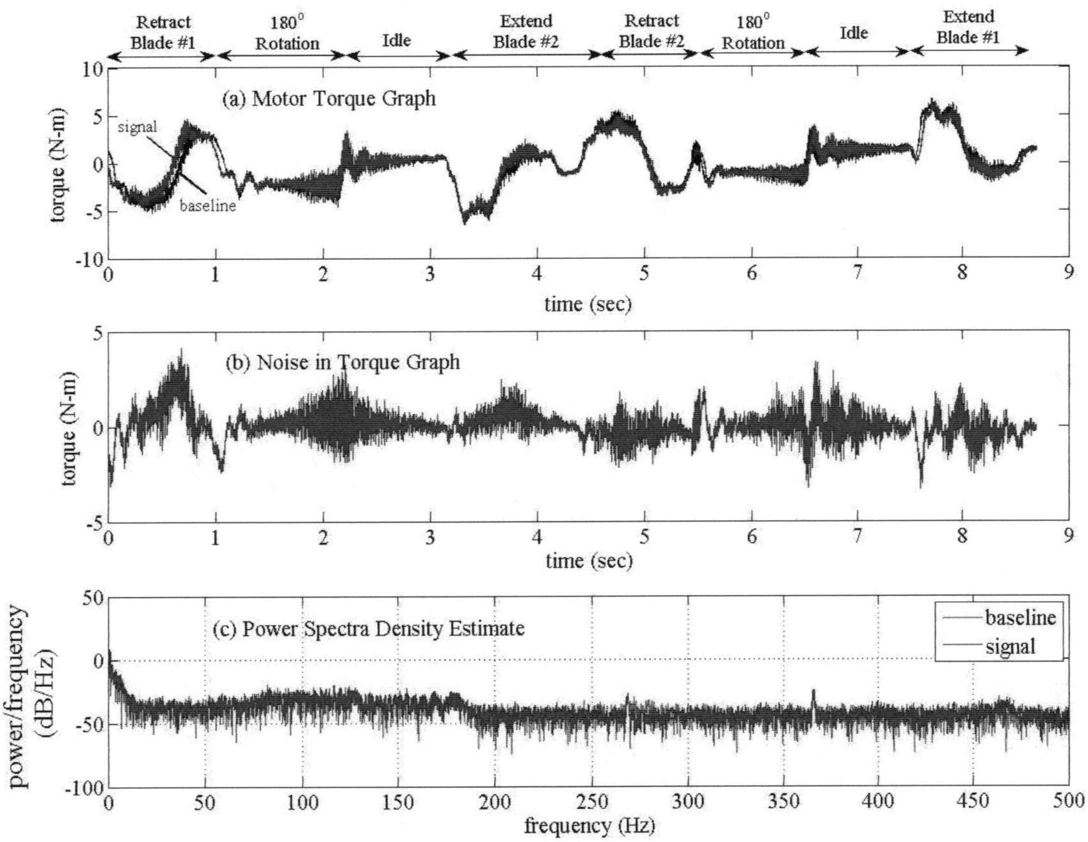

Figure 5: Motor torque response from robot 1 on January 15, 2010.

3.4 Robot Motor Degradation Measurement. Data of torque signal was collected from 4 robot assemblies, each with two motors in good operating condition. These robot assemblies were pulled from properly working lab systems. An identical sequence of voltage step inputs were repetitively applied to each motor. Voltage step inputs for one cycle followed the sequence: blade 1 retraction, robot rotation 180°, idle, blade 2 extension, blade 2 retraction, robot rotation 180°, idle, and blade 1 extension. Motors ran continuously from December 9, 2009 to February 5, 2010. Each motor output torque was measured on December 9, December 18, January 15, January 21, and February 5. The December 9 measurements, designated as baseline torques, served as desired received signals $y_o(t)$ in Equation (5). All subsequent measurements $y(t)$ on the same motor were compared to the corresponding $y_o(t)$. Before any calculations, a signal $y(t)$ was first correlated to its $y_o(t)$, to synchronize the signal alignments and samples in time. Figure 5a shows the motor torque signal $y(t)$ on January 15 from robot 1 (blue curve), and its baseline $y_o(t)$ (black curve). Fault noise in Figure 5b distills the disturbance signal contained in $y(t)$ generated by the fault. Power spectra of signal and noise are included as Figure 5c. Channel capacity C was estimated via Equation (4) and tabulated in Table 2. Dimensions of C in bits per second are relative and not included.

Table 2: Channel capacity C for motors A and B of four identical robots evaluated at four different times, and percent change of C at the last three times relative to the first time compared to ratings of the robot motions by engineers and technicians.

Date	Robot 1		Robot 2		Robot 3		Robot 4	
	Motor A	Motor B	Motor A	Motor B	Motor A	Motor B	Motor A	Motor B
12/18/09	2193	2164	1780	2326	1878	1647	2051	1679
1/15/10	1965	1784	1335	1481	1307	964	1383	989
1/21/10	2039	1827	1375	1466	1465	1072	1406	1005
2/5/10	1907	1985	1188	1340	1252	929	1475	1043
% change 12/18 - 1/15	10% BEST 1	18% 2	25% 3	36% WORST 8	30% 4	41% 6	33% 5	41% 7
% change 12/18 - 1/21	7% BEST 1	16% 2	23% 3	37% WORST 8	22% 4	35% 5	31% 6	40% 7
% change 12/18 - 2/5	13% BEST 2	8% 1	33% 4	42% WORST 8	33% 5	44% 7	28% 3	38% 6

For measurements subsequent to December 18, fractional changes in channel capacity C, relative to values associated with December 18 measurements were tabulated as percent changes of C in Table 2 for all motors. Inspection of the upper rows reveals a trend of diminishing channel capacity values over time. For example, for motor B of robot 2, C diminishes from 2,326 to 1,340 from 12/18/09 to 2/5/10. In subsequent rows, the percent change of channel capacity from December 18 to January 21 is displayed, along with a composite of human produced evaluations of motion performance by a team of engineers and technicians. The human evaluations rank-ordered the motors, and identified the best and worst performing motors. In general, the channel capacity estimates agreed well with human (team) assessments. Motor 'A' in robot 1 with the smallest channel capacity reduction, was ranked best (number 1 or 2) during testing. Motor 'B' in robot 2, with poorest performance, consistently showed the largest reduction of channel capacity, and was ranked WORST (number 8). This motor eventually developed a grinding noise, which prematurely ended testing of this robot. In general, the drop in the channel capacity values correlated with the human perceived amount of motor degradation. This application suggests that the channel capacity metric can quantify system degradation in industry settings. Fluctuations in channel capacity apparent in Table 2 for most of the motors at the beginning of tests are consistent with a product break-in period. An overall trend of decline in channel capacity indicates degradation. The majority of faults occur on the upper and lower vacuum bearings due to lubrication breakdown.

3.5 Robot Hub Assembly Motor Degradation Measurement. In the real world, multiple signals measured from the same system operating with satisfactory performance can result in variation of C values. Such variations are attributed mostly to intrinsic measurement noise present in sensors, data collection gear, and the system itself. Assuming a normal distribution, a probability density function is defined for C values at certain time instances during system performance. System degradation is determined through statistical evaluation of expected C values as described in this subsection.

On February 2011 current was measured from a switch of the 6 stator coils in motor B using a Tektronix A622 AC/DC current probe. Both upper and lower hubs (see Figure 3) were moved counter to each other, and each completed a full rotation. Each motor accelerated from rest at 120,000 steps/s^2, until a speed of 100,000 steps/s was attained. Motors maintained this speed until decelerated to rest. This exercise over one revolution comprised 491,520 steps. On February 15 ten current measurements, which exemplify good health, were collected from the switch in motor B. All measurements were correlated to the first measurement, to align signals in time.

Since the ten February 15 measurements represent an exemplar population, statistics were applied to define the baseline curve needed to compute channel capacity. After aligning the ten signals in time, an average of the ten signals at each time step produced a baseline for the current signal. Since all ten measurements taken under the same test conditions represent the same "good" motor performance, an average can construct a baseline signal $y_o(t)$ for motor B during its virgin February 15 condition.

Table 3: Channel capacity from 10 signals, motor B.

measurement	C (s^{-1})
signal 1	5616.2
signal 2	5565.4
signal 3	5749.5
signal 4	5400.1
signal 5	5420.6
signal 6	5526.7
signal 7	5407.4
signal 8	5579.9
signal 9	5311.5
signal 10	5599.6

Noise was defined as the difference between a measured signal $y(t)$ and the baseline $y_o(t)$. For each of the ten signals, using the aforementioned baseline $y_o(t)$, a channel capacity was evaluated via Equation (4) with noise defined by Equation (5). Channel capacity values calculated for each of the ten current signals ranged between 5,311 s^{-1} and 5,750 s^{-1}, see Table 3. This distribution of channel capacity has mean and standard deviation of 5,518 s^{-1} and 131 s^{-1} respectively. Statistically the channel capacity for a motor B current measurement from a healthy specimen should be expected within six standard deviations from the mean, from 5,125 s^{-1} to 5,911 s^{-1}. A C value inside this range represents the baseline and good health. Signal degradation is noted for average C values less than the six sigma minimum mean of 5,125 s^{-1}. Statistical confidence in a population average increases with the number of samples. The table lists the channel capacity values from the 10 measurements, which has average 5,518. This represents the sample average and not the population. Performance degradation considers the population average. Based on the measurement average over the sample size of 10 and a normal distribution, a population average 5,424 and 5,611 is expected with 95% confidence. Figure 6 shows the expected spread in the channel capacity value labeled as the 2/15/2011 curve to the right. System degradation is noted for channel capacity values less than 5,125. Channel capacity dimensions of s^{-1} have no real importance. Important are relative magnitudes.

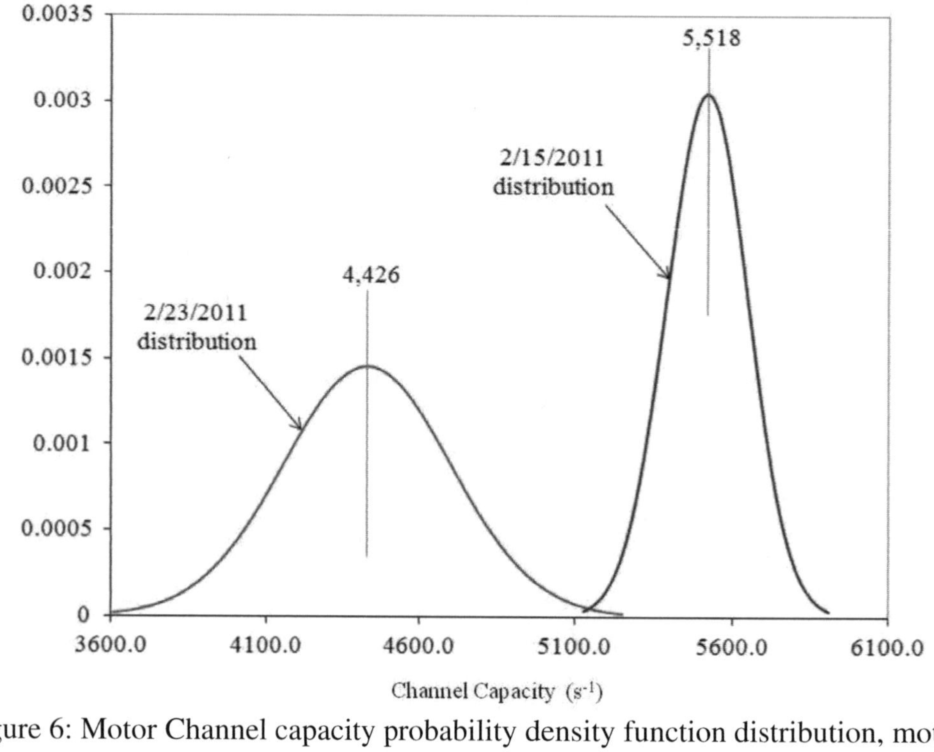

Figure 6: Motor Channel capacity probability density function distribution, motor B, 02/15/2011 and 02/23/2011.

For proper operation of the robot of Figure 3, the hub assembly of Figure 4 must maintain tolerances on parts. Heating of the hub assembly will cause parts to expand. Parts of different materials (the bearing races and vacuum barrier are steel, while hubs are aluminum) when heated expand at different rates. Thus heating will alter the bearing internal diametral clearance as parts expand. The vacuum hub bearings were built with a nominal internal diametral clearance of 0.0004 inches. Thermal analysis of the hub assembly predicts that the bearing internal clearance will start to decrease at a bearing temperature of 55°C. Calculations of the bearing's internal clearance versus temperature presented in Figure 7 suggests that the internal clearance will vanish near a bearing temperature of 58°C. This leaves little margin to operate the assembly at temperatures much higher than about 50°C.

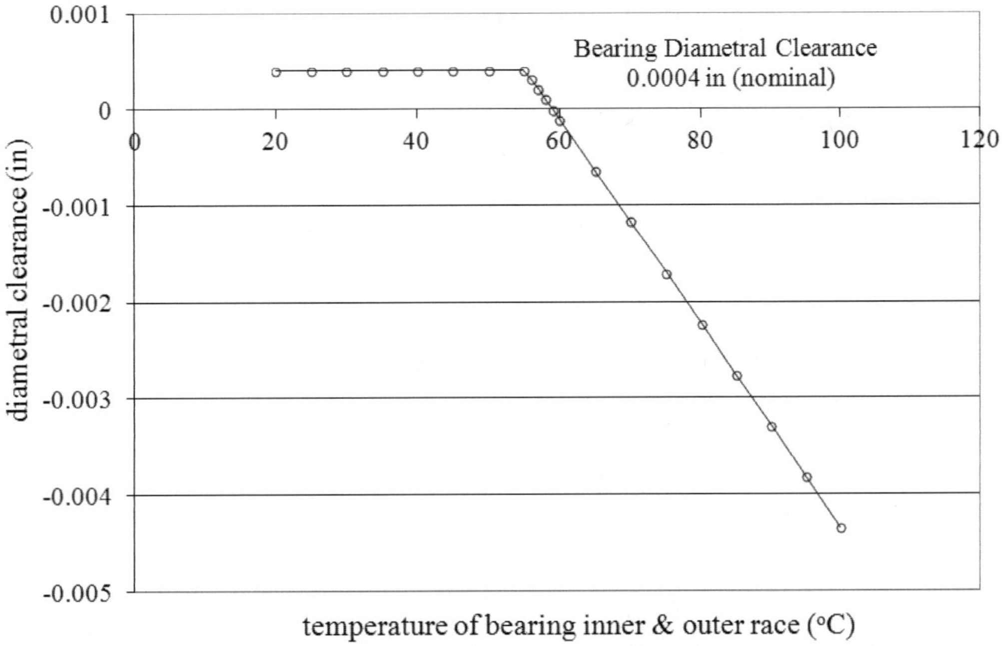

Figure 7: Internal diametral clearance of vacuum bearing versus temperature.

Here channel capacity applied to measured current signals from motors will indicate diminished clearance from thermal expansions. To study change in channel capacity of measured current signals, a heating element placed at the base of the vacuum barrier bowl held temperature to 85°C. Current measurements from motors were then taken, on February 15. The system was cycled with hubs rotating in opposite directions through a full revolution. A cycle rotated the upper hub counter-clock-wise one revolution, stopped then rotated back clock-wise to stop at the initial position. The same input signal profile as before was applied to the motors. Current measurements were taken on February 23 after the hub assembly completed about 38,400 cycles at 85°C temperature using measurement procedures explained earlier. Ten synchronized current measurements were used to calculate the signal channel capacity with respect to the baseline signal calculated as an average over the February 15 data. The new signals render channel capacity values ranging from 4,097 s^{-1} to 4,832 s^{-1} with a mean at 4,426 s^{-1}, and a standard deviation of 274 s^{-1}. The February 23 data reveals a lower channel capacity with a higher standard deviation. A lower channel capacity indicates system degradation, while the higher standard deviation indicates higher variation among the current signals. Using the student`s t-test [13] the channel capacity mean of 4,426 from February 23, compared to the 5,518 value from February 15, indicates a significant difference with a confidence level greater than 99%. The spread in the channel capacity value from the February 23 data is plotted in Figure 6, to the left of the spread from the 2/15/2011 data, indicating the significantly reduced level of channel capacity. Visual interpretation shows a small overlap in the distributions which is in accordance with results from the t-test interpretation.

4.0 Robot Hub Assembly Motor Machine Fault Detection. The channel capacity diagnostic concept is applied to detect failure in the robot motor hub assembly when the hub is part of a larger robot system. Because semi-conductor devices consist of dense integrated circuits with micron range deposition layers, particle contamination is a serious issue during microchip manufacture. Major sources of particle contaminants include air, production, facility, clean-room personnel, process material, and static electricity. Particles of various shapes range from 0.001 to 10,000 microns. Smoke particles are 12 times smaller while dust particles are 3 times smaller than the diameter of a human hair [14]. Particles that settle between metal lines formed on the wafer cause problems in device performance, reliability, and yield. Particle control in a micro-chip manufacturing site consists of a uniform gas flow through corresponding modules to a factory exhaust system where particles are monitored and counted. The transfer chamber serves as a wafer loading and unloading platform to several stations. Figure 8 shows a typical wafer processing system. Unprocessed wafers placed inside the Front Opening Unified Pod (FOUP), a plastic container to hold wafers, are then transported by the Factory Interface (FI) robot to the load-lock. The transfer chamber robot moves the wafer from the load-lock, through the transfer chamber to the various process chambers. After processing, the wafer is guided, through a reverse path to the FOUP. A door seals the wafer process environment.

Because the FI has its own particle control strategy, the possibility of a particle escaping into the transfer chamber is nil. Possible particle sources inside the transfer chamber include degradation of the slit valve elastomeric material on the door, escape of process gases from the process chambers, and robot degradation. Whenever particles are detected, the customer usually replaces the transfer chamber robot, believed to be the root cause. Failure analysis indicates about 80% of these replaced robots are fully functional. After robot replacement the customer completely wipes the transfer chamber interior to remove particles. Although the chamber is clean, the particle source might still be present in the chamber. The system must operate a few months for particle counts to become measurable. In robotic systems most of the particles are generated from excessive friction between moving parts caused by system degradation. The robot of Figure 4 has multiple bearings that can shed particles. To study ability of the channel capacity health measure to assess system health, in this section data will arise from simulations generated by a model. This enables study of the measure with multiple different faults that are easily varied. This section uses basic principles of communication theory to evaluate robot performance and (if necessary) diagnoses which robot hub needs repair.

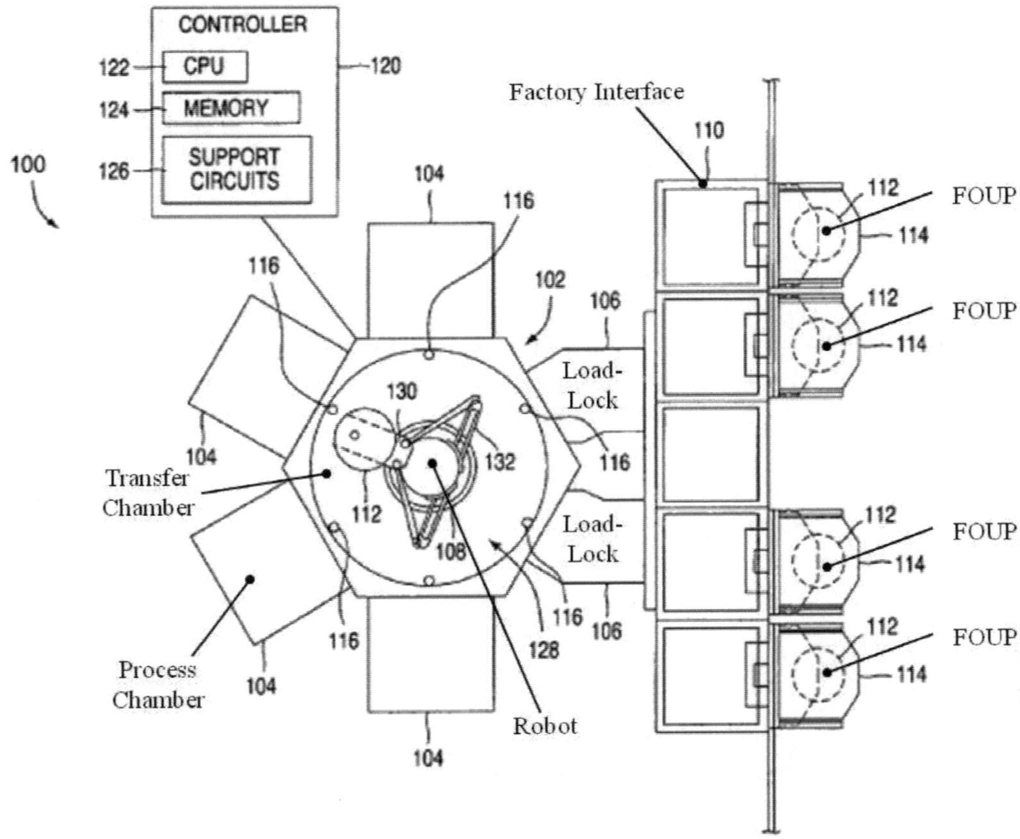

Figure 8: Wafer processing tool [15].

4.1 Sensitivity Analysis Via Measurements. Because both vacuum bearings of the robot depicted in Figure 4 are of the same type, pairs of known "good" bearings in pristine condition and known "bad" bearings from nonconformance cases were gathered. All experiments were performed on a common robot hub assembly with various vacuum bearing combinations. The test setup and apparatus was the same as in section 3.0 except the robot's motion profile (described in section 3.5) was modified to allow each motor to complete 5 revolutions. Experiments collected 7 torque measurements from each motor, in each of the four bearing conditions listed in Table 4.

Table 4: Design-of-Experiment bearing conditions.

	Upper Vacuum Bearing	Lower Vacuum Bearing
Experiment No. 1	GOOD	GOOD
Experiment No. 2	GOOD	BAD
Experiment No. 3	BAD	GOOD
Experiment No. 4	BAD	BAD

Experiments simultaneously rotated the upper hub clockwise, and the lower hub counterclockwise 5 complete revolutions. Torque data such as Figure 9 was read from the NSK driver at a sampling rate of 2,000 Hz. Before any analysis, all signals were correlated to synchronize alignments in time. Because of intrinsic noise in the signals, the average of all seven signals from the first experiment was defined as a motor torque baseline $y_o(t)$ expected for "good" bearings. The average signal still had intrinsic noise, but averaging makes this intrinsic noise in the averaged signal different from any of the samples, thus not biasing the calculations towards any of the measured signals. Subsequent signals were compared to this baseline signal. For each signal, a channel capacity was calculated via Equations (4) and (5). Variation in the channel capacity between the different signals was expected due to the appreciable intrinsic noise in the torque signals. Channel capacity values were quantified for each signal in each bearing condition experiment. An average ($\bar{C}$) and standard deviation (σ) for channel capacity were obtained for each motor. Upper Control Limit (UCL) and Lower Control Limit (LCL) establish the channel capacity spread 3σ from $\bar{C}$. Statistical information for the four bearing condition scenarios is presented in Table 5. Measurement noise from the NSK driver caused slight variations in samples over the same experiment. When both bearings were "good", channel capacity values greater than 10,000 appeared for both motors.

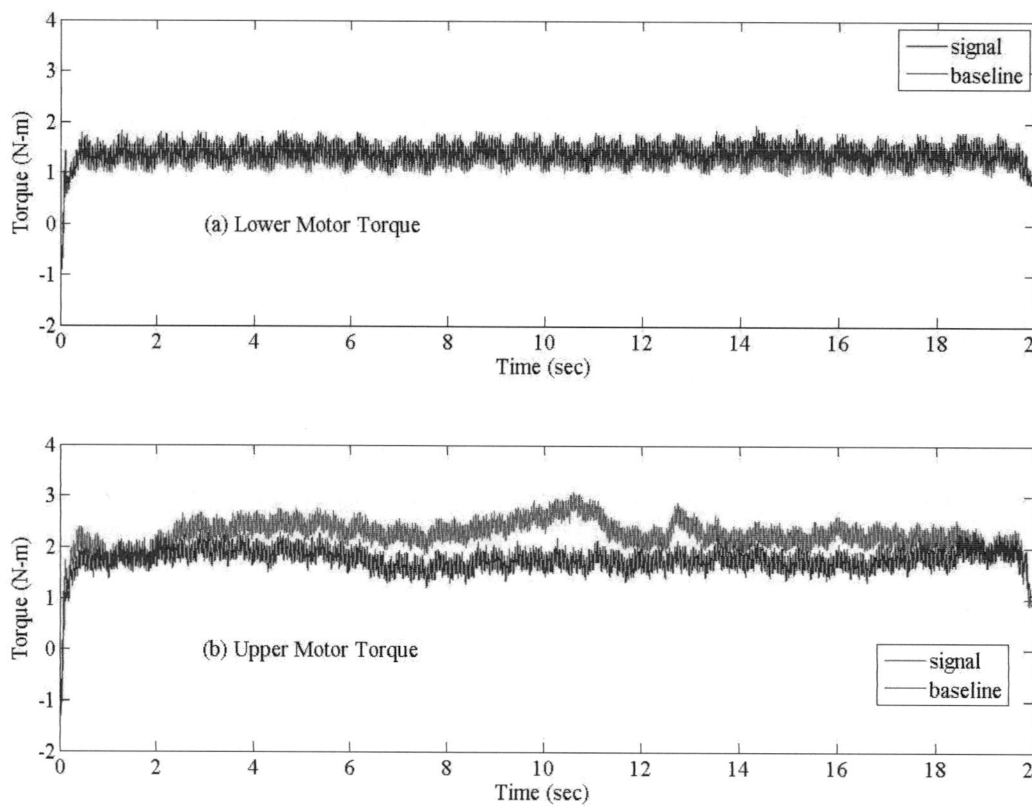

Figure 9: Motor torque from experiment 2.

Results with "good" upper bearing and "bad" lower bearing, experiment 2, show channel capacity values less than 10,000 for the lower motor, and much lower values around 4,000 for the upper motor (see Table 5). Oscillations at 120 Hz not accounted in the simulations appear embedded in the motor torque measurements, due to the 120 equally spaced teeth inside the motor. Bearing friction appears to have small effect on the driver noise output, approximately 1% decrease from the expected channel capacity value of the lower motor. This channel capacity change is less important compared to a 60% decrease calculated for the upper motor. Figure 9 depicts motor torque signals and corresponding baselines. Because bearing friction was not measured, exact comparison between simulated and measured results is not possible. However, a relative comparison is. A change in upper vacuum bearing friction affects the torque channel capacity value observed on both motors. Further, a change in lower vacuum bearing friction affects the channel capacity value of the upper motor, leaving the lower motor unaffected. A change in friction on both upper and lower vacuum bearings affects the upper motor friction more than the lower. These observations are in accordance with the measurement results of Table 5. Results with "bad" upper bearing and "good" lower bearing show significant change in the torque channel capacity on both motors. Motor torque channel capacity values further decreased with both motors "bad". Upper motor torque was most affected. The effects on channel capacity compound when both motors are "bad".

Table 5: Channel capacity variation measurements.

Vacuum Hub Bearing Condition		Channel Capacity of Motor Torque Signal (s^{-1})					
		Lower Motor			Upper Motor		
Upper	Lower	LCL	$\overline{C}$	UCL	LCL	$\overline{C}$	UCL
GOOD	GOOD	10,036	10,069	10,102	10,467	10,500	10,533
GOOD	BAD	9,919	9,951	9,983	4,181	4,224	4,267
BAD	GOOD	3,284	3,761	4,238	3,453	3,814	4,174
BAD	BAD	3,192	3,328	3,464	2,604	2,638	2,672

To quantify change among the four bearing condition scenarios, the noise change of the torque signals with respect to the baseline was established in terms of corresponding channel capacities. Starting with Equation (4), C values C_1 and C_2 for two signals were rearranged, to estimate the ratio of noise power levels

$$\frac{N_1}{N_2} = \frac{\left(2^{C_2/W}\right)-1}{\left(2^{C_1/W}\right)-1}. \quad (6)$$

In Equation (6), since C_1 and C_2 refer to the same health measure for the same signal, but for different bearing cases, and since except for total failure signals $y(t)$ are dominated by baseline components $y_o(t)$,[1] it was assumed that $S_{y_1} \approx S_{y_2}$ and $W = W_1 = W_2$. Treating C_2 as reference, Equation (6) compares noise power levels of different cases. Of interest for reference C_2 is when both lower and upper bearings are "good". Table 6

[1] For even the most degraded systems, baseline signal $y_o(t)$ usually constitutes over 90% of the energy in signal $y(t)$. This was concluded from many years experience and observations of power spectra [5]-[7], [10], [11].

presents a noise ratio measured in the motor torque signal for the corresponding vacuum hub bearing condition scenario compared to the reference where both lower and upper bearings are "good".

Table 6: Noise ratio values torque signal to baseline.

Vacuum Hub Bearing Condition		Noise Ratio (signal/baseline)	
Upper	Lower	Lower Motor	Upper Motor
GOOD	GOOD	1	1
GOOD	BAD	1	11
BAD	GOOD	12	13
BAD	BAD	15	25

The torque signal of the lower motor contains about the same noise as the reference "good"-"good" case when the lower hub bearing is bad, and 12 times the noise of the reference when the upper hub bearing is bad. The corresponding noise levels from the upper motor are 11 times and 13 times the reference respectively. This suggests a minimum change in lower motor torque when only the lower hub bearing is defective. A defect of this type affects the upper motor torque. Significant change in lower motor torque is noted when the upper hub bearing is defective. An upper hub bearing defect affects torque in both upper and lower motors equally. Motor torque requirements were significantly higher when both hub bearings were bad, mainly caused by the superposition of bearing effects on motor torque requirements.

5.0 Conclusions. The paper studied a new diagnostic health assessment concept wherein values of channel capacity can assess mechanical system health by quantifying signal degradation directly from the measurement and evaluate system behavior. The method is more sensitive to fault noise variation and less sensitive to random intrinsic noise, requires minimal computational effort since power of signal and noise is calculated directly from time domain signal information without needing knowledge of frequency content. Another application evaluated torque signatures from 4 robot assemblies, assessing effects of degradation at each robot motor, and identifying the worst performing robot. Another application showed how the channel capacity value can be used to diagnose robot health by comparing torque signals from 2 motors. By monitoring the channel capacity of the torque signal in both motors faulty vacuum bearings were identified. The utility and power of the channel capacity technique is its direct application to measurement data without mathematical models and their lengthy development times. Because of its simplicity, its effective computation, its robustness, and its sensitivity to intermittent faults, the channel capacity technique is ideal for industrial applications.

6.0 References:

[1] Starr A. G., Wynne R. J., Kennedy I., "Failure Analysis of Mature Robots in Automated Production", *Proceedings Institution of Mechanical Engineers*, Vol. 213, Part B, 813-824, 1999.

[2] Nyberg M., "Model-Based Diagnosis of an Automotive Engine Using Several Types of Fault Models", *IEEE Trans. On Control Systems Technology*, Vol. 10, No. 5, 2002.

[3] Venkatasubramanian V., Rengaswamy R., Yin K., Kavuri S., "A Review of Process Fault Detection and Diagnosis. Part I: Quantitative Model-Based Methods", *Computers and Chemical Engineering*, Vol. 27, pp. 293-311, 2003.

[4] Venkatasubramanian V., Rengaswamy R., Yin K., Kavuri S., "A Review of Process Fault Detection and Diagnosis. Part III: Process History Based Methods", *Computers and Chemical Engineering*, Vol. 27, pp. 327-346, 2003.

[5] Bryant M.D., Nakhaeinejad M., Choi J., "Model Based Diagnostics and Fault Assessment of Induction Motors with Incipient Faults", *Proceedings of the Society for Experimental Mechanics Series*, Vol. 8, pp. 439 - 449, 2011.

[6] Choi J., *Model Based Diagnostics of Motor and Pumps*, Ph.D. dissertation, University of Texas @ Austin, Austin, 2006.

[7] Costuros T., *Application of Communication Theory to Health Assessment, Degradation Quantification, and Robot Cause Diagnosis*, Ph.D. dissertation, University of Texas @ Austin, Austin, 2013.

[8] Shannon C. E., "Communication in the Presence of Noise", *Proceedings of the IRE*, Vol. 37, pp. 10 - 21, January 1949.

[9] F. Stremler, *Introduction to Communication Systems*, 2nd Edition, Addison - Wesley, December, 1982.

[10] Bryant M.D., Karlapalem L.C., Murphy J.T., and Ruizpalacios R., "Model Based Machine Diagnostics Using Shannon`s Information Theory", *Journal of Dynamic Systems, Measurement, and Control*, Vol. 128, No. 3, pp.584 - 591, 2006.

[11] Bryant M.D., Kim J., Lee S., Choi J., *Model-Based Machine Diagnostics and Prognostics Using Theory of Noise and Communications*, U.S. Patent 20040236450A1, Nov. 25, 2004.

[12] *Megatorque Motor System User's Manual*, NSK Ltd., Document Number: C20075-01, 2000.

[13] R. Langley, *Practical Statistics*, Dover Publications Inc., New York, N.Y., 1970.

[14] P. V. Zant, *Microchip Fabrication*, 3rd Edition, McGraw Hill, 1997.

[15] D. Cox, M. Freeman, *Auto-Diagnostic Method and Apparatus*, U.S. Patent 20050137751A1, 11/002732, 2005.

MACHINERY FAULT DIAGNOSIS USING POLAR WAVELET ANALYSIS

Lim Meng Hee and M. Salman Leong
Institute of Noise & Vibration
Universiti Teknologi Malaysia,
International campus, Jalan Semarak
54100 Kuala Lumpur
mhlim@utm.ic.my

Abstract: Most of the mechanical faults that occur in rotating machinery are repetitive in nature, of which the same forcing mechanism recur once per revolution. The ability to visualize these forcing mechanisms over a complete rotational cycle would provide crucial information about the characteristics of the faults and thus enables a more prudent fault diagnosis to be undertaken. The feasibility of using polar wavelet analysis to provide visualization of the cyclical forcing mechanism of faults in a rotor system is studied in this paper. An experimental study was conducted to simulate various machinery faults namely rotor eccentricity, stage rubbing and creep induced rubbing in a rotor system. Subsequently, polar wavelet analysis was applied to analyze the vibration response of the rotor. Experimental results showed that the characteristics of faults such as blade rubbing mechanism in the rotor could clearly be seen in the polar wavelet plots. Polar wavelet analysis is therefore deemed to be an excellence tool to visualize the cyclical force in a rotor and thus making fault diagnosis more intuitive and effective.

Key words: Polar; wavelet; fault diagnosis; eccentricity; rub.

Introduction: For many decades, vibration spectrum analysis has been widely used to detect and diagnose a wide range of machinery faults that are known to generate stationary signals such as rotor unbalance and misalignment. However, some machinery faults are also known to produce non-stationary or transient signals on top of the normal stationary signal such as in the cases of mechanical looseness and rubbing. Vibration spectrum analysis is known to be less effective in analyzing these type of signals which required the ability to display both the stationary and non-stationary signals in one single display. The ability of time-frequency analysis methods and in particular the wavelet analysis to represent both type of signals in one single display made it an ideal tool to diagnose transient machinery faults. Peng and Chu [1] presented a comprehensive review on the application of the wavelet analysis in machinery fault diagnosis and showed that one of the most prominent application of wavelet analysis is to detect changes in rotor vibration responses due to rubbing. Peng et al. [2] investigated the application of wavelet plot and reassigned wavelet plot in detecting the impact signals caused by rubbing. They observed that when rubbing occurred, an increase of magnitudes not only can be observed at the fundamental rotational frequency, but also at high frequency regions. Wang and Chu [3] reported a hybrid method of acoustic emission and wavelet analysis to determine the locations of rubbing in the rotor system. Peng et al. [4] used wavelet analysis and its phase maps to study the non-linear behaviours of rubbing that occurred in rotor systems. Al-Badour et al. [5] investigated stator-to-blade rubbing phenomena of a running machine and had found that the combination of continuous wavelet and wavelet packet transforms provides a better tool to monitor vibration signal of rubbing. Ma et al. [6] studied two different phenomenon of rotor rubbing namely the rotor rub coupled with crack and bearing rub coupled with oil-film instability. It was found that by combining spectrum cascade, reassigned wavelet plot, rotor trajectory, and frequency spectrum, some novel fault features usable for rubbing detection can be discovered from the vibration signals measured during run-up and run-down of a rotor. Beside this, the current authors, Lim and Leong [7, 8] had also conducted a series of experimental studies on the feasibility of wavelet analysis in detecting various types of blade faults (loose blades and blade rub) in an experimental rotor system. The application of polar wavelet analysis to detect faults was first introduced by Meltzera and Nguyen [9]. They have conducted an experimental study on faulty gearboxes under non-stationary rotational speed. The polar wavelet transform was then used to process the vibro-acoustics signals of the gear box. The polar wavelet analysis was reported to provide a better visualization of wavelet results for crack detection in gear.

This paper investigates the application of polar wavelet analysis in depicting the cyclical forcing mechanisms of rotor eccentricity, stage rubbing, and creep induced rubbing, which could then be useful for faults diagnosis and classification. Essentially, rotor eccentricity is a condition caused by the misalignment of shaft centerline and casing centerline, which could in turn cause by foundation sagging, misalignment, rotor bow, casing distortion, etc. This condition is a precursor to stage rubbing whereby the entire stage of blades rub against the rotor casing. For creep induced blade rubbing, only the elongated or creep blade is to be rubbed against the rotor casing. An illustration showing the mechanical condition of these faults are given in Figure 3 under the heading of Experimental Study.

Formulation of Polar Wavelet Analysis: Wavelet analysis is a very versatile signal analysis tool. There are many variations of wavelet analysis algorithms which could be used to analyze the signals in hand (e.g. continuous wavelet analysis, discrete wavelet analysis, stationary wavelet analysis). Polar wavelet analysis is an extension of continuous wavelet analysis plotted in polar coordinates system. The overall signal processing procedure of the polar wavelet analysis is illustrated in Figure 1 with the following explanations:

a) First, the measured vibration signal was divided into smaller segments in accordance to the time-gated tachometer signal. These individual segments represent the vibration signal of rotor in one complete shaft rotation. A total of 18 time-gated signal segments was created in 1s of vibration signal as seen in Figure 1(A).

b) These time-gated signals were then averaged to produce the Time Synchronized Averaged (TSA) signal whicht represents the average vibration signal of one complete shaft rotation (see Figure 1(B)).

c) Continuous Wavelet Transform (CWT) was then applied to the TSA signal (see Figure 1(C)). Coiflet was selected as the mother wavelet to analyse the TSA signals as the wavelet results produced with this mother wavelet were seen to be more symmetrical and easy to be interpreted. In brief, CWT is defined as the sum over all time of the signal *f(t)* multiplied by scaled (scale parameter, *a*) and the shifted (position parameter, *b*) versions of the wavelet function, ψ as showed in Equation (1):

$$C(a,b) = \int_{-\infty}^{\infty} f(t)\frac{1}{\sqrt{a}}\psi^{*}(\frac{t-b}{a})dt \quad (1)$$

Where * denotes the complex conjugate of wavelet function. The results of CWT are many wavelet coefficients, C in function of scale, a and position, b. CWT analysis was carried out from scale 1 to 32 with scale resolution of 0.1. The conventional results of CWT are wavelet coefficients in the form of *b* x *a* matrix plotted in Cartesian coordinates format.

e) Subsequently, the wavelet coefficients in Cartesian coordinates format were transforms into spherical polar coordinates format to produce the polar wavelet plot as seen in Figure 1(D). The spherical coordinates (radius *r*, inclination θ, azimuth φ) of a point can be obtained from its Cartesian coordinates (x, y, z) as showed in Equation (2)-(4). The inverse tangent denoted in φ was suitably defined taking into account the correct quadrant of (x, y).

$$r = \sqrt{x^2 + y^2 + z^2} \quad (2)$$

$$\theta = \arccos\left(\frac{z}{r}\right) \quad (3)$$

$$\varphi = \arctan\left(\frac{y}{x}\right) \quad (4)$$

f) Figure 2 illustrates a typical result of polar wavelet analysis. It can be seen that the pseudo-frequency of the polar wavelet plot range from 50 Hz (radius R1) to 1625 Hz (radius R4) in correspondence to wavelet scale of 1 to 32. The colour intensity scale in the polar wavelet plot indicates the amplitudes of the wavelet coefficients. The blade pass frequency (BPF) ring is also specifically indicated in the polar wavelet plot in dotted line (scale 7) to facilitate machinery fault diagnosis.

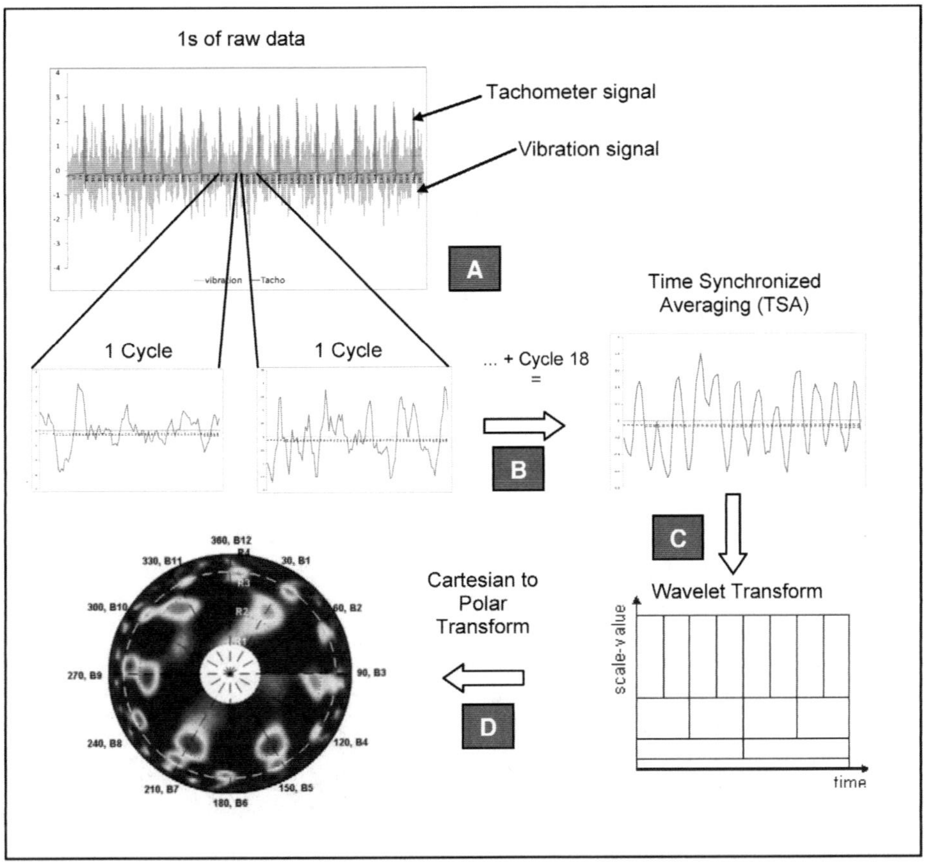

Figure 1: Overall signal processing procedure for polar wavelet analysis

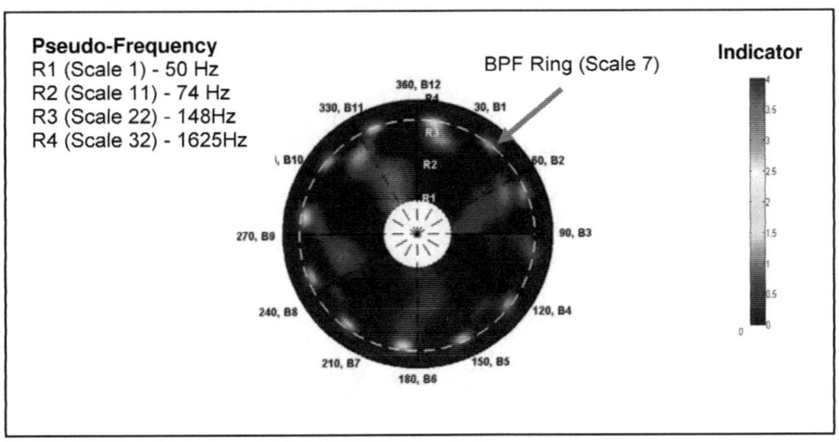

Figure 2: Pseudo-frequency and colour intensity scale of polar wavelet analysis

Experimental Study: The mechanical condition of faults that studied in this paper are showed in Figure 3. An experimental rotor system consists of casing and a rotor fitted with twelve number of blades was designed and fabricated as showed in Figure 4. In order to create creep induced rubbing condition, a "Top-head" (see Figure 5) made of aluminum sheet metal was used to extent the blade length and to act as the rubbing medium between the blade and the casing. Meanwhile, the initial blade tip clearance of 7mm was reduced to 2mm in vertical direction in order to create rotor eccentricity condition (or the impending stage rub). This was achieved by dropping the upper casing in vertical direction by removing some pieces of shim placed in between the upper and lower part of casing. Subsequently, stage rubbing condition was created by fitting Top-head onto each and every blade to rub against the uppermost spot of the casing. The operating speed of the experiment was set at 1200 rpm. Vibration and tachometer signals were measured simultaneously in this experiment at the sampling frequency of 5 kHz. With this sampling frequency, a total of 217 data points were sampled in one complete shaft rotation. The vibration signals were measured using the 786M accelerometer in tri-axes at upper casing. IMC C1 Data Acquisition Box (DAQ) was used to capture the vibration and tachometer signals simultaneously. The overall experimental set up is showed in Figure 6.

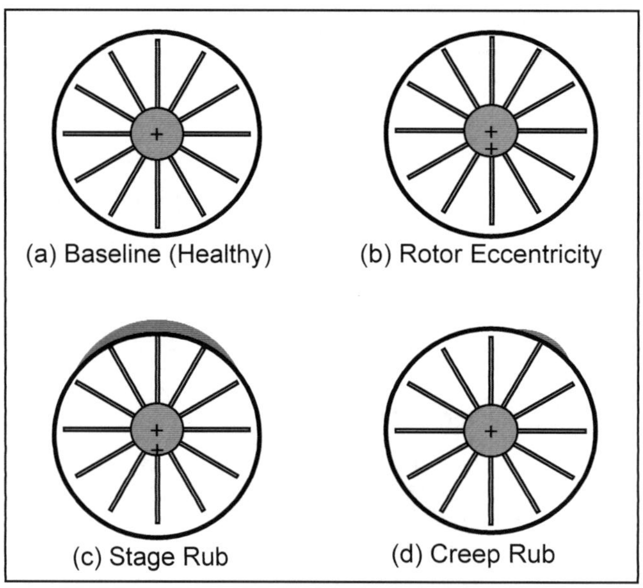

Figure 3: Experimental conditions created in the rotor system

Figure 4: Experimental rotor system consists of a rotor and twelve number of blades

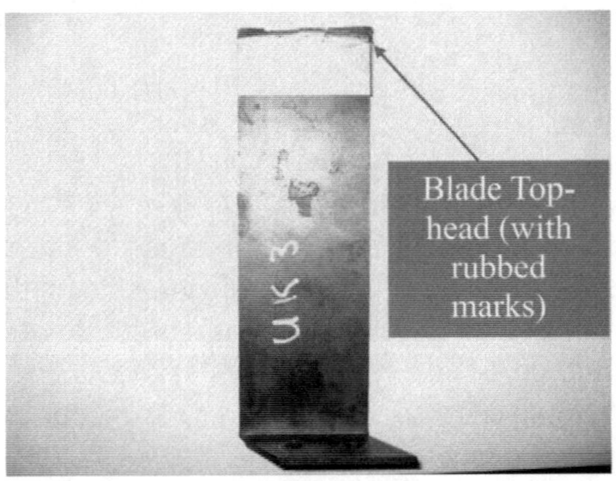

Figure 5: Blade fitted with "Top-head" with marks of rubbing

Figure 6: Overall experimental set up

Result and Discussion:
Vibration Spectrum Analysis

The vibration spectrums of baseline (healthy condition), rotor eccentricity, stage rubbing, and creep rubbing are showed in Figure 7. By comparing the vibration spectrum of rotor eccentricity to baseline condition, no significant difference in the pattern of vibration spectrum could be observed, except for the vibration amplitude of x2 and x4 components of rotor eccentricity which were seen only 0.1mm/s higher than the baseline condition. Hence, vibration spectrum analysis was seem to provide inconclusive evident about the occurrence of rotor eccentricity in the rotor. In comparison, the vibration spectrum of stage rubbing was seen to be significantly different from baseline condition. Multiple peaks of high frequency harmonics could be seen in the vibration spectrum, which implied the occurrence of blade rubbing as this phenomenon was not present in both the vibration spectrums of baseline and rotor eccentricity. In addition, the vibration amplitudes of x2 to x6 component were also seen to increase significantly from the baseline condition. For creep rubbing, some tiny frequency peaks could be seen scattered in the high frequency region of the vibration spectrum, which indicate the presence of minor blade rubbing. Besides this, the x1 component was seem to increase twofold from 0.3 mm/s to 0.7 mm/s. Based on these results, although vibration spectrum analysis was showed to be a viable tool to detect stage and creep rubbing, extra information which are useful for fault diagnosis and classification such as the quantity, the position, and the cyclical mechanism of fault in the rotor could not be obtained based on the vibration spectrum analysis.

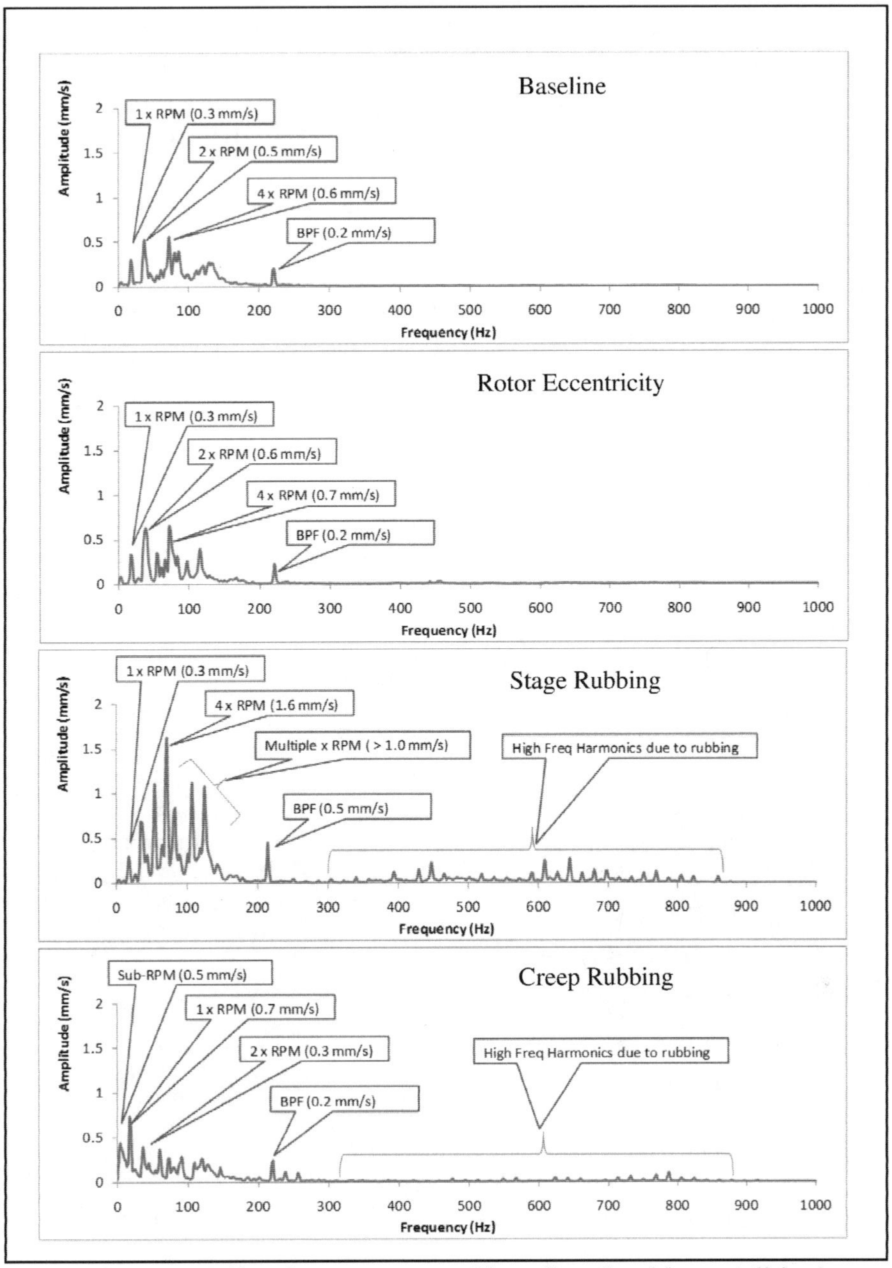

Figure 7: Vibration spectrum of baseline (healthy condition), rotor eccentricity, stage rubbing, and creep rubbing.

Polar Wavelet Analysis

The polar wavelet plot of baseline and rotor eccentricity are showed in Figure 8. In a glance, the polar wavelet plot of baseline condition was seen to be normal as there were no high intensity spots (with the magnitude higher than 2) to be observed. The fluctuation of blade pass frequency in one complete shaft rotation could also be seen, which were represented by the twelve number of relatively high intensity spots located along the BPF ring. By comparing the polar wavelet plot of baseline and rotor eccentricity, significant difference could be seen. There were four high intensity spots located along the R2 ring (approximately 74 Hz or x3 component) of polar wavelet plot of rotor eccentricity, which were not present in the baseline condition. This provides clue about the changes in rotor dynamics of the rotor system due to rotor eccentricity. In addition, the twelve numbers of high intensity spots at BPF ring, which represent the pressure pulsation of each running blade in the rotor (B1 to B12) were more visible in polar wavelet plot of rotor eccentricity as compared to baseline condition. This phenomenon consistent to the fact that the blade pressure pulsation increased in correspondence to the decrease in blade tip clearance. Figure 9 illustrates the comparison of polar wavelet plots of stage and creep rubbing. For stage rubbing, a surge in rubbing energy at BPF ring could easily be seen throughout one complete shaft rotation. The twelve numbers of very high intensity spots which represent the blade rubbing energy of each running blade could readily be seen. In comparison, when creep rubbing (blade position B1) was induced in the experiment, only single spot of high intensity could be observed, which was located at position B1 (or 30 degree) of BPF ring. The difference in pattern between the polar wavelet plot of creep and stage rubbing provides a clearer pictures about the cyclical mechanism of faults which is useful for faults classification. The capability of polar wavelet plot in visualizing the energy and fault mechanism in rotor system provides valuable insight for machinery fault diagnosis, which could not be accomplished using the conventional vibration spectrum analysis.

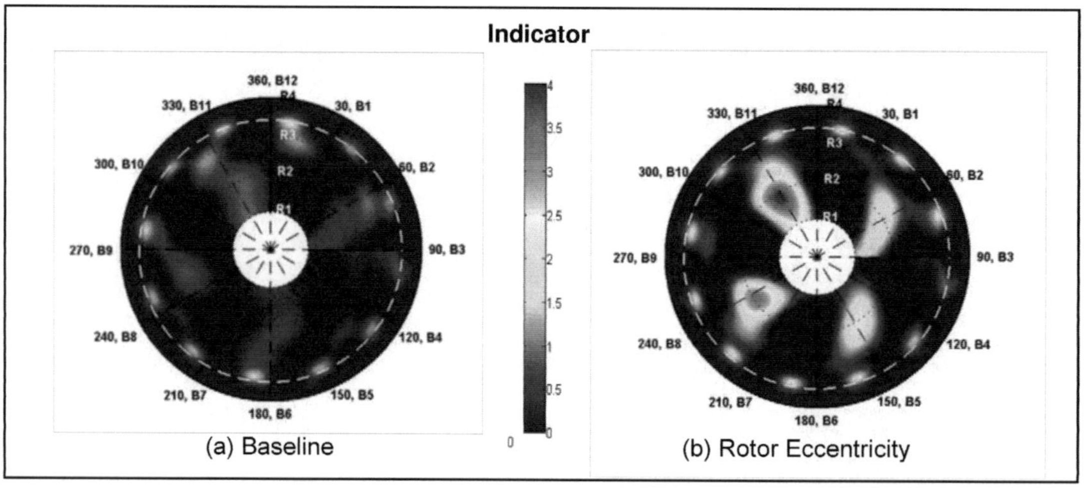

Figure 8: Polar wavelet analysis of baseline and rotor eccentricity

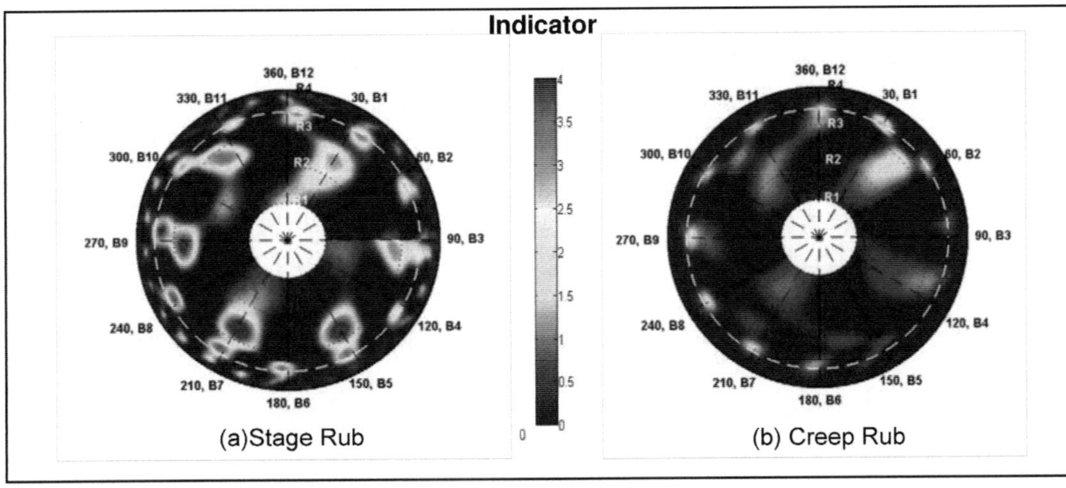

Figure 9: Polar wavelet analysis of stage and creep rubbing

Conclusion: The feasibility of using polar wavelet analysis to provide visualization of the forcing mechanism of faults in a rotor system is studied in this paper. Experiment results found that polar wavelet analysis provides a better representation than vibration spectrum analysis to envisage the forcing mechanism in rotor such as in the cases of rotor eccentricity, creep and stage rubbing. Polar wavelet analysis was also showed to be a more intuitive and effective method to identify the root cause of the faults such as in differentiating creep rubbing from stage rubbing. It is believed that this method could be extended to diagnose many other machinery faults of which its forcing mechanism are cyclical in nature such as in gear and bearing.

Acknowledgement:

This work is supported by Exploratory Research Grant Scheme (ERGS - R.K130000.7840.4L089) financed by the Ministry of Education Malaysia.

References:

[1] Z. K. Peng, F.L. Chu. 'Application of the wavelet transform in machine condition monitoring and fault diagnostics: a review with bibliography'. Mechanical Systems and Signal Processing, 18(2), pp. 199–221, 2004.

[2] Z.K. Peng, F.L. Chu, Peter W. Tse. 'Detection of the rubbing-caused impacts for rotor–stator fault diagnosis using reassigned plot'. Mechanical Systems and Signal Processing. 19(2), pp. 391–409, 2005.

[3] Q. Wang, F. Chu. 'Experimental determination of the rubbing location by means of acoustic emission and wavelet transform'. Journal of Sound and Vibration. 248(1), pp. 91–103, 2001.

[4] Z. K. Peng, Y. He, Q. Lu, F. Chu. 'Feature extraction of the rub-impact rotor system by means of wavelet analysis'. Journal of Sound and Vibration. 259(4), pp. 1000–1010, 2003.

[5] F. Al-Badour, M. Sunar, L. Cheded. 'Vibration analysis of rotating machinery using time–frequency analysis and wavelet techniques', Mechanical Systems and Signal Processing. 25. pp. 2083–2101, 2011.

[6] H. Ma, T. Yu, Q.K. Han, et al. 'Time-frequency features of two types of coupled rub-impact faults in rotor systems'. Journal of Sound and Vibration, 321, pp. 1109-1128, 2009.

[7] M. H. Lim, M.S. Leong., 'Diagnosis for loose blades in gas turbines using wavelet analysis'. ASME Journal of Engineering for Gas Turbines and Power, 127, pp 314-322, 2005.

[8] M. H. Lim and M. S. Leong., 'Reconstruction of Vital Blade Signal from Unsteady Casing Vibration', Advances in Mechanical Engineering, Volume 2014, Article ID 146983, 9 pages.

[9] G. Meltzera, Nguyen Phong Dien. 'Fault diagnosis in gears operating under non-stationary rotational speed using polar wavelet amplitude maps'. Mechanical Systems and Signal Processing. Volume 18, Issue 5, September, pp. 985–992, 2004.

A NEW ACOUSTIC EMISSION SENSOR BASED BEARING FAULT DIAGNOSTIC TECHNIQUE

Brandon Van Hecke, Yongzhi Qu, and David He
Department of Mechanical and Industrial Engineering
University of Illinois at Chicago, IL, 60607, USA
Telephone: 312-996-2387
bvanhe2@uic.edu

Abstract: Bearing fault diagnosis by quantifying acoustic emission (AE) data has been an area of interest for recent years due to the numerous advantages over vibration based techniques. However, most AE based methodologies to date are data-driven technologies. This research takes a new approach combining a heterodyne based frequency reduction technique, time synchronous resampling (TSR), and spectral averaging to process AE signals and extract condition indicators (CIs) for bearing fault diagnosis. The heterodyne technique allows the AE signal frequency to be shifted from several MHz to less than 50 kHz, which is comparable to that of vibration based techniques. Then, the digitized signal is band pass filtered to retain the information associated with the bearing defects. Finally, the tachometer signal is used to time synchronously resample the AE data to allow the computation of a spectral average and the extraction of CIs for bearing fault diagnosis. The presented technique is validated using the AE signals of seeded fault steel bearings on a bearing test rig. The result is an effective physics based approach validated to diagnose all four fault types: inner race, outer race, ball, and cage.

Keywords: Acoustic Emission Sensor; Bearing; Fault Diagnosis

1. Introduction: Most heavy machinery and mechanical devices that have moving components are likely to contain at least some type of bearing. Additionally, because bearings are typically implemented in non-ideal working environments, problems may surface that can result in failures that occur much earlier than intended. When bearings fail, it can result in significant downtime, high repair costs, and an accompanying decrease in production. Additionally, it has been reported that of the bearing failure modes, the majority are due to user related issues [1]. Thus, condition monitoring and bearing health analysis has been an area of interest for many years, which has resulted in numerous signal processing methods and algorithms that aid in the detection, classification, and prediction of bearing failures.

Currently, vibration analysis is the industry standard for bearing fault detection and diagnosis. However, acoustic emission (AE) based bearing fault diagnostic techniques have recently gained the attention of researchers and has resulted in many successful studies [2-4]. AE techniques have numerous advantages over vibration based approaches. Namely, AE techniques are capable of incipient fault detection. Also, they are sensitive to fault location, allowing its use for fault location detection. Additionally, AE sensors can retrieve information in frequencies that are much higher than vibration signals. In the context of bearing diagnostics, it has been shown that AE based techniques can detect faults earlier than other technologies [5,6]. However, there are a number of difficulties in the implementation of AE techniques. Namely, in comparison with vibration sensors, AE sensors usually require a much higher sampling rate (typically above 1MHz). Also, AE requires a significant storage and computational burden for large volumes of data. In this paper, a frequency reduction technique is implemented that down shifts the sampling rate to a level that is comparable to vibration methodologies while simultaneously reducing the computational burden. Lastly, the non-stationary behavior of AE signals makes traditional frequency analysis ineffective. This paper presents an AE based technique that can effectively and efficiently be used for bearing fault diagnosis.

Additionally, most AE based bearing fault diagnostic techniques utilize data driven technologies [7-9]. Although effective, data driven methodologies require significant quantities of data and are time consuming due to training of the models. In this paper, time synchronous resampling (TSR) and spectral averaging are used to develop an effective and efficient bearing fault diagnostic tool. First, a heterodyne based sampling frequency reduction technique is used to digitize the AE signal at a rate comparable to vibration methods. Then, this methodology utilizes band pass filtering to preserve the signal information related to the bearing defects. Then, the band pass filtered signal is time synchronously resampled using the zero crossing times (ZCT) of an incoming tachometer signal. Finally, features are extracted using a spectral averaging result to compute and evaluate the effectiveness of condition indicators (CIs) for bearing fault diagnosis.

The remainder of the paper is organized as follows. Section 2 provides a detailed explanation of the implemented methodology. In Section 3, the details of the seeded fault tests and experimental setup used to validate the methodology are discussed. Section 4 presents the bearing fault diagnosis results from the conducted seeded fault test and Section 5 concludes the paper.

2. The Methodology: Provided in Figure 1 is an overview of the presented methodology. First, a heterodyne based frequency reduction technique is used to acquire the AE signal at a rate comparable to vibration methods simultaneously with tachometer signal acquisition. Then, a band pass filter is chosen that contains the signal information related to the bearing defects. Next, the band pass filtered AE signal is time synchronously resampled using the tachometer signal as a trigger signal which allows a spectral average to be computed. Finally, features are extracted from the spectral average results for bearing fault diagnosis.

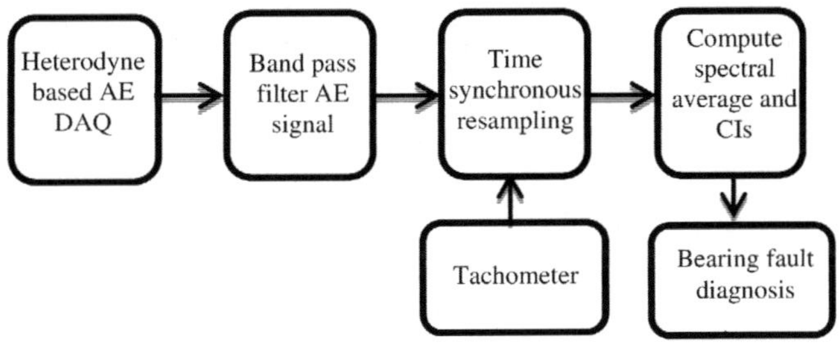

Figure 1: Overview of the Methodology

2.1 The Sampling Frequency Reduction Technique using Heterodyne: One disadvantage of AE based methodology is the significant computational burden. Because the frequency of the output signal from AE sensors is typically as high as several MHz, AE based techniques are usually accompanied with sampling rates as high as several to 10MHz. In this paper, a heterodyne based sampling frequency reduction technique is employed that down shifts the energy related to the signal so that a sampling rate comparable to vibration methods can be utilized. This is significant because less data needs to be collected and stored on the computer. Thus, the cost associated with data acquisition is reduced.

The idea has long been utilized in the field of communications. In radio, the carrier signal of a typical AM signal is often as high as several MHz. However, the audio signal that is modulated to that carrier signal is often as low as a few kHz. Through demodulation, the AM signal frequency is down shifted, allowing the audio to be sampled at a much lower rate. The result is a reduction in not only the sampling rate, but also the computational power required to process the data.

The AE signal demodulator implemented in this paper works similarly to a radio quadrature demodulator: shifting the carrier frequency to baseband, followed by low pass filtering. The technique applied here is called heterodyne. Mathematically, heterodyning is based on the trigonometric identity. For two signals with frequency f_1 and f_2 , respectively, it could be written as

$$sin(2\pi f_1 t)\, sin(2\pi f_2 t) = \frac{1}{2}\cos\,[2\pi(f_1 - f_2)] - \frac{1}{2}cos\,[2\pi(f_1 + f_2)] \quad (1)$$

where f_1 is the AE carrier frequency and f_2 is the demodulator's reference signal frequency.

For example, let $f_1 = 4$ Hz an $f_2 = 5$ Hz, and note $y_1 = sin(2\pi 4t)$ and $y_2 = sin(2\pi 5t)$. Then, their multiplication $Y = y_1 y_2$., is shown in Figure 2.

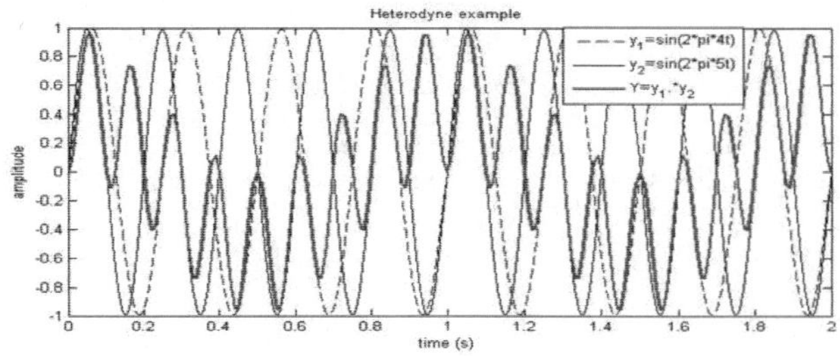

Figure 2: The Multiplication of Two Sinusoid Signals

Next, as shown in Figure 3, the modulated signal is low pass filtered to reject the high frequency image at frequency $(f_1 + f_2)$.

A detailed discussion of the heterodyne technique applied on the raw AE signal is given in the following. In general, amplitude modulation is the major modulation form for AE signals. Although, frequency modulation and phase modulation are potentially present in the AE signal, they are considered trivial and will not be discussed here. The amplitude modulation function is given in Eq. (2).

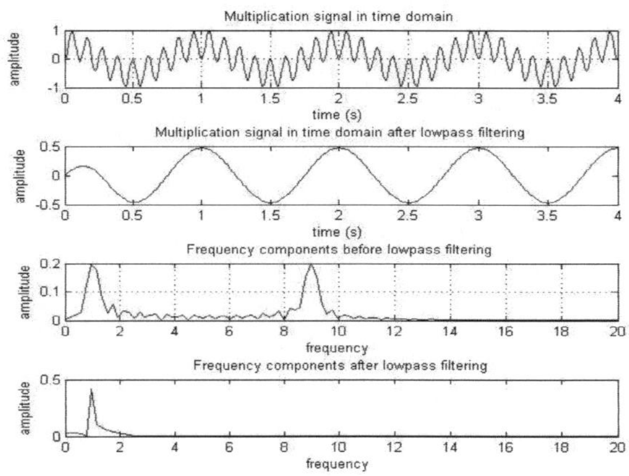

Figure 3: The Extraction of the Heterodyned Signal by Frequency Domain Filtering

$$U_a = (U_m + mx)\cos\omega_c t \tag{2}$$

where, U_a is the modulated signal, U_m is the carrier signal amplitude, ω_c is the carrier signal frequency, and m is the modulation coefficient, x is the signal of interest. With an amplitude X_m and frequency Ω, assume that x can be expressed as:

$$x = X_m cos\Omega t \tag{3}$$

Note that it is assumed that frequency Ω of signal x is normally much smaller than frequency ω_c of the carrier signal. Then, with the heterodyne technique, the modulated signal will be multiplied by a unit amplitude reference signal $\cos(\omega_c t)$. The result is U_o given as following:

$$U_o = (U_m + mx)\,cos(\omega_c t)\,cos(\omega_c t) = (U_m + mx)\left[\frac{1}{2} + \frac{1}{2}cos(2\omega_c t)\right] \tag{4}$$

Then, after substituting Eq. (3) into Eq. (4):

$$\begin{aligned} U_o &= \frac{1}{2}U_m + \frac{1}{2}mX_m cos\Omega t + \frac{1}{2}U_m cos(2\omega_c t) \\ &+ \frac{1}{4}mX_m[cos(2\omega_c + \Omega)t + cos(2\omega_c - \Omega)t] \end{aligned} \tag{5}$$

Since U_m does not contain any useful information related with the modulated signal, it could be set as 0, or removed by de-trending. From Eq. (5), it can be seen that only the part $\frac{1}{2}mX_m cos\Omega t$ which is the signal of interest will be retained after low pass filtering, while the high frequency components around frequency $2\omega_c$ will be removed.

By adding a demodulation step, it can achieve the purpose of reducing the signal frequency to 10s of kHz. This is close to the frequency range of general vibration signals. Thus, any data acquisition board with a low sampling rate should be able to sample the pre-processed AE data.

2.2 Band Pass Filter Selection: Once the AE bearing signal information is downshifted and sampling frequency reduced, the signal must be band pass filtered to retain the signal information related to the bearing defect. Generally, a properly selected band pass filter removes the high frequency components and low frequency contents associated with shaft imbalance [10]. Hence, when a band pass filter is executed properly, the information associated with the bearing fault is maintained.

In theory, as a bearing rotates at a constant speed, its AE signal can be characterized by a periodical property. In general, there are four fundamental defect frequencies to describe this motion. The four defect frequencies are the fundamental train frequency (FTF), ball spin frequency (BSF), ball pass frequency outer (BPFO), and ball pass frequency inner (BPFI), respectively representing the defect frequencies of the cage, ball, outer race, and inner race [11]. The defect frequencies are represented by the following:

$$FTF = (\omega/2)\left[1 - \left(D_e/D_p\right)\cos(ß)\right] \tag{6}$$

$$BSF = \left(\omega\, D_p/2D_e\right)\{1 - [\left(D_e/D_p\right)\cos(ß)]^2\} \tag{7}$$

$$BPFO = (\omega Z/2)\left[1 - \left(D_e/D_p\right)\cos(ß)\right] \tag{8}$$

$$BPFI = (\omega Z/2)[1 + (D_e/D_p)\cos\,(ß)] \tag{9}$$

where D_e is the diameter of the rolling elements , D_p is the pitch circle diameter , Z is the number of rolling elements , ß is the contact angle in degrees, and ω is the rotational speed of the shaft in revolutions per second.

The drawing of a 6205-2RS steel ball bearing is provided in Figure 4. Additionally, the parameters of the bearing and the calculated defect frequency multipliers are provided in Table 1 and Table 2 respectively. If the bearing rotates at a constant shaft speed, the fundamental defect frequencies can be computed by multiplying the defect frequency multipliers by the rotational shaft speed. In the absence of mechanical noise, the fault frequencies could be observed, leading to the selection of a filter band that contains the bearing fundamental defect frequencies. However, in practice, bearing signals are in the presence of mechanical noise from surrounding mechanical components, which eliminates the possibility of identifying the defect frequencies, and therefore, the ability to select a filter band that encompasses them. Thus, the filter band can be selected based on prior knowledge of the system or by utilizing some other criteria. If a correct filter band is selected, the useful signal information related to the fault can be effectively extracted. Consequently, the filter band that can extract the signal information must be determined. In this paper, the filter band is selected based on Shannon's entropy, which is defined as the average amount of information in a message. To select the proper filter band, different bands can be tested on the healthy bearing signals. After filtering, the entropy in the resulting signal is measured. Hence, the filtered signal that contained the highest level of information will be used to select the filter band to implement before further processing. Shannon's entropy is formally defined by:

$$Entropy = -\sum_{i=1}^{n} x_i^2 \ln(x_i^2) \tag{10}$$

where $Entropy$ is the Shannon's entropy, or average amount of information in the dataset x, and x_i is the ith element of the dataset x.

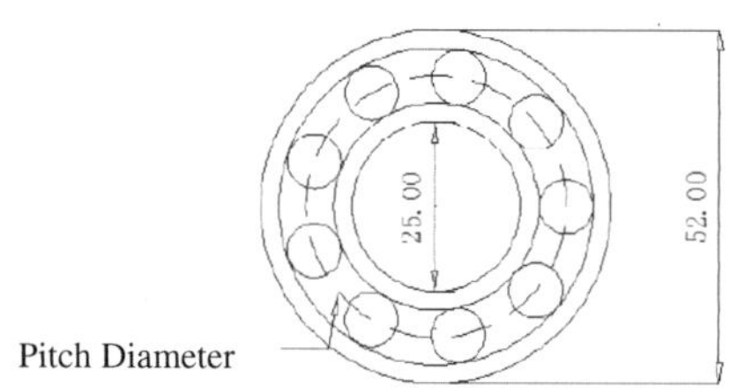

Figure 4: The Drawing of the 6205-2RS Ball Bearing

Table 1: The Parameters of the 6205-2RS Steel Ball Bearing

Parameter	Ball diameter	Pitch diameter	Inside diameter
Value	8 (mm)	39 (mm)	25 (mm)
Parameter	Outside diameter	Thickness	Number of balls
Value	52 (mm)	15 (mm)	9

Table 2: The 6205-2RS Steel Ball Bearing Fundamental Fault Multipliers

Fundamental Frequency Type	Inner Race	Outer Race	Cage	Ball
Fundamental Frequency Multipliers	5.415	3.585	0.398	4.714

2.3 Spectral Averaging of AE Signal: TSA is a proven technique for the extraction of periodic waveforms and has numerous applications to the tooth meshing vibrations for gear fault diagnosis [12,13]. Additionally, TSA has been effectively implemented in the processing of enveloped vibration signals for bearing fault diagnosis [14,15]. The concept is to compute the ensemble average of successive periods of a waveform of interest, resulting in a significant reduction of noise and an enhanced signal representing one period of the averaged waveform. Braun [16] formally expressed the TSA $y(nT)$ of a signal $x(t)$ sampled at interval nT by the following:

$$y(nT) = (1/N)\sum_{r=0}^{N-1} x(nT - rmT) \tag{11}$$

where the averaged period is denoted by mT. More details concerning TSA can be found in [12].

The successful implementation of TSA to quantify the vibration signals for bearing fault diagnosis has created interest to use it for the processing of AE signals. The successful application of TSA requires the knowledge of either the repetition frequency of interest, or a synchronous signal that is free of noise. Thus, two types of approaches have been presented in the literature, namely, the tachometer based TSA, and the tachometer less TSA. When compared to TSA using a tachometer, the tachometer less approach uses the repetition frequency of interest to estimate the angular information from the accelerometer signal. However, in the context of bearing fault diagnosis, this requires the computation of a trigger signal for each fault type. This research applies a tachometer TSA approach.

Although TSA has been widely applied to gear fault analysis [17-20], literature to date contains limited implementations for bearing fault diagnosis [14,15,21]. Additionally, these applications have all been based on the quantification of vibration signals, and none have reported the ability to diagnose cage fault, only presenting results for inner race, outer race, or ball fault diagnosis. Furthermore, these studies have focused on either multiple spalls of the same fault type or severity of damage and none have shown the ability to diagnose all four of the bearing fault types. Additionally, although it was shown that TSA can result in an improvement upon the envelope analysis technique [15], it still requires the need for multiple averages for each of the investigated bearing fault types. A number of difficulties need to be overcome to successfully apply TSA to quantify AE signals for bearing fault diagnosis. For example, the tachometer signal must be synchronous with all bearing fault types. Additionally, the large data volume and non-stationary behavior of the AE signals make the direct computation of TSA and real time condition monitoring unrealistic. In this paper, time synchronous resampling and spectral averaging is used to overcome these challenges and results in the extraction of features that are used for bearing fault diagnostics.

In addition to the time domain TSA approach, frequency domain TSA algorithms have been presented in [22]. One challenge in a time domain TSA approach for bearings is the requirement of four TSAs to account for each bearing fault type. Additionally, bearings could experience fluctuations in speed during industrial operation, and due to slippage, fault rates based on bearing geometries could not be exact. It has also been shown that by using a synchronous resampling technique based on shaft revolutions, potential fluctuations in shaft speed can be accounted for, and the effect of spectral smearing reduced [23]. Thus, by resampling to an even number of points between trigger signal revolutions, a better FFT result was obtained. In this paper, a similar time synchronous resampling approach is accomplished using the shaft ZCTs and is expressed in the following. Formally, the resampling process is achieved by interpolating the r number of data points in one shaft revolution, into L number of data points, such that:

$$L = 2^{Ceiling[log2(r)]} \tag{12}$$

where, L is the number of interpolated points between ZCTs, and r the average number of points between shaft crossings before resampling. Once L is determined, each segment contains an equal number of data points for FFT computations and the spectral average can be implemented.

The methodology presented in this paper combines the aforementioned TSR approach with spectral averaging to compute a single average which allows the extraction of effective condition indicators for bearing fault diagnosis. Thus, the computation of a single spectral average eliminates the necessity to compute multiple TSAs. Presented in Figure 5, is an overview of the spectral averaging approach. The computation of the spectral average requires the data to be sectioned so that the Fourier transform of each section can be computed and the ensemble average of the squared magnitude of the segments executed.

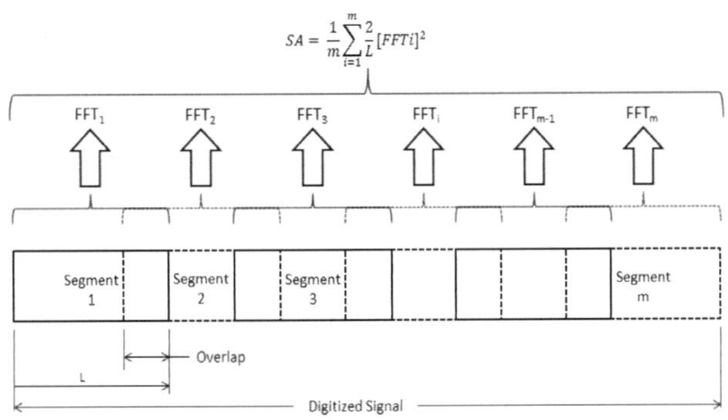

Figure 5: The Spectral Averaging Approach

Welch [24] formally expressed the implementation of this approach. By letting $Z(j), for\ j = 0,\dots, N-1$ be a sample from a stationary, stochastic sequence whose mean is zero and letting $Z(j)$ have spectral density $P(f), |f| \leq \frac{1}{2}$, where f is the normalized frequency. Then, if one takes several possibly overlapping segments of length L with starting points of the segments M points apart and letting $Z_1(j), j = 0,\dots, L-1$ be the first segment. Then,

$$Z_1(j) = Z(j), for\ j = 0,\dots, L-1. \tag{13}$$

Likewise,

$$Z_2(j) = Z(j+M), for\ j = 0,\dots, L-1. \tag{14}$$

And finally,

$$Z_K(j) = Z(j+(K-1)M), for\ j = 0,\dots, L-1. \tag{15}$$

The result is K segments, $Z_1(j)$, . . . , $Z_K(j)$, that cover the entire data sample $Z(j)$. Moreover, $(K-1)M + L = N$.

Then, for each segment of length L we obtain the Fourier transform. In other words, we choose a data window $W(j),\ for\ j = 0,\dots, L-1$, and take the Fourier transforms $F_1(n)$, . . . , $F_K(n)$, of the segments $Z_1(j)W(j),\dots, Z_K(j)W(j)$. This is formally expressed in Eq. (16):

$$F_k(n) = \frac{1}{L}\sum_{j=0}^{L-1} Z_k(j)W(j)e^{-2kijn/L} \tag{16}$$

where, $i = (-1)^{\frac{1}{2}}$. The result is K Fourier transforms $B_k(f_n)$, which correspond to the K number of segments:

$$B_k(f_n) = \frac{L}{S}|F_k(n)|^2, for\ k = 1, 2,\dots, K. \tag{17}$$

Where,

$$f_n = \frac{n}{L}, for\ n = 0,\dots, L/2. \tag{18}$$

And,

$$S = \frac{1}{L}\sum_{j=0}^{L-1} W^2(j). \tag{19}$$

Finally, the spectral average result, $SA(f_n)$, is obtained by taking the average of the K Fourier transforms:

$$SA(f_n) = \frac{1}{K}\sum_{k=1}^{K} B_k(f_n). \tag{20}$$

To implement the spectral average, the signal must first be sectioned. By using the shaft ZCTs to section and resample the AE signal, and then averaging the magnitude squared spectrums of the sections, the spectral average result is obtained. In other words, the duration between shaft revolutions is utilized as the sectioning function to average the AE signal. Moreover, the data segment length L, is determined by the number of data points between ZCTs of the shaft after resampling. By resampling the segments to L data points, an equal number of data points is used for FFT computations. Hence, this methodology uses the ZCTs of shaft rotations to section and resample the AE signals and compute the ensemble average of the squared magnitude Fourier transforms of the sections. This approach affords the ability to compute one spectral average and effectively diagnose any bearing fault that may occur, while eliminating the necessity to compute multiple TSAs. Moreover, the need to compute a single average also reduces the computational burden on the hardware, making on-line analysis and maintenance decision-making feasible. After the computation of the spectral averaging of the AE signals, various bearing fault features are extracted and evaluated.

2.4 Condition Indicators for Bearing Fault Diagnosis: There are numerous bearing fault condition indicators in the literature that quantify accelerometer signals to aid in bearing fault diagnosis. In addition, studies have developed effective CIs that accomplish bearing fault diagnosis through the quantification of AE signals [7,8]. A significant difference among the available CI feature extraction techniques is in the way the CIs are computed, i.e. features can be extracted from both the time and frequency domain. Additionally, features can be extracted from the raw signal or after a signal processing technique such as TSA or spectral averaging. This study uses the inverse Fourier transform of the spectral average result to extract effective CIs that allow the clear diagnosis of inner race, outer race, ball, and cage faults. The time domain signal used for CI computation is formally obtained by Eq. (21):

$$x = ifft[SA(f_n)] \tag{21}$$

Where x is the data set used for CI computations and $ifft[SA(f_n)]$ is the inverse Fourier transform of the spectral average result $SA(f_n)$ obtained by Eq. (20).

In this paper, several condition indicators were investigated to use for bearing fault diagnosis. The first is the root mean square (RMS) which is a statistical measure of the magnitude of a varying quantity. The RMS of a discretized signal is defined by the following:

$$x_{rms} = \sqrt{(1/N)\sum_{i=1}^{N} x_i^2} \tag{22}$$

where x_{rms} is the root mean square value of dataset x, x_i is the ith element of x, and N is the length of the dataset x.

The second is peak value which generally refers to the maximum value in the collected data. The peak value is calculated by the following:

$$x_{peak} = \frac{(x_{max} - x_{min})}{2} \tag{23}$$

where x_{peak} is the peak value of data set x, and x_{max} and x_{min} are the maximum and minimum values in the data set x.

The third is crest factor, which requires the aforementioned RMS and peak value. The crest factor can be obtained by the following:

$$CF = \frac{x_{peak}}{x_{rms}} \tag{24}$$

where CF is the crest factor, x_{peak} is the peak amplitude in the data x, and x_{rms} is the RMS of data x. For an incipient fault, the RMS value may not increase. However, the peak amplitude should increase which will result in an increase in crest factor.

Next, is kurtosis which describes the peakedness or smoothness of the dataset x. If the faulty bearing signal contains sharp peaks that are a result of high values generated from the fault, the distribution function should be sharper. Thus, the kurtosis of a faulty bearing signal should be higher than that of a healthy bearing signal. Kurtosis can be computed by the following:

$$Kurt = \frac{N\sum_{i=1}^{N}(x_i - \bar{x})^4}{[\sum_{i=1}^{N}(x_i - \bar{x})^2]^2} \tag{25}$$

where $Kurt$ is the kurtosis of dataset x, x_i is the ith element of x, $\bar{x}$ is the mean of dataset x, and N is the length of the dataset x.

Moreover, skewness is a measure of how asymmetric the data is around the sample mean. A negative value for skewness implies the data is spread out more to the left of the mean. Similarly, a positive value indicates the data is spread out more to the right of the mean. It is important to note that the skewness of the normal distribution or any perfectly symmetric distribution is zero. The skewness of the data set x is defined as:

$$Skew = \frac{(\frac{1}{N})\sum_{i=1}^{N}(x_i - \bar{x})^3}{[\sqrt{(\frac{1}{N})\sum_{i=1}^{N}(x_i - \bar{x})^2}]^3} \tag{26}$$

where $Skew$ is the skewness of data set x, x_i is the ith element of x, $\bar{x}$ is the mean of data set x, and N is the length of the data set x.

Additionally, the above CIs were evaluated using the energy operator (EO) of the time domain representation of spectral average result. The EO is a type of residual of the autocorrelation function [25]. In discretized form, the mathematical formula is given as:

$$EO[x_i] = x_i^2 - x_{i-1} \cdot x_{i+1} \tag{27}$$

Where $EO[x_i]$ is the ith element in the EO, and x_i is the ith element of dataset x.

3. Experimental Setup: This section covers the experimental setup used to validate the AE sensor based bearing fault diagnostic technique. Figure 6 shows the bearing test rig used to collect the AE data and conduct the steel bearing seeded fault test. A wide band (WD) type AE sensor was mounted on the face of the bearing housing using instant glue.

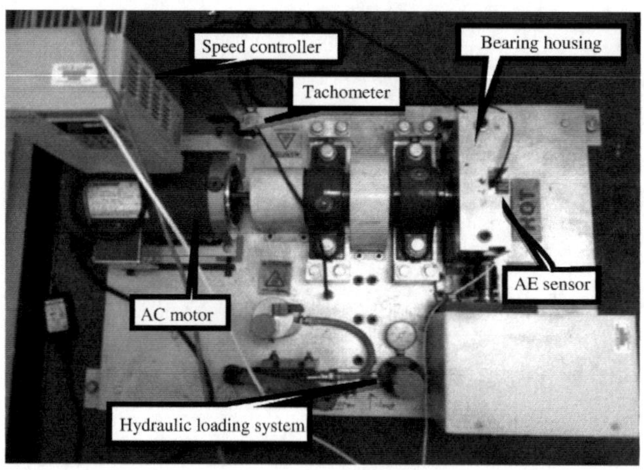

Figure 6: The Bearing Test Rig

Type 6205-2RS steel ball bearings were used for testing. Four fault types were simulated on steel bearings: inner and outer race faults, rolling element fault, and cage fault (see Figure 7). The inner and outer race faults were generated by scratching the steel race surfaces with a diamond tip grinding wheel bit to cover the ball contact surface. The ball fault damage was created by cutting the steel cage in one of the ball locations and then using the diamond tip grinding wheel bit to create a small dent in one of the steel balls. For the cage fault, the steel cage was cut in between two ball locations. For all seeded fault tests, the bearing seal and grease was removed and replaced following the creation of the fault.

Figure 7: The Steel Bearing Seeded Fault Tests

Figure 8 presents the demodulation board and power supply, along with the function generator and sampling device. The demodulation device performs the multiplication of the AE sensor signal with the reference signal output from the function generator which allows the aforementioned sampling frequency reduction technique to be implemented. It takes the two signals as inputs and the output is the multiplication of the signal inputs. After shifting the signal information to a lower frequency range, the output is fed to the sampling board while filtering out the high frequency component. To down shift the AE signal frequency, its carrier frequency had to be determined in order to introduce a reference signal for demodulation. Thus, the goal was to determine the central AE carrier frequency and set the reference signal frequency as close to it as possible.

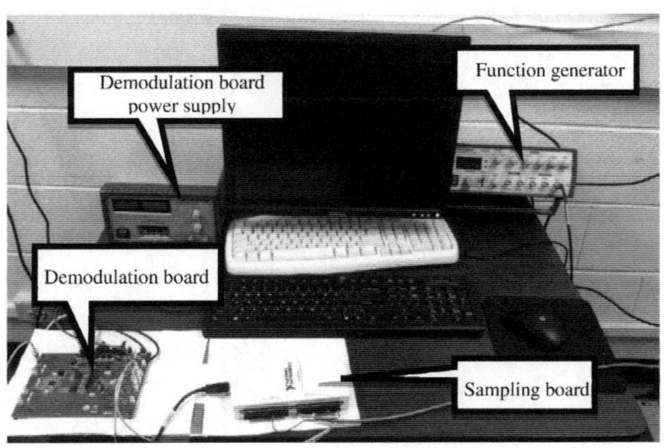

Figure 8: Demodulation and Sampling Devices

A sweep function, created by the function generator, was used to record the output of the system. After examining the energy envelope at different frequencies, an estimate of the frequency range of the output was found. It was determined that the central AE signal carrier frequency was 400 kHz and was used as the demodulation reference frequency.

NI Labview signal express was used for data acquisition, collecting the continuous AE signals at a sampling rate of 100 kHz. Both the healthy bearing and the seeded fault bearings were tested at three shaft speeds: 30 Hz, 45 Hz, and 60 Hz. At each speed, 5 samples were collected for a total of 15 samples of data per bearing type. It is important to note that for this study, only the aforementioned shaft speeds were tested while leaving low speed validation for future investigation. Additionally, no load was applied for the duration of the seeded fault tests. For consistency, the AE sensor was placed in the same axial location for all data acquisitions.

4. Results: This section presents the validation results for the seeded fault tests conducted on the bearing test rig. After heterodyne, the AE data was digitized at a sampling rate of 100 kHz. The AE signal was acquired simultaneously with the tachometer signal which was a record of the shaft rotation. Provided in Figure 9 is the tachometer and raw AE signal from the seeded inner race fault bearing recorded at a shaft speed of 30 Hz.

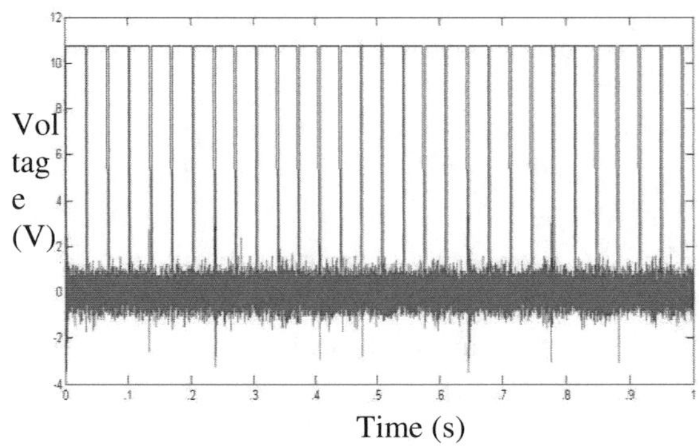

Figure 9: Inner Race Fault Raw AE and Tachometer Signal at 30 Hz Shaft Speed

Before the spectral average of the AE signal could be computed, the bearing signals were band pass filtered. Upon examining the frequency spectrums, it was observed that the fundamental fault frequencies could not be identified. Thus, the calculated fault frequencies could not be used as the basis for band pass filter selection. However, to choose a proper filter band, Shannon's entropy was computed on the resulting healthy bearing signals after the implementation of different filter bands. Thus, following this technique, a 5 kHz to 20 kHz filter band was used.

After selecting the filter band, the AE signal was band pass filtered, synchronously resampled, and spectrally averaged using the signal information contained between ZCTs of the shaft. Then, after the time domain representation of the spectral average was acquired, numerous CIs were computed and evaluated. Although a total of 10 CIs were tested, two were found to be effective: RMS and peak. Additionally, for comparison purposes and to exemplify the effectiveness of the selected filter band, results are presented for signals with and without the implementation of a band pass filter before additional processing. Figure 10 presents the RMS CI results for all collected data samples without the use of a band pass filter and Figure 11 presents the results with the implementation of the selected filter band. For both figures, samples 1-5 correspond to a shaft speed of 30 Hz, while samples 6-10 correspond to a shaft speed of 45 Hz, and samples 11-15 correspond to a shaft speed of 60 Hz.

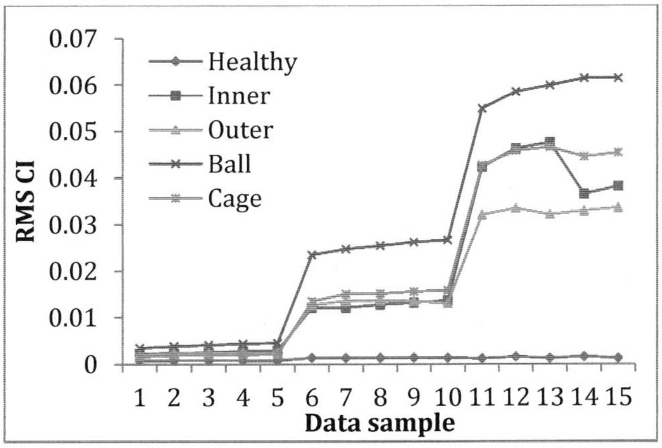

Figure 10: RMS by Sample Number without Filter

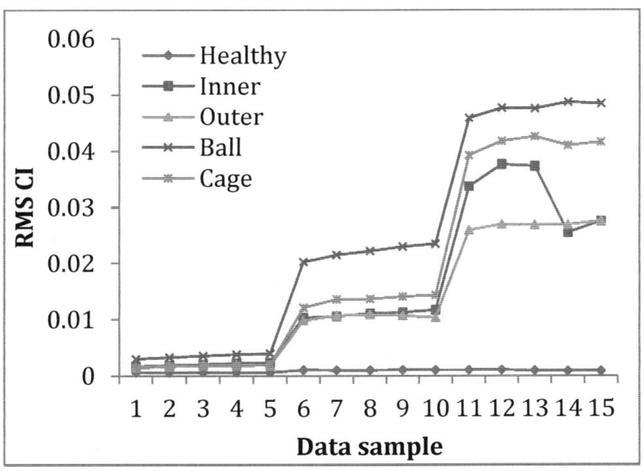

Figure 11: RMS by Sample Number with Filter

It should be noted that for the faulty bearing signals, the RMS CI value increases with shaft speed, regardless of fault type and whether or not a filter was utilized. Additionally, there is clear differentiation of any of the faulty bearing signals when compared to the healthy case for all acquisitions. However, it should be noted that with the use of a filter, greater separation is observed between the cage and inner race results for samples 11 – 13 at the 60 Hz shaft speed. Conversely, a decrease in separation was observed at samples 14 and 15. Moreover, at the 45 Hz shaft speed, the filtered result provided better separation of the cage fault signal from inner and outer race fault signals. The least amount of separation was observed at the 30 Hz shaft speed regardless of whether or not the filter was implemented. To determine the diagnostic ability of the methodology, the RMS CI values for all signals were averaged and plotted versus shaft speed. These results are presented in Figure 12 and Figure 13.

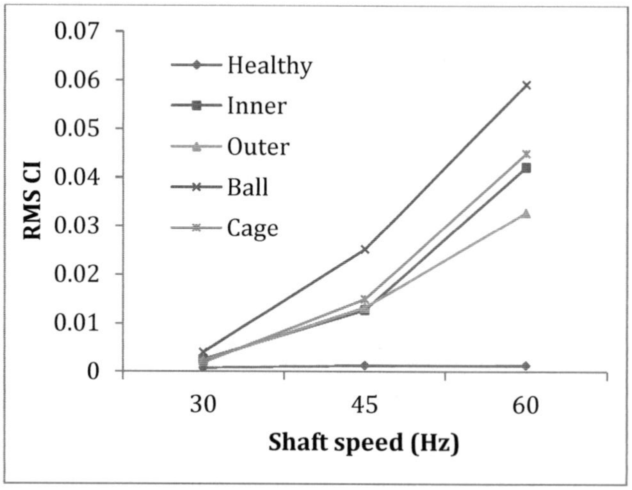

Figure 12: Average RMS by Shaft Speed without Filter

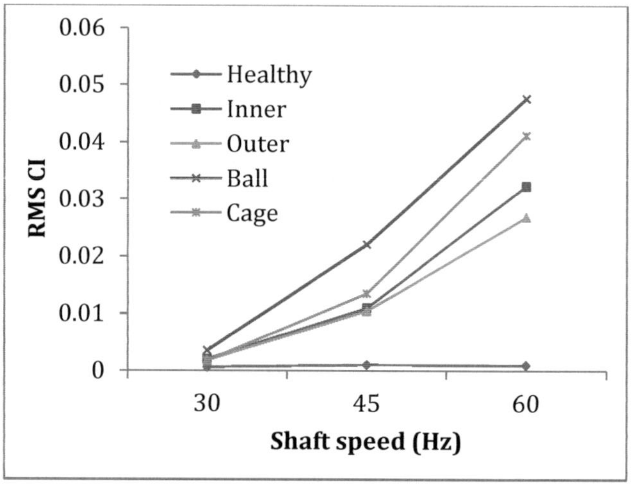

Figure 13: Average RMS by Shaft Speed with Filter

The results in Figures 12 and 13 indicate that RMS is an effective CI at the tested shaft speeds. All fault types can be differentiated from each other and the healthy bearing case. The averaged results further exemplify the previously mentioned observations. With the implementation of a band pass filter before spectral averaging, a greater separation of the results is observed at the 60 Hz shaft speed. Additionally, without filtering the inner and outer race average switches as the 45 Hz shaft speed which was not problematic for the case when a band pass filter was utilized. Similar results were observed for the peak CI and are shown in Figure 14 and Figure 15.

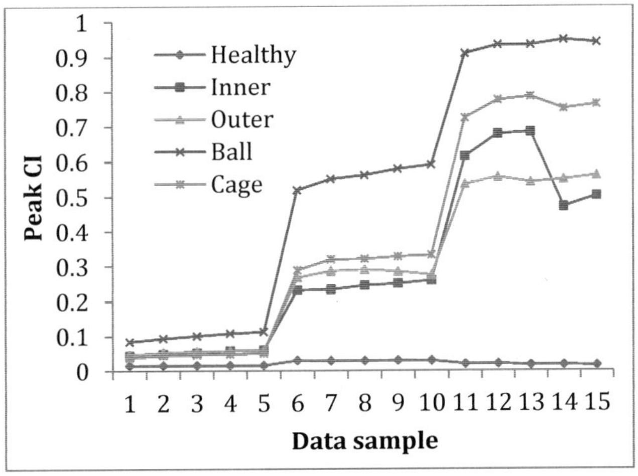

Figure 14: Peak by Sample Number without Filter

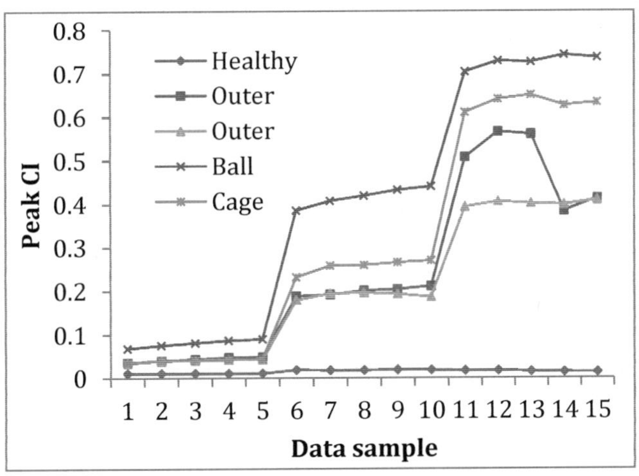

Figure 15: Peak by Sample Number with Filter

As shown above, the peak CI also effectively differentiates the faulty bearing signals from the healthy case. Although the results appear similar, the average of the peak value versus shaft speed provides a better view for comparison. The peak CI average results are presented in Figures 16 and 17.

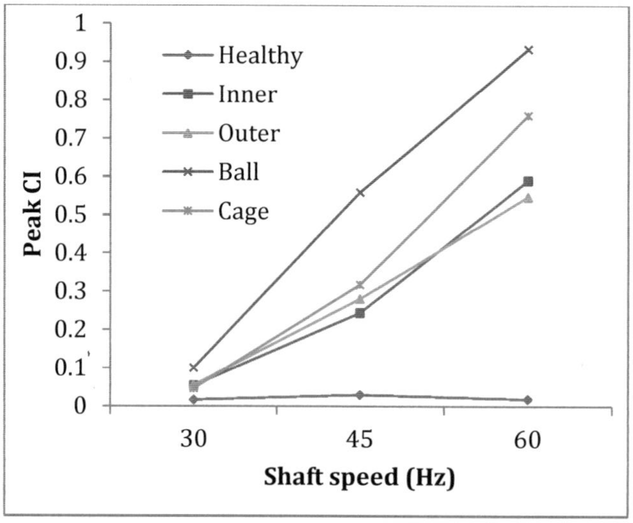

Figure 16: Average Peak by Shaft Speed without Filter

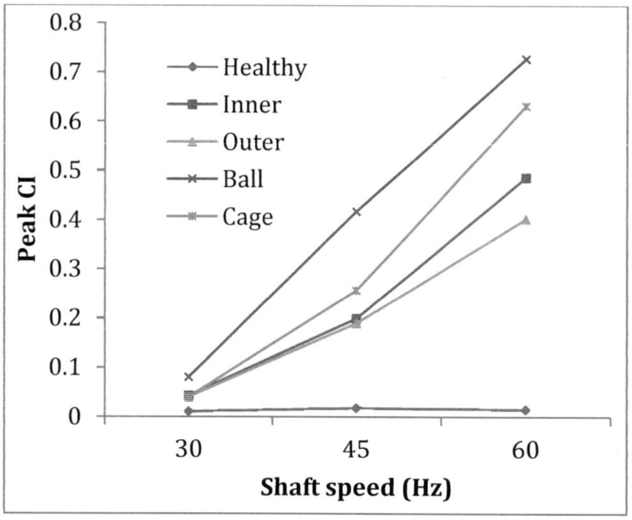

Figure 17: Average Peak by Shaft Speed with Filter

As with the average RMS CI results, the average peak CI also is effective in differentiating the faulty bearing cases from each other and the healthy case. Additionally, it should be noted that the separation of cage and inner race faults when using the peak CI is greater than when using RMS, suggesting that peak value is a more effective CI for bearing fault diagnosis using this methodology. Also, the previously mentioned switch of inner and outer race results at the 45 Hz shaft speed is distinct for the results obtained without filtering. As with RMS, when the filtered results are utilized, this phenomenon did not occur. Moreover, as with the previous case, the faulty results are clearly separable from the healthy case, with ball fault results indicating the largest level of separation. However, it should be noted that although the inner race, outer race and cage faulty results are differentiated from the healthy bearing case, the separation from each other cannot be seen as distinctly as with the 45 and 60 Hz shaft speeds. In summary, the RMS and peak features extracted from the spectral average results effectively differentiate all four fault types from each other and from the healthy bearing case, which has not been presented in the literature. Thus, the use of the time synchronous resampling based spectral averaging approach presented in this paper exemplifies the effectiveness of implementing an AE based approach. Additionally, the presented approach requires the computation of a single average, eliminating the need for multiple averages for each bearing component. Hence, the results in this paper indicate that the presented technique is an effective and efficient approach to AE based bearing fault diagnostics.

5. Conclusions: In this paper, a new AE based bearing fault diagnostic methodology was presented. This was accomplished through a combination of a heterodyne based frequency reduction technique, time synchronous resampling, and spectral averaging to extract features for bearing fault diagnosis. The heterodyne technique allows the AE signal to be digitized at a rate comparable to vibration based methods. Moreover, a band pass filter was implemented to retain the signal information related to the bearing defects. Then, after generating the filtered signals, time synchronous resampling and the spectral average were computed to allow the evaluation of numerous CIs. Current vibration based implementations of synchronous averaging for bearing fault diagnosis has lacked the ability to diagnose all four bearing fault types. Additionally, these implementations require the computation of multiple TSAs. Using the RMS and peak CIs, the results of this paper indicate that using an AE based approach allows such diagnosis to be accomplished. Thus, by implementing a time synchronous resampling based spectral averaging approach, bearing fault diagnosis is accomplished via the execution of a single spectral average.

References:

[1] Applied Industrial Technologies, 2009, "Lubricant failure = bearing failure", URL http://www.machinerylubrication.com/Magazine/Issue/Machinery%20Lubrication/1/2009.

[2] Morhain, A. and Mba, D., 2003, "Bearing defect diagnosis and acoustic emission", *Proceedings of the Institution of Mechanical Engineers, Part J: Journal of Engineering Tribology*, Vol. 217, No. 4, pp. 257 – 272.

[3] Mba, D., 2008, "The use of acoustic emission for estimation of bearing defect size", *Journal of Failure Analysis and Prevention*, Vol. 8, No. 2, pp. 188–192.

[4] Nienhaus, K., Boos, F.D., Garate, K., and Baltes, R., 2012, "Development of acoustic emission (AE) based defect parameters for slow rotating roller bearings", *Journal of Physics: Conference Series*, Vol. 364, No. 1, June 18–20, Huddersfield, UK.

[5] Yoshioka, T. and Fujiwara, T., 1982, "A new acoustic emission source locating system for the study of rolling contact fatigue", *Wear*, Vol. 81, No. 1, pp. 183 – 186.

[6] Mba, D. and Rao, R., 2006, "Development of acoustic emission technology for condition monitoring and diagnosis of rotating machines; bearings, pumps, gearboxes, engines and rotating structures", *The Shock and Vibration Digest*, Vol. 38, No. 1, pp. 3 – 16.

[7] He, D., Li, R., Zhu, J., and Zade, M., 2011a, "Data mining based full ceramic bearing fault diagnostic system using AE sensors", *IEEE Transactions on Neural Networks*, Vol. 22, No. 12, pp. 2022 – 2031.

[8] He, D., Li, R., Zade, M., and Zhu, J., 2011b, "Development and evaluation of ae based condition indicators for full ceramic bearing fault diagnosis", *IEEE International Conference on Prognostics and Health Management*, June 20-23, Denver, CO.

[9] Yoon, J. M., He, D., and Qiu, B., 2013, "Full ceramic bearing fault diagnosis using lamstar neural network", *IEEE PHM Conference Proceedings*, June 24-27, Gaithersburg, Maryland.

[10] Shiroishi, J., Li, Y., Liang, S., Kurfess, T., and Danyluk, S., 1997, "Bearing condition diagnosis via vibration and acoustic emission measurements", *Mechanical Systems and Signal Processing*, Vol. 11, No. 5, pp. 693 – 705.

[11] Felten, D., 2003, "Understanding bearing vibration frequencies", *Mechanical Field Service Department, L&S Electric, Inc., Schofield, Wisconsin*, pp. 1 – 3.

[12] McFadden, P. D., 1987, "A revised model for the extraction of periodic waveforms by time domain averaging", *Mechanical Systems and Signal Processing*, Vol. 1, No. 1, pp. 83 – 95.

[13] McFadden, P.D., 1991, "A technique for calculating the time domain averages of the vibration of the individual planet gears and the sun gear in an epicyclic gearbox", *Journal of Sound and Vibration*, Vol. 144, No. 1, pp. 163 – 172.

[14] McFadden, P. D. and Toozhy, M. M., 2000, "Application of synchronous averaging to vibration monitoring of rolling element bearings", *Mechanical Systems and Signal Processing*, Vol. 14, No. 6, pp. 891 – 906.

[15] Siegel, D., Al-Atat, H., Shauche, V., Liao, L., Snyder, J., and Lee, J., 2012, "Novel method for rolling element bearing health assessment – A tachometer-less synchronously averaged envelope feature extraction technique", *Mechanical Systems and Signal Processing*, Vol. 29, No. 1, pp. 362 – 376.

[16] Braun, S., 1975, "The extraction of periodic waveforms by time domain averaging", *Acustica*, Vol. 32, pp. 69 -77.

[17] Bonnardot, F., El Badaoui, M., Randall, R.B., Daniere, J, and Guillet, F., "Use of the acceleration signal of a gearbox in order to perform angular resampling (with limited speed fluctuation)", *Mechanical Systems and Signal Processing*, Vol. 19, No. 4, pp. 766 – 785.

[18] Bechhoefer, E., 2013, "An enhanced time synchronous averaging for rotating equipment analysis", *Proceedings for the joint conference: Machinery Failure Prevention Technology 2013 and International Instrumentation Symposium 2013*, May 13 – May 17, Cleveland, OH.

[19] Qu, Y., Bechhoefer, E., He, D., Zhu, J., 2013, "A new acoustic emission sensor based gear fault detection approach", *International Journal of Prognostics and Health Management*, Vol. 4, Sp. 2, pp. 1 – 14.

[20] Qu, Y., Zhu, J., He, D., Qiu, B., and Bechhoefer, E., 2013, "Time synchronous average based acoustic emission signal analysis on gear fault detection", *IEEE International Conference on Prognostics and Health Management*, June 24 – 27, Gaithersburg, MD.

[21] Christian, K. N., Mureithi, N., Lakis, A., and Thomas, M., 2007, "On the use of synchronous averaging, independent component analysis and support vector machines for bearing fault diagnosis", *First International Conference on Industrial Risk Engineering*, Dec. 17–19 Montreal, QC, Canada.

[22] Bechhoefer, E. and Kingsley, M., 2009, "A review of time synchronous average algorithms", *Annual Conference of the Prognostics and Health Management Society,* Sept 27 – Oct 1, San Diego, CA.

[23] Bechhoefer, E., Van Hecke, B., and He, D., 2013, "Processing for improved spectral analysis", *Annual Conference of the Prognostics and Health Management Society,* Oct 14 – 17, New Orleans, LA.

[24] Welch, P., 1967, "The use of fast fourier transform for the estimation of power spectra: a method based on time averaging over short, modified periodograms" *IEEE Transactions on Audio and Electroacoustics*, Vol. 15, No. 2, pp. 70 – 73.

[25] Teager, H. M., and Teager S. M., 1992, "Evidence for nonlinear sound production mechanisms in the vocal tract", *Speech Production and Speech Modeling Symposium, Time Frequency and Time-Scale Analysis*, Victoria, British Columbia, Canada, pp. 345 – 348.

STATISTICAL ANALYSIS OF THE VIBRATION OF MISTUNED CENTRIFUGAL IMPELLERS WITH CRACK DAMAGE

Shuai Wang, Yanyang Zi, Minquan Zhang and Zhengjia He
State Key Laboratory for Manufacturing and Systems Engineering
Xi'an Jiaotong University
Xi'an 710049, PR China
Telephone: +86 29 82663689
wangsh_106551@yeah.net

Abstract: Statistical analysis of the effects of random mistuning on the nonlinear vibration of a cracked impeller of centrifugal compressor is presented and used to calculate the statistical characteristics of the forced response and the distribution of frequencies. Based on a finite element model, a hybrid interface method of component mode synthesis (CMS) is employed to generate a reduced-order model (ROM) of the cracked impeller. The degrees of freedoms on the crack surfaces are retained in the ROM so that the crack breathing effects can be simulated. Random mistuning is modeled by introducing small variations in Young's modulus of each blade. Then, the resulting nonlinear equations are solved by applying multi-harmonic balance method. Probability distributions of the forced response and frequencies of the mistuned impeller are obtained by a numerical Monte Carlo simulation and are used to calculate means and standard deviations. The results show that as the crack length grows, resonant frequencies tend to decrease, which is a potential feature for crack detection.

Key words: Statistical analysis; centrifugal impeller; crack damage; component mode synthesis.

Introduction: Centrifugal compressors are important equipment in industrial production and are widely used in the fields of petrochemical, steel and electric power. The reliability of centrifugal compressors for the most part depends on the reliable operation of impellers [1, 2]. However, due to the high cycle fatigue damages caused by mistuning factors, such as manufacturing tolerances and property deterioration of components during service, an impeller tends to initiate a fatigue crack at the weak position. For a rotor of centrifugal compressor operating under high speed, a crack accident of blades may cause huge economic losses, or even threaten the life safety of field personnel. Thus, an efficient method for detection of crack of centrifugal impellers is urgently needed in the real industrial fields. However, a reliably technique for detection of cracks of impellers in operation remains a challenge at present. Then, investigations on the nonlinear vibration of cracked impellers will help to increase some in-depth understandings on the effects of cracks and will help to develop an industrial technique for crack detection.

Most of the previous research on the vibration of turbomachinery was focused on the bladed disks of axial compressors [3-7], and few investigations were reported to deal with the vibration analysis of centrifugal impellers. Methods for analysis of bladed disks can be divided into two categories. In the earlier stage, the models [4, 5] which simplified the blades as a mass-spring system or Euler-Bernoulli beams and treated the crack as local stiffness loss, were employed to analyze the vibration response and localization of cracked bladed disks. Then, models [6, 7] that were based on the finite element models of real structures and CMS methods were used to predict the forced response of mistuned bladed disks with crack damages. These methods can provide quantitatively predictions for structures and can describe the effects of crack breathing.

An impeller is fabricated via welding different components, including blades, cover and disk. The existence of cover makes the impeller inherently resistant to alternating forces and greatly increases the reliability. However, this feature makes the impeller insensitive to crack damages, and a crack may not lead to significant change in the vibration response unless the damage is severe. Then, a comprehensive study should be conducted to gain some insights into the effects of a crack on impellers.

The effects of mistuning on the nonlinear vibration of cracked impellers are not negligible. However, the mistuning pattern of a real impeller is almost impossible to be measured from the perspective of application. Then, the mistuning effects are taken into account by introducing a large number of random mistuning patterns, and a probabilistic technique is employed to investigate the statistical features of the response parameters of interest. Several papers [8, 9] were focused on the statistics of the maximum resonant response amplitude of mistuned bladed disks over a frequency range. In this paper, the statistical features of resonant frequencies of mistuned impellers with cracks are of interest.

In this paper, a statistical analysis of the effects of random mistuning on the nonlinear vibration of a cracked impeller of centrifugal compressor is presented. Firstly, based on a finite element model of an impeller, a hybrid interface method of CMS method is employed to generate a compact ROM. The degrees of freedoms (DOF) on the crack surfaces are retained in the ROM to simulate the nonlinear crack breathing effects. Mistuning is implemented by introducing small variation in the Young's modulus of each blade. Then, statistical analysis is performed using the Monte Carlo simulation to study the statistical characteristics of the maximum response amplitudes and the resonant frequencies. The bilinear frequency approximation method [6, 13, 14] was applied to calculate the resonant frequencies for cracked structures. The effects of crack and mistuning levels are considered. The results show that as the crack length grows, resonant frequencies tend to decrease, which is a potential feature for crack detection.

2 Reduced-order modeling: An impeller can be divided into one cover component, one disk component and several blades. The finite element model of an impeller is depicted in Figure 1. As can be seen in the figure, the fatigue crack studied in this paper is initiated at the weld toe on cover sides of the blade. An impeller contains numerous interfaces between the cover, blade and disk components, and there are a large number of DOFs on the interfaces. Then, a hybrid interface CMS method [10] is employed to generate a compact ROM while assuring high accuracy.

Based on the finite element model of an impeller, the governing equations of motion can be written as

$$\mathbf{M}_s\ddot{\mathbf{u}}_s(t)+\mathbf{C}_s\dot{\mathbf{u}}_s(t)+\mathbf{K}_s\mathbf{u}_s(t)=\mathbf{b}_s(t)+\mathbf{f}_s(\mathbf{u}_s) \quad (1)$$

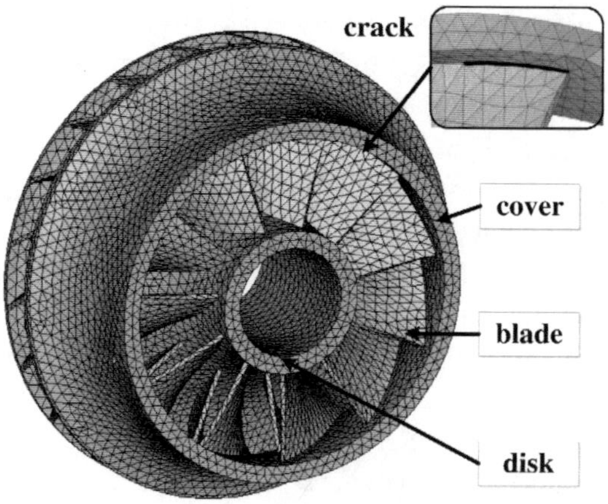

Figure 1: Finite element model of an impeller

where $\mathbf{M}_s$, $\mathbf{C}_s$ and $\mathbf{K}_s$ are the mass, damping and stiffness matrices of each component; $\mathbf{u}_s$ is the vector of nodal displacements; $\mathbf{b}_s$ is the vector of external forces acted on the blade component; and $\mathbf{f}_s(\mathbf{u}_s)$ the vector of interface reactions including the nonlinear forces caused by intermittent contact at the crack surfaces. The external force is assumed to be a traveling wave excitation in this paper, and the force on *n*th blade is represented as

$$\mathbf{b}_n(t) = \mathbf{B}\cos(\omega t - \varphi_n), \quad n = 1, 2, \cdots, N \tag{2}$$

where n is the number of blade; $\mathbf{B}$ is the forcing amplitude vector; ω is the angular frequency of excitation; and φ_n is the phase angle, defined as $\varphi_n = (n-1)2\pi C/N$, where C is the engine order of excitation.

In the hybrid CMS method, the Ritz vectors comprise a truncated set of normal modes Φ_s^k, attach modes Ψ_s^a and constraint modes Ψ_s^c. The displacement vector $\mathbf{u}_s$ can be projected onto a set of generalized coordinates, and the projection is expressed as

$$\mathbf{u}_s = \begin{Bmatrix} \mathbf{u}_{si} \\ \mathbf{u}_{sb} \end{Bmatrix} = \begin{bmatrix} \Phi_s^k & \Psi_s^a & \Psi_s^c \\ \mathbf{0} & \mathbf{0} & \mathbf{I} \end{bmatrix} = \Phi_s^H \mathbf{p}_s \tag{3}$$

where $\mathbf{u}_{si}$ and $\mathbf{u}_{sb}$ are the displacements vectors of internal DOFs and interface DOFs; and $\mathbf{I}$ is an identity matrix. By using this projection, the governing equations of each component can be written as

$$\mathbf{M}_s^H \ddot{\mathbf{p}}_s + \mathbf{C}_s^H \dot{\mathbf{p}}_s + \mathbf{K}_s^H \mathbf{p}_s = \mathbf{b}_s^H + \mathbf{f}_s^H(\mathbf{p}_s) \tag{4}$$

In the hybrid CMS method, components are synthesized by satisfying displacements compatibility and forces equilibrium conditions on the interface DOFs. The DOFs of the nodes on both crack surfaces are regarded as fixed interface DOFs, and the intermittent contact model employed by Saito [11] are used in this paper to simulate the crack breathing effects. Then, the nonlinear equations of motion for the impeller can be represented as

$$\mathbf{M}\ddot{\mathbf{p}} + \mathbf{C}\dot{\mathbf{p}} + \mathbf{K}\mathbf{p} = \mathbf{b} + \mathbf{f}_{nl}(\mathbf{p}) \tag{5}$$

where $\mathbf{M}$, $\mathbf{C}$ and $\mathbf{K}$ are mass, damping and stiffness matrices of the ROM; $\mathbf{p}$ denotes the retained generalized coordinates; $\mathbf{b}$ is the external excitations; and $\mathbf{f}_{nl}(\mathbf{p})$ is the nonlinear contact forces on the crack surfaces. Then, the obtained ROM is used in the following parts.

3 Statistical analysis of the vibration of mistuned impellers: In this section, statistical analysis of the vibration parameters of mistuned impellers with crack damages are presented, and the statistical features of the maximum response amplitudes and the resonant frequencies are of particular interest.

3.1 Effects of mistuning and crack: The random mistuning is taken into account by introducing small random variation in the Young's modulus of each blade. Then, the Young's modulus of the *n*th blade can be represented as

$$E_n = \left(1+\delta_n\right) E_b \tag{6}$$

where E_b is the nominal Young's modulus of a blade, and δ_n is a random variable following a uniform distribution with mean value 0 and standard deviation σ. The mistuning pattern can be easily implemented by changing the elements in the stiffness matrix of the ROM.

The finite element model used in this study is an industrial impeller shown in Figure 1 which consists of 11 blades. The material of the impeller is an alloy steel with Young's modulus $E = 214.6\,\text{GPa}$, density $\rho = 7850\ \text{kg/m}^3$ and Poisson's ratio $\nu = 0.3006$. The damping model is the Rayleigh damping, $\mathbf{C} = \beta\mathbf{K}$, where β is a damping coefficient. The modes with two nodal diameters (ND) shown in Figure 2 are of interest in this paper, which contain two equal natural frequencies and two cyclic modes for a tuned impeller.

Forced response analysis was conducted on the ROMs to study the effects of mistuning and crack. An impeller with a 20 mm crack was used as the representative model. The amplitude of the external excitation is 100 N with an engine order of 2, and the damping coefficient β is 5e-8. The nonlinear equations were solved by the HFT [12] method with the harmonic components of 0-9 harmonics, and the penalty coefficient was set to be $2\times10^9\ \text{N/m}$. These parameters are used in the subsequent analysis.

Figure 2: The 2ND mode shape of the tuned impeller without crack

Firstly, the effects of a 20 mm crack on the forced response of a tuned impeller are studied. The maximum displacement norm of the excitation node on the cracked blade in the frequency ranges of the 2ND modes is shown in Figure 3. As can be seen in the figure, the frequency response of the healthy impeller differs from that of the cracked structure, and the sole resonant peak split into two isolated peaks due to split of the two resonant frequencies. Due to the structural features of a covered impeller, strong structural coupling between the blades, cover and disk components makes it insensitive to crack damages and the maximum response amplitudes are not obviously increased compared with that of the structure without a crack.

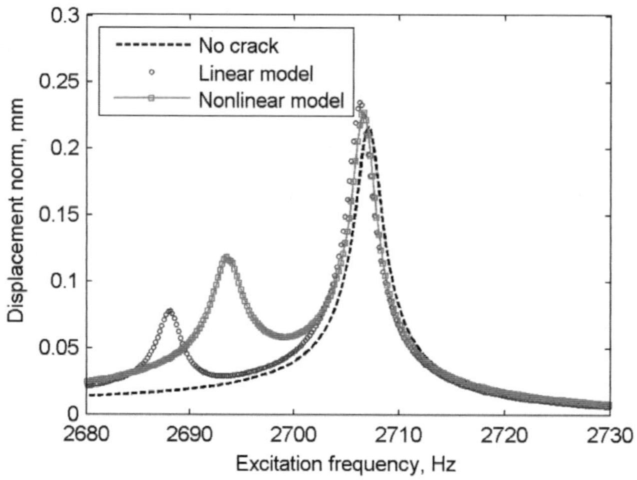

Figure 3: Frequency response of the impeller with a 20 mm crack of the 2ND mode.

Then, analysis of the effects of mistuning on the vibration of the impeller with a 20 mm crack is presented. By combining the mistuning and crack, similarities and differences of the two factors are pointed out. The mistuning pattern used for the blade component is taken from a uniform distribution of mean zero and standard deviation 2.0%. The mistuning parameters δ_n are listed in Table 1.

Table 1: Mistuning parameters of the blade component

Blade	δ_n (%)	Blade	δ_n (%)	Blade	δ_n (%)
1	-2.80	5	-2.35	9	-0.39
2	1.56	6	-0.57	10	2.32
3	1.79	7	-1.36	11	-1.80
4	2.09	8	1.70		

The frequency responses of the 2ND modes are shown in Figure 4. For the cracked impeller, two cases are considered, including a nonlinear model without mistuning except a crack and a model combining mistuning and the crack breathing effects which are referred to as the tuned model and the mistuned model. In the figure, a conclusion can be reached that mistuning changes the resonant frequencies but does not introduce additional resonant peaks. This characteristic will help to establish a quantitative relationship between resonant frequencies and the mistuning levels and the length of crack.

Due to the randomness of the mistuning for a real impeller, statistical analysis should be carried out to investigate the statistical characteristics of the forced response including the response amplitudes and the resonant frequencies of the cracked impeller which are presented in the following section.

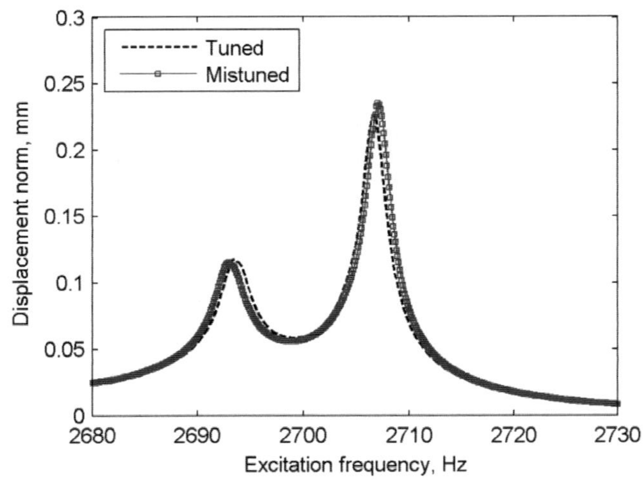

Figure 4: Frequency response of the impeller combining the effects of mistuning and a 20 mm crack in the frequency range of the 2ND modes.

3.2 Statistical characteristics of the response amplitudes: In this section, the statistical characteristics of the response amplitudes are investigated on an impeller with a 20 mm crack using the Monte Carlo simulation based on the obtained ROM.

Mistuning patterns were generated by a uniform distribution in the range $\left(-\sqrt{3}\sigma, \sqrt{3}\sigma\right)$, where σ is the standard deviation which is set to be 0.02 during the analysis, and the number of mistuning patterns considered is 20,000. Due to the huge demands for computation time when calculating the nonlinear resonant amplitudes, a linear model which did not incorporate the crack breathing effects was used to obtain the maximum response amplitudes of the mistuned cracked impeller. The amplitude magnification factor of a blade is obtained by dividing the maximum response amplitudes of the mistuned impeller by that of the tuned counterpart. The crack of the impeller is on the interfaces between the 11th blade and the cover component. The probability densities of the amplitude magnification factor of the 6th and 11th blade at the mistuning levels of 0.02 are shown in Figure 5 and Figure 6, respectively.

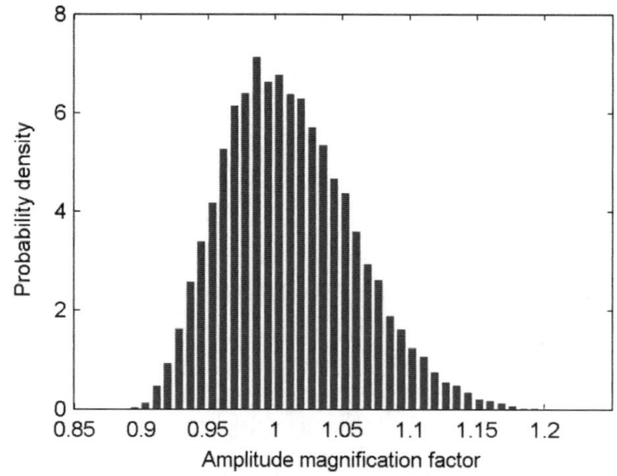

Figure 5: Probability density of the amplitude magnification factor of the 6th blade of the cracked impeller at the mistuning levels of 0.02.

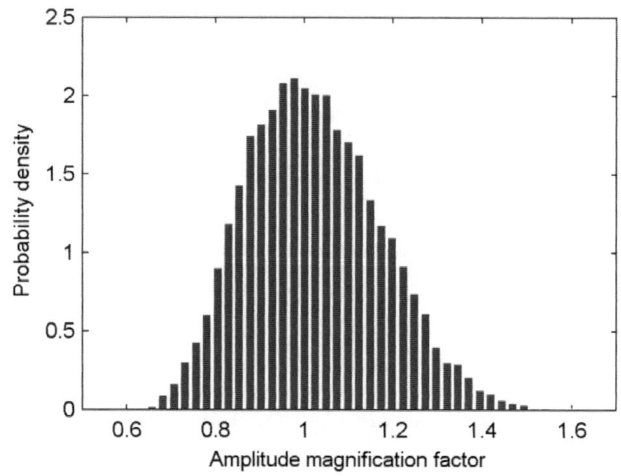

Figure 6: Probability density of the amplitude magnification factor of the 11th blade of the cracked impeller at the mistuning levels of 0.02.

As can be seen in the figures, for the 6th blade which is not the cracked one, the maximum magnification factor is less than 1.2, and that for the cracked blade can reach as large as 1.5. The lower bound of amplitude magnification factor of the cracked blade is smaller than the 6th blade. It can be concluded that mistuning can result in a much larger change to the response amplitudes of cracked blade. Then, if a cracked impeller is subjected to some mistuning patterns, the maximum response amplitude will be significantly increased, which will lead to high cycle fatigue earlier.

3.3 Statistical characteristics of the resonant frequencies: In this section, the statistical characteristics of the resonant frequencies are studied using the Monte Carlo simulation. The model used is the impeller with a 20 mm crack, and the number of mistuning patterns is 20,000.

The resonant frequency of a nonlinear system is defined as the frequency where response amplitude is the maximum within a certain frequency range. To decrease the demands for computing time, the bilinear frequency approximation is used to estimate the resonant frequencies of the mistuned impeller with a crack. An assumption is made that the cracked impeller has only two linear states which are referred to as state 1 and 2. State 1 is an open crack state, meaning that the DOF constraints of nodes on the crack surface are removed. State 2 corresponds to the closed crack state which constrains the relative perpendicular motion between the crack surfaces and the relative tangent motion is set free. The ith bilinear frequency ω_{bi} can be expressed as

$$\omega_{bi} = \frac{2\omega_{1i}\omega_{2i}}{\omega_{1i} + \omega_{2i}} \tag{7}$$

where ω_{1i} and ω_{2i} are the frequencies of the ith mode of the state 1 and 2. During the process of Monte Carlo simulation, mistuning patterns were also generated by a uniform distribution, and the standard deviation is 0.02 either. Probability density histograms of the 5th resonant frequency is shown in Figure 7, where red dashed line denotes the probability density of a normal distribution with a mean and deviation estimated by the obtained 20,000 frequency data.

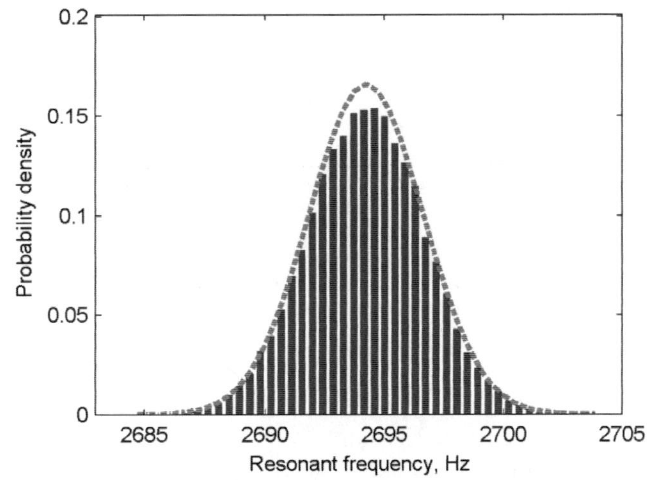

Figure 7: Probability density of the 5th resonant frequencies of the cracked impeller at the mistuning levels of 0.02.

Results in Figure 7 show that the 5th resonant frequency conforms to normal distribution. This conclusion can be applied to the other frequencies. As a common sense, the probability of a random variable following normal distribution within the range $(\mu-2\sigma, \mu+2\sigma)$ is 95.45%. The upper and lower boundaries of the resonant frequencies are derived from $\pm 2\sigma$ range. Then, analysis was conducted on the impeller with 13 different cracks ranging from 0 mm to 48 mm with an interval of 4 mm. The shifts of the 5th resonant frequency versus the length of crack at the mistuning levels of 0.01 and 0.02 are shown in Figure 8 and Figure 9, respectively.

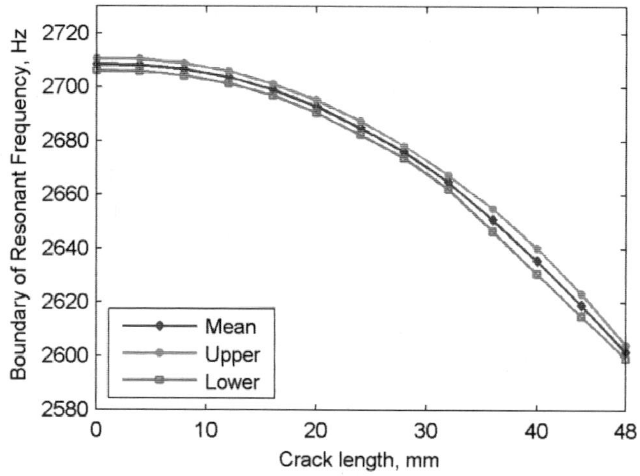

Figure 8: Boundary values of the 5th resonant frequency versus crack length at the mistuning levels of 0.01.

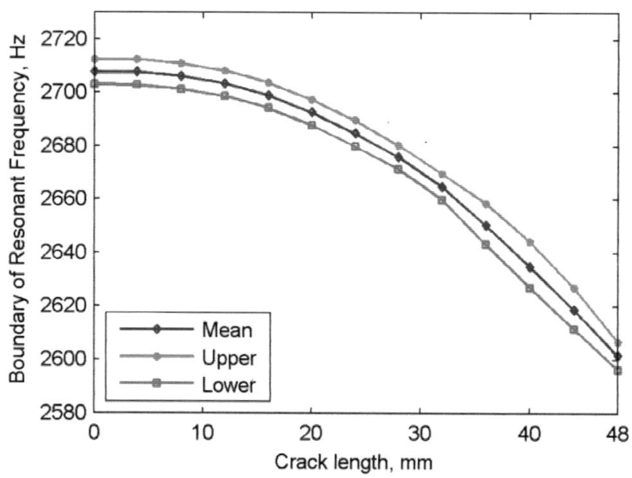

Figure 9: Boundary values of the 5th resonant frequency versus crack length at the mistuning levels of 0.02.

As can be seen in the figures, the most significant phenomenon is that the frequency range between the upper and the lower boundaries enlarges as the mistuning level increases. In addition, the boundary values shift monotonously as the crack length increase. Mistuning is a troublesome issue for crack detection. Firstly, the boundaries of resonant frequencies make it ambiguous that the diagnostic crack length becomes a range rather than a particular value. Moreover, for an impeller with large mistuning levels, the frequency range between the boundaries will be too wide to detect a crack quantitatively.

4 Detection of crack: The core issue for crack detection is to find sensitive prognostic indicators, and the resonant frequencies discussed in previous sections are potential indicators for quantitative detection of cracks. In the real situations, vibration features that reflect the inherent properties of the structure, such as resonant frequencies, will be included in the vibration signals. Though these features may be weak, the components of interest can be obtained and used to determine the length of crack by using proper signal processing methods for features extraction. A frequency based indicator which is referred to as frequency shift is discussed in this section. The ith frequency shift f_{sh}^{i} is defined as

$$f_{sh}^{i} = f_{r}^{i} - f_{0}^{i} \tag{8}$$

where f_{r}^{i} is the ith resonant frequency of a cracked impeller; f_{0}^{i} is the ith resonant frequency of the impeller without cracks. In practical applications, the mistuning pattern of an impeller is difficult to be tested. Then, the effects of mistuning are taken into account by introducing mistuning levels. A new well manufactured impeller tend to be have a small mistuning level, and the mistuning level should be relative large if the impeller is operated in a hostile environment for years. The shift of the fifth resonant frequency versus crack length at the mistuning level of 0.02 is used herein as an example to describe the application of the indicators. Quantitative relationship between the fifth resonant frequency and the crack length is shown in Figure 10.

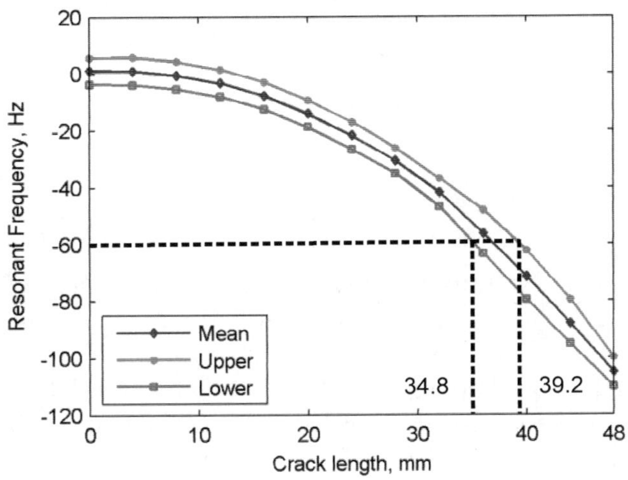

Figure 10: Detection of the length of crack via shift of the fifth resonant frequency: an illustration of crack detection when frequency is assumed to be -60 Hz.

As can be seen in the Figure, change of the frequency is so slow for a small crack that it is difficult to be detected by the frequency indicator. If a relatively large frequency shift is obtained from test data, the length of crack can be identified. For example, a frequency shift of the fifth resonant frequency is measured to be -60 Hz, then, crack length within the range (34.8, 39.2) can be identified with a 95.45% probability as illustrated in Figure 10. In the real applications, the curves of each resonant frequency are obtained beforehand, and the frequencies obtained from experimental data are used as the diagnostic indicators. The results show that the shift of resonant frequency is a valuable indicator for detection of a medium or large crack.

5 Conclusions: Statistical analysis of the vibration of mistuned centrifugal impellers with crack damages is presented in this paper and is used to investigate the statistical characteristics of the forced response and the resonant frequencies. Based on the results, some conclusions can be obtained. Firstly, both mistuning and cracks can lead to split of the two resonant frequencies with the same nodal diameters. Moreover, resonant frequencies of a cracked impeller follow normal distribution for random mistuning patterns generated by a uniform distribution. In addition, due to the boundaries caused by random mistuning, the diagnostic length of crack will be a range rather a certain value. The results in this paper will help to gain some insights into the statistical features of the vibration of cracked impellers and will provide some instructions for detection of cracks for centrifugal compressors in industrial applications.

References:

[1] F. Kushner, S.J. Richard, R.A. Strickland, Critical review of compressor impeller vibration parameters for failure prevention, in: Proceedings of the Twenty-Ninth Turbomachinery Symposium, 2000, pp. 103-112.

[2] M.P. Singh, B.K. Thakur, W.E. Sullivan, G. Donald, Resonance Identification for Impellers, in: Proceedings of the Thirty-Second Turbomachinery Symposium, 2003, pp. 59-70.

[3] M.P. Castanier, C. Pierre, Modeling and Analysis of Mistuned Bladed Disk Vibration: Current Status and Emerging Directions, Journal of Propulsion and Power, 22 (2006) 384-396.

[4] J.F. Hou, Cracking-induced mistuning in bladed disks, AIAA journal, 44 (2006) 2542-2546.

[5] B.-W. Huang, Effect of number of blades and distribution of cracks on vibration localization in a cracked pre-twisted blade system, International journal of mechanical sciences, 48 (2006) 1-10.

[6] A. Saito, M.P. Castanier, C. Pierre, O. Poudou, Efficient nonlinear vibration analysis of the forced response of rotating cracked blades, Journal of computational and nonlinear dynamics, 4 (2009).

[7] K. D'Souza, A. Saito, B.I. Epureanu, Reduced-Order Modeling for Nonlinear Analysis of Cracked Mistuned Multistage Bladed-Disk Systems, AIAA journal, 50 (2012) 304-312.

[8] S.-T. Wei, C. Pierre, Statistical analysis of the forced response of mistuned cyclic assemblies, AIAA journal, 28 (1990) 861-868.

[9] M. Bah, P. Nair, A. Bhaskar, A. Keane, Forced response statistics of mistuned bladed disks: a stochastic reduced basis approach, Journal of sound and vibration, 263 (2003) 377-397.

[10] R.H. MacNeal, A hybrid method of component mode synthesis, Computers & Structures, 1 (1971) 581-601.

[11] A. Saito, M.P. Castanier, C. Pierre, Effects of a cracked blade on mistuned turbine engine rotor vibration, Journal of vibration and acoustics, 131 (2009).

[12] O. Poudou, C. Pierre, Hybrid frequency-time domain methods for the analysis of complex structural systems with dry friction damping, in: Structural Dynamics and Materials Conference, Norfolk, VA, USA, 2003, pp. 111-124.

[13] S.W. Shaw, P. Holmes, A periodically forced piecewise linear oscillator, Journal of Sound and Vibration, 90 (1983) 129-155.

[14] S.-K. Hong, B.I. Epureanu, M.P. Castanier, D.J. Gorsich, Parametric reduced-order models for predicting the vibration response of complex structures with component damage and uncertainties, Journal of Sound and Vibration, 330 (2011) 1091-1110.

Calibration of Industrial Vibration Protection Systems

Mike Scott
Industrial Product Manager
Certified Category II Vibration Analyst

The Modal Shop
3149 E Kemper Rd
Cincinnati, OH 45241
513-351-9919
mscott@modalshop.com

MFPT Conference
Technology Solutions for Affordable Sustainment
May, 20, 2014
Virginia Beach, VA

MFPT

Presentation Overview

- Importance of Vibration Protection Systems
- Calibration Methods
  - Industrial Concerns
- Standards
- Types of Industrial Instrumentation
- Why Calibrate/Verify?
  - Shift of Accountability
  - Key Specifications
- In-Situ Calibration
- How To Calibrate
  - Accelerometers
  - Proximity Probes

TMS THE MODAL SHOP, INC. A PCB GROUP CO.
PCB PIEZOTRONICS

Why are Vibration Protection Systems Important?

- "Plant Shut Off by Quake Vibration not Power Failure"
  - Source: NBC News
  - **North Anna Nuclear Power Station** – Mineral, VA
  - Vibration causes shutdown due to 5.8 magnitude earthquake
  - First plant that "experienced an earthquake that exceeded its design parameters."
  - "First time an earthquake caused an automatic shutdown."
  - "The root cause team determined that this occurred as result of vibration in the reactor or the monitoring devices in the reactors, or both,"
    - Rick Zuercher, Dominion Resources
  - "Plant's systems reacted as they should have."
- **Safety**
- **Uptime**
- **Efficiency**
- **Quality**

TMS THE MODAL SHOP, INC. A PCB GROUP CO. PCB PIEZOTRONICS

Calibration Overview

- **What is calibration?**
  - Verifies performance over the intended range of use
  - Provides physical meaning to the electrical output
  - Typical calibration includes:
    - Reference sensitivity
    - Frequency response
    - Phase response
    - Output bias level or supply voltage
    - Transverse sensitivity
    - Resonant frequency
    - Time constant

TMS THE MODAL SHOP, INC. A PCB GROUP CO. PCB PIEZOTRONICS

Industrial vs. Laboratory Calibration

- **Industrial Concerns**
  - Trending
  - Overall vibration levels
  - Frequency response
    - Effect of magnetic mounting
  - Alert & alarm set points
  - Bearing fault methodology
  - Proximity probes
  - Heavier payloads
  - Cable connections & properly matched cabling
  - Various sensing technologies
  - Safety
  - Uptime & productivity
  - "Chasing ghosts"
  - Challenging conditions
  - Troubleshooting
- **Lab Concerns**
  - Transverse sensitivity
  - Phase
  - Precision at specific frequencies
  - Resonant frequency
  - Time constant
  - Bias voltage
  - Linearity
    - Shock sensors
  - 0 measured offset
    - DC sensors (MEMS)
  - Drift

TMS THE MODAL SHOP, INC. A PCB GROUP CO. PCB PIEZOTRONICS

Back-to-back calibration method

- **Test sensor mounts "back-to-back" with a reference accelerometer that has been previously calibrated by primary means or by a recognized laboratory**

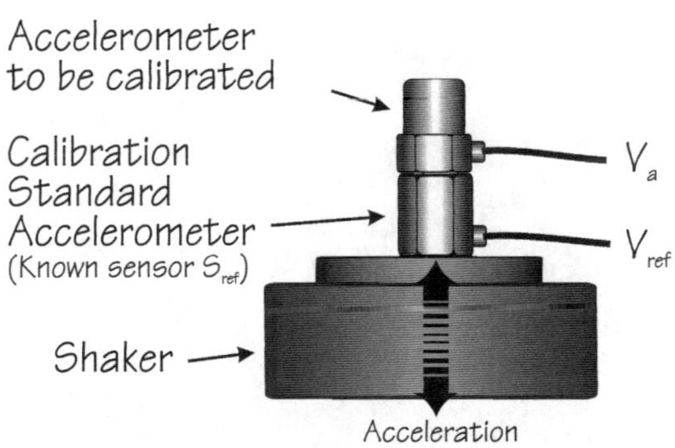

$$A_a = A_{ref}$$

$$\frac{V_a}{S_a} = \frac{V_{ref}}{S_{ref}}$$

$$S_a = S_{ref}\frac{V_a}{V_{ref}}$$

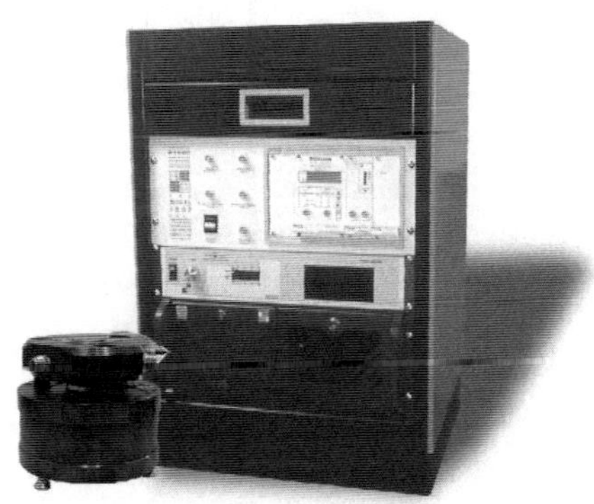

TMS THE MODAL SHOP, INC. A PCB GROUP CO. PCB PIEZOTRONICS

Calibration & Quality Standards

- What are the standards?
  - ISO 9001 Company Quality Management System
  - ISO 17025 Calibration Laboratory Quality System
    - Identifies the general requirements for the competence of calibration laboratories
    - Accreditation bodies include A2LA, ACLASS, NVLAP, etc.
  - ISO 16063 Methods for the Calibration of Vibration and Shock Transducers
    - 16063 Part 11. Primary vibration calibration by laser interferometry
    - 16063 Part 13. Primary shock calibration by laser interferometry
    - 16063 Part 21. Vibration calibration by comparison method
    - 16063 Part 22. Shock calibration by comparison method
  - API 670 Machinery Protection Systems
    - Section 7.6: Field Testing

TMS THE MODAL SHOP, INC. A PCB GROUP CO.   PCB PIEZOTRONICS

What Can Be Calibrated?

- Accelerometers
  - Amplitude linearity
  - Frequency response
- Loop powered, 4-20 mA velocity transmitters
  - Amplitude range/scale
  - Frequency band & filtering
- Proximity Probes
  - Static linear range
  - Dynamic linear range
  - Frequency response
  - Output vs. different materials

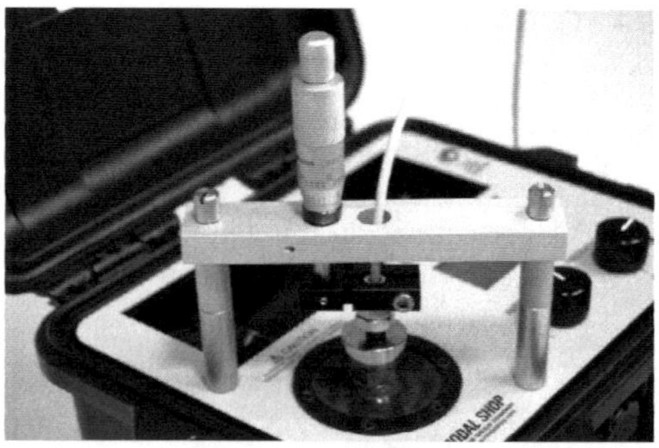

TMS THE MODAL SHOP, INC. A PCB GROUP CO.   PCB PIEZOTRONICS

What Can Be Calibrated?

- PLC, DCS & SCADA systems
  - Alert & Alarm set points
  - Time delays
- Electronic vibration switches
  - Alert & alarm set points (programming)
  - Time Delays
  - 4-20 mA retransmit range/scale
- Impact Detectors
  - Impact count
- Analyzers & Online Systems
  - FFT display
  - Overall values
  - Bearing fault methodology

TMS THE MODAL SHOP, INC. A PCB GROUP CO. PCB PIEZOTRONICS

Why Calibrate?

- Accelerometers
  - Route-based vibration monitoring programs count on a handful of sensors to provide data for various machines, applications & running speeds
  - Consistency: quarterly vibration readings are considered "frequent"
  - Bearing fault methodologies rely on sensor output at frequencies greater than 1 kHz
  - Sensitivity drift can mask vibration issues or prevent user from picking up high frequency, low amplitude bearing faults at stage three
- Loop powered, 4-20 mA velocity transmitters to PLC/DCS
  - Alert & alarm set points critical for uptime and safety
  - Ensure time delays are set properly to prevent false trips
- Proximity Probes
  - Ensure linearity
  - Comply with API 670 standard
  - Alert & alarm set points critical for uptime and safety
  - Mismatched cables can cause sensitivity errors
  - Measure output vs. alternate shaft materials

TMS THE MODAL SHOP, INC. A PCB GROUP CO. PCB PIEZOTRONICS

Why Calibrate?

- Electronic vibration switches
  - No self-test function on most models
- Analyzers & Online Systems
  - Ensure bearing fault detection methodology is working properly
- Impact Detectors
  - Understand & verify output
  - Check alert & alarm levels
- Industrial instrumentation endures extreme harsh conditions 24/7
- Rebuild after downtime is the most common time for machinery protection system malfunction
  - Cables routed to incorrect channels
- Troubleshooting
  - Ski Slope
  - Noise Issues
- Confidence, Safety, Assurance, Quality, Uptime

TMS THE MODAL SHOP, INC. A PCB GROUP CO. PCB PIEZOTRONICS

What do analyzer OEM's say?

- Emerson (CSI) →
- Commtest
  - "We recommend yearly performance checks for the instruments and a full calibration during the final year of warranty."
- SKF
  - "It makes good sense to check your vibration measurement equipment from time to time to ensure that your data trends are reliable."

Why Calibrate?

Environmental conditions, elapsed time, and type of application can all affect the stability of an instrument, and instruments from different manufacturers can show varying performance. Regular calibration of instruments helps maintain a high level of performance and safety in processes, as well as controlling emissions or maintaining other environmental regulations. Regular calibration will maintain the credibility of measurements and the quality of process measurements at a good-as-new level, enabling you to run your process with finer control.

EMERSON. Process Management

TMS THE MODAL SHOP, INC. A PCB GROUP CO. PCB PIEZOTRONICS

Quotes about industrial calibration

- **"Machinery Health Monitoring Depends on Accelerometers (Part 1)"**
  - Author: Wayne Tustin
  - Source: Reliability Web
  - "Sometimes, mechanical errors can result from other factors, like dropping the accelerometer…If that *does* occur, be sure to "red tag" that accelerometer. Don't use it until thoroughly recalibrated it over its entire frequency range…Single frequency checking is not sufficient"
- **"Why You Need Continuous Calibration"**
  - Author: Brian Graney
  - Source: Reliability Web
  - "Single point calibration is not a calibration at all. It is a sensitivity check at 100 Hz. An accelerometer is a dynamic sensor therefore it is required to have its sensitivity checked over its entire frequency range."
  - "If you do not check the sensitivity over the full range of the accelerometer you do not know if it is performing to specification."

TMS THE MODAL SHOP, INC. A PCB GROUP CO. PCB PIEZOTRONICS

Shift of Accountability

- **The price of lower cost trends**
  - Single point calibrations
  - Wider sensitivity tolerances
  - Loose frequency response requirements
  - Lack of ISO 17025 accredited manufacturers

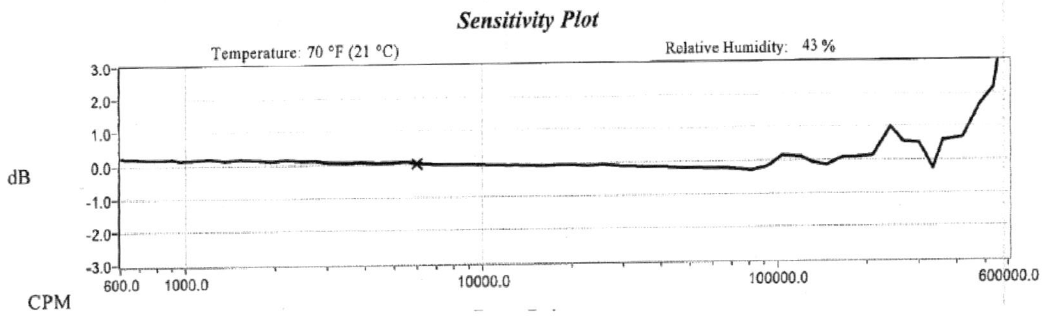

TMS THE MODAL SHOP, INC. A PCB GROUP CO. PCB PIEZOTRONICS

Key Accelerometer Specifications

- Sensitivity & Tolerance
  - At reference frequency: 100 Hz
- Frequency Range
  - Low end – determined by discharge time constant
  - High end – determined by resonant frequency
  - Tolerance: 5%, 10%, 3 dB?...what does 3 dB mean?
- Measurement Range
  - (Supply voltage – bias voltage – 1 VDC) / Sensitivity
- Settling Time
  - Caution: lower frequency response = longer settling time
- Temperature Range
- Accessories
  - Calibration Information
  - ISO 17025 accredited?

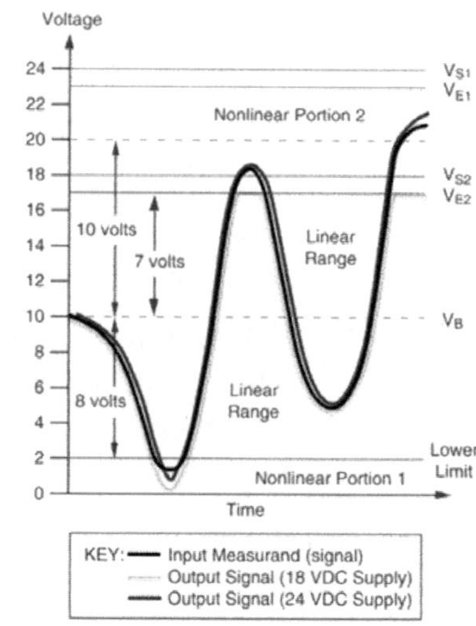

Accelerometer Frequency Response

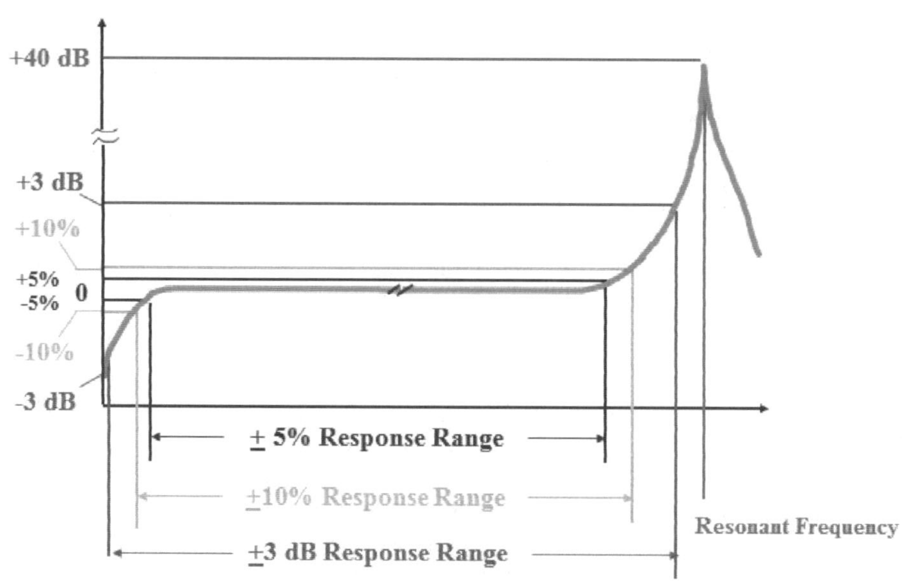

Traditional Transducer Calibration

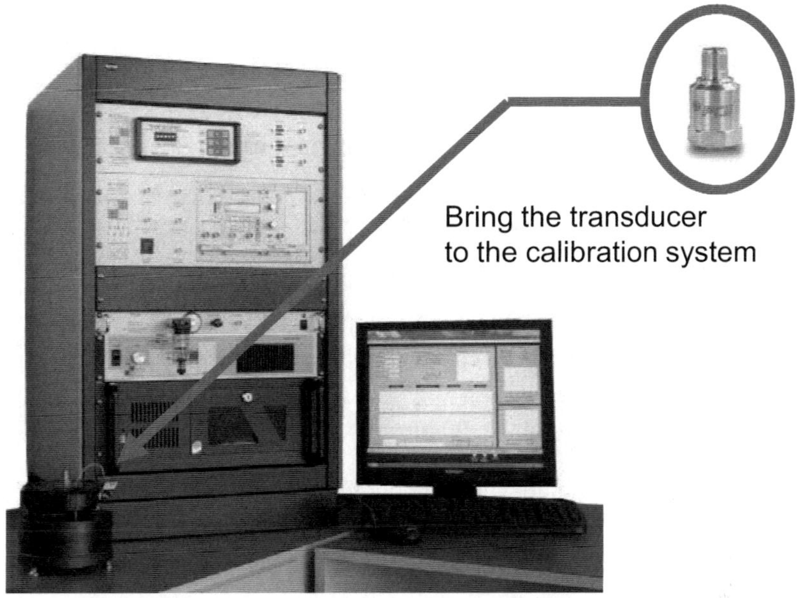

Bring the transducer to the calibration system

TMS THE MODAL SHOP, INC. A PCB GROUP CO. PCB PIEZOTRONICS

In-situ Calibration

Digital Control System

Plant Operator

Industrial Sensor

TMS THE MODAL SHOP, INC. A PCB GROUP CO. PCB PIEZOTRONICS

In-situ Calibration

- What is in-situ calibration?
  - Provides end-to-end confirmation of machinery protection vibration subsystem operation
  - In-situ means "in-place". The sensor is removed from the machine, but not from the measurement or monitoring chain.
  - Typical in-situ calibration includes:
    - Reference sensitivity check only- in the "units" being monitored (A,V,D)
    - Confirmation that alarm trip points are set
    - Confirmation that panel alarms are propagated through to Digital Control System or Turbine Supervisory Instrumentation which plant operators use to determine equipment health and make critical decisions about when to run and when to shut down the equipment

TMS THE MODAL SHOP, INC. A PCB GROUP CO. PCB PIEZOTRONICS

Calibrating Piezo Instrumentation

- Frequency Response Calibration
  - Performed on accelerometers and velocimeters
  - Hold amplitude steady
  - Measure sensitivity as frequency changes
  - Use critical frequencies for the application
    - 1x running speed
    - 2x running speed
    - BPFO & BPFI
    - Gear Mesh Frequency
    - RBPF
  - Else evenly spaced frequency points

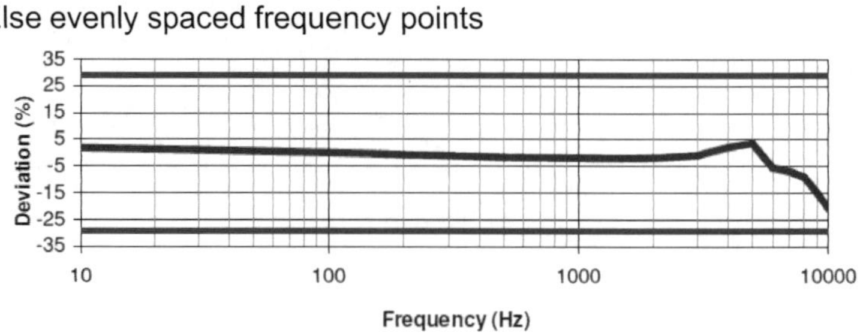

TMS THE MODAL SHOP, INC. A PCB GROUP CO. PCB PIEZOTRONICS

Calibrating Piezo Instrumentation

- Amplitude Linearity Calibration
  - Performed on 4-20 mA loop-powered vibration transmitters
  - Hold frequency steady
  - Measure sensitivity as amplitude changes
  - Can be done in acceleration, velocity or displacement
  - Piezoelectric sensors are extremely linear
    - Manufacturers do not always perform this test for each accelerometer

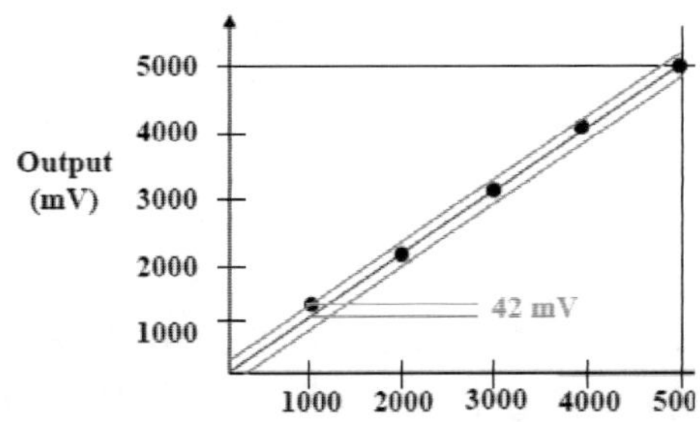

TMS THE MODAL SHOP, INC. A PCB GROUP CO.

PCB PIEZOTRONICS

Calibrating Proximity Probes

- Static Calibration
  - Measure DC voltage as distance changes
  - 10-90 mils typical range
  - Data points taken every 10 mils
- Dynamic Calibration
  - Measure AC voltage output as distance changes
  - Use 1x RPM frequency
  - Confirm alert & alarm set-points

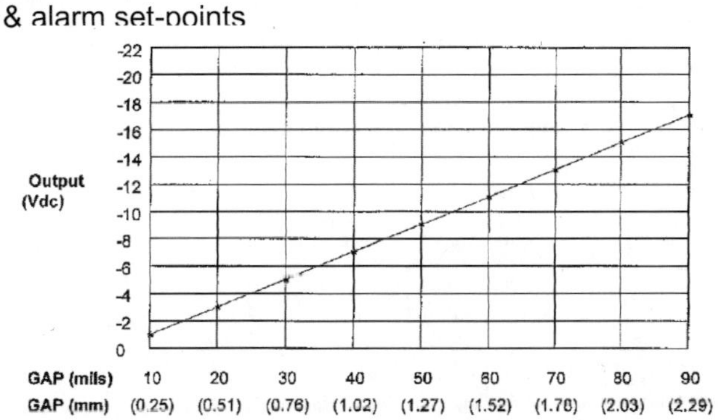

TMS THE MODAL SHOP, INC. A PCB GROUP CO.

PCB PIEZOTRONICS

In Conclusion

- Industrial machinery vibration protection systems are relied upon to ensure continued uptime and productivity while protecting the safety of those that work in and around the machines they monitor.
- In-situ calibration gives maintenance personnel the ability to ensure these systems are working properly
- Responsibility for the machinery protection system falls upon the plant personnel
- What is the cost of failure?
- What are the benefits of making quality measurements?

- Thank you! Questions?

TMS THE MODAL SHOP, INC.
A PCB GROUP CO.

PCB PIEZOTRONICS

Dealing with Storage, Aging, Value, and Visibility

Big Analog Data™ Lessons Learned from Power Generation Fleet Monitoring Applications

Preston Johnson
Principal Sales Engineer: Condition Monitoring Systems
National Instruments
preston.johnson@ni.com

ni.com

NATIONAL INSTRUMENTS

This presentation considers the current lessons learned in a fleetwide deployment of condition monitoring data acquisition systems. With fleet wide applications, there is an extreme amount of data from both high speed/dynamic sensors and from static, often process oriented, sensors.

NATIONAL INSTRUMENTS™

ni.com

Lets first take a look at the Volume of data as part of the fleet wide application.

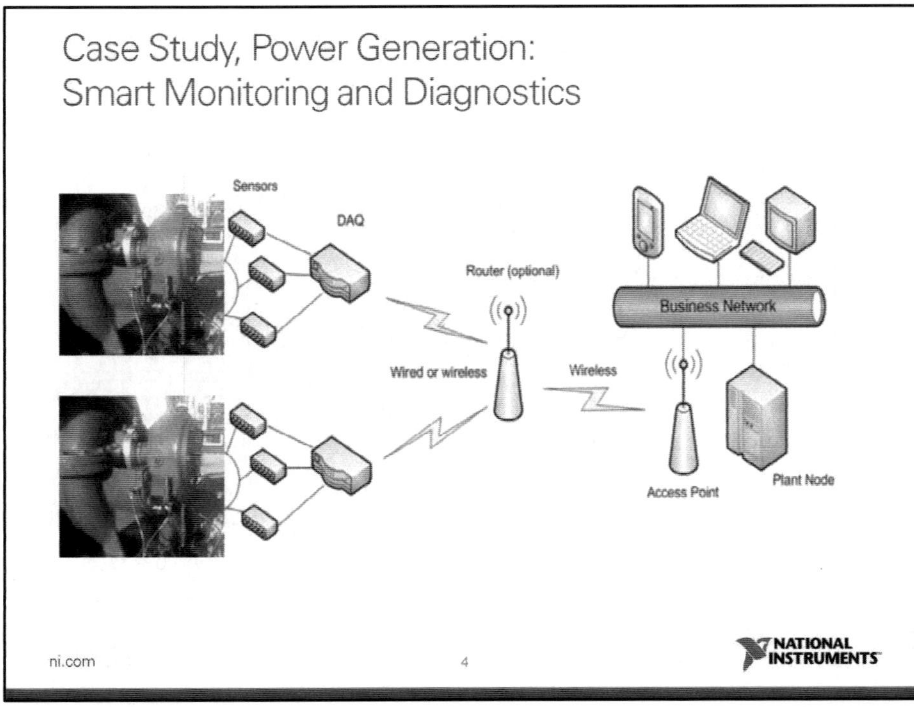

In many power generation plants, the operators are experiencing the effects of an aging workforce, aging rotating equipment, and new complex combined cycle gas turbine generation.

The experienced PdMs (predictive maintenance personnel) are spending up to 80% of time simply collecting data. Further, with Natural Gas combustion turbines and combined cycle plants, power producers are using new technologies in their plants, bringing new risks. All the while there is higher demand for reliability of GigaWatt electrical power supply.

The Electrical Power Research Institute (epri.com), sponsors research and technology innovation that both promotes and advances fleet wide condition monitoring of critical equipment such as turbines and generators, as well as supporting equipment including pumps, fans, motors, blowers, and so on.

To address the fleet wide condition monitoring application, permanent sensors are placed on machines, electrical apparatus, and even structural members where once, hand held measurements have been used.

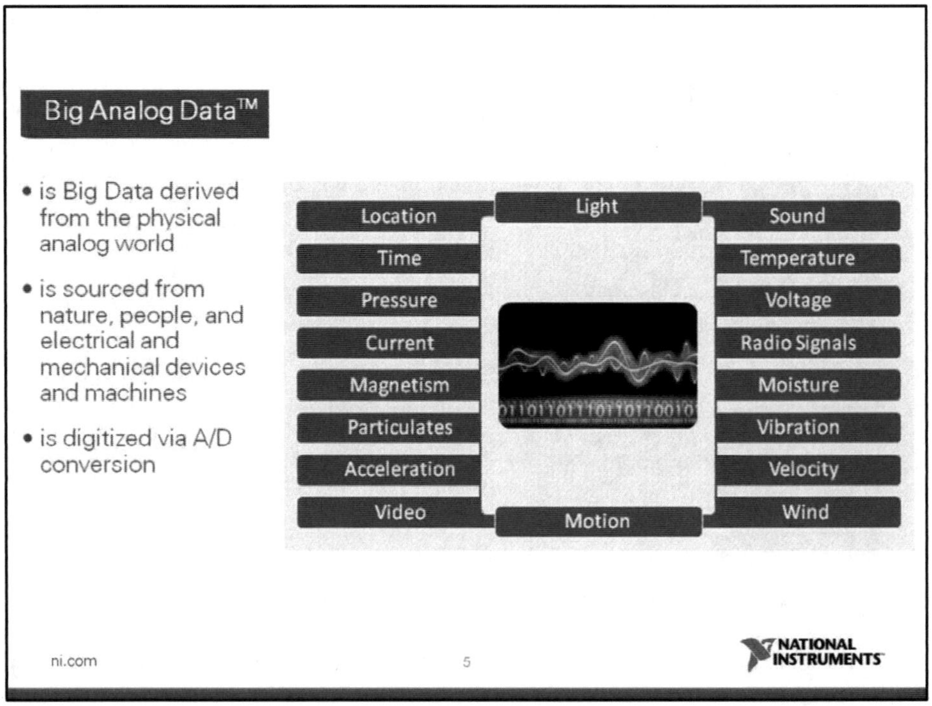

This is creating a Big Data problem, in particular of the Analog kind.

Big Analog Data™ is not social media data, it is not business CRM and Revenue Data, it is the digitization of the physical world around us. In the case of plant machinery, it is vibration, current, oil particulates, video, temperature, electromagnetic interference, and atmospheric conditions. Big Analog Data™ is the oldest, and the highest volume of all Big Data.

In general, Big Analog Data™ applications are those where sensors derive big data from the physical analog world. Many of these physical phenomenon are measured in our power plants today, being fossil, nuclear, or even wind.

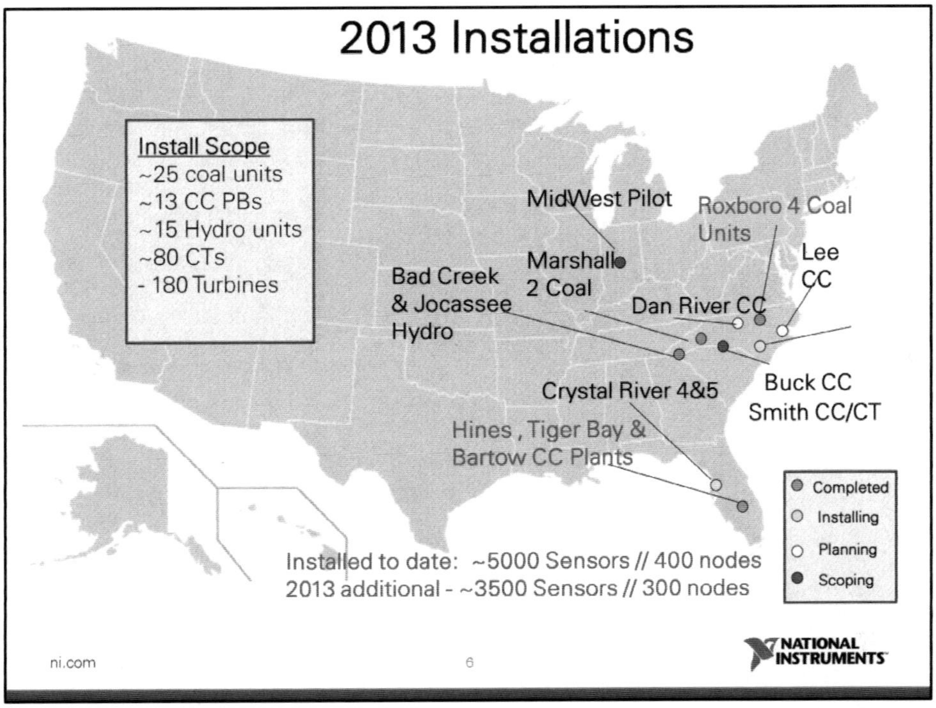

Here is a power generation company rolling out fleetwide monitoring at over 60 power generation plants (NIWeek 2013 Energy Summit). The addition of thousands of vibration, speed, temperature, and other sensors builds a tremendous volume of time series sensory data.

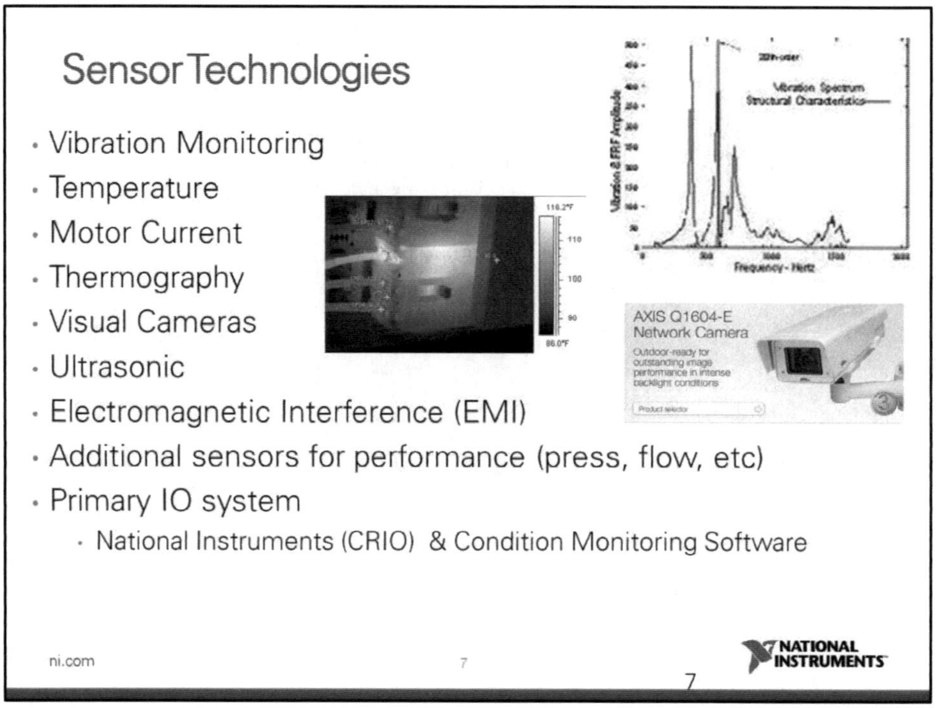

To get started, the power generation community is using traditional condition monitoring sensors, Vibration and Temperature. However, The power generation industry is rapidly expanding their automated sensory data collection to include all of the available technologies.

These technologies produce 10's of thousands of sensory data points per second. Multiply this by 10's of thousands of sensors and you quickly can imagine the big data challenge.

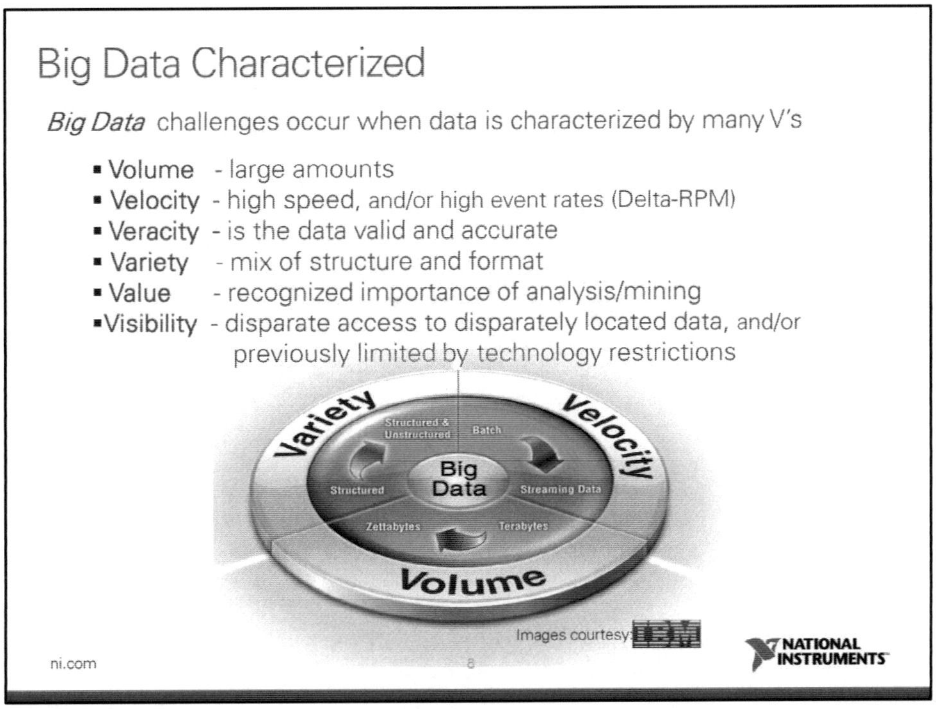

So lets define Big Analog Data™ characteristics.

Some people talk about the 3 V's of Big Data, some prefer 4. Myself, I think of it in terms of 6 V's. The first three most of you have probably heard of, but there are two more I like to talk about and that is Veracity, Visibility and Value. If you can't get at your data or get visibility into your data and analysis then it is worthless. And it is the data and analysis that bring us to our main objective VALUE.

So this is the challenge of our MFPT Big Data Management Focus Group. What are the best practices in acquiring, validating, managing, and gaining value from the sensory data collected in fleetwide applications?

Lets look at a quick illustration to underscore the amount of sensory data being collected in a fleetwide power generation condition monitoring application.

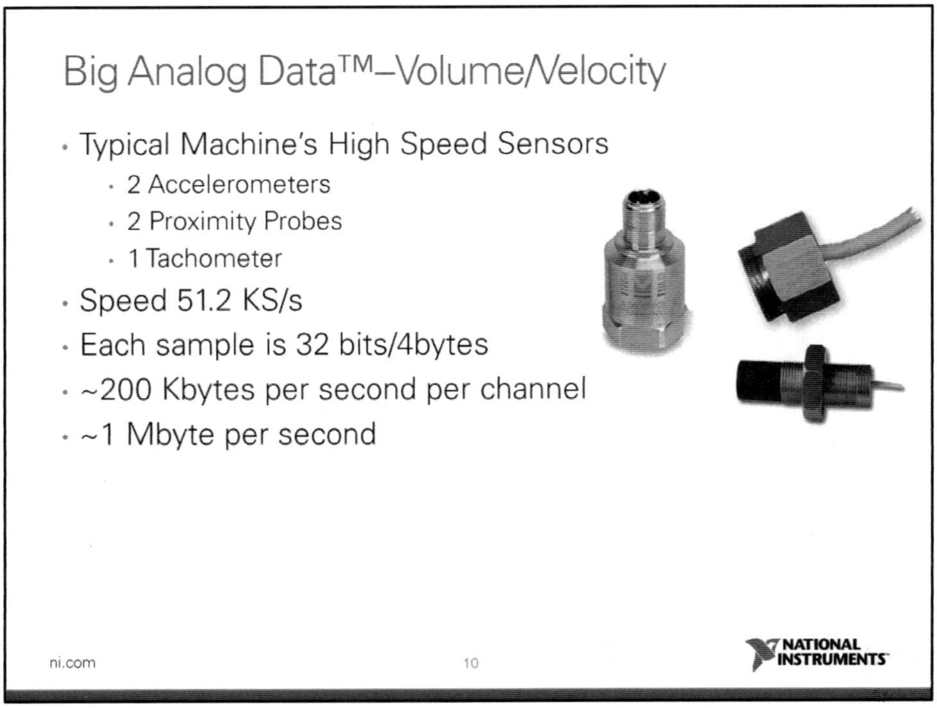

Consider a single machine.

This outlines a typical scenario in a permanently mounted, distributed monitoring application.

So what about Volume and Velocity (our first two V's).

From a Big Analog Data™ perspective in the power generation arena, vibration sensors are common. For a simple machine, we can easily generate 1MB of data per second.

How much data is this per week?

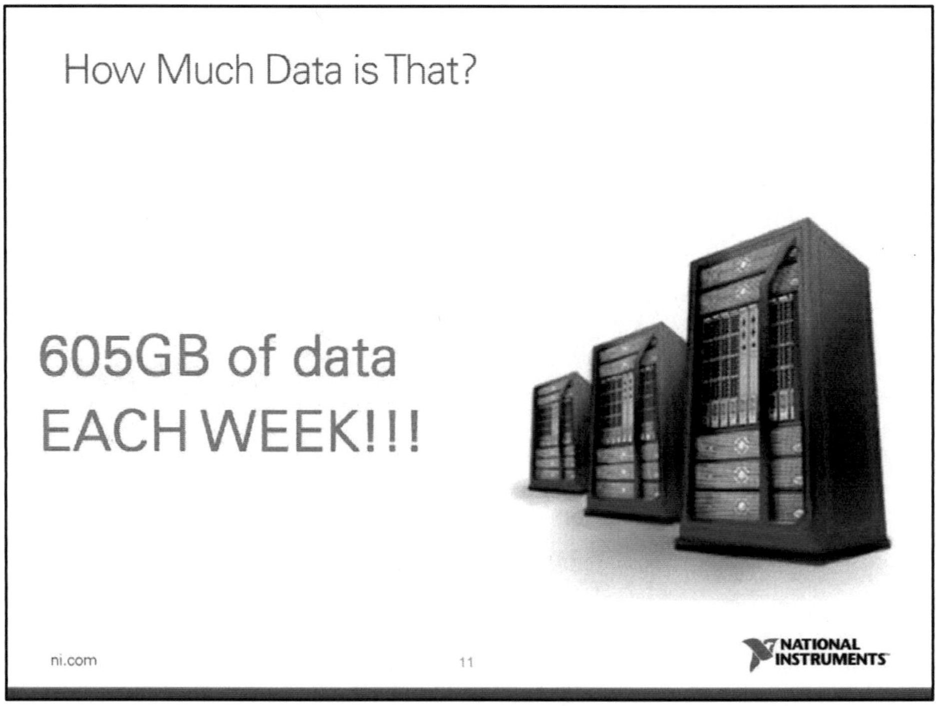

With in one week that one machine will produce 605Gb of data each week

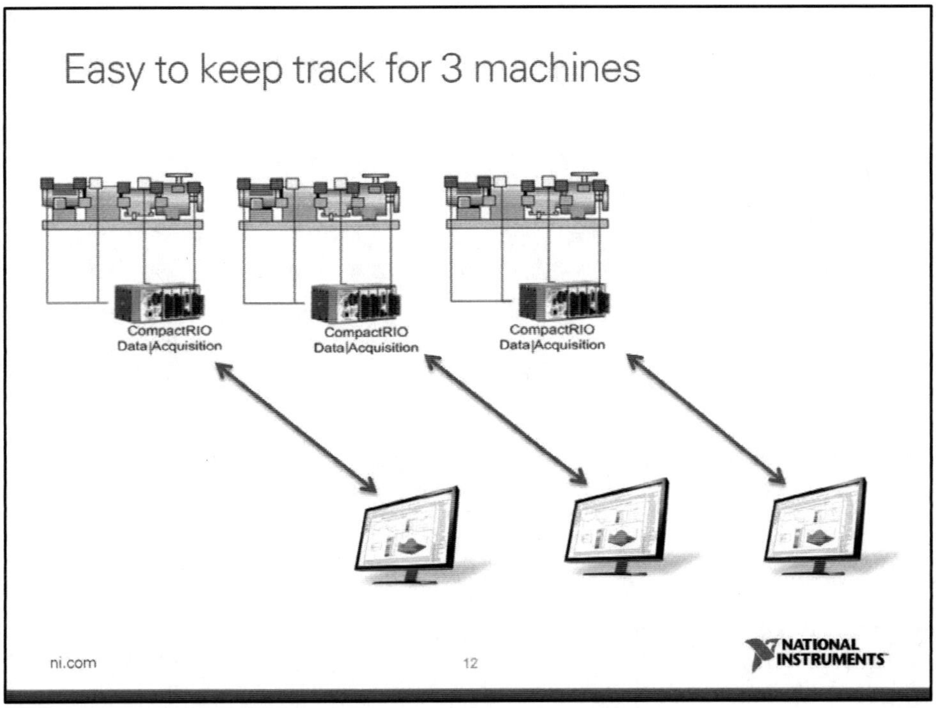

So not a big analog data(tm) problem for just three machines........

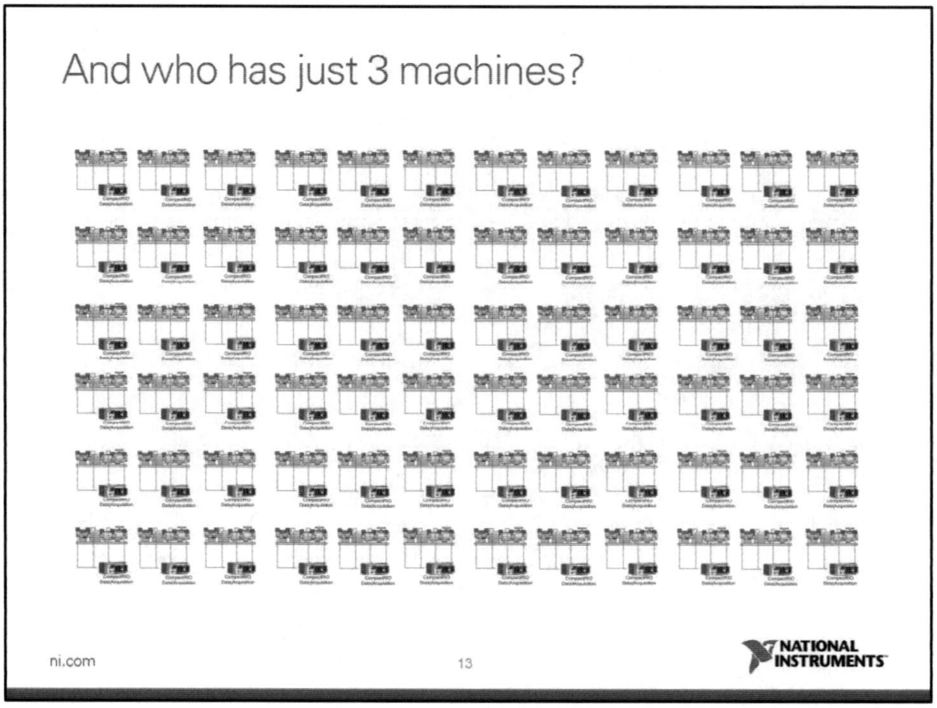

However, when working a complete fleet, the data can quickly become 100's of MB per second.

This is where the information technology department MUST be a part of the discussion in our solution.

There are, of course, intelligent ways in which the data can be monitored and captured in order to reduce the amount of data while still maintaining business value, YET we are still talking about a tremendous amount of data.

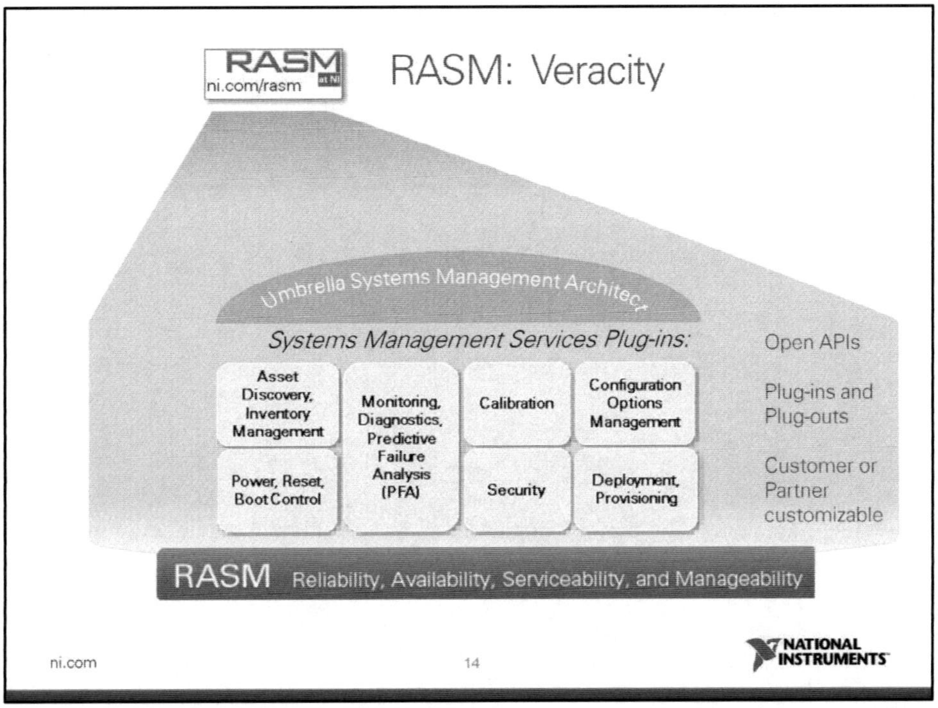

Oh, and with all that data, how can we be sure it is accurate and therefore useful?

The information system managing the data acquisition nodes must be able to verify data acquisition and sensor health.

In addition to managing the data, the network connected data acquisition nodes must also be managed.
Can they be discovered?
What self diagnostics and status information is available?
Is the device secure?
Are the sensory measurements valid / good?
How are configurations updated and deployed?

This is yet another information technology that needs to be addressed.

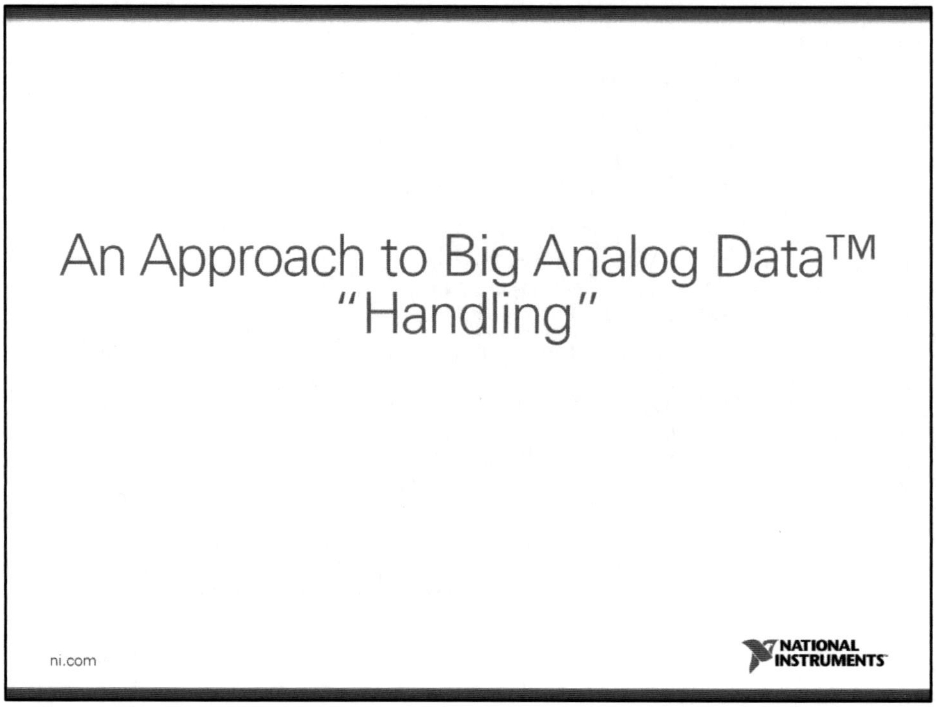
An Approach to Big Analog Data™
"Handling"
ni.com
NATIONAL
INSTRUMENTS

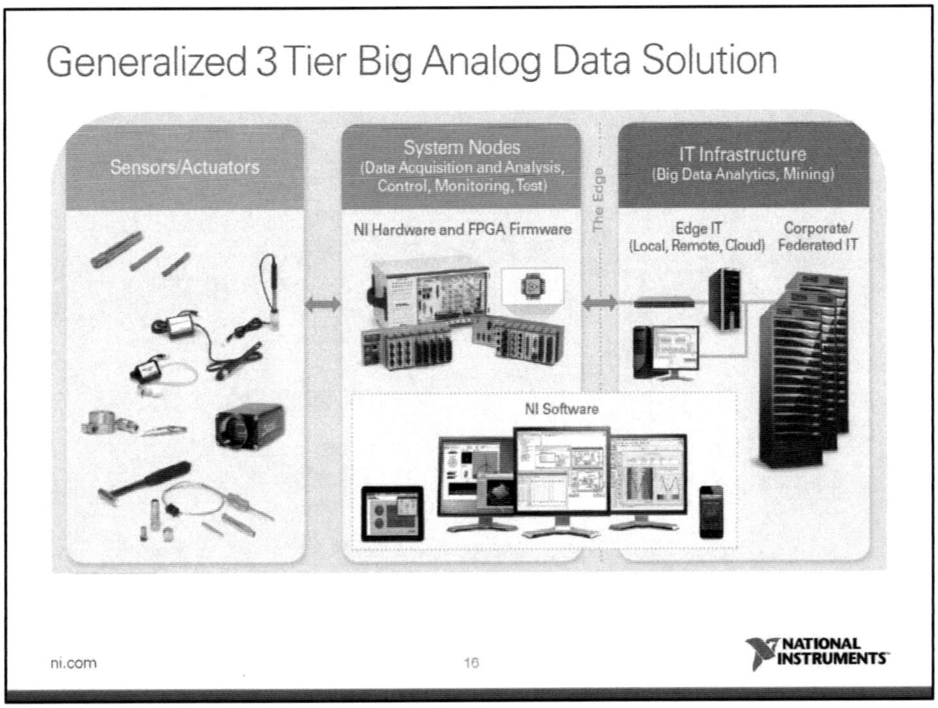

So lets look at a Big Analog Data™ Solution.

There are 1000's of sensors in a fleetwide monitoring application.

There are 100's of Data Acquisition and Analysis Nodes (DAAN's) working to capture, review, and record sensory data.

There are 10's of computer nodes in the IT infrastructure managing the DAANs and providing real-time trending, alarming, and visualization.

At the corporate level, aggregated sensory data is used to build best practices observations, normal operational signatures, fault signatures, and otherwise work to automate the diagnostics and predictive analytics used in operations of the plant.

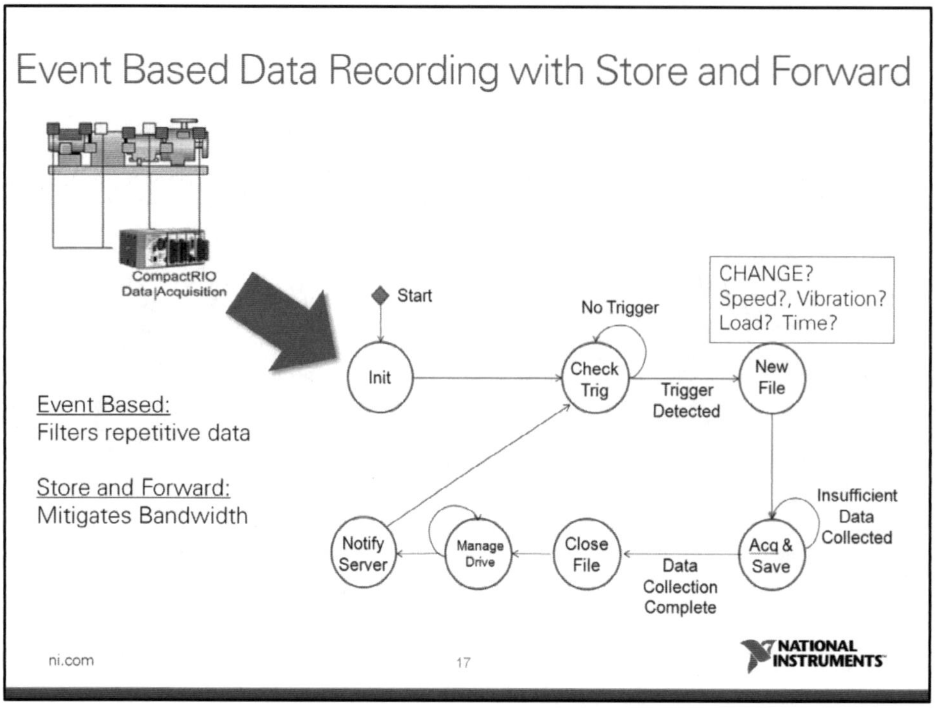

Lets start at the left side of the 3 Tier solution presented on the previous slide.

There are a wide range of sensors used in condition monitoring applications. These sensors generate Giga-bytes of data. So we must have a strategy to manage and perhaps filter the flow of data.

EVENT BASED RECORDING
Analog Sensors produce terabytes of data
Event based recording filters sensor data to that which is different (consider change in speed, load, vibration level, time, etc. as recording triggers)

STORE AND FORWARD
Don't get stuck behind low bandwidth or intermittent communications paths
Store data locally to the embedded solid state drive
Forward it when the communications path is ready

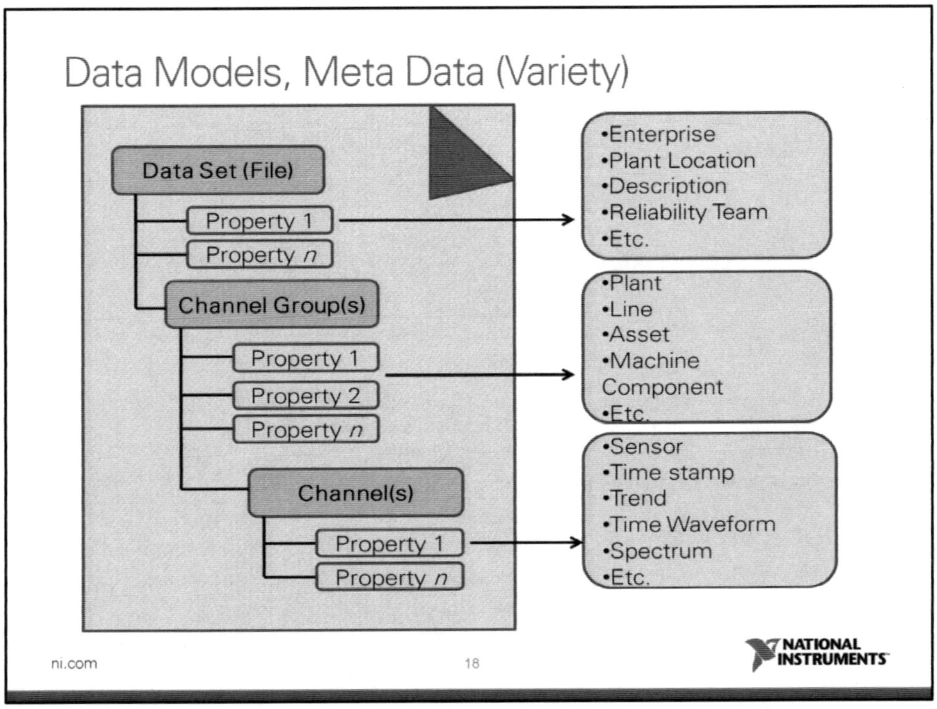

In addition, the sensory data must be properly labeled.

META DATA IS KEY

Enrich Data with descriptive meta data, at the recording source, in the data acquisition node.
This allows improved data mining during subsequent find and analyze operations.

This is very important in order to compare machines and components in order to find bad actors and develop patterns of degradation.

It also helps us to combine many types of data from a single event record into a single file. This helps address the variety of sensory data and condition indicators we have in our monitoring applications.

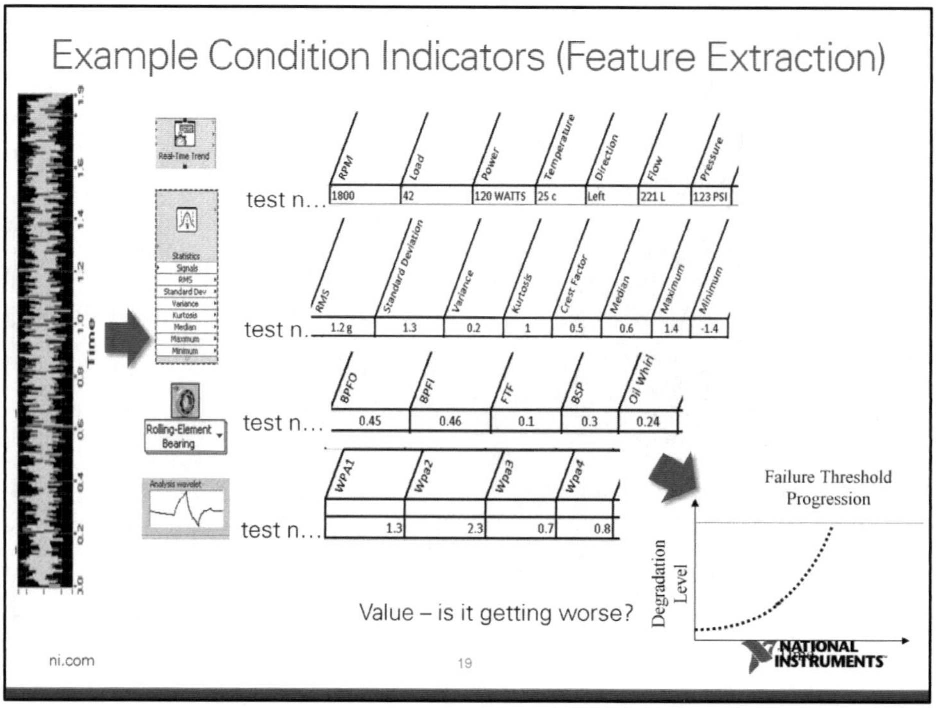

Another key process is extracting useful information from the sensory data. We can do this with signal processing and statistical analysis of the sensory data.

Sensor data is measured (the green waveform on the left) and features are extracted from these time waveforms. These features often provide indications of wear in specific components of the asset. These condition indicators can be trended and alarmed, to provide an early warning of a specific failure mode.

The collection of several features form patterns of normal and faulty behavior, and often of specific faults under specific operating conditions. These collections of features can be used for developing fault signatures and using predictive analytics to make predictions of future readiness of the asset for service.

Once we have derived all the condition indicators from the sensory readings, these can be recorded as meta data properties for the time waveform from which they were calculated. If enough features are calculated, it may be possible to reduce the size of the time waveform in the file, or remove it completely from the file.

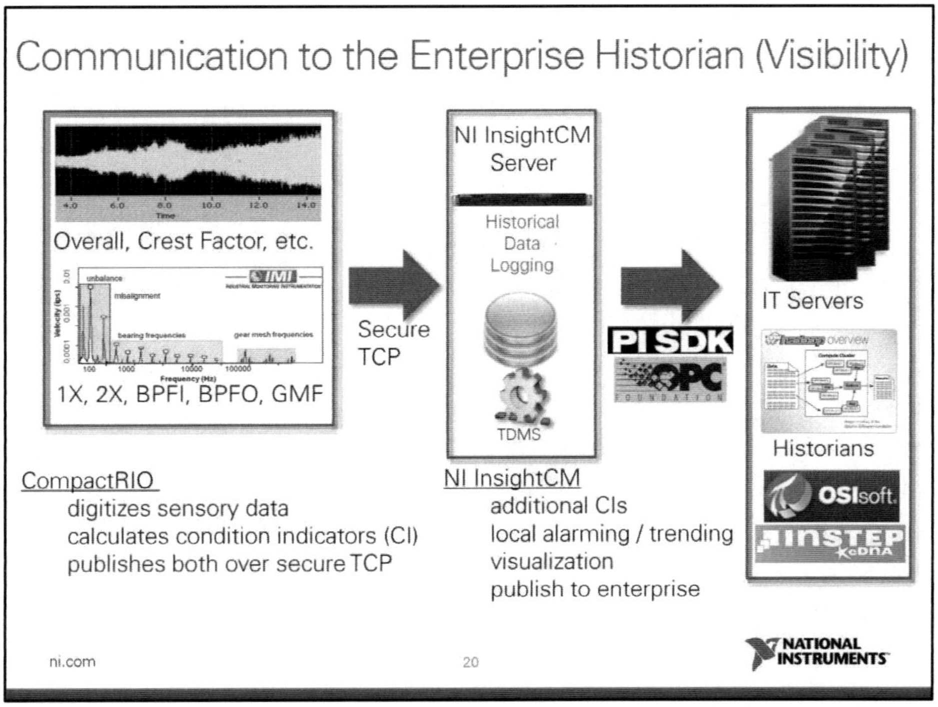

These condition indicators, whether calculated in the data acquisition device, or on the managing server, can be posted to plant historians.

Within the historian, the condition indicating features can be combined with process sensory information to produce advanced trending and alarming charts. Now it is possible to begin providing high level dashboards regarding the entire fleet, the plant, or even individual assets.

In some cases, there are anomaly detection tools on the market that operate from data historian systems. To feed these prognostics tools, the condition indicators / features are calculated and transferred to the historian. There other sensory, operational data, maintenance records and so on, are fused to provide a complete picture of the asset health, as well maintenance history and risks of failure.

The data historian then becomes a convenient form of fusion of many types of numerical data for off-line predictive analytics.

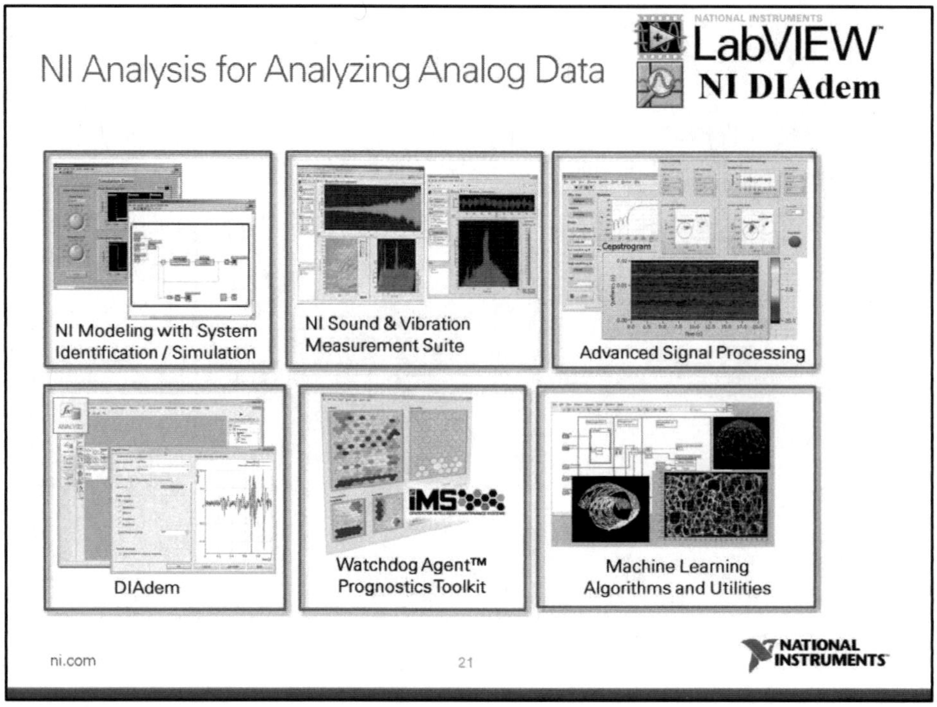

- Once Data arrives on the Host Computer, a number of additional and more advanced algorithms can go to work.

- Sound & Vibration Suite, extensive development throughout the years of analysis algorithms
- Through the Tools Network, NI offers the Watchdog Agent® Prognostics Toolkit for LabVIEW, NI also offers the
- Systems Identification toolkit for asset model development, and the
- Machine Learning toolkit for evaluation of large field sensory data sets.

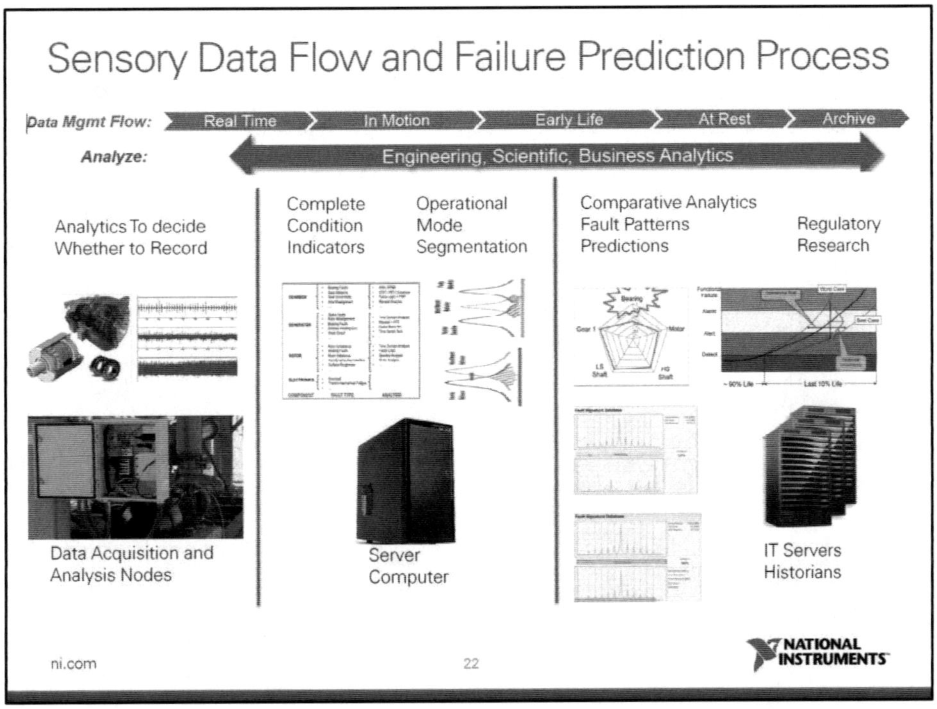

Here is a high level view of the overall Big Analog Data architecture, with more of a focus on the analysis used along the data flow.

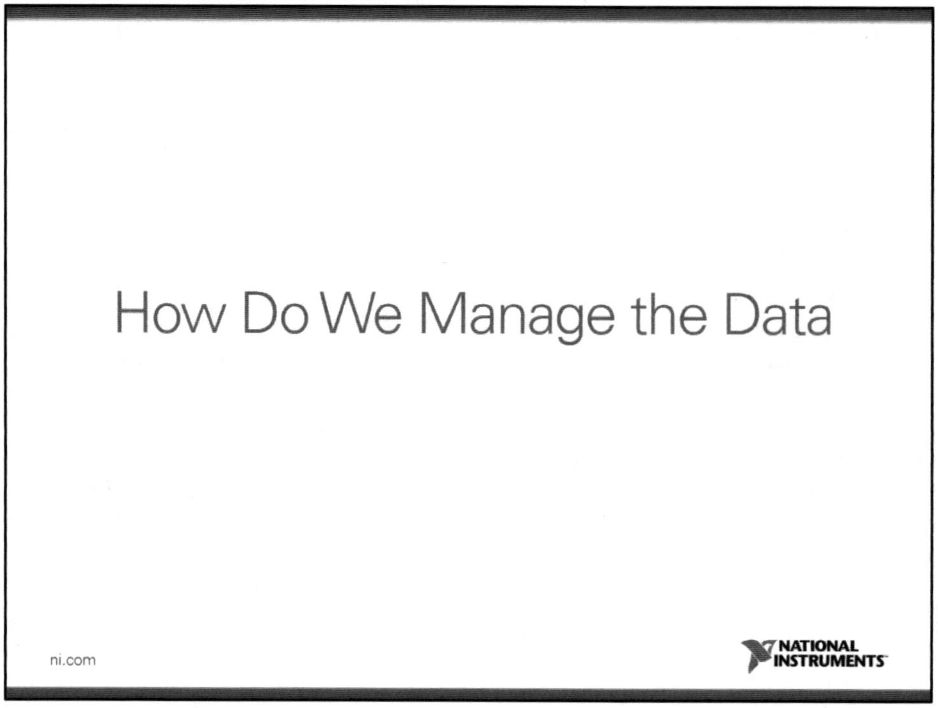

So now we have been introduced to the 6 V's of Big Data. Volume, Variety, Velocity, Veracity, Value, and Visibility.

Even though, we have filtered and reduced to some extend, some of the volume of data, the volume remains very large. How do we manage the data?

Data Aging and Archiving

- Storage is Cheap, but is it?
  - STORAGE:
    - we MUST Manage it
    - Procure it
    - Index it
- Keep the important data
  - based on its uniqueness from other data sets.
- Archiving
  - ensure an indexing system knows it is there

ni.com 24 NATIONAL INSTRUMENTS

Should we keep all the data? What strategies can we use to reduce the footprint and maintenance costs.

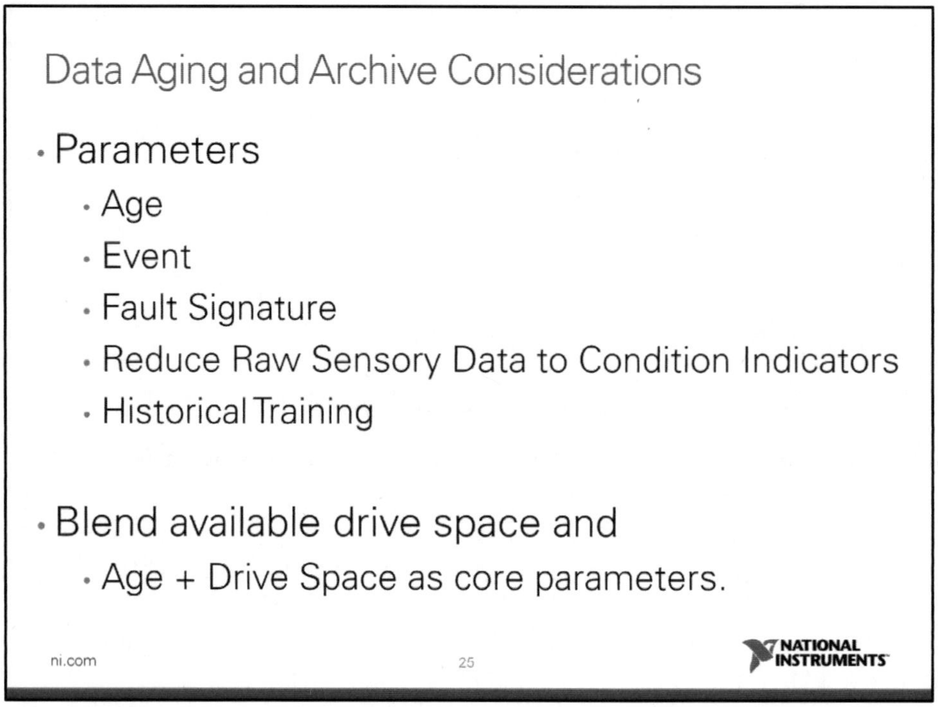

Aging considerations include
Age of the data
Occurrence of an event causing the recording
Is the data representative of a fault signature, and usable for pattern recognition
Can we use the data for training algorithms or people
Also consider the size of the data store

And with the data we keep, we must index it: (Manage the Index)
Optimize the fields needed for finding data
Do NOT index all fields, only the best properties for search
Do NOT undersize the data store
Use Server Class data stores
Use Solid State Drives where appropriate
Use Time Filters in query (last week, last 24Hrs, etc.)

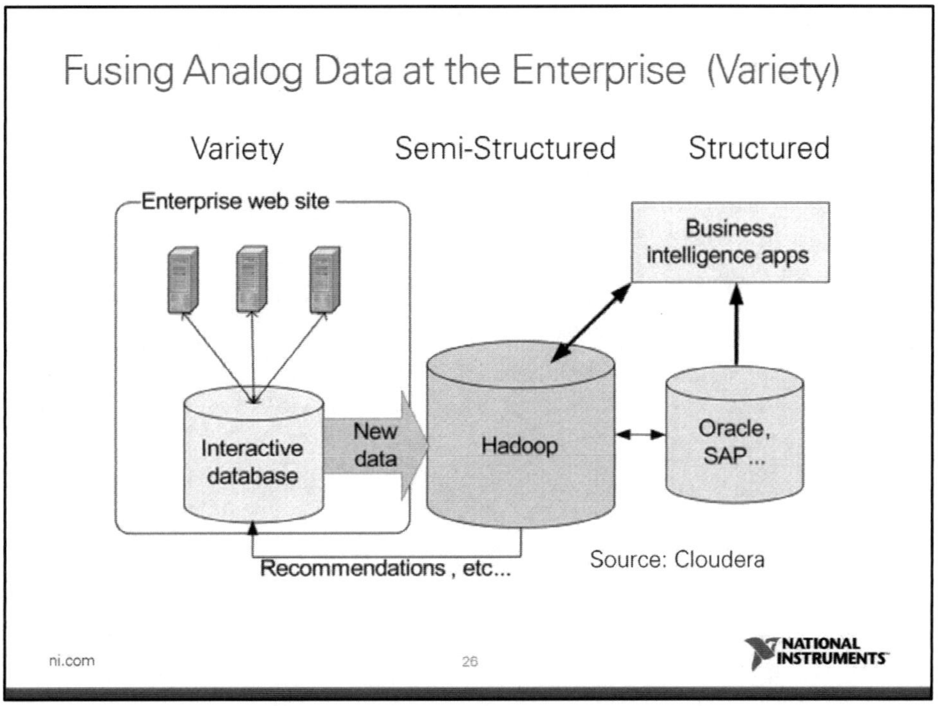

Now that we have our sensory data recorded (on event) and retained (on usefulness) Is it possible to fuse with less structured data?

Information technology suppliers including Dell Services, offer Hadoop Clusters that are able to organize and analyze unstructured data, including textual reports. Services are available to develop a Hadoop system strategy that may be an appropriate enhancement to our Predictive and Reliability problem.

We will see more of this in just a few moments.

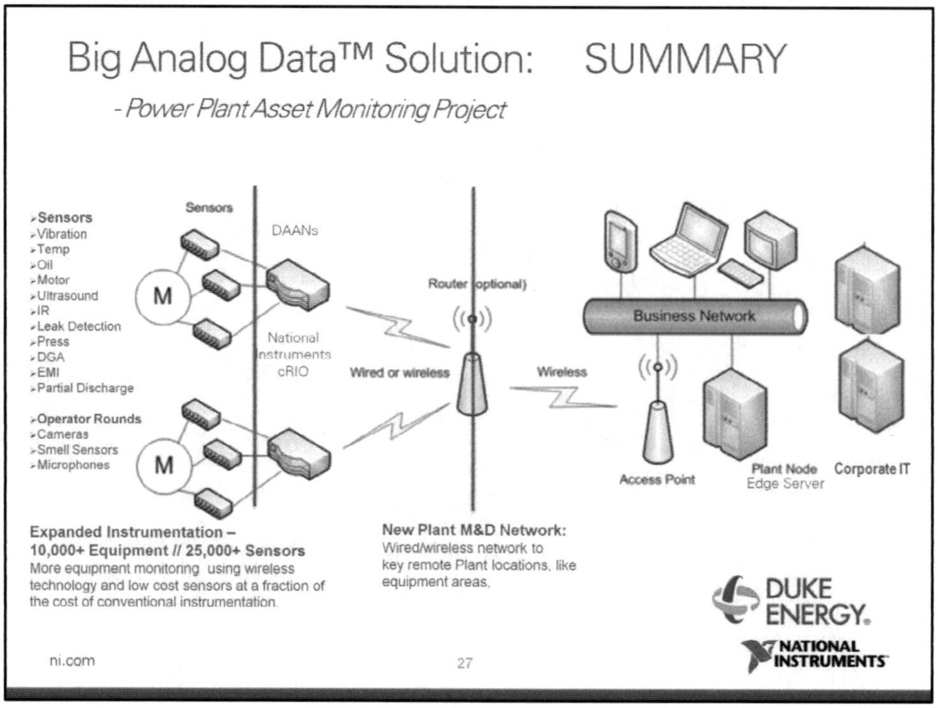

To help summarize, this graphic describes the fleet wide monitoring efforts now underway at several power generation plants in the Americas. Many data acquisition and analysis nodes are being deployed to initially capture vibration and temperature time waveforms based on time or speed events. Often, a business wireless tcp/ip network is used for data transfer to a business server in the PdM work area at the plant. This business server is accessible from other computers on the network for configuration and data analysis. This overall architecture is reflective of the Big Analog Data™ solution I showed earlier.

As we add more sensors, we have more analog data to work with. Yet intelligent analytics can reduce the data while "in-motion" to those features which indicate condition and degradation rates. With modern server technology, we have the resources at hand to fuse time waveform data and condition indicators with operational and maintenance notes to create the systems envisioned by EPRI, and its power generation members. With the right data, the right analytics, automation of diagnostics and prognostics yields advisory generation of the future.

All along the flow of Data, it is managed to address Volume, Velocity, Variety, Veracity, Value and Visibility. With the proper infrastructure in place, we can indeed grow our understanding of the condition of our assets, and set the stage for predicting their ability to perform their functions.

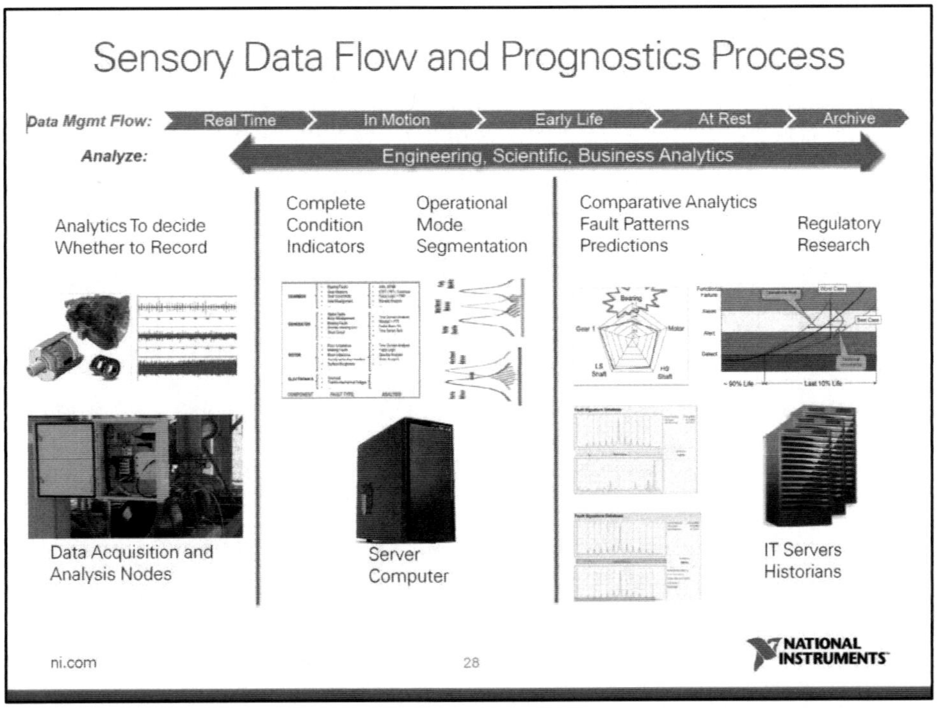

To summarize, condition monitoring and prognostics analytics play a role all along the life of the data.

At National Instruments, we welcome the opportunity to help you deploy an open, modular, and extensible system where you can get the value and visibility of your machinery and asset condition and utilization.

Dealing with Storage, Aging, Value, and Visibility

Big Analog Data™ Lessons Learned from Power Generation Fleet Monitoring Applications

Preston Johnson
Principal Sales Engineer: Condition Monitoring Systems
National Instruments
preston.johnson@ni.com

ni.com

NATIONAL INSTRUMENTS

AUTHOR INDEX

AUTHOR INDEX

MFPT Conference 2014

Technology Solutions for
Affordable Sustainment

Virginia Beach, Virginia, USA
20-22 May 2014

Volume 2 of 2

ISBN: 978-1-5108-8842-5

Printed from e-media with permission by:

Curran Associates, Inc.
57 Morehouse Lane
Red Hook, NY 12571

Some format issues inherent in the e-media version may also appear in this print version.

Printed by Curran Associates, Inc. (2019)

For permission requests, please contact Society for Machinery Failure Prevention Technology at the address below.

MFPT Society
715 Twining Road
Suite 135
Dresher PA 19025

Phone: (215) 346-2870
Fax: (215) 352-6902

https://mfpt.org

Additional copies of this publication are available from:

Curran Associates, Inc.
57 Morehouse Lane
Red Hook, NY 12571 USA
Phone: 845-758-0400
Fax: 845-758-2633
Email: curran@proceedings.com
Web: www.proceedings.com

TABLE OF CONTENTS

VOLUME 1

VOLUME 2

Dell Engineering Intelligence & Analytics

The Merger of Field Sensory Data (Big Analog Data™) with Control and Semi-Structured and Un-Structured Data Breeds New Opportunities for Data Management including Hadoop™ and Appplication Specific File Formats

Tik Sze, Dell Innovation Services

What is Dell Engineering Intelligence & Analytics?

Dell Engineering Intelligence & Analytics is an End–to–End Solution to Maximize Return on Investment of Assets & Resources

Equipment	Data	Actions
Data Generation	**Data Collection**	**Data Analytics**
Sensor Data • Pressure • Temperature • Flow Rates • Etc.	Gather Data • Analog to Digital Conversion • Collect "good" & "bad" measurements • Pre-determined aggregation, peak & low values	In Motion and At Rest Analysis • Predict events before they occur • Maintenance scheduling for highest ROI based on Risk Models • Real Time Monitoring & Alerting

Value Propositions

"RASM"

Reliability, Availability, Serviceability, and Manageability

Prognostics and Health Management (PHM) predicts when the assets may fail, before they fail

Smart maintenance allows preventive actions, just-in-time services, and automatic scheduling / prioritization

Automatic alarms and control to correct critical failures and minimize impact

Technology Stack

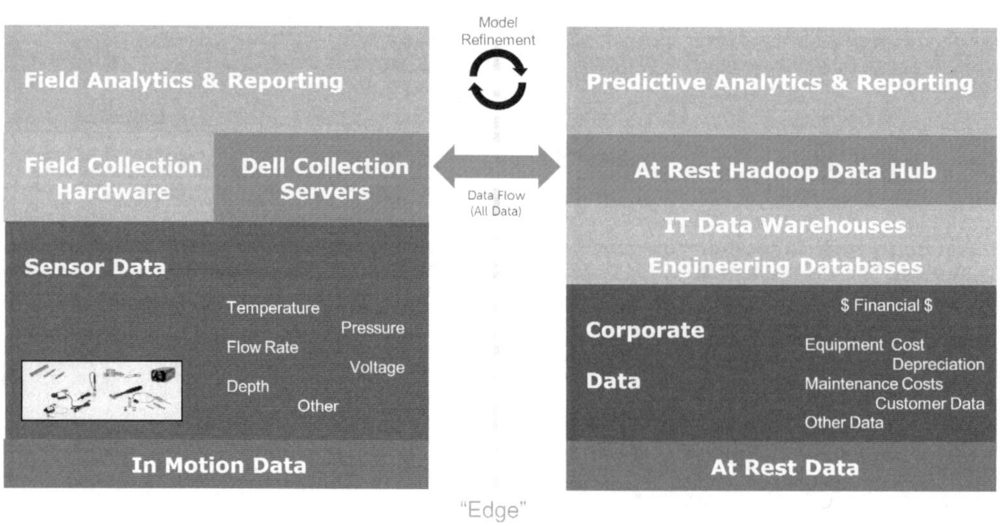

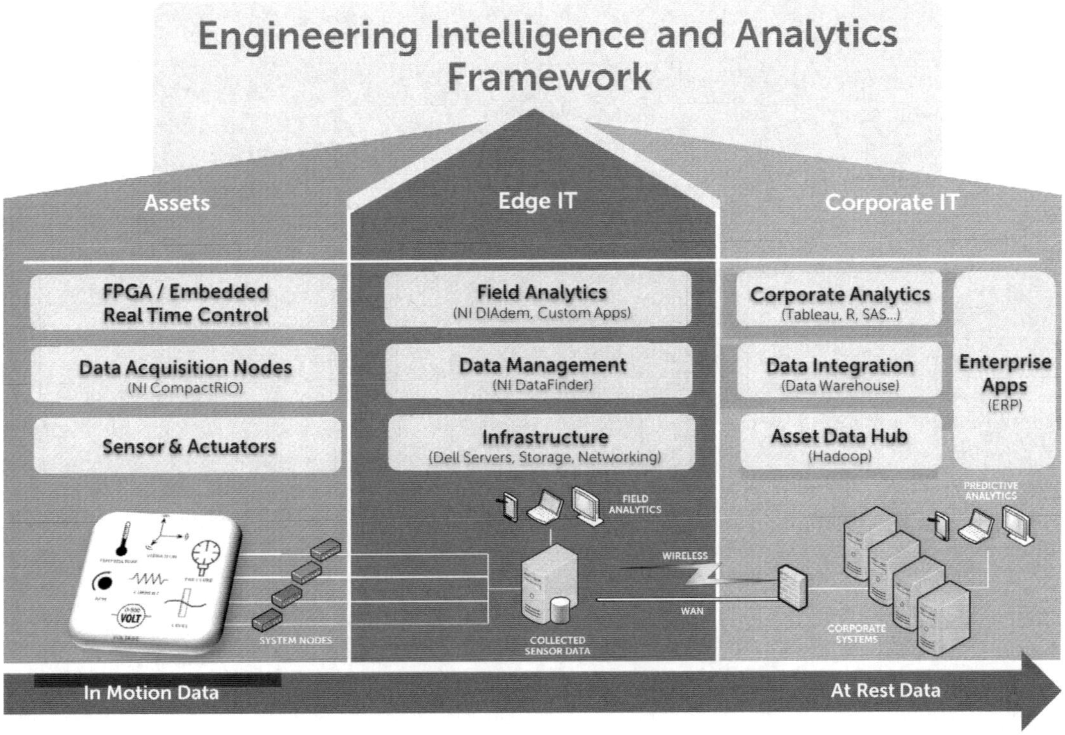

Assets

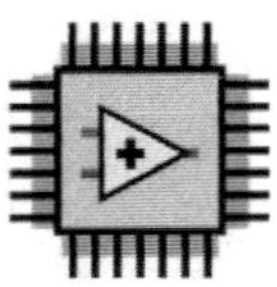

FPGA / Embedded Real Time Control

FPGA and embedded software running on RTOS provides fastest control and data processing closest to hardware

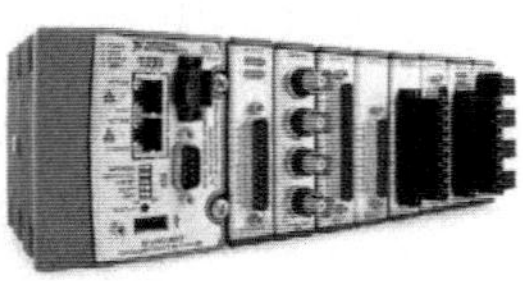

Data Acquisition Nodes
(NI CompactRIO)

Convert analog to digital with signal conditioning

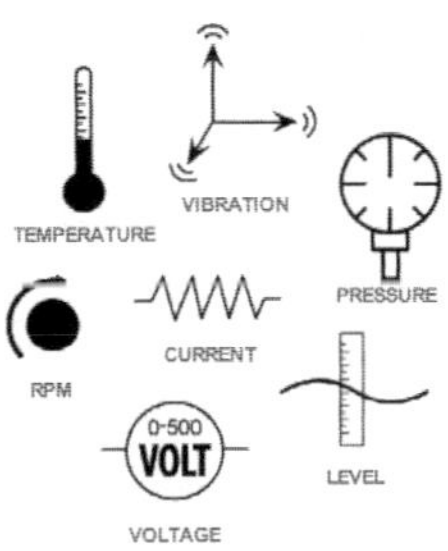

Sensor & Actuators

Collect analog data on physical conditions of assets, e.g. pressure, temperature, vibration...

Edge IT

Field Analytics
(NI DIAdem, Custom Apps)

Field analytics provides operational control and visualization

Data Management
(NI DataFinder)

Automatic indexing of large datasets based on industry standard TDM format

Infrastructure
(Dell Servers, Storage, Networking)

Reference architecture tested to work on large stream of data from hundreds of nodes

Corporate IT

Advanced Analytics & Visualization

Analytics provides insight into optimized operations & supply chain -> Action Plan: Maintenance schedule adjustments – adjustments to supply

Corporate Data

Data Management
(Integration with Hadoop Data Hub)

Integrate asset monitoring data with enterprise data – Asset Cost, Depreciation, Maintenance Schedules, Repair History

Data Hub
(Hadoop)

Hadoop data hub for asset data

.TDMS File Format from National Instruments Comparison Chart of Format Characteristics

	ASCII	Binary	XML	Database	TDMS
Exchangeable	✓		✓		✓
Small Disk Footprint		✓			✓
Searchable				✓	✓
Inherent Attributes			✓		✓
High-Speed Streaming		✓			✓
NI Platform Supported	✓	✓	✓	✓*	✓

*May require a toolkit or add-on module.

Table 1. The TDMS file format combines the benefits of several data storage options in one file format.

Elegant NI .TDMS File Format for Test Data

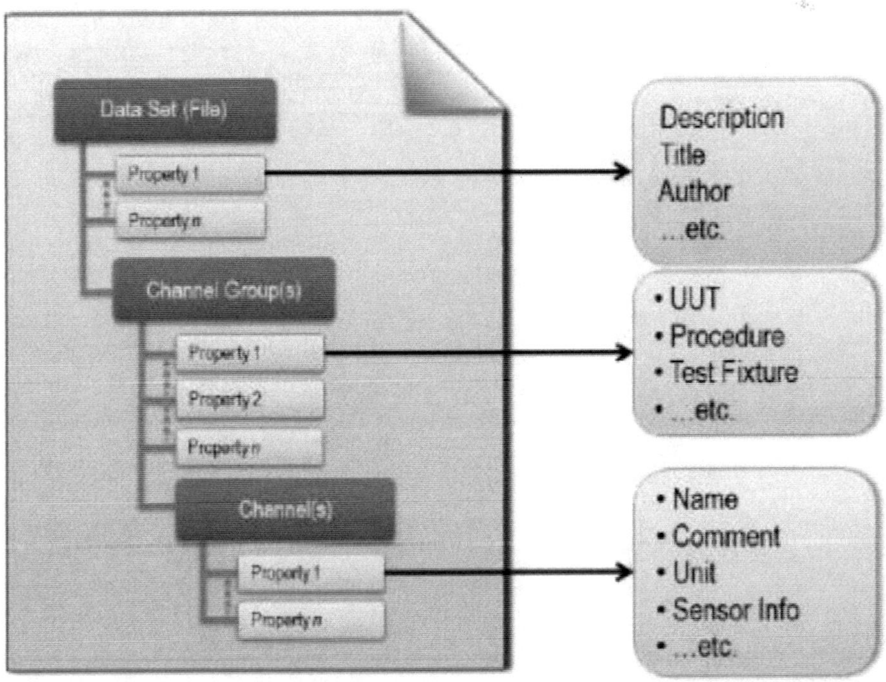

Data Tables in Text -> HDFS -> Hive -> Database Presentation (SAP HANA)

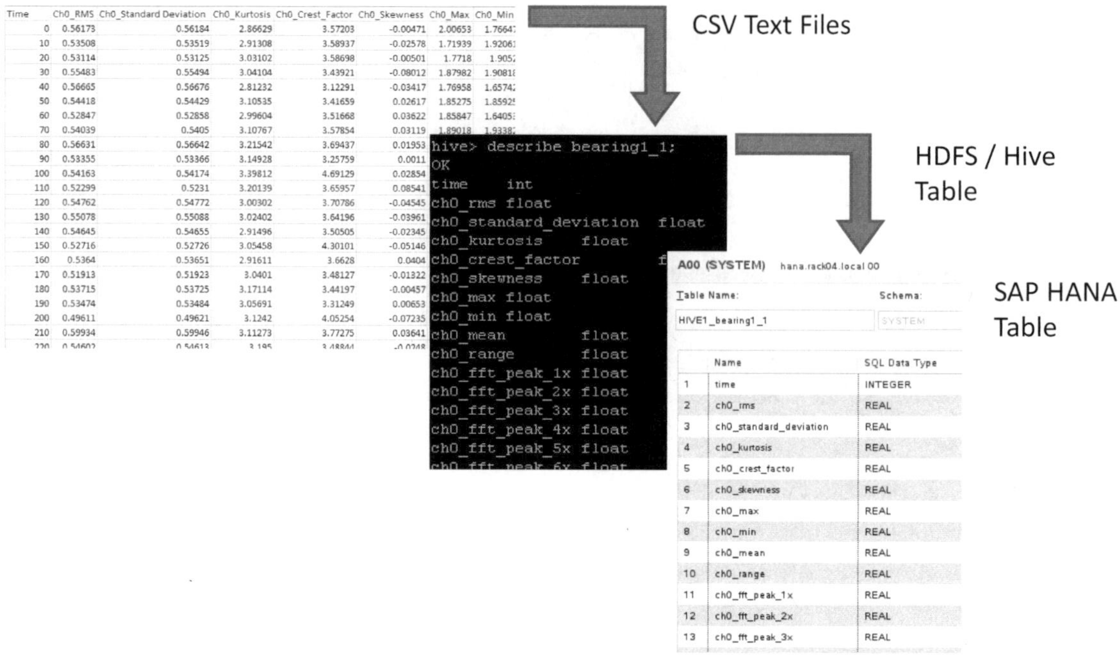

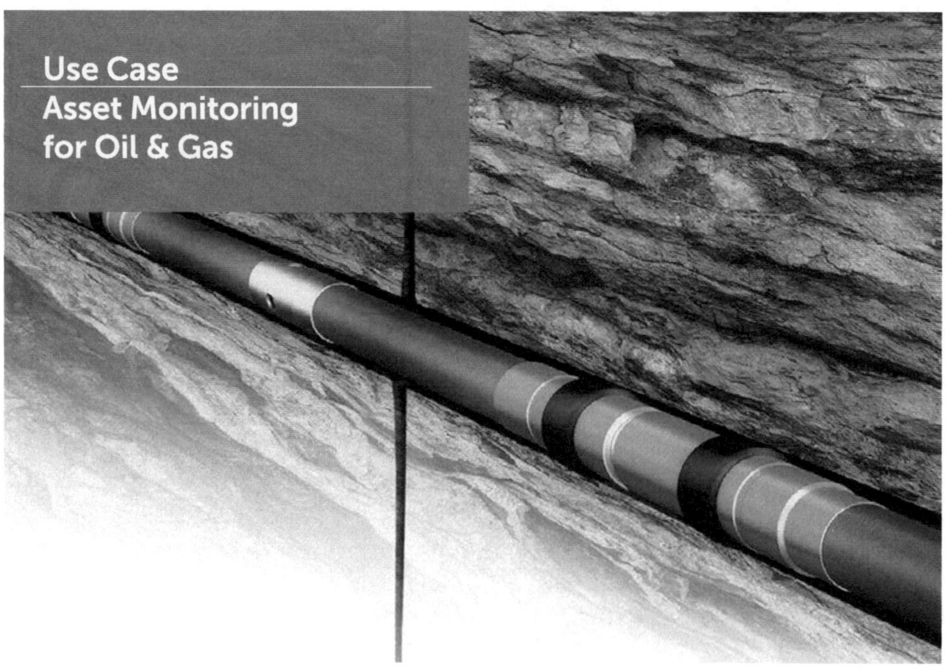

Operation View

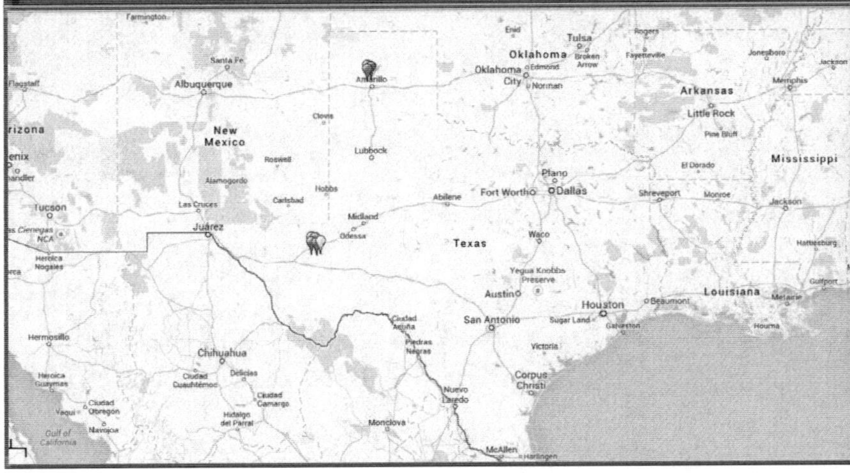

Site	Pump	Issue
RB Masterson C-1	1	Intake Manifold Pressure
RB Masterson B-1	6	Pump Oil Pressure
Essman C9-20	9	Crankcase Oil Temp
HSF State 3		
Sheldon State 3		
Sky King 47		
Beckham Caldwell 5-43		
RB Masterson A-2		

Maintenance View

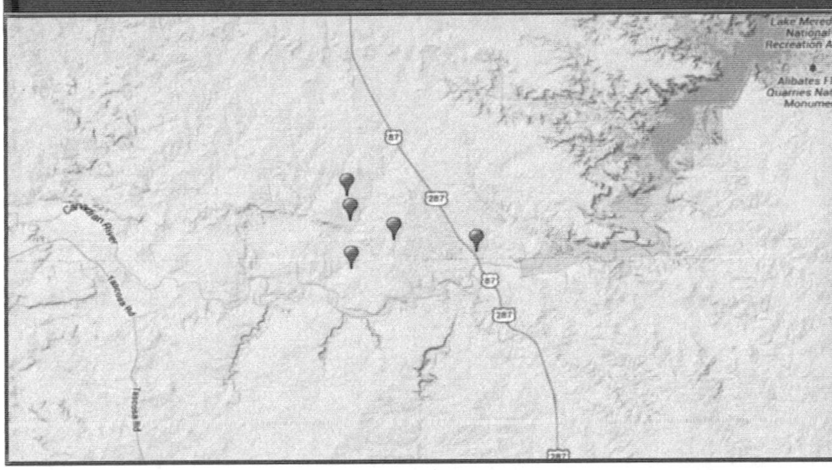

Site	Pump	Due
RB Masterson C-1	1	10/2/13
RB Masterson B-1	6	10/8/13
RB Masterson A-2		12/10/13
RB Masterson A-1		5/6/14
RB Masterson D-1		1/13/14

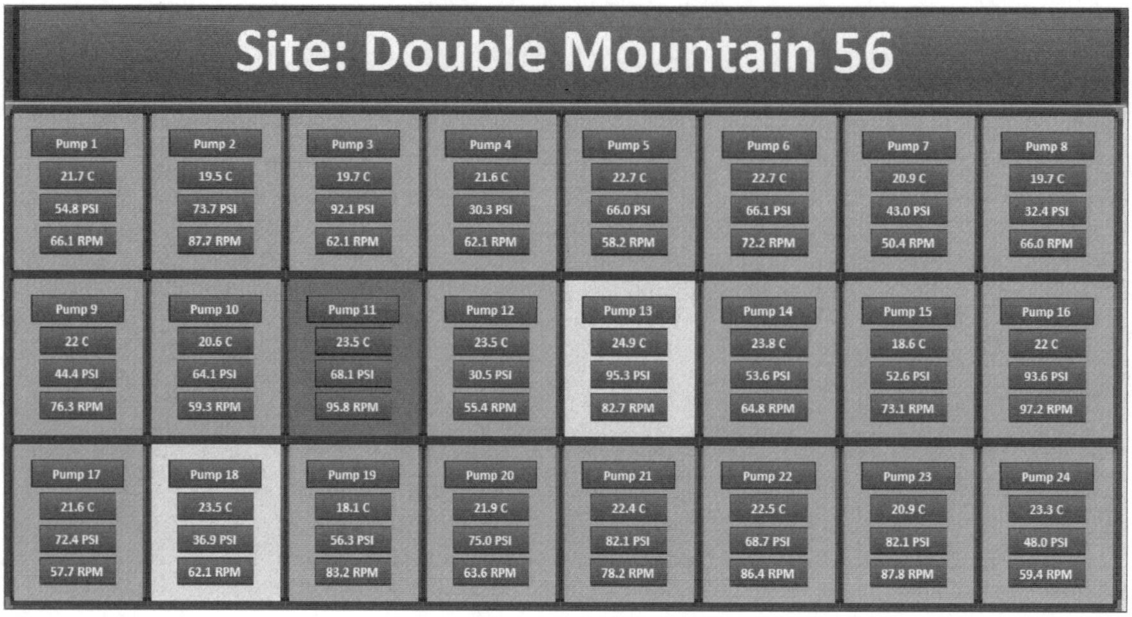

15 Business Intelligence & Analytics Services DELL

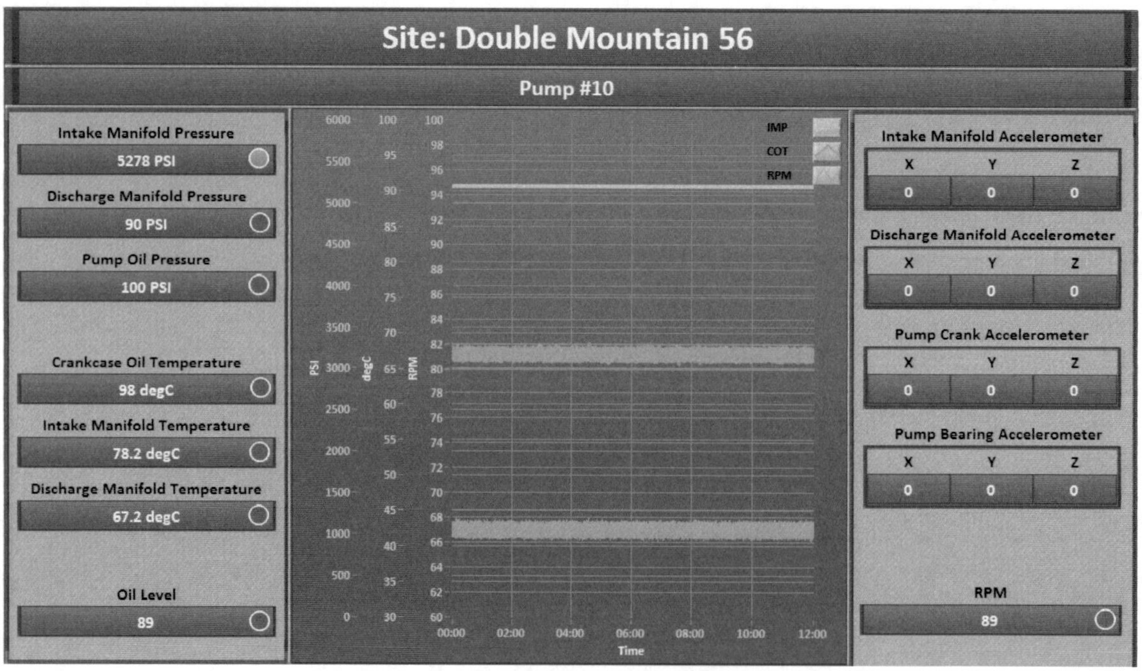

16 Business Intelligence & Analytics Services DELL

Thank you

PRACTICAL HEAT PUMPS IN COLD CLIMATES – AN ENABLING TECHNOLOGY

Thomas J. Walter
Mechanical Solutions, Inc.
431 New Karner Road
Suite 1C
Albany, NY, 12205-3868
Telephone: (518) 320-8552
tjw@mechsol.com

Abstract: The on-going development of a compact centrifugal compressor as a first stage or pre-compressor for cold climate operation of heat pumps is described. Positioned in the low pressure vapor portion of the refrigerant loop, this compressor will, on cold days, operate automatically to boost refrigerant pressure, in a manner similar to the way an automotive supercharger pressurizes air. The single stage motor driven centrifugal compressor runs on oil-free bearings. It is being configured to work in concert with a traditional heat pump compressor. The goal of this work is to enable air source heat pumps to efficiently extract heat from even the coldest ambient air without backup heat and without changing refrigerant type or the basic design of the positive displacement style compressors widely used in today's heat pumps. Doing so will allow air source heat pumps to operate effectively in virtually all major population centers. To date, a first article supercharger has been designed and built, and its performance mapped. Work to integrate it with a commercially available positive displacement (PD) heat pump compressor is underway.

Key words: Air source heat pumps, Centrifugal compressor; Gas foil bearings; High-speed motor-driven compressor; Refrigerant compressor; Vapor compression.

Introduction: Simply put, a heat pump is a device that uses a small amount of energy to move heat from one location to another. Heat pumps are typically used to pull heat out of the air or ground to heat a home or office building, but they can be reversed to cool a building. One of the biggest advantages of a heat pump over a standard heating ventilating and air conditioning (HVAC) unit is that there is no need to install separate systems to heat and cool a dwelling. Heat pumps also work extremely efficiently, because they simply transfer heat, rather than burn fuel to create it. This makes them more "green" than gas or oil-burning furnaces [1], and are far more efficient than directly heating with electricity.

Electrically driven heat pumps are an effective method of extracting heat from ambient air. As air temperature falls, however, heat pump performance falls off, essentially limiting their year-round usefulness to warmer climates. In colder climates heat pump-equipped heating systems typically have a secondary means to provide heat, usually electric strip heaters. This adds complexity, duplication, cost and pollution. An air source heat pump system that maintains both capacity and coefficient of performance (COP) in cold climates could have a significant energy-savings impact, particularly where natural gas is not available.

For the US Department of Energy [2], Mechanical Solutions, Inc. (MSI) is developing key elements of an integrated refrigerant compressor system that efficiently extracts heat from very cold ambients. The goal is a system that requires zero backup heat in even the coldest climates, in effect achieving a viable heat pump furnace. The core innovation is a second compressor, in the low pressure portion of the refrigerant loop that will, on cold days, boost the refrigerant pressure. The goal is a compact, highly efficient pre-compressor that works in concert with a traditional heat pump compressor. Once developed, the market objective is heat pumps that can be applied to virtually all major markets without changing the type of refrigerant employed or the basic design of existing heat pump compressors. As sources of electric power become cleaner, expanded heat pump use offers the ability to produce heat with minimal increase in greenhouse gases. A heat pump that can efficiently function as a furnace, with no requirement for backup heat, will revolutionize major portions of the HVAC industry. A recent presentation at an International Energy Agency (IEA) Heat Pump conference [3] describes the overall system and the thermodynamic role of the supercharger, which functions in a manner similar to automotive supercharger. The following sections describe technical aspects of the device, and test results to date.

Application: The DOE's performance targets for a cold climate air source heat pump includes a goal of efficient operation down to -13°F. Although heat pump performance begins to wane as temperatures fall below the mid-forties, a number of models are capable of heat extraction to the mid-teens. The objective of the supercharger therefore is to begin operation in this temperature range, then run faster as ambient temperature falls, reaching max speed at the coldest ambient. To be of optimum value, the supercharger's outlet pressure should be fairly constant and matched to a (PD suction) pressure where the PD compressor has peak efficiency.

Prototype Supercharger: The pre-compressor consists of a single stage centrifugal compressor driven by a permanent magnet motor. Gas foil bearings (no moving parts, oil-free) support the rotating components. It is driven by a variable frequency drive (VFD). Figure 1 shows the principal parts.

Based on the duty cycle described above and with R410a (a commonly used HFC refrigerant in heat pumps) as the working fluid, these conditions were selected for compressor operation: Inlet Total Pressure = 34 psia; Inlet Total Temperature = -16°F; Total Pressure Ratio ≈ 3.3; Mass Flow Rate = 24.8 lbm/min. Based on these correlations and loss models, a single stage compressor operating at a rotational speed of 100,000 rpm was selected. The compressor stage was modeled using a centrifugal impeller, vaned diffuser, and elliptical volute. This resulted in an impeller with 18 full height blades and zero partial height splitter blades, and a diffuser with 13 vanes. Based upon the analysis, the compressor was predicted to have a Total Adiabatic Efficiency = 0.72. Figure 2 shows the aero components.

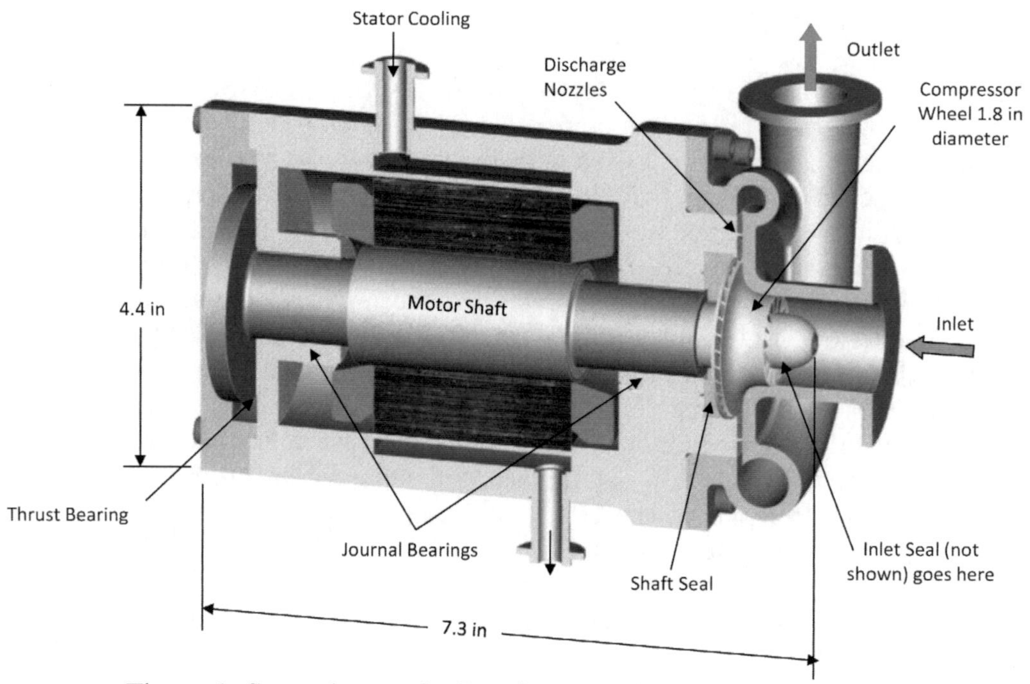

Figure 1: Supercharger for Pre-Compressing Refrigerant Vapors

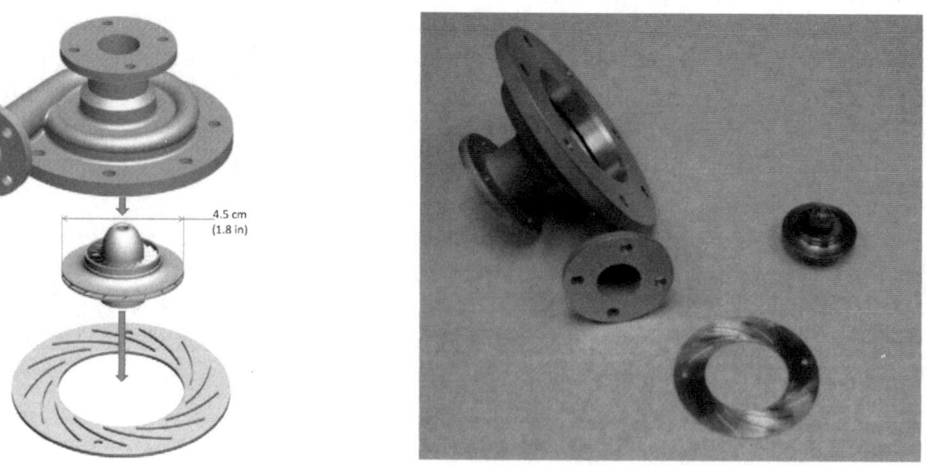

Figure 2: Final Design and Prototype Aero Parts - Volute Housing (Top) Impeller (Middle) and Diffuser (Bottom)

The impeller, shaft, and attachment bolt were analyzed using finite element-based software to evaluate operating stress and modal characteristics, with satisfactory results.

Manufacturing Methods for Compressor Aero Components: Historically, manufacturing prototypes for these components is expensive and time-consuming, with components made using traditional machining, casting, or a combination of both processes. This involves initial development cost and lead time to produce the first article components, which were prohibitive for this project. To address this issue, MSI internally funded an effort to identify and evaluate alternative manufacturing techniques that are faster for prototypes and less expensive. *Additive manufacturing* and *rapid investment casting* were identified as two advancing technologies that offered much promise

Additive Manufacturing: Also known as 3D printing, is growing in popularity and capability. Originally developed to create 3D shapes using thermoplastics and photopolymers, the technology is broadening and now includes a variety of metals. The process involves the use of a high energy beam, either laser or electron, which is focused on a bed of powder metal. Three processes currently used to fabricate metal components are Direct Metal Laser Sintering (DMLS), Selective Laser Melting (SLM), and Electron Beam Melting (EBM). Of the three processes, DMLS is the most commonly used and offers the best dimensional control and finish quality for small parts, and therefore was selected for manufacturing the impeller.

Rapid Investment Casting: Though the technology has been around for centuries, investment casting has recently been able to make improvements in reducing initial cost and lead times by taking advantage of additive manufacturing technology. The process involves building the pattern for the molds directly from a CAD file using a 3D printing process, such as Stereolithography (SLA) or Selective Laser Sintering (SLS). Lead times are reduced to days versus 8 to 10 weeks for standard investment castings. Parts are limited in size to those that can be built with the additive manufacturing process, but in some cases sections can be built and joined together to produce larger parts. Because of its size, the volute housing was manufactured using the rapid investment casting process.

Material Assessment: Although the DMLS process possesses many advantages, concerns exist regarding the strength and surface quality of the sintered power metal. For the impeller, a precipitation hardened stainless steel, 17-4PH heat treated to the H900 condition was selected for its high tensile strength. To evaluate the tensile properties of the sintered material, three tensile test specimens were fabricated and processed identically along with the impellers. These specimens were then tested to failure and the results compared to the specifications for wrought and cast material, as listed in Table 1. The results show the sintered material exceeded the minimum strength requirements for both wrought and cast material. The sintered material exhibits slightly lower ductility than wrought material, but better ductility than cast material. In addition, one of the finished impellers was successfully spin-tested to 115,000 rpm (15% over-speed) without failure.

Impeller Surface Finish: Ideally, surfaces exposed to high velocity gas flows should be as smooth as possible to minimize turbulence and friction losses. A surface finish of 32 µin or better is preferred. As processed, the DMLS material exhibits a surface finish of 130 µin. Accessible surfaces can be improved through bead blasting or polishing, however, the presence of the shroud prevents direct access to the blades. Alternate polishing methods were explored to improve the surface finish of these inaccessible areas, including electro-polishing, electro-plating, and extrusion honing. Of these, electro-polishing was determined to be the most likely to improve the surface finish most uniformly with minimal impact on dimensional tolerance. One of the impellers was sent to an electro-polishing vendor, who adjusted process parameters to optimize the process but with marginal results.

Table 1: Tensile Test Results Comparing DMLS Produced Samples to Other Forms

DMLS Data				
Specimen ID	Ultimate Tensile Strength, ksi	0.2% Offset Yield Strength, ksi	% Elongation	% Reduction in Area
SN1	208	186	5	11
SN2	210	197	7	8.5
SN3	211	193	9	15
Published Data (minimum values)				
Wrought	190	170	10	35
Investment Cast	180	160	4	12

While improvement was realized in some sectors, other areas were worsened, with the process creating localized pitting. The vendor attributed the varying result to the non-uniform nature of the sintered material. It was therefore decided to proceed with the as-manufactured surface finish and accept the aero performance degradation for initial testing. Methods of producing the impeller in two pieces (i.e., a separate shroud), enabling full access to the blade surfaces prior to attaching the shroud, can be explored later. As related below, initial testing revealed an adiabatic efficiency of 0.68 vs. the above-noted prediction of 0.72. We view this result as quite acceptable for a first prototype, with impeller surface finish and the type of seals implemented (see discussion below) deemed the principal contributors to the difference observed.

Volute Housing Fabrication: Initially we planned to make the volute housing from 17-4PH stainless steel, the same material as the impeller. However, the casting vendor ran into difficulties getting uniform flow through the thin-walled sections of the casting. Since the strength of 17-4PH is not required for the volute housing, the material was changed to A-356 aluminum, an alloy with superior casting properties. The result was a complete success, with the end product shown in Figure 2. To verify the aluminum casting had adequate strength, the housings were successfully hydro-tested to 150 psi (150% of the working pressure).

Diffuser Fabrication: We used traditional machining methods for this part, using Grade 304 stainless steel.

Motor and Controller: Two motor types (Induction and Permanent Magnet) were evaluated. A PM style was selected for the prototype since it results in a smaller, higher efficiency motor. Based on an assessment of parasitic losses, a 100,000 rpm 4-pole motor with nominal rating of 10 kW (13 hp) was selected. It contains stator laminations that are 0.007 in thick and copper windings, with a predicted efficiency of 96.9 percent. Figure 3 shows a completed stator and shaft. A commercially available digital drive variable frequency drive (VFD) which operates without the need for a Hall Effect sensor was selected.

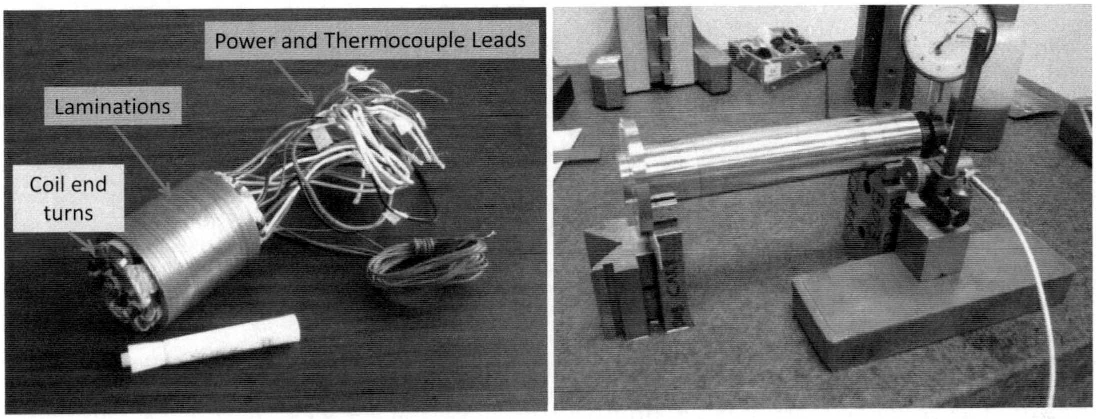

Figure 3: Motor Stator and Shaft

Bearings and Seals: A parametric design study was conducted to size the gas foil bearings. Load capacity was determined over a range of bearing lengths and diameters at 100,000 rpm. Once the final designs related above were complete, the bearing sizes were finalized as follows: Both radial bearings are 1.3 in. diameter x 1 in. long. This results in a more-than-adequate 13 lbf load capacity per bearing for the vertically mounted machine. The single acting thrust bearing has a 2.5 in. OD and 1.5 in. ID with 6 pads of approximately 60° arc angle per pad. The load capacity of this bearing is 55 lbf, which is ample to carry the weight of the rotor and anticipated compressor thrust load.

In small centrifugal compressors, sealing around the impeller is an especially key part of overall machine performance. As shown in Figure 1, there are two principal dynamic seals in the supercharger: At the impeller inlet, and on the shaft behind the impeller. For initial operation, labyrinth seals have been implemented at both locations. Labyrinth seals are commonly used in turbomachinery. They provide a close-clearance but non-contacting seal, so there is a small amount of flow (leakage) through the seal. In a large compressor such leakage is generally of no consequence. In small machines however, this leakage can be a significant portion of the total flow. At the shaft seal, a small amount of purposeful flow through the seal cools the motor and bearings by passing refrigerant through the bearings and motor-stator gap. As operational experience is gained, we will consider alternate inlet seal configurations that may moderately improve supercharger efficiency. A potential downside associated with some higher-performance seals is their potential to introduce small forces on the shaft that could degrade the dynamic stability of the supercharger's suspension system. It should be noted that all flow through the seals is retained within the sealed refrigerant system and merely recompressed; there is no "leakage" in the traditional sense. Figure 4 shows a set of bearings and seals.

Figure 4: One Set of Bearings with Thermocouples Attached (left) and Seals (right)

Rotordynamics: Compared to most other bearing types, gas foil bearings have lower stiffness and lower load carrying capability. As part of the design process, initial predictions were made to confirm no critical speeds within the anticipated operating speed range. With all key aspects of the rotor and bearings established, a final rotordynamic evaluation of the rotor was performed to predict critical speeds and their mode shapes. The main portion of the shaft is a high-strength alloy steel. The magnets are neodymium-iron-boron, segmented into three pieces axially and four pieces circumferentially. The magnets are bonded to the shaft with epoxy and held in compression with an Inconel sleeve which is installed with a heavy interference fit.

The analysis was performed using the journal bearing stiffness values generated by one of MSI's proprietary foil bearing codes. These values, the rotor geometry, and impeller mass properties, were input into the computer code RotorLab® [4] to determine rotor critical speeds and mode shapes. To be conservative, the motor magnet assembly was not included in the stiffness calculation, although its mass was included. Including a stiffness allocation would tend to raise the first bending critical speed (third overall critical speed).

The results of the rotordynamic analysis for the first three modes are summarized in Table 2 for running speeds of 50,000 and 100,000 rpm[1]. A conservative approach is to achieve a design wherein the rigid body modes are below the machine's minimum operating speed, and the bending modes are comfortably above the maximum operating speed. In this case, the rigid body modes were predicted to occur well below the supercharger's minimum operating speed of 85,000 rpm. The predicted first bending critical was well above the machine's 100,000 rpm max speed.

[1] The shift up in speed is the result of several effects, including increase in bearing stiffness as a function of speed, and gyroscopic stiffening.

As discussed below, Table 2 also shows the approximate critical speeds actually observed in the test data. For modes 1 and 2, the predicted vs. actual differences are attributed to lower bearing stiffnesses due to smaller shaft displacements and bearing loads than those used in the analysis. For mode 3, the difference is likely due to higher overall shaft stiffness, owing to a contribution from the magnet attachment method.

Table 2: Predicted and Actual Critical Speeds (in rpm) of the Rotor

Mode Number	Mode Shape	@ 50 krpm	@ 100 krpm	Observed During Test
1	Rigid Body	23,430	24,450	15,000
2	Rigid Body	28,800	31,400	18,000
3	Bending	143,720	146,970	154,000

Initial Testing and Results: An in-house vapor-only test loop to assess compressor performance was designed and built. In addition to flow meters, accumulators and instrumentation to measure performance and vibration, this loop includes heat exchangers to remove the compression heat added by the supercharger. A National Instruments LabVIEW™ data acquisition system gathers 30+ channels of process and dynamic data in real time.

Initial tests involved grooming the motor controller. This was followed by scoping runs with a vane-less impeller, dummy housing, and air not refrigerant in the loop, to confirm rotordynamic performance. For these tests, additional instrumentation was in place, including eddy current displacement sensors to observe shaft motion at all speeds.

Shaft displacements were quite low at all speeds, nominally 0.0002 inches zero-peak when compensated for runout.[2] Figure 5 shows shaft orbits at the lower (Compressor end) and upper (Thrust Bearing end) regions of the Supercharger at various speeds. The four-lobe pattern in the uncompensated orbits is attributed to A) the effect (magnetic influence or actual shaft motion) of the 4-pole motor, or (more likely) B) the effect of the shims (4 in number, 90 degrees apart) installed in the radial bearings, a common method used to preload gas foil bearings in vertically mounted machines. The "fat" orbits at the highest running speeds are attributed to data sampling rate limitations. An oscilloscope directly observing sensor outputs confirmed narrower orbits at all speeds.

Confirmation of Rotordynamics: The rigid body critical speeds occurred in the 15,000 rpm range, lower than predicted. As noted above, the discrepancy is attributed to the low shaft displacements, which results in lower bearing stiffnesses and therefore lower critical speeds. Figure 6 shows shaft displacements (uncompensated for runout) through the 15,000 rpm (250 Hz) range were comfortably low, indicating a good amount of damping in the bearings.

---

[2] The term runout refers to local-to-the-sensor surface and near-surface discontinuities, and small variations in the shaft's mechanical and electrical properties, which are detected by this type sensor, in addition to the dynamic movement of the shaft. "Compensated" means these effects have been vectorally subtracted from the overall data, leaving only the dynamic data, which is of primary interest here.

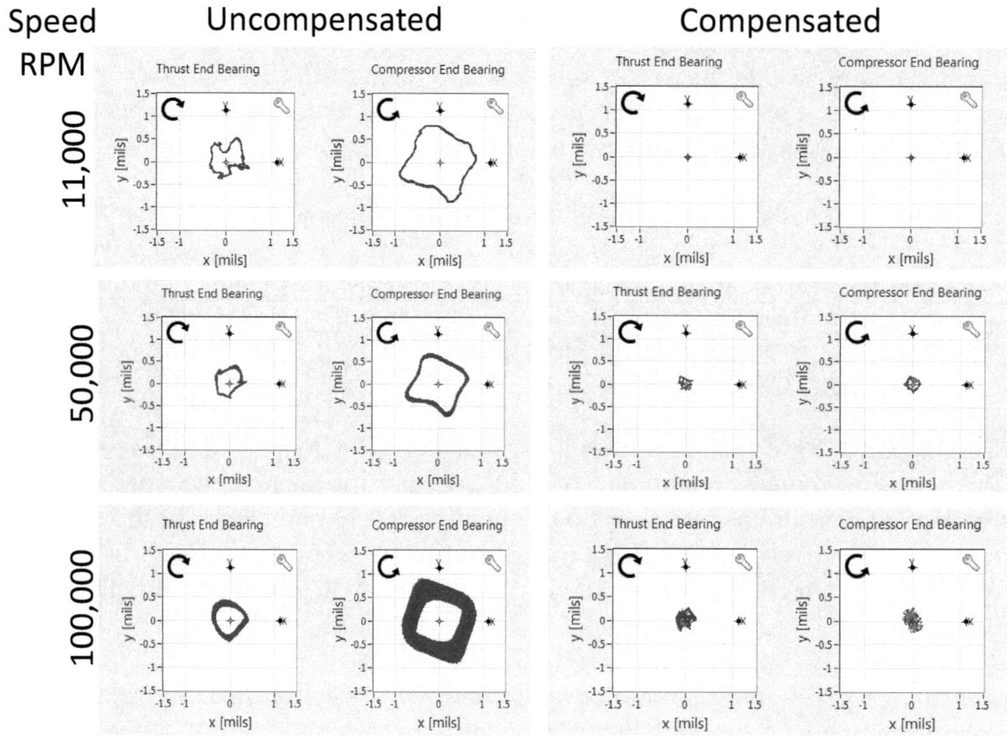

Figure 5: Shaft Orbits at Three Speeds, with (right side) and without (left) Compensation for Runout

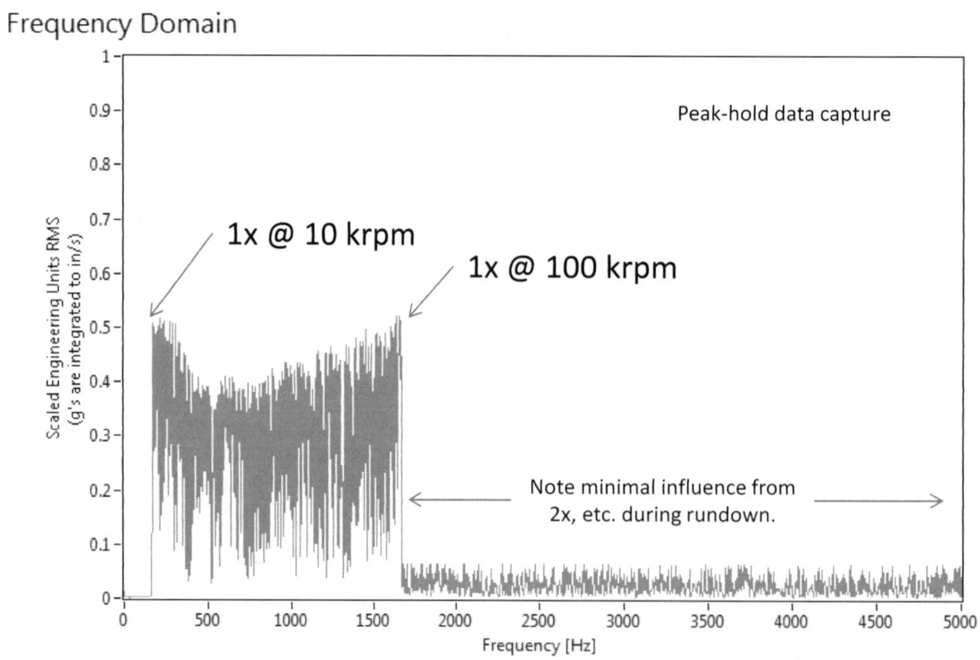

Figure 6: Compressor End Vibration, mils$_{rms}$, During Rundown from 100,000 rpm to 10,000 rpm; No Runout Compensation

Subsynchronous Vibration: At speeds above 90,000 rpm, a low level subsynchronous was observed, at a frequency equivalent to the rigid body criticals, see Figure 7. Once encountered, the subsynchronous did not increase significantly in amplitude as running speed was increased to max. Once present, the subsynchronous was discernable, at low amplitudes, until the running speed was dropped below 50,000 rpm. This phenomenon, sometimes labeled a controlled instability, is not unusual for shafts running on gas foil bearings.

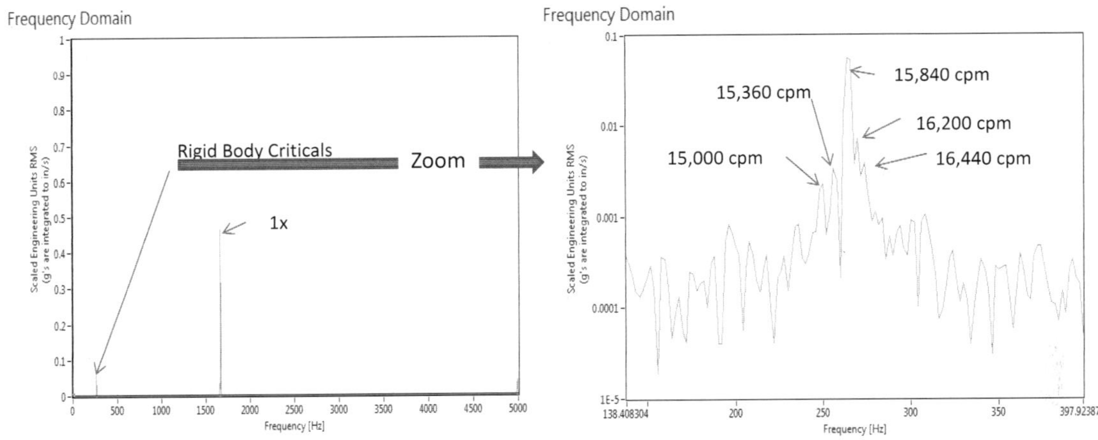

Figure 7: Compressor End Shaft Displacement, $mils_{rms}$ @ 100,000 rpm; No Runout Compensation

Damping Determination: An important aspect of the dynamics of high speed machinery is the amount of damping present at the bearings, as low damping can cause dynamic stability issues. Instability often results in excessive vibration and sometimes machine failure in short order. Gas foil bearings generally have lower damping than most other types of hydrodynamic bearings. Combining this aspect with the usual situation where a foil bearing-supported machine operates at speeds many items higher than the machine's rigid body critical speeds makes damping the most critical ingredient for successful machine operation.

For these reasons, we elected to experimentally determine the amount of damping in the supercharger's rotor-bearing system by implementing a commonly used "ring-down" method. Specifically, with the supercharger running, the motor housing was tapped with a hammer while observing lateral displacement of the spinning shaft at the shaft displacement sensors. Figure 8 shows a typical response, with these average results: Log Decrement = 0.64; damping ratio = 0.1; amplification factor = 5.

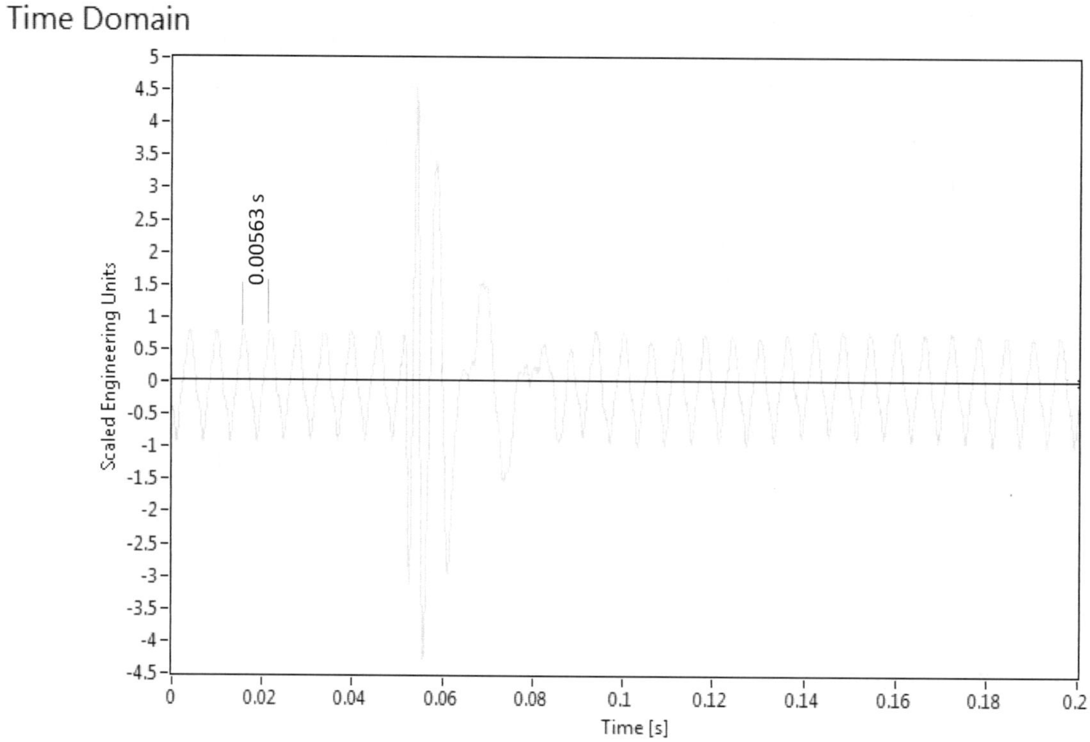

Figure 8: Compressor End Shaft Displacement, mils$_{rms}$ @ 10,000 rpm; No Runout Compensation

With these tests complete, the vane-less impeller was switched out for the bladed impeller, diffuser and shaft seal but not the volute or inlet seal. This configuration was run to max speed to verify no significant changes in vibration or rotordynamics, and none were observed. The test loop was then re-plumbed for refrigerant, and the supercharger's test housing switched out for the volute and inlet seal.

During the impeller change-out, the motor internals and bearings were inspected and found to be in excellent condition, with only minor wear-in marks noted. The supercharger was reassembled using the same bearings.

Aero Performance: Test runs to map as-built performance were conducted in late 2013 and early 2014. Figures 9 and 10 relate performance data from all of the runs conducted to date. Figure 9 shows the results as generally depicted for centrifugal compressors. For this style compressor, optimum operation is obtained when running through the center of the "islands" of constant efficiency. Figure 10 shows the corresponding head vs. flow coefficients, which can be used to assess how the compressor will perform in other applications. Both figures show encouraging data, with consistent results, reasonably shaped performance curves, and excellent data point clustering. Although somewhat lower than predicted, both head and flow coefficients are quite acceptable for a first prototype. As noted above, this initial build included purposeful simplifications regarding seal type and impeller finish. The Operating Line shown in Figure 9 represents current plans for supercharger operation when combined with a commercially available PD compressor. The region above the top end of the Operating Line is being reserved as margin to fine tune compressor interoperation, if needed.

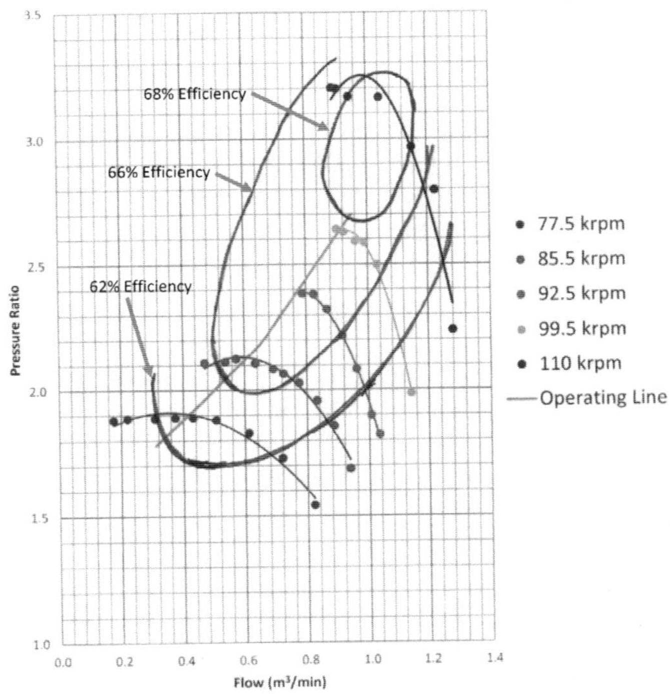

Figure 9: Compressor Performance Map for First Article, and anticipated Operating Line

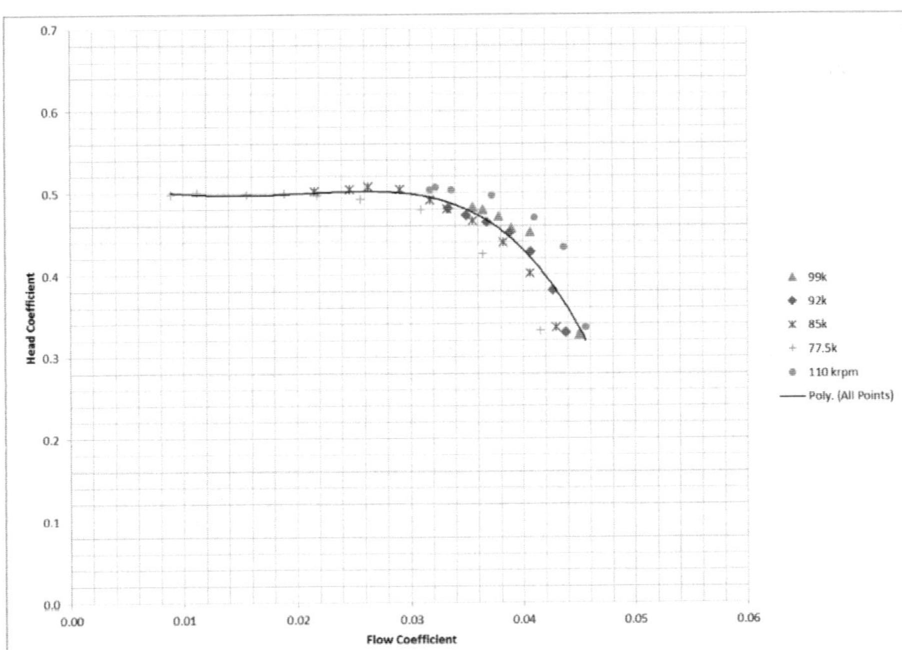

Figure 10: Head and Flow Coefficients for First Article Compressor

Other Performance Metrics: Motor housing vibration levels were typically 0.02 in/sec_{rms} or less. Acoustic (airborne) sound levels were not measured but observed as principally a low-level tone, likely associated with impeller vane or blade pass, at speeds above 50,000 rpm. Thermally, bearing losses were in line with predictions. Windage and stator losses were somewhat higher than expected.

Present Status and Plans: As presently configured, the prototype supercharger is suited for a heat pump in the 45 kW (150,000 Btu/hr) class, with a reasonable COP_h of 2.32. Work is underway to complete the DOE SBIR grant work scope by marrying the supercharger to a properly sized PD compressor and demonstrating effective solo and simultaneous operation of both compressors on our in-house test loop. Within the constraints of remaining schedule and budget, we are also implementing hardware modifications to reduce windage and motor losses.

Planned work includes discussions with potential commercializers, such as heat pump compressor original equipment manufacturers (OEMs), who bring built-in marketing and sales capabilities, along with service and support. The idea here is to integrate both compressors into a single package with overarching controller, for sale under their brand name.

Follow on work includes an anticipated grant from the State of New York to configure the compressors with associated heat exchangers, fans, and associated controls to fashion the components of a cold climate heat pump for demonstration to interested third parties, including heat pump OEMs, who have direct access to end users and are well suited to establish the bottom line economics needed for a viable cold climate heat pump [5]. This work is expected to commence in spring 2014.

Other potential uses for the refrigerant pre-compression concept described herein include industrial and commercial coolers, chillers and freezers as a compact and energy-efficient alternative to existing methods, with the potential for increased capacity or lower temperature operation. Outreach to suppliers of this type of equipment is planned as related system demonstration work continues.

Conclusion: Centrifugal compressors are used throughout industry because they have fewer rubbing parts, are relatively energy efficient, and give higher flow than a similarly sized reciprocating compressor or positive-displacement compressor. However, they cannot achieve the high compression ratio of reciprocating compressors without multiple stages [6]. In many respects, the application discussed herein makes optimum combined use of both compressor types. It is concluded that work conducted to date has verified the underlying design of a motor driven supercharger to improve the cold-day performance of air source heat pumps.

References:

[1] http://home.howstuffworks.com/home-improvement/heating-and-cooling/heat-pump.htm

[2] Small Business Innovative Research Grant DE-SC0006162 "Supercharger for Cold Climate Heat Pumps".

[3] Proceedings: 11th Biennial IEA Heat Pump Conference, May 11-16, 2014, Montreal, Canada.

[4] Rotating Machinery and Controls (ROMAC), University of Virginia.

[5] New York Energy Research and Development Authority Grant number 34903 "Supercharger for Cold Climate Heat Pumps".

[6] Wikipedia the free encyclopedia: http://en.wikipedia.org/wiki/Centrifugal_compressor

DESIGN OF A MODULAR, COST-EFFECTIVE EMBEDDED WIRELESS CONDITION MONITORING SYSTEM FOR ASSET MANAGEMENT OF DISTRIBUTED FLEETS OF LIGHT COMMERCIAL VEHICLES

Michael G. Lipsett Stephen C. Dwyer, James D. Yuen,
Nicolas A. Olmedo, Michael G. Mills
Department of Mechanical Engineering
University of Alberta
Edmonton, Alberta, Canada T6G 2G8
Telephone: (780) 492-9494
mlipsett@ualberta.ca

Abstract: In an effort to improve the return on investment for assets, it is becoming increasingly important for companies to develop and implement techniques for more efficient asset management, particularly through condition monitoring. Improved reliability and productivity levels for technological assets such as fleet vehicles and material handling systems in factory and warehouse environments are key metrics for competitiveness. On a system such as a forklift fleet, distributed on-demand condition monitoring, combined with model-based fault analysis and prediction techniques, can raise productivity through more efficient maintenance schedules and reduced downtime. A small, customizable embedded sensor system is described for forklifts, with wireless connectivity and automatic status reporting for a set of condition indicators. Technical specifications are based on information flow for providing asset tracking of location, hours used, and machine condition. The system will be capable of logging and transmitting sensor data over a cellular network to a centralized server and database system capable of providing improved model-based condition monitoring reports to fleet owners and operators. An embedded system has been designed, prototyped, and demonstrated in field service using cell modems and a web-based reporting tool.

Key words: Diagnostics; fleet monitoring; embedded systems; cellular; asset management; vehicles.

Introduction: For a company operating a distributed set of assets, such as a large fleet of vehicles, ensuring they are being properly monitored and maintained is crucial to productivity and profitability [1]. An Alberta-based company manufactures onboard sensor systems which track the usage of light commercial vehicles (forklifts). This company wanted to improve the technology to allow for automatic data collection and analysis of machine condition across a large fleet of vehicles, with uplink to a central computer system for reporting and decision-making, specifically a prototype system to automatically log the number of hours that a forklift is operating, based on when the engine is on. Hour meters are commonly installed on light commercial equipment, but collecting the operating information is laborious and often not captured accurately. Much of the maintenance that is done on a scheduled calendar basis is excessively conservative.

As well, there is little information collected on machine condition, and so predictive maintenance programs have not been implemented [2]. Enabling technologies now exist to provide low-bandwidth information from mobile equipment simply and inexpensively [3,4].

A generalized system architecture would allow for a range of performance and condition monitoring applications for remote or distributed assets, from mobile equipment [5-8] to plant components that are not connected to control systems [9-12]. A University of Alberta project team created a general framework for system development, and an implementation suitable for industrial applications, using embedded systems with wireless connectivity [13,14]. The bandwidth requirements depend on the sensing application and the amount of onboard processing [15-17]. For mobile equipment, information can be used for condition monitoring or for operational decision making [18-22], especially when equipped with a positioning system [23,24]. This applies to commercial vehicles, mining equipment, and railcars, to name a few [25,26].

The project comprised system specification development, initial prototype demonstration, revisions to specifications, and development and testing of two prototypes on vehicles. The alpha prototype design and preliminary test results are presented, covering hardware, firmware, and software. A working alpha prototype was delivered in less than ten months; and a number of areas for development have been identified to bring the hour meter unit to a production ready product.

Functionality: For proof of concept, an alpha prototype system was designed to implement the following features:

- Hour meter count based on a (digital) on/off input from the vehicle
- Status feedback with simple LED to indicate local system heartbeat
- GPS location updates with receiver integrated in a COTS certified cellular modem
- Periodic reporting of hour meter, cell reception, and GPS data pushed from the onboard device to a central server
- Event reporting to track times when a vehicle is turned off and on (based on hour meter input)
- Low power standby mode for the embedded device when the vehicle/hour meter is off to reduce power draw from the vehicle battery
- Wakeup when vehicle turns on or when it is time for another periodic update
- Periodic update rate adjustment through communication with the server.

The flow of operation of the embedded device is shown in Figure 1.

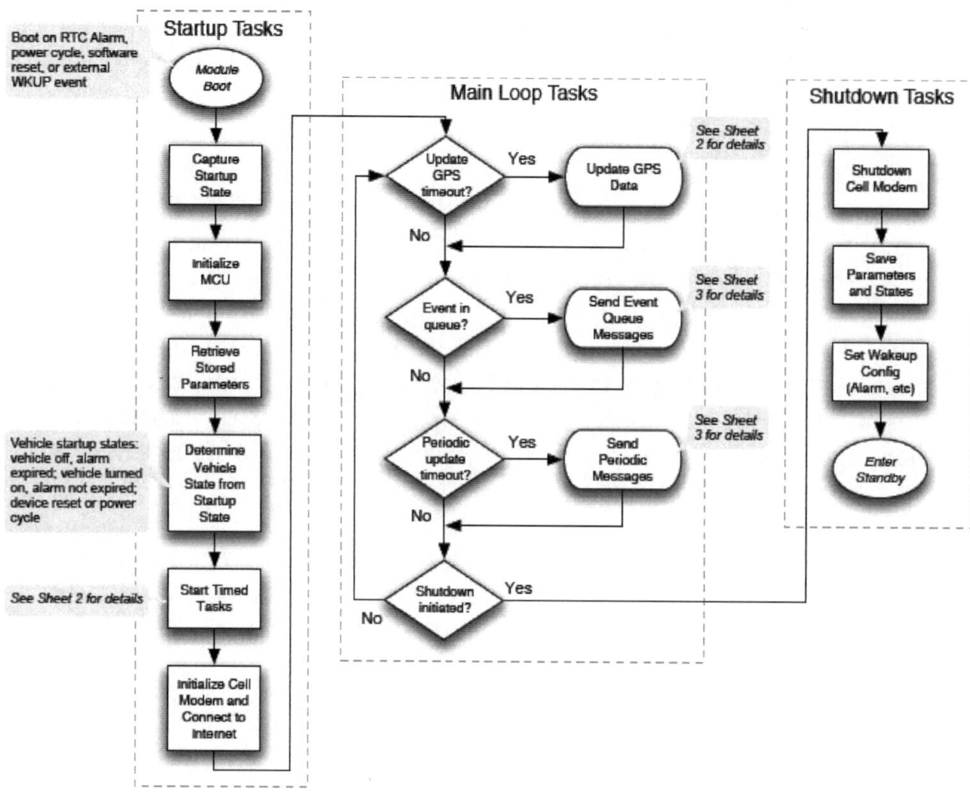

Figure 1: Software Flow of Activities

System Design: A general system architecture has been designed, and a prototype system has been developed and tested for a cellular-network-connected embedded device to monitor simple features from a vehicle. The application of first interest was a forklift-hour meter. A block diagram for the embedded onboard device running on vehicle power is shown in Figure 2.

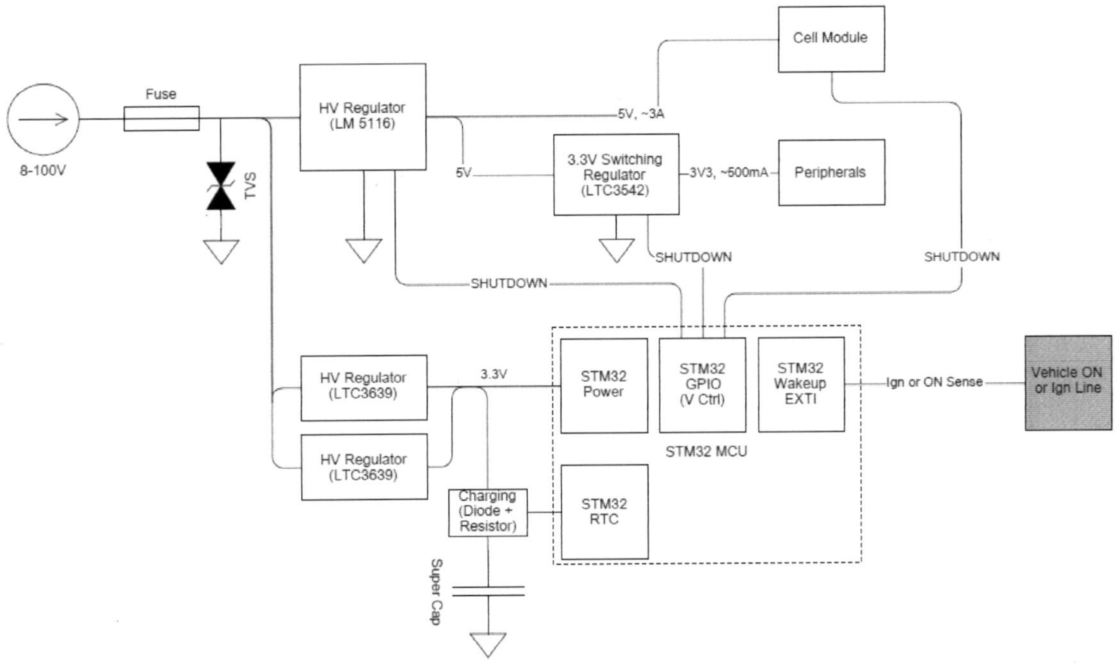

Figure 2: Alpha Prototype Embedded Onboard Device System Block Diagram

Hardware and Firmware: The core functionality of the device side of the cellular hour meter monitoring system is primarily driven by the device firmware, which complements the hardware to provide all required features. Firmware is written in C, cross-compiled on a host machine with gcc, and targeted against the Cortex-M3 based microcontroller on the alpha prototype, the STM32F103. The cellular module used in the system was the HSPA910.

Firmware files, along with other software source, documentation and hardware design files, are version controlled using the *Git* source code management system. In general, the code is organized by firmware, which each firmware calling on its own configuration or supporting source files. Supporting source files are specific to the project and help implement features critical or unique to the functionality of the hour meter device.
Specifically, the code implementing the communications protocol with the server is implemented in such supporting source files. Each firmware also calls on local (to the project) configuration and source libraries. Local libraries are generic, and not specific to the project, for functionality such as low-level device drivers and utilities. In some cases, firmware makes use of additional third party libraries to implement functionality, in particular, the libopencm3 open source low-level library for Cortex-M processors is heavily integrated into the system.

Firmware was developed incrementally according to feature. Features are identified and broken into reusable library functions, or kept as part of firmware-specific source code. For example, state machines were created to include all the necessary states for the cellular module. After initial code development, unit testing firmware is written to debug and verify the feature. Such unit testing is typically accomplished using development hardware that is well suited to this task. Some features that are specific to the hour meter unit itself are directly tested on the device hardware. Finally, all features are integrated into the appropriate demonstration or main device firmware.

The following tools and libraries are used in the development of firmware:
gcc-arm-embedded - gcc open source toolchain targeted against the ARM Cortex-M
make for building source automatically
libopencm3 to provide convenient low level access to the STM32 MCU
git for source code management
Olimex open source development hardware
Black Magic Probe open-source JTAG/SWD debugger
gdb as part of this debugging tool, and
Saleae Logic logic analyzer for low level timing and communications debugging

An effort was made to use free, open source tools in development as much as possible. In general these tools come at a greatly reduced cost and in many cases are equal if not better than commercial development suites. While many of the tools are open source, care was taken not to use any source libraries or tools that compromised the capacity of the hour meter to remain proprietary with closed source code.

After initial alpha testing, improvements were identified. Periodic polling of hour meter state should be done instead of wakeups triggered by a rising edge. (This change is needed because if the rising edge occurs in a few tiny slices of time during shutdown, then there is a slight chance the edge will be missed, and the device will go to sleep instead of immediately waking back up to continue counting hours.) Another recommended change is to improve cellular AT command interaction and parsing to reduce wait times when unnecessary and handle additional errors or states as required. There are some desirable features that may be easily included in a future revision to improve device functionality. Input (battery) voltage monitoring and reporting will allow for improved status monitoring and power management of the embedded hardware. Improved non-volatile data storage will improve reliability. Improved separation between data sampling and data reporting will allow different sampling rates versus reporting rates (for example, sampling and storing GPS data once per minute, but only reporting all of this data once per day), thereby reducing power and cell data consumption. An expanded set of data collection and associated event reporting will make the device more versatile.

Initial firmware testing was carried locally using traditional debug tools. Then, a basic debug server was developed that allowed the embedded device to communicate with the server in a manner very similar to the full server system, with additional test and debug information available. An embedded device simulator was also developed to aid in testing both the debug server and the full server system. After testing the firmware with the debug server, final testing was completed with the full server system. Current firmware has been bench tested under normal operating conditions, and tested inside vehicles under normal operating conditions, for up to 48 hours consecutively. Further testing is required under extreme or abnormal operating conditions. This should be carried out both in the lab and on a test vehicle.

Software: The server software is divided into two major components, the device server and web server. This layout is shown schematically in Figure 3.

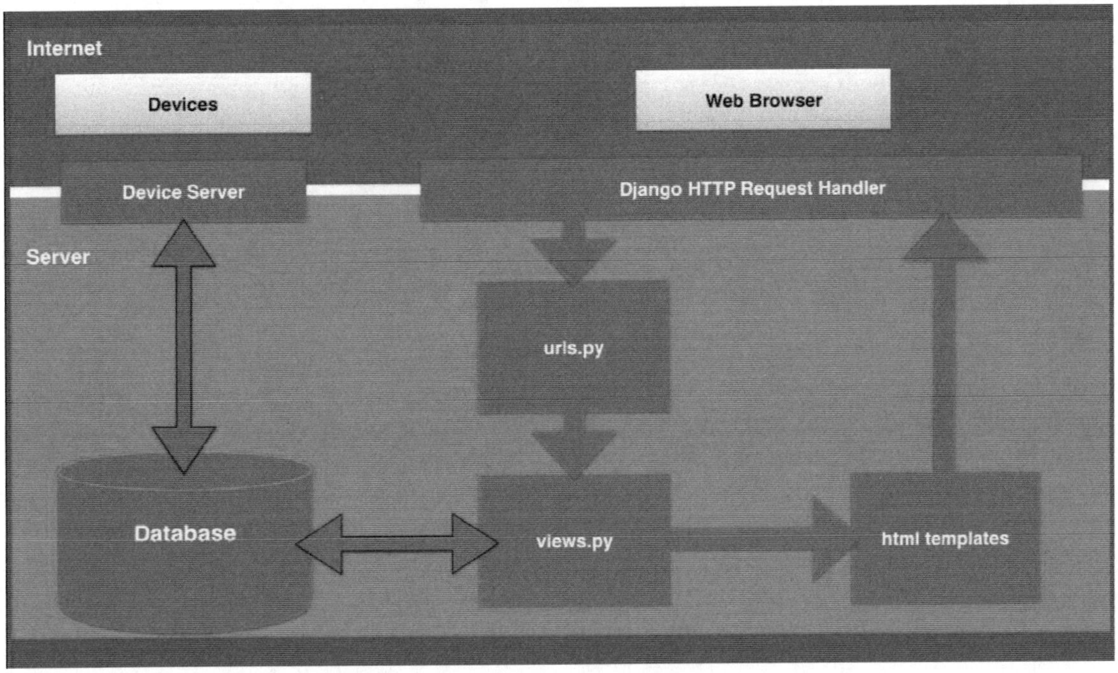

Figure 3: Schematic Structure of Software with Two Servers

Devices on board vehicles communicate with the device server which then saves data into the database. The web application receives http requests and filters them based on the url structure using a script *urls.py*. The url requests are then passed onto the *view.py* module which uses the database and html templates to build the response pages for the user. Both servers are run on a Debian 6.0 system.

The device server receives data from devices in the field, saves the data to the main database and responds to the device with an ACK or NACK message. This application must run with a global IP address to allow devices to directly connect to it.

The web server was built using the django web framework which is implemented in the Python programming language. Django is a free and open source web application framework which follows a model-view-controller architectural pattern. Django's primary goal is to ease the creation of complex, database-driven websites. Django emphasizes reusability and modularity of components, rapid development, and the principle of don't repeat yourself. Python is used throughout, even for settings, files, and data models. Django also provides an optional administrative create, read, update and delete interface that is generated dynamically through introspection and configured via admin models.

Currently the Django built-in web server is used to run the application and the database engine used is SQLite. This was deemed to be adequate for a demonstration prototype; but this part of the system would need to be upgraded in a production version.

A web interface is an important part of the system. It allows for flexible access to information on a range of devices, including smartphones. The home page of the web interface shows a list of vehicle groups and the vehicles in each group. This page also provides access to the administration interface and can be customized for customers as required. The vehicle page shows the details about the vehicle in question. A map shows its current location along with a history of vehicle movements. The vehicle's current state is shown as the last event that the vehicle sent. All pages are easily customized. The vehicle page as implemented for testing and demonstration is illustrated in Figure 4. The figure shows a screenshot of the vehicle page for a vehicle being tracked for hours and location as it drove in an urban area.

Figure 4: Web Interface Vehicle Page Sample Screenshot

Administrator pages provide administration of the entire system, including users, permissions, devices, events, periodic updates, device groups, and histories of changes made to the system. An AJAX-style interface is appropriate to reduce the effort required by the end user to make page updates.

An analytics page shows a graph of when and how long the vehicle was in use. There is of course a significant opportunity to customize the display and variety of analytics for tracking vehicle usage, condition indicators that may indicate an impending failure, etc.

A production system will require a more robust and large scale platform to run the web site and device communications. In a production environment the web app code will likely run on an Apache server with a MySQL database backend or equivalent. Also required will be to lease or buy data hosting services with proper data lines, power, backups and maintenance personnel. In the original design the website was designed to be a monolithic entity serving all the users and admins from a central web application. It may be better to administer services for a given company's fleet on a separate web server (which may be virtual). In this way, each application lives independently of the others and can easily be added or removed as needed. Third-party visualization tools such as Google Maps would require licensing.

Testing and Design Compliance: The prototype design was compared to the design specification matrix created at the beginning of the project. Overall the alpha prototype was in compliance with the specifications for a networked embedded system on a light vehicle, with these notable exceptions:

- Report Generation: Although user report generation has not yet been implemented, all the data required to do so is already stored in the database. Implementing report generation is now a relatively minor task and can be done quickly after a report template is agreed upon.
- Product Life: Further testing is required to ensure a product lifetime of 5 years under expected operating conditions.
- Connection Security: Encryption was not implemented in the Alpha prototype, but is staged to be added in the next revision.
- Power Draw: Fully complies with an average of 1.5~W during transmit and 1.3~mW in shutdown.
- Embedded Environment: Further testing is required under expected operating conditions.

Because the embedded device runs on vehicle electrical system, the device must not be affected by voltage transients in the vehicle. These transients generally occur during starting, when the current draw can pull the supply voltage to less than 30% of its nominal voltage. Features were included for voltage regulation, spike suppression, and provision of sufficient current to the device to not power down during low voltage events. There are two types of voltage sense circuits on the alpha board: voltage dividers to scale supply voltages for measurement with the microcontroller unit analog-to-digital converter, and the optocoupled digital sense lines for hour meter state control. The voltage dividers were measured by applying a known input voltage. The voltage dividers were selected to ensure a factor of safety beyond the nominal operating voltage. The optocoupler maximum voltage is determined by the current limiting resistors in series with the emitter.

On-board power supplies were tested independently (isolated with jumpers), except for the 3.3V peripheral supply, whose input was connected to the output from the 5V supply. A known voltage sourced from a laboratory-grade power supply was applied to the input, a known load was applied to the output (using a lab programmable DC load), and the resulting output voltage was measured with an oscilloscope. Noise was also estimated based on oscilloscope output. Both prototype devices were able to remain operational during engine start events.

Conclusions: A simple but effective system has been demonstrated for monitoring hour meters remotely over a cellular network, with a simple web-based user interface for generating reports. In the first stage of the project, system specifications were developed and a basic architecture was designed for a distributed wireless fleet monitoring system. The initial implementation considered hour metering only, but location and machine health status indicators were also added. Next, a preliminary proof-of-concept prototype was prepared and demonstrated. Reviewing the initial specifications after completing the proof of concept prototype introduced some changes into the system requirements.
A complete embedded prototype was designed and developed. Two prototypes were built and underwent basic testing. Firmware implementing device functionality was written in parallel with hardware development. Firmware was initially tested on lab development hardware, and later verified on the alpha prototype. A database was developed on a server that received information from the onboard units through cellular modems. A web client accessed the database and generated simple reports.

The next steps in development are to take the working alpha stage prototype and produce a beta product for testing at actual sites where forklifts are operating. Firmware will be tested to ensure that it performs well under the full range of conditions without missing events, even if there is a loss of power to the system. The hour meter counting strategy will be critiqued, with more polling to avoid missed state changes. Parsing of AT commands will be implemented to check for errors, and message structure will be modified and moved to HTTPS. A more robust cell driver will reduce the possibility of dropped messages. Additional features that need to be added include an over-the-air boot loader for remote firmware upgrades, data encryption, and stress testing.

Additional hardware testing remains to be done to ensure antenna reliability and suppress a wider range of possible power input spikes. The next stage of development would involve building a beta prototype for limited deployment and long term testing on a forklift truck fleet, with reduced power consumption, smaller board size and reduced component count. A ruggedized case and connector set, and other features for durability. After testing of the beta prototype, additional steps will be required to move into hardware production. Several development steps are also needed for a commercial deployment of the software, including certification for a custom cell implementation. The software, communication system, and database must each be scalable for more features, multiple companies, and large fleets, with security features for each company's information. Any third-party software will have to be licensed.

This technology offers promise to deliver reasonable quality operational and condition monitoring information for a variety of distributed, remote assets, such as light commercial vehicles, dedicated process facilities that lack sensing and communication infrastructure (pipelines, railway), and environmental monitoring stations. The architecture even allows for actuation, which would be useful for closing valves, taking samples, and other simple actions.

Acknowledgments: Funding support was provided by Mitacs Inc. through an Accelerate Cluster internship program.

References:

[1] D. Wikoff. Billions of dollars to be saved with reliability excellence. TMS Annual Meeting, pages 187 – 191, 2009.

[2] T. Schmidberger, A. Horch, A. Fay, and R. Drath. Rule based engineering of asset management system functionality. Proceedings. 5th Vienna Symposium on Mathematical Modelling, page 226, 2006.

[3] J. Muller, U. Epple, M. Wollschlaeger, and C. Diedrich. An efficient information server for advanced plant asset management. EFTA 2003. 2003 IEEE Conference on Emerging Technologies and Factory Automation. Proceedings (Cat. No.03TH8696), vol.2:66 – 73, 2003.

[4] B. L. Johnson and K. Bailey. Leveraging innovative technology delivers benefits in start up and daily operations. ISA TECH/EXPO Technology Update Conference Proceedings, 403:119 – 124, 2000.

[5] J. Peng, R. Jiang, and J. Xu. Field reliability and maintenance analysis of a heavy engineering vehicle. Appl. Mech. Mater. (Switzerland), 130-134:3875 – 80, 2012.

[6] G. B. Anderson and R. S. McWilliams. Vehicle health monitoring system development and deployment. American Society of Mechanical Engineers, Rail Transportation Division (Publication) RTD, 26:143 – 148, 2003.

[7] J. Kim, K.-S. Lee, and J.-G. Oh. A study on the wireless onboard monitoring system for railroad vehicle axle bearings using the saw sensor. Lecture Notes of the Institute for Computer Sciences, Social- Informatics and Telecommunications Engineering, 57:52 – 58, 2011.

[8] N. Meyendorf, B. Frankenstein, and L. Schubert. Structural health monitoring for aircraft, ground transportation vehicles, wind turbines and pipes-prognosis. Emerging Technologies in Non-Destructive Testing V - Proceedings of the 5th Conference on Emerging Technologies in NDT, pages 15 – 22, 2012.

[9] S. Singh and A. Singh. Condition monitoring of reciprocating air compressor using ir thermography. Indian J. Pure Appl. Phys. (India), 49(1):35 – 41, 2011.

[10] A. Webster. Condition monitoring process for critical rotating equipment within the oil and gas industry. Institution of Mechanical Engineers - 10th European Fluid Machinery Congress: Advances in the Optimisation, Design and Maintenance of Process Machinery, pages 201 – 210, 2008.

[11] M. Pinelli, P. R. Spina, and M. Venturini. Gas turbine health state determination: Methodology approach and field application. International Journal of Rotating Machinery, 2012.

[12] C. da Costa, M. Mathias, and M. Kashiwagi. Development of an instrumentation system embedded on fpga for real time measurement of mechanical vibrations in rotating machinery. Proceedings of the 2012 International Symposium on Instrumentation & Measurement, Sensor Network and Automation (IMSNA), vol.1:60 – 4, 2012.

[13] C. Ferreira, P. Ventura, R. Morais, A. Valente, C. Neves, and M. Reis. Sensing methodologies to determine automotive damper condition under vehicle normal operation. Sens. Actuators, A Phys. (Switzerland), 156(1):237 – 44, 11 2009.

[14] S. Scott, A. Kovacs, L. Gupta, J. Katz, F. Sadeghi, and D. Peroulis. Wireless temperature microsensors integrated on bearings for health monitoring applications.

2011 IEEE 24th International Conference on Micro Electro Mechanical Systems (MEMS 2011), pages 660 – 3, 2011.

[15] I. Bravo-lmaz, A. Garcia-Arribas, E. Gorritxategi, M. Hernaiz, A. Arnaiz, J. Terradillos, and J. Barandi- aran. Magnetoelastic viscosity sensor for lubricant oil condition monitoring. Key Eng. Mater. (Switzerland), 495:71 – 4, 2012.

[16] M. Brouwer, L. Gupta, F. Sadeghi, D. Peroulis, and D. Adams. High temperature dynamic viscosity sensor for engine oil applications. Sens. Actuators, A Phys. (Switzerland), 173(1):102 – 7, 1 2012.

[17] M. Antunano and S. Dessureault. Development of a real-time fleet cost tool as part of an integrated remote mine control centre. 35th APCOM Symposium - Application of Computers and Operations Research in the Minerals Industry, Proceedings, pages 789 – 793, 2011.

[18] R. Subtil, D. Silva, and J. Alves. A practical approach to truck dispatch for open pit mines. 35th APCOM Symposium - Application of Computers and Operations Research in the Minerals Industry, Proceedings, pages 765 – 777, 2011.

[19] S. Ercelebi and A. Bascetin. Optimization of shovel-truck system for surface mining. SIAMM - Journal of The South African Institute of Mining and Metallurgy, 109(7):433 – 439, 2009.

[20] H. P. Simao, J. Day, A. P. George, T. Gifford, J. Nienow, and W. B. Powell. An approximate dynamic programming algorithm for large-scale fleet management: A case application. Transportation Science, 43(2):178 – 197, 2009.

[21] M. Asawa and W. E. Stark. Cellular digital packet data for intelligent vehicle highway systems. IVHS journal Print, 2(4):341 – 358, 1995.

[22] G. Phillips-Wren. Intelligent decision support to assist real-time collaboration. 2008 International Symposium on Collaborative Technologies and Systems, CTS'08, page 375, 2008.

[23] X. Ma, E. D. McCormack, and Y. Wang. Processing commercial global positioning system data to develop a web-based truck performance measures program. Transportation Research Record, (2246):92 – 100, 2011.

[24] H. Kargupta, K. Sarkar, and M. Gilligan. Minefleet: An overview of a widely adopted distributed vehicle performance data mining system. Proceedings of the ACM SIGKDD International Conference on Knowledge Discovery and Data Mining, pages 37 – 46, 2010.

[25] R. W. Ngigi, C. Pislaru, A. Ball, and F. Gu. Modern techniques for condition monitoring of railway vehicle dynamics. Journal of Physics: Conference Series, 364(1):012016, 2012.

[26] R. V. Anzabi, D. S. Nobes, and M. G. Lipsett. Haul truck tire dynamics due to tire condition. Journal of Physics: Conference Series, 364(1):012005, 2012.

Impact Technologies

A Sikorsky Innovations Company

Formal Validation and Verification of Flight Critical System Software using Modeling and Simulation

Michael Roemer, Jianhua Ge, and Sudarshan Bharadwaj

May 10, 2013

FORMAL V&V MOTIVATION

- Models already developed during design stages are often not utilized throughout product life cycle
- Traditional V&V methods accomplished through testing are typically very time-consuming and expensive
- Full coverage of potential code problems is often difficult to accomplish

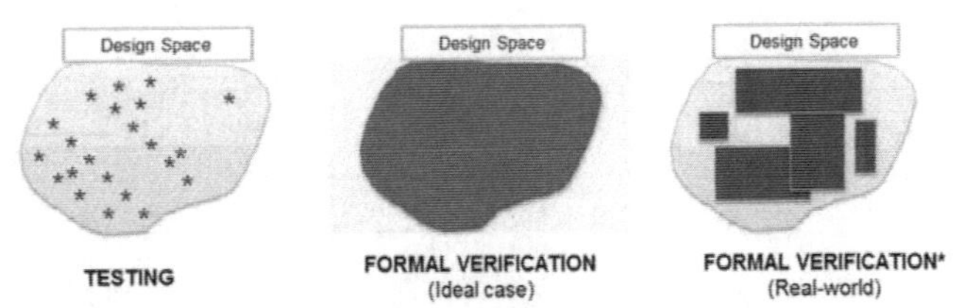

- Point coverage through simulation
- Complete coverage of design space (formal proof)
- Real-world application of formal verification

* Source: Erik Seligman, "Formal Verification – An Overview"

Formal Verification is the act of proving or disproving the correctness of intended algorithms underlying a system with respect to a specification or property, using mathematical and computer science proof/logic strategies.

MODEL-BASED DESIGN AND V&V

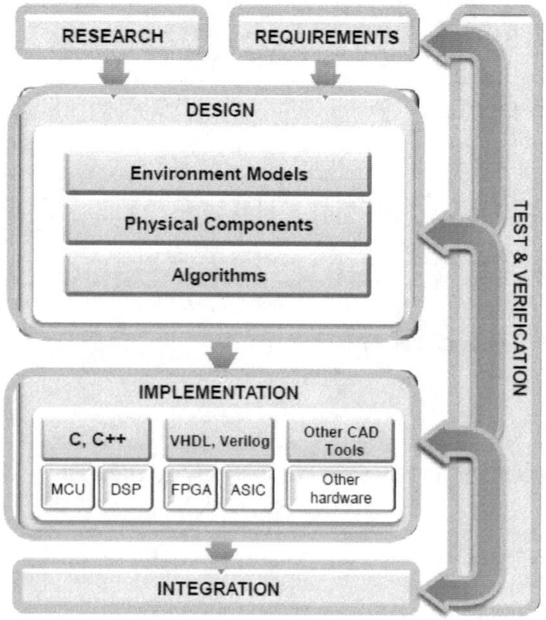

Manage Requirement Changes

Assess Design Robustness

Exploring Design Space

Design Reuse

Automatic Code Generation

Safety-Critical Certification

Impact Technologies
A Sikorsky Innovations Company

This page contains no technical data controlled by the ITAR or EAR

MODEL BASED DESIGN AND V&V

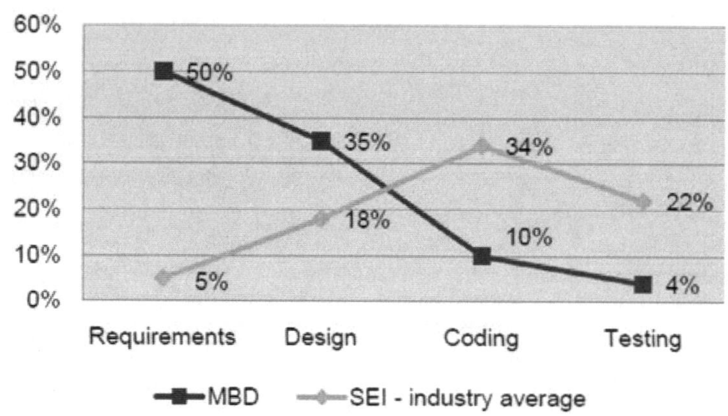

Number of bugs founds shifts to earlier in development phase

SEI: Software Engineering Institute

Impact Technologies
A Sikorsky Innovations Company

This page contains no technical data controlled by the ITAR or EAR

DEVELOPMENT AND MODEL-BASED V&V

- Building a Model from Requirements
  - Simulink Model Development with Requirements Hyperlinks
- Traceability of a Model to Requirements
  - Using the Requirements Management Interface
  - The Requirements Report
- Conformance to Modeling Standards
  - Using the Model Advisor – DO-178C
  - Customizing the Model Advisor
  - Model Advisor Report
- Verification of the Model against Requirements
  - Simulink Design Verifier – Model Coverage
  - System Test Report
  - Property Proving within Simulink blocks

Impact Technologies
A Sikorsky Innovations Company

This page contains no technical data controlled by the ITAR or EAR

5

MBD & DO-178C PROCESS FLOW

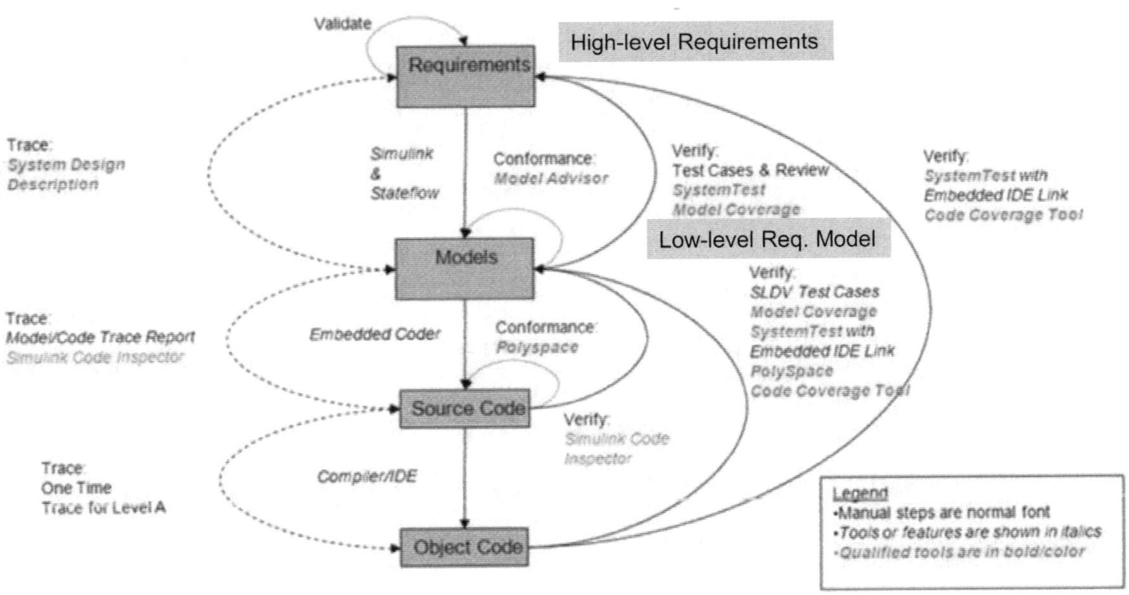

Impact Technologies
A Sikorsky Innovations Company

This page contains no technical data controlled by the ITAR or EAR

6

FORMAL VERIFICATION METHODS

- Static analysis (source code inspection process – automated)
- Model checking (explicit model checking, symbolic model checking, model coverage)
- Theorem and property proving
- Dynamic analysis (run against test/use cases – typical for validation aspects)

1. NASA Formal Methods Symposium, 2009, 2010, 2011, 2012
2. S. A. Jacklin, M. R. Lowry, J. M. Schuman, P. P. Gupta, J. T. Bosworth, E. Zavala, J. W. Kelly, K. J. Hayhurst, C. M. Belcastro, C. M. Belcastro, Verification, Validation, and Certification Challenges for Adaptive Flight-Critical Control System Software, NASA, 2004.

S/W TOOLS FOR FORMAL VERIFICATION

- Simulink Design Verifier by MathWorks (approach: static code analysis uses Polyspace)
- Temporal Logic of Actions (TLA) TLA+ by Microsoft (Approach: Model Checking)
- VeriSoft by Bell Laboratories (Approach: Model Checking)
- Slam and Static Driver Verifier (Microsoft)
- SPIN, MALPAS Static Analysis Toolset
- JPF, tools within NASA (specifically ARC, JPL, LaRC)
- Many more....

Impact Technologies
A Sikorsky Innovations Company
This page contains no technical data controlled by the ITAR or EAR

SIMULINK AND MODEL-BASED V&V

Formal Methods Integrated into Design Process

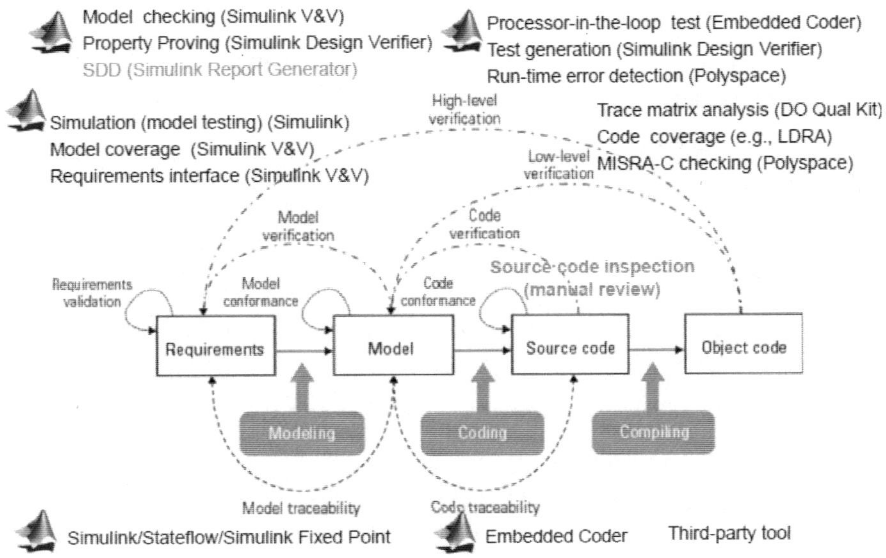

Reference: T. Erkkinen and P. Mathapati, Verification and Validation of Safety Critical Systems, 2011 Mathworks.

SIMULINK V&V TOOLBOX

- Compliance checking for MAAB style guidelines and high-integrity system design guidelines (DO-178B and IEC-61508)
- Model Advisor Configuration Editor, including custom check authoring, model coverage analysis
- Requirements Management Interface for traceability of model objects, code, and tests to requirements documents
- Automatic test-harness generation for subsystems
- Component testing via simulation, software-in-the-loop (SIL), and processor-in-the-loop (PIL)
- Programmable scripting interface for automating compliance checking, requirements traceability analysis, and component testing

Mathworks, *User Menu of Simulink Verification and Validation Toolbox*, 2011.

SIMULINK DESIGN VERIFIER

- Polyspace® (static code analysis) and Prover Plug-In formal analysis engines
- Detection of dead logic, integer and fixed-point overflows, division by zero, and violations of design properties
- Blocks and functions for modeling functional and safety requirements
- Test vector generation from functional requirements and model coverage
- Objectives, including condition, decision, and modified condition/decision
- Property proving, with generation of violation examples for analysis and debugging
- Fixed-point and floating-point model support

Mathworks, *Simulink Design Verifier User Menu*, 2011.

SIMULINK DESIGN VERIFIER

- Simulink Design Verifier (SDV) performs exhaustive formal analysis of Simulink models and generated code to confirm their correctness with respect to given properties. The user can specify these properties directly in Simulink as assertions. The assertions are specified in the form of prove objectives.
- To prove these objectives, Simulink Design Verifier searches for all possible values of a Simulink function in order to find a simulation that satisfies an objective.
- Prover Plug-In models problems as propositional logic formulas and applies proof methods to decide whether the formulas evaluate to true for all assignments, or not. This problem is solved by using a reduction and absurdum argument, where a combination of different analyses are used to check whether the negation of the formula is unsatisfable.

PROPERTY PROVING PROCESS

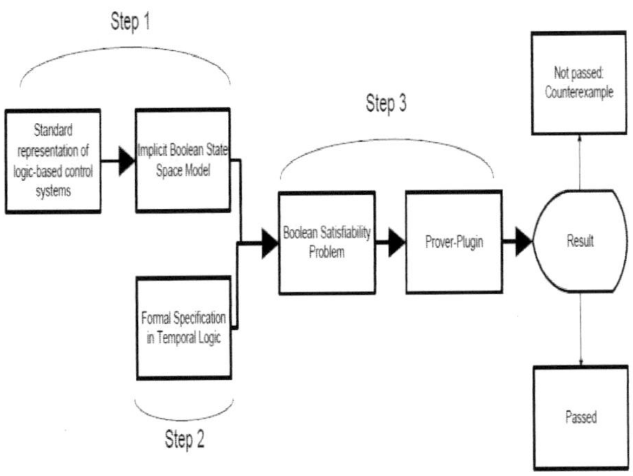

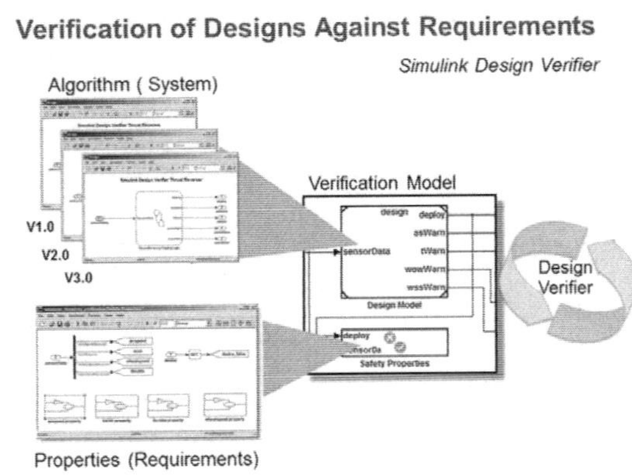

* Stålmarck's Proof Procedure which was patented in 1992

Impact Technologies
A Sikorsky Innovations Company

This page contains no technical data controlled by the ITAR or EAR

PERFORMING FORMAL METHODS

- English Language Requirements
- Requirement Translation into Logic (CNL)
- Coding the Requirements within Model
- Assess Model Coverage and Standards (i.e. DO 178C)
- Assess Design Errors in Model/Code (Abstract Interpretation)
- Checking Model Compatibility with Formal Requirements
- Updating Model with Proof Objectives and Assertions
- Exercise Prove Properties

Property Proving

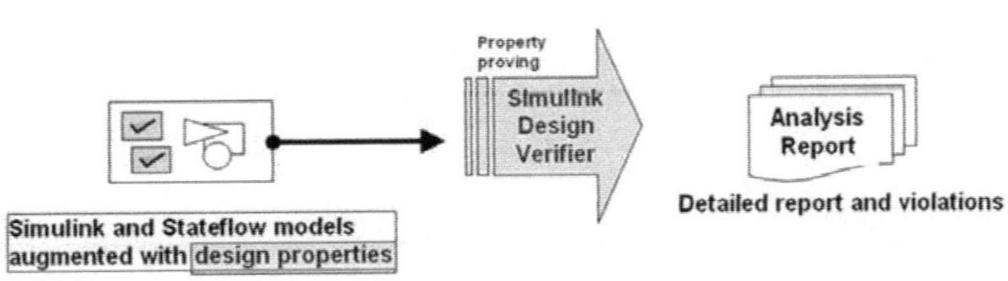

Impact Technologies
A Sikorsky Innovations Company

Use Case

Flight Control Actuator PHM Algorithm

- Model Advisor and DO-178C
- Actuator PHM Code
- Model Coverage Analysis
- Requirements-based Verification

FLIGHT CONTROL ACTUATOR PHM

TWO-STAGE KALMAN FILTER FOR ACTUATOR FAULTS

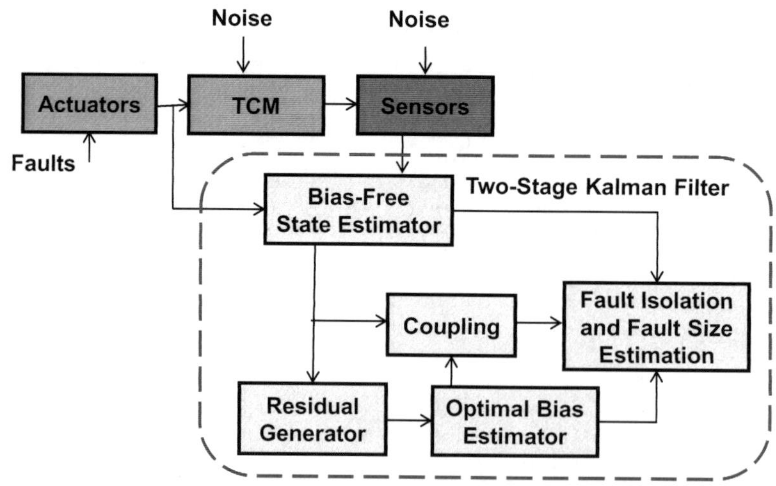

The computational complexity due to high state dimension is alleviated by decomposing an *n+6* dimension Kalman Filter as an *n* dimension Bias-free State Estimator and a *six* dimension Optimal Bias Estimator for six motors.

ESTIMATING DEGRADATION VIA LOSS OF EFFECTIVENESS

We model the vehicle dynamics in linear discrete-time state variable form as given by:

$$x_{k+1} = A_d x_k + B_d u_k + w_k^X \quad (1)$$

$$y_k = C x_k + v_k \quad (2)$$

w_k^X and v_k are zero mean uncorrelated random sequences with covariance matrices Q^x and R, respectively. We assume that the initial conditions x_0 is a random Gaussian vector with $E\{x_0\} = \bar{x}_0$ and $E\{(x_0 - \bar{x}_0)(x_0 - \bar{x}_0)^T\} = P_0^x > 0$.

The loss of control effectiveness can be modeled as follows:

$$\gamma_k = \begin{bmatrix} \gamma_{1_k} \\ \gamma_{2_k} \\ \vdots \\ \gamma_{6_k} \end{bmatrix}, \; 0 \le \gamma_{i_k} \le 1; \; \text{i=1,...,6} \quad (3)$$

Where $\gamma_{i_k} = 0$ and $\gamma_{i_k} = 1$ means that the *i*th actuator is completely healthy or fully damaged, respectively.

ESTIMATING DEGRADATION

To demonstrate actuator faults explicitly for the convenience of fault parameters estimation, the state equation in Eq. (1) is modified as

$$x_{k+1} = A_d x_k + B_d u_k - B_d U_k \gamma_k + w_k^x \quad (4)$$

where $U_k = diag[u_k^1 \;\; u_k^2 \;\; \ldots \;\; u_k^6]$. We want to determine the amount of loss in the effectiveness, γ_k, in a recursive form, so that the control allocation routine can be reconfigured. The effectiveness factors are modeled as the following form:

$$\gamma_{k+1} = \gamma_k + w_k^\gamma \quad (5)$$

where w_k^γ is zero-mean white Gaussian noise vectors with covariance Q^γ. The γ_k term can also be thought of as a health parameter related to the health condition of the corresponding actuators.

SIMULINK LATERAL MODEL WITH LQR

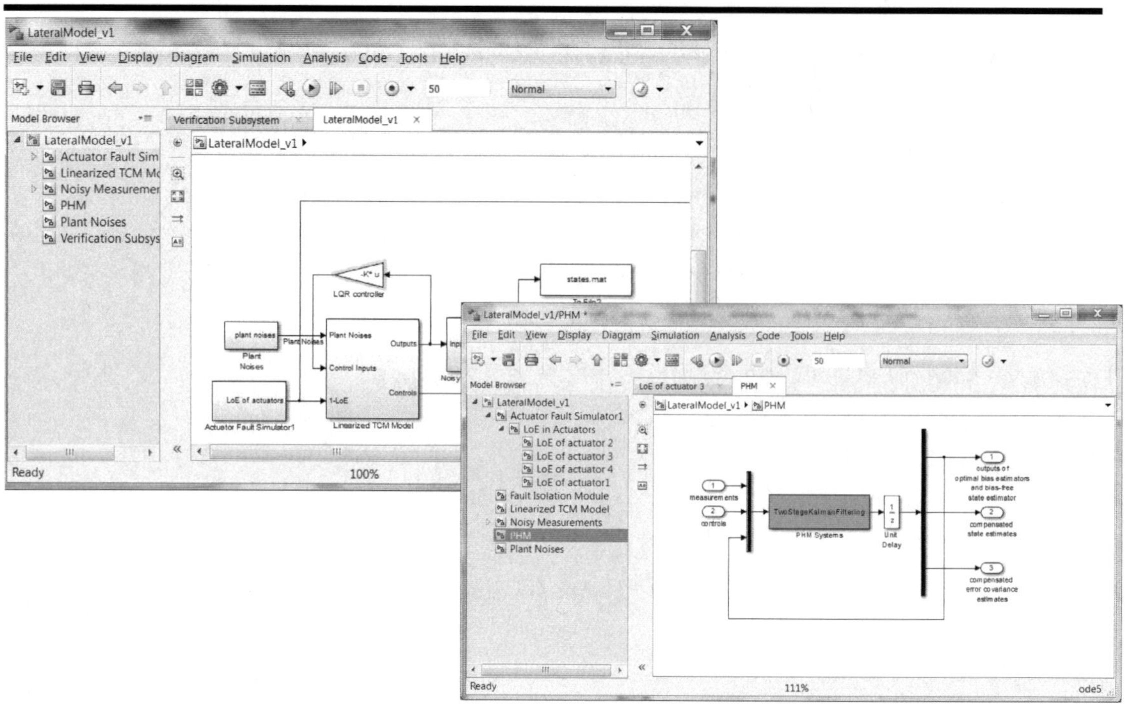

Model Advisor Using Modeling Standards for DO-178C

MODEL ADVISOR – CODE SELECTION

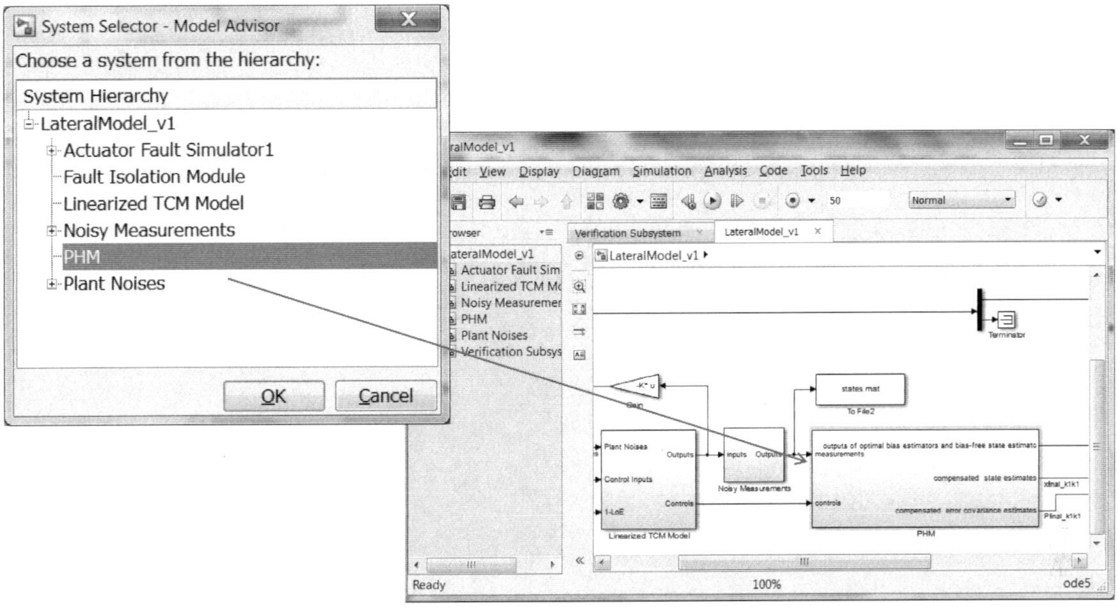

CHECKING DO-178C STANDARD

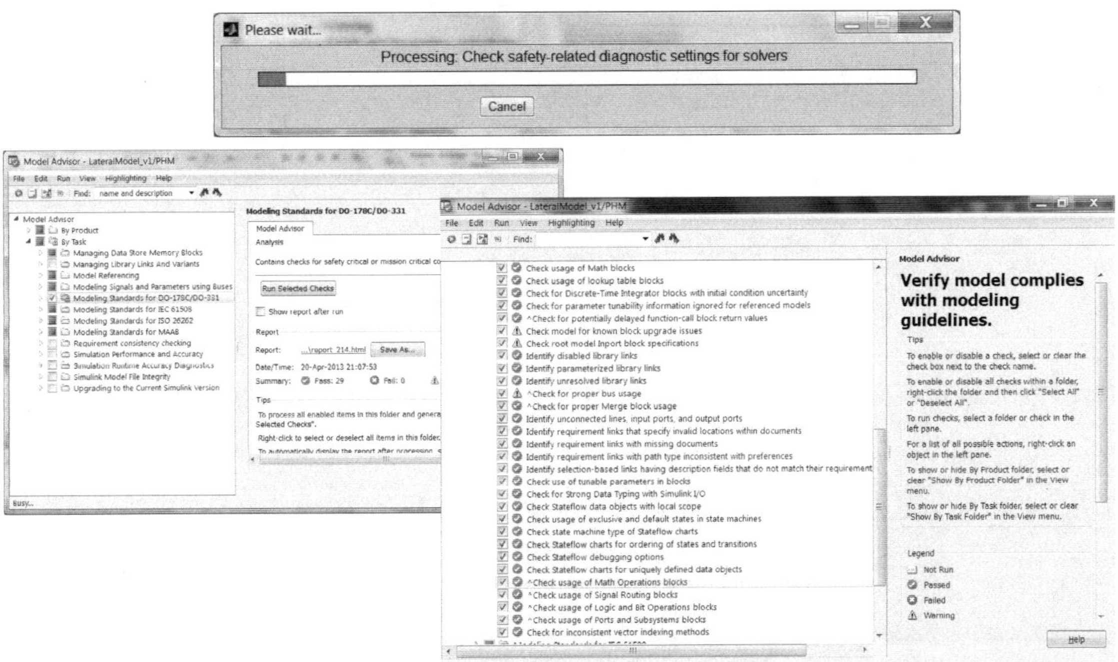

Coverage Analysis

COVERAGE SETTING

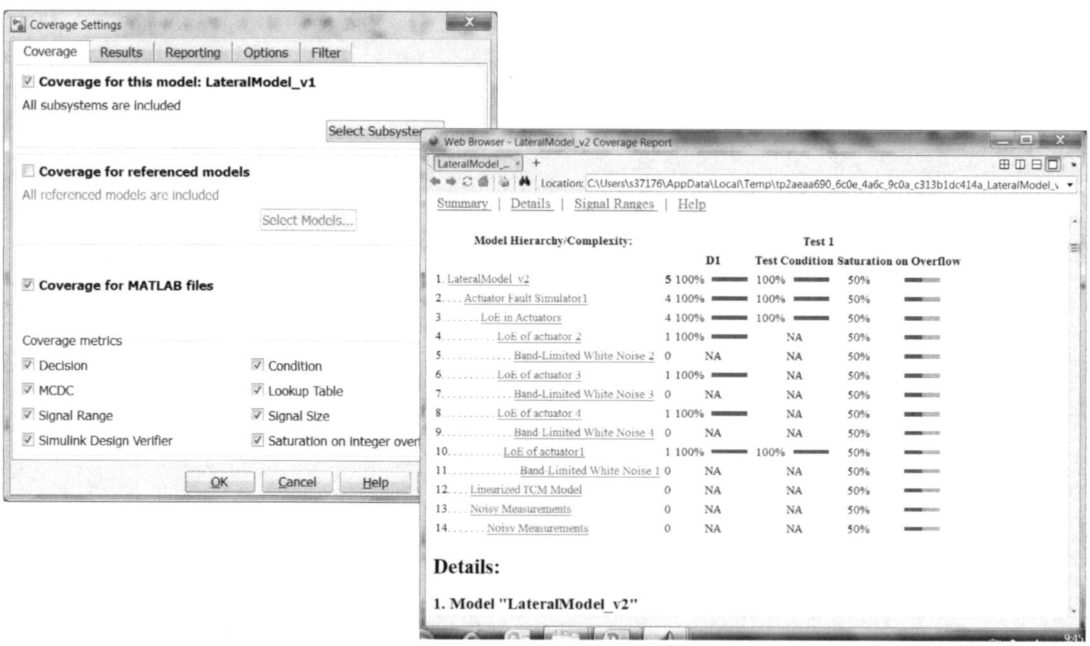

Requirements-Based Verification: Property Proving

Impact Technologies
A Sikorsky Innovations Company

This page contains no technical data controlled by the ITAR or EAR

REQUIREMENTS AND TRACEABILITY

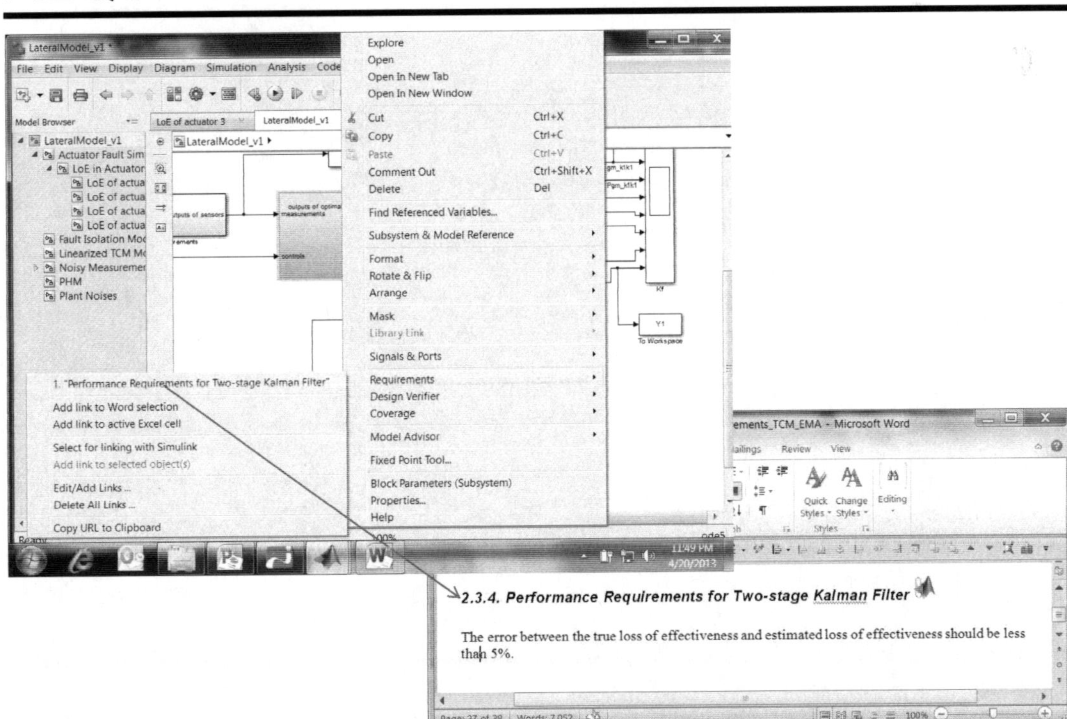

Impact Technologies
A Sikorsky Innovations Company

This page contains no technical data controlled by the ITAR or EAR

VERIFICATION SUBSYSTEM

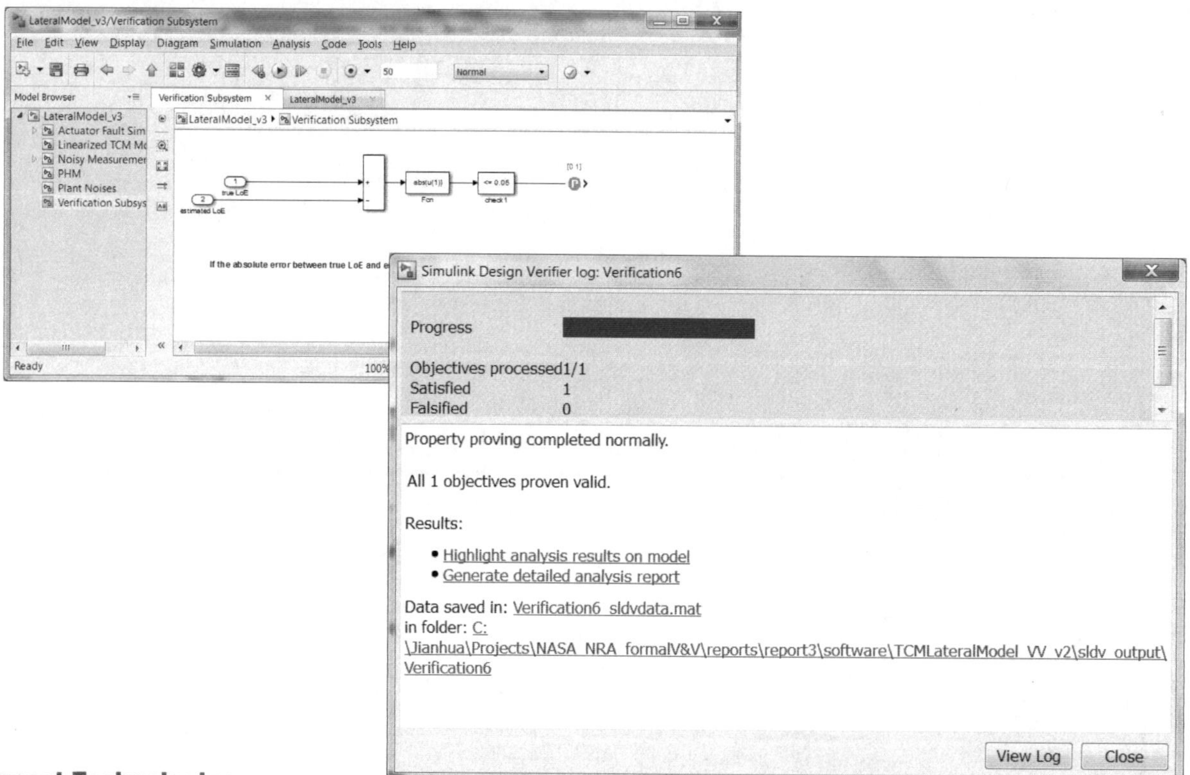

DATA FUSION TOOL FOR SPIRAL BEVEL GEAR CONDITION INDICATOR DATA

Paula J. Dempsey
National Aeronautics and Space Administration
Glenn Research Center
21000 Brook Park Rd.
Cleveland, Ohio 44135
Telephone: (216) 433-3398
Paula.J.Dempsey@nasa.gov

Lance J. Antolick, Jeremy S. Branning, and Josiah Thomas
RMCI Inc.
1525 Perimeter Parkway
Huntsville, Alabama 35806

Abstract: Tests were performed on two spiral bevel gear sets in the NASA Glenn Spiral Bevel Gear Fatigue Test Rig to simulate the fielded failures of spiral bevel gears installed in a helicopter. Gear sets were tested until damage initiated and progressed on two or more gear or pinion teeth. During testing, gear health monitoring data was collected with two different health monitoring systems. Operational parameters were measured with a third data acquisition system. Tooth damage progression was documented with photographs taken at inspection intervals throughout the test. A software tool was developed for fusing the operational data and the vibration based gear condition indicator (CI) data collected from the two health monitoring systems. Results of this study illustrate the benefits of combining the data from all three systems to indicate progression of damage for spiral bevel gears. The tool also enabled evaluation of the effectiveness of each CI with respect to operational conditions and fault mode.

Keywords: Data fusion; diagnostics; gears; health monitoring

Background: Helicopter transmission health is important to aircraft health and safety because helicopters depend on the power train for propulsion, lift and control. Health and Usage Monitoring Systems (HUMS) capable of predicting impending transmission component failure for "on-condition" maintenance have the potential to decrease operating and maintenance costs and increase safety and aircraft availability. These systems employ "condition indicators" (CIs), typically calculated from vibration signatures, that are generated when a faulted component interacts with its operational environment. The CI must be correlated to a known failure mode to reliably detect the health of the system.

One method used to provide evidence that a HUMS can replace current maintenance practices required for "on-condition" maintenance is the performance of seeded fault tests. Seeded fault tests, within this paper, refer to initiating a known fault into a component to accelerate damage while monitoring its progression. The fault can be seeded by several different methods. The fault can be initiated by machining a fault into the component prior to test. A component with an existing fault that was removed from a helicopter can be installed in a test rig for testing of its progression. The fault can also be forced to naturally occur through test design and operational conditions. For this analysis, the third method was used.

Damage progression "seeded fault" tests were performed in the NASA Glenn Spiral Bevel Gear Fatigue Test Rig to simulate failures of spiral bevel gears installed in a helicopter. Gear sets were tested until damage occurred on two or more gear or pinion teeth. Gear vibration, oil debris, torque and speed data were recorded with a research data acquisition system. Vibration and speed data were also recorded and processed with a helicopter HUMS system. A facility data acquisition system was used to monitor and record the operational conditions of the test rig. Tooth damage progression was documented with photographs of the gear teeth at periodic intervals throughout the test.

The objective of this work was to demonstrate the benefit of a software tool used to fuse the data generated from the three data acquisitions systems and the damage progression photos. The tool enabled correlation of the CI data with operational conditions the gear sets were exposed to during damage initiation and progression. Use of this tool will promote a better understanding of why gear vibration based CIs trend clearly with damage under some conditions but trend poorly for others, by determining which operational parameters and fault modes impact the effectiveness of each CI to indicate a gear fault.

Test Facility: Tests were performed in the Spiral Bevel Gear Fatigue Test Rig at NASA Glenn Research Center. A detailed description of this test facility is provided in references [1] and [2]. The Spiral Bevel Gear Fatigue Test Rig is illustrated in Figure 1. The facility operates as a closed-loop torque regenerative system, where the drive motor only needs enough power to overcome friction losses within the system. Two sets of spiral bevel gears are installed in the test rig and tested simultaneously. Facing the gearboxes, per Figure 1, the left gear set (pinion/gear) is referenced as left and the right gear set (pinion/gear) is referenced as right within the paper.

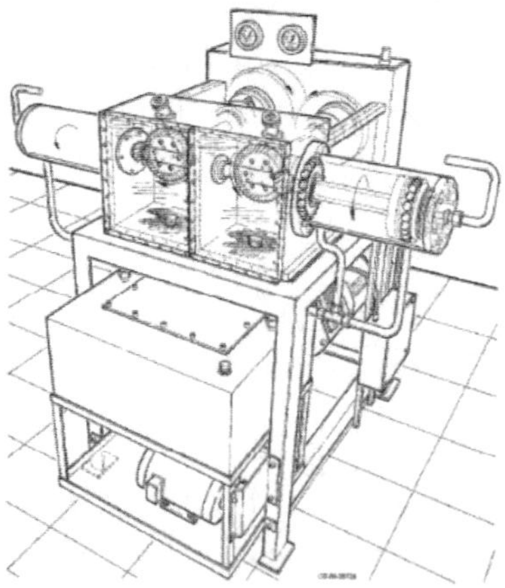

Figure 1: Spiral Bevel Gear Fatigue Test Rig.

Qualified helicopter transmission oil, AEROSHELL Turbine Oil 555, was used in the test rig during testing. Both gear sets are lubricated with oil jets pumped from an oil reservoir. The lubrication from the gearbox then exits the gearbox and flows through an inductance type in-line oil debris sensor, then past a magnetic chip detector. A strainer and a 3 μm filter are located downstream of the oil debris (OD) sensor to capture any debris before returning to the sensor and gearbox.

Data Acquisition and Instrumentation: During these tests, three data acquisition systems were used. Vibration, oil debris, torque and speed data were collected once every minute with the NASA Glenn research data acquisition system, referred to as the Mechanical Diagnostic System Software (MDSS). Vibration and speed data were collected from a second set of sensors with a helicopter HUMS referred to as the Modern Signal Processing Unit (MSPU). Operational parameters were collected with a third system referred to as the Daytronic. These operational parameters included torque, speed, and right and left gearbox oil temperatures.

A non-contact rotary transformer shaft mounted torque sensor was used to measure torque during testing. Thermocouples were used to measure inlet and outlet oil temperatures. An inductance type oil debris sensor was used to measure the ferrous debris generated during fatigue damage to the gear teeth. The MDSS records the particle counts measured by the oil debris sensor, their approximate size and an approximate mass. The sensor measures the number of particles and their approximate size based on user defined particle size ranges or bins. Based on the bin configuration, the average particle size for each bin is used to calculate the cumulative mass by assuming the average particle size as a diameter spherical in shape and multiplying it by the density of steel. Chip indications from the magnetic chip detector, when the gap was closed with debris, were also measured.

For the MDSS system, accelerometers were installed on the right and left side of the test rig housing. Accelerometer frequency range is 0.7 to 20 KHz with a resonant frequency of ≥70 KHz. The MDSS accelerometers were mounted on the housing, radially and vertically with respect to the pinion, as shown in Figure 2. Facing the gearboxes, the left gear set (pinion/gear) and right gear set (pinion/gear) accelerometers were referenced as such in the MDSS system. Speed was measured with optical tachometers mounted on the left pinion shaft and left gear shaft to produce a separate once-per-rev tach pulse for the pinion and gears. Time synchronous averaging (TSA) of the vibration data collected from the left and right accelerometer is performed in the MDSS system for the pinions via the pinion tach pulse and the gears via the gear tach pulse.

For the MSPU system, accelerometers are also installed on the right and left side of the test rig. Accelerometer frequency range is 0.5 to 5 KHz with a resonant frequency of 26 KHz. A magnetic tachometer is installed on the right pinion and measures pinion pulses per tooth pass. This is used to calculate the TSA for both the pinion and the gear. The gear ratio is used to process the data at the correct speed for the gear. The MSPU accelerometers were mounted on the housing, in close proximity to the MDSS accelerometers, radially and vertically with respect to the pinion, as shown in Figure 2.

Figure 2: Location of Accelerometers.

Vibration data was collected at sample rates that provided sufficient vibration data for calculating time synchronous averaged data (TSA). TSA refers to techniques for averaging vibration signals over several revolutions of the shaft, in the time domain, to improve signal-to-noise ratio [3]. Using a once-per-revolution signal or tachometer, the vibration signal is interpolated into a fixed number of points per shaft revolution. Vibration signals synchronous with the shaft speed intensify relative to non-periodic signals which become weaker.

Since helicopter gears generate vibration signals synchronous with gear rotational speed, most current helicopter gear CIs are calculated from TSA data. Many CIs are based on statistical measurements of vibration energy. Signal processing techniques used to extract useful information to calculate a gear CI from the vibration signal are discussed in detail in reference [4]. Some gear CIs are calculated directly from the TSA signal, such as Root Mean Square (RMS). Some are calculated from the TSA converted to the frequency domain, such as Sideband Index (SI). Some convert the TSA signal to the frequency domain, filter specific frequencies, convert it back to the time domain, then calculate a statistical parameter from this data [3 to 5].

Figure 3 illustrates the information used to calculate the TSA for the right gear and pinion. Using the sample rate of 200 KHz for 1 sec duration and the speed of both shafts, the number of TSA revolutions averaged for each acquisition is determined. To calculate the TSA, the accelerometer data is divided into segments equivalent to 1 revolution of the shaft. Each segment is then linearly interpolated into equal numbers of points that have been rounded down to a power of two. A power of two is used because it eases the future use of the Fast Fourier Transform (FFT) to transform the TSA to the frequency domain. Per Figure 3, the right accelerometer raw data is plotted in the top plot. The two lowest plots are the TSA signal calculated from the 1/rev and vibration data for the right gear and pinion. Pulses from the 41 tooth gear and the 19 tooth pinion can easily be seen within these two plots.

TSA Info	Pinion	Gear
Test Gear Teeth	19	41
Max Speed (RPM)	7553	3500
Max Speed (Hz)	126	58
Gear Mesh (Hz)	2392	2392
Sample rate	200000	200000
Sample Duration (sec)	1	1
Revs per Acq	126	58
TSA int. points/rev	1024	2048

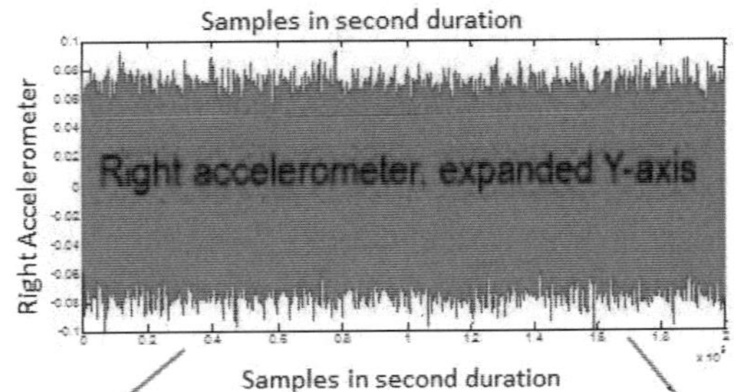

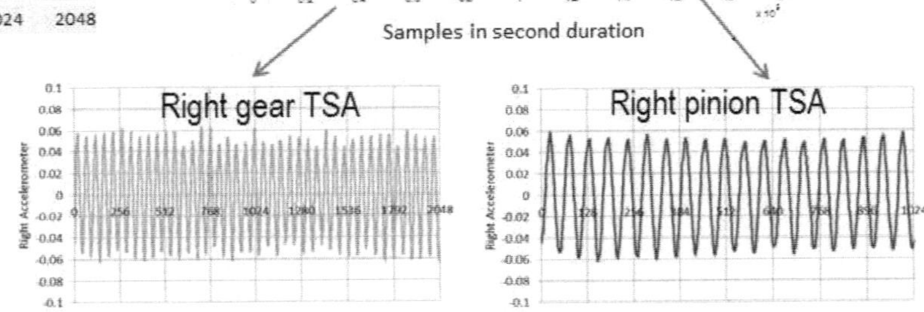

Figure 3: Information Used to Calculate TSA.

Test Gear Sets: The gears were made from a steel alloy CEVM 9310, carburized, hardened and ground. The test gears are the gear sets (pinion/gear) installed on the left side of the rig. The slave gears are the gear sets installed on the right side of the rig. The right gear sets are superfinished to decrease wear rates and increase their life. The gear sets have a 6.4 in. diametral pitch, 20° pressure angle, 25° spiral angle, 0.94 in. face width and a 2.15 gear ratio. The gears have 41 teeth and the pinions have 19 teeth. The test gears were designed to operate at a gear speed of 3500 rpm, gear torque of 8000 in.-lbs., pinion speed of 7553 rpm, pinion torque of 3707 in.-lbs. and 240 to 265 °F inlet oil temperatures with an American Gear Manufacturers Association (AGMA) calculated contact stress of 237 ksi.

Test Description: For this study, two left gear sets were tested at a gear speed of 3500 rpm and pinion speed of 7553 rpm. At the start of each test, a run-in was performed for a minimum of 1 hour at 4000 in.-lbs. gear torque/3500 rpm gear speed and 1854 in.-lbs. pinion torque/7553 rpm pinion speed. Then the gear torque was increased to 8000 in.-lbs. and pinion torque was increased to 3707 in.-lbs. for the remainder of the test. Contact cycles accumulated at a rate of 210,000 per hour for the gear and 453,180 per hour for the pinion. At completion of the run-in, inspection photos were taken of the left and right gear and pinion teeth. Inspection photos were also taken periodically throughout the test to document damage progression to the gear teeth.

The failure modes to be investigated for this study were defined by class (contact fatigue), general mode (macro pitting) and degree (progressive) per American Gear Manufacturers Association (AGMA) standards for gear wear terminology [6]. Using [6] for tooth damage terminology, a numbering scheme [7] was developed to streamline the identification of gear damage. Table 1 illustrates the types of damage observed during these tests. The tests ran until macro pitting/spalling larger than 1 mm in diameter covered a significant area of two or more gear or pinion tooth surfaces. Definitions for tooth surface pitting modes [6] are summarized as follows:

- Initial—Pits less than 1 mm in diameter.

- Progressive—Pits in different shapes/sizes greater than 1 mm in diameter.
- Flake—Pits that are shallow thin flakes.
- Spalling—Pits that cover tooth contact surfaces that exceed progressive pitting.

During test 2, scuffing occurred prior to contact fatigue damage. Scuffing causes transfer of metal from one tooth surface to another. This failure class is also listed in Table 1.

Table 1: Numbering Scheme for Nomenclature of Gear Failure Modes [6].

Class	General Mode	Specific Mode or Degree
2.0 Scuffing	2.1 Scuffing	2.1.1 Mild
		2.1.2 Moderate
		2.1.3 Severe
4.0 Contact Fatigue	4.1 Subcase Fatigue	
	4.2 Micropitting	
	4.3 Macropitting	4.3.1 Initial
		4.3.2 Progressive
		4.3.3 Flake
		4.3.4 Spall

Software Tool: Many factors can affect the gear CI's ability to respond to tooth damage through vibration response. The response of the accelerometer to a specific fault can depend on the sensor specifications, the signal processing of the raw signal, mounting and its location. The CI method of calculation, operational conditions and type of failure mode can also affect its response.

Assessing whether a change in any particular condition indicator was due solely to a change in damage level, a change in operating condition or some combination of both was a challenge due to three data acquisitions systems with different measured parameters and acquisition rates. To assess the relationship between the different parameters, a software tool was developed to fuse the three systems. The tool provided a means to analyze the data generated during gear damage initiation and progression from the three data acquisitions systems and the damage progression photos. This software tool enables fusing the information from the three data acquisition systems (MDSS, MSPU and Daytronic) and the damage modes. The data can be plotted and analyzed separately or correlated together. The tool also provides a means for inputting damage levels for further correlation. Summarized statistical parameters for CIs and operational conditions can be calculated and exported into tables. A screen shot of the tool is shown in Figure 4.

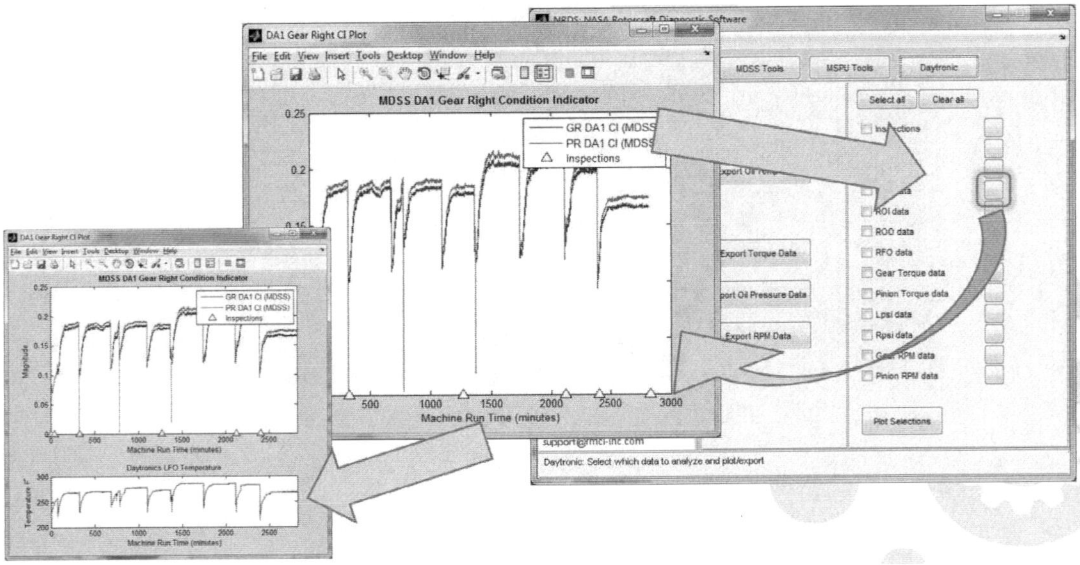

Figure 4: Data Fusion Tool.

Tool Demonstration: Summaries of the tools capabilities on the Spiral Bevel Gear Fatigue Test Rig test data will be discussed in the following sections. These summaries will include plots of CIs from both the MDSS and MSPU systems, debris generated and gear torque. Note that the CIs for each component are identified as GL (gear left), GR (gear right), PL (pinion left) and PR (pinion right). Torque was measured in inch-pounds and debris generated was measured in milligrams. Tooth damage photos for the gear and pinion during inspection intervals were converted to a damage scale per Table 2. User defined damage scales of zero through three identify gear tooth condition. The damage scales are followed by a user defined letter that indicates the classification of the damage. The scales are used to provide a damage level indication on the x-axis of the plots via green, yellow or red triangles. For the tests to be discussed, 2a indicated macropitting on one tooth, 3a indicated macropitting initially observed on two or more teeth, and 3b indicated the pitting area increased on the tooth surface between inspection intervals.

Table 2: Damage Scales.

Damage Scale	Gear Condition	Indication
0	New	●
1[a-z]	Minor Wear	▲
2[a-z]	Minor Damage	△
3[a-z]	Significant Damage	▲

For test 1, contact fatigue damage was first observed on left pinion tooth 10 at run time 2119 min. Damage was observed on a second tooth, left pinion tooth 13, at run time 2402 min. The size of the spall increased on left pinion tooth 13 at run time 2402 min. Photographs of damage progression on the two pinion teeth are shown in Figure 5. Table 3 lists the inspection intervals and damage scales observed during testing. The CIs evaluated will be discussed in the following section.

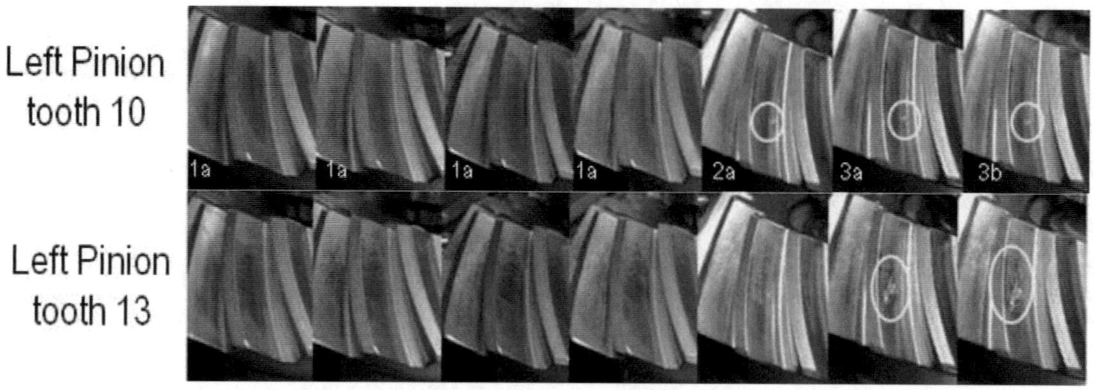

Figure 5: Left Pinion Tooth Damage.

Table 3: Test 1 Inspection Intervals and Observed Damage Scales.

Date	Run Time (min.)	Damage Scale
2013-03-21	1	1a
2013-03-22	76	1a
2013-03-26	324	1a
2013-04-02	1370	1a
2013-04-11	2119	2a
2013-04-15	2402	3a
2013-04-17	2833	3b

Condition Indicators: FM4 is one CI used to indicate gear tooth damage. Figure 6 is a block diagram of the steps required to calculate FM4 (Figure of Merit 4), a common vibration algorithm used in commercial HUMS [3].

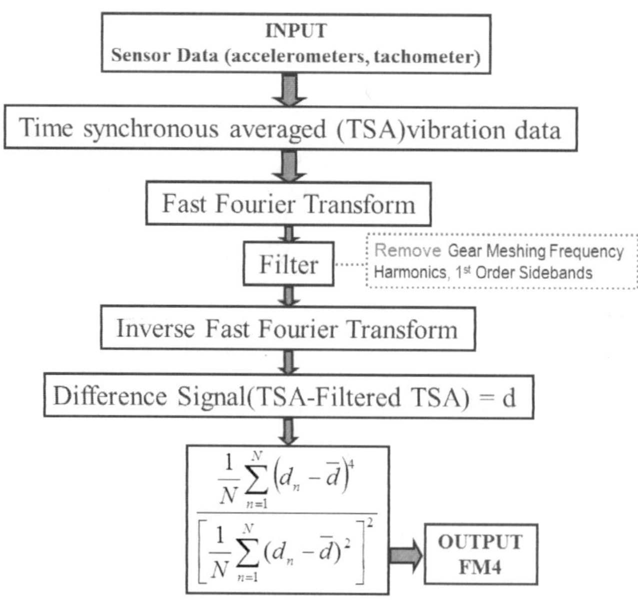

Figure 6: FM4 Calculation.

Figure 7 contains plots of FM4, oil debris, gear torque, gear speed, and left and right side of gearbox oil output temperatures. FM4 values for the left pinion increased significantly when damage was observed on two left pinion teeth. Oil debris mass also increased as damage progressed on the pinion teeth. FM4 did not appear to be significantly affected by the varying operational parameters (gear torque, gear speed and output oil temperatures).

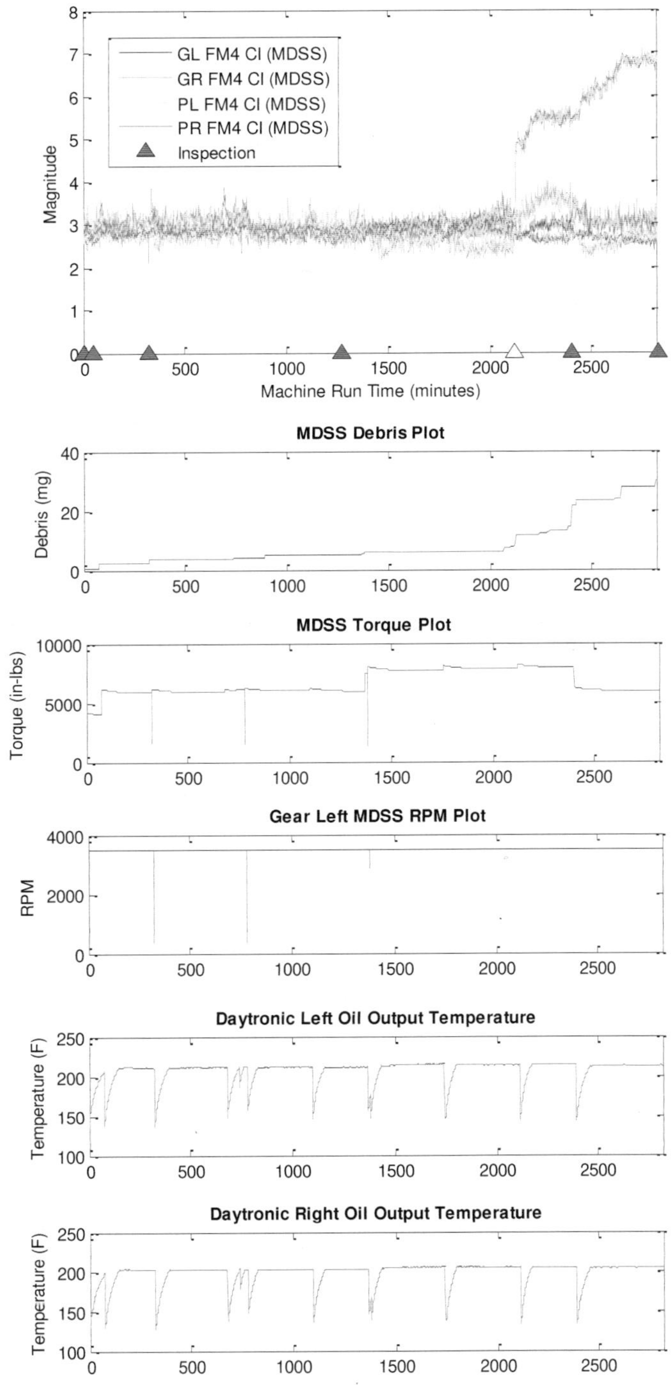

Figure 7: FM4 and Operational Parameters.

RMS or DA1 is another CI used to indicate gear tooth damage. Figure 8 is a block diagram of the steps required to calculate RMS.

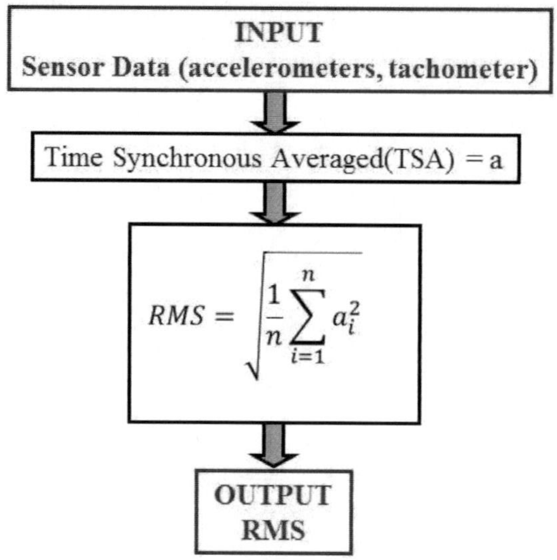

Figure 8: RMS Calculation.

Figure 9 contains plots of RMS, oil debris, gear torque, gear speed, and left and right oil output temperatures. RMS values for the *right* side were significantly higher than the left and were sensitive to torque and oil outlet temperature transients. RMS values for the *left* side were sensitive to torque. However, once damage was observed on two pinion teeth, the RMS value for the left pinion did not drop when torque decreased.

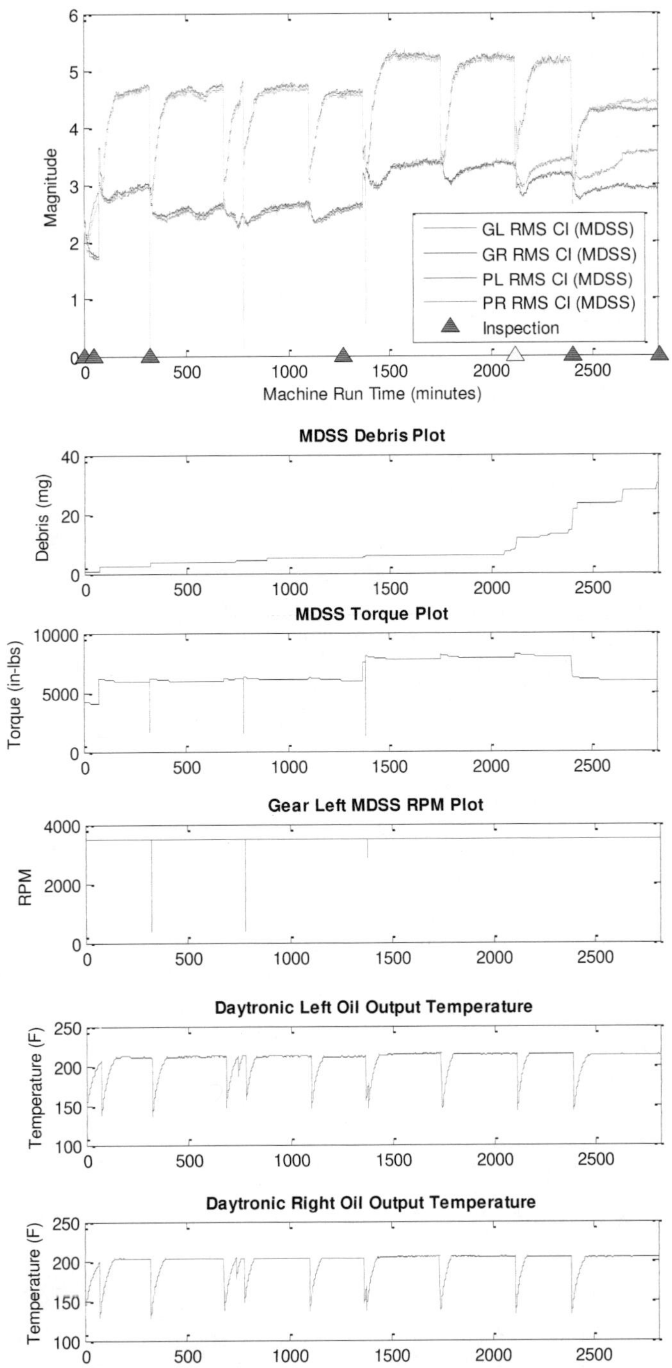

Figure 9: RMS and Operational Parameters.

Sideband index (SI) is another CI used to indicate gear tooth damage. SI is a frequency domain based CI. The CI value is an average value of sideband amplitudes about the fundamental gear mesh frequency. All gears generate a dominant gear mesh (GM) frequency in the vibration signature due to each tooth impacting the gear it is driving as the pinion and gear mesh. The gear (or pinion) mesh frequency is equal to the number of teeth multiplied by its speed. The number of sidebands included in the calculation of the sideband CI can vary with different health monitoring systems. Averages of ±1 (SI1) and ±3 (SI3) were used for this analysis. Figure 10 is a block diagram of the steps required to calculate SI1.

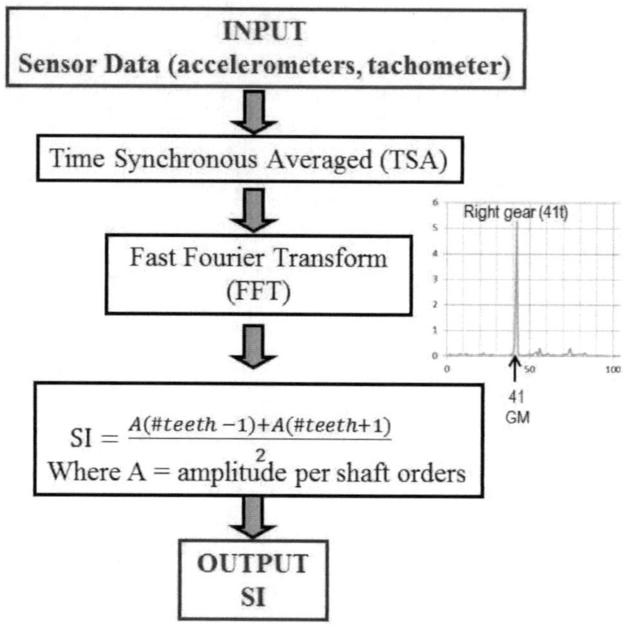

Figure 10: SI1 calculation.

Figure 11 contains plots of the average of ±1 sideband (SI1), oil debris, gear torque, gear speed, and left and right oil outlet temperatures. SI1 values for the *right* side were sensitive to torque and oil outlet temperature transients. SI1 values for the *left* side were not sensitive to torque. SI1 values for the left pinion increased significantly when damage was observed on two left pinion teeth.

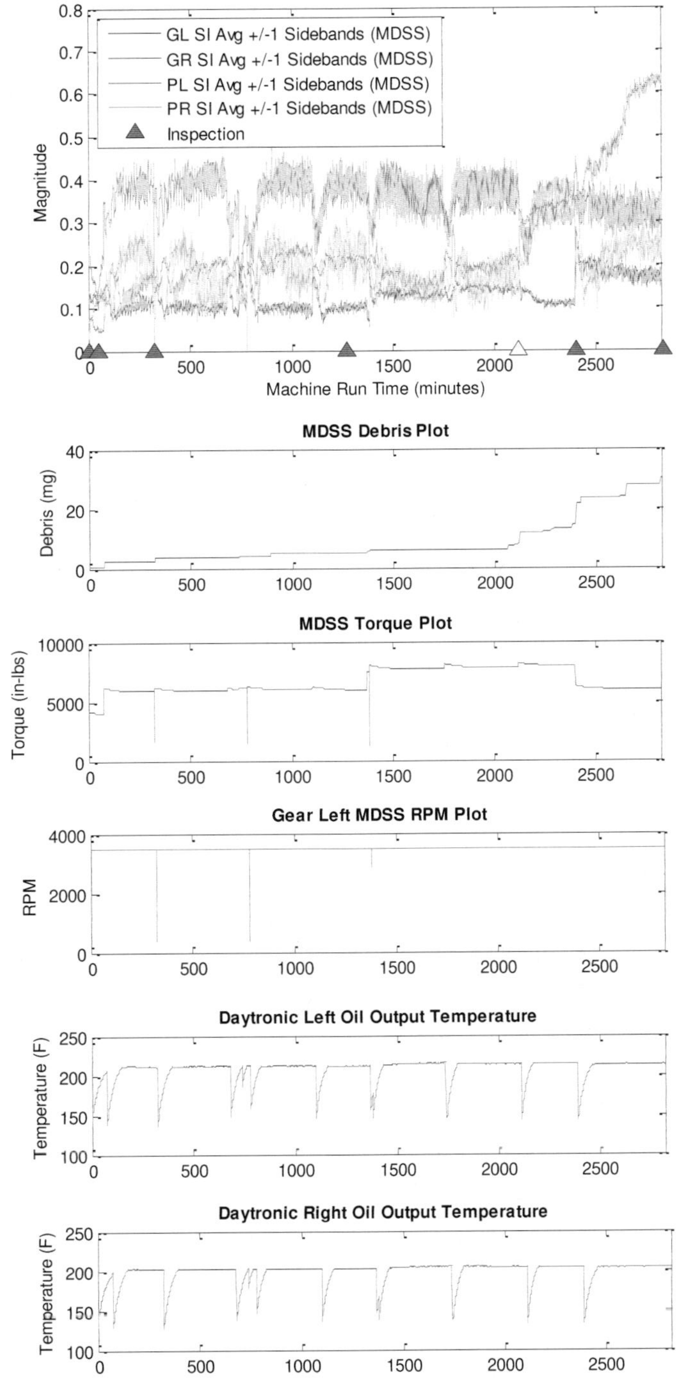

Figure 11: SI1 and Operational Parameters.

CIs calculated separately from the two different health monitoring systems were compared to determine if system differences affected CI response. Figures 12 and 13 contain plots of the average of ±1 sideband (SI1) and average of ±3 sidebands (SI3) for the MDSS and MSPU systems. Figure 14 contains plots of RMS, referred to as DA1, in the MDSS system for the MDSS and MSPU systems. Figure 15 contains plots of FM4 for the MDSS and MSPU systems. Although data was collected more frequently with the MDSS system, both systems trended well together.

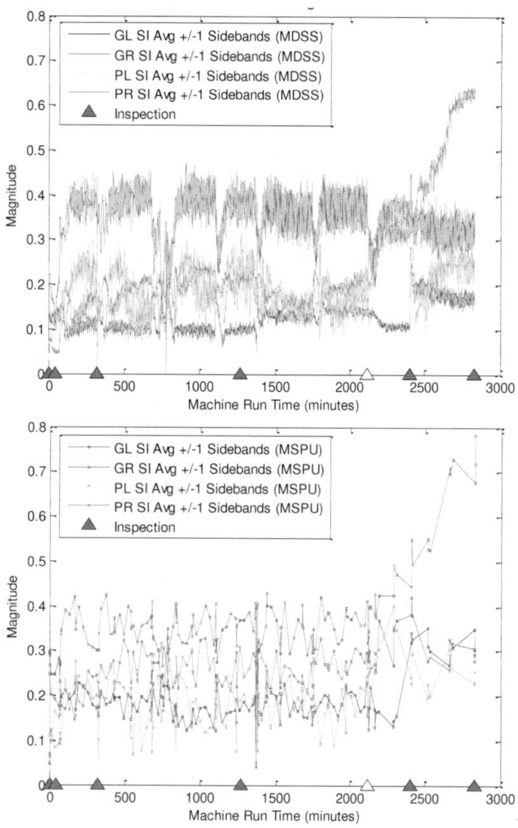

Figure 12: MDSS and MSPU SI1.

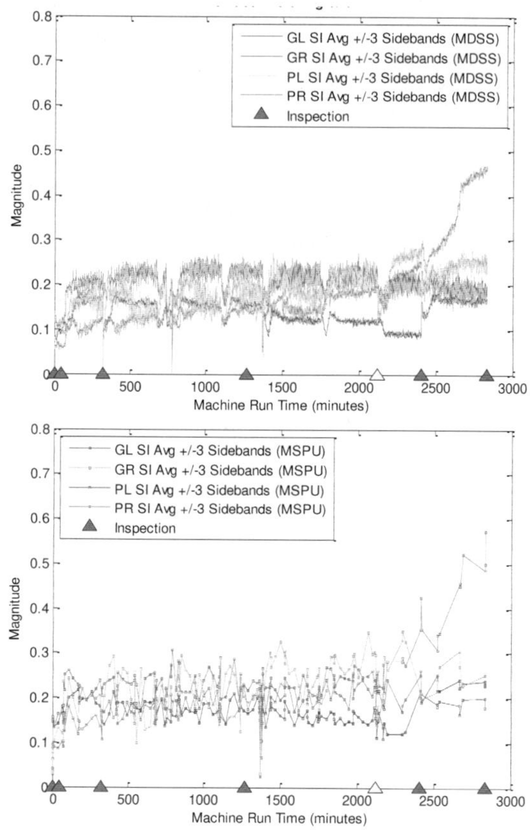

Figure 13: MDSS and MSPU SI3.

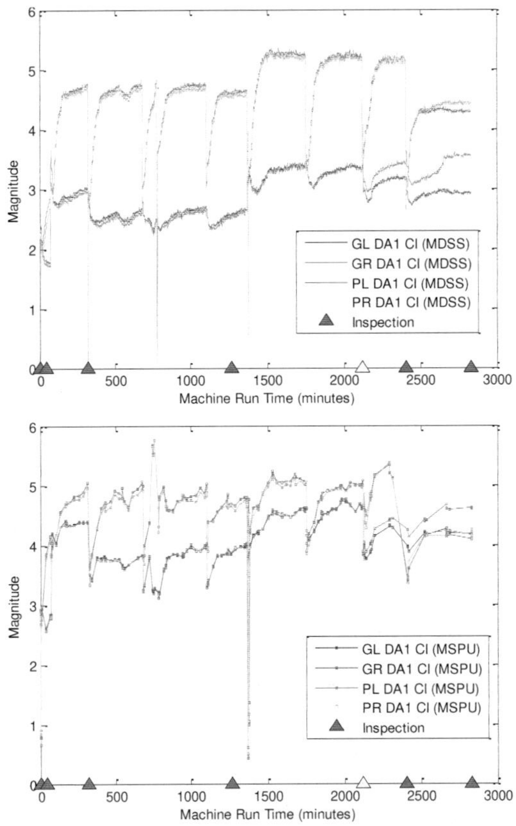

Figure 14: MDSS and MSPU DA1.

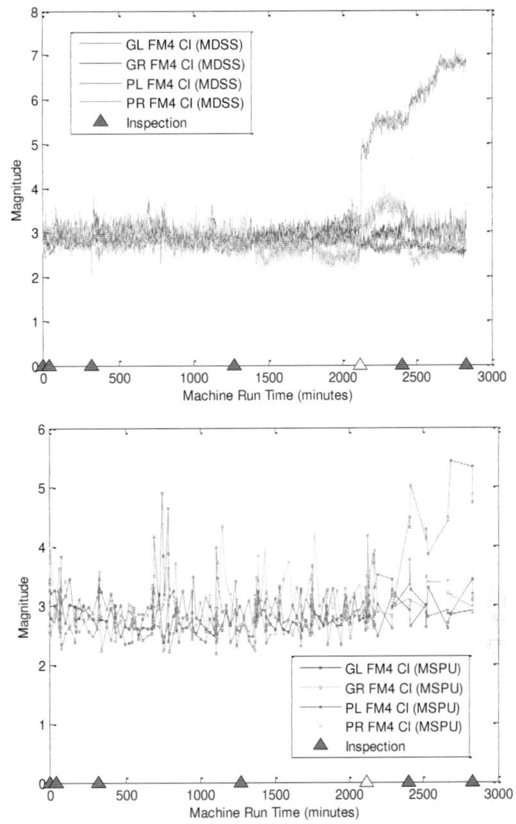

Figure 15: MDSS and MSPU FM4.

The software tool also generates tables of statistical parameters within each inspection interval for comparison. Table 4 summarizes the mean, standard deviation and RMS values calculated from MDSS within each inspection interval for FM4, RMS, SI1, and SI3. The cells highlighted identify the timeframe when damage was observed on the left pinion teeth (PL). During the inspection interval when damage was observed on two or more pinion teeth, run time 2120 to 2403 min, mean values for FM4 and SI1 were higher for the left pinion than the other three components monitored.

Table 4: RMS, FM4 and SI Statistical Parameters for Test 1.

MDSS RMS	Mean	STDEV	RMS	Mean	STDEV	RMS	Mean	STDEV	RMS	Mean	STDEV	RMS
Inspection	GL	GL	GL	GR	GR	GR	PL	PL	PL	PR	PR	PR
1 - 43	2.05	0.22	2.06	2.26	0.14	2.26	2.00	0.23	2.01	2.09	0.15	2.09
43 - 325	2.77	0.42	2.80	4.15	0.78	4.22	2.71	0.42	2.74	4.08	0.81	4.16
325 - 1266	2.53	0.13	2.53	4.40	0.49	4.43	2.48	0.13	2.48	4.35	0.48	4.37
1266 - 2120	3.18	0.26	3.19	4.93	0.53	4.96	3.16	0.27	3.17	4.87	0.54	4.90
2120 - 2403	3.08	0.13	3.08	4.85	0.51	4.88	3.28	0.15	3.28	4.83	0.52	4.86
2403 - 2833	2.91	0.07	2.91	4.19	0.30	4.20	3.35	0.19	3.36	4.28	0.31	4.29
MDSS FM4	Mean	STDEV	RMS	Mean	STDEV	RMS	Mean	STDEV	RMS	Mean	STDEV	RMS
Inspection	GL	GL	GL	GR	GR	GR	PL	PL	PL	PR	PR	PR
1 - 43	2.83	0.11	2.83	2.68	0.10	2.68	2.95	0.16	2.96	2.74	0.20	2.75
43 - 325	2.80	0.10	2.80	2.88	0.21	2.88	2.96	0.17	2.97	2.91	0.19	2.92
325 - 1266	2.88	0.09	2.88	2.91	0.22	2.92	2.90	0.22	2.91	2.87	0.20	2.88
1266 - 2120	2.97	0.14	2.97	2.84	0.13	2.84	2.92	0.23	2.92	2.61	0.24	2.62
2120 - 2403	2.91	0.13	2.91	2.67	0.10	2.67	5.29	0.40	5.31	3.47	0.23	3.48
2403 - 2833	2.65	0.10	2.65	3.07	0.20	3.07	6.37	0.46	6.39	2.62	0.19	2.63
MDSS +1 SI	Mean	STDEV	RMS	Mean	STDEV	RMS	Mean	STDEV	RMS	Mean	STDEV	RMS
Inspection	GL	GL	GL	GR	GR	GR	PL	PL	PL	PR	PR	PR
1 - 43	0.1265	0.0079	0.1268	0.1799	0.0324	0.1827	0.0649	0.0102	0.0656	0.0844	0.0278	0.0888
43 - 325	0.1098	0.0202	0.1116	0.3466	0.0768	0.3550	0.1370	0.0420	0.1433	0.1933	0.0548	0.2009
325 - 1266	0.1064	0.0230	0.1089	0.3625	0.0584	0.3671	0.1978	0.0269	0.1996	0.1801	0.0519	0.1875
1266 - 2120	0.1311	0.0175	0.1323	0.3646	0.0427	0.3670	0.1900	0.0267	0.1918	0.1709	0.0388	0.1752
2120 - 2403	0.1164	0.0179	0.1177	0.3261	0.0477	0.3295	0.3265	0.0254	0.3275	0.2991	0.0596	0.3049
2403 - 2833	0.1831	0.0215	0.1843	0.3275	0.0350	0.3293	0.5100	0.0967	0.5191	0.2190	0.0394	0.2225
MDSS +3 SI	Mean	STDEV	RMS	Mean	STDEV	RMS	Mean	STDEV	RMS	Mean	STDEV	RMS
Inspection	GL	GL	GL	GR	GR	GR	PL	PL	PL	PR	PR	PR
1 - 43	0.0871	0.0076	0.0874	0.1134	0.0108	0.1139	0.0635	0.0043	0.0637	0.0970	0.0130	0.0978
43 - 325	0.1486	0.0271	0.1511	0.1914	0.0319	0.1940	0.1020	0.0193	0.1038	0.1504	0.0307	0.1535
325 - 1266	0.1494	0.0189	0.1506	0.2162	0.0260	0.2177	0.1284	0.0166	0.1295	0.1619	0.0324	0.1651
1266 - 2120	0.1251	0.0163	0.1261	0.2168	0.0214	0.2179	0.1610	0.0217	0.1625	0.1972	0.0272	0.1991
2120 - 2403	0.0952	0.0116	0.0959	0.1979	0.0207	0.1990	0.2151	0.0254	0.2166	0.2381	0.0436	0.2421
2403 - 2833	0.1645	0.0118	0.1649	0.1961	0.0159	0.1967	0.3585	0.0768	0.3666	0.2364	0.0202	0.2373

For test 2, a different failure mode occurred when full load was applied after run-in. This caused scuffing to occur on the left pinion and gear teeth. Scuffing is not a fatigue failure mode. This failure mode causes the transfer of metal between the meshing teeth and does not cause debris to be generated. Photographs of the damage are shown in Figure 16. The first two photos are at test start and after run-in. The scuffing occurred when full load was applied at reading 70. This accelerated the contact fatigue failure mode as shown in the photographs of left pinion teeth 2, 4, 8, 14, 18, and 19. A representative photo of all left gear teeth damage is shown for left gear tooth 3. At reading 214, macro pitting was observed on six pinion teeth.

Plots of RMS for all four components and S1I and SI3 averages for the left pinion and gear are shown in Figure 17. RMS was a good indicator of the left pinion damage and trended as damage increased. Note the increase in debris at reading 180 indicates the pitting initiation on the pinion teeth. Both gear sideband indexes slightly increased with scuffing, then slightly decreased with pitting.

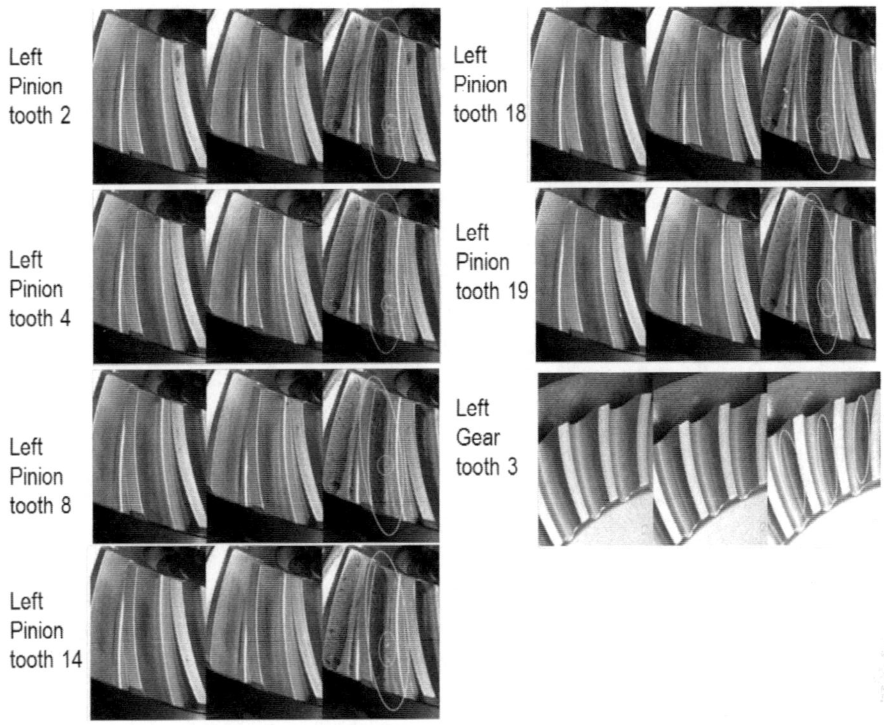

Figure 16: Pinion and Gear Teeth Damage.

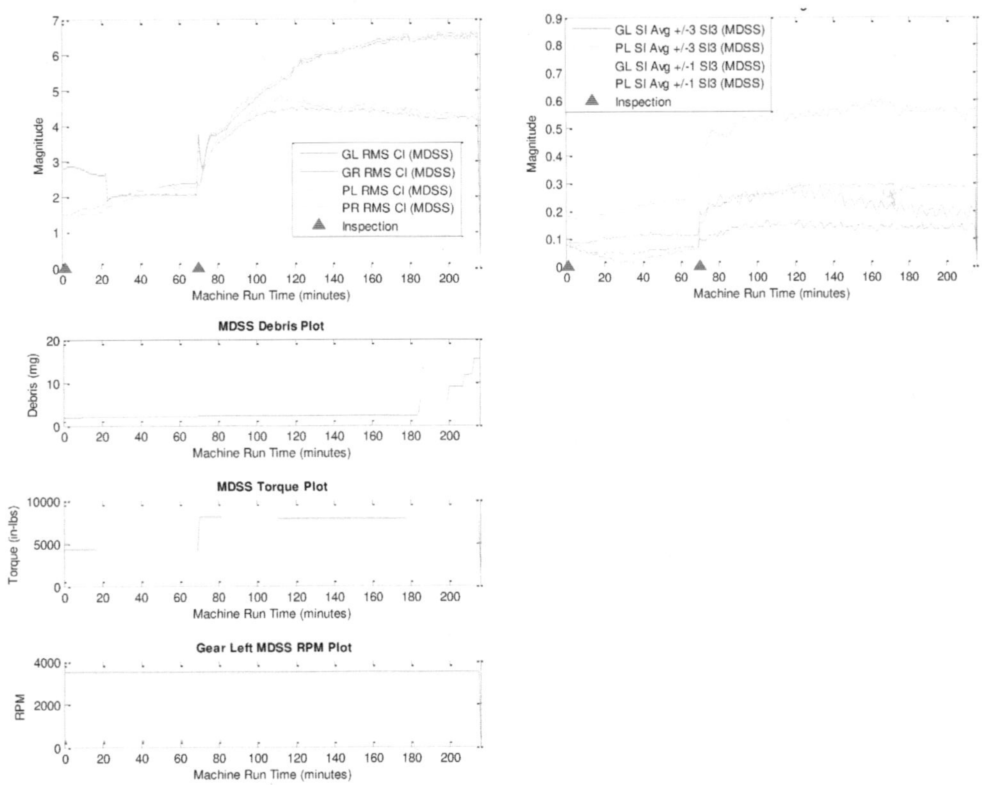

Figure 17: MDSS RMS and S1I and SI3 Averages.

Correlation Analysis: To evaluate the relationship between CIs and operational conditions, the software tool allows the user to select parameters to correlate using the Pearson Correlation Coefficients (r). Correlation coefficients measure the strength and direction of the linear relationship between two parameters [8]. Correlation coefficients are calculated by dividing the covariance of the two variables by the product of their standard deviations. Its value ranges between –1 and +1. A perfect linear relationship between two parameters will have a correlation coefficient of 1 or –1. Positive values indicate both parameters increase. Negative values indicate as one parameter increases, the other decreases. A value close to zero indicates no linear relationship between the two parameters. Hypothesis tests can be used to assess the significance of the relationship between the two parameters; however, a good rule of thumb is that values greater than 0.8 indicate a strong correlation and values less than 0.5 indicate a weak correlation.

When comparing data from several different data acquisition systems collected at varying acquisition intervals, a method to align the timestamps and fill in the gaps of data for the systems recorded at a lower acquisition rates must be defined. This is required when applying statistical methods or data mining techniques to correlate relationships between variables across the systems. The MDSS and Daytronic systems acquired data once per minute. The MSPU system acquired data once per 30 min. Since the MDSS, MSPU and Daytronic systems collected data at varying acquisitions intervals, the software tool provided three methods to align timestamps and fill in data gaps: (1)Nearest Neighbor; (2) Linear Interpolation, and (3) Filtered Spline. The linear interpolation method was used in the presented analysis. For this method, data points from the smaller data set are truncated to fall within the time range of the larger data set and timestamps from the smaller data set are used to linearly interpolate new values for the larger data set.

Figure 18 illustrates the linear interpolation method applied to the left pinion (PL) average S1I for the MDSS and MSPU data set previously plotted in Figure 12. The plot on the top left is the actual data. The plot on the right is the linear interpolated data with a correlation coefficient of 0.94. The bottom plot compares the left pinion MSPU SI1 values to the MDSS SI1 values. Figures 19 to 24 are plots of CIs versus oil outlet temperatures and torques. The r values for the two plotted parameters are shown in red.

Figure 19 compares the RMS values for the right gear and right pinion to the right oil outlet (ROO) temperatures measured on the right side of the test rig. RMS values for the left gear and pinion are compared to the left oil outlet (LOO) temperatures. The plots indicate a strong positive correlation with right oil outlet temperatures and right gear and pinion RMS values (0.78 and 0.79). Figure 20 compares the RMS values for the right gear, right pinion, left gear and left pinion to the applied torque. All four RMS values correlate to positive torque values. However, the left gear indicates a significantly stronger correlation (0.82) than the other three RMS values.

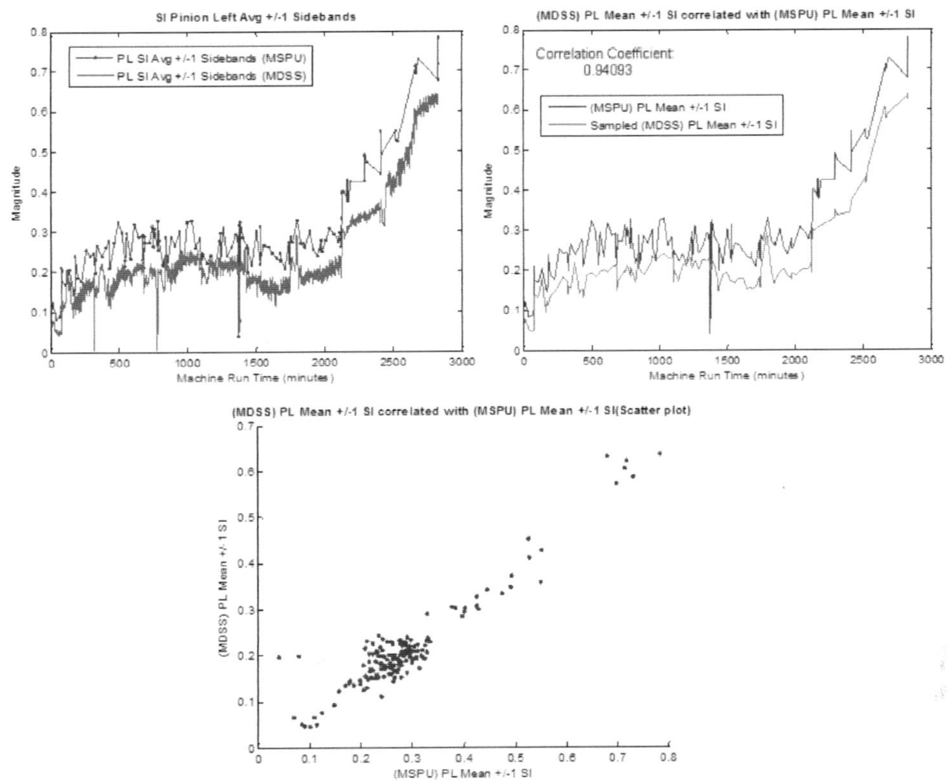

Figure 18: Linear Interpolation of MDSS and MSPU SI1.

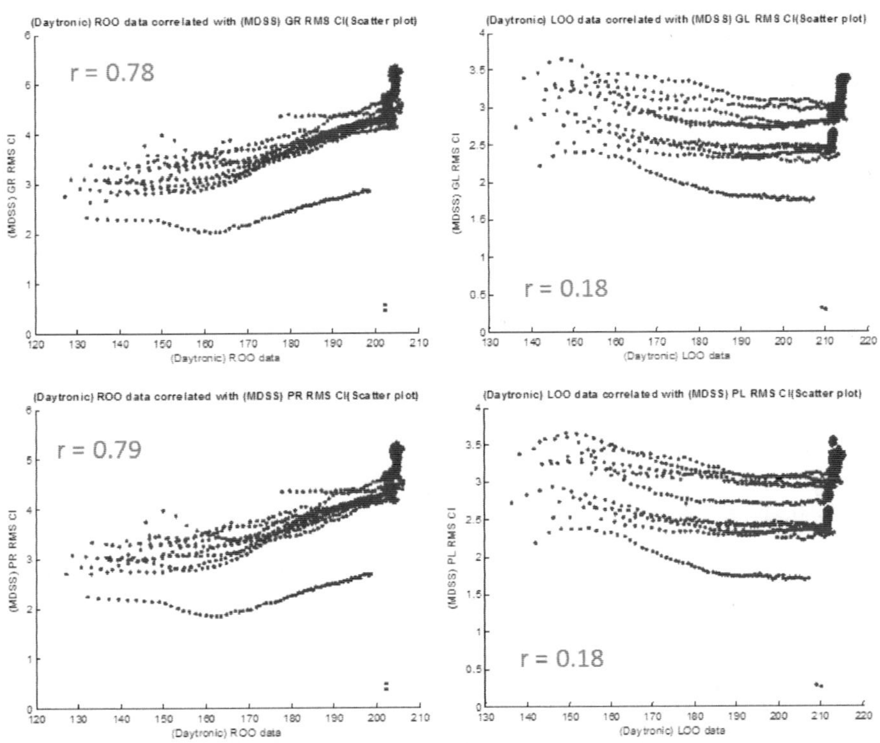

Figure 19: Correlation Between RMS and Outlet Oil Temperatures.

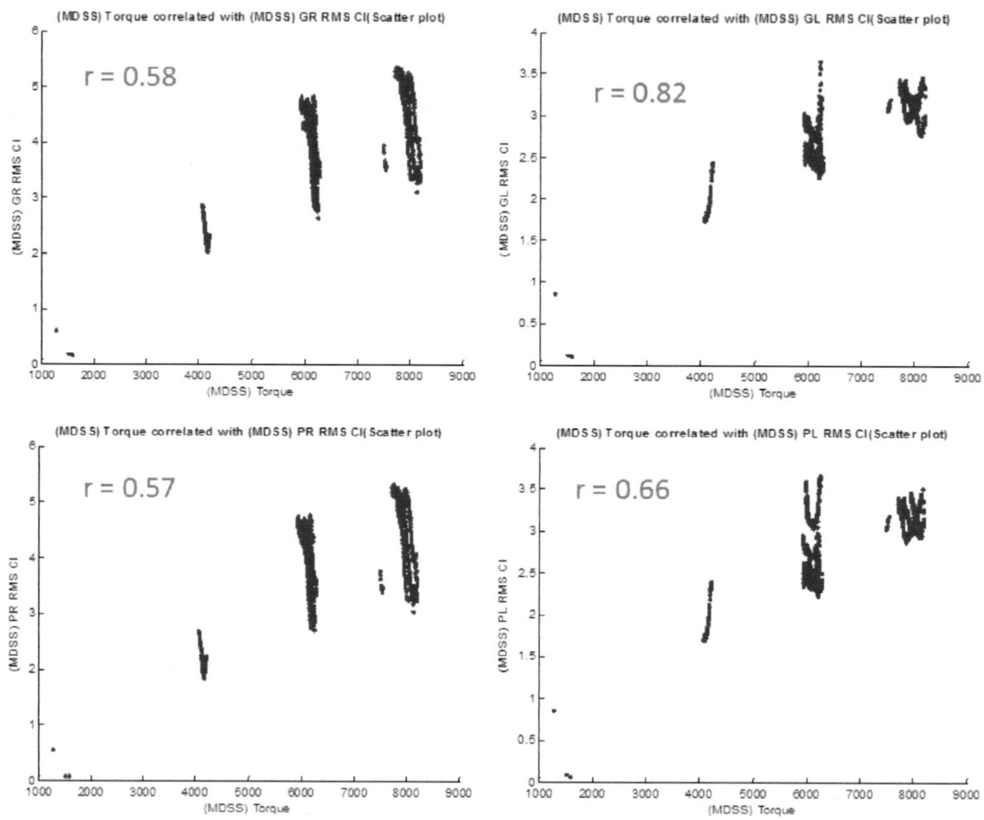

Figure 20: Correlation Between RMS and Torques.

Figure 21 compares the FM4 values for the right gear and right pinion to the oil outlet temperatures measured on the right side of the test rig. FM4 values for the left gear and pinion are compared to the left oil outlet temperatures. The plots indicate a weak correlation between FM4 and oil outlet temperatures for all four components. Figure 22 compares the FM4 values for the right gear, right pinion, left gear and left pinion to the applied torque. All four FM4 values indicate a weak correlation to torque. Note the plot of FM4 for the left pinion in the bottom right hand corner. The separation of FM4 values at 6000 and 8000 in.-lbs. are an indication of the larger FM4 values due to pinion tooth damage.

Figure 23 compares the SI1 values for the right gear and right pinion to the oil outlet temperatures measured on the right side of the test rig. SI1 values for the left gear and pinion are compared to the left oil outlet temperatures. All four SI1 values indicate a weak correlation with oil outlet temperatures. Figure 24 compares the SI1 values for the right gear, right pinion, left gear and left pinion to the applied torque. All four SI1 values indicate a weak correlation with torque.

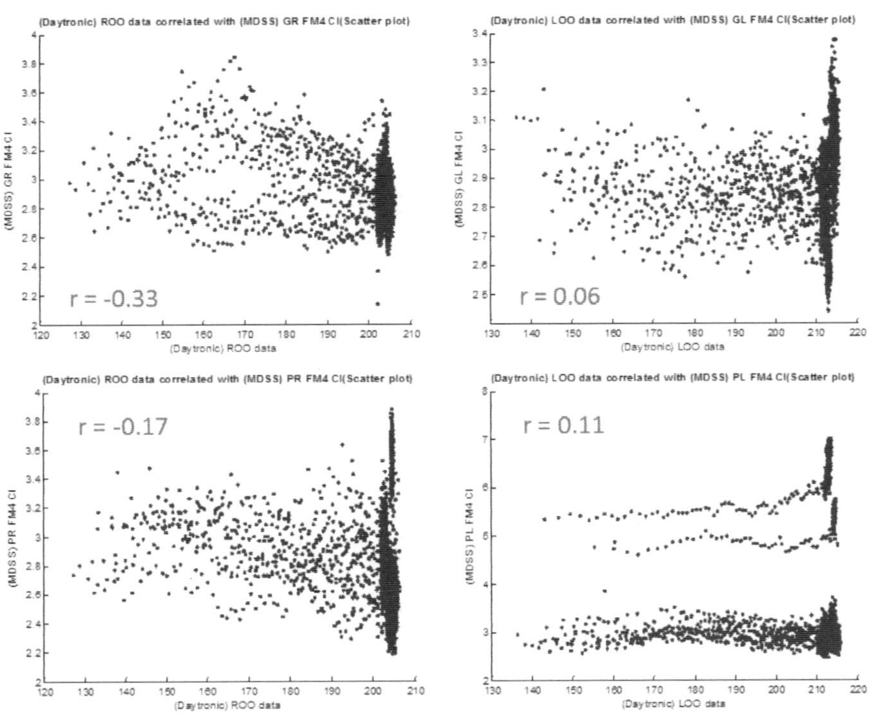

Figure 21. Correlation Between FM4 and Outlet Oil Temperatures.

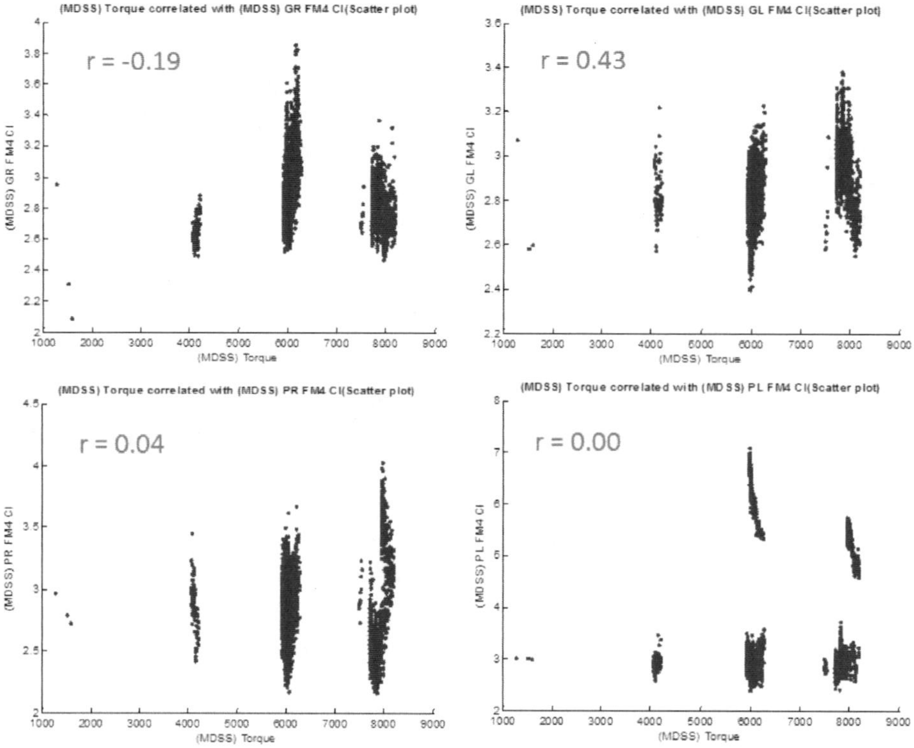

Figure 22: Correlation Between FM4 and Torques.

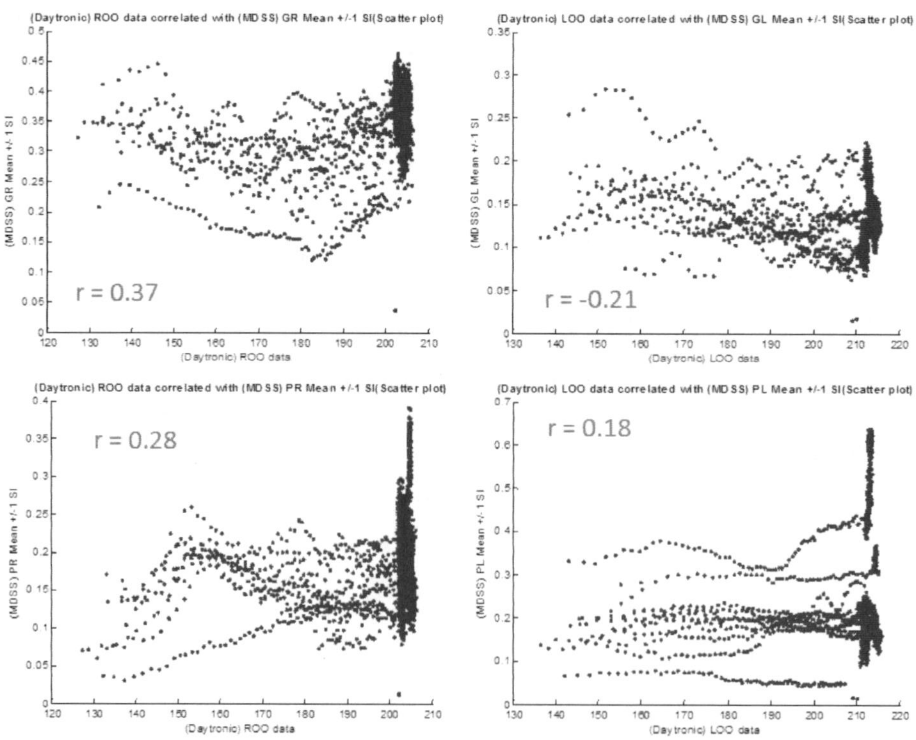

Figure 23: Correlation Between ±1 Average Sideband (SI1) and Outlet Oil Temperatures.

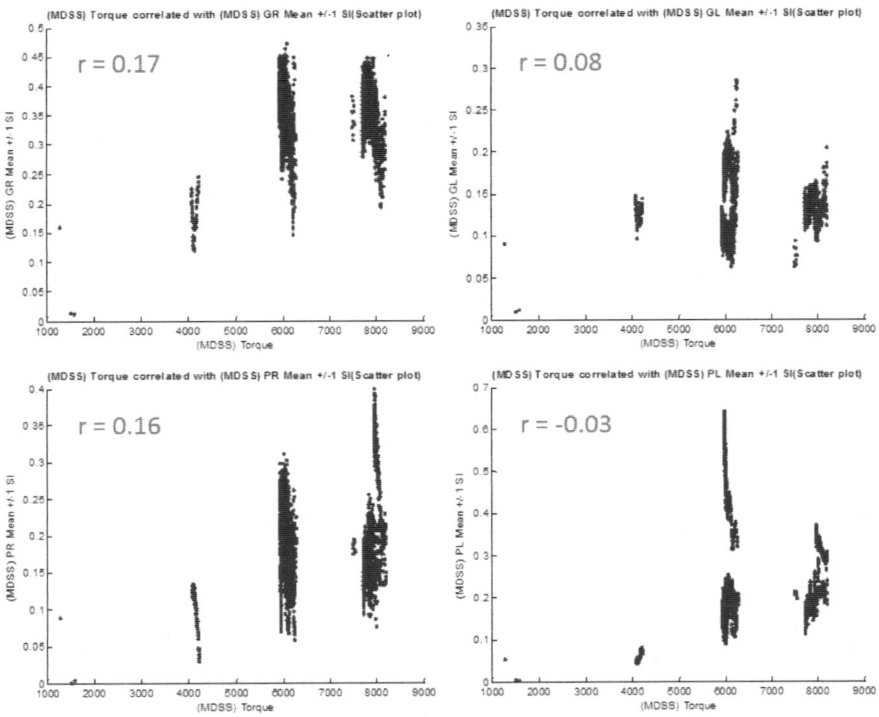

Figure 24. Correlation Between ±1 Average Sideband (SI1) and Torques.

Correlation analysis determined that some CIs were sensitive to operating conditions during test 1. However, the correlation data was not separated into specific failure modes and levels of damage. Sensitivity of a gear CI to environmental conditions may increase or decrease based on damage. Non-linear correlations may also exist between CIs, damage modes and operating conditions that will require further analysis with different methods. Further studies are required to investigate these effects. Additional analysis methods and capabilities are planned for the tool that enables grouping the data into comparable damage levels for analysis.

Summary: The objective of this work was to demonstrate the benefit of a software tool used to fuse the data generated from three data acquisitions systems and damage progression photos during spiral bevel gear damage progression tests in the NASA Glenn Spiral Bevel Gear Fatigue Test Rig. The response of gear CIs RMS, FM4 and average sideband indexes were compared to gear tooth damage progression, gearbox oil temperatures and applied torque. The tool enabled comparison of CI performance for different damage modes. CI data was also correlated with the operational conditions that the gear sets were exposed to during damage initiation and progression. Results found sensitivities of some CIs to damage modes, levels of damage and environmental conditions that must be taken into consideration when developing reliable gear health monitoring tools. Lessons learned during this preliminary investigation will be used to define additional analysis methods to be incorporated into the data fusion tool.

References

[1] Handschuh, R.F.: Thermal Behavior of Spiral Bevel Gears. NASA TM-106518, 1995.
[2] Handschuh, R.F.: Testing of Face-Milled Spiral Bevel Gears at High-Speed and Load. NASA/TM—2001-210743, 2001.
[3] Stewart, R.M. 1977. Some useful data analysis techniques for gearbox diagnostics. Machine Health Monitoring Group, Institute of Sound and Vibration Research, University of Southampton, Report MHM/R/10/77, July 1977.
[4] Zakrajsek, J.J. 1989. An investigation of gear mesh failure prediction techniques. NASA TM–102340, AVSCOM TM 89–C-005.
[5] Zakrajsek, J.J., Handschuh, R.F., and Decker, H.J. 1994. Application of fault detection techniques to spiral bevel gear fatigue data. Proceedings of the 48th Meeting of the Mechanical Failures Prevention Group. Office of Naval Research, Arlington, Virginia, pp. 93–104.
[6] ANSI/AGMA 1010-E95. Appearance of Gear Teeth – Terminology of Wear and Failure.
[7] Delgado, Irebert; Dempsey, Paula; Antolick, Lance and Wade, Dan: Continued Evaluation of Gear Condition Indicator Performance on Rotorcraft Fleet. Airworthiness, CBM, and HUMS Specialists' Meeting, February 11-13, 2013, Huntsville, Alabama.
[8] Triola, Mario. Elementary Statistics. 6th edition. Addison-Wesley 1995.

Acknowledgements: The authors would like to thank the NASA Rotary Wing Project and Traci Stadtmueller and Paul Swindell of the FAA Technical Center for their assistance and support of this work.

ROTORDYNAMIC MODEL DEVELOPMENT OF A SPIRAL BEVEL GEAR TEST RIG DRIVE TRAIN FOR GEAR HEALTH MONITORING

Michael J. Hurrell
Sierra Lobo, Inc.
102 Pinnacle Drive
Fremont, Ohio 43420
Telephone: (216) 433-9335
michael.j.hurrell@nasa.gov

Jerzy T. Sawicki
Department of Mechanical Engineering
Center for Rotating Machinery Dynamics and Control
Cleveland State University
Cleveland, Ohio 44115-2214
j.sawicki@csuohio.edu

Paula J. Dempsey
National Aeronautics and Space Administration Glenn Research Center
21000 Brook Park Road
Cleveland, Ohio 44135
paula.j.dempsey@nasa.gov

Abstract: The objective of this analysis was to show the importance of understanding the dynamic characteristics of the gear train driveline for implementation of a gear health monitoring system. A rotordynamic model of the NASA Glenn Spiral Bevel Gear Fatigue Test Rig was developed. The modeling considered the uncoupled lateral, axial and torsional vibration of the test rig driveline. The model results were then compared to vibration response measurements conducted using the rig gear health monitoring system at various rig operating conditions. In addition, the vibration response measurements were used to tune model parameters such as bearing preload and fastener torque. The results showed a good correlation between model and measurement. Of particular interest to gear health monitoring system implementation, is a double impact indication in the time synchronous average signal that was observed at the resonance conditions predicted by the rotordynamic model. The model will be used in further studies of gear train health monitoring and gear mesh dynamic modeling of the rig drive train including coupling between lateral, axial and torsional vibration.

Key words: Axial vibration; gear health monitoring; lateral vibration; rotordynamics; spiral bevel gear; time synchronous average; torsional vibration

Introduction: Health monitoring systems and tools that provide accurate fault detection and condition assessment of drive trains will benefit industries that perform condition based maintenance on these dynamic rotating components by improving safety, decreasing maintenance costs, and increasing system availability. These capabilities include detection of failures, damage severity and prediction of remaining useful life.

Many health monitoring systems employ vibration signatures generated from fault patterns produced when damaged components interact with their environment. For gear sets, these signatures relate to changes in dynamics at the gear mesh due to tooth damage. Typically, data from "healthy" transmission components define healthy vibration signatures and significant changes in these are used to indicate damage. Implementing a model based approach does not require validation data from all modes of damage and will identify system dynamic characteristics under varying operational conditions that can affect vibration response. Developing a dynamic model of system response under varying operational conditions can minimize experimental testing required to validate the health monitoring system. Correlating the model to respond to changes in dynamic response at fault initiation and progression can be used to predict system remaining useful life.

The objective of this investigation was to demonstrate the benefits of developing a system rotordynamic model to understand the interaction of the component within its environment and the resulting health monitoring vibration signatures. Understanding and modeling the system has the potential to predict performance of the health monitoring system during system operation.

Test Facility: The testing reported herein was conducted on the NASA Glenn Research Center (GRC) Spiral Bevel Gear Fatigue Rig. A sketch of the test rig is shown in Figure 1. The test rig drive train components are shown in Figure 2. A detailed description of this test facility can be found in references [1] and [2].

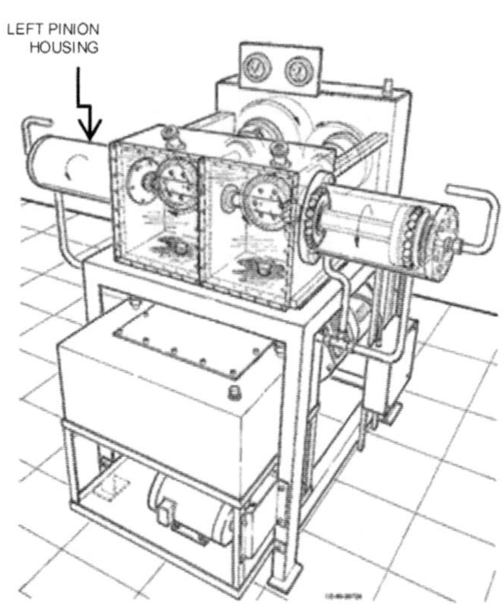

Figure 1: Spiral Bevel Gear Fatigue Rig.

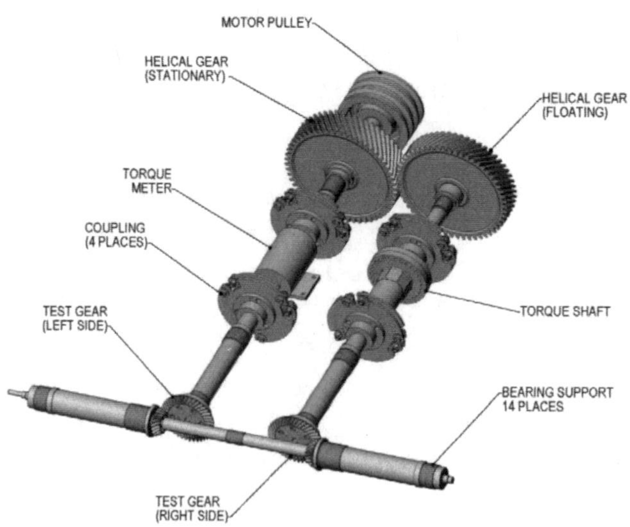

Figure 2: Rig Drive Train

For this investigation, the NASA Glenn Spiral Bevel Gear Fatigue Rig is being used to evaluate current gear health monitoring systems and techniques. In addition to developing gear health monitoring tools, the rig has been used for spiral bevel gear fatigue/endurance testing to study the effects of gear material, gear tooth design, and lubrication on the fatigue strength of gears.

As shown in Figure 2, the test rig is designed as a closed-loop arrangement where the torque is locked into the loop via a split shaft. Driveline rotation is provided by the drive motor through a v-belt connection to an axially stationary helical gear. Additional torque is created by way of a thrust piston that forces an axially floating helical gear into mesh. Two sets of spiral bevel gears are installed in the test rig and are tested simultaneously. The left side spiral bevel gear set is operated in the normal speed reducer mode where the pinion drives the gear. In contrast, the right side gear set acts as a speed increaser. However the gear sets on both the test and slave sides (right and left) of the rig operate with contact on the concave side of the pinion and convex side of the gear. An inline torque/speed sensor is used to monitor the load and speed of the test side gear output shaft [2].

Vibration, torque and speed data were measured and collected once every minute with the NASA Glenn research data acquisition system, referred to as Mechanical Diagnostic System Software (MDSS). Vibration data was collected from an accelerometer mounted on the left pinion housing, radially and vertically with respect to the pinion, as shown in Figure 3. Speed was measured with an optical tachometer mounted on the left pinion shaft. Vibration and speed data were collected at sample rates that provided sufficient vibration data for calculating time synchronous averaged data (TSA). TSA refers to techniques for averaging vibration signals over several revolutions of the shaft, to improve signal-to-noise ratio [3]. Failure specific condition indicators (CI) are then calculated using the TSA. Each CI is tailored to indicate specific failure modes such as surface fatigue (pitting, spalling), scoring, etc. Time synchronous averaging of the vibration data collected from the left accelerometer and pinion tachometer pulse is performed in the MDSS system.

Figure 3: Left Accelerometer Location

The gears were made from a steel alloy CEVM 9310, carburized, hardened and ground. The gear sets have a 6.4 in. diametral pitch, 20° pressure angle, 25° spiral angle, 0.94 in. face width and a 2.15 gear ratio. The gear has 41 teeth and the pinion has 19 teeth.

Rotordynamic Analysis: In order to understand the dynamic characteristics of the test rig and how the vibration signal and health monitoring techniques are affected, an uncoupled rotordynamic analysis was performed for each of the three types of vibration; lateral, axial, and torsional. An open source Matlab®-based rotordynamic finite element analysis (FEA) code [4] was used for the analysis presented herein. The code was modified and extended to include the lateral analysis of co-axial rotors and harmonic excitation at multiples of the rotor speed. The original code was limited to the lateral vibration of rotors. Axial and torsional capabilities were added to the original code for this analysis.

The MDSS accelerometer is uniaxial with the measurement direction aligned radially to the pinion shaft, or in the lateral direction. This is a typical configuration for a vibration based health monitoring system. An analysis coupling the three types of vibration was not considered. If strong coupling occurs between the lateral vibration and the axial and/or torsional vibration, then the axial and/or torsional modes will be evident in the lateral vibration signature. Evidence of this would warrant a coupled analysis.

Each type of vibration was analyzed using a finite element analysis referred to as a shaft-line rotordynamic FEA. A shaft-line analysis considers nodal locations along the rotor centerline and the degrees of freedom are displacements (translations and rotations) at the nodal locations [4]. The coordinate system and degrees of freedom are shown in Figure 4.

Shaft-line modelling is ideal for axisymmetric rotors such as the rotors that constitute the test rig drive train.

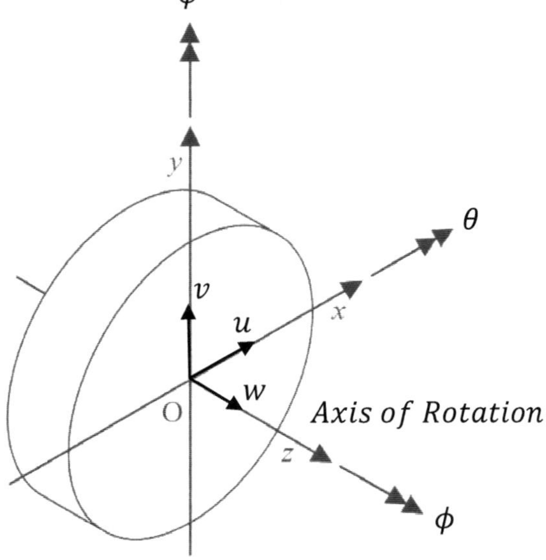

Figure 4: Coordinate System and Degrees of Freedom.

Lateral Vibration Model: A lateral analysis considers the two translations (u, v) and two rotations (θ, ψ) in the $x - y$ plane (see Figure 4). The resonance frequencies of lateral vibration of a rotor depend on the speed of rotation through the gyroscopic effects and speed dependent bearing properties. The speed dependence of bearing properties is of a concern mainly when the bearings are of the fluid film type. In addition, the lateral vibrations of a rotor are strongly coupled to the supporting structure. However, because of the lateral flexibility of shaft couplings, shaft assemblies can most often be analyzed laterally in isolation from the complete drive train at least as a first approximation [4]. Given these assumptions, and the fact that the health monitoring accelerometer is located on the left pinion housing, the left pinion housing assembly was modeled for lateral vibration in isolation from the rest of the rig drive train.

Considering the left pinion housing vibration, the housing assembly cross-section and shaft-line FEA locations are shown in Figure 5. The pinion shaft consists of an inner shaft in tension which holds the pinion to the outer shaft. The outer shaft locates the bearing surfaces for the three angular contact bearings that support the rotating components in the pinion housing. The pinion housing is bolted to the rig main housing through a mounting flange. The front end of the pinion housing has a pilot diameter that has a slip fit with the mating bore on the test rig. This fit locates the pinion housing and pinion with respect to the gear independent of the pinion housing mounting bolts and provides proper alignment of the gear set.

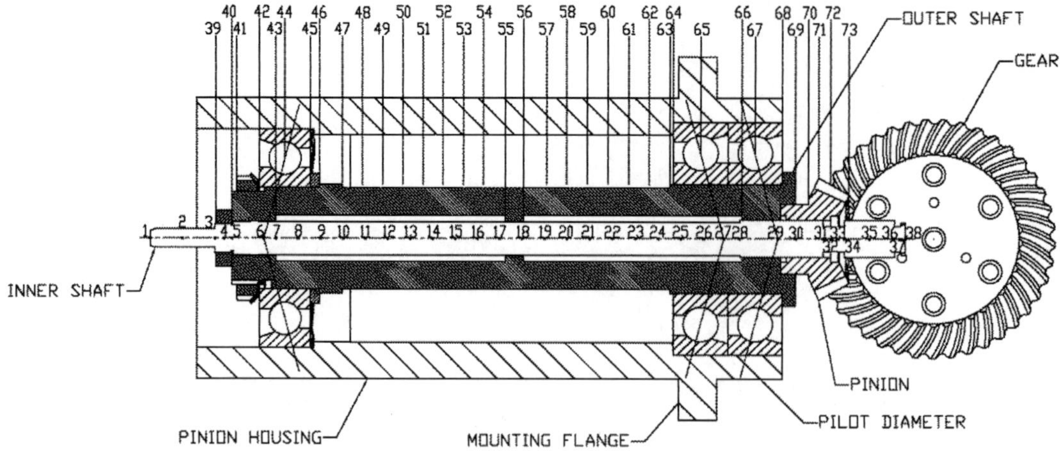

Figure 5: Left Pinion Housing Cross-section with Nodal Locations for Lateral and Axial Models.

The shaft elements were modelled as Timoshenko beam elements with shear, rotary inertia and gyroscopic affects included [5-6]. A disk element was added to the right end of the shaft assembly at node 30 to simulate the inertia of the pinion. The inner and outer shafts were modelled as coaxial shafts with two interfaces at node 5 and node 30. The inner shaft was modeled with an axial load between node 4 and 29 to simulate the tension on the inner shaft due to the shaft end hex nut torque of 200 ft-lbs. As a first approximation, the bearings at node 44, node 65 and node 67 were modeled as rigid short bearings. The lateral FEA model consisted of 73 nodes and 72 elements.

Axial Vibration Model: The axial analysis considers deflection w along the z-axis (see Figure 4). Similar to the consideration of the lateral flexibility of shaft couplings that led to examining the pinion housing assembly in isolation from the rest of the rig drive train, the axial vibration model considered the pinion housing assembly only. In general, shaft couplings are axially, as well as laterally, flexible. Nodal locations for the axial model were similar to the nodal locations for the lateral model as shown in Figure 5.

The axial model considered the effects of the axial support of the angular contact bearings and the wave spring stack that provides preload to the angular contact bearing located at node 44. The shaft elements were modelled with axial stiffness and mass. In addition, a disk was included at node 30 to simulate the inertia of the pinion. The axial FEA model consisted of 38 nodes and 37 elements.

Torsional Vibration Model: The torsional analysis considers rotation ϕ about the z-axis (see Figure 4). Typically shaft couplings are laterally and axially flexible to allow for lateral and axial misalignment but are torsionally stiff to provide the necessary driving torque transmission from one shaft assembly to the next. For this reason, it was necessary to consider all rotors transmitting torque in the drive train. As a result, the torsional analysis was performed on the complete drive train.

The model considered eleven rotational displacements and eleven inertias as shown in Figure 6. The shaft elements were modelled with torsional stiffness and rotational inertia due to the radial distribution of mass in the shaft. The two helical gears, the two spiral bevel gears, the two spiral bevel pinions and the shaft couplings were modeled as rotary inertias.

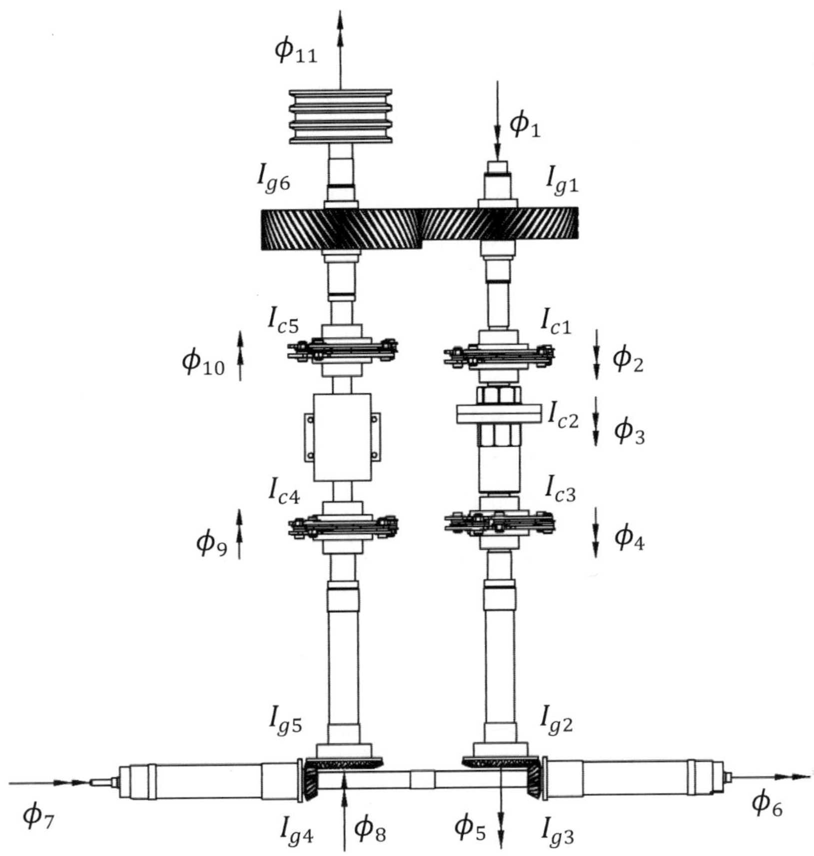

Figure 6: Top View of Rig Drive Train with Displacements and Inertias for the Rig Torsional Model.

Vibration Response Measurements: Testing consisted of two runs performed on two separate days. The test rig was allowed to warm to normal operating temperature before each data acquisition. Gear torque was maintained at 8000 in-lbs. throughout the testing. Gear speed was varied for run #1 per the schedule listed in Table 1. Table 2 lists the sequence of speeds for run #2. Each speed condition listed was maintained for approximately 3 to 5 minutes duration to allow for the MDSS to capture at least one data set.

Pinion housing vibration, gear shaft speed, gear torque and test box oil debris generation were recorded during the tests using the MDSS. Pinion shaft speed is calculated by multiplying the measured gear speed by the gear ratio

$$Gear\ Ratio = \frac{\#\ Gear\ Teeth}{\#\ Pinion\ Teeth} = \frac{41}{19}. \tag{1}$$

Table 1: Speed Schedule for Run #1.

Run-time (min)	Gear Torque (in-lbs)	Gear Speed (RPM)	Pinion Speed (RPM)
161	8000	3500	7553
166	8000	3300	7121
170	8000	3200	6905
174	8000	3100	6689
178	8000	3000	6474
181	8000	2900	6258
185	8000	2800	6042
189	8000	2700	5826
192	8000	2600	5611
196	8000	2500	5395
199	8000	2400	5179
204	8000	2300	4963
208	8000	2400	5179
211	8000	2500	5395
215	8000	2600	5611
219	8000	2700	5826
223	8000	2800	6042
226	8000	2900	6258
230	8000	3000	6474
233	8000	3100	6689
237	8000	3200	6905
242	8000	3300	7121

Table 2: Speed Schedule for Run #2.

Run-time (min)	Gear Torque (in-lbs)	Gear Speed (RPM)	Pinion Speed (RPM)
348	8000	2800	6042
352	8000	2750	5934
355	8000	2726.27	5883
358	8000	2700	5826
362	8000	2500	5395
366	8000	2450	5287
369	8000	2400	5179
372	8000	2375	5125
377	8000	2358	5088
381	8000	2325	5017
384	8000	2300	4963
388	8000	2325	5017
392	8000	2358	5088
395	8000	2375	5125
399	8000	2400	5179
402	8000	2450	5287
406	8000	2500	5395
410	8000	2700	5826
413	8000	2726.27	5883
417	8000	2750	5934
420	8000	2800	6042

The test speeds were chosen as a result of the rig resonance conditions determined by the lateral rotordynamic analysis. The results from this analysis and the torsional and axial analysis will now be discussed.

Results and Discussion: A Campbell plot, sometimes called an interference plot, displays the natural frequencies from a given rotordynamic analysis versus shaft speed. Load lines representing possible excitations are also plotted. The intersection of the natural frequencies (solid lines) and the load lines (dashed lines) represent possible resonance conditions and are called critical speeds if they occur within the normal operating speed range. Load lines corresponding to 1X, 2X, and 19X the pinion shaft speed were considered. The load line corresponding to 1X rotational speed represents loading due to rotor imbalance. Loading due to misalignment is usually represented by a 2X load line [7]. Gear mesh (GM) excitation is represented by the 19X load line. The spiral bevel pinion has 19 teeth.

The Campbell diagram resulting from the torsional rotordynamic analysis and axial rotordynamic analysis are shown in Figure 7 and Figure 8, respectively.

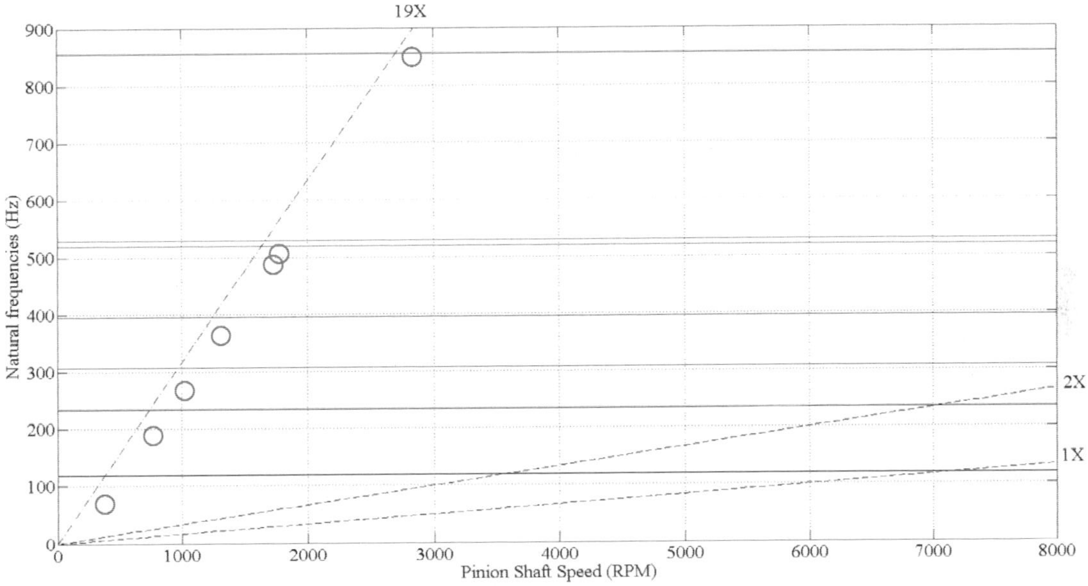

Figure 7: Campbell Diagram for Drive Train Torsional Vibration.

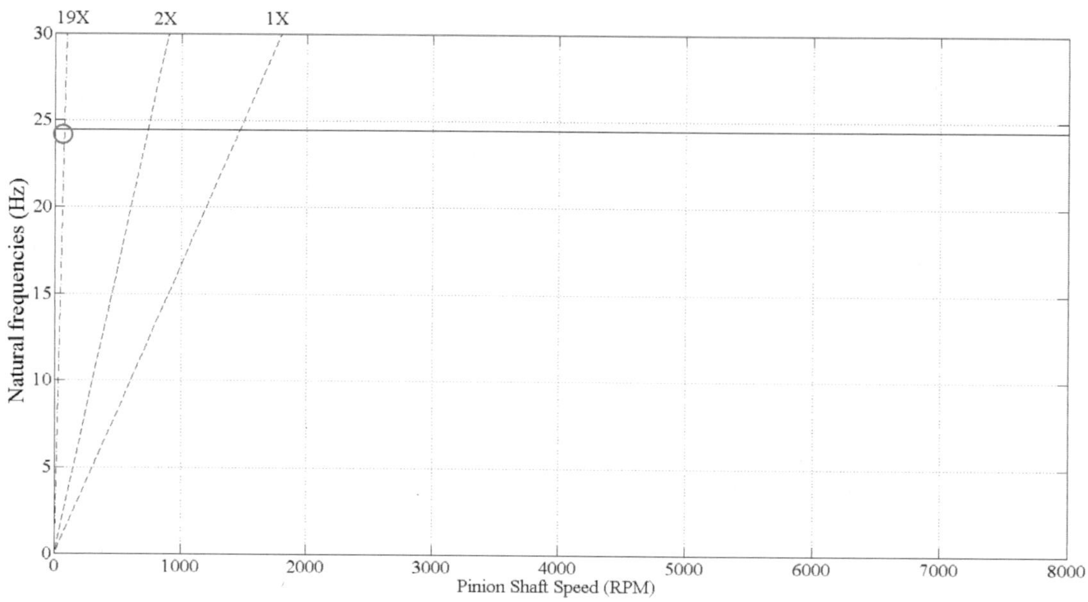

Figure 8: Campbell Diagram for Pinion Shaft Assembly Axial Vibration.

The analysis that follows will focus on loading due to gear mesh excitation since some frequencies used for health monitoring of gear trains correspond to the gear mesh frequency and sidebands. Excitations at 1X and 2X the pinion shaft speed are shown on the Campbell diagrams for reference only.

The torsional analysis indicates seven resonances (circled in red) corresponding to the first seven modes of vibration given gear mesh excitation. These resonances occur below a pinion shaft speed of 3,000 rpm. Since the normal operating speed for the pinion is 7,553 rpm, these resonances are not considered to be critical speeds since they occur well below the operating speed. Likewise, the axial analysis shows a resonance condition at the first mode of vibration at a pinion shaft speed of 77 rpm, well below the normal operating speed. This mode is similar to a rigid body mode and is a result of the relatively weak stiffness of the wave spring stack used to preload the pinion shaft-end angular contact bearing. The next resonance condition (not shown) occurs at 15,524 rpm well above the normal operating speed. Both Campbell diagrams show no potential critical speeds near the operating speed range.

The Campbell diagram for the lateral rotordynamic analysis is given in Figure 9. The lateral analysis shows eight resonance conditions below a pinion speed of 6,000 rpm. Note that the natural frequencies occur in complex conjugate pairs as a result of the gyroscopic nature of the rotor. As the rotational speed of the rotor increases the natural frequency pairs diverge. The largest of the pair is the forward mode and is the mode excited by the gear mesh excitation [4]. In contrast to both the torsional and axial analysis, the lateral analysis shows two potential critical speeds at 5088 rpm and 5883 rpm pinion speed.

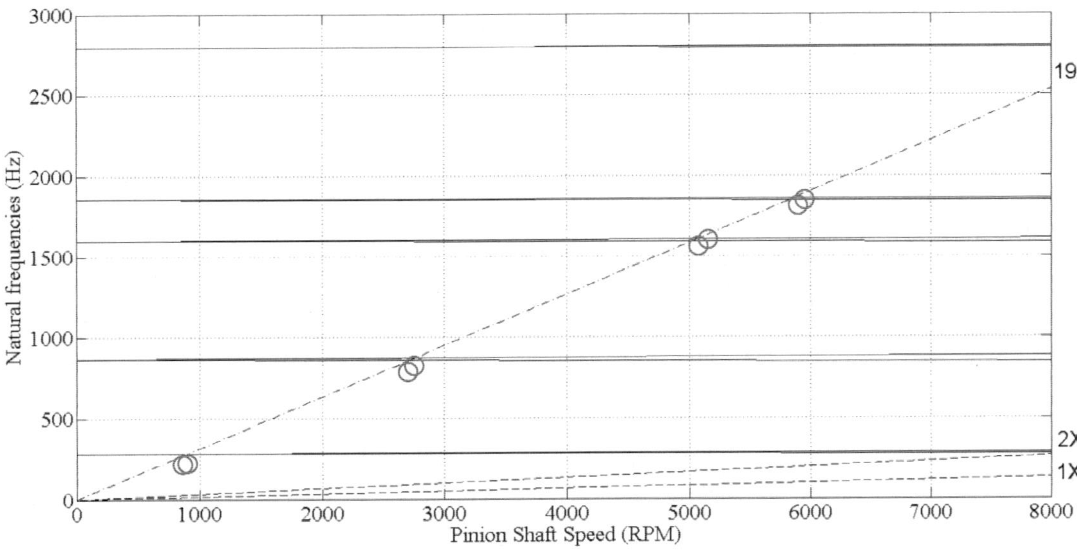

Figure 9: Campbell Diagram for Pinion Shaft Assembly Lateral Vibration.

A forced response analysis was conducted to understand the nature of the response along the pinion shaft length. The forcing was assumed to be a harmonic of the gear mesh frequency. This force was applied to the FEA model at node 72 to simulate gear contact.

The response at various node locations along the pinion shaft is shown in Figure 10. The peaks in the response correspond to the resonance conditions shown in the Campbell diagram at 914 rpm, 2783 rpm, 5088 rpm and 5883 rpm pinion speed. Of particular interest are the two potential critical speeds that occur at approximately 5088 rpm and 5883 rpm. These speeds were chosen for establishing the speed schedule for the vibration response testing shown in Tables 1 and 2.

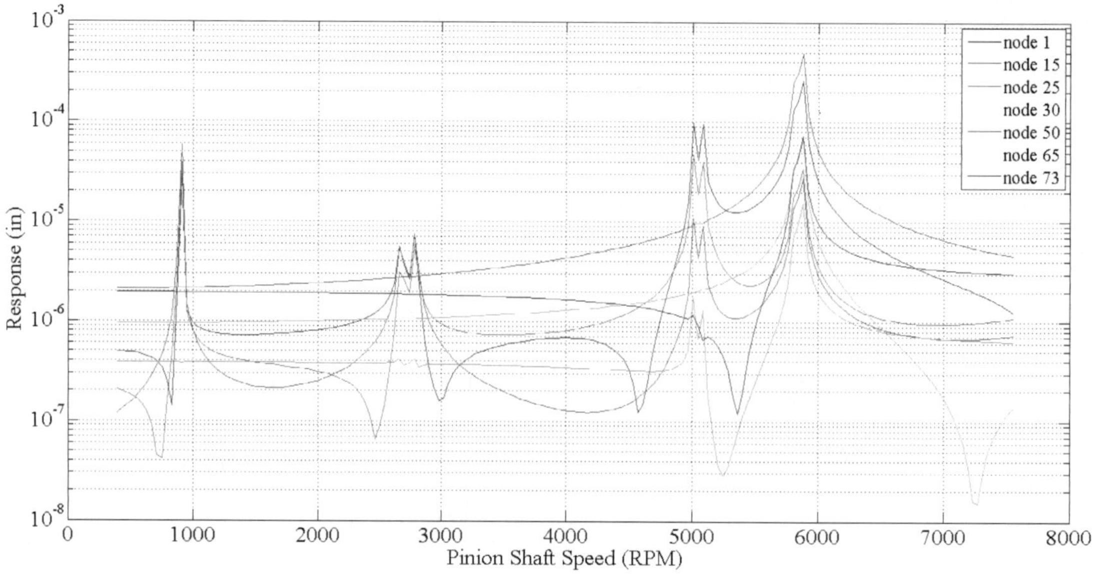

Figure 10: Forced Response at Gear Mesh Frequency for Pinion Shaft Assembly Transverse Vibration.

The results from the first run of the vibration response tests are given in Table 3. Comparisons between the different test conditions were made using the amplitude of the gear mesh frequencies. All gears generate a dominant gear mesh frequency in the vibration signature due to each tooth impacting against the gear it is driving as the teeth mesh. The gear (or pinion) mesh frequency is equal to the number of teeth multiplied by its speed. It is obtained by converting the TSA data, the average of the vibration response over 1 pinion revolution, into the frequency domain. Once converted to the frequency domain, the gear mesh is found at the gear mesh frequency equal to number of teeth multiplied by shaft speed. This is the response magnitude of the first gear mesh harmonic, designated GM1. The second gear mesh harmonic, 2X shaft speed multiplied by the number of pinion teeth, is referred to as GM2 and the third gear mesh harmonic, 3X shaft speed multiplied by the number of pinion teeth, is referred to as GM3.

The trend from the data suggests that the second harmonic dominates at, or near, a critical speed. To illustrate this, the time synchronous average (TSA) and Fast Fourier Transform (FFT) of the vibration signal from the pinion housing accelerometer corresponding to a pinion shaft speed of 6474 rpm is shown in Figure 11. The TSA and FFT at this speed are as expected with a TSA showing 19 peaks and the dominant response at the first harmonic of the gear mesh frequency. However, Figure 12 shows the TSA and FFT for a pinion shaft speed of 5179 rpm. At this speed the TSA has 38 peaks and the dominant response occurs at the second harmonic of the gear mesh frequency. Similar results were observed during run #2. These results indicate that a double impact of the mating gear teeth occurs at or near a resonance condition.

In comparing the analysis with the vibration response measurements, the observed resonance at 5179 rpm is slightly higher than the predicted critical speed of 5088 rpm. A dominant response of the second harmonic of the gear mesh frequency was observed at a pinion speed of 6042 rpm close to the predicted critical speed of 5883 rpm. The difference between the predicted critical speeds and the observed critical speeds may be the result of variation of damping used in the model versus actual damping. The number of degrees of freedom used in the model is an approximation for the continuous nature of the actual rotor and will also affect the predicted critical speeds.

Table 3: Speed Schedule for Run #2.

Pinion Speed (RPM)	GM1	GM2	GM3	Comments
7553	1.75		0.50	
7121	2.25	0.75	0.25	
6905	4.00	1.80		
6689	5.00			
6474	6.00			
6258	3.00	1.00		
6042	2.50	2.25		
5826	3.75	3.25		
5611	4.00	4.50		GM2 > GM1
5395	4.25	5.25		GM2 > GM1
5179	3.00	6.25		GM2 > GM1
4963		3.50		GM2 > GM1
5179	3.00	6.50		GM2 > GM1
5395	4.00	5.25		GM2 > GM1
5611	4.00	5.00		GM2 > GM1
5826	5.00	2.50		
6042	2.50	2.75		GM2 > GM1
6258	3.00	1.50		
6474	5.25			
6689	5.00			
6905	3.75	1.50		
7121	2.25	1.00		

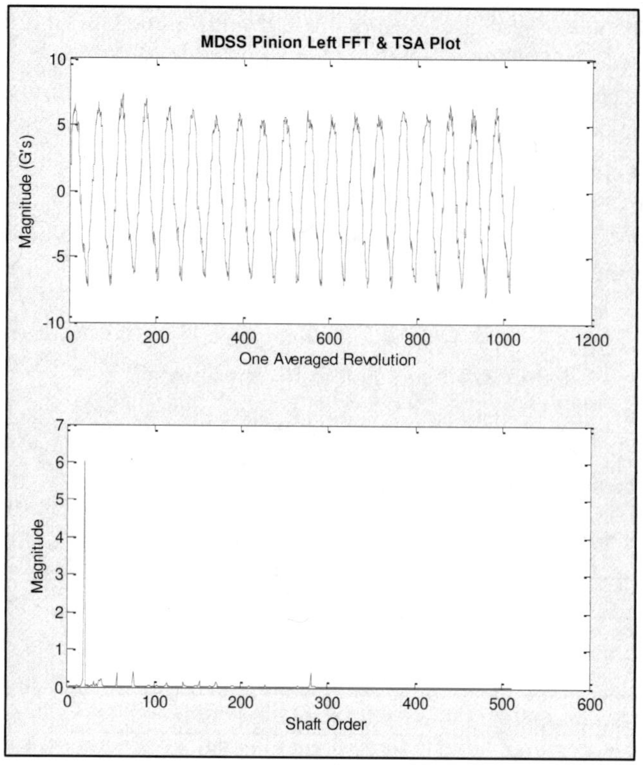

Figure 11: TSA Plot and FFT for Left Pinion at 6474 rpm, 8000 in.-lbs Gear Torque.

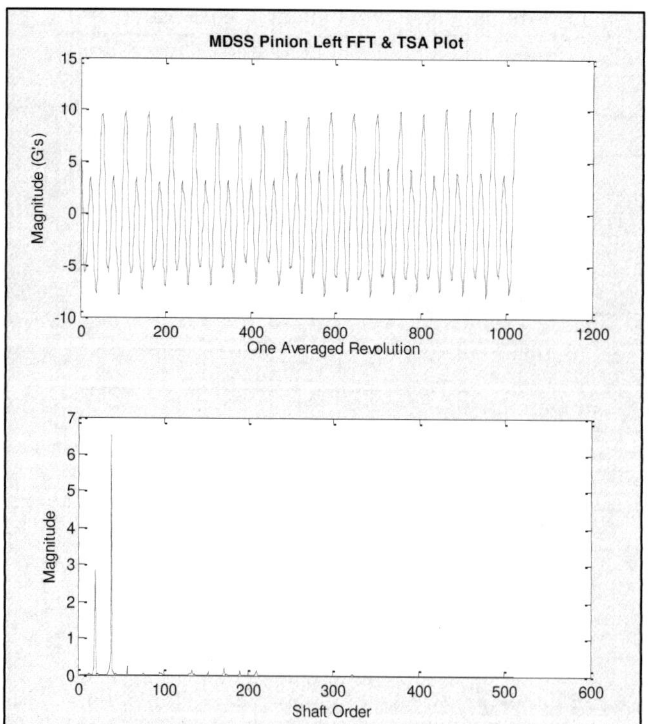

Figure 12: TSA Plot and FFT for Left Pinion at 5179 rpm, 8000 in.-lbs Gear Torque.

Conclusions: This investigation demonstrated the benefits of developing a system rotordynamic model to predict system resonances that can affect the response of vibration signatures used to predict system health.

Implementation of a gear health monitoring system should include evaluation of the drive train dynamics and identification of potential resonance conditions and corresponding critical speeds. The model predicted resonance conditions that affected the response of the time synchronous average of the vibration signal used for health monitoring. At critical speeds, the second harmonic of the gear mesh frequency was found to dominate the response. Physically, the TSA exhibited a double impact indication. Clearly, the dynamics of the rotor assembly has an influence on the gear mesh dynamics and resulting health monitoring vibration signature.

A drive train rotordynamic model can aid in the development of a health monitoring system that will provide reliable system response within the operating conditions of interest. If a more complex model can predict the output response based on the system and operating conditions, it will reduce the number of tests required for condition based maintenance validation and improve the design of the health monitoring system. A more sophisticated model incorporating gear mesh dynamics may lead to a better understanding of the physics of CI response.

References:

[1] Handschuh, R. F: Thermal Behavior of Spiral Bevel Gears. NASA TM-106518, 1995.

[2] Handschuh, R. F.: Testing of Face-Milled Spiral Bevel Gears at High-Speed and Load. NASA/TM—2001-210743, 2001.

[3] Stewart, R.M. 1977. Some useful data analysis techniques for gearbox diagnostics. Machine Health Monitoring Group, Institute of Sound and Vibration Research, University of Southhampton, Report MHM/R/10/77, July 1977.

[4] Friswell M. I, Penny J. E. T., Garvey S. D., Lees A. W. Dynamics of Rotating Machines, Cambridge University Press, Cambridge, UK (2010).

[5] Timoshenko, S. P., 1921, On the correction factor for shear of the differential equation for transverse vibrations of bars of uniform cross-section, Philosophical Magazine, p. 744.

[6] Timoshenko, S. P., 1922, On the transverse vibrations of bars of uniform cross-section, Philosophical Magazine, p. 125.

[7] Corbo M. A., Malanoski S. B. "Practical Design Against Torsional Vibration" Turbomachinary Symposium 25th, Houston TX, USA (1996).

DYNAMIC CONDITION BASED FEATURE EXTRACTION STRATEGY FOR MACHINE HEALTH MONITORING APPLICATIONS

Hung-An Kao, Chao Jin, Zongchang Liu, Shanhu Yang and Zhe Shi
NSF I/UCRC Center for Intelligent Maintenance Systems (IMS)
University of Cincinnati
Cincinnati, OH, USA 45221-0072
Telephone: (513) 556-3412
kaohn@mail.uc.edu

Abstract: Machine health condition monitoring and fault diagnosis are two important tasks for the optimization of factory up time, diagnosis of component failure modes, and eventually support predictive maintenance decisions. That is why research to prognostics and health management (PHM) is being conducted world wide in different industrial applications. Within these applications, data acquisition is a key to collect machining data and understand machine conditions. The traditional data acquisition devices deal with controller data, sensors and other manufacturing systems at the same time. The feature extraction strategy is static once the PHM system being designed and established. However, which features to extract and when to extract them are highly related to real-time machine/component conditions. Therefore, this paper presents a framework for dynamic condition based feature extraction strategy for machine health monitoring applications. A knowledge base for the relationship of features and corresponding failure modes is developed first using quality function deployment (QFD) format. Then, a decision tree for prioritizing feature extraction order will be built based on the calculation of Bayesian probabilities. Because of the hierarchy and priority, the DAQ system can dynamically extract critical features based on the real-time condition of monitored machine systems.

Key words: Condition based monitoring; data acquisition; decision tree; fault diagnosis; feature extraction; prognostics and health management; quality function deployment

Introduction: Prognostics and health management (PHM) is an emerging engineering discipline that evaluates the reliability of a system within its life cycle in order to detect beforehand any upcoming failures and reduce risks. Due to such importance, PHM has gained more and more attention in both academia and industry. In order to develop and implement efficient and reliable PHM solutions to machining systems, there are two urgent needs: 1) to develop a dynamic feature extraction strategy that intelligently select optimal features based on failure modes and available sensors; 2) to balance between local computational resource of the hardware and data transmission bandwidth.

To improve the feasibility of PHM, enabling machines to communicate with a PHM system, which consists of a set of analytic tools, is somehow a critical technology that needs to be solved first. The communication between machines and the system may include collecting and transferring data/features from machining process and connecting

to a platform for analytical service. Currently, there are various data resources, including machine controllers (CNC, PLC, etc.), add-on sensors (accelerometers, thermocouples, etc.), and e-manufacturing systems (ERP, MES, CAD/CAM, etc.).
Since these data sets are essential inputs needed by PHM analytic tools, data acquisition systems (DAQ), either embedded in controller or externally connected, are responsible for extracting all the data, preprocessed and transferred to a designated central system. To assure the accuracy and reliability of the PHM results, the whole DAQ process is required to be continuous and precise. However, the limitations of transferring bandwidth and computational barrier have put a great challenge to PHM tasks with respect to current manufacturing environment.
To solve the above-mentioned problem, a mechanism so called 'Dynamic Condition based Feature Extraction Strategy' is proposed in this paper, which in the system features can be prioritized into at least two tiers according to their criticality for fault detection, extracting effort, and also based on user constrains. The framework is shown in Figure 1.

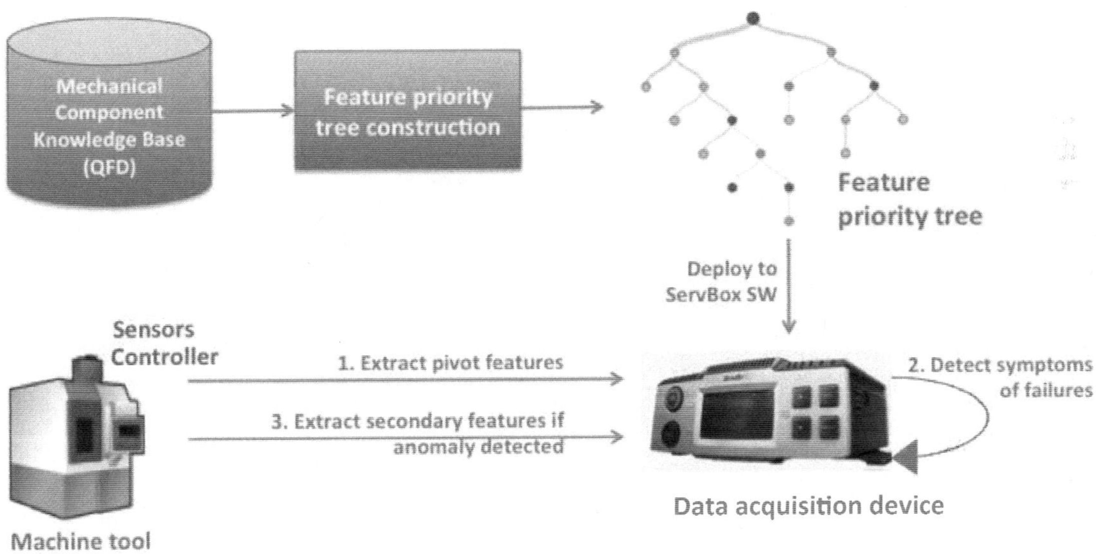

Figure 1: Dynamic condition based feature extraction strategy

The system includes 3 main function units:
1) A knowledge base for critical machine components, their failure modes, and related data resources/features. It is constructed through expertise interviewing and literature survey and based on quality function deployment (QFD).
2) A decision tree for feature priority will be constructed automatically based on Bayesian probability learning from machine condition and prior knowledge. The execution flow defines most critical features at the first level (so called Pivot Features) and secondary features in lower levels.
3) In routine monitoring tasks, under normal conditions, DAQ only calculates Pivot Features and analyzes them to make initial judgments of whether anomaly happens. If any abnormal condition is detected, DAQ will further go through lower levels, calculate and evaluate secondary features, and eventually infer root cause of the problem.

By incorporating the proposed mechanism, DAQ will be able to operate with minimal computational resource under normal conditions. The overall procedure is presented in Figure 2. Besides, the PHM system can react and perform diagnosis promptly when anomaly happens. With such improvement, the future DAQ and PHM system together will be able to perform real-time health diagnosis and prognosis in a more bandwidth and computational efficient way.

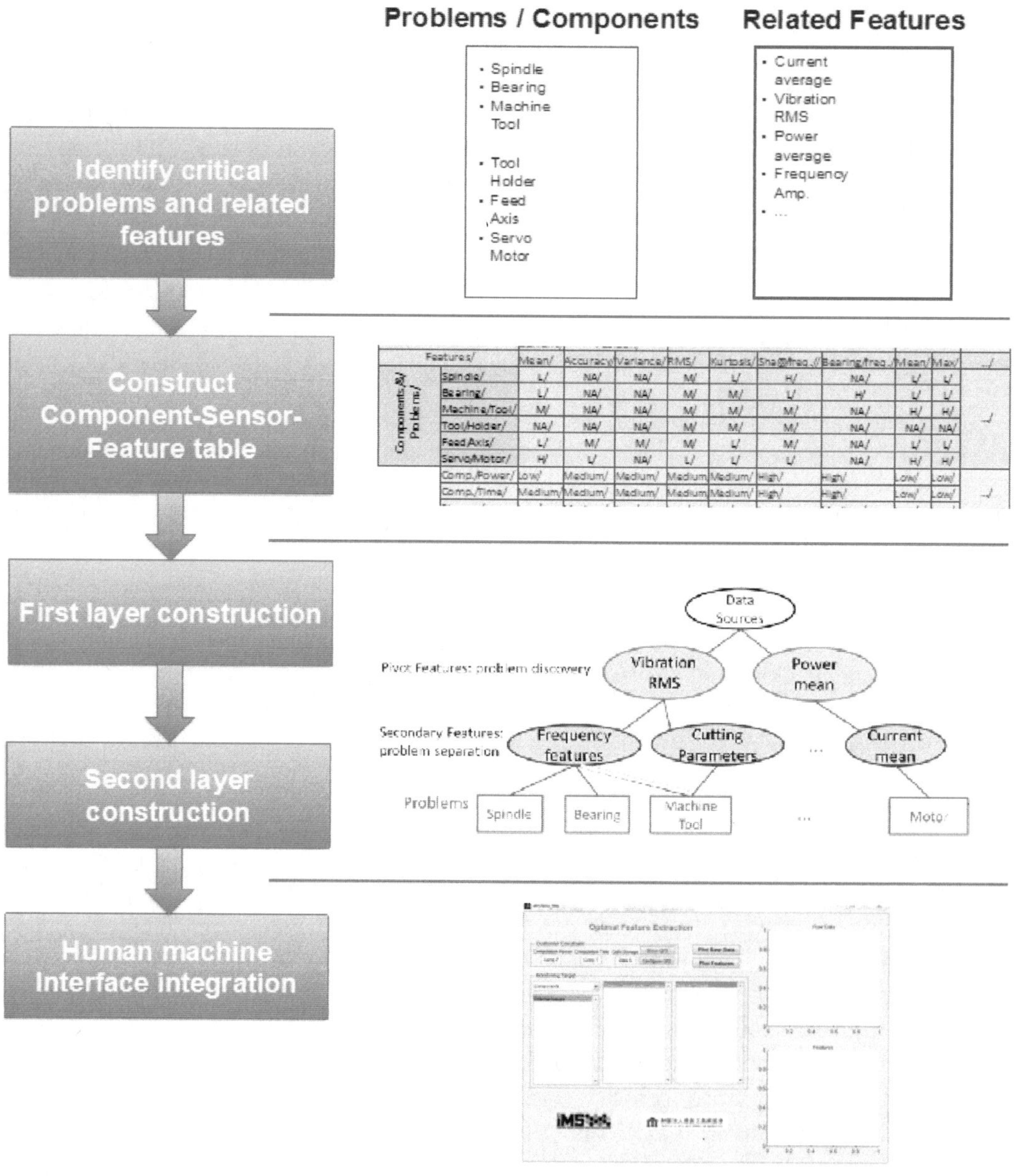

Figure 2: Overall procedure for machine health condition monitoring applications

The paper is organized as follows. Section 2 presents previous research efforts that are related to design of machine PHM framework and input variable selection strategy.

Section 3 describes the process of knowledge base development using QFD. Section 4 focuses on the decision tree learning strategy based on Bayesian theory. Section 5 describes the data set used to validate the proposed framework and experimental results are reported in this section. Section 5 concludes the paper with some perspectives.

Related work: With respect to machine condition monitoring design process, there is already ongoing research in a few of these key areas, such as developing common prognostic and diagnostic evaluation metrics [1], data quality [2] and signal validation methods [3], selecting the suitable hardware platform [4], and substantial guidance on which analysis tool to use for a given application [5].

As mentioned in Introduction section, advanced sensing technologies are now widely used on factory automation for machine condition based monitoring. Using a redundant or improper data input will decrease the efficiency of machine health assessment and fault detection. Even worse, it may convey wrong interpretations and hence eventually mislead important decision-making process. How to decide an optimal data acquisition and feature extraction strategy has started to serve as an important issue. Traditional and most common approach highly relies on expertise knowledge and experience; hence, some previous research summarized critical sensors for specific systems based on literature survey and industrial experience. Lee et al. (2013) [10] listed a set of critical sensors including related features and PHM tools for rotary systems. Hameed et al. (2009) [9] reviewed condition based monitoring and fault detection system for wind turbines and known sensing technologies are summarized, including vibration, acoustics, oil quality, etc. Besides these academia references, when deploying sensors for condition based monitoring process design, expertise and experience on how to detect degradation and faults are also included for sensor decision. However, it does not consider important features, such as the capability of fault detection, detection speed, cost and transmission bandwidth, etc., through a consistent quantitative method [8].

In order to improve PHM process, optimal input variable selection has added new dimensions to solving existing problems of prognostics and diagnostics. Santi et al. (2005) [8] developed a systematic sensor selection strategy for rocket engine system. In the selection process, a merit value will be calculated for each potential sensor suites. Genetic algorithm will then be applied to select optimal sensor suite. Zhao et al. (2002) [7] proposed an information-driven approach to dynamically select optimal sensors for tracking application. Factors including the information utility of sensory data and communication cost are considered in the selection process. Wang et al. (2004) [6] followed the idea of information utility to select sensors, which can yield greatest reduction in the entropy of the target location distribution. Isler and Bajcsy (2005) [11] also developed a sensor selection strategy on minimizing the error in estimating the position of a target. However, the above-mentioned research focuses on areas different from factory automation, which would need more focus on how to quantify feature contribution with machine failure modes and eventually develop a methodology for optimizing the performance.

Knowledge base development using QFD: A typical embedded system with data acquisition capability usually consists of processor, memory, I/O and operating system. Because of its single task characteristic, it has small size processor and comprised operation system. The embedded system attracts intensive research interests in recent years. Among those research, majority works focus on minimizing system's response time and accelerating process speed in order to have higher efficiency. For embedded system based mechanical system fault diagnosis, how to extract useful information from numerous signals efficiently is one of the major challenges. Here, the 'efficiently' means that the system with limited computational resource is able to analyze signals effectively, extract feature vector which can represent system operational performance precisely, and eventually make condition-based monitoring (CBM) feasible and reliable. In real practice, more factors need to be considered in order to get comprehensive and systematic analysis results. Each factor may represent some characteristics or performance indicators of the overall system.

With the growth of sensing technologies, more and more mature sensors have been implemented into machine condition monitoring, which leads to the growth in numbers of variables. However, when analyzing multi-variable problem via statistical approaches, when the number of variables increases, higher computational resources are required. If some independent variables, which can detect faults earlier, can be extracted in early stage (instead of extracting all features at the same time), the computational costs should be minimized. Hence, the first step in the proposed framework is to develop a knowledge base so that the relationship among features and failure modes can be represented.

To quantify how well the features can detect certain machine component failures, QFD is utilized to express all of the relationship and later will be served as a support tool for building decision tree. Note that although these components are quite different either in geometry or in functionality, their health conditions are all highly related to machine tool overall operational performance.

Currently, most common used signals in condition monitoring are: voltage, current, rotation speed, flow rate, pressure, vibration, etc. Before actually acquiring data, specific measurement signal must be selected. The measurement signal value change should have certain relationship with the performance of component under monitoring. Further analysis work will figure out the relationship and utilize it for fault diagnosis purpose. Next step is to choose sensor type, sensor location, and sampling rate. All these factors are necessary in developing a comprehensive expert knowledge base for machine fault diagnosis. To demonstrate the feasibility of the proposed framework, in this paper we select seven critical machine components, including tool holder, feed axis, drive motor, bearing, gear, shaft, and cutting tool, and conduct literature survey on which features can detect which component failures. Table 1 shows the literature review table for feed axis fault diagnosis feature priority. The table provides feature priority for different feed axis failure modes. The priority ranks are based on three judgment criterions: 1) Effectiveness for fault diagnosis; 2) Availability of measurement signal; 3) Measurement cost. In Table 1, high (H) stands for very suitable and recommend using, and the priority decrease from high (H) to medium (M) to low (L). Similar tables are also developed for the other six components based on previous research work [12-20]. Table 2 shows the final knowledge base for mechanical system fault diagnosis, which is developed based on all seven tables. The feature priority ranks for most common failure modes are included and the next section will show how feature priority decision tree is constructed based on Table 2.

Table 1: Feed axis literature review for fault diagnosis feature priority

Problems		Sensors & Signals & Measurement Frequencies										
		Acoustic Emission [19]					Vibration [19]					
		Time Domain		Frequency Domain			Time Domain	Frequency Domain				
		Statistical	Sound Pressure Level [12]	High-pass/ Band-pass filtering [12]	Time-Frequency Analysis	FFT	Statistical	Fourier Transform	Envelope Analysis [13][14]	Wavelet/Mel-frequency cepstral coefficients [15]	Fractal dimension	EMD [13] [20]
Feed Axis System: Ball Screw Component Level	Ball screw groove/balls wear	H	H	H	H	H	H	H	H	H		
	Bent screw											
	Ball Screw Preload Loss						H	H	H	H		H
	Ball Nut Misalignment								L	M		
	Lubrication [17]											
Feed Axis System: Bearing Component Level	Bearing inner race/outer race wear/ball wear						H		H	H	H	H
	Bearing misalignment						H					
Concerns & Comments for Each Feature				The peak frequency range is at 11 to 12kHz in the frequency spectrum				Requires high sampling rate to be effective.				

Table 2: Mechanical system fault diagnosis expert knowledge base

Signals & Features / Failure Modes		Sensors & Signals & Measurement Frequencies													
		Vibration					Current, Power of Spindle Motor		Acoustic Emission			Tempera-ture	Cutting Force, Torque		Hydr-aulic pressure
		Time Dom-ain	Frequency Domain				Time dom-ain	Freq-uency domain	Time Dom-ain	Frequency Domain		Time Domain	Time Domain		Time Domain
		Stat-istics	FT	Enve-lope / filter-ing	Time-freq-uency analysis	EMD	Aver-age	FT	Stat-istics	FT / filter-ing	Time-freq-uency analysis	Max / average	Y-axis varia-tion	Resid-ual varia-tion	Varia-tion
Bear-ing (Spin-dle, Feed Axis, Motor)	Inner/outer race/ball wear	M	H	H	H	H	M		H	H	H				
	Bearing misalign-ment	M					M		H						

Spindle	Unbalance	M	M		M		M								
Motor [21-22]	Stator Eccentricity	M	H		H				L			M			
	Insulation						H	H				H			
	Broken/bent shaft	M	H		M		M					M			
Gears	Manufacturing error	H	H	M	H	H			H	H	H	M	L	L	
	Tooth missing	H	H	M	H	H			H	H	H	M	L	L	
	Tooth pitting/spall	H	H	M	H	H			H	H	H	M	L	L	
	Gear crack	H	H	M	H	H			H	H	H	M	L	L	
	Gear fatigue/wear	H	H	M	H	H			H	H	H	M	L	L	
Shaft [23-25]	Unbalance	H	H	L	H	M	L	L	L	L	L	L	L	L	
	Bend	H	H	H	M	L	L	L	L	L	L	L	L	L	
	Crack	H	H	H	L	L	L	L	H	H	M	L	L	L	
	Misalign-ment	H	H	H	M	M	L	L	L	L	L	L	L	L	
	Rub	H	L	L	L	L	M	L	M	M	L	M	M	M	
Feed Axis Ball Screw	Wear-out	M	M	H			L		M	M		M			
	Bent screw	M	H		L		L								
	Preload Loss	M	M	M											
	Ball Nut Misalign-ment	M	M	M			L								
	Lubrication failure	M	M				L		M	M		H			
Tool Holder Unbalance [36-38]		M	H	M	L		L	H					H		
Mach-ine Tool [26-35]	Normal tool wear	M	H	H	L		H	H	H	M	M	H	H		
	Chipped tooth	M	H	H	L		H	H	H	M	M	H	H		H
	Broken tooth	M	H	M	L		H	H	H	M	M	H	H	M	H

Decision tree learning based on Bayesian theory: After the knowledge base library is constructed, the features need to be ranked based on their ability in reflecting various faults in order to realize optimized feature extraction. On the one hand, the features should be effective to capture the signature of the concerned fault types. On the other hand, the feature extraction process should be efficient.

As a classic algorithm in machine learning, decision tree has the ability to automatically construct rules to divide the variable space for prediction. Since the tree structure only makes decisions like "Yes/No", the efficiency of computation would rarely be an issue. Due to these advantages of a decision tree, the layered tree structure is adopted for optimal feature extraction with some modifications, which is shown in Figure 3.

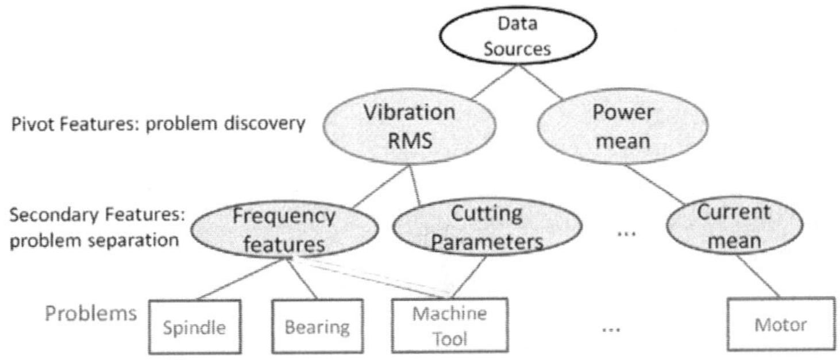

Figure 3: Decision tree structure for prioritized optimal feature extraction

In order to prioritize the feature extraction sequence, a metric is defined on the basis of Bayesian probability theory. The basic form of Bayesian probability is

$$p(\mathrm{y}=\mathrm{c} \mid \mathrm{x}, \theta) \propto \mathrm{p}(\mathrm{x} \mid \mathrm{y}=\mathrm{c}, \theta)\, \mathrm{p}(\mathrm{y}=\mathrm{c} \mid \theta) \tag{1}$$

We define the likelihood, prior probability, and posterior probability as follows:

- Likelihood: $\mathrm{p}(\mathrm{F}=\mathrm{f}_i \mid \mathrm{P}=\mathrm{p}_j)$, the probability that feature f_i can "find" fault p_i on condition that p_i happens
- Prior probability: $\mathrm{p}(\mathrm{P}=\mathrm{p}_j)$, the probability of fault p_i happening in general (expert knowledge/experience)
- Posterior probability: $p(\mathrm{P}=\mathrm{p}_j \mid F=f_i)$, the probability that fault p_i exists on condition that f_i is "abnormal" (exceeding threshold, etc.)

Thus, the score for feature f_i is obtained as

$$p(\mathrm{P}=\mathrm{p}_j \mid F=f_i)=\frac{\mathrm{p}(\mathrm{F}=f_i \mid \mathrm{P}=\mathrm{p}_j)\,\mathrm{p}(\mathrm{P}=\mathrm{p}_j)}{\sum_{j=1}^{N} \mathrm{p}(\mathrm{F}=f_i \mid \mathrm{P}=\mathrm{p}_j)\,\mathrm{p}(\mathrm{P}=\mathrm{p}_j)} \tag{2}$$

Assume that a mechanical system is consisted of J components or potential faults, and that there are in total I features to be known to be extracted, then the overall score s_{ji} of Feature i to Fault j can be calculated by Equation (3).

$$s_{ji}=\begin{pmatrix} \text{Importance}_{\text{ji}} \\ \text{Comp.Power}_{\text{i}} \\ \text{Comp. Time}_i \\ \text{Data Storage}_i \end{pmatrix}^T \cdot \begin{pmatrix} w_1 \\ w_2 \\ w_3 \\ w_4 \end{pmatrix} \tag{3}$$

where *Importance* corresponds to the knowledge base value, and Comp. Power (computational power) and other two items are acquired from customer survey on hardware condition. Considering the tree structure described in Figure 3 again, we find that the purpose of the first layer features are for health assessment, which means that features at this layer should be able to reflect faults or problems in a wide range, but is not necessary to identify which problem it is. Therefore, the first layer is looking for features with higher probability in finding the problem given there are one or some. This matches with the definition of likelihood that is defined earlier, and can be calculated as shown in Equation (4).

$$l_{ji} = \mathrm{p}(\mathrm{F} = f_i \,|\, \mathrm{P} = \mathrm{p}_j) = \frac{s_{ji}}{\sum_{j=1}^{N} s_{ji}} \tag{4}$$

The features likelihood matrix is obtained in Table 3, and the feature likelihood score is calculated by summing up each column.

$$L_n = \sum_{j=1}^{N} l_{ji} \tag{5}$$

Table 3: Feature likelihood score matrix

	Feature 1	Feature 2	Feature 3	...	Feature i
Problem 1	l_{11}	l_{12}	l_{13}	...	l_{1i}
Problem 2	l_{21}	l_{22}	l_{23}		l_{2i}
...		...		...	
Problem j	l_{j1}	l_{j2}	l_{j3}	...	l_{ji}

For the secondary layer features, since the purpose is to roughly diagnose the faults or problems, and they will only be used after the first layer features exceed the threshold, the features then need to identify the which fault it is. The question this time becomes that what would be the probability of fault or problem p_i happening given that feature f_i is abnormal, which matches the definition for posterior probability.

$$v_{ji} = p(\mathrm{P} = \mathrm{p}_j \,|\, F = f_i) = \frac{\mathrm{p}(\mathrm{F} = f_i \,|\, \mathrm{P} = \mathrm{p}_j)\,\mathrm{p}(\mathrm{P} = \mathrm{p}_j)}{\sum_{j=1}^{N} \mathrm{p}(\mathrm{F} = f_i \,|\, \mathrm{P} = \mathrm{p}_j)\,\mathrm{p}(\mathrm{P} = \mathrm{p}_j)} = \frac{l_{ji}\,\mathrm{p}(\mathrm{P} = \mathrm{p}_j)}{\sum_{j=1}^{N} l_{ji}\,\mathrm{p}(\mathrm{P} = \mathrm{p}_j)} \tag{6}$$

The feature posterior probability matrix is thus obtained and the feature posterior score is

$$V_n = \sum_{j=1}^{N} v_{ji} \tag{7}$$

Table 4: Feature posterior probability score matrix

	Feature 1	Feature 2	Feature 3	...	Feature i
Problem 1	v_{11}	v_{12}	v_{13}	...	v_{1i}
Problem 2	v_{21}	v_{22}	v_{23}		v_{2i}
...		...		...	
Problem j	v_{j1}	v_{j2}	v_{j3}		v_{ji}

Here because the feature posterior probability score is calculated based on the first layer features, and the first layer feature cannot always cover all the faults, we argue that the sum value will be 1 most of the cases. The implementation will be explained later.

The first layer features are extracted based on the feature likelihood score matrix. The rule for selecting the first layer features are that they need to cover and alarm all the problems, and that the computation should be minimized. All the features available in respect to the faults or problems to be monitored are prioritized first according to their likelihood score, and then from the top, the features are selected until all the faults are able (the value in QFD table is higher or equal to "M") to be alerted. The detailed flowchart is shown in Figure 4.

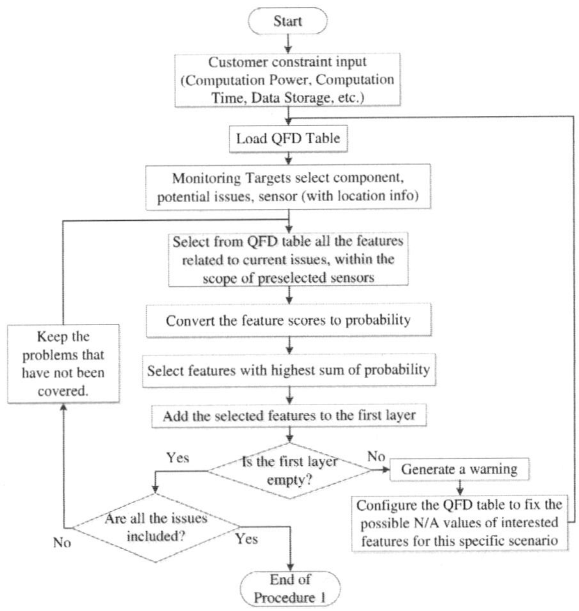

Figure 4: Flowchart of first layer prioritized optimal feature extraction

After one or more features in the first layer exceed the predefined threshold, the features will be ranked under each first layer feature. Basically, the first layer feature will serve as a "root", and secondary "branch" features will be prioritized considering the already triggered root feature. The branch features will be chosen by avoiding repeating the root feature, features that cover all the problems, and features that have already been calculated. Ranking is based on the posterior probability score matrix. The flowchart is shown in Figure 5. Note that it might be possible that the score of a feature sums up to 1. If this scenario applies to all the problems, secondary features need to be ranked by the posterior probability score problem-wise.

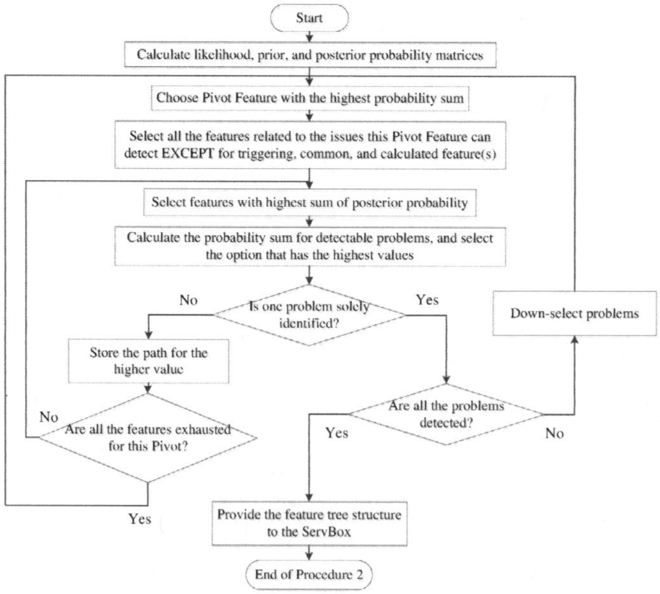

Figure 5: Flowchart of first layer prioritized optimal feature extraction

Experiments and results: Condition based monitoring and prognostics for rotary machinery has been a domain of interest for several decades. Being the key component for motion transmission, gearbox system has been identified as most critical component for rotary machinery as its failure can lead to downtime and damage to other components. Each smaller components of gearbox, such as gears, shafts, and bearings, have their own degradation mechanisms and failure modes. With each component have its own characteristics and fault signatures, prognostics and diagnosis can be performed with the utilization of the proper signal processing and feature extraction techniques [10].

The PHM 2009 Data Challenge focused on developing diagnostic techniques for gearbox system. A comprehensive collection of faults was induced on the gearbox and vibration signals were collected while the gearbox is operating under different regimes. The schematic of the gearbox for data collection is shown in Figure 6. Two accelerometers were mounted on the casing at the input shaft side between idler and output shaft, and the output side between input shaft and idler shaft. Validation of the proposed dynamic condition based feature extraction approach will be performed with a portion of this open source data.

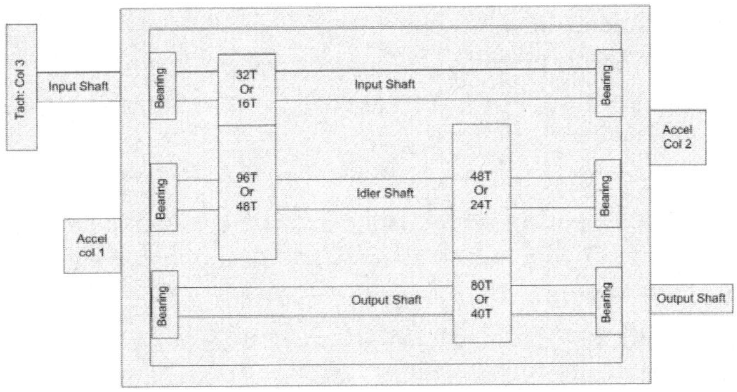

Figure 6: Schematic of gearbox of data collection (PHM 2009 Data Challenge)

The wining method for PHM 2009 Data Challenge that was proposed by D. Siegel extracted more than 200 Features from time domain statistics, frequency spectrum, time synchronize averaged signal, wavelet decomposition, envelop spectrum among others [39]. Those features have been identified to be effective for fault classification. However, the computational load for the process of signal processing and feature extraction is too high for high-speed data stream.

The knowledge about gearbox faults related features summarized from [14, 19, 39-45] has lead to the QFD table in Table 2. According to the QFD table, the structure of the decision tree for feature extraction is constructed as Figure 7. The 'Pivot Features' for health assessment include RMS of vibration at input shaft and output shaft, FFT magnitude at 5X input shaft, and the bearing related frequency peaks at envelop frequency spectrum. The health condition of the system can be quantified by confidence value given by Equation (8), where (X_f, N_f) is the Euclidean distance between the pivot feature vectors of health condition and current condition.

$$CV = \exp\left(-D\left(X_f, N_f\right)\right) \quad (8)$$

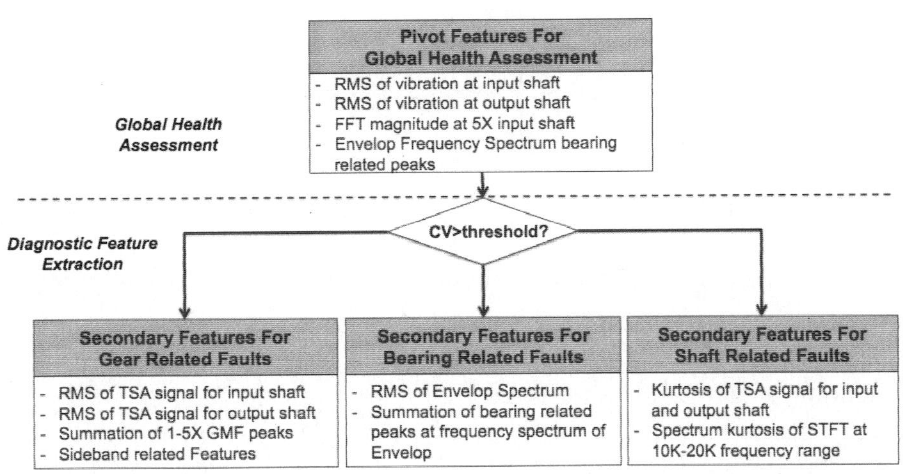

Figure 7: Decision tress for feature extraction generated from QFD table

If the CV is larger than threshold, the algorithm will trigger the diagnostic feature extraction for fault diagnosis and allocation. The secondary features for each component are used to compute the likelihood of fault existing on certain component based on logistic regression model.

Model training and fault diagnostic results: Eight sets of experiment data with different health conditions as shown in Table 3 were used for model training and testing. Each condition has two tests, and one test has 4 seconds of data. The data of each test were segmented by a moving window with length of 1.5s and increment of 0.1s, therefor dividing each test condition into 50 segments of data. 20 segments for each health condition were used as training data, while the rest as testing data. The preparation of training data for health assessment and fault diagnosis models are presented in Table 4.

Table 5: Feature posterior probability score matrix

	Gear Conditions				Bearing Conditions						Shaft Conditions	
Case	32T	96T	48T	80T	IS:IS	ID:IS	OS:IS	IS:OS	ID:OS	OS:OS	Input	Output
Spur 1	Healthy	Healthy	Healthy	Healthy	Healthy	Healthy	Healthy	Healthy	Healthy	Healthy	Healthy	Healthy
Spur 2	Broken Tooth	Healthy	Eccentric	Healthy	Healthy	Healthy	Healthy	Healthy	Healthy	Healthy	Healthy	Healthy
Spur 3	Healthy	Healthy	Eccentric	Healthy	Healthy	Healthy	Healthy	Healthy	Healthy	Healthy	Healthy	Healthy
Spur 4	Healthy	Healthy	Eccentric	Broken Tooth	Ball	Healthy	Healthy	Healthy	Healthy	Healthy	Healthy	Healthy
Spur 5	Broken Tooth	Healthy	Eccentric	Broken Tooth	Inner	Ball	Outter	Healthy	Healthy	Healthy	Healthy	Healthy
Spur 6	Healthy	Healthy	Healthy	Broken Tooth	Inner	Ball	Outter	Healthy	Healthy	Healthy	Imbalance	Healthy
Spur 7	Healthy	Healthy	Healthy	Healthy	Inner	Healthy	Healthy	Healthy	Healthy	Healthy	Healthy	Keyway Sheared
Spur 8	Healthy	Healthy	Healthy	Healthy	Healthy	Ball	Outter	Healthy	Healthy	Healthy	Imbalance	Healthy
IS: Input Shaft; ID: Idler Shaft; OS: Output Shaft;												
IS: Input Side; OS: Output Side. (OS:IS means output shaft input side)												

Table 6: Feature posterior probability score matrix

	Input Features	Training Data
Health Assessment Model	Pivot Features	Baseline (1)：Spur 1 Compare (0): Spur 5,6,7
Gear Fault Diagnostic Model	Gear related secondary features	Baseline (1)：Spur 2,3 Compare (0): Spur 7,8
Shaft Fault Diagnostic Model	Shaft related secondary features	Baseline (1)：Spur 7,8 Compare (0): Spur 5,6
Bearing Fault Diagnostic Model	Bearing related secondary features	Baseline (1)：Spur 6,8 Compare (0): Spur 2，3，4

For model validation, testing data were combined together to simulate continuous data stream from the gearbox. Health assessment and fault diagnostic feature extraction were performed based on the decision tree in Figure 7. The results for health value and likelihood of component failure are shown in Figure 8. The model has achieved 100%

accuracy for health assessment, 100% accuracy for gear and bearing fault diagnosis, and 100% detection rate (with 12% false alarms) for shaft fault diagnosis. In terms of data processing speed, the algorithms consumed on average 0.24 seconds to process 1.5 length of data under healthy conditions, and 0.92 seconds under faulty conditions when diagnostic feature extraction was triggered. The results are shown in Figure 9. The testing results well demonstrated the capability of the proposed method in optimization of computational recourses while maintaining fault classification accuracy. Compared with the results from traditional DAQ and feature extraction methods (shown in Figure 10), it proves that the proposed systems can improve the computational time by about 1.46 seconds (85.9%) in common DAQ process and 0.78 seconds (45.9%) once faults have been detected.

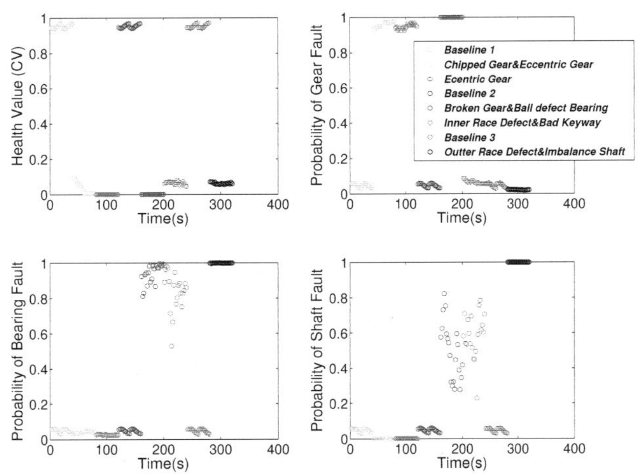

Figure 8: Health assessment and fault diagnosis results

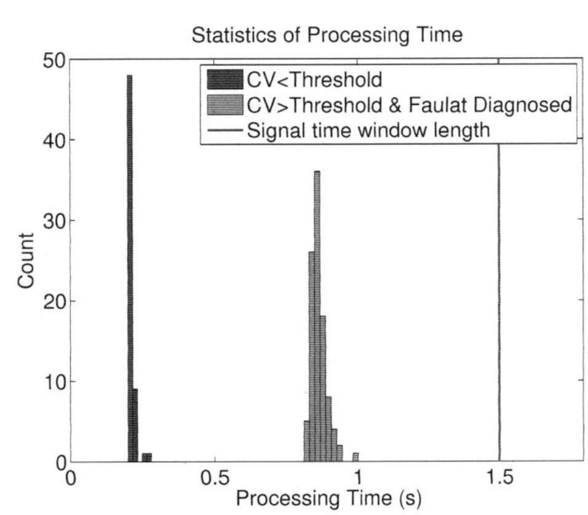

Figure 9: Data processing time distribution

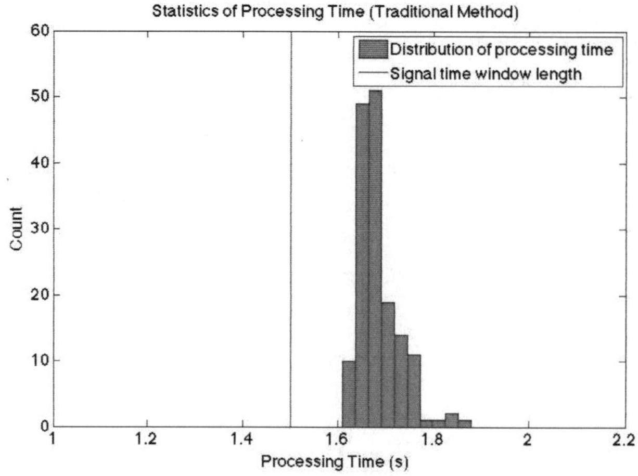

Figure 10: Data processing time distribution

Conclusion: A strategy to optimize the performance of machine data acquisition systems, using a decision tree-based approach with a QFD knowledge base, has been presented in this article. It includes a general process for feature extraction task considering dynamic machine conditions. To realize the relevance of features and corresponding failure modes of critical machine components, several literature surveys have been conducted and a knowledge base has been developed in QFD format. With this QFD table, a feature extraction tree can be built through the calculation of defined Bayesian probabilities and then be utilized on dynamic feature extraction procedure. Under this scenario, only pivot features will be extracted and analyzed in routine DAQ process. Once anomaly occurs (pivot feature analysis results in fault detected), then the DAQ would be triggered to extract corresponding secondary features. Because of the hierarchy and priority, the DAQ system can dynamically extract critical features based on the real-time condition of monitored machine systems.

Evaluations on real data set have shown that the proposed dynamic condition based feature extraction strategy can optimize the computational time by about 1.46 seconds (85.9%) in common DAQ process and about 0.78 seconds (45.9%) once faults have been detected. It can still remain a 100% detection rate and 12% false alarm rate. By the proposed system, feature extraction and fault detection are executed at the same time. Each feature extracted and analyzed will update the results of fault detection and also support the decision of triggering secondary feature extraction. Hence, this framework assures no redundancy of feature extraction and finds the optimal set of features for fault diagnosis and has better computational resource utilization than traditional methodologies.

Acknowledgement: This study is conducted under the "Project of Development of Industrial Platform for Value-Added Service via Intelligent Sensing Technology (1/4)" of the Institute for Information Industry which is subsidized by the Ministry of Economy Affairs of the Republic of China.

References:

[1] Saxena, A., Celaya, J., Saha, B., Saha, S. and Goebel, K., Metrics for Offline Evaluation of Prognostic Performance, 2010, International Journal of Prognostics and Health Management, Volume 1.

[2] Chen, Y. and Lee, J., Data Quality Evaluation and Improvement for Prognostic Modeling using Visual Assessment based Data Partitioning Method, 2013, Computers in Industry, Volume 64, Issue 3, pp. 214-225.

[3] Jablonski, A. and Barszcz, T., Validation of Vibration Measurements for Heavy Duty Machinery Diagnostics, 2013, Mechanical Systems and Signal Processing, Volume 38, pp. 248-263.

[4] Liao, L. and Lee, J., Design of a Reconfigurable Prognostics Platform for Machine Tools, 2009, Expert Systems with Applications, Volume 37, Pages 240-252.

[5] Siegel, D. and Lee, J., Reconfigurable Informatics Platform for Rapid Prognostic Design and Implementation: Tools and Case Studies, 2013, Proceedings of the Machine Failure Prevention Technology (MFPT) Conference.

[6] Wang, Hanbiao, et al., Entropy-based Sensor Selection Heuristic for Target Localization, 2004, Proceedings of the 3rd international symposium on Information processing in sensor networks. ACM.

[7] Zhao, Feng, Jaewon Shin and James Reich, Information-driven Dynamic Sensor Collaboration, 2002, Signal Processing Magazine, IEEE 19.2, Pages 61-72.

[8] Santi, L. Michael, T. Shane Sowers and Robert B. Aguilar, Optimal Sensor Selection for Health Monitoring Systems, 2005, National Aeronautics and Space Administration, Glenn Research Center.

[9] Hameed, Z., et al, Condition Monitoring and Fault Detection of Wind Turbines and Related Algorithms: A Review, 2009, Renewable and Sustainable energy reviews 13.1, Pages 1-39.

[10] Jay Lee, Fangji We, Wenyu Zhao, etc., Prognostics and Health Management Design for Rotary Machinery Systems - Reviews, Methodology and Applications, 2014, Mechanical Systems and Signal Processing, Pages 314-334.

[11] Isler, Volkan and Ruzena Bajcsy, The Sensor Selection Problem for Bounded Uncertainty Sensing Models, 2005, Proceedings of the 4th international symposium on Information processing in sensor networks. IEEE Press.

[12] Teruo Igarashi, Yasushi Tokunaga and Kazushige Ookuma, Studies on the Sound and Vibration of a Ball Screw: Sound Characteristics of a Ball Screw, 1988, JSME international journal. Ser. 3, Vibration, control engineering, engineering for industry 31(4), Pages 732-738.

[13] Huang, Norden E., et al, The Empirical Mode Decomposition and the Hilbert Spectrum for Nonlinear and Non-stationary Time Series Analysis, 1998, Proceedings of the Royal Society of London. Series A: Mathematical, Physical and Engineering Sciences 454.1971, Pages 903-995.

[14] Peng, Z. K., Peter W. Tse and F. L. Chu, A Comparison Study of Improved Hilbert–Huang Transform and Wavelet Transform: Application to Fault Diagnosis for Rolling Bearing, 2005, Mechanical systems and signal processing 19.5, Pages 974-988.

[15] Nelwamondo, Fulufhelo V., Tshilidzi Marwala and Unathi Mahola, Early Classifications of Bearing Faults using Hidden Markov Models, Gaussian Mixture

Models, Mel-frequency Cepstral Coefficients and Fractals, 2006, International Journal of Innovative Computing, Information and Control 2.6, Pages 1281-1299.

[16] Schoen, Randy R., et al., Motor Bearing Damage Detection using Stator Current Monitoring, 1995, Industry Applications, IEEE Transactions on 31.6, Pages 1274-1279.

[17] Studt, P., Influence of Lubricating Oil Additives on Friction of Ceramics under Conditions of Boundary Lubrication, 1987, Wear 115.1, Pages 185-191.

[18] Garinei, A. and R. Marsili, A New Diagnostic Technique for Ball Screw Actuators, 2012, Measurement.

[19] Byington, Carl S., et al., Recent Case Studies in Bearing Fault Detection and Prognosis, 2006, Aerospace Conference, 2006 IEEE.

[20] Sridhar Rahulram, Electromechanical Actuator Bearing Fault Detection using Empirically Extracted Features, 2012, Diss. Rochester Institute of Technology.

[21] A. Korde, On-Line Condition Monitoring of Motor Using Electrical Signature Analysis, 2002, presented at the Recent Advances in Condition-Based Plant Maintenance, A seminar organized by the Indian Institute of Plant Engineers, Mumbai.

[22] Thomson, W. T. and Fenger, M., Current Signature Analysis to Detect Induction Motor Faults, 2001, Industry Applications Magazine, IEEE, 7(4), Pages 26-34.

[23] Zhang, Bang Cheng, et al, Research on Fault Diagnosis for Bearing of CNC Machine Tools, 2013, Advanced Materials Research 605, Pages 713-717.

[24] Law, Leh-Sung, Jong Hyun Kim, Willey YH Liew and Sun-Kyu Lee, An Approach Based on Wavelet Packet Decomposition and Hilbert–Huang Transform (WPD–HHT) for Spindle Bearings Condition Monitoring, 2012, Mechanical Systems and Signal Processing.

[25] Jing Tian, Michael Pecht and Changning Li, Diagnosis of Rolling Element Bearing Fault in Bearing-gearbox Union System using Wavelet Packet Correlation Analysis, 2012, Machinery Failure Prevention Technology.

[26] Kuljanic E, Sortino M and Totis G, Application of Wavelet Transform of AE Signal for Tool Condition Monitoring in Face Milling, 2006, 39th CIRP Int. Sem. on Manuf. Systems, Ljubljana, Pages 39–44.

[27] Axinte D and Gindy N, Tool Condition Monitoring in Broaching, 2003, Wear 254(3-4), Pages 370–382.

[28] Axinte D and Gindy N, Assessment of the Effectiveness of a Spindle Power Signal for Tool Condition Monitoring in Machining Processes, 2004, International Journal of Production Research 42, Pages 2679–2691.

[29] Teti R and Baciu IL, Neural Network Processing of Audible Sound Signal Parameters for Sensor Monitoring of Tool Conditions, 2004, 4th CIRP Int. Sem. on ICME, Sorrento, 30 June–2 July, Pages 385–390.

[30] Lee JM, Choi DK, Kim J and Chu CN, Real-time Tool Breakage Detection for NC Milling Process, 1995, CIRP Annals 44(1),Pages 59–62.

[31] Ryabov O, Mori K and Kasashima N, An In-process Direct Monitoring Method for Milling Tool Failures Using a Laser Sensor, 1996, CIRP Annals 45(1), Pages 97–100.

[32] Lee DJ, Kim SH and Ahn JH, Breakage Detection of Small-diameter Tap Using Vision System in High-speed Tapping Machine with Open Architecture Controller, 2004, KSME International Journal 18(7), Pages 1055–1061.

[33] Kim H, Ahn J, Kim S and Takata S, Real Time Drill Wear Estimation Based on Spindle Motor Power, 2002, Journal of Materials Processing Technology 124(3), Pages 267– 273.
[34] Arrazola PJ, Arriola I, Davies MA, Cooke AL and Dutterer BS, The Effect of Machinability on Thermal Fields in Orthogonal Cutting of AISI 4140 Steel, 2008, CIRP Annals 57(1), Pages 65–68.
[35] Gandarias E, Dimov S, Pham DT, Ivanov A, Popov K, Lizarralde R and Arrazola PJ, New Methods for Tool Failure Detection in Micromilling, 2006, Proceedings of IMechE 220(2), Pages 137–144.
[36] Snr, Dimla and E. Dimla, Sensor Signals for Tool-wear Monitoring in Metal Cutting Operations—A Review of Methods, 2000, International Journal of Machine Tools and Manufacture 40.8, Pages 1073-1098.
[37] Jemielniak, K. and O. Otman, Tool Failure Detection Based on Analysis of Acoustic Emission Signals, 1998, Journal of Materials Processing Technology 76.1, Pages 192-197.
[38] König, W., K. Kutzner and U. Schehl, Tool Monitoring of Small Drills with Acoustic Emission, 1992, International Journal of Machine Tools and Manufacture32.4, Pages 487-493.
[39] Hassan Al-Atat, David Siegel and Jay Lee, A Systematic Methodology for Gearbox Health Assessment and Fault Classification, 2011, International Journal for PHM, Volume 2.
[40] Eric Bechhoefer and Michael Kingsley, A Review of Time Synchronous Average Algorithms, 2009, Annual Conference of the Prognostics and Health Management.
[41] David Siegel and Hassan Al-Atat, Novel Method for Rolling Element Bearing Health Assessment—A Tachometer-less Synchronously Averaged Envelop Feature Extraction Technique, 2012, Mechanical System and Signal Processing.
[42] David Siegel, Canh Ly and Jay Lee, Evaluation of Vibration-Based Health Assessment and Diagnostic Techniques for Helicopter Bearing Components, 2011, AIAC14 Fourteenth Australian International Aerospace Congress, HUMS 2011.
[43] Antoni and R.B. Randall, The Spectral Kurtosis: Application to the Vibratory Surveillance and Diagnostics of Rotating Machines, 2006, Mechanical Systems and Signal Processing, vol. 20, Pages 308-331.
[44] K. Choy, S. Huang, J.J. Zakrajsek, R.F. Handschuh and D.P. Townsend, Vibration Signature Analysis of a Faulted Gear Transmission System, 2004, NASA Technical Memorandum, NASA TM-106623/ARL-TR 475/AIAA-94-2937.
[45] Vecer, M. Kreidl and R. Smid, Condition Indicators for Gearbox Condition Monitoring Systems, 2005, Acata Polytechnica, vol. 45, No. 6.

INVESTIGATION AND EVALUATION OF CONDITION INDICATORS, VARIABLE SELECTION, AND HEALTH INDICATION METHODS AND ALGORITHMS FOR ROTORCRAFT GEAR COMPONENTS

David Siegel and Jay Lee
Center for Intelligent Maintenance Systems
University of Cincinnati, P.O. Box 210072
Cincinnati, OH 45221-0072
Telephone: (513) 556-3412
siegeldn@mail.uc.edu

Paula Dempsey
NASA Glenn Research Center
21000 Brookpark Road, MS 23-3
Cleveland, Ohio 44135
Telephone: 216-433-3398
paula.j.dempsey@nasa.gov

Abstract: Providing better fleet-wide availability, reliability, and maintainability, are some of the potential benefits of a rotorcraft health and usage monitoring system. Despite several advances in research and development on drivetrain condition monitoring and also many deployed rotorcraft health and usage monitoring systems, there is a still a significant need to further validate and improve upon the diagnostic and prognostic algorithms for drivetrain components. This study focuses on performing an investigation on gear condition indicators, methods for ranking and selecting the condition indicators, and also health indication algorithms that fuse multiple gear condition indicators into a single health index. For investigating the gear health monitoring algorithms, a spiral-bevel gear fatigue test-rig is used to conduct four run-to-failure experiments. Vibration measurements along with a tachometer pulse are collected throughout the life testing, and visual inspection and oil-debris measurements are used to compare the vibration symptom response with the actual gear health condition. Time domain statistics, time synchronous average spectrum indicators, amplitude modulation, residual signal, auto-regressive residual signal, and second order cyclostationary indicators are evaluated in this study. Three different variable ranking methods based on the Fisher criterion, the area under the receiver operating characteristic curve and the monotonic criteria were weighted together to rank and select the best performing condition indicators. The top performing condition indicators were from the residual and amplitude modulation signal, and consisted of the residual signal RMS, residual signal peak to peak value, the sideband level indicator, and the amplitude modulation peak to peak value. A principal component based health indicator provided the most consistent health trend when compared with the distribution overlap and auto-associative neural network methods. The future work will consider further validation of the algorithms with additional run-to-failure tests, and a more in depth study on the influence of load and speed on the condition indicators.

Key words: Health Indicators, Gear Condition Indicators, Variable Selection

1. Introduction: Rotorcraft health and usage monitoring systems (HUMS) aim to provide better fleet wide availability and reliability, since maintenance can be performed based on the condition of the asset as opposed to a time based schedule. For the rotorcraft drivetrain components, the HUMS use vibration signatures and advanced processing techniques to infer the gear, shaft, and bearing health conditions [1]. Although there are commercially available rotorcraft HUMS systems that are currently being used to monitor fleets of rotorcraft [2]; there is still a need to further advance the science on rotorcraft condition monitoring for the drivetrain components. This study focuses on algorithms for monitoring the gear component health, in which one of the key objectives is to quantitatively rank the performance of different gear condition indicators. This research work also looks at health indicator methods that fuse multiple gear condition indicators into a single health index. The outline of the paper is the following: Section 2 discusses the gear condition indicator algorithms, the variable selection methods, and the health indicator algorithms; Section 3 describes the experimental setup and testing; Section 4 highlights the results obtained from this study; and conclusions are provided in Section 5.

2. Health Monitoring Algorithms for Rotorcraft Gear Components: Assessing the health condition of drivetrain components such as gears, involves a series of processing steps. The typical first step involves a set of signal processing methods that enhance the vibration signature and transform the data into appropriate domains that can better reveal the interesting patterns or fault symptoms in the data. For gear components, there is a wealth of established signal processing techniques and condition indicators; section 2.1 describes the most widely used gear condition indicator algorithms. Considering the vast amount of different indicators that have been proposed for gear components, it is important to consider which indicators provide the best discrimination between a normal and damaged gear component, and also which ones provide the best long term trend of an impending failure. Section 2.2 describes a set of quantitative methods for ranking and comparing the gear condition indicators and ultimately selecting the best indicators for diagnostic and prognostic purposes. Considering that there is a multitude of different gear failure modes, it is usually advantageous to combine multiple condition indicators when assessing the gear health condition. Section 2.3 discusses three different health monitoring algorithms that fuse multiple condition indicators into a single health index. The experimental testing (Section 3) and the results from these evaluated algorithms (Section 4) are provided in the subsequent sections and are used to recommend which processing methods provide the best diagnostic and prognostic assessment.

2.1 Condition Indicator Techniques: There are a variety of signal processing algorithms that are used to extract information from the gear vibration signature, in which six of the more commonly used methods are listed in Table 1. Time domain statistics such as the root mean square (RMS) , peak to peak, and kurtosis values from the raw signal can offer some overall indication of the gear health condition, but are less robust and offer less diagnostic information then the more advanced processing methods. Using the frequency domain spectrum without any pre-processing enhancement is also a straightforward approach that can be used to assess the gear condition based on the sideband patterns and magnitudes. However the use of time synchronous averaging (TSA) for ensemble averaging and filtering out vibration that is not synchronous with the shaft can provide a more reliable signal for frequency domain analysis and sideband level and magnitude indicators [3]. In fact, the most widely used methods are based on further processing of the TSA signal. The first two methods outlined in Table 2 can also be applied to the TSA data. Further processing of the TSA signal by including a few sidebands around a gear mesh frequency, and taking the envelope of the filtered signal can be used to analyze the amplitude modulation patterns in the gear vibration signature [4]. In addition, departures from the regular gear meshing pattern can be considered by filtering out the gear mesh frequency, the harmonics of the gear mesh frequency, and first order sidebands; this filtered signal is named the "residual" signal and the widely used FM4 condition indicator is based on this processing method (kurtosis of the residual signal) [5]. An auto-regressive model can also be used to learn the regular gear meshing patterns, in which the regular patterns can be subtracted out of the signal, and the resultant signal (AR-residual) can be further processed to infer the gear health condition. The AR residual signal method involves more processing steps and baseline data for learning the AR model coefficients, however Wang [6] indicated that it had better diagnostic results for his experiments.

Table 1: Listing of Signal Processing Methods for Gear Condition Monitoring

	Method	**Processing Description**	**Rationale**
1	Time Domain Statistics	Statistics such as root-mean square, peak to peak, kurtosis	For overall vibration condition and statistical characteristics
2	Frequency Domain Spectrum	Fast Fourier Transform of raw vibration signal, observe gear sideband magnitude/patterns	Can be used to extract peak information at sidebands to characterize gear health
3	Time Synchronous Averaging (TSA)	Uses tachometer signal to resample data in angular domain and ensemble average the vibration signals for one rotation	Enhances signal to noise ratio for shaft/gear vibration and allows for additional processing of gear condition indicators
4	Amplitude Modulation Signal	After TSA, filter around gear mesh frequency (plus/minus a few sidebands) and extract envelope (Hilbert Transform)	Amplitude modulation is a common signature for some gear faults (eccentricity, crack on a single gear tooth)
5	Classic Residual Signal	After TSA, filter out gear mesh frequency, its harmonics, and first order sidebands to calculate residual signal	Removes regular meshing pattern; FM4 and other widely known condition indicators are based on this approach
6	Auto-Regressive (AR) – Residual Signal	After TSA, auto-regressive model is fit to a baseline data pattern and used to calculate residuals	Another method for focusing on departures from regular gear mesh vibration signature

In terms of the mathematical details with respect to the time synchronous averaging method including the different interpolation methods, and alternative filtering or frequency domain methods, the interested reader can review the work by Braun [3] and Bechhoefer [7]. After performing the TSA calculation, the amplitude modulation, phase modulation, and residual signal can be derived from the TSA signal using the appropriate filtering and processing steps. An example flow chart that describes how one can obtain the amplitude modulation gear signal is provided in Figure 1; which consists of filtering the TSA signal around the gear mesh frequency (2-5 sidebands) and using the Hilbert transform to extract the envelope of the filtered signal [4]. The classical residual signal method is also one of the more commonly used techniques, and the processing flow chart for this method is provided in Figure 2. The processing steps consist of filtering out the gear mesh frequency (GMF), harmonics of the GMF, and first order sidebands, and then converting this signal back to the time domain to obtain the residual signal.

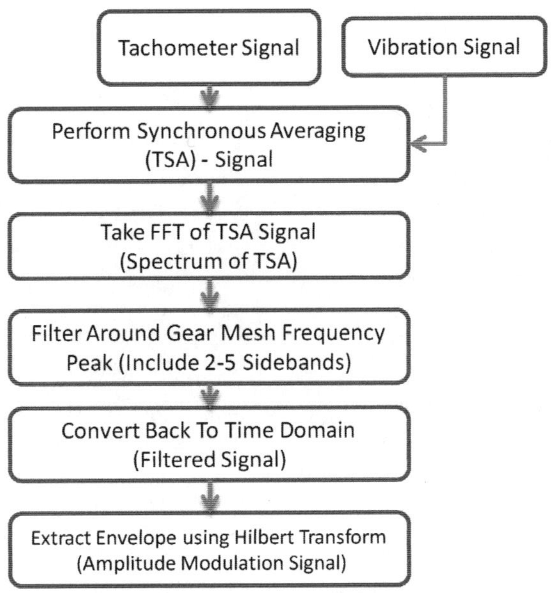

Figure 1: Narrowband Amplitude Modulation Signal Flow Chart

Variations on the residual signal processing method include how many sidebands and GMF harmonics to remove in the filtering operation, and also the use of auto-regressive models to remove the regular meshing pattern. The processing steps for the auto-regressive residual algorithm are shown in Figure 3, in which the auto-regressive (AR) model coefficients are learned from baseline vibration TSA signals [6]. For selecting the AR model order, the Akaike information criterion (AIC) criterion is commonly used to balance the residual error with the model complexity. After saving the model coefficients, the AR model provides a predicted gear mesh vibration response which can be subtracted from the TSA signal. The AR residual signal would ideally be Gaussian noise for a system that is free of gear defects, but abnormal gear vibration fault patterns would remain in the residual signal if there is damage on one or more of the gear teeth.

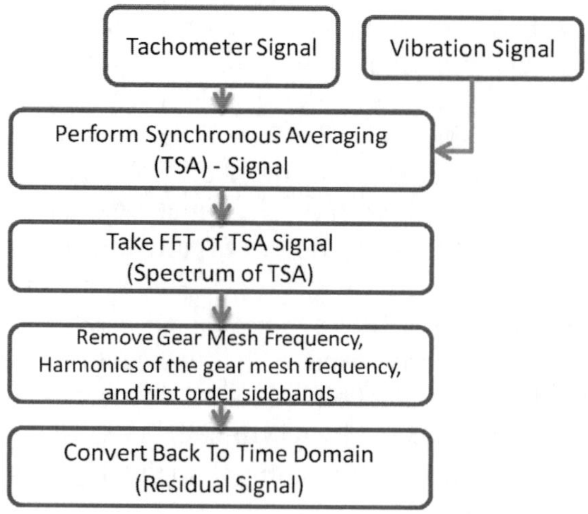

Figure 2: Classic Residual Gear Signal Processing Method

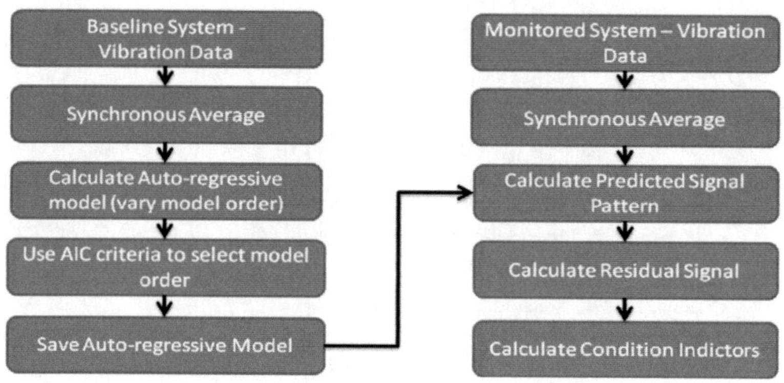

Figure 3: Auto-Regressive Residual Processing Method for Gear Components

After using the various signal processing tools to enhance the vibration signature, statistical indexes of the processed signal are calculated. These include the root mean square (RMS), the peak to peak value, and the normalized fourth statistical moment (kurtosis). Formulas for calculating the RMS, peak to peak, and kurtosis are provided in Equation 1-3, in which x(t) is the sampled vibration waveform, N is the number of samples in the analysis block, and σ is the standard deviation of the vibration signal.

$$RMS = \sqrt{\frac{1}{N}\sum_{i=1}^{N} x_i^{\ 2}} \tag{1}$$

$$Peak\_to\_Peak = \max(x(t)) - \min(x(t)) \tag{2}$$

$$Kurtosis = \frac{\frac{1}{N}\sum_{i=1}^{N}(x_i - \bar{x})^4}{\sigma^4} \quad (3)$$

These three statistics were calculated from the raw time domain signal, the amplitude modulation signal, the classic residual gear signal, and the AR residual signal.

For the frequency domain indicators, the sideband level (SBL) and sideband ratio indicators can be calculated using Equation 4 and 5, in which (Sb_{a1}) and (Sb_{a2}) are the magnitude of the first order sidebands and GMF_{peak} is the magnitude of the gear mesh frequency peak [8].

$$SBL = \frac{1}{2}\left(SB_{a1} + SB_{a2}\right) \quad (4)$$

$$SBR = \frac{SB_{a1} + SB_{a2}}{GMF_{peak}} \quad (5)$$

Table 2: Tabular Listing of Condition Indicators Used in this Study

	Method	Processing Description	Type of Indicators	Number of Indicators (n)
1	Time Domain Statistics	Raw Time Signal	RMS, Peak to Peak, Kurtosis	3 for each accelerometer (6)
2	TSA Spectrum Indicators	Synchronous Averaging, FFT	Sideband Level, Sideband Ratio	2 for each gear (8)
3	Amplitude Modulation	TSA, Narrowband Filtering, Envelope	AM Signal RMS, Peak to Peak, Kurtosis	3 for each gear (12)
4	Classic Residual Gear	TSA, Removing GMF and harmonics and first order sidebands	Residual RMS, Peak to Peak, Kurtosis	3 for each gear (12)
5	AR Residual	AR Model, Residual Signal	Residual RMS, Peak to Peak, Kurtosis	3 for each gear (12)
6	FMO	TSA and Ratio Calculation	Ratio of TSA peak to peak to Sum of GMF and Harmonic Peaks	1 for each gear (4)
7	Second Order Cyclostationary Index	TSA, Random Residual Signal (Raw-TSA)	Ratio of Residual Signal Harmonic Sum by Total Signal Power	1 for each gear (4)

A tabular listing of the condition indicators used in this study are provided in Table 2. It should be noted that the TSA spectrum, amplitude modulation, residual signal and FMO indicators can be calculated for each individual gear. This is because the TSA signal can be formulated for each gear shaft speed, and these filtering methods can focus on a set of sideband frequencies that are only related to a particular gear of interest. The second order cyclostationary metric, which is based on comparing the strength of the random signal (after removal of the TSA signal) to the periodic signal (TSA signal) was also calculated since a previous study from the same data set indicated its potential diagnostic ability [9]. In total, 29 gear indicators were considered for each gear pair (58 in total since there are 4 gears). These 29 gear indicators are evaluated using variable ranking methods in order to select the ones with the best diagnostic and prognostic performance.

2.2 Variable Selection Methods: Not all of the 29 gear condition indicators have the same ability to diagnose or prognose the gear health state and one should consider ways for quantitative ranking the indicators so the top performing ones can be selected. A flow chart for variable selection that considers three different ranking methods is provided in Figure 4, in which two diagnostic metrics (Fisher criterion and area under the curve) are weighted together with a prognostic metric (monotonic criteria). Considering that there are four data sets, the ranking method averages the scores from the four data sets when ranking the indicators. Also, for the diagnostic metrics, one must define a set of data samples that is considered normal and a set of samples when the gear is believed to have significant damage. In this study, visual inspection of the oil debris trend is used for assigning the portions in each run-to-failure test that are considered normal and degraded.

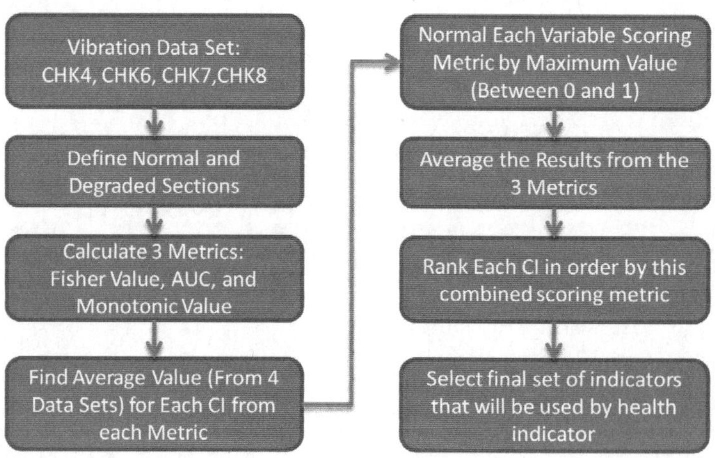

Figure 4: Combined Variable Selection Method

The first of the three metrics for variable selection is the Fisher criterion, which is a measure of separation between the normal and degraded health condition using a particular condition indicator. The formula for calculating the Fisher Criterion (f_k) is shown in Equation (6), in which $\mu_{i,k}$ and $\mu_{j,k}$ are the means and $\sigma_{i,k}$ and $\sigma_{j,k}$ are the variances for the normal and degraded health conditions for the kth condition indicator [10]. This metric would be calculated for all 29 condition indicators and the ones with the highest fisher values are best at separating the normal and degraded gear health states.

$$f_k = \frac{\left|\mu_{i,k} - \mu_{j,k}\right|^2}{\sigma_{i,k}^2 + \sigma_{j,k}^2} \tag{6}$$

Another diagnostic measure is the area under the curve (AUC) of the receiver operating characteristic curve (ROC). The ROC curve provides a false positive and true positive rate as the threshold is varied and allows one to consider the performance of different condition indicators in a quantitative manner [11]. If one were to consider the AUC of the ROC curve as shown in Figure 5, this provides an index for comparing different condition indicator ROC curves, in that it is bounded between 0 and 1, with a perfect indicator having a value of 1. The AUC of the ROC curve is used to also rank the diagnostic ability for each of the 29 condition indicators.

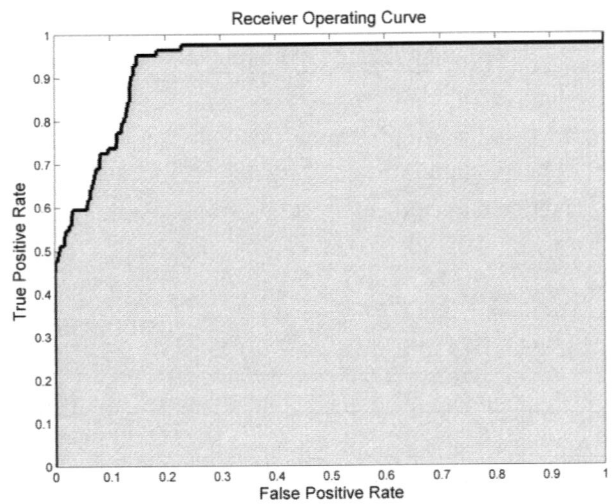

Figure 5: Visual Example of Area Under the Curve of the Receiver Operating Curve

One of the ideal characteristics for a condition indicator is to have a consistent and always increasing trend. Coble [12] proposed a metric for measuring how monotonic a condition indicator is, in which the equation is provided in Equation 7. Effectively it considers the number of times the derivative is positive ($\#pos\,d/dx$) and negative ($\#neg\,d/dx$), in that the more monotonic a trend is, the less often the derivative will change signs. The monotonic index is also bounded between 0 and 1, and can be used to rank the 29 indicators on how monotonic of a trend each one provides. For carrying out the calculation, it is common to use a smoothing function and localized linear regression; in this study an exponential moving window (smoothing value of 0.1) and a 80 point local linear regression method was used.

$$\text{Monotonicity} = mean\left(\left|\frac{\#pos\,d/dx}{n-1} - \frac{\#neg\,d/dx}{n-1}\right|\right) \tag{7}$$

<u>2.3 Health Indicator Algorithms:</u> Combining multiple condition indicators into a single health index can offer more coverage for the various gear failure modes and also provide a more robust assessment of the gear health condition. Three different health indicator methods were considered in this study, in which a brief description of those methods are provided in Table 3. All three methods are based on comparing the current behavior to the baseline behavior, in which one approach is based on comparing statistical distributions, while the other two approaches are based on linear and non-linear principal component analysis. The motivation for including algorithms that only require baseline data for training, is that it is usually difficult in practice to have a significant amount of historical data from systems with damaged gears and thus supervised classification type methods are difficult to train and implement.

Table 3: Overview of Health Indicator Methods Considered in this Study

	Method	**Processing Description**	**Rationale**
1	Principal Component Analysis – T^2 Health Metric	Multivariate distance from baseline calculation using principal component analysis	Combines multiple indicators and has statistical limits for thresholds
2	Auto-Associative Neural Network – Square Prediction Error Metric	Similar to PCA, but a non-linear version; also requires baseline data for training	Considers that the indicators could have a non-linear correlation relationship.
3	Distribution Overlap Method	Compares current multivariate distribution with baseline distribution	As gears degrade over time, CI distributions will move further and further from baseline condition

The first health indicator method considered is based on principal component analysis, in which the processing flow chart for this multivariate health index is shown in Figure 6. Effectively it is using the baseline data for determining the principal component coordinates, and is then using those coordinates to project the monitored data. Principal component analysis has two different health metrics (square prediction error-SPE) and the T^2 metric. The SPE metric is better for finding changes in correlation while the T^2 is looking for changes in magnitude from the baseline condition. From the author's personal experience, the T^2 is works better for vibration based monitoring applications.

The T^2 monitoring statistic is effectively a distance from normal calculation (Mahalanobis distance) using the data projected into the principal component coordinates. The formula for T^2 is provided in Equation (8) in which $\{u\}_{1xr}$ is the projected data and $[\Sigma]^{-1}$ is the inverse of the eigenvalue matrix [13].

$$T^2 = \{u\}_{1xr} [\Sigma]^{-1}_{rxr} \{u\}^T_{1xr} \tag{8}$$

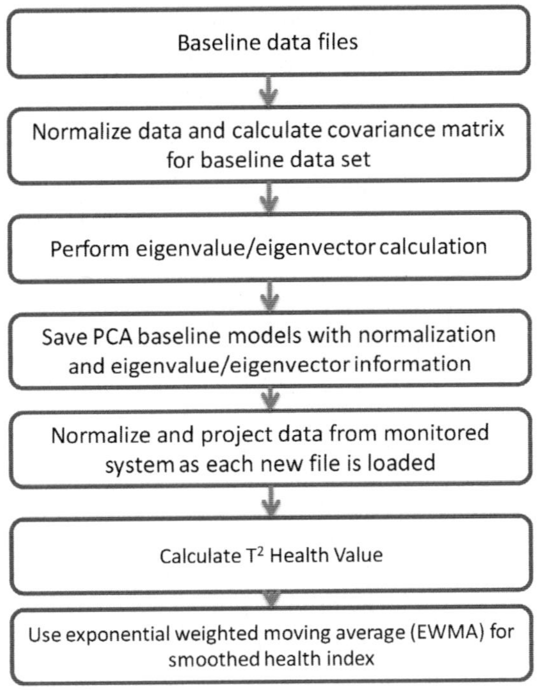

Figure 6: Principal Component Analysis Health Index Method Flow Chart

The distribution overlap health indicator method is based on comparing the baseline condition indicator distribution H(x), with the condition indicator distribution G(x), from the monitored gearbox. Mathematically, the overlap between the two distributions can be calculated using the L2 distance, in which the formula for the L2 distance is provided in Equation (9). The L2 distance is finding the difference between the two distributions, squaring that difference, and then summing the results through an integration calculation.

$$\|H(x)-G(x)\|_{L2} = \sqrt{\int (H(x)-G(x))^2 dx} \tag{9}$$

In order to bind the health indicator value (called Confidence value - CV) between 0 and 1, a normalized version of the L2 distance is used and is provided in Equation (10). Effectively, the normalized L2 distance is a measure of similarity between two distributions, in which a value of 0 would imply no overlap or similarity, and value of 1 would occur if the two distributions were identical.

$$CV = 1 - \frac{\|H(x)-G(x)\|_{L2}}{\sqrt{\int \left((H(x))^2 + (G(x))^2\right) dx}} \tag{10}$$

Although the integration expression, which is a product of two Gaussians appears intractable, this expression reduces to a constant that is based on the covariance matrix (Σ_a and Σ_b) and means (μ_a and μ_b) from both distributions. The equation for calculating this constant is provided in Equation (11) and this simplification implies that only matrix calculations are needed for carrying out this distribution overlap health monitoring method [14]. The constant expression in Equation (11) is simply comparing the means and covariance matrices from the two distributions, and would be a very small number if the means between the two distributions are very distant.

$$z_c = \left[\det\left(2\pi(\Sigma_a + \Sigma_b)\right)\right]^{-1/2} \exp\left((\mu_a - \mu_b)^T (\Sigma_a + \Sigma_b)^{-1} (\mu_a - \mu_b)\right) \quad (11)$$

A non-linear principal component analysis based health indicator method was also considered during this study and is based on auto-associative neural networks. An example auto-associative network structure is provided in Figure 7, and effectively is a feed-forward neural network with three hidden layers. Four transfer functions are used, in which the mapping and de-mapping layer use non-linear functions, and linear functions are used in the bottleneck (middle layer) and output layer. The network is trained with the same inputs and targets, and is based on learning the non-linear correlation relationship between the different condition indicators [15]. After training the network with baseline data, predicted values and residual values (E) can be calculated using Equation (12). In addition, the health index shown in Equation (13) is simply the summation of the residuals squared (Squared Prediction Error -SPE), in which one could use n condition indicators in this calculation.

$$\{E\}_{1xn} = \{x\}_{1xn} - \{\hat{x}\}_{1xn} \quad (12)$$

$$SPE = \sum_{i=1}^{n} (E_i)^2 \quad (13)$$

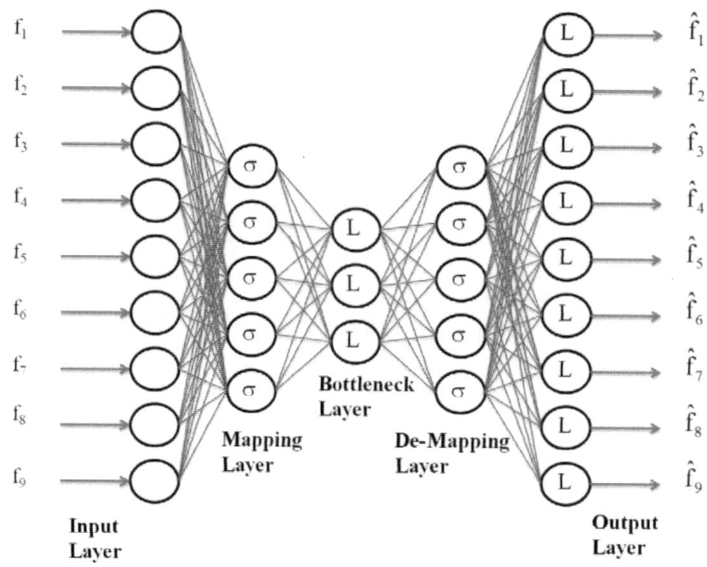

Figure 7: Auto-Associative Neural Network Example Structure - Used to Calculate Predicted Condition Indicator Values and Residual-Based Health Value

3. Experimental Setup and Testing: Tests were performed in the Spiral Bevel Gear Fatigue Test Rig at NASA Glenn Research Center. A detailed description of this test facility is provided in references [16] and [17]. The Spiral Bevel Gear Fatigue Test Rig is illustrated in Figure 8. The facility operates as a closed-loop torque regenerative system, where the drive motor only needs enough power to overcome friction losses within the system. Two sets of spiral bevel gears are installed in the test rig and tested simultaneously. Facing the gearboxes, per Figure 8, the left gear set (pinion/gear) is referenced as left and the right gear set (pinion/gear) is referenced as right within the paper.

The aerospace quality gear sets tested had 5.14 in. (13.06 cm) diametral pitch, 35° spiral angle, 1 in. (2.54 cm) face width, 90° shaft angle, and 22.5° pressure angle. The gear had 36 teeth and the pinion had 12 teeth. A photo of the gear set is shown in Figure 8. Qualified helicopter transmission oil, was used in the test rig during testing. Both gear sets are lubricated with oil jets pumped from an oil reservoir. The lubrication from the gearbox then exits the gearbox and flows through an inductance type in-line oil debris sensor, then past a magnetic chip detector. A strainer and a 3 μm filter are located downstream of the oil debris (OD) sensor to capture any debris before returning to the sensor and gearbox.

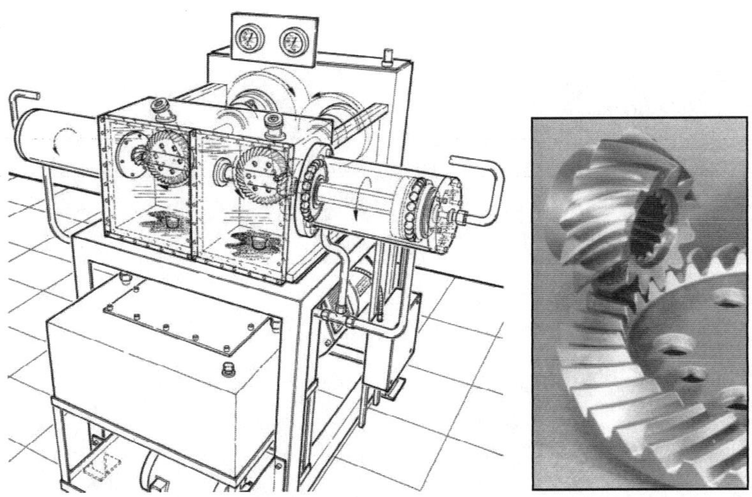

Figure 8: Spiral Bevel Gear Fatigue Test Rig.

A non-contact rotary transformer shaft mounted torque sensor was used to measure torque during testing. An inductance type oil debris sensor was used to measure the ferrous debris generated during fatigue damage to the gear teeth. The sensor measures the number of particles and their approximate size based on user defined particle size ranges or bins. Based on the bin configuration, the average particle size for each bin is used to calculate the cumulative mass by assuming the average particle size as a diameter spherical in shape and multiplying it by the density of steel. A detailed discussion of the oil debris data collected during these experiments can be found in reference [18]. Chip indications from the magnetic chip detector, when the gap was closed with debris, were also measured. During testing, the data from the chip detector, oil debris sensor and torque sensor were collected every minute using the Mechanical Diagnostics System

Software (MDSS) data acquisition system, a LabView based, NASA Glenn developed, research data acquisition system.

The vibration data was collected from accelerometers installed on the right and left side of the test rig housing. Accelerometer frequency range is 0.7 to 20 KHz with a resonant frequency of ≥70 KHz. The accelerometers were mounted on the housing, radially and vertically with respect to the pinion. Speed was measured with an optical tachometer mounted on the left gear shaft. This is used to calculate the TSA for both the pinion and the gear. The gear ratio is used to process the data at the correct speed for the pinion.

Table 4. Summary of Tests

Test	Sample rate (KHz) Sample time (sec)	Run time (min)	Description of Damage
1 (CHK4)	300 0.3	2578	Right pinion and gear teeth damage
2 (CHK6)	300 0.3	3302	Left pinion scoring on three teeth
3 (CHK7)	150 1	1057	Left pinion pitting on three teeth Right pinion pitting on four teeth
4 (CHK8)	150 1	1057	Left pinion pitting on two teeth

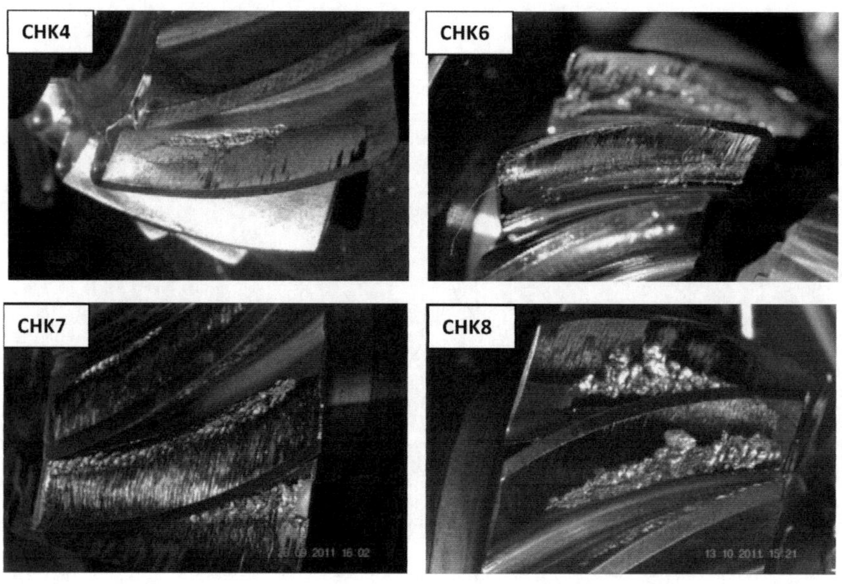

Figure 9. Photographs of tooth damage

Torque on the gear shaft during testing was on average 8000 in.-lbs, 239 ksi gear contact stress and 3500 rpm speed. The failure modes to be investigated for this study were defined by class (contact fatigue), general mode (macro pitting) and degree (progressive) per American Gear Manufacturers Association (AGMA) standards for gear wear terminology [19]. The tests ran until pits larger than 1 mm in diameter covered a significant area of gear or pinion tooth surfaces. Table 4 is a summary of the sample rates, duration, run-time and description of damage at the end of the test for the four experiments. A representative photo of the damage observed during testing is shown in figure 9.

4. Case Study and Health Monitoring Results:

4.1 Signal Processing Methods and Condition Indicators: Example results applying the various signal processing techniques to test CHK8 are used to illustrate the gear fault patterns and how the condition indicators are used to extract this signature information for assessing the gear health condition. An example frequency domain spectrum is provided in Figure 10, in which the TSA processing was used to resample the data into the angular domain and the spectrum is shown in terms of pinion shaft orders. The spectrum is shown for one of the run-to-failure tests, in which the left pinion gear had pitting damage on two of its teeth at the end of the test. The top plot in Figure 10 consists of a vibration acquisition during the early portion of the run-to-failure test, when there was no damage on any of the gear teeth; thus the regular vibration spectrum consists of the gear mesh frequency (12X) and harmonics but the first order sideband magnitudes are quite low (11X and 13X). This is in sharp contrast to the middle and bottom graphs, in which the sideband magnitudes are much larger and in some cases, are nearly the same magnitude as the gear mesh frequency. From this visual observation, one would expect that the sideband level condition indicator would work well for this data set.

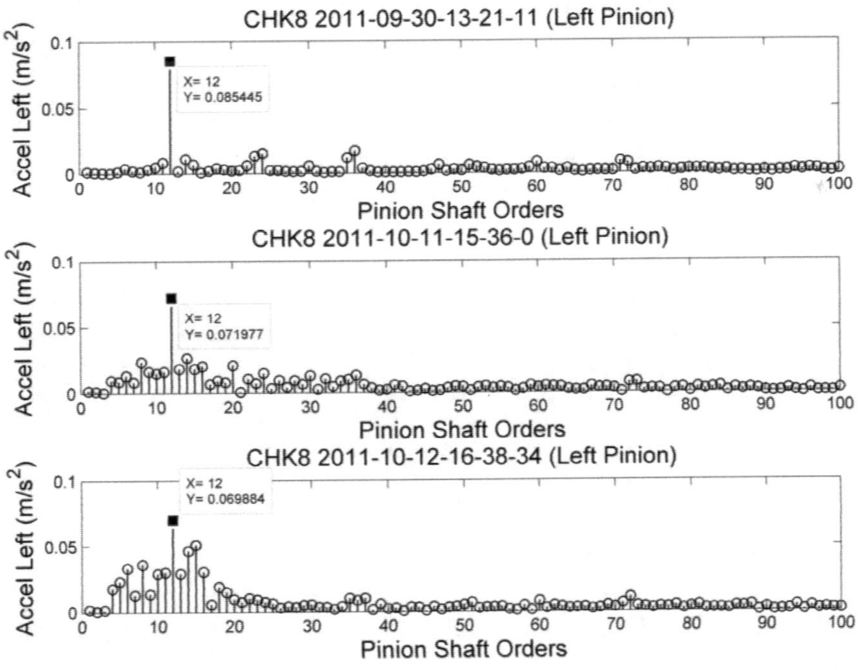

Figure 10: Spectrum Plot From Three Time Periods (Acquisition 101, 949, and 1050); the three acquisitions represent normal, degraded, and near failure.

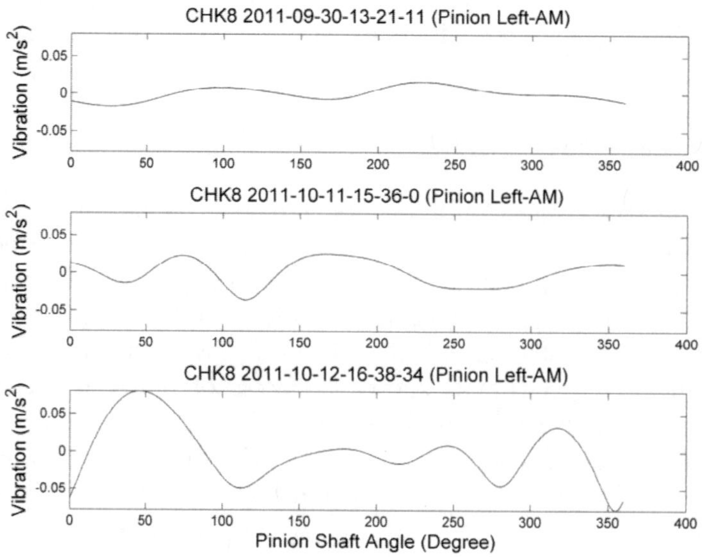

Figure 11: Amplitude Modulation Signal Plot (Shown for Acquisitions 101, 949, and 1050); Left Pinion Had Pitting Damage on Two Teeth

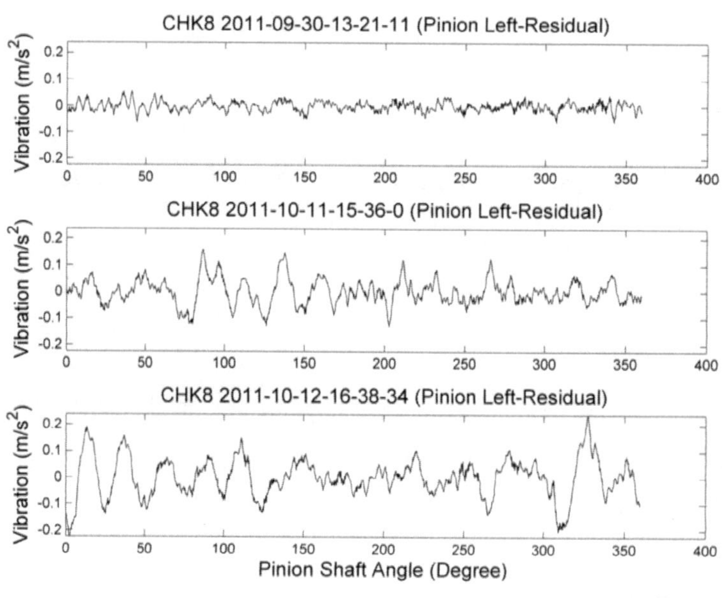

Figure 12: Residual Signal Plot (Shown for Acquisitions 101, 949, and 1050)

The amplitude modulation signal is plotted in Figure 11, and one can also observe a significantly different pattern and amplitude modulation magnitude when comparing the acquisition at the beginning of the test, and also when there was pitting damage on the left pinion. The peak to peak magnitude of the amplitude modulation signal could also be an effective condition indicator for inferring the gear health condition for this data set.

The same data files are also processed using the classical residual signal method, in which the graphs are presented in Figure 12. There is a clear difference in the residual signal pattern and magnitude if you compare the top plot (normal condition) with the middle and bottom plots (pitting damage on the left pinion). The RMS level of the residual signal could also be a good indicator for assessing the gear health state.

Prior to considering the condition indicator trends, an initial data quality check was also performed to make sure each vibration acquisition and waveform was free of any measurement errors or anomalies in the recording of the data file. Some recent work on the area of vibration signal validation has been published, and the consecutive sample check that was described by Jablonski et al. [20] was used to remove any potential files that had suspicions data quality. Also, a few acquisitions were collected at different rotational speeds, and those were also not included when evaluating the condition indicator trends. An example trend of the RMS, the consecutive value waveform check, and the cleaned condition indicator trend are provided in Figure 13, in which one can observe that the low outlier values are removed using this data quality check.

Example trends of three condition indicators are provided in Figure 14; the sideband level, the amplitude modulation peak to peak, and the residual RMS trends are shown on the top, middle, and bottom plots respectively. All three indicators have a clear and increasing trend as the pitting damage worsens for the left pinion. For this experimental data set, it appears that the amplitude modulation peak to peak indicator has the most consistent and clear trend of the three indicators. The aspect of condition indicator evaluation and ranking is explored in more detail in Section 4.2.

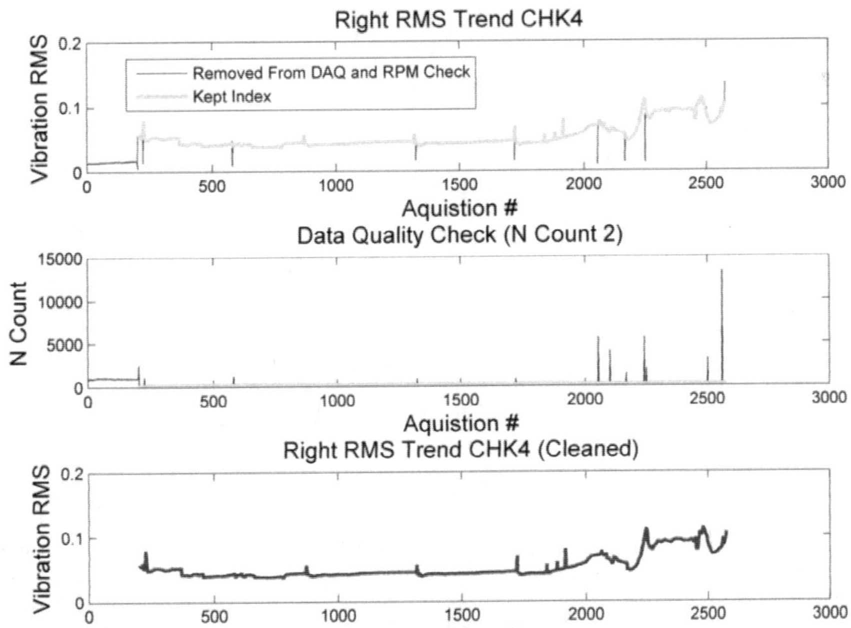

Figure 13: Data Quality Check and Outlier Removal- Based on Consecutive Sample Vibration Waveform Check

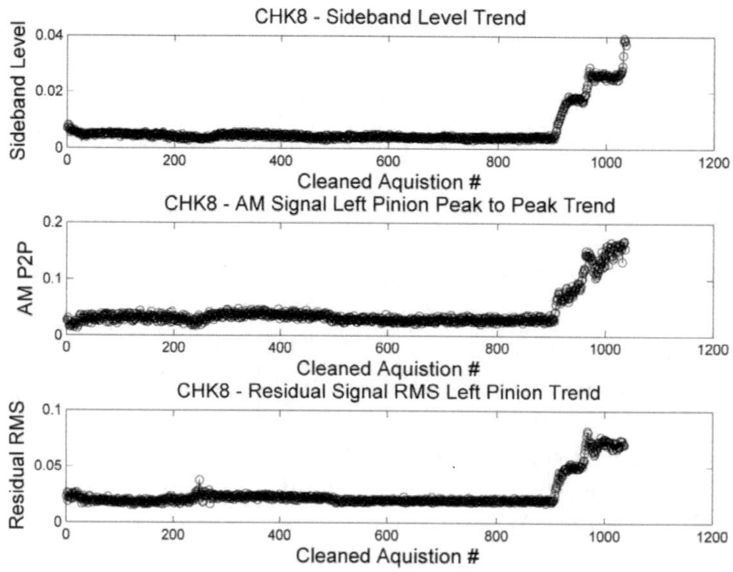

Figure 14: Example Trends for the CHK8 Gearbox Experiment; Sideband Level, Amplitude Modulation Peak to Peak, and Residual Signal RMS Trends

4.2 Variable Ranking Results: One aspect of this study was to evaluate and quantitatively rank the condition indicators diagnostic and prognostic performance for gear condition monitoring. The Fisher criterion, monotonic ranking method, and the area under the curve (AUC) of the receiver operating curve were all used to individually rank the condition indicators using the four run-to-failure data sets. Although the individual rankings offer some insight on the condition indicator performance, it is perhaps more useful to focus on the weighted result; since this considers multiple diagnostic and prognostic performance metrics. The weighted results from the variable selection routine are presented in Table 5, in which the top 10 condition indicators are listed. From this table, one can observe that the residual signal RMS and peak to peak indicators are the best two, while the raw RMS, sideband level, and amplitude modulation signal peak to peak indicators also performed well. Although the raw accelerometer RMS performed well, from the author's experience, the RMS value is sensitive to damage to other mechanical components (bearing and shafts) and would not be suited for only detecting gear related problems. For selecting the final set of condition indicators to use for further processing, gear and pinion indicators are grouped together and the top four are selected (8 condition indicators in total). Considering the top four ranked condition indicators in Table 5 and excluding the accelerometer RMS, the final selected condition indicators are the residual signal RMS and peak to peak value, the sideband level, and the amplitude modulation peak to peak value. This final list of 8 condition indicators is summarized in Table 6, and highlights that the residual and amplitude modulation processing methods appear to be quite effective for assessing the gear health condition over time. This is an encouraging result, in that these are some of the most established condition indicators for gear condition monitoring, and this validates much of the prior work done in this field.

Table 5: Ranking of Condition Indicators (Weighted Average Ranking) from Fisher Criterion, AUC Metric, and Monotonic Score)

#	Top 10 -Weighted Method	Final Score
1	'Residual Signal Pinion RMS'	0.94
2	'Residual Signal Pinion Peak to Peak'	0.91
3	'Residual Signal Gear RMS'	0.87
4	'Residual Signal Gear Peak to Peak'	0.82
5	'Accel RMS'	0.76
6	'Sideband Level Pinion'	0.69
7	'AM Signal Pinion Peak to Peak'	0.68
8	'AR Residual Signal Pinion RMS'	0.67
9	'AM Signal Pinion RMS'	0.67
10	'AR Residual Signal Left Gear RMS'	0.66

Table 6: Final Selection of Condition Indicators Used for Further Processing

#	Final Selection (8 Indicators)
1,2	Residual Signal RMS (Pinion/Gear)
3,4	Residual Signal Peak to Peak (Pinion/Gear)
5,6	Sideband Level (Pinion/Gear)
7,8	AM Signal Peak to Peak (Pinion/Gear)

4.3 Health Indicator Results: The comparison of the health indicator methods are based on looking at the correlation of the health indicator values with the oil-debris measurement. The oil debris measurement is not typically measured on a deployed rotorcraft health and usage monitoring system, however for this test-rig, it provides a measure of ground truth on the actual gear health condition. Thus, a health indicator trend that is very well correlated to the oil-debris measurement is more desirable then a health trend that is less correlated with the oil-debris values.

Based on this rationale, the health trend for each of the four run-to-failure gear experiments were compared with the oil-debris measurements and a correlation value was calculated. The results for the PCA based health indicator are provided in Figure 15 for the four data sets, in which the correlation values between the health trend and the oil-debris trend are very high for all four data sets. A smoothed health trend was also calculated using an exponential weighted moving average (smoothing constant of 0.2) and one can observe that the moving averaged provided a slightly better trend that was more correlated with the oil-debris measurement. Although the correlation values are all above 0.8, it is also interesting to note that the health value trend for the third and fourth test are more well correlated with the oil-debris trend then the previous two tests. For the last two run-to-failure tests, the sampling parameters were adjusted and consisted of acquiring the waveform at a lower sampling rate (150KHz compared to 300KHz) but for a longer duration (1 second compared to 0.3 seconds); the vibration health indicator results appear better when the waveform is acquired for a 1 second duration.

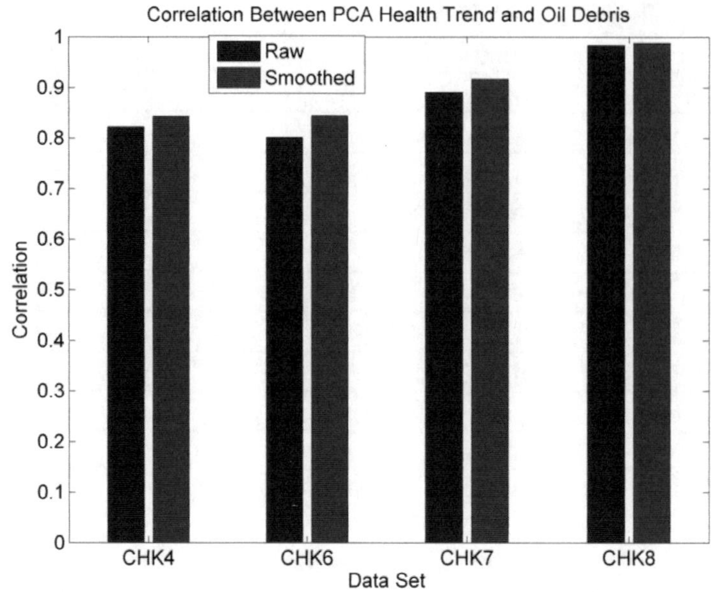

Figure 15: Correlation Between PCA Health Indicator and Oil-Debris Measurement

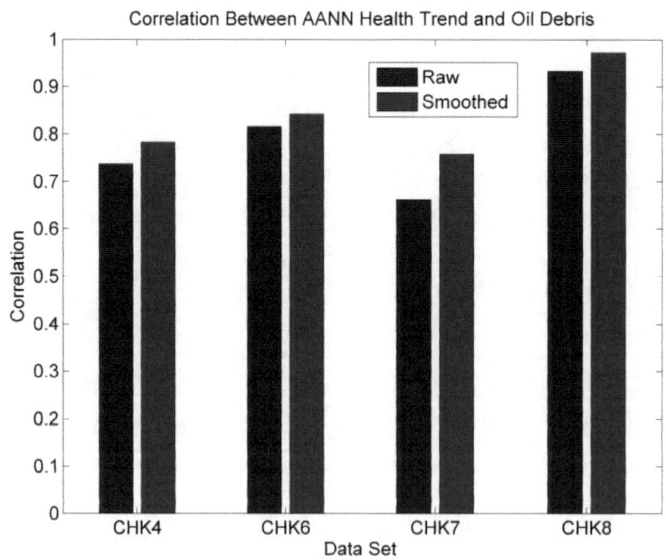

Figure 16: Correlation Between AANN Health Trend and Oil Debris Measurement

The health indicator results using the auto-associative neural network (AANN) method were slightly worse then what was obtained for the linear principal component health indicator method. The correlation between the AANN health values and the oil-debris measurement values are provided in Figure 16 for each of the four tests. Similar observations for the AANN health indicator were also observed, in that the best health trend was obtained for the last run-to-failure test (CHK8) and also the smoothing function offered some slightly improvement in the health trend. The results for the distribution overlap method were less encouraging and the method in general appeared to sensitive to changes in the condition indicators. The overlap in the distribution between the current situation and the baseline situation would become small in magnitude very early into the run-to-failure test, and the health trend would show too dramatic of a trend. This provided low correlation values between the distribution overlap health index and the oil-debris measurement as illustrated in Figure 17.

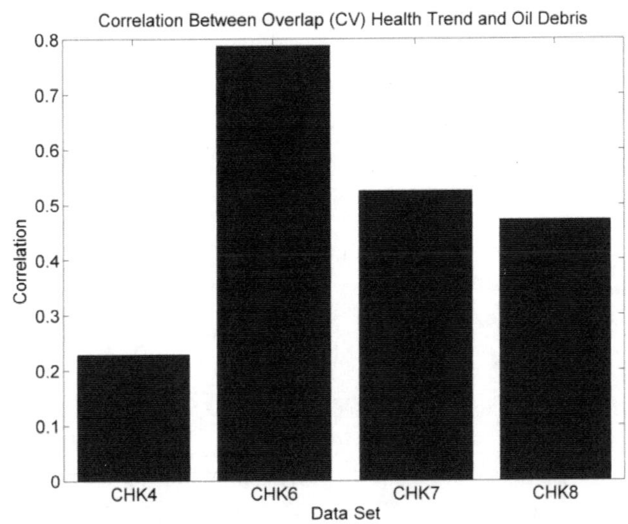

Figure17: Correlation Between Overlap Health Value and Oil Debris Measurement

The health monitoring results for each of the four data sets are shown using the principal component analysis health indicator method, since the PCA method had the best results from the three health indicator methods that were considered in this study. The smoothed health indicator trend for the CHK4 data set is provided in Figure 18, in which the health values above the threshold of 1 are highlighted in red. Although there is some fluctuation in the health value, the health value is above the limit when the oil-debris measurement starts to increase. This implies that the vibration indicators are sensitive to the early stages of gear pitting damage. In addition, it implies that the PCA method has a reliable threshold that can accurately determine when a problem exists but is not too sensitive to generate erroneous false alarms. Although the fluctuations in the health value are a bit of a concern, additional testing would be needed to determine if this is due to less desirable data acquisition settings, such as only acquiring the vibration waveform for 0.3 seconds.

The health trend for the CHK6 data set, shown in Figure 19, shows a sharp rise in the vibration based health value. However, this has very good agreement with the oil-debris measurement that also has a very rapid increase. The CHK6 run-to-failure experiment resulted in scoring damage to the 3 teeth on the left pinion and one postulate is that this particular failure mode is more sudden and causes a more rapid change in the vibration signature. A more thorough investigation on the vibration symptom response to gear scoring would be needed to validate this hypothesis on this failure mode being more sudden. It does appear that a quick detection of gear scoring is more practical then considering prognostic methods, since this failure mode does not have a long term trend.

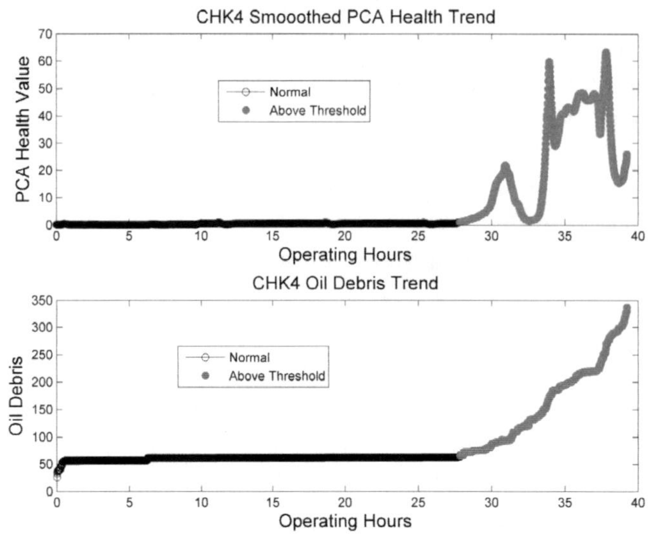

Figure 18: PCA Health Indicator Trend (CHK4 Data Set)

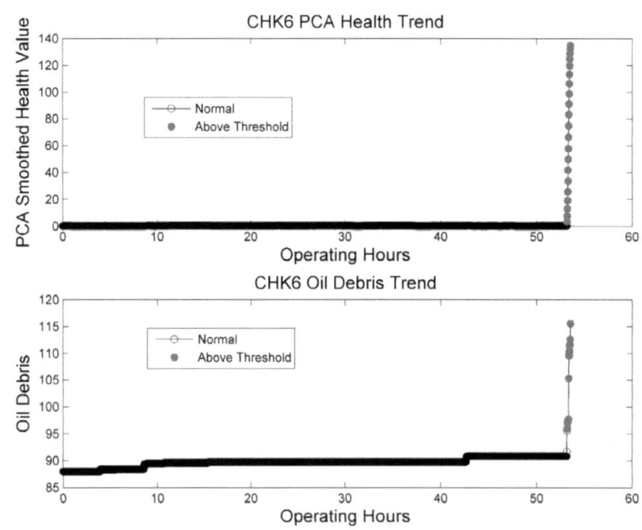

Figure19: PCA Health Indicator Trend (CHK6 Data Set)

The results for the CHK7 and CHK8 data set are shown in Figure 20 and Figure 21 respectively. The CHK7 experimented resulted in pitting damage to the right pinion, while at the end of the CHK8 test, the left pinion had pitting damage on two of the teeth. The results for the CHK7 data set show a clear and increasing health trend that is in good agreement with the oil-debris measurement. In addition, the health values are above the threshold when the oil-debris values start to increase; this implies that the vibration indicators have good sensitivity and that the threshold is set appropriately. The health monitoring results for the CHK8 data set are even more encouraging and the health trend and oil-debris measurement have nearly the same pattern and trend. With a correlation value above 0.95, this shows that the vibration based health indicator is in very good agreement with the oil-debris measurement for this experiment.

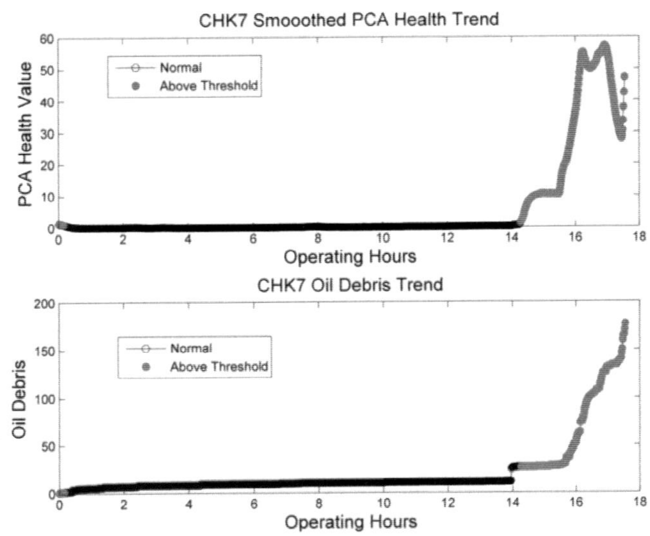

Figure 20: PCA Health Indicator Trend (CHK7 Data Set)

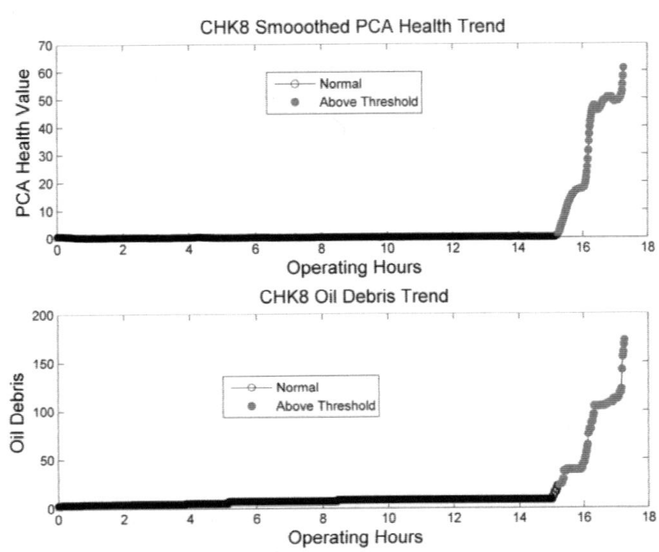

Figure 21: PCA Health Indicator Trend (CHK8 Data Set)

5. Conclusion: The study evaluated a multitude of different gear condition indicator methods for rotorcraft drivetrain condition monitoring, in which the amplitude modulation, residual signal, and sideband indicators had the best performance. The importance of variable selection methods was also revisited in this study, in which it was observed that multiple diagnostic and prognostic ranking methods should be used to select the indicators that have the best performance. Lastly, three different health indicator methods were evaluated and the linear principal component analysis health indicator method offered the most consistent health trend of the three algorithms. Additional run-to-failure gear testing will be performed to further validate these methods. Also, further study will consider health indicator methods based on fuzzy logic and gear failure prediction algorithms.

Acknowledgments:

The work is supported by the NSF Industry/University Cooperative Research Center on Intelligent Maintenance Systems (IMS) at the University of Cincinnati as well as its company members.

References:

[1] P.D. Samuel, D.J. Pines, 2005, A review of vibration-based techniques for helicopter transmission diagnostics, Journal of Sound and Vibration 282, Pages 475-508.
[2] E. Bechhoefer and E. Mayhew, 2006, Mechanical diagnostics system engineering in IMS HUMS, In IEEE Aerospace Conference, Pages 1-8.
[3] S. Braun, 2011, The synchronous (time domain) average revisited, Mechanical Systems and Signal Processing 25, Pages 1087-1102.
[4] P.D. McFadden, 1986, Detecting fatigue cracks in gears by amplitude and phase modulation of the meshing vibration, ASME Journal of Vibration, Acoustics, Stress, and Reliability in Design 108, Pages 165-170.
[5] R.M. Stewart, 1977, Some useful data analysis techniques for gearbox diagnostics. Machine Health Monitoring Group, Institute of Sound and Vibration Research, University of Southampton, Report MHM/R/10/77, July 1977.
[6] W. Wang and A.K. Wong, 2002, Autoregressive model-based gear fault diagnosis, Journal of Vibration and Acoustics 124, Pages 172-179.
[7] E. Bechhoefer and M. Kingsley, 2009, A review of time synchronous average algorithms, Annual Conference of the Prognostics and Health Management Society.
[8] F. Combet, L. Gelman, Novel adaptation of the demodulation technique for gear damage detection to the variable amplitude of mesh harmonics, Mechanical Systems and Signal Processing 25, Pages 839-845.
[9] J. Isom, Y. Chen, Z. Chaudhry and P. Dempsey, Performance Evaluation of a Second-Order Indicator of Cyclostationarity for Gear Condition Monitoring, MFPT 2013 Conference, May13-17, 2013, Cleveland, Ohio.
[10] R. Duda and P. Hart, 2001, Pattern Classification, Wiley , Pages 114-121.
[11] M. Zweig and G. Campbell, 1993, Receiver-operating characteristics (ROC) plots: a fundamental evaluation tool in clinical medicine, Clinical Chemistry 39, Pages 561-577.
[12] J. Coble, 2010, Merging data sources to predict remaining useful life- an automated method to identify prognostic parameters, PhD Dissertation, The University of Tennessee.
[13] A.J. Bayba, D.N. Siegel, K. Tom, and C. Ly, 2011, Approaches to health monitoring of the CAT 7 diesel engine, ARL-TR-5677, Army Research Lab, Sensors and Electron Devices Directorate.

[14] E. Lapira, D. Brisset, H.D. Ardakani, D. Siegel, and J. Lee, 2012, Wind turbine performance assessment using multi-regime modeling approach, Renewable Energy 45, Pages 86-95.

[15] M.A. Kramer, 1991, Nonlinear principal component analysis using autoassociative neural networks, AIChe Journal 37, Pages 233-243.

[16] Handschuh, R.F., Thermal Behavior of Spiral Bevel Gears, NASA TM-106518, 1995.

[17] Handschuh, R.F., Testing of Face-Milled Spiral Bevel Gears at High-Speed and Load, NASA/TM—2001-210743, 2001.

[18] Dempsey, P.J., Tuck, W.R. and Showalter, S., Comparison of an Inductance In-Line Oil Debris Sensor and Magnetic Plug Oil Debris Sensor Proceedings of the American Helicopter Society 68th Annual Forum, Ft. Worth, Texas, May 1–3, 2012.

[19] ANSI/AGMA 1010-E95, Appearance of Gear Teeth – Terminology of Wear and Failure.

[20] A. Jablonski, T. Barszcz, and M. Bielecka, 2011, Automatic validation of vibration signals in wind farm distributed monitoring systems, Measurement 44, Pages 1954-1967.

Spectro Scientific
Confidence in knowing

Fluid-Based Portable Machinery Assessment Tools

MAY 21, 2014

MFPT 2014 Virginia Beach, VA

Outline

Miniaturized Systems for Machinery Assessment with Fluid Analysis

- Introduction
- Current Status
- Application examples
- Future Outlook
- Conclusions

Spectro Scientific
Confidence in knowing

Introduction

Spectro Scientific
Confidence in knowing

Fluid Analysis

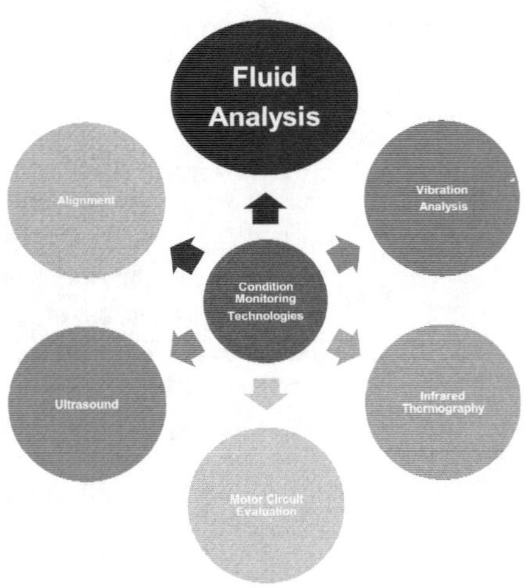

fluid analysis is part of a comprehensive Predictive Maintenance Program

Spectro Scientific
Confidence in knowing

System Definition

- Probe the status of fluids in a machinery system (lube, hydraulic, fuel, …)
- Self-contained, provide near real-time assessment
- Built for field use by a single operator (hand-carry, handheld, …)
- Operates at the machinery with no machinery mods

Spectro Scientific
Confidence in knowing

Status Indicators

- Wear metals in fluids – abnormal levels can indicate machinery is damaged (eg. Fe, Cu)
- Fluid contamination – indicates presence of materials (eg. wrong fluid) which can rapidly degrade machinery
- Fluid Condition – indicates the lubricant is no longer able to provide machinery needed lubrication (eg. acid/base number)

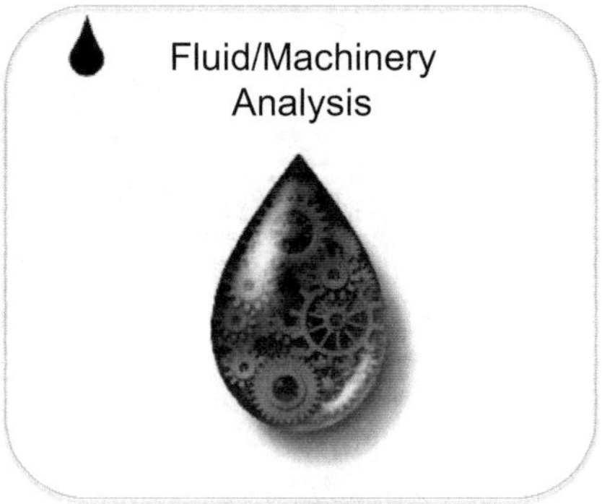

Spectro Scientific
Confidence in knowing

Example Tools

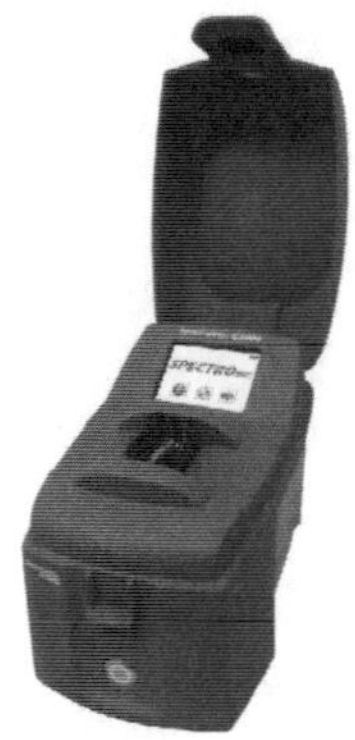

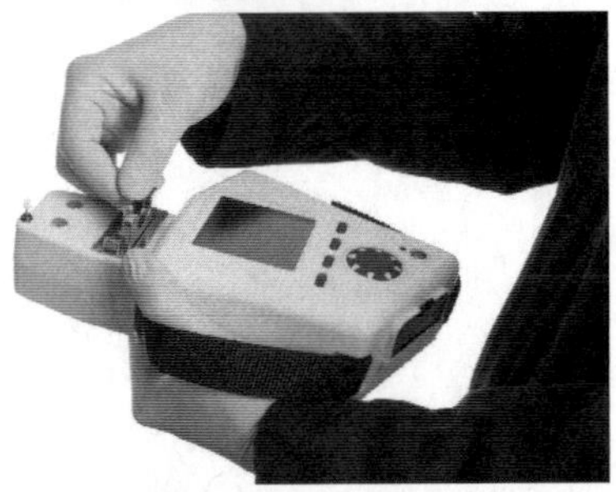

- Handheld or hand-carry, battery-operated
- Self-contained
- Instant answers
- Built for field use

Spectro Scientific
Confidence in knowing

Why is it different from miniature lab equipment?

- need to include recommendations and actions with each result
- the equipment should be self-guiding; use out-of-box
- data management needs to be figured out ahead of time– don't assume someone has a PC
- clarify asset-based vs. sample-based upfront
- automate – don't make the user have to select analysis guidelines

Spectro Scientific
Confidence in knowing

Current Status

Spectro Scientific
Confidence in knowing

Portable Tools – Motivation

- Labs still process hundreds of millions of oil analysis samples per annum
- Comprehensive online technology still emerging
- Field tools can provide a bridge between the technologies
- Prognostics is a must (RUL)
- In both the DoD and commercial sectors, economics continues to drive maximum equipment uptime efforts

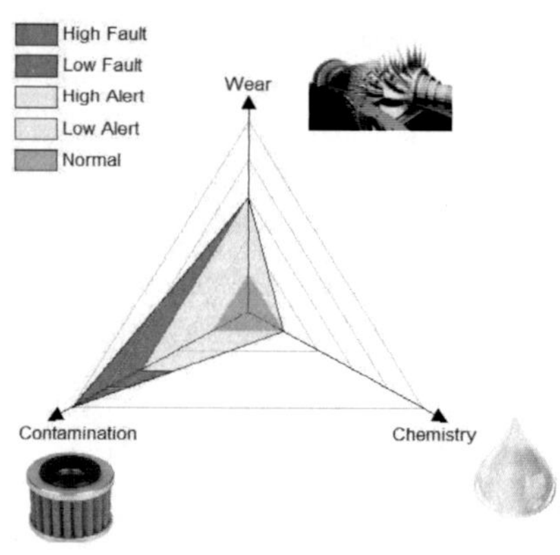

Spectro Scientific
Confidence in knowing

Portable Tools – Current Status

- getting to be near the entire lab test suite
- standards are being established (eg. ASTM D7889)
- infrared spectrometers, particle counters, elemental analyzers
- no longer specialized equipment but lab-to-field
- seeding best practices across industries (eg. incoming lubricant QC)

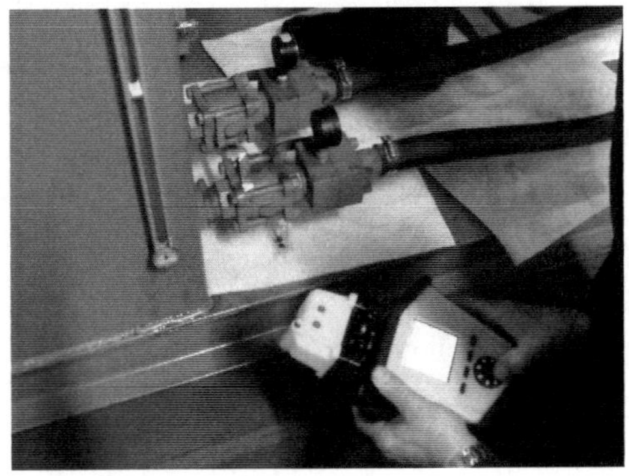

Spectro Scientific
Confidence in knowing

Portable Tools – Observations

- data reachback/integration is essential (command and control)
- establishing quantitative, significant units of measure is essential
- you must understand why/if your data will be different from the laboratory upfront (eg. water content)
- expanding an existing capability is very desirable
- What establishes ROI can be very different depending on program:
  - No existing program on use and downtime
  - Existing program on use-case
  - Existing program on analytical capability

Spectro Scientific
Confidence in knowing

Application Examples

Spectro Scientific
Confidence in knowing

DoD – Naval Facility

- Large fleet of heavy equipment – previous time-based maintenance
- RUL of equipment and oil is essential
- Significant extension of oil changes by real-time monitoring
- By monitoring all (wear metals, particulate, oil condition) properties issues with equipment components can be rapidly identified

Spectro Scientific
Confidence in knowing

DoD – Military Sealift Command

- Ships monitor a range of lube and hydraulic oils
- Asset-based software allows equipment status to be directly monitored – no setup, labor-saving
- Complete elimination of chemicals for the analysis
- Monitors wide range of fluids: (T)BN, Viscosity, Water
- Integration with SAMM on-ship database

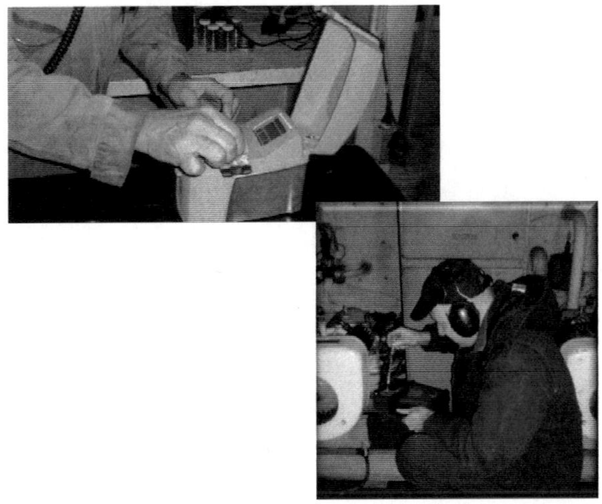

Spectro Scientific
Confidence in knowing

Commercial – Route-Based Oil Analysis

- Large machinery supervisory database implants an analysis route into handheld equipment
- Operator follows route and analyzes oil of machinery
- Resynch with supervisory database at end of route
- Eliminates need to draw sample bottles for analysis

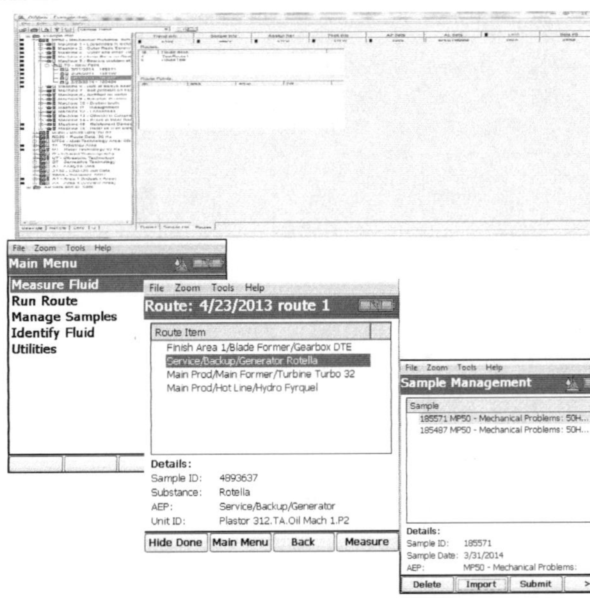

Spectro Scientific
Confidence in knowing

Future Outlook and Conclusions

Spectro Scientific
Confidence in knowing

Portable Tools – Future Outlook

- continued miniaturization (always) – allows new uses and applications
- more sophisticated reachback – figure out how to provide continuous updates without infringing on proprietary data and systems
- ordering of lab, field, and online analysis – roles will become better-defined
- application-specific solutions will become the norm
- in-system feedback – how field equipment is linked to other field equipment and feedback systems

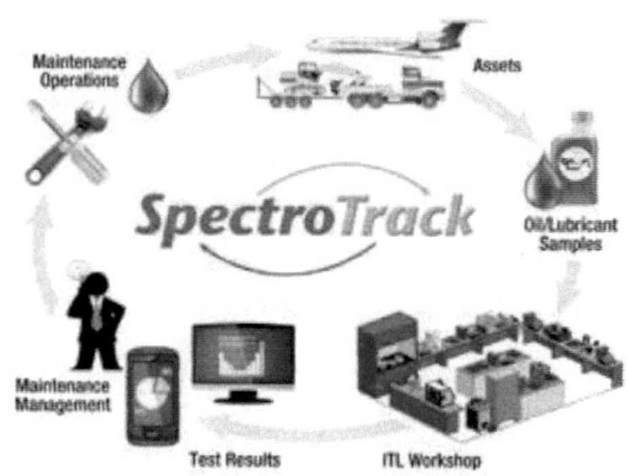

Spectro Scientific
Confidence in knowing

Conclusions

- Portable machinery assessment tools for fluids analysis are a reality today in the DoD and commercial sectors:
  - Provide near real-time answers
  - Differentiated from lab equipment but provide the same quantitative answers
  - Path to future condition monitoring architectures
- Easy integration into existing data architectures is a leading requirement
- Portable tools enable condition-based maintenance on-the-fly

Spectro Scientific
Confidence in knowing

Acknowledgements

- Office of Secretary of Defense for Maintenance Policy and Programs
- US Army Special Forces
- United States Military Sealift Command
- United States Coast Guard
- United States Air Force
- United States Navy
- United States Army
- Commercial Technologies for Maintenance Activities
- Emerson Process Management
- MFPT

Spectro Scientific
Confidence in knowing

Non-intrusive Load Monitoring of Electric Motor Driven Machine Equipment

This material is based upon work supported by the Air Force under Contract No FA9101-11-C-0032. Cleared for Public Release. Distribution is unlimited.

Any opinions, findings and conclusions or recommendations expressed in this material are those of the authors and do not reflect the views of the Air Force.

LUNA

Jacob Siegel and Nathan K. Brown
Tuesday, May 21, 2014

Outline

- Motivation
- Technical Approach
- System Overview
- Experimental Study
- Prototype System Development
- Conclusions

2

Motivation

- Increase the service life and reliability of rotating machines
- Improve efficiency of routine maintenance, troubleshooting, and diagnostics
- Decrease daily workload in facilities with large number of installed machines
- Reduce operation and maintenance costs

3

Technical Approach

- Build widely applicable, general purpose health monitoring platform technology
- Nonintrusive motor load estimation method
  - Correlate motor stray flux with mechanical output torque
  - Achieve or exceed 10% full scale accuracy
- Secure wireless communication
  - Provides convenient configuration and data access
  - Eliminates need for cabling
- Low-power
  - Thermoelectric generators considered to extend service life

4

System Overview

- Sensor node communicates with base station over closed Wi-Fi network
- SBC receives wireless data, translates to analog signal for existing data collection infrastructure (configurable for given application)
- Processing at node reduces transmission bandwidth and associated energy expenditure (RMS coil/accel, avg temp). Embedded models translate flux amplitude into estimate of mechanical motor load.

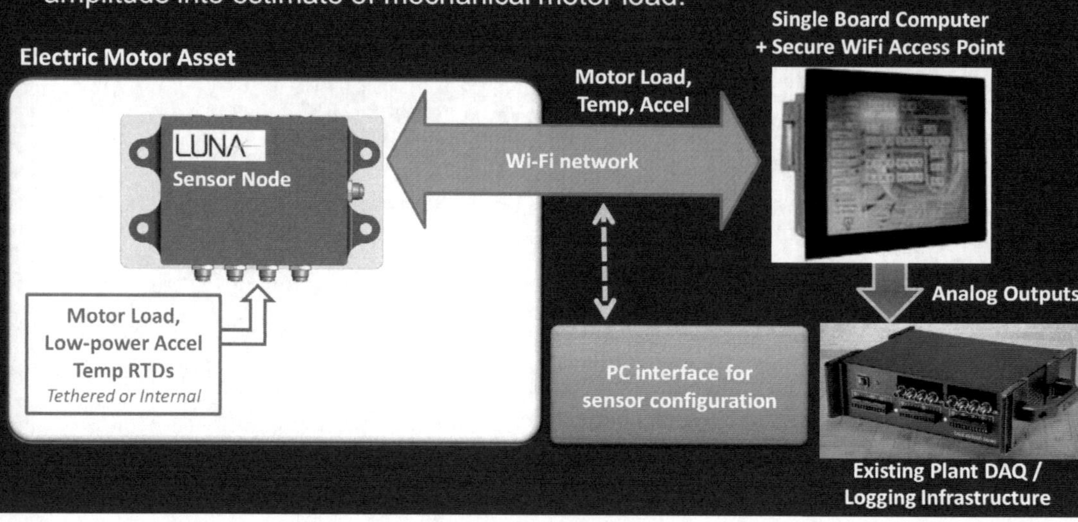

5

Candidate Stray Flux Transducers

- Hall-effect flux density sensor
  - Active, relatively low power
  - Small measurement area, localized

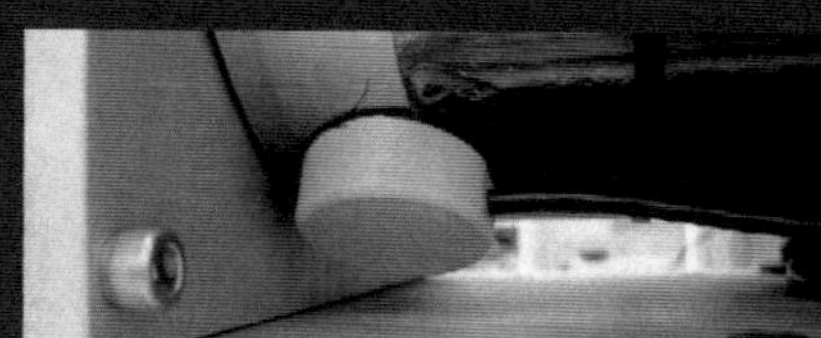

- Faraday induction coil
  $V_{induced} = -N \, \Delta(B\,A) / \Delta t$
  - For AC flux measurement only
  - Integrates field over large area, less localized
  - Higher sensitivity by increasing area and/or winding turns

- Small induction coil array
  - Same principle as above, but more conformal packaging

6

Induction Coil Measurement

- Correlate raw induction coil data to practical load measurement
- Root mean square (RMS) of coil voltage observed to trend predictably with motor load
- RMS method is less prone to noise than peak detection and spectral methods
- RMS is a cheap computational process for embedded systems (i.e., low power)

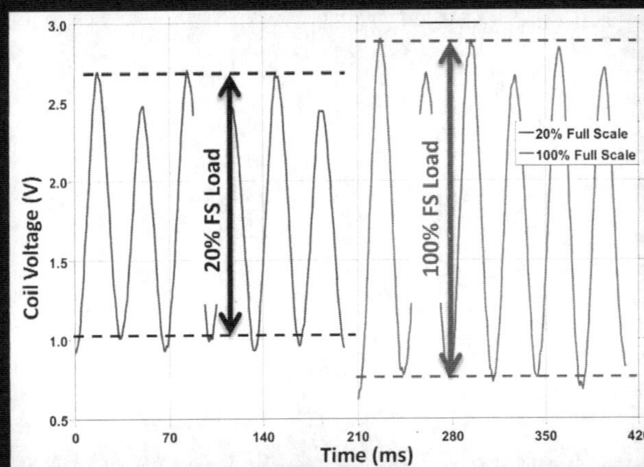

RMS amplitude response for:
5 hp induction motor
Large coil transducer
30 Hz excitation

$V_{induced} = -N\ \Delta(B\ A) / \Delta t$

7

Flux Transducer Downselection

- Exercised motor across speed and load range
  - VFD Frequency: 30–60 Hz
  - Gen Power: 0–430 W
  - Random step ordering
  - Coil V sampled at 5/10 kHz
- Surface response map
  - Calibration curves obtained between coil V and generator power for each motor speed
  - Response across all speeds forms calibration map
  - Response deviation from straight line fits indicates measurement repeatability

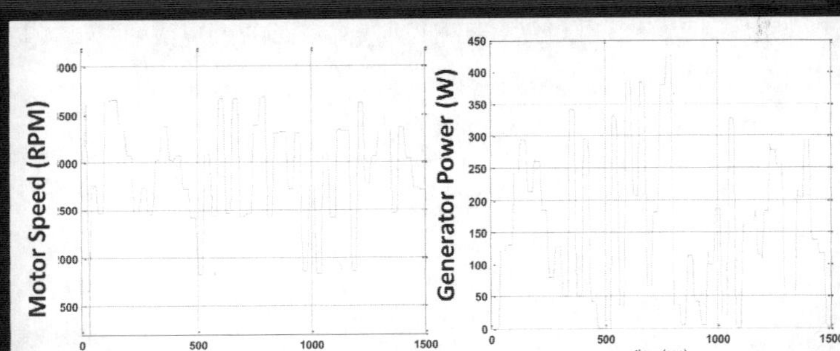

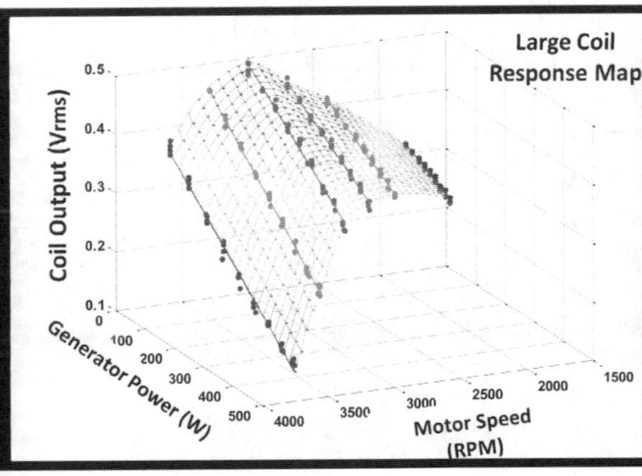

8

Flux Transducer Downselection

- Compare flux sensitivity for 3 candidates
- Large coil: 5x sensitivity of coil array
  - more turns, larger area, less cancellation
- Hall sensor: lowest sensitivity, opposite trend and very localized response
- Load estimate error based on std dev of data spread across RPM, Power range
- 10 kHz sampling rate: less variability
  - Better rejection of high frequency VFD noise
- Large coil produces most consistent results with 95% CI of +/- 3.2% full scale power
- ***Large coil selected for further development***
- Test stand load uncertainty, room for improvement with direct load measurement

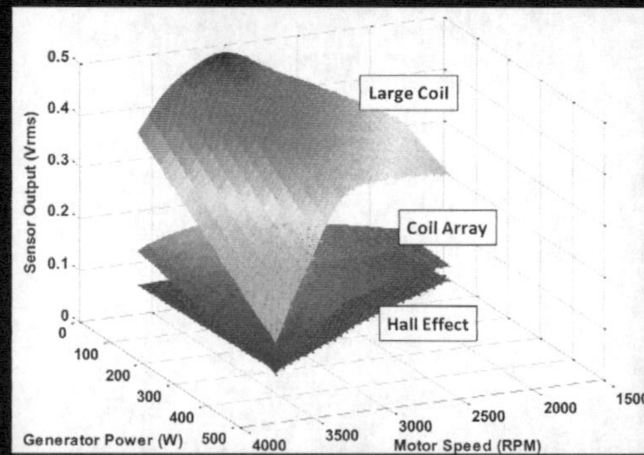

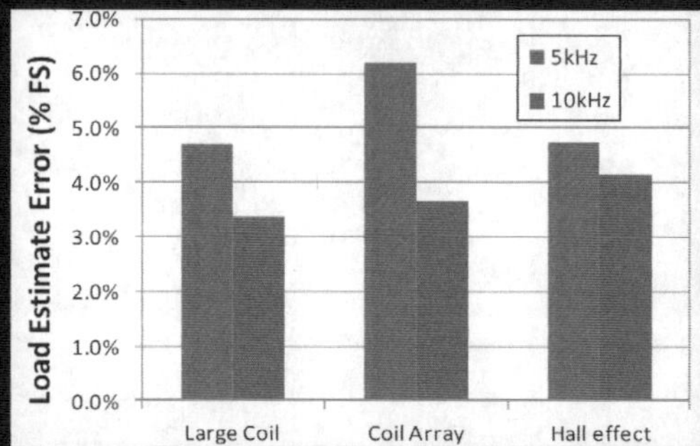

9

Phase II Motor Test Platform

- Addresses shortcomings of earlier test stand
- *5 HP motor capacity*
  - Induction and synchronous machines
- *Direct torque measure ground truth (0.05% FS accuracy)*
- *Fully automated load control*
  - Labview controls alternator field voltage to adjust torque
- Automated speed control
- Soft-start module for line-power operation

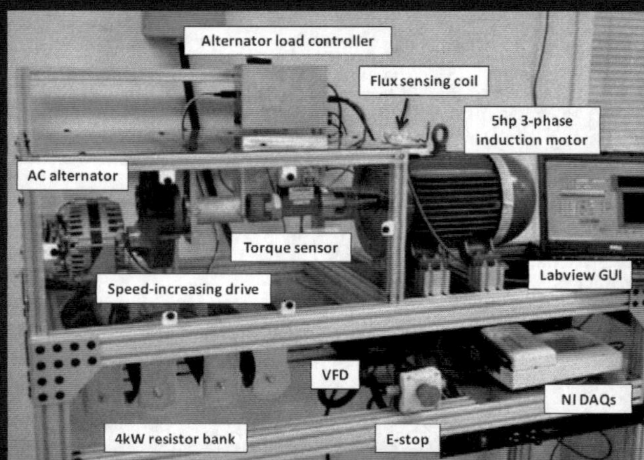

Interface T4 – 30 Nm rotary torque sensor

5 HP 3-phase synchronous motor

10

Phase II Motor Test Platform

- Rotary Torque Sensor Calibration
  - Ensure stray flux is correlated against accurate assessment of motor output torque
  - Performed *in-situ* random loading profile
    - Dead load weights placed at varying distances
    - Applied Torque: 0 – 24 Nm

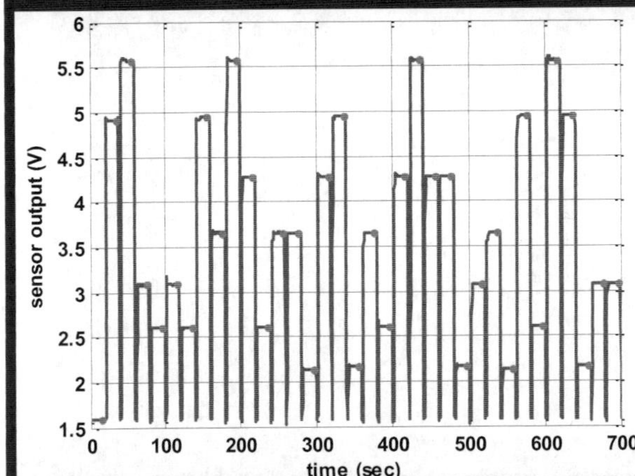

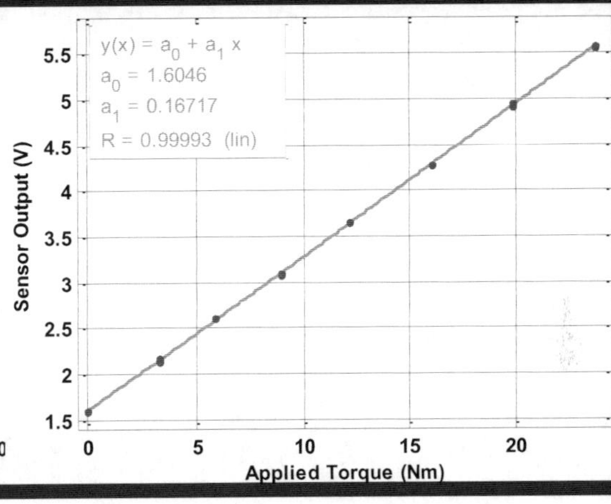

11

Stray Flux Characterization

- **With new test stand, observe stray flux sensitivity to motor speed and torque**
- Motor: 5 hp induction
- Speed: 30, 40, 50, 60 Hz
- Torque: 10-100% FS cycles, 3X

Results:

- 10% TQ steps readily resolved, good repeatability
- Speed has profound impact on stray flux output
- Coil output 5X lower than 1hp
  - 1 hp motor: much thinner case, aluminum endplates = less shielding
  - Amplify signal or use more windings to utilize A/D range

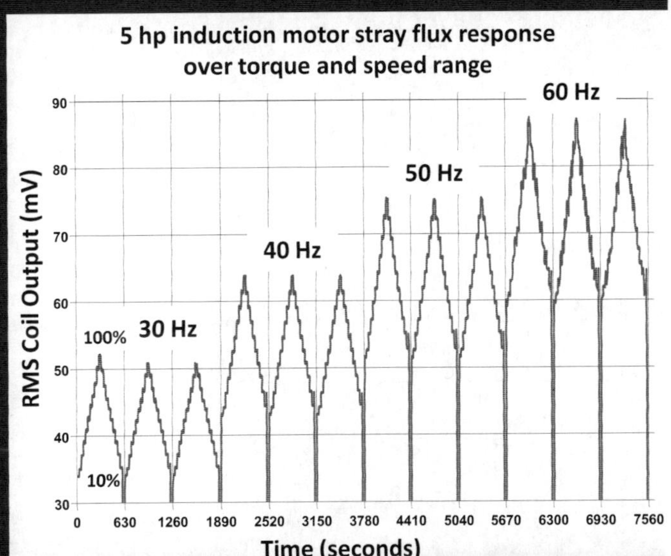

12

Stray Flux Characterization

- **Correlating coil output to motor torque, speed** with 95% CI bands (±2σ)

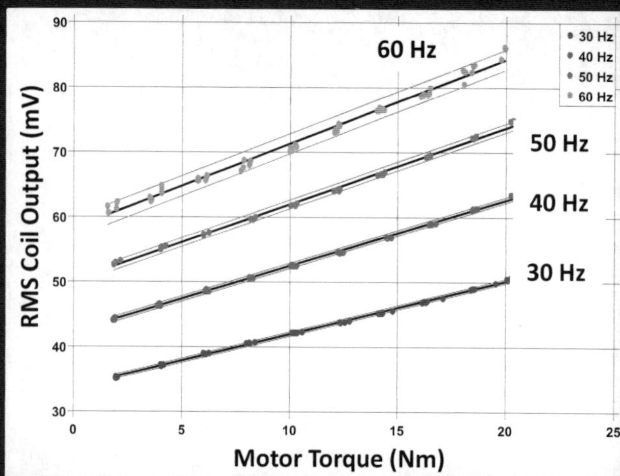

- **More linear than 1 hp data, equal spacing btw speeds**
- **Improved variability due to direct torque measurement**
- **30-50 Hz data < 3% FS error**
- **60 Hz response indicates significantly higher spread**

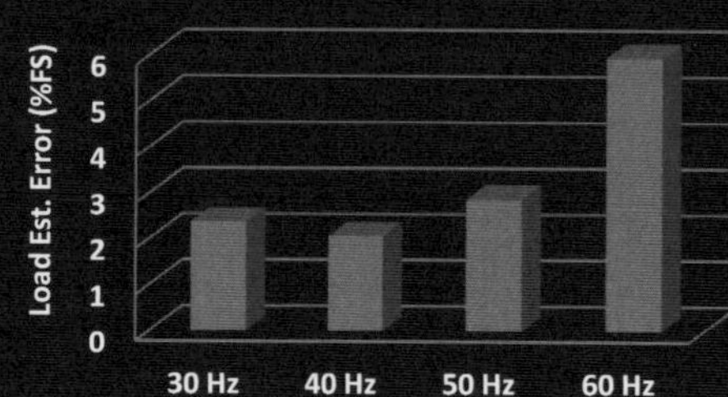

13

Stray Flux Characterization

- **Observe stray flux sensitivity to motor speed and torque**
- Motor: 5 hp synchronous
- Speed: 30, 40, 50, 60 Hz
- Torque: 5-35% FS cycles, 3X
  - Synch unable to be maintained at higher loads

Results:

- 10% TQ steps readily resolved, good repeatability
- Hysteresis observed between cycles
  - Test stand heating decreases the applied load
- Coil output 30X higher than 5hp induction machine
  - Synchronous motor: aluminum casing = less shielding
  - Amplify/attenuate signal during initial calibration

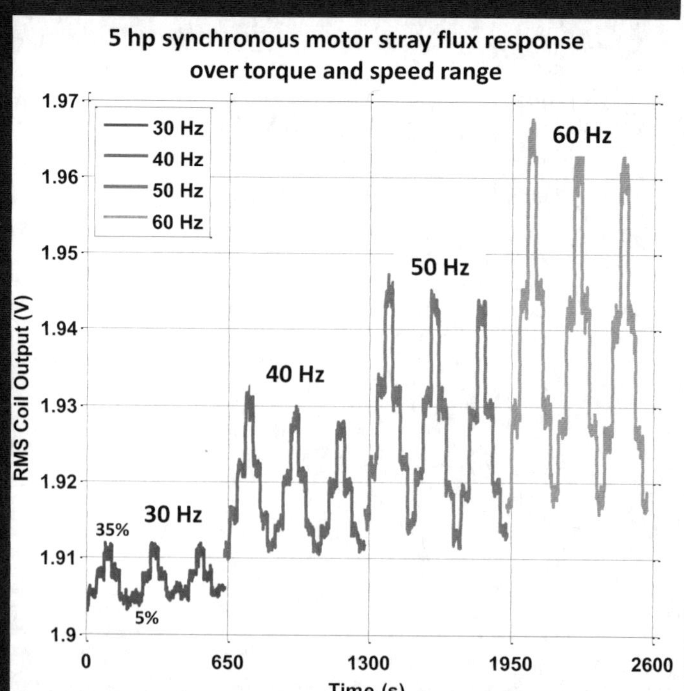

Stray Flux Characterization

- **Correlating coil output to motor torque, speed** with 95% CI bands (±2σ)

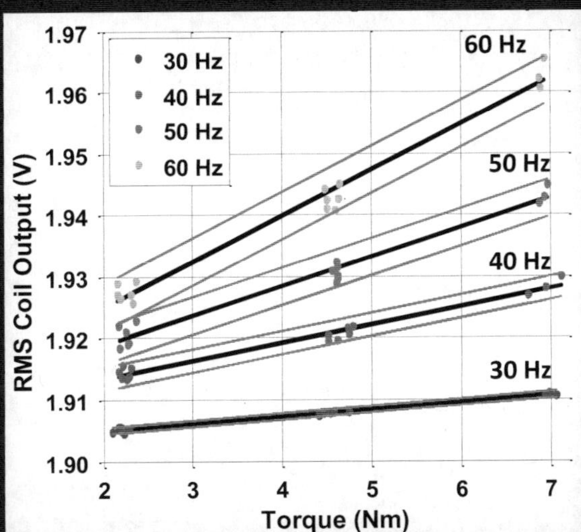

- **Linear trending similar to 5HP synchronous machine**
- **More variation in slope between speeds**
- **30-60 Hz data < 3.5% FS error**

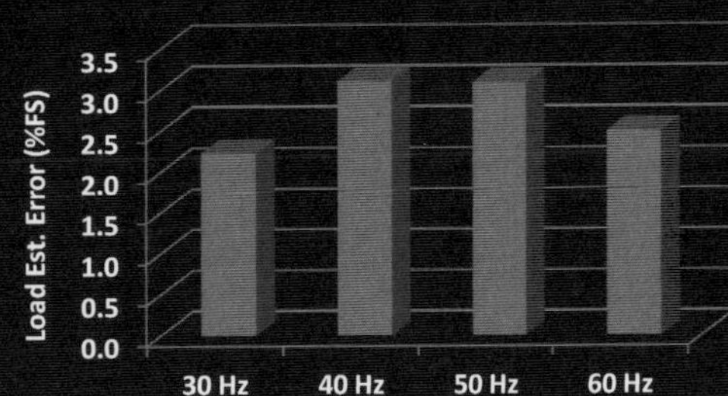

15

Stray Flux Characterization

- 5HP synchronous motor stray flux location study
  - Determine how observed stray flux field varies due to induction coil sensor location
  - (8) positions investigated, (5) locations on the top and (3) on the front side of the motor
  - Highest output observed at position #3, slightly left of top dead center
  - Observed RMS coil output varied ~80 mV (or 4.0% of the 1.98 FS Vrms observed)
  - Sensor placement precision is necessary when remounting to ensure data comparability

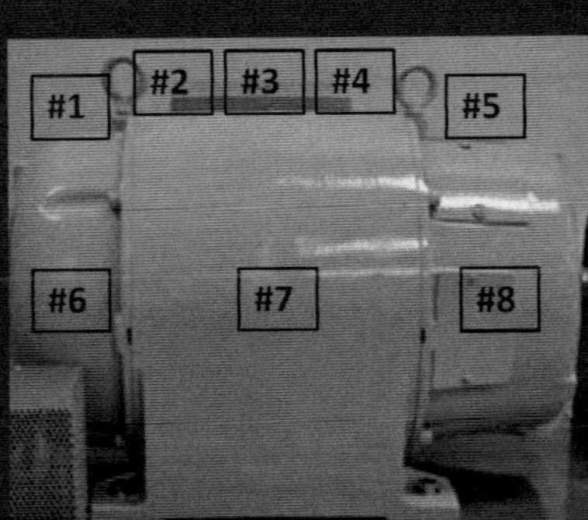

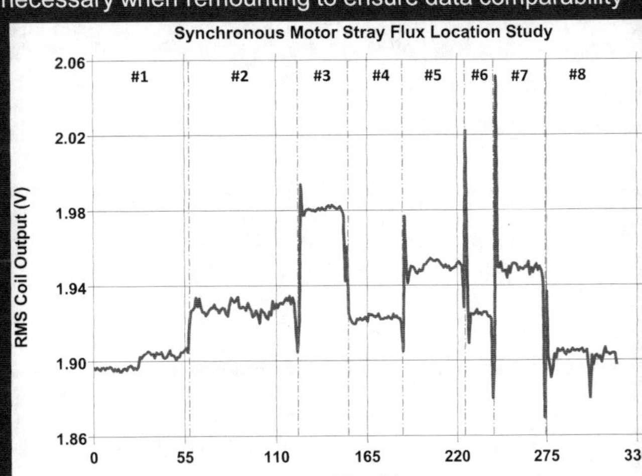

16

Experimental Conclusions

- Air core RF coil selected for stray flux measurement
  - Low-power, high sensitivity, large sensing area
- Non-intrusive load estimation shown feasible
  - Applicable to induction and synchronous type machines both VFD-driven and line-driven
  - Coil voltage RMS calculation correlates well with load and speed
  - Exceeded 10% full scale accuracy
- Scalability of the method demonstrated on varying size and casings of in-house machines
  - Amplify/attenuate signal or use more windings to fully utilize A/D range
    - Performed during initial calibration

17

Prototype System Development - Node

Sensor Node

- Overall dimensions: 3.86" x 2.60" x 2.16" (L x W x H)
- Embedded processing and non-volatile flash memory
- (4) tethered transducers: (1) induction coil, (1) accelerometer, (2) temperature sensors
- Magnetic mounting

Faraday Induction Coil (RMS)

- In-house custom build
- 1230 turns of 34 AWG magnet wire
- Flux to voltage sensitivity: 0.1 V/Gauss

Accelerometer (RMS)

- Sensitivity: 100 mV/g
- Full Scale: 20 g
- RMS in lieu of spectral analysis for power savings

Temperature Sensor

- Machine surface temperature
- 2-wire RTD

Ø 2.77

1.01

Ø 1.30

1.06

Dimensions in inches

Ø 1.45

0.35

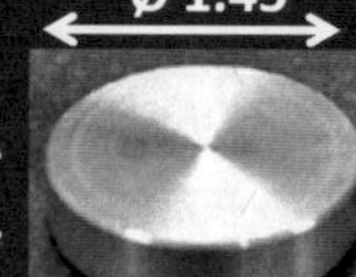

18

Prototype System Development – Base Station

Base Station

- Overall dimensions: 7.4" x 7.4" x 2.7" (L x W x H)
- Houses win7 SSD single board computer
- Wi-Fi wireless with AES 256 encryption
- RS-485/USB wired interface
- BNC analog outputs for interfacing with AEDC existing infrastructure

GUI

- Configure nodes (e.g., wired/wireless, duty cycle, sensing duration streaming/autonomous data logging)
- Monitor and review up to (3) sensor nodes simultaneously

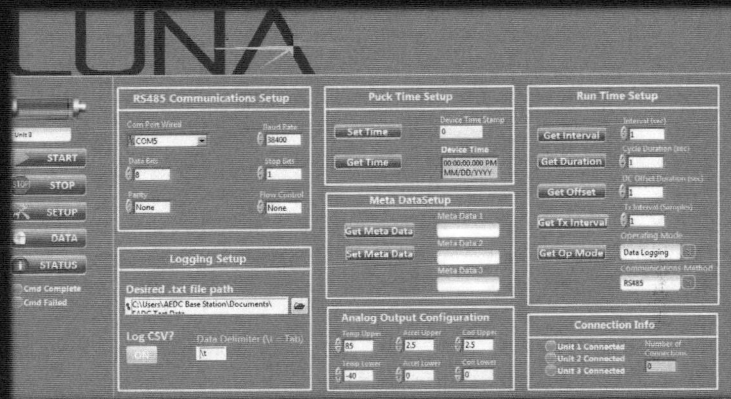

19

Prototype System Development - Evaluation

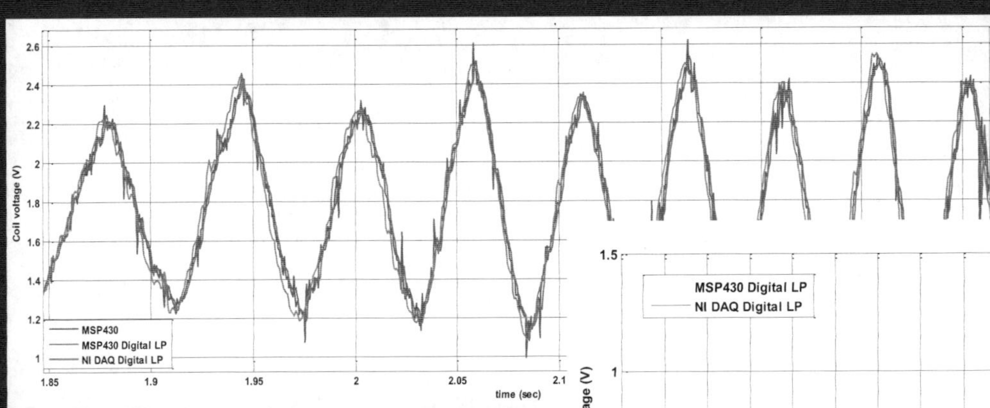

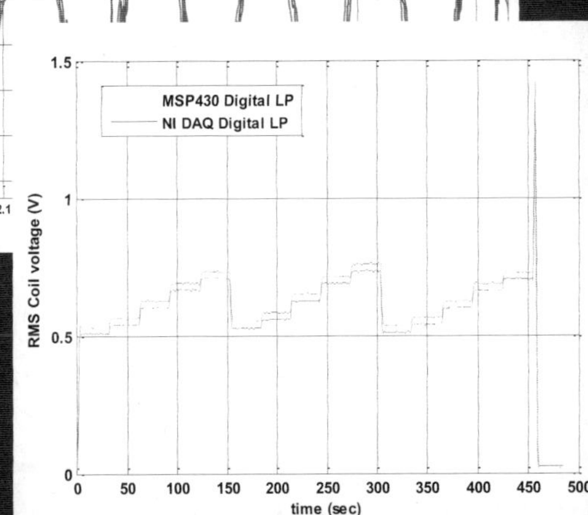

- Comparing NI DAQ and MSP430 acquired coil signal
- RMS voltage trends as expected with applied torque and speed

20

Deployable System Development – On site

- Motor survey at USAF Arnold Engineering Development Complex
  - Baldor 250 HP AC induction motor, 1790 RPM, 60 Hz line-driven, TEFC
  - Support system development
- Demonstrated ability to monitor motor start and stop times
  - Coil Position #1: motor fins, Coil Position #2: motor end plate
  - Change in coil position affects stray flux output

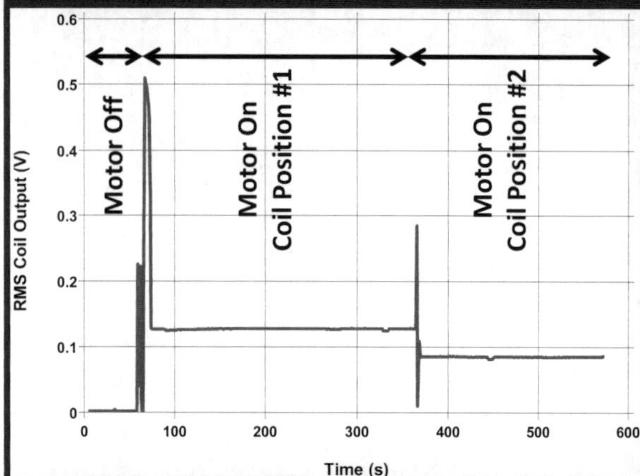

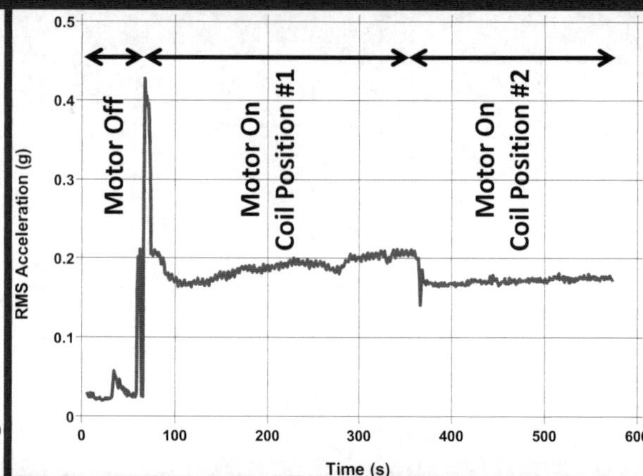

21

Prototype System Development – On site

- Plant monitoring system deployed at USAF Arnold Engineering Development Complex
  - Determine applicability of non-intrusive load estimation methodology on large-scale AEDC machines
- Demonstrated ability to monitor motor start and stop times
  - Coil output trends well with motor current draw
  - Coil output RMS level varies with machine; consistent A/D gain for all tests

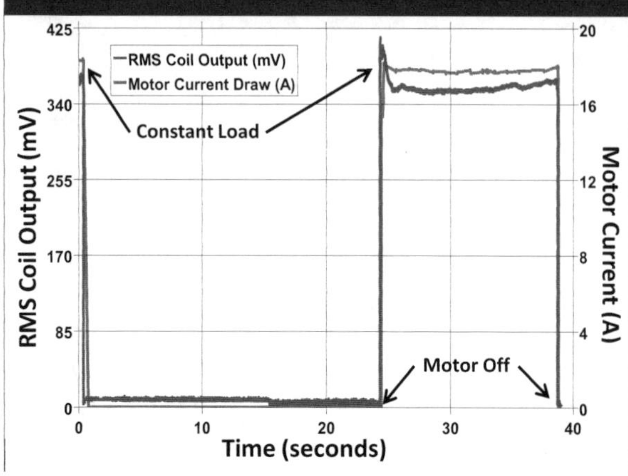

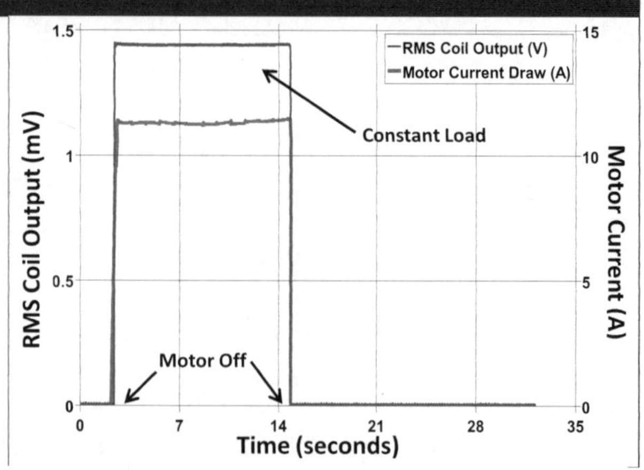

Conclusions

- Testing demonstrated the ability to non-intrusively record motor start/stop times and determine percent motor load better than 10% full scale accuracy
- Induction coil provides robust means of non-intrusively determining motor load
- Prototype system was developed to survey large-scale AEDC machines
  - Deployable sensor node successfully tested at AEDC prior to delivery
  - Deployable sensor node performs embedded signal processing and records data to on-board flash memory
  - Autonomous measurement via tethered transducers: stray flux, acceleration, and temperature

23

Contact Information

- **Jake Siegel**, Research Engineer
  siegelj@lunainc.com

- **Nathan K. Brown**, Mechanical Team Lead
  nbrown@lunainc.com

24

ADAPTIVE TUNABLE Q-FACTOR WAVELET TRANSFORM FOR FAULT FEATURE EXTRACTION OF GEARBOX BASED ON VIBRATION SIGNALS

Wangpeng He, Yanyang Zi, Zhiguo Wan, Shuilong He and Zhengjia He
State Key Laboratory for Manufacturing Systems Engineering, Xi'an Jiaotong University
Xianning West Road, Xi'an, Shaanxi, 710049, P.R. China
No.28
Telephone: (86) 15829301428
hewangpeng2007@stu.xjtu.edu.cn

Abstract: Fault diagnosis of gearbox is of great importance to avoid catastrophic accidents. Feature extraction has always been a key problem for fault diagnosis. In this paper, a novel fault feature extraction method called adaptive tunable Q-factor wavelet transform for gearbox fault diagnosis is proposed. The proposed adaptive method is implemented using the tunable Q-factor wavelet transform (TQWT). Kurtosis as an effective index of impulses is adopted to choose the optimal TQWT basis. The new method can obtain the optimal Q-factor according to the maximum of kurtosis. Thus, the Q-factor of the TQWT can match the oscillatory behavior of signals optimally without artificially specified. The interested fault feature is extracted by the single branch reconstruction of optimal subband. The proposed method is applied to vibration signals analysis of a bevel gear with a scratch defect from an antenna transmission chain and a gearbox from an electric locomotive. The processed results demonstrate that the proposed method can extract weak fault features of gearbox efficiently.

Key words: Fault diagnosis; gearbox; tunable Q-factor wavelet transform; vibration signals

Introduction:

Gearboxes are widely used in industrial applications to transmit speed and torque. An unexpected failure of the gearbox may cause significant economic losses or even result in catastrophic personal casualties. Therefore, successful and efficient detection of early faults developed in gearboxes is of significant importance.

The widespread application of gearboxes in industry requires advanced technologies to efficiently and effectively monitor their health status. Typical faults of gears include pitting, chipping, and more seriously, crack [1]. Vibration signal analysis has been the most popular approach to the fault diagnosis of rotating machinery [2,3]. The modern development of vibration signal processing techniques pays significant attention in enhancing their capacity in analyzing the non-stationary and transient nature of the measured vibration data. When a localized defect is induced on the gearbox, periodic impulses with the rotation frequency of the defected gear will be generated in vibration signals [4]. The gear fault signals are varied in oscillatory nature and transient characteristics. Moreover, these useful fault features are often immersed in heavy background noise. Thus, it is necessary to develop an effective feature extraction technique for gearboxes fault diagnosis based on signals measured from vibration sensors.

The last two decades have seen the rapid advancement of many signal processing methods such as Wigner distribution [5], short-time Fourier transform [6], cyclostationary analysis [7], empirical mode decomposition (EMD) [8], spectral kurtosis (SK) [9-11], and wavelet transform [12]. Wavelet transform (WT), evolving from Fourier transform, has made great progress in the theory and applications of machinery fault diagnosis [13-15]. A remarkable advantage of WT is its multi-resolution analysis capability, which is extremely suitable for detecting transient vibration signatures. WT converts a signal in time domain into time-frequency domain through a series of convolution operations between the analyzed signal and a wavelet basis. The essence of WT is a similar measurement between the signal and the wavelet basis functions. Thus, the meaningful fault feature extraction of WT is focused on the construction and selection of a proper wavelet basis. Among the varieties of wavelet analysis, DWT has been widely applied in mechanical fault diagnosis. However, the effectiveness of DWT is weakened by a few inevitable limitations [16]. First, DWT is sensitive to translation of the input signal, which causes waveform distortion in extracting impulse features. Second, DWT has fixed dyadic frequency domain partition manner, which may separate useful features into adjacent subbands. Third, wavelet functions of DWT are of low oscillation, which weakens its ability in matching high oscillatory vibration features.

Ideally, the Q-factor (defined as the ratio of its center frequency to its bandwidth) of a wavelet should be chosen to match the oscillatory behavior of the signal under analysis. Specifically, the WT with a low Q-factor is suitable for processing piecewise smooth signals, such as a scan line of an image. On the other hand, a relatively high Q-factor is useful for processing oscillatory signals, such as audio or physical vibration measurements [17]. Nowadays, conventional wavelet transform is commonly known as constant-Q transform. Therefore, it has the difficulty in tuning Q-factor to match the oscillatory behavior of the signal under analysis.

Motivated by the need for adjusting Q-factor of a wavelet according to the oscillatory behavior of signals, Selesnick recently developed a wavelet transform, denoted as the tunable Q-factor wavelet transform (TQWT), for which the Q-factor is easily tunable [18,19]. The flexible ability of the TQWT in tuning the Q-factor makes it suitable for revealing fault-related information from non-stationary vibration signals sampled on machines [20-22]. However, the Q-factor needs to be specified artificially, which means that the Q-factor cannot be tuned adaptively according to the oscillatory behavior of the signal under analysis.

To overcome the drawbacks, an adaptive TQWT-based feature extraction technique for gearbox is proposed in this paper. The proposed adaptive method is based on the maximization of the kurtosis values computed from the multiple wavelet subband signals generated by this TQWT basis. Here, the kurtosis is used as an indicator of selecting the best filter bank. The useful fault features can be extracted adaptively for the purpose of fault diagnosis. Moreover, the proposed method based on TQWT is applied to incipient fault diagnosis of gearbox in engineering applications, and the satisfactory diagnosis results validate its effectiveness and feasibility. Due to its operation simplicity in extracting fault features from vibration signals, the proposed TQWT-based feature extraction technique has potential to become an effective tool in fault diagnosis of gearbox.

The rest of the paper is organized as follows. The summary of TQWT is introduced in Section 2. In Section 3, the proposed adaptive feature extraction technique is presented. In Section 4 applies the proposed method to fault diagnosis of gearboxes in an antenna transmission chain and an electric locomotive for further validation of its effectiveness. Finally, conclusions are summarized in Section 5.

2. Summary of the TQWT theory:

Recently, a wavelet transform, namely the TQWT, has been developed for discrete-time signals for which the Q-factor is easily and continuously adjustable [18]. The TQWT is implemented using perfect reconstruction over-sampled filter banks with real-valued sampling factors. The associated analysis and synthesis filter banks for the TQWT is shown in Fig. 1 (where LPS and HPS represent low-pass scaling and high-pass scaling, respectively).

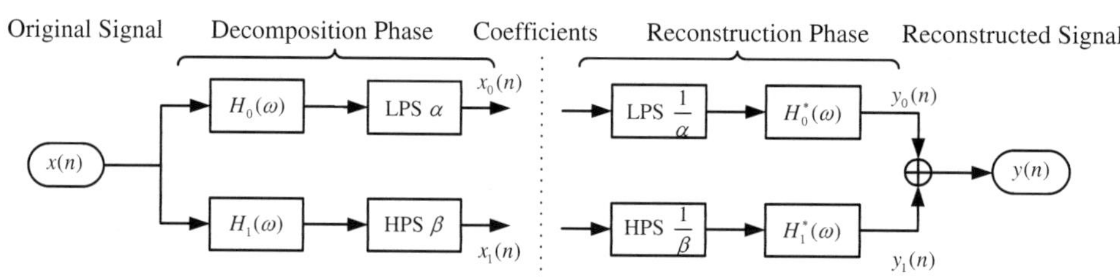

Figure 1: Block diagram of analysis and synthesis filter banks for the implementation of the TQWT.

For the purpose of perfect reconstruction, the frequency responses $H_i(w)$, $i = 0,1$ must be chosen so that the reconstruction signal $y(n)$ equals the input signal $x(n)$. Specifically, the frequency responses of $H_0(w)$ and $H_1(w)$ are defined as

$$H_0(w) = \begin{cases} 1, & |w| \leqslant (1-b)p \\ q\left(\dfrac{w + (b-1)p}{a+b-1}\right), & (1-b)p \leqslant |w| < ap \\ 0, & ap \leqslant |w| \leqslant p \end{cases} \quad (1)$$

$$H_1(w) = \begin{cases} 0, & |w| \le (1-b)\pi \\ q\left(\dfrac{a\pi - w}{a+b-1}\right), & (1-b)\pi \le |w| < a\pi \\ 1, & a\pi \le |w| \le \pi \end{cases} \quad (2)$$

where

$$0 < a < 1,\ 0 < b\ £\ 1,\ a + b > 1,$$

$$q(w) = \frac{1}{2}(1 + \cos w)\sqrt{2 - \cos w} \quad \text{for } |w|\ £\ \pi . \tag{3}$$

The transition function $q(w)$, originating from the Daubechies filter with two vanishing moments, is used to construct the transition bands of $H_0(w)$ and $H_1(w)$. The variables ¶ and b are the LPS parameters and the HPS parameter, respectively. The filter bank parameters ¶ and b should be properly chosen so as to achieve a wavelet transform with the desired Q-factor and oversampling *r*. Here, the relationship between (¶ , b) and (Q, r) can be expressed as

$$b = \frac{2}{Q + 1}, \qquad ¶ = 1 - \frac{b}{r}. \tag{4}$$

The TQWT is specified by three parameters: the Q-factor *Q*, the redundancy *r*, and the number of stages (or decomposition levels) *J*. Generally, *Q* is a measure of the number of oscillations the wavelet exhibits. According to the definition of the Q-factor, the bandwidth varies inversely to the Q-factor for a given center frequency. Therefore, a higher Q-factor has a better frequency resolution in comparison with a lower one. For Q, a value of 1.0 or greater can be specified. The parameter *r* is the redundancy of the TQWT when it is computed using infinitely many levels. The specified value of *r* must be greater than 1, and a value of 3 or greater is generally recommended. Therefore, the value of *r* is specified to be 3.0 in this research. The parameter *J* denotes the number of filter banks. There will be *J*+1 subbands: the high-pass filter output signal of each filter bank, and the low-pass filter output signal of the final filter bank.

Fig. 2 illustrates the TQWT for two different sets of TQWT parameters. By increasing the value of Q, the wavelet function becomes more oscillatory, as shown in Fig. 2. Obviously, it can be observed that for Q=1.0 the wavelet consists of few oscillatory cycles compared with the wavelet for Q=4.0. Meanwhile, the frequency responses of the band-pass filters constituting the TQWT can be easily observed in Fig. 2. For a low Q-factor, the band-pass filters are quite wide. In contrast, for a high Q-factor, the filters are narrower. Therefore, a high Q-factor wavelet transform requires more levels to cover the same frequency range as a low Q-factor transform. Moreover, the TQWT can be implemented efficiently using radix-2 FFTs.

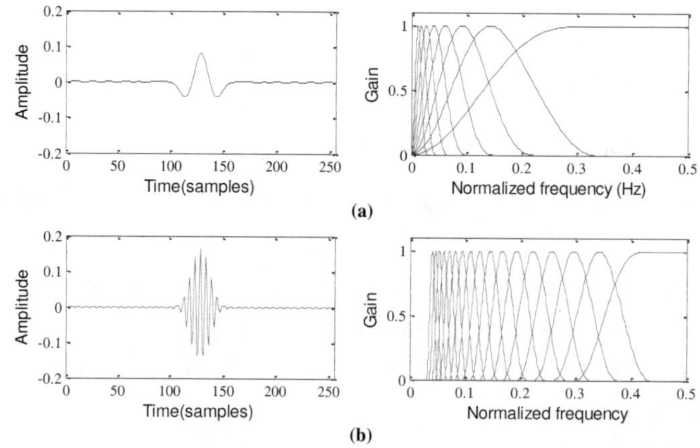

Figure 2: TQWT wavelets and frequency responses (at level 6) for two Q-factors: (a) Q = 1, r = 3, J = 8; and (b) Q = 4, r = 3, J =17.

3. Adaptive feature extraction method based on TQWT:

In this section, an adaptive TQWT for fault feature extraction is proposed. The procedure of the proposed method is illustrated in Fig. 3. The proposed method is composed of two major steps:

(1) Selection of an optimal Q-factor of the TQWT;
(2) Fault feature extraction by using the selected optimal TQWT basis.

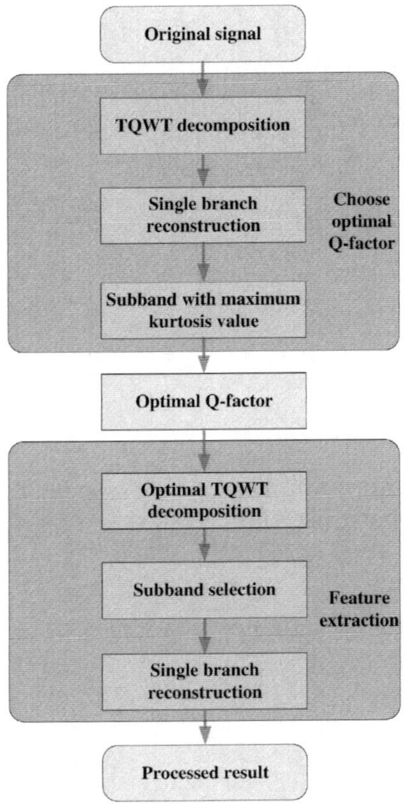

Figure 3: The procedure of the proposed adaptive feature extraction method based on the TQWT.

Note that a value of 4.0 or greater of the Q-factor is large enough for the precise division of frequency band of mechanical vibration signals. Therefore, the lower and upper bounds of the Q-factor are set to be 1.0 and 5.0 respectively in this article. The detailed descriptions of these steps are presented in the following subsection.

3.1 Selection of an optimal Q-factor of the TQWT:

As mentioned earlier in section 1, the similarity between the wavelet basis and the fault features plays an important role in its effectiveness. Therefore, an appropriate wavelet is particularly important for feature extraction in machinery fault diagnosis. In order to address this problem, an adaptive method to obtain a TQWT base related to fault feature is proposed. This optimal TQWT basis selection method is based on the maximum of kurtosis values, which is defined as

$$Kurt(x) = \frac{E(x - m)^4}{s^4} \tag{5}$$

where μ is the mean of x, σ is the standard deviation of x, and $E(t)$ represents the expected value of the quantity t. This equation is the operator that calculates the kurtosis value of the input time series x. The wavelet subband with the greatest kurtosis value is defined as the ''sensitive wavelet subband''

Generally, the steps for the selection of an optimal Q are as following:

(1) Decompose the input vibration signal by using a TQWT dictionary, which is constructed by a set of TQWT bases with different Q-factors. It should be noted that Q is varied from 1.0 to 5.0 with a step of 0.5;
(2) For a certain Q, reconstruct the signal with single subband and subsequently calculate the kurtosis values Kurt(j), (j=1, 2...J+1). Similarly, this process is utilized to other Q-factors bases;
(3) Select the optimal Q-factor based on maximization of kurtosis values among candidate TQWT bases.

3.2 Adaptive fault feature extraction:

After the selection of Q-factor is completed, we get the optimized TQWT basis determined by parameters (Q, r, J), then the fault feature can be extracted by the following steps:

(1) Decompose the input vibration signal by using the optimal TQWT basis;
(2) Calculate the kurtosis values and select the optimal subband;
(3) The selected subband is reconstructed to reveal the fault features hidden in the original signal.

4. Engineering application:

In this section, the proposed method is applied to process vibration signals of a bevel gear with a scratch defect from an antenna transmission chain and a gearbox from an electric locomotive.

4.1 Fault feature extraction of a bevel gear:

Gearboxes play an important role in the transmission train of a satellite communication antenna or a telemetry, tracking, and command antenna for aerospace industry. A satellite communication antenna consists of three axes: the azimuth axis, the pitching axis and the crossing axis. Each transmission chain consists of an induction motor, a planetary gearbox and a two-stage gearbox.

The fault detection and diagnosis on gearboxes are gaining importance to prevent malfunction and failure. However, the vibration signals of the antenna transmission are complicated and are submerged in heavy background noise due to its complexity of the structure and the severe working conditions. Therefore, it is necessary to extract the hidden useful fault features through advanced signal processing methods. Hence, this section presents an application of a bevel gearbox fault diagnosis to validate the effectiveness of the proposed method.

A bevel gear with a scratch defect is displayed in Fig. 4. The failure bull gear was mounted on the pitching axis. Accelerometers for vibration detection were installed on the output shaft housing of the planetary gearbox. The transmission ratio of the planetary gearbox is 32. The vibration signal was collected using a YE6267 data acquisition system. The sampling frequency is 5 kHz and each record of signal is of length 32768. The expected characteristic frequencies of the gearbox is listed in Table 1.

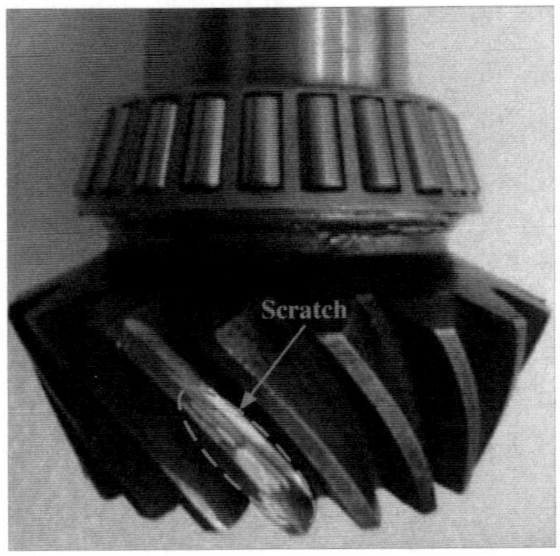

Figure 4: The tested bevel gear with a scratch defect.

Table 1: The expected characteristic frequencies of the gearbox.

Rotating speed of motor [r/m]	20
Rotating frequency of input shaft of the gearbox [Hz]	0.625
Meshing frequency of the last stage bull gear shaft [Hz]	5.9375

The time domain waveform and the Fourier spectrum of a record vibration signal are shown in Fig. 5. It can be noticed from the figure that the characteristic frequency is difficult to be distinguished from both the time domain waveform and its spectrum due to the heavy background noise.

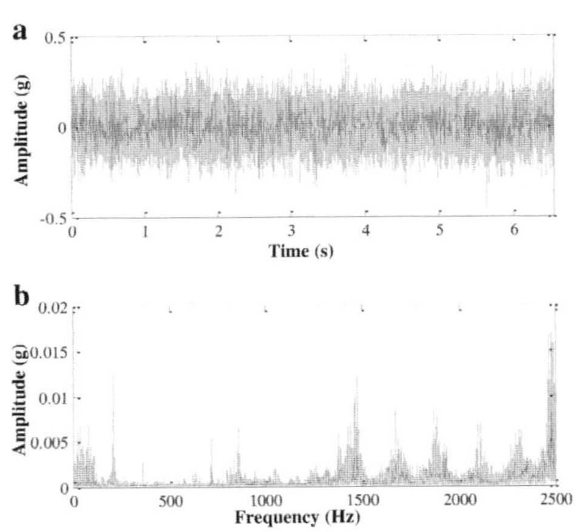

Figure 5: Measured signal and its Fourier spectrum (a) time domain wave and (b) Fourier spectrum.

To further investigate the useful fault features hidden in the vibration signal, the proposed adaptive method is applied. Reconstruction signal of selected optimal subband is plotted in Fig. 6. Fig. 6 shows that there are repeated strong impulses with the periodic interval of 0.178 s (5.628 Hz), which is much closed to the meshing frequency of the last stage bull gear shaft. In addition, there are several big envelopes with the occurrence period of 1.659 s (0.603 Hz), as shown in Fig. 6. This occurrence rate is much closed to the rotating frequency of input shaft of the gearbox. Thus, the fault feature of bevel gear with a scratch defect is successfully extracted.

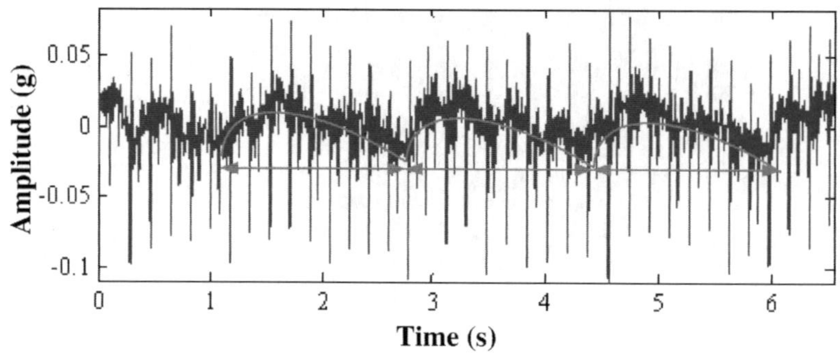

Figure 6: The selected optimal subband by the proposed adaptive feature extraction method.

4.2 Fault feature extraction of an electric locomotive gearbox:

In this engineering application, the proposed method is employed to vibration signal analysis of an electric locomotive to extract hidden periodical impulses. The gearbox of electric locomotive is a single-stage transmission and accelerometers were mounted on the outer case of the gearbox. The structural parameters of the gearbox and some expected characteristic frequencies are listed in Table 2. The signals were sampled at 12.8 kHz.

Table 2: The expected characteristic frequencies of transmission system of the gearbox.

Number of teeth [Z2/Z1]	87/20
Rotating freq. of low speed shaft [Hz]	4.47
Rotating freq. of high speed shaft [Hz]	19.44

A record of signal and its Fourier spectrum are shown in Fig. 7. The length of the signal is 16384. It is evident that the signal is contaminated by a large amount of background noises, and no meaningful fault symptoms could be found. The frequency components are abundant and evident sidebands relevant to gearbox faults can hardly be identified in the spectrum.

For the purpose of fault diagnosis, the proposed method is adopted to analyze the signal. Reconstruction signal of selected optimal subband is shown in Fig. 8. Repeated impulses are clearly observed in the result. The averaged spacing of the impulses is 0.2003 s (4.992 Hz), much closed to the rotating frequency of the bull gear. Based on the above information, the bull gear was suspected of suffering tooth damage.

After being disconnected from the testing platform, an incipient tooth crack was found on the bull gear, as shown in Fig. 9. The gear fault coincides with the diagnostic conclusion of the proposed method. Hence, the analyzed results show that the proposed method could successfully identify the root crack.

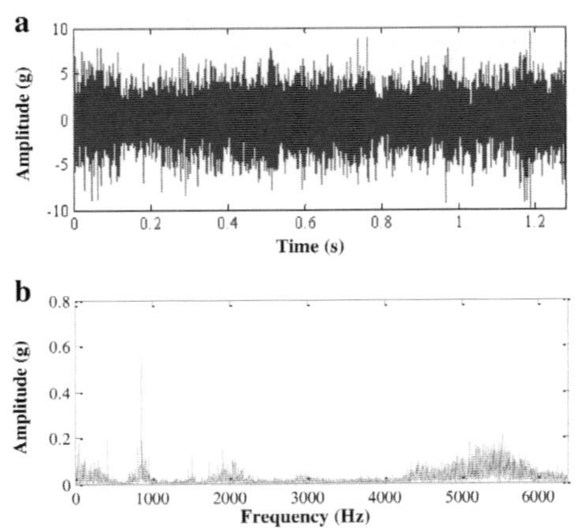

Figure 7: Measured signal and its Fourier spectrum (a) time domain wave and (b) Fourier spectrum.

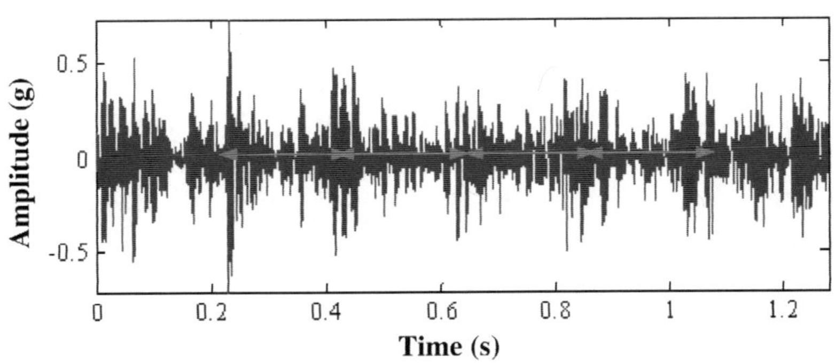

Figure 8: The selected optimal subband by the proposed adaptive feature extraction method.

Figure 9: The crack fault of the bull gear.

4.3 Discussion on limitations of the proposed method:

This paper has proposed an adaptive feature extraction technique for gearbox fault diagnosis. The effectiveness of the proposed method is verified by the engineering applications. However, some problems still exist in the proposed method.

(1) The choosing of Q-factor still needs to be optimized. In this study, just several Q-factors are available to match the oscillatory behavior of the signal under analysis. In fact, some optimization algorithms should be used to adapt the Q-factor more effectively and efficiently.

(2) The proposed technique requires more computation burden compared with the original TQWT. However, for the TQWT can be efficiently implemented using radix-2 FFTs, the total computational cost is still low and acceptable for engineering applications.

(3) The proposed method is more suitable for vibration data measured from rotating machinery with constant or stable rotation speed. To solve this problem, the vibration signals should be collected under the condition that the rotation speed of the machinery is roughly constant or at least varies slowly.

5. Conclusion:

In this paper, a new adaptive fault feature extraction technique for gearboxes is proposed based on TQWT. The proposed method can select suitable Q-factor to match the oscillatory behavior of the analyzed signal according to the useful fault features. The adaptive criterion is based on the maximization of the kurtosis values. The effectiveness of the proposed method has been verified by two engineering cases. The processing results show that the proposed method can successfully extract useful features in fault diagnosis of gearboxes. Future work would focus on applying this proposed method to bearing fault diagnosis.

Acknowledgments

This research is supported financially by the National Natural Science Foundation of China (Grant No. 51275384), the Key Project of National Natural Science Foundation of China (Grant No. 51035007), the Important National Science and Technology Specific Projects (Grant No. 2010ZX04014-016) and the National Basic Research Program of China ("973" Program) (Grant No. 2009CB724405).

References:

[1] X. Fan, M.J. Zuo, Gearbox fault detection using Hilbert and wavelet packet transform, Mech. Syst. Sig. Process. 20 (4) (2006) 966-982.
[2] W.J. Wang, P.D. McFadden, Application of wavelets to gearbox vibration signals for fault detection, J. Sound Vib. 192 (5) (1996) 927–939.
[3] R. Yan, R.X. Gao, Approximate entropy as a diagnostic tool for machine health monitoring, Mech. Syst. Sig. Process. 21 (2) (2007) 824-839.
[4] B. Liu, S. Riemenschneider, Y. Xu, Gearbox fault diagnosis using empirical mode decomposition and Hilbert spectrum, Mech. Syst. Sig. Process. 20 (3) (2006) 718-734.
[5] N. Baydar, A. Ball, A comparative study of acoustic and vibration signals in detection of gear failures using Wigner–Ville distribution, Mech. Syst. Signal Process. 15 (6) (20 01) 1091– 1107.
[6] H. Xie, J. Lin, Y.G. Lei, et al., Fast-varying AM–FM components extraction based on an adaptive STFT, Digital Signal Process. 22 (4) (2012) 664–670.
[7] J. Antoni, Cyclostationarity by examples, Mech. Syst. Signal Process. 23 (4) (2009) 987–1036.
[8] N.E. Huang, Z. Shen, S.R. Long, The empirical mode decomposition and Hilbert spectrum for nonlinear and non-stationary time series analysis, Proc.R. Soc. London Ser. A 454 (1998) 903–995.
[9] J. Antoni, The spectral kurtosis: a useful tool for characterizing non-stationary signals, Mech. Syst. Signal Process., 20 (2) (2006) 282–307.
[10] J. Antoni, The spectral kurtosis: Application to the vibratory surveillance and diagnostics of rotating machines, Mech. Syst. Signal Process., 20 (2) (2006) 308–331.
[11] J. Antoni, Fast computation of the kurtogram for the detection of transient faults, Mech. Syst. Sig. Process. 21 (1) (2007) 108–124.
[12] Lin J, Zuo M J, Gearbox fault diagnosis using adaptive wavelet filter, Mech. Syst. Sig. Process. 17 (6) (2003) 1259-1269.
[13] R. Yan, R. X. Gao, Harmonic wavelet-based data filtering for enhanced machine defect identification, J. Sound Vib. 329 (15) (2010) 3203–3217.
[14] R. Yan, R. X. Gao, Energy-based feature extraction for defect diagnosis in rotary machines, IEEE Trans. Instrum. Meas. 58 (9) (2009) 3130–3139.
[15] Z. Peng, F. Chu, Application of the wavelet transform in machine condition monitoring and fault diagnostics: A review with bibliography, Mech. Syst. Signal Process., 18 (2) (2004) 199–221.

[16] B. Chen, Z. Zhang, C. Sun, B. Li, Y. Zi, Z. He, Fault feature extraction of gearbox by using overcomplete rational dilation discrete wavelet transform on signals measured from vibration sensors, Mech. Syst. Signal Process. 33 (2012) 275– 298.

[17] I. Bayram, I. W. Selesnick, Frequency-domain design of overcomplete rational-dilation wavelet transforms, IEEE Trans. Signal Process. 57 (8) (2009) 2957–2972.

[18] I. W. Selesnick, Wavelet transform with tunable Q-factor, IEEE Trans. Signal Process. 59 (8) (2011) 3560–3575.

[19] I. W. Selesnick, Sparse signal representations using the tunable Q-factor wavelet transform, Proc. SPIE. 81381U-81381U-15 (2011).

[20] G. Cai, X. Chen, Z. He, Sparsity-enabled signal decomposition using tunable Q-factor wavelet transform for fault feature extraction of gearbox, Mech. Syst. Signal Process. (2013).

[21] W. He, Y. Zi, B. Chen, S. Wang, Z. He, Tunable Q-factor wavelet transform denoising with neighboring coefficients and its application to rotating machinery fault diagnosis, Sci. China Tech. Sci. 56 (8) (2013) 1956–1965.

[22] J. Luo, J. Yu, M. Ming, A kurtosis-guided adaptive demodulation technique for bearing fault detection based on tunable-Q wavelet transform, Meas. Sci. Technol. 24 (5) (2013).

INTRINSIC METHOD FOR MONITORING INDUSTRIAL APPLICATION FOR VIRTUAL DIAGNOSTICS

Markus Schütz, Marc Hilbert, Domenic Boos, Christian Bernet, Ralph Baltes and Karl Nienhaus
Institute for Mining and Metallurgical Machinery (IMR), RWTH Aachen University, Wüllnerstraße 2, 52056 Aachen, Germany
Telephone: +49 241-8096878
mschuetz@imr.rwth-aachen.de

Claudia Sequeira
Universidade do Algarve, Campus da Penha, 8005-139 Faro, Portugal
Telephone: +351 289-800100
cdsequeira@ualg.pt

Abstract: The Internet of Things (IoT) refers to a network and virtual representations of individual objects (e.g. machine part) in an internet-like structure. This representation enables the comparison of the behavior of the real world object to a virtual counterpart. On the virtual counterpart the reactions of loads or an estimation of remaining life time among other aspects can be simulated. To realize those industrial visions each object's needs to have cheap and reliable systems that monitor and record the object conditions during the full lifetime. Only with these data a mapping of the virtual object will be possible and, therefore, an estimation of the remaining life can be provided.

This paper will describe a method that covers these challenges on two sample applications. 1. Condition monitoring of a wind turbine epicyclic gear, especially the planet meshing and the planet bearing using a measurement system mounted on the planet carrier. 2. Condition monitoring of a single centered chain strand built in an armored face conveyor, by use of a special measurement chain link.

Key words: Diagnostics; Prognostics; Condition monitoring; Internet of Things; Epicyclic gears; Wind turbine; Armored face conveyor; Measurement system; Chain link;

Introduction: The Internet of things (IoT) refers to a network of intelligent nodes. Each node is able to sense its condition and information which might influence that node. The term IoT originally comes from a vision of global infrastructure and was inspired by the success of RFID. The first applications of IoT were done by tracking objects in warehouses in order to allow an optimized logistic management. [1]

In the last years the development of smart objects capable of sensing information, pre-processing them and the communication within a large network increased rapidly. The focus was mainly on collecting information and optimizing the operation to achieve a more efficient process management.

Industry 4.0 is a term of the German government which was introduced in 2012. It represents a new form of industrialization, integrating IoT in factories as basic technology to improve methods of self-optimization, self-configuration and self-diagnosis of increasingly complex systems. [2]

This paper discusses the extension of the IoT towards industrial applications for condition monitoring (CM). CM is defined as a process of detecting material or geometric changes of a structural system including changes of the process conditions [3]. State of the art is to collect information for CM with sensors mounted as close as possible to the structural system and process the sensor information in a central node using fault diagnosis and identification (FDI) technology like plausibility test, software/analytics redundancy schemes and signal processing [4]. This method limits the amount of processed information in two ways: First, in most applications it is rather sophisticated to place a single sensor close to the location where the structural change of the system occurs and at the same time connecting it to a node for information processing. Second, the information transferred from the measurement point to the node for information processing is limited by the transmission capacity (e.g. data rate).

To overcome such limitations the authors believe that intelligent nodes that are able to preprocess measured information and, therefore, are able to be more loosely connected to a central node, allow it to place the sensor close to the assumed structural change and, thereby, collect more process information. Being able to preprocess information and not be limited to the available rate of information that can be transferred to the processing node, it is possible to collect more information to gain a comprehensive idea about the structural condition of the object of interest. In the future, this will also make it possible to let intelligent measurement nodes communicate with each other directly (and not over a central node) and increase the information content of an overall system. To realize those industrial visions each structured object needs to have cheap and reliable systems capable to measure and preprocess the information during the full lifetime.

Using preprocessed information of measured nodes a virtual object based on a model-based approach can be designed and used for the prediction of the condition of its real counterpart. The preprocessed measurement information will give a data mapping of the current state of the structured object under the current and past conditions. Having such a virtual object allows predicting the future condition using a variety of possible boundary conditions without interrupting the utilizations. Furthermore, the comparison of the behavior of the virtual object and the real object allows model-based FDI technologies to monitor the current condition of the real object. A classic example of condition prediction is predicting the remaining life time and will be given in this paper.

Measurement System: A broad range of online measuring systems for CM is available on the market, e.g. [5]. For most common applications (e.g. measuring force, torque, vibration) these systems are sufficient. But for special applications, such as measurements on rotating high speed shafts, under harsh environmental conditions or in small machine components, conventional systems reach their limits. This is where the research project i-MaSS (integrated Machine Sensor System) offers a solution. By an interdisciplinary consortium consisting of mechanical, electrical and mineral resources engineers from industry and university a new measurement system, especially for these applications, is currently developed.

Figure 1: Design of the developed measurement system.

A first state of the hardware, shown in figure 1, is able to connect to up to four sensors (e.g. strain gauge, accelerometer, temperature sensor, etc.). The required energy is provided by a lithium-ion battery. To reduce the energy consumption an intelligent energy management system is integrated. Depending on the measurement setup the power supply of the sensor and parts of the measurement system are continuous or pulsed. In times of shutdown a sleep mode can be activated automatically. Therefore, energy can be saved and the period of application can be extended. Data preprocessing and transmission is conducted on the main circuit board. In case strain gauges are used, the data preprocessing also includes a conversion of the electrical raw signal into mechanical component load or stress. This is achievable in the case of known or unknown main direction of tension. If required, a real time classification of load time functions is possible. Therefor the Markov algorithm [6] is implemented (two dimensional counting method). This "smart" feature is especially for long duration measurements and can also be used for remaining life time estimation. Furthermore, energy consumption during data transmission sinks because of the small memory size of a Markov matrix compared to load time signal. On the secondary circuit board a micro secure digital (SD) card is installed. All measurement data, load time function or matrix, can be stored. The data transfer to the PC is realized via wireless low power protocol.

The use of this measuring system is supposed to be applied under harsh environmental conditions, like underground mining. Hence, the whole system is designed with respect to the ATEX (Athmosphères Explosibles) directive [7].

Application Epicyclical Gears [12]**:** Present and future wind turbines are equipped with gearboxes which have to fulfill highest requirements. The requirement is to transmit variable powers under extreme environmental conditions while at the same time providing a life span up to 20 years. This can only be met with the newest technologies. For reliable design and operation of wind turbines, an explicit knowledge of gearboxes is necessary. Vibration measurement on gearboxes is an important tool for the verification of design and early fault detection. Due to their high-power density, epicyclic gears are often chosen for gearboxes of wind turbines. Currently used typical gear types are designed based on the principle "one planetary gear stage and two spur gear stages" or "two planetary gear stages and one spur gear stage". Planetary stages in configuration with a fixed ring gear are used almost exclusively. As described in [8], gearing concepts being developed demand a mechanical power split because of an increasing torque at the gearbox input due to constantly increasing rotor diameters. This task can be solved by a parallel connection of two planetary stages or differential planetary stages. An example of a differential gear for wind turbines of the multi-megawatt classes is Bosch Rexroth´s REDULUS GPV-D.

Condition monitoring of gearboxes, especially the planet meshing and the planet bearing, is challenging. The changing position of the planet (by a fixed ring gear) to the fixed sensor (accelerometer) leads to a time-variant transfer function between the vibration source and the accelerometer. The measurement of vibrations at the gearbox housing or ring gear, which is state of the art, provides unsatisfying results. If the accelerometer is attached to the cage near the bearing (figure 2), the known diagnostic algorithms for bearing monitoring can be applied. To detect faults in the planet bearing as early as possible, the accelerometer has to be placed close to the acoustic source. This guarantees a sufficiently high signal-to-noise ratio between the bearing frequency or meshing frequency and noise. Furthermore, it guarantees the time-variant transfer function between vibration source and accelerometer.

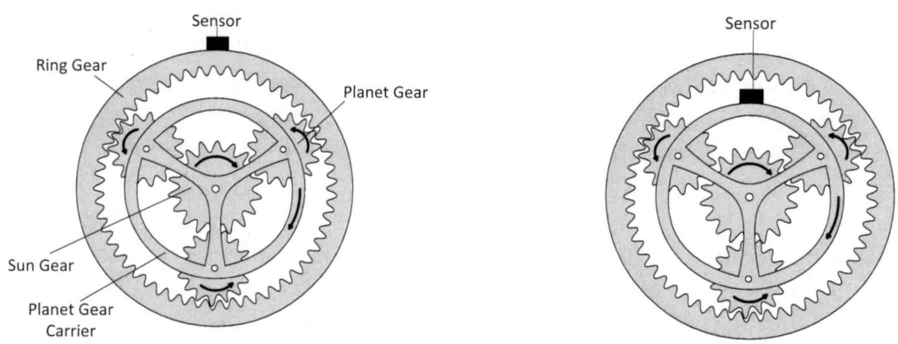

Figure 2: Design of an epicyclic gear with attachment of the sensor to the outside of the ring gear (left) and planet gear carrier (right). [12]

In the last decades, a number of articles were published describing the measurement of the condition of epicyclic gearboxes using vibration analysis with focus on the detection of damage on tooth flank and planet gear carrier, especially for epicyclic gearboxes on helicopters. There are only a few publications that deal with damaged bearings in epicyclic gearboxes. The state of the art detecting damages on rolling bearings of the planet gear can be found e.g.in [9][10].

The complex and time-variant transfer function is the main problem. Damage in a planet bearing produces several transfer paths of the signal to the accelerometer. Figure 3 depicts the different transfer paths which result in different transfer functions $G(t)$. The vibration excitation, which, for example, occurs on the inner ring of the planet bearing, is transferred (transfer function $G_1(t)$) to the accelerometer by the rolling elements of the bearing, the outer ring of the planet bearing, the planet gear, the mesh between planet gear and ring gear, the ring gear and the housing, respectively. Additionally, the signal transfer is possible using other transfer functions as shown in figure 3. For the transfer of the signal, the following factors are important [11]: the lubrication of the planet bearing and the planet mesh, the gap between planet bearing and planet mesh, the rigidity of the planet mesh and the gear structure and the position of the planet gear respectively to the sensor.

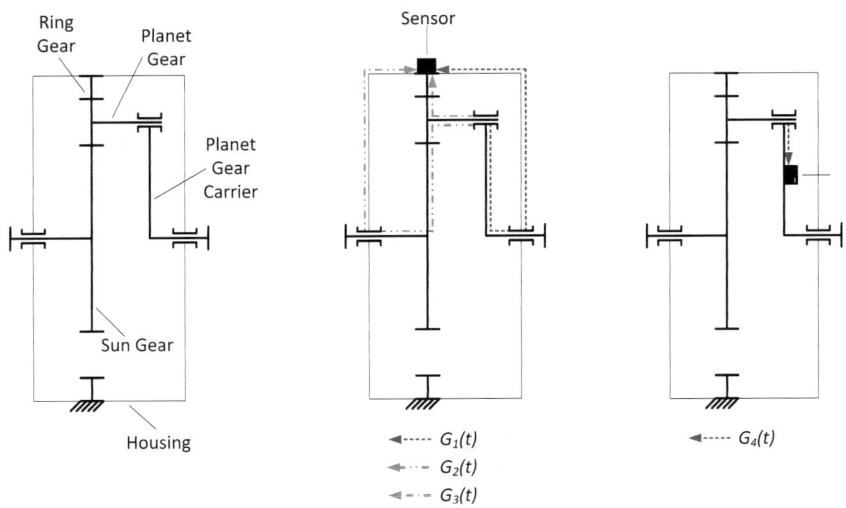

Figure 3: Schematic setup of an epicyclic gearbox with fixed ring gear (left). Different transfer functions *G(t)* of the vibration signal of a damaged rolling bearing with the mounting of the accelerometer on the exterior of the ring gear (center) and the planet gear carrier (right). [12]

This leads to a substantial time-varying attenuation of the vibration signal. Despite extensive analysis, damage detection of rolling bearings of planet gearboxes with accelerometers on the planet gear housing is very difficult.

The goal of the presented method is to measure vibrations inside the gearbox with a standard acceleration sensor and transmit the data to a receiver wireless. Because the sensor and the transducer are installed on the rotating machine part, they can be placed close to the vibration source. Measuring the vibration directly on the planet gear carrier, as presented in this paper, enables the use of conventional analysis methods to monitor the bearings of the planets. The mentioned interfering factors must no longer be considered and a defined time-invariant transfer function ($G_4(t)$) between vibration source and accelerometer is guaranteed (which is less damped than all other transfer functions).

The installation of the sensor system inside the gearbox is challenging regarding the wireless data connection. Because of the properties of the epicyclic gear, there is only visual contact between sensor system and data receiver for short periods of time in irregular time intervals. The visual contact is also disturbed by lubrication which can contain ferromagnetic particles and, therefore, can disturb the wireless connection. It could be shown with a connection test in a 2.5 MW wind turbine gearbox (with two epicyclic gear stages and one spur gear stage) that a wireless connection between a sensor system, installed on a planet gear carrier, and a receiver, installed on inspection opening, is possible. The sensor system was installed on the planet gear carrier with magnets and the with lubrication filled gearbox. During the successful data transmission the planet gear carrier was rotating so that it was immersed in lubrication for part of the time.

Application to armored face conveyor: Armored face conveyors (AFC) are deployed in longwall mining operations. It is placed between the powered roof support and the coalface. A shearer loader located above cuts coal out of the coal seam and loads the AFC. Due to scraper chains the coal is removed from the coal face to the main gate. Such systems can be up to 470 m long and transport up to 5000 t coal and dead waste rock per hour. The following picture shows a simplified computer-aided design (CAD) model of a 20 m long experimental conveyor. [13]

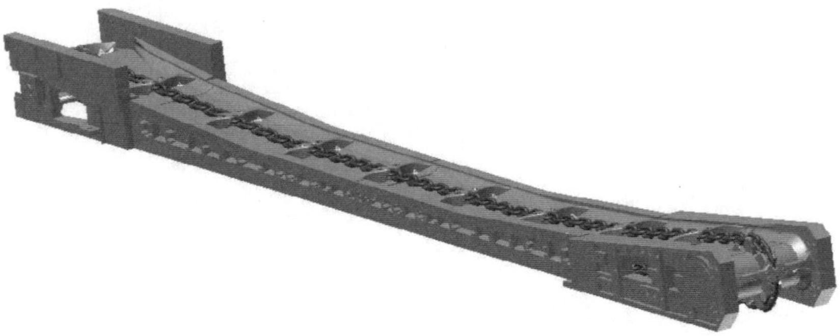

Figure 4: CAD model of an AFC.

The transport route from the coal face to the main gate is not redundant. If the AFC has a breakdown the whole coal production in the affected longwall comes to a shutdown. Chain and scraper are the major components in the transmission system. They are highly stressed and vulnerable for failure. Typically malfunctions are chain break, jumped out scraper and increased chain wear and fatigue due to wrong adjusted preload force [14]. For preventing these scenarios an intelligent monitoring system is necessary. Furthermore, the safety at work will be increased, because unplanned repair works of AFC are associated with safety hazards for the miners. Several conventional measurement systems for AFC have been developed for the last 30 years. They are usually big, not consumption optimized, not included to chain or scraper and can only be used for few production circles. Hence, there is still a brisk interest in finding an adequate solution. In Australia, one of the biggest mining countries in the world, projects for development of such systems have been supported for more than 10 years [15] [16].

The authors accomplished different preliminary studies concerning force measurement in chain links. One research project included a temporary force measuring method for round steel chain with a wire diameter > 30 mm. [17]. This approach to measure tensile force has been taken up in i-MaSS project, again. By means of strain gauges, specific produced chain links and our new developed measuring system a possibility is given to get operating data out of the production process. Figure 5 depicts a measurement chain link, left chucked in a testing machine, right as a CAD model. Centered in the chain link is a bore hole. Inside a jig is positioned, which includes the strain gauge, the measurement system and the battery.

Figure 5a: Measurement chain link with jig (Measurement system is for testing purposes located outside).

Figure 5b: CAD model with jig.

First real bench tests were executed at an experiment conveyor. For these experiments four measurement chain links were calibrated and included to the single center chain strand. The AFC has a head drive, so that the theoretical force distribution in tight strand and slack strand is as in figure 6a.

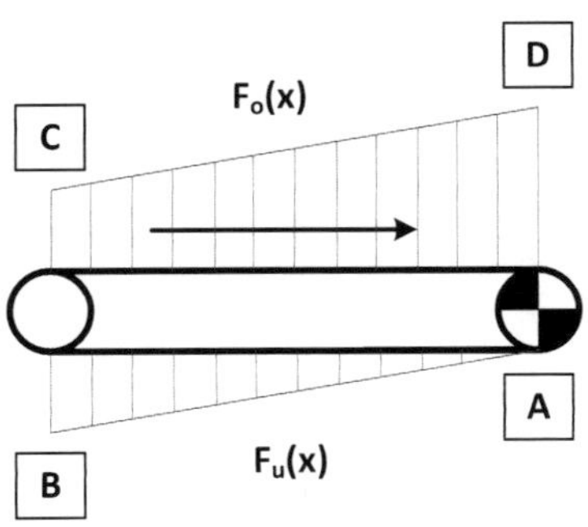

Figure 6a: Theoretical force distribution. [18]

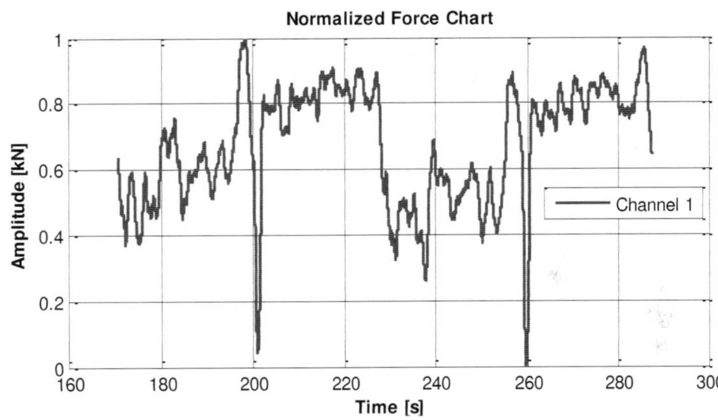

Figure 6b: Normalized measurements result of two circulations.

At point A, located in slack strand, the force is on the lowest level. On the way to Point B the force increases slowly, represented by formula $F_u(x)$. The redirection of force to the tight strand is done by a sprocket, located in point B and C. Afterwards the force increases continuously, represented by formula $F_o(x)$, until the next sprocket at the main drive is reached. Then, the force is redirected a second time, to point B and the procedure begins again. This theoretical force progression can also be seen in figure 6b, the measurement data of one chain link during two circulations in idle state of the AFC. At time 170 s the chain link is on point A in the slack strand, comparison to figure 6a. Next, the force increases slowly until the sprocket at the rear head is reached. A peak which is caused by the entering of the chain link into the sprocket can be seen at 200 s. After that, the measurement chain link is in the tight strand and the force increases continuously again. At time 230 s the chain link reaches point D, the sprocket at the main drive. Then a force drop occurs, which is caused by the redirection to point A.

The sampling rate of the force signal was 100 Hz. However, many undulations therein are present. These undulations are no noise, but result from the surface condition of pan chute and scraper. Due to the age of the experiment conveyor the surface is partially rusted. Therefore, the coefficient of friction has different values which lead to undulations of the chain strand during circulation.

Summary: The presented approach represents state-of-the-art of the intrinsic systems for condition monitoring. It provides a possible set-up for several applications, but in particular for wind energy turbines and armored face conveyors. The system provides a platform for further developments in condition monitoring and its integration in the IoT with the described features. The introduced system overcomes the two mentioned limitation of the common systems by: First, Preprocessing information at the measurement point. Second, providing additional information for a virtual representation of the machine component in the IoT. This enables to use that specific information for self-configuration and self-diagnosis for the overall system. For the wind turbine gearbox a vision could be to use the acquired information to estimate the remaining lifetime of on the virtual representation of the planet carrier under different simulated scenarios (e.g. of load, speed). The second example depicted an AFC which was used to monitor chain forces during operation over a defined period of time. With this information, a better and adaptive adjustment of preload force during operation can be realized in the future. This would increase the lifetime of the machine and improve the safety condition. Similar to the application wind turbine gearbox, this could be done using a virtual counterpart to predict the remaining lifetime of the AFC e.g. the most highly stressed component chain strand. By way of example, the appearance of chain breaks can be estimated more easily, so that maintenance works can be placed in planned maintenance shifts.

Since the development is public it will be possible to do research on the device and use it for developing methods and algorithms to integrate the system in the overall CM.

Acknowledgement: The i-MaSS project is funded by the "NRW-EU Ziel 2" program and the "Europäischen Fonds für regionale Entwicklung" (EFRE).

EUROPÄISCHE UNION
Investition in unsere Zukunft
Europäischer Fonds
für regionale Entwicklung

Ziel2.NRW
Regionale Wettbewerbsfähigkeit und Beschäftigung

References:

[1] Kortuem, G., Kawsar, F., Fitton, D., & Sundramoorthy, V. (2010). Smart objects as building blocks for the internet of things. Internet Computing, IEEE, 14(1), 44-51.

[2] Sendler, U. (2013). Industrie 4.0 - Beherrschung der industriellen Komplexität mit SysLM, Springer-Verlag Berlin Heidelberg, ISBN 978-3-642-36916-2

[3] Farrar, C. R., Worden, K. (2012). Condition Monitoring: A Machine Learning Perspective. John Wiley & Sons, Ltd, ISBN: 9781119994336

[4] Ding S. X. (2008). Model-based Fault Diagnosis Techniques, Springer-Verlag, Berlin Heidelberg, Germany

[5] MANNER Sensortelemetry (2014). Imagebrochure, www.sensortelemetrie.de/en

[6] Haibach, E. (2006). Betriebsfestigkeit: Verfahren und Daten zur Bauteilberechnung. Springer-Verlag, Berlin Heidelberg New York, ISBN: 978-3-540-29363-7

[7] DIN EN 60079-11 (08.2007). Explosionsfähige Atmospäre – Teil 11: Geräteschutz durch Eigensicherheit "i". Beuth Verlag GmbH, Berlin, Germany

[8] Grzybowski, R., Steingröver, K. (2011). Anforderungen an zukünftige Hauptgetriebe für Windenergieanlagen aus Sicht der Zertifizierung. Dresdner Maschinenelemente Kolloquium p. 429-439

[9] Kolerus, J. (2011) Zustandsüberwachung von Maschinen, das Lehr- und Arbeitsbuch für den Praktiker. 5. Ausgabe, Expert-Verl., Renningen

[10] Feng, Z., Zuo, M. J., (2012) Vibration signal models for fault diagnosis of planetary gearboxes. Journal of Sound and Vibration, vol. 331, S. 4919 – 4939

[11] Shan, J. (2001) Zustandsdiagnose von Planetengetrieben bei stationärem und instationärem Betrieb mit Hilfe der Schwingungsanalyse. PhD Thesis, RWTH Aachen, Aachener Beiträge zur angewandten Rechentechnik, Band 45, Verlag Mainz, Aachen

[12] Hilbert, M., Schütz, M., Boos, D., Bernet, C., Baltes, R., Bartnitzki, T., Kleine-Rüschkamp, L., Küch, C., Jacek, A., Nienhaus, K. (2013). i-MaSS: Approach for Vibration Measurements on Rotating Machine Parts in Gearboxes of Wind Turbines. 26th International Congress on Condition Monitoring and Diagnostic Engineering (COMADEM 2013), 11–13 June 2013, Helsinki, Finland.

[13] Caterpillar Inc. (2011). High-Capacity AFC Systems, www.cat.com

[14] Flöte, K., Grätz, A., Konrad, G., Wölfle, M. (2000). DMT Berichte aus Forschung und Entwicklung 87 Verbesserung der Arbeitssicherheit durch neue Komponenten und Zustandsüberwachung für Kettenkratzerförderer, Bochum, ISSN: 0941-4282

[15] CRC Mining (2005). Annual Report 2004-05, Kenmore East QLD, Australia, www.crcmining.com.au

[16] CRC Mining (2013). Annual Report 2012-13 Transforming Mining, Kenmore East QLD, Australia, www.crcmining.com.au

[17] Pierburg, L. (2009). Entwicklung und Anwendung eines neuartigen Konzepts zur Aufnahme von Bauteilbeanspruchungen für Betriebsfestigkeitsanalysen. PhD Thesis, RWTH Aachen

[18] Sann, B. (1988). Die Mechanik des Förderers. Westfalia Lünen, Lünen

Bibliography:

[1] Markus Schütz received the Dipl.-Ing. degree in mechanical engineering at the Dresden Technical University, Germany, in 2009. Since 2010 he is a PhD candidate at the Institute for Mining and Metallurgical Machinery at RWTH Aachen University. His focus is on field and test measurements, signal analysis and signal evaluation of heavy machinery.

[2] Marc Hilbert received the Dipl.-Ing. degree in mechatronic engineering in the Dresden Technical University, Germany, in 2009. From 2009 to 2011, he was with the System Department Operation and Control, Nordex Energy GmbH, as a Development Engineer for Condition Monitoring Systems and Automation. Since 2011 he is a PhD

candidate at the RWTH Aachen University. His research is related to monitoring and diagnostics of heavy machinery.
[3] Claudia Sequeira finished her PhD in mechanical engineering, “Mechanical Vibrations as a tool to improve maintenance in wind turbines”, in the New University of Lisbon - Faculty of Science and Technology, Portugal, in 2013. Since 2008, develops research and consultant activities in maintenance applied at wind turbines in Iberwind, one of the leading wind energy companies in Portugal. From 2003 she is professor in Mechanical Engineering Department at the University of Algarve, Institute of Engineering, Portugal.

A STUDY ON COMPARING ACOUSTIC EMISSION AND VIBRATION SENSORS FOR GEARBOX FAULT DIAGNOSTICS

Yongzhi Qu, Brandon Van Hecke, David He, Jae Yoon
Department of Mechanical and Industrial Engineering
University of Illinois at Chicago, Chicago, IL, 60607, USA
Telephone: 312-996-2387
davidhe@uic.edu

Eric Bechhoefer
Green Power Monitoring Systems, LLC, Vermont, 05753, USA

Junda Zhu
NRG Systems, Hinesburg, VT, 05461, USA

Abstract: In recent years, acoustic emission (AE) sensors and AE based techniques have been developed and tested for gearbox fault diagnosis. In general, AE based techniques require much higher sampling rate than vibration analysis based techniques for gearbox fault diagnosis. Therefore, it is questionable if an AE based technique would give a better or at least the same performance as the vibration analysis based techniques using the same sampling rate. To answer the question, this paper presents a comparative study for gearbox tooth damage level diagnostics using AE and vibration measurements, the first known attempt to compare the gearbox fault diagnostic performance of AE and vibration analysis based approaches using the same sampling rate. Partial tooth cut faults are seeded in a gearbox test rig and experimentally tested. Results have shown that AE based approach has the potential to differentiate gear tooth damage levels in comparison with vibration based approach. While vibration signals are easily affected by mechanical resonance, the AE signals show more stable performance.

Key Words: Fault Diagnosis; Gearbox; Acoustic Emission; Vibration

1. Introduction: Gearboxes are used in almost all of the transmission systems and power systems. Wind turbine systems and helicopter impetus systems are two typical systems that rely heavily on gearboxes. According to the existing literature, gear failure plays a critical role in the overall failure modes of a gearbox and transmission system. As reported in [1], approximately 59% of the failure modes involved gear failure on wind turbines. In another report [2], it was shown that, among all of the failures of helicopter transmission systems, 19.1% of them included gear failure. The use of health and usage monitoring system (HUMS) is mandatory for helicopter operating service. HUMS allows better and cost effective maintenance practice because it can give indications and warnings prior to collateral damages. Various sensing techniques have been applied to gearbox fault diagnostics, such as vibration, acoustic emission (AE), oil debris, and so on.

Currently, vibration analysis is still the most widely used technique in industry. Acoustic emission has been investigated as a potential alternative for machinery fault detection and diagnosis.

Vibration analysis is a well-developed technique for machinery fault diagnosis. A large number of research papers on vibration based fault detection and diagnosis have been published. When there is force variation in a gearbox, the component will generate a vibration. This vibration is then transmitted to the surrounding structure, and therefore noise and vibration will be generated on the gearbox [3]. Transmission error (TE) is generally considered to be the primary excitation mechanism for gear noise and vibration. According to [4], a transmission error is defined as "the difference between the actual position of the output gear and the position it would occupy if the gear drive were perfectly conjugate". Vibration signal analysis is an important tool to experimentally investigate gear vibration because gears generate vibration at specific frequencies, which are related to the number of teeth and the rotational speed of the gear shaft.

AE is defined as transient elastic waves within a material caused by deformation and the release of localized stress energy [5]. Even though AE has been studied as a potential tool for machine fault diagnosis for a long time, the source and characteristics of AE signals, especially in machine fault detection, cannot be fully understood. Initially, burst type AE signals were used for fault detection in structural health monitoring. The AE bursts are believed to be fault related. While this might hold a ground truth for static structural fault detection, it has never been proved for rotating machines. For bearings, it has been proposed that asperity contact was the primary sources of AE signals [6]. For gears, similar studies have not been performed systematically yet. The relationship between AE signals and asperity contact under elastohydrodynamic lubrication regime has been studied, which is synonymous with gears [7][8]. The authors in [7] and [8] identified asperity contact as a significant source of AE signals but did not investigate other sources in detail, such as the gear dynamics, backlash and so on. It is generally accepted that an increase in meshing stress would generate larger amplitude AE responses [9]. In this paper, AE signals are postulated to be mostly related to the interaction and impact of teeth during tooth meshing. The impact on the surface of the tooth causes material deformation and is followed by the stress energy release, which will then cause transient elastic waves.

AE signals are not relatively affected by structural resonance and could be more sensitive to early fault activities [10]. When an unknown fault starts to form in the machinery, energy loss actions such as impacts, friction, and crushing generate sound wave activity that spans a broad range of frequencies [11]. AE sensors could catch frequencies that are much higher than those in vibration signals and therefore their use enables the technicians to detect inchoate faults before any damage occurs. Also, by quantitative methods, one could monitor the fault evolution process from the very beginning. Compared with vibration analysis, AE signals have the potential to detect small abnormal friction, initial cracking and so on. There are some possible explanations for this. The first one, as discussed above, is that AE emitted by very small defects occurs in frequency ranges that are higher than the operational ranges of vibration sensors and therefore might not be caught by vibration sensors. The second explanation is that when there is only a small crack or surface wear in the machinery, it is not severe enough to change the structural vibration. The vibration signal, which measures the second derivative of the displacement, may still remain the same, and thus unable to detect the incipient fault.

Many studies on AE and vibration based gear fault detection have been reported. Ogbonnah [12] applied a wavelet analysis method to gear fault diagnosis and prognosis using AE sensors. A linear relationship between AE amplitude, gearbox running time, and pit progression has been shown in the study. It has been shown that the wavelet analysis method offers good prognosis for the pitting progression as well as the pitting rate. In an early study which applied AE technique to the analysis of fatigue crack growth in a carburized gear tooth [13], AE energy rate was found to be proportional to the stress intensity factor range and crack growth rate. Another comparative study using AE, vibration and spectrometric oil samples for spur gear pitting fault detection was reported in [7]. As an experimental study, it was found that based on the raw signal root mean square (RMS) levels, the AE technique was more sensitive for fault detection purpose. However, in their experiments, the AE sensors were attached directly on the faulty gear inside the gearbox, which is infeasible for most of real applications. Baydar and Ball [14] used the smoothed pseudo-Wigner–Ville distribution to compare the results from acoustic signals and vibration signals. They simulated three types of progressing local faults, broken tooth, gear crack and localized wear. Their results suggested that acoustic signals are more effective for the early detection of faults and may provide a powerful tool to indicate the various types of progressing faults in gearboxes. However, the acoustic signal presented in their paper was collected by a microphone, which was not exactly acoustic emission. Acoustic emission signal is the elastic stress wave generated inside a solid material, typically metal, due to energy release. Acoustic signal refers to the sound signal which reaches in the air and can be collected by a microphone. Acoustic signal is different from acoustic emission signal in the way that acoustic signal generally lies in the audible range (20 Hz ~ 20 kHz), while acoustic emission frequency lies in the high frequency range (1 kHz ~ 1 MHz).

Used as a ground reference, reliable AE signals of healthy cases have been acquired by many researchers as an important pre-requisite for the success of AE based fault detection. In a very recent study on wind turbine condition based monitoring, a design of a new continuous condition monitoring system with automated warnings based on a combination of vibrational and AE analysis was reported in [15]. The authors tried to determine a ground reference for the healthy turbine. The vibrational and AE signatures for a healthy wind turbine gearbox and generator were obtained as a function of wind speed and turbine power. They listed a number of limitations in current research of AE on rolling elements diagnostics. First of all, the measurements are mostly performed in laboratory test rig other than field service conditions. Second, the signal to noise ratio is low due to short time data collection. Third, classification algorithms such as pattern recognition could possible cancel the coherent elements of the noise but not the random or quasi random components. Thus, they proposed that in order to address the above limitations, it is needed for future AE work using much longer monitoring times and repeated measurement on actual defect rolling elements in service to compensate the random noise and instrument performance error. In another AE based gear diagnosis paper [16], an energy-based condition indicator were introduced for monitoring and diagnosis for any machine operating conditions in spite of speed and load variations. A feature called energy index (EI) was proposed to measure the statistical relative energy levels of segments in a time domain signal over a cycle. The proposed technique was validated by comparison with some of the existing methods using the same AE data for early fault detection. The proposed method was also tested with vibration data. When applied to AE signals, it was able to effectively detect the early fault. However, in their research, AE signals were sampled at a high rate of 1 MHz, which hindered them from doing time synchronous averaging due to the large data volume. They used an alternative method of plotting the result of each revolution together to get a visual data graph of the results. Also, their work aimed to evaluate AE and vibration for fault detection purpose other than fault level diagnostics.

In general, AE based techniques require much higher sampling rate than vibration analysis based techniques for gearbox fault diagnosis. In a recent paper, Qu *et al.* [17] proposed a new AE based gearbox fault diagnostic approach. Their proposed approach combines a heterodyne based frequency reduction technique with time synchronous average (TSA) and spectral kurtosis (SK) to collect AE signals with a sampling rate that is comparable to that of vibration sensors and process AE sensor signals and extract features as condition indictors for gearbox fault detection. They have showed that the proposed AE sensor based approach gave good gear fault diagnostic results. However, it is questionable if an AE based technique would give a better or at least the same performance as the vibration analysis based techniques using the same sampling rate. To answer the question, this paper presents a comparative study for gearbox tooth damage level diagnosis using AE and vibration measurements, the first known attempt to compare the gearbox fault diagnostic performance of AE and vibration analysis based approaches using the same sampling rate. Partial tooth cut faults are seeded in a gearbox test rig and experimentally tested.

2. Gear Mechanics Background: Backlash, Contact Ratio, and Tooth Cut: Gear conjugating involves several kinds of stress, among which two basic stress are: contact stress and root bending stress [18][19]. Excessive contact stress causes surface pitting/wear, while the latter one causes tooth breakage or tooth root crack. The process of gear mating is complex. Numerous papers studied the numerical models and performed extensive experiments to identify the dominant cause of gear noise. The major factors contributing to the gear noise include tooth stiffness, manufacture error, tooth profile, backlash, asperity ratio, and so on. These factors determine the smoothness of the gear meshing and therefore cause the vibration and acoustic emission signals.

Since gears with partial tooth cut were utilized in this research, one could infer that backlash and contact ratio are major factors that affect sensor measurement. These two factors will be the primary concern of this paper. Backlash, in the context of gears and gear trains, is the amount of clearance between mated gear teeth. It is the gap that can be seen when the direction of movement is reversed and the slack or lost motion is taken up before the reversal of motion is complete. The presence of backslash has a significant effect on impact dynamics of meshing gear teeth-pair. Backlash is one of the most common faults found in any geared systems. Backlash is essential for the gear transmission in the sense that too little backlash may result in interference between the teeth while excessive backlash would cause looseness during gear mating. Generally, the greater transmission accuracy required, the smaller backlash needed. The presence of excessive backlash could result in larger impact energy than normal. The initial contact can be modeled as an impact phenomenon when backlash exists between mating gears. Gear impact is generally approximated by a linear model although some limitations have already been established for the model [20].

Several recent gear mechanics studies regarding backlash would be briefly reviewed here. De La Cruz and Rahnejat [21] described a model based on classic Hertzian impact for gear mating under medium to heavy load. They modeled the instantaneous geometry of the contact and prevailing kinematics of contiguous surface for helical teeth pairs. In paper [22], dynamic response of a gear pair system with periodic stiffness characteristics and backlash were analyzed. The study provided an explanation for intermittent chaos and other complicated dynamic behaviors observed in previous experimental studies. Impact-induced vibration analysis was performed theoretically and experimentally in [23]. The results showed that remote impact of meshing transmission teeth through backlash induced high frequency components in drivetrain vibrational response.

Tuma [24] gave an overall review on the practical techniques and procedures employed to quiet gearboxes and transmission units. He concluded that sufficiently rigid housing, shafts and gears, and high contact ratio gears are important to reduce gearbox noise.

As reported in [20], vibration analysis could be used to establish a backlash model. Disturbances, such as impacts, rolling, sliding and friction produce elastic deformations which absorb the energy during gear meshing. These elastic deformations causes elastic wave which transmit the energy from the gear to the gear housing. From the aspect of sensors mounted on the gear house, the typical propagation path is gears, shaft, bearing and housing [25]. The elastic wave released by material deformations is exactly the direct source of AE signals. On the other hand, a vibration sensor measures the velocity, which is the second derivation of the micro-displacement.

In order to understand the actual effect of the varying tooth conditions on the gear meshing activity, it is important to take a brief look at the gear profile before and after the tooth cut. The schematic of two gears meshing is shown in Figure 1.

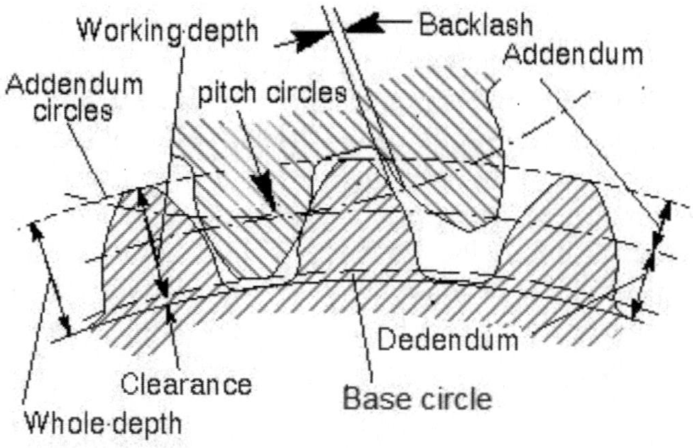

Figure 1: Schematic diagram of two gears meshing

The termilogies shown in Figure 1 are defined as follows:

(1) Pitch circle: The circle through the pitch point having its center at the axis of the gear. The pitch circle of a gear is an imaginary circle which passes through the point where the teeth touch when one gear meshes with another.
(2) Base circle: The circle from which an involute tooth curve is generated or developed.
(3) Addendum: The radial distance between the pitch circle and the outside diameter or the height of the tooth above the pitch.
(4) Dedendum: The radial distance from the pitch circle to the bottom of the tooth space.

Contact ratio is defined as the number of angular pitches through which a tooth surface rotates from the beginning to the end of contact. In a simple way, it can be defined as a measure of the average number of pairs of teeth in contact during the period in which a tooth comes and goes out of contact with the mating gear. It can be calculated as:

$$Contact\ ratio = \frac{\sqrt{r_{a1}^2 - r_{b1}^2} + \sqrt{r_{a2}^2 - r_{b2}^2} - Csin\emptyset}{P_c cos\emptyset} \quad (1)$$

where, r_{a1} and r_{b1} are addendum radius (distance from the tops of the teeth of a gear to the gear center) and base radius (distance from the base circle to the gear center) for the pinion gear center, and r_{a2} and r_{b2} are addendum radius and base radius from the pairing gear center, respectively; C is the gear axis center distance; $\emptyset$ is the angel of the pressure line; P_c is the circular pitch of the pinion gear. Circular pitch is length of the arc of the pitch circle between the centers or other corresponding points of adjacent teeth. For more detail of the concept and calculation, refer to [26].

From Eq. (1), it can be inferred that as the tooth cut gets deeper, the term $r_{a1}^2 - r_{b1}^2$ keeps decreasing until it becomes 0 when the tooth cut reaches the base circle. In this experiment study, we can draw the conclusion that due to the tooth cut, the local contact ratio of the gear keeps decreasing in the order of healthy, 25%, 50% and 100%. As the contact ratio decreases, the amount of meshing looseness increased, which is expected to generate large gear noise.

Based on the diametral pitch system, addendum a and dedendum b can be calculated as:

$$a = \frac{1}{P_d} \tag{2}$$

$$b = \frac{1.157}{P_d} \tag{3}$$

where, P_d is the diameter pitch; a is the addendum and b is the dedendum as shown in Figure 1.

From Eq. (2) and Eq. (3), it is clear to see that in the case of 50% tooth cut, the depth of tooth loss would be greater than the addendum depth. In other words, the remaining tooth depth is somewhere between the pitch circle to base circle. For a gear with a 50% tooth cut, when the gear mates with another, the tooth would lose the initial contact point until the point on the pitch circle and a little over, which will cause a larger backlash than normal condition. But the remaining tooth is still above the base circle, which would make the tooth able to provide support in the next mating cycle. Based the above analysis, it can be inferred that the gear with 50% tooth loss will have a larger contact ratio and smaller backlash compared to 100% tooth loss. Similarly, 50% tooth loss will have a larger backlash and smaller contact ratio compared with 25% tooth cut. For a comparison between a gear with 25% tooth cut and a healthy gear, the 25% tooth cut gear will have smaller contact ratio because it loses the tooth tip which is essential for contact ratio. Thus, one would expect that a 25% tooth cut gear will have larger backlash than healthy gear.

In summary, it is shown that the contact ratio decreases in the order of healthy, 25% tooth cut, 50% tooth cut, and 100% tooth cut. An increasing backlash is also likely to be associated, at least presented in 25%, 50% and 100% tooth loss. For the interest of backlash analysis, more details can be found in [20].

3. Gearbox Fault Diagnosis using AE and Vibration Sensors: The purpose of this paper is to investigate and compare the diagnostic performance of AE sensor and vibration sensor based techniques on a set of seeded gear tooth cut fault test data using the same sampling rate. Therefore, both the diagnostic techniques using AE sensors and vibration sensors in this paper are explained in this section.

3.1 AE based Gear Fault Diagnosis

3.1.1 The Heterodyne Technique: To explain the AE based gear fault diagnostic approach using a comparison, the traditional AE signal processing procedure is first presented in Figure 2.

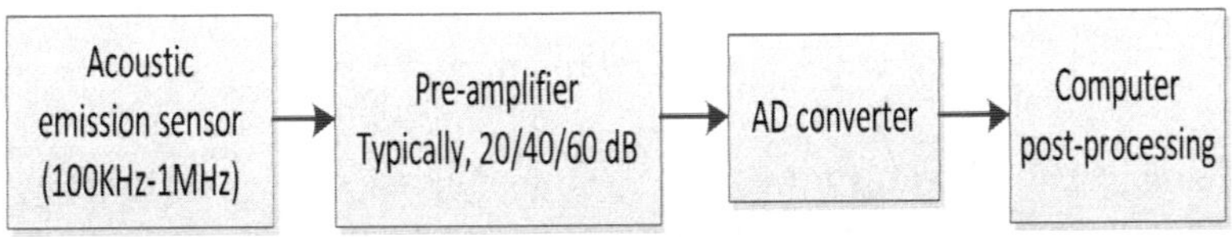

Figure 2: Traditional AE signal acquisition and preprocessing procedure

In a traditional AE signal processing procedure, all of the data is collected and stored to computer without any signal processing. There are two disadvantages associated with this procedure. First, it increases the data acquisition cost. Second, it relies on the computer to process the resulting large data set. A heterodyne based frequency reduction technique has been proposed in a previous research [17]. For the purpose of completeness, the basic principles are introduced below.

For rotating machinery, a periodic displacement (which may only cause a small acceleration) can be indicative of a fault. The displacement will cause a dislocation associated with the AE signature. The information contained in the AE signature is not related to the AE signature itself, but the modulation rate of the signature. This load information is recovered through a demodulation process. The demodulation process is similar to information retrieval in an amplitude/phase modulated radio frequency signal. The carrier signal of a typical AM radio signal is several MHz, while the information modulated onto that signal is audio signal of a couple of kHz. After demodulating the carrier using an analog signal conditioning circuit, the acquisition system can then be sampled at audio frequency (10s of kHz). This signal processing can then be performed at lower cost with an analog circuit than: a high speed analog to digital converter, and the associated computation power required to process the large data set resulting from the high sample rate.

The AE signal demodulator implemented in this paper work similarly to a radio quadrature demodulator: shifting the carrier frequency to baseband, followed by low pass filtering. The technique is called heterodyne. Mathematically, heterodyning is based on the trigonometric identity. For two signals with frequency f_1 and f_2, respectively, it could be written as

$$\begin{aligned} &sin(2\pi * f_1 * t) * sin(2\pi * f_2 * t) \\ &= \frac{1}{2}\cos[2\pi(f_1 - f_2)t] - \frac{1}{2}\cos[2\pi(f_1 + f_2)t] \end{aligned} \tag{4}$$

where, f_1 is the carrier frequency, f_2 is the demodulator's reference input signal frequency. This process could be explained with a simple example.

A detailed discussion of the heterodyne technique applied on the raw AE signal is given in the following. In general, amplitude modulation is the major modulation form for AE signal. Although, frequency modulation and phase modulation could also present in the AE signal potentially, they are considered trivial and will not be discussed here. The amplitude modulation function is given in Eq. (5).

$$U_a = (U_m + mx)co\,s\,\omega_c t \tag{5}$$

where, U_m is the carrier signal amplitude, ω_c is the carrier signal frequency, m is the modulation coefficient. x is the modulated signal, note as

$$x = X_m cos\Omega t \tag{6}$$

Then, with heterodyne technique, the modulated signal will be multiplied with a unit amplitude reference signal cos $(\omega_c t)$. The result is given in the following.

For the amplitude modulation signal,

$$U_o = (U_m + mx)\cos\omega_c t\cos\omega_c t = (U_m + mx)\left[\frac{1}{2} + \frac{1}{2}cos(2\omega_c t)\right] \tag{7}$$

By substituting Eq. (6) into Eq. (7), one gets:

$$\begin{aligned} U_o &= \frac{1}{2}U_m + \frac{1}{2}mX_m cos\Omega t + \frac{1}{2}U_m cos(2\omega_c t) \\ &+ \frac{1}{4}mX_m[cos(2\omega_c + \Omega)t + cos(2\omega_c - \Omega) \end{aligned} \tag{8}$$

Since U_m does not contain any useful information related with the modulated signal, it could be set as 0, or removed by de-trending. According to Eq. (8), it can be seen that only the modulated signal will be left after low pass filtering, where the high frequency components around frequency $2\omega_c$ will be removed.

The diagram of the proposed down sampling system using heterodyne is shown in Figure 3.

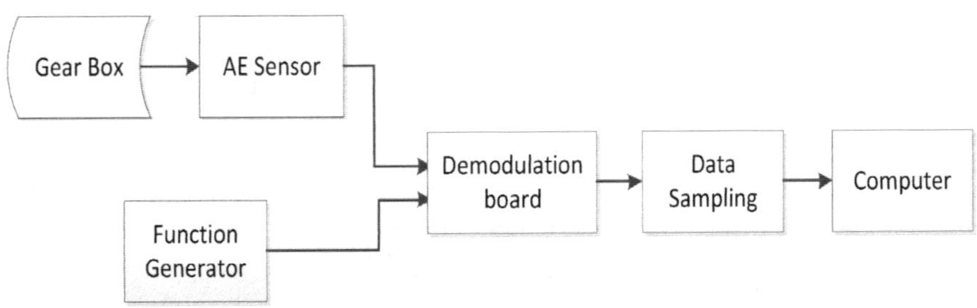

Figure 3: The AE signal acquisition and preprocessing procedure

By adding a demodulation step, it could achieve the purpose of reducing the signal frequency to 10s of kHz. This is close to the frequency range of general vibration signals. Any data acquisition board with a low sampling rate could be able to sample the pre-processed AE data.

3.1.2 Time Synchronous Averaging: TSA has been widely used in processing the vibration signals for rotating machine fault diagnosis [27-30]. The idea of TSA is to use the ensemble average of a raw signal over certain number of revolutions in order to enhance signal of interest with less noise from other sources. For a function *x*(*t*), digitized at a sampling interval *nT*, resulting in sampling in samples *x*(*nT*). Denoting the averaged period by *mT*, TSA is given as [30]:

$$y(nT) = \frac{1}{N}\sum_{r=0}^{N-1} x(nT - rmT) \tag{9}$$

More details about TSA could be found in [27].

The successful application of TSA on vibration signal analysis provides opportunities for processing AE signals. Basically, two types of TSA algorithms are available in literature, i.e., TSA with tachometer, and tachometer less TSA. In comparison with TSA with tachometer, tachometer less TSA needs to estimate the angular information from the vibration data. For slow speed variation cases, time domain feature like gear meshing information could be used. However, tachometer less TSA will introduce more phase reference error and thus have less accuracy than TSA with tachometer. In this work TSA with tachometer is applied.

Despite of the popular application of TSA to vibration signal analysis, application of TSA to AE signal processing for gear fault diagnosis has not been reported in the literature. The complicated feature and large data volume of AE signals make it unrealistic to perform TSA algorithm directly on AE data with an on-line condition monitoring system. In this paper, the authors explore the application of TSA to AE signal analysis.

TSA enables the direct comparison of the vibration/acoustic signals produced by each tooth on the same gear over one revolution. TSA for gear diagnosis generally computes the vibration/acoustic signals of a single shaft revolution. After TSA is calculated, basically all kind of fault detection condition indicators can be evaluated on the TSA signal.

3.1.3 AE Condition Indicators: There are many condition indicators available in literature. Most of the condition indicators deal with the data distribution, such as peakiness, amplitude level, deviation from the mean and so on. A brief introduction of the condition indicators applied on AE signals is given below.

RMS: The root mean square for a discretized sampled signal is defined as:

$$x_{rms} = \sqrt{\frac{1}{N}\sum_{i=1}^{N}(x_i{}^2)} \tag{10}$$

where x_{rms} is the root mean square value of dataset x, x_i is the ith element of x, N is the length of dataset x.

P2P: Peak to peak value of a dataset x is defined as:

$$P2P = \frac{Max(x) - Min(x)}{2} \tag{11}$$

where, $Max(x)$ is the maximum value of x, $Min(x)$ is the minimum value of x.

Kurtosis: kurtosis describe how peaky or how smooth of the amplitude of dataset x. If a signal contains sharp peaks with high values generated by a fault in the gearbox, it is expected that its distribution function will be sharper. Thus, the kurtosis of the fault signal should be higher than that of the healthy signal. The function of kurtosis is given below:

$$Kurt = \frac{N\sum_{i=1}^{N}(x_i - \bar{x})^4}{(\sum_{i=1}^{N}(x_i - \bar{x})^2)^2} \quad (12)$$

where *Kurt* is the kurtosis of dataset x, x_i is the ith element of x, N is the length of dataset x.

It is worth mentioning that for any normal distribution, the kurtosis value is 3. This could be easily verified by the moment generating function.

In addition to the condition indicators computed using the AE TSA signals directly, RMS of the residual signal is also computed. A residual signal is generally defined as a synchronous averaged signal with the shaft, gear mesh, and their associated harmonic frequencies removed.

3.2 Vibration based Gear Fault Diagnosis: For vibration signal, a similar process flow is applied except heterodyne technique which is not needed for vibration signal. After TSA, the condition indicators tested on vibration are different from those tested on AE signals. Basically, the vibration signal frequency is more closely related to gearbox rotational frequency and mechanical interaction. In addition to RMS and P2P, two physical meshing behavior related condition indicators: FM0 and SLF are introduced next.

FM0: FM0 is the zero-order figure of merit. It is a global indicator that will react to changes in the whole frequency range of the average and identifies major abnormal behaviors with regard to meshing pattern. FM0 is defined as the ratio of peak to peak amplitude (PPA) of the TSA signal to the sum of amplitudes of gear mesh frequency and its harmonics. An increase in peak to peak level is generally observed in case of major tooth faults, such as tooth breakage without significant change in the mesh frequency, which will results in increase of FM0 value [31]. FM0 will increase if a periodic signal contains a local increase in amplitude. Mathematically, it is expressed as following:

$$FM0 = \frac{P2P_{TSA}}{\sum_{i=1}^{n} A(f_i)} \quad (13)$$

where FM0 is the zero-order figure of merit; $P2P_{TSA}$ is the peak to peak value of the vibration signal of TSA in the time domain; $A(f_i)$ is the amplitude of the ith harmonic of the gear meshing frequency.

SLF: SLF stands for the sideband level factor. It is the sum of the first order sideband amplitudes of the fundamental gear meshing frequency normalized by the RMS of the synchronous time average [32]. SLF is a good indicator of single tooth damage or gear shaft damage. The formula for SLF calculated is given as,

$$SLF = \frac{R_{I,-1}(x) + R_{I,+1}(x)}{RMS(x)} \tag{14}$$

where x is the vibration signal TSA, $R_{I,-1}(x)$ is the amplitude of the first order left-hand side sideband, $R_{I,+1}(x)$ is the amplitude of first order of right-hand side sideband. $RMS(x)$ is the RMS of x.

4. Experimental Setup: In order to compare the gearbox fault diagnostic performance of the AE and vibration sensors, tests with gear tooth cut seeded faults were conducted on a notational two stage split torque gearbox (STG). In a STG, there are several identical intermediate gear pairs which could split the torque evenly. Also, the intermediate gear pair could offer a larger transmission ratio. Both the input side and output side of the STG use parallel shaft layout. All of the gears inside are spur gears. On the input side, the input driving gear is a 40 teeth gear which drives three input driving gear with 72 teeth each. On the output side, three output driving gears with 48 teeth drive a 64 teeth gear. A 3HP three phase induction AC motor with a maximum speed of 3600 rpm is used to drive the notational gearbox. To accommodate for shaft misalignment and reduce the vibration transmission, a disc type coupling is utilized to transmit the torque from the motor to the driving shaft. A magnetic loading system is controlled by a power supply and the load can be adjusted by changing the output current of the power supply. Figure 4 shows the structure of the notational split torque gearbox. The test rig and sensor locations are shown in Figure 5.

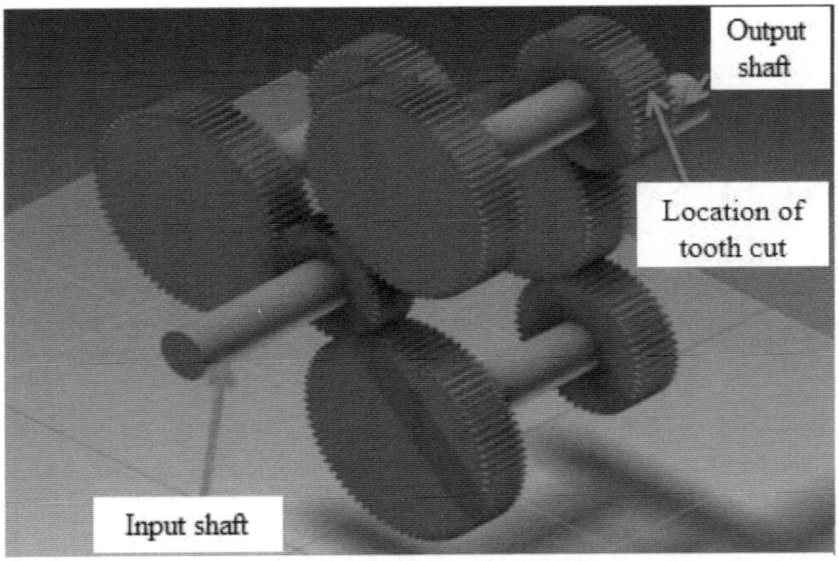

Figure 4: The structure of the notational split torque gearbox

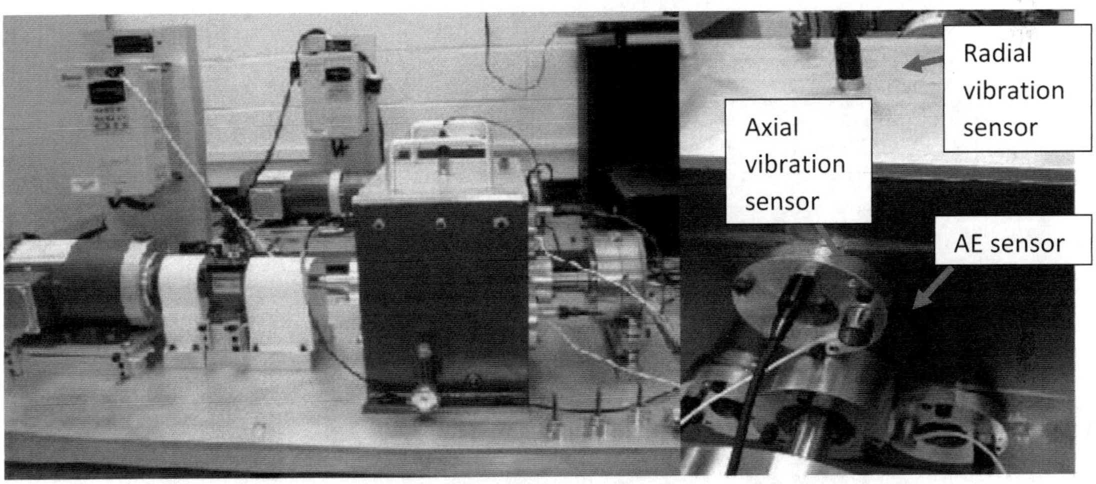

Figure 5: The notational split torque gearbox and sensor locations

As a speed reduction gearbox, the input side and the output side have a 2.4 times speed reduction ratio. Based in the input speed tested in the experiments, the corresponding output shaft speed and intermediate shaft (faulty gear shaft) speed is provided in Table 1.

Table 1: Output shaft speed corresponding to input shaft speed

Input shaft speed (Hz)	10	20	30	40	50	60
Faulty gear shaft frequency (Hz)	5.56	11.1	16.7	22.2	27.8	33.3
Output shaft speed (Hz)	4.17	8.33	12.5	16.7	20.8	25

For the faulty gearbox, one of the intermediate gears with 48 teeth on the output side was damaged by cutting a gear tooth with certain percentages. As shown in Figure 6, 25%, 50%, and 100% tooth cut were created, respectively. The fault created here is relatively large, but the tooth cut did not cause any severe failure to the gearbox because of the split torque feature. The slack or lose of motion due to the tooth cut or tooth missing on one of the driving pinion gears can be compensated by the other two driving pinion gears.

Figure 6: Seeded tooth cut faults

One AE sensor and two accelerometers were mounted on the gearbox. The AE sensor was attached to the gear housing using adhesives as shown in Figure 5. One accelerometer was mounted on the gearbox housing in the axial direction and the other one was mounted on top of gearbox housing in the radial direction (see Figure 5). The signals from all of the three sensors were collected simultaneously during the test runs. In addition, tachometer signals were collected along with vibration and AE signals.

For AE data acquisition, a true differential wideband sensor with high sensitivity and bandwidth wideband differential sensor was used. It has a good frequency response over the range of 100 – 900 kHz. Differential sensors offer a lower noise output from a pre-amplifier. The frequency response of the accelerometers was from 0.5 Hz– 10 kHz. The heterodyne process was accomplished by a hardware demodulation. A demodulation board and a sampling device were used. The demodulation board performed the multiplication of sensor signals and reference signals. The demodulation board is an analog device and much more affordable than a high sampling rate data acquisition board. It takes two inputs, one from the AE sensor, and the other from function generator as a reference signal. The output current of this circuit is an accurate multiplication of the base currents of the both inputs. According to Eq. (4), it could convert the signals to baseband and twice the carrier frequency. The frequency of reference signal was obtained as 400 kHz. In searching for the optimized reference frequency, a chirp function with a range of 50 kHz - 1050 kHz was selected to cover the whole sensor response range. The chirp function started with an initial frequency of 50 kHz and chirp up at a rate of 139.89 kHz/s. The output of the demodulation board goes to the sampling board and the high frequency component is filtered out.

For signal acquisition, Labview signal express software was used. During the experiments, continuous AE signals were collected. The data sampling rate was set to 100 kHz for both vibration and AE signals in order to make a fair comparison. A light torque load was applied to the gearbox during the test. The gearbox was run with 6 different input shaft speed staring from 10 Hz then increased with a 10 Hz increment to 60 Hz. For each speed, 5 data sets were collected. In order to get a good TSA results, signals over approximately 200 revolutions were recorded.

5. Results and Discussions: In this section, the diagnostic results of the gear seeded cut fault tests using both AE and vibration sensors are provided and discussed.

5.1 Results of AE signal analysis: After heterodyning, TSA was performed on the signals first to get the time synchronous average signals using tachometer signal as the phase reference. Then the AE signal condition indicators were calculated on the TSA signals. Three condition indicators as introduced in Section 3.1.3, were computed using the AE TSA signals: RMS, P2P and kurtosis. In addition, RMS values of the residual signals were also computed for comparison.

Figure 7 shows the RMS plot of AE TSA signals. The data set numbers were arranged from 10 Hz – 60 Hz, 5 data sets for each speed. Other plots in the following context are arranged in the same manner. It can be seen from Figure 6 that the RMS provided a good trend for the energy level when speed increase. For different level of tooth cut, it offers clear separation. This indicators that AE signal are very sensitive to the gear meshing impact due to both speed and level of severity of tooth fault.

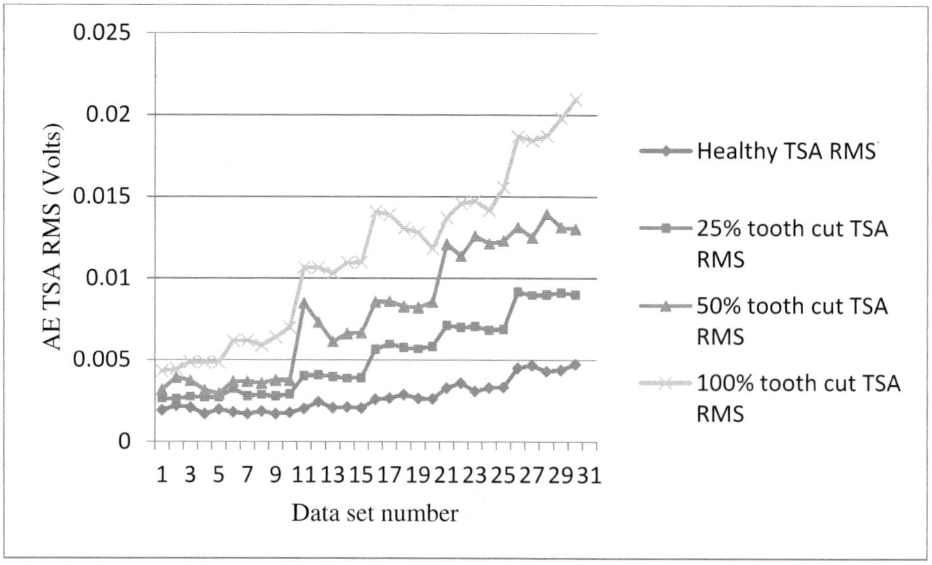

Figure 7: RMS of AE TSA signals

Figure 8 shows the plot for residual RMS of AE TSA signals. Residual RMS provides similar separation as the TSA RMS but reduces the degree of fluctuation. It also increases the fault detectability between healthy and 25% tooth loss case since the separation between the two is more clear. As residual signals normally contain the fault features except gear meshing and harmonics, it could be more effective than the TSA signal itself.

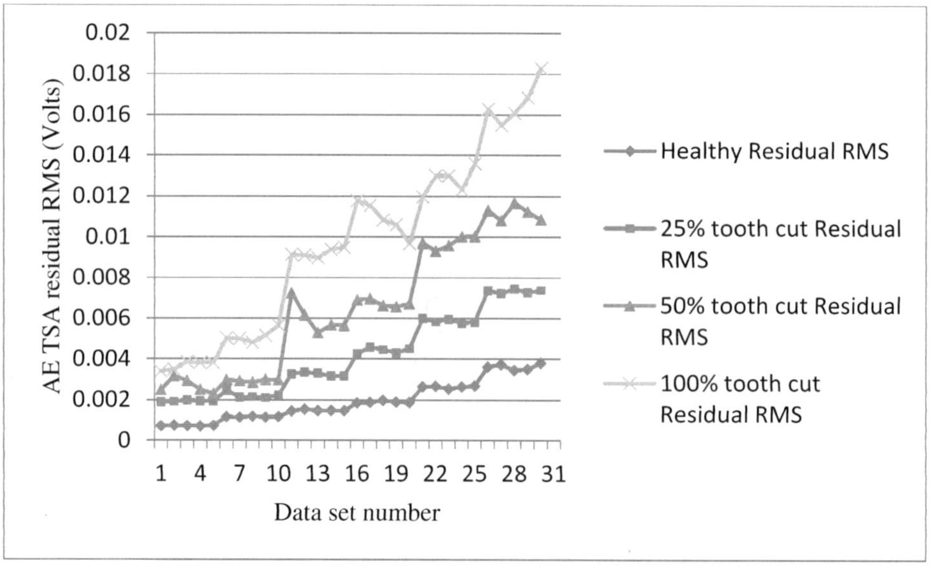

Figure 8: RMS of AE TSA residual signals

Figure 9 shows the plot for P2P of AE TSA signals. As can be seen from here, P2P generally follows the trends, but it contains some more fluctuation compared with RMS.

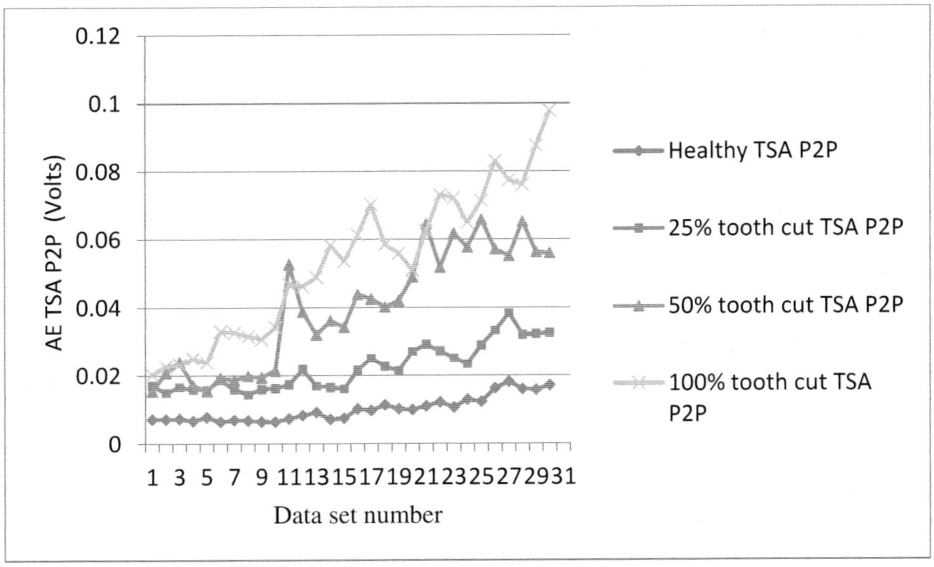

Figure 9: P2P of AE TSA signals

Figure 10 shows the plot for kurtosis of AE TSA signals. Although kurtosis is not able to distinguish fault levels, it acts as a good condition indicator for fault detection. Kurtosis measures the peakiness of the signals. It is a non-quantitative parameter, which means it is independent of the magnitude of the signal. For any Gaussian distribution, the value of kurtosis is calculated as 3. Since kurtosis is not affected by the speed, it is useful for making fault detection decisions.

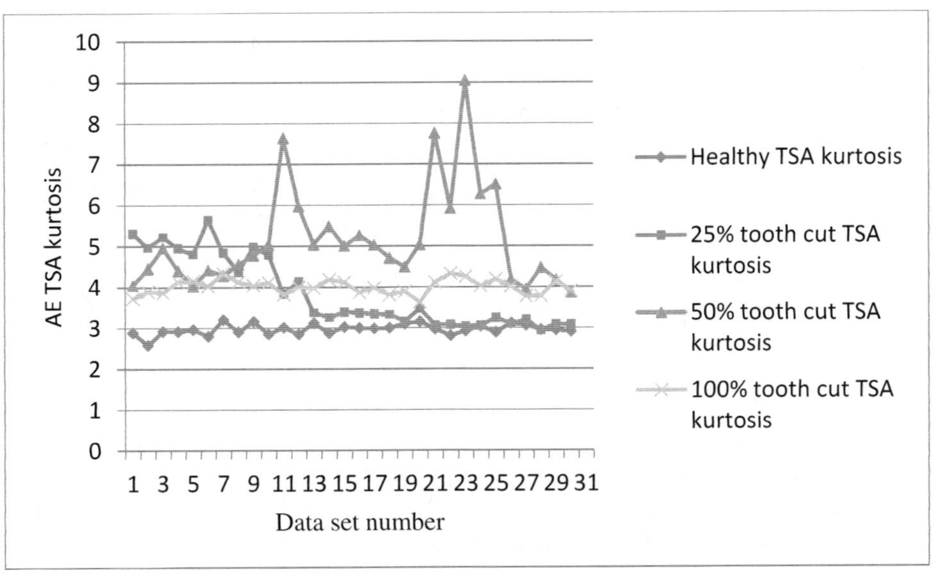

Figure 10: Kurtosis of AE TSA signals

5.2 Results of vibration signal analysis: Vibration sensors, typically accelerometers, measure the velocity, which is the second derivation of displacement. The frequency response range of vibration signals is much lower than that of AE signals. Therefore a vibration signal has the advantage of representing the mechanically behaviors more closely. But it also has the disadvantage of being easily effected by mechanical resonance. Similar to AE signal processing, TSA was performed on raw vibration signals first, then the condition indicators were computed. A total of 4 condition indicators: RMS, P2P, FM0 and SLF were computed for vibration signals.

During the experiments, both axial and radial direction vibration signals were collected and analyzed. Figure 11 to Figure 14 show the results from the axial direction vibration sensor. Figure 15 to Figure 18 give the results from the radial vibration.

Figure 11 shows the RMS plots of the vibration TSA signals. It can be seen that in the low speed range below 30 Hz, the vibration RMS does not give any indication of the fault. In the high speed range above 30 Hz, vibration RMS with tooth faults increases significantly and provides good indication for fault detection. However, vibration RMS is not sensitivity to the level of tooth cut as the vibration RMS for 100% cut is lower than that of 50% and 25% tooth cut.

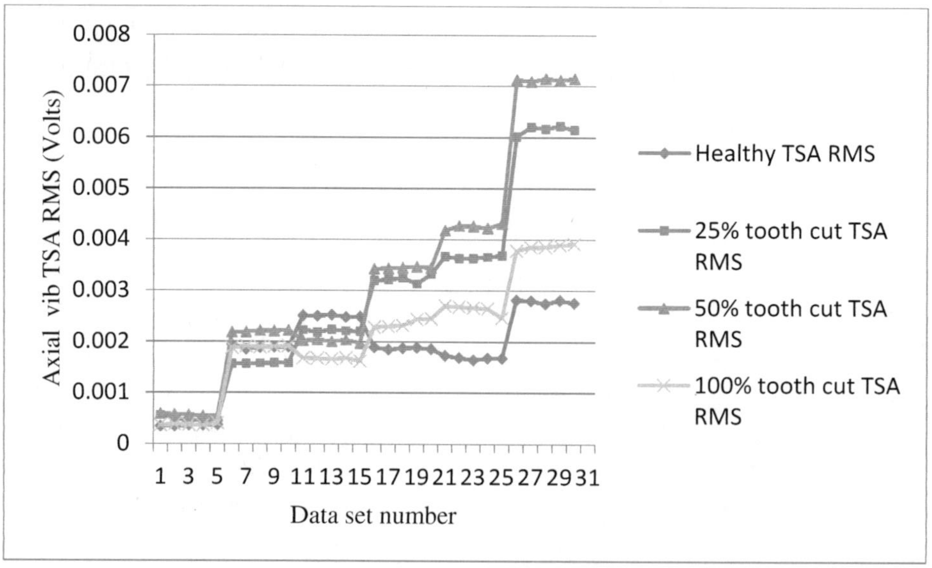

Figure 11: RMS of axial vibration TSA signals

Figure 12 gives the P2P plots of vibration TSA. P2P values of the faulty signals are mostly higher than the healthy counterpart except at 10 Hz input speed. Similar to RMS, P2P shows potential capability for fault detection but not for fault level diagnostics.

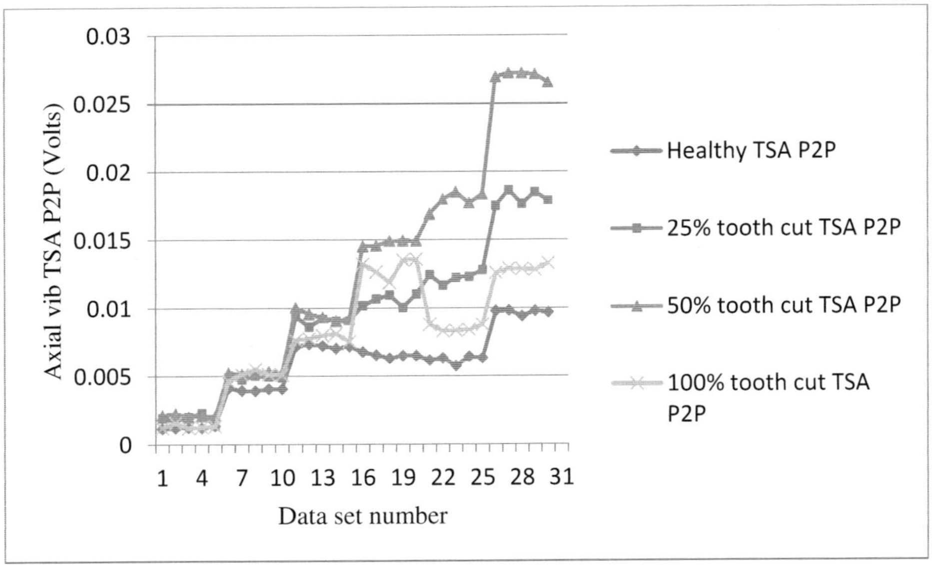

Figure 12: P2P of the axial vibration TSA signals

Figure 13 shows the FM0 of the axial vibration signals. FM0 could detect the anomalies in most of the cases. However, it has a lot of fluctuation at different speeds. Again it is not effective for damage level separation.

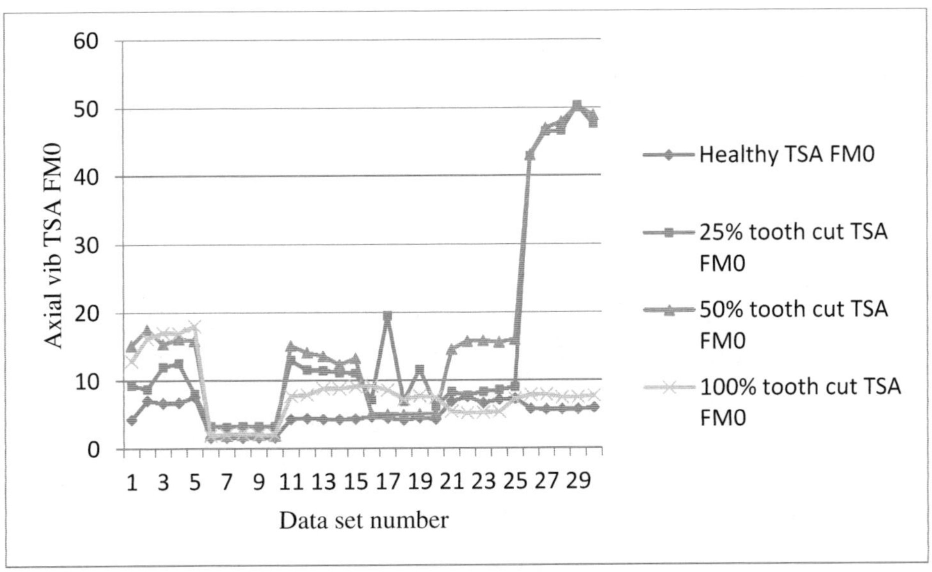

Figure 13: FM0 of the axial vibration TSA signals

Figure 14 shows the condition indicator SLF plots of the axial sensors. It can be seen that at 10 Hz, 30 Hz, and 60 Hz, the healthy SLF are all lower than faulty ones. But for the other speeds, some of the faulty signals SLFs are lower than healthy. This result shows that SLF of axial vibration is not effective for fault detection.

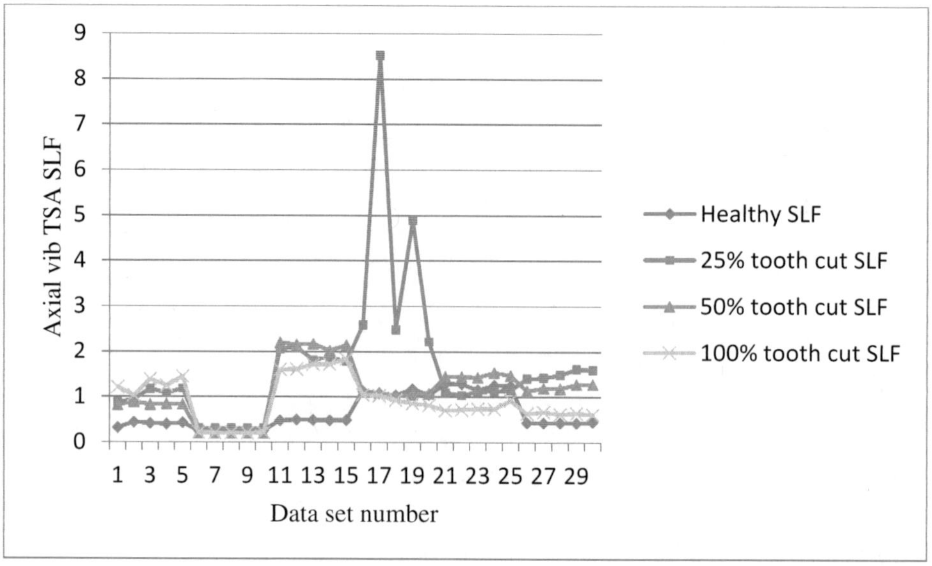

Figure 14: SLF of axial vibration TSA signals

Figure 15 shows the RMS plots of the radial vibration TSA signals. It can be seen from the plot that the radial vibration signal are seriously affected by the mechanical resonance, especially at 30Hz. Basically, the RMS of radial vibration does not give a good indication for gear tooth cut fault and the level of the cut.

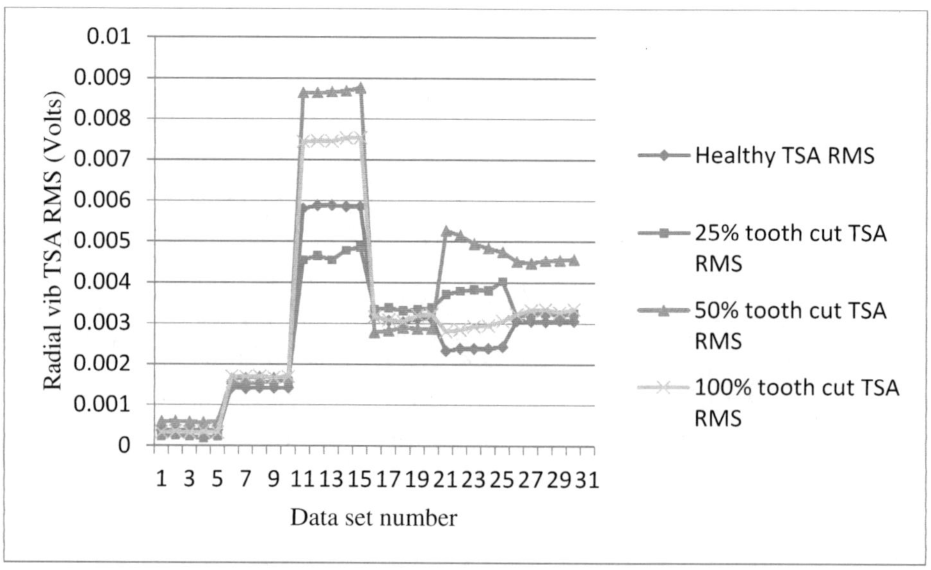

Figure 15: RMS of radial vibration TSA signals

Similarly, the P2P of the radial vibration does not give a good indication neither, as shown in Figure 16.

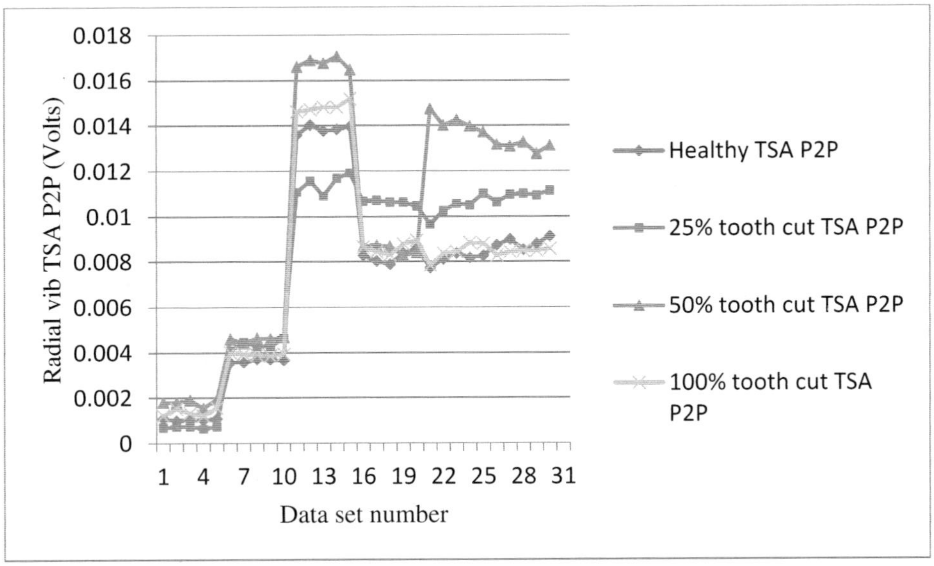

Figure 16: P2P of radial vibration TSA signals

Figure 17 shows the condition indicator FM0 of the radial vibration sensors. It can be see that except at 20Hz and 30Hz, the FM0 of the faulty signals are all higher than the healthy ones. At 30Hz, FM0 for all cases dropped to a very low level and overlapped each other. At the same time, RMS and P2P from figure 15 and figure 16 show that the peaks and energy levels for all cases increased significantly. It is highly likely to be caused by mechanical resonance. Other than that, the healthy FM0 is relatively stable with the bound of approximately 10, while the faulty signals can go as high as 40, which makes FM0 a good condition indicator for tooth cut fault detection purpose.

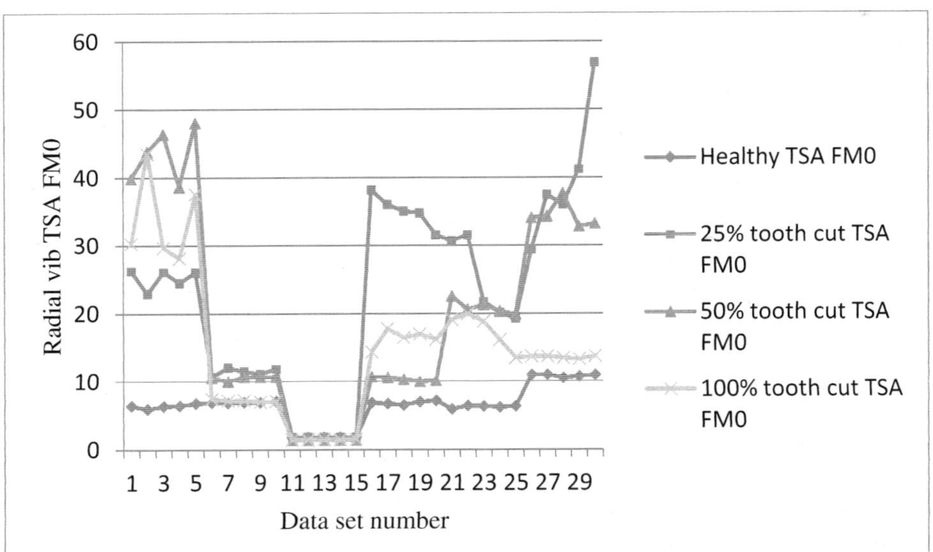

Figure 17: FM0 of radial vibration TSA signals

Figure 18 shows the condition indicator of SLF. SLF works similarly as FM0, it could separate the healthy signals from the faulty one clearly in the low speed at 10 Hz as well as high speed 30 Hz and above. But it fails to distinguish the fault at 20 Hz and 30 Hz.

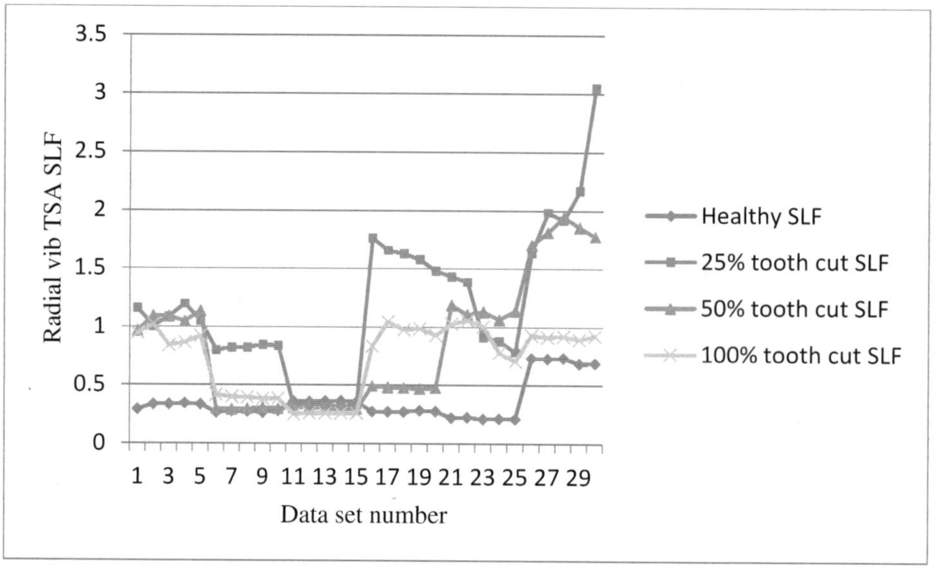

Figure 18: SLF of vibration TSA signals

In summary, it can be seen that for axial vibration sensor mounted on the bearing house, the RMS and P2P show good fault detection potential. FM0 and SLF of the axial sensor work in most cases but not stable. On the other side, for the radial sensor mounted on the top of the gearbox, RMS and P2P fail to work, while FM0 and SLF work for fault detection purpose. Compared with AE results, none of the vibration condition indicators could detect the tooth cut level. The vibration signals are highly affected by background noise or mechanical resonance, making it unstable in performance. AE RMS and P2P show a roughly linear relationship against shaft speed. They could clearly indicate the tooth cut levels for diagnostics. Also, kurtosis of AE signals offers another effective index for fault detection.

It is also needed to point out that vibration signals offer better frequency domain resolution as both FM0 and SLF are calculated based on gear meshing frequency. AE offers better time domain features related to energy levels, such as RMS and P2P.

As explained in Section 2, the tooth cut fault is the direct cause of larger backlash and reduction in contact ratio. Both the large backlash and low contract ratio introduce more looseness during gear meshing and therefore cause higher impact and gear noise. From this perspective, it can be inferred that AE sensors are much more sensitive to impact energy. Vibration measured by accelerometer is the velocity signal, which is less sensitive to direct impact energy.

6. Conclusions: Previous research has showed that AE sensor based approach using a sampling rate that is comparable to that of vibration analysis gave good gear fault diagnostic results. However, it is questionable if an AE based technique would give a better or at least the same performance as the vibration analysis based techniques using the same sampling rate. To answer the question, this paper presented a comparative study for gearbox tooth damage level diagnostics using AE and vibration measurements. Partial tooth cut faults were seeded in a gearbox and experimentally tested. Three different levels of tooth cut fault were artificially created and tested on a notational split torque gearbox. For AE based gear fault diagnostic approach, a hardware frequency convertor based on heterodyne technique was used for AE data collection. Both the AE signals and vibration signals were collected with the same sampling rate of 100 kHz. Time synchronous averaging was applied to both types of signals. Condition indicators were then calculated respectively for AE and vibration signals. Experimental results were provided and explained. Based on the experimental results, several conclusions can be drawn:

(1) Acoustic emission is more sensitive to the energy level in rotating mechanical system. The energy level could be caused by speed variation or introduced by the presence of different level of faults.

(2) Acoustic emission is insensitive to mechanical background noise and mechanical resonance. Therefore, acoustic emission has the potential to provide better condition indicator during subtle situations.

(3) Vibration signals provided better frequency resolution to mechanical related behavior such as gear meshing in the low frequency range.

References:

[1] Link H., LaCava W., van Dam J., McNiff B., Sheng S., Wallen R., McDade M., Lambert S., Butterfield S., and Oyague F., 2011, *Gearbox Reliability Collaborative Project Report: Findings from Phase 1 and Phase 2 Testing*, NREL Technical Report: NREL/TP-5000-51885.

[2] Astridge D. G., 1989, "Helicopter Transmissions - Design for Safety and Reliability", *Proceedings of the Institution of Mechanical Engineers, Part G: Journal of Aerospace Engineering*, Vol. 203, No.2, pp. 123-138.

[3] Smith J. D. 2003, *Gear Noise and Vibration*, Marcel Dekker Inc., New York, NY.

[4] Åkerblom M., 2001, *Gear Noise and Vibration - A Literature Survey*, Maskinkonstruktion, Stckholm, Sweden.

[5] Mathews J. R., 1983, *Acoustic Emission*, Gordon and Breach Science Publishers Inc., New York, NY.

[6] Al-Ghamd A. M. and Mba D., 2006 "A Comparative Experimental Study on the Use of AE and Vibration Analysis for Bearing Defect Identification and Estimation of Defect Size", *Mechanical Systems and Signal Processing*, Vol. 20, No. 7, pp. 1537 - 1571.

[7] Tan C. K. and Mba D., 2005, "Experimentally Established Correlation between AE Activity, Load, Speed, and Asperity Contact of Spur Gears under Partial Elastohydrodynamic Lubrication". *Proceedings of the Institute of Mechanical Engineers, Part J: Journal of Engineering Tribology,* Vol. 219, No. 6, pp. 401 – 409.

[8] Tan C. K. and Mba D., 2006, "A Correlation between Acoustic Emission and Asperity Contact of Spur Gears under Partial Elastohydrodynamic Lubrication", *International Journal of COMADEM*, Vol. 9, No. 1, pp. 9 - 14.

[9] *Non-destructive testing handbook*, Volume 5 – Acoustic Emissions, Chapter 1, American Society of Non-destructive Testing, 1987, Eds, R. K. Miller and P. McIntire.

[10] Holroyd T., 2000, *Acoustic Emission & Ultrasonics Monitoring Handbook*, Coxmoor Publishing Co, Oxford, U.K. .

[11] Loutas T. H., Roulias D., Pauly E., and Kostopoulos V., 2012, "The Combined Use of Vibration, AE and Oil Debris On-line Monitoring Towards A More Effective Condition Monitoring of Rotating Machinery", *Mechanical Systems and Signal Processing*, Vol. 25, No. 4, pp. 1339–1352.

[12] Ogbonnah V., 2007, *Condition Monitoring of Gear Failure with AE*, Master's Degree Thesis, Department of Mechanical Engineering, Blekinge Institute of Technology, Karlskrona, Sweden.

[13] Tomoya M., Katsumi I., and Masana K.,1994, "Acoustic Emission during Fatigue Crack Growth in Carburized Gear Tooth" *Transactions of the Japan Society of Mechanical Engineers, Part C*, Vol. 60, No. 575, pp. 2456 - 2461.

[14] Baydar N. and Ball A., 2001, "A Comparative Study of Acoustic and Vibration Signals in Detection of Gear Failures using Wigner–Ville Distribution", *Mechanical Systems and Signal Processing*, Vol. 15, No. 6, pp. 1091 – 1107.

[15] Soua S., Lieshout P., Perera A., Gan T. H., and Bridge B., 2013, "Determination of the Combined Vibrational and AE Signature of A Wind Turbine Gearbox and Generator Shaft in Service as A Pre-requisite for Effective Condition Monitoring", *Renewable Energy*, Vol. 51, No.3, pp. 175 – 181.

[16] Al-Balushi K. R., and Samanta B., 2002, "Gear Fault Diagnosis using Energy-based Features of AE Signals", *Proceedings of the Institution of Mechanical Engineers, Part I: Journal of Systems and Control Engineering*, Vol. 216, No. 3, pp. 249 - 263.

[17] Qu Y., Bechhoefer E., He D., and Zhu J., 2013, "A New Acoustic Emission Sensor Based Gear Fault Detection Approach", *International Journal of Prognostics and Health Management*, Vol. 4, Special Issue on Wind Turbine PHM, pp. 1 – 14.

[18] Li S., 2013, "Effects of Centrifugal Load on Tooth Contact Stresses and Bending Stresses of Thin-rimmed Spur Gears with Inclined Webs", *Mechanism and Machine Theory*, Vol. 59, No.1, pp. 34 – 47.

[19] Tiwari S. K. and Joshi U. K., 2012 "Stress Analysis of Mating Involute Spur Gear Teeth", *International Journal of Engineering Research & Technology (IJERT)* , Vol. 1, No. 9, pp. 2173 -2180, ISSN: 2278-0181

[20] Sarkar N., Ellisb R. E., and Mooreb T. N., 1997 "Backlash Detection in Geared Mechanisms: Modeling, Simulation, and Experimentation", *Mechanical Systems and Signal Processing*, Vol. 11, No. 3, pp. 391–408.

[21] De la Cruz M. and Rahnejat H., 2008, "Impact Dynamic Behaviour of Meshing Loaded Teeth in Transmission Drive Rattle". *Proceedings of the Sixth EUROMECH Nonlinear Dynamics Conference*, pp. 418 - 430.

[22] Theodossiades S. and Natsiavas S., 2000, "Nonlinear Dynamics of Gear-pair Systems with Periodic Stiffness and Backlash". *Journal of Sound and Vibration*, Vol. 229, No. 2, pp. 287 - 310.

[23] Gnanakumarr M., Theodossiades S., Rahnejat H., and Menday M., 2005, "Impact-induced Vibration in Vehicular Driveline Systems: Theoretical and Experimental Investigations". *Proceedings of the Institution of Mechanical Engineers, Part K: Multibody Dynamics*, Vol. 219, Special Issue, pp. 1 – 12.

[24] Tuma J., 2009, "Gearbox Noise and Vibration Prediction and Control". *International Journal of Acoustics and Vibration*, Vol. 14, No. 2, pp. 99 - 110.

[25] Ognjanović M. and Snežana Ćirić Kostić, 2012, "Gear Unit Housing Effect on the Noise Generation Caused by Gear Teeth Impacts", *Journal of Mechanical Engineering*, Vol. 58, No. 5, pp. 327-337

[26] Budynas R. G. and Nisbett J. K., 2011, *Shigley's Mechanical Engineering Design*, 9th Edition, McGraw Hill, New York, NY.

[27] McFadden P.D., 1987, "A Revised Model for the Extraction of Periodic Waveforms by Time Domain Averaging", *Mechanical Systems and Signal Processing*, Vol.1, No.1, pp. 83 – 95.

[28] McFadden P. D. and Toozhy M. M., 2000, "Application of Synchronous Averaging to Vibration Monitoring of Rolling Element Bearings", *Mechanical Systems and Signal Processing*, Vol. 14, No. 6, pp.891–906.

[29] Bonnardot F., Badaoui M. El, Randall R.B., Daniere J., and Guillet F., 2005, "Use of the Acceleration Signal of A Gearbox in order to Perform Angular Resampling (with Limited Speed Fluctuation) ", *Mechanical Systems and Signal Processing*, Vol. 19, No. 4, pp. 766 - 785.

[30] Braun S., 1975, "The Exaction of Periodic Waveforms by Time Domain Averaging", *Acustica*, Vol. 32, No.2, pp. 69 - 77.

[31] Nooli P. K., 2011, *A Versatile and Computationally Efficient Condition Indicator for AH-64 Rotorcraft Gearboxes*, Master thesis, Department of Mechanical Engineering, University of South Carolina, Columbia, SC.

[32] Antolick L. J., Branning J. S., Wade D. R., and Dempsey P. J., 2010, "Evaluation of Gear Condition Indicator Performance on Rotorcraft Fleet, *Proceedings of the American Helicopter Society 66th Annual Forum,* Phoenix, AZ, May 11 – 13.

FAULT DETECTION OF ROTATING MACHINERY USING A HYBRID HIDDEN MARKOV MODEL

Ayan Sadhu and Sriram Narasimhan
University of Waterloo
200 University Avenue West,
Waterloo, ON N2L 3X4, Canada
Telephone: 519 888 4567 (x33252)
asadhu@uwaterloo.ca

Abstract: In this paper, a generalized hybrid hidden Markov model (HMM) is proposed to perform multi-state fault detection of rotating machinery. Time-frequency signatures extracted from wavelet packet transformation based Teager Kaiser energy operator are first used to construct de-noised damage features of the raw signal, and then GMM-based HMM is undertaken for fault detection. This approach improves the traditional HMM over a more accurate state estimation and simpler codebook preparation under multiple failures and operating conditions. The proposed method is finally validated using bearing data of Case Western Reserve University. This research provides an attractive solution to practising engineers to undertake timely intervention and maintenance scheduling of rotating machineries in several industrial applications.

Key words: Condition-based monitoring; fault detection, hidden Markov model; wavelet packet transform; Teager Kaiser operator; Gaussian mixing model; bearing failure

Introduction: Sudden failure of machining equipment in manufacturing and processing industries is of paramount importance due to their repair cost and costs associated with business interruption. Over last few decades, maintenance of rotating machinery has experienced extensive research activity in both the industry and academia. Different strategic approaches have evolved over time, which can be categorized as: (a) *breakdown* or *run-to-failure* maintenance, (b) *preventive* or *time-based* maintenance and (c) *predictive* or *condition-based* maintenance [1]. Out of these, condition-based maintenance (CBM) leads to a solution where maintenance is undertaken whenever and wherever needed; based on the condition of the system rather than on a pre-scheduled agenda. In this mode of scheduling, a machine component is continuously monitored and repaired when a failure is likely to occur. In this paper, time-frequency decomposition based robust hidden Markov model is proposed for diagnosis of faults in gearboxes.

Gearbox is one of the key components of rotating machinery used in modern industrial applications. Unexpected failures of gearboxes may endanger machine's normal operation and productivity, and thus cause significant economic losses. Fault diagnosis of gearboxes therefore plays a valuable role mitigating catastrophic failure. However, a typical gearbox is a complicated dynamical system consisting of several pairs of meshing gears mounted on rotating shafts, while the shafts themselves are supported by bearings and fixtures. Subsequently a wide range of faults can potentially occur in a gearbox, e.g., eccentric or cracked gears, tooth failure, bearing malfunction or misalignment. When a gear has a local fault, the vibration signal consists of amplitude and phase modulated signals containing gear meshing frequencies with sidebands [2]. Unlike gear faults, bearing faults manifest themselves as recurring impulses, which are cyclo-stationary [3]. When the vibration signal with gear and/or bearing failure is masked with measurement noise, early fault detection at relatively low levels of deterioration or damage becomes a challenge. In this paper, wavelet packet transformation-based Teager Kaiser operator is first employed as a de-noising feature selection tool, followed by a Gaussian mixing model-based robust hidden Markov model for fault detection and diagnosis.

Traditionally, vibration-based fault detection techniques are classified into four categories: time-domain, frequency-domain, time-frequency domain, and pattern recognition analysis [1]. While time-domain methods involve features that are sensitive to the impulsive nature of the signal, frequency domain methods like high frequency resonance technique (HFRT) require expensive computations and impact tests to find the characteristic frequency of vibration signal [4]. Furthermore, these methods are not suitable for non-stationary signals. Time-frequency decomposition algorithms [5] overcome drawbacks in both time and frequency domain methods, and have garnered significant attention in the diagnosis of gearboxes. These methods include wavelet transform [6, 7], empirical mode decomposition (EMD) and its variants [8, 9, 10], cyclo-stationarity analysis [11], blind source separation (BSS) [12, 13, 14], and many other time-frequency decompositions [15]. On the other hand, pattern recognition methods including Gaussian Mixing model (GMM) [16], Hidden Markov model (HMM) [17, 18, 19] or semi-Markov model [20], and symbolic analysis [21] have also been investigated in the context of fault diagnosis. Amongst these, HMMs offer a unified mathematical structure representing the dynamic relationship between observation and severity of faults.

In their seminal work [18] showed how data and expert knowledge of a mechanical system can be used to train HMMs in order to detect faults in a system. This HMM-based diagnosis scheme was further explored in the area of metal cutting tools [19, 22], where the issues associated with selection of number of states, sensor fusion, and preparation of codebook were underscored. In [23], linear prediction model coefficients are used as features in a HMM model while dealing with single or multiple faults. A combined BSS-based factorial HMM is proposed as a fault classifier for the speed-up and speed-down processes in rotating machinery [24]. In another work, matching pursuit decomposition is first used to extract time-frequency damage features of the signal and then these features are used as observation sequences in HMM for damage detection in laminated composites [25]. In [26], the authors proposed an online HMM-based fault classification technique using features that are functions of energies of time-frequency distribution.

Most previous studies of fault diagnosis using HMM assume that the number of health states is known *a priori*. The discretization of states and observations (i.e. preparation of codebook) make the task of characterizing multiple faults under different operating conditions difficult [19]. Moreover, when the features are relatively insensitive to faults and signal is buried in measurement noise, HMM faces significant challenges [22] in identifying correct physical states of the system as well as in preparing the codebook. Therefore there is a need for more robust and generalized HMM models for fault diagnostics under multiple failure modes and noisy measurements. To address these issues, wavelet packet transformation is first employed to decompose raw vibration signals into sparse amplitude and frequency modulated signals. Out of these, low energy signals are discarded and the de-noised signal is reconstructed back using the higher energy packet coefficients. Next, Teager Kaiser Energy operator is calculated on the de-noised signal, and the amplitude and frequency modulated signals are obtained. Condition indicators are then computed from the modulated signals that represent the observations of HMM. GMM is then utilized to estimate the unknown number of states of the HMM and to prepare a simpler codebook to discretize the states and observations. Finally HMMs are constructed to represent faulty training data and are used for fault diagnosis.

The paper is organized as follows. A brief background of GMM, HMM, wavelet packet transform and Teager Kaiser Operator are first presented, followed by the details of the proposed method. Finally, the proposed method is validated using vibration data obtained from the Bearing Data Center at the Case Western Reserve University.

Background: Before going into the details of the proposed method, a brief background on GMM, HMM, and wavelet packet transform are presented in this section.

(a) Gaussian Mixture Model: A Gaussian Mixture Model (GMM) [27] is a powerful clustering method, where each cluster is represented by a Gaussian distribution. The entire data set is therefore expressed as a weighted sum of Gaussian density components:

$$p(\boldsymbol{x}|\boldsymbol{\lambda}) = \sum_{i=1}^{M} w_i f(\boldsymbol{x}|\boldsymbol{\mu}_i, \boldsymbol{\Sigma}_i) \tag{1}$$

where $\boldsymbol{x}$ is an n-dimensional continuous data or feature vector, $\boldsymbol{w_i}$ is the weighting coefficient of each GMM. $\boldsymbol{f(x|\mu_i ,\Sigma_i)}$ is the component Gaussian density that can be represented as:

$$f(\boldsymbol{x}|\boldsymbol{\mu}_i, \boldsymbol{\Sigma}_i) = \frac{1}{(2\pi)^{n/2}|\boldsymbol{\Sigma}_i|^{1/2}} \exp\{-\frac{1}{2}(\boldsymbol{x}-\boldsymbol{\mu}_i)^T \boldsymbol{\Sigma}_i^{-1}(\boldsymbol{x}-\boldsymbol{\mu}_i)\} \tag{2}$$

where the mixing coefficients satisfies the constraint. The model parameters are symbolized as $\boldsymbol{\lambda} = \{w_i , \mu_i , \Sigma_i\}$. These parameters can be estimated by maximizing the likelihood function (i.e. the probability of the observed data). Consider the observations $\mathbf{X} = \{\mathbf{x_1}, \mathbf{x_2}, ..., \mathbf{x_p}\}$. By assuming independence among the observations, their joint probability can be written as:

$$p(\boldsymbol{X}|\boldsymbol{\lambda}) = \prod_{j=1}^{p} p(\boldsymbol{x}_j|\boldsymbol{\lambda}) \tag{3}$$

Then log-likelihood function is then represented as:

$$L(\boldsymbol{\lambda}) = \sum_{j=1}^{p} \log p(\boldsymbol{x}_j|\boldsymbol{\lambda}) = \sum_{j=1}^{p} \log \sum_{i=1}^{M} w_i f(\boldsymbol{x}_j|\boldsymbol{\lambda}_i) \tag{4}$$

The likelihood function is then maximized by taking its derivatives w.r.t λ_k [27],

$$\frac{\partial L}{\partial \lambda_k} = \sum_{j=1}^{p} \frac{w_k}{\sum_{i=1}^{M} w_i f(\boldsymbol{x}_j|\boldsymbol{\lambda}_i)} \frac{\partial f(\boldsymbol{x}_j|\boldsymbol{\lambda}_k)}{\partial \lambda_k}$$
$$= \sum_{j=1}^{p} \frac{w_k f(\boldsymbol{x}_j|\boldsymbol{\lambda}_k)}{\sum_{i=1}^{M} w_k f(\boldsymbol{x}_i|\boldsymbol{\lambda}_k)} \frac{\partial \log f(\boldsymbol{x}_j|\boldsymbol{\lambda}_k)}{\partial \lambda_k}$$

The above computation is accomplished using the iterative Expectation-Maximization (EM) algorithm. Figure 1 shows the performance of clustering of two Gaussian features using GMM. It is seen that GMM is able to separate the individual Gaussian components clearly as evident in two distinct clusters for the case of two Gaussians comprising the mixtures.

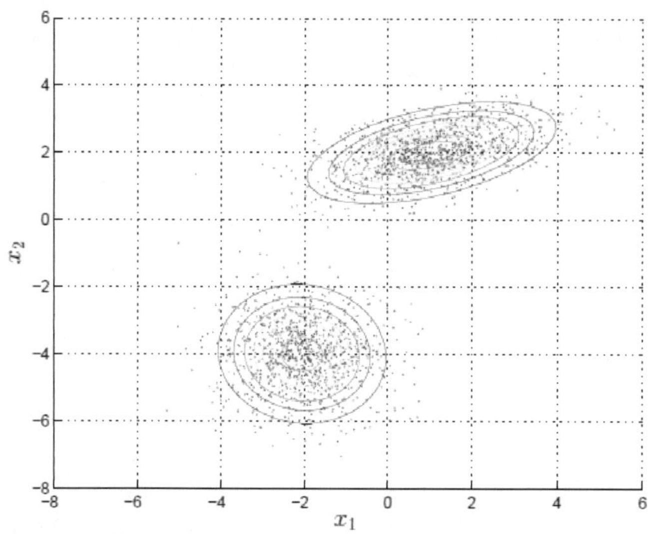

Figure 1: Clustering of two Gaussian features using GMM

(b) Hidden Markov Model: The basic idea of Hidden Markov model (HMM) [17] stems from classical Bayesian theory. In fact, HMM is a doubly stochastic Markov model where both states and observations are random in nature. Before going into the details of HMM, let us consider a system with n discrete states. According to Baye's rule, the joint probability of these states can be expressed using chain rule:

$$P(q_t, q_{t-1}, .., q_1) = P(q_t|q_{t-1}, q_{t-2}, .., q_1) \times P(q_{t-1}|q_{t-2}, q_{t-3}, .., q_1) \times .. \times P(q_2|q_1) \times P(q_1|q_0) \quad (5)$$

$$P(q_t, q_{t-1}, .., q_1) = \prod_{l=1}^{t} P(q_l|q_{l-1}, q_{l-2}, .., q_1) \quad (6)$$

Assuming Markov properties for this process, the future state depends only on its current state, not on previous states. Therefore Equation 6 can be simplified as:

$$P(q_t, q_{t-1}, ..., q_1) = \prod_{l=1}^{t} P(q_l|q_{l-1}) \quad (7)$$

This forms the basic equation of a discrete Markov process. A Markov process can be classified into two classes: (a) Markov chain, and (b) Hidden Markov model (HMM). Unlike in a Markov chain, states are not uniquely observed in HMM and are indirectly observed as a probabilistic function of states. Using simpler notation, the probability terms of Equation 7 can be expressed as:

$$P(q_l = s_j | q_{l-1} = s_i) = a_{ij} \tag{8}$$

where, a_{ij} is the transition probability between states. The transition probability of the entire system with n states can be expressed in a matrix form:

$$A = \begin{bmatrix} a_{11} & a_{12} & .. & .. & a_{1n} \\ a_{21} & a_{22} & .. & .. & a_{2n} \\ . & . & . & . & . \\ . & . & . & . & . \\ a_{n1} & a_{n2} & .. & .. & a_{nn} \end{bmatrix} \tag{9}$$

where each of the elements of A satisfies the stochastic constraints $a_{ij} \geq 0$ and $\sum a_{ij} = 1$, $i = 1, 2, ..., n$ with associated initial probability matrix $I = \{\pi_i\}$, $\pi_i = P(q_t = s_i)$. In HMM, each observation is associated with states through observation probability matrix, $B = \{b_i(k)\}$, $b_i(k) = P(x_k | q_t = s_i)$; $1 \leq k \leq m$. The HMM model λ is comprised of above three terms, i.e. $\lambda = \{I, A, B\}$. Basic idea is to select the model that best explains the data (e.g. likelihood function) or maximizes the posterior probability. Since we do not know the state sequence; numerical computation is expensive as the total number of computation contains $(2n-1)N^T$ multiplications and $(N^T - 1)$ additions. Therefore, it is computed numerically [17]:

- Compute probability of a given observation sequence $P(X|\lambda)$ using forward and backward procedure.
- Optimal state estimation that explains best the given observation using Viterbi algorithm.
- Adjust model parameter to maximize $P(X|\lambda)$ using Baum-Welch algorithm that gives maximum likelihood estimate of the parameters of an underlying distribution from a given set.

Figure 2 shows a simulated 3-state HMM model where A and B matrices are also shown. In this example, the three states correspond to three health conditions (good, alarming and faulty) and three observations correspond to low, medium and higher values of condition indicators respectively. Using this HMM model, state and observation sequences are estimated and shown in Figure 3. These state and observation sequences form the test bed for HMM, from which the model parameters (λ) are estimated and then used for the fault detection.

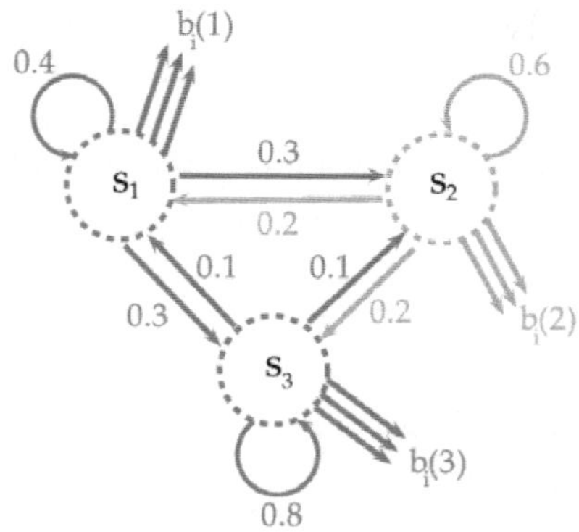

$$A = \begin{bmatrix} 0.4 & 0.3 & 0.3 \\ 0.2 & 0.6 & 0.2 \\ 0.1 & 0.1 & 0.8 \end{bmatrix}$$

$$B = \begin{bmatrix} 0.7 & 0.2 & 0.1 \\ 0.15 & 0.75 & 0.1 \\ 0.2 & 0.2 & 0.6 \end{bmatrix}$$

Figure 2: Simulated HMM model

(c) Wavelet packet transforms

One of the fundamental building blocks in the proposed algorithm is the redundant version of the discrete Wavelet transform (WT), called the Stationary Wavelet Packet Transform (SWPT) [28, 29]. In this study, SWPT is used as a de-noising tool to extract nonstationary vibration signatures associated with gears and bearings while the

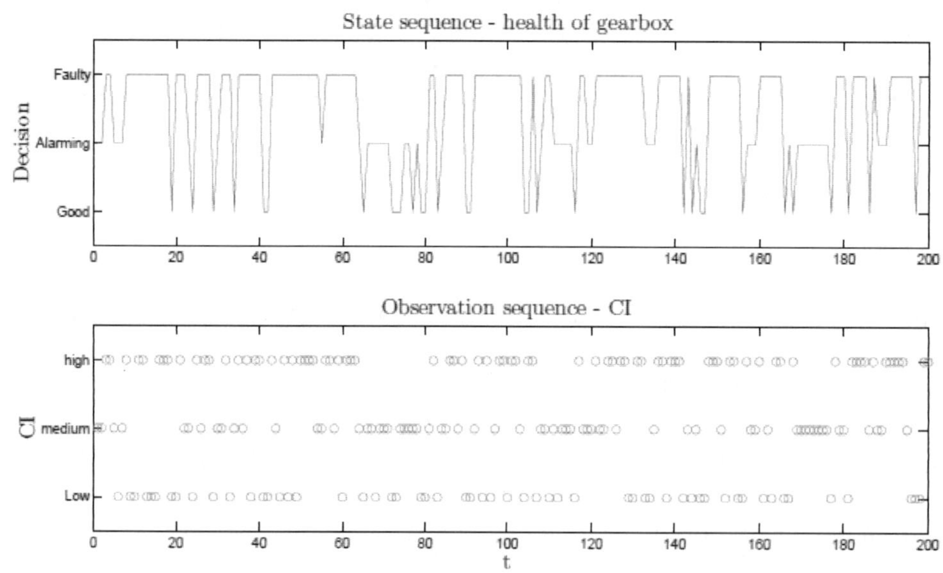

Figure 3: Simulated state and observation sequences

unnecessary contributions from noise are rejected. This is accomplished in a multi-resolution framework [28]. The WT of a signal *y(t)* is a linear transform defined as:

$$w_k^j(y) = \frac{1}{\sqrt{j}} \int_{-\infty}^{\infty} y(t)\, \psi_k^{j*}(t)\, dt \tag{10}$$

Where the function ψ is commonly known as the mother wavelet and * stands for complex conjugation. The indices *j* and *k* denote the scale and translation parameters, respectively. Thus, WT decomposes a signal *y(t)* using basis functions that are simply scaled and translated versions of the mother wavelet [28, 29]. The wavelet packet transform (WPT) [30] is an extension of WT that offers a complete level-by-level decomposition, where both the details as well as the approximations are filtered. As a result, the information in both the low and high frequency sub-bands are retained and processed via filter-banks. A wavelet packet is a triple-index function, and a WPT basis is defined as [30]:

$$\psi_k^{j,v}(t) = \frac{1}{2^j}\psi^v\left(\frac{t-k}{2^j}\right) \tag{11}$$

Where, $\psi^{1,0}(t) = \varphi(t)$ and $\psi^{1,1}(t) = \psi(t)$ represent the scaling (father) and wavelet (mother) functions, respectively. For a given scale level j, a binary tree as shown in Figure 4 can be formed using the basis functions, $\psi_k^{j,v}(t)$, whose nodes are indicated by the scale level, j and the frequency parameter v=0, 1, 2, …, (2^j-1). The wavelet packet coefficients at each node (j, v) are written as:

$$w_k^{j,v}(y) \triangleq \left\langle y(t) \,\middle|\, 2^{-j/2}\psi^v\left(\frac{t-k}{2^j}\right)\right\rangle = \frac{1}{2^{j/2}}\int_{-\infty}^{\infty} y(t)\psi^v\left(\frac{t-k}{2^j}\right)dt \tag{12}$$

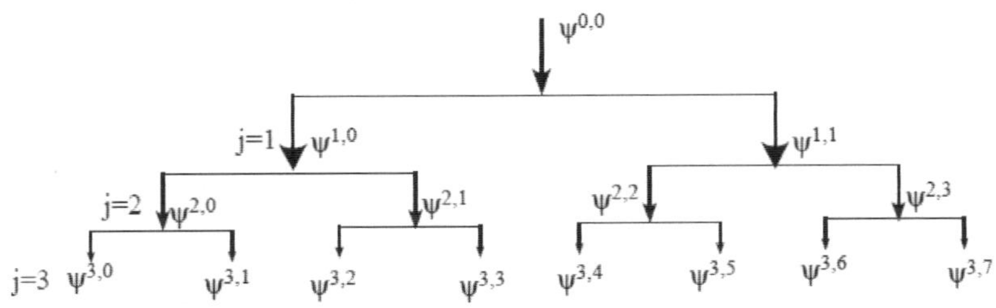

Figure 4: SWPT binary tree indexing of $\psi^{j,v}$

For the *j-th* level of decomposition, the original signal *y(t)* can be expressed as a summation of all the wavelet packet component signals, $y^{j,v}$

$$y(t) = \sum_{v=0}^{2^j-1} y^{j,v} = \sum_{v=0}^{2^j-1}\sum_k w_k^{j,v}\psi_k^{j,v}(t) \tag{13}$$

Thus, the WPT is a generalization of SWT such that each frequency band of the wavelet spectrum is further sub-divided into finer frequency bands, repeatedly.

(d) Teager Kaiser Operator

Teager Kaiser (TK) operator is a nonlinear energy operator, defined as [31]:

$$\Psi(x(t)) = \dot{x}^2(t) - x(t)\ddot{x}^2(t) \tag{14}$$

It can be shown that TK operator of an amplitude and frequency modulated signal yields its amplitude and frequency modulation respectively (see Figure 5). Such an amplitude envelope of a complex modulated signal facilitates fault detection by suppressing noise and intensifying local features while cancelling the global trend of the signal as explained in next section.

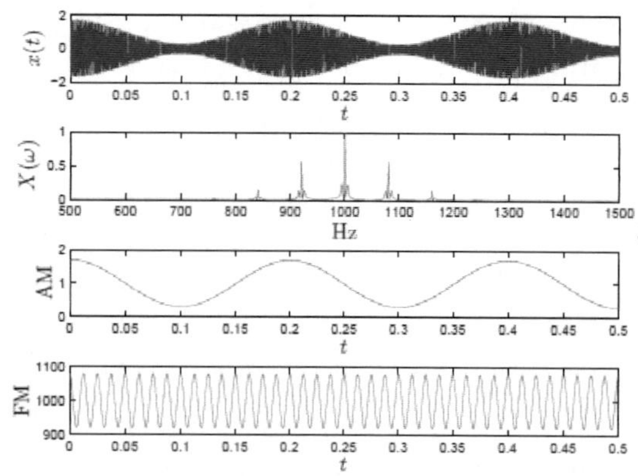

Figure 5: Taeger Kaiser operator

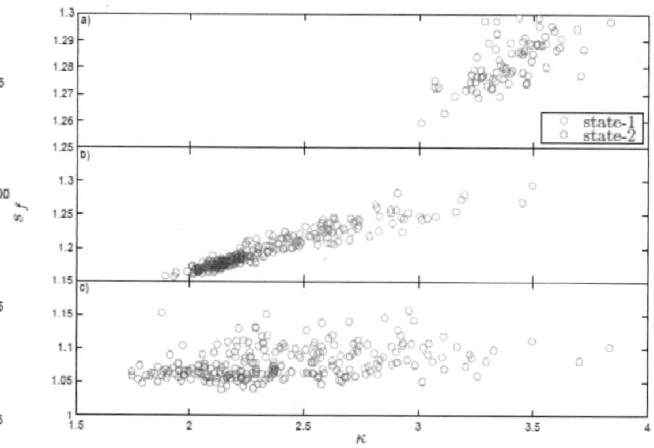

Figure 6: (a) Raw data (b) WPT (c) WPT-TK

Proposed method: The basic idea of the proposed method is revealed in Figure 6, where two different AM-FM signals with the same carrier frequencies but different modulated frequencies are considered as state-1 and state 2. Figure 6(a) shows scatter plot of the condition indicators (CIs) of the raw signals, where s_f and κ are respectively the shape factor and kurtosis of the signal. It can be seen that clusters of state-1 and state-2 are not clearly separated with a definite decision boundary. When the signals are de-noised using WPT (see Figure 6(b)), the separation of two clusters corresponding to two different states are relatively separable. However, when CIs of the amplitude modulation (using TK operator) of the de-noised signal are used, the separability of the clusters is further improved (see Figure 6(c)). This gives the motivation of the proposed method as illustrated in Figure 7, where HMM is employed on the time-frequency domain features with improved separability.

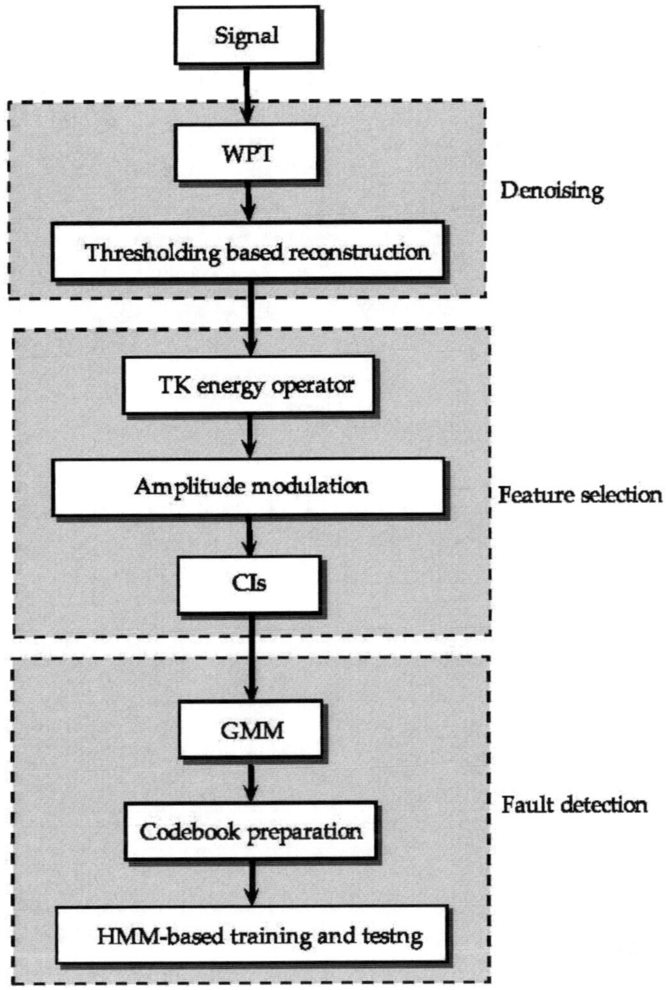

Figure 7: Flowchart of the proposed fault detection scheme

Case study: The test data used in this study was obtained from the Case Western Reserve University bearing data center (http://www.eecs.case.edu/laboratory/bearing). The experimental setup as shown in Figure 8, comprises of a 3-phase induction motor with a torque transducer, fan-end (FE) and drive-end (DE) bearing, and a dynamometer. Both the DE and FE bearing (deep grove ball bearing) data are available and are used in this analysis. The specification details of bearings are tabulated in Table 1.

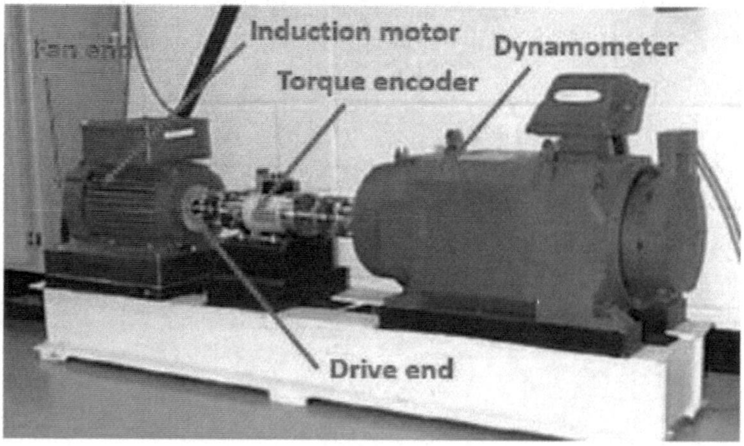

Figure 8: Test rig – Case Western Reserve University Bearing data center

Table 1: Specifications of bearing

Parameters (inch)	DE Bearing	FE Bearing
Inside diameter	0.9843	0.6693
Outside diameter	2.0472	1.5748
Thickness	0.5906	0.4726
Ball diameter	0.3126	0.2656
Pitch diameter	1.537	1.122

With this information, the fault characteristic frequencies of the bearing are estimated [32], which are shown for DE bearing in Table 2. Single point faults are introduced in inner raceway (IR), outer raceway (OR) and ball elements (BP) in each bearing using electron-discharge machining with fault diameters of 7, 14, and 21 mils respectively. Vibration data are collected at 12 kHz and for different fault sizes with load varying from 0 to 3 HP with 1 HP increments. The corresponding speed varies from 1797 RPM (no-load) to 1730 RPM (3 HP load). Accelerometers were placed at both drive-end and fan-end of the motor housing. The data for the outer race fault are taken with the fault position centered at the 6 o'clock position with respect to the load zone. 30% of these data is used to train the HMM and rest is used for the testing purpose that produced 40 samples of test data.

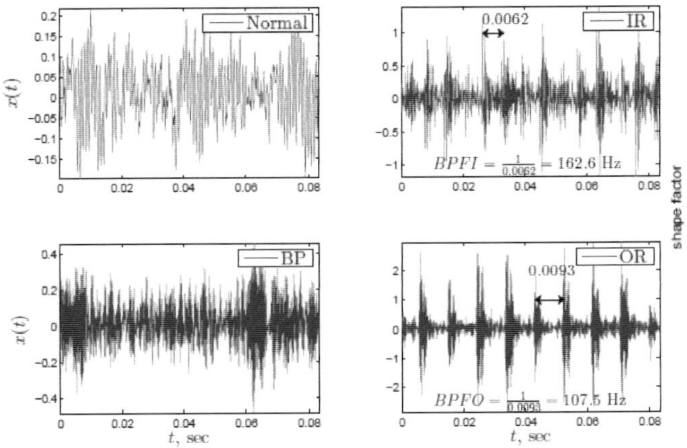

Figure 9: DE signal: 7 mils fault under 0 HP

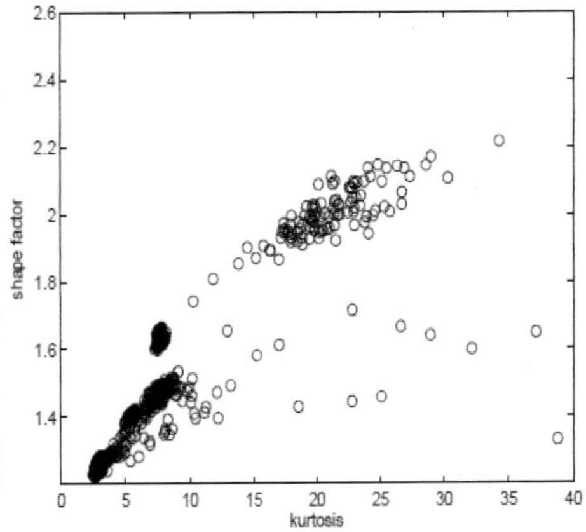

Figure 10: Combination of all failures in DE and FE bearings

Figure 9 shows the acceleration at DE motor housing where vibration data under normal condition is compared with 7 mils of faults at IR, BP and OR failure cases. Due to simulated faults, the intervals between impulse trains are observed in IR and OR failure cases. For example, under IR failure case, the approximate interval between two successive impulses is observed as 0.0062 sec yielding BPFI of 162.6 Hz that matches the theoretical BPFI (i.e. 162.2 Hz in Table 2).

Table 2: Theoretical fault characteristic frequencies of DE bearing

Load (HP)	f_r (RPM)	IR (5.41* f_r)	BP (4.71* f_r)	IR (3.58* f_r)
0	1797	162.2	141.2	107.4
1	1772	159.9	139.2	105.9
2	1750	157.9	137.5	104.6
3	1730	156.1	135.9	103.4

Frequency-domain spectral information reveals such fault characteristic frequencies and their multiples in the vibration data. However when the acceleration data are buried in measurement noise [3], it is difficult to extract such fault frequencies directly from raw data. In this paper, time-domain features (e.g. kurtosis and shape factors) of the de-noised signal using WPT-TK operation are used for fault detection, instead. Figure 10 represents a combination of all the fault cases. It is noted that the example here is used only for demonstration purposes, primarily to demonstrate the methodology.

Before training the data with HMM, GMM is first employed to estimate the number of states in the signals of individual faulty signals. In order to choose the optimal number of states, AIC and BIC criteria are used. As shown in Figure 11, model order of three results in a relatively stable AIC value. In fact, the number of true states is 3 (i.e., normal, 7 mils and 21 mils of faults). Figure 12 shows the GMM modeling of the inner race failure data. It also gives the mean and variances of the corresponding clusters from which observation clusters are obtained. This operation is repeated for other clusters and then a combined observation set is obtained. Finally, the codebook is prepared based on thresholding these observations.

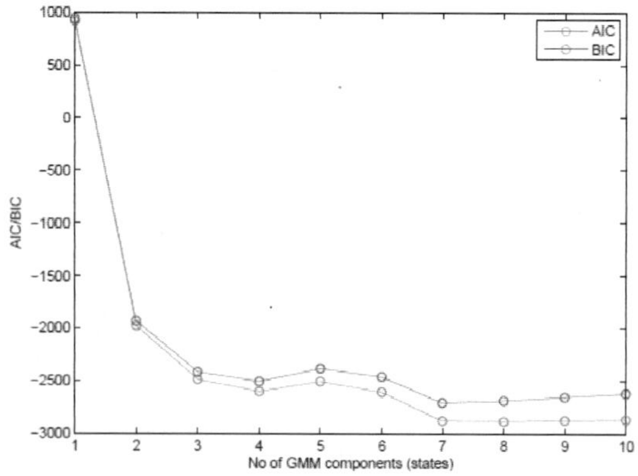

Figure 11: AIC and BIC of GMM model

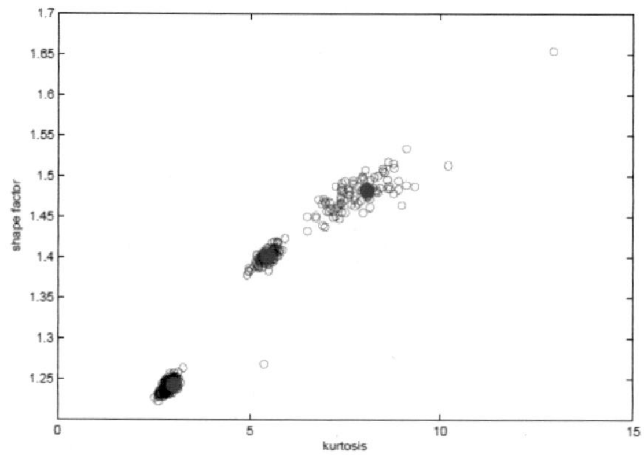

Figure 12: Outer race failure

Figure 13 and 14 show the discretized state and observation sequences of inner race and outer race data data, respectively, using the combined codebook from which the HMM parameters are calculated. Once the HMMs are estimated for all the individual fault cases, the test data can be used to estimate its unknown states. The data corresponding to

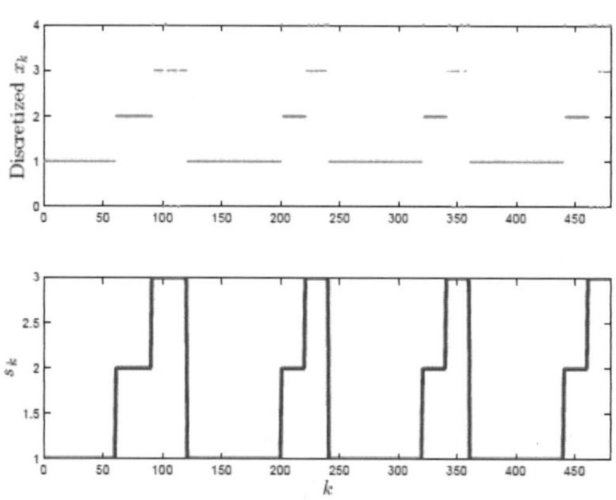

Figure 13: State/observation of IR model

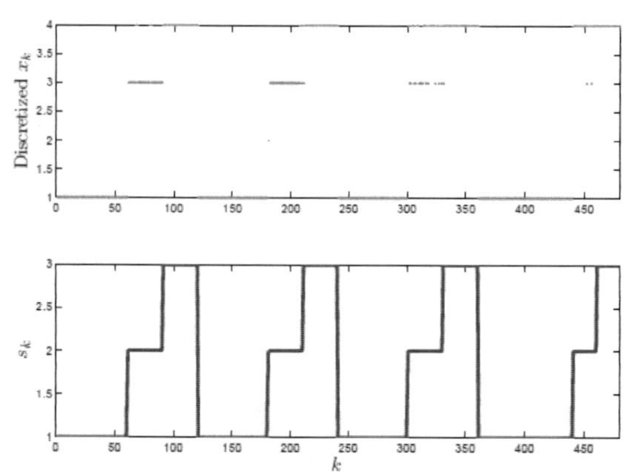

Figure 14: State/observation of OR model

an unknown health state can be identified based on the trained HMM model which results in the highest likelihood of the state. Table 3 shows the fault detection result of a sample

Table 3: Results of a sample test data using hybrid HMM

Test	IR HMM	OR HMM	IR-OR HMM
IR	**-300**	-940	-1130
OR	-690	**-424**	-870
IR-OR	-926	-478	**-227**

Table 4: Confusion matrix for 40 test data without and with measurement noise (the data in bracket denotes the results under SNR=10) using hybrid HMM

Test	IR HMM	OR HMM	IR-OR HMM
IR	39 (38)	0 (1)	1 (1)
OR	1 (1)	38 (37)	1 (2)
IR-OR	0 (2)	0 (1)	40 (37)

test data using hybrid HMM. Similar exercise is repeated for other 40 sets of data with and without measurement noise (SNR=10). Table 4 shows the fault detection result of the test data using HMM model, in terms of a confusion matrix. Results indicate more than 90% success rate even with noisy measurements containing SNR=10. The example shown here is an illustrative example to introduce the main theory in the proposed method. Currently, the authors are working with large set of interconnected gear motors to test and verify the performance of this method.

Conclusion: In this paper, an improved hybrid HMM is proposed to perform multi-state fault detection. Wavelet packet transformation based energy operator is first used to construct de-noised features from the raw signal, and then GMM-based HMM is undertaken for fault detection. The proposed method enables a generalized and robust HMM approach where state estimation and codebook preparation are accurate and simpler compared to the traditional HMM methods under measurement noise. The proposed method will be further exploited for condition based maintenance of a critical facility in Toronto airport.

References

[1] A Jardine, D Lin, and D Banjevic, A review on machinery diagnostics and prognostics implementing condition-based maintenance, Mechanical Systems and Signal Processing, 2006, 20, 1483-1510.

[2] W Wang, F Ismail, and M F Golnaraghi, Assessment of gear damage monitoring techniques using vibration measurements, Mechanical Systems and Signal Processing, 2001, 15(5), 905-922.

[3] R B Randall and J Antoni, Rolling element bearing diagnostics: A tutorial, Mechanical Systems and Signal Processing, 2011, 25(2), 485-520.

[4] V Purushothama and S Narayanana, Multi-fault diagnosis of rolling bearing elements using wavelet analysis and hidden Markov model based fault recognition, NDT&E International, 2005, 38, 654-664.

[5] Z Feng, M Liang, and F Chu, Recent advances in time-frequency analysis methods for machinery fault diagnosis: A review with application examples, Mechanical Systems and Signal Processing, 2013, 38(1), 165-205.

[6] Z Peng and F Chu, Application of the wavelet transform in machine condition monitoring and fault diagnostics: a review with bibliography, Mechanical Systems and Signal Processing, 2004, 18(2).

[7] R Yan, R X Gao, and X Chen, Wavelets for fault diagnosis of rotary machines: A review with applications, Signal Processing, 2013.

[8] Y Lei, J Lin, Z He, and M J Zuo, A review on empirical mode decomposition in

fault diagnosis of rotating machinery, Mechanical Systems and Signal Processing, 2013, 108-126.
[9] H Jiang, C Li, and H Li (2013), An improved EEMD with multi-wavelet packet for rotating machinery multi-fault diagnosis. Mechanical Systems and Signal Processing, 36(2), 225-239.
[10] X Zhao, T Patel, and M J Zuo, Multivariate EMD and full spectrum based condition monitoring for rotating machinery. Mechanical Systems and Signal Processing, 2012, 27(0), 712-728.
[11] Z Zhu, Z Feng, and F Kong. Cyclostationarity analysis for gearbox condition monitoring: Approaches and effectiveness. Mechanical Systems and Signal Processing, 2005, 19(3), 467-482.
[12] G Gelle and M Colas, Blind source separation: a tool for rotating machine monitoring by vibrations analysis?, Journal of Sound and vibration, 2001, 248(5), 865-885.
[13] V H Nguyen, and J C Golinval, Fault detection based on kernel principal component analysis. Engineering Structures, 2010, 32(11), 3683-3691.
[14] V H Nguyen, C Rutten, and J C Golinval, Fault diagnosis in industrial systems based on blind source separation techniques using one single vibration sensor. Shock and Vibration, 2012, 19, 795-801.
[15] C Li and M Liang, Time-frequency signal analysis for gearbox fault diagnosis using a generalized synchrosqueezing transform, Mechanical Systems and Signal Processing, 2012, 26(0), 205-217.
[16] F Nelwamondo, T Marwala, and U Mahola, Early classifications of bearing faults using hidden Markov models, Gaussian mixture models, mel-frequency cepstral coefficients and fractals, 2005.
[17] L R Rabinar, A tutorial on hidden Markov models and selected applications in speech recognition. Proc. of IEEE, 1989, 77(2), 257-287.
[18] C Bunks and D Mccarthy, Condition-based maintenance of machines using hidden Markov models, Mechanical Systems and Signal Processing, 2000, 14(4), 597-612.
[19] T Boutros and M. Liang, Detection and diagnosis of bearing and cutting tool faults using hidden Markov models, Mechanical Systems and Signal Processing, 2011, 25(6), 2102-2124.
[20] Q Liu, M Dong, and Y Peng, A novel method for online health prognosis of equipment based on hidden semi-Markov model using sequential monte carlo methods. Mechanical Systems and Signal Processing, 2012, 32(0), 331-348.
[21] S C Chin, A Ray, and V Rajagopalan, Symbolic time series analysis for anomaly detection: a comparative evaluation, Signal Processing, 2005, 85, 1859-1868.
[22] P Baruah and R B Chinnam, HMMs for diagnostics and prognostics in machining processes, International Journal of Production Research, 2005, 43(6), 1275-1293.
[23] H Ocak and K A Loparo, HMM-based fault detection and diagnosis scheme for rolling element bearings. Journal of vibration and acoustics, ASME, 2005, 127(4), 299-306.
[24] Z Li, Z Wu, Y He, and C Fulei, Hidden Markov model-based fault diagnostics method in speed-up and speed-down process for rotating machinery, Mechanical Systems and Signal Processing, 2005, 19, 329-339.

[25] W Zhou, N Kovvali, W Reynolds, A Papandreou-Suppappola, A Chattopadhyay, and D Cochran, On the use of hidden Markov modeling and time-frequency features for damage classification in composite structures, Journal of Intelligent Material Systems and Structures, 2009, 20(11), 1271-1288.

[26] Q Miao and V Makis, Condition monitoring and classification of rotating machinery using wavelets and hidden Markov models, Mechanical Systems and Signal Processing, 2007, 21, 840855.

[27] G McLachlan, Mixture Models, Marcel Dekker, New York, NY, 1998.

[28] M Wickerhauser, Adapted Wavelet Analysis from Theory to Software, A. K. Peters, Ltd., Wellesley, Massachusetts, 1994.

[29] S G Mallat, A wavelet tour of signal processing, Academic Press, San Diego, 1998.

[30] J C Pesquet, H Krim, and H Carfantan, Time-invariant orthonormal wavelet representations. IEEE Transactions in Signal Processing, 1996, 44(8), 1964-1970.

[31] J F Kaiser, Some useful properties of Teager's energy operators. Proc. IEEE ICASSP, 1994, 3, 149-152.

[32] I Howard. A review of rolling element bearing vibration: Detection, diagnosis and prognosis. DSTO-AMRL report, DSTO-RR-00113, 1994, 115-121.

PERFORMING DIAGNOSTICS ON THE SPALLATION NEUTRON SOURCE VISION BEAM LINE TO ELIMINATE HIGH VIBRATION LEVELS AND PROVIDE SUSTAINABLE OPERATION

Blake Van Hoy
Oak Ridge National Laboratory
P.O. Box 2008 - 6174
Oak Ridge, Tennessee 37831-6174
Telephone: (865) 576-0340
vanhoybw@ornl.gov.

Abstract: The Spallation Neutron Source (SNS) at the Oak Ridge National Laboratory (ORNL) provides a range of neutron energies for a variety of experiments. The neutrons proceed down beam lines in the experiment hall, which houses a variety of experiments and test instruments. Each beam line has one or more neutron choppers which filter the neutron beam based on neutron energy by using a rotating neutron absorbing material passing through the neutron beam.

Excessive vibration of the Vision beam line, believed to be caused by the T0 chopper, prevented the Vision beam line from operating at full capacity. This problem was expected to be specific to the installation at the Vision instrument and not the T0 chopper itself. Vibrations of the T0 chopper were 3 times higher installed at Vision than they were when the T0 chopper was installed in the chopper shop. To determine the cause of the high vibration, dynamic testing was performed. Twenty-seven channels of data, both accelerometers and motor current, were collected during drive up, drive down, coast down, and steady-state conditions; resonance testing and motor current signature analysis were also performed.

The data was analyzed for traditional mechanical/machinery issues such as misalignment and imbalance using time series analysis, frequency domain analysis, and operating deflection shape analysis. The analysis showed that the chopper base plate was experiencing an amplified response to the excitation provided by the T0 neutron chopper. The amplified response was diagnosed to be caused by higher than expected base plate flexibility, possibly due to improper grouting or loose floor anchors.

Based on this diagnosis, a decision was made to dismantle the beam line chopper and remount the base plate. Neutron activation of the beam line components make modifications to the beam line especially expensive and time consuming due to the radiation handling requirements, so this decision had significant financial and schedule implications. It was found that the base plate was indeed loose because of improper grouting during its initial installation.

The base plate was modified by splitting it into multiple sections, isolating the T0 chopper from the rest of the beam line, and each section was then reinstalled and re-grouted. After these modifications, the vibration levels were reduced by a factor of 30. The reduction in vibration level was sufficient to allow the Vision beam line to operate at full capacity for the first time since its completed construction date.

Key words: Diagnostics; operation deflection shape; vibration; Spallation Neutron Source; chopper; looseness; motor current signature analysis

Introduction: The Oak Ridge National Laboratory's Spallation Neutron Source (SNS) is a $1.4 billion facility for neutron interaction and scattering studies with 19 separate beam line experiments (Figure 1). The SNS is the world's most powerful pulsed neutron facility with over 800 kilowatts of power in each pulse and full power of 1.4 megawatts of power in each pulse expected within the next four years.

The beam pulses operate at 60 Hz and as such have several beamline devices that condition the neutron pulse for the desired experiments at each station in the experiment building. A typical beamline device used to condition the neutron pulses are neutron choppers.

Figure 1 - Aerial view of the ORNL SNS facility

The T0 chopper is one of multiple choppers in each beam line that determine the duration and energy of the desired pulse of neutrons entering the experiment at the end of the beam line. The chopper blade is synchronized to arrive in the path of neutrons within a few micro-seconds of the intended neutron energy for each pulse. The blade consists of a 350 pound rotating mass of high density material at a specific location on the chopper wheel that effectively filters the neutron beam to meet that particular experiment's requirements. The chopper is directly driven with a variable speed drive controller that synchronizes the chopper's rotation to the beam pulse. The chopper blade is the large (red) mass shown in Figure 2 as the bolted on piece at the bottom of the wheel.

A beamline would typically have other choppers as well, such as bandwidth choppers, frame overlap choppers, or Fermi choppers to more finely control the energy content of the neutrons reaching the experiments.

Figure 2 - The T0 chopper

The choppers are critical to the uptime and availability of the beam lines. One operational goal of a T0 chopper is to obtain vibration levels and bearing life that will allow the machine to operate 10,000 hours or more at 3,600 RPM of continuous duty before any shutdown maintenance is performed.

The Vision Beamline consists of several pieces of equipment that are in line and closely mounted together on the same baseplate. Figure 3 shows a design rendering of the beamline equipment. The bright yellow unit is the T0 chopper and the pastel green unit to the right is a bandwidth chopper support. The neutrons run from left to right in this schematic.

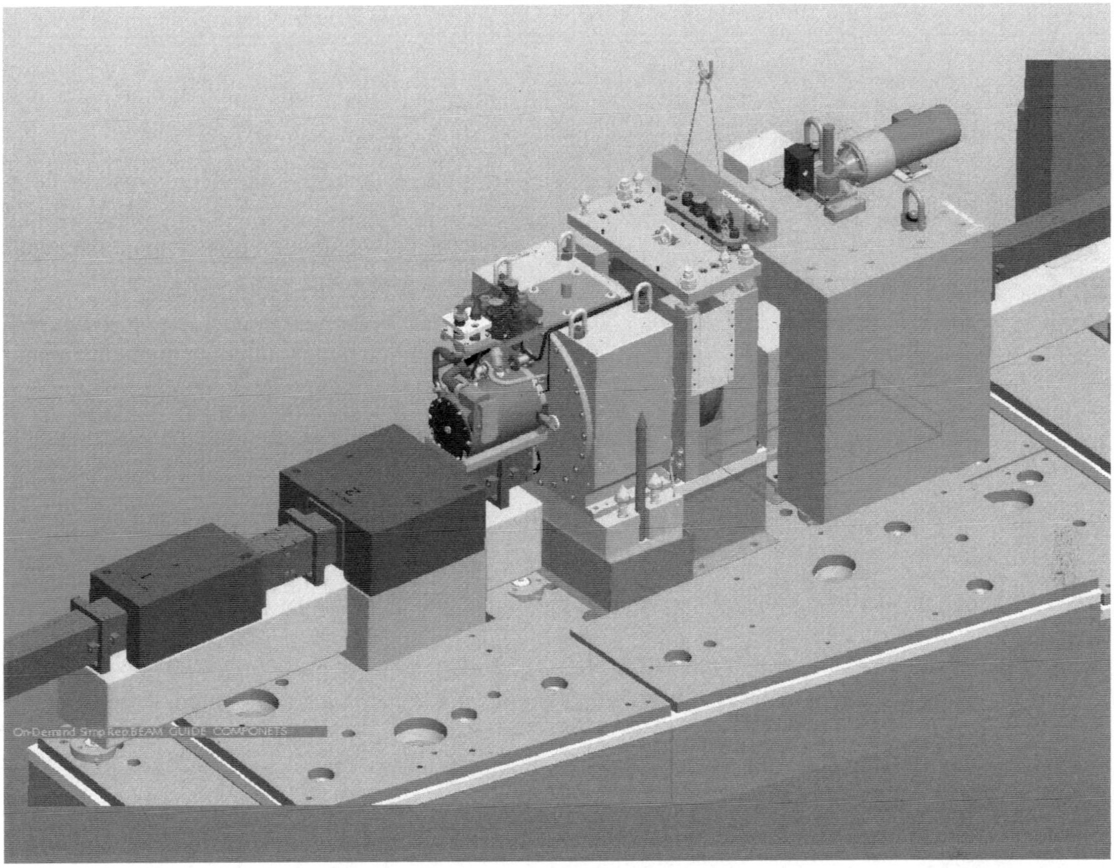

Figure 3 - The Vision Beamline

Problem: The T0 chopper on the Vision Beamline was exhibiting very high levels of vibration at running speed. At 80% of running speed or less the vibration levels were acceptable. The vibration appeared to be speed dependent and because the chopper must be synchronous to the linear accelerator pulses it was impossible to operate at the reduced speed. The vibration generated by the T0 chopper was so large that the bandwidth chopper immediately downstream had to be disabled. This bandwidth chopper runs on SKF magnetic bearings with software limits; these limits were being exceeded and the hard stops were being contacted, caused by the vibration transmitted from the T0 chopper. The inability to run the bandwidth chopper kept the experiment line from being able to meet some of it major project milestones.

Figure 4 shows typical time domain signatures during acceleration to operating speed. The four traces were all at the four corners of the common baseplate.

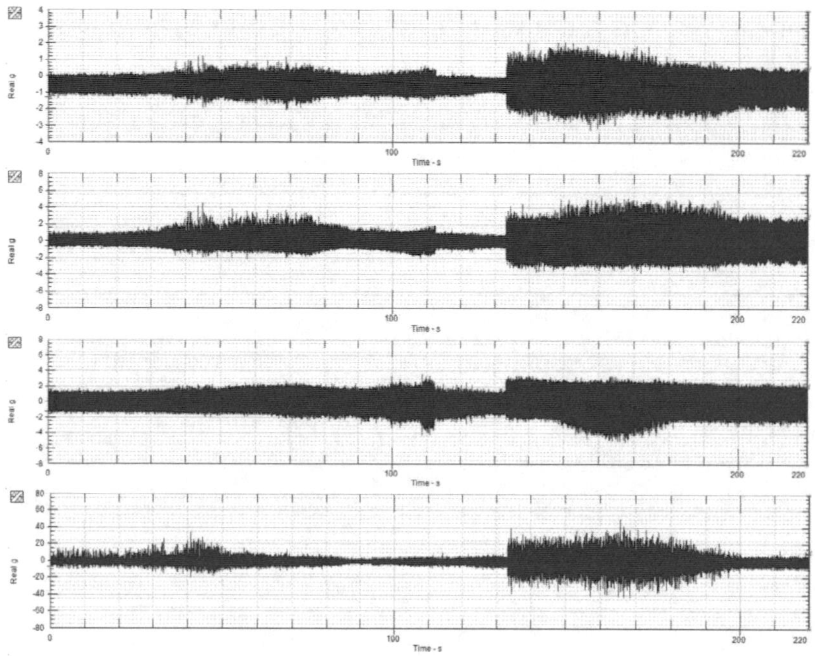

Figure 4 – Time domain response measured during acceleration to operating speed

The acceleration plots of all the measurement points showed the same increasing trend with a sudden amplitude change near operating speed. The area where the measurements were made had limited accessibility. Mounting the sensors and running the cables was challenging. Figure 5 shows the top view after the beam line had been opened and the radiation shielding removed for access.

Figure 5 – Top view of T0 chopper (on the far right)

Diagnostics: The investigation into the cause of the problem used all of the standard machine diagnostics techniques. Unbalance was eliminated because the Chopper Shop at

SNS had certifications from the manufacturer that its rotating components had been balanced to an ISO 1940 G-0.4 level prior to assembly. The unit also ran smoothly in the SNS chopper shop test bed.

Bad or loose bearings were also ruled out because new high precision re-designed bearings had been installed and had seen very few service hours. In addition, there were no bearing frequencies of significant amplitude present in the frequency domain plots.

A modal analysis was conducted with impact excitation on the entire structure and no resonant frequencies near operating speed were shown to be of any concern.

Misalignment was ruled out as the system was direct drive with a very structurally sound machine mount and there were no misalignment frequency components or characteristics in the frequency domain plots.

The only hint of any problem other than amplitude was the high frequency content as the system approached running speed. See the data in Figure 6.

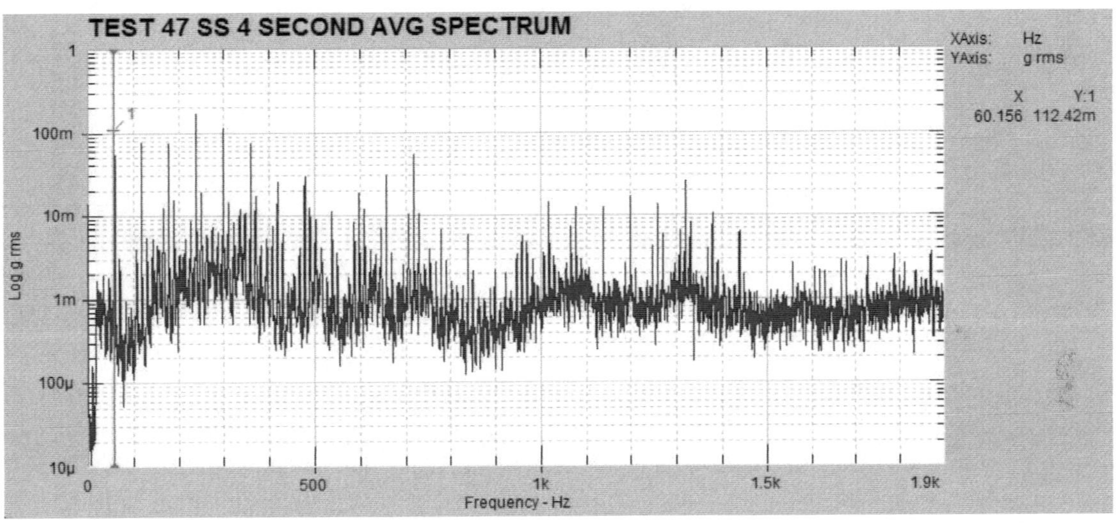

Figure 6 – Frequency domain plot at operating speed

Harmonics in a mechanical structure out to 20 times the running speed are not normal and usually are electrical in nature. Based on this data, motor current signature analysis was employed to check for any unusual electrical or variable frequency drive faults. This again yielded no definitive conclusions regarding the problem or a possible solution.

In all over 27 measurement points and 30 GBytes of data were acquired under different boundary conditions and different operating modalities such as drive up, drive down, coast down, long term steady state with temperature equilibrium, etc.

With no obvious cause and a significant amount of data acquired, several operating deflection shape (ODS) models were run. With the measurement points of the four corners of the base, the top two points on the top of the chopper and the motor as the driving point, the animation finally yielded insight into the cause of the high vibration levels. It was found that the vibration levels on the bandwidth chopper were much higher than those on the T0 chopper, *even with the bandwidth chopper not running*. This unexpected behavior could be explained by a cantilever arrangement of the base which could result in larger vibrations further out on the cantilever. Figure 7 shows the extreme right hand displacement of the base in the ODS analysis. This base of the figure rocks right to left in the animation in phase and equal amplitude with the top and the motor.

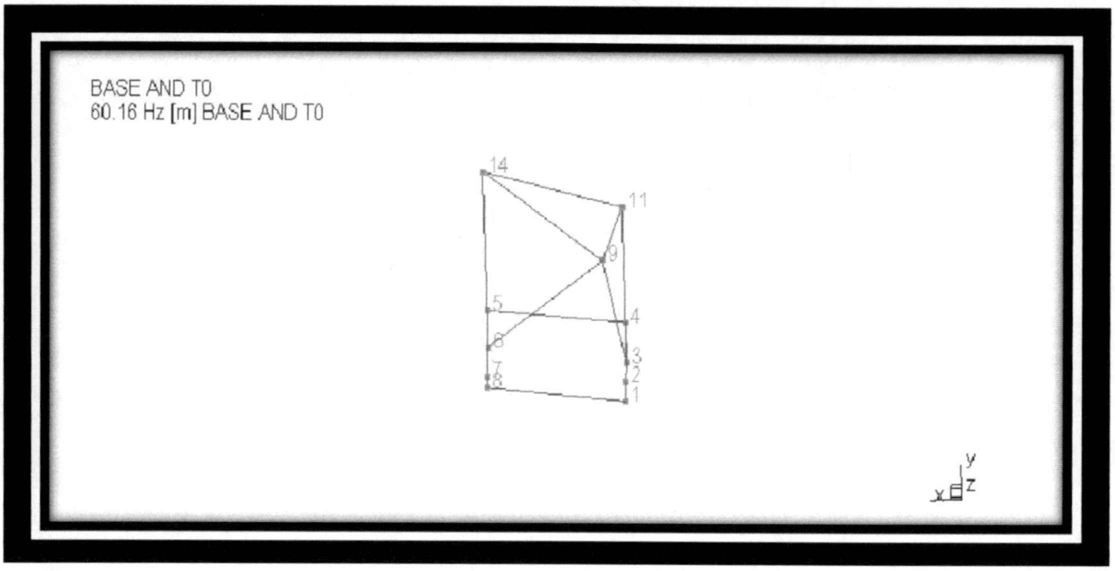

Figure 7 – ODS of T0 and base plate

With this animation and the large number of higher harmonics, it was concluded that looseness of the base plate was the cause of the high vibration levels. This looseness results in a reduction of the stiffness of the mounting system and, as the chopper gets to full speed and generates its maximum unbalance force, that force it sufficient to overcome the stiffness resulting in a dramatic increase in vibration levels.

Fixing the Root Cause Problem: Based on the analysis conclusion, the SNS engineers re-designed the base plates for each piece of equipment in the beam line. Weep holes, extra-thin grout, and a new installation procedure were all brought to bear to assure bonding. The existing equipment was removed from the beam line and the activated base plate was drilled in consecutive holes so the plate could be extracted without generating dust or fumes from cutting with a saw or a torch.

The base plate was removed approximately 10 days after the completion of testing; Figures 8 and 9 show the de-bonded portions of the common base plate. Figure 10 shows the installation of the new base plate.

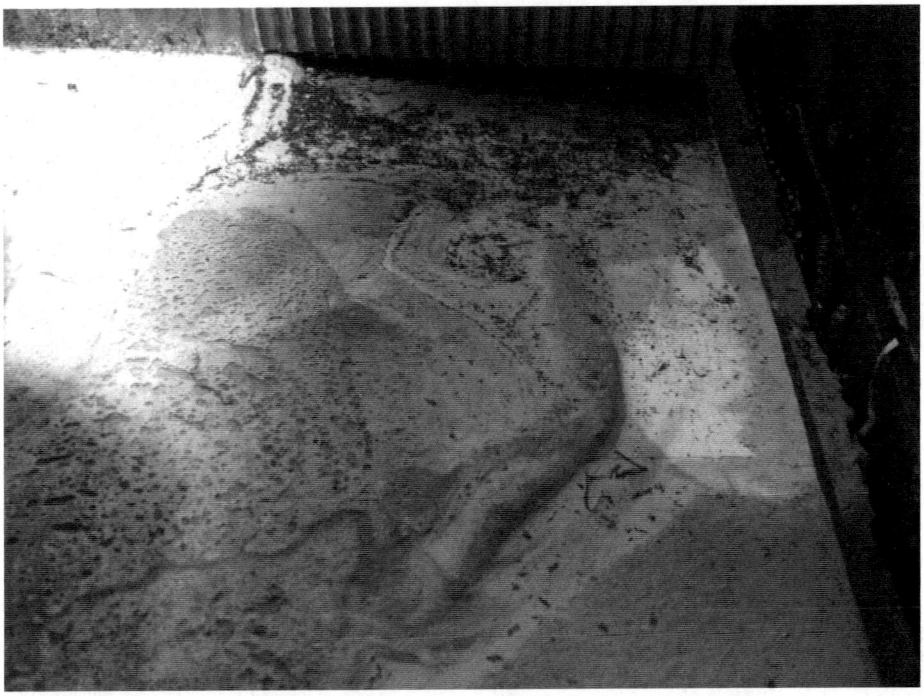

Figure 8 – De-bonded base plate

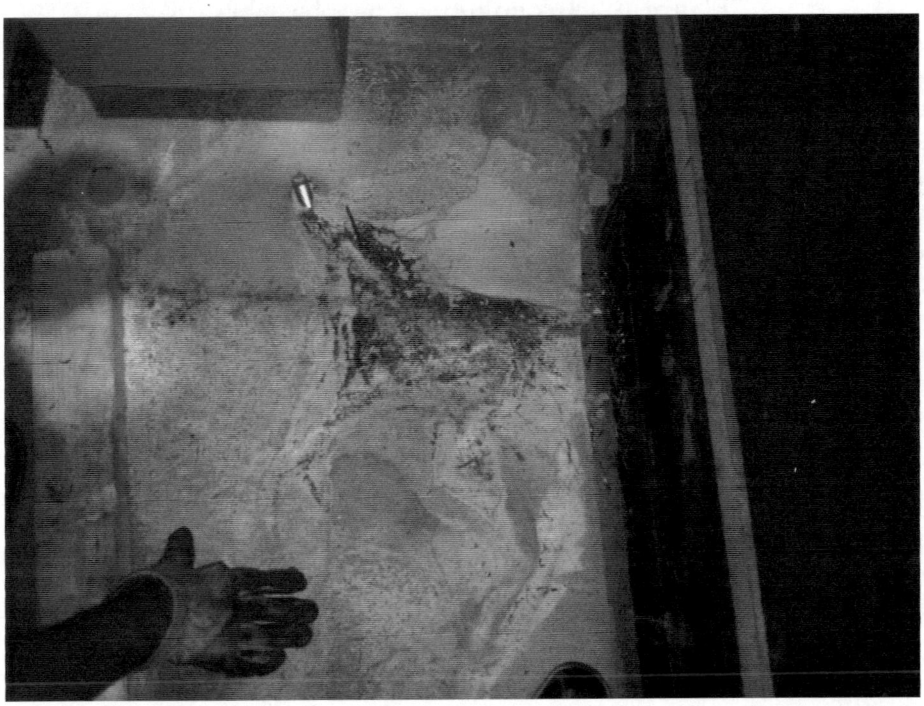

Figure 9 – Total base plate area showing area not bonded

Figure 10 – Installation of new base plate

The new base plate system used separate plates for each piece of equipment to minimize cross coupling between components, larger and more anchor bolts were used to attach each individual plate to the concrete slab, and weep holes were included in the base plates to vent the air during grouting to assure full contact with each base plate.

Results: After installation and alignment the grout was allowed to set to full strength, and then the system was prepared for operation. The first set of runs gave the equipment 4 hours to reach thermal equilibrium and the steady state running yielded a maximum vibration level that was 30 times less than the original readings. The new readings correspond to a newly commissioned machine per ISO 10816-3. Figure 11 shows a data comparison.

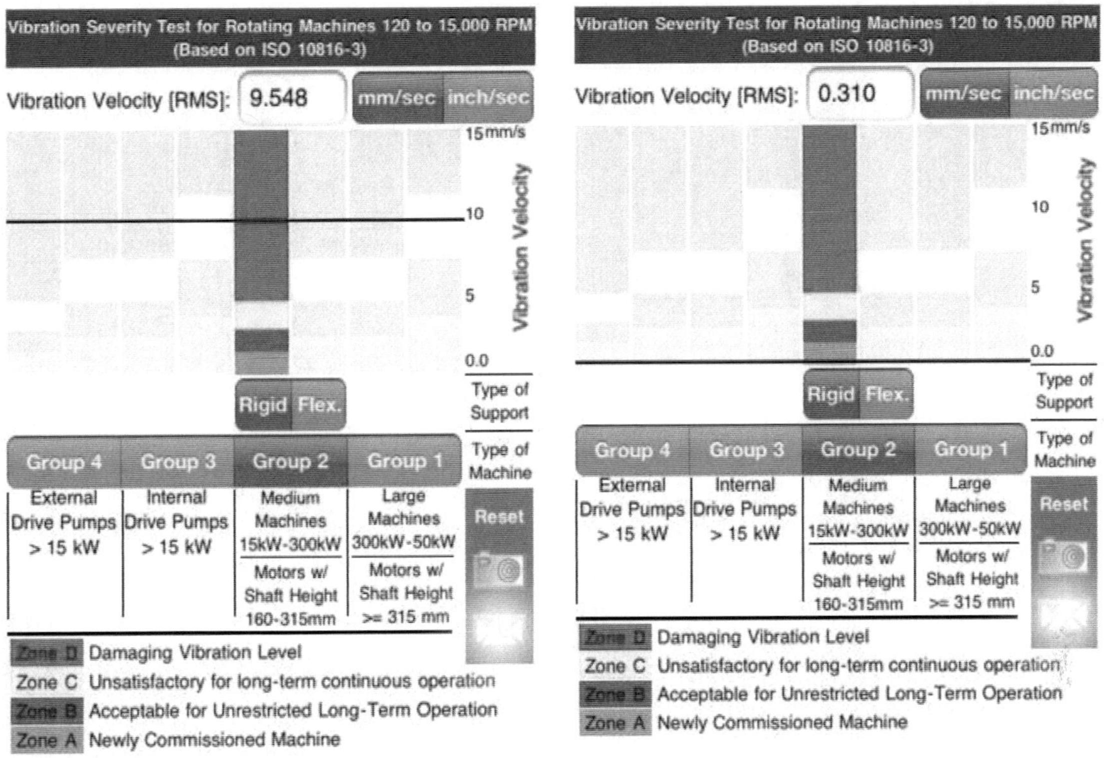

Figure 11 – Comparison of before and after overall vibration velocity

Another useful comparison of effectiveness of the repair is to use the ISO 1940 balance quality chart to compare the after-repair vibration readings with other common machinery. Per ISO 1940, the starting value of 9.5 mm/sec relates to a balance quality of G-40 for crankshafts of trucks and locomotive engines. The final value of 0.31 mm/sec is the value of G 1.0 expected for precision items such as phonographs and precision grinding machine spindles. Figure 12 shows this chart with the before and after values plotted.

The bandwidth chopper with the magnetic bearings has a limit of 150 micrometers displacement on the hard stops. The bearings will not spin or come up on their magnetic field if their displacement is at or above this limit to keep them from sustaining permanent damage. Figure 13 shows the recorded displacement prior to the installation of the new base plate. Notice that the inboard and outboard bearing plots show their displacement and that these are maxed out against the hard stops for one bearing.

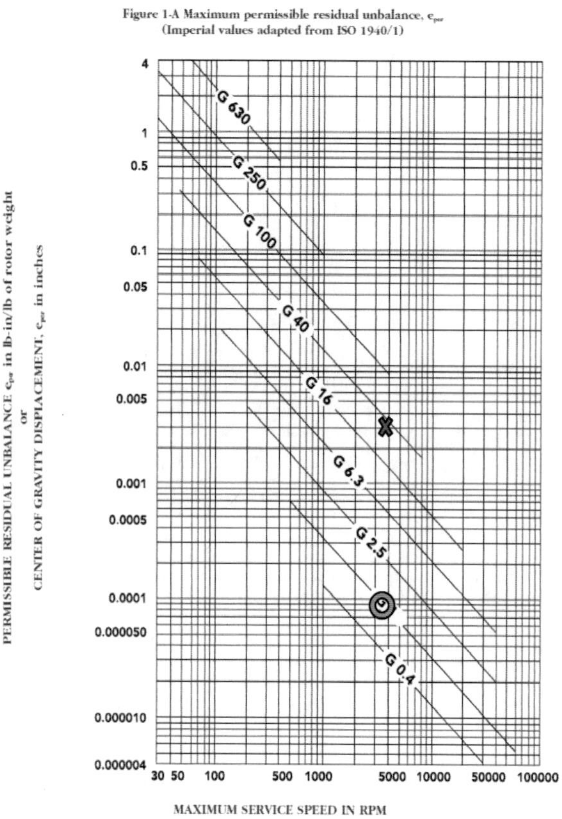

- Original values at 60 Hz run speed. A vibration of 9.5 mm/sec (0.0028 in) is close to a G40 classification. G40 corresponds to crankshafts of trucks and locomotives.
- At the re-designed and re-worked values of 0.31 mm/sec (0.00009 in). G1 corresponds to phonographs and precision grinding machines.

Figure 12 – ISO 1940 balance quality chart

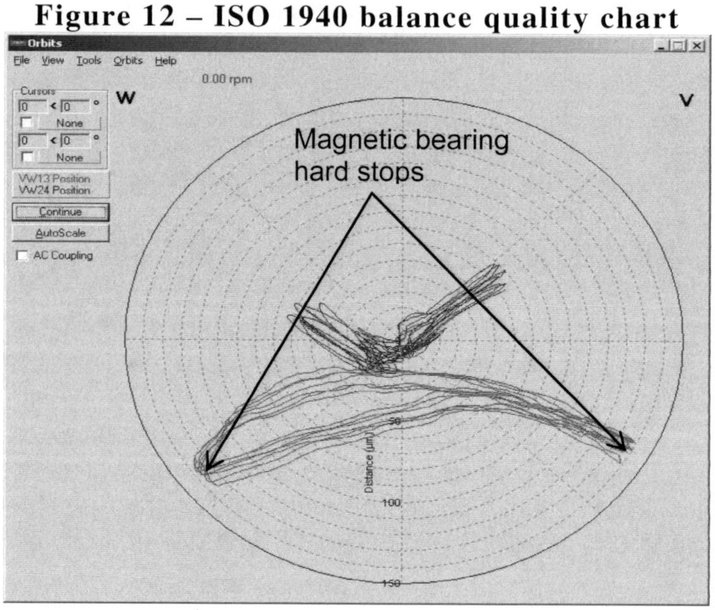

Figure 13 – Un-levitated magnetic bearing displacements with old base plate

After installation of the new base plates and setup of the equipment the un-levitated displacement of the magnetic bearings is shown in Figure 14.

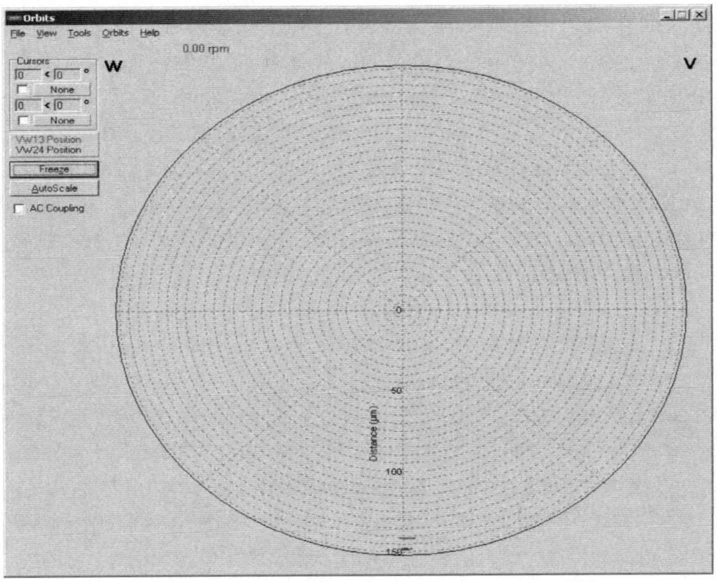

Figure 14 – Un-levitated magnetic bearing displacements after installation of new base plate.

The magnetic bearing displacement with the T0 running and the bandwidth chopper levitated is shown in Figure 15.

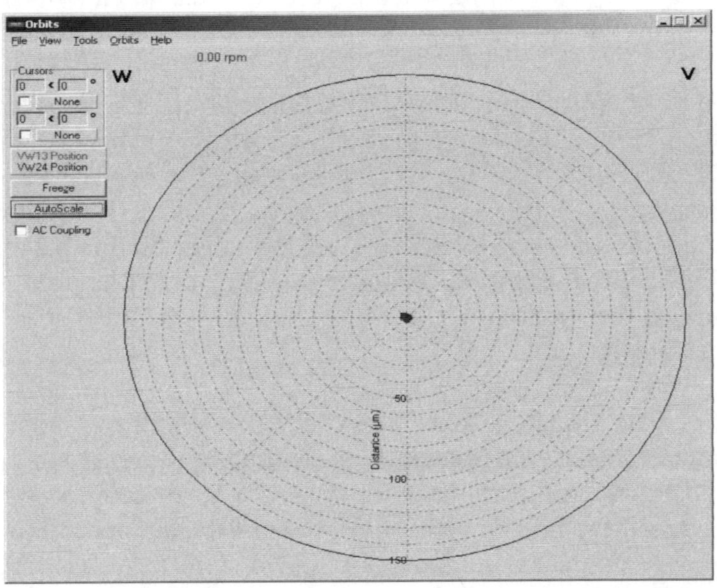

Figure 15 – Levitated magnetic bearing bandwidth chopper after new base plate installation,T0 chopper running.

Additional Results: Action reports and further investigation involving other beam line choppers have resulted in some lessons learned being added to the conclusions of this effort. Since the conclusion of this investigation, another T0 chopper has been built and delivered and its vibration performance was poorer than expected. The chopper shop has obtained a precision balance machine since this effort was concluded. Using the precision balance machine on the new T0 chopper has shown that even though the certification report from the manufacturer indicated that a G 0.4 balance quality was achieved; in fact those components were not balanced to a G 0.4 quality (or even a G 1.0 quality!). So the unbalance force was much greater than expected or specified for this unit. This result suggests the very real possibility that higher imbalance levels were present in the previous chopper units.

Recently, the Vision T0 chopper readings increased to 4 times the current running value, which itself was twice the original base values presented in this paper. Upon further investigation, the decision was made to disassemble the chopper sub assembly and inspect for looseness or possibly cracked or destroyed grout under the base plate. All four bolts were loose or not torqued properly, the specified lock washers were not used in the bolt up assembly, and two of the bolts were a couple of inches shorter than the length specified. When repaired, the overall vibration values decreased below the original values stated in this paper.

Conclusions: Some very basic conclusions can be made from this investigation:

- Base stiffness is absolutely essential to ensuring the proper life of a rotating machine.
- Looseness of any sort in a rotating machine will decrease the life drastically.
- Certification reports should always be verified.

This Vision beamline T0 chopper has reached 9,000 hours of operation. Even though the repaired mount resulted in vibration levels to decrease below the original values, it was determined to be a good decision for the Operations Group to replace the T0 chopper now, while it was accessible. This would save manpower costs six months from now when the Vision T0 chopper was scheduled to be replaced. The Vision beamline has met its goal for availability and experimental beam time. Sustainability means designing, building, and maintaining for the long term. This equipment has met its goal and was likely capable of exceeding it.

Acknowledgement: The Spallation Neutron Source chopper shop mechanics, supervisors and engineers were instrumental in facilitating the testing, evaluation, equipment details, and operational parameters of this testing. In addition figures 13, 14, and 15 were processed and furnished by Bill McHargue of the chopper shop engineering team. The SNS engineers on the design team also added significant contributions to the effort as well as furnishing the documentation for figures 2 and 3.

EXPERIMENTAL INVESTIGATION AND ANALYSIS OF SOME COMMON FAULTS FOR WIND TURBINE

Wenguang Yang, Shaohua Li and Dongxiang Jiang
State Key Laboratory of Control and Simulation of Power System and Generation Equipments,
Department of Thermal Engineering, Tsinghua University,
Beijing, 100084, China
Telephone: (86) 6279-4046
jiangdx@mail.tsinghua.edu.cn

Abstract: This paper focus on the fault characteristic by experimental research. A test rig is built for the direct driving stall fix-pitch horizontal axis wind turbine. 8 signals are collected by the experiment system, including wind speed, rotor rotational speed, shaft torque, the output electric power, vertical and horizontal vibration displacement of main shaft, vibration acceleration of the front bearing pedestal and the back bearing pedestal. 11 common wind turbine faults are researched by experiment. These faults are rotor mass imbalance, pneumatic asymmetry, airfoil changed, yaw fault, shaft horizontal misalignment, shaft vertical misalignment, front bearing pedestal looseness, rear bearing pedestal looseness, front bearing outer ring fault, front bearing inner ring fault, and front bearing roller fault. Several signal analysis methods, which are waveform analysis, orbit analysis, amplitude domain analysis, frequency domain analysis and envelope spectrum analysis, are adopted to extract characteristics from vibration signals. The influence on output power and vibration characteristics of these faults is presented by comparing with normal operation. The findings can be used to guide wind turbine fault diagnosis in engineering applications.

Keywords: Wind turbine; fault experiment; signal analysis

1. Introduction: Wind turbines are one of the fastest growing sources of power generation in the world wide, which we can find in the World Wind Energy Report (see [1]). Wind turbines are working in the outdoor so its work environment is very poor, such as the time-varying wind field, high humidity and sandstorm etc. This situation leads to the high failure rate of wind turbines, which can be find in this report (see [2]). On the other hand, wind turbines are usually installed in remote areas. Once a failure occurs, the maintenance is inconvenience and the downtime is long, which makes the maintenance cost and breakdown loss extremely high. So operation and maintenance (O&M) is a major issue with wind power system, which can also be seen in these papers [3-7]. Wind turbine fault diagnosis is a promising technique for solving this problem and improving the safety.

The research of fault diagnosis can be roughly divided into three parts, the study of fault characteristics and mechanism based on theory and experiment, the research of diagnosis algorithms, such as signal analysis, feature selection, pattern classification etc., and the development of diagnosis system (see [8]). Fault characteristics research is the foundation of the study of fault diagnosis algorithm and diagnosis system. It mainly focus on the common faults of the key components, and one of its goals is to build a model to describe the response of the whole machine under all conditions when these faults happened. For wind turbine, blades and rotor, gearbox, tower and generator are its most precious and hard to maintain components. A lot of research works have been done in these directions respectively. Sheibani, M., Akbari, A. A., Imani, B. M. (see [9]) investigated damage detection in blade root joints of a 100 kW wind turbine using frequency tracking. Jiang, D. X.; Huang, Q., Hong, L. Y. (see [10]) studied rotor unbalance from theory and experiment. Adams, D.; White, J., Rumsey, M. (see [11]) study the operational response under various wind states as well as the effects of three types of damage in the blade, low speed shaft and yaw joint. Tsai, J. I., Leu, G. C., Liao, G. C. (see [12]) researched the torque vibrations excited by balanced network faults on the mechanism of wind turbine generators. Ali, B., Sundaresan, M. J., Schulz, M. J. (see [13]) studied the early detection of local buckling in structural members. Icing fault research based on theory and experiment can be find here [14, 15]. Each of these research are focus on one or two faults based on theory or experiment. In this paper, we study the fault characteristics of direct-drive wind turbine by experimental research and 11 faults are studied at the same time.

Signal analysis is necessary since the raw signals from the sensors contain noise and fault symptoms are implicit. Developing new algorithms and choosing appropriate analysis methods both deserve studying. Signals can be divided into vibration signal and state parameter signal. Vibration signals usually contain more information. For wind turbine vibration signals include mechanical vibration, electric vibration and acoustic emission (AE). Acceleration signal are widely used for gearbox and bearing fault diagnosis, related work can be find in these papers [16-18]. Gong, X. etc. in [19] studied the incipient bearing fault detection via wind generator stator current, and they make a contribution on imbalance fault detection of direct-drive wind turbines using generator current signals (see [20]). Acoustic emission is a relatively new and promising technology. Boczar, T. in [21] presented comparative analysis results of the size of acoustic oscillations emitted by the work of wind turbines of three various rated power values of 150 kW, 0.5 MW and 1 MW, respectively. Soua, S. in [22] studied wind turbine condition monitoring by combining vibration and acoustic emission signature of gearbox and generator shaft. Zhao, X. G. in [23] used acoustic emission for wind turbine blade fault diagnosis.

There are many kinds of vibration signal analysis methods, include statistical analysis and parametric model and correlation analysis in time domain, such as orbit, frequency domain analysis, envelope analysis, time-frequency domain analysis, and adaptive analysis etc. The time-frequency analysis methods used for non-stationary signal include Short Time Fourier Transformation (STFT), wavelet transformation etc. Guo, Y. P. in [16] researched a new way to use the discrete wavelet decomposition for wind turbine gearbox fault diagnosis. Zhang, X. Y. in [17] presented a High Frequency Resonance (HFR) method to implement bearing fault detection and fault diagnosis. Chen, Z. G. in [18] used information entropy to extract characteristic parameter from vibration signal to judge the gear box state. The representative adaptive decomposition methods include empirical mode decomposition (EMD), intrinsic time-scale decomposition method (ITD). Feng, Z. P. in [24] proposed a simple and effective method to diagnose planetary gearbox faults based on ensemble empirical mode decomposition and energy separation. Liu, H. in [25] gave a research on wind turbine fault diagnosis based on EMD. Xueli An in [26] studied the application of ITD method to wind turbine bearing fault diagnosis. In our experiment research, we mainly focus on mechanical vibration, include two shaft displacement signal and two bearing acceleration signal. Waveform and shaft orbit analysis, statistical analysis in time domain, spectrum analysis, marginal spectrum analysis and envelop analysis are adopted.

2. Wind Turbine Fault Test Rig and Fault Experiment: A test rig is built to simulate the direct driving stall fix-pitch horizontal axis wind turbine. The structure diagram is depicted in Figure 1.

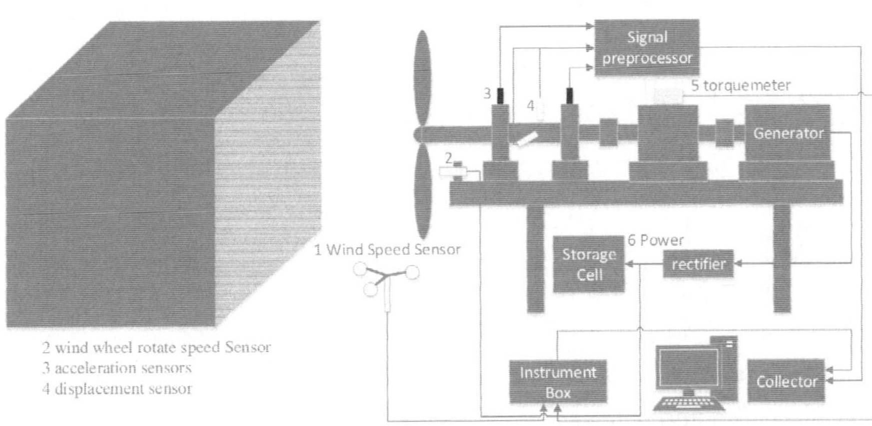

Figure 1: Hardware Structure Diagram of Test Rig.

Table 1: The Faults Simulated and Its Simulation Methods.

Fault Modes	Identi-Fies	Simulation Methods
Rotor Mass Imbalance	F1	Append a mass block to a blade at the low pressure side near the tip.
Pneumatic Asymmetry	F2	Change the airfoil characteristics of a blade by pasting papers on the surface close to the tip.
Airfoil Changed	F3	Change the airfoil characteristics of all blades by pasting papers on the surface close to the tip.
Yaw Fault	F4	Change the angle between the shaft axis direction and the wind direction from zero to a value, 6 degree is taken in our experiment.
Shaft Horizontal Misalignment	F5	Make the horizontal position of the front bearing pedestal offset from its normal position, 2mm is taken.
Shaft Vertical Misalignment	F6	Make the horizontal position of the front bearing pedestal offset from its normal position, 2mm is taken.
Front Bearing Pedestal Looseness	F7	Unscrew the bolts connecting front bearing pedestal with the base.
Rear Bearing Pedestal Looseness	F8	Unscrew the bolts connecting bear bearing pedestal with the base.
Front Bearing Outer Ring Fault	F9	Make a crack of a certain width and depth along axial direction on the bearing outer ring.
Front Bearing Inner Ring Fault	F10	Make a crack of a certain width and depth along axial direction on the bearing inner ring.
Front Bearing Roller Fault	F11	Make a crack of a certain width and depth on a bearing roller.

Wind is produced by a wind tunnel under laboratory conditions. The wind tunnel equipped an axial flow fan, whose nominal power is 15KW, can generate wind with speed range from 0 to 15m/s. The rotor consists of three blades, the diameter of the rotor is 1.44m. The type of the bearing is 22206, and the rated speed is 10000rpm. The type of the generator is FF-300, and the rated power is 300W and the rated speed 400rpm. There are two storage cell, whose type is HGL100-12, capacity is 100AH, and voltage is 12V. The shaft is supported by two deep groove ball bearings. The shaft and generator are connected by flexible coupling. The experimental test rig installed 8 sensors, the parameters be collected include: wind speed, rotor rotational speed, shaft torque, the output electric power, vertical and horizontal vibration of the shaft, vibration acceleration of front bearing block and rear bearing block. The sampling frequency of the four vibration sensors was 20KHZ, which was more than twice of the highest frequency component in our study. More details about the test rig can be find in this paper [27].

The components threatened by failure in wind turbine includes blades, shaft bearing, gearbox, tower, etc. Considering the characteristics of the test rig, 11 kinds of faults were simulated by experiment. For each fault, only one severity was considered, and the wind speed range is from 4 to 12m/s. These faults and its simulation methods are listed in Table 1, and the real shots and schematic of some faults are shown in Figure 2.

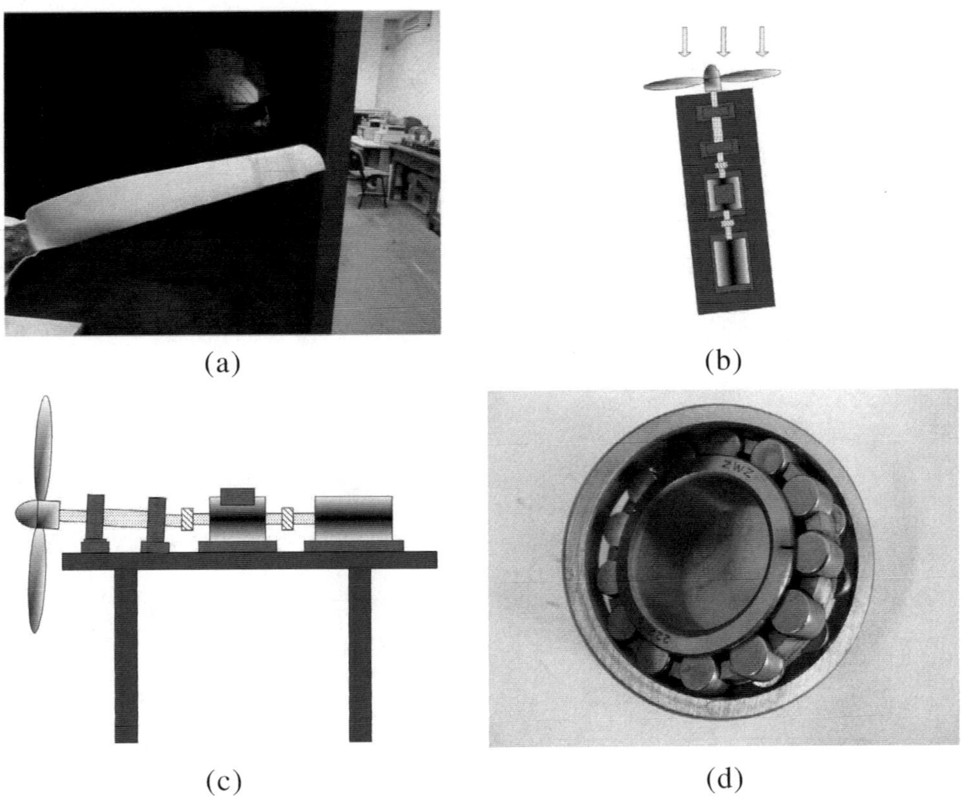

Figure 2: Real shots and schematics of some faults. (a) Mass unbalance of rotor; (b) Yaw fault; (c) Shaft vertical misalignment; (d) Bearing inner ring fault.

3. Signal Analysis Methods: In our experiment, we focus on mechanical vibration, include two shaft displacement signal and two bearing acceleration signal. The statistical analysis in time domain, shaft orbit, spectrum analysis and envelop analysis are adopted. For a data set consist of n data, $x_1, x_2, \cdots x_n$. The common statistical parameters can be expressed as following formulas listed in Table 2.

Table 2. The Calculation Formulas of Statistical Parameters

Peak-to-Peak Value	$X_{PP} = \max(x_i) - \min(x_i)\ (i = 1,2,\cdots,N)$	(1)
Peak Value	$X_{\max} = \max(\lvert x_i \rvert)\ (i = 1,2,\cdots,N)$	(2)
Effective Value	$X_{rms} = \sqrt{\frac{1}{N}\sum_{i=1}^{N} x_i^2}$	(3)
Root Amplitude	$X_r = \left[\frac{1}{N}\sum_{i=1}^{N}\sqrt{\lvert x_i \rvert}\right]^2$	(4)
The Average Absolute Amplitude	$\lvert \overline{X} \rvert = \frac{1}{N}\sum_{i=1}^{N}\lvert x_i \rvert$	(5)
Skewness	$SK = \frac{1}{N}\sum_{i=1}^{N} x_i^3 \cdot \frac{1}{X_{rms}^3}$	(6)
Kurtosis	$K = \frac{1}{N}\sum_{i=1}^{N} x_i^4 \cdot \frac{1}{X_{rms}^4}$	(7)
Corrugation Factor	$S_f = X_{rms}/\lvert \overline{X} \rvert$	(8)
Peak Factor	$C_f = X_{max}/X_{rms}$	(9)
Impulse Factor	$I_f = X_{max}/\lvert \overline{X} \rvert$	(10)
Margin Factor	$CL_f = X_{max}/X_r$	(11)

Spectrum analysis is one of the most important vibration signal analysis methods, which is calculated by Fourier transform. For a periodic signal $f(t)$ with cycle time T, the Fourier transform can be expressed as,

$$\boldsymbol{f(t) = \sum_{n=-\infty}^{\infty} F(n) e^{j2\pi nt/T}}, \tag{12}$$

$$\boldsymbol{F(n) = \frac{1}{T}\int_0^T f(t) e^{-\frac{j2\pi nt}{T}} dt}. \tag{13}$$

The Fourier transform of any signal is defined as,

$$F(\omega) = \mathcal{F}\big(f(t)\big) = \int_{-\infty}^{\infty} f(t)\, e^{-i\omega t} dt, \tag{14}$$

$$f(t) = \mathcal{F}^{-1}\big(F(\omega)\big) = \frac{1}{2\pi}\int_{-\infty}^{\infty} F(\omega)\, e^{i\omega t} dt. \tag{15}$$

Envelope spectrum analysis can extract impulse component. Envelope can be calculated by Hilbert Transform, which is a part of Hilbert-Huang Transform (HHT).
The Hilbert transform of signal $x(t)$ is defined as

$$\bar{x}(t) = \frac{1}{\pi}\int_{-\infty}^{+\infty} \frac{x(t)}{t-\tau} d\tau. \tag{16}$$

The complex envelope h(t) is defined as,

$$h(t) = x(t) + i\bar{x}(t) = a(t)e^{i\theta(t)}, \tag{17}$$

where,

$$a(t) = \sqrt{x^2(t) + \bar{x}^2(t)},$$

$$\theta(t) = arctan\,(\frac{\bar{x}(t)}{x(t)}).$$

The instantaneous amplitude $a(t)$ is also called the envelope of original signal $x(t)$. The instantaneous frequency $\omega(t)$ is defined as following,

$$\omega(t) = \frac{d\theta(t)}{dt}. \tag{18}$$

The instantaneous amplitude and frequency comprise a frequency-time distribution. This distribution is a line in three-dimensional space, not a surface. It is designated as the Hilbert spectrum, H(w,t). The marginal spectrum can be defined as,

$$h(\omega) = \int_0^T H(\omega, t)dt. \tag{19}$$

Impulse frequency can be obtained by marginal spectrum analysis, spectrum analysis of envelope amplitude signal and spectrum analysis of instantaneous frequency signal.

3. Experiment Data Analysis:

3.1 Output Power Comparison: The drive train faults have less effect on power generation, but it can bring damage to the structure. Figure 3 shows the output power of the direct-drive wind turbine in normal operation, and four fault mode, rotor mass imbalance, pneumatic asymmetry, airfoil changed, and yaw fault. The mean wind speed ranges from

4m/s to 12m/s. From this analysis it can be seen that the rotor fault have significant effect on power generated, especially when airfoil of all the blades changed. Slightly yaw fault does not reduce the total output power.

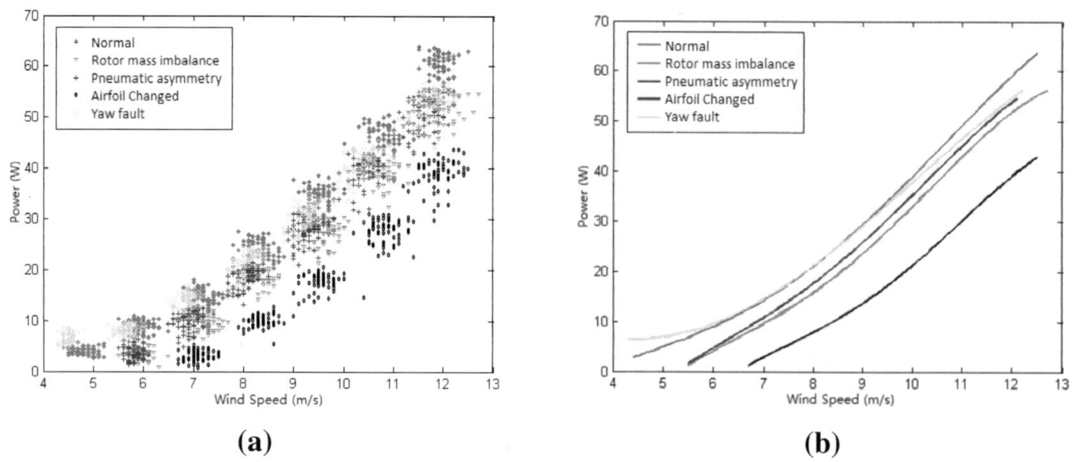

Figure 3: Power Generation of Different Fault Mode: (a) Raw Data; (b) The Fitted Curve.

3.2 Shaft Displacement Signal Analysis:

(1) Waveform and Orbit Analysis: The waveform distortion of shaft displacement signal and the irregular and chaotic of shaft orbit can be used directly to determine whether there is impulse force on the shaft. Figure 4 and Figure 5 present an example of waveform and orbit in normal operation and in shaft vertical misalignment fault, where wind speed is around 12m/s.

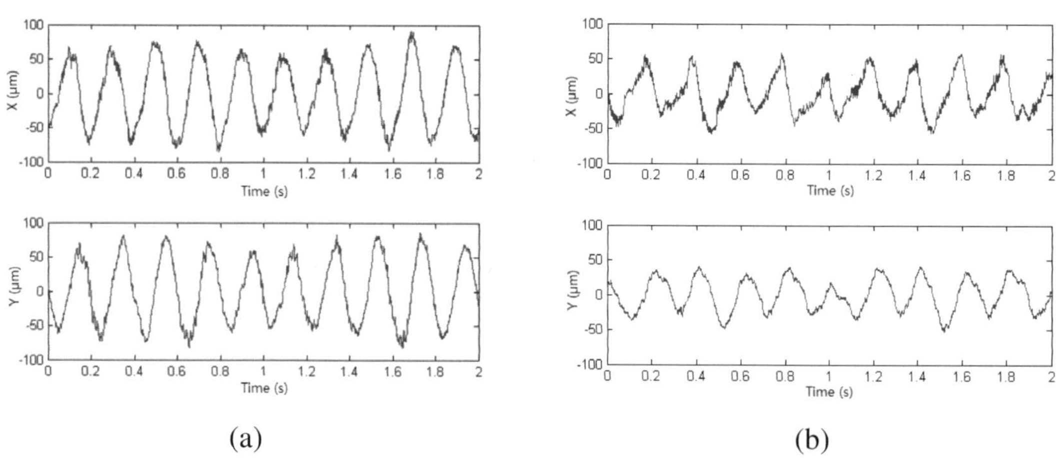

Figure 4: Shaft Displacement Waveform of Different Fault Mode: (a) Normal Operation; (b) Shaft Vertical Misalignment.

From Figure 4, it can be seen that the vibration wave of normal state is clean and close to sine, while the waveform changed a lot when shaft vertical misalignment occurs. Figure 5 tells that the orbit in normal condition is similar to a circle, while it becomes chaos in shaft vertical misalignment fault. Since there are fluctuations in wind speed, the orbit and waveform in our normal condition experiment if different from the situation under fixed rotation speed. Besides, yaw fault, shaft vertical misalignment, front bearing pedestal looseness and rear bearing pedestal have obvious effective on waveform and orbit.

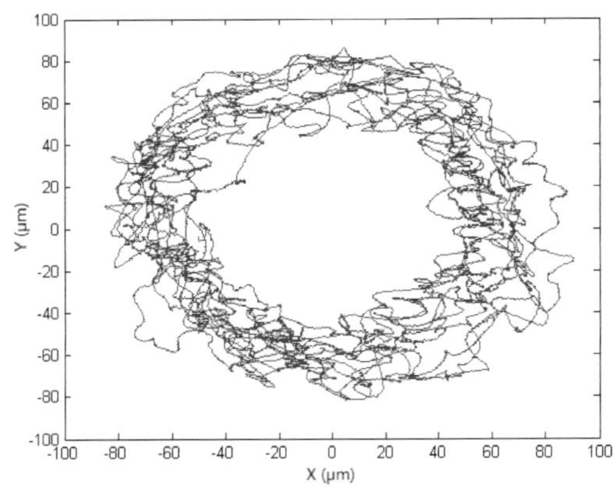

(a)

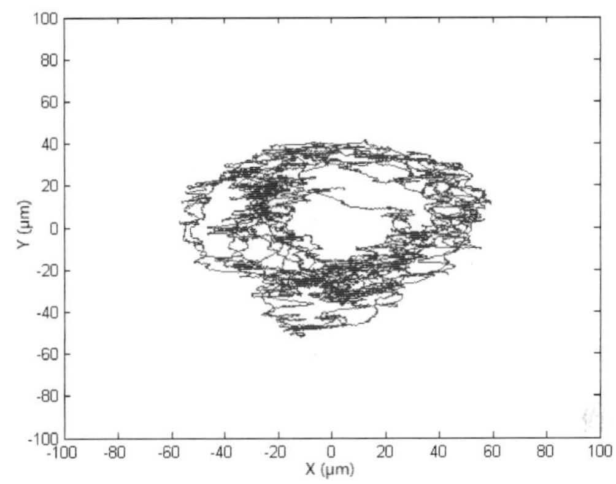

(b)

Figure 5: Shaft Orbit of Different Fault Mode: (a) Normal Operation; (b) Shaft Vertical Misalignment.

(2) Statistical Parameters Analysis: The statistical parameters of vibration signal in time domain are effective indications for fault diagnosis. Seven statistical parameters are chosen in our analysis, they are peak-to-peak value (PP value), skewness, kurtosis, corrugation factor, peak factor, impulse factor, and margin factor. The calculation formulas is given in Section 3. We get 28 statistical parameters in total since there are 4 vibration sensors in our test rig. Table 3 listed the parameters of shaft displacement in x direction, when the wind speed is about 12m/s.

The analysis finds that when front and rear bearing pedestal looseness occurs, the peak-to-peak value in x direction increase significantly. Kurtosis , peak factor, impulse factor, margin factor of yaw fault, shaft vertical misalignment and front and rear bearing pedestal looseness is larger than normal condition.

Table 3: statistical parameters of shaft displacement in x direction

State	PP value	Skewness	kurtosis	Corrugation factor	Peak factor	Impulse factor	Margin factor
Normal	173.5	0.02	1.65	1.12	1.95	2.19	2.41
F1	183.1	0.13	1.70	1.13	2.11	2.38	2.63
F2	165.8	0.10	1.76	1.14	2.08	2.36	2.62
F3	176.9	0.04	1.65	1.12	1.95	2.19	2.41
F4	115.7	0.00	2.03	1.17	2.43	2.85	3.22
F5	112.9	-0.10	1.72	1.13	2.14	2.41	2.65
F6	127.1	0.26	2.13	1.18	2.52	2.97	3.37
F7	192.9	0.15	1.85	1.15	2.25	2.58	2.87
F8	215.6	-0.10	1.77	1.14	2.09	2.39	2.66
F9	220.6	-0.15	1.84	1.14	2.40	2.74	3.03
F10	107.9	-0.14	1.88	1.15	2.50	2.86	3.18
F11	194.5	-0.17	1.82	1.14	2.23	2.55	2.83

(3) Spectrum Analysis: Figure 6 shows the spectrum of displacement signal under different fault state, where the wind speed is 275rpm and the rotational frequency is 4.58HZ. Under this experiment condition, the wind speed is different from each other in our experiment, so it's meaningless to compare the value of amplitude between different fault states. We should focus on the differences among the amplitudes of the same vibration spectrum. The spectrum can be normalized.

The analysis results include, the main component of the spectrum under normal operation is the rotational frequency, while harmonic and sub-frequency harmonic become obvious under some fault state. 1/2 rotational frequency increases significantly when shaft vertical misalignment, front bearing pedestal looseness and shaft bearing inner ring fault happened. Double rotational frequency increases significantly when shaft vertical misalignment and shaft bearing outer ring fault happened. Tripling rotational frequency increases significantly with shaft bearing outer ring fault.

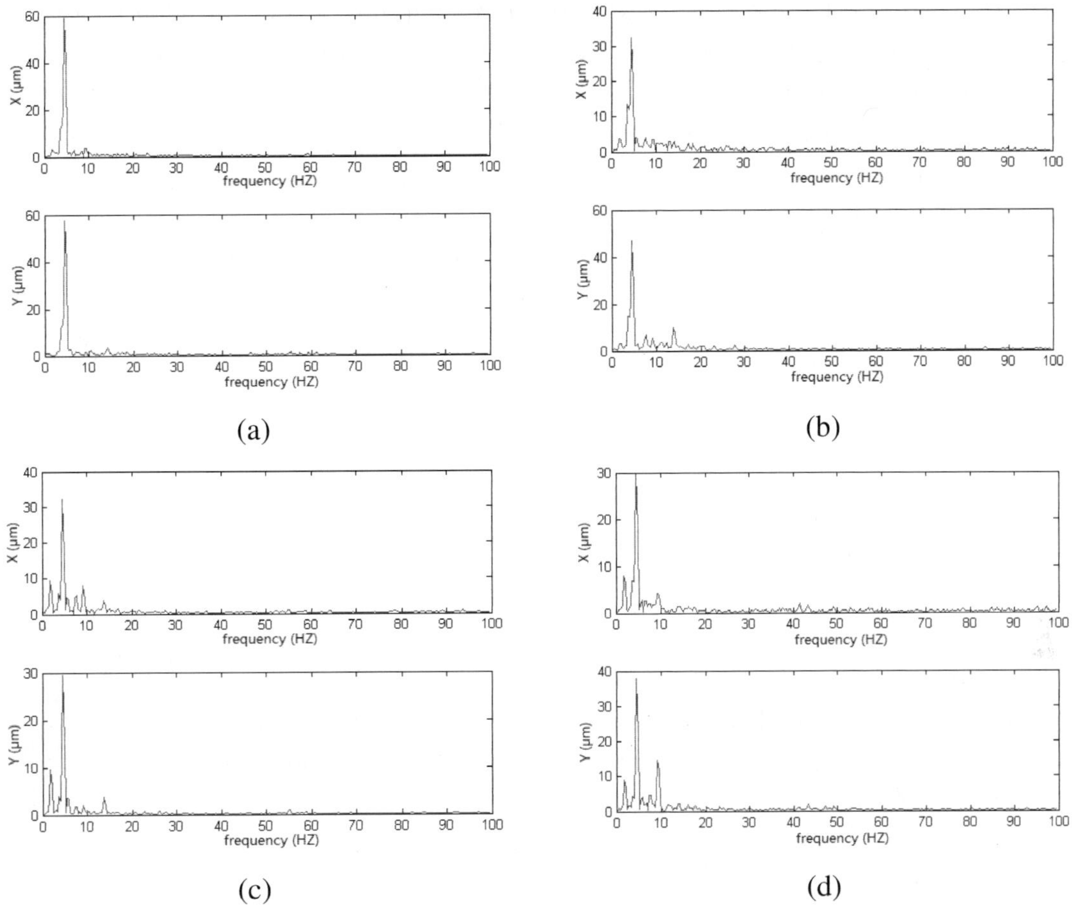

Figure 6: Spectrum Of Displacement Signal Under Different Fault State: (a) Normal Operation; (b) Yaw Fault; (c) Shaft Vertical Misalignment; (d) Front Bearing Inner Ring Fault.

(4) Marginal Spectrum Analysis: Hilbert-Huang Transform are good at complex non-stationary signal processing. The corresponding algorithm is given in Section 3. Figure 7, Figure 8 and Figure 9 shows the marginal spectrum of the shaft displacement vibration signals under normal operation, front bearing pedestal looseness and rear bearing pedestal looseness respectively. In these figure, the corresponding rotational speed is around and 266 rpm, and the rotational frequency is 4.14. The decomposition IMFs using EEMD are omitted.

With marginal spectrum, we can find that the main energy peak appears in the rotational frequencies, about 4.14HZ; when front bearing pedestal become looseness, the rotational frequency, 1/2 rotational frequency, 3/2 rotational frequency and two times rotational frequency are obvious in x-direction vibration; 1/2 rotational frequency, 3/2 rotational frequency are smaller in the marginal spectrum in y-direction. Under the fault of rear bearing pedestal looseness, the rotational frequency and 1/2 rotational frequency can be seen clearly, and some high-frequency components can also be found, but they are slight. The symptom of front bearing pedestal looseness is stronger than rear bearing pedestal looseness due to the cantilever structure of the wind turbine rotor, which means the front pedestal carries bigger load.

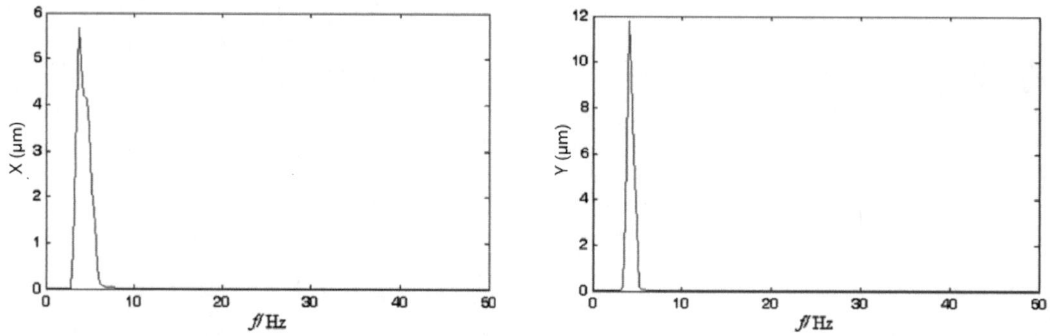

(a) (b)

Figure 7: Marginal Spectrum of Displacement Vibration Signal in Normal Operation: (a) Marginal Spectrum in X-Direction; (b) Marginal Spectrum in Y-Direction.

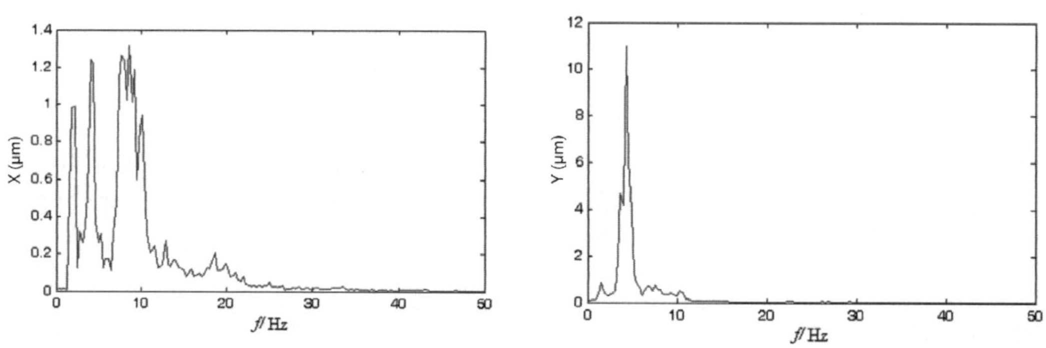

(a) (b)

Figure 8: Marginal Spectrum of Displacement Vibration Signal with Front Bearing Pedestal Looseness: (a) Marginal Spectrum in X-Direction; (b) Marginal Spectrum in Y-Direction.

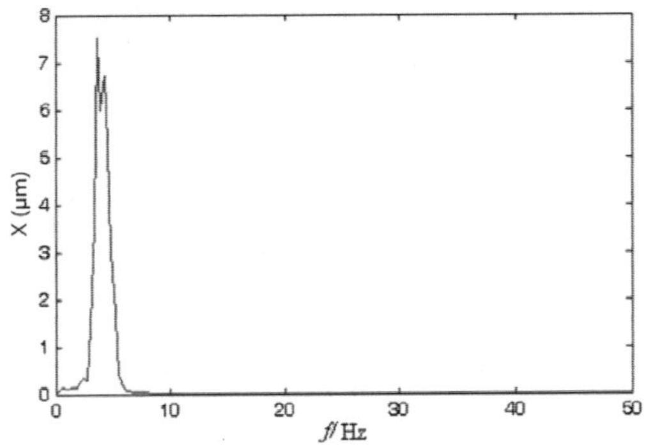

(a)

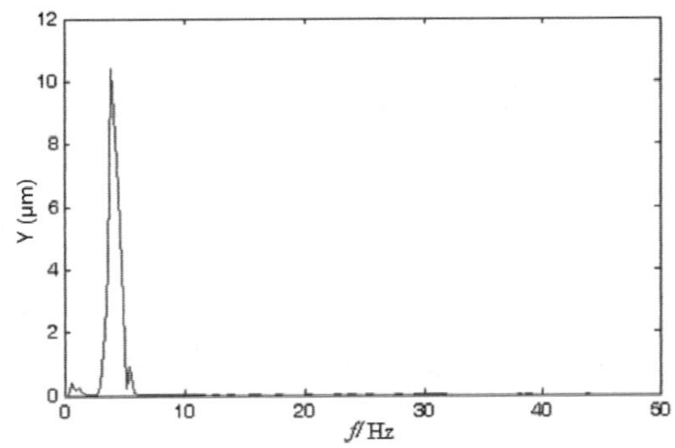

(b)

Figure 9: Marginal Spectrum of Displacement Vibration Signal with Rear Bearing Pedestal Looseness: (a) Marginal Spectrum in X-Direction; (b) Marginal Spectrum in Y-Direction.

3.3 Bearing Acceleration Signal Analysis:

(1) Waveform Analysis: Figure 10 shows the waveform of acceleration signal in time domain in normal operation and three fault modes respectively, where the wind speed is about 12m/s. These faults modes are front bearing outer ring fault state, front bearing inner ring fault and front bearing roller fault.

It can be found that, when the bearing has fault, the vibration amplitudes of front bearing acceleration and rear bearing acceleration increase significantly. The impulse components can be seen directly. This symptom of inner ring fault and outer ring fault is quite obvious, while the roller fault is slightly. One bearing's faults have effect on the acceleration signal of the other bearing, the impulse components of rear bearing acceleration signal increase when there is fault on front bearing. Furthermore, the acceleration amplitude also increase when there are bearing pedestal looseness faults.

(2) Statistical Parameters Analysis: Same as the displacement vibration signals, seven statistical parameters are chosen in our analysis. This analysis reveals that, for front bearing pedestal looseness fault, the peak to peak value of front vibration acceleration increase; when front bearing has fault, kurtosis, corrugation factor, peak factor, impulse factor and margin factor of front bearing vibration acceleration increase obviously. These statistical parameters of rear bearing vibration acceleration signal changed slightly when these faults happen.

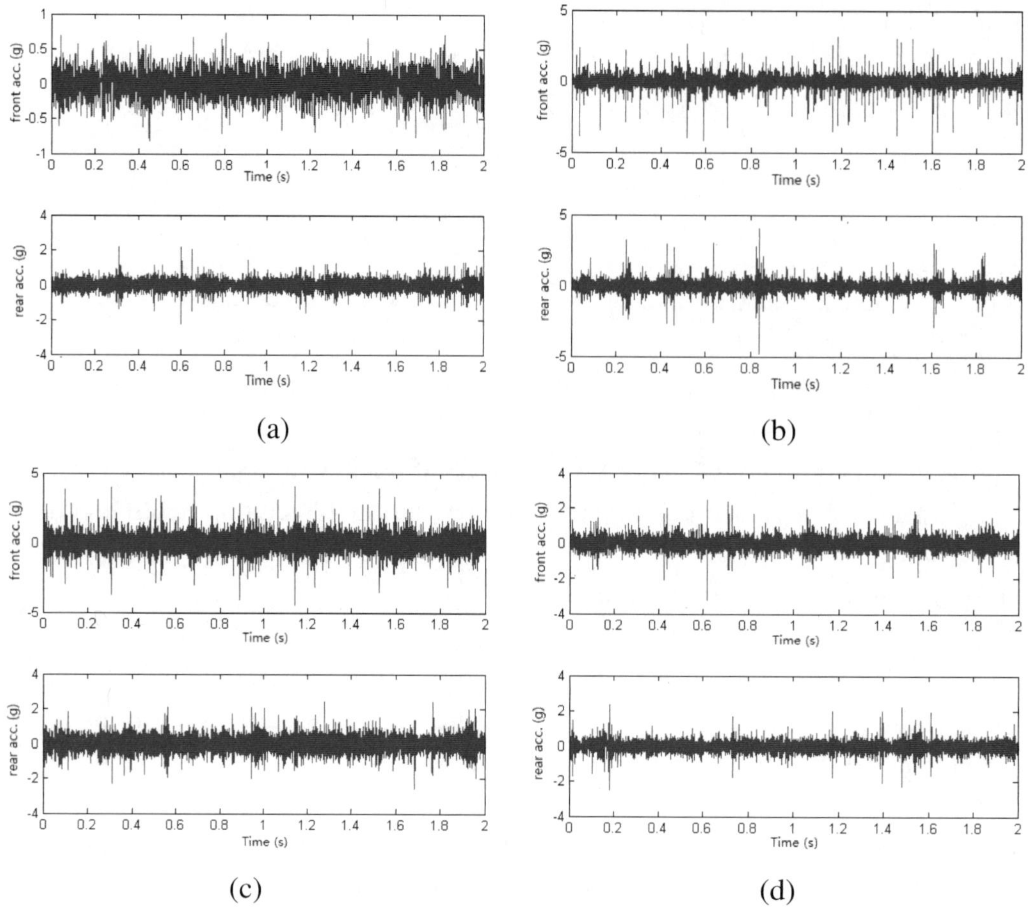

Figure 10: The Waveform of Acceleration Signal In Time Domain with Wind Speed About 12m/s: (a) Normal Operation; (b) Front Bearing Outer Ring Fault; (c) Front Bearing Inner Ring Fault; (d) Front Bearing Roller Fault.

(3) Spectrum analysis: Figure 11 shows the spectrum of acceleration signal with rotational speed around 275rpm in normal operation and several fault modes, include pneumatic asymmetry, front bearing pedestal looseness and front bearing roller fault.

Through the spectrum analysis of acceleration signal, we find the third harmonic of acceleration signal is a quite powerful fault indicator. The third-harmonic amplitude of front bearing acceleration increases obviously when the fault of pneumatic asymmetry, front bearing pedestal looseness, rear bearing pedestal looseness or front bearing roller fault occurs. The third-harmonic amplitude of rear bearing acceleration increases significantly when shaft vertical misalignment or bearing outer ring fault occurs.

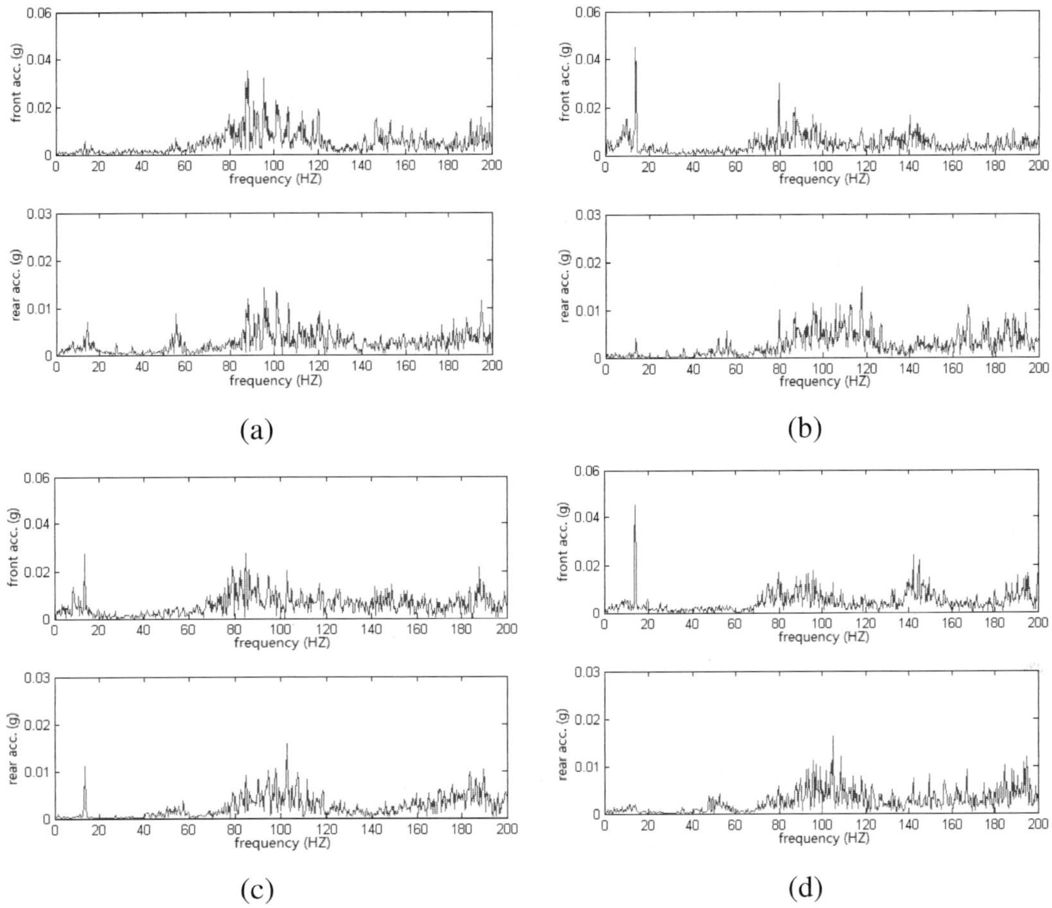

Figure 11: Spectrum of Acceleration Signal with Rotational Speed about 275rpm: (a) Normal Operation; (b) Pneumatic Asymmetry; (c) Front Bearing Pedestal Looseness; (d) Front Bearing Roller Fault.

(4) Envelope Spectrum Analysis: Acceleration signal is much more complex and sensitive than displacement signal. Envelop analysis is a valid method for acceleration signal analysis. The corresponding algorithm is given in Section 3. For its envelop spectrum, besides rotational frequency component, harmonic component and sub-frequency harmonic component, the component of characteristic frequencies of bearing outer ring fault, inner ring fault and bearing roller fault also need attention. These characteristic frequencies are very useful for bearing fault diagnosis. They can be calculated using following formulas listed in Table 4.

Table 4. The Calculation Formulas of Characteristic Frequencies

Characteristic Frequency of Bearing Outer Ring Fault	$f_o = \frac{z}{2}\left(1 - \frac{d}{D}\cos\alpha\right)f$	(20)
Characteristic Frequency of Bearing Inner Ring Fault	$f_i = \frac{z}{2}\left(1 + \frac{d}{D}\cos\alpha\right)f$	(21)
Characteristic Frequency of Bearing Roller Fault	$f_b = \frac{D}{2d}\left(1 - \frac{d^2}{D^2}\cos^2\alpha\right)f$	(22)

In the table, f is the rotational frequency, d is the diameter of roller, D is the pitch diameter, z is number of rollers, α is the contact angle, and N is rotational speed whose unit is revolutions per minute (rpm).

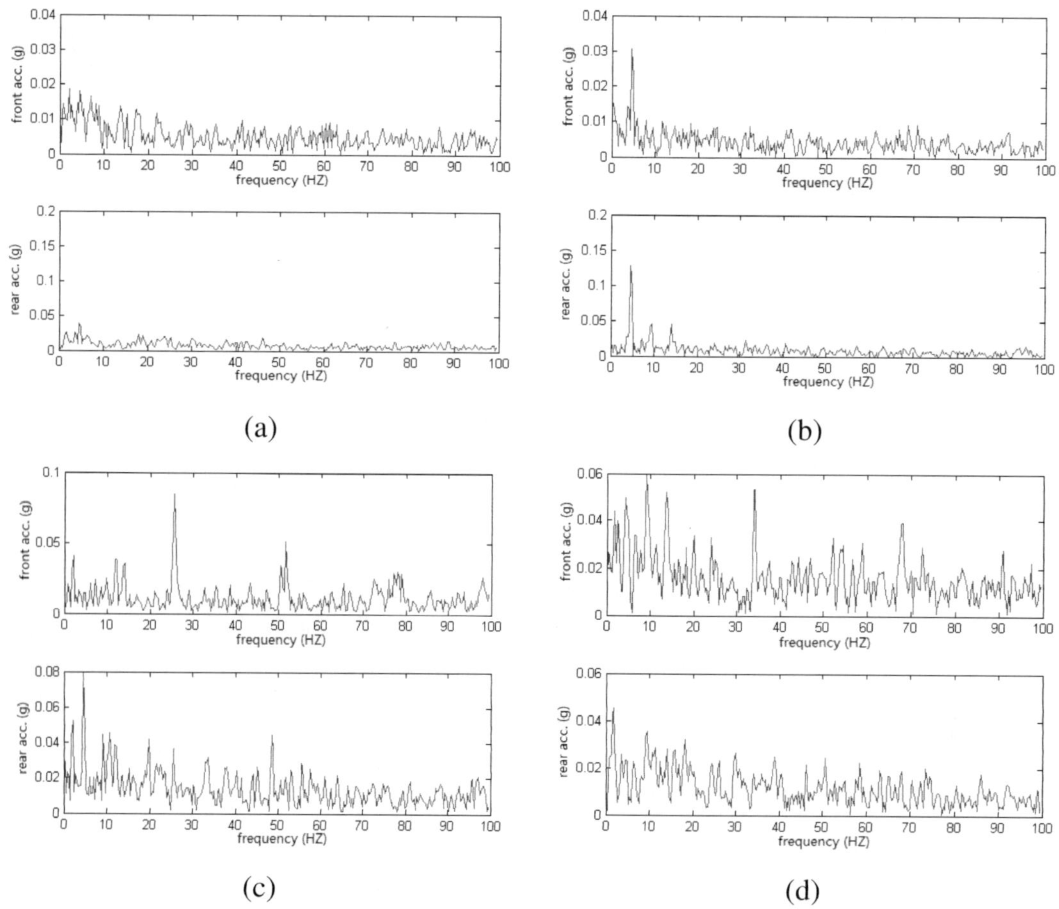

Figure 12: Envelop Spectrum of Acceleration Signal with Rotational Speed about 275rmp. (a) Normal Condition; (b) Rear Bearing Pedestal Looseness; (c) Front Bearing Inner Ring Fault; (d) Front Bearing Outer Ring Fault.

Figure 12 shows the envelop spectrum of acceleration signal in normal operation and three fault mode, rear bearing pedestal looseness, front bearing inner ring fault and front bearing outer ring fault, of which the rotational speed is 275rmp. The corresponding characteristic frequencies of bearing outer ring fault, bearing inner ring fault, bearing roller fault are 24.61Hz, 34.97Hz and 12.51Hz respectively.

It can be seen that, when rear bearing pedestal looseness fault happens, the rotational frequency component in envelop spectrum of front bearing acceleration signal increases significantly; while in the envelop spectrum of rear bearing acceleration signal, the amplitudes at rotational frequency, second harmonic frequency and third harmonic frequency increase obviously. When front bearing outer ring fault occurs, the component at characteristic frequency of bearing outer ring fault becomes outstand. When bearing inner ring fault occurs, the component at characteristic frequency of bearing outer ring fault becomes outstand. However, when bearing roller fault happens, the component at characteristic frequency of bearing roller fault changes slightly.

4. Conclusion: A test rig for the direct driving stall fix-pitch horizontal axis wind turbine is built. 11 common wind turbine faults are researched by experiment with wind speed range from 4m/s to 12m/s. 8 signals are collected. 5 methods are adopted for signal analysis. The results of experiment data analysis reveal the behavior of these faults.

When rotor imbalance occurs, kurtosis, corrugation factor, peak factor, impulse factor and margin factor of shaft displacement in x direction increase. Rotational frequency amplitudes of displacement signals increase slightly since the imbalance mass is small in our experiment. When the fault of pneumatic asymmetry occurs, the third harmonic component of front bearing acceleration vibration signal increases significantly. When airfoil changed, the output power decrease significantly. When yaw fault happens, the orbit become chaotic, and kurtosis, corrugation factor, peak factor, impulse factor and margin factor for both displacement in x direction and y direction increase. If the shaft has horizontal misalignment, kurtosis, corrugation factor, peak factor, impulse factor and margin factor of shaft displacement in y direction will increase. If the shaft has vertical misalignment, the orbit become chaotic, these statistical parameters of shaft displacement in Y direction kurtosis increase, include corrugation factor, peak factor, impulse factor and margin factor increase, and the 1/2 rotational frequency component and doubling rotational frequency component become outstand. When the fault of front bearing pedestal looseness occurs, the high-frequency, 3/2 rotational frequency and 1/2 rotational frequency components appear and are obvious. When the fault of rear bearing pedestal looseness occurs, the high-frequency and 1/2 rotational frequency can also be found, but they are slight; and the rotational frequency component is dominating. When the bearing has fault, the impulse components of both front and rear bearing acceleration signal become obvious. When bearing outer ring fault occurs, the component at characteristic frequency of bearing outer ring fault becomes outstand. When bearing inner ring fault occurs, the component at characteristic frequency of bearing outer ring fault becomes outstand. When bearing roller fault happens, the component at characteristic frequency of bearing roller fault changes slightly.

Further experiment studies are needed in order to furthering discovery and fully understand the behavior of each fault. For each fault, experiment for several severities should be conducted and more signal analysis methods need to be taken to get more effective symptom parameters. Besides, models describing how these parameters are influenced by the fault severity and the operation parameters should be built.

Acknowledgements: This work was supported by National Natural Science Foundation of China (Grant 51174273).

References:

[1] W.W.E. Association, World Wind Energy Report 2012, 2013.

[2] C.W.I. Forum, Summary of Wind Turbine Accident data to 31 March 2013, 2013.

[3] A brief status on condition monitoring and fault diagnosis in wind energy conversion systemsrev.

[4] A review of recent advances in wind turbine condition monitoring and fault diagnosis.

[5] Condition Monitoring and Fault Diagnosis in Wind Energy Conversion Systems A Review.

[6] Condition monitoring and fault detection of wind turbinesand related algorithms A review.

[7] F. Marquez, A.M. Tobias, J. Perez, M. Papaelias, Condition monitoring of wind turbines: Techniques and methods, Renew Energ, 46(2012)169-178.

[8] W. Guobiao, H.E. Zhengjia, C. Xuefeng, L. Yinan, Basic Research on Machinery Fault Diagnosis—What is the Prescription, JOURNAL OF MECHANICAL ENGINEERING, 49(2013)63-72.

[9] M. Sheibani, A.A. Akbari, B.M. Imani, Investigation of damage detection in blade root joints of a 100 kW wind turbine using frequency tracking, JOURNAL OF VIBROENGINEERING, 15(2013)364-373.

[10] D.X. Jiang, Q. Huang, L.Y. Hong, Theoretical and Experimental Study on Wind Wheel Unbalance for A Wind Turbine, 2009 WORLD NON-GRID-CONNECTED WIND POWER AND ENERGY CONFERENCE, (2009)351-355.

[11] D. Adams, J. White, M. Rumsey, C. Farrar, Structural health monitoring of wind turbines: method and application to a HAWT, WIND ENERGY, 14(2011)603-623.

[12] J.I. Tsai, G.C. Leu, G.C. Liao, W.C. Wu, C.Y. Hsueh, R.C. Wu, Torque Vibrations on the Mechanism of Wind Turbine Generators Excited by Balanced Network Faults, 2012 ASIA-PACIFIC POWER AND ENERGY ENGINEERING CONFERENCE (APPEEC), (2012).

[13] B. Ali, M.J. Sundaresan, M.J. Schulz, D. Hughes, Early detection of local buckling in structural members, Smart Structures and Materials 2005: Smart Structures and Integrated Systems, 5764(2005)494-500.

[14] M.H. Zhao, D.X. Jiang, S.H. Li, Research on Fault Mechanism of Icing of Wind Turbine Blades, 2009 WORLD NON-GRID-CONNECTED WIND POWER AND ENERGY CONFERENCE, (2009)356-359.

[15] X. Wang, E.L. Bibeau, G.F. Naterer, Experimental investigation of energy losses due to icing of a wind turbine, CHALLENGES OF POWER ENGINEERING AND ENVIRONMENT, VOLS 1 AND 2, (2007)1143-1147.

[16] Y.P. Guo, W.J. Yan, Z.J. Bao, Gear Fault Diagnosis of Wind Turbine Based on Discrete Wavelet Transform, 2010 8TH WORLD CONGRESS ON INTELLIGENT CONTROL AND AUTOMATION (WCICA), (2010)5804-5808.

[17] X.Y. Zhang, S.G. Yang, X.L. Li, Wind turbine Bearing Condition Monitoring Based on High Frequency Resonance method, 2011 INTERNATIONAL CONFERENCE ON ELECTRONICS, COMMUNICATIONS AND CONTROL (ICECC), (2011)1792-1795.

[18] Z.G. Chen, Fault Diagnosis of Gear Box Based on Information Entropy, 2010 CHINESE CONTROL AND DECISION CONFERENCE, VOLS 1-5, (2010)1239-1242.

[19] X. Gong, W. Qiao, W. Zhou, Incipient Bearing Fault Detection via Wind Generator Stator Current and Wavelet Filter, IECON 2010 - 36TH ANNUAL CONFERENCE ON IEEE INDUSTRIAL ELECTRONICS SOCIETY, (2010).

[20] X. Gong, W. Qiao, Imbalance Fault Detection of Direct-Drive Wind Turbines Using Generator Current Signals, Ieee T Energy Conver, 27(2012)468-476.

[21] T. Boczar, Comparative analysis of vibroacoustic oscillations of various power wind turbines, PRZEGLAD ELEKTROTECHNICZNY, 86(2010)131-133.

[22] S. Soua, P. Van Lieshout, A. Perera, T.H. Gan, B. Bridge, Determination of the combined vibrational and acoustic emission signature of a wind turbine gearbox and generator shaft in service as a pre-requisite for effective condition monitoring, Renew Energ, 51(2013)175-181.

[23] X.G. Zhao, C.Z. Chen, B. Zhou, X.H. Liu, Study on Monitoring Damage about Material of Blade in Wind Turbine Based on Acoustic Emission, ADVANCED MANUFACTURING SYSTEMS, PTS 1-3, 201-203(2011)2753-2758.

[24] Z.P. Feng, M. Liang, Y. Zhang, S.M. Hou, Fault diagnosis for wind turbine planetary gearboxes via demodulation analysis based on ensemble empirical mode decomposition and energy separation, Renew Energ, 47(2012)112-126.

[25] H. Liu, C. Wang, W.J. Yan, Research on Fault Diagnosis of Drive Train in Wind Turbine Based on EMD and LSSVM, RENEWABLE AND SUSTAINABLE ENERGY II, PTS 1-4, 512-515(2012)763-770.
[26] X. An, D. Jiang, J. Chen, C. Liu, Application of the intrinsic time-scale decomposition method to fault diagnosis of wind turbine bearing, J Vib Control, 18(2012)240-245.
[27] D.X. Jiang, Q. Huang, L.Y. Hong, Test System of a Small Wind Turbine under Laboratory Conditions, 2009 WORLD NON-GRID-CONNECTED WIND POWER AND ENERGY CONFERENCE, (2009)283-286.

MECHANICAL WAVEGUIDE VIBRATION AND TEMPERATURE SENSORS FOR MONITORING GAS TURBINE ENGINES

Chris Larsen, Richard Roth, and Oleg Lobkis
Etegent Technologies
5050 Section Ave
Suite 110
Cincinnati, OH 45312
Telephone: (513) 631-0579
chris.larsen@etegent.com

Abstract: Solid, mechanical waveguide sensors have been demonstrated for health-monitoring in the gas turbine environment. These sensors can be made of any material which conducts ultrasound; examples made of stainless steel, high-temperature aerospace alloys, and ceramics are presented here. Mechanical waveguides can be configured to measure temperature, pressure, heat flux, strain, and vibration. This paper presents results for two different waveguide-based sensors – a vibration sensor, and a temperature sensor. The waveguide vibration sensor was demonstrated for monitoring main-shaft bearings in a Rolls-Royce 501KB5+ industrial turbine engine. A high-temperature version was demonstrated in the exhaust of this engine for monitoring the #4 bearing at the turbine outlet; this requires survivability in the approximately 760°C (1400°F) exhaust stream. Additionally, a waveguide temperature sensor was demonstrated for turbine inlet temperature measurement in the same industrial turbine engine. This sensor provides accurate temperature measurement in combustion gases at over 1100°C (2000°F).

Key words: Diagnostics; prognostics; health management; gas turbine; bearings; temperature; ultrasound, turbine

Introduction: Solid, mechanical waveguide-based sensors are a mechanical counterpart to fiber-optic sensors: fiber optic sensors encode information in light which is transmitted by an optical fiber; whereas mechanical waveguides encode information in ultrasonic-mechanical strain energy which is transmitted by a solid, mechanical waveguide. The advantage of mechanical waveguides is that they can be made of any material which conducts ultrasound without significant attenuation. This enables waveguide sensors to be made of a wide variety of materials which are appropriate for the intended environment, such as high-temperature alloys, refractory metals, ceramics, and ceramic-matrix-composites (CMC).

Waveguide Sensor Background: Ultrasonic waveguides have been used for non-destructive testing (NDT) and process measurement in hot or harsh environments for many years [1]. One of the simplest examples of this is using a metal rod to isolate an ultrasonic transducer from a hot object for NDT or acoustic emission testing [2]. The metal rod is simply a conduit to transmit ultrasonic stress waves from the hot surface to the piezoelectric ultrasonic transducer. Figure 1 illustrates the concept of ultrasound propagating through a cylindrical waveguide.

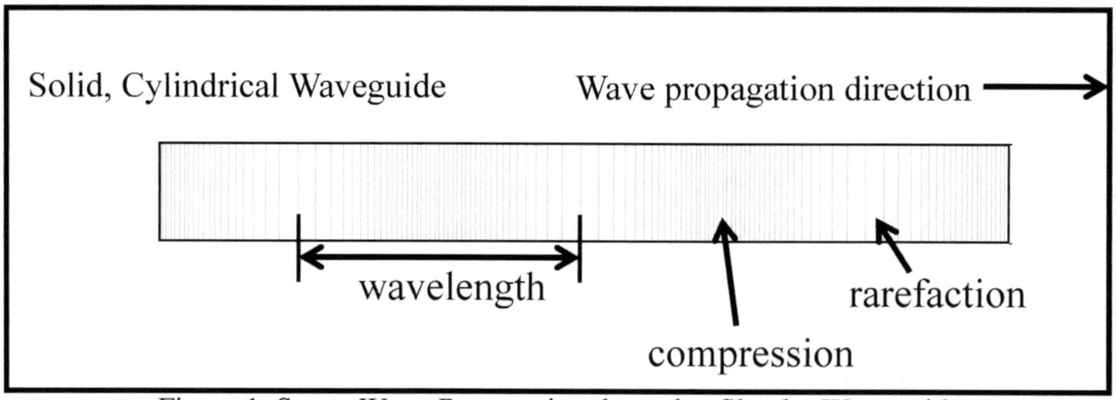

Figure 1: Stress-Wave Propagation through a Slender Waveguide

Waveguides can be used to construct a number of different sensor types. They can be broken up into two fundamental classifications, active and passive. Passive waveguides simply transmit the vibration signal of interest from the sensing location to a benign measurement location, but they are not able to measure static quantities. Active waveguide sensors utilize an ultrasonic transducer to actively interrogate the sensing head, enabling sensors to measure static or quasi-static quantities (e.g. temperature, pressure, or strain).

Both waveguide technologies can be implemented to measure a variety of different physical quantities. To date, passive waveguides have been demonstrated for vibration and dynamic pressure measurement. Active waveguide sensor prototypes have been demonstrated for pressure, strain and temperature measurement. This paper presents the most recent test results for two different waveguide sensors for monitoring a Rolls-Royce 501KB5+ industrial turbine engine. A passive waveguide vibration sensor was demonstrated in the turbine section for monitoring the #4 roller bearing. An active waveguide temperature sensor was demonstrated for turbine inlet temperature measurement in the same engine.

Broadband Waveguide Vibration Sensor: Mechanical waveguides can be designed to transmit stress wave energy over long distances with minimal attenuation or change [3]. Stress waves are generated by any dynamic force (e.g. pressure, vibration) and are transmitted from the sensing end of the waveguide to the measurement end. That is, a thin, metallic wire can be attached to a surface (with appropriate considerations given to the attachment design), and vibration from the surface will travel down the wire where it can be measured by an appropriately-designed piezoelectric element at the other end. They key advantage of mechanical waveguides is that the piezoelectric element does not have to be placed directly on the object to be monitored; instead it can be placed wherever it is most convenient - the vibration is mechanically transmitted from the object through the waveguide to the remotely-mounted piezoelectric element. Both the traditional measurement approach and the advantages of the waveguide are presented in Figure 2.

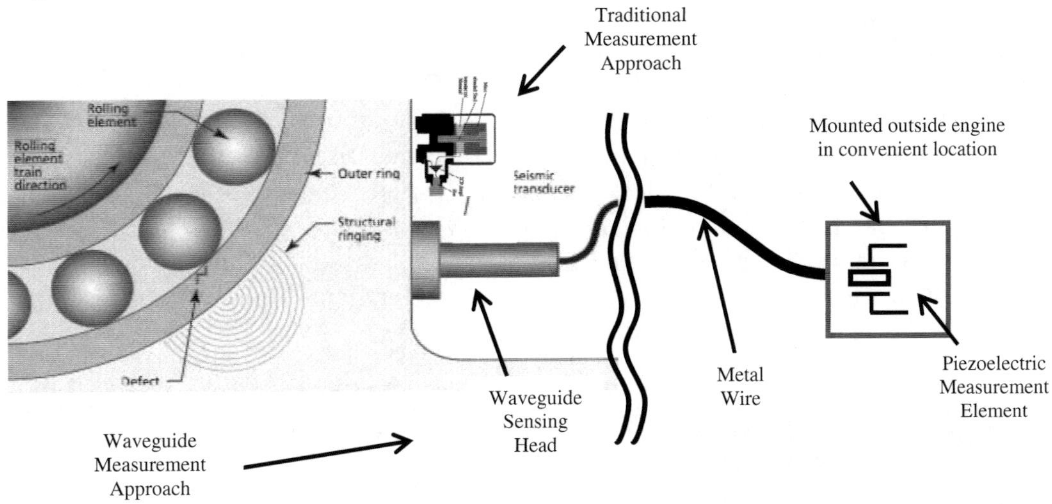

Figure 2: Traditional Vibration Measurement Requires Piezoelectric Element Directly on Structure – Waveguide Mechanically Transmits Vibration to Convenient Location for Piezoelectric Element

The waveguide vibration sensor's ability to mechanically transmit vibration makes it an excellent device for vibration monitoring turbine engine main-shaft bearings. Turbine engine main-shaft bearings are located deep inside the engine, where it is impractical to place accelerometers, since a sensor failure would require a costly teardown. In addition, the bearings are in a harsh environment, with high temperatures and high vibration levels. This environment along with the service requirement forces accelerometers to be mounted on the engine case. Due to the long vibration transfer path, attenuation, and noise and vibration from many other sources, identifying failing turbine engine main-shaft bearings with case-mounted accelerometers is generally impractical. The waveguide vibration sensor, however, can be permanently mounted inside the engine, since the portion inside the engine is a simple, rugged, metal device, and the piezoelectric portion is outside the engine where it can be serviced if necessary. In addition, the waveguide is insensitive to the vibration transmission path dynamics and noise sources present in the engine.

As was discussed in the authors' 2013 MFPT paper [4], the waveguide vibration sensor was successfully demonstrated on a T63 turbine engine with seeded inner and outer race defects as well as on a Rolls-Royce 501KB5+ industrial turbine engine with a seeded outer race defect. The waveguide vibration sensor provided 50% better signal-to-noise (SNR) of the defect frequency on the T63 and 200% better SNR on the 501KB5+.

Very High Temperature Waveguide Vibration Sensor: For the T63 and the 501KB5+ engines, the environment where the waveguide vibration sensor was installed was approximately 260°C (500°F). To monitor the main-shaft bearings in the turbine section requires even higher temperature resistance. Case-mounted accelerometers are not practical in the turbine section due to the high temperatures; they require additional isolation (e.g., a long pedestal to separate the sensor from the hot surface) or placement in cooler sections which are even more distant from turbine section main-shaft bearings. All of these factors make turbine-section main-shaft bearings even more difficult to monitor; this is an application which is well-suited to the waveguide.

A waveguide vibration sensor was constructed of a high-temperature aerospace alloy and installed at the #4 roller bearing location in the Rolls-Royce 501KB5+ industrial gas turbine. This waveguide was mounted to a strut in the exhaust close to the #4 roller bearing. The exhaust gas temperature of this engine is typically above 540°C (1000°F) and can exceed 760°C (1400°F) at maximum load – the alloy used for this waveguide was chosen for its strength and corrosion resistance at temperatures over 1100°C (2000°F). Figure 3 shows the waveguide mounted in the engine.

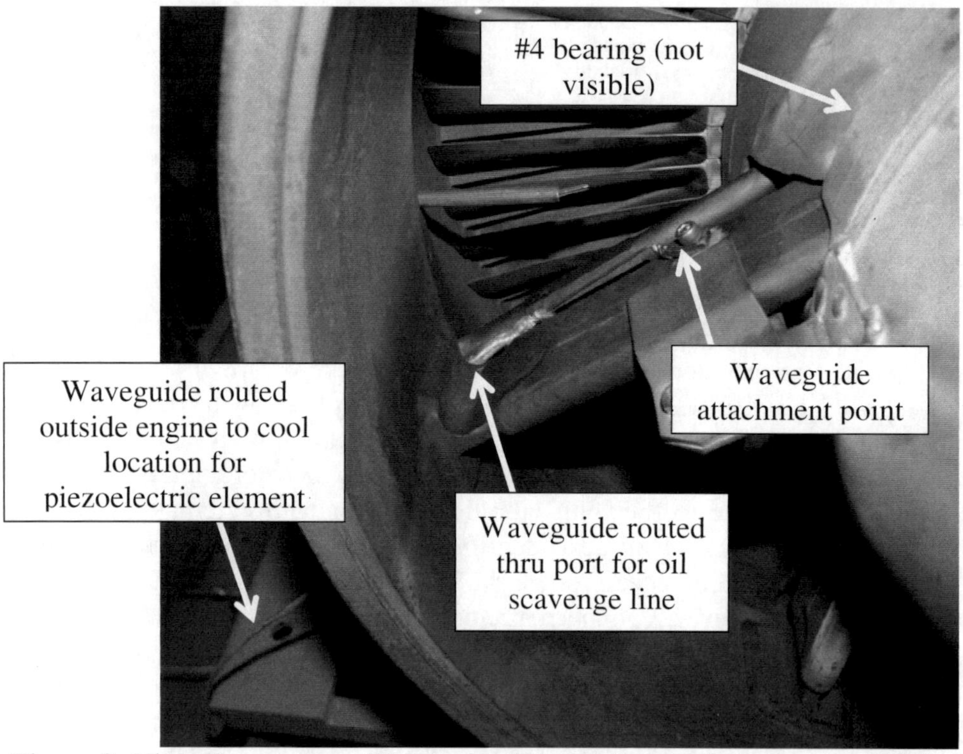

Figure 3: High-Temperature Waveguide Mounted in Rolls-Royce 501KB5+ Industrial Turbine Exhaust

The engine was run with a healthy bearing, and data was recorded for later comparison to planned seeded-defect testing. Figure 4 shows some sample data from the very high temperature waveguide vibration sensor. Blade-pass frequencies and other content are visible – this data will provide a good baseline for later seeded-defect testing, which is planned for late in 2014.

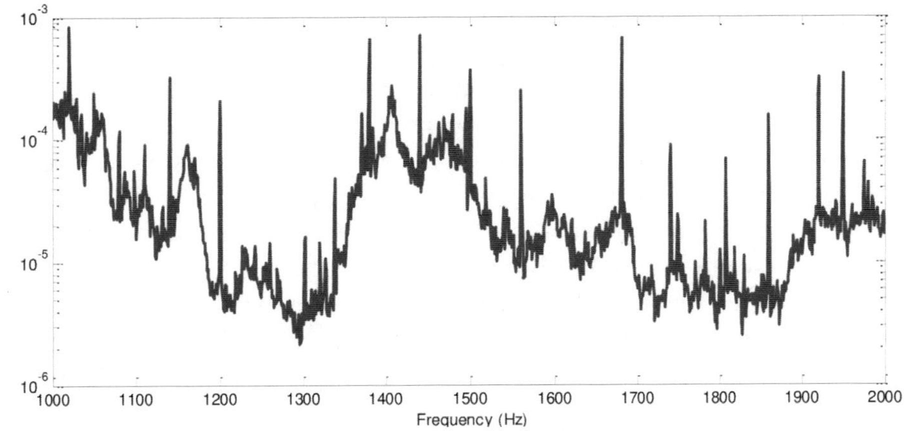

Figure 4: High-Temperature Waveguide Power Spectral Density (Zoom of 1 – 2 kHz)

Ultrasonic Thermometry: Since mechanical waveguides are well-suited to high-temperature environments, they can be configured to make temperature measurements where thermocouples cannot survive long-term or are not practical.

The speed of sound in any medium is temperature dependent due to the changing elastic modulus; therefore, the measurement of ultrasonic velocity between two points can be used as a temperature measurement with the appropriate calibration. The simplest means of measuring ultrasonic velocity is to measure the difference in propagation time between two ultrasonic reflection points. For example, a rod with a notch as shown in Figure 5 will reflect ultrasound both at the notch and at the end; the measurement of the velocity between these two points provides (with the appropriate calibration) provides the *average* temperature between the points. This measurement is completely insensitive to the temperature anywhere else along the waveguide.

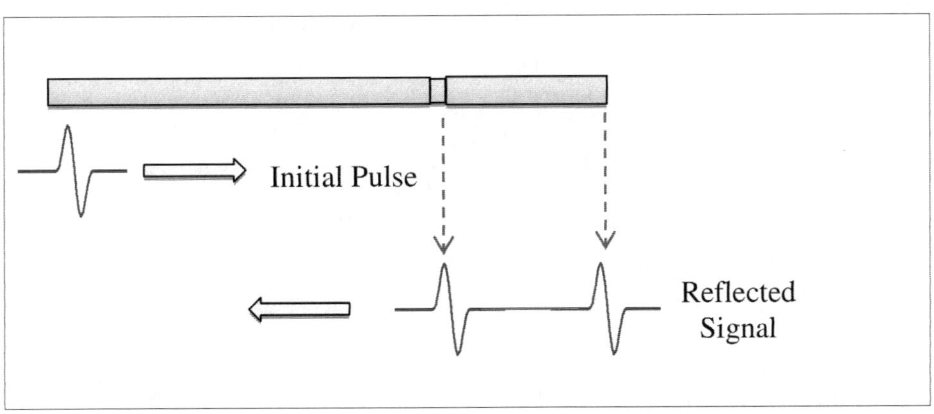

Figure 5: Reflections of Ultrasound for Velocity and Temperature Measurement

As mentioned previously, the rod pictured in Figure 5 can be made of any material which conducts ultrasound. Nickel-based superalloys, refractory metals, and ceramics are all suitable for use as ultrasonic waveguides.

Ultrasonic turbine inlet Temperature probe: The turbine inlet temperature has a great effect on efficiency, emissions, and the life of the turbine. Modern engines have turbine inlet temperatures exceeding the maximum temperature at which thermocouples can reliably operate and therefore thermocouples cannot be used to directly make this measurement. This requires *inferring* turbine inlet temperature by using a model and other available measurements such as turbine outlet temperature, compressor inlet temperatures, and shaft speed. Combustor issues, seal degradation, wear of the compressor or turbine sections, and other factors can cause error in the inferred temperature which degrades the control effectiveness – either the actual temperature is lower than inferred, and more power could be safely generated or the actual temperature is higher than inferred and the turbine section will have a shortened life. A safety margin built into the measurements and the temperature-estimation model reduces the maximum available power. Additionally, when one or more combustors are burning hotter than the others, the *average* turbine inlet temperature may be within specification; however, the first-stage turbine nozzles behind an excessively-hot combustor will have shortened life. Also, combustors burning hotter than specifications dictate generally produce more nitrogen-oxide emissions. Ideally, the turbine inlet temperatures would be measured directly, enabling more precise control to achieve maximum engine efficiency, performance, and life, while minimizing emissions.

Etegent has demonstrated an ultrasonic waveguide temperature sensor at the turbine inlet location on a Rolls-Royce 501KB5+ industrial turbine engine. This engine is generally controlled to maintain an inferred turbine inlet temperature of approximately 1060°C (1940°F). An appropriate high-temperature alloy was selected, and a sensor was constructed with a stagnation tube and mounting plate, as shown in Figure 6.

Figure 6: Ultrasonic Waveguide Turbine Inlet Temperature Probe Demonstrated on Rolls-Royce 501KB5+ Industrial Turbine Engine

This engine has ports for reference thermocouples in turbine inlet which can be used to setup the control system or for diagnostics – these thermocouples are not used for normal operation because they have a very limited lifetime in this environment. Reference thermocouples were used for this test for comparison to the Etegent waveguide temperature sensor.

Figure 7 shows the prototype ultrasonic waveguide turbine inlet temperature sensor installed in the engine along with a reference thermocouple in two adjacent measurement locations.

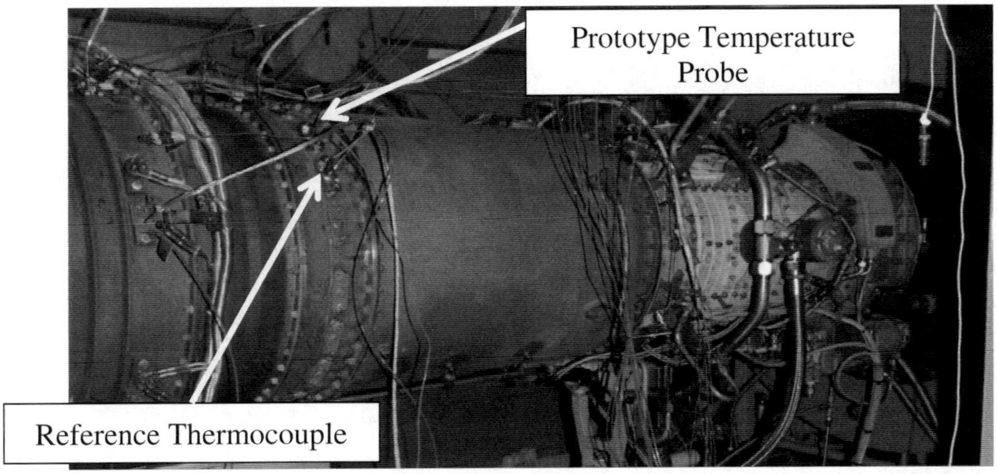

Figure 7: Ultrasonic Waveguide Turbine Inlet Temperature Probe in Rolls-Royce 501KB5+ Industrial Turbine Engine

The engine was operated at nominal speed (14,400 RPM) with no-load, and then the engine was loaded at 500 kW, 1 MW, 1.5 MW, 2 MW, 2.5 MW, and 3.0 MW in the first ramp period (from 0 to 27 minutes). Then the engine was brought back to idle for a few minutes and the operating speed of the turbine was increased to 14,700 RPM. The engine was then loaded at 1.5 MW and then 3.0 MW. Temperatures were recorded with the prototype waveguide temperature sensor and the reference thermocouple; Figure 8 shows the waveguide sensor's recorded temperature along with the reference thermocouple. The temperature measurement compare very well between the reference thermocouple and the Etegent waveguide sensor.

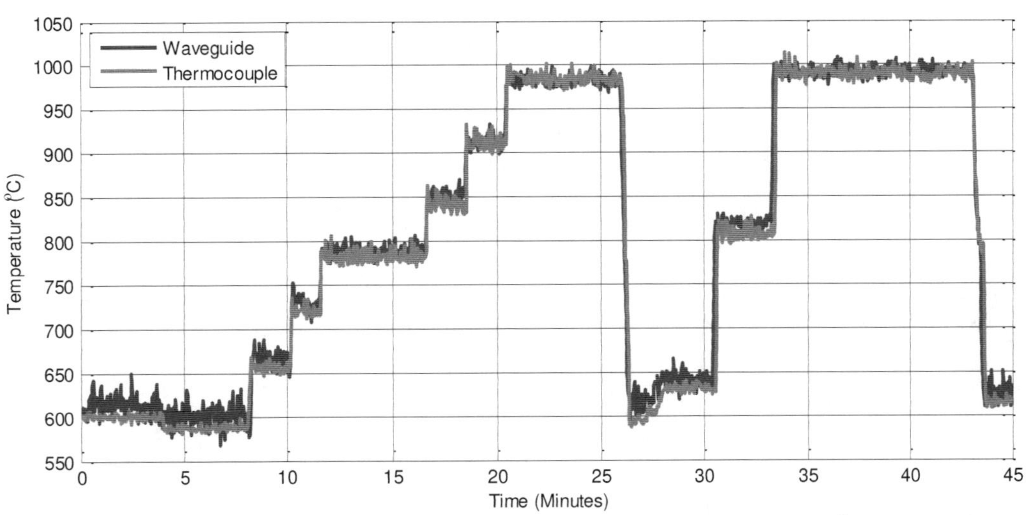

Figure 8: Waveguide temperature measurements on-engine

This test demonstrated the capability of a waveguide-based temperature sensor to provide accurate temperature measurements in the harsh turbine environment. Based on laboratory kiln and torch testing as well as published data for the high-temperature alloy, it is believed that this sensor should last many thousands of hours in this engine.

Future Work: More seeded defect testing is planned for the 501KB5+ engine with waveguide vibration sensors. A seeded defect will be run in separate tests in the #2 bearing and the #4 bearing. These tests will be run long enough for the defect to propagate; this will provide valuable data for the development of health-monitoring algorithms.

Additional testing is planned for the temperature sensor as well. Ceramic versions of the temperature sensor have been fabricated and tested in the lab; calibration procedures are being established for these sensors to 2000°C (3600°F) and beyond.

Acknowledgments: Waveguide vibration sensor development is currently funded by an Air Force small business innovation research (SBIR) Phase II award: AF121-173, "Engine Health Management of Mechanical Systems for High Performance Turbine Engines" (contract No. FA8650-14-C-2435, technical point of contact (TPOC): Chris Klenke). Waveguide temperature sensor development was funded by two NAVAIR Phase II SBIR awards: N093-185, "Robust Pressure Transducer for Propulsion Control Systems" (contract No. N68335-11-C-0203, TPOC: Jessica LaFond), and N101-013, "Low Cost, Dual Purpose Engine Control and Diagnostic Sensors" (contract No. N68335-11-C-0385, TPOC: Eric Carney).

Conclusions: Mechanical waveguide-based sensors have been demonstrated for monitoring both main-shaft bearings and turbine inlet temperature in a Rolls-Royce 501KB5+ industrial gas turbine engine at varying operating conditions. These sensors offer great flexibility in material selection, enabling measurements in environments which were previously difficult if not impossible to monitor.

References:

[1] L. C. Lynnworth, Ultrasonic Measurements for Process Control: Theory, Technoques, Applications, Academic Press, 1989.

[2] I. Neill, I. Oppenheim and D. Greve, "A Wire-Guided Transducer for Acoustic Emission Sensing," in *Proc. SPIE 6529, 652913*, 2007.

[3] D. Zakharov, S. Pichkov and V. Shemyakin, "Acoustic Emission Signal Attenuation in the Waveguides Used in Underwater AE Testing," in *ECNDT 2010 - Acoustic Emission*, Moscow, Russia, 2010.

[4] C. Larsen and N. Branch, "Waveguide Vibration Sensors for Aerospace Health Monitoring," in *MFPT*, Dayton, OH, 2013.

SMART SENSORS FOR TECHNICAL HEALTH ASSESSMENT OF SUBSEA OIL AND GAS EQUIPMENT

Pratichi Vaidya
Norwegian University of Science and Technology
Norway
Telephone: (47) 45490287
pratichivaidya@yahoo.com

Abstract: Life extension is becoming an important and highly discussed issue in the subsea oil and gas industry. Decision-making related to life extension is a multidisciplinary exercise and the primary factor that influences the decision is the remaining useful life. Models being used by nuclear and avionics industry for remaining useful life prediction are based on the operational data. In subsea oil and gas industry a lack of sufficient operational data does not make it possible to use similar models. Data acquisition from subsea equipment is a unique challenge due to remoteness of installations, water depth, and environmental conditions. Attempts for developing new techniques are being made by suppliers of sensors, subsea instrument, and subsea equipment. The challenge however is not only limited to technical enhancements of individual components but also to the need of interoperability among them. The paper gives an overview of the ongoing work on the development of smart devices and the common communication standards which will aid data transfer within the sea and from deep sea to a topside control room. Acknowledging that this is a big leap forward towards determination of remaining useful life, the paper especially focusses on the steps towards standardization, taken as a joint effort by subsea suppliers, contractors and operators. The paper explains how this technology will be useful in subsea oil and gas industry for technical health assessment.

Key words: Remaining useful life, Subsea sensors, IWIS, SIIS, SEAFOM

1 Introduction: Offshore deep-water oil and gas fields are increasingly being developed with subsea production systems. By subsea production systems, we mean production (Xmas) trees, templates, manifolds, pumps, separators, and associated equipment that are placed on the seabed. A subsea production system is usually tailor-made for a specific application and for a specified service life. New methods for enhanced oil recovery and tie-in with neighbouring wells often lead to an actual service life that is significantly longer than the initially planned service life. The question then is: Can we predict and trust that the existing subsea production system will survive and perform satisfactorily till the end of the extended service period? Or, should the current system be discarded and replaced by a new system. To answer this question, it becomes essential to assess the technical health of a system and to predict the remaining useful life (RUL). The factors influencing RUL are shown in Figure 1. The technical health concept is defined by Vaidya and Rausand [1] as an assessment or judgement of the state of the item at a specified time say t based on measured or observed factors related to the technical condition of the item, the operational history of the system up to time t, and the design quality of the system. Detailed explanation of all the influencing factors of RUL is covered in a separate paper by the authors [2]. The aim of this paper is to explain how it will be possible to obtain the operational field data with the help of smart sensors to address the issue of RUL.

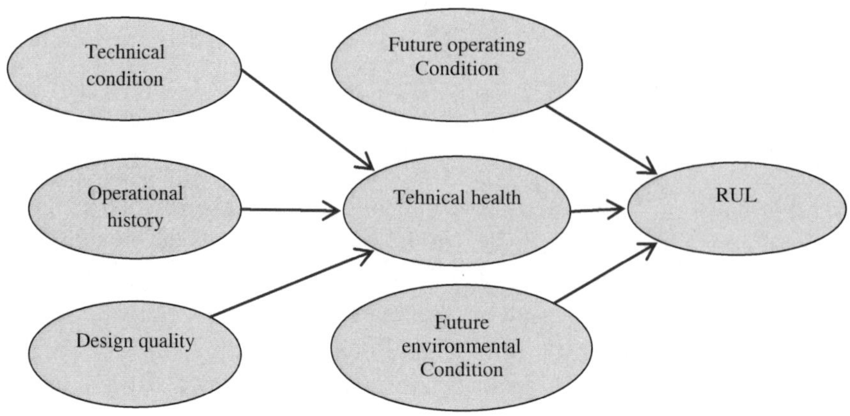

Figure 1: Factors influencing RUL

The quantity and quality of operational data is dependent on the data acquisition process. Design and functioning of data acquisition system is governed by the hardware, firmware, communication protocols and the corresponding software. The paper presents the efforts being taken by the community of operators, contractors, subsea equipment suppliers and sensor suppliers to standardize the interfaces which will result in seamless transfer of monitored subsea data to Integrated Operation Centres for further analysis.
The paper is organized as follows: Section 1 is an introduction. The section 2 describes the specifications developed as a joint effort within the industry for addressing interoperability and long distance communication. Section 3 is a discussion about data and standardization process. Conclusions are covered in section 4.

2 Communication Specifications: This section describes the three specifications developed by subsea industry as a joint industrial effort. These specifications are:

1. Intelligent Well Interface Specification (IWIS) for interface between down-hole system and subsea control module
2. Subsea Instrument Interface Specification (SIIS) covering subsea sensors and subsea control module
3. Subsea Fibre Optic Monitoring (SEAFOM) for use of fibre optics based sensing devices.

2.1 Intelligent Well Interface: Digital communication era in subsea industry started in the early 90s, with the deployment of digital down-hole gauges to measure pressure and temperature in reservoir. There was however no agreed way to interface these devices with a control system. The suppliers developed various tailored interfaces based on input from oil companies and in turn the companies picked up the bill and the schedule impact for the resulting interface development. With the advent of a more advanced down-hole technology where multiple monitoring and control devices would be placed in a well, it was clear that the industry needed a common standard to bring together existing down hole gauging systems and the emerging intelligent well control systems.

In an intelligent well, the down-hole instrumentation, well head monitoring system and control elements are connected to a single control unit. This generated the need for a standardized system with a compatible power supply and data communication protocol so that the operators do not have to spend a high cost for developing or adapting interfaces either at the start or later when new devices are introduced. This resulted in formation of a committee [3] for Intelligent Wells Interface Standard (IWIS) consisting of the Operators, the subsea equipment suppliers and the Intelligent Well / down-hole equipment suppliers. The committee worked towards integration of down-hole power and communication architectures, subsea control systems and topside by providing recommended specifications and standards for power and communication architectures and other hardware requirements. This is shown pictorially in Figure 2.

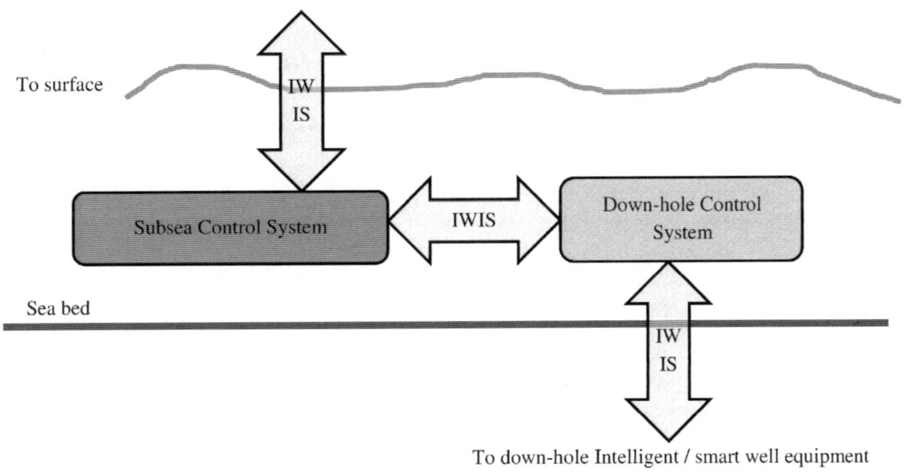

Figure 2: IWIS Interface for Physical Size, communications, Electrical Power & Testing

The IWIS Recommended Practice was finalized in Sept 2007. It was further updated in 2011 and is referenced as IWIS-RP-A2, IWIS recommended practice, in ISO-13628-6 called Design and operation of subsea production systems - Part 6: Subsea production control systems [4] [5]. There are three physical configurations for interface options supported by the standard as explained in Figure 3.

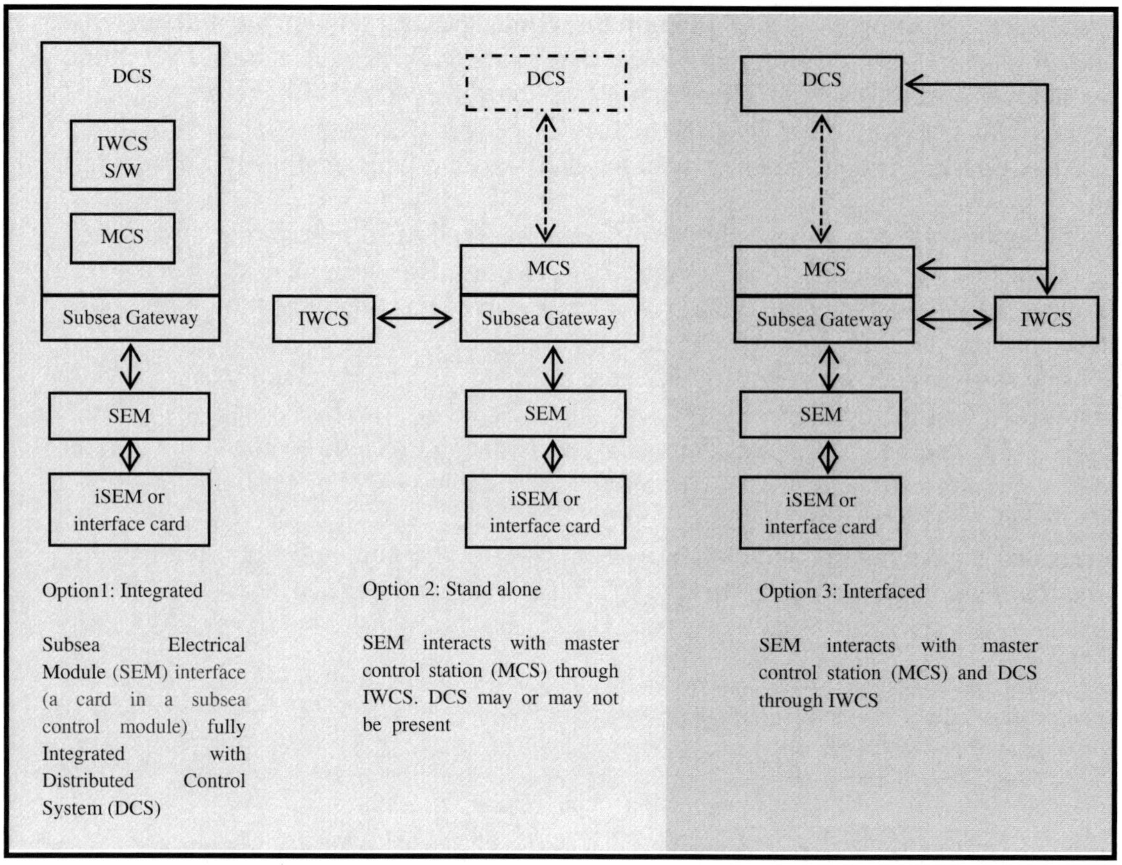

Figure 3: System data flow diagrams for the interface options

2.2 Subsea Instrumentation Interface: During IWIS work, representatives from the oil companies noted that along with a focus on down-hole equipment and its integration with the distributed control system at the topside or a central office, a parallel standardization activity is needed for interfacing subsea equipment sensors with themselves and with subsea control systems. This led to the formation of a SIIS group. The scope of the SIIS initiative is distinct from IWIS and it is limited to subsea production system instruments interfacing directly with the Subsea Control Module.

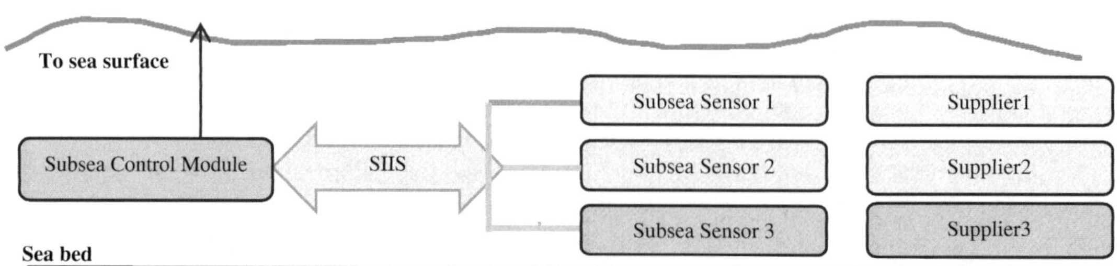

Figure 4: SIIS Interface for Physical/electrical connection, field bus standard & Testing

SIIS group initially established five sensor interface levels. Later, levels 2 and 3 were combined and merged to level 4, making total of three levels as explained in Table 1 [6].

Table 1: Three interface levels within SIIS

Level 1	Analog Devices	4-20 mA sensors, two wire loop powered analog output sensing devices
Level 2	Digital Serial Devices	Complex sensors, instruments which can measure different variables, communication through 'open standard' protocols
Level 3	Ethernet TCP/IP devices	Intelligent Seabed Devices (ISD), direct communication access between ISD and Surface Application Systems (not through Subsea Production Control System)

At present, the entire work is focused on the development of sensors and an open standard communication protocol for the level 2 interface. The committee has selected CAN bus and CANOpen protocol to be run over an ISO 11898-3 (fault tolerant) physical layer and suppliers are developing CAN based Level 2 devices [7]. Originally, the CAN bus was developed by BOSCH for the automobile industry. It specifies a maximum signalling rate of 1 megabit per second. Unlike a traditional network such as USB or Ethernet, CAN does not send large blocks of data from one point to another under the supervision of a central bus master. In a CAN network, many short messages like temperature, pressure or frequency are broadcast to entire network with a high reliability. CAN is a message based and not an address based protocol and hence is suitable for applications where data is needed by more than one location. The protocol provides data consistency throughout the system. Fault tolerance is another major benefit of CAN. Faulty nodes get dropped from the bus and the entire network does not go down because of any single faulty node. The bandwidth is always available for the transmission of critical messages. Thus fault tolerant feature enables a system to continue its intended operation, possibly at a reduced level, rather than failing completely. This is particularly important in subsea environment where it is difficult or impossible to access the subsea equipment during certain times for example Artic regions where ice may impede surface based maintenance for up to six months of the year.

2.3 Subsea Fibre Optic Monitoring: In modern subsea developments for enhanced oil recovery, few systems like multiphase booster pumping and separation system are shifted subsea to improve efficiency. The electrical noise environment both around this equipment, and surrounding the power umbilical that supply them, create difficult conditions for conventional means of communication. In this situation, inherent noise immunity of fibre optic communication has made it popular in offshore oil industry. Optical fibre by its very nature offers a secure communication medium, as it is impossible to "tap" an optical fibre without causing a detectable power loss, unlike a copper conductor which can be "read" from the field surrounding it [8]. The fibre itself can be used as a distributed sensor, using either the Brillouin or Raman scattering effects, inherent in all fibre, to measure temperature and strain over fibre lengths of up to 30 km [9].

Considering these facts, an international joint industry project namely Subsea Fibre optic monitoring (SEAFOM) was formed [10]. It promotes the growth of fibre-optical sensing in subsea applications. Membership of the group is open to operators, contractors, connector suppliers, sensor suppliers and umbilical suppliers who work towards standardizing functional requirements. Typical fibre optic subsea applications are Distributed Temperature Sensing, In-well Flow Metering, Distributed Pressure Sensing, Acoustic Sensing (fibre hydrophones), Seismic Sensing, Fixed point Pressure/Temperature Sensing, Strain Sensing (smart structures)

3.0 Discussions: As the oil and gas reserves are diminishing, the attention of the oil and gas industry is going towards the marginal fields. The most common method to make such pockets productive is by having a tie back to the existing facilities, creating a need to use existing fields beyond their design life. Marginal field recovery is the driving factor that is pushing the subsea industry towards condition based maintenance and life extension. Technical health models are used to predict RUL. This prediction is the main factor that influences the life extension decision. There are various modelling approaches as described by Jardine et.al [11]. The statistical reliability based approach needs the past historical failure data of the same or the similar system to determine time to failure. The life cycle load based prediction approach considers assembly, storage, handling, and the operational load related to the equipment [12]. Model based approach uses data related to environmental conditions such as temperature, humidity, vibration etc. and operating conditions such as stress, strain, current, power etc. [13]. Artificial Intelligence based approach need extensive data for analysis as well as for training the model [2]. Lack of adequate operational data is the main handicap at present towards RUL prediction in subsea industry. The lack of data can be supplemented by expert knowledge and Bayesian technique [14] [15] but for the accuracy and acceptability of RUL, the need of real time operational data cannot be undermined. We can say that unlike nuclear and aviation industries, assessment of technical health and RUL are still only on 'wish-list' and not yet in practice in subsea oil and gas industry.

To make the problem complex, the same factors that lead to the lack of data namely water depth, remoteness, geographically diverse locations, and hostile environment are the factors that demand tailor made sensors and instruments for subsea operations. These factors lead to a situation that the device developers resort to custom made solutions. Sensors from different suppliers could be intrinsically different from each other depending on the basic principles of physics used for their design for example, acoustics, electromagnetics or optics. There are some specific challenges related to subsea sensors as opposed to top-side sensors. Radio waves are not suitable for subsea usage because of extremely limited propagation. Acoustic telemetry is gaining grounds in subsea applications but off the shelf acoustic modems are not suitable and need tailoring with respect to power and operating range. Propagation delay is the main factor that needs consideration in acoustic sensors. Each type of sensor involves transmission of information and has its unique advantages and limitations with respect to power requirement, data size, bandwidth, propagation in media and information exchange. Physical size, electrical power and communication compatibility of such devices for end to end data transfer from subsea to topside has been an important unresolved issue in subsea industry.

In our view the efficacy of the open standards towards RUL modelling can be summarized as follows:

1 JIP work has helped in reducing the lead time for introduction of new sensors and hardware, meaning thereby that new sensors can be installed in the existing (old) fields. For example, to measure strain at specific locations
2 The new hardware can be introduced at low cost, thus giving flexibility to RUL evaluation team to demand for surveillance of additional parameters if deemed necessary. For example to measure the actual pipe thickness
3 The standards have promised high scanning speed and fast transmission which will enable monitoring of vibration data which is one of the main parameters for technical health assessment of rotating equipment
4 Standardization of interfaces and compatibility between devices have reduced the technical risk involved in such high technology work

It is expected that the suppliers of the down-hole equipment and the sensors follow the standards namely IWIS, SIIS and SEAFOM and enable the requisite data which can help to create operational history as shown in Figure 5.

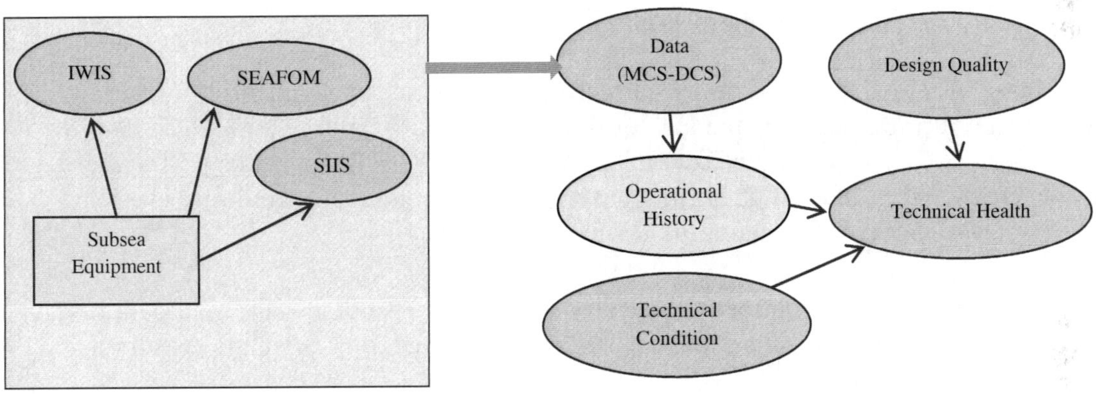

Figure 5: Data flow - from smart sensors to operational history

As of now the knowledge about the standards is limited to the initial set of partners who are members of Joint Industry Program (JIP) for these standardization programmes [3]. The specialized suppliers from aviation and nuclear industries which are more advanced in the field of prognosis are still not fully involved because the standardization is still ongoing.

The problem of *operational history* has three dimensions namely scientific/technological, social and economic. The issue is not limited to intricacies of field theory, design, fabrication etc. but extends to social networks and cost implications. If a subsea production system is viewed as a Large Technical System (LTS) as described by Thomas Hughes, the historian of technology, then we can expect that the models of technological determinism and social determinism [16] would be acting simultaneously. In due course of time, the socio- technical system will gain the *technological momentum.* This will hopefully lead to a situation where standards will drive the interface and sensor developments.

Within the field of institutional economy, some scholars have studied standards as a part of more general phenomena that are labelled as "self-reinforcing mechanisms" [17]. Self-reinforcing mechanisms appear when the value of a particular product or technology for individual adopters increases as the number of adopters increase. The basic mechanism is that the large installed base attracts complementary production and makes the standard cumulative more attractive [18]. A larger base with more complementary products also increases the credibility of the standard as shown in Figure 6.

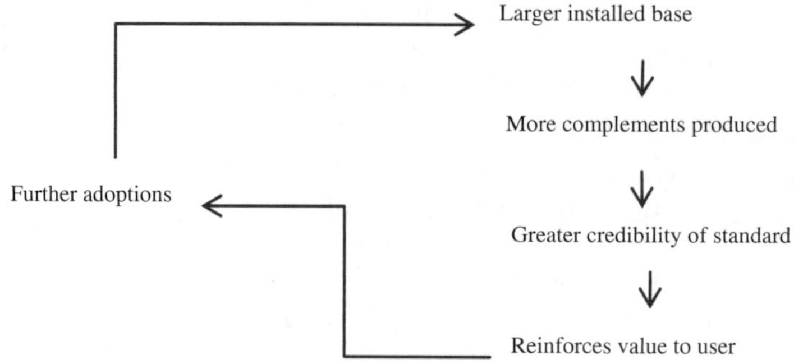

Figure 6: Standards reinforcement mechanism [17]

Together these make a standard more attractive to new users. This brings in more adoptions which further increases the size of the installed base, and so on [19]. It can be expected that the self-reinforcing effects of the installed base will be achieved for these subsea standards. From that point in time onwards, operational data would be available and prognosis would become meaningful in subsea industry.

Conclusion: The paper summarizes the efforts of the subsea community of operators, contractors and suppliers who are collaborating for standardization of smart sensors and data communication.

- Historically, the service and operator companies have been free to choose interfaces as there were no standards for subsea. This has led to a plethora of communication protocols across the industry when interfacing to external sensors causing issues related to compatibility and lack of data for analysis
- It is recommended to comply with IWIS specification as it aids to reduce lead times for hardware, allows wider choice of suppliers and hence possibility of making subsea data available
- With SIIS specification, service companies and sensor suppliers will not have to focus on designing new interfaces, qualify and test different interfaces to their sensors; but instead can focus their energy on improving the products
- The oil and gas operators will be able to save time in not having to select a protocol, but being able to quote the standards and being able to provide better data service
- SEAFOM will result in secure data flow from subsea to topside

References:

[1] Vaidya, P & Rausand, M. 2009. Life extension of machinery in the oil and gas industry. *Proceedings of European Safety and Reliability Conference ESREL,* Praha

[2] Vaidya,P &Rausand, M. 2010 Technical healh of a system – in the context of condition based maintenance *Proceedings of European Safety and Reliability Conference ESREL,* Crete Greece

[3] OTM Consultants : www.otmnet.com

[4] ISO 13628-6:2006 Petroleum and natural gas industries – Design and operation of subsea production systems- Part6: Subsea production control system

[5] IWIS Recommended Practise, IWIS –RP-A2, prepared for IWIS panel by OTM Consulting Ltd, April 2011

[6] Saul, D. (2006) Subsea instrumentation interface standardization in the offshore oil and gas industry International CAN Conference (iCC), Stockholm

[7] ISO 11898-3: 2006 Road vehicles- Controller are network (CAN) - Part3: Low –speed, fault tolerant, medium –dependent interface

[8] Wright, P.J (2000). Optical fiber's gigabit bandwidth, 200Km range attractive for subsea work, *Offshore Magazine*:1-8

[9] Udd, E (2006). *Fiber Optic Sensors: An Introduction for Engineers and Scientists,* Wiley-Interscience, New York, NY, 2nd Edition,

[10] http://www.seafom.com/

[11] Jardine, A. K. S., Daming, L. & Banjevik, D.(2006). A review on machinery diagnostics and prognostics implementing condition-based maintenance. *Mechanical systems and signal processing,* 20: 1483-2510.

[12] Kumar, S.& Pecht, M (2010). Modelling Approaches for Prognostics and Health Management of Electronics, *International Journal of Performability Engineering*, 6(5):467-476

[13] Farrar, C. R. and N. A. J. Lieven (2006). Damage prognosis: The future of structural health monitoring. *Philosophical Transactions of the Royal Society 365*, 623–632.

[14] Zuashkiani, A., Banjevic, D. and Jardine, A.K.S, Estimating Parameters of Proportional Hazards Model Based on Expert Knowledge and Statistical Analysis, Journal of the Operations Research Society (2009) 60, 1621-1636

[15] Vaidya, P. (2010). Prognosis – Subsea oil and gas industry, Annual Conference of the Prognostics and Health Management Society, (PHM), Portland USA

[16] Hughes, T.P (1994). *Technological Momentum, Does Technology Drive History?* The Dilemma of Technological Determinism edited by Merritt Roe Smith and Leo Marx, MIT

[17] Grindley, P. (1995) *Standards, strategy, and policy: Cases and stories*, Oxford: Oxford University Press

[18] Hanseth, O. (2000) *The economics of Standards*, From Control to Drift, edited by C. Ciborra, Oxford Univ. Press, 56-70

[19] Ciborra C. U., Braa K., Cordella A., Dahlbom B., Failla A., & Hanseth O. (2001). *From Control to Drift: The Dynamics of Corporate Information Infrastructures*, Oxford University Press

MFPT SENSORS TECHNOLOGY FOCUS GROUP MEETING

Chris Nemarich
Sensors Focus Group Chair
sensors@mfpt.org

MFPT 2014
Virginia Beach, VA
May 20-22, 2014

MFPT SENSORS TECHNOLOGY

MFPT Technical Committee Chairs are charged with promoting their technical area of expertise and are the MFPT Society's driving force in moving forward the state-of-the-art. They shall accomplish this by creating, growing and leading a group of interested individuals (the technical committee) who are linked through their involvement in MFPT activities.

MFPT Focus Groups

- Sensors Technology
- Data Management
- Signal Analysis
-
- Failure Analysis
- Diagnostics & Prognostics
- Systems Engineering

http://www.mfpt.org/FocusGroups/FocusGroups.htm

MFPT SENSORS TECHNOLOGY

- **MFPT technical committee chairs shall promote the aims of MFPT through:**

  a. Maintaining a Technical Committee Roster
  b. Providing MFPT HQ with names and contact details of individuals potentially interested in MFPT activities
  c. Developing and maintaining aims and objective for their technical committee.
  d. Communicating routinely with their technical committee on technical matters
  e. Soliciting papers, tutorials and other items of interest from their technical committee members, and others, for inclusion in the MFPT meeting.
  f. Obtaining or writing articles for the Society Newsletters.
  g. Encouraging committee members to attend the MFPT annual conference and to promote the Society and the annual conference to others.
  h. Holding a committee meeting at the annual conference to recruit new members and further develop and promote the committee's aims and objectives.
  i. Acting as MFPT ambassador / publicist in the workplace and community

MFPT SENSORS TECHNOLOGY

- Communication
  - Official MFPT email has been setup

    sensors@mfpt.org
  - Create a "Sensors Technology" page under www.mfpt.org website.
  - Post useful links, presentations, requests for information.
  - Interest in a Members Forum for members and guests? LinkedIn?

MFPT SENSORS TECHNOLOGY

- News articles of interest to Focus Group
  - New Products
  - News articles
  - Standards and practices
  - Application notes
  - Case Histories
  - Useful links to products / working groups / resources
  - Q&A

MFPT

MFPT SENSORS TECHNOLOGY

Focus Group Goals

- Solicit Papers for MFPT 2015 Sensor Sessions
  - Topic: Systems Engineering approach to CBM
- Schedule and hold periodic focus group meetings in by phone conference or by internet meeting
- Grow the Sensors Focus Group membership, become active in addressing needs of MFPT communities.

MFPT

MFPT SENSORS TECHNOLOGY

- Paper areas for 2015
- Issues / Topics of interest
  - Smart sensors
  - Wireless sensors / networks
  - Energy harvesting
  - Accelerometers
  - High Temperature / harsh environment
  - Corrosion
  - Remote interrogation
  - Sensor Reliability
  - Low Cost Sensors
  - Acoustic / Ultrasonic

" Advanced Sensors and Measurement Technologies for Improved Gas Turbine Engine Operation"

Development engine testing in the future

PIWG Propulsion Instrumentation Working Group

Harvey Niska
Honeywell Aerospace Co
PIWG CoChair
MFPT - Virginia Beach, VA
May 2014

Test Cell Measurement Technologies

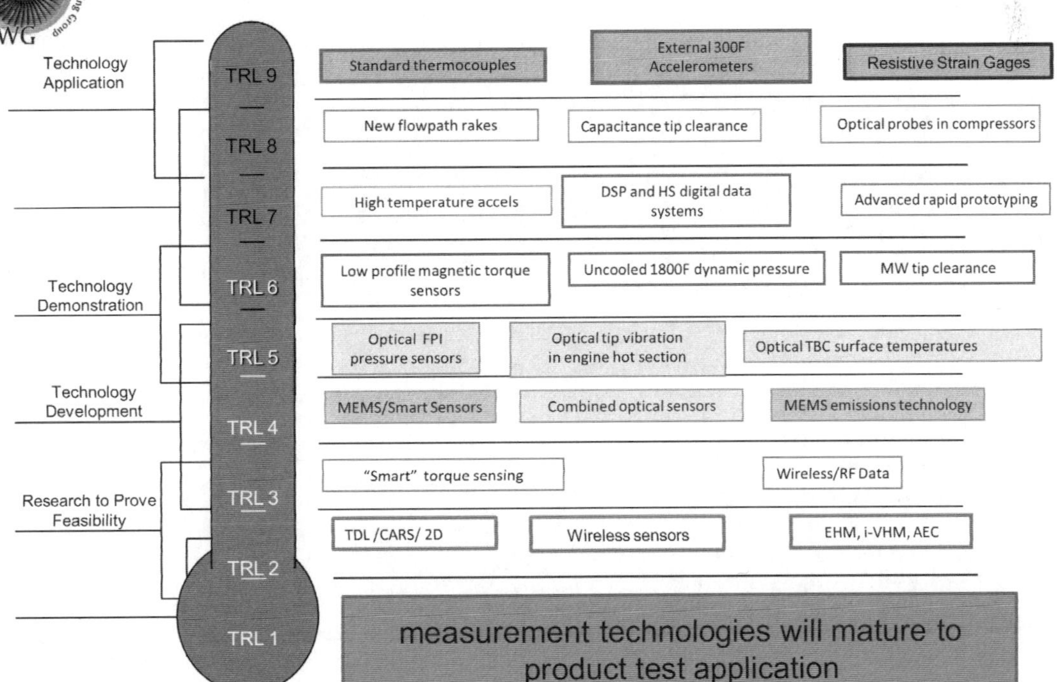

PIWG Mission

- Identify common sensor and measurement needs and requirements for gas turbine engines
- Create and implement a common strategy and vision to guide future research and development
- Provide a consensus voice to represent the gas turbine test cell instrumentation industry
- Identify and pursue instrumentation development efforts

www.PIWG.org

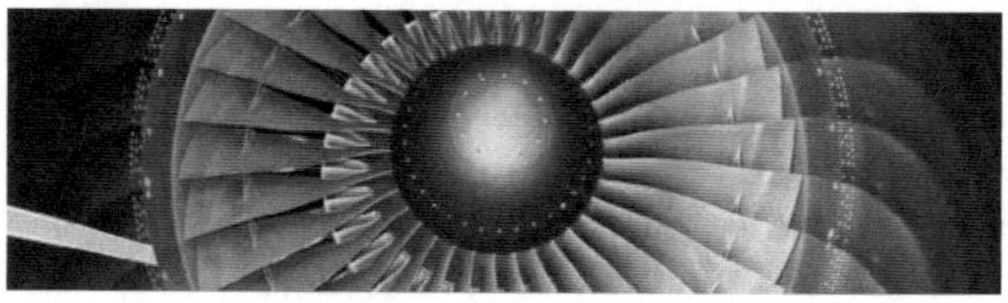

Gas Turbine Engine Measurement Focus

Current Needs

- **Hot section measurements**
- **Wireless data transmission**
- **Optical sensors and methods**

THE focus will be on transition

- **CBM, PHM, EHM , HUMS**
- **Active engine controls**

Customer "CTQ's"

- Lower cost of sensors and data acquisition
- Reduce product development cycle time
- Reduce SFC
- Improved performance
- Reduce TMF and HCF
- Extend MTBO - hot section overhaul
- Fewer "in flight" engine mishaps

Future Measurements and Payoffs

- Cost of turbine engine sensors – minimally invasive
- CBM and HUMS – accelerometers and thermocouples
- Flowpath measurements will have the highest payoff
- CBM – inserted, in wall, structural
- Customer defines the requirements
  - Military and government
  - Commercial

What Is driving change but also creating opportunities?

- Changing landscape across the globe
  - Greenhouse gases and climactic issues
  - Fossil fuel impact

- Regulatory requirements – EU and US
- Industry goals and objectives
  - Improve performance / efficiency
  - Reduce specific fuel consumption (SFC)
  - Reduce noise and exhaust emissions
  - Increase hardware life (MTBF)
  - Reduce downtime
  - Extend hot section overhaul intervals

    CBM, HUMS, VHM, iVHM, EHM, and AEC

Aerospace Engine - Customer Priorities Survey

- Price
- Weight
- Reliability
- Life – MTBF
- Maintainability - MTBO
- Fuel consumption - SFC
- Reduced acoustic noise emissions
- Reduced exhaust species emissions

Gas Turbine Engine “Domain”

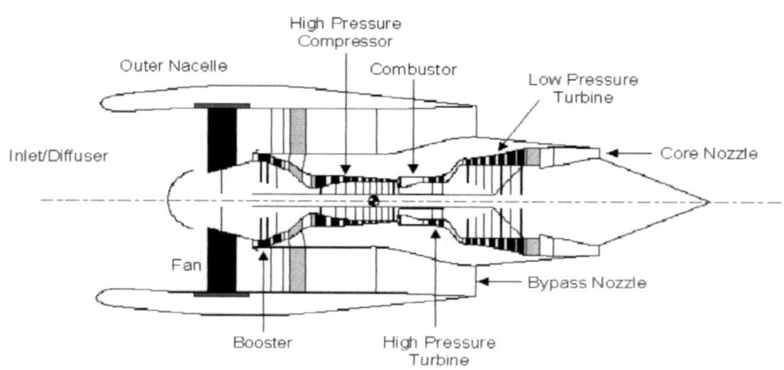

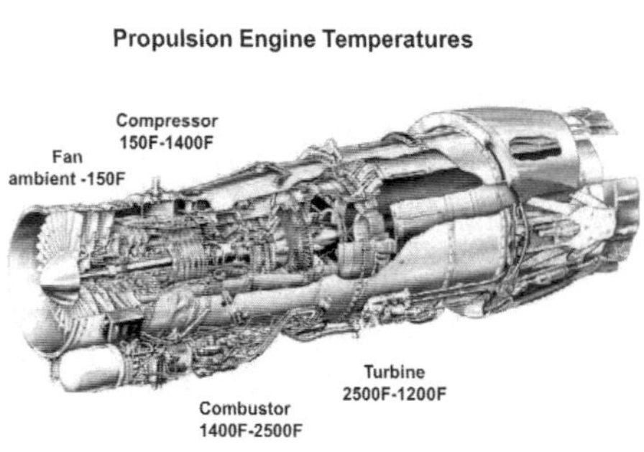

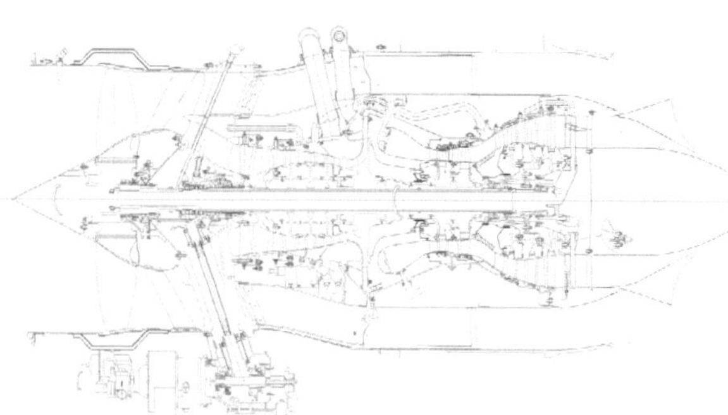

Land Power Generation Gas Turbines

Military –Land and Sea based

F135 Engine Tests

Turboshaft, Turbofan, and Turboprop

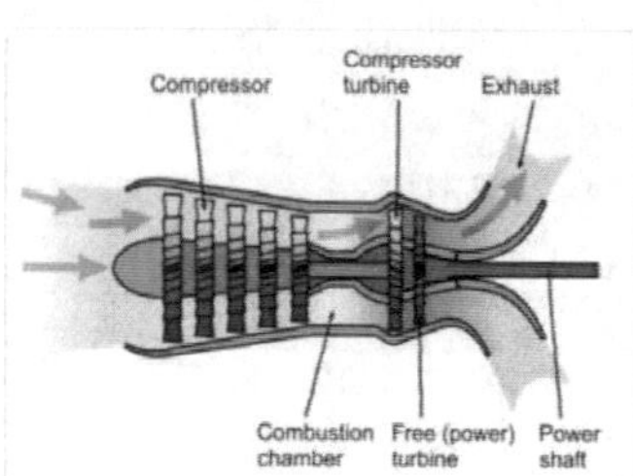

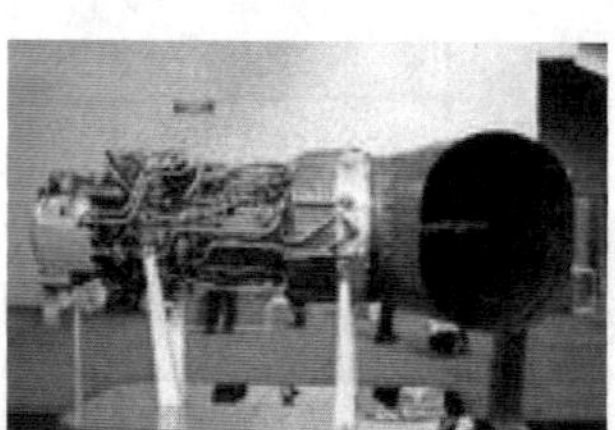

PIWG Propulsion Instrumentation Working Group

USAF Class A & B Mishaps - Component Areas

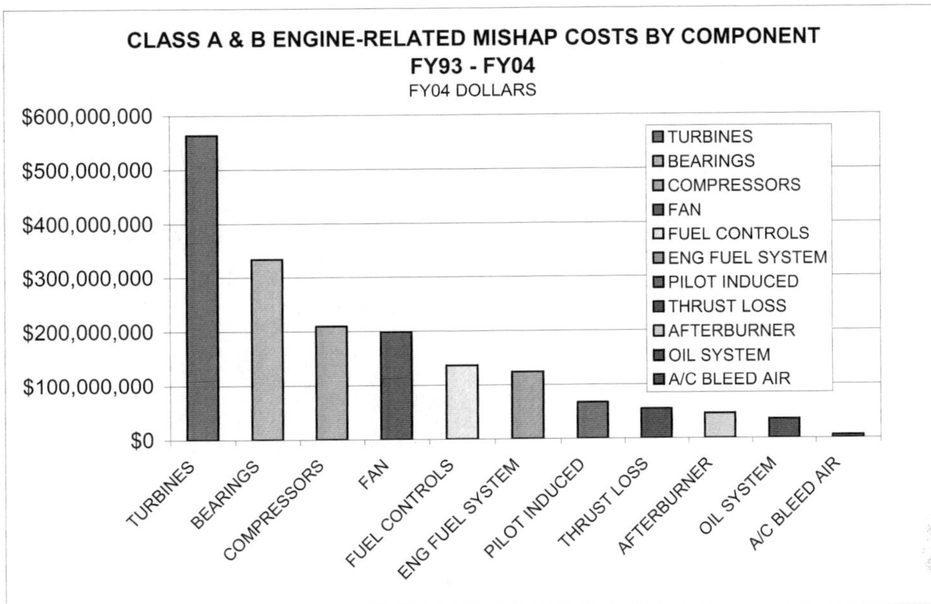

Approved for public release; distribution is unlimited.

Gas Turbine Engine Hot Section Conditions

- T3 compressor discharge flow
  750C (1382F) 40 Bar (600 psi)
- T4 high pressure combustor
  discharge/turbine inlet
  1600C/2100C (2912F /3812F)
- HPT blades and stators surface
  1100C (2012F)
- HPT thermal barrier coating (TBC)
  1300C (2372F)

Standard Instrumentation Used on Development Engines

"80/20 rule"

- Thermocouples
- Pressure tubing and transducers
- Accelerometers

- Resistive strain gages
- Proximity sensors
- Capacitive clearance
- Optical based sensors
- Flow sensors
- Other . . . .

OEMs use common instrumentation installation methods

Standard Pressure & Temperature Installations

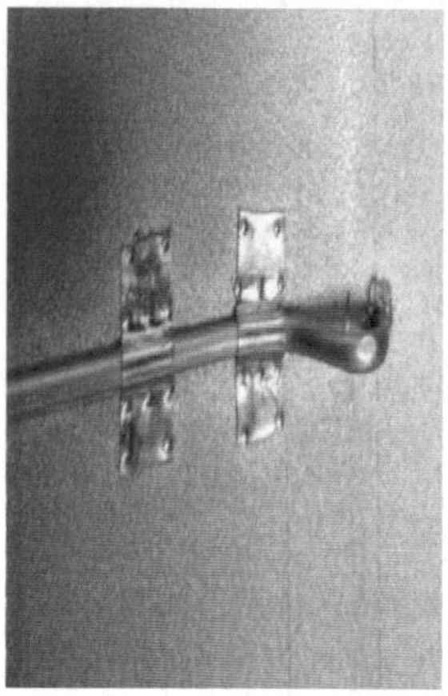

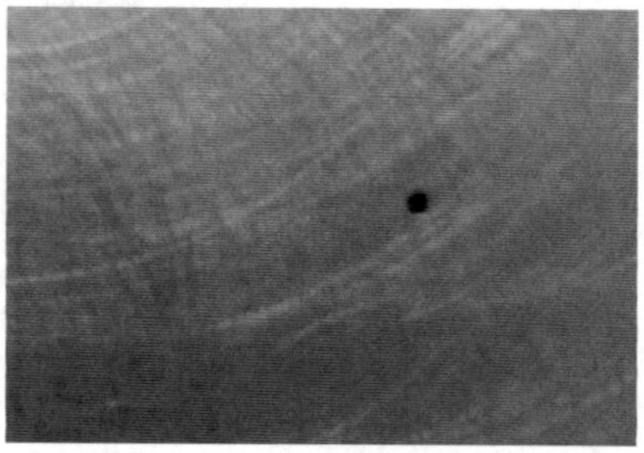

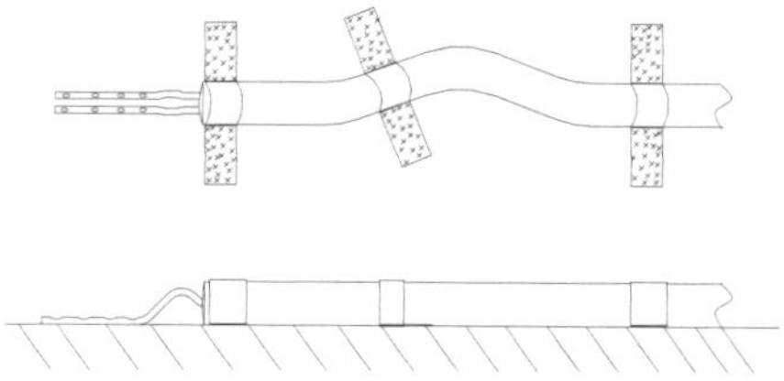

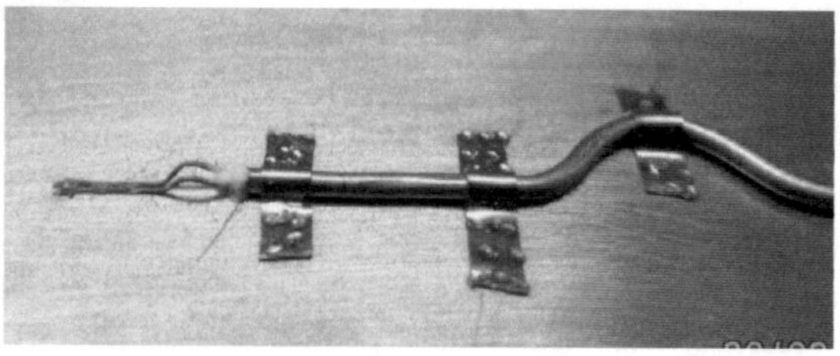

Standard Strain Gage Methods

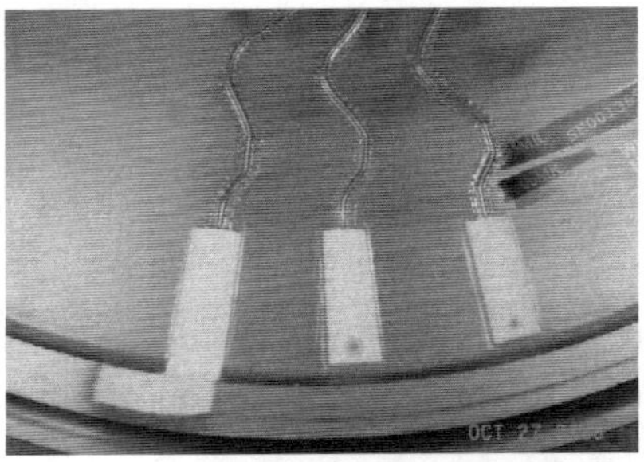

Dynamic Rotating Strain Gages

Flowpath Performance Measurements

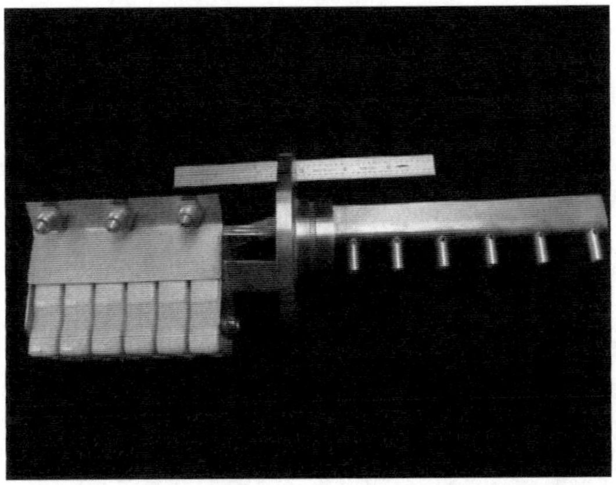

Thermal Paints and Surface Temperature Mapping

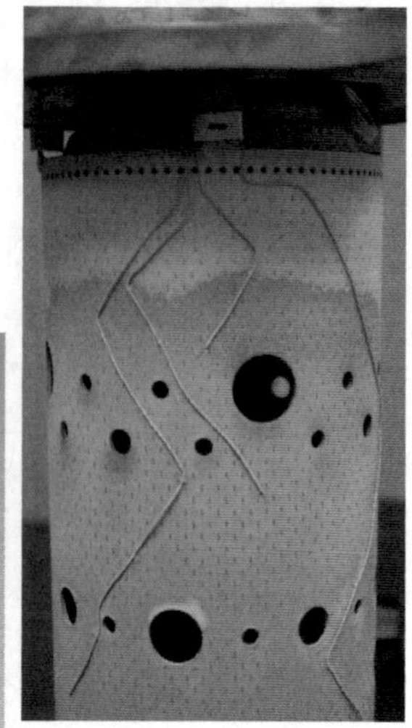

PIWG Is Addressing Key New Technologies!

- BTT/BTC and NSMS maturity
- Thermographic phosphors for hot section TBC measurements
- Dynamic pressure sensors for use in the engine flowpath
- Engine hot section flow path temperature mapping

PIWG has created task teams to address these 4 areas

PIWG has Identified Key Technology Enablers

- High temperature electronics
- High temperature optical materials
- High temperature optical fibers
- High temperature metal materials
- High temperature ceramic materials
- High temperature dielectric materials and application methods
- Improved material vibratory capability
- High temperature flexible soft wire
- High temperature attachment methods and ceramic cements
- Connector and transition technology
- Thin film coatings, compliant layers, and dielectrics
- Sensor composition
- Electronic circuit design
- Laser system maturity
- Probe sensor head configuration and design
- Miniaturization and small packaging
- Data acquisition
- Software and algorithms
- Data and signal processing

NSMS/BTT Development

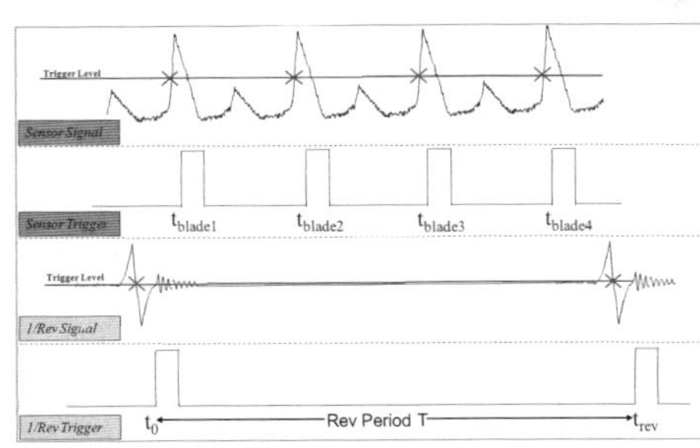

System can be used in conjunction with strain gages
All blades monitored allow for lowering the Goodman limits
X probes support N modes

BTT/BTC and NSMS "Options"

- Capacitance blade tip sensor
- Optical sensors
- Microwave sensors
- Eddy current sensors
- Radar sensors

- Demonstrate and transition to EHM
- Performance and reliability in the engine hot section

Suppliers / vendors are and will be "intimately" involved

Thermographic Phosphors - Temperature

- New thermal barrier coatings are being developed and deployed into propulsion and power engines

- Yttria stabilized 7% zirconia (7YSZ) is the current standard
- Additives improve the TBC performance - low K TBC

- HPT1 and aft . . . .
- Measure deterioration and delamination
- Measure surface temperatures for mapping
- Isolate hot spots and check cooling effectiveness

Optical Probe Access - Insertion and Seal

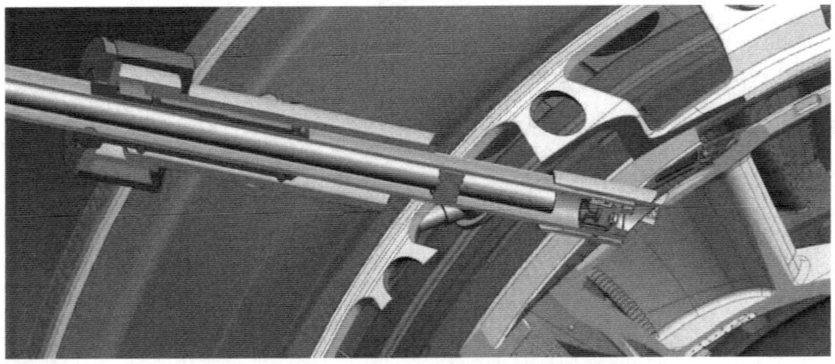

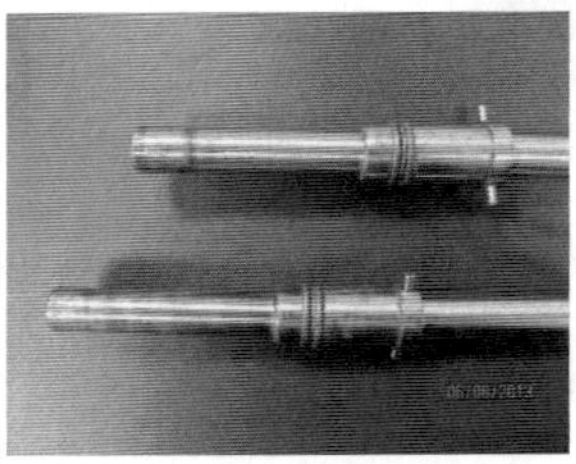

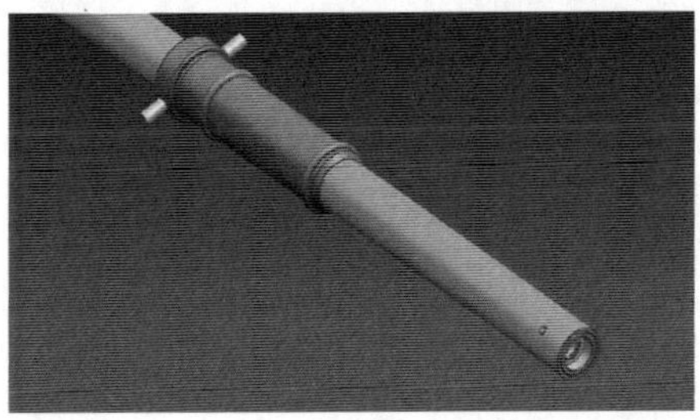

Dynamic Pressure Sensors

- New technologies include;
  - Optical Fabry Perot Interferometry (FPI) sensors
  - Piezo resistive based sensors using SOI and SiC
  - MEMS devices and assemblies

- Oxsensis LTD (UK)
- Prime Photonics
- Kulite Semiconductor
- Sporian Microsystems
- Chiral Optics

- All work is at suppliers externally funded and via consortia
- Goals are; higher frequency response, high temperature, miniaturization, and improved accuracy

Dynamic Pressure and Temperature Transducer

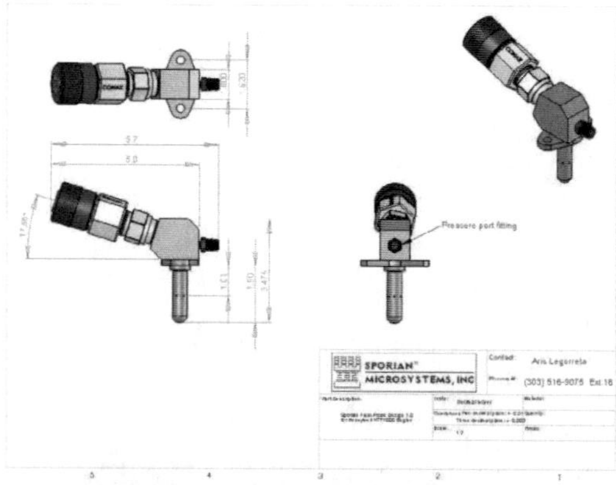

Sporian Microsystems
MEMS Tip Sensor
Pressure and temperature

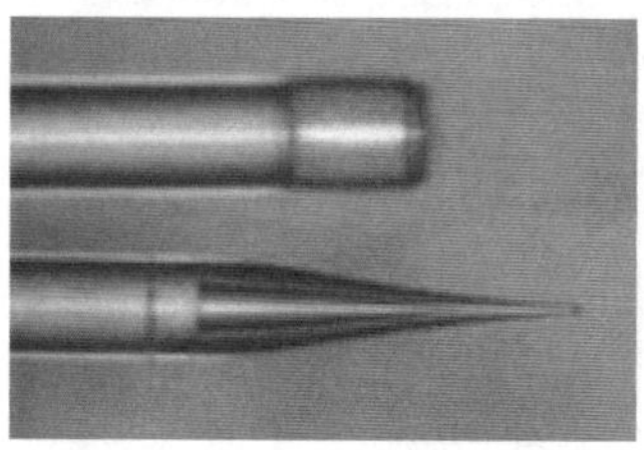

Prime Photonics
Optical FPI Sensor
Pressure ranges from: 5 psi - 30,000psi
Temperature range: 0 - 750° F
Frequency Response: >>100 kHz

Hot Section Flowpath Temperature Mapping

PIWG Propulsion Instrumentation Working Group

- Current technology uses discrete point measurements using precious metal thermocouples to assess temperatures in the hot section
- Full field measurements would interrogate the flowpath using in wall sensors or measurement systems
- Prime interest is to measure T 4.0 at the combustor exit
- This is the highest priority measurement for the OEM community
- The combustor discharge has the harshest P&T conditions in the engine

EHM and Condition Based Maintenance

- Accelerometers – measure vibration
- Baseline normal engine performance signature
- Deviations indicate maintenance may be required

- Opportunities for new optical sensors
- Temperature measurements – thermal mechanical fatigue (TMF)
  - Combustor
  - High pressure turbine
  - Low pressure turbine
- Hot section flowpath temperatures
- Compressor blade tip timing – vibration signature
- Fan – optically visible – blade tip vibration

PIWG Collaboration provides energy toward success!

Innovation is a team activity and happens throughout the product life cycle

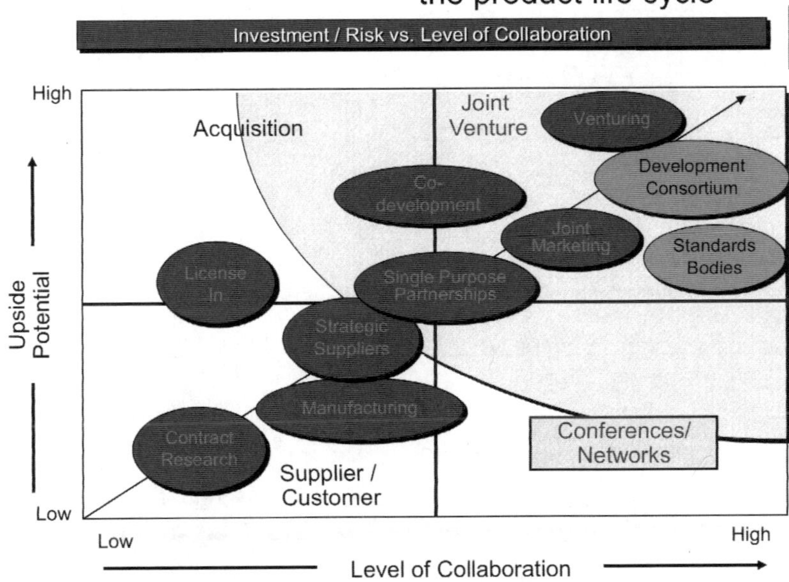

Increasing management complexity & risk

Benefits of Collaboration

- **Acquiring Technology**
- **Influencing OEMS**
- **Co-developing products**
- **Building strategic suppliers**
- **License-in potential**
- **Risk sharing/capital investments**
- **Monitoring development**
- **Feasibility**
- **Cost reduction/efficiency**
- **Build/obtain competencies**

ISA TCI Standards Committees

- ISA 107.X – Advanced Measurement Techniques for Gas Turbine Engines
  - Develop standards, recommended practices and technical reports on measurement techniques that are vital in the safe and reliable operation of gas turbine engines
- Five Subcommittees
  - 107.1 - Nonintrusive Stress Measurement System/Blade Tip Timing *Bryan Hayes, AEDC*
  - 107.2 – Temperature Measurement using Thermographic Phosphors *Steve Allison, EMCO*
  - 107.3 Blade Tip Clearance *Paul Seccombe, Pentair (UK)*
  - 107.4 Wireless Data Transmission in the Test Cell Environment *Dan Sexton, GE Global Research*
  - 107.5 Dynamic Pressure Measurement *Adam Hurst, Kulite (US)*

Measurement Technologies

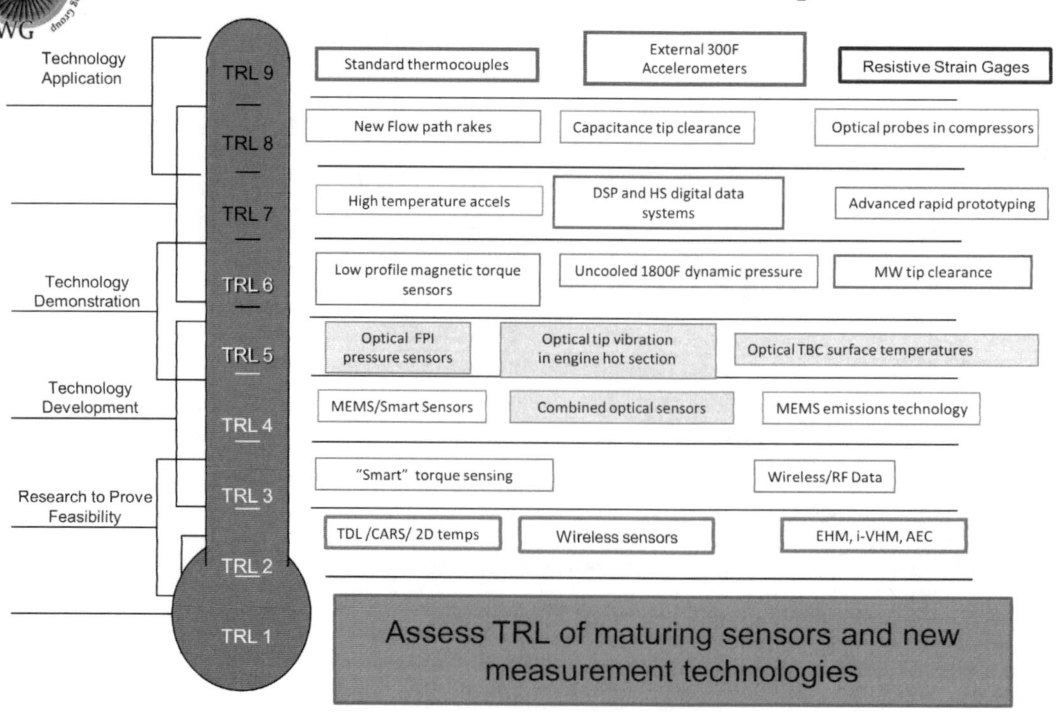

The University of New South Wales

MFPT 2014

PHASE DEMODULATION BASED COMPUTED ORDER TRACKING – A TUTORIAL

R. B. Randall

School of Mechanical and Manufacturing Engineering

The University of New South Wales
Sydney 2052, Australia

The University of New South Wales

MFPT 2014

Background

- **Order tracking means processing signals so that their spectra are converted from a basis of frequency to harmonic order of a reference shaft**
- **This means that time signals are converted from time to rotations, or periods**
- **Sampling of the signals must be changed from equal steps in time to equal steps in rotation angle**
- **This can be done for small variations in speed, or for very large variations such as run-ups amd run-downs**

ORDER TRACKING
for variable speed machine signals

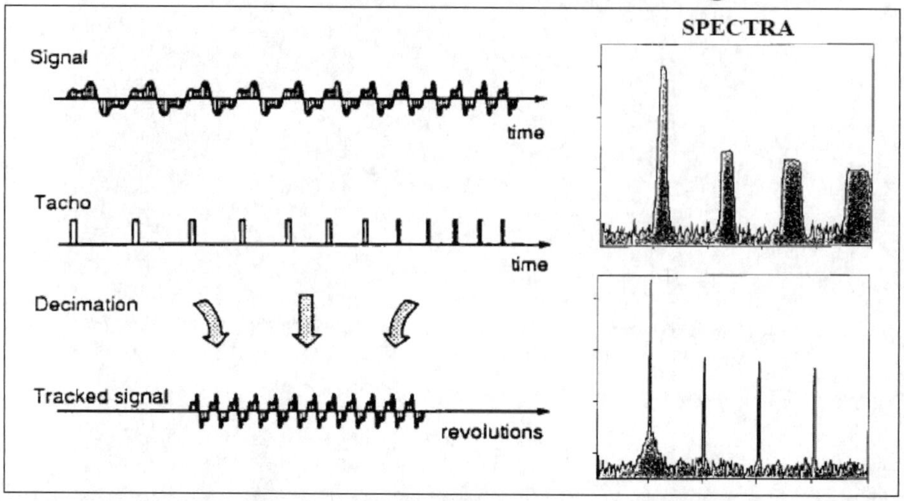

May also be required to correct for small speed variations before averaging

ORDER TRACKING APPLIED TO SIGNALS FROM A MINING SHOVEL GEARBOX

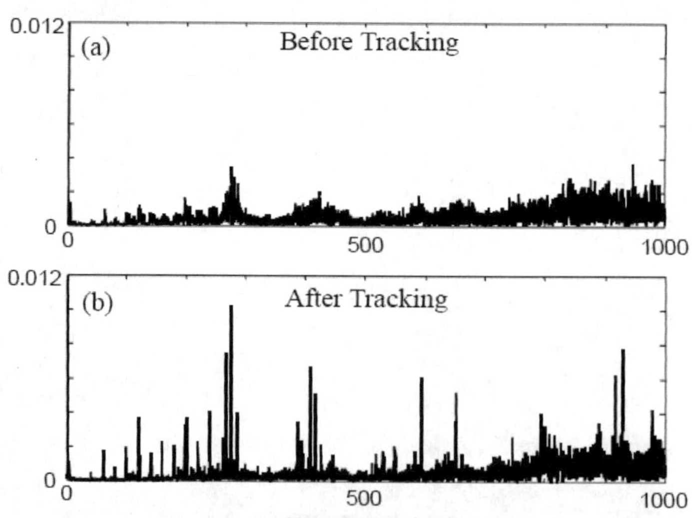

The University of New South Wales

MFPT 2014

Order Tracking Methods

1. **External sampling controlled by phase-locked loop tuned to tacho signal (N samples per tacho period) – Loses all information about time, and requires a speed related tracking lowpass filter for antialiasing**
2. **Computed order tracking where reference signal is digitised at same time as measurement signals and used to resample at constant intervals of rotation angle**
   a) Polynomial interpolation of various orders to find resample times and the values of the signals at those resample times
   b) Phase demodulation of reference signal to obtain correct phase/time map (to any resolution), and then polynomial (usually cubic spline) interpolation of signals
3. **Computed order tracking using a reference signal extracted from the measurement signal**

The University of New South Wales

MFPT 2014

Phase Demodulation Based Order Tracking

- **Order tracking involves two separate stages of interpolation**
  1. Generation of a phase vs time map with fine resolution. Samples of phase are typically only available once per rev from a tacho signal, and must be interpolated for intermediate values
  2. Interpolation of originally sampled signals from equal spacings in time to equal spacings in rotation angle
- **If the spectrum of a given harmonic of the reference signal is completely isolated in the frequency domain it can be windowed out with a rectangular window and demodulated to give a phase/time map with any desired resolution. Interpolation of the signal will then be more accurate**

The University of New South Wales

MFPT 2014

Phase Demodulation based on Hilbert Transform

- The Hilbert transform represents the relationship between the real and imaginary parts of the Fourier transform of a one-sided signal
- Thus the inverse transform of a one-sided spectrum is a complex time signal whose real and imaginary parts are related by a Hilbert transform
- The amplitude of this complex time signal represents the amplitude modulation of the signal, and its phase represents the phase modulation signal
- The Hilbert transform can be calculated based on Fourier transforms, so some basic theory is required

The University of New South Wales

MFPT 2014

FOURIER SERIES

For any periodic signal *g(t)* of period *T* for which

$$g(t) = g(t+T)$$

it can be shown that:

$$g(t) = \frac{a_0}{2} + \sum_{k=1}^{\infty} a_k \cos(k\omega_0 t) + \sum_{k=1}^{\infty} b_k \sin(k\omega_0 t) \qquad (1)$$

where ω_0 is the fundamental angular frequency in rad/s $= \frac{2\pi}{T}$

The University of New South Wales

MFPT 2014

FOURIER SERIES

Thus, the total component at frequency $\omega_k (= k\omega_0)$

$$= a_k \cos(\omega_k t) + b_k \sin(\omega_k t) \qquad (3)$$

which can be written as:

$$C_k \cos(\omega_k t + \phi_k) \qquad (4)$$

where $C_k = \sqrt{a_k^{\,2} + b_k^{\,2}}$ and

$$\phi_k = \tan^{-1}(b_k / a_k)$$

The University of New South Wales

MFPT 2014

FOURIER SERIES

This can again be represented as:

$$\frac{C_k}{2}\left[\exp\{j(\omega_k t + \phi_k)\} + \exp\{-j(\omega_k t + \phi_k)\}\right] \qquad (5)$$

which can be interpreted as two rotating vectors, each of length $C_k / 2$ one rotating at angular frequency ω_k with initial phase ϕ_k and the other rotating at angular frequency $-\omega_k$ with initial phase $-\phi_k$

The University of New South Wales

MFPT 2014

EXPRESSION OF A SINUSOID AS A SUM OF ROTATING PHASORS

$$C_k \cos(\omega_k t + \phi_k) = \frac{C_k}{2}\left[\exp\{j(\omega_k t + \phi_k)\} + \exp\{-j(\omega_k t + \phi_k)\}\right]$$

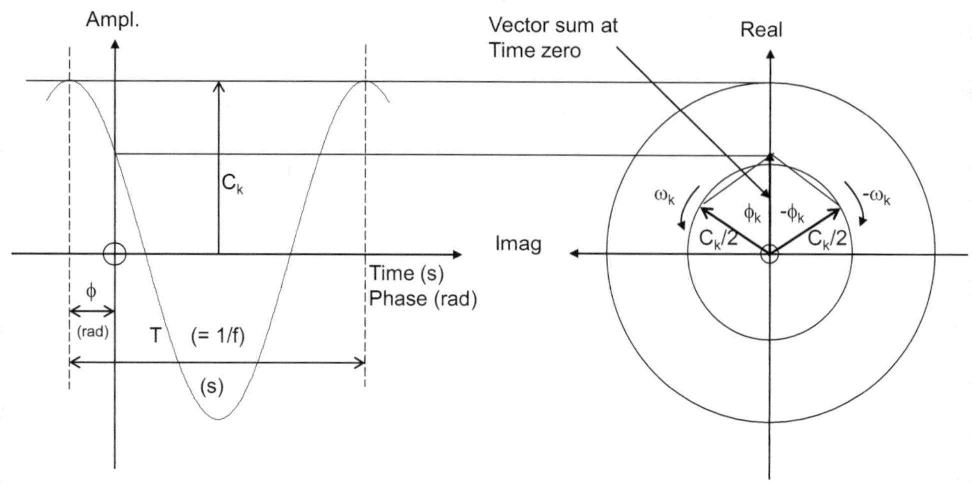

The University of New South Wales

MFPT 2014

ALTERNATIVE REPRESENTATION

as a projection on the real axis of a 1-sided spectrum The time signal with a 1-sided spectrum is complex and is known as an analytic signal

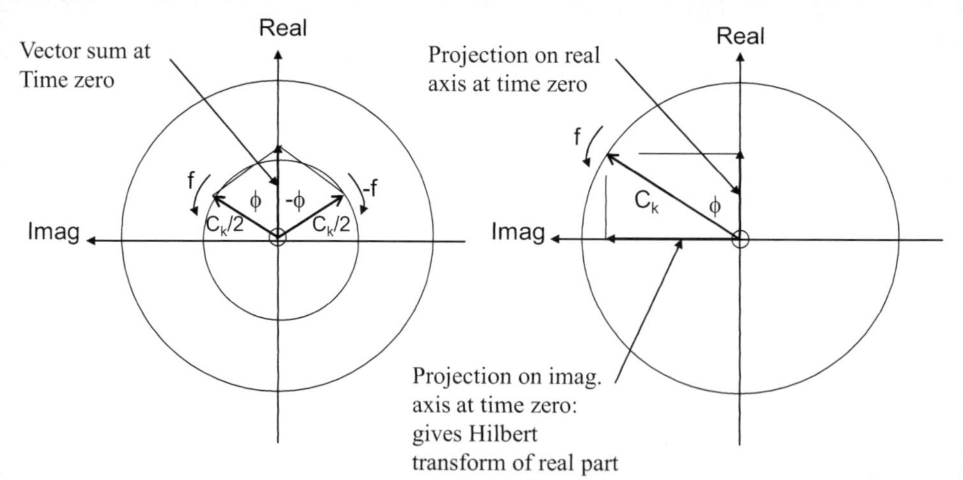

Real and imaginary parts related by a Hilbert transform

The University of New South Wales

MFPT 2014

HILBERT TRANSFORM

Relationship between the real and imaginary parts of the Fourier transform of a one-sided function

$$x(t) = x_e(t) + x_o(t) \qquad \Im\{x(t)\} = X(f)$$

$$x_e(t) \qquad \Im\{x_e(t)\} = \mathrm{Re}[X(f)]$$

$$x_o(t) \qquad \Im\{x_o(t)\} = \mathrm{Im}[X(f)]$$

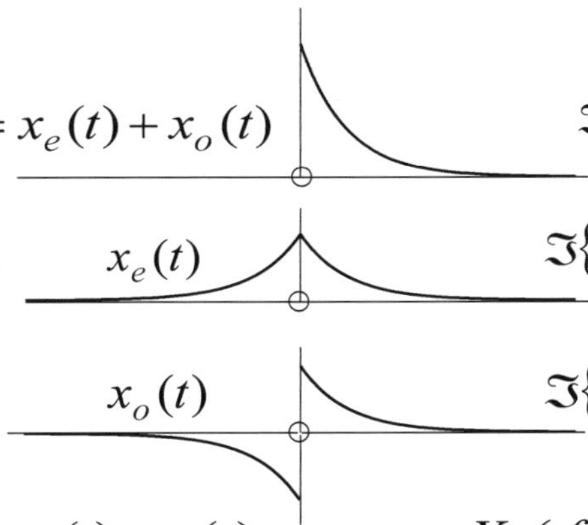

$$x_o(t) = x_e(t) \cdot \mathrm{sgn}(t) \qquad X_I(f) = X_R(f) * \Im\{\mathrm{sgn}(t)\}$$

Fourier transform of a sign function is 1/πf, which is convolved with $X_R(f)$

The University of New South Wales

MFPT 2014

AMPLITUDE MODULATION

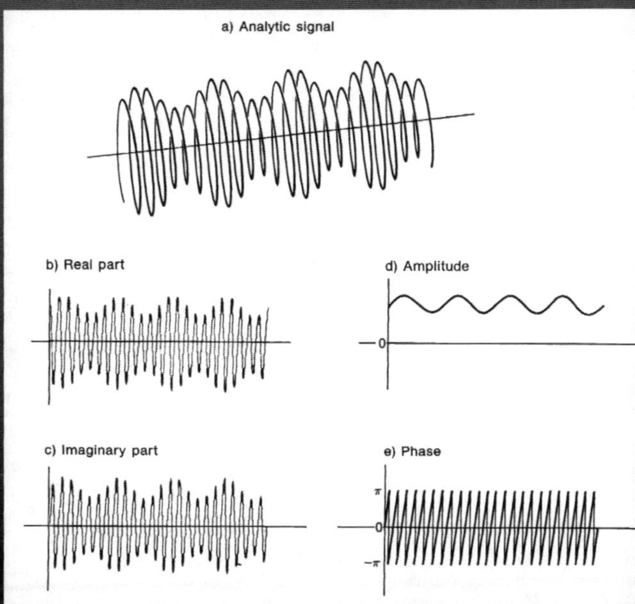

The University of New South Wales

MFPT 2014

AMPLITUDE MODULATION AS A SUM OF THREE CONSTANT VECTORS OR AS A SINGLE VARYING VECTOR

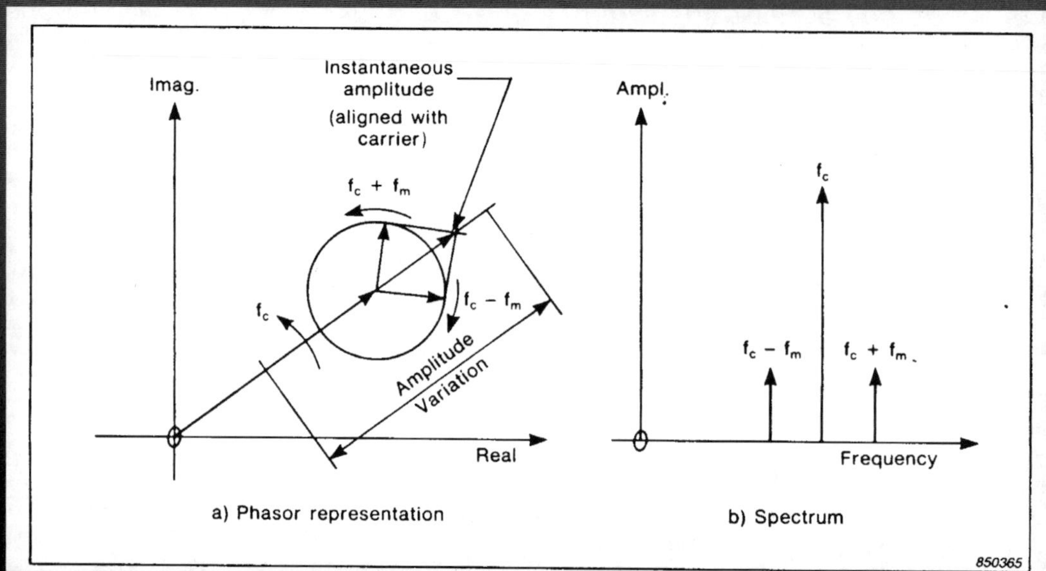

The University of New South Wales

MFPT 2014

PHASE (FREQUENCY) MODULATION

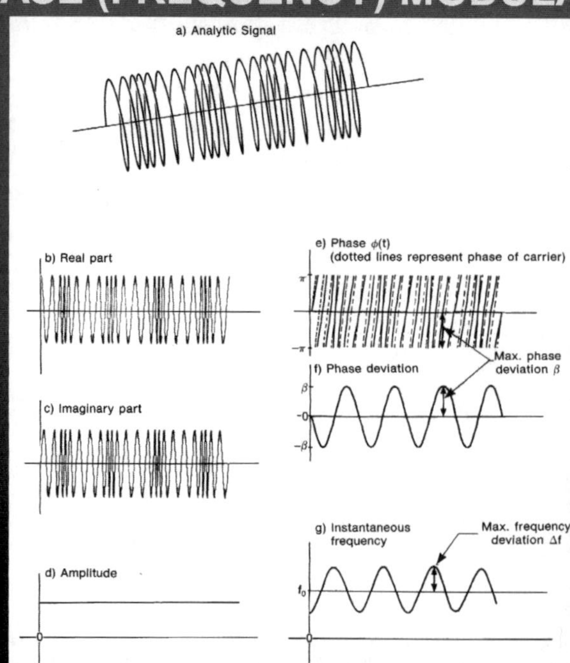

The University of New South Wales

MFPT 2014

PHASE RELATIONSHIPS OF SIDEBANDS FOR AMPLITUDE AND PHASE MODULATION

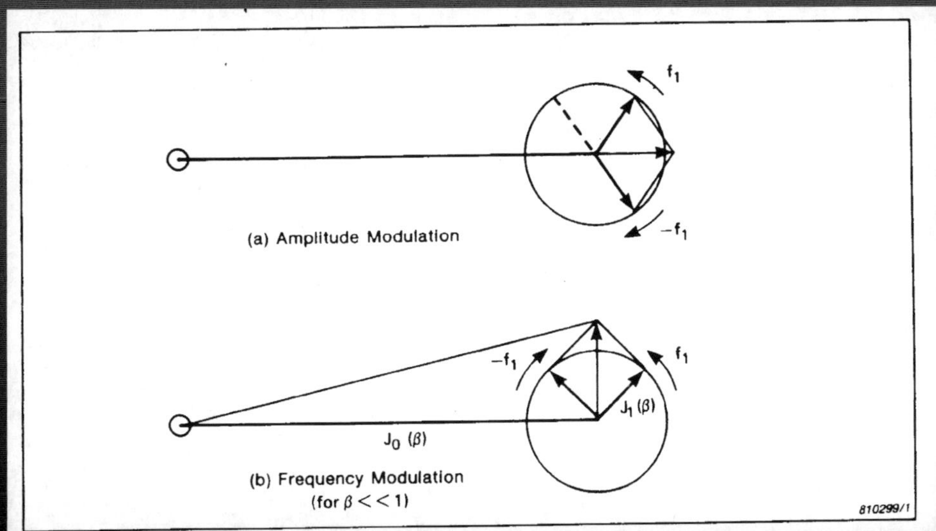

For small phase deviations only one pair of sidebands is necessary, but when hypotenuse changes significantly, more sidebands are required

The University of New South Wales

MFPT 2014

Sideband spectrum for single frequency modulation with larger phase deviation

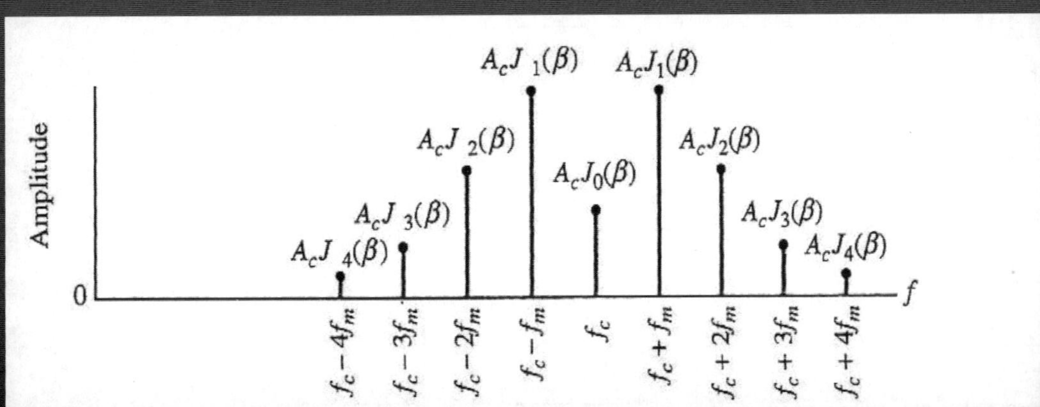

Sidebands defined by Bessel functions. Theoretically they extend to infinity, but only a limited number are significant

The University of New South Wales

MFPT 2014

Sidebands for Single Frequency Modulation

with modulation index β

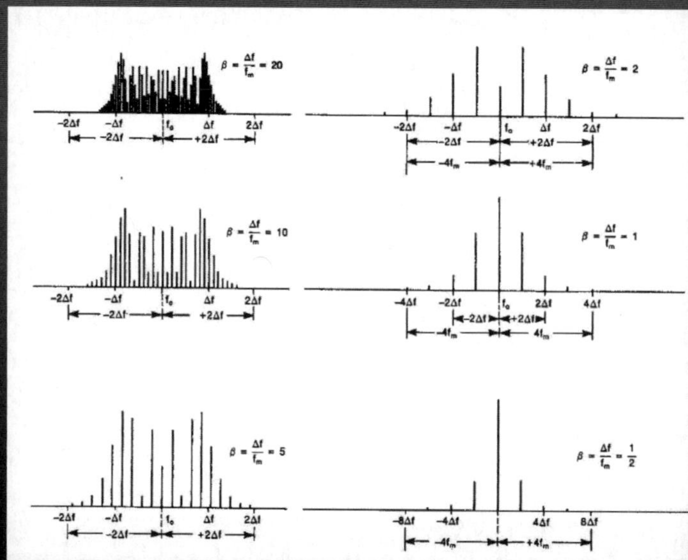

Governed by frequency sweep   Governed by modulation frequency

NB Modulation index β equals max phase deviation from carrier

β = 1 (radian) requires 2 pairs of sidebands (second pair to counter amplitude variation)

β << 1 only requires one pair of sidebands

The University of New South Wales

MFPT 2014

FFT ZOOM PROCESS

NB Can also be used to demodulate carrier frequency f_k

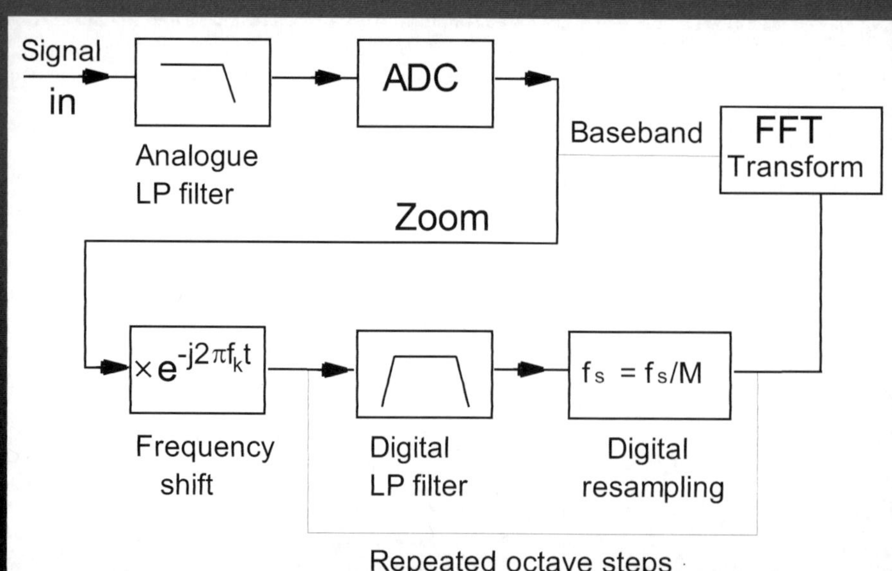

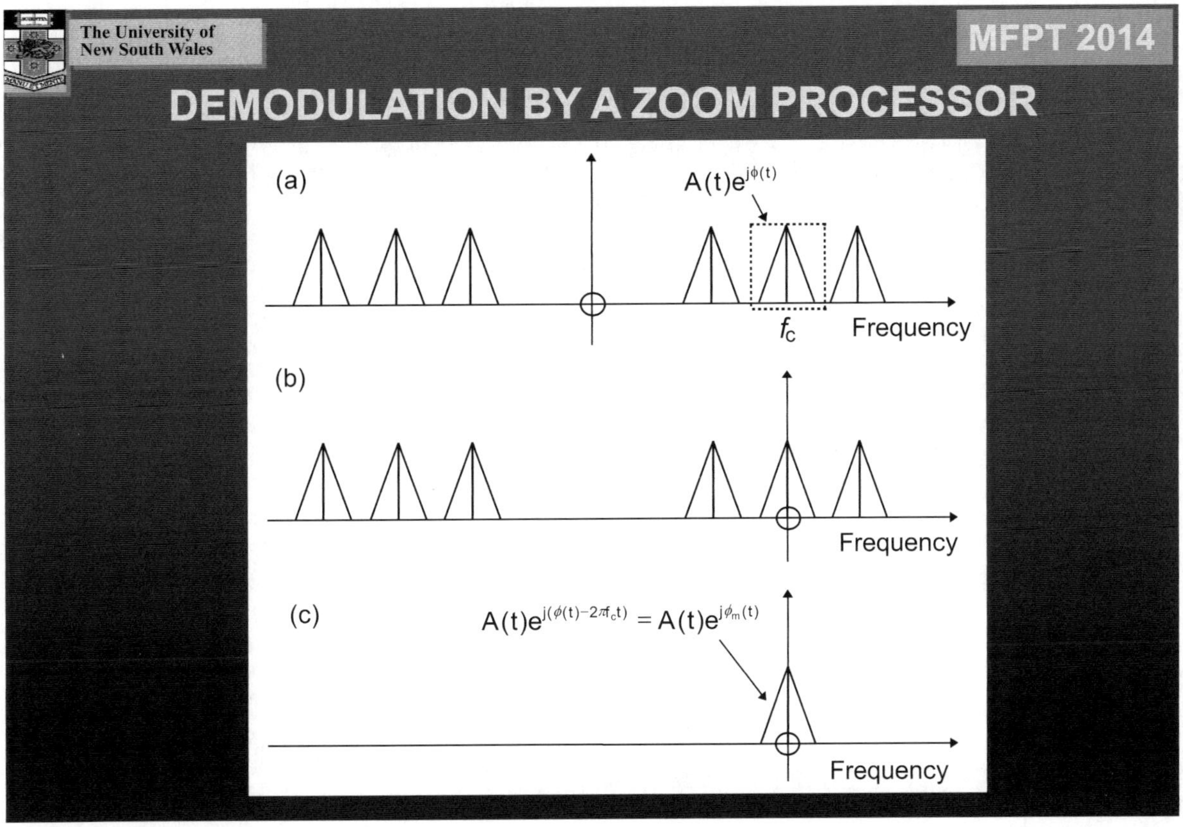
The University of New South Wales
MFPT 2014
DEMODULATION BY A ZOOM PROCESSOR
(a)
$A(t)e^{j\phi(t)}$
f_c
Frequency
(b)
Frequency
(c)
$A(t)e^{j(\phi(t)-2\pi f_c t)} = A(t)e^{j\phi_m(t)}$
Frequency

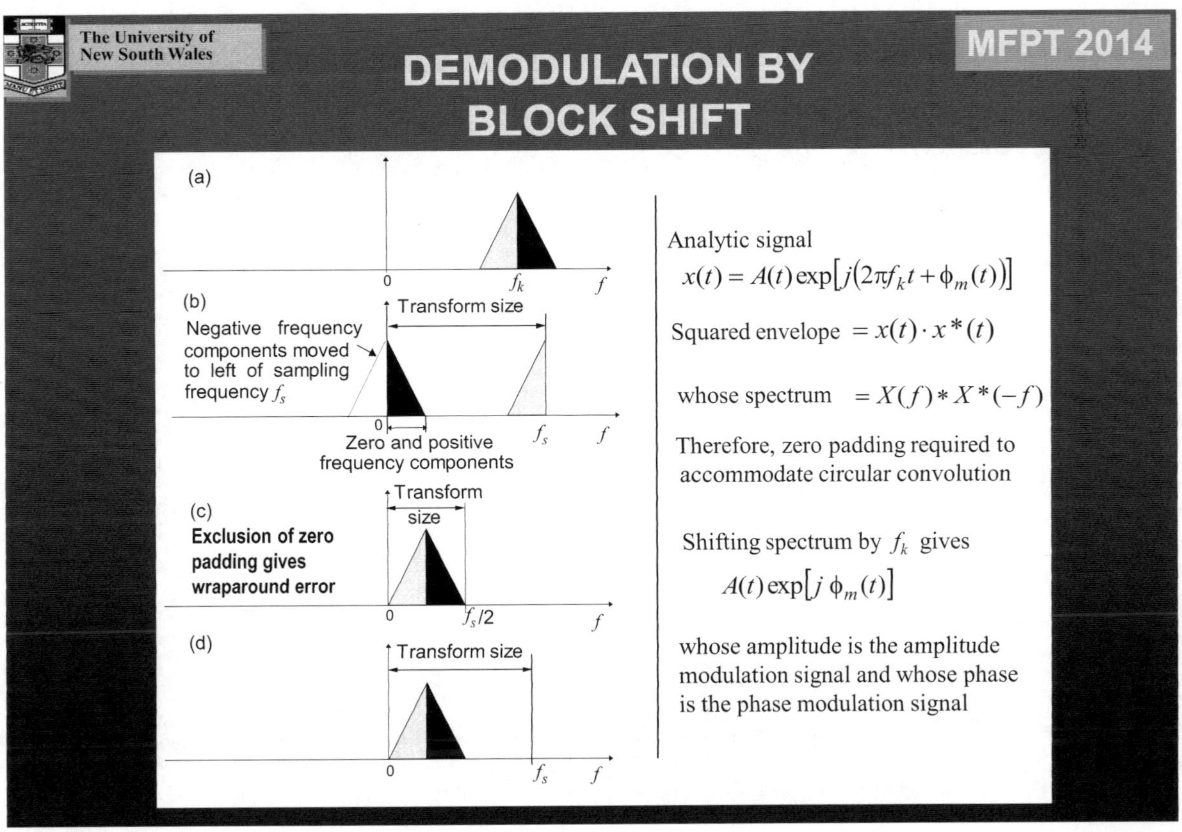
The University of New South Wales
MFPT 2014
DEMODULATION BY BLOCK SHIFT
(a)
0
f_k
f
(b)
Transform size
Negative frequency components moved to left of sampling frequency f_s
0
Zero and positive frequency components
f_s
f
(c)
Exclusion of zero padding gives wraparound error
Transform size
0
$f_s/2$
f
(d)
Transform size
0
f_s
f
Analytic signal
$x(t) = A(t)\exp[j(2\pi f_k t + \phi_m(t))]$
Squared envelope $= x(t) \cdot x^*(t)$
whose spectrum $= X(f) * X^*(-f)$
Therefore, zero padding required to accommodate circular convolution
Shifting spectrum by f_k gives
$A(t)\exp[j\,\phi_m(t)]$
whose amplitude is the amplitude modulation signal and whose phase is the phase modulation signal

The University of New South Wales

MFPT 2014

Angular resampling with phase/time map

Rectangular window in frequency domain corresponds to sinx/x interpolation in time domain

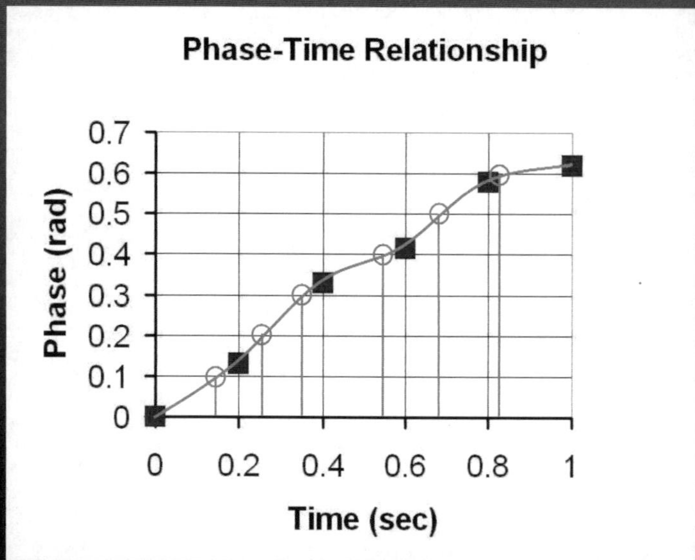

Phase demodulation gives phase/time map uniformly sampled in time (arbitrary resolution)

Use interpolation to resample at equal phase intervals

The University of New South Wales

MFPT 2014

Phase vs Time Map by "Hilbert transform" methods

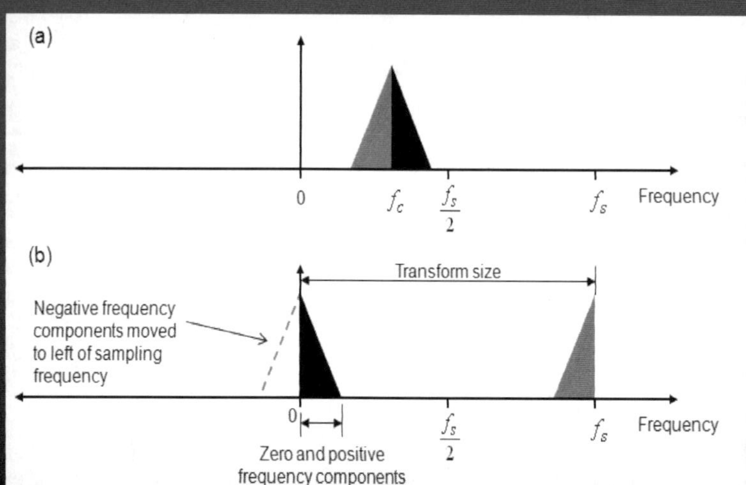

Select band to be demodulated

Shift carrier frequency to zero and inverse transform to a (complex) time signal

Phase of complex time signal is phase modulation signal, but modulo 2π. Must be unwrapped to continuous function of time. This requires removal of carrier component, but it can be added back in afterwards.

The University of New South Wales

MFPT 2014

RESAMPLING BY INTERPOLATION

Choose nearest sample

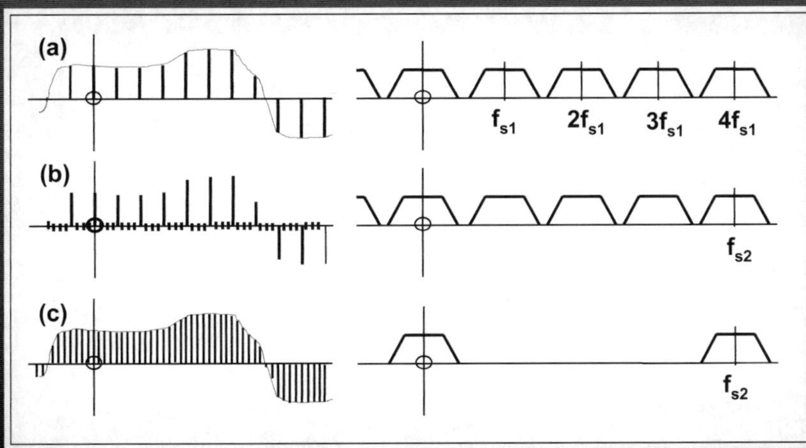

Original spectrum

Sampling frequency increased by adding zeros

Interpolation by lowpass filtration

The above applies to realtime operation in time domain but is equivalent to Matlab INTERPFT in frequency domain.

By postprocessing, the resolution can be increased arbitrarily by increasing sampling frequency with no loss of information

The University of New South Wales

MFPT 2014

POLYNOMIAL INTERPOLATION

From P. McFadden (1989) "Interpolation Techniques for Time Domain Averaging of Gear Vibration". *Mechanical Systems and Signal* Processing 3(1), pp 87-97

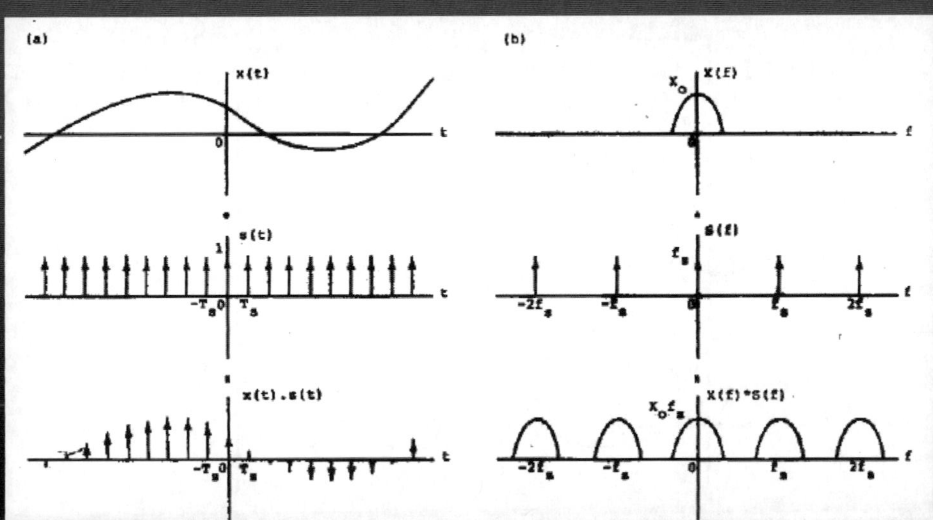

Sampling of original signal

The University of New South Wales

MFPT 2014

POLYNOMIAL INTERPOLATION

Courtesy P. McFadden

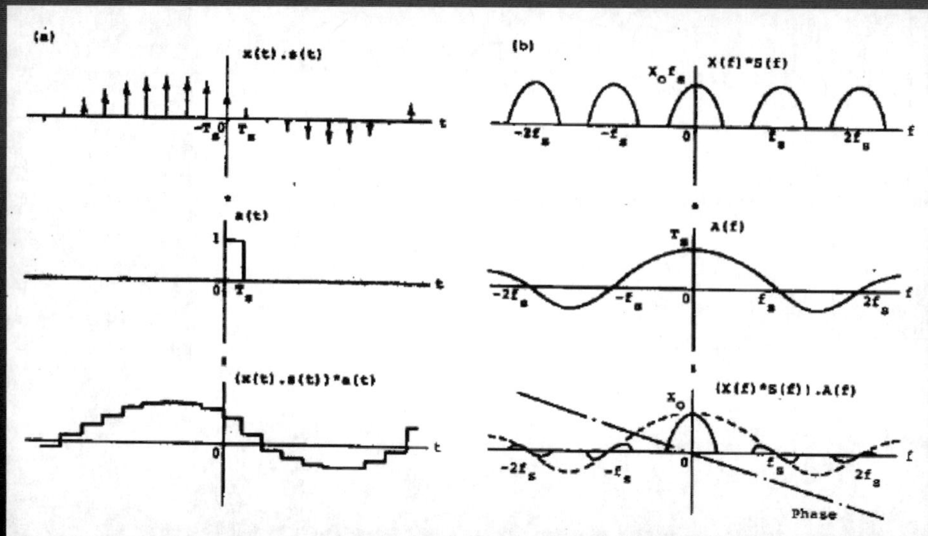

Sample and hold interpolation (convolution with rectangle)

The University of New South Wales

MFPT 2014

POLYNOMIAL INTERPOLATION

Courtesy P. McFadden

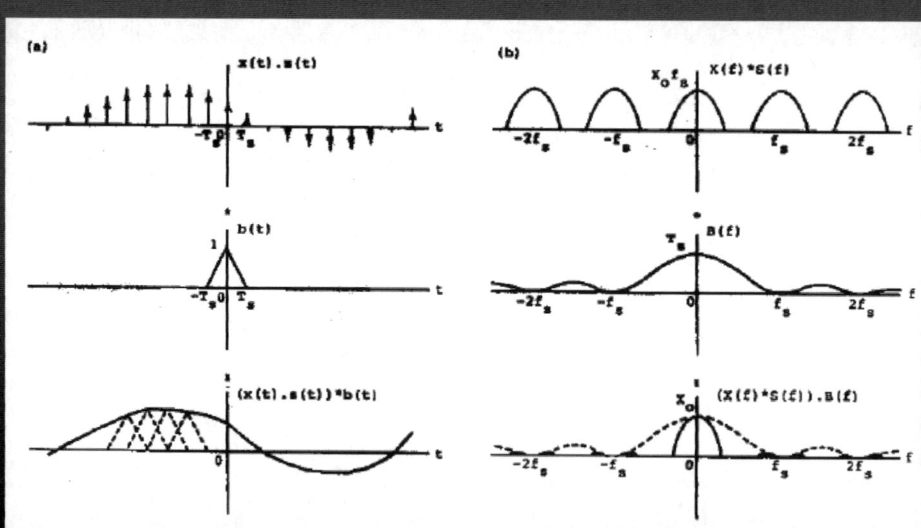

Linear interpolation (convolution with triangle)

The University of New South Wales

MFPT 2014

POLYNOMIAL INTERPOLATION

Courtesy P. McFadden

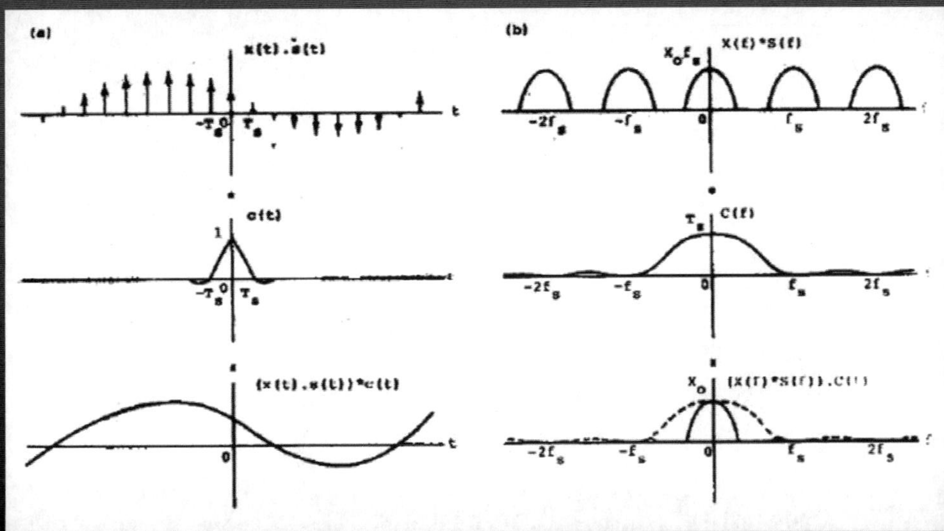

Cubic spline interpolation (convolution with cubic function)

The University of New South Wales

MFPT 2014

POLYNOMIAL INTERPOLATION

Courtesy P. McFadden

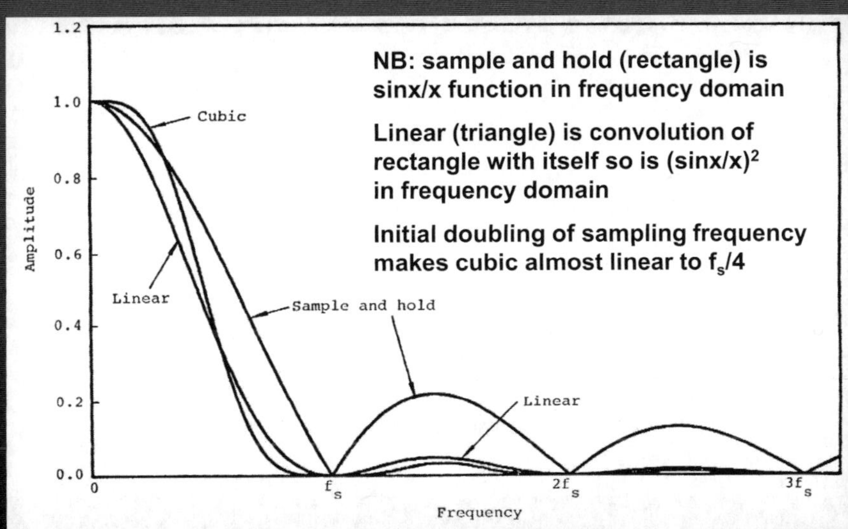

Cubic interpolation has least in-band effect and smallest sidelobes

The University of New South Wales

MFPT 2014

Limiting speed variation for a carrier with multiple harmonics

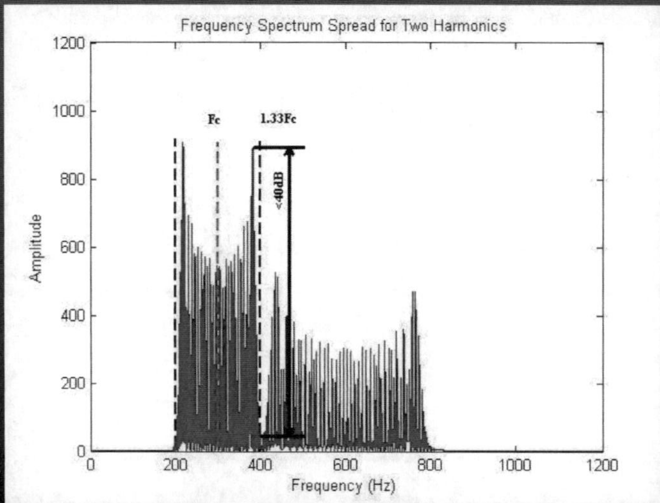

Note similarity with PDF for a sinusoid

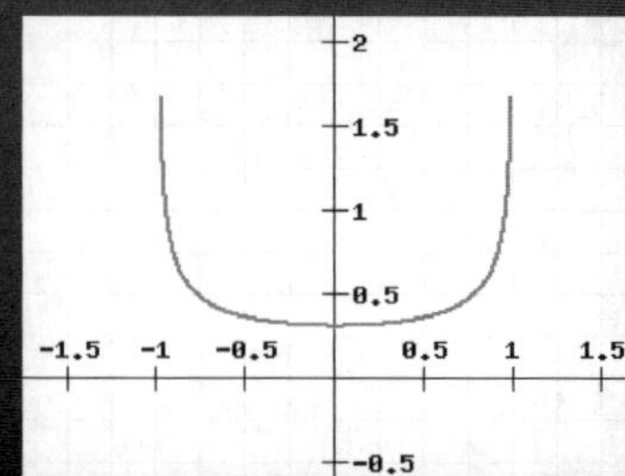

Maximum deviation of first harmonic theoretically 33% for slow variation
(Spread approximately prop. to centre frequency)

The University of New South Wales

MFPT 2014

Overlap of higher order sidebands

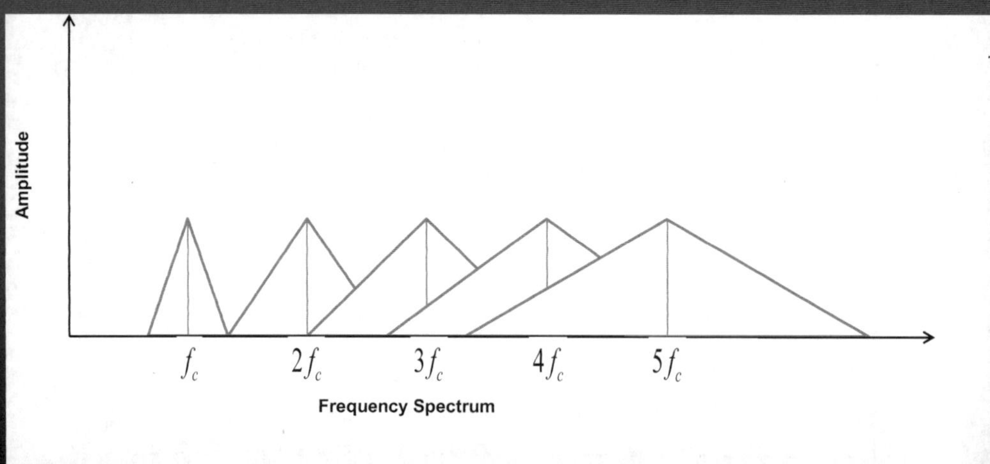

Higher order sidebands do not overlap if spread proportional to centre frequency

The University of New South Wales

MFPT 2014

Order Tracking for Machine Diagnostics

Three main areas of use

1. To make the phase-locked signals (gears) periodic in the rotation angle domain. This usually requires limited speed variations (< 5%) to limit amplitude modulation and allow time synchronous averaging (TSA). Process can be iterated using progressively higher harmonics
2. To make bearing pulse repetition "frequencies" almost constant and similar applications. Can accommodate larger speed variations up to ±30% which can be achieved by frequency demodulating a first harmonic
3. To track the orders over a wide speed range such as run-ups and run-downs. Can be used to track variation in individual orders with speed, and perhaps find special conditions for further investigation. Requires overlapping windowed segments in order to use phase demodulation based order tracking

The University of New South Wales

MFPT 2014

Example of correction for small speed variation of wind turbine gearbox

(squirrel cage induction generator)

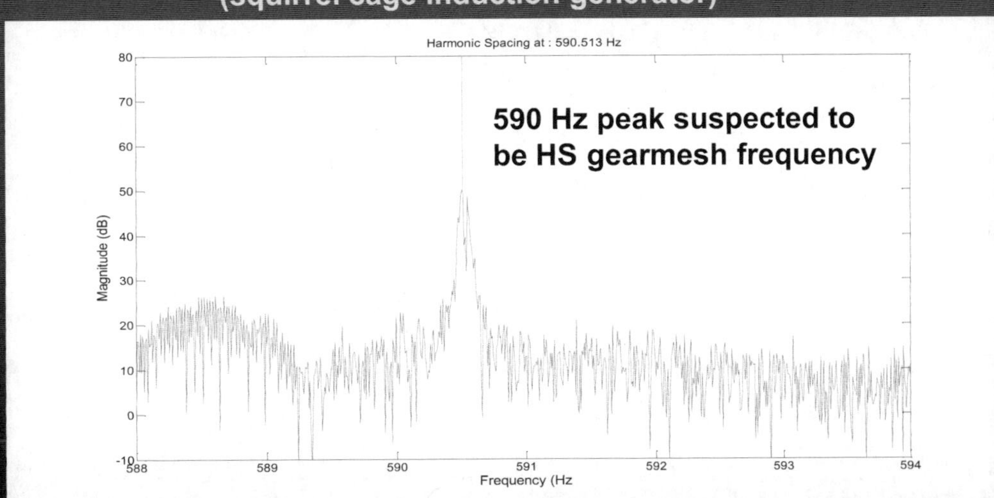

Zoom on 590.5 Hz peak before tracking.
Double peak resolved into one after order tracking

The University of New South Wales

MFPT 2014

Harmonic cursor on suspected shaft speed

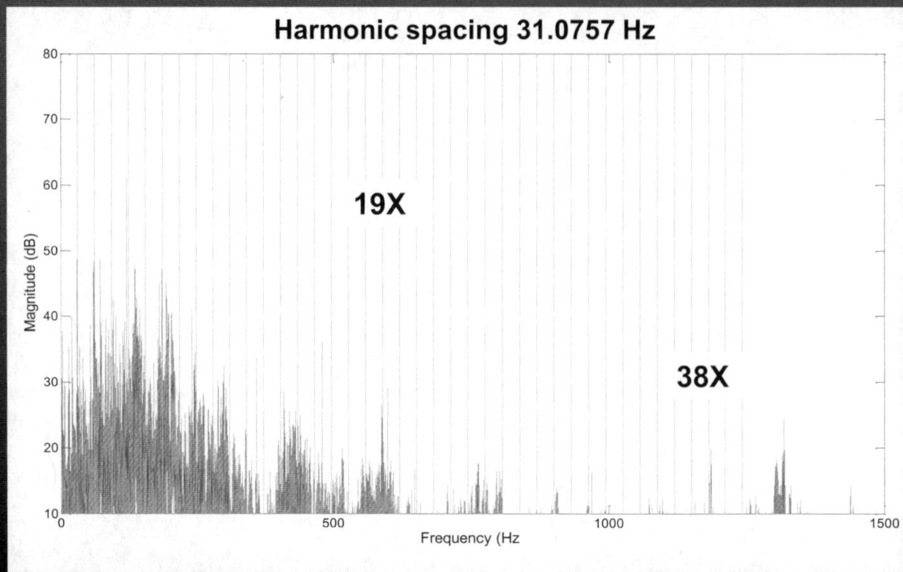

After order tracking “gearmesh” shown to be harmonic of “shaft speed”

The University of New South Wales

MFPT 2014

USE OF GEARMESH SIGNAL TO ESTIMATE LOAD

- Amplitude of gearmesh signal responds directly to the load because of increased tooth deflection – can be obtained by amplitude demodulation
- The instantaneous frequency of the gearmesh signal is directly coupled to the generator speed – can be obtained by frequency demodulation
- Deviation of the instantaneous generator speed from synchronous gives the slip frequency, which is also a measure of load

The University of New South Wales

MFPT 2014

DEMODULATION OF GEARMESH SIGNAL USING HILBERT TRANSFORM

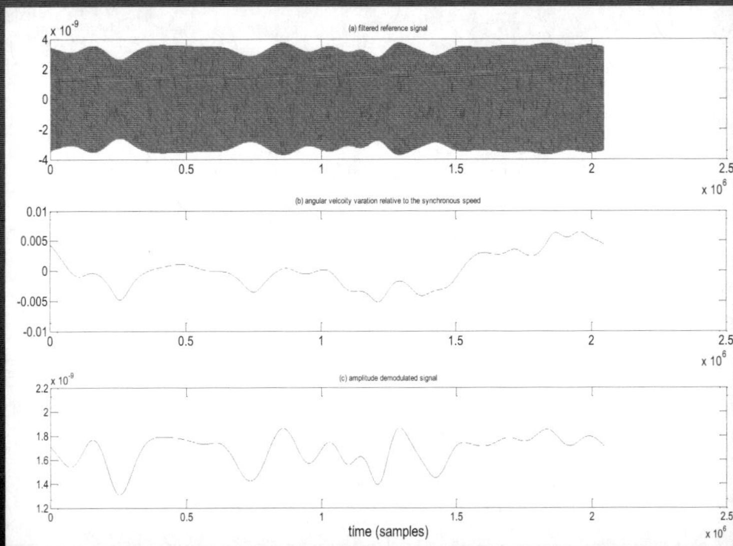

Gearmesh signal used as pseudo-encoder

Frequency modulation signal indicates slip frequency

Amplitude modulation signal indicates tooth load

Frequency modulation given by derivative of phase modulation

The University of New South Wales

MFPT 2014

Potential to use vibration measurement as a measure of load

- Possible for stall controlled turbines with a squirrel cage induction generator
- Not possible for synchronous generators because load gives a phase deviation rather than frequency deviation, but for constant speed, amplitude modulation could possibly be used
- It appears that so-called “doubly fed induction generators” with a wound rotor used for pitch controlled turbines are actually working as synchronous generators, so slip frequency cannot be used. Amplitude modulation would be very difficult to use because speed variation gives amplitude modulation without change in load

The University of New South Wales

MFPT 2014

Correction for larger speed variation

- A tacho signal is usually a series of pulses, so each harmonic is a potential carrier frequency
- Since speed variations give a time deviation Δt for each period, this corresponds to a phase deviation proportional to the harmonic order and best resolution is obtained by demodulating the highest possible harmonic order
- The limitation is given by overlap of modulation sidebands around each harmonic (Bedrosian condition)
- Maximum frequency deviation is about ±30% for first harmonic and most machine signals contain a strong first harmonic
- Deviation is also limited by the rate at which the frequency is changing

The University of New South Wales

MFPT 2014

Limitation given by rate of change

Max. frequency deviation vs modulation frequency (both relative to carrier frequency

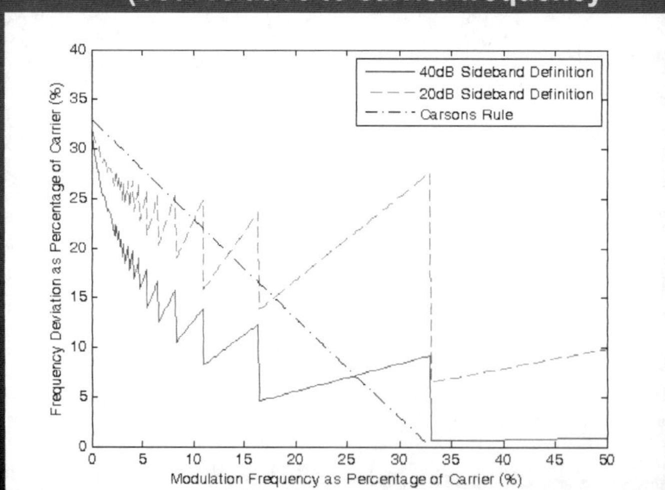

Sudden steps when new pair of sidebands to be included (criterion dynamic range 40 dB or 20 dB)

Weaker criterion given by Carson's rule

The University of New South Wales

MFPT 2014

Selection of section with small speed variation – wind turbine driving doubly fed induction geneator

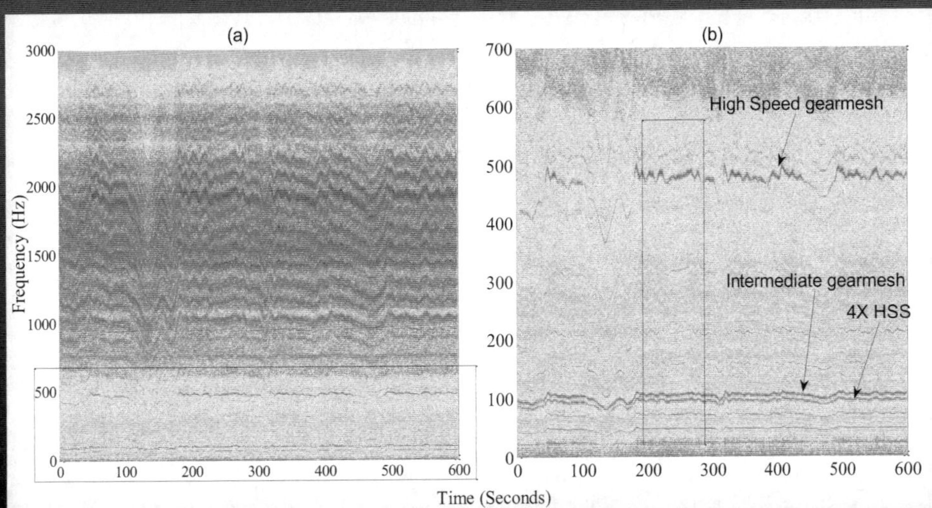

Maximum speed variation ±30%, but section selected where variation < 5%

The University of New South Wales

MFPT 2014

2 Stage iterated order analysis using different order harmonics of response signal

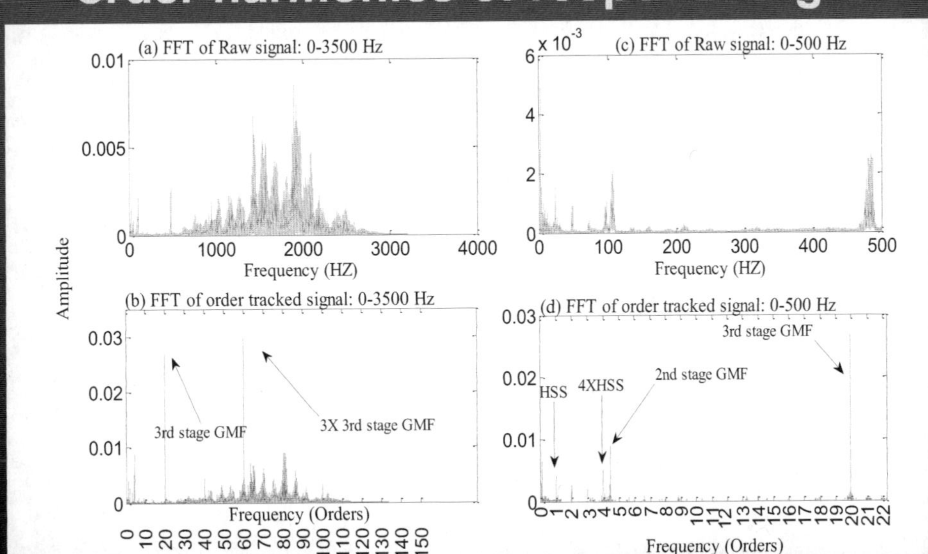

First iteration used HSS, and second iteration used 3rd stage GMF

The University of New South Wales

MFPT 2014

Use of finely tuneable cursor to identify fundamental frequency very accurately

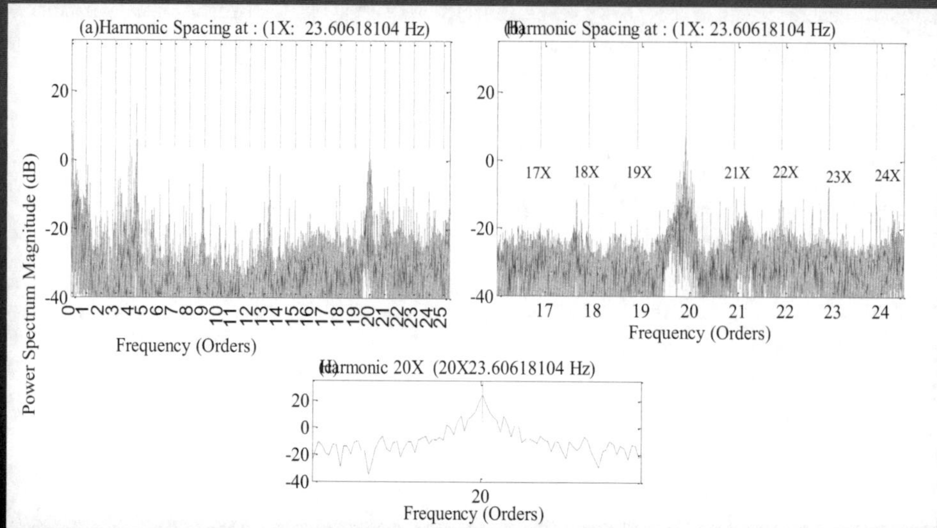

Establishes that 3rd stage GM is 20th harmonic of HSS

The University of New South Wales

MFPT 2014

Find another harmonic series with same GM frequency

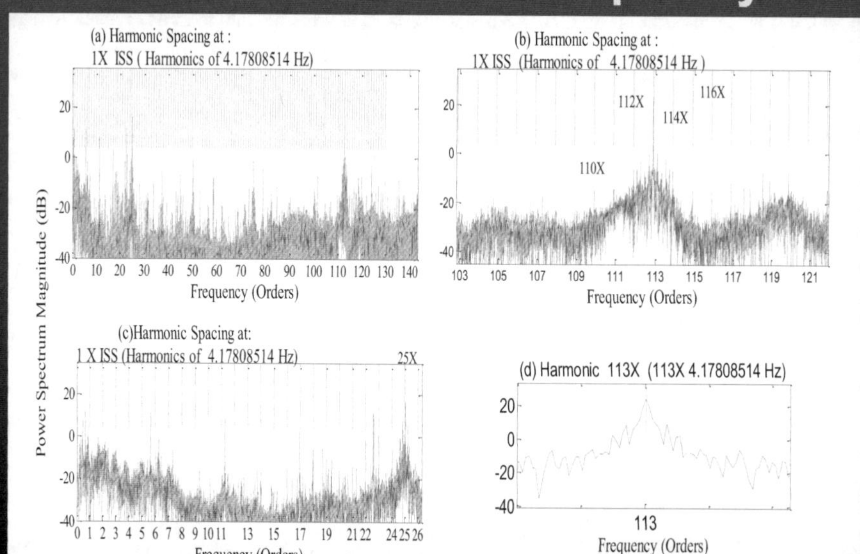

Same GM frequency is 113th harmonic of another series (ISS)

The University of New South Wales

MFPT 2014

Establishing gearmesh frequency as first common harmonic

	HSS Orders	Frequency (Hz)	ISS Orders	Frequency (Hz)
HSS Mean Fundamental Frequency (Hz) 23.60618104	17	401.30507768	105	438.69893970
	18	424.91125872	106	442.87702484
	19	448.51743976	107	447.05510998
ISS Mean Fundamental Frequency (Hz) 4.17808514	20	472.12362080	108	451.23319512
	21	495.72980184	109	455.41128026
	22	519.33598288	110	459.58936540
	23	542.94216392	111	463.76745054
			112	467.94553568
			113	472.12362082
			114	476.30170596
			115	480.47979110
			116	484.65787624
			117	488.83596138
			118	493.01404652
			119	497.19213166
			120	501.37021680

Common frequency much closer than 1/(m×n) showing that tooth numbers are 20 and 113 and that the design is "hunting tooth"

For a "hunting tooth" gearset (no common factors) the gearmesh frequency is the first which is a harmonic of both shaft speeds.

The closest they can come elsewhere is 1/(m×n) where m and n are the numbers of teeth on the two gears

The University of New South Wales

MFPT 2014

Application 2 using simple test rig with larger speed variation

Time signals

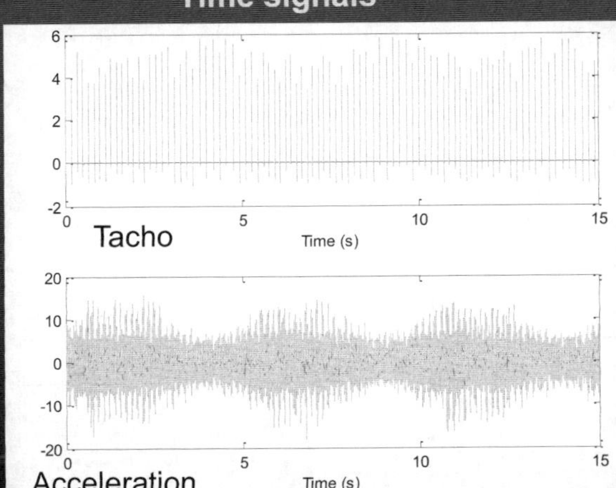

Spectra

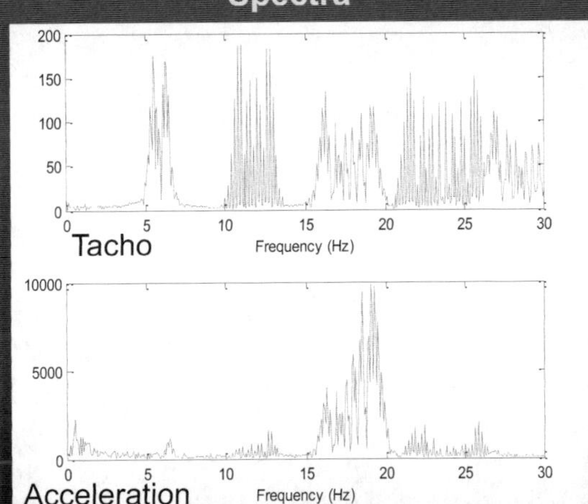

Third harmonic is the highest that is separated for ±10% speed variation
It is the lowest which is strong enough in the acceleration signal.
Note amplitude modulation because gearmesh passes through resonances

The University of New South Wales

MFPT 2014

Order tracking using third harmonic

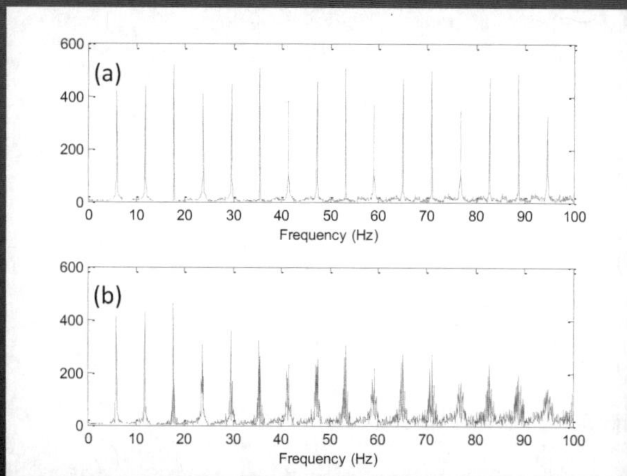

Order tracked tacho signal using
(a) Tacho signal (b) Acceleration signal
Even (b) is adequate for bearing diagnostics but could be improved by iteration

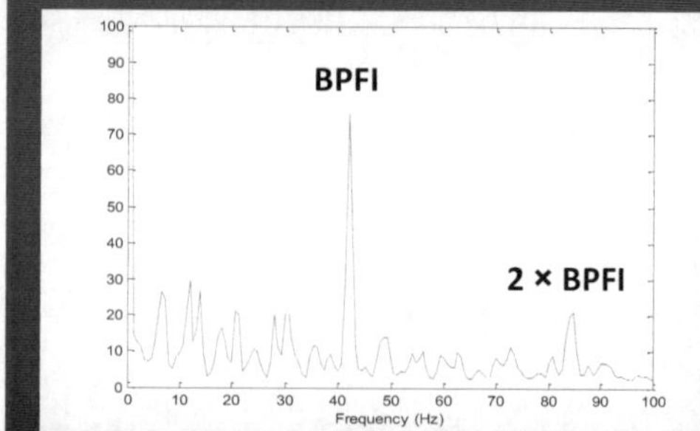

(Squared) envelope analysis of acceleration signal for inner race bearing fault using autonomous order tracking

First harmonic of tacho signal was used for ±25% speed variation with similar result

The University of New South Wales

MFPT 2014

Order Tracking for a Wider Speed Range
eg Run-up or Run-down

- Divide up signal into a series of overlapping windowed segments in each of which the speed range is < 2:1
- Allowable frequency range will depend on sweep rate (need to find equivalent modulation frequency for a given sweep rate)
- If each segment is windowed at each end by half Hanning windows, they can be separately processed and added back together afterwards
- The final order tracked signal will have a maximum valid order defined by highest valid frequency at highest speed.
- For lower speed segments, the signal will have to be lowpass filtered at progressively lower cutoff frequencies to avoid aliasing
- This is most simply done by eliminating frequencies above this limit in the frequency domain and transforming back to time domain

The University of New South Wales

MFPT 2014

Typical Windowed Segments

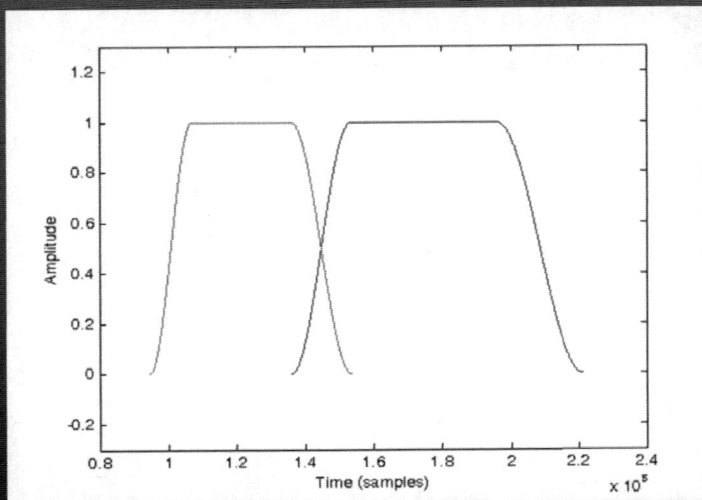

Matching half Hanning windows in overlap zone. Add up to 1 everywhere. Windows distorted by transformation from time to angle domain, but still add up to 1 everywhere. Suggested break points where windows overlap to equilibrate phase in adjacent segments

The University of New South Wales

MFPT 2014

Data used for Demonstration

Gearbox casing (without internals) excited by a force signal, applied through a shaker, with ten harmonics of equal amplitude, with fundamental frequency increasing from zero to 200 Hz over 30s. Simulates a typical run-up situation

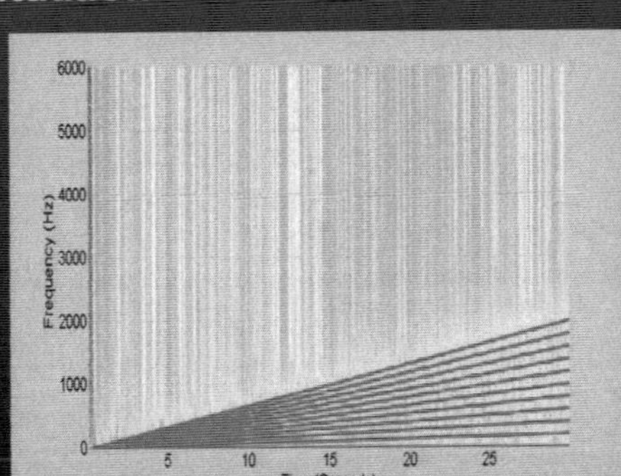

Force signal

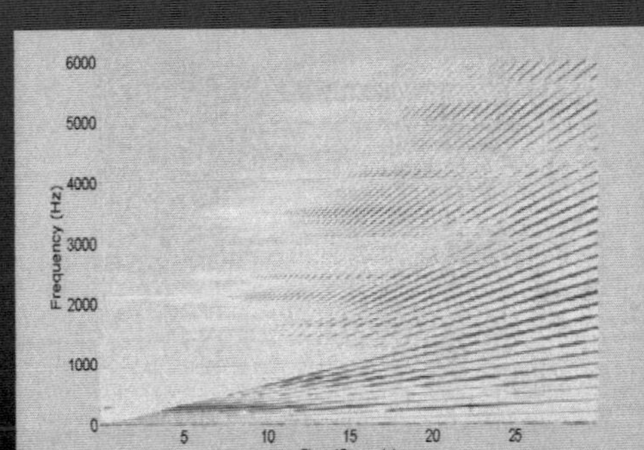

Response signal

First harmonic of force signal used as reference signal to decide break points and give phase vs time map for order tracking for each segment

The University of New South Wales

MFPT 2014

Equivalent Modulation Frequency

for a given sweep rate W Hz/s

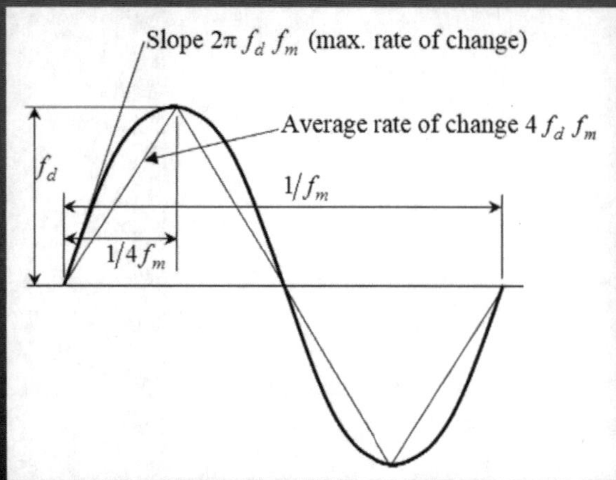

A given modulation frequency has a maximum sweep rate (conservative), or an average sweep rate as shown.

Average sweep rate used here and shown to be satisfactory

$$W = f_d \times 4 \times f_{meq} \quad \text{from which} \quad f_{meq} = \frac{W}{f_d \times 4} \quad \text{and} \quad \frac{f_{meq}}{f_c} = \frac{W}{f_c^2 \times \left(\dfrac{f_d}{f_c}\right) \times 4}$$

The University of New South Wales

MFPT 2014

Segments for demonstration Example

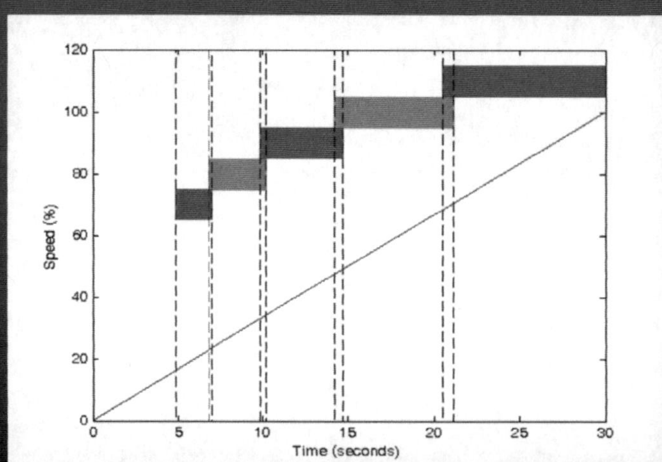

Permissible frequency deviation found to lie between 25% and 30%. 18% used here for demonstration. Overlap chosen to be ±5% (of length of lower speed segment in each case).

The University of New South Wales

MFPT 2014

Typical Windowed Segments in Time Domain (4th Segment)

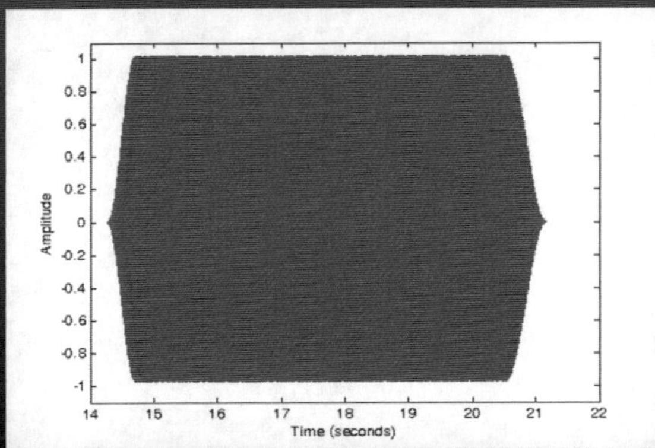

Reference (tacho) signal

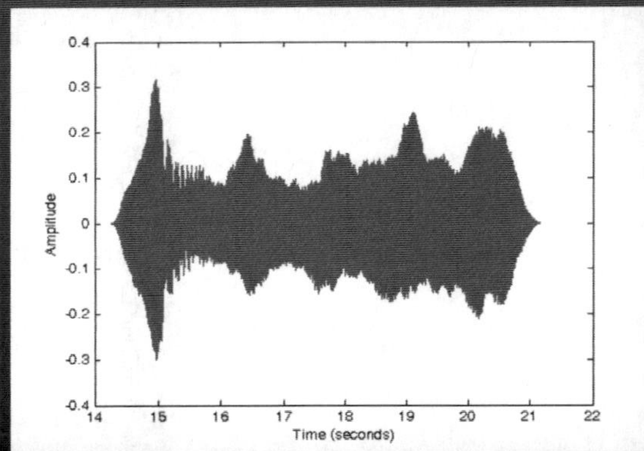

Response signal

The University of New South Wales

MFPT 2014

Spectra of Typical Segments of Tacho Signal

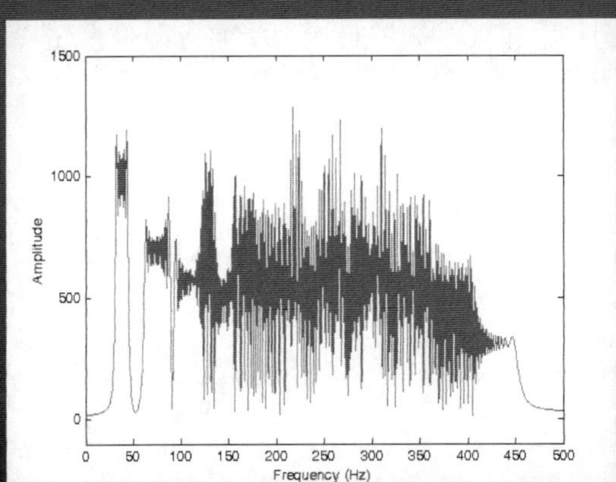

Segment 1

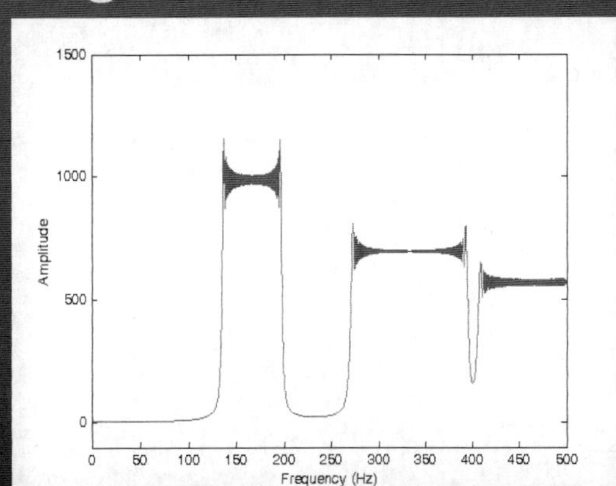

Segment 5

First harmonic windowed out with a rectangular window and phase demodulated to give phase/time map for order tracking of that segment

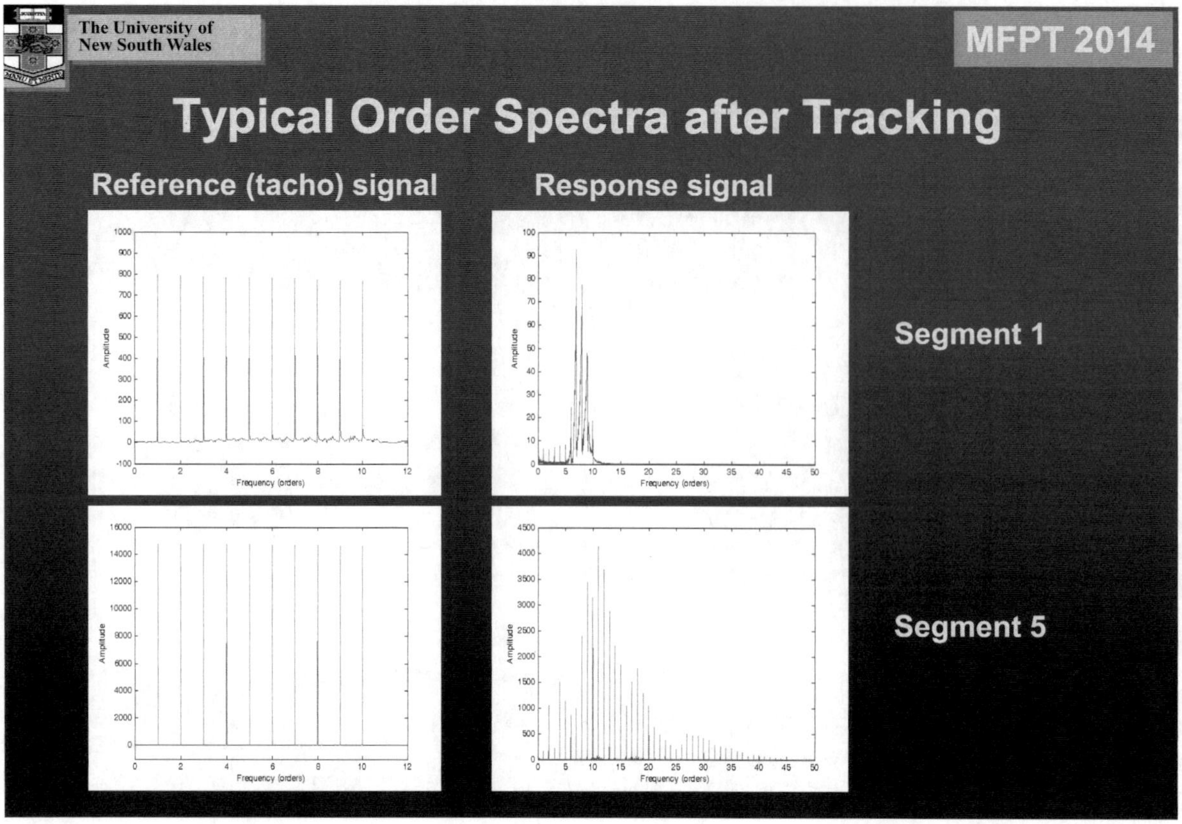
The University of New South Wales
MFPT 2014
Typical Order Spectra after Tracking
Reference (tacho) signal
Response signal
Segment 1
Segment 5
Amplitude
Frequency (orders)

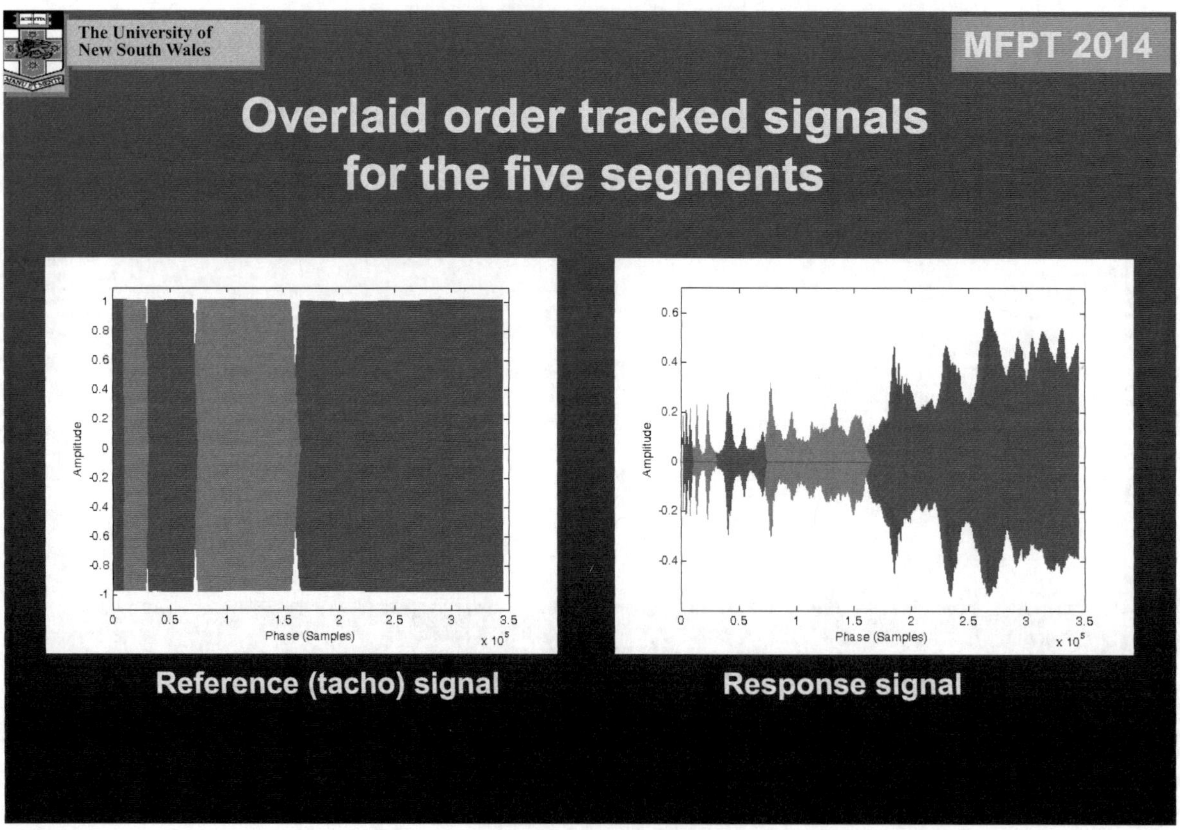
The University of New South Wales
MFPT 2014
Overlaid order tracked signals for the five segments
Amplitude
Phase (Samples)
x 10^5
Reference (tacho) signal
Response signal

The University of New South Wales

MFPT 2014

Recombined signals

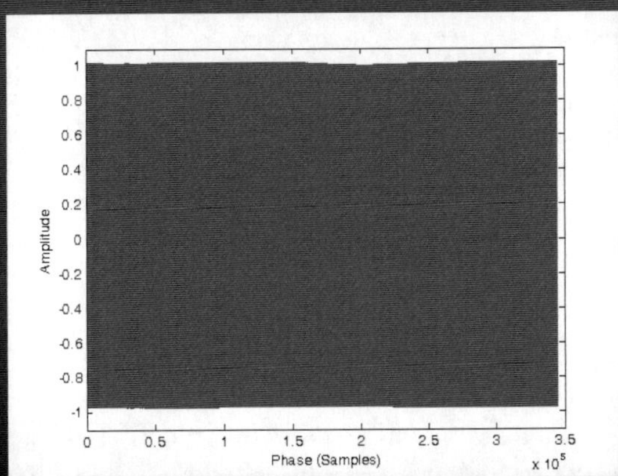

Reference (tacho) signal

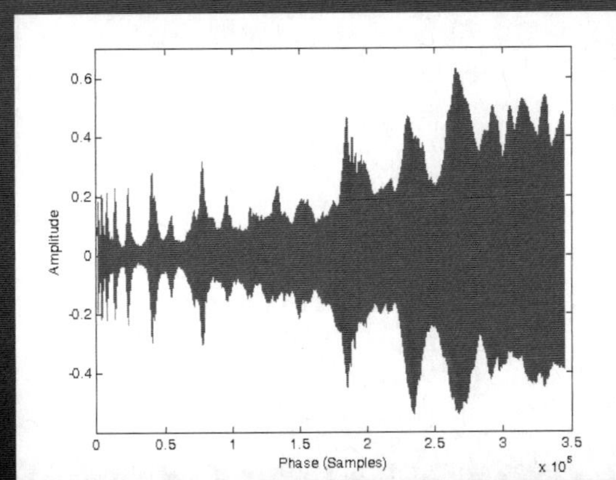

Response signal

The University of New South Wales

MFPT 2014

Order Spectra of the Re-combined Signals

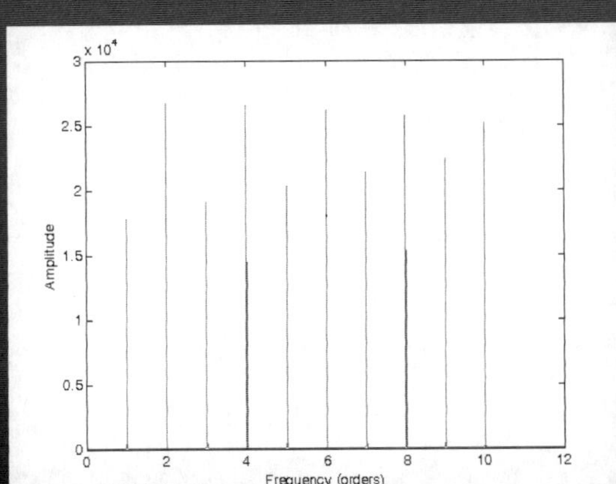

Reference (tacho) signal

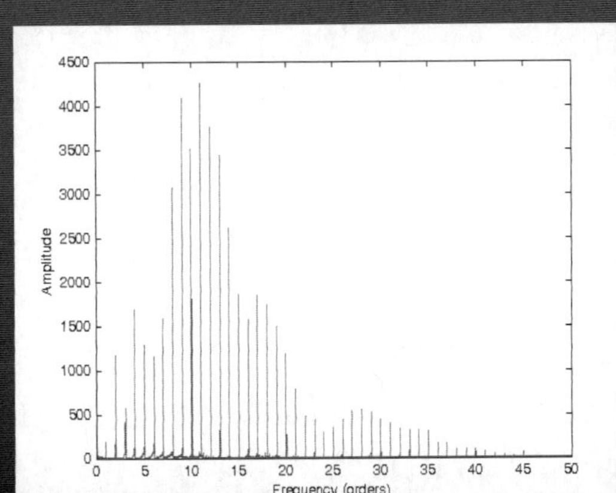

Response signal

The University of New South Wales

MFPT 2014

Spectrograms for the order tracked signals

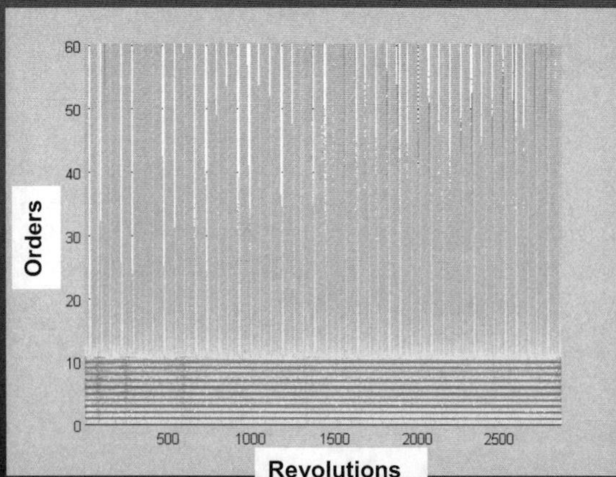

Reference (tacho) signal

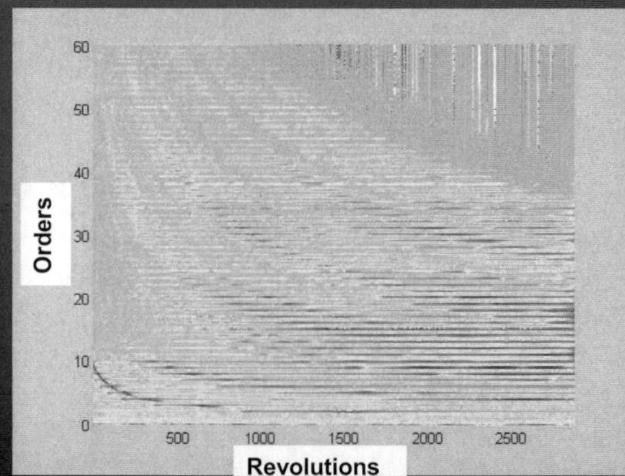

Response signal

Nyquist frequency for the original signals was 8192 Hz (>40 orders), of which 6400 Hz (order 32) was valid below the lowpass filter. Sampling frequency of order tracked signal must be > 80 samples per period. Here 120 samples were chosen, allowing display of 60 (32 fully valid).

The University of New South Wales

MFPT 2014

Limitation on use of signal itself as reference

- It is assumed that the harmonics of the response signal are integer multiples of the current running speed at each instant
- This will not be the case if the response time of the mechanical system is of the order of the length of the analysis records analysed, as the current spectrum may reflect the speed a short time prior to the current response
- This phenomenon was studied by Borghesani in the paper "Order tracking for discrete-random separation in variable speed conditions", MSSP 30, (2012), 1–22.
- It is unlikely to be a problem with for example wind turbines, where the inertia is so large that the speed cannot change very rapidly

The University of New South Wales

MFPT 2014

CONCLUSIONS

- Phase demodulation based method of order tracking has advantages in obtaining a phase/time map with any resolution
- Polynomial interpolation still used to resample signals at the resample points determined from the map, but potentially with finer resolution
- Phase demodulation method can be iterated several times using higher harmonics to improve result
- Method can use an external reference such as a once-per-rev tacho , a multi-pulse-per-rev shaft encoder, or the signal itself as reference
- Amplitude modulation from traversing resonances can restrict some applications (eg TSA of gear signals), but up to ±30% can be used for bearings
- Phase demodulation based method of order tracking, normally restricted to maximum speed range of 2:1, can be extended to analyze run-up or run-down signals over a much wider frequency range
- Reference signal extracted from the response cannot be used unless the response time of the system is short with respect to the rate of speed change

CURSORY DEVELOPMENT OF AN NDI METHOD FOR INTERNAL COATINGS ON ARMY GUN BARRELS

Reinhold Ludwig
Gene Bogdanov
Center for Imaging and Sensing
Department of Electrical and Computer Engineering
Worcester Polytechnic Institute

Victor Champagne
William de Rosset
Marc Pepi
US Army Research Laboratory
Weapons and Materials Research Directorate
Aberdeen Proving Ground, Maryland

Abstract: The Army has researched the concept of utilizing cold spray coating technology as an alternative to the conventional electroplated chromium coating within 20mm gun barrels. This project was undertaken as part of the Army's Toxic Metal Reduction program to remove hexavalent chromium from the manufacturing process, and involves the creation of tantalum-10% tungsten "donor tubes" for explosive bonding to the inner gun barrel. A major concern of any internal coating process (such as the interior of a gun barrel) is the integrity of the final product, since it is not a location that is easily inspected. As such, a nondestructive inspection method was developed, allowing successful evaluation of the interior gun barrel coating. The research involved a high-frequency ultrasonic inspection whereby the elastic waves were transmitted from the exterior. The equipment consisted of contact and immersion probes with center frequencies of ~10 MHz connected to a pulser/receiver system with variable duty cycles. The analog time-amplitude signals were acquired with a high-speed (2 GHz) digitizing oscilloscope and evaluated/stored in a PC. The test arrangement was used to measure the average liner thickness as a function of distance along the barrel. Results comparing well-bonded and delaminated samples were also achieved.

1. Objective:
Monitor the thickness and delamination of a cold-sprayed liner inside a gun barrel via traditional ultrasonic testing (UT) technology.

2. Approach:
Utilize ultrasonic testing (UT) with 5 MHz and 10 MHz transducers for:

- performing contact measurements on the external surface of the gun barrel
- performing immersion measurements from inside the gun barrel

3. INTRODUCTION

3.1 Gun Tube

For this study the interior coating of a medium caliber cannon was investigated. The interior of this gun, as well as most other firearms, was traditionally plated with hexavalent chromium. Historically, chromium plating provided adequate protection of the steel substrate in terms of wear and erosion, and its characteristic cracking held lubrication, providing an almost self-lubricating mechanism. However, in recent years, with the advent of more potent propellants, chromium plating has been shown to be detrimental to the underlying steel substrate. These propellants cause increased erosion of the chromium, as well as chemical attack of the steel barrel when ultimately exposed by the worn away protective coating. The erosion damage of chromium-plated tubes is accelerated by chromium spallation, which allows hot gas attack and/or melting of the unprotected substrate steel [1]. In fact, reference [1] found that chemical attack is the precursor to erosion damage in the steel gun barrel. For this reason, as well as the benefit of removing hexavalent chromium from the process, the Army has been investigating alternative protective coatings for the interior of gun barrels. This work has resulted in the testing and evaluation of four refractory metals that have potential as possible liner materials [2]. These materials included tantalum-10% tungsten (Ta-10W), Ta-5W-2Mo, Ta-4Mo and Stellite 25, all of which have higher melting points than chromium. Reference [3] lists BioDur CCM (a cobalt, chromium and molybdenum alloy) as another possible candidate. These materials were prepared as liners that could subsequently be explosively bonded, or applied using the GLEEM (Gun Liner Emplacement with an Elastomeric Material) process, to the interior of the gun barrel. Each of these materials was machined to the final tube dimensions necessary for explosive bonding. However, due to the cost of machining these refractory materials, cold spray technology was also explored. Tubes of Ta-10W were prepared utilizing the cold spray process for explosive bonding. To show that the explosive bond was of high quality, adhesive bond tests and metallography were performed on sections of a coated gun [2]. However, these processes are destructive and require a coated gun to be sectioned, which can be a very expensive means of assuring the coatings will perform in-service. A nondestructive method of inspecting this liner material is also needed after repeated firings, to ensure no cracking or delamination had occurred.

3.2 Gun Barrel Coating Inspection

For the nondestructive inspection, a number of techniques were explored, including eddy currents [5], electric conductivity, X-ray, immersion and contact ultrasonics. After assessing the relative merits and cost of each approach, a high-frequency contact ultrasonic technique was adopted.

Ultrasonic inspection is a nondestructive method in which beams of high-frequency acoustic energy are introduced into the material under evaluation in order to detect surface and subsurface flaws and to measure the thickness of the material or the distance to a flaw [4]. According to reference [4] and our own experience, the following are some of the advantages and disadvantages of using this inspection method:

- Advantages
    - Superior penetrating power, which permits detection of flaws deep in the part.
    - High sensitivity, permitting detection of extremely small flaws.
    - Greater accuracy than other nondestructive methods in determining the positions of internal flaws, estimating their sizes, and characterizing them in terms of nature, orientation and shape.
    - Typically only one surface needs to be accessible.
- Disadvantages
    - Manual operation requires careful attention by experienced technicians.
    - Technical knowledge is required for development of inspection procedures.
    - Specimens with high surface roughness are difficult to inspect.
    - Discontinuities that are present in a shallow layer immediately beneath the surface may not be detectable.

For this study, an ultrasonic (UT) test system consisting of a pulser/receiver, a LeCroy digitizing scope, and a PC equipped with Matlab for signal evaluation was utilized. The 10-MHz immersion UT probe was integrated into a 0.24" steel rod, which could be inserted into the gun barrel. The ultrasound propagated through a liquid coupling medium (water, oil, glycerin) into the coating layer, interacted with the layers, and was subsequently received by the same 10-MHz immersion probe acting as a receiver. Furthermore, a hand-held 10-MHz UT contact probe was used for all external measurements. The equipment used for immersion is shown in Fig. 1, while a schematic of contact testing is shown in Fig. 2. The immersion probe was integrated into a steel rod, which ultimately allowed for rotation and translation of the transducer probe (for the interior of gun barrels). The probe operated in pulse/receive mode, and was hand-held.

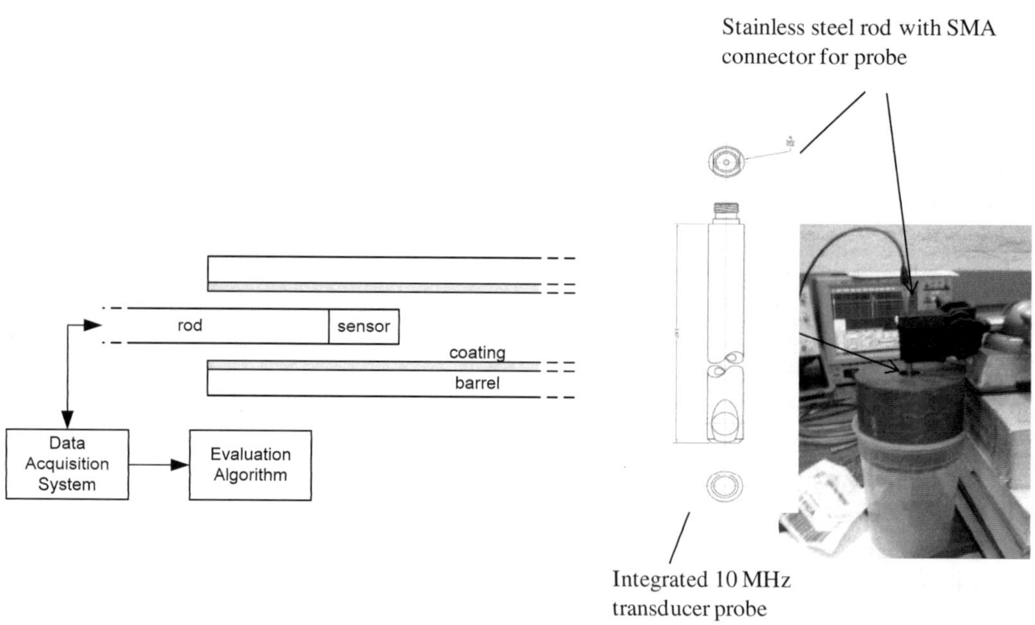

Fig. 1: a) Schematic of immersion probe, b) immersion probe set up.

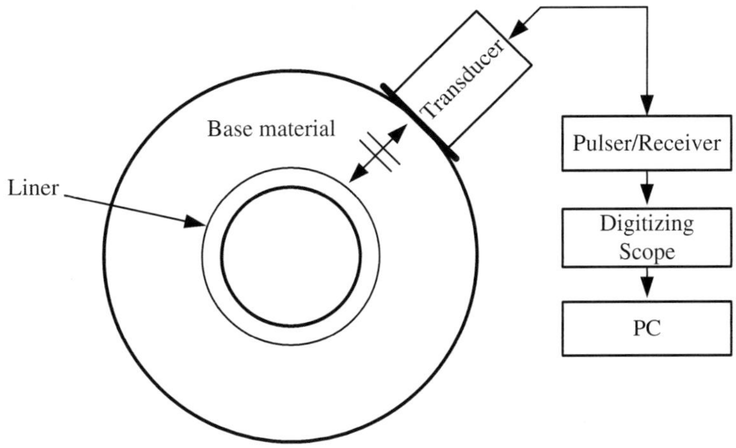

Fig. 2: Generic test arrangement: a high-frequency UT contact probe connected to a pulser/receiver insonifies the liner from the exterior. A high-speed digitizing oscilloscope acquires the time-amplitude signals before transferring to a PC for display and storage.

3.3 Test Samples

Three samples were subjected to testing:

1) Stainless steel (SS) pipe with 1" by 1" inner dimensions and 0.125" wall thickness,
2) 3" gun barrel section (smooth bore) with Ta-10W liner,
3) 4" gun barrel section (rifled) with a Stellite 25 liner.

These samples are shown in Figs. 3 a), b) and c).

Smooth bore
2.3mm liner

Rifled
1.7mm max liner

SS reference tube:
Rectangular 304
stainless steel tube
31.7x31.7mm outer
2.8mm wall

a) b) c)

Fig. 3: a) Smooth bore, b) rifled 20 mm gun barrel, and c) SS reference tube.

3.4 UT Testing Objective

The objective was to obtain UT signals in the time/frequency domain that allowed reliable detection of: a) the liner thickness, and b) the condition of the liner (delamination, cracking, porosity, etc.)

3.5 Measurements

Preliminary testing with two different probes (10-MHz immersion and 10-MHz contact transducers) was conducted. Figs. 4 a), b) and c) show UT immersion testing of a stainless steel tube with known thickness of 3.2mm, using the contact method, and after immersion in oil and glycerin. Glycerin provided stronger back wall echoes, and the frequency of the signal was reduced (wider echoes). As the time-domain A-scan data in Fig. 3 indicates, the time separation between the first and second echo, and between the second and third echo, is almost exactly 1 µs. Based on a sound velocity of 5,960 m/s (302 austenitic stainless steel), a thickness of approximately 3mm was obtained. It was determined that the 10 MHz contact transducer yielded the best results.

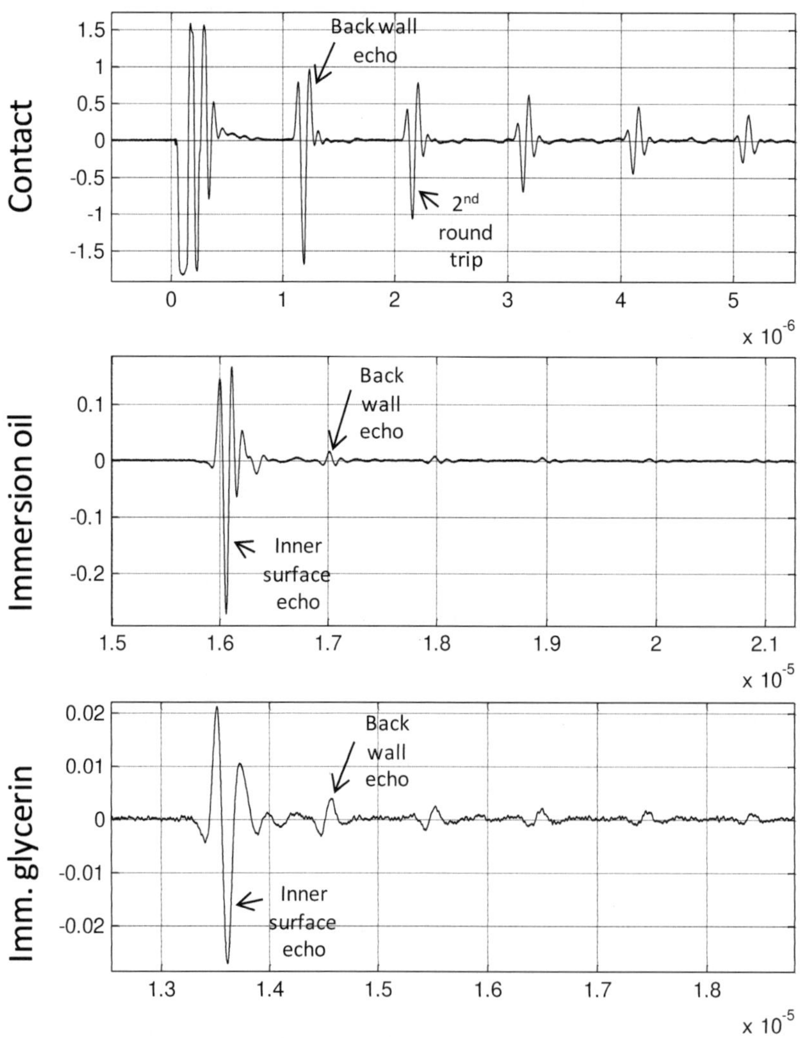

Fig. 4: Time-domain A-scans for the stainless steel reference using a) contact, b) oil immersion, and c) glycerin immersion.

The 10-MHz immersion probe was inserted into the barrel at various depths and measurements were taken and displayed for both the smooth bore (Fig. 5) and rifled (Fig. 6) specimens. The 10MHz contact transducer yielded strong liner-barrel and bore echoes for the smooth barrel, whereas the immersion transducer yielded weak liner-barrel echoes.

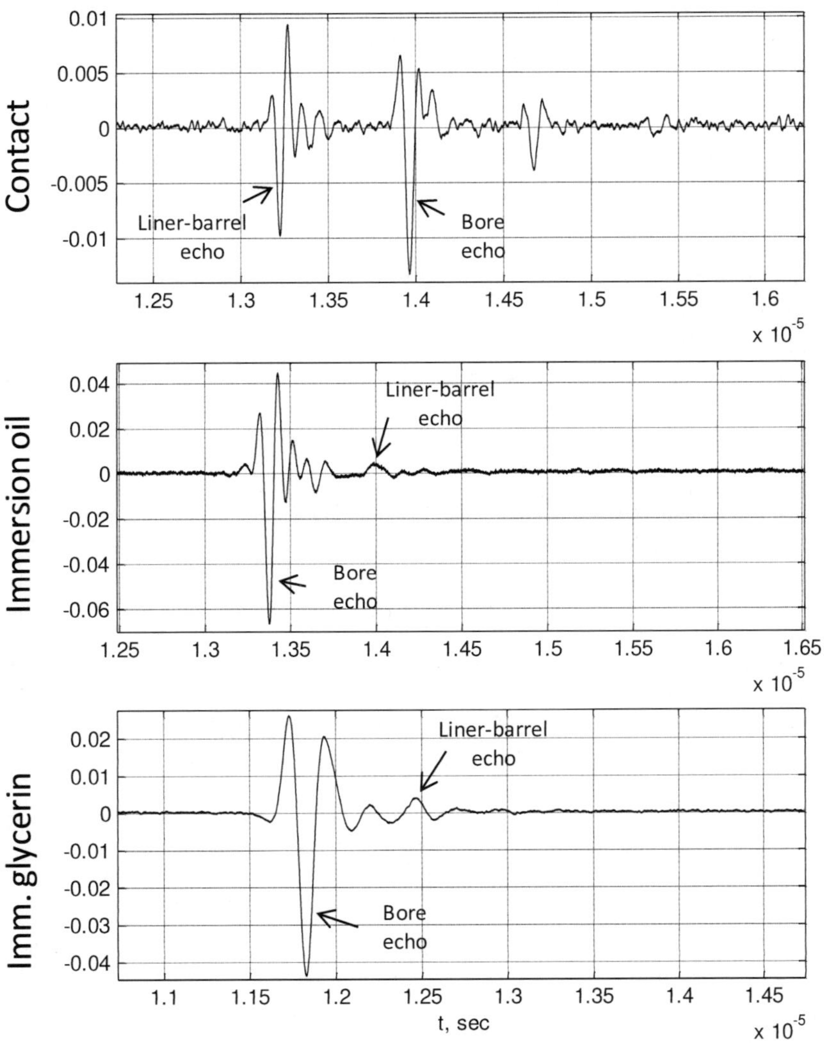

Fig. 5: Time-domain A-scans for the smooth barrel sample using a) contact, b) oil immersion, and c) glycerin immersion.

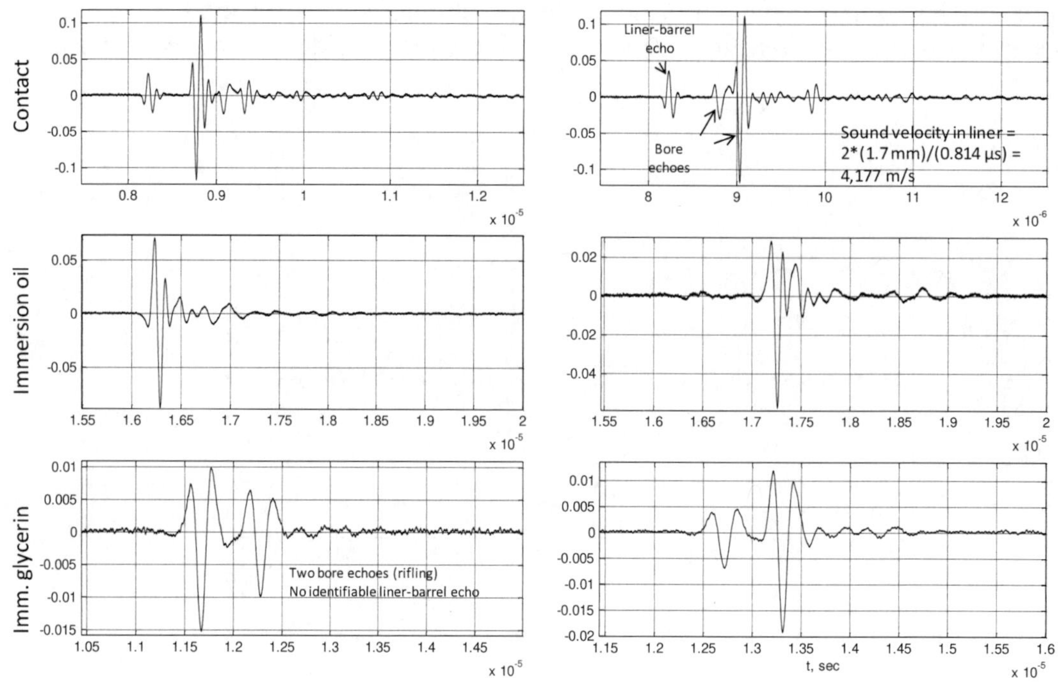

Fig. 6: Time-domain A-scans for the rifled barrel sample using a) contact, b) oil immersion, and c) glycerin immersion.

To confirm that the contact transducer possessed a ~10 MHz center frequency, the time-domain data of Fig. 7 was gated and converted into the frequency domain. Fig. 15 shows that the transducer has a center frequency slightly below 10 MHz.

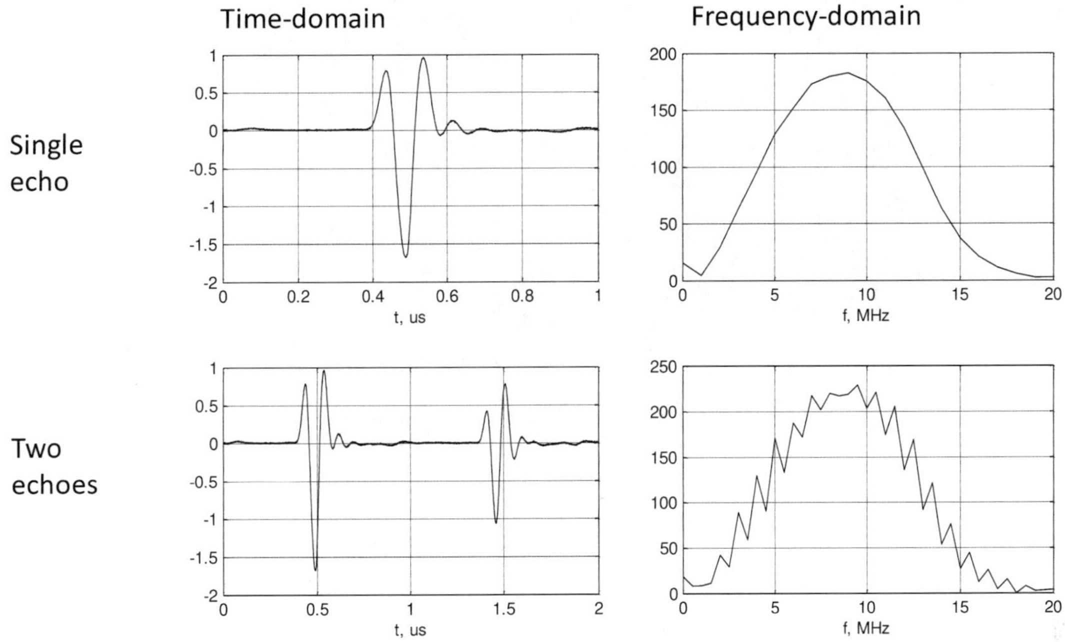

Fig. 7: Time-domain A-scans (left) and frequency-domain curves (right).

Next, a 10 MHz contact probe was used on a flat sample that represented typical coating variation that could be seen on the gun barrel liners. The coating was intact, i.e., good adhesion with the substrate. The sample had a thicker coating on one end, tapering to a thinner coating on the other end. Fig. 8 shows the 10 MHz contact transducer on the sample. Fig. 9 shows the time-domain scans from the thick and thin coating sides.

In addition, a sample with a "delaminated" coating was used to compare results to the intact coating sample. This sample was fabricated via cold spray technology with known delaminations (poor adhesion). Fig. 10 shows the 10 MHz contact transducer on the delaminated sample, while Fig. 11 shows the time-domain scans from the delaminated region, the middle of the sample, and the thin, intact coating.

Fig. 8: 10 MHz contact transducer on the tapered coating sample.

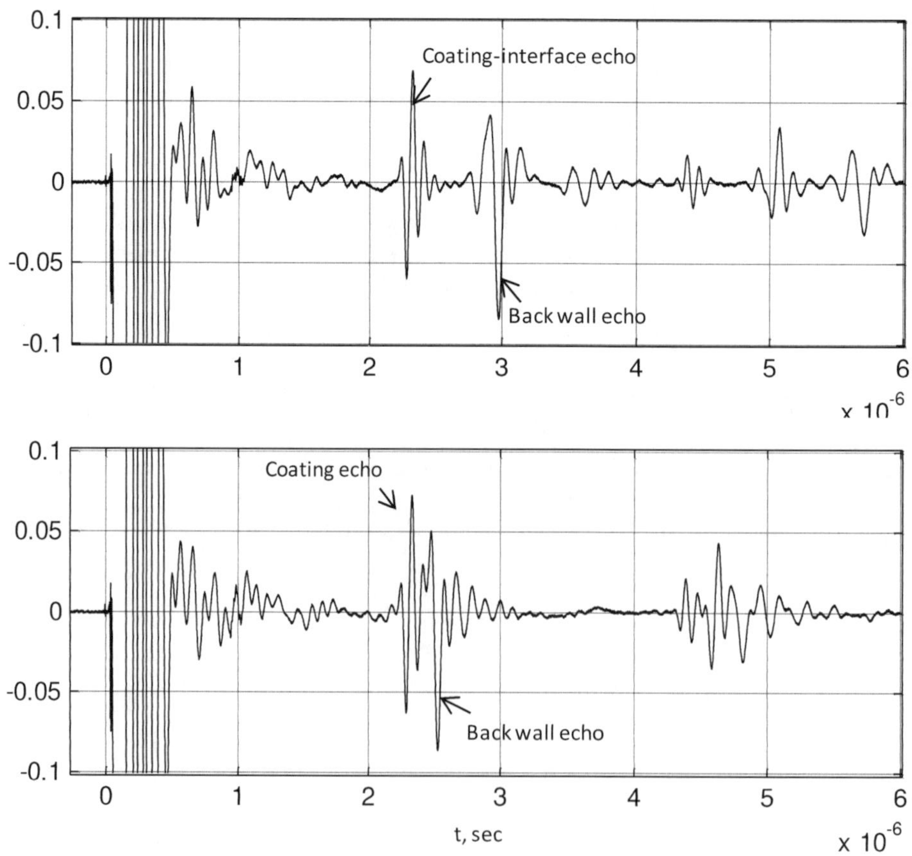

Fig. 9 shows the time-domain scans from the a) thick and b) thin coating regions.

Fig. 10: 10 MHz contact transducer on the delaminated coating sample.

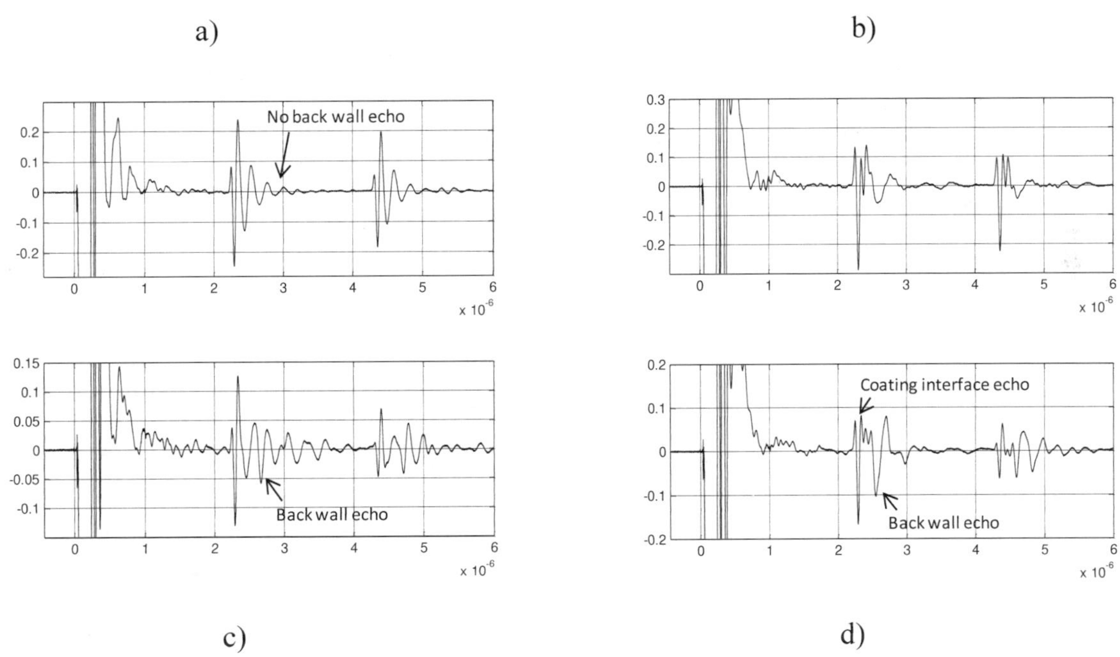

Fig. 11: Time-domain scans from the a) and b) delaminated, c) middle and, d) thin coating (intact).

4. DISCUSSION

Preliminary UT results with a 10 MHz probe indicate that high-frequency contact testing from the exterior of the gun tube provides sufficient time-domain information to resolve the thickness of the liner. Specifically, the liner thickness d_{th} can be determined from:

$$d_{th} = \frac{1}{2} t_E \cdot u_p \qquad (1)$$

where t_E is the round-trip time delay between echoes, and u_p is the previously mentioned longitudinal phase velocity of 5,920 m/s in the stainless steel base material. The time delay between echoes of approximately 0.8 μs, predicts a liner thickness of 2.4 mm. This compares well with the known thicknesses of roughly 2.3mm (smooth bore) and 1.7mm (max., rifled bore). Obviously, additional work with controlled samples is needed to determine these numbers more precisely. Also, higher frequency ultrasound should improve resolution. Increasing the center frequency of the probes should not pose a major problem since the base material exhibits very little attenuation.

Delamination between the liner and the steel barrel is detectable because the acoustic impedance mismatch at the liner-steel interface is maximized, resulting in larger reflected waves from a delaminated than from an intact interface. This also results in an inability to detect the back wall echo behind the delaminated region since the acoustic wave is completely reflected and there is no acoustic energy transmitted through the rest of the part. A comparison of the well-bonded and delaminated signals in Figure 11 shows this difference clearly, and can be utilized for nondestructive assessment of bonding integrity.

It appears that the acoustic testing methodology presents one of the more advantageous NDI opportunities in terms of capabilities and cost. Furthermore, transducers of higher frequency settings (20 MHz, even 50 MHz) are currently being explored as they offer higher resolution, well below 1mm. Ongoing research with flat samples and interfacial layers has shown that 0.3 mm (0.15mm) thick liners can be detected with 20 MHz (50 MHz) probes.

5. CONCLUSIONS

This paper demonstrates that high frequency ultrasonic testing with 10 MHz contact transducers in the time-domain allows for the testing of gun tube liners. In addition to the approach of insonifying the liner from the interior of the tube with immersion transducers, an exterior test arrangement in pulse/receive mode was also used. This was possible since the elastic wave attenuation in the base material proved very low. Low material attenuation allowed the deployment of high frequency transducers well beyond the conventional 1 – 5 MHz probes. In fact, ongoing research has demonstrated that 20 MHz, even 50MHz, transducers can be utilized; this should yield thickness resolutions below 300 microns. The UT approach is cost-effective and easy to deploy in the field. Moreover, because of the large difference in acoustic impedance between the liner medium and air, it is well-suited for detecting delaminations: a delaminated liner location results in a large reflected signal at the interface and an inability to transmit additional acoustic energy to the back wall.

6. REFERENCES:

[1] Cote, P.J. and Rickard, C., "Gas–Metal Reaction Products in the Erosion of Chromium-Plated Gun Bores", *US Army Armament Research, Development, and Engineering Center, Benet Laboratories, Watervliet, NY 12189, USA* 13 October 1999, Wear 241-2000. 17–25.

[2] Miller, M.D., Campo, F., Troiano, E., Smith, S and de Rosset, W., "Explosive Bonding of Refractory Metal Liners", Proceedings of the 27th Annual Army Science Conference, Orlando, FL, December 2010.

[3] de Rosset, W., "Cost Estimate for Gun Liner Emplacement", ARL-CR-0677, August 2011.

[4] Posakony, G.J., "Ultrasonic Inspection", Metals Handbook, Desk Edition, American Society for Metals, 1985, pp. 33·32 – 33·40.

[5] Nondestructive Testing – Eddy Current, CT-6-5, Second Edition, General Dynamics, Convair Division, 1979, p. 7-39 & 7-58.

BENDING FATIGUE LIFE COMPARISON BETWEEN DMLS AND COLD-ROLLED NICKEL ALLOY 718

Onome Scott-Emuakpor, Jeremy Schwartz, Tommy George, and Casey Holycross
Air Force Research Laboratory
Aerospace Systems Directorate
1950 Fifth Street
Building 18D
WPAFB, OH 45433
Telephone: (937) 255-6810
onome.scott-emuakpor@wpafb.af.mil

Joseph Slater
Wright State University
3640 Colonel Glenn Highway
Dayton, OH 45435

Abstract: Bending fatigue life behaviors of direct metal laser sintering (DMLS) and cold-rolled Nickel (Ni) Alloy 718 have been compared in this study. The work was motivated by the possibility of using DMLS manufacturing to improve functionality of hot section components in gas turbine engines. In other words, by using DMLS, features like cooling passages with enhanced functionality can be created in turbine blades and heat exchangers, leading to larger flight envelopes and improved specific fuel consumption (SFC). The work in this document focuses on assessing high cycle fatigue (HCF) behavior of DMLS Ni Alloy 718. DMLS specimens from two different suppliers were fatigued via a vibration-based bending technique, and compared to published rotating bending behavior of cold-rolled Inconel 718 (Ni Alloy 718 composition). Tensile material properties of the DMLS specimens were observed as well. The comparisons conclude that fatigue and tensile properties may be sensitive to DMLS process parameters and post-sintering treatments, thus showing a need for process control optimization. Nonetheless, fatigue performance of DMLS Ni Alloy 718 compares well with rotating bending fatigue of cold-rolled Inconel 718.

Keywords: Additive manufacturing; Direct metal laser sintering; Nickel Alloy 718; Gas turbine engine; Fatigue; Tensile

Introduction: Improved functionality of hot section components such as cooled turbine blades and heat exchangers is critical to improve the efficiency of gas turbine engines. These components can be enhanced with novel design concepts; however, conventional manufacturing approaches are limited in their ability to maintain component dimensions within tolerances and structural integrity within design allowables. An alternative to conventional gas turbine engine manufacturing practices is additive manufacturing (AM). The allure of AM is the ability to three-dimensionally print components, which allows for

Nomenclature:

Alloy#	Nickel composition identification number
E	Elastic modulus
G	Gravitational constant (9.81 m/s^2)
H	Rolling marks or DMLS build plane normal to load
Roll_Dogbone	Rolling direction of tensile specimen
Roll_Plate	Rolling direction associated with the plate stock
Specimen#	Fatigued specimen number
Stock#	Bulk stock number
V	Rolling marks or DMLS build plane parallel to load
N_{fail}	Number of cycles to failure
N_{step}	Number of cycles per step
Ra	Roughness Average
$\mu\varepsilon_a$	Microstrain amplitude
σ_a	Stress amplitude
σ_{fail}	Stress amplitude during failed step
σ_{pr}	Stress amplitude during previous step

intricate cooling features for small-scaled turbine blades and minimized heat-affected zones for efficient heat exchangers. In other words, AM is an enabling technology that could allow gas turbine engines to operate at higher temperatures, which would provide more versatile flight envelopes and improved SFC.

Direct Metal Laser Sintering (DMLS) is one AM process that has garnered recent interest from the gas turbine engine community. DMLS works by sintering metal powder, layer by layer, on a bedplate until a part is completed [1, 2]. Monotonic material properties of DMLS manufactured specimens have been favorably compared to the wrought material properties of a few aerospace alloys [3-6]. This qualifies the use of DMLS to manufacture some non-rotating and non-flowpath components with standard geometries. However, acceptance of DMLS manufactured parts such as integrally bladed rotors (IBRs), combustor liners, heat exchangers, or cooled blades will be reliant on full characterization of aeromechanics induced issues such as fatigue crack growth, low cycle fatigue (LCF) and high cycle fatigue (HCF) failure [7]. Therefore, extensive knowledge on fatigue behavior is necessary to qualify the use of DMLS built rotating and/or flowpath components in gas turbine engines.

The work in this manuscript compares the fatigue behaviors of DMLS and cold-rolled Ni Alloy 718 as a step for future qualification of the AM process for HCF induced engine components. DMLS fatigue behavior was determined with a vibration-based bending fatigue testing approach, and the results were compared to rotating bending fatigue life behavior of cold-rolled Inconel Alloy 718 [8]. Vibration-based bending fatigue testing was chosen for this study to simulate the type of surface stress loading seen in a turbine blade at resonance. DMLS fatigue specimens were manufactured by two vendors: Vendor

I plates were heat treated, hot isostatic pressed (HIP), and stress relieved; Vendor II plates were only stress relieved. The effects of the different DMLS processes were observed in the strength and fatigue comparisons, where Vendor I showed high strength properties but lower fatigue performance than both cold-rolled and Vendor II. Future microscopy is planned to understand and support the results of the comparisons.

DMLS Synopsis: DMLS is a three-dimensional printing process that sinters (or melt, depending on the laser power) layers of 20 micron thick metal powder with a 200-400W Ytterbium-fiber laser until a part is completed [1, 2]. Sintering is conducted on a steel bedplate inside a sealed chamber with argon gas between 40° – 100°C. The DMLS process is automated, thus machines can be unattended while part builds are underway [2]. Despite the precision of the DMLS manufacturing capability, surface finish modifications are still necessary to reduce the roughness of manufactured components [1].

There are a number of benefits to manufacturing with DMLS as opposed to conventional methods. One benefit is the reduction in material waste and subtractive tooling wear during manufacturing [2]. This along with part fabrication occurring in a controlled, automated environment means that DMLS requires less production time and labor than conventional methods [2]. Another benefit of using DMLS is the capability to manufacture components that have small and/or complex geometric features or that require assembling multiple parts.

Background studies have been conducted on a few DMLS alloys to validate and improve the microstructure characteristics and mechanical properties [3-6]. Two key disadvantages found in early studies were the presence of inhomogeneous residual stresses and porosity. Both of these issues favor crack nucleation in dynamic and quasi-static (monotonic) loading [3, 4]. Homogeneity of residual stresses and the reduction of porosity have been addressed for some materials by controlling the temperature of the bedplate and incorporating post-sintering treatments such as shot peening, surface polishing, aging heat treat, stress relief, and HIP [3-5].

Vibration-Based Method and Test Setup: Experimental HCF failure results for this study were acquired using a vibration-based testing approach referred to as the George fatigue method [9, 10]. The purpose of this method is to experimentally acquire either uniaxial or biaxial fatigue life results at loading conditions similar to those that could cause HCF failure in turbine engine blades and vanes. Testing for this method is carried out by supplying a dynamic base excitation from an electrodynamic shaker to a cantilever specimen at a resonant frequency. The resonant frequencies required to generate either a biaxial or uniaxial bending load are based on the material composition and the geometry of the specimen. These geometries were determined using a geometric design procedure, which optimized the location of the maximum von Mises stress amplitude (the fatigue zone) so that it was not along the clamped-edge of the specimen [9].

The work in this document focuses only on the uniaxial bending fatigue specimen pictured in Figure 1. The dimensions of the specimen are 114.3 x 165.1 x 3.2 mm with a square test-section area of 114.3 x 114.3 mm. In order to attain a maximum stress region away from the clamped edge of this particular geometry, a chord-wise bending (or two-stripe) mode is excited at a frequency of roughly 1600 Hz. Figure 2 shows the vibratory analysis of this mode, which includes contour images of von Mises stresss and displacements (mode shapes).

The instrumentation used to measure data during vibration-based uniaxial fatigue tests are an Endevco 2271A accelerometer, a Micro-Measurements CEA-06-062UW-350 uniaxial strain gage, and a non-contact Polytec OFV 505 single-point laser vibrometer. The location for the test instrumentations can be seen in Figures 1 and 3. The accelerometer is used to control the acceleration force of a 90 kN Unholtz-Dickie electrodynamic shaker (Figure 3), the strain gage is placed at the fatigue zone of the test specimen to acquire strain amplitude, and the laser vibrometer is positioned near the two-stripe nodal line of the vibrating specimen to measure velocity during testing [9]. The strain gage used in this experiment is not fatigue rated for the strain amplitudes and cycles accumulated during vibration-based bending. The fatigue rating for this strain gage is 1500 $\mu\varepsilon_a$ for 10^5 cycles, whereas most tests were conducted between 1700 - 2700 $\mu\varepsilon_a$ for $2x10^5$ - 10^7 cycles. Therefore, in order to determine strain amplitude values throughout the duration of a test, the velocity results from the laser vibrometer are linearly correlated with strain results from the gage at sufficiently low strain amplitudes (below 1500 $\mu\varepsilon_a$) and low loading cycles (50 Hz sinusoidal sweeps around resonance for 60 sec durations). The velocity is then used to identify strain/stress amplitudes during fatigue testing.

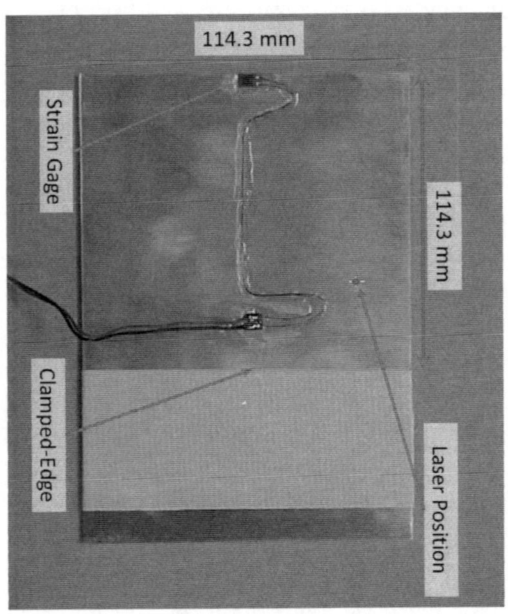

Figure 1. Vibration-based uniaxial bending fatigue specimen

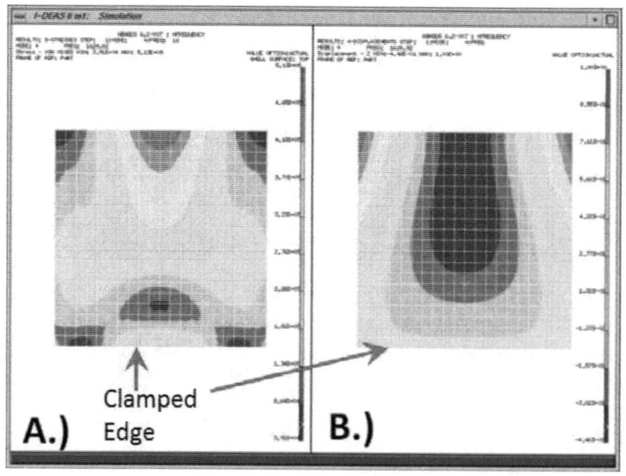

Figure 2. Vibratory analysis of the chord-wise mode on a square plate: A.) von Mises stress, B.) mode shape

Failure during a vibration-based test, which is defined as crack initiation, is determined by monitoring the frequency change with respect to time. Furthermore, observation of the shaker acceleration force versus time is also monitored to support the occurrence of failure. Figure 4 illustrates a plot of operating frequency and acceleration force over time during a fatigue test on an Inconel Alloy 625 specimen [11]. Failure can be determined from this data as the instance the natural frequency, which is tracked by the driving frequency, shifts more than 0.1%. This frequency shift is due to a change in the stiffness of the plate as a crack appears, and can be associated with crack lengths as small as 2-3 mm. Damping in the plate also increases with the presence of a crack, thus more force (i.e. higher shaker head acceleration) is required to maintain the desired velocity response. Note, in Figure 4, the shaker acceleration limit is notched at 120 G despite having a rating of approximately 133 G for the armature payload of the vibration-based setup; the data spike above and below 120 G is due to the noise from velocity control.

Fatigue testing to failure is conducted two ways: sinusoidal amplitude dwell or the step test procedure. Sinusoidal amplitude dwell, referred to as dwell for the remainder of this manuscript, is conducted by controlling the strain amplitude, via the velocity amplitude, at a constant value throughout the duration of the test until failure occurs. The step test procedure, which was utilized more than dwell, is a series of dwell steps for a specified number of cycles (N_{step}), each step having consecutively increasing sinusoidal dwell amplitudes, where the failure stress is calculated by interpolating between the failure and previous steps [12, 13]. The initial step in the series of N_{step} cycles is subjected to a constant stress amplitude level well below the anticipated failure stress amplitude. In the George fatigue method, the stress for the initial step is chosen at roughly 50% of the anticipated failure stress; however, there is not a specific guideline for the choice of this value. When, or if, failure does not occur after N_{step} cycles, the stress amplitude is increased by 5% of the anticipated failure stress, and the specimen is subjected to the new stress amplitude for another N_{step} cycles. These steps are conducted until failure occurs. When failure does occur, the failure stress amplitude (σ_a) is calculated by Equation (1), where σ_{pr} is the stress amplitude during the previous un-failed step, σ_{fail} is the stress amplitude during the failed step, and N_{fail} is the accumulated cycles to failure during the last step. Therefore, σ_a correlates to the failure stress amplitude for the controlled N_{step} cycles to failure.

$$\sigma_a = \sigma_{pr} + \frac{N_{fail}}{N_{step}}\left(\sigma_{fail} - \sigma_{pr}\right) \tag{1}$$

Figure 3. Vibration-based test setup and instrumentation

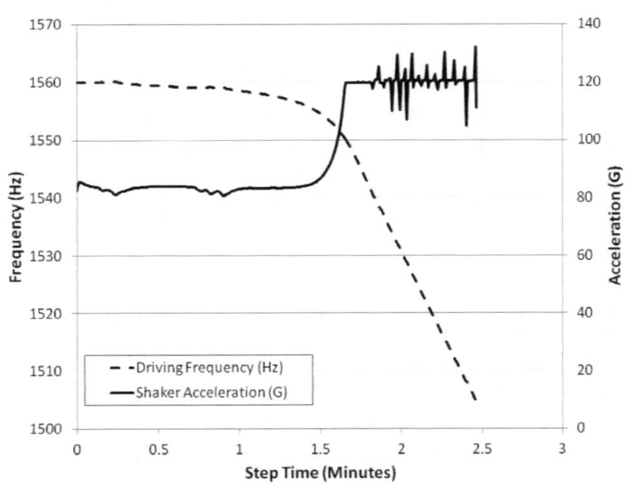

Figure 4. Sinusoidal resonant dwell time vs. frequency and acceleration during vibration-based test: Inconel Alloy 625 [11]

DMLS Manufacturing Process: The Ni Alloy 718 DMLS specimens used in this study were fabricated by two different vendors. Each vendor, designated as Vendor I and Vendor II, used different DMLS process parameters and post-sintering treatments. The DMLS process parameters, e.g. cooling rates, laser power, and atmosphere gas and temperature, were not specified by either vendor as this information is considered proprietary to the manufacturers. Each vendor, however, specified differences in post-sintering treatments. Vendor I plates were solution heat treated, HIP, and stress relieved, and Vendor II plates were only stress relieved. The as-built surface roughness for each DMLS plate varied between Ra values of 5.08 and 12.7 microns; however, each plate was polished with a Merit fine-grit buffing disk to Ra values between 1.27 and 5.08 microns. Differences in the tensile properties and fatigue performance of each DMLS vendor are observed in the following Sections.

Monotonic Tensile Results: Tensile material properties were determined by pulling dogbone/coupon specimens monotonically until separation. Monotonic tensile tests were conducted with a Mechanical Test Systems (MTS) 312.21 load frame. The load frame is equipped with a 100 kN load cell, a linear variable differential transformer (LVDT) for displacement control, and an extensometer for strain measurement. The displacement loading rate and the specimen geometry used for monotonic tests were determined based on ASTM standard E 8-09 recommendations [14]. The displacement rate for each tensile test was 102 μm per second, and the geometry of the tensile dogbone specimen is shown in Figure 5.

In order to efficiently utilize available DMLS material, and to correlate the strengths of specimens to respective fatigue lives, monotonic tensile data was acquired from dogbones cut out from fatigued plates. Four fatigue tested plates from each DMLS Vendor were analyzed, and six dogbone specimens with the geometry in Figure 5 were cut from each plate. For half of the eight DMLS plates, dogbone specimens were cut longitudinal with the build direction of the plate, and the other four were cut transverse to the build direction. The build direction shown in Figure 6 was chosen under the assumption that it provided the worst case fatigue life results for the vibration-based specimen. As shown in Fig. 6, specimens that were cut longitudinal and transverse are respectively denoted by *H* and *V* to identify the sintering plane orientation with respect to the loading axis. This data was compared against two pristine (non-vibration affected) data points provided by Vendor I, and MIL-HDBK-5J minus three sigma (referred to as min) material property values for long-transverse sheet/plate specimens with thicknesses between 0.254 and 25.4 mm [15].

Since the pristine DMLS tensile data was received from Vendor I and not generated under the same controlled procedure as the fatigue affected data, cold-rolled Inconel alloys were investigated to understand tensile property variation between fatigue affected and pristine dogbones from the same Ni-based stock. The cold-rolled materials investigated were Inconel Alloy 625 and unhardened Inconel Alloy 718. These alloys were chosen due to the similarities in chemical composition to DMLS Ni Alloy 718. The unhardened 718 specimens came from one 610 x 610 x 3.2 mm plate stock, and the 625 specimens came from two 610 x 305 x 3.2 mm plate stocks referred to as Stock I and

Stock II. Each cold-rolled specimen was cut either along or transverse to the rolling marks of the bulk stocks, and are respectively denoted as *V* and *H* to identify the rolling mark orientation with respect to the loading axis.

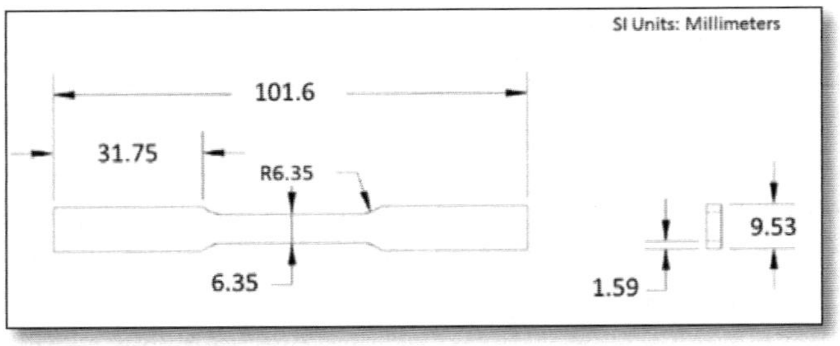

Figure 5. Monotonic tensile specimen with ASTM E 8-09 recommendations

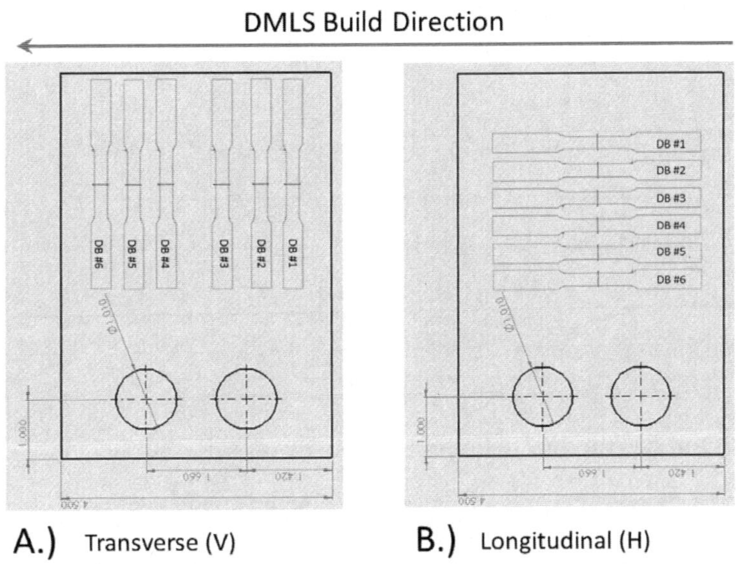

Figure 6. Diagram for dogbone cutting configuration of DMLS plate specimens

Tables 1 and 2 show the average results for ultimate tensile and yield strength, the elastic modulus (E), and the total percent elongation before separation of vibration affected and pristine cold-rolled Inconel Alloy 625 and unhardened Inconel Alloy 718. Specimen labels for each Table are defined as follows: for Table 1, *Alloy#* ***Roll_Plate Specimen#*** *-* ***R****oll_Dogbone;* and for Table 2, *Alloy#* ***Roll****_Plate+Stock#* ***Specimen#*** - ***Roll****_Dogbone.* Parameters *V* and *H* in both tables indicate all dogbones cut longitudinal and transverse to the roll direction, respectively. These labels are defined in the Nomenclature Section.

For Table 1, the rows of the vibration affected results are categorized by the fatigued plate labels; *718 H #3 – V* is a plate with chord-wise rolling direction and dogbones cut longitudinal to the roll, and *718 V #1 – H* is a plate with span-wise rolling direction and dogbones cut transverse to the roll. Table 2 also categorizes the vibration affected properties by the fatigue plate labels; 625 *V1 #3 – H* is a plate from Stock I with span-wise rolling direction and dogbones cut transverse to the roll, and 625 *H2 #1 – V* is a plate from Stock II with chord-wise rolling direction and dogbones cut longitudinal to the roll. Since rolling direction will affect tensile properties, the variation for the combined pristine specimen properties (i.e. all pristine *H* and *V* specimens combined) were used for overall comparisons to simulate unforeseen manufacturer uncertainties not present in the studied material stocks. Results show a 0.3-1% increase in strength properties due to vibration on 625 specimens in the *V* direction, and a 0.4-0.7% decrease in strength properties for specimens in the *H* direction. The 718 comparison shows the opposite results, where *V* specimens shows roughly 1.5% decrease and *H* specimens show a 0.4-1.3% increase in strength properties due to the effects of vibration. These results do not indicate any trends based on cold-rolling direction or accumulated fatigue cycles. Results also show that the vibration affected tensile properties of each individual specimen are within one standard deviation of the total pristine results in Tables 1 and 2. Figures 7 and 8 respectively show the yield and tensile strength comparisons against a single standard deviation (one sigma) bound for cold rolled Inconel Alloy 718. The plots on the Figures illustrate no particular trends in strength properties along the span of each specimen as well.

Table 1. Inconel Alloy 718 tensile property results

Category	Yield (MPa)	Ultimate (MPa)	E (GPa)	Elongation %	Samples #
Total	495	930	187	48	10
V	507	943	181	48	5
H	484	916	192	49	5
718 H #3 – V	489	929	175	48	3
718 V #1 – H	490	920	190	49	6

Table 2. Inconel Alloy 625 tensile property results

Category	Yield (MPa)	Ultimate (MPa)	E (GPa)	Elongation %	Samples #
MIL_HDBK-5J	*481*	*942*	*205*	*30*	*n/a*
Total	491	914	194	*50*	12
Stock I – H	499	932	201	*50*	4
Stock I – V	506	919	191	*51*	2
Stock II – H	477	902	191	*52*	4
Stock II – V	489	898	189	*47*	2
Stock I – Total	502	928	197	*50*	6
Stock II – Total	481	901	190	*50*	6
625 V1 #3 – H	496	928	197	*51*	6
625 H2 #1 – V	494	904	194	*54*	6

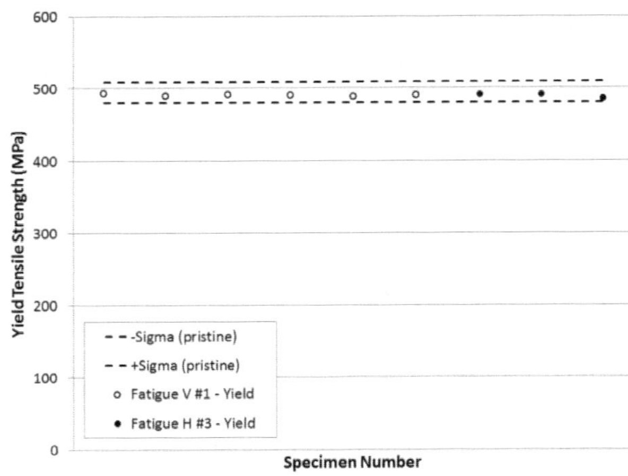

Figure 7. Yield tensile strength comparisons for unhardened Inconel Alloy 718

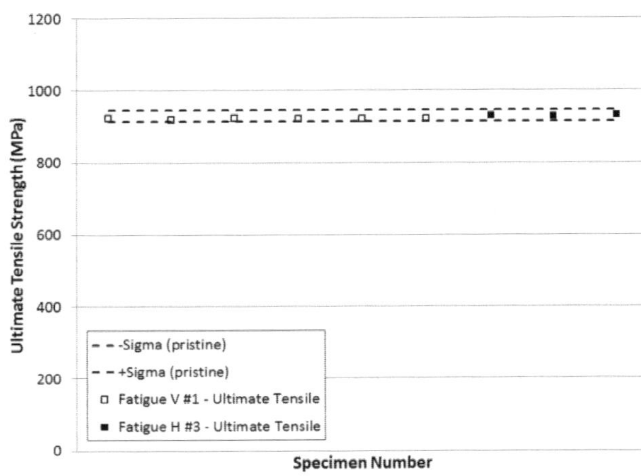

Figure 8. Ultimate tensile strength comparisons for unhardened Inconel Alloy 718

Results for the DMLS Nickel Alloy 718 comparisons are shown in Table 3, where the final eight rows represent the vibration affected tensile properties. The average results for ultimate tensile and yield strength, the elastic modulus, and the total percent elongation before separation for vibration affected DMLS specimens are compared with two pristine data points provided by Vendor I. Also, the Table shows minus three sigma (referred to as min) material property values for long-transverse sheet/plate specimens with thicknesses between 0.254 and 25.4 mm [15]. Both vendors show that the *V* specimens have stronger tensile properties than the *H* specimens; however, this discrepancy is more noticeable in Vendor II than in Vendor I. Overall, Vendor I shows higher strength properties than both Vendor II and the published minimum data, but the ultimate tensile strength data for Vendor I is 1-3% less than the reference pristine data points from Vendor I. Other comparisons for the yield and ultimate tensile strength properties are shown in Figures 9 and 10, respectively. These Figures conclude that Vendor II strength properties do not meet the minus three sigma material strength limits of the MIL-HDBK-5J published data.

Table 3. DMLS Nickel Alloy 718 tensile property comparison

Category	Yield (MPa)	Ultimate (MPa)	E (GPa)	Elongation %	Samples #
MIL_HDBK-5J – min	*1034*	*1240*	*203*	*12*	*n/a*
Vendor I Baseline – V	*1068*	*1344*	*n/a*	*27*	*n/a*
Vendor I Baseline – H	*1034*	*1309*	*n/a*	*27*	*n/a*
Vendor I #1 – H	1070	1316	195	18	6
Vendor I #2 – V	1082	1331	199	20	6
Vendor I #5 – V	1071	1322	198	20	6
Vendor I #6 – H	1059	1293	191	17	6
Vendor II #2 – H	789	1059	174	31	6
Vendor II #3 – V	868	1162	182	26	6
Vendor II #4 – H	787	1034	185	31	6
Vendor II #5 – V	854	1148	172	26	6

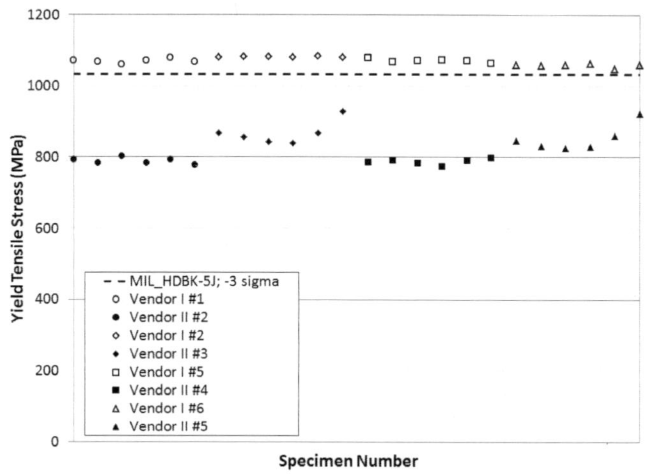

Figure 9. Yield tensile strength comparisons for DMLS Nickel Alloy 718

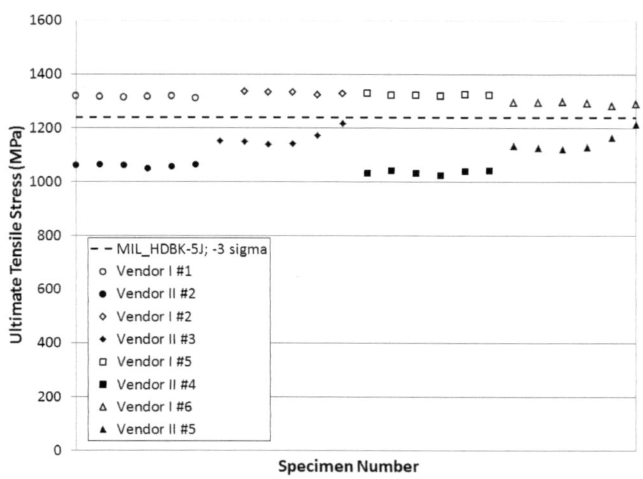

Figure 10. Ultimate tensile strength comparisons for DMLS Nickel Alloy 718

Bending Fatigue Life Comparison: The George Fatigue Method was used to characterize bending fatigue behavior for DMLS Ni Alloy 718. Using the 90 kN electrodynamic shaker shown in Figure 3, fatigue life data was acquired from 24 DMLS plates, 12 per vendor, with build directions normal to the axis of the anticipated crack path. Based on Vendor I pristine data and other supporting strength properties from Table 3, this build direction corresponds to a worse case failure scenario. Four of the 12 tests conducted for each Vendor batch were fatigued to failure with dwell tests, and the rest were fatigue using the step test method discussed in the *Vibration-Based Method and Test Setup* Section. Previous research shows that though the step test has a lower life performance than dwell, the results are comparable with respect to normal fatigue data scatter [11, 16, 17]. The comparisons between dwell and step test methods are provided in Figures 11 and 12 for Vendor I and II, respectively [18]. Observing the regression correlation of the experimental data, the comparison in Fig. 11 shows more deviation between the step test and the dwell results than the comparison in Fig. 12. However, based on 80% prediction intervals from the dwell regression correlations of each Vendor, the scatter of the step test data is acceptable to combine with dwell test data for comparison against cold-rolled Inconel Alloy 718 fatigue performance [18].

The step test and the dwell results were combined for both Vendors I and II, and the data was compared against published rotating bending, cold-rolled data from Special Metals Corporation in Fig. 13 [8]. The comparisons show that the fatigue performances of both Vendors agree within a reasonable scatter with the published rotating bending results. Also, despite Vendor II showing weaker and more inconsistent strength properties than Vendor I, the comparison shows that Vendor II has slightly better fatigue performance than Vendor I.

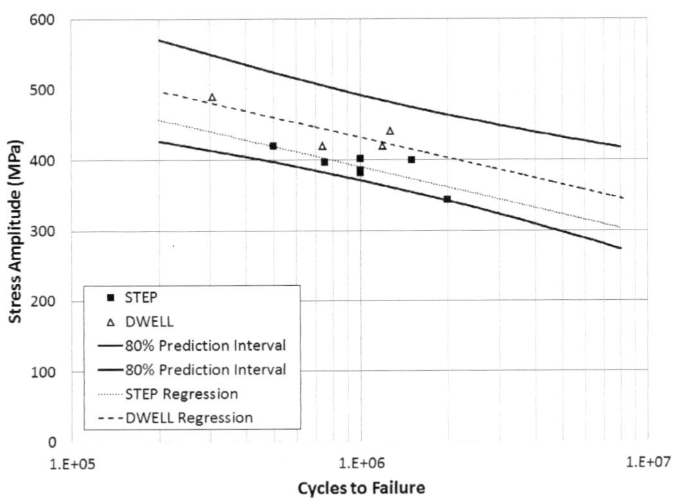

Figure 11. Step test comparison against sinusoidal amplitude dwell: Ni Alloy 718, Vendor I

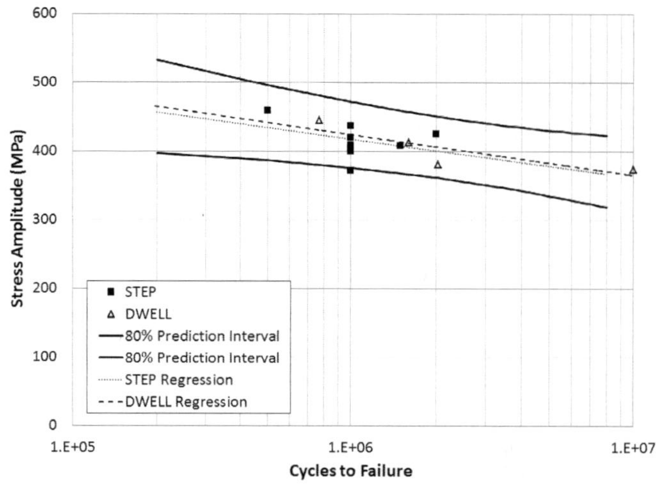

Figure 12. Step test comparison against sinusoidal amplitude dwell: Ni Alloy 718, Vendor II

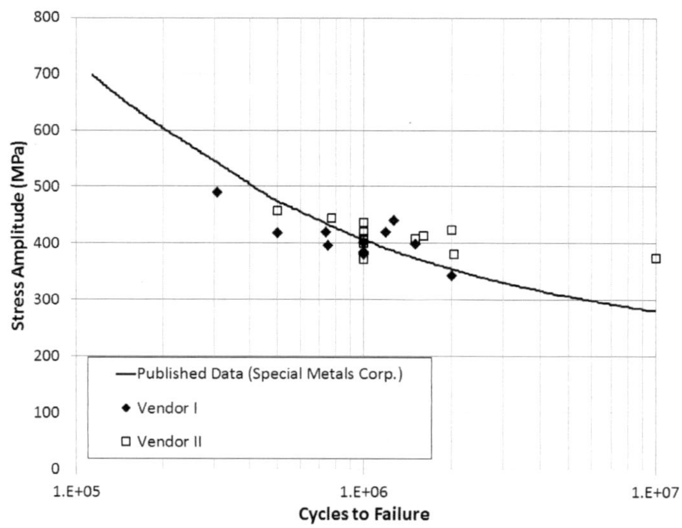

Figure 13. Fatigue life comparison: DMLS Ni Alloy 718 versus Inconel 718 [17]

Conclusion: DMLS and cold-rolled Ni Alloy 718 tensile strength and bending fatigue performances were compared in the presented work. The DMLS specimens were provided by two vendors with different post-sintering treatments: Vendor I plates were heat treated, HIP, and stress relieved; Vendor II plates were stress relieved. Tensile strength DMLS properties, which were acquired from vibration affect bending fatigue plates, were compared against published cold-rolled plate data as well as provided DMLS data. DMLS bending fatigue, acquired via a vibration-based test method, was compared against published rotating-bending fatigue results.

Along with the DMLS tensile comparison, cold-rolled Inconel alloys 625 and 718 was investigated to understand tensile property variation between fatigue affected and pristine dogbones from the same stock. The cold-rolled comparisons showed conflicting results for tensile strength properties with respect to the effects of rolling direction and fatigue cycles. However, the variations of the vibration affected tensile properties are all within one standard deviation of the pristine data. The strength comparisons for DMLS show mechanical properties are sensitive to post-sintering treatments and could be sensitive to process parameters. More manufacturing and microscopy information is required to fully support these claims.

Bending fatigue performance comparison of Vendor I, Vendor II, and cold-rolled 718 showed good visual agreement between $2x10^5$ and $2x10^6$ cycles to failure. Vendor II showed better fatigue performance than the other data sets; rotating-bending cold-rolled 718 showed second best, and Vendor I, though marginal, showed the weakest performance. As with the tensile strength comparison, the discrepancy in the fatigue comparisons shows a need to better understand process parameters and conduct microscopy.

Acknowledgements: The authors would like to thank the following organizations for funding, support, and collaboration: the Turbine Engine Fatigue Facility (TEFF) of the United States Air Force Research Laboratory (AFRL), Universal Technology Corporation (UTC), the Department of Mechanical Engineering at Wright State University, and the Department of Mechanical and Aerospace Engineering at The Ohio State University. Also, the authors would like to acknowledge UTC contractor Tyler Kennedy for contributing to laboratory testing.

References:

[1] Lind, J-E., Hanninen, J., Kotila, J., Nyrhila, O., and Syvanen, T., "Rapid Manufacturing with Direct Metal Laser Sintering," Materials Research Society Symposium proceedings, 2002, Vol. 758.

[2] "Additive Manufacturing – Direct Laser Sintering of Titanium and other Metals," April 5, 2010, Next Big Future, <http://nextbigfuture.com/2010/04/additive-manufacturing-direct-laser.html>.

[3] Sanz, C., Garcia Navas, V., Gonzalo, O, and Vansteenkiste, G., "Study of Surface Integrity of Rapid Manufacturing Parts after Different Thermal and Finishing Treatments," 1st CIRP Conference on Surface Integrity proceedings, 2011, Vol. 19, pp. 294 – 299.

[4] Zaeh, M., and Ott, M., "Investigation on Heat Regulation of Additive Manufacturing Processes for Metal Structures," CIRP Annals – Manufacturing Technology, 2011, Vol. 60, Issue 1, pp. 259-262.

[5] Santos, E., Masanari, S., Osakada, K., and Laoui, T., "Rapid Manufacturing of Metal Components by Laser Forming," International Journal of Machine Tools and Manufacturing, 2006, Vol. 46, Issue 12-13, pp. 1459-1468.

[6] Murr, L, Quinones, S., Gaytan, S., Lopez, M., Rodela, A., Martinez, E-Y., Hernandez, D., Martinez, E., Medina, F., and Wicker, R., "Microstructure and Mechanical Behavior of Ti-6Al-4V Produced by Rapid-Layer Manufacturing for Biomedical Applications," Journal of the Mechanical Behavior of Biomedical Materials, 2009, Vol. 2, Issue 1, pp. 20-32.

[7] Engine Structural Integrity Program (ENSIP), MIL HDBK-1783B (USAF), 15 Feb 2002.

[8] Special Metals Corporation, "Inconel Alloy 718," New Hartford, NY, September 7, 2007, Publication No. SMC-045.

[9] George, T., Seidt, J., Shen, M.–H.H., Cross, C., and Nicholas, T., "Development of a Novel Vibration-Based Fatigue Testing Methodology," International Journal of Fatigue, 2004, Vol. 26, Issue 5, pp. 477-486.

[10] George, T., Shen, M.-H.H., Cross, C., and Nicholas, T., "A New Multiaxial Fatigue Testing Method for Variable Amplitude Loading and Stress Ratio," Journal of Engineering for Gas Turbines and Power, ASME, 2006, Vol. 128, pp. 857-864.

[11] Scott-Emuakpor, O., Schwartz, J., Holycross, C., George, T., Cross, C., and Shen, M.-H.H., "In-situ Study on Coaxing during Vibration-Based Bending Fatigue of Inconel 625 and 718," ASME/IGTI Turbo Expo, San Antonio, TX, 2013, Paper number: GT2013-94233.

[12] Nicholas, T., and Maxwell, D., "Mean Stress Effects on the High Cycle Fatigue Limit Stress in Ti-6Al-4V," Fatigue and Fracture Mechanics: 33rd Volume, ASTM STP 1417, W.G. Reuter and R.S. Piascik, Eds., American Society for Testing and Materials, West Conshohocken, PA, 2003, pp. 476-492.

[13] Nicholas, T., *High Cycle Fatigue – A Mechanics of Materials Perspective*, Oxford, UK, Elsevier, 2006.

[14] American Society for Testing and Materials, "E 8M - 09: Standard Test Methods for Tension Testing of Metallic Materials," *ASTM Book of Standards,* 2009; Vol. 03.01, ASTM International, West Conshohocken, PA.

[15] Metallic Materials and Elements for Aerospace Vehicle Structures, MIL-HDBK-5J (USAF), 31 Jan 2003.

[16] Bellows, R., Muju, S., and Nicholas, T., "Validation of the Step Test Method for Generating Haigh Diagrams for Ti-6Al-4V," International Journal of Fatigue, 1999, Vol. 21, Issue 7, pp. 687-697.

[17] Scott-Emuakpor, O., George, T., Cross, C., Wertz, J., and Shen, M.-H.H., "A New Distortion Energy-Based Equivalent Stress for Multiaxial Fatigue Life Prediction," International Journal of Non-Linear Mechanics, 2012, Vol. 47, No. 3, pp. 29-37.

[18] American Society for Testing and Materials, "E739 - 10: Standard Practice for Statistical Analysis of Linear or Linearized Stress-Life (S-N) and Strain-Life (ε-N) Fatigue Data," *ASTM Book of Standards,* 2010; Vol. 03.01, ASTM International, West Conshohocken, PA.

Mitigating Technical Risk when Applying Novel Materials & Processes

21 May 2014

Bob Ware
Structural Failure Analyst
Materials & Manufacturing Directorate
Air Force Research Laboratory

Outline

- Background / Rationale
- List of Recent Examples
- Two Historical Examples
  - Project Azorian
  - Liberty Ship Fractures
- Guidelines offered M&P Substitution
  - AFLCMC/EZ Structures Bulletin EZ-SB-13-001

 2

Background / Rationale

- Emerging Materials & Process (M&P) solutions hold the promise of cost reductions and performance improvements.
  - Many initial implementations are selected out of desperation, because existing approaches are no longer authorized by the EPA or current materials and processes do not meet margins
- Advocates of the new M&P solutions are often biased and may not be aware of service requirements.
  - Cost savings in one area are offset by unquantified risks in another.

DISTRIBUTION A: Approved for public release; distribution unlimited - 88ABW-2013-3860 8/27/2013 3

Recent Examples

- *High Velocity Oxy Fuel (HVOF) for Hard Chrome*
- *Cold Spray for hydraulic tube chafing protection*
- *Laser Depainting for Plastic Media Blasting*
- *Zinc Nickel for Cadmium Plating*
- *Laser Shock Peening (LSP) for residual stress*
- *High Speed Machining (HSM) over conventional*
- *Citrus Cleaners*
- *Diegme Fuel Additive for deicing*
- *Ultra High-Strength Steels*
- *Friction Stir Welding (FSW)*
- *Laser Cutting of Aluminum Structure*
- *Metal Additive materials*
- *Bearing Refurbishment*

The rate of change in M&P seems to be increasing

DISTRIBUTION A: Approved for public release; distribution unlimited - 88ABW-2013-3860 8/27/2013 4

Historical Examples

Our predecessors also dealt with extreme performance and schedule requirements

- **Project Azorian, Raising a Soviet Sub**
- **Liberty Ship Fractures**

DISTRIBUTION A: Approved for public release; distribution unlimited - 88ABW-2013-3860 8/27/2013 5

Project Azorian Background

- **Russian Nuclear Sub K-129 sinks in 3 miles' depth in the Northwest Pacific in 1968**
  - **3 nuclear missiles**
  - **Torpedoes**
  - **Codebooks**
- **CIA initiated Project Azorian to recover it**

Image used with permission.

Sources:
Nova episode "Submarines, Secrets, and Spies", aired January 19, 1999, online transcript
Images from Project Azorian online Trailer

DISTRIBUTION A: Approved for public release; distribution unlimited - 88ABW-2013-3860 8/27/2013 6

Project Azorian
Glomar Explorer

Howard Hughes constructed the Glomar Explorer. The reported purpose was undersea mining.

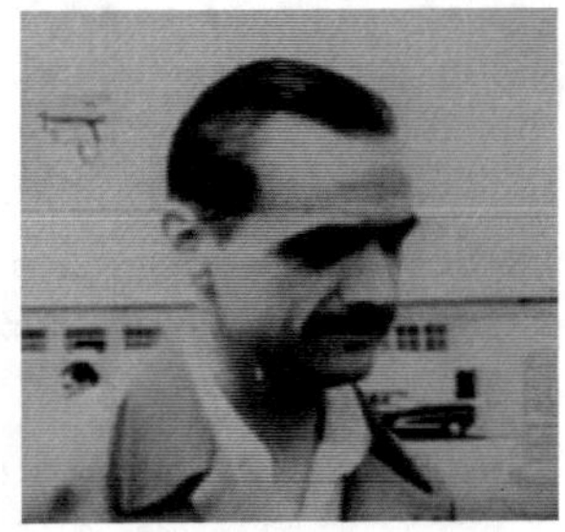

Nova and National Geographic advanced the undersea mining cover story.

In 1974, the recovery operation proceeded.

Project Azorian
Glomar Explorer

The Glomar Explorer had multiple thrusters to hold her position in high winds and seas.

The middle of the ship was a single giant room with floor that retracted, allowing access to the ocean below. Three hundred 60-foot steel pipes were strung together to lower a huge claw down the Soviet sub.

Together, the claw, pipe and flooded sub weighed 15 million pounds.

Project Azorian
'the Claws'

Seven custom claws made from recently developed high-strength Maraging Steel were custom designed to grip the nose section of the of the Sub.

DISTRIBUTION A: Approved for public release; distribution unlimited - 88ABW-2013-3860 8/27/2013 9

Project Azorian
Aftermath

Three of seven of these claws fractured during the lift operation, only a small section was recovered, *without code books or missiles.*

Despite tremendous expenditures, the mission goals were not achieved.

A Soviet Naval Attaché advised the US that a return to the recovery site would prompt a declaration of war.

In 1993, a film of the funeral service held for the Soviet seamen was given to Boris Yeltsin. This was the only official acknowledgment of the operation.

DISTRIBUTION A: Approved for public release; distribution unlimited - 88ABW-2013-3860 8/27/2013 10

Project Azorian, What Happened?

Polmar's and White's film suggests the undersea temperatures were below the steel's ductile to brittle transition temperature (DBTT).

Current ASMH handbook data shows significant reductions in toughness at lower temperatures.

It seems likely project engineers did not adequately test the material in realistic conditions.

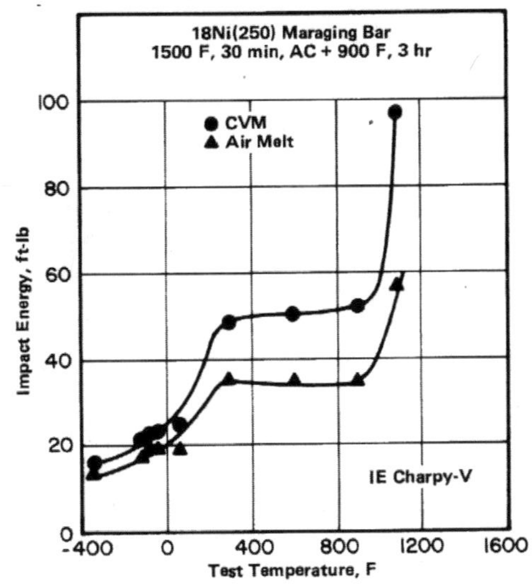

FIGURE 3.0334. EFFECT OF TEST TEMPERATURE ON THE IMPACT ENERGY OF ANNEALED AND AGED BAR (6)

Source: Aerospace Structural Metals Handbook, Section 1220 18Ni (250 Grade) Maraging Steel

Liberty Ship Production

Approximately 2700 Liberty Ships were built during WWII for sealift across the Atlantic.

Production was accelerated by using prefabricated all-welded construction to a standard design, rather than riveted panels.

Sources:
Fracture - some maritime examples, Douglas Wright, University of Western Australia;
Wikipedia;

Liberty Ship Fractures

There were nearly 1,500 instances of significant brittle fractures.

Several ships broke in half without warning.

Fracture of the SS John P. Gaines resulted in the loss of 10 lives.

In January 1943, the one-day old T2 tanker SS Schenectady fractured at the dock after seatrials. The vessel jack- knifed and the center portion rose so that no water entered.

Liberty Ship Fractures, What Happened?

Among other causes investigated, several were related to design and materials:

- The predominantly welded hull construction allowed cracks to run for long distances unimpeded.
- Some areas had super-positioned stress concentrators. For example, one square corner of a hatch coincided with a welded seam.
- The grade of steel suffered from a relatively high ductile to brittle temperature (DBTT).

Guidelines to Mitigate Risk

- Identify Unique Limitations
- Test Limitations Directly in Realistic Environments
- Select Initial Applications Carefully
- Establish Appropriate Process Control

Guidelines, Identify Unique Limitations

- With an open mind, evaluate the shortcomings of the new process.
  - They may be very different from the shortcomings of the original process.
  - If a proponent does not inform you of any process limitations, there are two possibilities
    - There are no limitations, or
    - They haven't looked.

Don't let familiarity with the original process blind you to the limitations of the new process.

Example: Chrome Electroplating vs High Velocity Oxy Fuel (HVOF) coating

- EPA requirements have led us away from electrolytic hard chrome
  to High Velocity Oxy Fuel (HVOF) WC-Co.
- These are very different processes.

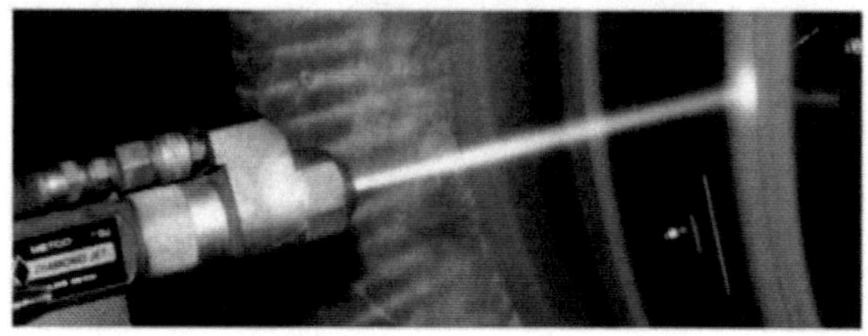

Guidelines, Identify Unique Limitations

Chrome has a constant hardness throughout, with microcracks.

These cracks allow some strain tolerance, and act as lubricant reservoirs.

Hydrogen embrittlement relief is a very real requirement.

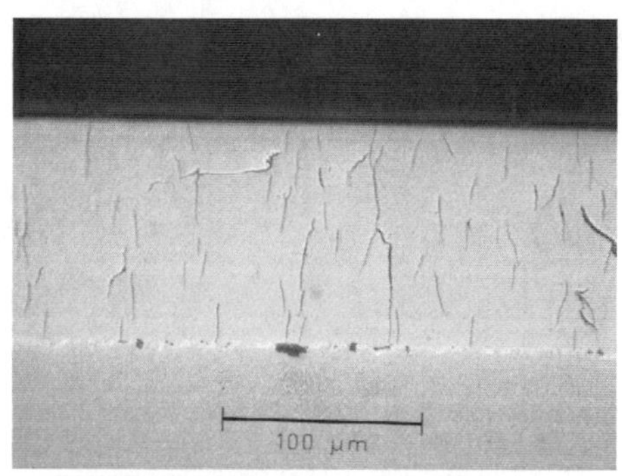

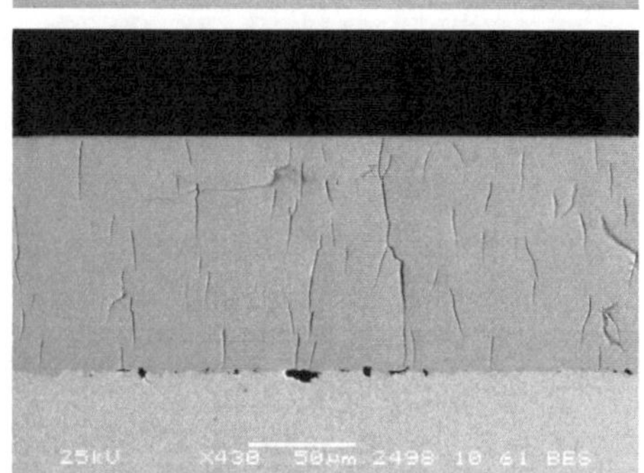

Guidelines,
Identify Unique Limitations

HVOF is a thermal spray process, depositing hard WC particles in a ductile cobalt binder.

This coating requires superfinishing to minimize seal wear.

WC-Co-Cr, in cross section (unground)

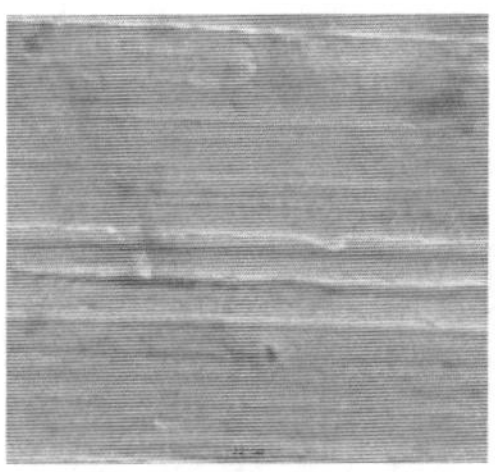

Chrome, 8 μin

WC-Co, 8 μin

Guidelines,
Test Limitations Directly

- **Test Limitations Directly in Realistic Environments**
  - Quantify the limits of the new process i.e. Find the edge of the cliff.

Modelling cannot reveal unknown unknowns

  - Don't depend on modeling to identify what you don't know about new M&P.
  - Test early to verify model assumptions.

Example: Brittleness of the Azorian Claws at cold temperatures was not foreseen.

Guidelines, Selecting Initial Applications

- **Select Initial Applications Carefully**
  - Don't use up all of the projected benefit; Leave margin to allow a fallback position to a proven technology.

  Example: Boeing's application of HVOF to Landing Gear vs Pratt & Whitney's initial application of Laser Shock Peening.

Try not to get boxed into a corner.

Guidelines, Process Control

- **Establish Appropriate Process Control**
  - **Particularly when the performance requirements are still undetermined.**

AFRL's early testing of HVOF coupons revealed wide differences in fatigue performance depending on Gun Vendor /and Processing.

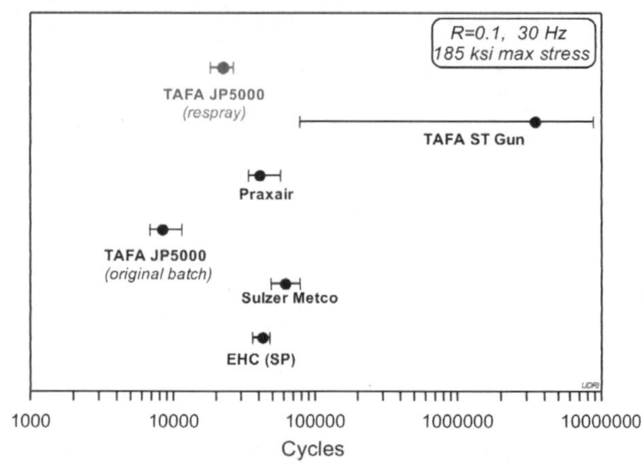

- Optimize early to avoid the need for retesting.

Guidelines to Mitigate Risk

- **Identify Unique Limitations**
- **Test Limitations Directly in Realistic Environments**
- **Select Initial Applications Carefully**
- **Establish Appropriate Process Control**

Guidelines, Technical Risk

- **Balancing Risk against potential savings and performance improvements.**

"I'd rather have 10 problems I understand than 1 that I don't"

Wayne Plack, SM-ALC

- **If we do not attempt technology insertions on low risk / low payoff applications, we put off learning lessons and overcoming barriers.**

EZ-SB-13-001, Guidelines for Material, Product Form, and Process Substitutions

- **Risk Level is based on component criticality and similarity to existing configuration**
- **Engineers should consult with AFRL/RX on potential substitutions characterized as "high" risk**

Substitution Type ► / ▼Part Type▼	Same Alloy System, Same Heat Treatment (e.g., 7XXX-TXYZ Al for 7ABC-TXYZ Al)	Same Alloy System, Different Heat Treatment (e.g., 7XXX-TXYZ Al for 7XXX-TABC Al)	Different Alloy, Same Base (e.g., 2XXX Al for 7XXX Al)	Different Base (e.g., Al for Ti)	Different Class (e.g., Composite for Metal)	Hybrid Material (e.g., Metal-Composite Laminate for Metal)
Safety-of-Flight*	M	H	H	H	NR	NR
Durability Critical	L	M	M	H	H	NR
All Others	L	L	L	M	H	H

KEY	Complexity / Risk
L	Low
M	Moderate
H	High
NR	NOT RECOMMENDED without extensive testing and AFRL/RX support

* Includes component types designated as Critical Safety Items (CSI), Safety Critical, and Mission Critical

Replacing ► / ▼With▼	Forging	Extrusion	Machined Plate	Casting	Welded Structure	Built-Up Structure
Forging	L	L	L	L	L	M
Extrusion	M	L	L	M	L	M
Machined Plate	H	H	L	M	M	H
Casting	H	H	H	L	M	H
Welded Structure	H	H	H	H	L	H
Additive Mfg.*	NR	NR	NR	NR	NR	NR

KEY	Complexity / Risk
L	Low
M	Modèrate
H	High
NR	NOT RECOMMENDED without extensive testing and AFRL/RX support

* Additive Manufacturing includes, but is not limited to, laser additive manufacturing, electron beam additive manufacturing, cold spray processes, etc.

DISTRIBUTION A: Approved for public release; distribution unlimited - 88ABW-2013-3860 8/27/2013 25

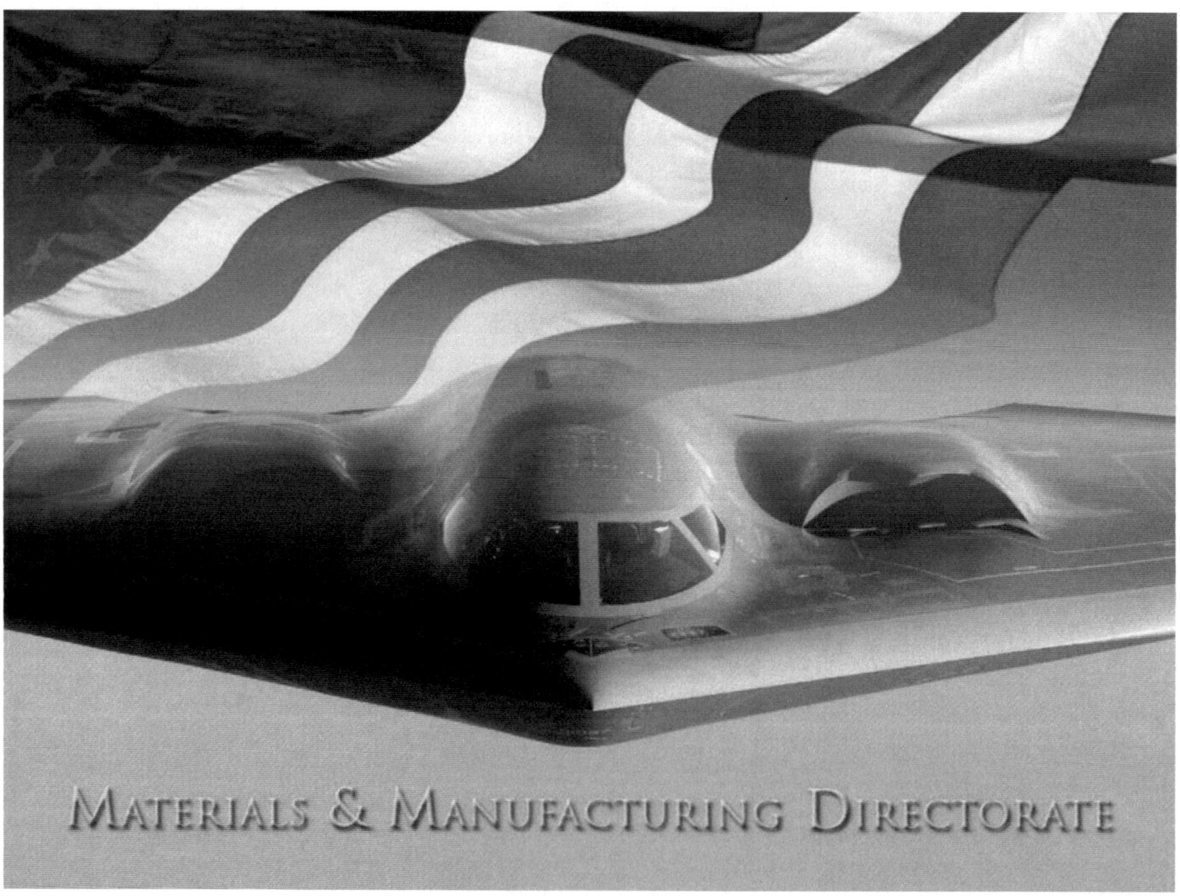

Materials & Manufacturing Directorate, Systems Support Division (AFRL / RXS)

Physical and organizational location fosters widespread collaboration:

- RXS Teams at WPAFB:
  - Structural Failure Analysis
  - Electronic Failure Analysis
  - Coatings Technology Integration Office
  - Adhesives Composites & Elastomers
  - Non Destructive Inspection
- RXS Field Offices at Air Logistics Centers:
  - Corrosion Prevention & Control;
  - Metals Technology (Robins)
  - Non Destructive Inspection (Tinker)
  - Advanced Composites (Hill)
- RXS personnel Collocated within WPAFB Program Offices

  B-2, C-17, JSF, F-22, Engines, Global Hawk, MQ-9
- Other RX Divisions, including:
  - Structural Materials (RXC)
  - Functional Materials (RXA)
  - Manufacturing Technology (RXM)
- Other AFRL Directorates at WPAFB:
  - Aerospace Systems (/ RQ)
  - Sensors (/ RY)
  - 711th Human Performance Wing (711 HPW)

We enable 2 way feedback between field issues and R & D

 27

Structural Failure Analysis Section Mission

- **Provide <u>Quick Reaction</u> Materials & Process (M&P) Support**
  - **Failure Analysis**
  - **Material Characterization**
  - **Process Reviews**
- **Monitor And <u>Understand Changing Needs</u> in the Field**
  - **Keep Research Focused on Relevant Issues**
- **<u>Transition</u> Emerging M&P Technologies to the Warfighter**

Prevent Failure Recurrence, Transition Lessons Learned

 28

Section Customers / Workload

- **Accidents, Safety Issues**
  - **Air Force Safety Center**
- **Production Issues**
  - **System Program Offices**
- **USAF Field Failures**
  - **Using Commands**
- **Other Gov't Agencies**
  - **Coast Guard, NTSB FAA, FBI, DOJ**

~ 30-40 Analyses / Year Including Several Mishaps

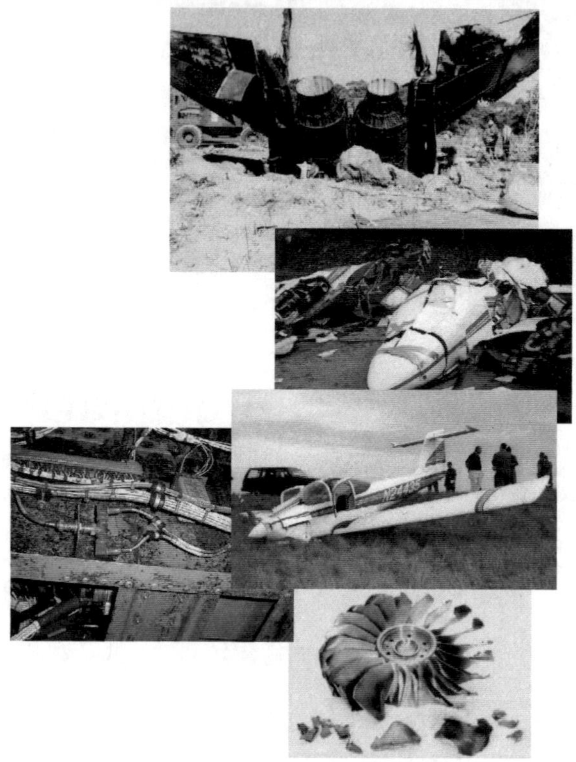

Using Reliability Analysis and Maintenance Process Simulation to Improve Decision Making

MG Lipsett
University of Alberta
May 21, 2014
MFPT 2014, Virginia Beach

Faculty of ENGINEERING
University of Alberta

This Is All About Understanding Risk

- Definitions
- Understanding elements of risk for a system of interest
  - Technology, process, people
- Qualitative models (risk assessment, HAZOP)
- Quantitative risk models
- Simulating effects of a failure, decision, or technology change
- Verifying and validating models
  - Training and testing the model
  - Estimating probabilities
  - Estimating costs and benefits
  - Monetizing non-economic risk, and discounting over time
- Examples, cautions, and summary

Faculty of ENGINEERING
University of Alberta

The Role of Modeling in Continuous Improvement

Understand flow and use of information in detecting failure and making maintenance decisions

- Characterize systems comprising flows and transformations (activities)
- Includes materials, energy, personnel, finances, and information
- Represent states and activities (including decisions) for all types of flows

Capture knowledge of how decisions are made and effect of decisions (and identify best practices)

Identify and manage uncertainties and associated risks

- Continual improvement in estimates of probabilities and costs
- Inputs, processes, and outputs

Make more effective individual decisions in context

Improve maintenance strategies for more effective organizational decision making

Faculty of ENGINEERING
University of Alberta

Sensitivity, Uncertainty, and Risk

We do sensitivity analysis to produce a set of results that brackets the possible range of results of our decisions

This range is possible because of our uncertainty with respect to inputs to the decision (such as what is really happening, costs to solve problems, timing, etc.)

This is a simple way to model the ***consequences*** of variability on the expected financial result (such as IRR or NPV)

There is another aspect to this analysis, namely, what is the ***probability*** of certain inputs being different from what our good judgment tells us they should be

Faculty of ENGINEERING
University of Alberta

Defining the System

The system includes all of the physical entities that can be affected, including people, and the environment.

All relevant events that can happen and their consequences need to be included.

Consequences can be far removed from the root cause of the event.

Faculty of ENGINEERING
University of Alberta

Technology, Process, People

The system is not merely the physical equipment that has failure modes

The system includes processes for operation, monitoring & inspection, parts procurement, maintenance, IT, documentation, finance,...

And people are also critical to a safe and reliable system. Their training, skill level, compliance to procedures, morale, and ability to solve novel problems all affect reliability.

Faculty of ENGINEERING
University of Alberta

Classical Reliability Analysis

Analytical reliability analysis uses the calculus of transforming probability distributions

In the simplest case, a system that is perfectly reliable has a probability of performing its intended process for its intended lifetime T, described by

$$R(T) = 1.$$

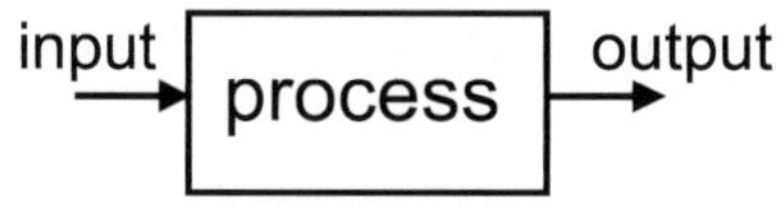

Ah, would that this were so….

Faculty of ENGINEERING
University of Alberta

System Reliability Calculation

For two identical, independent system elements in parallel, the reliability is twice what it would be for only one element: $R(T)_{parallel} = R(T) + R(T) = 2\ R(T)$.

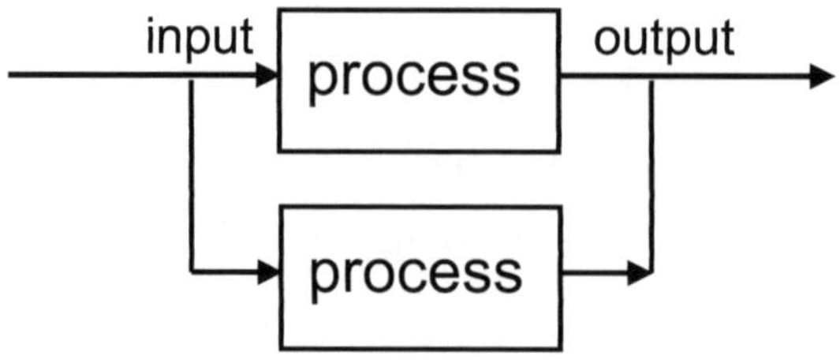

Faculty of ENGINEERING
University of Alberta

System Reliability Calculation

For two identical, independent system elements in series, the reliability is only half what it would be for only one element: $R(T)_{series} = R(T) \times R(T)$

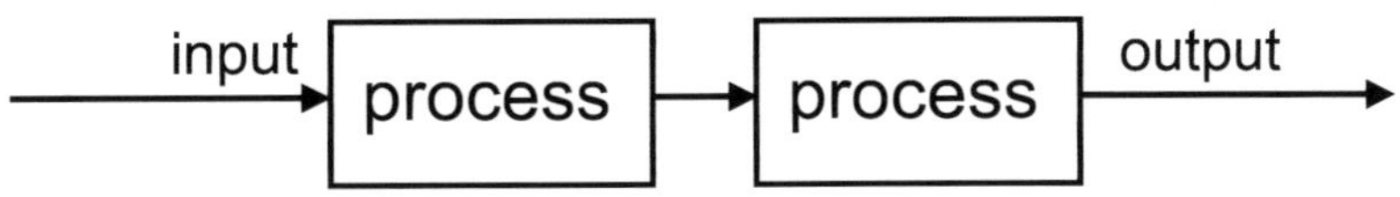

Faculty of ENGINEERING
University of Alberta

Reality of System Reliability Calculation

Our systems of interest may comprise hundreds (or thousands) of elements, and none of them may be identical

There may be underlying causes of failure (unreliability) that affect multiple elements, and so they are not really independent

Probabilities may not be a single number, but rather a distribution function describing a random process.

These functions may be very difficult to describe in analytical functions, making Jacobian transformation of their distribution functions to find the overall reliability extremely complicated.

Faculty of ENGINEERING
University of Alberta

Risk Associated with Event or Decision

And we also have to consider the consequence associated with a particular factor can be expressed as a cost (or a benefit, if the outcome is positive)

If the probability of occurrence is also known (analytically, or from data, or estimation), then we can determine the ***Risk***

Risk = Probability x Consequence

For a ***set*** of n independent probabilities and associated costs (or benefits), the expected outcome can be estimated as

$$Outcome = \sum_{i=1}^{n} P(i) \times C(i)$$

Faculty of ENGINEERING
University of Alberta

Expected Outcome for a Set of Probabilities

A utility is planning to install a new condition monitoring system for its plant. There are some uncertainties about each technology choice, which will affect the useful life of each. The probabilities are as follows:

	Probabilities	
Useful Life	Option A	Option B
10	0.1	0.05
20	0.5	0.25
30	0.3	0.5
40	0.1	0.2

Option A will cost \$1.5 M for the capital installation. Option B will cost \$3.0 M. Neglecting operating costs, which has the more attractive cost in \$/yr?

Faculty of ENGINEERING
University of Alberta

Expected Outcome for a Set of Probabilities

Useful Life	Probabilities Option A	Option B
10	0.1	0.05
20	0.5	0.25
30	0.3	0.5
40	0.1	0.2

Expected life A = .10(10) + .50(20) + .30 (30) + .10(40)
= 24 years

Expected life B = .05(10) + .25(20) + .50 (30) + .20(40)
= 28.5 years

Cost per year:

Expected cost A = 1,500,000/24 = $62,500/yr

Expected cost B = 3,000,000/28.5 = $105,300/yr

Faculty of ENGINEERING
University of Alberta

Uncertainty & Risk Example

- Consider that you can make an investment in a condition monitoring system that you expect will cost $1 million and provide $200k in benefits annually for 10 years, which will give an internal rate of return of 15.1%
  - If the minimum acceptable return is less than 15.1%, you will want to invest in that project
- Now consider that there is some degree of uncertainty:
  - Optimistically, the project could cost only $900k and provide $220k in annual benefits, with a probability of 30% ("P 30")
  - The realistic case is estimated to have a probability is 60% ("P 90")
  - Pessimistically, the project could cost as $1.1 million and provide only $180k in annual benefits with a probability of 10% ("P 99")
- Now will you invest in the project?

Faculty of ENGINEERING
University of Alberta

Uncertainty and Risk Example

We weight the uncertainties by their respective probability to determine the effect on the decision factor:

	Probabilities	Expected outcome
IRR most likely	**0.6**	**15.1 %**
IRR optimistic	**0.3**	**18.5 %**
IRR pessimistic	**0.1**	**10.1 %**
Expected IRR is:		

Faculty of ENGINEERING
University of Alberta

Uncertainty and Risk Example

We weight the uncertainties by their respective probability to determine the effect on the decision factor:

	Probabilities	Expected outcome
IRR most likely	**0.6**	**15.1 %**
IRR optimistic	**0.3**	**18.5 %**
IRR pessimistic	**0.1**	**10.1 %**
Expected IRR is:		**15.6 %**

(weighted by probabilities)

This is *highly* simplified (& linearized); but gives a pretty good estimate.

You would still go ahead with the project, if MARR < 15.6%

Faculty of ENGINEERING
University of Alberta

Dealing with Risk

There are three basic ways to deal with investment risks:

Contingency:

Include an allowance to cover for what you don't know. This is standard for engineering and construction contracts.

Return:

The higher the perceived risk, the higher the hurdle rate. For example, what percent of output is pre-sold? If low, require a higher return. This means if the investment is risky, the reward must be much higher.

Mitigation:

Identify the risk you can't accept, and develop a plan to buy your way out of that risk, by paying someone else to assume the liability

Faculty of ENGINEERING
University of Alberta

Some Pitfalls in Investment Analysis

Can not handle "base business" capital replacement analysis well

- What does one take as the incremental income?

The math is more sophisticated than the forecasts, which often be influenced by office politics and/or a judgment call from a senior person

The accuracy of the forecast is frequently much less than the precision of the analytical tools employed

- A deceptive confidence can ensue
- ROI analysis is best for relative investment tests with fixed outputs

Faculty of ENGINEERING
University of Alberta

Why Decisions Are Difficult

Many factors affect decisions

- Technical Issues
- Financial Issues
- Management Issues

People generally have a poor perception of risk

Factors should be weighted to represent their relative importance

- Not every factor is equally important
- Appropriate relative weighting is difficult to do

Faculty of ENGINEERING
University of Alberta

Decision Analysis Methods

Structured methods to determine actions to manage risk

Not necessarily minimize risk, just manage it!!

Depends on good corporate information on costs, probabilities, and dependencies

- Sometimes quantified
- More often qualitative

An analytical hierarchy process allows the different aspects of a decision to be ranked by their importance

Weighted decision making to understand risk and how different decision options can reduce risk to an acceptable level

A facilitation process is used to ensure all scenarios are explored and considered consistently

- Kepner-Tregoe, others

Faculty of ENGINEERING
University of Alberta

Basic Decision Analysis Process

- Gather experts and decision-makers
- Describe situation to the group
- Identify issues relevant to the situation
- Identify alternatives
  - Usually established in advance (to have good supporting info available)
  - Sometimes brainstorming is done during the process (free development of alternatives) followed by reality check and selection of a small set of realistic alternatives for analysis & decision
- Select important factors and relative weighting factors
- For each option or event, estimate value for each factor in terms of costs and costs & benefits

Faculty of ENGINEERING
University of Alberta

Basic Decision Analysis Process

- Gather experts and decision-makers
- Describe situation to the group
- Identify issues relevant to the situation
- Identify alternatives
  - Usually established in ... available)
  - Sometimes ... ess (free development of ... k and selection of a small set of rea... lysis & decision
- Select ... factors and relative weighting factors
- For each option or event, estimate value for each factor in terms of costs and costs & benefits

Faculty of ENGINEERING
University of Alberta

Usually use gut feel for loss functions - which is the hard part.
When time to decision is short, or in the early concept stage,
brainstorming is part of the process to develop options

Risk Assessment

Risk assessment is a form of decision analysis

When doing risk assessment, define nature of ***uncontrolled*** risks in terms of standard factors for potential cost multiplied by estimated probability (using techniques such as HAZOP, FMEA, etc.)

Brainstorm potential alternatives (controls)

Select appropriate controls and estimate costs & benefits

Estimate residual risk once controls in place (cost – benefit)

Identify people to take actions to mitigate risks

Faculty of ENGINEERING
University of Alberta

Estimating Probability of an Event

The general answer is “it depends”

Faculty of ENGINEERING
University of Alberta

Estimating Probability of an Event

The general answer is "it depends"

An expert who is familiar with the failure history of components and systems in the operating situation can usually give a good guess estimate of the failure rate. This is an unconditional probability for all conditions, assuming no changes in the plant or how it is run.

Super experts can differentiate amongst failure rates due to severe operating conditions

Faculty of ENGINEERING
University of Alberta

Estimating Probability of an Event

The general answer is "it depends"

An expert who is familiar with the failure history of components and systems in the operating situation can usually give a good guess estimate of the failure rate. This is an unconditional probability for all conditions, assuming no changes in the plant or how it is run.

Super experts can differentiate amongst failure rates due to severe operating conditions

Most of us are not super experts

Faculty of ENGINEERING
University of Alberta

Estimating Costs of an Event

Costs have to be comprehensive and represent mutually exclusive options

(meaning that each situation stands on its own)

Costs of failure may include consequential damage, extra labour & material costs, opportunity cost of lost production (but only if production is bottleneck) and penalties

Faculty of ENGINEERING
University of Alberta

The Importance of a Good Facilitator

Potential pitfalls in this process:

- Dopey estimates of probabilities
  - People generally have poor perception of risk
- Overly optimistic estimates on costs for risk controls
  - People generally don't do good cost estimates for complex actions
  - People sometimes confuse conditional probabilities and unconditional probabilities
- Jumping to an answer
  - People do not usually take time to consider enough options
- Identifying an action that doesn't really control the risk
  - Lack of understanding of dependencies

Faculty of ENGINEERING
University of Alberta

Risk Assessment

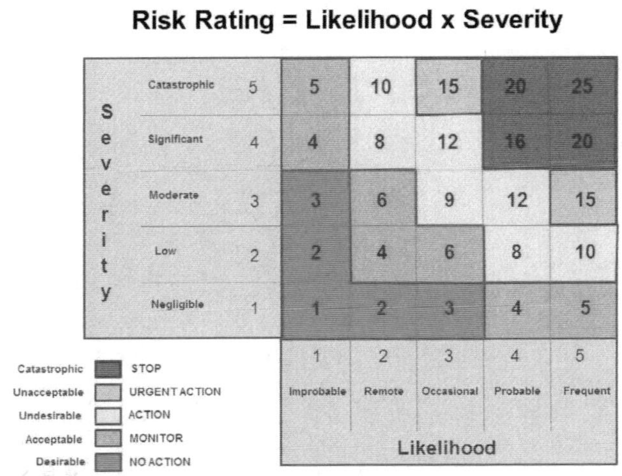

High uncontrolled risks require controls to bring the risk to an acceptable level

Common mistake: using overall unconditional probability

Faculty of ENGINEERING
University of Alberta

Conditional Risk

Faculty of ENGINEERING
University of Alberta

Conditional Risk

Faculty of ENGINEERING
University of Alberta

Conditional Risk

Risk is ***always*** conditional,
because the probability is affected by the context

The set of conditional risks for all the possible system states adds up to the overall conditional risk.

Faculty of ENGINEERING
University of Alberta

Conditional Risk & Minimum Risk

Each conditional risk is defined as:

$$R(\alpha_i|\mathbf{x}) = \sum_{j=1}^{c} \lambda(\alpha_i|w_j)\, P(w_j|\mathbf{x})$$

for $i = 1,\ldots, a$, where

a is number of possible actions α

c is number of states of the system

$\lambda(\alpha_i|w_j)$ is the loss function for taking an action α_i when the state of the system is w_j

$P(w_j|\mathbf{x})$ is the posterior probability that the state w_j given that the feature vector $\mathbf{x}$ has been measured

Select the action for which R is minimum (Bayes Risk):

$$R^* = \min \{ R(\alpha_i|\mathbf{x}),\ i = 1,\ldots, a \}.$$

Faculty of ENGINEERING
University of Alberta

Conditional Risk Is Hard to Calculate

<discuss>

Faculty of ENGINEERING
University of Alberta

Quantitative Risk Modeling

Sensitivity analysis is often used to determine amount contingency, or to decide whether to make an investment

- Uncertainty can be expressed as a range or a probability
- Scenarios (most likely / best /worst case) give a range, a coarse indication of relationship between an parameter and its effect (Tornado Diagram)
- Conditional probabilities are needed
- Actuarial data is not always available (maintenance histories, etc)
- But uncertainties in inputs can be used to estimate the uncertainties in what the system will do
- More precise scenario analysis can be done using simulations that test the probabilities
- This allows the effect of a change (such as implementing a condition monitoring system, or a new maintenance technique) to be assessed

Faculty of ENGINEERING
University of Alberta

Discrete-Event Systems

We usually think about continuous technological systems, in which variables are functions of continuous time and have a continuous range of values (also called analog systems), related by constitutive relationships based on physics

The state trajectory is the solution of a set of differential equations.

In many engineering systems, however, the behaviour can be described without having to model the underlying physics. This is done by defining a finite set of discrete states and the conditions by which they change.

- Activities in information technologies (digital computers, networks,…)
- Activities in projects, manufacturing, and services

Changes in state occur discontinuously

- In time & in the number of allowable states (finite states)

Faculty of ENGINEERING
University of Alberta

Discrete-Event System

A discrete-event system has only certain possible reactions to inputs, based on a set of prescribed conditions

The system may be a single element, or an entire supply chain

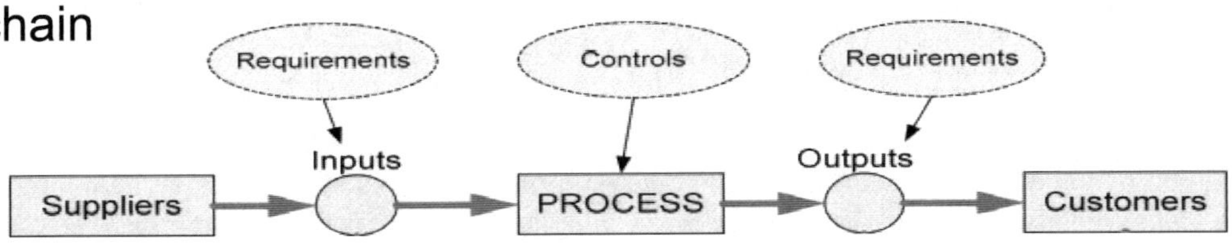

This diagram illustrates a SIPOC system representation, used in business operations modeling

Faculty of ENGINEERING
University of Alberta

Discrete-System Element

An element in a discrete-event system has only certain possible reactions to inputs (based on rules or conditions) and available resources

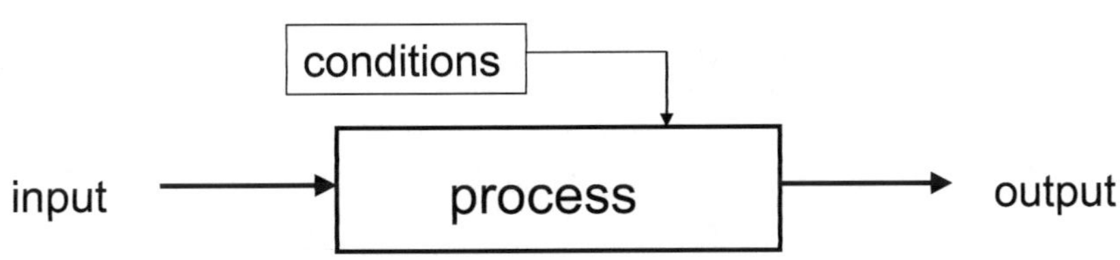

Faculty of ENGINEERING
University of Alberta

Discrete-System Element

As well, the element has only a finite number of possible inputs

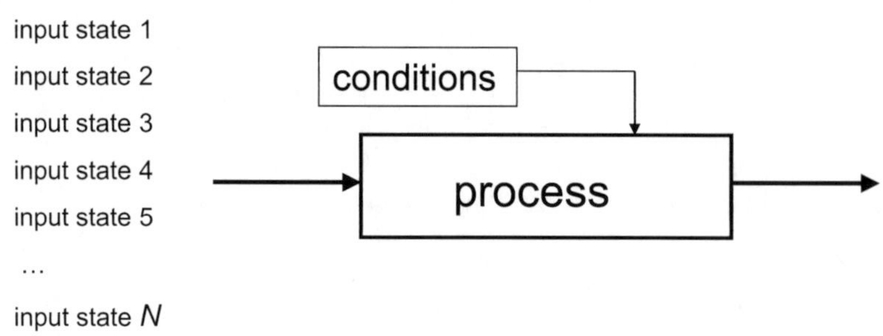

Faculty of ENGINEERING
University of Alberta

Discrete-System Element

As well, the element has only a finite number of possible inputs, & so it only has a finite number of possible outputs

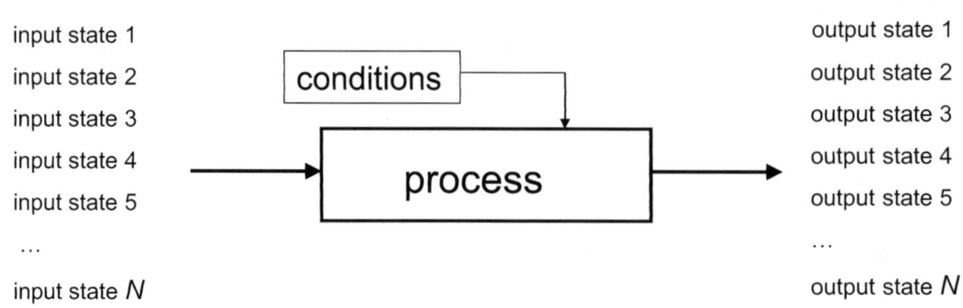

Faculty of ENGINEERING
University of Alberta

General Modeling of Discrete-Event Systems

Inputs - finite number of defined states

Processing - wait or "hold" until conditions are satisfied

Outputs (when conditions met) – finite number of states

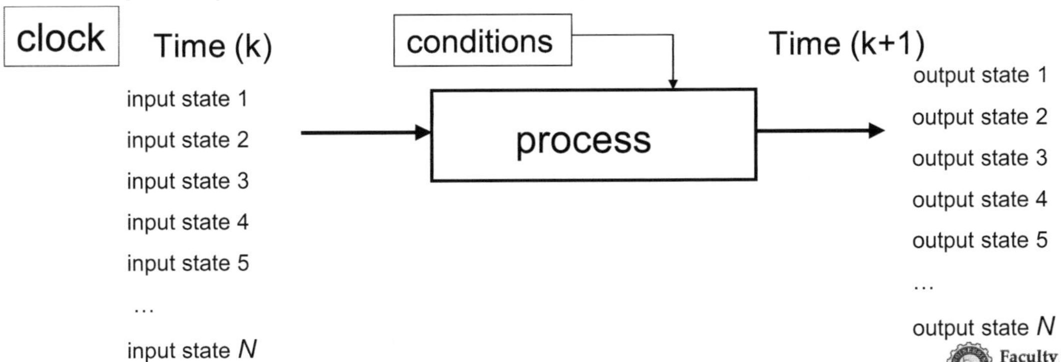

Faculty of ENGINEERING
University of Alberta

General Structured Analysis

Before constructing a computer simulation, a process has to be analyzed, to understand what all of the events are, how they work, and how they connect

Similar to setting up a set of governing equations for a physical system

Approaches include: event-based, activity-based, process-based, and "three-phase"

Essentially structured methods for creating flowcharts for stepwise processes (such as computer code execution)

- including decision choices, iteration, and possibly concurrency

Methods include

- Activity diagram (Universal Markup Language)
- Business Process Modeling Notation (BPMN)
- Event-driven Process Chain (EPC)
- Petri Net (Software packages allow this to be done graphically)

Faculty of ENGINEERING
University of Alberta

Structured Analysis of a System

Each step (activity) in the system is depicted in block diagram to show input, output, controls (such as decision criteria) and resources

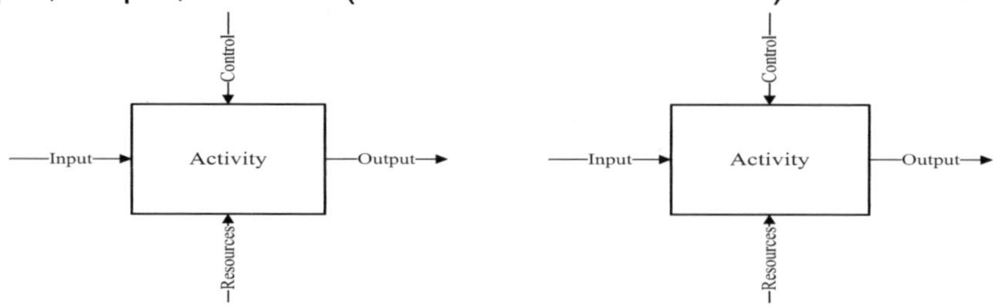

The blocks are then arranged in the order in which the activities occur to create a flowchart of input-output relationships

Also called an event tree

Faculty of ENGINEERING
University of Alberta

Monte Carlo Simulation

Monte Carlo simulation is a probabilistic analysis of a system

- Inputs are probabilities (distributions or functions)
- Process is the system dynamics
  (or the business process, or the investment analysis procedure, etc.)
- Output is the system response (or analysis result),
  also expressed probabilistically
- Conditional probabilities multiplied by their consequences (costs) gives the overall risk associated with the system

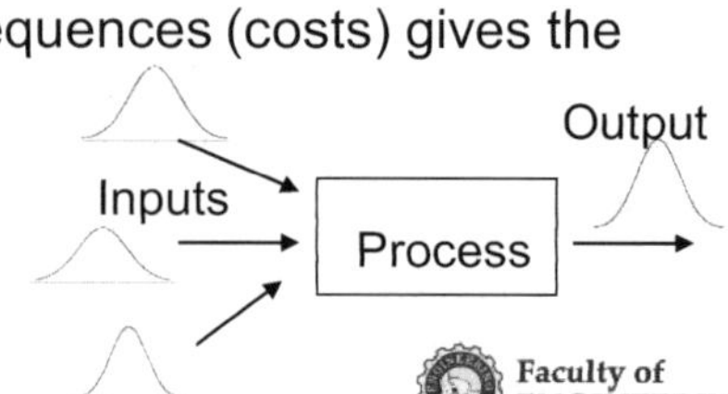

Faculty of ENGINEERING
University of Alberta

Monte Carlo Method for Forecasting

Forecasting is prediction with uncertainty

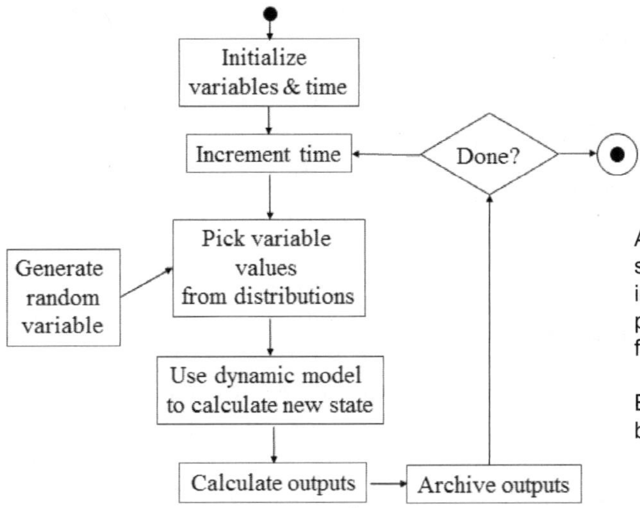

A time-driven Monte Carlo simulation steps through increments of time, just like a propagation simulations for solving differential equations.

Event-driven simulation is similar, but with variable time steps

Faculty of ENGINEERING
University of Alberta

Some Simulation Packages

Matlab SimEvents

AweSIM

Ideas/Extend

Weibull ++

<not at all a complete list>

Even Excel can be used as a simulation tool

Faculty of ENGINEERING
University of Alberta

Steps for System Analysis

- Identification of activities
- Interconnection of activities with each other
- Identification of hazards
- Determination of consequences
- Input probabilities
- Calculate output probabilities for different events
- Caclulate risk for each scenario
- For long time-scale scenarios, calculate discounted cash flow

Faculty of ENGINEERING
University of Alberta

Steps in Developing & Deploying a UAV-Based Inspection System

Evaluating whether using a UAV for inspection adds risk to the overall pipeline project

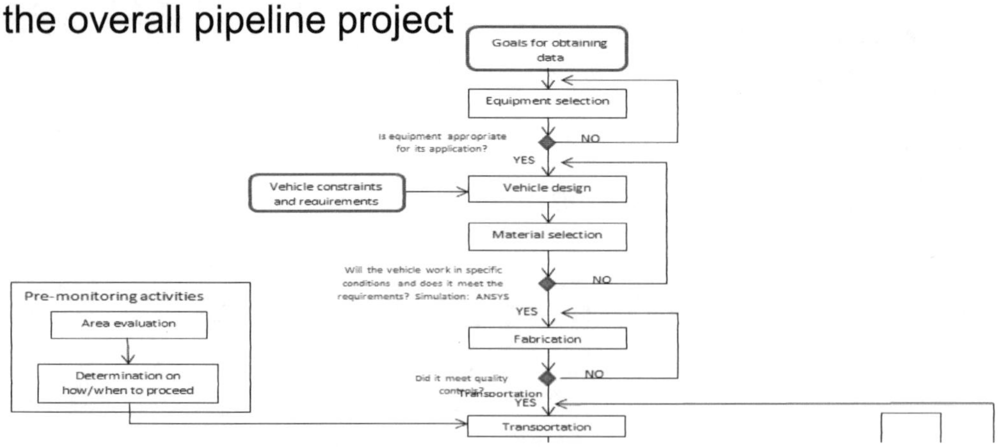

Deployment

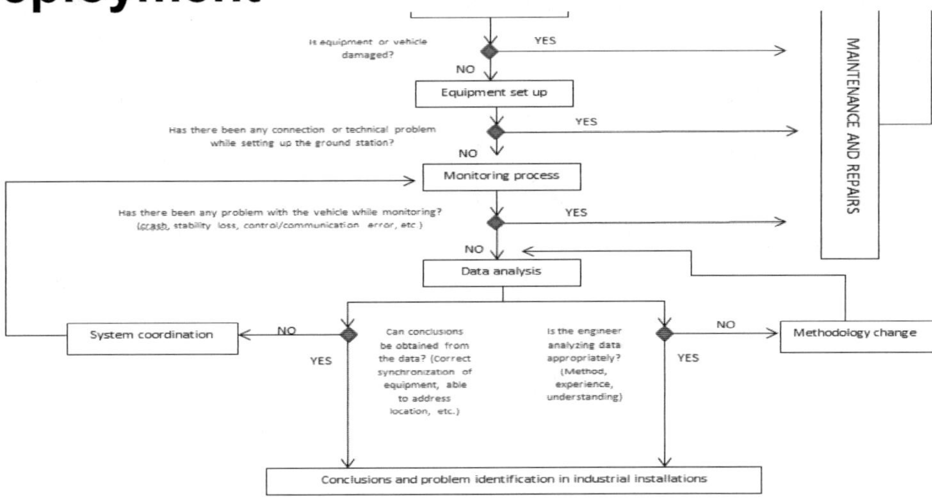

Faculty of ENGINEERING
University of Alberta

Summary of Activities and Sub-Activities

<u>1. Equipment selection</u>
1. Decision
   - 1.1 Ground Station
   - 1.2 Vehicle
   - 1.3 Improvements

<u>2. Vehicle design</u>
1. Aerodynamics
2. Equipment-vehicle compatibility
3. Improvements

<u>3. Material selection</u>
1. Decision
2. Improvement

<u>4. Fabrication</u>
1. Components
2. Assembly

<u>5. Transportation</u>
1. Move

<u>6. Equipment set up</u>
1. Ground Station installation
2. Connections vehicle-Ground Station

Faculty of ENGINEERING
University of Alberta

Summary of Activities and Sub-Activities

7. Pre-monitoring

1. Calibration
2. Process evaluation/Operability (Can it be carried out?)
3. Location evaluation (obstacles)

8. Monitoring process

1. Technical problem
   - 1.1 Control system
   - 1.2Communication system
2. Vehicle
   - 2.1 Functionality
   - 2.2 Preparation activities (i.e. clear memory card, charge, etc.)
   - 2.3 Bearing conditions (i.e. weather)
3. Equipment
   - 3.1 Sensors
   - 3.2 Spectrometer
   - 3.3 Camera
   - 3.4 GPS
   - 3.5 Machinery (i.e. for digging, taking samples etc.)
4. Weather conditions (stability loose)

9. Analysis

1. Computer
2. Data synchronization
3. Determination of facts
4. Methodology

Safety activities

Faculty of ENGINEERING
University of Alberta

Operational Phase and Consequences

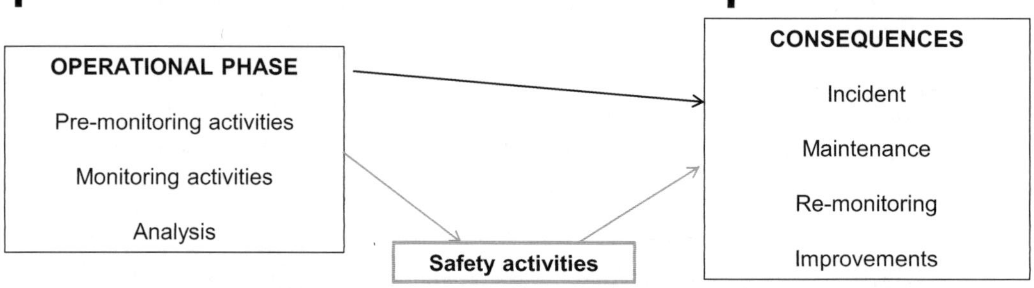

Incident	Re-monitoring	Maintenance	Improvement
Operability	Calibration	Station	
Control system	Preparation activity	Methodology	
Communication system	Equipment	Damages:	Functionality
Conditions	Synchronization	Vehicle	
Spot evaluation	Determination	Equipment	

Faculty of ENGINEERING
University of Alberta

Sub-Activities Connection Diagram

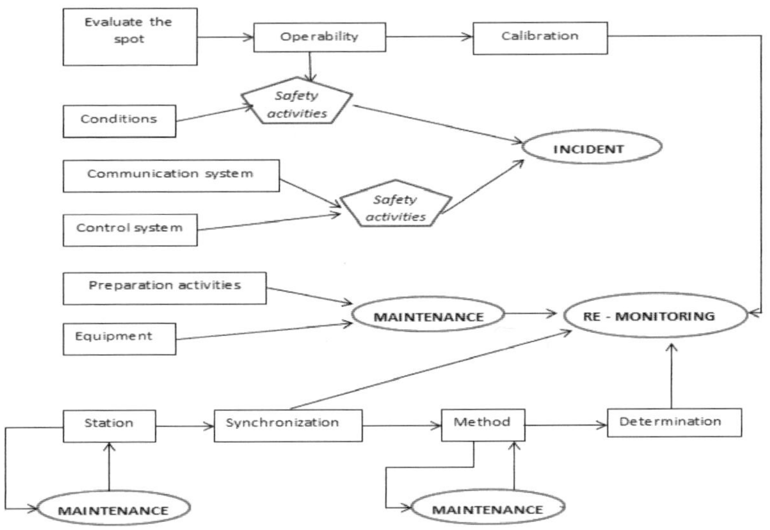

In the simplest case, each sub-activity has a single failure mode (& consequence)

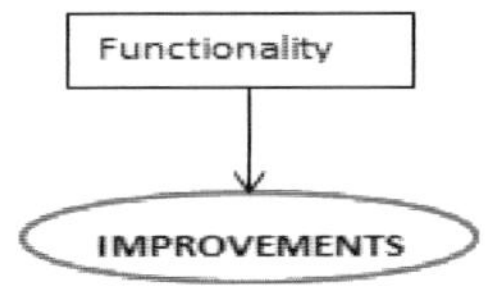

Faculty of ENGINEERING
University of Alberta

Calculations Using Probabilities in Excel

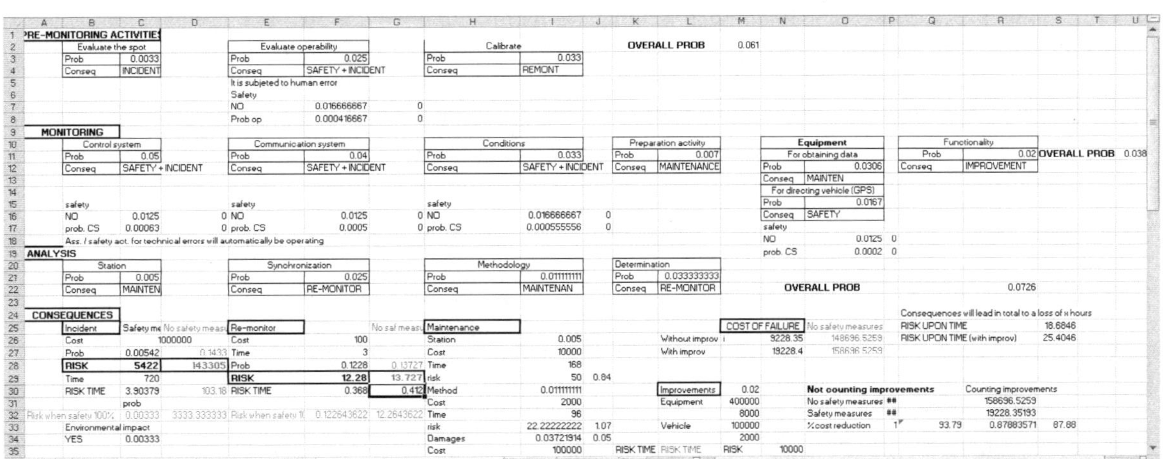

Faculty of ENGINEERING
University of Alberta

Assumptions for a UAV

Used at least once a week
Development phase and operational phase
Canadian installations
Continuous improvements
Safety activities
When unavailable, lost time has a cost, due to contracting replacement services (risk mitigation)

Faculty of ENGINEERING
University of Alberta

Sensitivity Analysis

Safety measures reduce risk

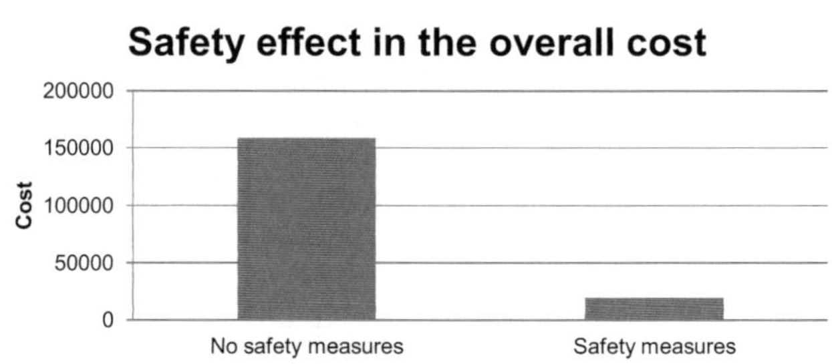

Overall cost of failure in the system	
No safety measure ($)	159 k
Safety measures exist ($)	19.2 k
Δ (No safety-safety) ($)	139 k
% cost reduction	88 %

Faculty of ENGINEERING
University of Alberta

Best Case & Worst Case Scenarios

Comparing no safety measures to full safety measures

Consequences		Probabilities	Risk ($)
Incident	No safety measures (0%)	14.3%	143 k
	Safety measures (100%)	0.33%	3.3 k
Re-monitoring	No safety measures (0%)	13.7%	13
	Safety measures (100%)	12.26%	12
Maintenance (Damages)	No safety measures (0%)	5.31%	5.3 k
	Safety measures (100%)	3.70%	3.7 k

Consequences	Δ ($)	Reduction (%)
Incident	140 k	98 %
Re-monitoring	2 k	11 %
Maintenance (Damages)	1. 7 k	31 %

Faculty of ENGINEERING
University of Alberta

Monetizing Time

Failure in the system = time lost

Translate time into cost

⬇

Cost of contracting a replacement, lost revenue, legal penalties,...

Faculty of ENGINEERING
University of Alberta

Time into Cost

CONSEQUENCES	Incident	Maintenance	Station	Methodology	Damages	Improvement	Vehicle	Equipment
Time for activity	1 month		7 days	4 days	14 days		14 days	14 days
Times of monitoring	5		2	1	3		3	3
Cost of contracting ($)	40.0 k							
Overall risk ($)	11.2 k							

Overall Cost

Overall cost	
Cost most likely ($)	19.2 k
Cost in time ($)	11.2 k
Total cost ($)	30.4 k

Faculty of ENGINEERING
University of Alberta

Base Case:
Assumptions for Piloted System

Same basic equipment as for unmanned systems
Need of a professional pilot
Safety measures applied by the pilot
Communication system failing would not result in an incident
More stable vehicle (because of pilot who can react better than an autonomous control system)
Higher operating costs
Transportation costs apply for maintenance or improvements

(Recall that we need mutually exclusive options)

Faculty of ENGINEERING
University of Alberta

Cost of Consequences (Piloted System)

Consequences	Cost ($)
Incident	95 k
Re-monitoring	39 k
Maintenance	9 k
Improvement	14 k
Overall cost	118 k

Cost in time

CONSEQUENCES	Incident	Maintenance	*Station*	*Methodology*	*Damages*	Improvement	*Vehicle*	*Equipment*
Time for activity	45 days		7 days	4 days	21 days		21 days	14 days
Times of monitoring	7		2	1	4		4	3
Cost of contracting ($)	40 k							
Overall risk ($)	22.5 k							

Faculty of ENGINEERING
University of Alberta

Expected Benefits of UAVs for Distributed Asset Inspection

Safety (no risk to pilot), provided that UAV does not increase risk to the public (through crashes, which could injure people and cause property damage)

Improved information capture and audit trail of inspection results

Increased flexibility (likely able to fly in more extreme conditions)

Reduced cost (likely, because of reduced expert labour and reduced cost of infrastructure; but requires analysis)

Reduced energy consumption and emissions (smaller aircraft)

Faculty of ENGINEERING
University of Alberta

Risk Comparison between Systems

	Manned system	Unmanned system
Cost ($)	118 k	20 k
Cost related to time ($)	22.5 k	11.2 k
Overall risk	140 k	30.4 k
Risk reduction ($)	110 k	
Times unmanned system cost	4.6	

(does not include non-monetary benefits of improved environmental performance)

Faculty of
ENGINEERING
University of Alberta

Using Simulations to Understand the Effect of a Decision

A good simulation of how technology, processes, and people interact can allow the effect of an event - or a change in the system - to be predicted.

Simulations allow for the effects of uncertainty on outcomes to be assessed.

Faculty of
ENGINEERING
University of Alberta

Example: Discounted Cash Flow

Costs and benefits of a system usually occur over time.

If the time is long, or the interest rate is high, then future values need to be corrected to their net present worth (NPV)

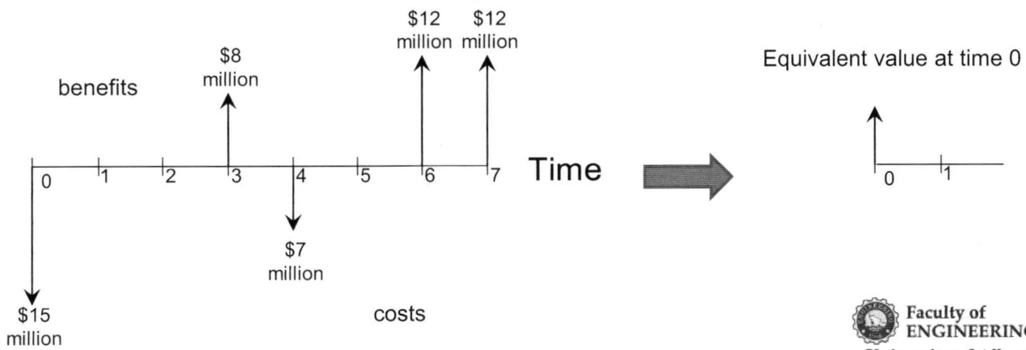

Faculty of ENGINEERING
University of Alberta

Probabilistic Discounted Cash Flow

When there is uncertainty, each revenue & cost may be a distribution (usually getting more uncertain with time)

Monte Carlo simulation can yield a good estimate of the probability of the NPV

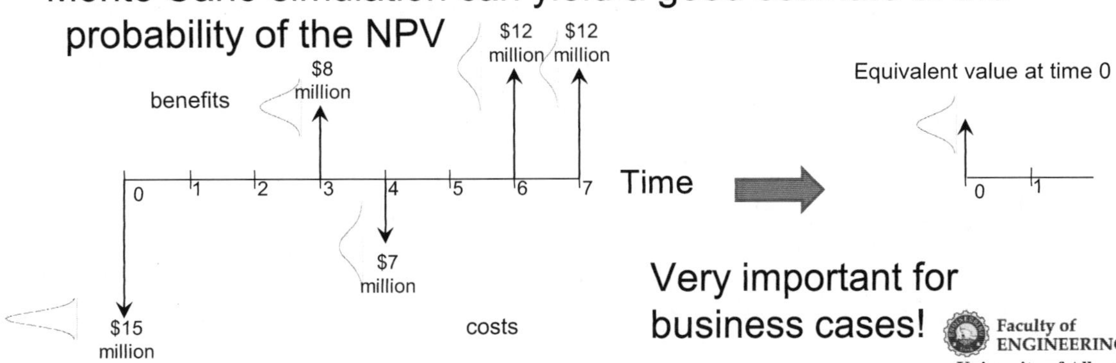

Very important for business cases!

Faculty of ENGINEERING
University of Alberta

Probabilistic Discounted Cash Flow

When there is uncertainty, each revenue and cost may be a distribution (usually getting more uncertain with time)

Monte Carlo simulation can yield a good estimate of the probability of the net present worth

Very important for business cases!

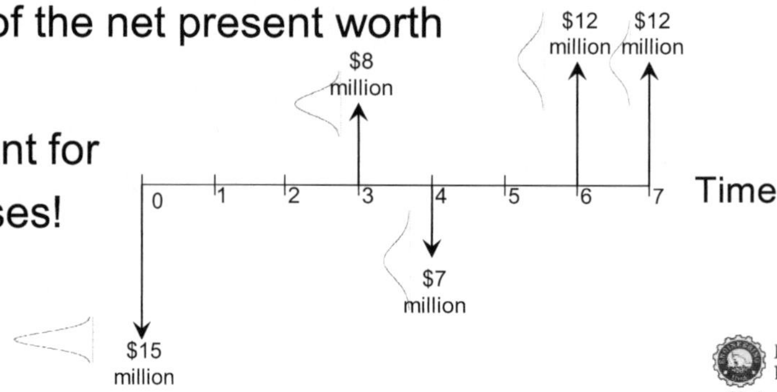

Faculty of ENGINEERING
University of Alberta

Problems with Obtaining Data

For new technologies, there will be no historical data

Few precedents for relevant data

Variety of equipment options (not necessarily completely defined at the basic design stage)

In that case, simplifications & approximations are necessary.

For the UAV system, a probability was assumed based on UAV experience in the lab, and estimated mean values were used.

Faculty of ENGINEERING
University of Alberta

Input Data Modeling (When Data Exist)

General method:

1. Sample independence
2. Analysis of variance
3. Type of distribution
   - Empirical vs. theoretical
   - Discrete vs. continuous
4. Parameter estimation
5. Distribution selection

$S \sim lognormal\ (\alpha = 0.873, \beta = 62.818)$

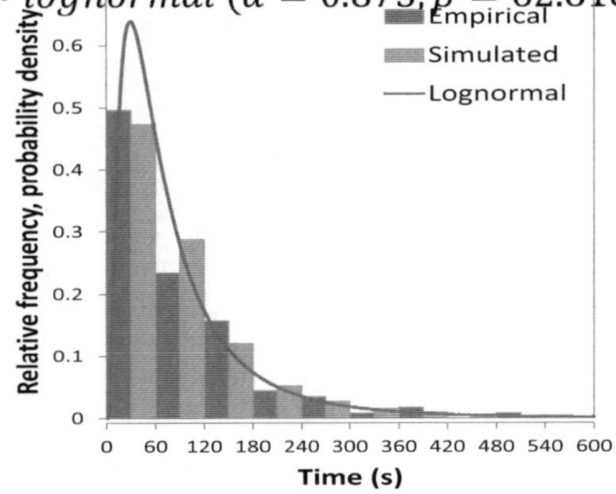

Faculty of ENGINEERING
University of Alberta

Distribution Modeling

A distribution is a convenient way to express the data using an analytical function with only a few parameters. To find an appropriate distribution:

1. Collect Data of Events
2. Select a Candidate Distribution (Normal, Weibull, Exponential, ...)
3. Estimate Parameters of Distribution (maximum likelihood method)
4. Conduct Goodness-of-fit Test (chi-squared test)
5. Choose distribution with best goodness-of-fit

Faculty of ENGINEERING
University of Alberta

Example: Haul Truck Tire Lifetimes

- ***Chosen distribution for the data set***

  3-parameter Weibull distribution, defined by:

$$f(t) = \frac{\beta}{\alpha}\left(\frac{t-\gamma}{\alpha}\right)^{\beta-1} \exp\left(-\left(\frac{t-\gamma}{\alpha}\right)^{\beta}\right)$$

  α is a scale parameter
  β is a scale parameter
  γ is a location parameter

- ***Estimating Parameters***

  Maximum Likelihood method results:

  $\alpha = 5399.0033; \quad \beta = 2.5219; \quad \gamma = 34.6234$

- ***Goodness-of-fit test***

  Chi-Squared test results showed that the chosen distribution fitted to the data set well

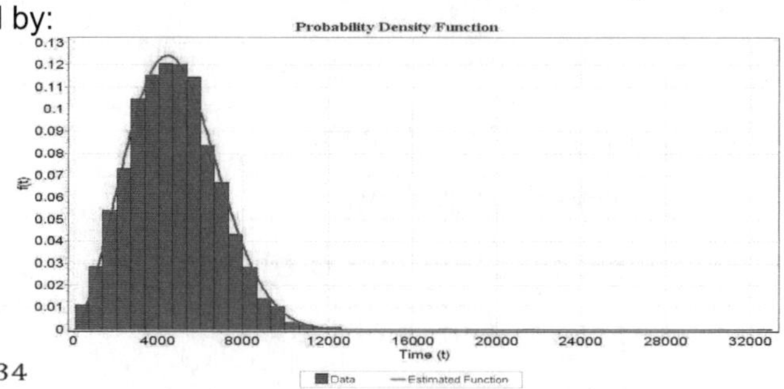

Fitted distribution to all failure data (3-parameter Weibull)

Faculty of ENGINEERING
University of Alberta

Modeling Maintenance Decisions

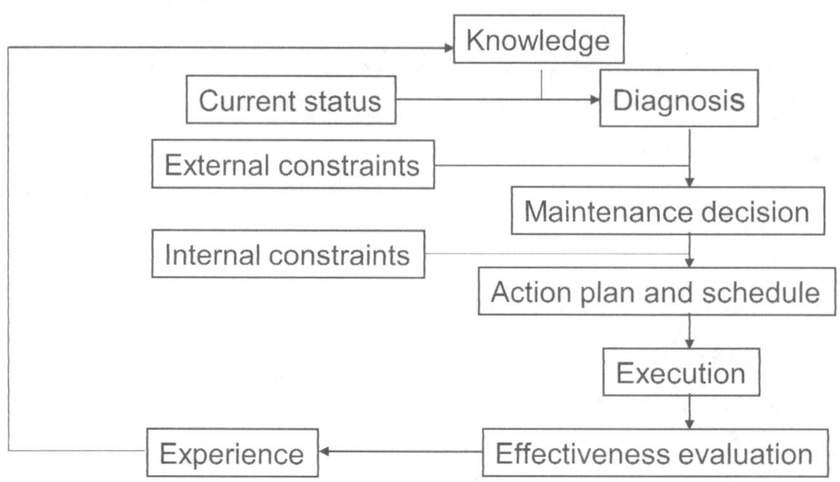

Faculty of ENGINEERING
University of Alberta

Risks When Doing Maintenance

Is there an impending failure?

How long until it is a problem?

What is the expected impact of taking action?

An information system allows for informed decision making based on the best available information about the situation

That is, knowing what the state of the system is, thus allowing for the appropriate conditional risk to be understood

Faculty of ENGINEERING
University of Alberta

Improving Decision-Making

Improved flow of information for decisions
- Removing bottlenecks
- Understanding uncertainty and managing risk
- Monitoring consistency of maintenance decisions and activities

Structured understanding of maintenance
- Model-based analysis for decision support
  - Anomaly Detection (Is something wrong)
  - Diagnostics (what is wrong)
  - Prognostics (how long will it remain in a functional state)
  - Planning and Scheduling (what needs to be done and when can it happen)
  - System Reliability (Interferences and bottlenecks in business processes related to system being in a reliable state, including external factors)
- Consistent application of best practices

Faculty of ENGINEERING
University of Alberta

Model Verification

Verification is a reality check: is the model appropriate?

Have someone else check the "narrative" of the model

Test the model under known conditions where you know what to expect from the system (e.g. benchmark cases)

Use appropriate inputs (including random inputs)

Look for known outputs

Validation is confirmation that you are simulating the system of interest correctly with a sufficiently accurate model and an appropriate solver (test model with *different* data)

Faculty of ENGINEERING
University of Alberta

Mine Mobile Equipment Example

We used simulations to understand how machine condition monitoring and maintenance management practices affect a surface mining operation

We verified the model by interviewing an expert (the maintenance manager, who was formerly a maintenance engineer, thus knowing technology, processes, and people issues)

We evaluated the quality of the simulation using transient event-risk trees and an analytical hierarchy process (AHP) for decision making

Faculty of ENGINEERING
University of Alberta

Application Case

Management and maintenance of hauler trucks for surface mining operations, e.g. oil-sand production in Alberta

Truck-shovel mining methods operate under unfavorable conditions

Faults and failures related to engine, suspension struts, and tires

Typical maintenance program: visual and manual inspection and regular planned maintenance processes

77

Faculty of ENGINEERING
University of Alberta

Application Case

Organization intends to increase production by reducing machine downtime, and decrease costs by improving procedures and reducing the instances of field events

To achieve this, the organization is considering a CMMIS

78

Faculty of ENGINEERING
University of Alberta

CMMIS

An information system focused on procedures and practice of handling the reliability and availability of equipment and systems

It provides advanced machine monitoring, management, and maintenance functionalities

79

Faculty of ENGINEERING
University of Alberta

Existing System Assessment

Machines operate based on schedule

Until planned service time or until they fail or cause incident

Faults may occur than can lead to events

Faults may be detected by operator, inspector, mechanic, or CMS (if available)

Inspector: determines machine condition

Operator and inspector can make decisions on the field

Extra costs for field failures and incidents

Technicians and tools limited

80

Faculty of ENGINEERING
University of Alberta

Existing System Assessment

Routine inspection and planned maintenance

Procedure relies on technician expertise

Basic documentation of procedure and job reports

Integrated CMMIS does not exist

Machines can be equipped with CMS

WWAN exists in the field

81

Faculty of ENGINEERING
University of Alberta

Hypothetical (Alternative) CMMIS

Offline CMMIS

- Onboard sensors and CMS
- Manually downloaded from data logger
- Data used in desktop RCM software

Real-time CMMIS

- Wireless sensors
- CMS "cloud" application

82

Faculty of ENGINEERING
University of Alberta

Alternative Systems

BDM system (base)

PM system (existing practice, no CMMIS)

Offline CMMIS

Real-time CMMIS

83

Metrics Selection

Financial:
- Production loss risk
- Direct cost of failure risk

Environmental:
- Production loss risk from the environmental incidents
- Direct cost risk associated with environmental incidents

Safety:
- Production loss risk from the safety incidents
- Direct cost risk associated with safety incidents

Operational excellence:
- Up-time (Availability)

Risk awareness:
- Operators' risk awareness of machine faults

Asset and maintenance management:
- Lean service

Information management and flow:
- Communication time

84

Case Model 1

10 machines

One failure mode: tire failure

One generic safety incident mode and one environmental

Faults may occur which result in failure after 48 h

No resource limitations and queuing

No operator or managerial decision making

Inspector decision making for diagnosis

Machines repaired to "like new" state

85

Faculty of ENGINEERING University of Alberta

Transient Event-Risk Tree Method

TERT development:

Define risk scenarios: "good", "bad", "ugly", "expected"

Calculate event probabilities: $F(t) = 1 - e^{-(\frac{t-\gamma}{\alpha})^{\beta}}$

Calculate scenario probabilities: $\mathcal{F}(x,t) = \binom{n}{x} F(t)^{x}(1 - F(t))^{(n-x)}$

Calculate severity for events: $S_k(x)$

Calculate risk at various times: $R_k(x,t) = \mathcal{F}_k(x,t) \times S_k(x)$

86

Faculty of ENGINEERING University of Alberta

Event Cumulative Distribution Functions

Tire Faults

Uniform:

Normal:

Weibull:

Safety incident

Uniform:

Environmental incident

Uniform:

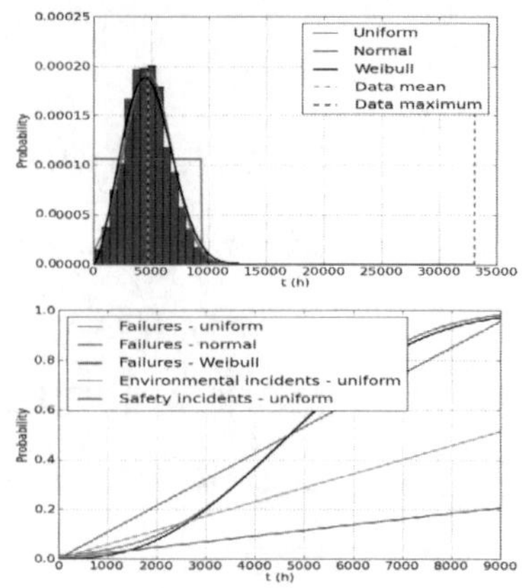

87

Faculty of ENGINEERING
University of Alberta

Base System

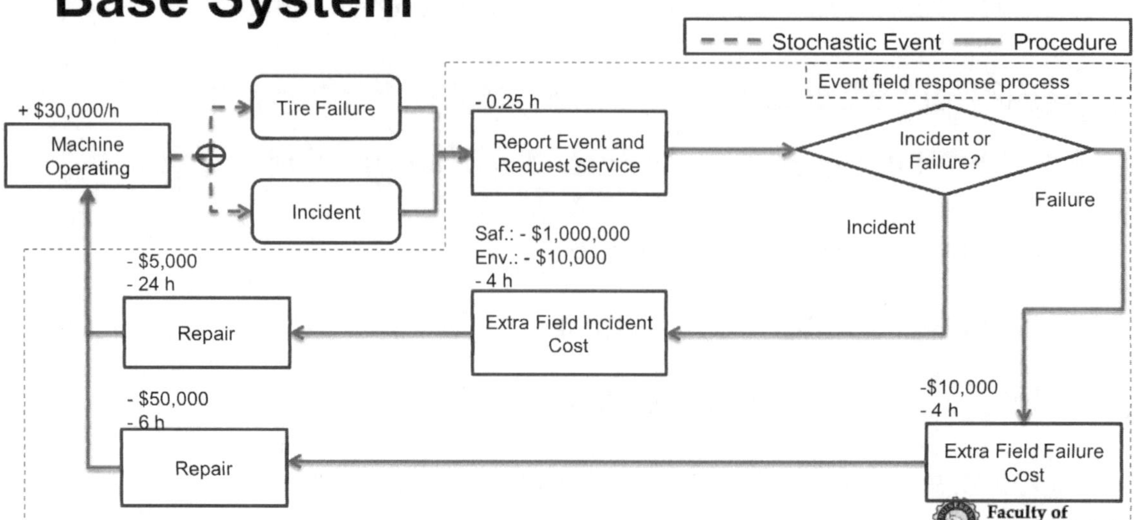

88

Faculty of ENGINEERING
University of Alberta

Base System Risk Plots

"Ugly" scenario: risk value if all of the machines fail prior to t

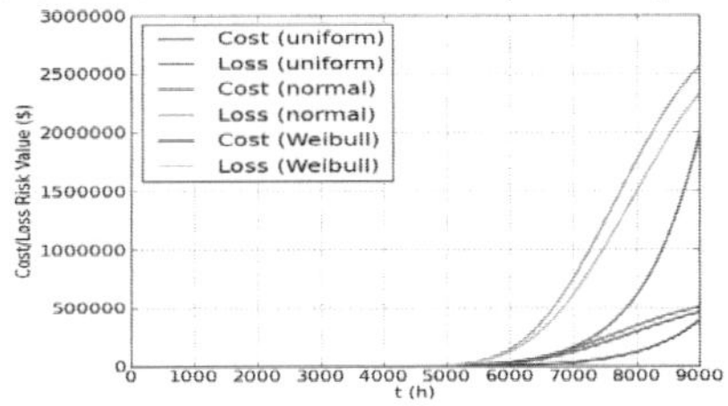

"Expected": total average risk values of machine failures prior to t

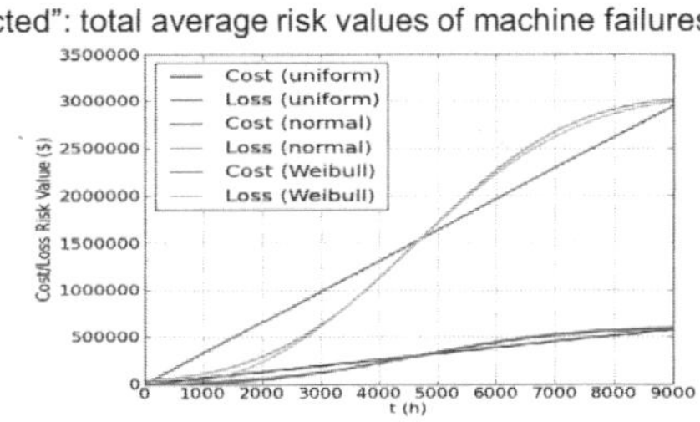

"Bad " scenario: risk value if a minimum of 5 machines fail prior to t

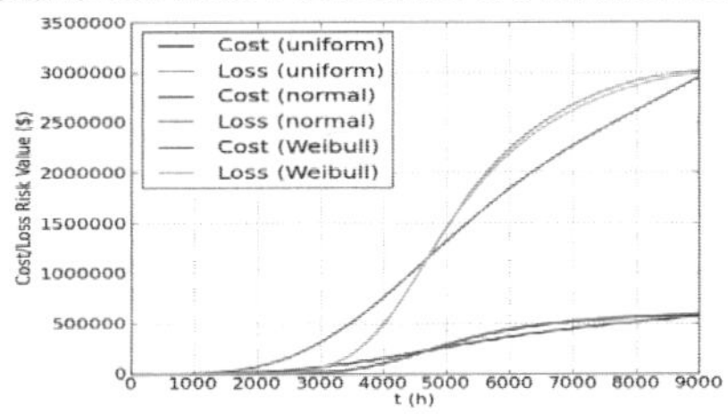

All scenarios: risk value of machine environmental incidents

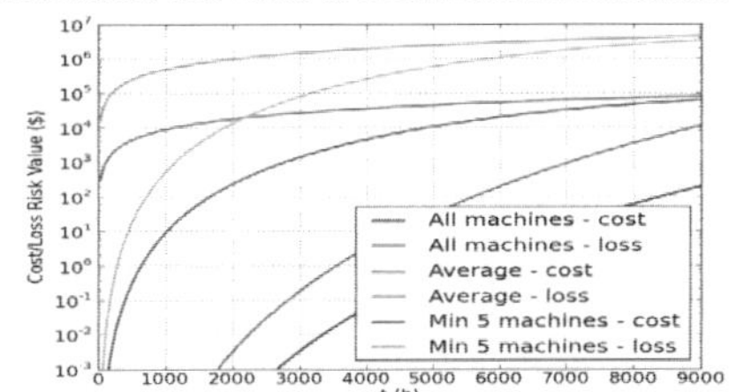

ıf
ERING
Alberta

Existing System

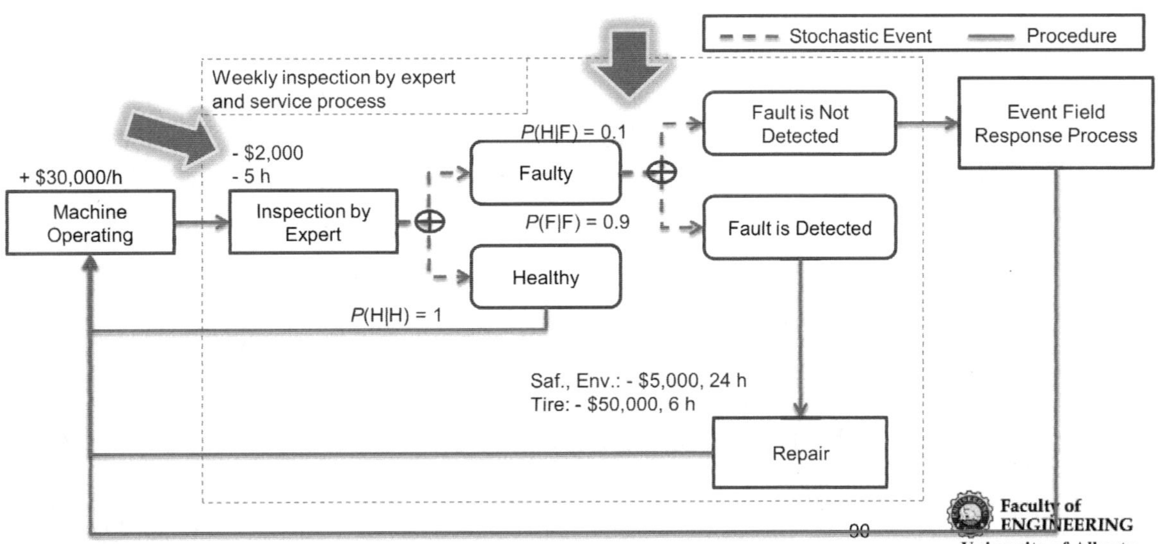

90

Faculty of ENGINEERING
University of Alberta

Existing System Risk Plots

"Ugly" scenario: risk value if all of the machines fail prior to t

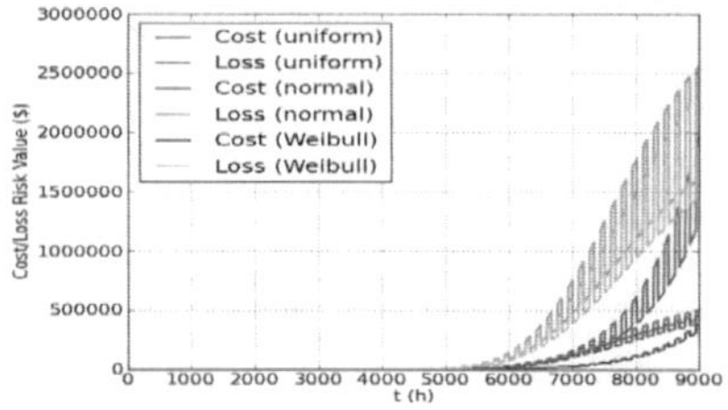

"Expected": total average risk values of machine failures prior to t

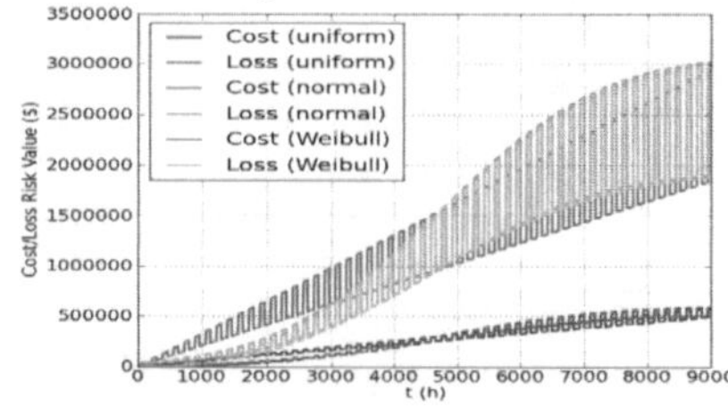

"Bad " scenario: risk value if a minimum of 5 machines fail prior to t

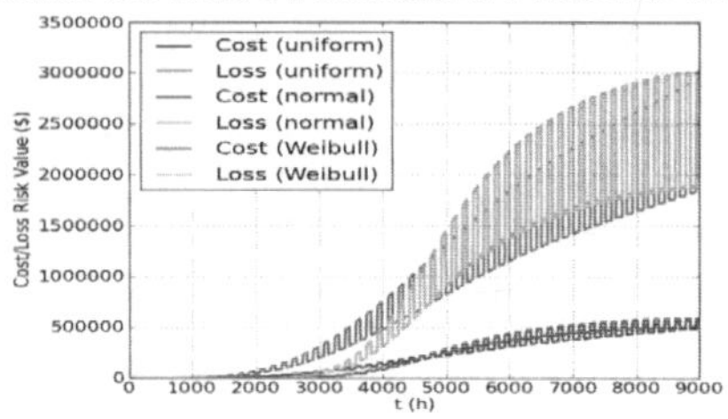

All scenarios: risk value of machine environmental incidents

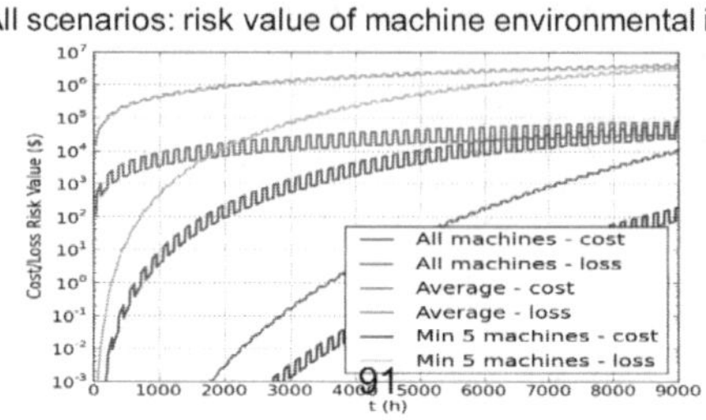

91

of
ERING
Alberta

Offline CMMIS

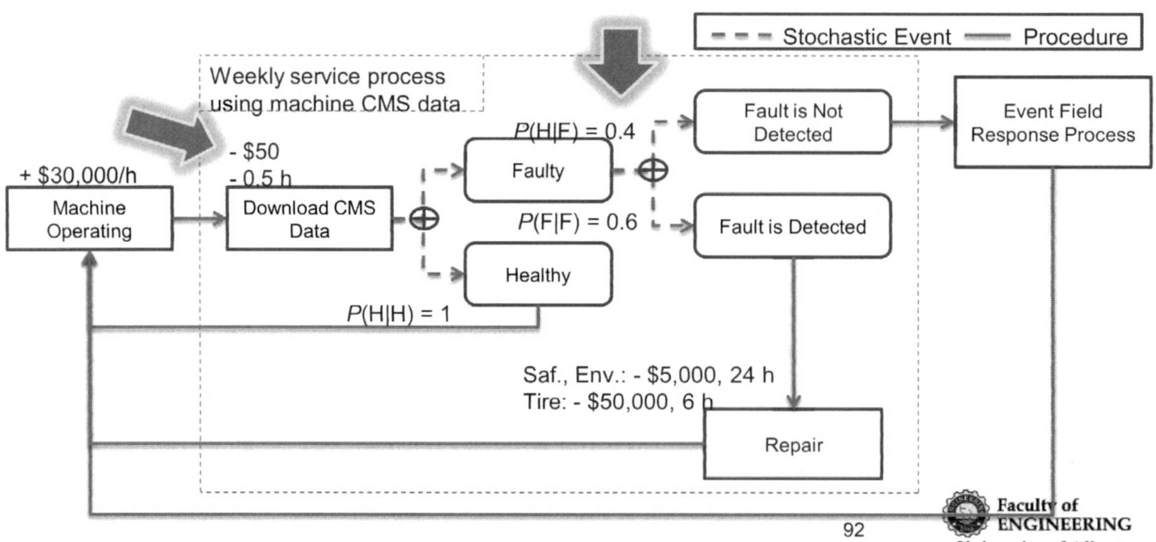

92

Faculty of
ENGINEERING
University of Alberta

Offline CMMIS Risk Plots

"Ugly" scenario: risk value if all of the machines fail prior to t

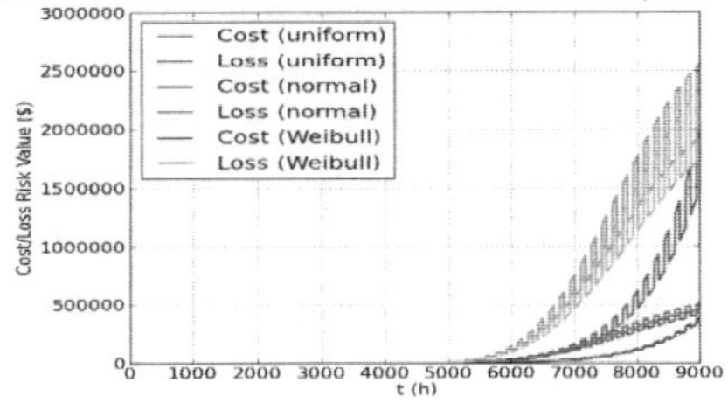

"Expected": total average risk values of machine failures prior to t

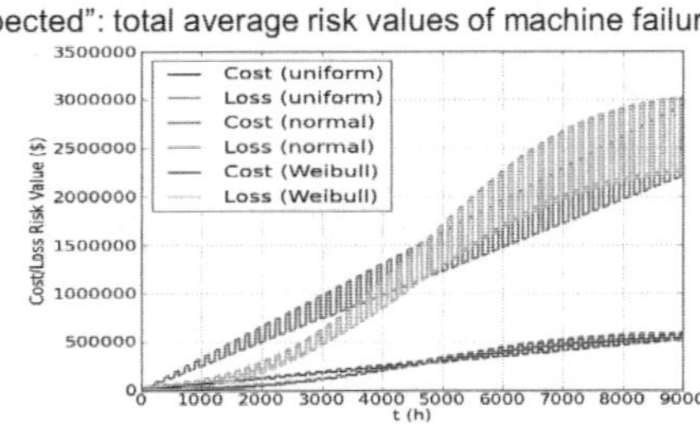

"Bad " scenario: risk value if a minimum of 5 machines fail prior to t

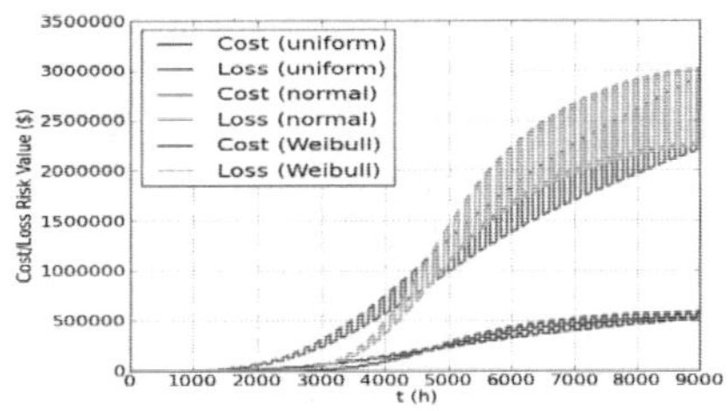

All scenarios: risk value of machine environmental incidents

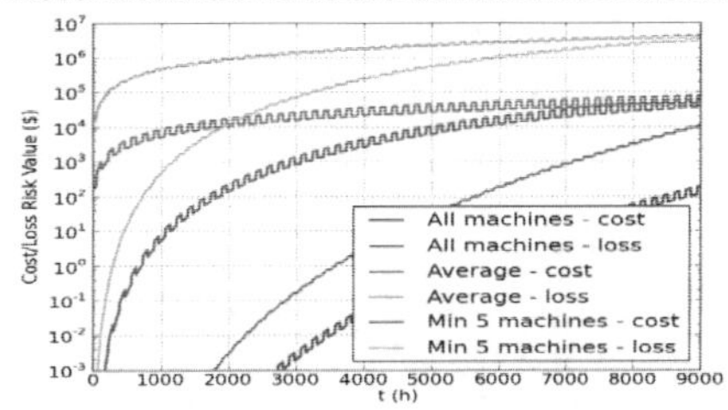

of
EERING
f Alberta

Real-time CMMIS

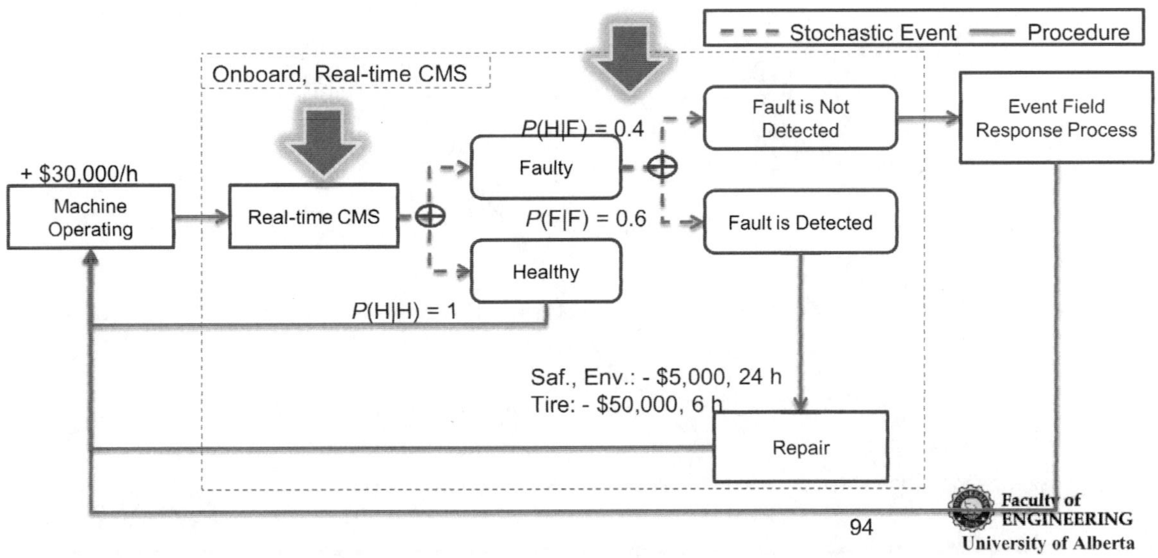

94

Faculty of
ENGINEERING
University of Alberta

Real-time CMMIS Risk Plots

"Ugly" scenario: risk value if all of the machines fail prior to t

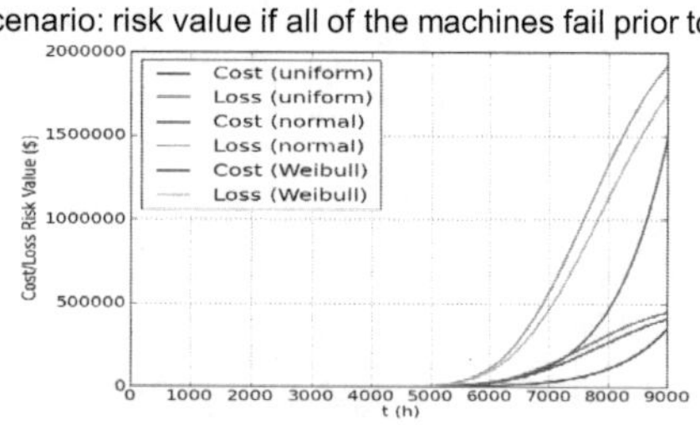

"Expected": total average risk values of machine failures prior to t

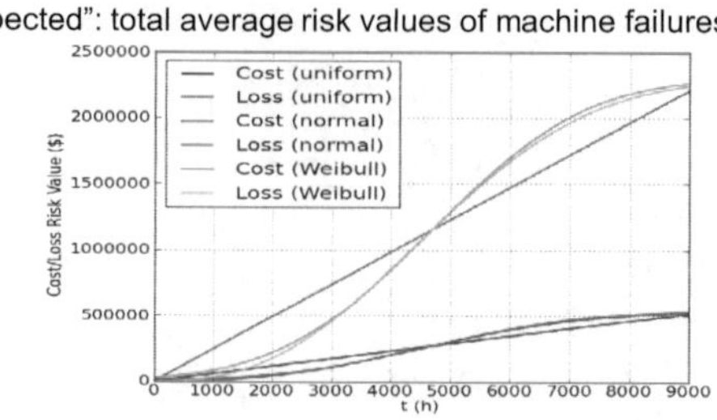

"Bad " scenario: risk value if a minimum of 5 machines fail prior to t

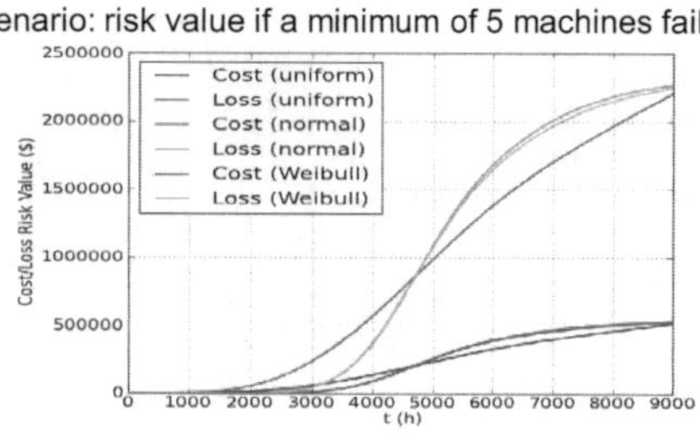

All scenarios: risk value of machine environmental incidents

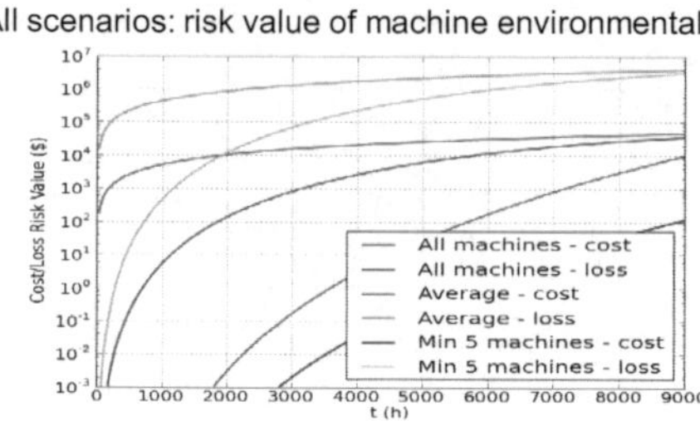

of
ERING
Alberta

System Comparison Risk Plots

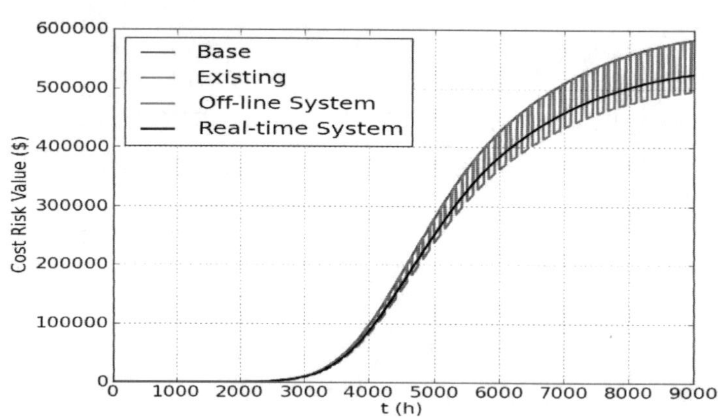

"Ugly" scenario: cost risk value if all of the machines fail prior to t (Weibull distribution)

96

Faculty of
ENGINEERING
University of Alberta

System Alternative Selection using the Analytic Hierarchy Process (AHP)

- Scale-based ranking method for multi-criteria decision making
- Can include many criteria, sub-criteria, and alternatives
- Both subjective and measured concepts can be used as long as they are quantitatively scale

Pair-wise comparison of criteria, sub-criteria and alternatives → Comparison Matrices → Eigenvector calculation → Ranking

97

Faculty of ENGINEERING
University of Alberta

Ranking of Performance Criteria & Metrics

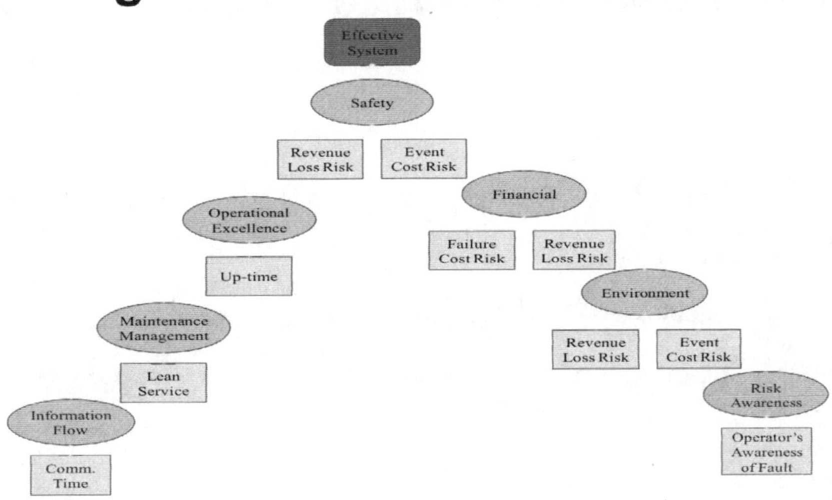

98

Faculty of ENGINEERING
University of Alberta

Pair-wise Ranking of Performance Criteria

	Safety	Financial	Operational Excellence	Environment	Maintenance Management	Risk Awareness	Information Flow
Safety	1	5	7	9	9	9	9
Financial	1/5	1	5	3	9	9	9
Operational Excellence	1/7	1/5	1	3	5	7	7
Environment	1/9	1/3	1/3	1	5	3	3
Maintenance Management	1/9	1/9	1/5	1/5	1	3	3
Risk Awareness	1/9	1/9	1/7	1/3	1/3	1	1
Information Flow	1/9	1/9	1/7	1/3	1/3	1	1

Safety	0.489
Financial	0.237
Operational Excellence	0.120
Environment	0.071
Maintenance Management	0.037
Risk Awareness	0.025
Information Flow	0.021

1: equal importance
3: slightly better
5: strongly better
7: very strongly better
9: extremely better

99

Faculty of ENGINEERING
University of Alberta

Total Ranking of Alternatives

Scenarios	Safety (0.489)		Financial (0.237)		Operational Excellence (0.120)	Environment (0.071)		Maintenance Management (0.037)	Risk Awareness (0.025)	Information Flow (0.021)	Total Rank
	Revenue Loss (0.167)	Event Cost (0.833)	Revenue Loss (0.875)	Event Cost (0.125)	Up-time (1.000)	Revenue Loss (0.833)	Event Cost (0.167)	Lean Service (1.000)	Operator Awareness (1.000)	Comm. Time (1.000)	
Base	0.18	0.14	0.14	0.16	0.12	0.18	0.15	0.03	0.08	0.05	0.14
Existing	0.28	0.29	0.25	0.29	0.04	0.28	0.29	0.08	0.08	0.05	0.23
Off-line	0.24	0.22	0.23	0.23	0.22	0.24	0.22	0.27	0.08	0.19	0.22
Real-time	0.3	0.35	0.38	0.32	0.62	0.30	0.34	0.62	0.76	0.71	0.41

100

Faculty of ENGINEERING
University of Alberta

Final System Ranking

	Value Ratio	Total Cost (TCO $k)	Opportunity Loss (Production Loss $M)
Base	1.00	0	0
Existing	1.42	1,040	78
Offline CMMIS	1.38	$26 + SC_1$	8
Real-time CMMIS	2.33	SC_2	0

Profitability index = Value created / TCO

Similar results uniform or normal distributions for tire failures

101

Faculty of ENGINEERING
University of Alberta

Summary

Risk always depends on the conditions of the system

Consequences can be in a different part of the system

Qualitative risk models are usually good enough

Quantitative models require time and care, but are less prone to judgment errors

Discrete-event simulations show how uncertainty affects outcomes

Risk models can be used for decision support, capture institutional knowledge, and support best practices in reliability and maintenance.

Faculty of ENGINEERING
University of Alberta

CHARACTERIZATION OF TAPE ADHESION TO CHEMICAL AGENT RESISTANT COATINGS

Thomas A. Considine, Thomas E. Braswell, and Joseph P. Labukas
4600 Deer Creek Loop
Aberdeen Proving Ground, MD 21005
Telephone: 410-306-2564
thomas.a.considine.ctr@mail.mil

Abstract: Advanced military coating technologies have incorporated chemical agent resistance, desirable mechanical properties, and corrosion mitigating properties into CARC systems currently in use. The performance of these coatings is evaluated using MIL-DTL-53072[1]. During the coating application phase and in the field, a tape pull off test is required by MIL-DTL-53072 to gauge adhesive strength of a coating to a primer or substrate. We have investigated the chemical and physical interactions of a variety of tapes used for the verification of coating adhesion with ASTM D3359[2] on several substrates using tensile pull-testing, infrared-spectroscopy, and contact-angle measurements. A correlation between tape adhesion and surface wetting characteristics has been established. Tapes meeting the minimum performance parameters of 80 inch ounce-force over the selected CARC systems were indentified.

Keywords: CARC; Coatings; Coating Adhesion; ASTM D3359; Cross Hatch; Surface Wetting; Tape Adhesion

Introduction: Under regulation AR 750-1[3], Army based ground equipment must have a full Chemical Agent Resistant Coating (CARC) system as defined in MIL-DTL-53072, "CHEMICAL AGENT RESISTANT COATING (CARC) SYSTEM APPLICATION PROCEDURES AND QUALITY CONTROL INSPECTION". Several performance test requirements for CARC are specified in MIL-DTL-53072. One such performance requirement is adhesion testing of the system in accordance with ATSM D3359 Methods A and B, "Standard Test Methods for Measuring Adhesion by Tape Test". The primary test for coating adhesion is the cross-hatch tape adhesion method described in ASTM D3359 Method B. Applicators of CARC at OEM's, depots, and suppliers can provide objective quality evidence (OQE) that the integrity of the CARC adhesion is within the established limits of MIL-DTL-53072.

Recently, reports from the users in the field indicate that there is a significant fault with the tape adhesion testing used on Army assets[4,5]. More specifically, many of the tapes that are currently used in dry adhesion testing of CARCs reportedly do not adhere well to CARC topcoats containing polymer-beads used for gloss reduction[6,7]. In prior revisions of ASTM D3359 a single tape was specified for use in adhesion testing. Without adequate tape adhesion to the CARC surface, false positive results of satisfactory CARC adhesion to the substrate can easily result in unacceptable fielded materiel, thereby exposing the warfighter to significant risk of IR detection and a coating system that is no longer resistant to chemical agents. To emphasize the problem more clearly, a schematic representation of a tape pull-off test on identical substrates with different tapes is shown in Figure 1. The performance specification, MIL-DTL-53072 does not specify the type of tape that should be used for adhesion testing or the required peel strength needed to verify the coating adhesion. In order to mitigate this risk, it would require the closing of the technology gap by identifying suitable tapes for dry adhesion testing and developing a fundamental understanding of their adhesive properties with respect to various CARC surfaces.

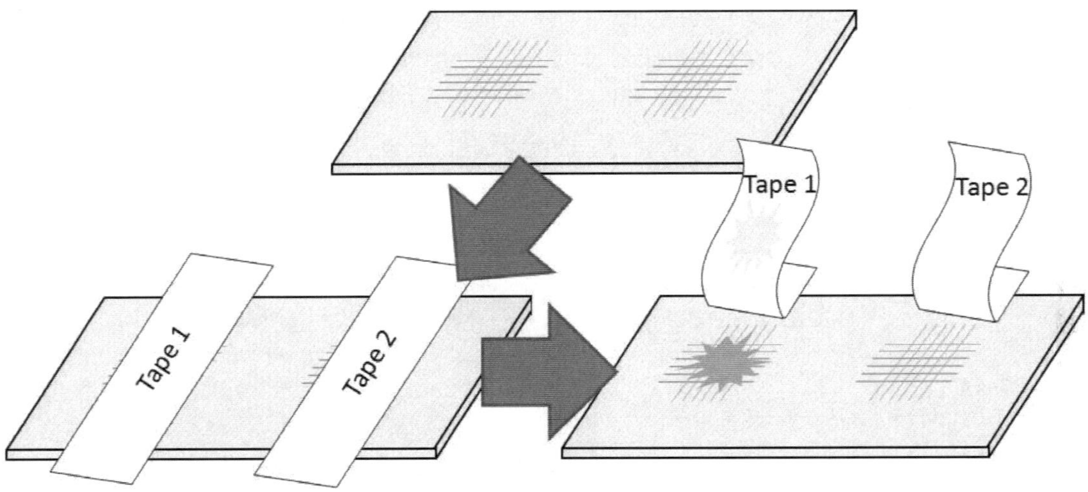

Figure 1: Diagram displaying differences in tapes used in cross hatch adhesion

In this research, the adhesion of several commercially available tapes to various CARC surfaces was evaluated. These tapes were selected based on interviews with several OEMs and military coating applicators as to which are commonly used in the field for adhesion testing[8,9,10]. The CARC formulations were varied to investigate the relationship between the tapes and the different types of CARC's available for use by OEM's and depots. More specifically, each tape was tested on water dispersible and solvent borne polyurethane topcoats in the two most common colors (Green 383[11] and 686A Tan[12]) that were prepared and provided by four different paint manufacturers. Attenuated total reflectance Fourier transform infrared spectroscopy (ATR-FTIR) was used to investigate chemical differences in the adhesives used for each tape. Advancing water contact angles were measured to provide insight on variations in the surface energies of various CARCs and the adhesive backing on the tapes.

Performance criterion for tape adhesion has been set by the CARC Commodity Manager at 80 inch ounce-force using IAW ASTM D 3330, Peel Adhesion of Pressure Sensitive Tape, which used a 180 degree pull angle and 1-inch wide tape [13] Tapes that meet or exceed the 80 inch ounce-force minimum on CARC systems will be recognized by ARL for qualifying the adhesion strength at interfaces within the CARC system. Typically, adhesion throughout the various interfaces in this system is verified using IAW ASTM D 3359.

That the governing specification for tape adhesion testing has become less precise with respect to what tapes are acceptable for use which has created an opening for many different types and brands of tapes with varied adhesive compositions. These variations in tape adhesive composition can lead to differences in adhesion to a substrate. CARC formulations within the same specification vary between manufacturers and because of this, not all CARCs have similar surface energies. We hypothesize that the wide variation in surface energy of CARCs affects adhesion between various tapes and the substrate.

Experimental:

CARC coated substrates were tested as received from the manufacturers. Table 1 summarizes the manufacturer, specification, type, color, and extender.

Table 1: CARC Information

Manufacturer	Spec	Type	Color	Extender
Hentzen	MIL-DTL-64159A	I	Green 383 686A Tan	Silica Particles/Flake
	MIL-DTL-64159B	I	Green 383 686A Tan	Polymer Beads
	MIL-DTL-53039C	I	Green 383 686A Tan	Silica Particles/Flake
		IV	Green 383 686A Tan	Polymer Beads
NCP	MIL-DTL-64159B	I	Green 383 686A Tan	Polymer Beads
	MIL-DTL-53039C	I	Green 383 686A Tan	Silica Particles/Flake
		II	Green 383 686A Tan	Silica Particles/Flake
	MIL-DTL-53039D	IV	Green 383 686A Tan	Polymer Beads
Sherwin Williams	MIL-DTL-64159A	I	Green 383 686A Tan	Silica Particles/Flake
	MIL-DTL-64159B	I	Green 383 686A Tan	Polymer Beads
	MIL-DTL-53039C	II	Green 383 686A Tan	Silica Particles/Flake
	MIL-DTL-53039D	IV	Green 383 686A Tan	Polymer Beads
Spectrum	MIL-DTL-64159B	I	Green 383 686A Tan	Polymer Beads

Tapes were used as received from the manufacturers. Table 2 summarizes the manufacturer and type of tape used in this work.

Table 2: Tape Information

Manufacturer	Product Name	Adhesive Type	Cost/roll
3M	250 Flatback Masking Tape	Rubber	~$55
3M	396 Super Bond Polyester Film Tape	Rubber Resin	~$13
Intertape	LA-26 Polyester/Rope-Fiber Laminate Tape	Thermosetting Rubber	~$16
SEMicro	Cross Hatch Tape (CHT)	Synthetic Rubber	~$35

An Instron model #8871 load frame mechanical test system was used to measure pull-off forces of the different tapes on various CARC surfaces. The Instron was equipped with a +/- 5kN dynamic load cell, a 90° panel clamping apparatus, and a pulling arm which gripped and performed the 180° tape pull. The Instron was connected to a personal computer (PC). This PC served as both digital controller and data logger. A picture of the experimental set-up is shown in Figure 2. The tape was pressed onto the CARC surface using an applicator in accordance with ASTM D3359. A suitable length of tape was used to cross the surface of the test panel and double back 180° up into the test fixture. A photo of this can be seen in Figure 3. Air bubbles were removed from the tape/surface interface by applying an even pressure with the backside of a plastic spoon until no air pockets could be seen. A common plastic spoon was selected and a limit of 10 seconds was established to remove air pockets so as to make every application as consistent as possible and to establish an accommodating reference for future field applications. The tape covered sample was clamped into place on the panel mount at the base of the Instron. After the test panel was properly secured, and tape attached to the tape pull apparatus, the tape was pulled from the surface over 25mm at a rate of 125mm/s. A total of 18 pulls were collected for each coating specification, type, color, and vendor tested. All data was recorded utilizing the Instron WaveMatrix[14] software suite and output to excel files.

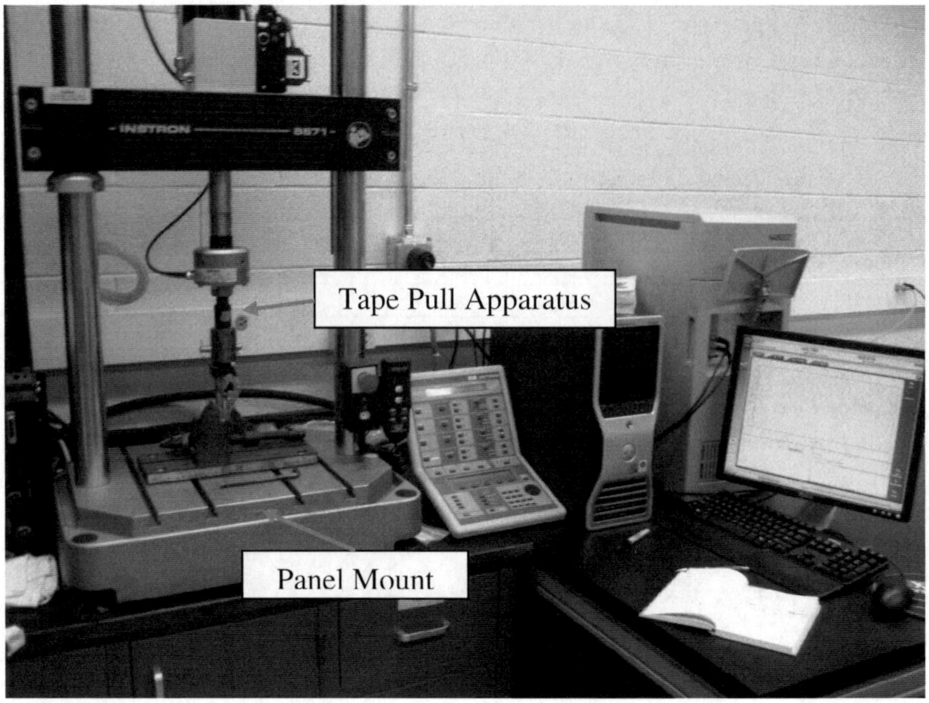

Figure 2: Instron load frame, panel mount, and tape pull apparatus

Figure 3: Close-up of panel mount and tape pull apparatus.

A Nexus 870 (Thermo-Nicolet) FTIR spectrophotometer was fitted with a Gateway ATR accessory (Specac). The accessory was allowed to purge with dry nitrogen for 10 minutes prior to collecting a background spectrum of the ZnSe crystal that was mounted in the accessory. Tape was applied onto the crystal and air bubbles were removed with finger pressure. The entire assembly was placed onto the ATR accessory through the sliding window in the FTIR sample compartment. The accessory compartment was purged with N_2 for 10 min prior to collecting the spectrum. The spectra were the summation of 32 scans collected between 650 and 4000 cm^{-1} at a 2 cm^{-1} resolution.

Advancing contact angles were measured using a Rame-Hart Model 290 Automated Contact Angle Goniometer. Water was used as the probe liquid and was purified before use to 18 MΩ cm using a Milipore water purification system. A droplet of water was placed on the surface and the volume of the water droplet was increased until the 3-phase contact line advanced. After an equilibrium angle was observed (i.e.- the drop edge stopped advancing), a measurement was recorded. Five drops were measured at random locations across the sample surfaces.

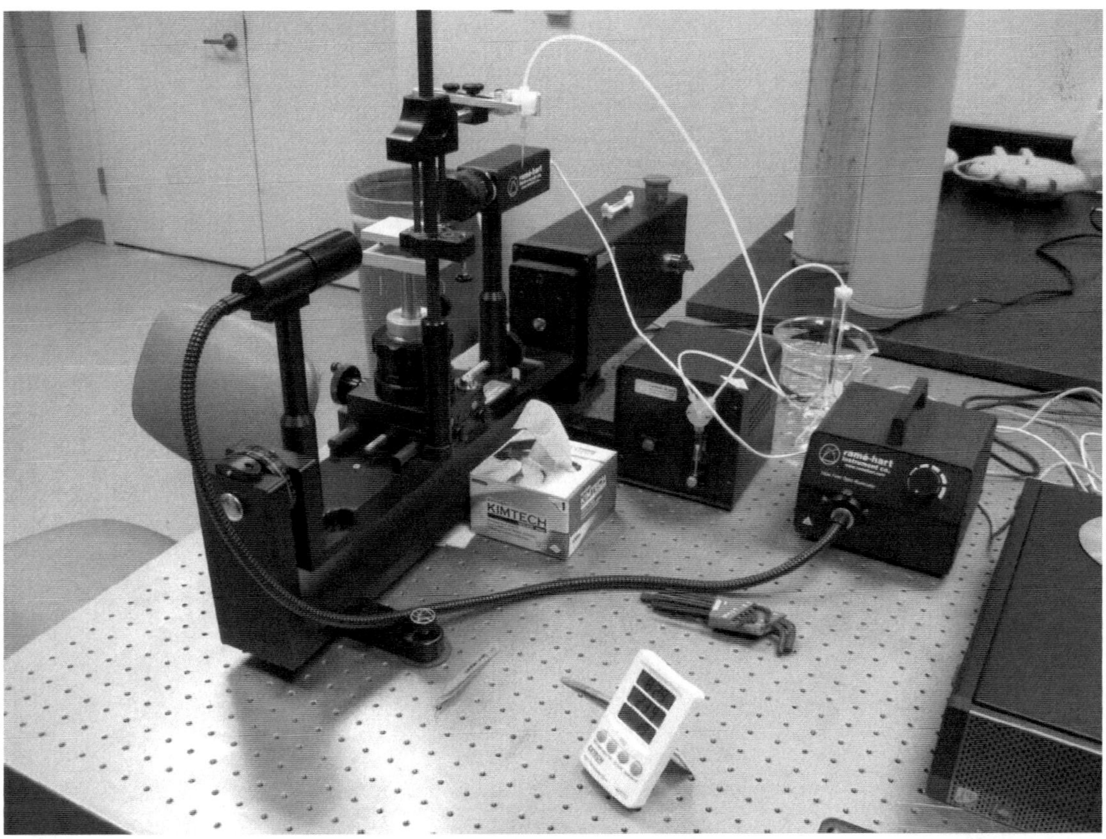

Figure 4: Rame-Hart Model 290 Automated Contact Angle Goniometer

Results and Discussion

Table 3: Average Pull-off Adhesion Strengths for Tapes on Hentzen Coatings

		Green 383	686a Tan
	Tape ID	Adhesive Torque (inch oz-F)	
64159A	3M 250	116.4343418	115.5536483
	3M 396	209.9502469	131.2387126
	CHT	75.30919954	98.0246156
	LA-26	170.2137478	172.8680479
64159B	3M 250	118.3693817	142.2385995
	3M 396	102.3657472	162.7249841
	CHT	80.23762291	100.8733565
	LA-26	165.0913277	193.604993
53039C TI	3M 250	143.5390277	137.2690834
	3M 396	195.0037631	191.8542441
	CHT	154.3933372	156.001803
	LA-26	200.3762747	198.6664471
53039C TIV	3M 250	131.3566168	136.2262631
	3M 396	190.6538985	186.2936355
	CHT	126.095947	122.2181742
	LA-26	178.2005384	170.9017335

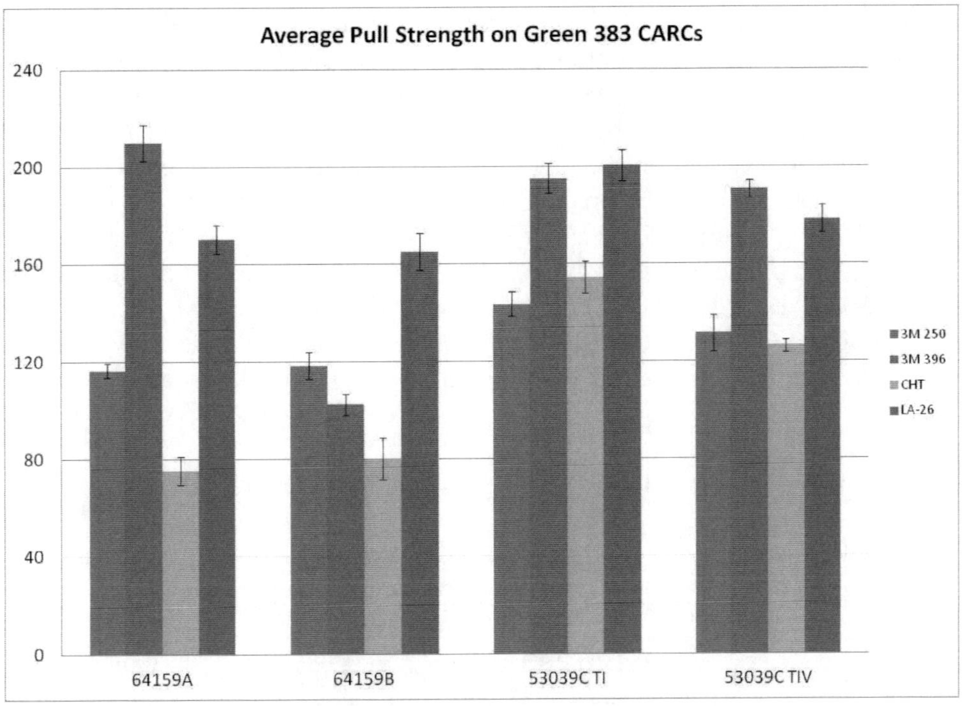

Figure 5: Average Pull-off Strength across Hentzen Green 383 CARCs

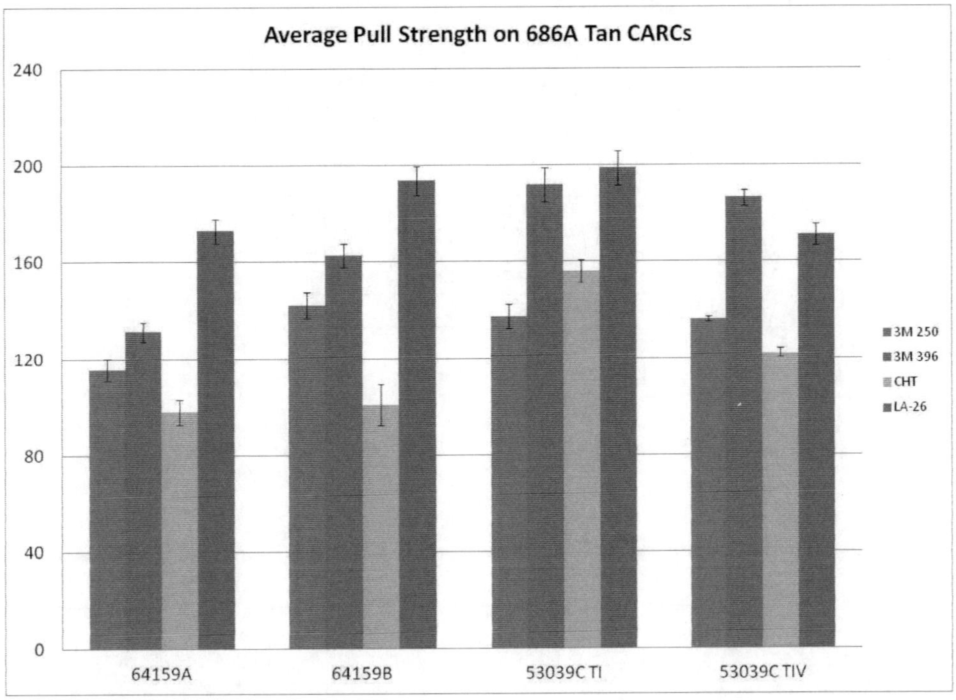

Figure 6: Average Pull-off Strength across Hentzen 686A Tan CARCs

All tapes tested met the established performance criteria of 80 inch ounce-force on all Hentzen coatings with the exception of Cross Hatch Tape over the Green 383 MIL-DTL-64159A. This coating is a silacious flattened waterborne CARC which had an average adhesive strength of 75.3 inch ounce-force, though standard deviation would meet the 80 inch ounce-force threshold. During testing it was noted that the CHT had issues with some adhesive residues left behind on the green silacious waterborne Hentzen coating, at first delaminating from the surface before the Instron was able to initiate the test. The baseline 3M 250 Flatback Masking Tape adhered similarly to each Hentzen coating. The 3M 396 Super Bond Film Tape and the LA-26 Intertape had higher adhesion values than the baseline 3M 250 on each Hentzen coating with the exception of 3M 396 having a lower adhesion strength on the Green 383 Hentzen MIL-DTL-64159B beaded waterborne CARC. The CHT typically had the lowest adhesion strengths of all the tapes tested on the Hentzen coatings with the exception of the type I solvent-borne CARC.

Table 4: Average Pull-off Adhesion Strengths for Tapes on NCP Coatings

		Green 383	686a Tan
	Tape ID	Adhesive Torque (inch oz-F)	
64159A	3M 250	132.7503765	120.9349896
	3M 396	158.3875335	195.7080301
	CHT	192.4183165	173.1846228
	LA-26	119.2778943	88.94703369
64159B	3M 250	125.0854736	99.3807805
	3M 396	202.6268211	208.8347205
	CHT	172.1919435	189.4485205
	LA-26	115.1765149	120.2415306
53039C TI	3M 250	140.4451796	112.6309499
	3M 396	144.2195001	155.3683186
	CHT	177.3376463	163.4457609
	LA-26	111.9850009	90.36489506
53039C TIV	3M 250	98.90985427	83.13689632
	3M 396	63.41534713	77.779158
	CHT	106.9569749	91.40965219
	LA-26	41.97781413	45.33741022

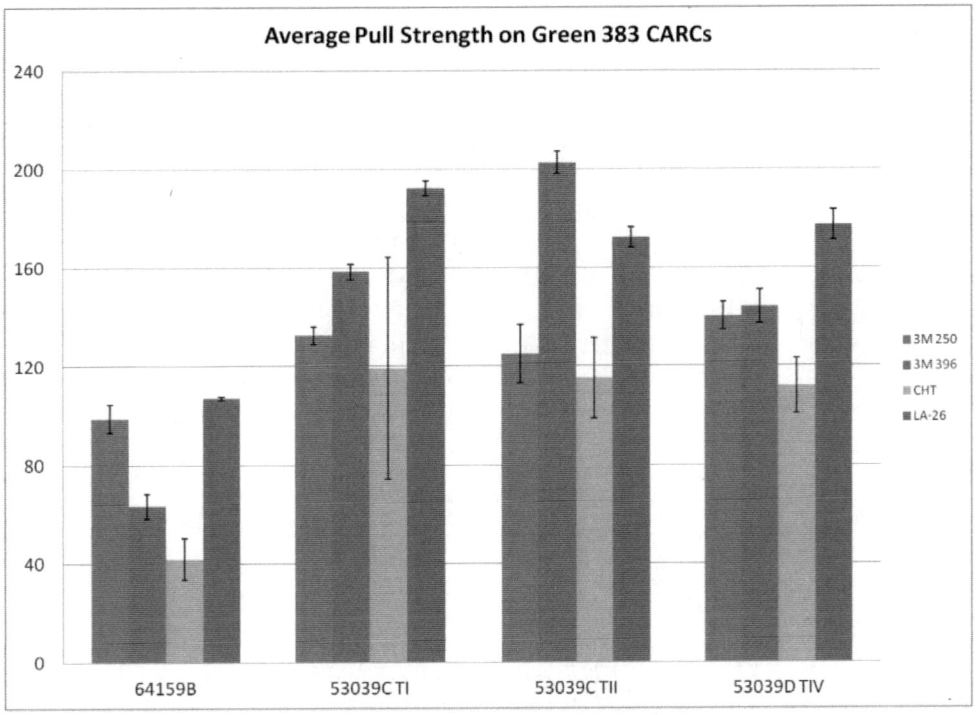

Figure 7: Average Pull-off Strength across NCP Green 383 CARCs

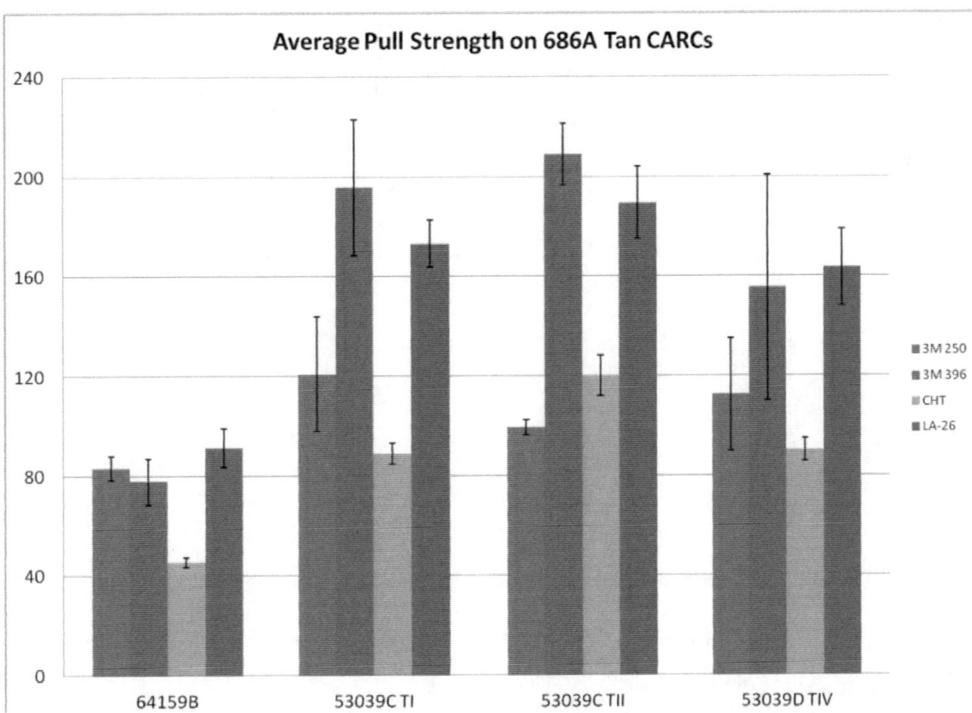

Figure 8: Average Pull-off Strength across NCP 686A Tan CARCs

All tapes tested met the established performance criteria of 80 ounce-force inch on all Hentzen beaded waterborne CARCs with the exception of 3M 396 and CHT. In the case of the 396 on the 686A Tan version of the beaded waterborne CARC, the standard deviation was greater than the margin by which the failure occurred. The standard deviation of the other failures mentioned above were less than the margin of failure. The 396 adhered strongly to the other coatings in the matrix, occasionally in excess of 200 inch ounce-force . The LA-26 adhered strongly to these other coatings as well, and performed adequately on the beaded waterborne CARCs. The baseline 3M 250 performed similarly across all coatings, and with the lower end of standard deviation could have been considered a failure on the beaded waterborne CARCs.

Table 5: Average Pull-off Adhesion Strengths for Tapes on Sherwin Williams Coatings

		Green 383	686a Tan
	Tape ID	Adhesive Torque (inch oz-F)	
64159A	3M 250	121.924166	126.2662192
	3M 396	143.1628696	121.4726765
	CHT	99.86779499	92.92666367
	LA-26	161.7851218	154.8465372
64159B	3M 250	98.1758155	86.83773152
	3M 396	68.95000138	52.87354516
	CHT	27.07132894	13.84469814
	LA-26	125.8186814	129.0627346
53039C TI	3M 250	144.8774794	129.4325893
	3M 396	188.2608935	160.4740078
	CHT	137.1784956	114.8251067
	LA-26	183.5808988	138.7905794
53039C TIV	3M 250	128.1307735	119.04685
	3M 396	152.6848133	149.5709988
	CHT	110.9296079	104.1712998
	LA-26	149.5745858	138.7905794

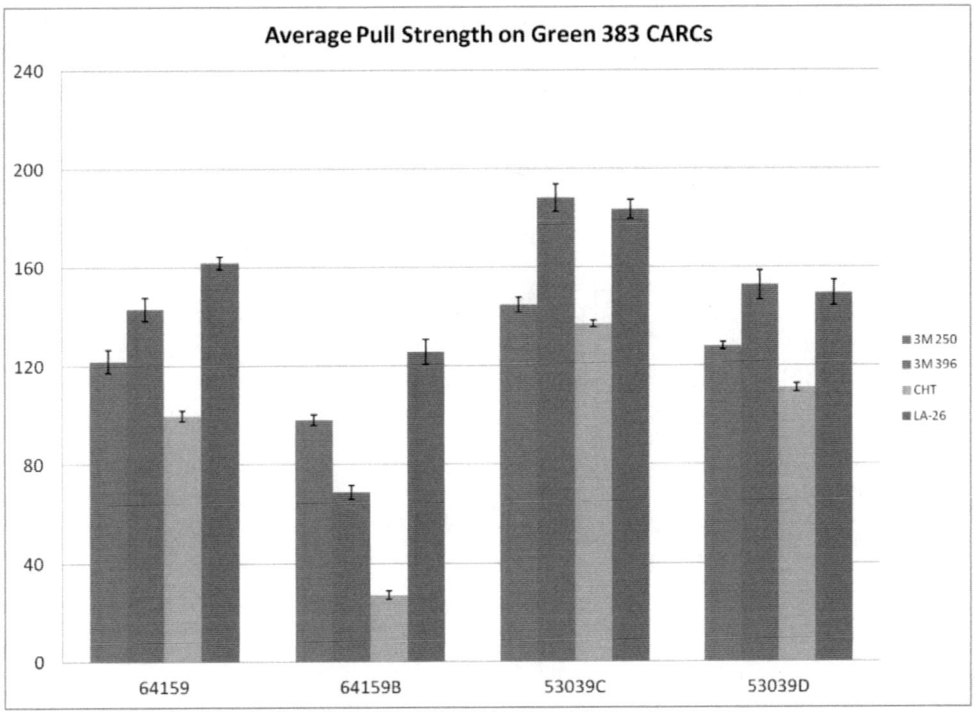

Figure 9: Average Pull-off Strength across Sherwin Williams 383 Green CARCs

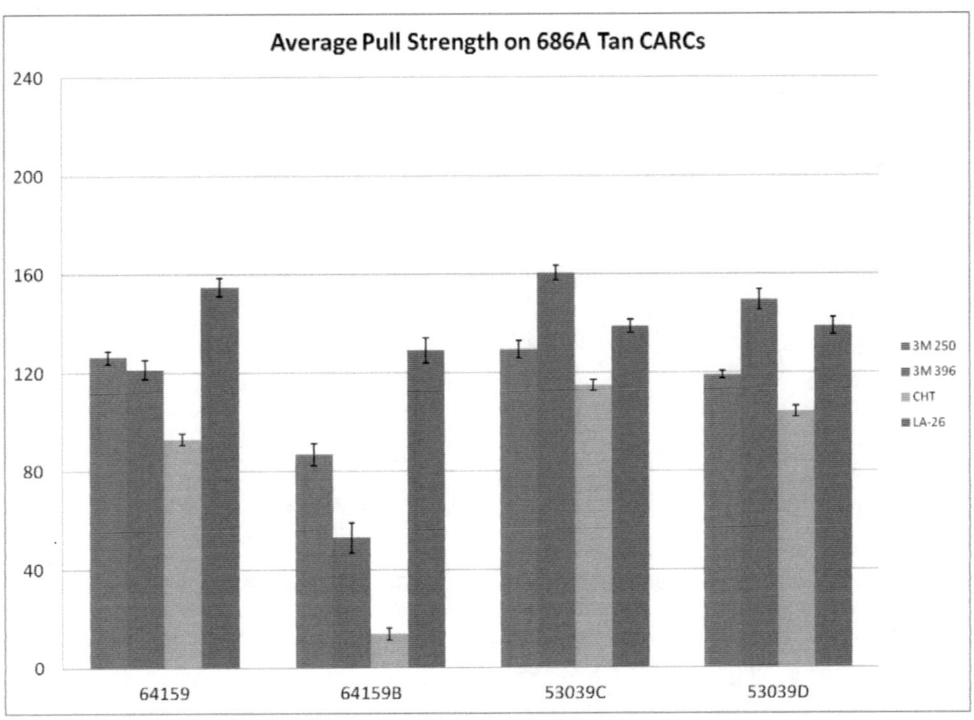

Figure 10: Average Pull-off Strength across Sherwin Williams 383 Green CARCs

The baseline 3M 250 met and exceeded the established performance criteria of 80 inch ounce-force on each of the Sherwin Williams coatings, as did the LA-26. The 396 and CHT met/exceeded 80 inch ounce-force on each of the Sherwin Williams coatings except for the beaded waterborne CARCs. The standard deviations do not exceed the margin of failure in the case of either the 396 or the CHT.

Table 6: Average Pull-off Adhesion Strengths for Tapes on Spectrum Coatings

	Green 383	686a Tan
3M 250	126.12747	134.24676
3M 396	98.256272	77.959094
CHT	73.289886	62.05658
LA-26	165.85749	171.37503

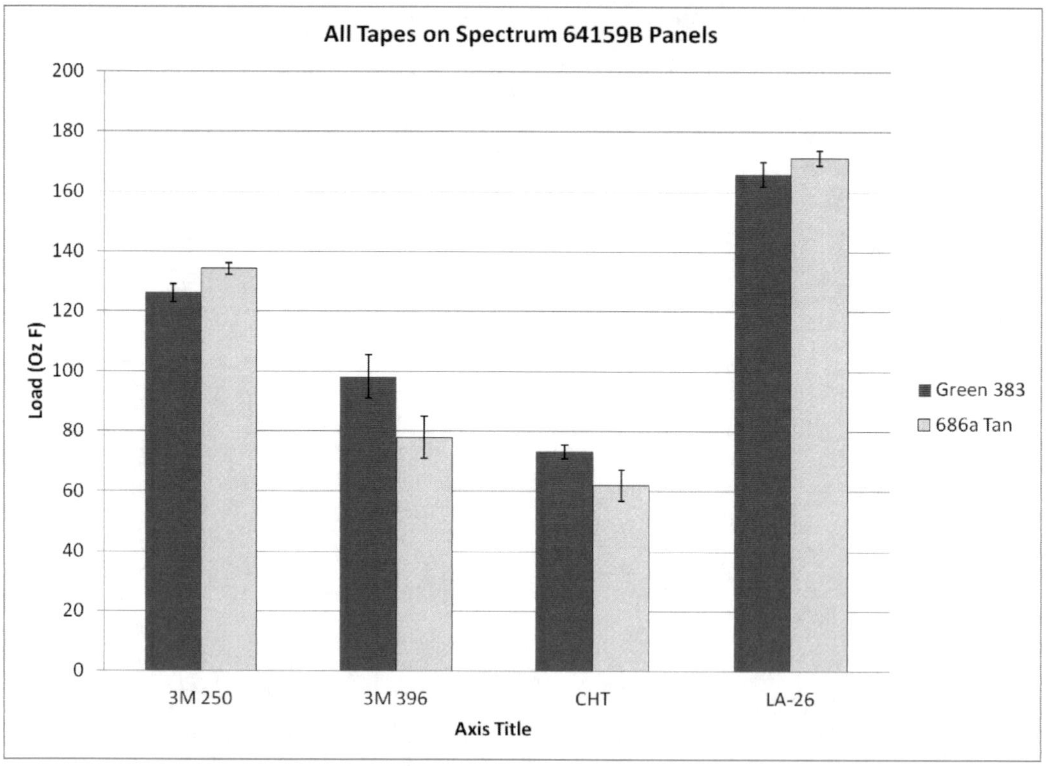

Figure 11: Average Pull-off Adhesion Strength across Spectrum 64159B CARCs

Table 7: Water Break Test Results

Manufacturer	Sample	Coating	250	396	CHT	LA-26
Sherwin Williams	64159A	Green 383	PASS	PASS	PASS	FAIL
		686A Tan	PASS	PASS	PASS	FAIL
	64159B	Green 383	PASS	PASS	PASS	PASS
		686A Tan	PASS	PASS	PASS	PASS
	53039C	Green 383	PASS	PASS	PASS	PASS
		686A Tan	PASS	PASS	PASS	PASS
	53039D	Green 383	PASS	PASS	PASS	PASS
		686A Tan	PASS	PASS	PASS	PASS
Spectrum	64159B	Green 383	PASS	PASS	PASS	PASS
		686A Tan	PASS	PASS	PASS	PASS
Hentzen	64159A	Green 383	PASS	FAIL	PASS	PASS
		686A Tan	PASS	PASS	PASS	PASS
	64159B	Green 383	PASS	PASS	PASS	PASS
		686A Tan	PASS	PASS	PASS	PASS
	53039C I	Green 383	FAIL	FAIL	Marginal	FAIL
		686A Tan	FAIL	FAIL	FAIL	FAIL
	53039C IV	Green 383	PASS	PASS	PASS	FAIL
		686A Tan	PASS	FAIL	PASS	PASS
NCP	64159B	Green 383	PASS	PASS	PASS	PASS
		686A Tan	PASS	PASS	PASS	PASS
	53039C I	Green 383	PASS	PASS	PASS	PASS
		686A Tan	PASS	PASS	PASS	PASS
	53039C II	Green 383	PASS	PASS	PASS	PASS
		686A Tan	PASS	PASS	PASS	PASS
	53039D IV	Green 383	PASS	PASS	PASS	PASS
		686A Tan	FAIL	FAIL	FAIL	FAIL

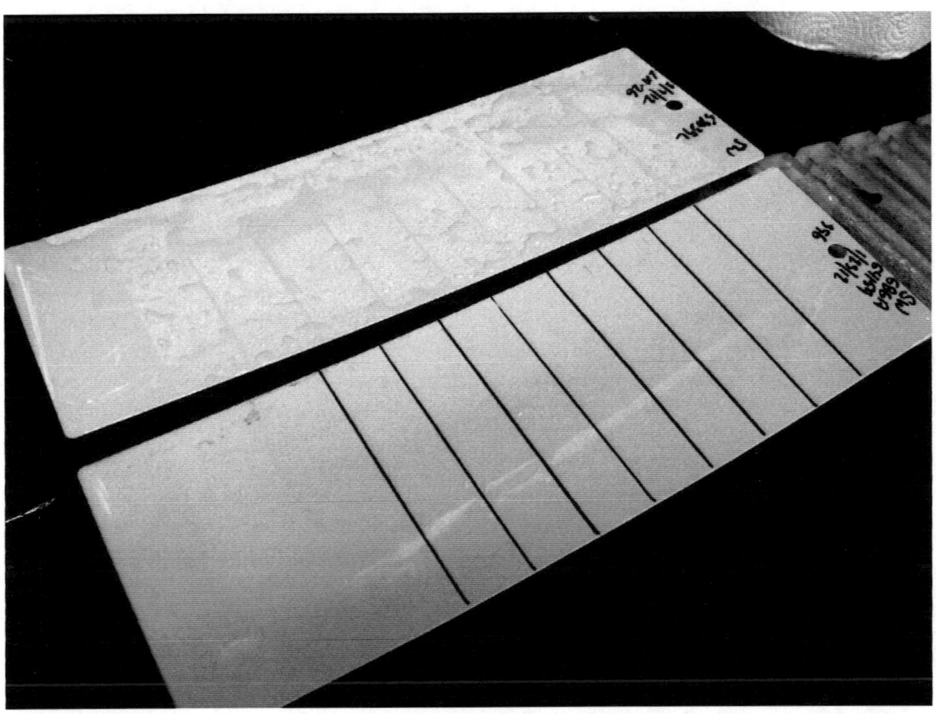

Figure 12: Example of Failed Water Break Test (Top) and Passed Water Break Test (Bottom)

The water break testing was done by running DI water over the test panels after the tape testing had been completed. For the purposes of this test, water that beaded up over the areas where tape had been applied to the surface was considered a failure and can be seen on the upper panel in Figure 11. The lower panel in Figure 11 in which you can see the water evenly spread out over the panel surface is an example of a pass. The instances in which failure occurs are attributed to adhesive residue left on the surface of the panel. These results could affect field application whereby these tapes would be used for masking during painting operations. There were few failures in the water break testing and were limited to LA-26 over Sherwin Williams' silacious waterborne CARCs, 396 over Hentzen's silacious waterborne (green only), every tape over Hentzen's silacious solvent-borne, 396 over Hentzen's beaded solventborne, and every tape over NCP's 53039D beaded solventborne (tan only).

Surface properties such as roughness, chemical composition, and homogeneity are well known to influence the wettability of surfaces. CARC topcoats used by the Army can exhibit a wide range of surface properties that result from subtle differences in formulation,, from variations in application parameters, and even ambient conditions, andGenerally, systems formulated to be water dispersible produced films that were slightly hydrophilic (water contact angle < 90°) whereas systems formulated with organic solvents that are not miscible with water cured to yield polymer films that were hydrophobic (water contact angle > 90°) (Table 8).

Table 8: Average water contact angle of various CARC systems.

	advancing contact angle of water (degrees)	
	waterborne polyurethane	*solventborne polyurethane*
silica extender	80 +/- 5	112 +/- 8
polymeric bead extender	79 +/- 9	106 +/- 4

The results in Table 8 also suggest that the type of extender has little or no impact on the wettability of the virgin polyurethane surfaces. A comparison of wetability for 686 tan and 383 green pigmented polyurethane coated samples obtained from several different vendors is shown in figure 12.

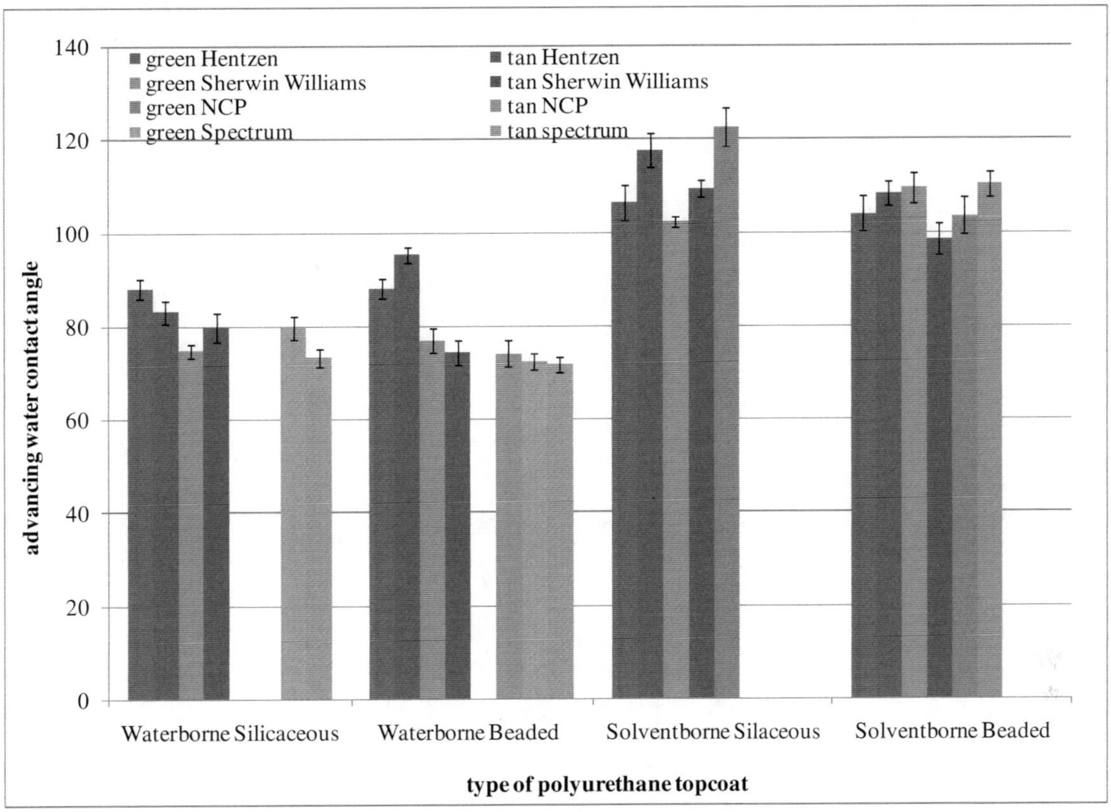

Figure 13: Advancing water contact angles of polyurethane topcoats formulated in tan and green from four different vendors.

There is little or no difference in the water contact angle measured on samples differing only in pigment composition. Although this observation suggests that pigment does not affect surface properties, additives used to disperse the pigments and extenders may preferentially migrate to the surface of the coating. Similarly, surfactants utilized for increasing the solubility of urethane precursors in waterborne formulations may reside at the surface preferentially. A more detailed and systematic study on a model system should provide clarity regarding surface segregation of particular additives used in formulation. Our research clearly shows waterborne polyurethanes have lower contact angles (higher surface free energies) than their solvent borne counterparts.

The relationship between wettability and work of adhesion is described by the Young-Dupré equation:

$$W_{SVL} = \gamma_{LV}(1 + \cos \theta), \quad (1)$$

where W_{SVL} is commonly referred to as the work of adhesion, γ_{LV} is the interfacial free energy at the liquid-vapor interface, and θ is the contact angle at the three-phase contact line. From this equation we expect the work of adhesion between a liquid and solid to decrease as the contact angle increases as shown in Figure 13.[15]

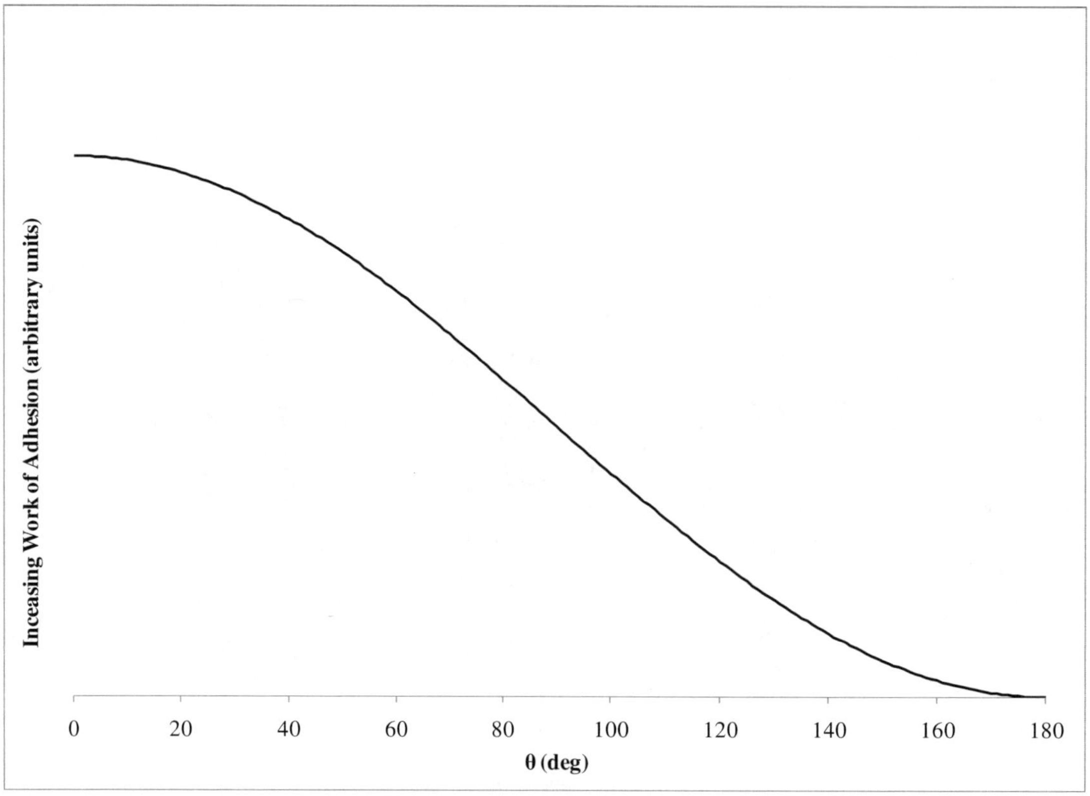

Figure 14: Theoretical behavior of the work of adhesion with respect to contact angle based on the Young-Dupré equation.

Extending this expected behavior to a solid-solid system such as an elastomer/polyurethane interface is more complicated. The Johnson-Kendall-Roberts (JKR)[16] and the Derjaguin-Muller-Toporov (DMT)[1718] theories are typically used to describe the work of adhesion for polymer-solid systems. Although these relationships will be useful in fundamental studies that model our system, they are beyond the scope of this report and are mentioned here to highlight the relationship between work of adhesion, pull-off-force, and contact angle. In our research the lack of a clear correlation between contact angle and pull-off strength is most likely due to the complexity of the systems studied.

Attenuated total reflectance-fourier transform infrared spectroscopy was used to identify chemical differences in the adhesive of the tapes used for pull-off tests. Figure 14 shows the region of the spectrum where –CH , –OH, and -NH stretching vibrations occur. Small stretches observed at ~3280 cm^{-1} for the adhesive on the 3M® 250 tape and 3330 $cm^{-1}$for the LA-26 tape indicate –OH or –NH groups within the adhesive or adsorbed water. Hydroxyl, amine, or amide functionality within an adhesive can hydrogen bond with free chain ends of the urethane and with the oxide surfaces of pigment particles and hydroxyl groups may have enough mobility to chemically react with any residual isocyanate groups present at the polyurethane surface. The likelihood that these hydroxyl groups would contribute significantly to tape pull-off forces is low due to the relatively low intensity of the normally strong –OH stretch. It is common, however, for –NH groups of secondary amines to be quite weak. Other evidence that supports the presence of –NH groups is seen in figure IRB (*vide infra*). Evidence for the presence of vinyl groups in all of the tape samples analyzed is seen by the –CH stretch just above 3000cm^{-1}. Although vinyl functionality may contribute to an increase in the overall adhesion to a substrate that contains thiol groups, the low intensity of this stretch complicated quantification of the differences in –C=C concentration between the various tapes.

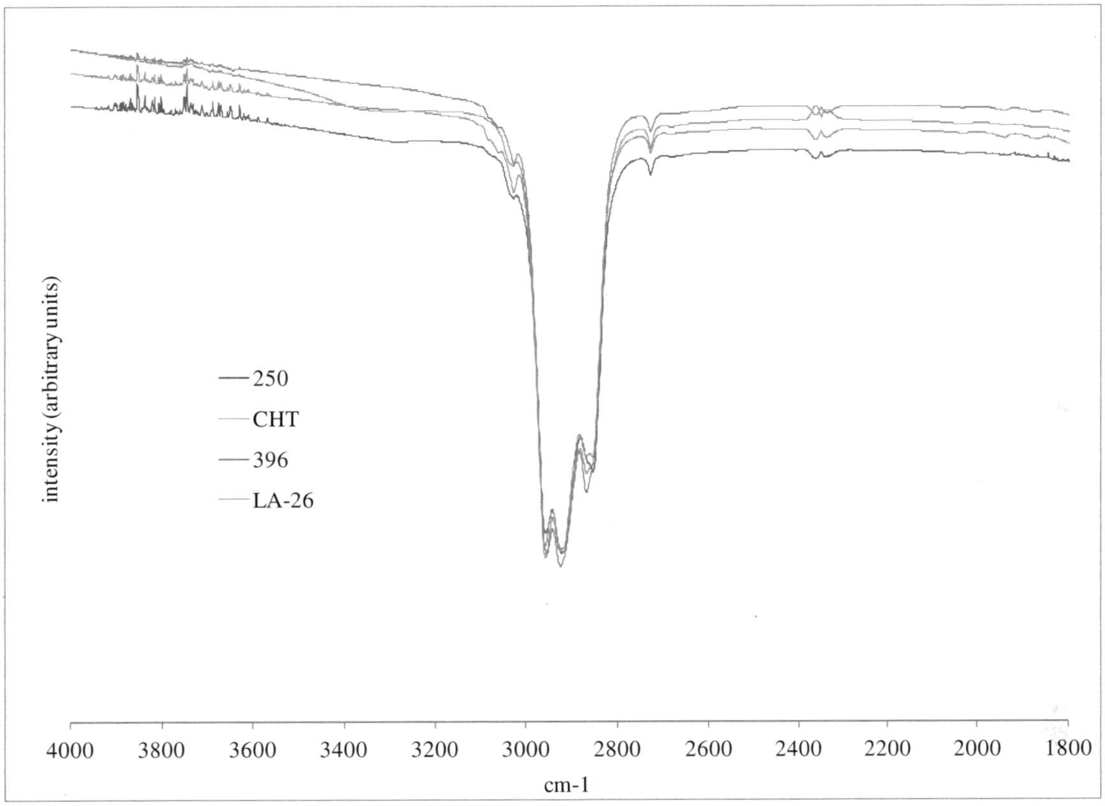

Figure 15: ATR-FTIR spectra of various tapes in the region of 4000-1800 cm^{-1}

Spectra from 650 – 2000 cm^{-1} for of each tape are seen in figure 15. Once again there is evidence that the 3M® 250 and the LA-26 tape contain chemical functional groups consistent with hydrogen bonding groups such as the – NH bend of an amide (~1585 cm^{-1}), an amine (1218 and 1236 cm^{-1}) for LA-26 and 3M® 250 respectively, and a strong stretch at 911 (3M® 250) and a weaker one at 918 cm^{-1}(LA-26) that is indicative of the double bond of an acrylate functional group. Several smaller peaks are present in spectra of the tapes that are difficult to confidently assign functionality to without complementary analytical techniques.

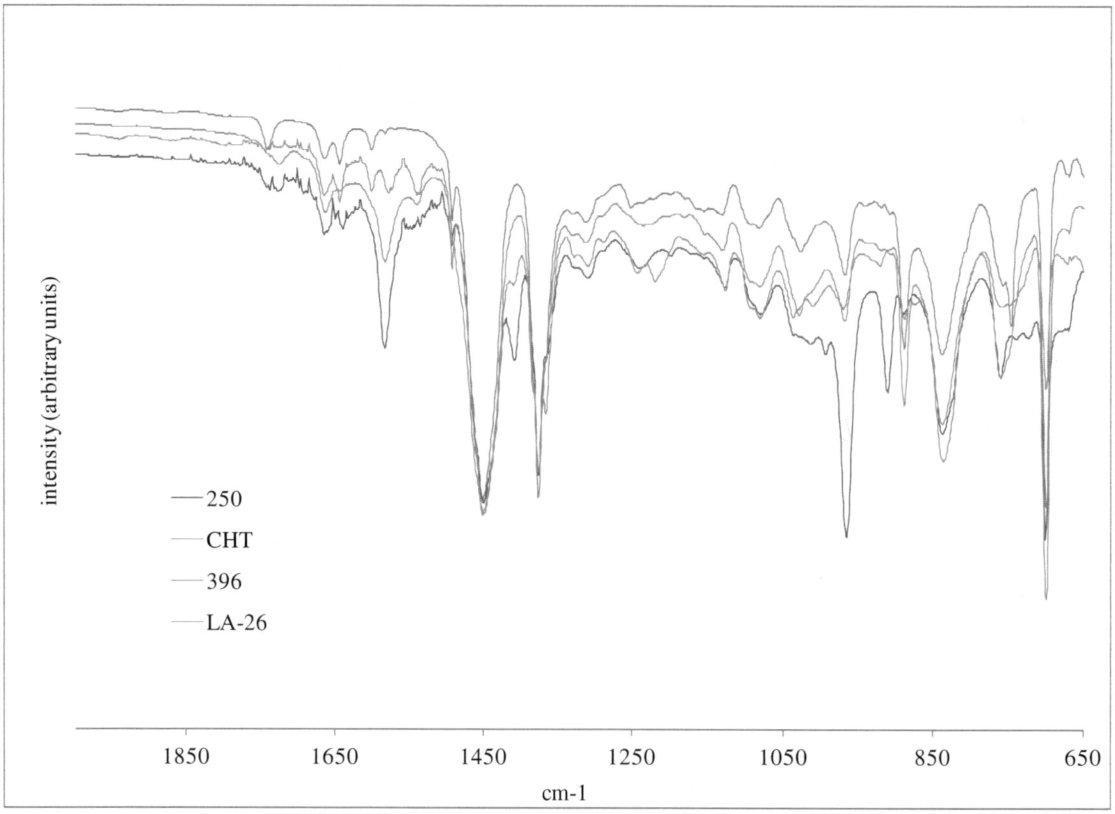

Figure 16: ATR-FTIR spectra of various tapes in the region of 650-1950 cm^{-1}

Conclusions

1) Waterborne formulations have lower advancing contact angles (higher surface free eneries) than solvent borne counterparts.
2) There is no clear difference in the wettability of CARC that contains polymeric beads and a CARC system that contains silica as an extender. The caveat here is that each system has unique additives (dispersants, surfactants, solvents) that allow for a final film with specific properties and these variables may be the underlying cause of the similarities in contact angles.
3) Green and tan pigments do not affect the wettablilty of the cured film. The same caveat that is described in #2 applies here.
4) Tapes that contain functional groups such as hydroxyl, amine, amide, or vinyl typically correlate with higher pull off forces on most samples.
5) A more fundamental study is needed to develop a more fundamental understanding.

References

[1] MIL-DTL-53072. *Chemical Agent Resistant Coating (CARC) System Application Procedures And Quality Control Inspection.* **2003**

[2] ASTM D3359-09e2. "Standard Test Methods for Measuring Adhesion by Tape Test". *Annu. Book ASTM Stand.* **2009**.

[3] AR-750-1. "Army Materiel Maintenance Policy". *Army Regulations.* **2007**.

[4] J. Adams and G. Somerville, telephone conversation with author, December 2009

[5] J. Mort, telephone conversation with author, December 2010

[6] M. Flory, telephone conversation with author, May 2010

[7] L. Knisley, telephone conversation with author, December 2010

[8] P. Do, telephone conversation with author, December 2010

[9] B. Tucker, telephone conversation with author, December 2010

[10] T. Tuttle, telephone conversation with author, December 2010

[11] FED-STD-595/34094. *Colors Used in Government Procurement/Green, Flat or Lusterless.* **1984.**

[12] FED-STD-595/33446. *Colors Used in Government Procurement/Yellow, Flat or Lusterless.* **1984.**

[13] ASTM D3330/D3330M-04(2010). "Standard Test Method for Peel Adhesion of Pressure-Sensitive Tape". *Aunnu. Book ASTM Stand.* **2010**.

[14] *Wave Matrix*. Computer software. *WaveMatrix Dynamic Testing Software*. Vers.

1.3.232.

[15] Frederick M. Fowkes. Contact Angle, Wettability, and Adhesion, Volume 43

AMERICAN CHEMICAL SOCIETY, 1964.

[16] K. L. Johnson and K. Kendall and A. D. Roberts, Surface energy and the contact of elastic solids, Proc. R. Soc. London A 324 (1971) 301-313

[17] Derjaguin, BV and Muller, VM and Toporov, Y.P., 1975, *Effect of contact deformations on the adhesion of particles*, Journal of Colloid and Interface Science, 53(2), pp. 314-326

[18] Muller, VM and Derjaguin, BV and Toporov, Y.P., 1983, *On two methods of calculation of the force of sticking of an elastic sphere to a rigid plane*, Colloids and Surfaces, 7(3), pp. 251-259.

A MEASUREMENT VALIDATION ALGORITHM USED IN TEST CELL

Aurélie GOUBY
Snecma
Rond-Point René Ravaud - Réau
77550 Moissy-Cramayel CEDEX - France
Phone: +33 1 60 59 42 53
aurelie.gouby@snecma.fr

Abstract: Aircraft engines are systematically tested throughout development phases to evaluate their performance. During a test, several measurements on engine and on the bench itself are acquired and then stored in a dedicated database during the test. Many factors can cause an abnormal measure is not detected on time; rerun a test can be very expensive. Prevent measurement errors is thus the priority during an engine test. In the following article, we present a measurement validation algorithm used in batch mode, during the test with a short delay or at the end of each test. The algorithm consists in modeling the measurement by building a mathematical model. An anomaly Z-score is then computed with model's residues and compared to an abnormality threshold learned on real data. The user will be alerted about the measurement value if the Z-score exceeds this abnormality threshold. Through this monitoring system, if a measurement problem occurs, the engine test owner can then be reactive and rerun the acquisition or change the sensor if faulty.

Key words: Anomaly; Alarm; Engine; Measure; Monitoring; Quality; Score; Test; Tube; Validation

Introduction: Test benches are used for each new production engines (Figure 1) or for new engines during development phases to evaluate the performance. This test procedure is very expensive and the bill can quickly become steep if sensor is faulty and prevents the nominal progress of the test. The goal is to reduce the risk of failure and alerting at the earliest by ensuring the measurements are correctly acquired.

Figure 1: Engine installed in a production test cell.

During the test, the engine and the test cell are monitored through the acquisition of hundreds of real-time measurements. This is a SPC (Statistic Process Control) tool that is able to record each sensor at different acquisition frequencies and presents graphs with alert bounds in real time. In the monitoring room, screens enable test engineers to visualize some important engine parameters. However, all measurements cannot be controlled in real time during the test. During the test or at the end of the test, two types of data are stored in a dedicated database:

• The transient data: there are low frequency (up to 100Hz) signals like performance measurement (pressures, temperatures, flows, ...) and high frequency dynamic measurements like tip timing, accelerometers, strain gauges and microphones (up to 200kHz).

• Stabilized data: that's a point acquired when the engine and the conditions are stabilized.

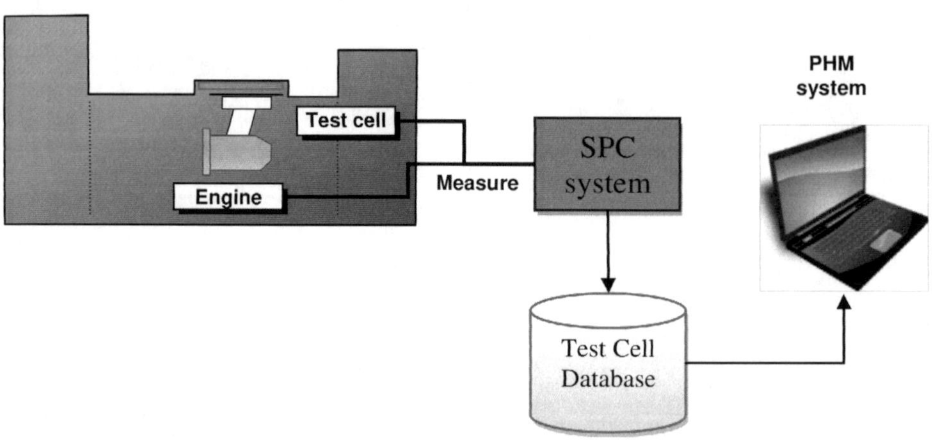

Figure 2 : PHM System in test cell with a direct link to database

The PHM (Prognostic and Health Monitoring) system has access to the cell database (Figure 2); hence, it's possible to analyze the data of the test that has just taken place and provide a diagnostic few minutes after data download in the database.

Context of application: The health-monitoring algorithms are developed by Snecma under the SAMANTA (Snecma Algorithm Maturation ANd Test Application) platform which was previously described in [1]. This environment industrializes blocs of mathematical processing tools in graphical units. PHM engineers are able to use each mathematical module to design their own specific solutions.

In order to help test people to use PHM system, a supervisor with the SAMANTA platform has been set up in the test cell. A direct link between the database and the platform has been created to perform the database analysis as quickly as possible. The developed algorithm presented in this paper analyzes in a first step the stabilized set of measures stored in cell database.

Before connecting the algorithm in test cell, it's important to ensure that computation results have a minimum of reliability: one doesn't want to interrupt an expensive test experiment scheduled for bad reasons. Hence, two performance indicators are defined: the PFA (Probability of False Alarm) and the POD (Probability Of Detection). It's sometimes difficult to compute these indicators if there are no available abnormal labeled data (which is not always easy to get in test context).
The PFA indicator is given by the following conditional probability:

$$PFA = \mathrm{P}(Healthy|Detected) \quad (1)$$

if one writes $\mathrm{P}(Detected)$ the probability that an abnormality is detected by the algorithm and $\mathrm{P}(Healthy)$ the probability that se measure is healthy, then the false alarm rate is just the probability that the measure is healthy while an abnormality is detected. This probability should not be confused with $\alpha = \mathrm{P}(Detected|Healthy)$ wich is the first species statistic error that one needs to define the test rejection domain.
The probability of detection is given by the following equation:

$$POD = \mathrm{P}(Detected|Faulty) \quad (2)$$

It's usually called in statistics the "test power".
Obviously, the ideal monitoring method minimizes the rate of false alarms while maximizing the rate of correct detection.

Monitoring algorithm: The sensor monitoring algorithm developed for this project is divided into two phases: learning and execution.

In following, pressure monitoring in test cell is taken as an example to illustrate the algorithm principle. As explained in the previous chapter, monitored measurements are stabilized sets of measures acquired during the test; therefore the temporal dimension cannot be used here. To overcome this problem, the measurement is monitored relatively to context parameter as the rotor speed for pressure.

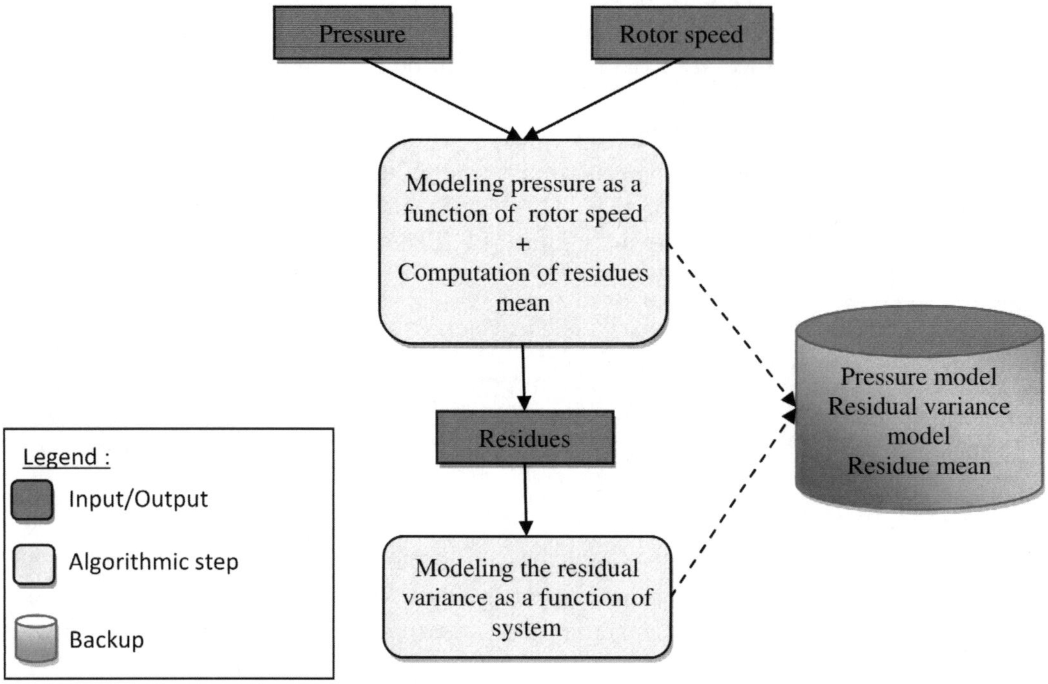

Figure 3: Input/Output and Algorithmic steps of monitoring algorithm for learning phase

This first step consists in modeling the cloud of points tagged as healthy by a curve. It is assumed that one have n observations. Bench pressure, denoted $Y = (y_1, \dots, y_n)$, is estimated by rotor speed, denoted $X = (x_1, \dots, x_n)$, with relation that may be linear, polynomial, exponential, and logarithmic, according to the cloud shape. The choice of transformation is given by the engine experts who know the physical relationship between the parameters or is computed automatically with a penalty method like Lasso algorithm described in [2].
Thus we obtain:

$$\widehat{y_i} = f(x_i) \text{ for } i = 1, \dots, n \tag{3}$$

where $\widehat{Y} = (\widehat{y_1}, \dots, \widehat{y_n})$ is the estimator of pressure Y and f the regression function. The regression residuals are given by the following formula:

$$res_i = y_i - \widehat{y_i} \text{ for } i = 1, \dots, n \tag{4}$$

Of these residues, the mean is computed using the following formula:

$$mean = \frac{1}{n}\sum_{i=1}^{n} res_i \tag{5}$$

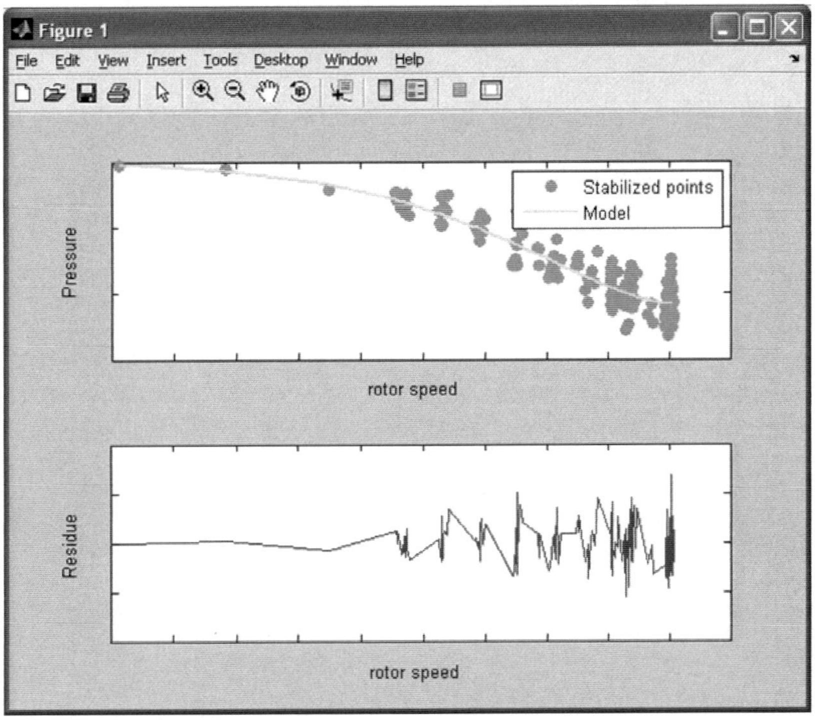

Figure 4 : Stabilized Points Scatter Plot and the associated regression model (top subplot). Regression residuals according to rotor speed (bottom subplot).

Figure 4 bottom plot shows that there is dependence between residuals variance and the explanatory parameter. The higher the rotor speed, the more the variance of the pressure measurement increases. This may be due to the sensor precision. Therefore, it's important to take into account this specificity in the detection algorithm.

Once created the regression model, the residual variance is estimated on a sliding window of size w (where w is a parameter of the algorithm) using the following formula:

$$var_j = \frac{1}{w-2} \sum_{i=j}^{j+w} (res_i)^2 \text{ for } j = 1, \dots, (i + w - 1) \qquad (6)$$

After estimating the variance, the purpose is to express it as a function of rotor speed. In the same way as before, one seeks the function g such that:

$$\widehat{var}(x_j) = g(x_j) \text{ for } j = 1, \dots, (i + w - 1) \qquad (7)$$

It is then possible to create a confidence tube around the curve by giving a threshold number of standard deviation θ. This tube takes into account the uncertainly sensor measurement and the data dispersion at high rotor speed. The upper and lower envelopes of the tube after the following equations:

$$env_{sup} = f(x) + \theta \times \sqrt{\widehat{var}(x)} \qquad (8)$$

$$env_{inf} = f(x) - \theta \times \sqrt{\widehat{var}}(x) \qquad (9)$$

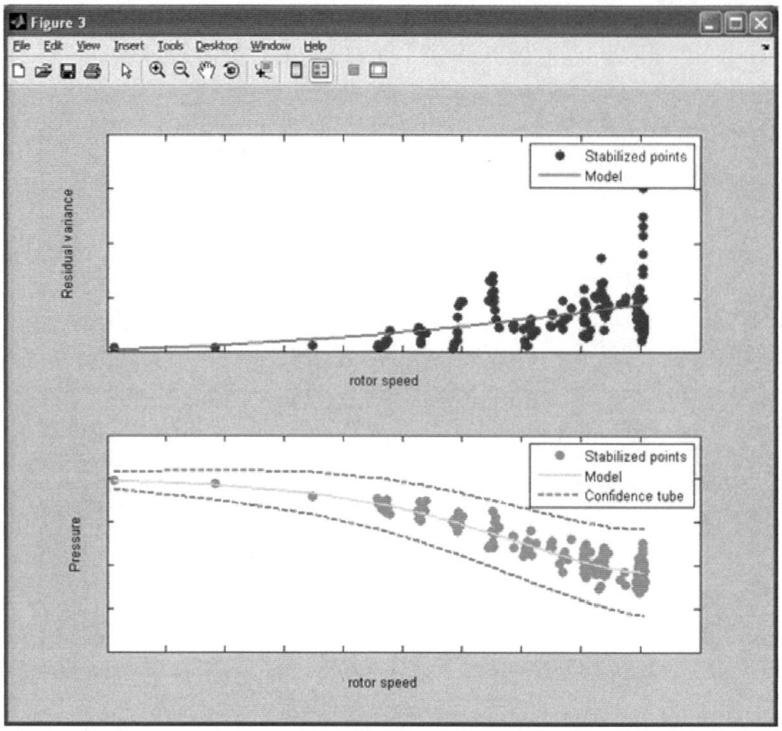

Figure 5 : Modelisation of Residual Variance by Polynomial Regression (top subplot). Confidence tube (3-sigma) around stabilized points (bottom subplot).

By default, two alert thresholds are set in learning phase: a threshold of three standard deviations and a threshold of six standard deviations because in normal operation condition the probability to observe something outside 3σ is less than $3 \times 10^{-3}$and the probability to observe data outside 6σ is lower than 2×10^{-9}. However, these thresholds are configurable by test people and can be adjusted according to false alarm rate and objective detection rate.

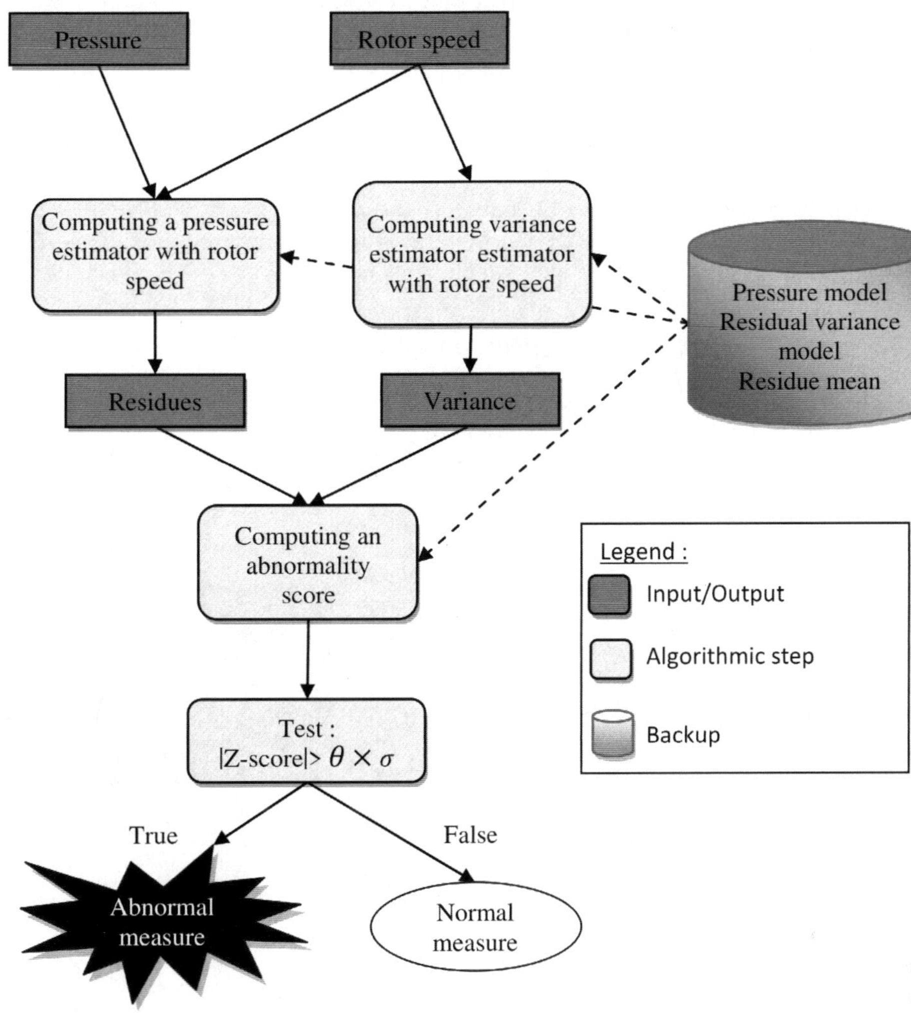

Figure 6: Input/Output and algorithmic steps of monitoring algorithm for running phase

At runtime, new stabilized points are analyzed one by one. Is noted $P = (x_{exec}, y_{exec})$ a new point which coordinates $x_{exec}$and y_{exec} are the values of this point (respectively the rotor speed and the pressure). The first step is therefore to apply equation (1) to obtain an estimate of the pressure value (denoted $\hat{y}_{exec}$) of the relevant point.

$$\hat{y}_{exec} = f(x_{exec}) \tag{10}$$

One can thus calculate the model residuals:

$$Res_{exec} = y_{exec} - \hat{y}_{exec} \tag{11}$$

The next step consists in applying equation (5) to obtain an estimate of the variance for the associated rotor speed:

$$\widehat{var}_{exec} = g(x_{exec}) \tag{12}$$

Once the pressure and the variance are estimated, a Z-score can be calculated using the residues and the mean (computed during the learning phase):

$$Z_{score} = \frac{Res_{exec} - mean}{\widehat{var}_{exec}} \tag{13}$$

Once the Z-score calculated, it's compared to the anomaly threshold which has been used to define the confidence tube in learning phase. If the Z-score exceeds this threshold, then the point is outside the tube and is therefore considered abnormal because outside the normal model dispersion.

Results: This algorithm has been tested on data from two Snecma engines types. The following examples given come from development test.

The first example is based on a engine 1 test. The measurement error is detected here was due to failure of fuel density.

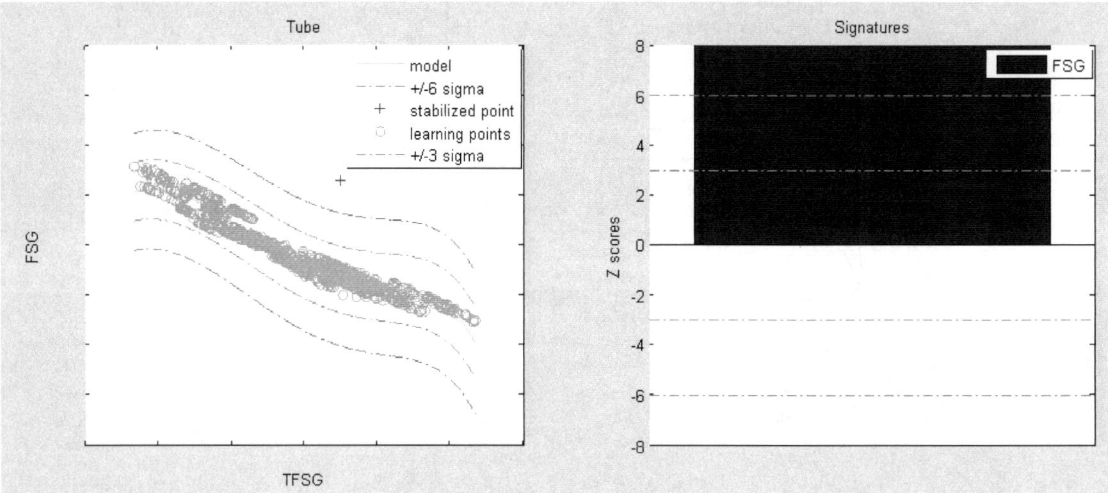

Figure 7: Detection of abnormal fuel density measurement (left subplot). Signature of failure (right subplot).

FSG parameter is the fuel density and TFSG is fuel temperature. One clearly can see on Figure **7** that the stabilized point is outside the red 6-sigma confidence tube. The failure signature is given by Z-score value of the stabilized point. It can be used if one seeks to classify the type of failure but that's not the subject in this paper.

The second example highlights the failure of a pressure sensor during a engine 2 test.

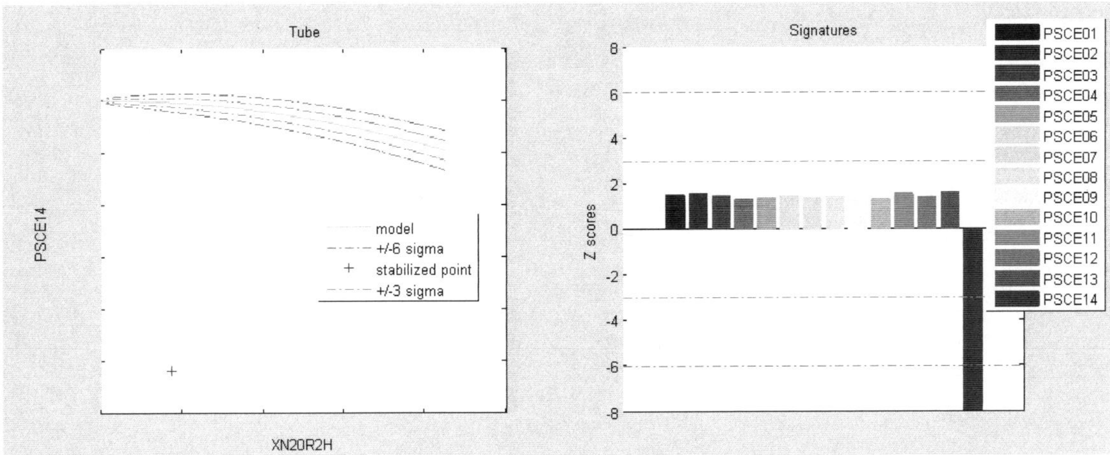

Figure 8: Detection of abnormal PSCE14 pressure sensor (left subplot). Signature of failure (right subplot).

All the PSCE parameters in Figure **8** represent pressure sensor installed in different cell parts and XN20R2H is the rotor speed. In this example, the stabilized point is far from normal curve and its confidence tube. One also notes on signatures plot that's PSCE14 measure is the only faulty one among all other measures of pressure. A signature interpretation method is given in [3]

Conclusion: The monitoring algorithm presented in this paper is a simple method to validate a measure (as a point). It doesn't need an a priori on the relationship between physical parameters acquired since the "normal curves" are learned from real observations and alarm algorithm thresholds are easily configurable by the user.
The first tests of this function are encouraging since the lowest POD obtained is 84.5% and the highest PFA is 4.17% with 6-sigma threshold. These results should be confirmed on a wide test campaign.

The next step will be to upgrade this monitoring function to monitor the real-time measurements in the test cell.

References:

[1] J Lacaille, A Maturation Environment to Develop and Manage Health Monitoring Algorithms, 2009, PHM Society Conference, San Diego, CA
[2] J Lacaille & E Côme, Sudden Change Detection in Turbofan Engine Behavior, 2011, 8th International Conference on CMMFPT, Cardiff, UK
[3] J Lacaille, Standardized Failure Signature for a Turbofan Engine, 2009, IEEE Aerospace Conference, Big Sky, MT

Bibliography:

[1] J Lacaille, V Gerez & R Zouari, 2010, An Adaptive Anomaly Dtector used in Turbofan Test Celss, PHM Society Conference, Portland, OR

[2] J Lacaille & V Gerez, 2011, Online Abnormality Diagnosis for real-time Implemntation on Turbofan Engines and Test Cells, PHM Society Conference, Montreal, Canada

[3] J Lacaille & V Gerez, 2012, A Batch Detection Algorithm Installed on a Test Bench, PHM Society Conference

FAILURE ANALYSIS OF FAN MILL IN THERMAL POWER PLANT

Chao Liu[1], Dongxiang Jiang[2,*], Fulei Chu[1]
1, Department of Mechanical Engineering, Tsinghua University, Beijing 100084, China
2, State Key Laboratory of Control and Simulation of Power System and Generation Equipments, Department of Thermal Engineering, Tsinghua University, Beijing 100084, P. R. China
chaoliu13@tsinghua.edu.cn, jiangdx@tsinghua.edu.cn

Abstract: Fan mill is an important device in thermal plants, which supply coal power to boiler. Crack failures recently occurred in pulverizing wheels of fan mills in a power plant in China. The cracks happens in quite a few pulverizing wheels and they affect the normal operation. To find out the root cause of the cracks, Finite Element Analysis (FEA) of the pulverizing wheel is carried out in this work. The assembly with real dimensions is modelled to calculate the concentrated stress and strain with centrifugal load, driving load, and operational temperature taken into account. Contact effects are included in the finite element model as well as nonlinear material property. Based on the analysis, the crack causes are determined: (i) the initial stress exceeds the yield limit in normal operational condition, resulting low safety coefficient; (ii) the stress increases significantly in the worn pulverizing wheel, where the risk of failure is greatly increased.

Key words: Cracks; Pulverizing wheel, Thermal power plant, Failure analysis

Introduction: Coal mill is widely applied in thermal power plants. Due to different characteristics of coal, types of coal mills are applied including ball mill, disk mill, roller mill, hammer swing sledge mill and fan mill [1]. Among different types of coal mills, fan mill presents advantages of compact structure, less construction cost, less installed area and less electricity consumption, which attracts wide application in thermal power plants with lignite coal fueled.

Cracks in pulverizing wheel of fan mill are recently observed in a power plant of China, which are shown in Figure 1. The cracks occur in the dovetail grooves of the front disk and the back disk, where the beaters are installed. Cracks are observed in several locations of the pulverizing wheel, and they all lie in the dovetail area with beaters installed.

After detecting fan mills in four units, cracks are observed in more than half of them. The operation time of different fan mills varies in a long time span, where the shortest one is about 6700 hours and the longest one is about 25000 hours. The crack depth reaches 20 percent of the thickness of the dovetail groove in severe ones. Crack failures in these fan mills greatly decrease the safety and influence the normal operation of the power plant.

---

* Corresponding author. Tel.: 0086-1062794046; Fax: 0086-1062770209.

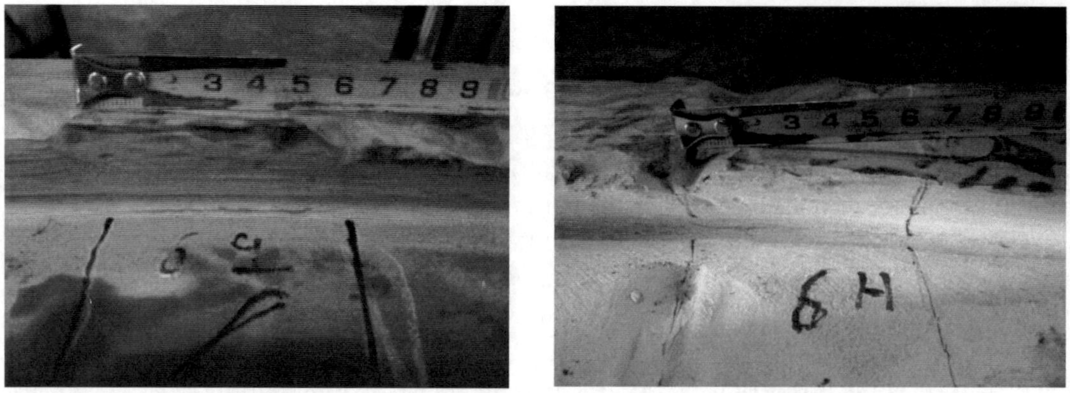

(a) Crack in dovetail groove of back disk

(b) Crack in dovetail groove of front disk

Figure 1: Cracks observed in pulverizing wheel of fan mill

Similar crack accidents emerged in fan mill [2]. Cracks are observed in the beaters of the coal mill, where the crack depth is less than 15mm with the crack ratio of 7 percent. The operation time is 1658 hours. Metallurgical analysis is carried out and the crack cause is determined as the defect of the material with too much carbon and lower teeming temperature than the standard. Cracks also occurred in shafts and other locations of fan mill [3–5].

Cracks are detected in more than a half of the pulverizing wheels, which greatly influence the normal operation of the steam turbine unit. Failure analysis of cracks are necessary to determine root causes. This work carried out root causes analysis of the cracked pulverizing wheel with finite element method (FEM), where the real dimensions of pulverizing wheel are applied. The analysis includes: (I) stress and strain analysis of the pulverizing wheel in the design condition; (II) stress and strain comparison with the worn condition measured from the worn pulverizing wheel; (III) comparisons of stress and strain between the design condition and the worn condition.

Finite Element Modeling of Pulverizing Wheel: The structure of the pulverizing wheel is shown in Figure 2a. The assembled pulverizing wheel contains a front disk, a back disk, twelve beaters (twelve inner beaters and twelve outer beaters), twelve connecting supports and accessories.

Bolts are applied to fix the connecting support with the back disk and the front disk. The rotating beater is limited by the dovetail grooves on the front disk and the back disk, which is not fixed by bolts or other components in operation. The back disk is installed on a shaft which is connected to the output of the drive chain.

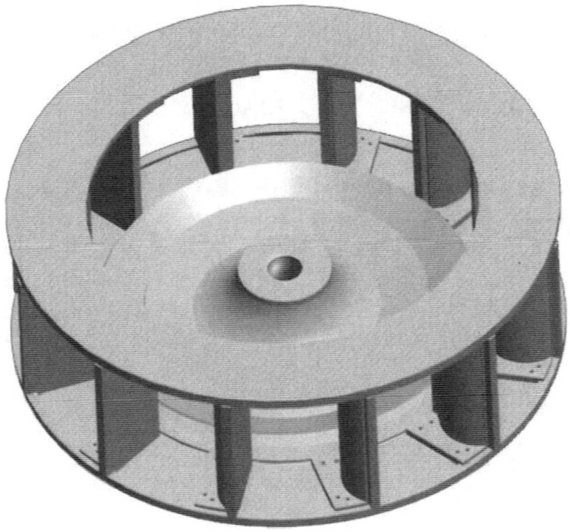

Figure 2: The geometric structure of pulverizing wheel in fan mill

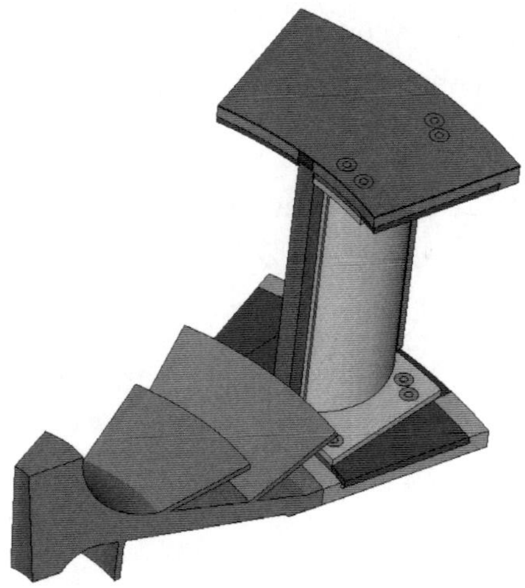

Figure 3:Model of 1/12 pulverizing wheel of fan mill

In order to reduce the computational scale, 1/12 of the pulverizing wheel is applied in this work based on St. Venant principle. Due to its periodic symmetric characteristics, 1/12 section can be expanded to the entire model. In modeling the structure with FEM, real dimensions of the back disk, the front disk, the beater and the connecting support are applied, to acquire the accurate behavior of the structure. To get the status of bolts in the operating condition, the behavior of the contact surface between the connecting support and the back disk is simulated with contact model.

Hexahedron element is applied in meshing the pulverizing model, meshes between contact surfaces are consistent to improve the solution efficiency and accuracy. Model of 1/12 section are shown in Fig. 3.

(1) Design condition
Dovetail groove is the location with stress concentration and its dimensions are critical to get accurate local stress and strain. The Dimensions of design condition of pulverizing wheel is listed in Table 1.

Table 1: Worn condition of dovetail groove in pulverizing wheel

Location		Dimension	
		Design condition	Worn condition
Back disk	Thickness	70	64
	Width	70	30
Front disk	Thickness	65	58
	Width	70	58

(2) Worn condition
Wearing continually occurs during the operation of fan mill, as the lumpy coal is broken up by the collision with the beater. Although the material used in the pulverizing wheel is

wear-resistant, wearing is cumulated in a long-time operation. The dovetail groove lies in the outside of the pulverizing wheel. The dimensions of dovetail groove is decreased during the wearing. The worn dimensions of the dovetail are listed in Table. 1.

Operational Loads and Material Property: The operation condition of the pulverizing wheel includes rotating speed, operation temperature and input power, which are listed in Table. 2.

Table 2: Operational loads of pulverizing wheel

Item	Content
Rotating speed	490 r/min
Input power	1400kW (rated)
Torque of the driven rotor	$2.73 \times 10^4 N \cdot m$
Operation temperature	200-300°C
Tightening force of bolts	$960 N \cdot m$

The rotating speed is the rated speed in the design condition. The driven power is inputted by the driven chain connecting the motor and the pulverizing wheel. Due to the principle of the fan mill, the lumpy coal is sent to the pulverizing wheel with the flue gas of boiler. The temperature of flue gas is 550-560 ℃ before it is mixed with coal. The flue gas dries the coal and increases the operation temperature of fan mill. According to the field measurement, the temperature of the pulverizing wheel is 200-300 ℃ in normal condition.

Concentrated stress exists in dovetail groove due to its geometry, and the maximal stress probably exceeds the yield limit of the material. The elastoplastic property is therefore necessary to determine the stress and strain in the stress concentration location. The material property in 20℃, 200℃ and 300℃ are tested. The material property at 20℃ is shown in Figure 4.

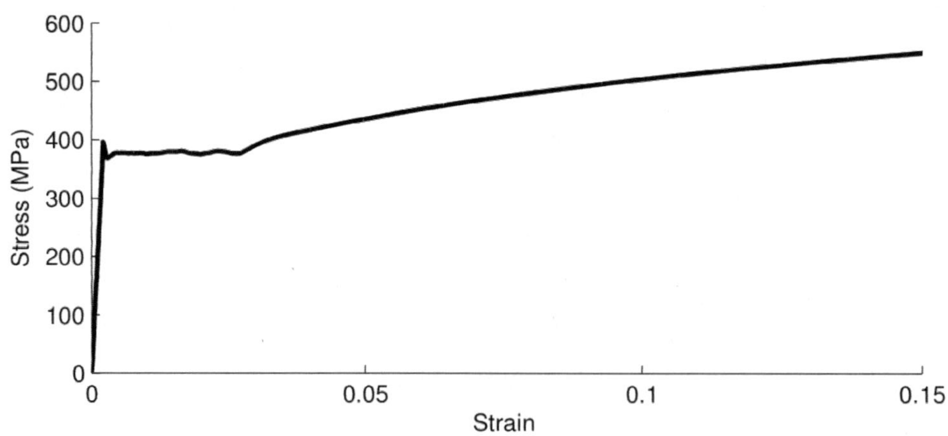

Figure 4: Material property at 20 ℃

The tested material is from the back disk and the front disk. As different materials are applied in different components of pulverizing wheel, the test material property is applied in the back disk and the front disk, and material property of the other components are based on handbook of metal materials [6].

Results and Discussion: The stress distribution at the dovetail groove of the front disk and the back disk in design condition is acquired and shown in Figure 5, where loads at rated power are applied and the material property at 20 ℃ is used. The strain distribution at the dovetail grooves is shown in Figure 6.

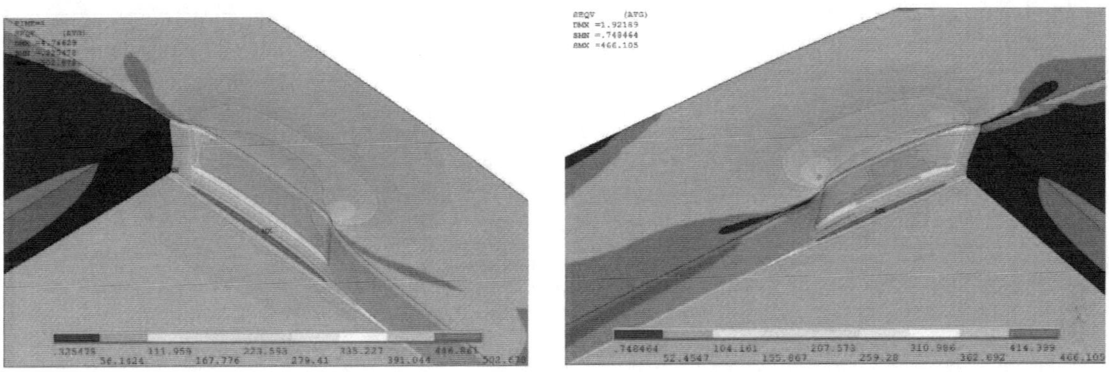

(a) Dovetail groove of front disk (b) Dovetail groove of back disk

Figure 5: Von Mises stress distribution of dovetail groove in design condition

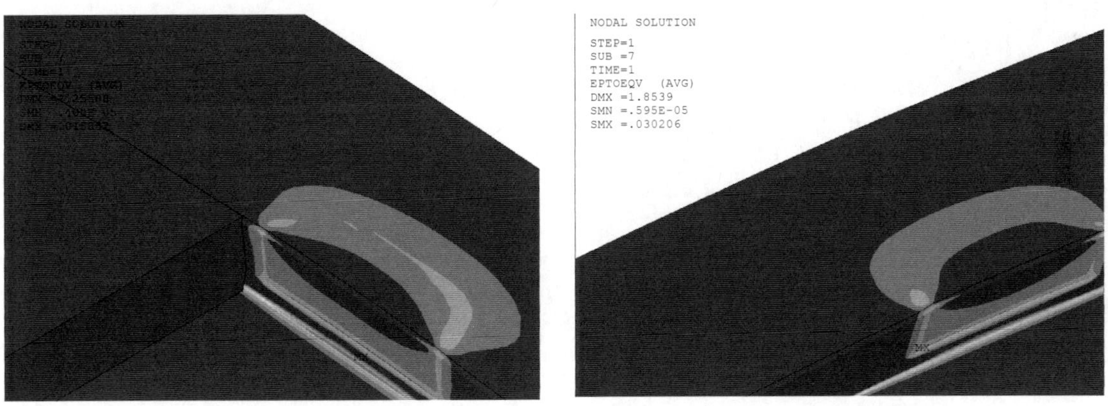

(a) Dovetail groove of front disk (b) Dovetail groove of back disk

Figure 6: Von Mises strain distribution of dovetail groove in design condition

The stress concentration is severe at the bottom of the dovetail groove in the front disk and the back disk. The maximal Von Mises stress of the front disk is 502.68MPa, the maximal Von Mises stress of the back disk is 466.10MPa. The location with maximal Von Mises stress is consistent with the crack position. Based on the material property of the front disk and the back disk, the maximal stress in the bottom of the dovetail groove exceeds the yield limit. The area of the contact surface between the front (back) disk and the outer beater is relatively small, and it results in high stress in the contact area in the range of 250MPa –350MPa.

The stress distribution at the dovetail groove of the front disk and the back disk in worn condition is shown in Figure 7, where loads at rated condition are applied and the material property at 20 ℃ is used. The strain distribution at the dovetail grooves is shown in Figure 8.

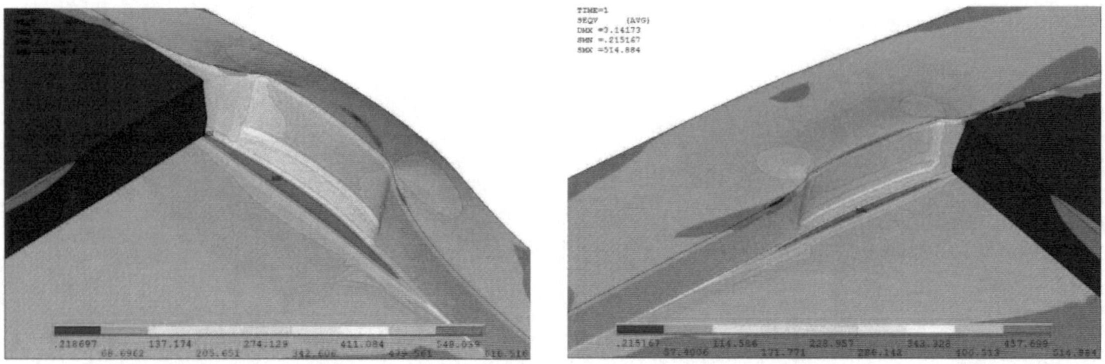

(a) Dovetail groove of front disk (b) Dovetail groove of back disk

Figure 7: Von Mises stress distribution of dovetail groove in worn condition

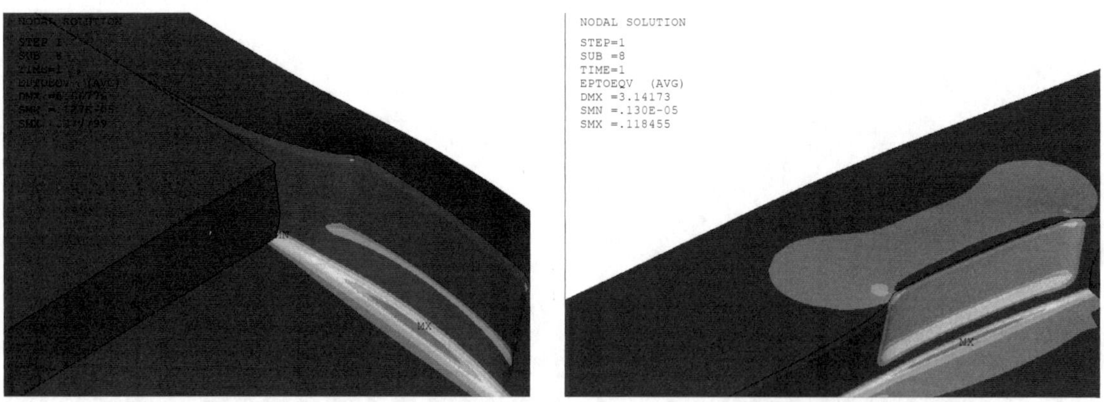

(a) Dovetail groove of front disk (b) Dovetail groove of back disk

Figure 8: Von Mises strain distribution of dovetail groove in worn condition

The location of stress concentration is identical with that in the design condition, while the maximal stresses of the front disk and the back disk are significantly increased. The maximal Von Mises stress of the front disk is 616.52MPa, and the maximal Von Mises stress of the back disk is 514.88MPa. The maximal stress of the front disk is close to the strength limit.

To compare the stress and strain in different operation temperatures, the maximal stress and strain of the design condition are listed in Table. 3.

Table 3: Stress and strain of design condition in different operation temperatures

Temperature	Location	Maximal stress		Maximal strain	
		Von Mises	Shear	Von Mises	Shear
20°C	Back disk	466.10	266.30	0.07856	0.07629
	Front disk	502.68	288.06	0.11490	0.11244
200°C	Back disk	350.76	192.85	0.05435	0.05230
	Front disk	343.79	195.71	0.04704	0.04540
300°C	Back disk	361.11	205.02	0.03182	0.02967
	Front disk	359.60	204.67	0.03078	0.02864

From Table. 3, maximal stress and strain of the front disk and the back disk are different in different temperatures. The maximal stresses at 200 ℃ are less than that at 20 ℃ and 300 ℃. It means that the maximal stress firstly decreases and then increases with the increasing temperature. The maximal strain of the front disk and the back disk decrease with the increasing temperature, which is different from the tendency of maximal stress. The maximal stain of the front disk is larger than that of the back disk when the material property at 20 ℃ is applied. However, the relation is inverse in the operation temperature of 200 ℃ or 300 ℃, where the maximal stain of the back disk is larger than that of the front disk.

The maximal stress and strain of worn condition are listed in Table. 4.

Table 4: Stress and strain of worn condition in different operation temperatures

Temperature	Location	Maximal stress		Maximal strain	
		Von Mises	Shear	Von Mises	Shear
20°C	Back disk	514.88	291.46	0.11846	0.11598
	Front disk	616.52	347.54	0.37780	0.37484
200°C	Back disk	366.17	208.01	0.04250	0.04054
	Front disk	486.88	273.01	0.31650	0.31403
300°C	Back disk	383.61	217.21	0.02955	0.02742
	Front disk	509.67	290.17	0.23260	0.22981

From Table. 4, the maximal stresses increase in the worn condition, and the front disk is more obvious. The maximal stress of the front disk is high and it is close to the strength limit of the material in all temperatures. It's therefore concluded that the maximal stress increases significantly in the worn condition, which greatly reduce the safety margin of the component and increase the risk of failure. The maximal strain is increased in the front disk, while it is decreased in some cases in the back disk.

To compare the differences of stress and strain in design condition and worn condition, the changes of worn condition are listed in Table. 5.

Table 5: Changes of stress / strain of worn condition comparing with design condition (%)

Temperature	Location	Maximal stress		Maximal strain	
		Von Mises	Shear	Von Mises	Shear
20°C	Back disk	10.47	9.45	50.78	52.03
	Front disk	22.65	20.65	228.81	233.37
200°C	Back disk	4.40	7.86	-21.81	-22.49
	Front disk	41.62	39.49	572.83	591.72
300°C	Back disk	6.23	5.95	-7.16	-7.58
	Front disk	41.73	41.78	655.69	702.43

Changes between the worn condition and the design condition are compared. The maximal stress of the front disk presents larger increase than that of the back disk, which is more obvious in the situation of high operation temperature. Von Mises stress and shear stress present same tendency in the design condition and the worn condition. The maximal strain at 200 ℃ and 300 ℃ presents different trends in the front disk and the back disk. The maximal strain of the back disk decreases in worn condition, while it increases significantly in the front disk. As different characteristics presented in different temperatures, stress and strain analysis at the operation temperature is necessary to get the accurate behavior of the local dovetail groove.

Considering the operation temperature of the pulverizing wheel, the stress and strain at 200 ℃ and 300 ℃ are applied to determine root cause of crack failures.

To determine the causes of crack failures in dovetail groove of pulverizing wheel, the following aspects are analyzed:

(1) Stress concentration
Stress concentration is severe in the bottom of the dovetail groove of the front disk and the back disk. Due to the geometric structure and dimensions of the dovetail groove, there exists stress concentration at the bottom the transition section with small fillet. The results of finite element analysis show that the maximal stress locates in the fillet area of the dovetail groove, which is consistent with the crack position in reality.

(2) High stress in design condition
The maximal stresses of the front disk are 343.79MPa and 359.60MPa with the operation temperature of 200 ℃ and 300 ℃, respectively. The maximal stresses of the back disk are 350.76MPa and 361.11MPa with the operation temperature of 200 ℃ and 300 ℃, respectively. The yield limit of the material in 200 ℃–300 ℃ is about 300MPa. The maximal stresses of the front disk and the back disk exceed the yield limit, which make the margin is small. And plastic strain emerges in the dovetail groove during operation.

(3) Significant increase of maximal stress and strain in worn condition
As the operating characteristics of the fan mill, wearing lasts in the operation. From the analysis of the maximal worn condition, the maximal stresses of the front disk are

486.88MPa and 509.67MPa with the operation temperature of 200 ℃ and 300 ℃, respectively. The maximal stresses of the back disk are 366.17MPa and 383.61MPa with the operation temperature of 200 ℃ and 300 ℃, respectively. The strength limit of the material is about 500MPa and 520Mpa in 200 ℃ and 300 ℃, respectively. It means that the stresses of the front disk are approaching to the strength limit of the material, which greatly increase the risk of crack failure.

Conclusion: Crack failures occurred in 600MW steam turbine units of a power plant in China are analyzed in this work. Stress and strain of pulverizing wheel in design condition and worn condition are calculated to find out the root causes of crack failures.

Based on the analysis, the crack causes are determined including: (i) the initial stress and strain exceed the yield limit under the normal operation condition, which means the design stress is high and the safety coefficient is low; (ii) the stress and strain increase significantly in worn condition, resulting in greatly increased risk of failure. Both causes play important effects in the crack failures of the pulverizing wheel.

References:

[1] R. G. Zalosh, Review of coal pulverizer fire and explosion incidents, in: Industrial Dust Explosions: Symposium on Industrial Dust Explosions: Pittsburgh, Pennsylvania, 10-12 June 1986, volume 958, ASTM International, 1987, p. 191.

[2] G. Zhang, T. Zhanyuan, Exploration on crack cause of coal mill beaters, Northeast Electric Power Technology 22 (2001) 24–26.

[3] A. Pavan, K. Vikrant, M. Swamy, G. Jayaraman, Root cause analysis of bowl-mill pinion shaft failures, Case Studies in Engineering Failure Analysis 1 (2013) 103–109.

[4] N. Parida, S. Tarafder, S. K. Das, P. Kumar, G. Das, V. Ranganath, D. Bhattacharya, Failure analysis of coal pulverizer mill shaft, Engineering Failure Analysis 10 (2003) 733–744.

[5] B. Ravi Kumar, D. Bhattacharya, S. K. Das, S. Ghosh Chowdhury, Premature fatigue failure of a spring due to quench cracks, Engineering Failure Analysis 7 (2000) 377–384.

[6] B. Wen, Handbook of Metal Material, Publishing House of Electronics Industry, Beijing, 2013.

ADAPTIVE WAVELET DECOMPOSITION

Dr. Suri Ganeriwala & Dr. Budhaditya Hazra
SpectraQuest Inc.
Richmond, Virginia, USA

SpectraQuest, Inc.
8227 Hermitage Road, Richmond, VA 23228 U.S.A.
TEL +1-804-261-3300/FAX +1-804-261-3303
www.spectraquest.com

MFPT-2014
Virginia Beach, Virginia

Why Wavelet and what is it?

- Fourier expansion cannot be used for non-stationary signals as it can tell about frequencies but nothing about time
- To overcome this Short time Fourier transform (STFT) was developed
- STFT can either give good resolution of tome or frequency not both
- Wavelet transform was developed to overcome this resolution problem
- This done via contraction (time resolution) and expansion (frequency resolution)

SpectraQuest, Inc. MFPT-2014

Introduction

Issues with Wavelet Decompositions(WDs)

- Non-adaptive basis
- Once chosen, cannot be altered
- Wavelet packets are attractive but have a fixed ratio of decomposition (e.g. dyadic)
- Adaptability of basis is the key requirement which is missing in most of the traditional WDs

SpectraQuest, Inc. MFPT-2014

Issues with WDs

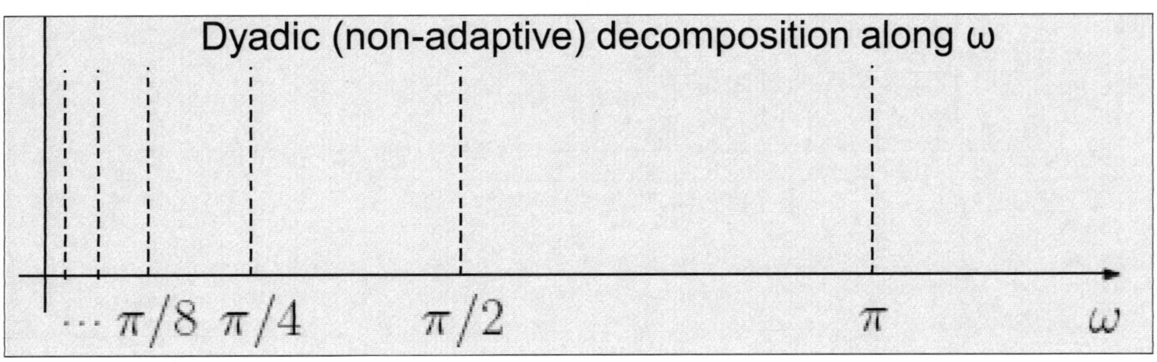

SpectraQuest, Inc. MFPT-2014

Adaptive Wavelet Decomposition

Motivation of AWD

- Fourier transform based approach to construct a bank of band pass filters in [0, π]
- Fourier support [0, π] is segmented into **N** contiguous segments

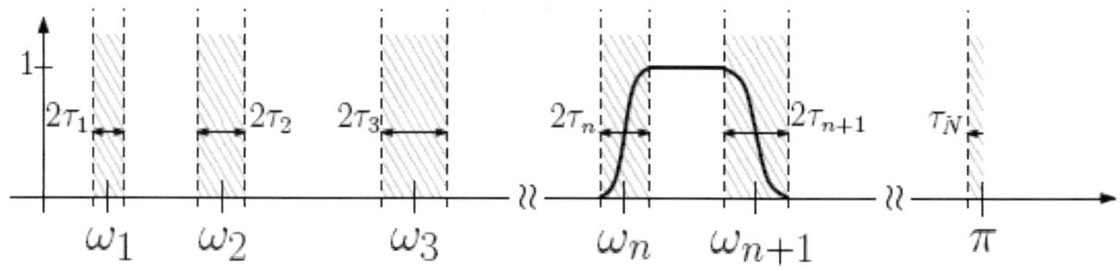

Basic Functions

$$\hat{\phi}_n(\omega) = \begin{cases} 1 & \text{if } |\omega| \le \omega_n - \tau_n \\ \cos\left[\frac{\pi}{2}\beta\left(\frac{1}{2\tau_n}(|\omega| - \omega_n + \tau_n)\right)\right] & \\ & \text{if } \omega_n - \tau_n \le |\omega| \le \omega_n + \tau_n \\ 0 & \text{otherwise} \end{cases}$$

Like SCALING FUNCTIONS

$$\hat{\psi}_n(\omega) = \begin{cases} 1 & \text{if } \omega_n + \tau_n \le |\omega| \le \omega_{n+1} - \tau_{n+1} \\ \cos\left[\frac{\pi}{2}\beta\left(\frac{1}{2\tau_{n+1}}(|\omega| - \omega_{n+1} + \tau_{n+1})\right)\right] & \\ & \text{if } \omega_{n+1} - \tau_{n+1} \le |\omega| \le \omega_{n+1} + \tau_{n+1} \\ \sin\left[\frac{\pi}{2}\beta\left(\frac{1}{2\tau_n}(|\omega| - \omega_n + \tau_n)\right)\right] & \\ & \text{if } \omega_n - \tau_n \le |\omega| \le \omega_n + \tau_n \\ 0 & \text{otherwise.} \end{cases}$$

Like WAVELET FUNCTIONS

$$\beta(x) = \begin{cases} 0 & \text{if } x \le 0 \\ 1 & \text{if } x \ge 1 \end{cases} \text{ and } \beta(x) + \beta(1-x) = 1 \quad \forall x \in [0,1].$$

Segmentation of Fourier Spectrum

- Separate different portions of the spectrum which correspond to modes; centered around a specific frequency
- For N modes, there are N-1 boundaries
- Boundaries: How to find them ?
  - ❖ **Detect the local maxima and sort them in decreasing order**

SpectraQuest, Inc. MFPT-2014

Segmentation of Fourier transform

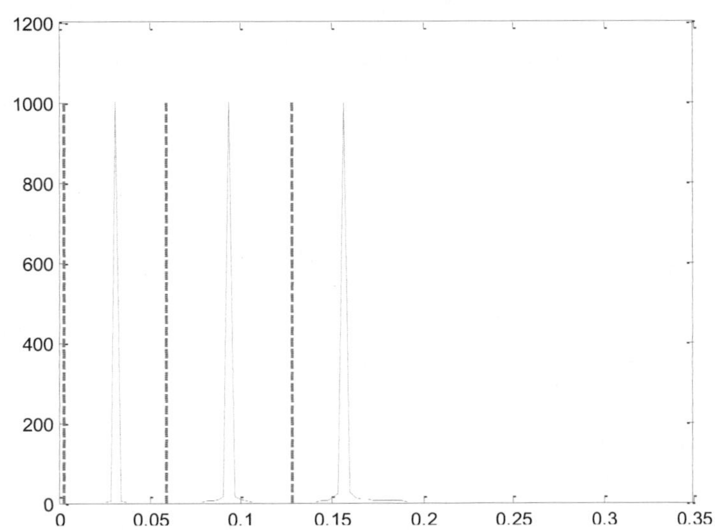

- Find the local maxima
- Take support boundaries as the middle between successive maxima

SpectraQuest, Inc. MFPT-2014

AWD: Definitions

AWT

$$\mathcal{W}_f^{\mathcal{E}}(n,t) = \langle f, \psi_n \rangle = \int f(\tau)\overline{\psi_n(\tau - t)}d\tau$$
$$= \left(\hat{f}(\omega)\overline{\hat{\psi}_n(\omega)}\right)^{\vee},$$

$$f(t) = \mathcal{W}_f^{\mathcal{E}}(0,t) \star \phi_1(t) + \sum_{n=1}^{N} \mathcal{W}_f^{\mathcal{E}}(n,t) \star \psi_n(t)$$ ➔ Reconstruction

$$f_k(t) = \mathcal{W}_f^{\mathcal{E}}(k,t) \star \psi_k(t)$$ ➔ **Modes**

AWT: Algorithmic Details

Input: **f, N** (number of components ≡ no of scales)

- Fourier transform of the signal [X(f)]
- Boundary detection: Compute the local maxima of X(f) on [0 π] and find the set $\{\omega_n\}$
- Build the filter bank
- Filter the signal to get each component (IMF)

Boundary Detection

- Pre-processing:
  - ➢ Remove trend (local and global)
  - ➢ Morphological mathematics
- Local maxima
- Local minima between 2 consecutive maxima
- Histogram segmentation

Generate Filter-banks & Recover Each Component

- Decomposition the segment [ω_n, ω_{n+1}]
- Generate 1D filter-bank

$$\hat{\psi}_n(\omega) = \begin{cases} 1 & \text{if } \omega_n + \tau_n \leq |\omega| \leq \omega_{n+1} - \tau_{n+1} \\ \cos\left[\frac{\pi}{2}\beta\left(\frac{1}{2\tau_{n+1}}(|\omega| - \omega_{n+1} + \tau_{n+1})\right)\right] & \\ & \text{if } \omega_{n+1} - \tau_{n+1} \leq |\omega| \leq \omega_{n+1} + \tau_{n+1} \\ \sin\left[\frac{\pi}{2}\beta\left(\frac{1}{2\tau_n}(|\omega| - \omega_n + \tau_n)\right)\right] & \\ & \text{if } \omega_n - \tau_n \leq |\omega| \leq \omega_n + \tau_n \\ 0 & \text{otherwise.} \end{cases}$$

- Filter the signal to extract each component

Case studies: Case-1

Original Components

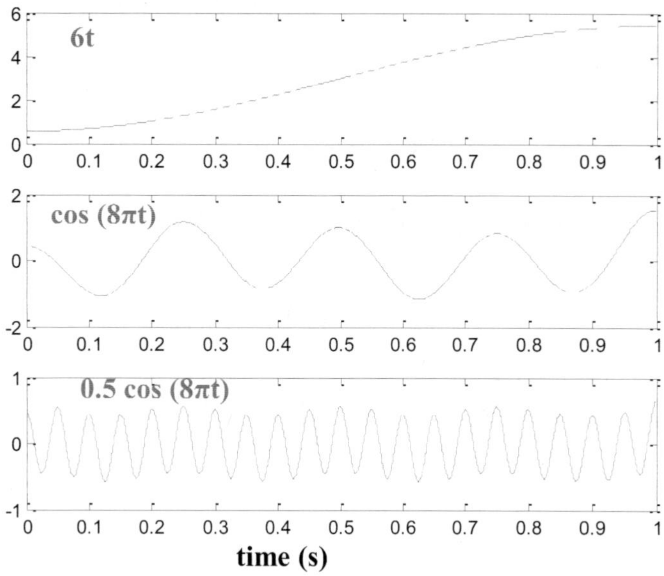

$$f_{c1}(t) = 6t$$
$$f_{c2}(t) = \cos(8\pi t)$$
$$f_{c3}(t) = 0.5\cos(40\pi t)$$

Case-1: Decomposed Components & Reconstruction

Recovered components

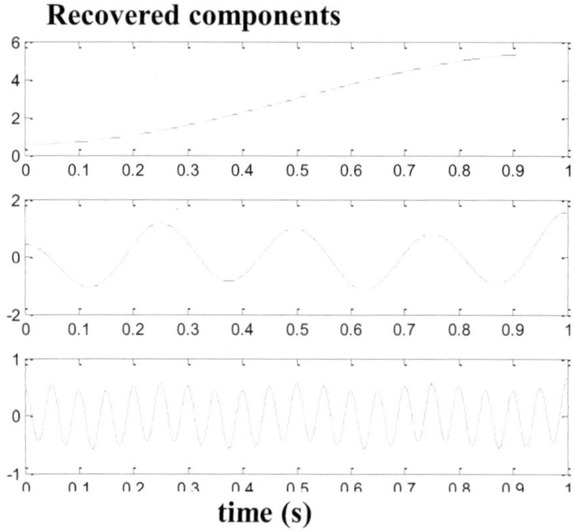

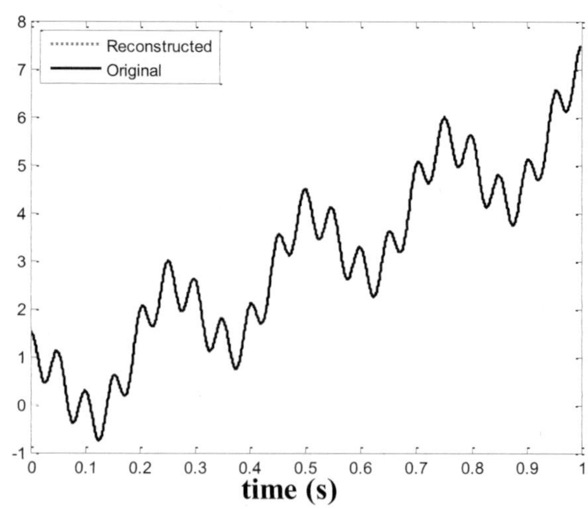

Case-1: Noisy Signal & Its Denoising

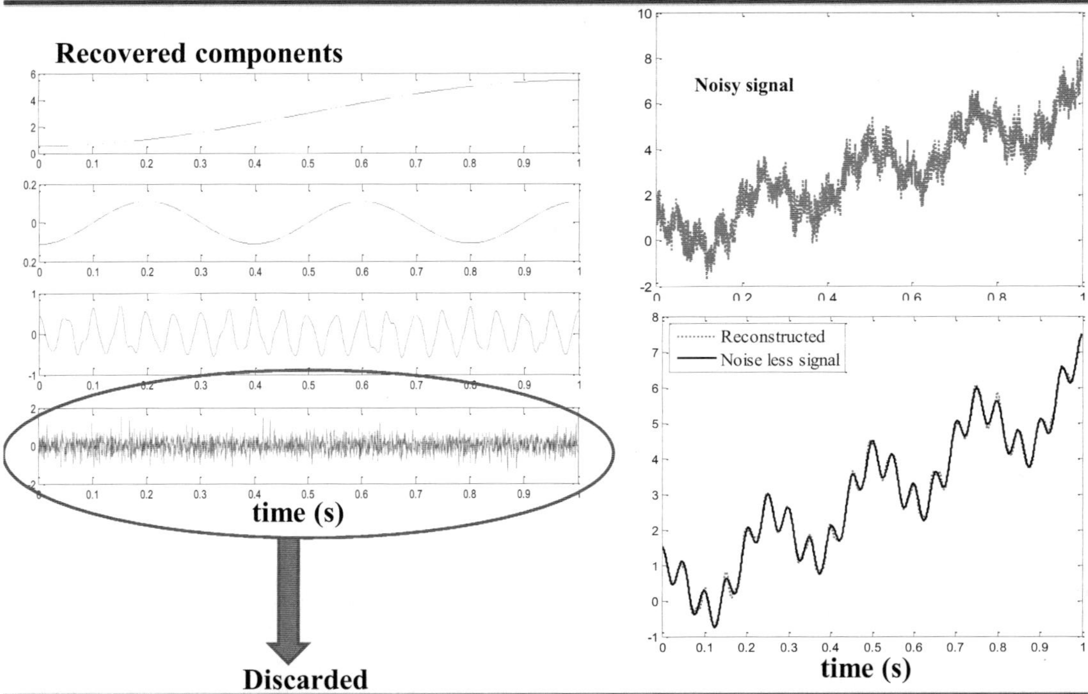

SpectraQuest, Inc. MFPT-2014

Case-2: Sine plus AM-FM

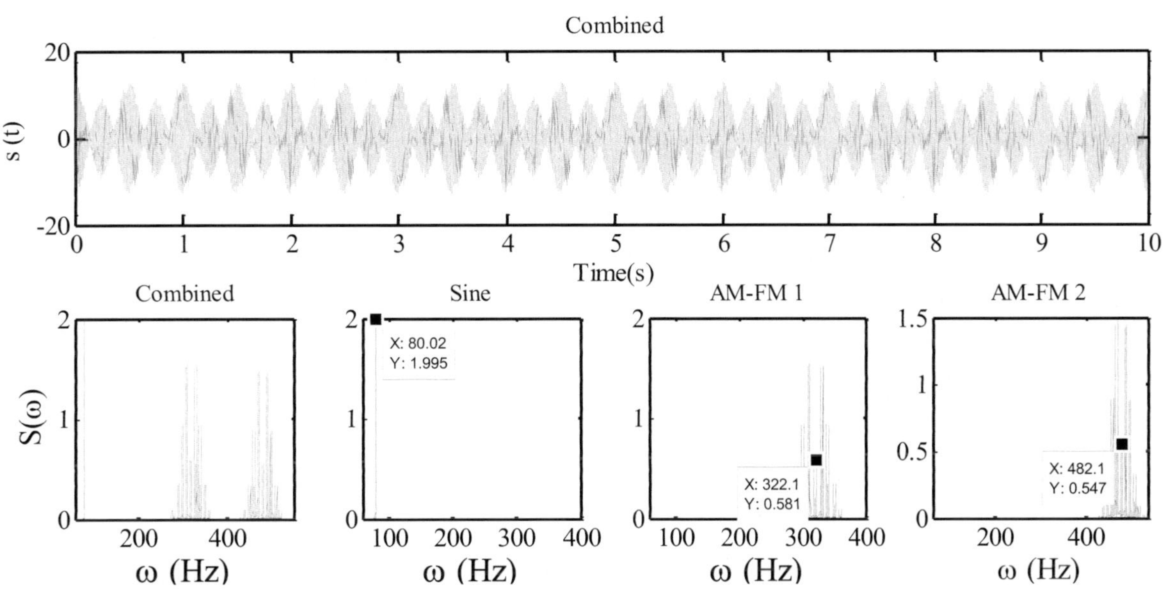

SpectraQuest, Inc. MFPT-2014

Partitioning of Fourier Spectrum

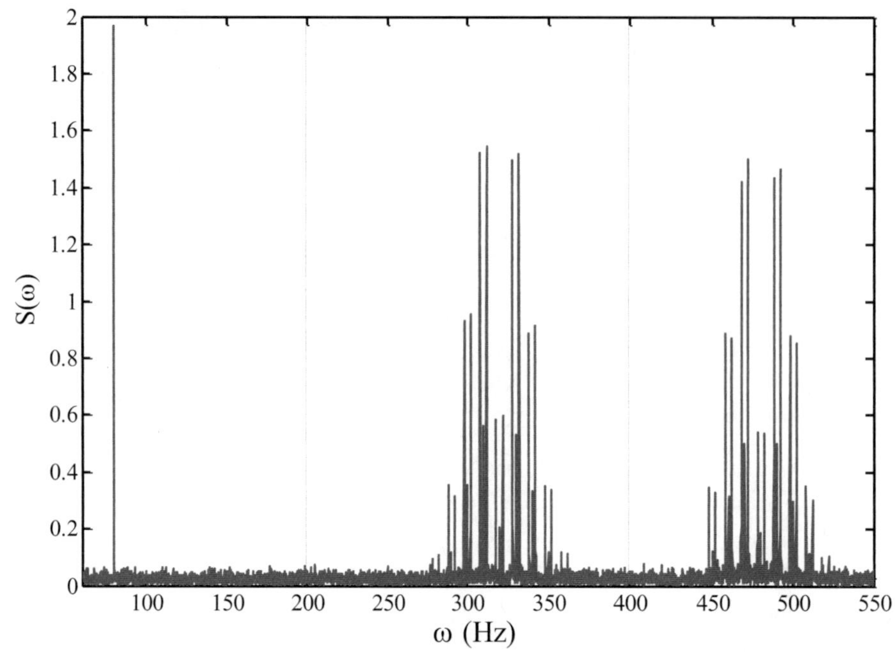

Qi SpectraQuest, Inc.

MFPT-2014

Recovered components for 20% noise

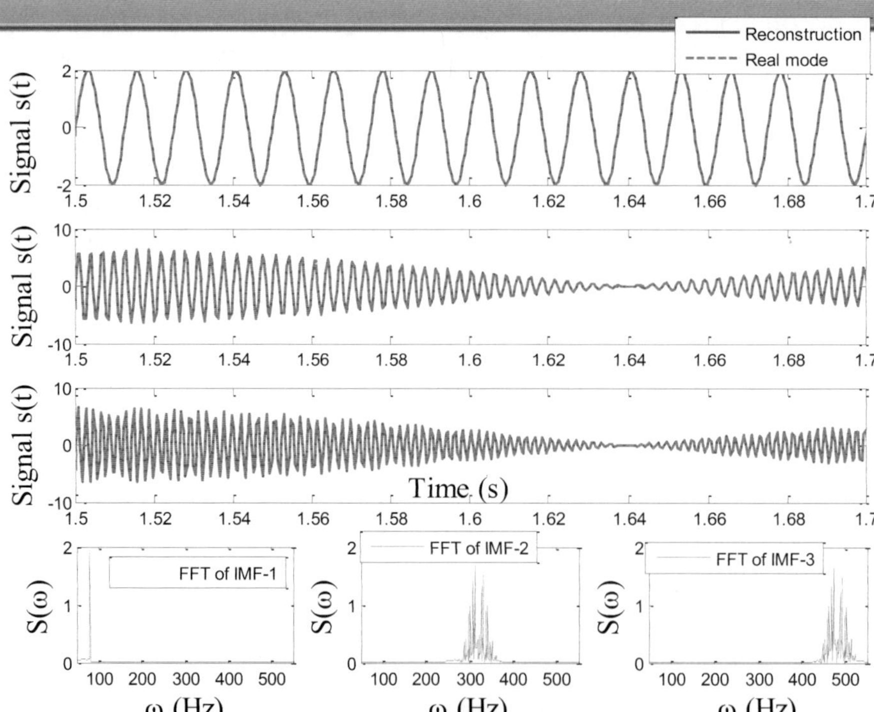

Qi SpectraQuest, Inc.

MFPT-2014

Filter-bank Property

8 channel white noise data is used and 9 components were extracted using AWD &EMD

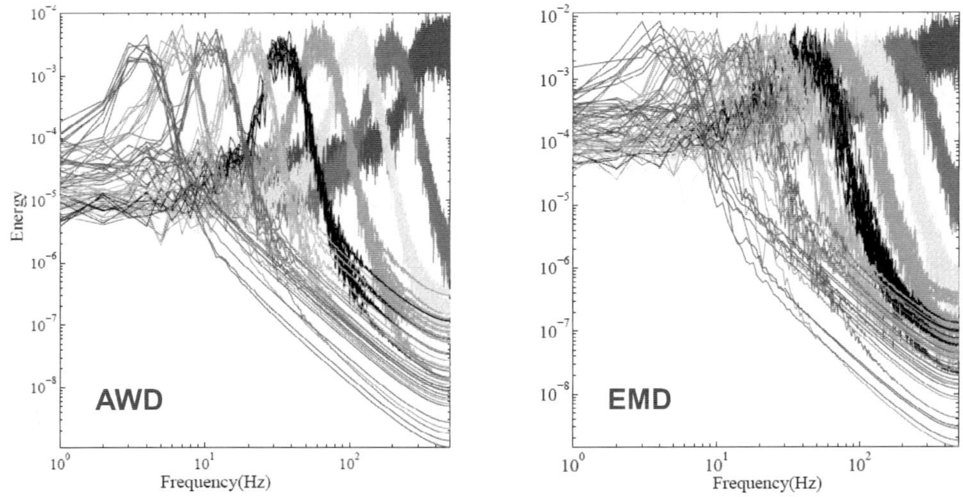

Recovered components using EMD

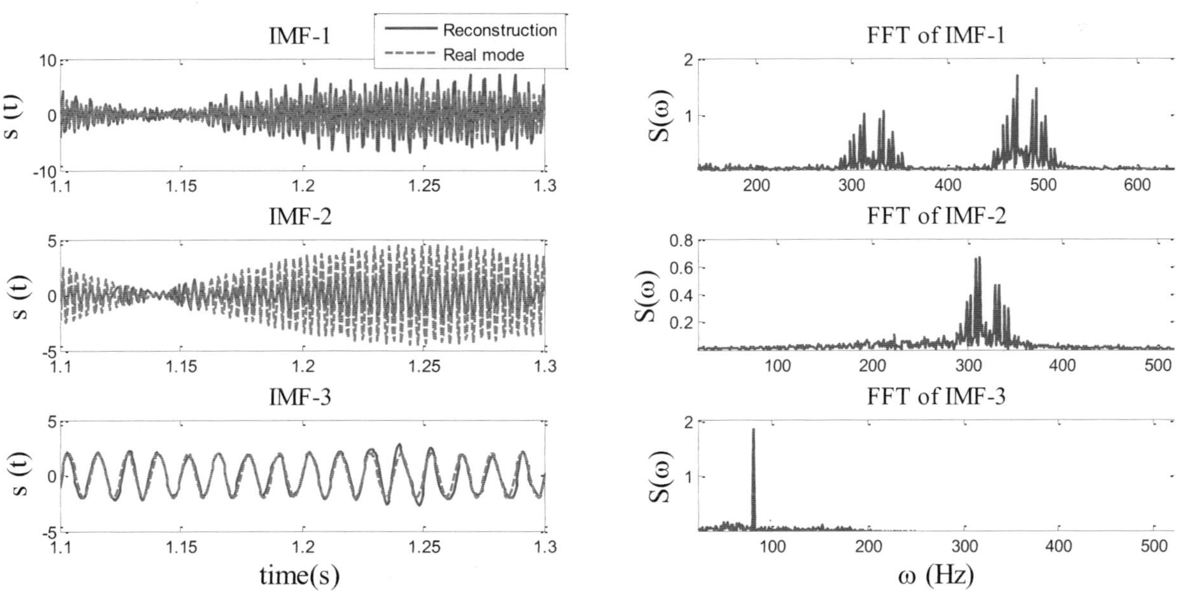

Results using field data

- Helicopter gearbox data
- Seeded fault test in Trenton, New Jersey
- Data: Input and output accelerometers and Tachometer
- Sampled at 100 KHz
- Only 3 out of the 36 datasets are made publicly available
- 3 data sets correspond to 3 health states: new, middle age and worn out
- Website:http://qsun.eng.ua.edu/cpw_web_new/gear1.htm
- Data partitioned into windows of 65536 samples each giving total 21 windows of data for spectral analysis

Results: Fourier Spectra

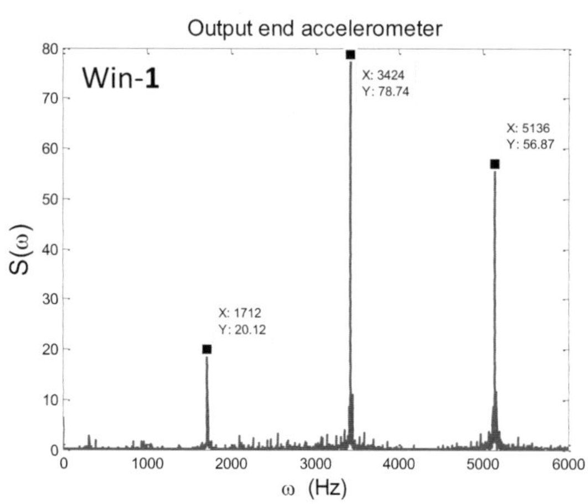

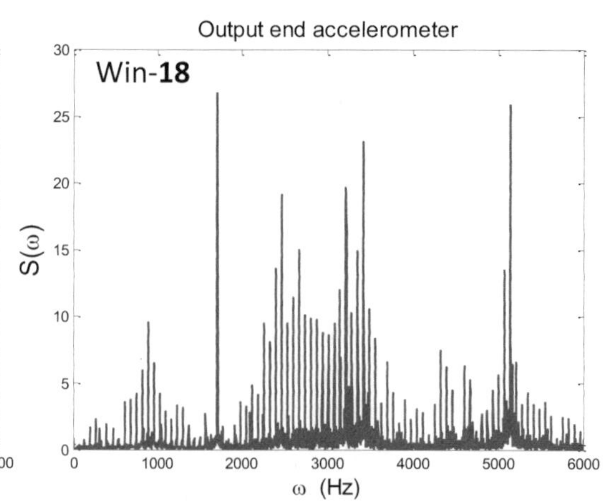

Results: Decomposed components

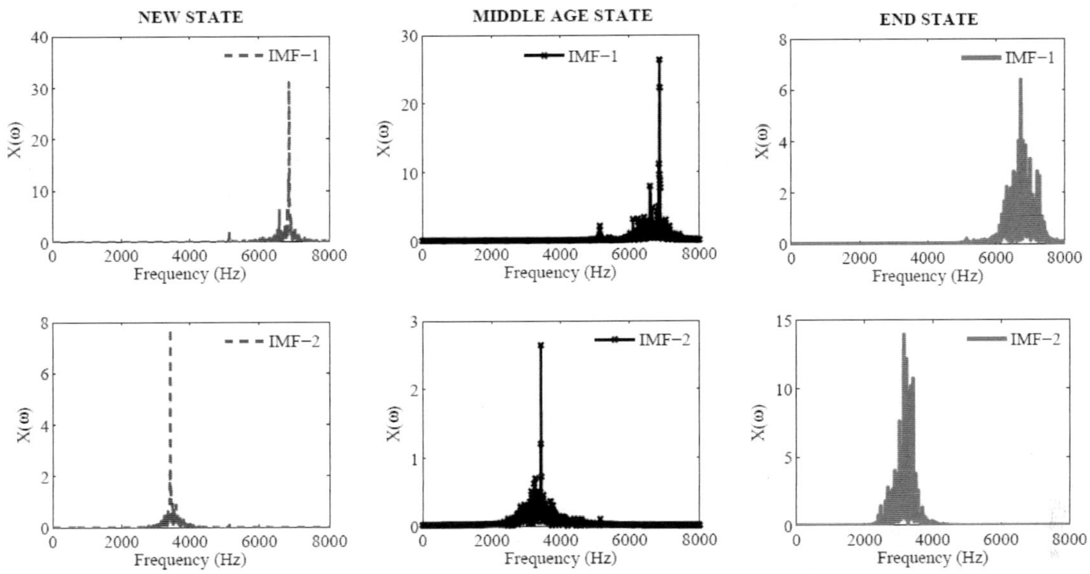

Results: Condition indicators (CI)

Variance	$\sigma^2(s(t)) = \frac{1}{N}\sum_{i=1}^{N}(s(t_i) - \mu)^2$
Kurtosis	$\kappa(s(t)) = \frac{1}{N}\sum_{i=1}^{N}\frac{(s(t_i)-\mu)^4}{\sigma^4}$
Energy operator(EOP)	$\chi = s(i+1)^2 - s(i)^2$
EOP variance and EOP kurtosis	$\sigma^2(\chi); \kappa(\chi)$
Crest factor	$CF = \frac{\max(s) - \min(s)}{\sigma}$

Results: Recursive CI's

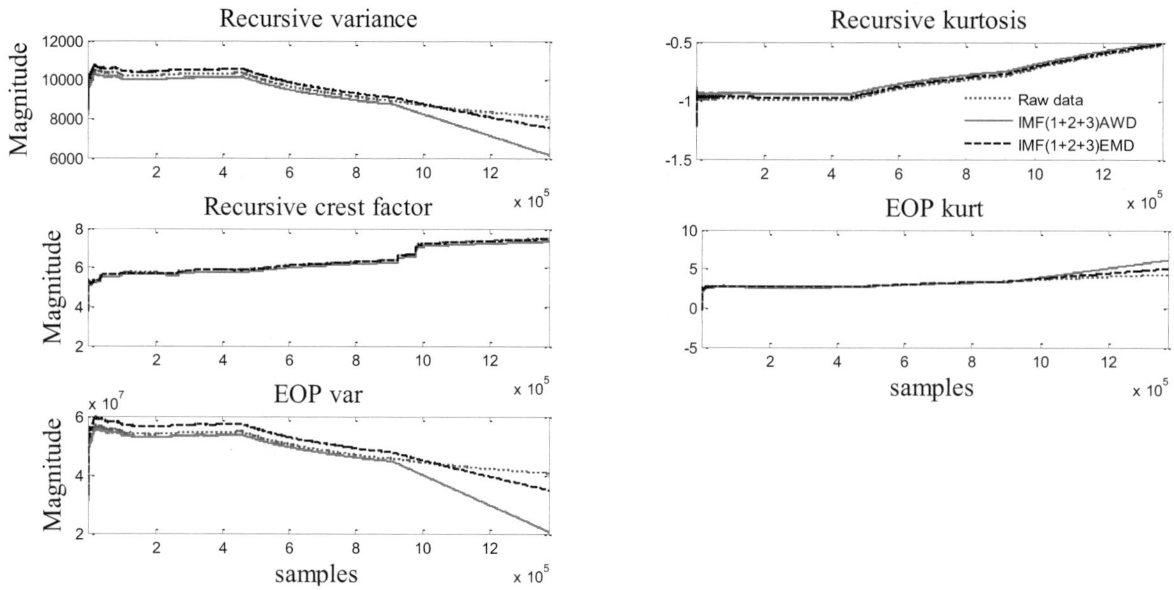

MFPT-2014

Conclusions

- Adaptive wavelet decomposition (AWD) has been introduced in the context of rotating machinery diagnosis
- AWD provides an improved decomposition capability and filter-bank structure than EMD
- AWD is also able to extract the harmonic, non-stationary and time-varying components from rotating machinery signals
- The proposed method provides better detectability compared to the cases where only raw-data is used

MFPT-2014

A COMPARISON OF DIFFERENT TECHNIQUES FOR ROLLING ELEMENT BEARING FAULT DIAGNOSIS

Suri Ganeriwala, Ph.D.
&
Liang Xu

SpectraQuest Inc.
Richmond, Virginia, USA
suri@spectraquest.com

SpectraQuest, Inc.
8227 Hermitage Road, Richmond, VA 23228 U.S.A.
TEL +1-804-261-3300/FAX +1-804-261-3303
www.spectraquest.com

MFPT-2014
Virginia Beach, Virginia

Issues with Bearing Diagnostics

Often bearing signals are massed by near by high amplitude signals and are hard to detect even using enveloping

- Bearings signals are impulsive and weak giving very small fault frequencies values (low signal to noise ratio)
- The random slip/frequency variation (1-2%) means that high "harmonics" in the spectrum of the raw signal are Smeared
- Direct frequency analysis gives little information
- Resonances enhance high frequencies response
- Frequency analysis of the envelope solves this problem, but does not work all the times

Rolling Element Bearing Faults

- Ball damage
- Inner race defect
- Outer race defect
- Cage damage

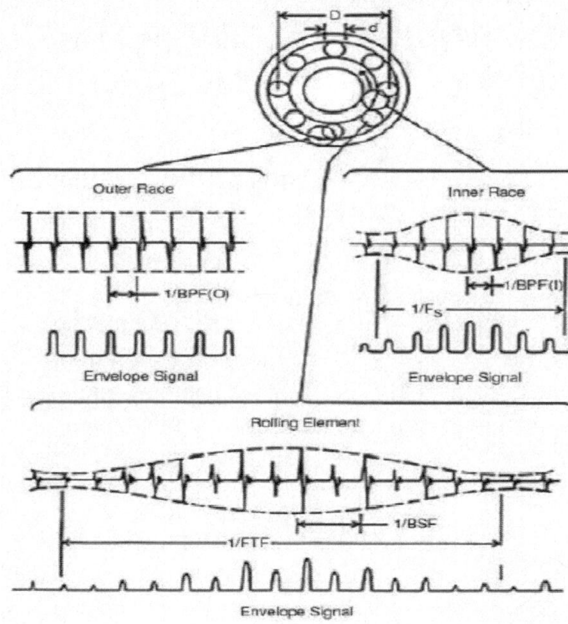

Rolling Element Bearing Faults

ball passing frequency outer race $BPFO = \frac{nf_r}{2}\left\{1 - \frac{d}{D}\cos\phi\right\}$

ball passing frequency inner race $BPFI = \frac{nf_r}{2}\left\{1 + \frac{d}{D}\cos\phi\right\}$

fundamental train frequency $FTF = \frac{f_r}{2}\left\{1 - \frac{d}{D}\cos\phi\right\}$

Ball spin frequency $BSF = \frac{D}{2d}\left\{1 - \left(\frac{d}{D}\cos\phi\right)^2\right\}$

D = pitch dia; d = ball dia; ϕ = contact angle;
n = no. of balls

Bearing Defects

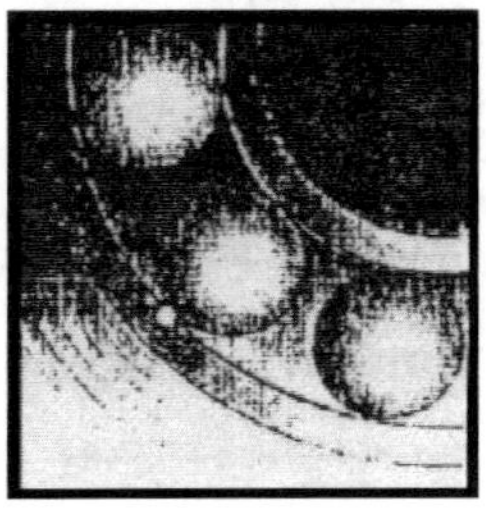

BPFO

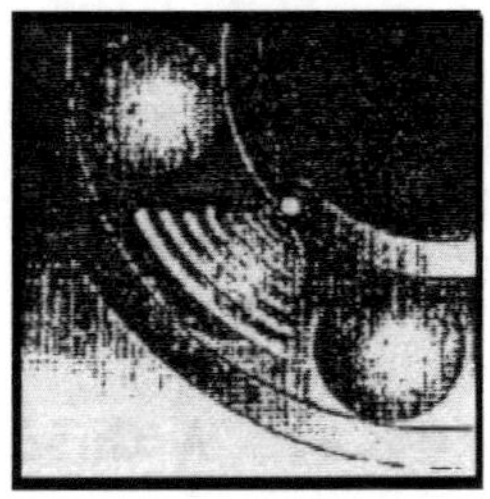

BPFI

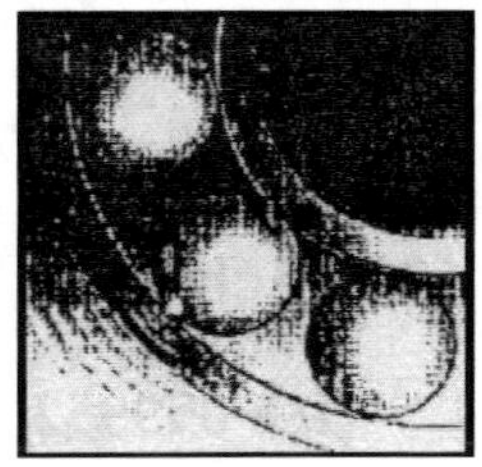

BSF

5

Time Domain Impact Response

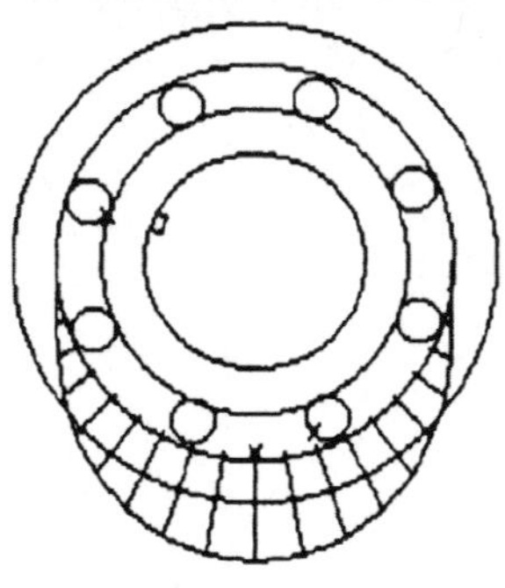

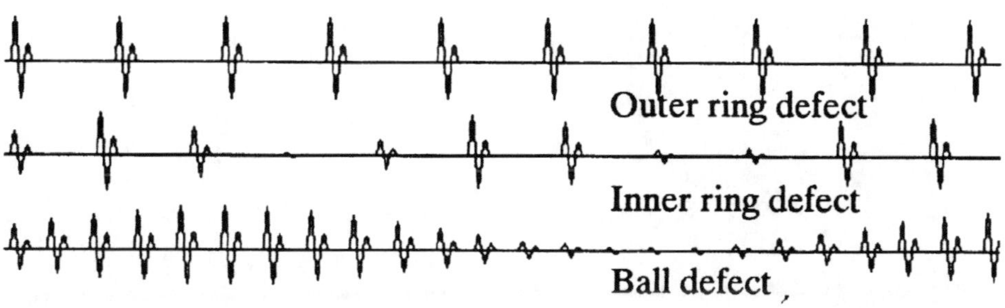

6

Illustration of Sidebands

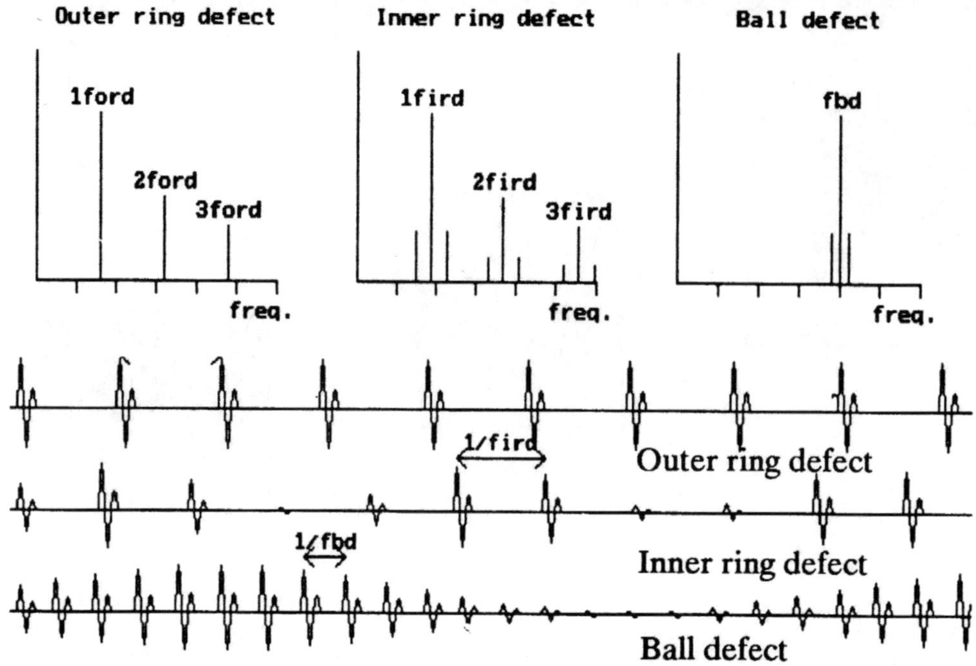

Rolling Element Bearing Faults Analysis

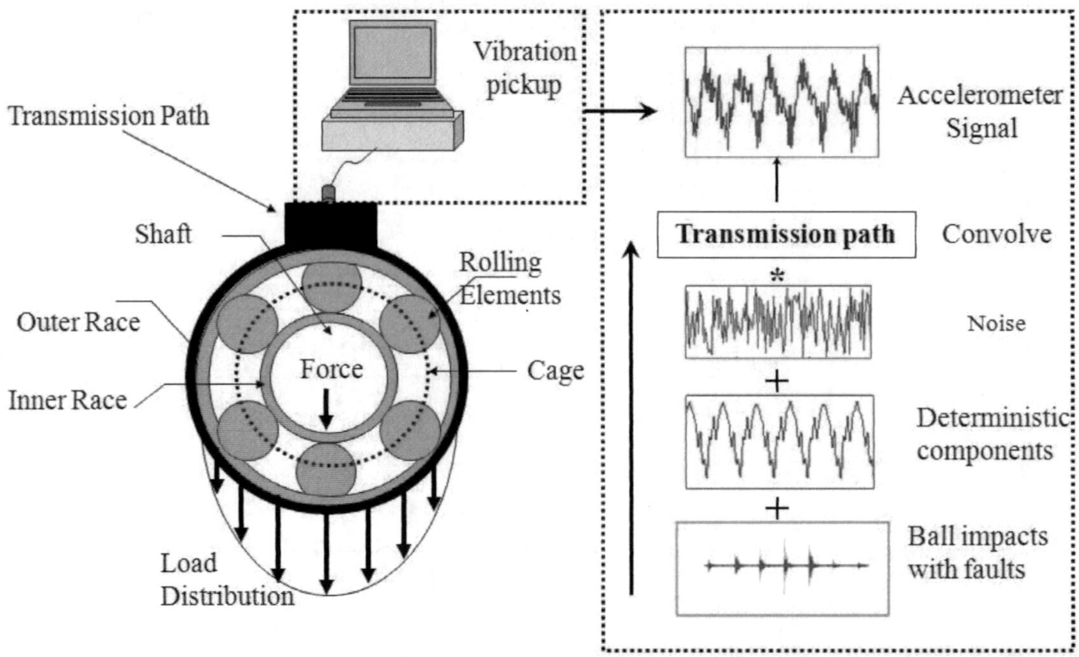

Signal Pre-processing stages

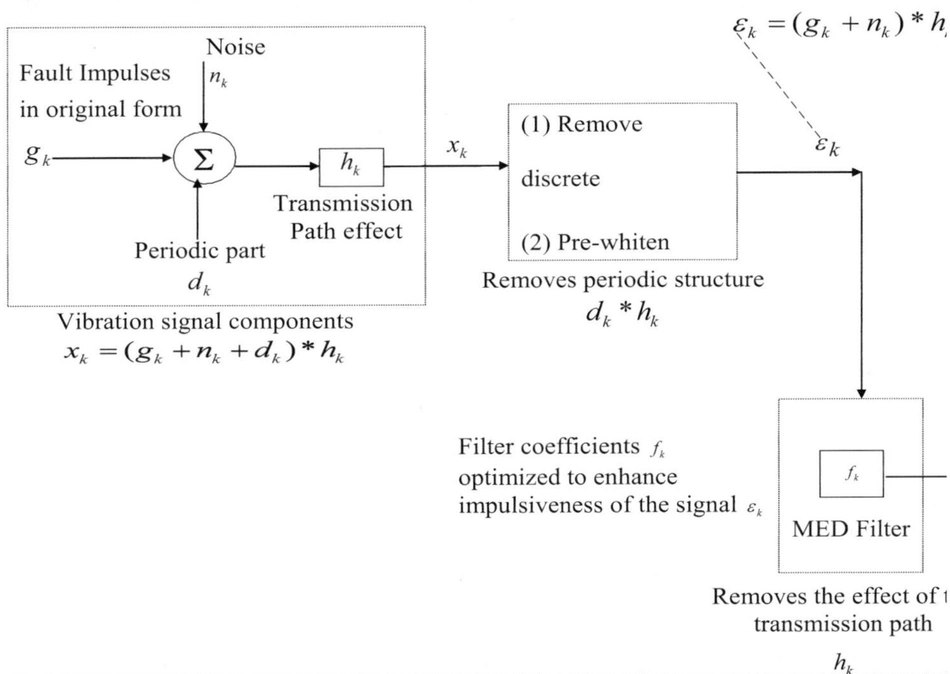

Detailed Processing Approach

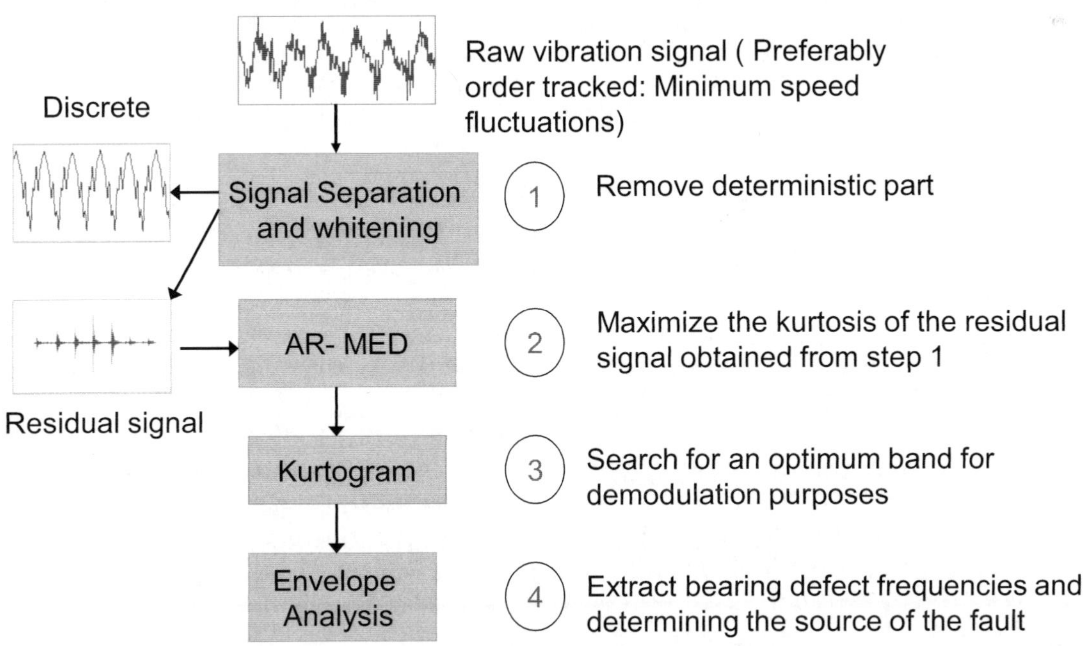

New Techniques

Signal Separation:

- Time Synchronous Averaging
- Self-adaptive Noise Cancellation (SANC)
- Discrete Random Separation

Signal Enhancement:

- Auto Regression Modeling
- Minimum Entropy Deconvolution

Determining Optimum Band for Enveloping

- Spectral Kurtosis

New Efficient Technique, use full spectrum enveloping

- Cepstrum Pre-whitening

SQi SpectraQuest, Inc. MFPT-2014

Signal Separation into Deterministic (Discrete) and Random

- Time synchronous averages (TSA) for each shaft in the system are found and subtracted from the total signal to give a residual signal that contains bearing signals and other nonstationarities in the system.
- Discrete Random Separation (DRS).
  - This utilizes the signal its self by generating a linear transfer function (using FFT methods) between the signal and a delayed version
  - In order to generate the linear transfer function and implement the DRS, the analyst has to select a suitable time delay and decide on the filter length.
  - The minimal filter length should be inversely proportional to the minimal frequency spacing between two discrete components in the power spectrum, which are to be enhanced independently
  - The minimal time delay must be longer than the autocorrelation length (memory) of the random signal to be filtered out.

SQi SpectraQuest, Inc. MFPT-2014

Signal Enhancing Using Autoregressive Models(AR)

- In an Autoregressive model (AR), the value Y for sample number n is expressed as a linear combination of previous p elements, i.e

$$Y_n=\underline{a2}Y_{n-1}+\underline{a3}Y_{n-2}+\underline{a4}Y_{n-3}+..........+\underline{a(p+1)}Y_{n-p}$$

- Order *p can be selected to maximize impulsiveness of residual signal*

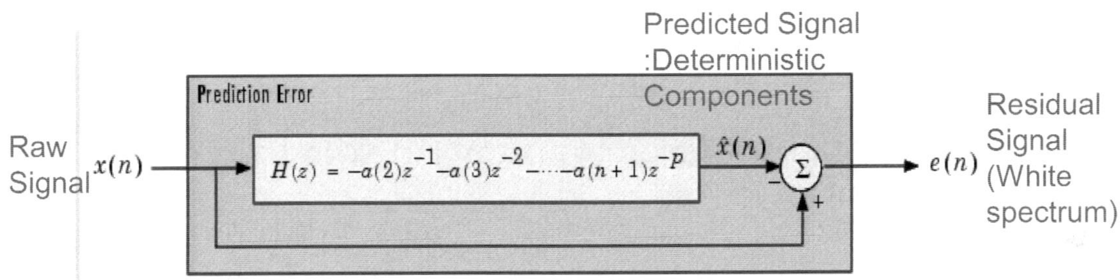

Minimum Entropy Deconvolution (MED) Technique for signal enhancement

- The MED technique aims at deconvolving the effect of the transmission path from the signal and clarifies the impulses.
- It was originally developed by (Wiggins, 1978) to aid extraction of reflectivity information in seismic data.
- MED is an iterative technique which searches for an optimum set of filter coefficients (Initial assumption as a delta function) that recover the output signal (of an inverse filter) with the maximum value of kurtosis.
- MED has been successfully used on Bearings in the cases of limited sampling frequency where impulses were overlapped and in cases of transducers mounted far from bearing.

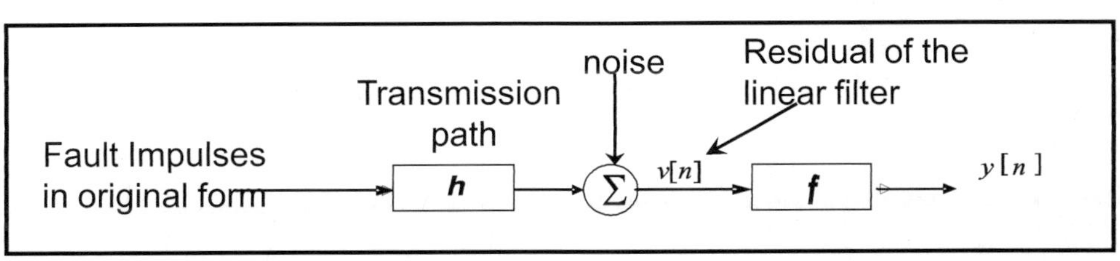

Spectral Kurtosis (SK)

1. Kurtosis (time Kurtosis) is a statistical parameter that is sensitive to spikiness of the signal.

2. Definition has been expanded to monitor the spikiness for every single frequency. "using a joint time frequency representation "

$$K_Y = \langle E\{Y(t)^4\}\rangle / \langle E\{Y(t)^2\}\rangle^2 - 3$$

3. First used in detecting transients in sonar signals (Dwyer 1984)

Wavelet Kurtogram to select best bandwidth of highest impulsiveness (highest kurtosis value)

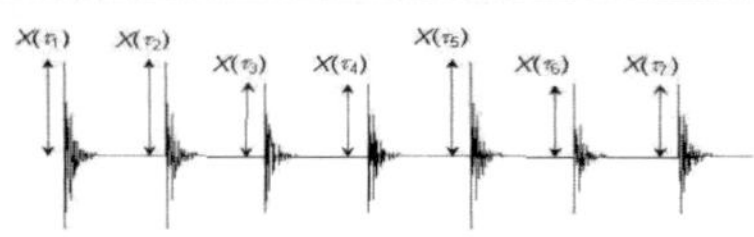

3 filters/Octave covering 4 octaves

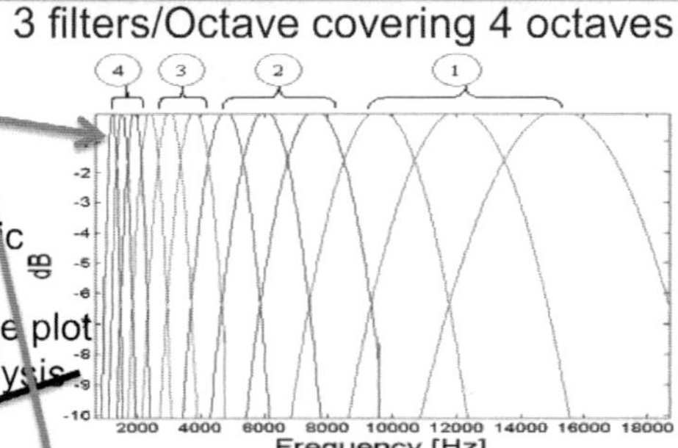

1. Signal Passed filtered using different filter characteristic
2. Kurtosis of each filtered signal is calculated
3. Kurtosis of each filtered signal is presented in an image plot
4. Filtered with highest kurtosis is used for envelope analysis

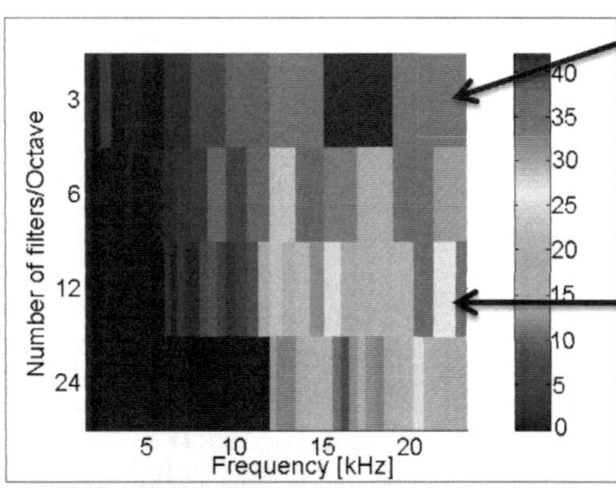

12 filters/Octave covering 1 octave

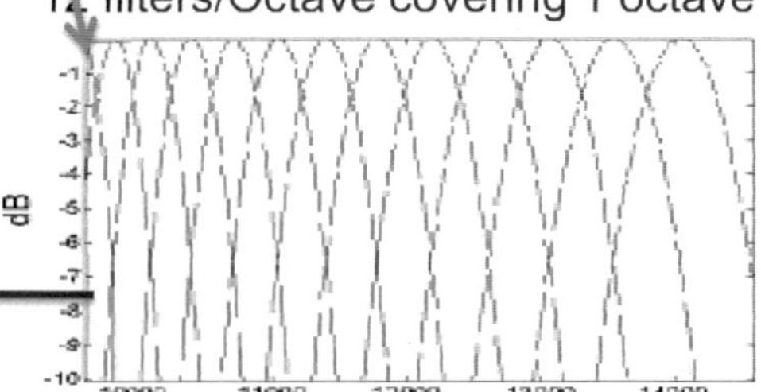

Frequency

Cepstrum whitening and full bandwidth enveloping

Cepstrum whitening involves:

- Transforming the signal into the cepstrum domain
- All discrete frequency components are set a very low value and phase of original data is saved
- To get back to the time domain, original phase information is restored before inverse transforming.
- Envelope analysis now performed over the entire band

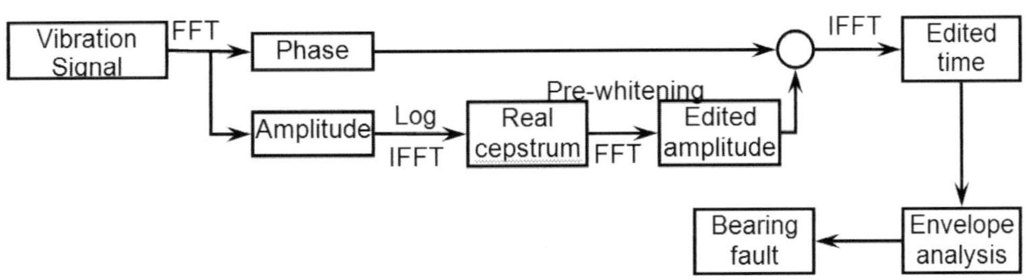

Experimental Setup

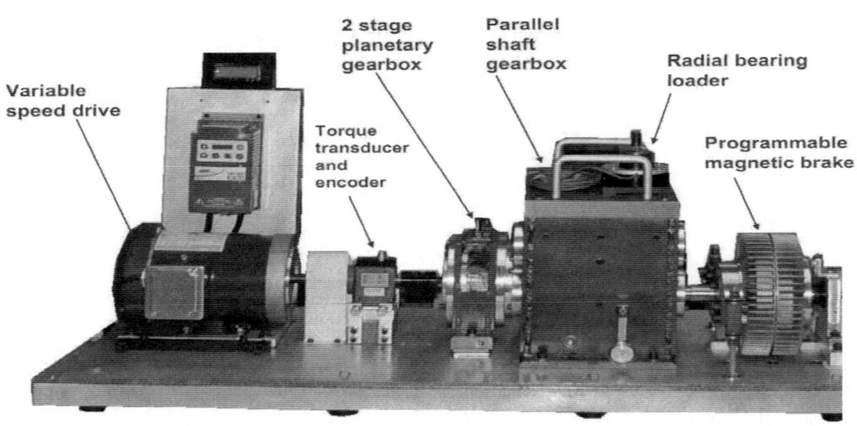

- Both torsional and radial loadings can be applied to study damage signature or propagation in gears and/or bearings.
- Torsional load is applied via a programmable magnetic particle brake and user-defined speed profiles.
- Radial load is applied to a shaft in the parallel gearbox.

Typical Data

Bearing with outer race fault

Bearing with inner race fault

Bearing with ball fault

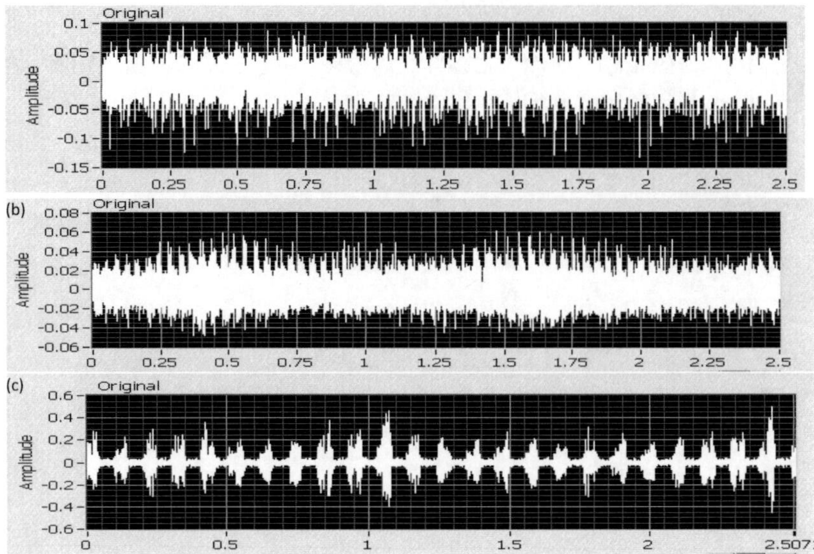

Using signal resampling based on the gears number of teeth

- Synchronous signal Extraction (No Gear fault case)
- Note high frequency content in intermediate and low shafts. See next sli(

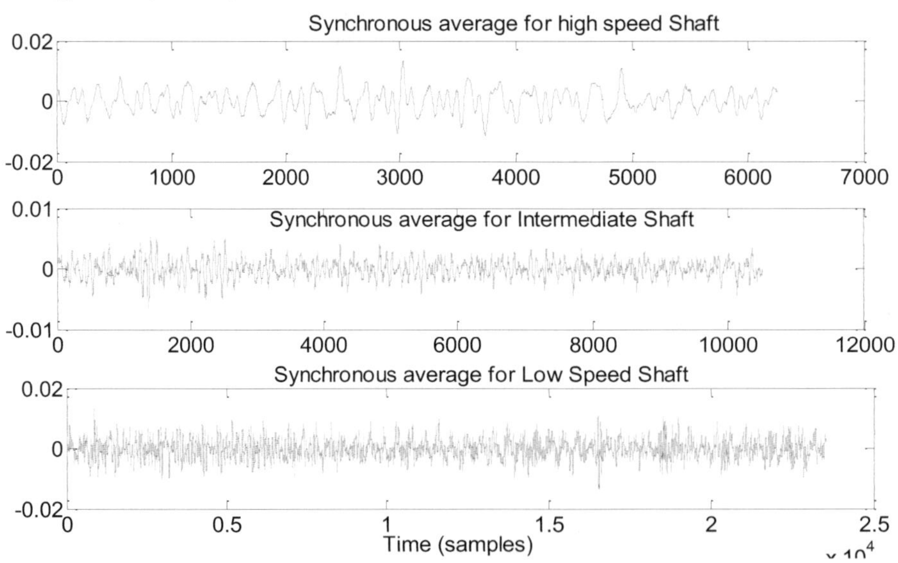

Raw signal compared to Residual after subtracting shaft synchronous components

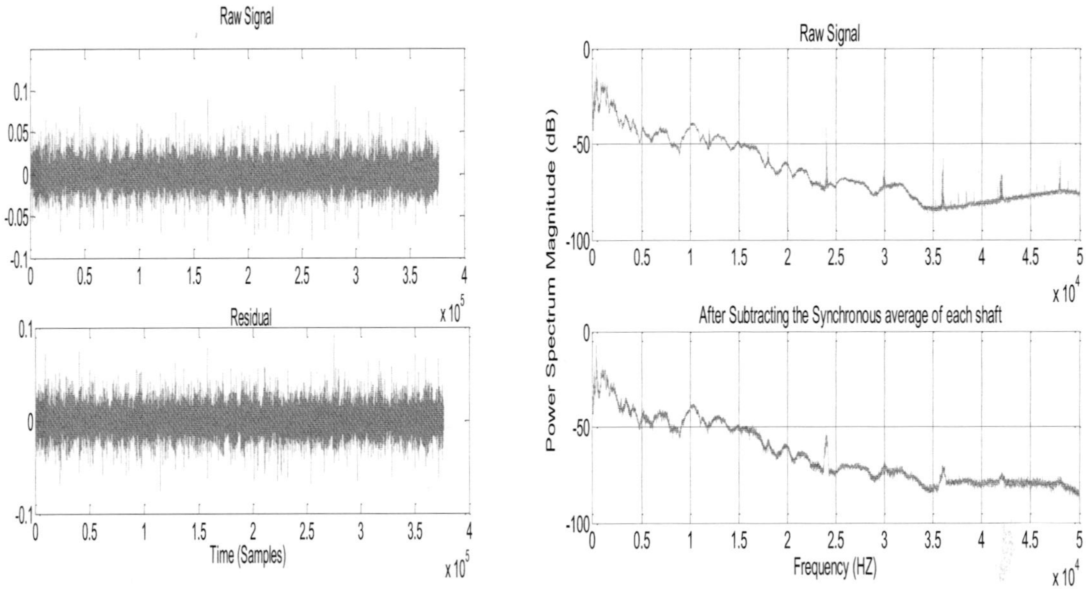

DRS : Time domain signal

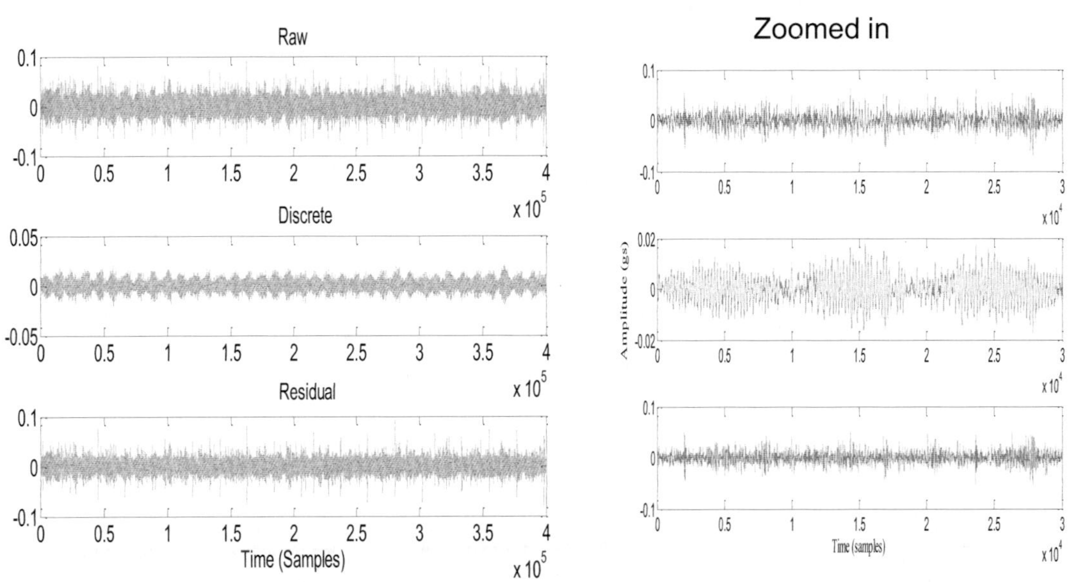

DRS Frequency Content

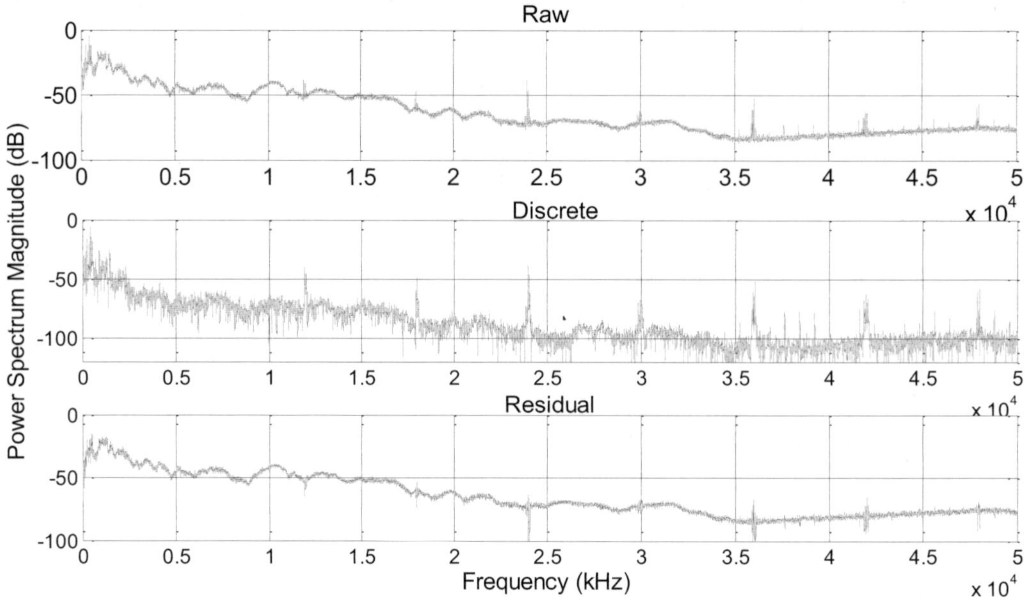

Signal Enhancement: AR-MED Filtration

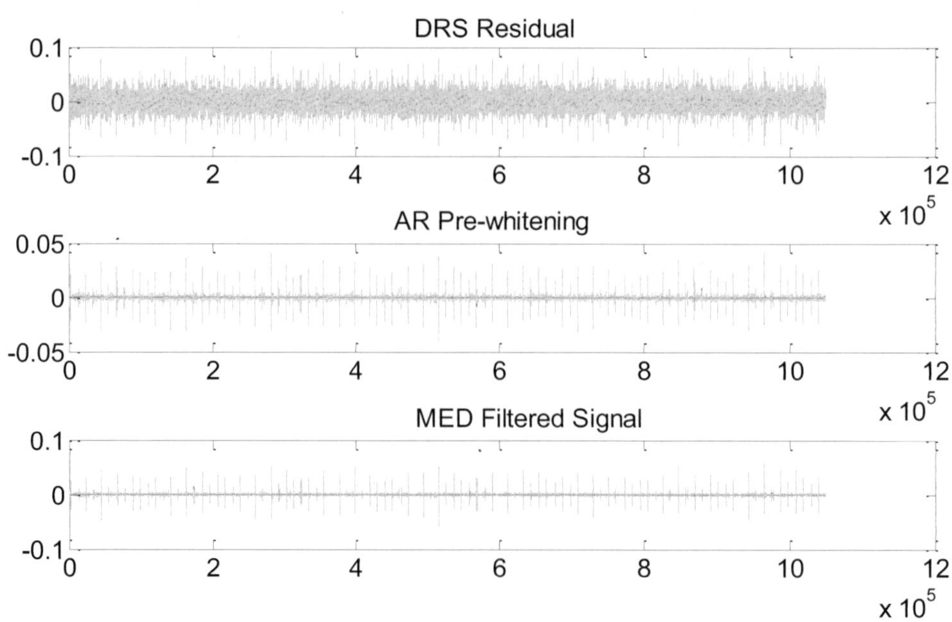

Signal Enhancement: AR-MED Filtration (Zoomed in)

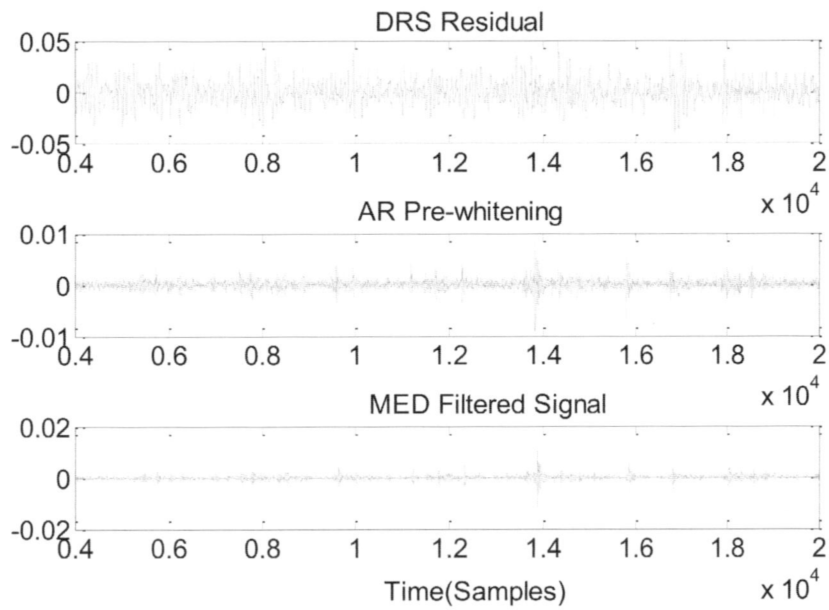

Wavelet Kurtogram

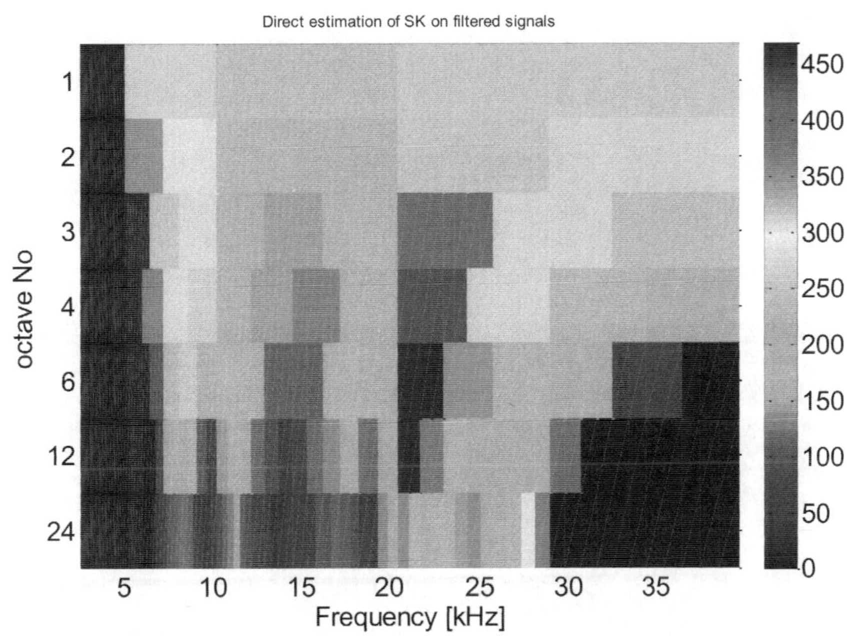

Squared enveloped based on SK

- BPFI at 52.52 Hz (*9*9.57*0.6*) Very strong modulation with Intermediate shaft sidebands

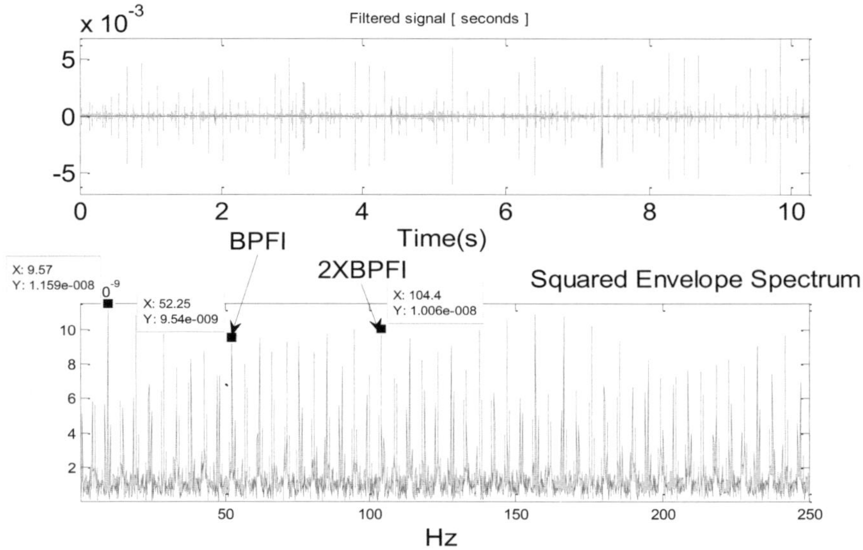

Cepstrum Pre-whitening-Inner race

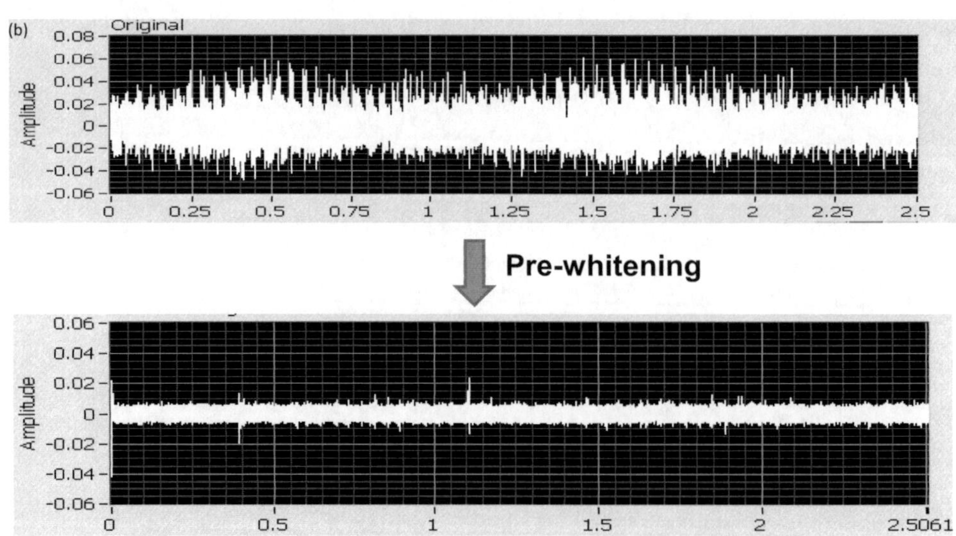

Bearing with inner race fault

Cepstrum Pre-whitening-Inner race

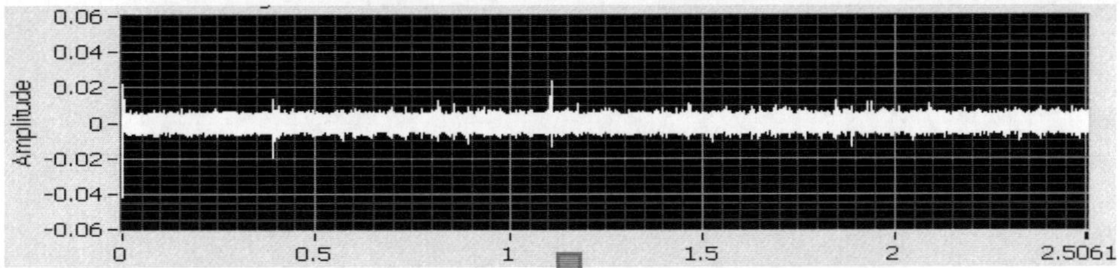

Envelope analysis

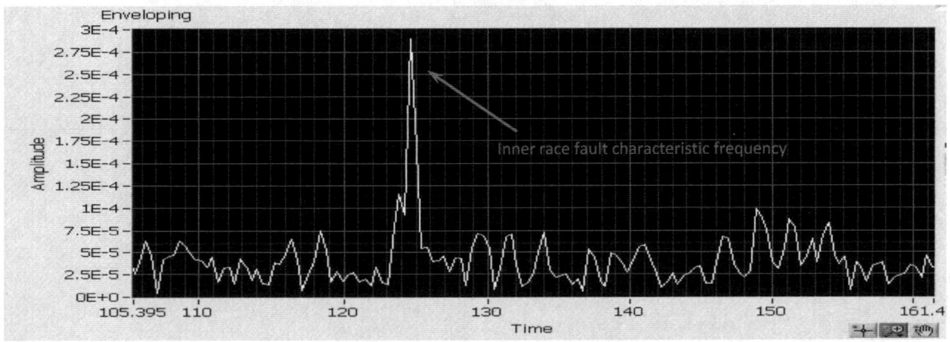

Bearing with inner race fault

Conclusions

- ❖ Bearing signals are non-stationary somewhat random and small (low signal to noise ration)
- ❖ A systematic Step-by-Step procedure has been outline to analyze rolling element bearing faults
- ❖ It is important to remove deterministic components (sinusoidal) and enhance impulsiveness of the random part
- ❖ Apply spectra Kurtosis to determine the optimum band for applying envelop
- ❖ Cepstrum pre-whitening is simple and has promise

Acknowledgement

My sincere thanks goes to:

Dr. Nader Sawahli and Jun Yang with data analysis
And
MFPT organization for organizing this conference

Comprehensive and Systematic Study of Shaft/Coupling Misalignment Signature using Vibration Analysis

Mate Toth and Suri Ganeriwala
SpectraQuest, Inc.
Richmond, VA 23228, USA

MFPT – May 22, 2014

SpectraQuest, Inc.

Outline

- Motivation
- What is misalignment?
- Investigated parameters
- Measurement setup
- Results
- Conclusions

SpectraQuest, Inc.

What is misalignment?

- The center lines of two shafts do no coincide.

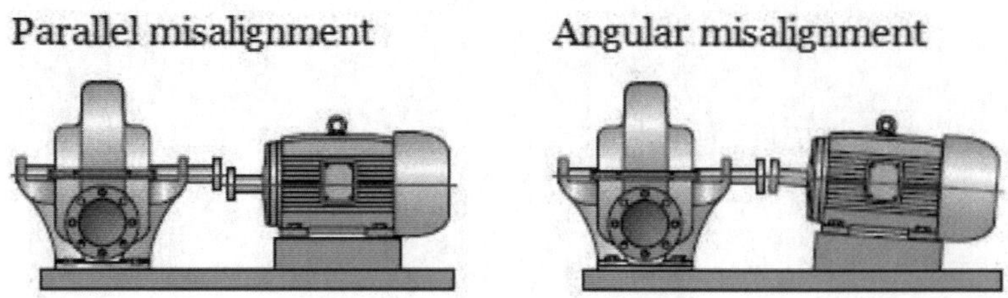

$$\underbrace{\text{Strain} \rightarrow \text{Stress} \rightarrow \text{Force}}_{\text{varying}} \Rightarrow \text{Vibration}$$

SpectraQuest. Inc.

Investigated parameters

Coupling type	Rubber	
	Spiral	
	Rigid	
Shaft diameter	½"	
	¾"	
	1"	
Shaft speed	1750 rpm	
	3880 rpm	
	5170 rpm	
Misalignment type & level	Parallel	5 mils
		15 mils
		25 mils
	Angular	0.03°
		0.10°
		0.17°

SpectraQuest. Inc.

Investigated parameters

$$V = FF \times FRF$$

$$FF = CS \times M$$

$$V = CS \times M \times FRF$$

- V : Vibration
- FF : Forcing Function
- FRF : Frequency Response Function
- CS : Coupling Stiffness
- M : Misalignment

SpectraQuest, Inc.

Investigated parameters

$$V = CS \times M \times FRF$$

Coupling type

Misalignment type & level

Shaft diameter

Shaft speed?

$$V = FF \times FRF$$

SpectraQuest, Inc.

$$V = FF \times FRF$$

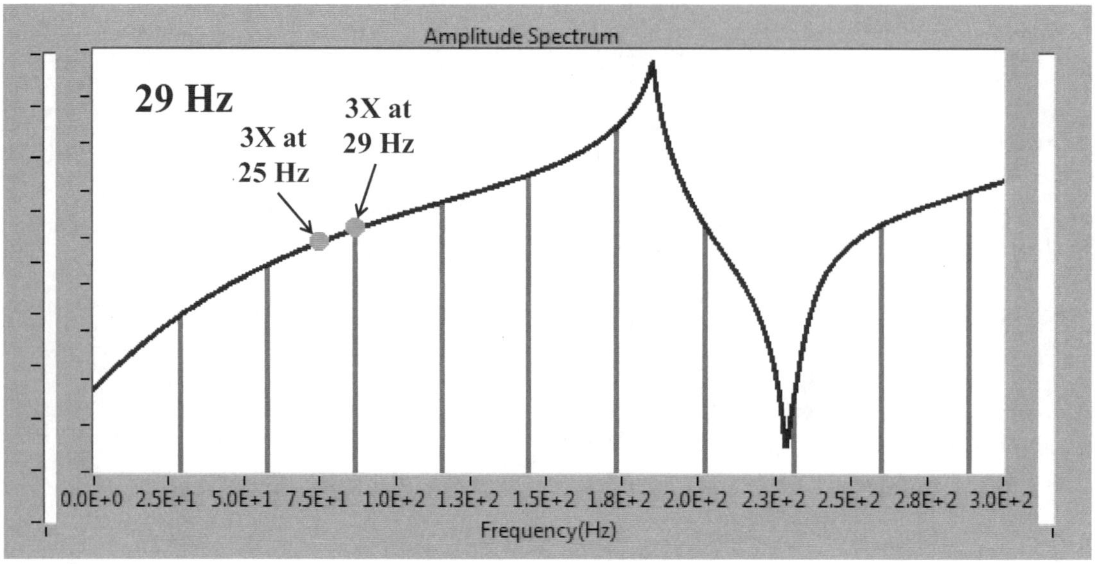

SpectraQuest, Inc.

Investigated parameters

$$V = CS \times M \times FRF$$

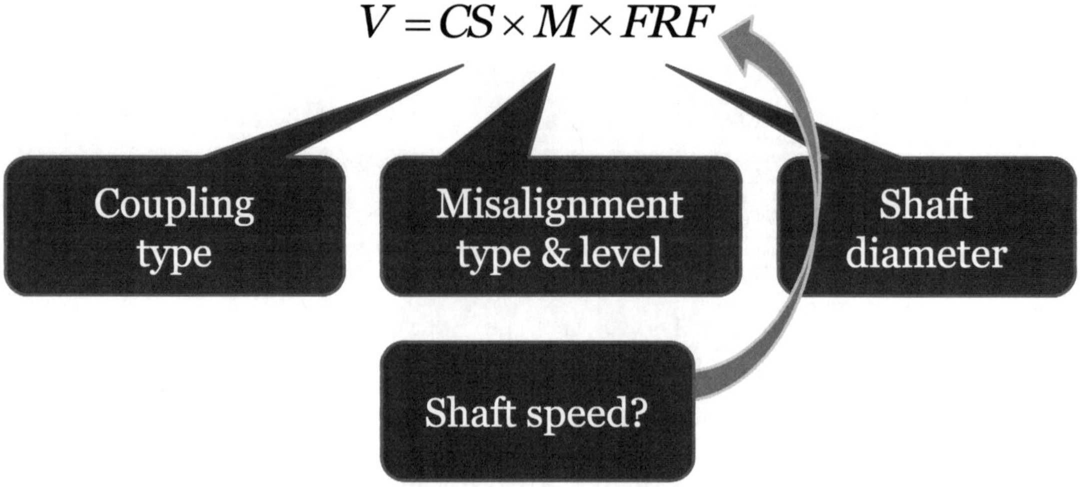

$$V = FF \times FRF$$

The shaft speed samples the FRF.

SpectraQuest, Inc.

Measurement setup

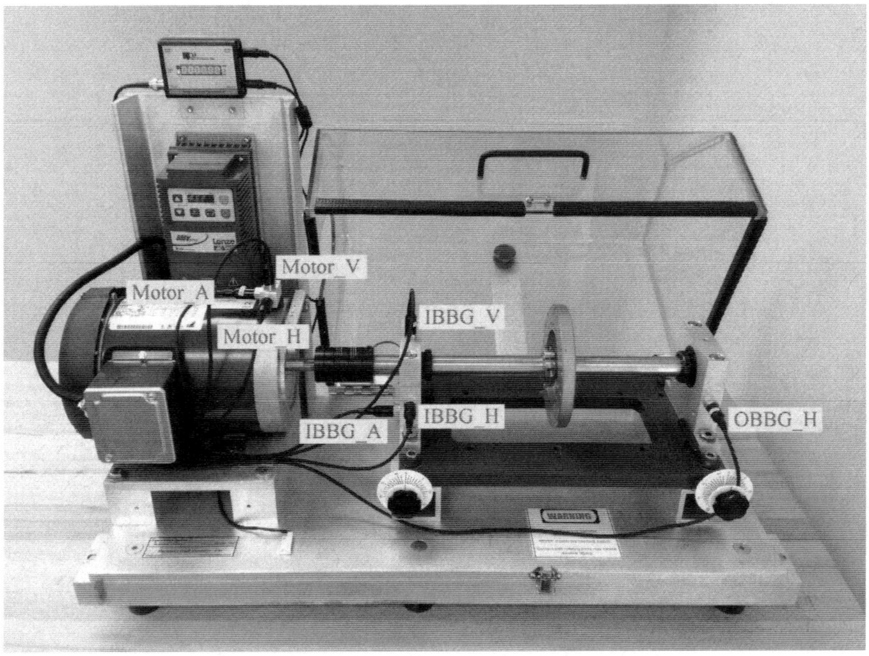

SpectraQuest. Inc.

Results: Coupling type

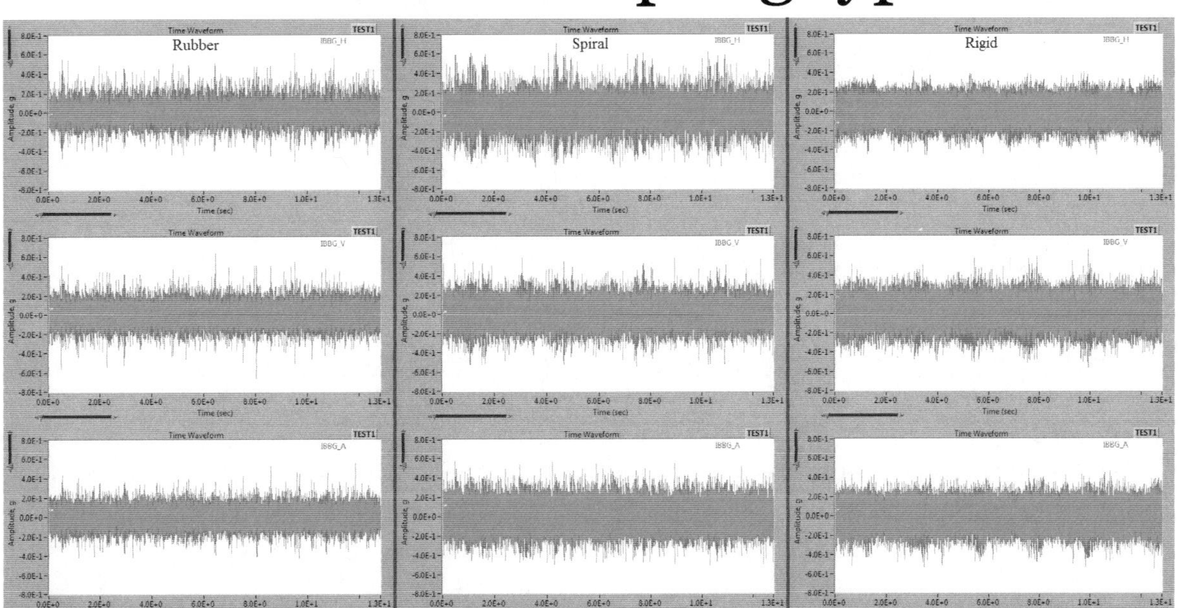

SpectraQuest. Inc.

Angular – 0.10°, 1750 rpm, ¾" shaft

Results: Coupling type

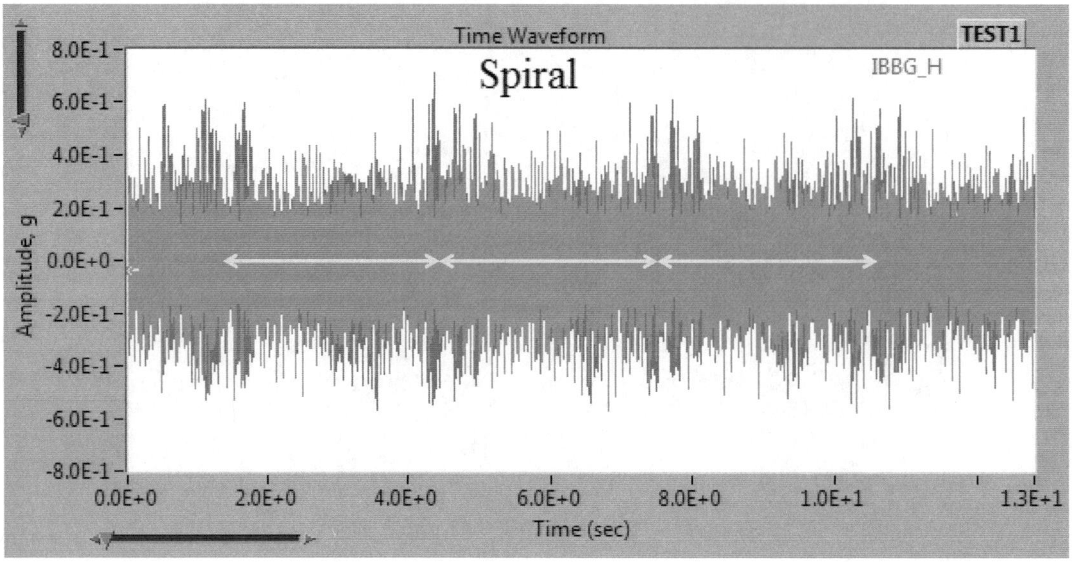

$\Delta t \approx 3$ s $\rightarrow$ $\Delta f \approx 1/3$ Hz

Results: Coupling type

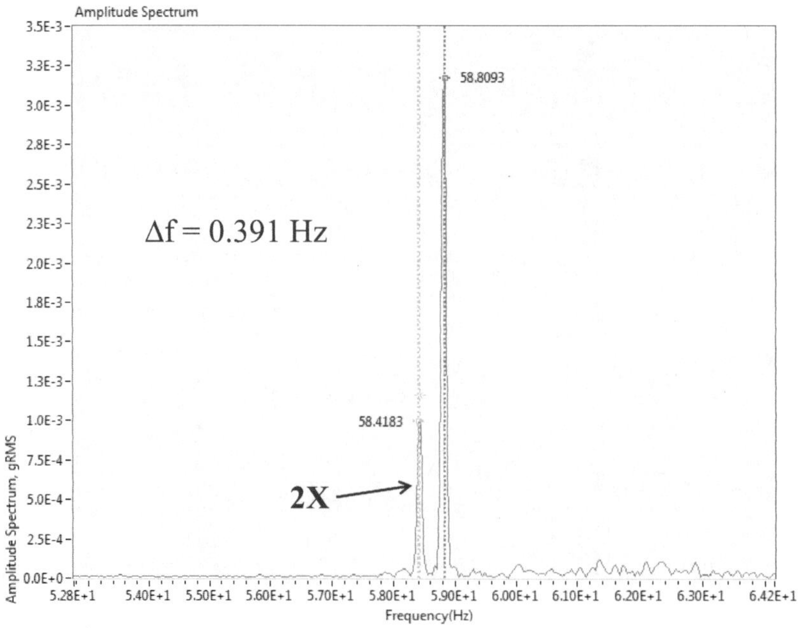

Results: Coupling type

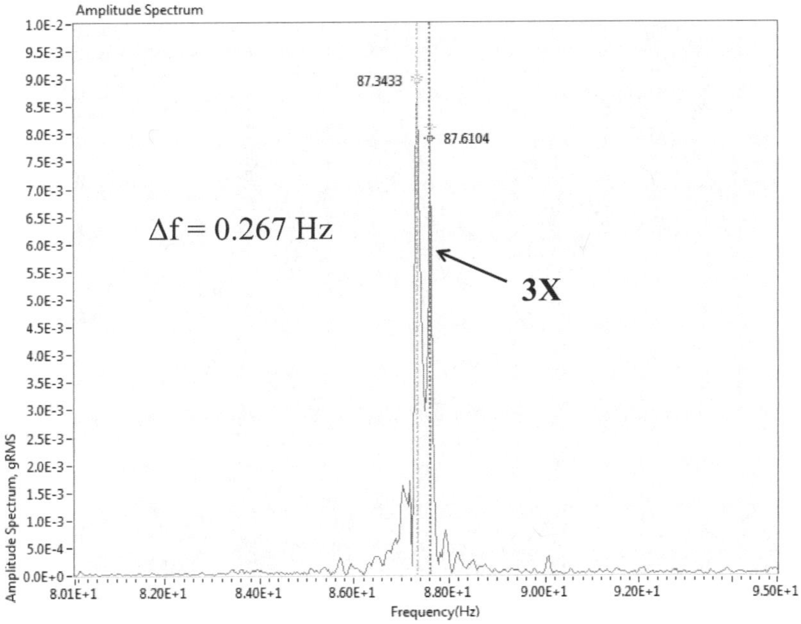

SpectraQuest. Inc.

Results: Coupling type

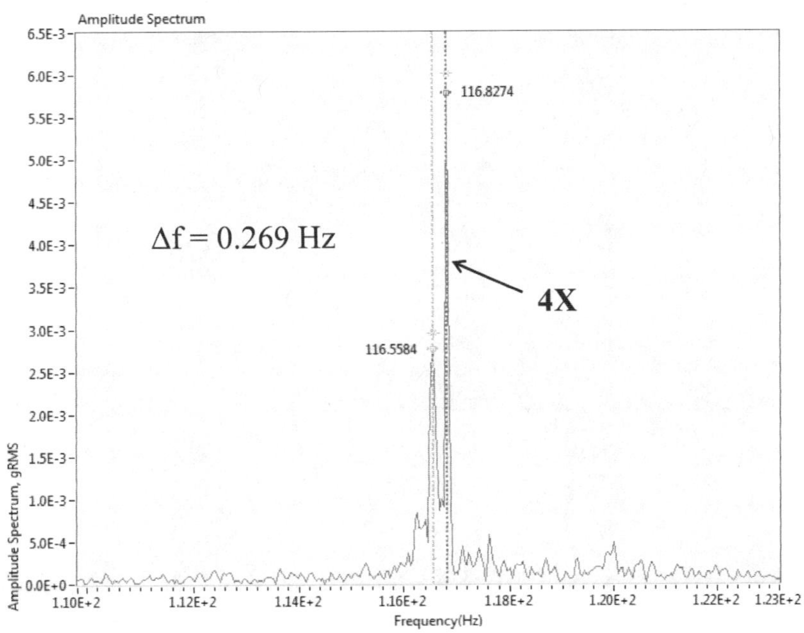

SpectraQuest. Inc.

Results: Coupling type

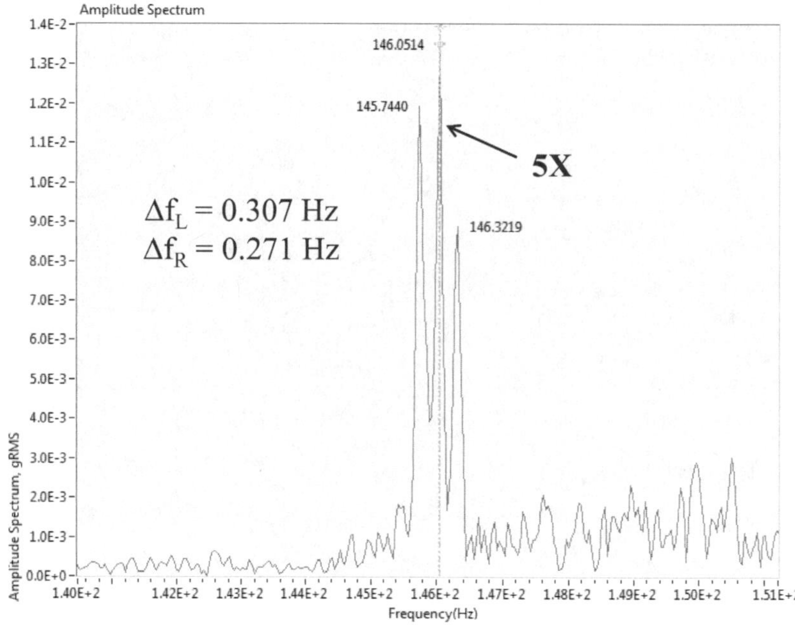

Results: Coupling type

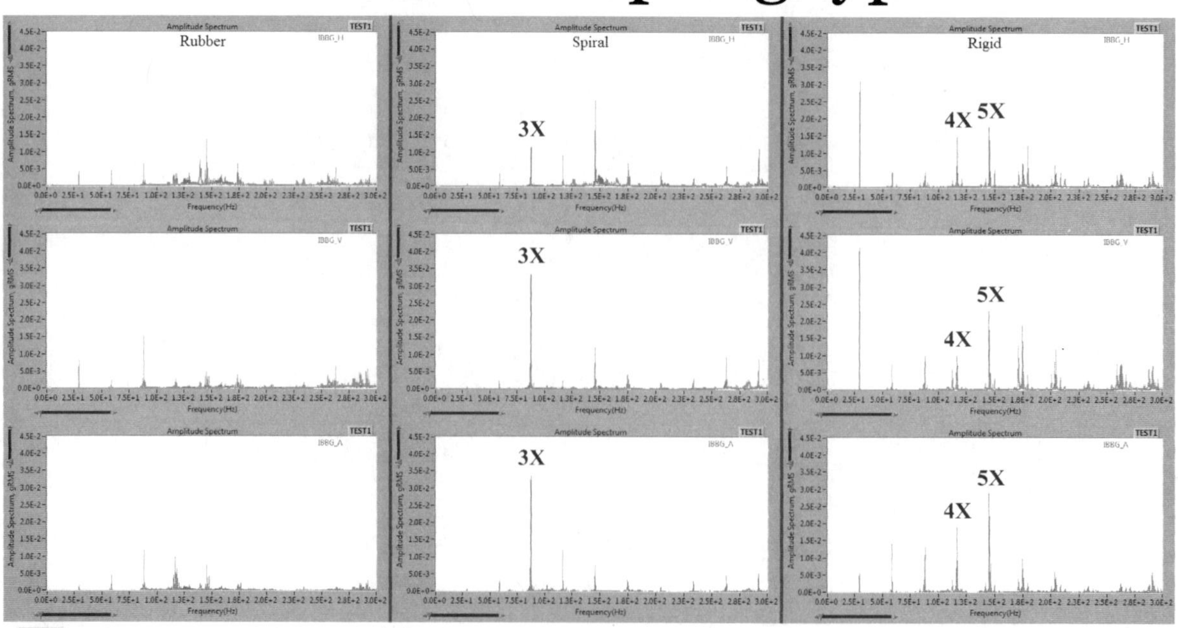

Angular – 0.10°, 1750 rpm, ¾" shaft

Results: Misalignment level

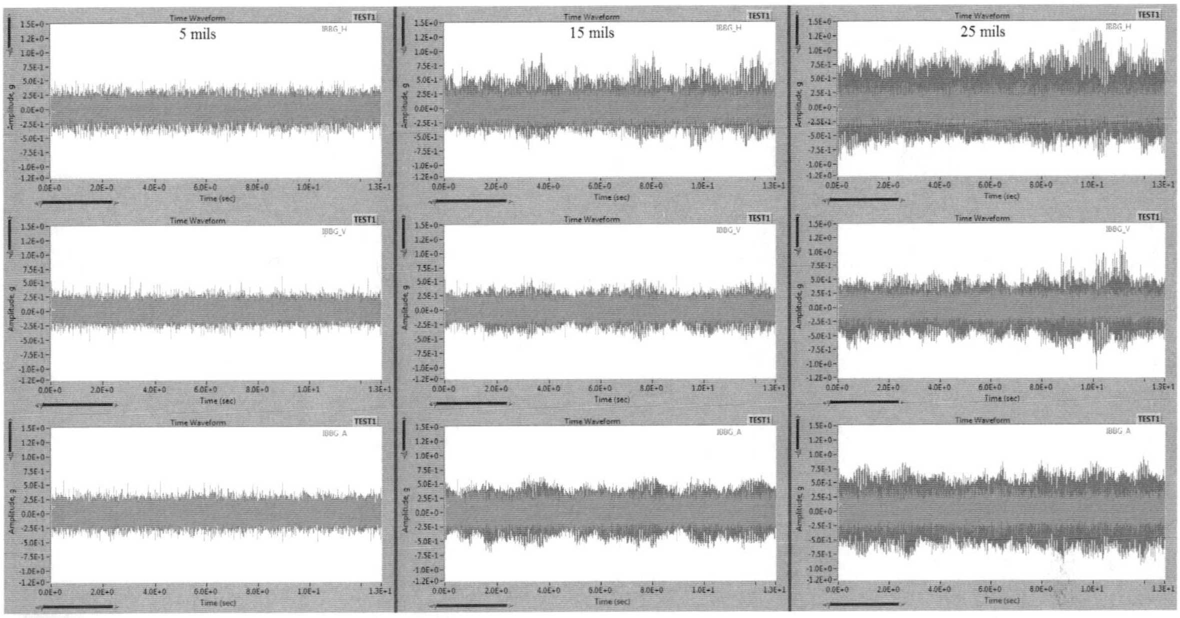

Spiral coupling, 1750 rpm, ¾" shaft

Results: Misalignment level

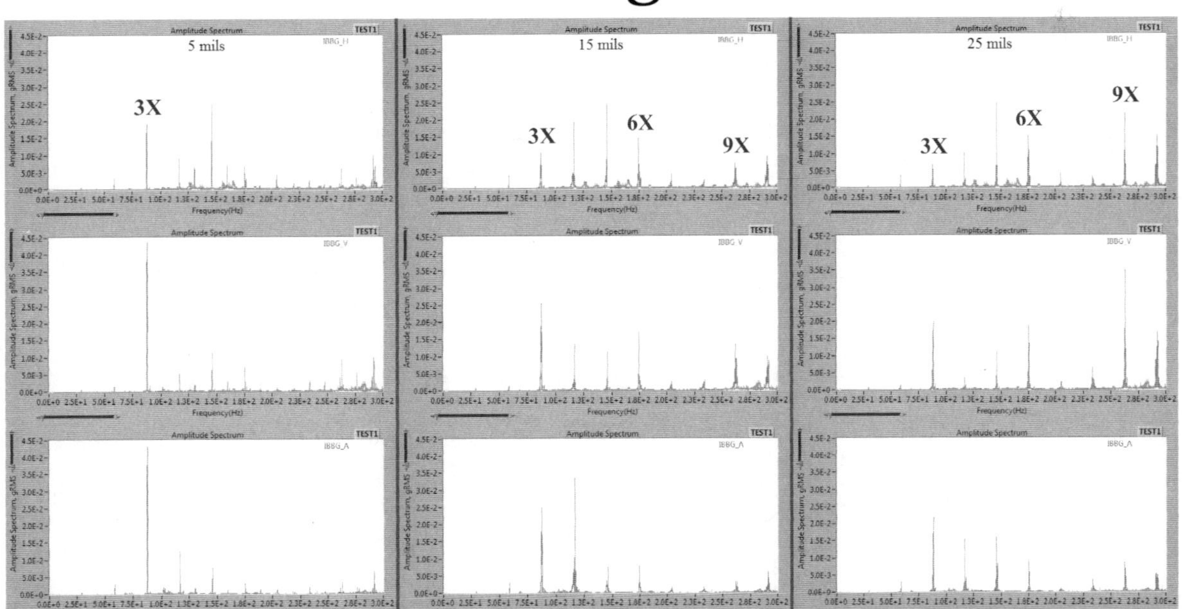

Spiral coupling, 1750 rpm, ¾" shaft

Results: Shaft speed

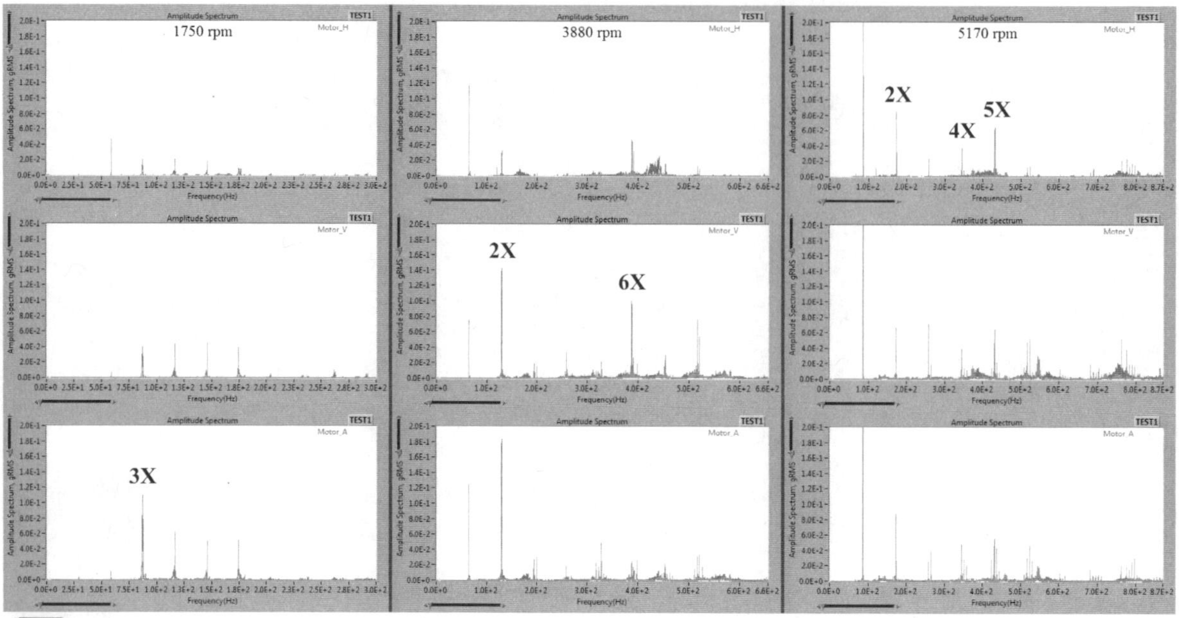

SpectraQuest, Inc.

Parallel – 15 mils, rigid coupling, ¾” shaft

Results: Shaft diameter

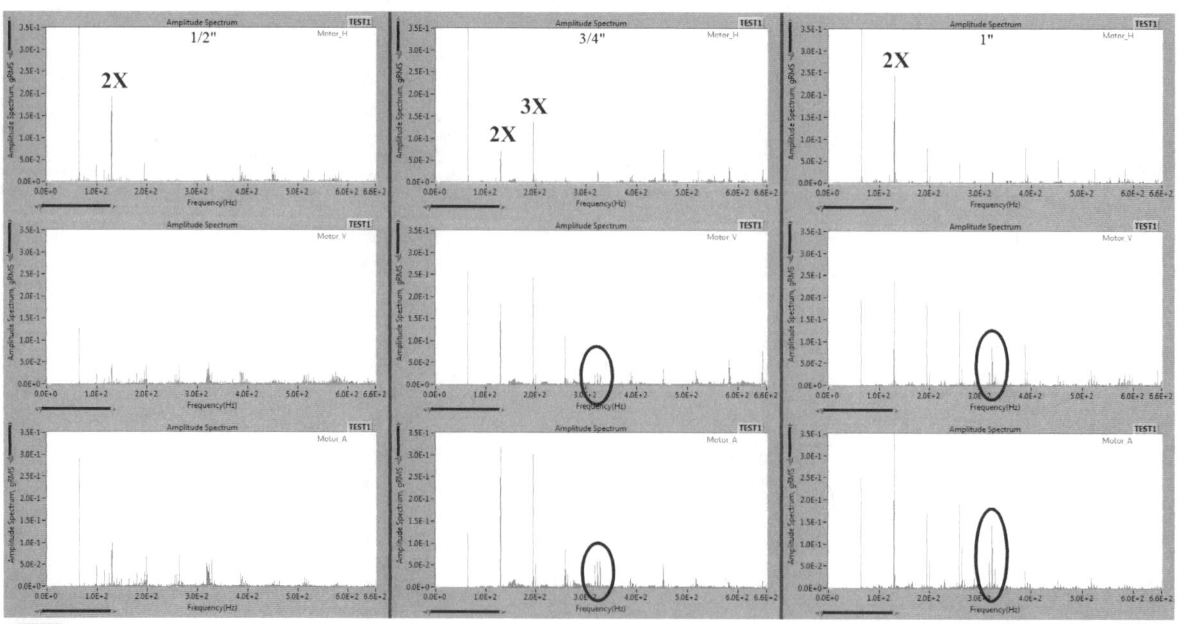

SpectraQuest, Inc.

Parallel – 15 mils, rigid coupling, 3880 rpm

Results: Shaft diameter

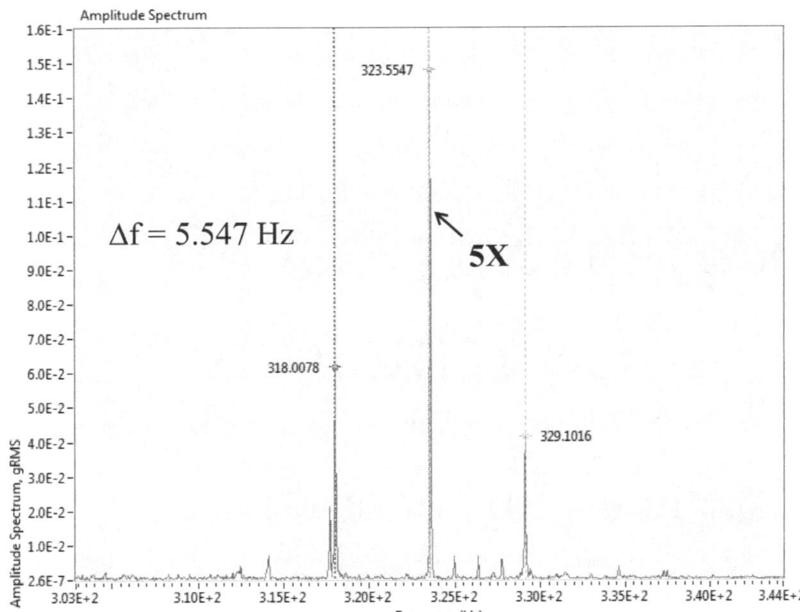

SpectraQuest, Inc.

Conclusions

- Effect of **coupling type**
  - Defines the shape of the forcing function.
- Effect of **misalignment type & level**
  - Type: Selects the direction of the coupling stiffness.
  - Level: Defines the magnitude of the forcing function.

ORIGIN

- Effect of **shaft speed**
  - Amplifies and attenuates harmonic peaks.
- Effect of **shaft diameter**
  - Changes the FRF → amplifies and attenuates harmonic peaks even at the same shaft speed.

TRANSFER

SpectraQuest, Inc.

Conclusions

- Inconsistencies, no general rules
- Peaks at harmonics of the shaft speed
- Amplitude modulation
- Sidebands
- Significant axial vibrations

- Baseline vibrations + CS + FRF
- Model (nonlinearities, rotor dynamic effects)

SpectraQuest, Inc.

THE GBS GROUP

MFPT 2014 CONFERENCE

Technology Solutions for Affordable Sustainment

Sheraton Virginia Beach Oceanfront Hotel
May 20–22, 2014

MFPT The Society for Machinery Failure Prevention Technology

Strain Gauge Tutorial

MAIN PROPULSION SHAFTING ANALYSIS FOR U.S. NAVY ASSETS USING STRAIN GAGE TECHNOLOGY

Presented By: The GBS Group/Don Thompson
Mechanical Engineering Consultant

MFPT 2014 CONFERENCE

Technology Solutions for Affordable Sustainment

Sheraton Virginia Beach Oceanfront Hotel
May 20–22, 2014

MFPT The Society for Machinery Failure Prevention Technology

- Introduction
  - Strain Gauge Technology
    - What is Strain
    - How does it work
    - Wheat-Stone Bridge Theory
    - Simple Beam Theory
    - Strain Gauge Applications
    - Strain Gauge Uses
- Main Propulsion Shafting Alignment
- Propulsion System Components & Key Terms
- Strain Gauge Advantages & Disadvantages
- Computer Modeling Considerations Using Strain Gauges
  - Thermal Growth
  - Bearing wear
- GBS Points of Contact

THE GBS GROUP

MFPT 2014 CONFERENCE

Technology Solutions for Affordable Sustainment

Sheraton Virginia Beach Oceanfront Hotel
May 20–22, 2014

MFPT
The Society for Machinery Failure Prevention Technology

INTRODUCTION:

Reliability of a system can be defined as the probability that the system in question will operate with out failure for a certain period of time under conditions previously established.

Reliability centered maintenance (RCM) focuses on identifying what should be done to assure the functions of a system are consistently carried out in a cost-effective and reliable way for a pre-determined time Fully reliable being defined as a ship that meets all underway commitments with no shafting restrictions.

Strain Gauge Analysis is a condition base maintenance tool used to determine the material readiness of the propulsion shafting system.

THE GBS GROUP

MFPT 2014 CONFERENCE

Technology Solutions for Affordable Sustainment

Sheraton Virginia Beach Oceanfront Hotel
May 20–22, 2014

MFPT
The Society for Machinery Failure Prevention Technology

STRAIN GAUGE TECHNOLOGY

- Strain gauge technology has been used by the US Navy to more accurately measure bearing vertical and horizontal loads on main propulsion shafting systems.
- This technology became crucial when the Spruance Class ships were introduced to the fleet as the new Anti-Submarine Warfare (ASW) platforms. The Main Reduction Gear Housing foundation was fitted with sound isolation mounts vice hard mounted to reduce noise. Subsequently making the conventional Jack Load Cell method for measuring shaft bearing loads ineffective.

THE GBS GROUP

WHAT IS STRAIN ?

- Strain is the amount of deformation of a body due to an applied force. More specifically, strain (e) is defined as the fractional change in length.
- Strain can be positive (tensile) or negative (compressive). Although dimensionless, strain is sometimes expressed in units such as in./in. or mm/mm.
- In practice, the magnitude of measured strain is very small. Therefore, strain is often expressed as micro-strain (me), which is e x 10-6.
- One micro-strain is the strain producing a deformation of one part per million.

THE GBS GROUP

MFPT 2014 CONFERENCE
Technology Solutions for Affordable Sustainment
Sheraton Virginia Beach Oceanfront Hotel
May 20–22, 2014
MFPT
The Society for Machinery Failure Prevention Technology

THE STRAIN GAGE: HOW DOES IT WORK?

- A strain gauge is a device used to measure strain (deformation) on an object.
- The most widely used gage is the bonded metallic strain gage.
- The metallic strain gage consists of a very fine wire or, more commonly, metallic foil arranged in a grid pattern. The grid pattern maximizes the amount of metallic wire or foil subject to strain in the parallel direction. The grid is bonded to a thin backing, called the carrier, which is attached directly to the test specimen

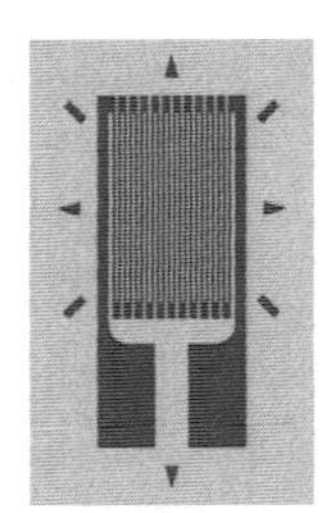

THE GBS GROUP

MFPT 2014 CONFERENCE

Technology Solutions for Affordable Sustainment

Sheraton Virginia Beach Oceanfront Hotel
May 20–22, 2014

MFPT
The Society for Machinery Failure Prevention Technology

STRAIN GAUGES: HOW DOES IT WORK

A small voltage is applied to the input leads of the gauge. As shown to the right, as the object is deformed either in compression or tension the change in resistance is measured as a micro-strain.

In many instances this measurement is converted to weight (lbs) or force (ft/lbs).

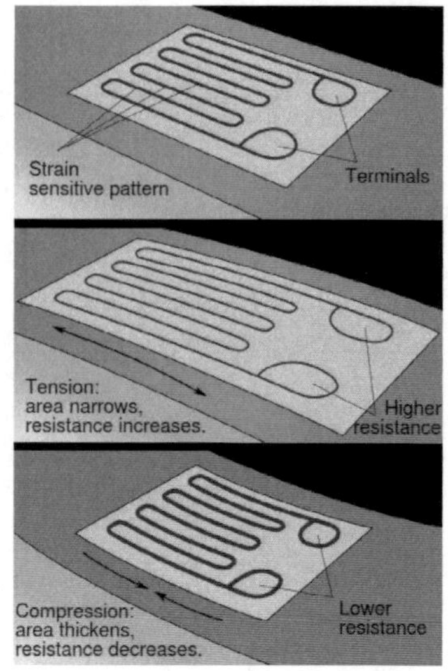

THE GBS GROUP

MFPT 2014 CONFERENCE

Technology Solutions for Affordable Sustainment

Sheraton Virginia Beach Oceanfront Hotel
May 20–22, 2014

MFPT
The Society for Machinery Failure Prevention Technology

WHEATSTONE BRIDGE

- Wheatstone Bridge is an electrical circuit where (4) resistance are connected to form a closed loop. Unknown resistance can be determined by balancing (2) legs of a bridge circuit. One leg includes the unknown component.
- Strain Gauge application a Half Bridge Strain Gauge circuit is used.

THE GBS GROUP

Half-bridge strain gauge circuit

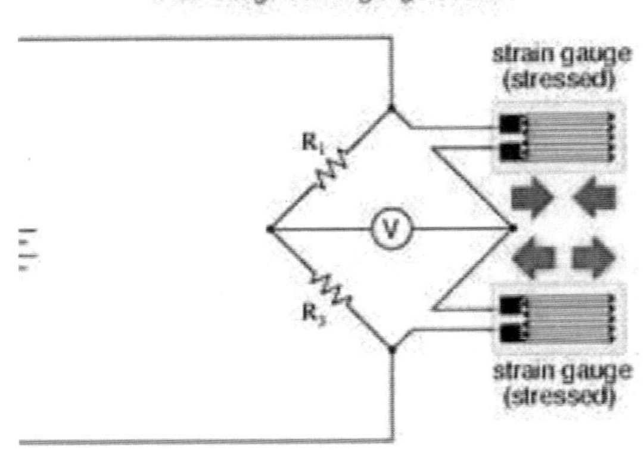

Wheatstone Bridge

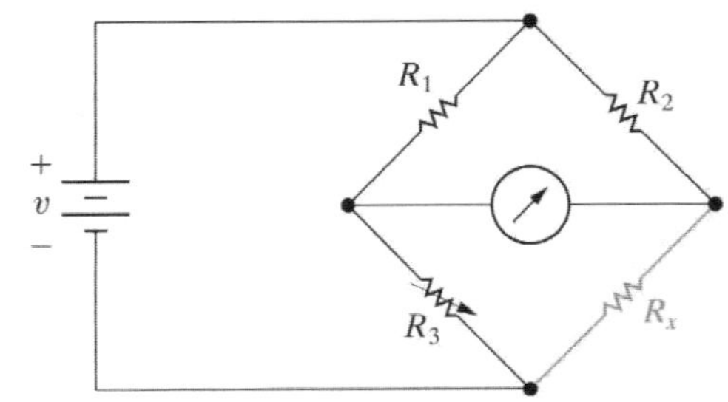

ECE 201 Circuit Theory I 1

MFPT 2014 CONFERENCE
Technology Solutions for Affordable Sustainment
Sheraton Virginia Beach Oceanfront Hotel
May 20–22, 2014
MFPT
The Society for Machinery Failure Prevention Technology

SIMPLE BEAM

- Bernoulli Beam Theory provides the means of determining the deflection as well as load carrying characteristics of a beam .
- Propulsion Main Shafts are treated as "Beams"
- Calculations regarding support bearing loads and any deformation as a result of external forces and temperature changes can be calculated

Beam Diagram of Shaft

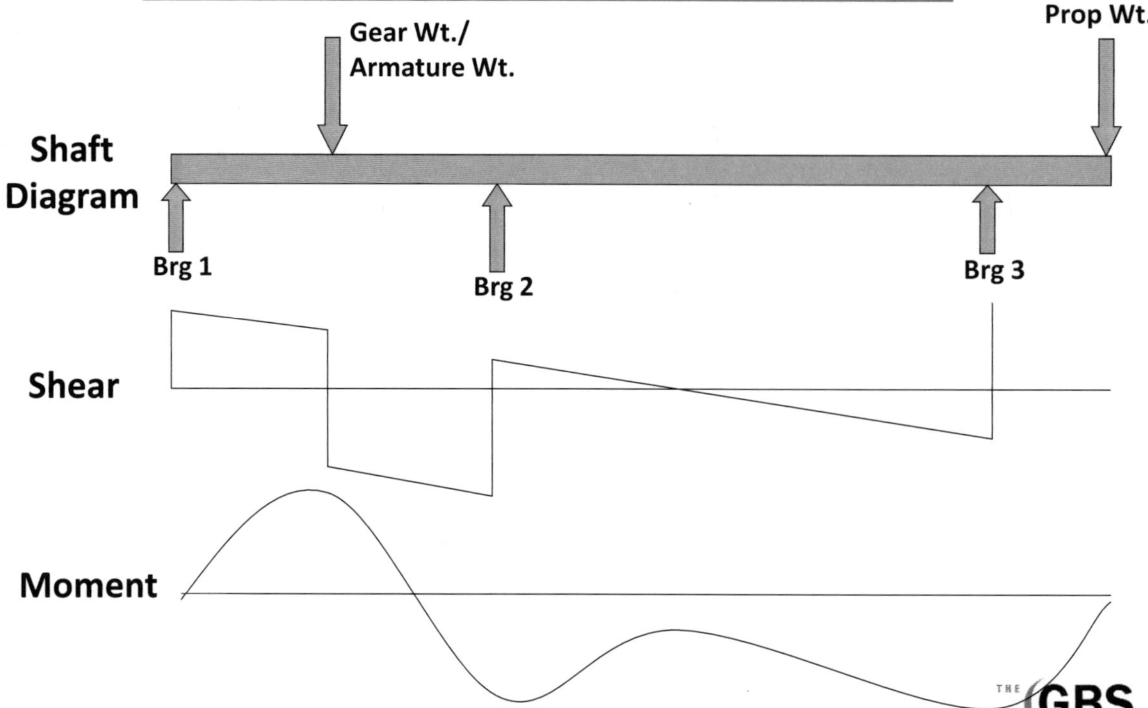

THE GBS GROUP

MFPT 2014 CONFERENCE
Technology Solutions for Affordable Sustainment
Sheraton Virginia Beach Oceanfront Hotel
May 20–22, 2014
MFPT
The Society for Machinery Failure Prevention Technology

STRAIN GAUGE APPLICATION

>Stress and Strain Measurement

The basic function of a strain gauge is to measure strain, and in turn stress. Strain is simply the change in length of a component divided by the original length of the component while experiencing an applied load. Stress is calculated by multiplying the strain by the modulus of elasticity for the specific material being measured.

>Load Measurement

Strain gauges are integral elements of load cell transducers, which measure the magnitude of an applied tensile or compressive force. The strain gauges are installed on precisely machined steel components where the exact cross section is known, and therefore the strain measurement can be converted to a load measurement output.

>Pressure Measurement

A pressure sensor uses the same strain gauge configuration and installation as a load cell, except that the sensor measures the applied load over a discrete area, or pressure.

THE GBS GROUP

Strain gauge application cont

>Torsion Measurement

Strain gauges are used to measure the torsion forces in a rotating shaft to ensure that the shaft is not being overloaded. Shaft overload can lead to fatigue over time or fracture, resulting in a system failure. In this case, either strain gauges are wired through a slip ring to transmit over the rotating connection or data is transmitted wirelessly to a remote receiver.

> Shear Strain Measurement

A combination of strain gauges can measure shear strain in a component. In this case, two single-axis strain gauges are installed with a 90-degree angle between them. The two strain measurements can be used to calculate the shear strain in a component and can be converted to shear stress.

THE GBS GROUP

MFPT 2014 CONFERENCE
Technology Solutions for Affordable Sustainment
Sheraton Virginia Beach Oceanfront Hotel
May 20–22, 2014
MFPT
The Society for Machinery Failure Prevention Technology

Strain Gauge Uses

Military

- Naval Shafting Analysis
- Army Gun Recoil Stresses
- Air Force Wing Stresses

Civil Engineering

- Bridge Monitoring
- Rail track monitoring
- Building movements

THE GBS GROUP

MFPT 2014 CONFERENCE

Technology Solutions for Affordable Sustainment

Sheraton Virginia Beach Oceanfront Hotel
May 20–22, 2014

MFPT
The Society for Machinery Failure Prevention Technology

When to use Strain Gauges

- On Main Propulsion Systems Using Resilient Mounted Main Reduction Gears
- Pre Dry Dock Evaluations
- Post Dry Dock Evaluations
- Extensive Damage/Repairs to Structural Areas Adjacent to Bearing Supports/Foundations
- When Extensive Bearing Side Loading is Suspected

THE GBS GROUP

15

MFPT 2014 CONFERENCE

Technology Solutions for Affordable Sustainment

Sheraton Virginia Beach Oceanfront Hotel
May 20–22, 2014

MFPT
The Society for Machinery Failure Prevention Technology

- In addition to US Navy & Commercial & Maritime Navy uses
- Strain Gauge application has also been used in some seemingly unlikely applications

- The Hutson-Athens Lighthouse has strain gauges attached to some of the brick external walls to measure and trend foundation settling.

THE GBS GROUP

MFPT 2014 CONFERENCE
Technology Solutions for Affordable Sustainment

Sheraton Virginia Beach Oceanfront Hotel
May 20–22, 2014

MFPT
The Society for Machinery Failure Prevention Technology

THE GBS GROUP

MFPT 2014 CONFERENCE
Technology Solutions for Affordable Sustainment

Sheraton Virginia Beach Oceanfront Hotel
May 20–22, 2014

MFPT
The Society for Machinery Failure Prevention Technology

THE GBS GROUP

MAIN PROPULSION SHAFTING ALIGNMENT

- DEFINITION
- IMPORTANCE
- WHEN TO PERFORM

THE GBS GROUP

MFPT 2014 CONFERENCE
Technology Solutions for Affordable Sustainment
Sheraton Virginia Beach Oceanfront Hotel
May 20–22, 2014
MFPT
The Society for Machinery Failure Prevention Technology

SHAFTING ALIGNMENT DEFINITION

Main Propulsion Shafting Alignment is defined as:

- The propulsion shafting is mated to the main reduction gear assembly in such a way that the low speed gear tooth contacts are not adversely affected
- The location of the main propulsion shaft bearings provide no excessive bearing loads, no excessive shaft vibration, no excessive bearing temperatures or shaft torques

- A properly aligned shafting system should provide satisfactory long term operation of the propulsion plant for a wide range of operational conditions.

THE GBS GROUP

MFPT 2014 CONFERENCE

Technology Solutions for Affordable Sustainment

Sheraton Virginia Beach Oceanfront Hotel
May 20–22, 2014

MFPT
The Society for Machinery Failure Prevention Technology

IMPORTANCE

- An adequate shafting system and bearing alignment provide the following operational and maintenance advantages:
  - Minimizes shaft vibration due to a lightly loaded or unloaded bearing(s) in the system.
  - Minimizes bending stresses in the shafting system which may result in a shaft failure.
  - Minimizes internal misalignment of main reduction gear elements which can lead to excessive gear tooth wear (pitting) or tooth failure.
  - Minimizes overload of shafting system bearings which can result in premature bearing failures.
  - Ensures proper wear of waterborne bearings

THE GBS GROUP

Gear Shaft Alignment

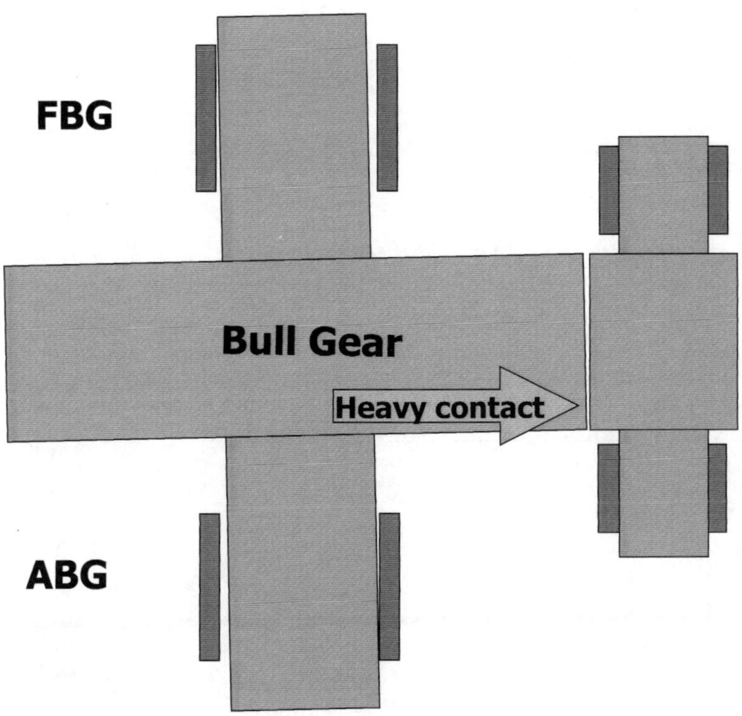

Vertical or horizontal misalignment causing skewed bull gear shaft. In this example the tooth contacts would be heavy aft. Skew in the other direction would cause heavier tooth contacts on the forward ends of the helices.

Note: misalignment of the pinion will cause the same condition. Multiple pinion systems help to determine cause of poor contacts. Otherwise bearing reaction measurements are required.

THE GBS GROUP

WHEN TO PERFORM ALIGNMENT

- If Significant Misalignment Is Suspected Due To:
  - Excessive Bearing Temperature
  - Abnormal MRG Tooth Contacts
  - Ship Grounding
  - Repeated Bearing Failures
  - Unusual Bearing Wear Pattern
  - Unexplained System Vibration
- After Major Ship Repairs or Modifications That May Affect Shaft Alignment
  - Re-staving Waterborne Bearings
  - Shaft Removal/Repair
  - Significant Welding or Structural Work in Vicinity of Propulsion Plant
- During Major Ship Overhaul Periods
  - Perform Final Alignment Check at the end of Overhaul with Ship Waterborne and Normal Draft
  - Perform check prior to and/or during Dry-dock

THE GBS GROUP

MFPT 2014 CONFERENCE
Technology Solutions for Affordable Sustainment
Sheraton Virginia Beach Oceanfront Hotel
May 20–22, 2014
MFPT
The Society for Machinery Failure Prevention Technology

PROPULSION SYSTEM COMPONENTS AND KEY TERMS

- Propeller
- Main Strut/Bearing & Intermediate Strut/Bearing
- Tail Shaft
- Intermediate Shaft
- Stern Tube Shaft
- Line Shaft
- CPP Components
- Line Shaft Bearings
- Main Reduction Gear & Prime Mover

THE GBS GROUP

MFPT 2014 CONFERENCE
Technology Solutions for Affordable Sustainment
Sheraton Virginia Beach Oceanfront Hotel
May 20–22, 2014

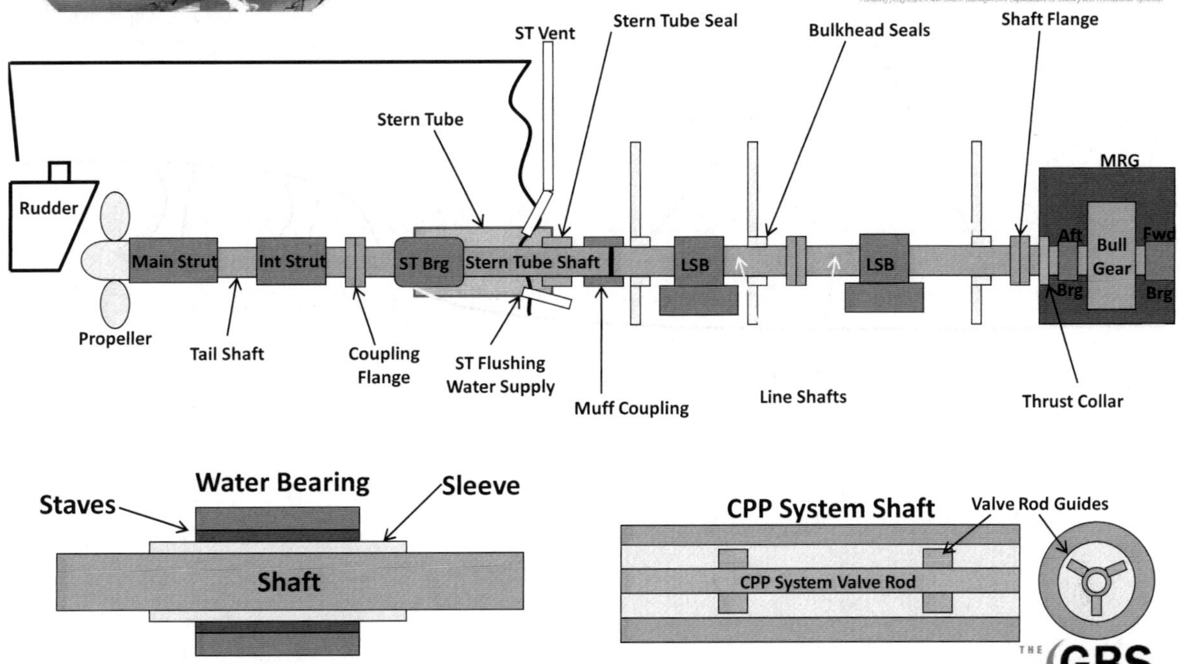

MFPT 2014 CONFERENCE
Technology Solutions for Affordable Sustainment
Sheraton Virginia Beach Oceanfront Hotel
May 20–22, 2014

STRAIN GAUGE ADVANTAGES/DISADVANTAGES

- Advantages
  - Measures vertical and horizontal bearing loads
  - Measures bull gear bearing loads without intrusion into reduction gear
  - Capability of measuring resiliently mounted bearing loads
- Disadvantages
  - Requires shafting model and validation of model
  - Provides no indication of bearing clearance
  - Cannot measure shaft run out/load variation
  - Requires a higher level of technical training

MFPT 2014 CONFERENCE
Technology Solutions for Affordable Sustainment
Sheraton Virginia Beach Oceanfront Hotel
May 20–22, 2014

MFPT
The Society for Machinery Failure Prevention Technology

STRAIN GAUGE PREPERATION

- Model Development
  - A shafting model is required to be developed and validated to use Strain Gage method. The model is basically a beam diagram of a shafting system. The following is required to develop a quality model.
    - Shafting dimensions from detailed drawings
    - Bearing locations from builders drawings
    - Concentrated weights such as gear or armature weight, thrust collar weight and propeller weight
    - Shafting material properties such as density and elasticity
  - Determining gage locations is critical for model

THE GBS GROUP

MFPT 2014 CONFERENCE
Technology Solutions for Affordable Sustainment
Sheraton Virginia Beach Oceanfront Hotel
May 20–22, 2014

MFPT
The Society for Machinery Failure Prevention Technology

STRAIN GAUGE PREPERATION

Strain gauges are attached to the shaft with a special glue. The type of glue depends on the required lifetime of the measurement system. For short term measurements (up to some weeks) cyan acrylic glue is appropriate, for long lasting installation epoxy glue is required.

The preparation of the surface where the strain gauge is to be glued is important. The surface must be smoothed (e.g. with very fine sand paper), cleaned with solvents, the solvent traces must then be removed and the strain gauge must be glued immediately after this to avoid oxidation or pollution of the prepared area. If these steps are not followed the strain gauge binding to the surface may be unreliable and unpredictable measurement errors may be generated.

THE GBS GROUP

Strain Gage Toolbox

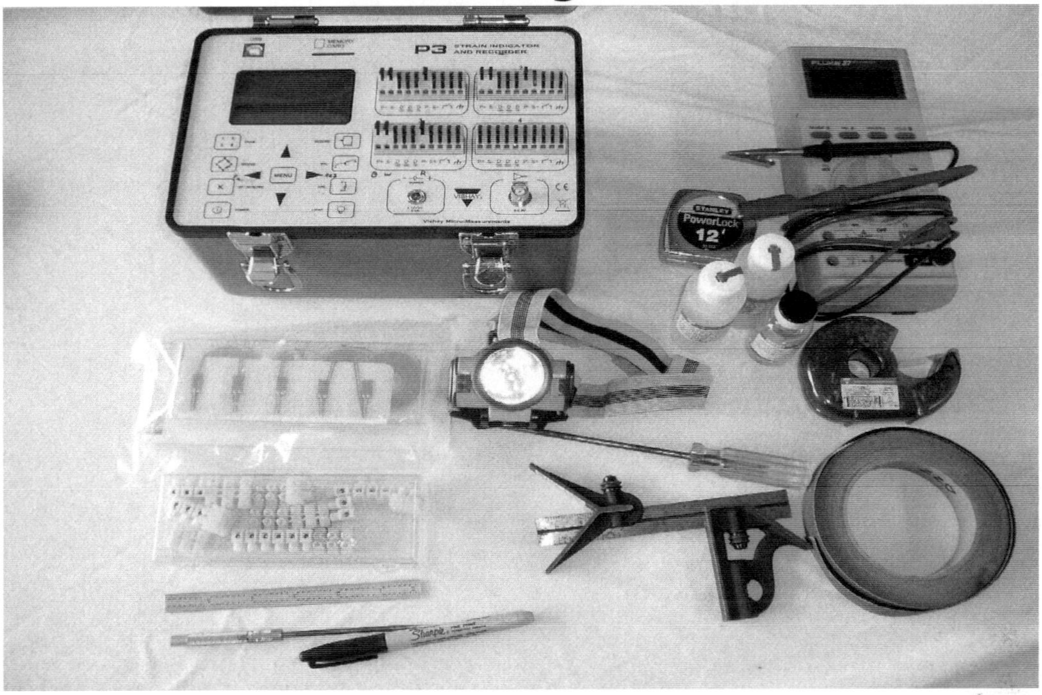

Figure 4.1 STRAIN GAGE NUMBERING DIAGRAM

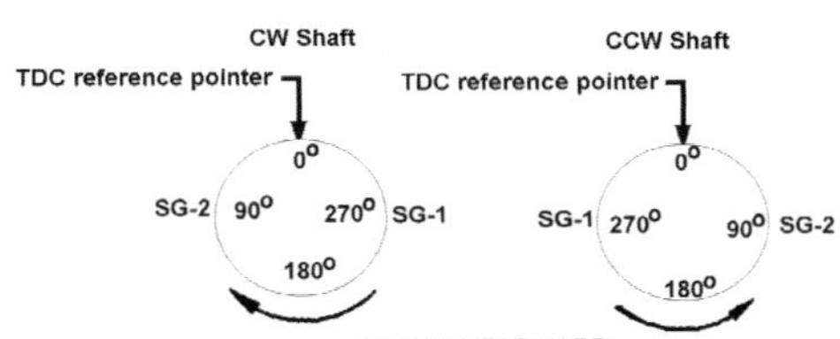

Figure 4.2 STRAIN GAGE WIRING DIAGRAM

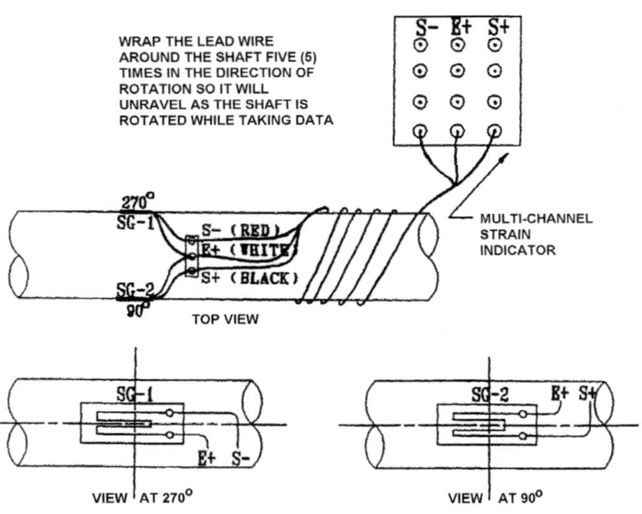

Gathering Data

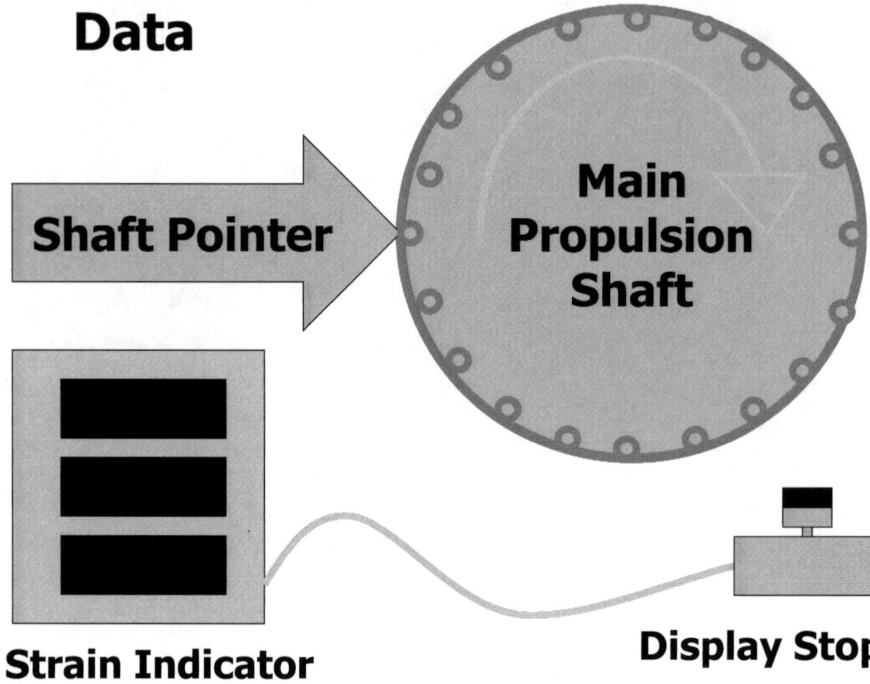

MFPT 2014 CONFERENCE

Technology Solutions for Affordable Sustainment

Sheraton Virginia Beach Oceanfront Hotel
May 20–22, 2014

MFPT

The Society for Machinery Failure Prevention Technology

- When to use Strain Gauges

- On Main Propulsion Systems Using Resilient Mounted Main Reduction Gears
- Pre Dry Dock Evaluations
- Post Dry Dock Evaluations
- Extensive Damage/Repairs to Structural Areas Adjacent to Bearing Supports/Foundations
- When Extensive Bearing Side Loading is Suspected

THE GBS GROUP

32

Thermal growth and bearing wear greatly affect long term computer modeling using strain gauge technology

Thermal growth affects main reduction gear to line shaft alignment and associated bearing loads

Bearing wear affects over all bearing loads.

THE GBS GROUP

Thermal Growth Effects

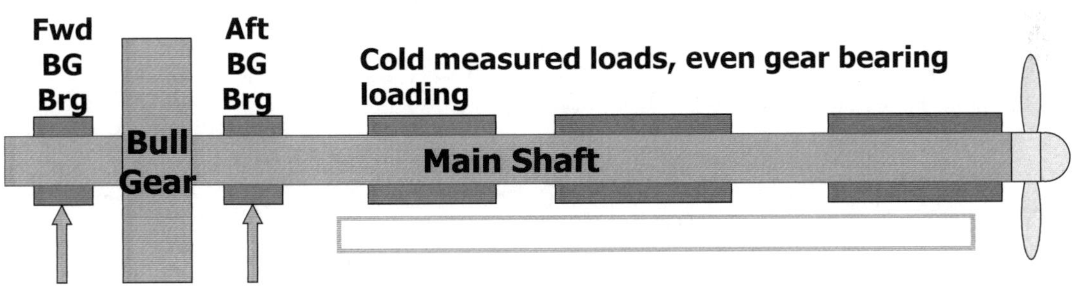

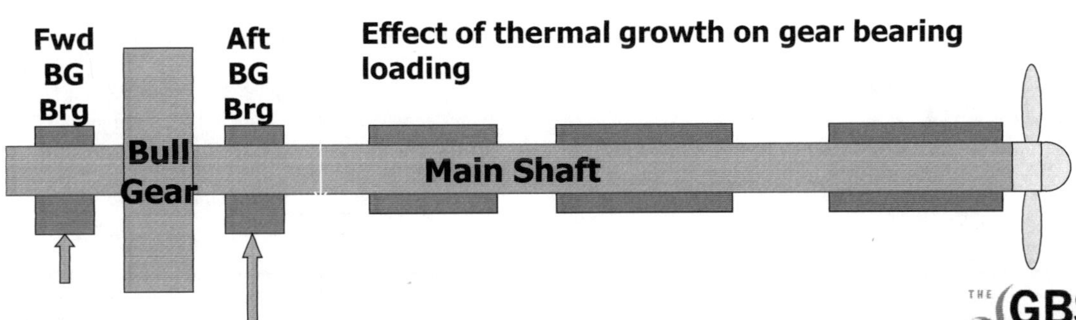

THE GBS GROUP

Bearing Wear Effects

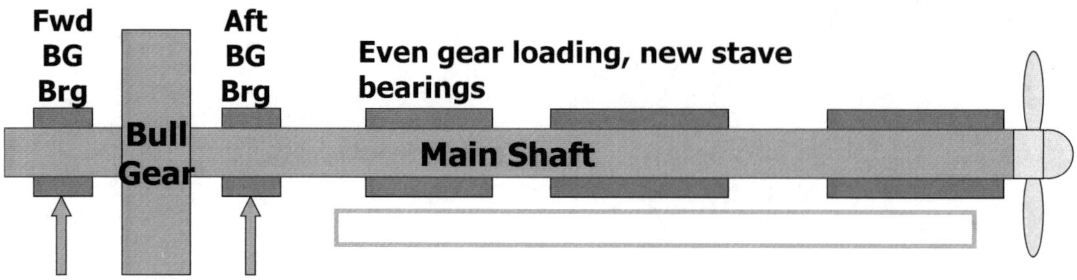

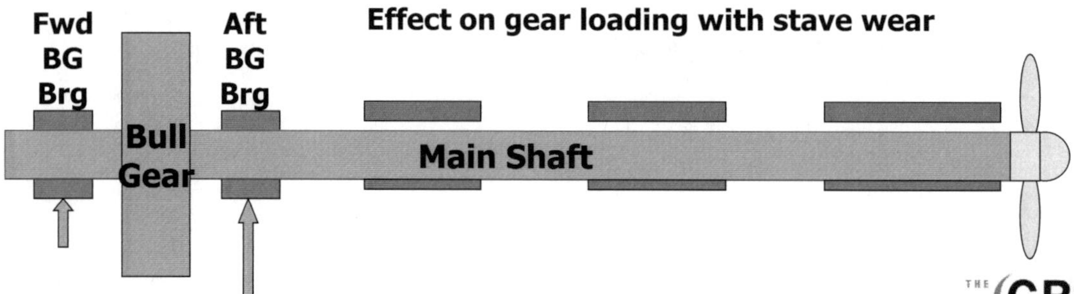

Bearing Wear Effects

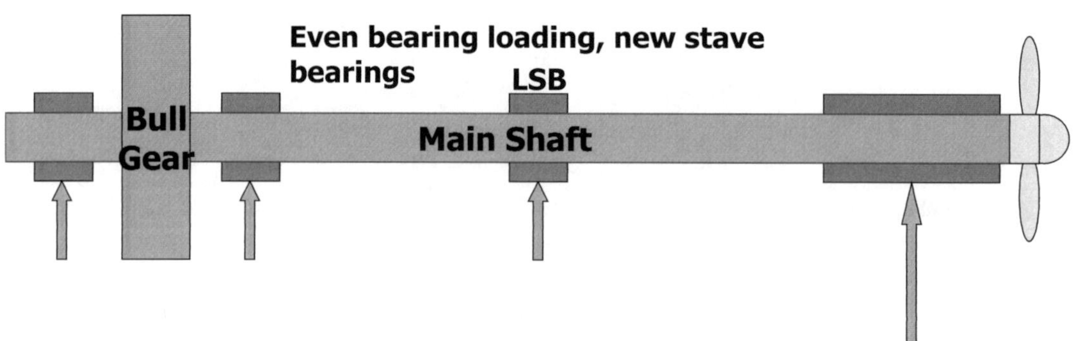

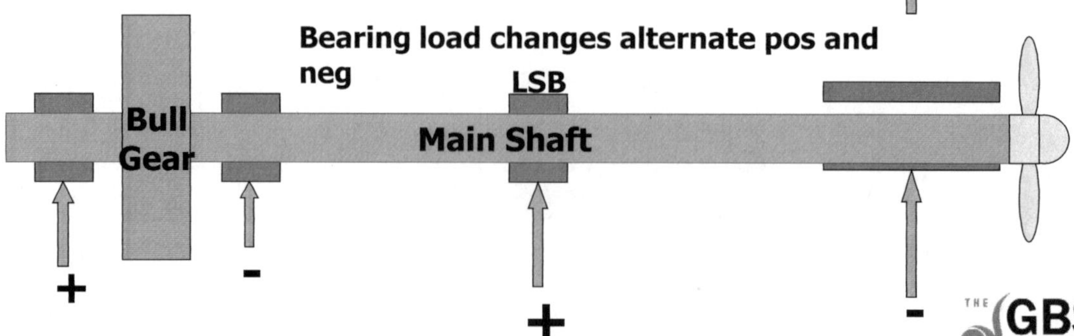

Solving the Toughest Challenges

Please visit our website at:

www.theGBSgroup.us

or contact:

Pete Martinez PE
VP The GBS Group/Maritime Division
petemartinez@theGBSgroup.us
(757) 965- 4274

Don Thompson
Mechanical Engineering Consultant
donthompson@theGBSgroup.us
(757) 636-3190

GBS Corporate Headquarters
397 Little Neck Road
Building 3300 South, Suite 204
Virginia Beach, VA 23452
Phone: 757.965.4274
Fax: 757.965.4518

AUTHOR INDEX

AUTHOR INDEX